Geophysical Monograph Series
American Geophysical Union

*Geophysical Monograph Series*
A. F. Spilhaus, Jr., Managing Editor

1 **Antarctica in the International Geophysical Year,** A. P. Crary, L. M. Gould, E. O. Hulburt, Hugh Odishaw, and Waldo E. Smith (editors)

2 **Geophysics and the IGY,** Hugh Odishaw and Stanley Ruttenburg (editors)

3 **Atmospheric Chemistry of Chlorine and Sulfur Compounds,** James P. Lodge, Jr. (editor)

4 **Contemporary Geodesy,** Charles A. Whitten and Kenneth H. Drummond (editors)

5 **Physics of Precipitation,** Helmut Wieckmann (editor)

6 **The Crust of the Pacific Basin,** Gordon A. Macdonald and Hisashi Kuno (editors)

7 **Antarctic Research: The Matthew Fontaine Maury Memorial Symposium,** H. Wexler, M. J. Rubin, and J. E. Caskey, Jr. (editors)

8 **Terrestrial Heat Flow,** William H. K. Lee (editor)

9 **Gravity Anomalies: Unsurveyed Areas,** Hyman Orlin (editor)

10 **The Earth Beneath the Continents: A Volume in Honor of Merle A. Tuve,** John S. Steinhart and T. Jefferson Smith (editors)

11 **Isotope Techniques in the Hydrologic Cycle,** Glenn E. Stout (editor)

12 **The Crust and Upper Mantle of the Pacific Area,** Leon Knopoff, Charles L. Drake, and Pembroke J. Hart (editors)

13 **The Earth's Crust and Upper Mantle,** Pembroke J. Hart (editor)

14 **The Structure and Physical Properties of the Earth's Crust,** John G. Heacock (editor)

15 **The Use of Artificial Satellites for Geodesy,** Soren W. Henriksen, Armando Mancini, and Bernard H. Chovitz (editors)

16 **Flow and Fracture of Rocks,** H. C. Heard, I. Y. Borg, N. L. Carter, and C. B. Raleigh (editors)

17 **Man-Made Lakes: Their Problems and Environmental Effects,** William C. Ackermann, Gilbert F. White, and E. B. Worthington (editors)

18 **The Upper Atmosphere in Motion: A Selection of Papers With Annotation,** C. O. Hines and Colleagues

19 **The Geophysics of the Pacific Ocean Basin and Its Margin: A Volume in Honor of George P. Woollard,** George H. Sutton, Murli H. Manghnani, and Ralph Moberly (editors)

# The Geophysics of the Pacific Ocean Basin and Its Margin

**geophysical monograph 19**

# *The Geophysics of the Pacific Ocean Basin and Its Margin*

A Volume in Honor of George P. Woollard

**GEORGE H. SUTTON**
**MURLI H. MANGHNANI**
**RALPH MOBERLY**
**editors**
**ETHEL U. McAFEE**
**associate editor**

**American Geophysical Union**
**Washington, D.C.**
**1976**

1909 K Street, N.W.
Washington, D.C. 20006

*Library of Congress Cataloging in Publication Data*

Main entry under title:

The geophysics of the Pacific Ocean basin and its margin.

(Geophysical monograph ; 19)
Papers presented at a symposium held in Honolulu, Dec. 8–10, 1974.
Bibliography: p.
Includes index.
1. Geophysics—Pacific area—Congresses.
2. Woollard, George Prior, 1908– I. Woollard, George Prior, 1908– II. Sutton, George H., 1927– III. Manghnani, Murli H., 1936– IV. Moberly, Ralph, 1929– V. Series: American Geophysical Union. Geophysical monograph ; 19.
QE350.4.G46 551′.09182′3 75-38816
ISBN 0-87590-019-4

Typeset by
Innovative Media, Inc.
Honolulu, Hawaii

Printed by
LithoCrafters, Inc.
Ann Arbor, Michigan

# Foreword

This volume is dedicated to Professor George P. Woollard, Director of the Hawaii Institute of Geophysics, University of Hawaii, in recognition of his contributions to science and scientific organizations at the local, national, and international level.

The research contributions of Professor Woollard have been wide in scope and prolific in nature. His outstanding contributions to fundamental geophysics are attested to by the award of the William Bowie Medal of the American Geophysical Union in 1973. Perhaps the greatest contributions have been in the areas of standardization of gravity observations throughout the world and of geological interpretation of the gravitational field through interdisciplinary geophysical studies. In 1939, he established the first transcontinental gravity and magnetic traverse across the United States to determine the relationship between these data and regional geologic features. Later, he and his students extended these studies over the entire United States. He was instrumental in launching comparable studies in Canada, Mexico, Central and South America, Africa, India, Spain, Australia, and Antarctica, and in establishing international pendulum-gravity standardization ranges. He was closely associated with the first U.S. program of explosion seismic refraction surveys of the continental margin and deep-sea floor, and later instituted one of the first university refraction programs to study the continental crust and upper mantle. His research in marine acoustics at Woods Hole Oceanographic Institution in World War II remains largely classified.

During his career, Dr. Woollard has developed two centers for geophysical research: the Geophysical and Polar Research Center at the University of Wisconsin, and the Hawaii Institute of Geophysics at the University of Hawaii.

Almost forty students have worked for advanced degrees under Professor Woollard's guidance. These scientists now hold responsible positions in universities, government, and private industry. Many other students have benefited from his membership on dissertation and thesis committees, and still others, both undergraduate and graduate at Princeton, Rutgers, Wisconsin, and Hawaii, have found easy access to his warm and sympathetic counseling.

George Woollard's own training is as strong and varied as his interest in the earth sciences. At the Georgia Institute of Technology he studied civil engineering (BS, 1932) and geological engineering (MS, 1934). At Princeton University's Department of Geology his work was in mineral deposits (MA, 1935) and structural geology (PhD, 1937). Later, he held a post-doctoral position at Lehigh University (1938–1939), and Guggenheim fellowships (1940–41 and 1947–48). He has been awarded honorary degrees by Uppsala University, Sweden, and the University of Wisconsin.

A past president of the American Geophysical Union and a former member of the Executive Council of the Geological Society of America and of the Board of Directors of the American Geological Institute, George P. Woollard has been chairman or member of numerous standing and special committees of the National Academy of Sciences, National Aeronautical and Space Administration, Advanced Research Projects Agency, the State of Hawaii, and the International Union of Geodesy and Geophysics.

# George P. Woollard

On behalf of the Latin American scientific community, it is my great honor and privilege to introduce the remarkable scientific personality of my colleague and friend, Dr. George P. Woollard. We who have had the good fortune to be associated with projects under his leadership are convinced that he is one of the outstanding earth scientists of our time. His colleagues have always identified him with the highest qualities of a scientist, a professor, and a man.

We have gathered this evening to show him our deep appreciation, which has continued to grow over the years, and to honor a universal man, who with tremendous energy has dedicated himself to new frontiers of discovery—to explore, to detect, and to define the physical processes hidden in our Mother Earth. He has always been eager to execute projects with the means and tools at his disposal, giving the opportunity to his partners to work freely and with initiative to study the physical phenomena of the Earth. The more we know about the Earth and reach some of our objectives, the more we are presented with new problems to challenge our minds and to make our lives more fascinating. Thanks to the continuing evolution of new problems, we may say that the scientist of today enjoys his life more than any other scientist has in history. A principal figure is George who, I am sure, takes great satisfaction in seeing around him outstanding scientists of diverse backgrounds who come from different places on earth not only to congratulate him but also to discuss with him some geophysical problems of today. We in turn have come to gain more knowledge from him, and from others at this international symposium.

Now, I will attempt to recount briefly George's Latin American work and its significance to our countries. We, as Latin Americans, have been fortunate because George really has been and is fascinated with the Spanish-speaking Americas, and he has, notwithstanding the language barrier, integrated with us. He has built up good teamwork, with which several of us here tonight are identified. With his talent and thought, he developed a valuable new strategy, a new scientific policy. This policy is not confined to pure science alone; it incorporates a humanistic philosophy in accordance with the spirit and wishes of the Latin American scientists. George Woollard initiated the strategy of scientific cooperation that exists between us today. Even government officials have accepted this new strategy little by little, year by year; and we see, with enormous satisfaction, its influence on research policies. We consider that this breakthrough has eliminated a lot of prejudices that previously existed in conducting scientific research.

It has been 25 years since George with his collaborators initiated gravity measurements in the Americas and worldwide, from pole to pole, including reoccupations, with better and more accurate equipment. This work has served to build a fertile and practical scientific comprehension. Due to his intelligent action, vision, and talent, a series of meetings in Mexico, Panama, Buenos Aires, and Lima, and in the USA and Canada, has resulted in a healthy environment for better understanding, and has led to high-level decisions that have effectively interrelated groups of people of different aims and origins.

George P. Woollard's name is a synonym for integration, comprehension, cooperation, collaboration, respect, and favorable results for the people of Latin America. When we refer to scientific development we therefore think of him as an architect who has built, is building, and will continue to build a variety of scientific programs in the earth sciences which will continue to have impact on future studies of the most exciting features on the Earth, such as the Nazca plate, the Andes, the continental shields, and others. Without a doubt, the scientific achievements in understanding these features are due to a high degree to the brilliant career of this American scientist who, at his silver anniversary of continuous relationship with Latin America, has already accomplished numerous goals.

Finally, on behalf of my colleagues, I want to tell you, George, that we have great admiration for your outstanding contributions to mankind. Our sincere and lively appreciation and thanks to you, your beloved and distinguished wife, and your magnificent children.

Thank you.

—ING. REYNALDO P. SALGUEIRO
Instituto Pan Americano de
Geografía e Historia
La Paz, Bolivia
(Given at the luau, 9 December 1974.)

# Preface

The papers included in this volume were originally presented at the International Woollard Symposium held at the Kahala Hilton Hotel in Honolulu, Hawaii, December 8 through December 10, 1974. The symposium honored Professor George P. Woollard on the occasion of his 66th birthday. It was cosponsored by the University of Hawaii, where Professor Woollard is Director of the University of Hawaii's Institute of Geophysics, by the U.S. Geodynamics Committee, by the Inter-Union Commission on Geodynamics of the International Council of Scientific Unions, and by the following Sponsoring and Organizing Committee:

| | |
|---|---|
| Philip H. Abelson | Everett Dale Jackson |
| L. Thomas Aldrich | Gordon A. Macdonald |
| Orson L. Anderson | Henry W. Menard, Jr. |
| William E. Bonini | Robert P. Meyer |
| Edward C. Bullard | Julio Monges-Caldera |
| Ramon R. Cabre, S. J. | Jack E. Oliver |
| R. M. Demenitskaya | Frank Press |
| Charles L. Drake | Russell W. Raitt |
| John I. Ewing | Donald A. Rice |
| Maurice Ewing (Deceased) | George H. Sutton |
| Richard A. Geyer | Manik Talwani |
| Alberto A. Giesecke | G. B. Udintsev |
| Wytze Gorter (Chancellor, University of Hawaii at Manoa) | Seiya Uyeda |
| | Tjeerd H. van Andel |
| | J. Tuzo Wilson |
| Anton L. Hales | J. Lamar Worzel |
| John Brackett Hersey | |

The Organizing Committee set the theme and scope of the meeting: the symposium was to explore the edge of knowledge, in those fields of science that during the past decade have been the principal concern of Professor Woollard and his associates and students, of the Pacific Ocean basin and its margin. A Program Committee was charged with inviting papers from internationally known scientists, for the following sessions:

Gravity and Geodesy
Seismology
Magnetism
Marine Geology and Tectonics
Volcanology and Petrology
Tectonophysics

Each of these six sessions commenced with a paper giving a general view of that field of science, with reference to the Pacific. The papers that followed described a selection of significant areas of current research. The majority of the 37 papers presented at the symposium are published in this monograph; the others are represented by abstracts.

Members of the Program Committee chaired the sessions. More importantly, they reviewed the manuscripts in their sessions. In nearly every instance, they obtained additional reviews from geophysicists outside the Program Committee. The work of the editors of this volume has thus been eased substantially by the Committee, whom we thank:

| | |
|---|---|
| O. L. Anderson | J. E. Oliver |
| C. L. Drake | M. Talwani |
| J. I. Ewing | G. B. Udintsev |
| E. D. Jackson | S. Uyeda |
| G. A. Macdonald | Tj. H. van Andel |

The success of the December 1974 meeting was due to the knowledge and enthusiasm of all the participants. It is a special pleasure to acknowledge general support for the symposium by the National Science Foundation, the Office of Naval Research, the Office of Marine Coordinator of the State of Hawaii, and the Rama Watumull Foundation of Hawaii. In addition, the National Science Foundation supported the travel expenses of fifty pre-doctoral and post-doctoral students. We editors, and Gordon A. Macdonald and John C. Rose, our co-workers on the Local Committee, thank the management and staff of the Kahala Hilton Hotel for their hospitality and thoughtfulness. The low cost of this volume has been made possible through generous grants-in-aid provided by the Office of Naval Research and the State of Hawaii toward the publication of the volume, and through the editorial services provided by the staff of the Publications Office of the Hawaii Institute of Geophysics. We are grateful to the AGU publications staff; and to Mrs. Ethel U. McAfee, our associate editor, and Mrs. Rita P. Pujalet, her assistant, for transforming manuscript into this book.

GEORGE H. SUTTON
MURLI H. MANGHNANI
RALPH MOBERLY

August 1, 1975

# Contents

## 4. MARINE GEOLOGY AND TECTONICS



## 5. VOLCANOLOGY AND PETROLOGY



## 6. TECTONOPHYSICS

# 1. GRAVITY AND GEODESY

## *1. Gravity Investigations of the Subduction Zone*[1]

J. LAMAR WORZEL

*Geophysics Laboratory, Marine Science Institute, The University of Texas, Galveston, Texas 77550*

The concept of island arcs being arcuate structures with a deep ocean trench seaward along the convex side of the arc, and associated Benioff zones dipping about 45° landward from the deep-ocean trench, does not describe present-day features. This circumstance is undoubtedly the result of describing these features at a time when data were inadequate. Island arcs are global features and hence must be considered on a global scale. Since I cannot improve on the description of these features which I wrote in the Colston Symposium Volume [*Worzel,* 1965*a*], I quote the following from that work:

"Probably the most important fact is that each trench is located at or near the margin of an island ridge or a continent. Some of these trenches are markedly arcuate as, for example, the Marianas, the Aleutian, the Puerto Rico and the South Sandwich Trenches. Others are less markedly arcuate as, for example, the Java, the Ryukyu, the Kurile-Kamchatka and the Middle America Trenches. Still others are nearly linear as, for example, the Tonga, the Philippines, the Chile and the Peru Trenches. There are several trenches that are markedly angular as, for example, the New Britain–Solomons Trench, the connections of the Yap Trench to the Marianas Trench, the northern end of the Peru–Chile Trench and the Banda Sea Trench. In general, one side of a trench faces a deep oceanic basin, but that is not invariably the case as, for example, the Banda Sea Trench, the New Britain–Solomons Trench and the Celebes Trench. On one side of most trenches there is an island ridge or a continent. There are many examples of continental margins and island ridges without trenches. Examples of continental margins without trenches are the eastern coast of the United States and South America, and the southern coast of Asia. Many island-chains, as well developed as those where there are deep sea-trenches, do not exhibit any trench, for example the Tuamotu Archipelago, the Hawaiian Chain, the Marshall Island Chain, the islands of the Palmer Peninsula, the South Shetland Islands and many others. It is concluded that deep trenches have been formed along the margins of continents or island-chains, that their shape is determined by the nearby coastal region and that trenches do not form at the margin of every continent or island ridge."

So it is clear that convex, concave, angular, and straight coastal margins of the continents and of island ridges with associated trenches all exist.

Figure 1 shows the structure section for the Puerto Rico trench derived by *Talwani et al.* [1959*a*] on the basis of seismic and gravity information. The gravity fits the transition zone, in this case for an island ridge. The seismic data confirm the structure determined by the gravity data where seismic data have been observed. The curve at the bottom shows the exceedingly fine fit of the calculated gravity curve to the observed values of that section.

Figure 2 is a section from the paper by *Talwani et al.* [1961] showing a similar structure for the Tonga trench. The dashed gravity curve at the bottom fits the observed pendulum observation points in this area and corresponds with the dashed curves of the structure section above, which is seen to fit the seismic refraction data of *Raitt et al.* [1955] quite satisfactorily. Once again a downdropped block satisfies the gravity curve and all of the seismic data quite satisfactorily.

Figure 3 shows very simplified models of the transition zone between a continent and ocean crust or between an island ridge and an ocean crust with a downdropped block in a part of the transition zone [*Worzel,* 1965*a*]. These are three separate suggested crustal cases which create

[1]Contribution No. 69 of the Galveston Geophysics Laboratory of the University of Texas Marine Science Institute.

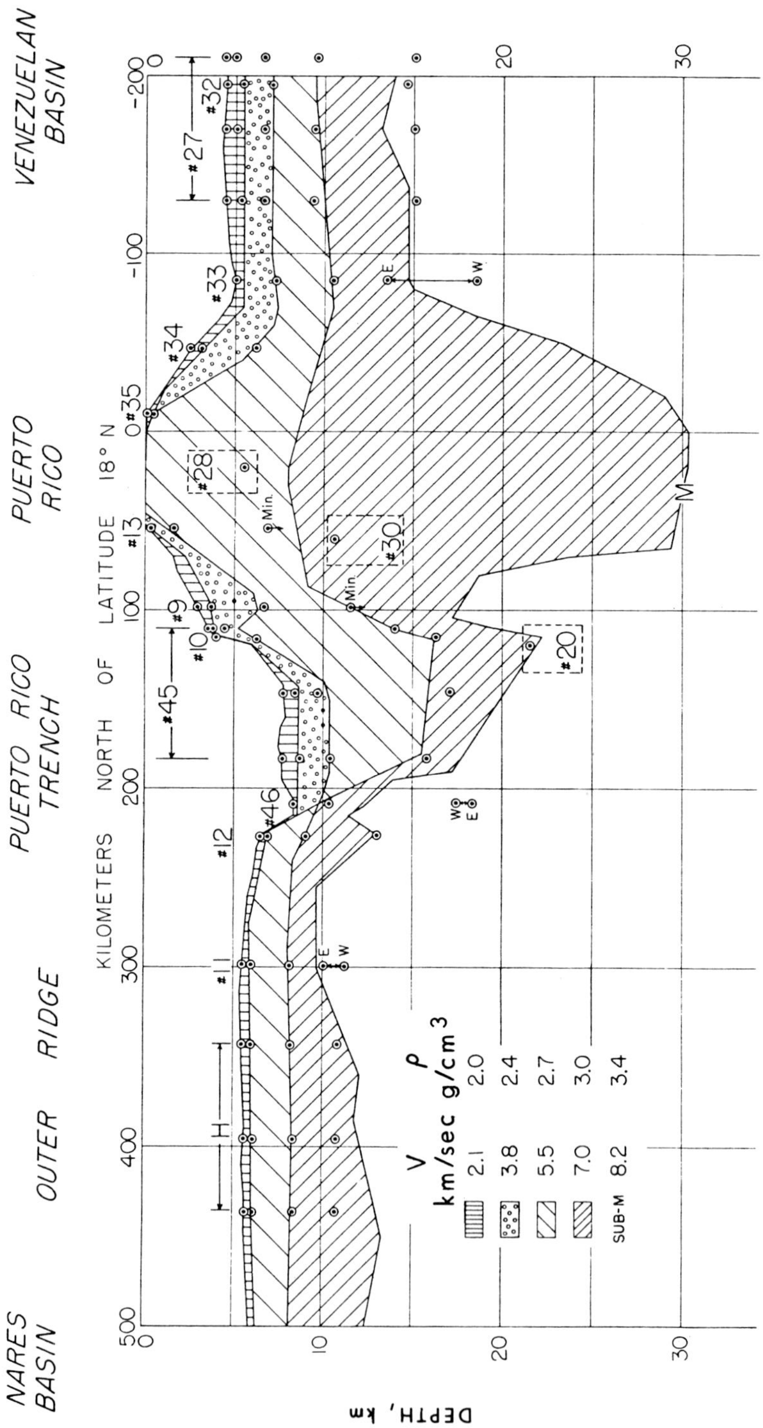
VENEZUELAN BASIN
PUERTO RICO
PUERTO RICO TRENCH
OUTER RIDGE
NARES BASIN
KILOMETERS NORTH OF LATITUDE 18° N
DEPTH, km
V km/sec
ρ g/cm³
2.1 2.0
3.8 2.4
5.5 2.7
7.0 3.0
SUB-M 8.2 3.4
M
Min.
E
W
H
#9
#10
#11
#12
#13
#20
#27
#28
#30
#32
#33
#34
#35
#45
#46

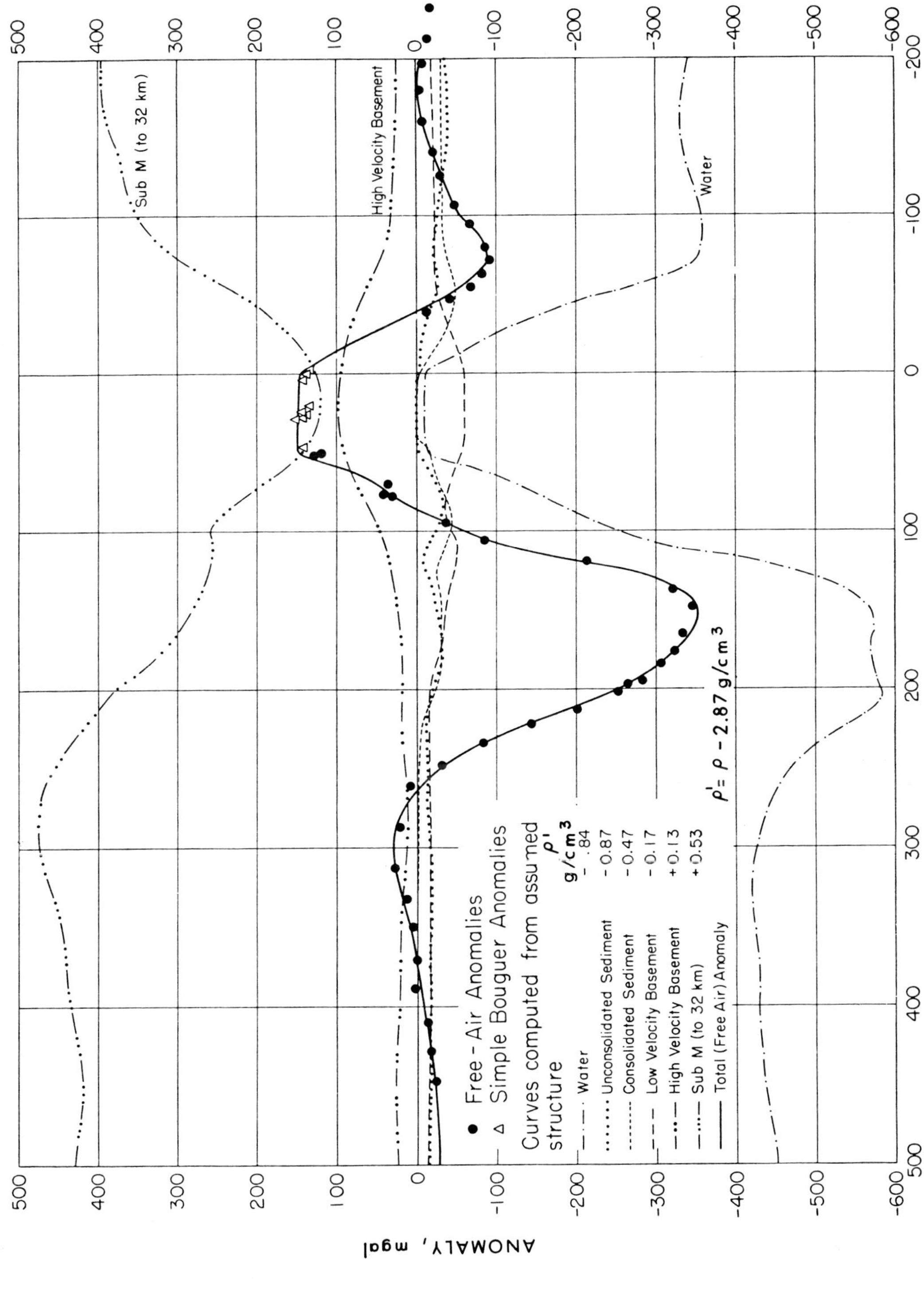


Fig. 1. Structure of the Puerto Rico trench region derived to fit the seismic refraction and gravity data. The solid curve in the bottom of the figure is the gravity anomaly computed for the section shown above, and the points are the observed data [after *Talwani et al.*, 1959*a*].

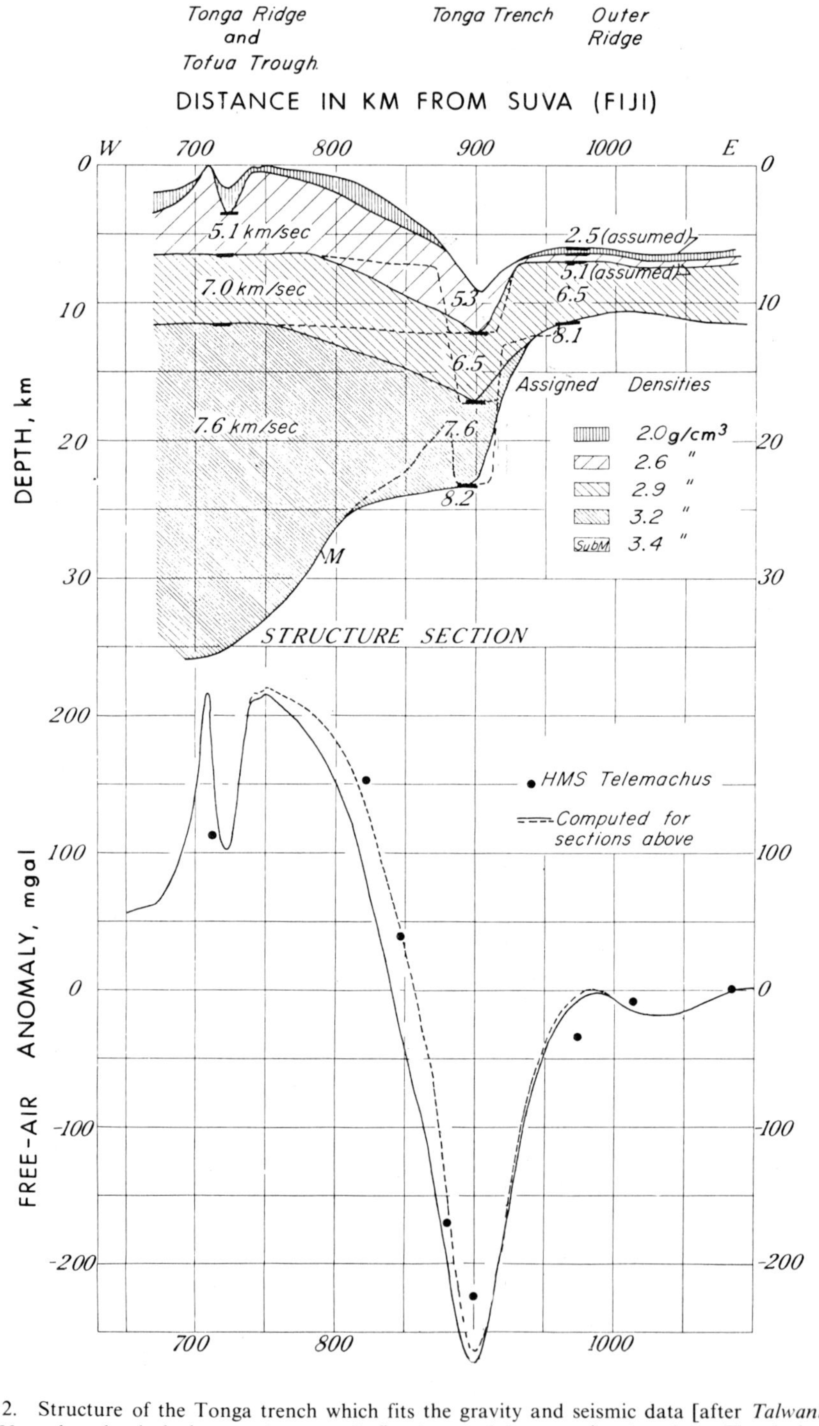


Fig. 2. Structure of the Tonga trench which fits the gravity and seismic data [after *Talwani et al.*, 1961]. Note that the dashed curve (computed) fits the pendulum gravity data best. The dashed layer boundaries in the structure section at the top correspond with the dashed computed gravity curve.

gravity anomaly curves that are quite similar to those observed in the various trenches.

Figure 4 shows a comparison of a gravity profile computed using a mean crustal density (the solid line) to fit the observed gravity curve on a profile crossing the Aleutian trench adjacent to Unimak Island [*Worzel,* 1965*b*] with model 5 (the dashed curves) shown in Figure 3. Note especially the upward bulge of the mantle-crust boundary required to fit the positive anomaly in the gravity curve near kilometer 290. The very close fit of this section to the simplified crustal model of the trench is impressive.

Figure 5 shows the case of the Puerto Rico trench of Figure 1 compared to model 4B which is a slight adaptation of model 4 of Figure 3. The adaptation is a 3-km rather than 2-km depression of the transitional crustal block. Once again the simplified model quite satisfactorily fits the structure determined from gravity and seismic data in the trench.

Figure 6 shows model 7 compared with the Tonga trench structure section of Figure 2. The downdropped section has been modified from a model 6 concept to a narrow portion of the outer part of the transition zone. Note that the fit of the data is quite satisfactory with the observed data in the Tonga trench case.

It can be seen from these sections that the large, negative gravity anomaly associated with the trench is fully accounted for by a downdropped crustal section that fits the observed seismic data.

Figure 7 shows a crustal section from *Tsuboi* [1957]. It is evident that a similar downdropped block in the trench region, resulting from the altogether different form of analysis used by Tsuboi, explains the gravity data that were observed in the Vening-Meinesz profile 17.

*Grow* [1973] reported on the crustal and upper mantle structure of the Aleutian arc near the island of Amlia, 350 miles west of our section near Unimak shown in Figure 4. Our structure section calculated from gravity data and topography alone has been superimposed on their Figure 7 for comparison (Figure 8). The heavy line at the bottom is the bottom of the crustal structure, and the heavy line at the top is the topography on our section. The narrower lines are the section reported by Grow. In our section we assumed that the island ridge structure was continuous behind the islands themselves as it is in our section, the Bering Shelf. The upper heavy line near the bottom on the right side of the figure shows the variation that would be required in our section in order to allow for the water depth indicated in the section by Grow. Allowances for the sedimentary section indicated by Grow would have made the crust mantle boundary fit almost perfectly. This simplified gravity calculation is quite a good fit to the crustal structure which has been given in more detail by the refraction data reported by *Grow* [1973]. In fact wherever there are actual seismic data our section fits them within about a kilometer, except in the Bering Sea region where the large accumulation of sediment would cause a significant change from the simplified gravity computations. Note that all of the volcanoes occur in the region just above the point where the upward deflection of the mantle crust boundary occurs on our structure.

Figure 9 shows Grow's interpretation of a subduction zone with our Unimak simplified crustal section from 350 miles east superimposed on it. Both sections fit the gravity observations quite satisfactorily. The gravity anomalies for both sections are substantially the same (Figure 10), with the anomaly beneath the terrace sediments being about 30 mgal less and that beneath the island ridge about 30 mgal greater for Grow's section. These small differences could readily be explained by the 350-mile geographic separation of the two sections. For Grow to fit the anomalies, Grow needs to assume mantle densities of .05 g/cm$^3$ lower for the mantle from 80 km to about 260 km deep, to introduce reduced density mantle beneath the island ridge to depths of at least 30 km, and to assume a basalt to eclogite transformation within layer 3 at a depth of 30 km. The consensus of most authors is that such a transformation would not occur until depths of at least 60 km were reached. It seems unjustified to make arbitrary density changes within the mantle without any data base, if the anomalies can be adequately fitted by crustal structures within the constraints of the existing data.

At the bottom of Figure 11 is a line drawing of a section obtained by *Beck and Lehner* [1974] from multi-channel seismic data in the Java trench region. It has been suggested by several workers that the oceanic basement surface indicated in Beck and Lehner's drawing represents proof of the downgoing limb of the lithosphere in the subduction zone. There is, however, an alternative hypothesis. If we consider that the sediment on top of the oceanic basement surface was deposited on that surface, and that isostatic adjustment took place, we can reconstruct the original surface by removing the sediment and returning the basement surface to isostatic equilibrium, as has been done at the top of Figure 11. As can

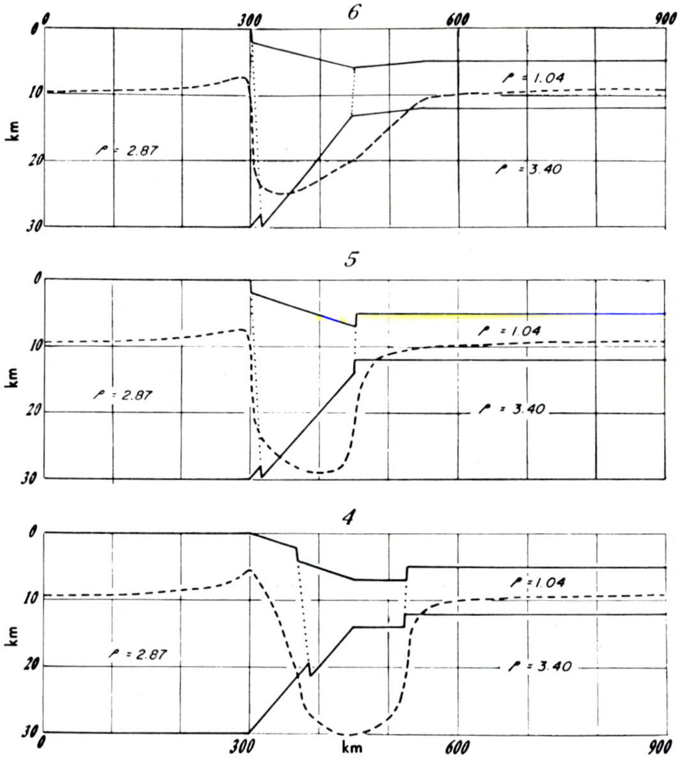


Fig. 3. Suggested crustal sections in transition zones at continental and island margins which produce gravity anomalies similar to those observed in trenches. The downdropped block is displaced downward by 2 km in each case [after *Worzel,* 1965*a*].

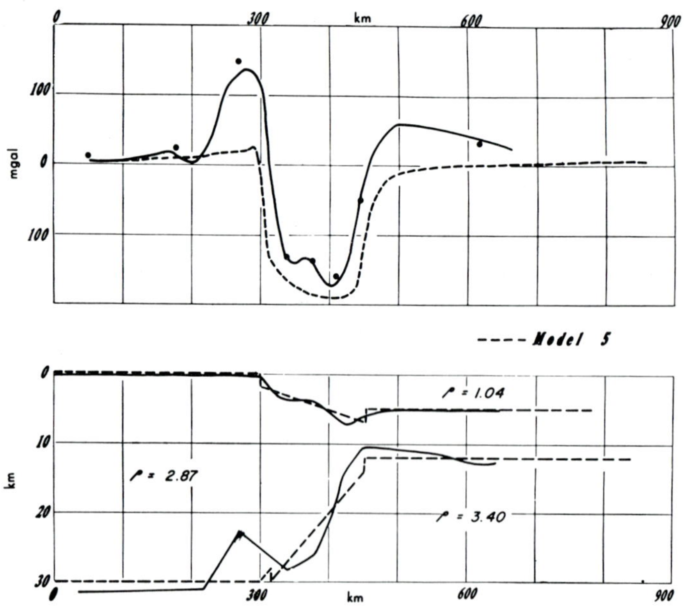


Fig. 4. Comparison of the model 5 crustal section shown in Figure 3 with the Aleutian trench section near Unimak Island. The dashed lines are those for the model and the solid lines are those derived from the observed gravity profile. The points are the pendulum gravity observations [after *Worzel,* 1965*a*].

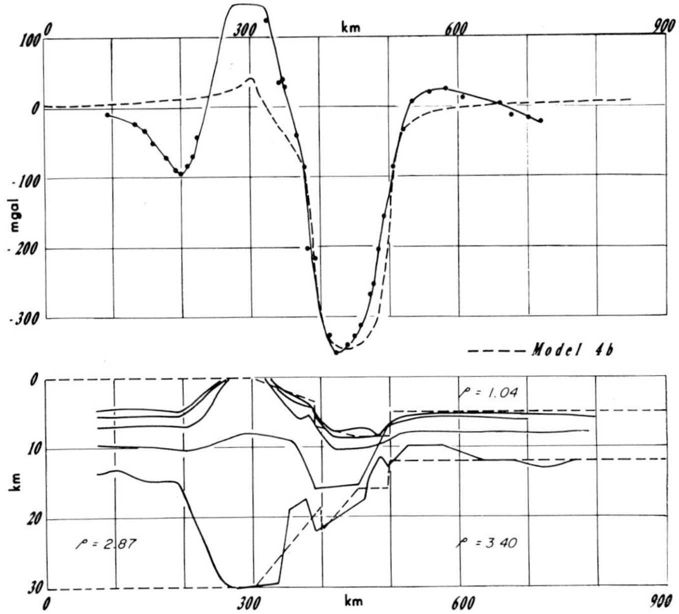


Fig. 5. Comparison of Puerto Rico trench section with model 4B, a variation of model 4 of Figure 3. The modification is an increase of the displacement of the downdropped block beneath the trench from 2 km to 3 km [after *Worzel,* 1965*a*].

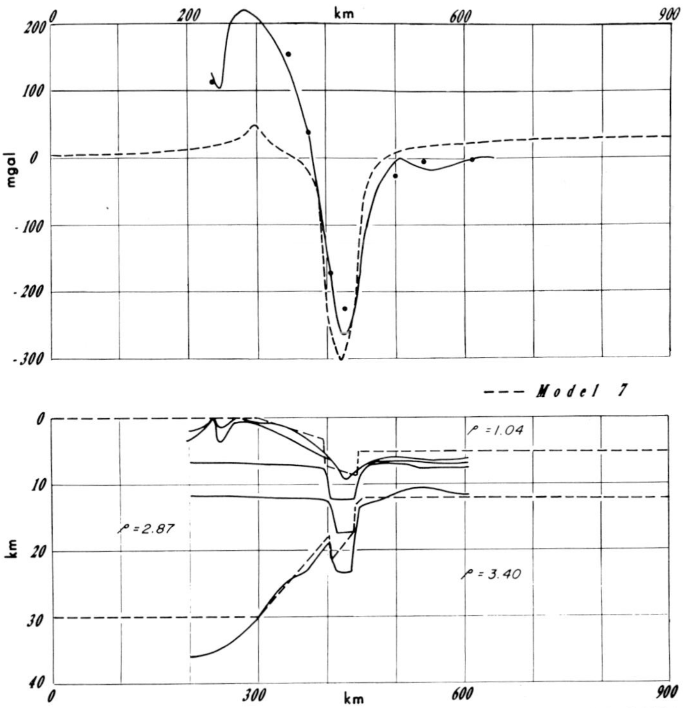


Fig. 6. Comparison of the Tonga trench section of Figure 2 with model 7. Model 7 is a modification of model 6 of Figure 3 where the downdropped crustal block in the trench is made narrower and placed at the seaward portion of the transition zone. The gravity curve (solid line) at the top is calculated to fit the crustal section (solid lines) at the bottom (for seismic velocities and densities of the layers of the solid line at the bottom see *Worzel* [1965*a*]).

be seen, this returns the basement surface to a nearly horizontal surface. Therefore the alternative explanation—that this is simply a loading of the crust by sediment deposition with near-contemporaneous isostatic adjustment—is just as good an interpretation of the observed section. Once again it appears that our consideration of multiple hypotheses first proposed by T. C. Chamberlin in 1897 is not receiving the consideration it is due. In a later epoch, *Bowen* [1948] called attention to this neglect by the earth-science community in his paper entitled: The granite problem and the method of multiple prejudices, published in *The Origin of Granite,* Memoir 28 of the Geological Society of America. It is well to reread this and refresh our memories from time

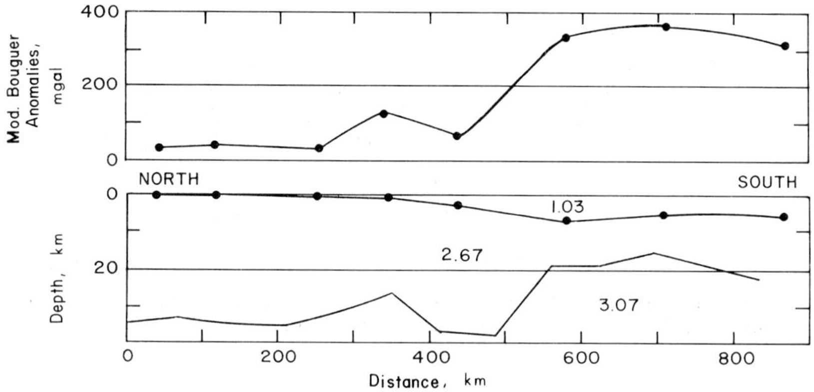


Fig. 7. Crustal section computed to fit Vening Meinesz profile 17 in the Dutch East Indies [after *Tsuboi,* 1957].

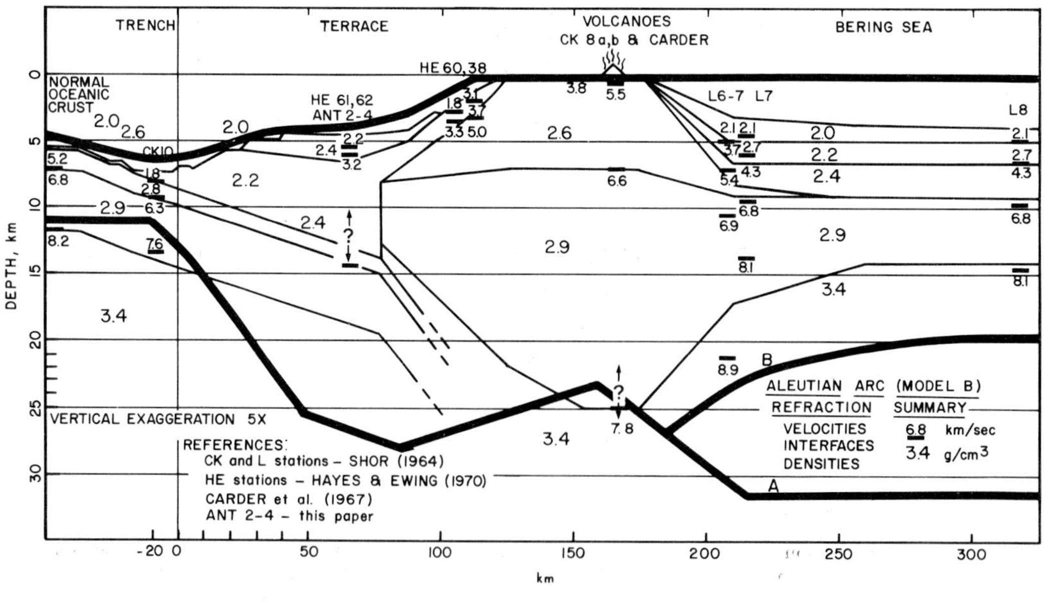


Fig. 8. Crustal section derived from seismic refraction data near Amlia. Our section near Unimak 350 miles to the east is superimposed (heavy lines) [adapted from *Grow,* 1973].

to time.

In Figure 12 by averaging the best data available on continents and ocean basins, *Worzel* [1974] has recently determined the average crustal structure for a truly basin-type oceanic crust and the low elevation continental crustal structure. On the right-hand side I have calculated what I call the equivalent ocean crust, that is, one without sediments, or water considered as part of the crust and with the two remaining crustal layers averaged into one layer of uniform density, 2.86 g/cm$^3$. This equivalent crust will be used below.

Figure 13A shows the density versus depth curves published by *Dorman* [1969], *Dziewonski* [1971*a*], *Press* [1968, 1970*a*], and *Wang* [1970] for the oceanic region. *Dorman* [1969] summarized the data from the shear velocity derived from Love and Rayleigh wave observations. Densities were compared for his velocity depth curve from the *Nafe and Drake* [1963*a*] velocity–density relationship. This portion of the curve is substantially from F. J. Birch's data on metamorphic and igneous rocks. The other authors determined their curves by inversion from Love and Rayleigh wave data, from free oscillations of the earth, and from the mass and moment of inertia of the earth. *Press* [1968, 1970*a*] gave a range of curves as his

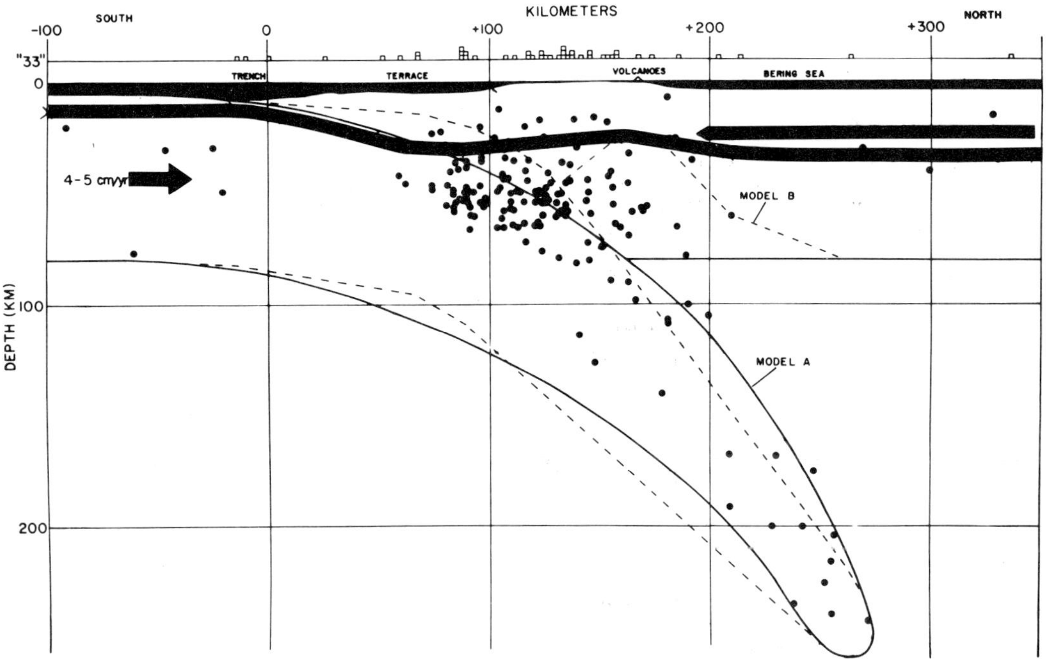


Fig. 9. Proposed subduction zone of lithosphere into the asthenosphere. Our superimposed crustal section is shown by the heavy lines. Note that the whole gravity anomaly is accounted for by the crustal section [adapted from *Grow*, 1973].

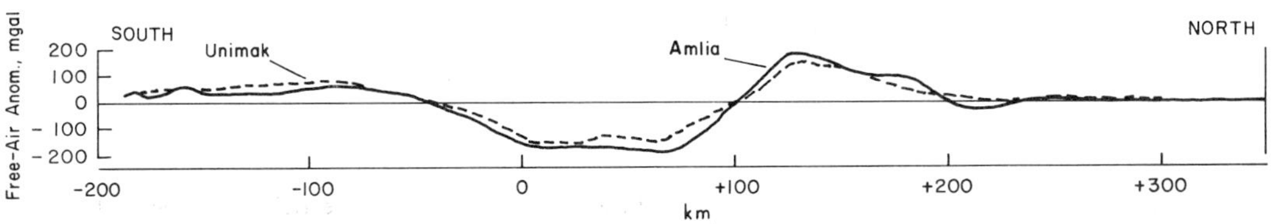


Fig. 10. Gravity profiles for the sections shown in Figure 9.

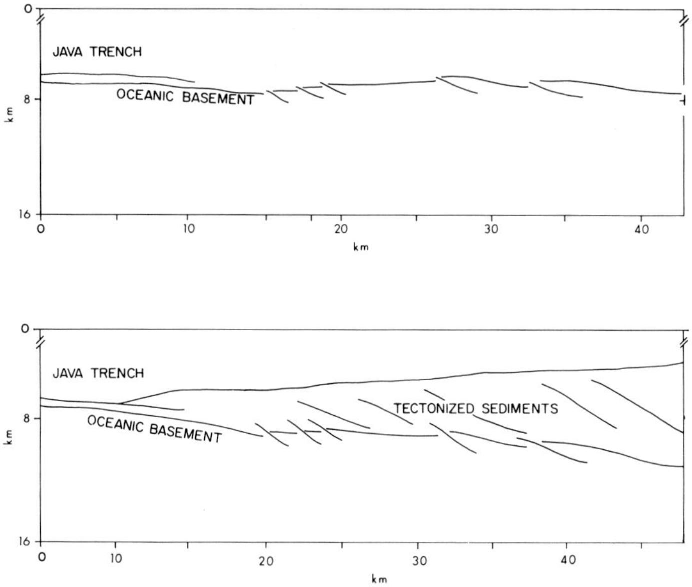


Fig. 11. (Bottom) Line drawing of a multichannel seismic section across the Java trench [after *Beck and Lehner,* 1973]. (Top) The same section with the tectonized sediments removed and returned to isostatic equilibrium. Densities (g/cm$^3$) used were 1.03 for the water layer, 2.30 for the tectonized sediments, 2.84 for the oceanic basement, and 3.40 for the mantle.

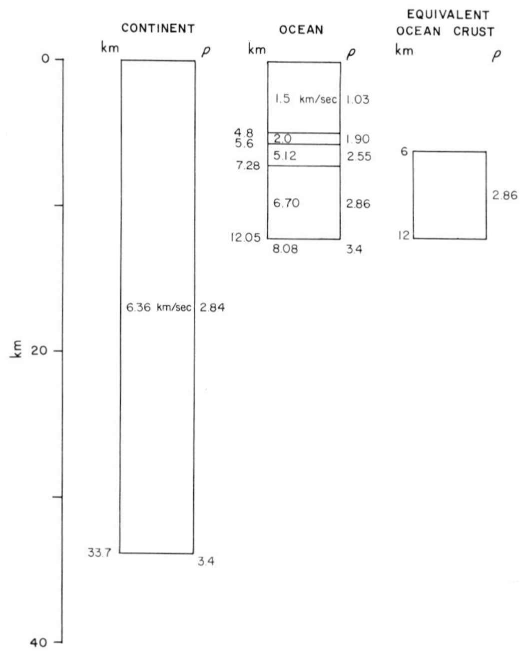


Fig. 12. Standard continent and ocean crusts [after *Worzel,* 1974]. At the right is the equivalent ocean crust, i.e., one of uniform density and of the same mass as the two deepest layers of the ocean section.

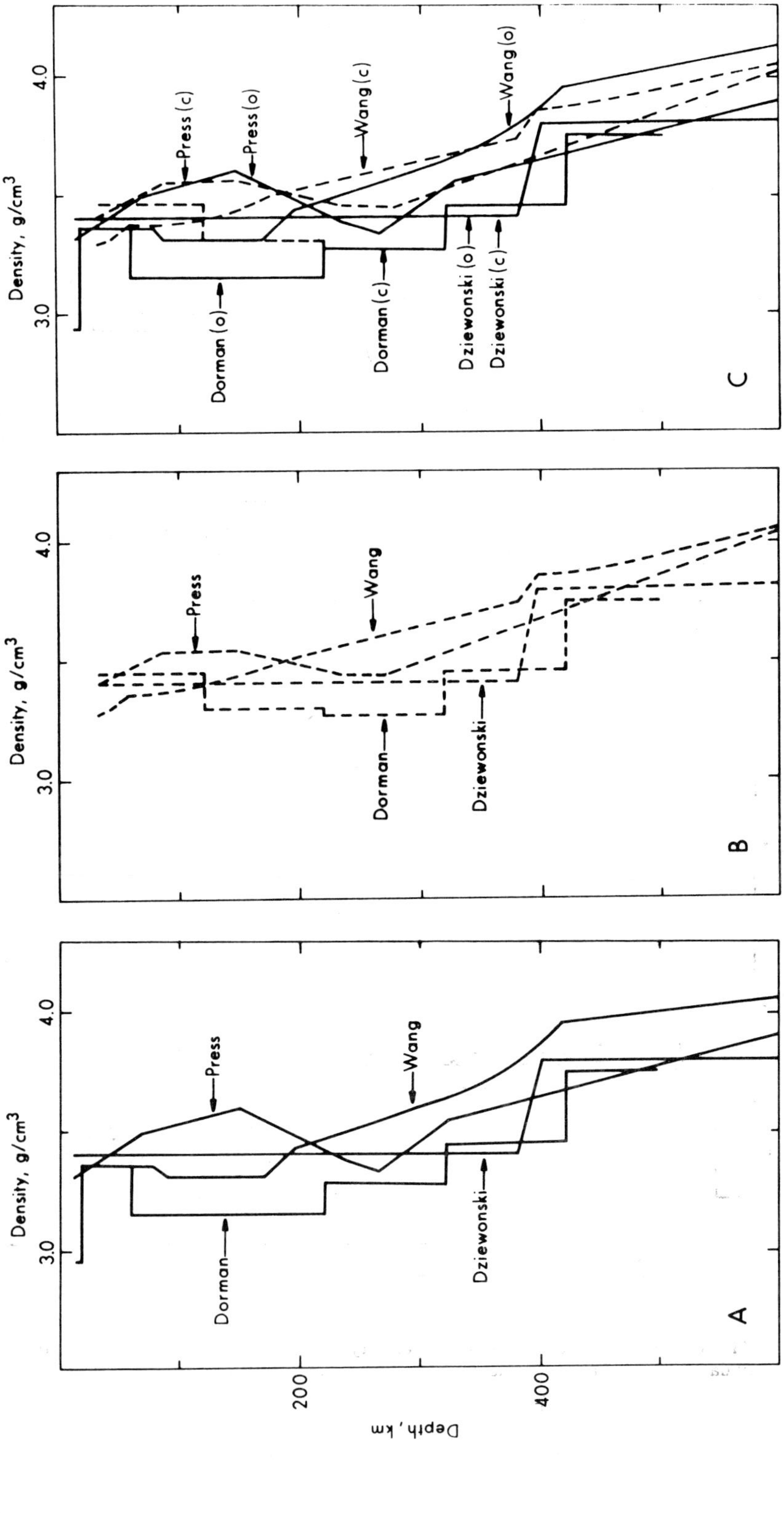


Fig. 13. A. Density depth curves for oceans obtained principally from earthquake seismology. B. Density depth curves for near-sea-level continental areas obtained principally from earthquake seismology. C. The curves of (A) normal ocean and (B) normal continent plotted together.

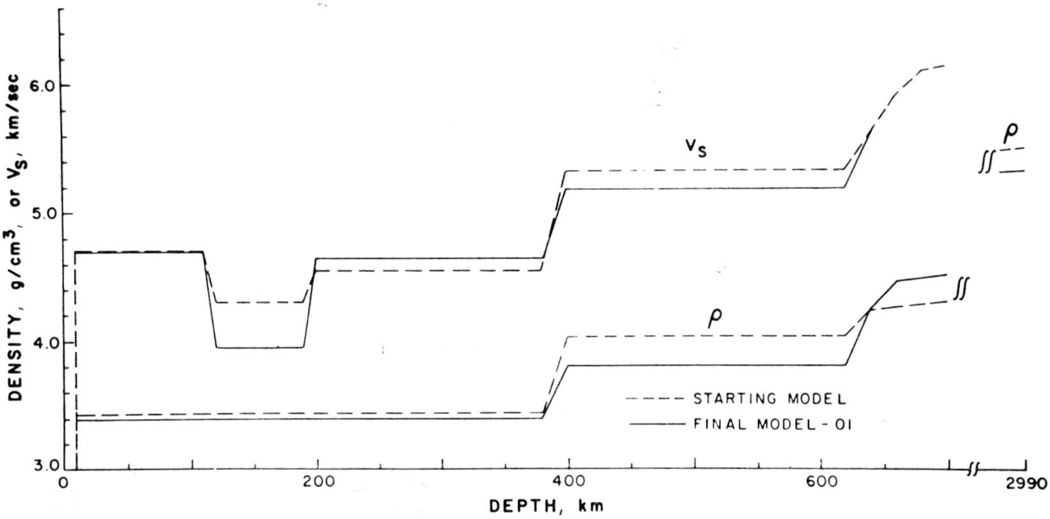


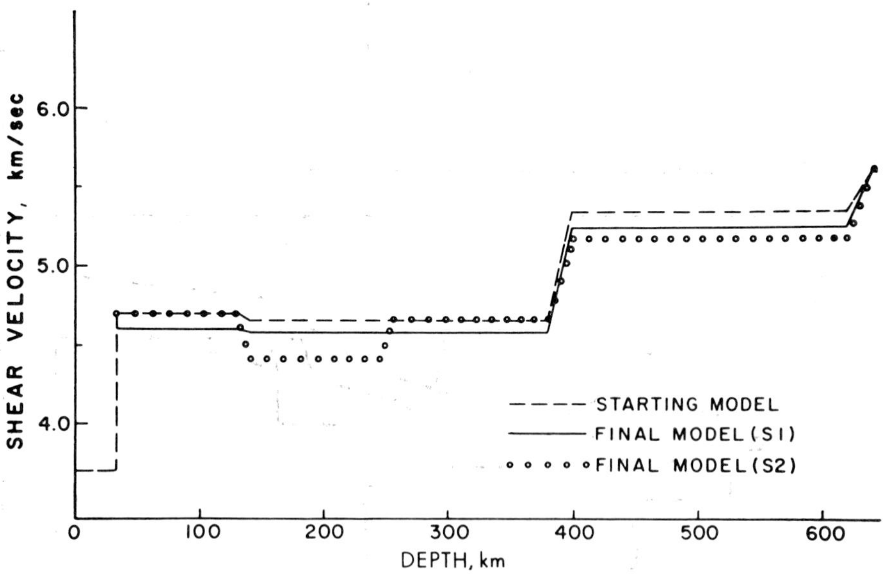


Fig. 14. (Top) Fit of shear velocity data and density with depth for oceanic data. (Bottom) Fit of shear velocity data with depth for continental data [after *Dziewonski,* 1971*a*].

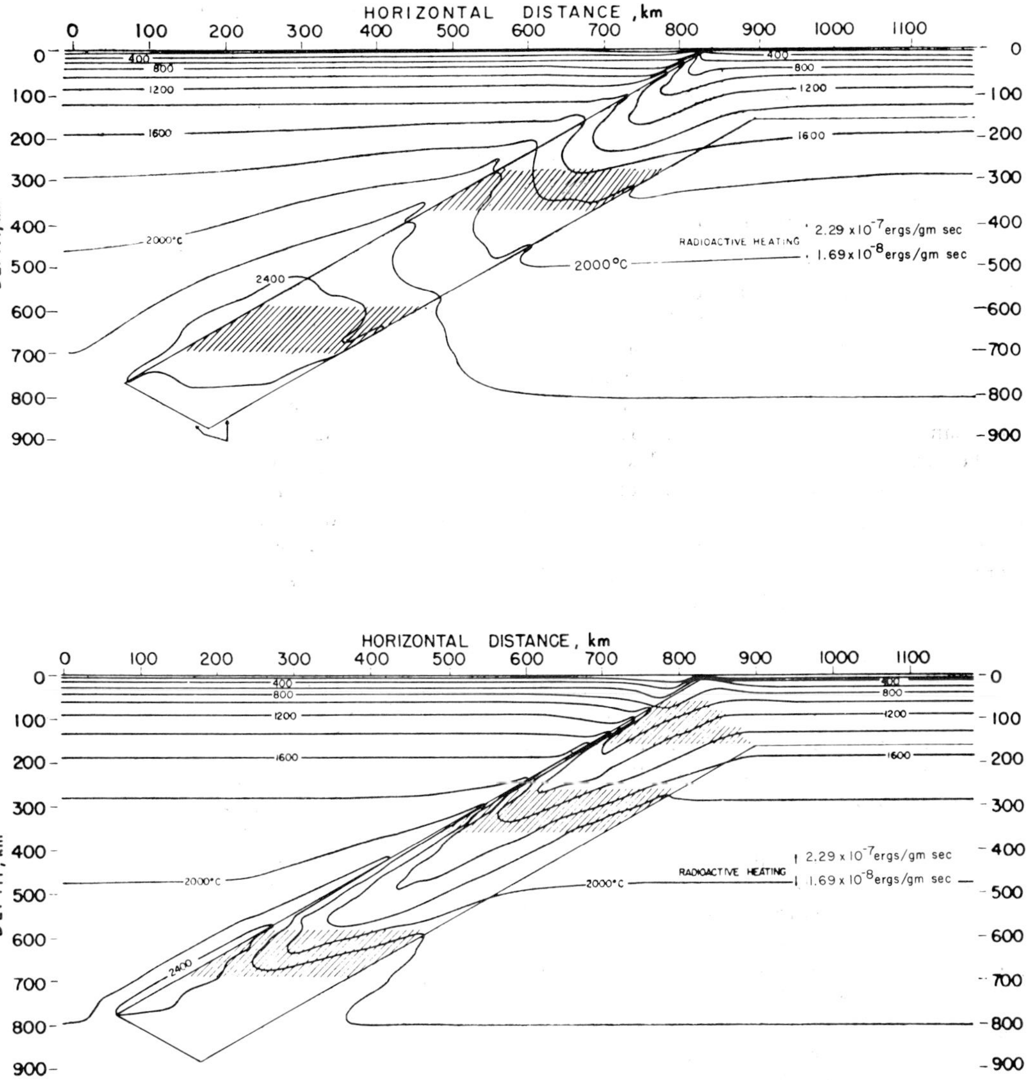


Fig. 15. (Top) Temperature-depth section for subducting lithosphere for spreading rate of 1 cm/yr. (Bottom) Temperature-depth section for subducting lithosphere for spreading rate of 8 cm/yr [after *Minear and Toksöz,* 1970].

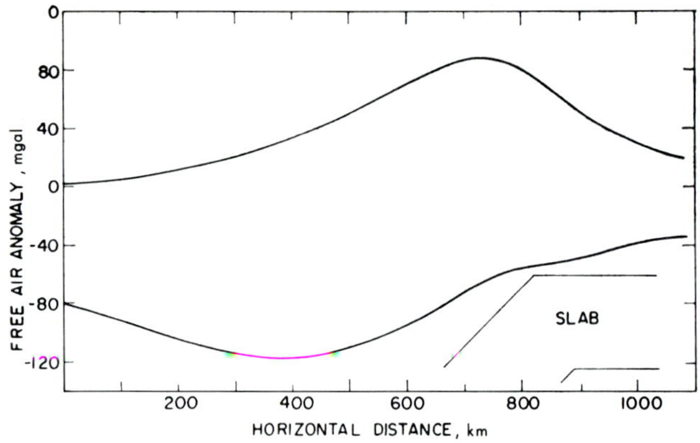


Fig. 16. Free-air gravity anomalies calculated for the cases of Figure 15. The bottom curve is for the 1 cm/yr spreading rate, the top for the 8 cm/yr spreading rate [after *Minear and Toksöz,* 1970].

Monte Carlo solutions for each of several regions. The curve I present here is the mean of the solutions which he obtained for the oceanic region.

Figure 13B shows the similar analysis for the subcontinental crust. *Press's* [1970*a*] solution is the mean of the several curves that he obtained for the shield regions. *Wang's* [1970] is pure path data (that is, data that include only one type of crustal structure in the seismic wave path), and *Dziewonski's* [1971*a*] curve is for the minimum number of variables necessary to fit the data. For further delineation of these terms I refer the reader to the original papers.

Figure 13C shows all of the curves plotted together. The minor differences between the oceans and the continents shown by any one author is impressive. Obviously there are much larger differences from author to author for the two different regions than there is within a single author's paper.

In order to make computations of the gravity effects of a subducting zone it is necessary to assume a density structure for the lithosphere and upper asthenosphere. Since Dziewonski's curve is closest to the mean of all of the solutions shown in Figure 13C, we have chosen to use his density depth curves.

Figure 14 is a figure adapted from *Dziewonski's* [1971*a*] Figure 2 and Figure 5. At the top are shown the shear velocity and density versus depth of the oceanic data of the best mean structure that fits the data adequately. The final model is the 01 model that Dziewonski chose. At the bottom are similar data for the continental case using the same density depth curve as the model 01 shown at the top. In the S1 model, the shear velocity of the overlying crust was held at 4.59 km/sec, and in the S2 model it was held at 4.70 km/sec. Both S1 and S2 fit the data with little choice between them.

Figure 15, adapted from Figures 10 and 11 of *Minear and Toksöz* [1970], shows calculated isotherms for lithosphere movements of 1 cm/yr at the top and 8 cm/yr at the bottom. For the assumption that the whole lithosphere was mantle material, they computed the resulting gravity curves shown in Figure 16. From these it is seen

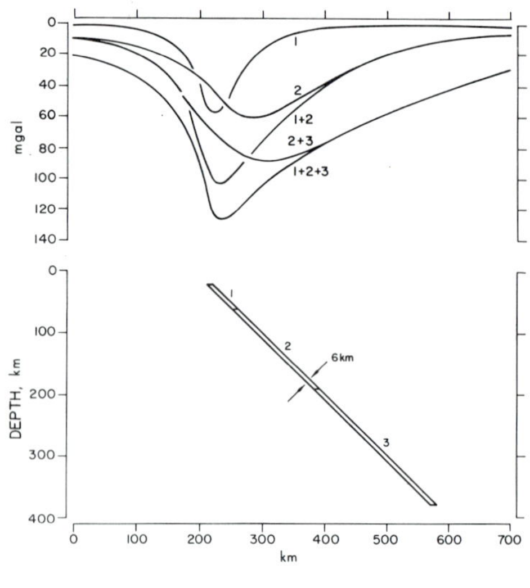


Fig. 17. Gravity anomalies that would be observed for various depths of subducting oceanic crust into the upper mantle.

that a gravity anomaly should be observed from just the changes of temperature of the mantle material, that is, for the 1 cm/yr spreading rate, a gravity anomaly of the order of −120 mgal should be observed about 400 km landward of the seaward trench wall and that a negative anomaly in excess of −40 mgal should extend at least 1000 km across the structure. For the 8 cm/yr rate, the maximum anomaly is about +85 mgal, peaking about 150 km shoreward from the seaward wall of the oceanic trench and remaining in excess of +20 mgal over a region of 600 km in width. Since this model neglects the oceanic crust which has been displaced into the mantle, we have calculated the gravitational effect of the crust under a number of different circumstances which are shown in Figure 17. The curves are calculated using the density differential of 0.54 obtained as a difference from our mean crustal density of 2.86 $g/cm^3$ and that for the mantle obtained from Dziewonski of 3.40 $g/cm^3$. Curve 1 is for the block of crustal material extending downward at a 45-degree angle from 20-km depth (depth of the base of the crust at the trench inner wall) to a depth of 60 km. It has been the consensus of most writers that the depth of the rigid crust of the earth is 60 km. Note Grow chose to calculate his gravity anomalies as if the basalt-eclogite transformation took place at 30 km while stating "Green and Ringwood (1967) experimental studies indicate that basalts invert into the high density mineral assemblages of eclogite at pressures between 10 and 20 kb (30–60 km depth)." He anticipates that the effects of tectonic 'overpressure,' low temperature, and shearing would cause the transformation at shallow depth.

Curve 2 is for block 2 which extends from 60 to 190 km, about the depth of the low-velocity layer (Dziewonski model S2). Curve 1 + 2 is the sum of these two blocks. Curve 1 + 2 + 3 is the curve if the crust should persist to a depth of 380 km without change of phase or melting and dissemination into the asthenosphere. Negative gravity anomalies up to 60 to 100 mgal extending over ranges of hundreds of kilometers centered at about 50 km landward from the inner trench wall would be observed from this source alone.

Since in the region of the deep trenches the gravity anomalies observed are in close agreement with those that would be observed from the crustal structure determined by seismic reflection and refraction data, the postulated subduction zone would require negative gravity anomalies of the order of 60 to 100 mgal, in some cases as large as −200 mgal, over a distance of 600 to 800 km. Such extensive, large negative gravity anomalies have not yet been observed. It must therefore be concluded that either the postulated subduction zone does not exist, or the concept has to be modified to include a substantial mass of high-density mantle material paralleling and very close to the subducting lithospheric layer.

# 2. *Gravity Field of the Northwest Pacific Ocean Basin and Its Margin*[1]

A. B. WATTS, M. TALWANI, AND J. R. COCHRAN

*Lamont-Doherty Geological Observatory and Department of Geological Sciences, Columbia University, New York, New York 10027*

*Abstract.* Surface-ship and pendulum gravity measurements have been combined in new free-air gravity anomaly maps of parts of the northwest Pacific Ocean basin and its margin. The most prominent features of these maps are the broad belts of positive anomalies which occur seaward of deep-sea trenches and landward of island arcs at the margins of the northwest Pacific basin and the broad belts of positive and negative anomalies which flank the eastern end of the Hawaiian ridge in the interior of the basin. The positive anomalies seaward of trenches can be easily isolated from the longer wavelength regional gravity field of the northwest Pacific upon which they are superimposed. These studies show that (1) positive free-air gravity anomalies of about +40 to +50 mgal occur seaward of trenches and (2) negative residual gravity anomalies of about −10 mgal occur over the Pacific basin, except over Hawaii and the Gulf of Alaska where they are positive.

The development of improved versions of spring-type gravimeters [*Graf and Shulze,* 1961] has greatly enhanced the measurement of gravity at sea. Gravity measurements can now be routinely obtained on research vessels in a wide variety of sea conditions. By combining surface-ship measurements with earlier measurements obtained with pendulums [*Vening Meinesz,* 1941, 1948; *Worzel,* 1965*b*] it is now possible to construct gravity anomaly maps for large regions of the world's ocean basins. These maps form an important basis for geological, geophysical, and geodetic studies.

Studies of the Earth's gravity field have long been recognized as one of the principal means of deducing Earth structure in oceanic regions. In the past it was customary to use gravity anomalies, with other geophysical data, to deduce crustal sections across geological features [e.g., *Talwani et al.,* 1961; *Worzel,* 1965*b*]. Crustal sections are important because they give an idea of the mass distribution which can explain observed gravity anomalies.

Since the development of the concept of plate tectonics there have been two additional approaches to the interpretation of marine gravity anomalies that have proved useful. The first approach, based on early studies by *Gunn* [1943] and *Vening Meinesz* [1941], is to use relatively short-wavelength gravity anomalies over volcanic islands to deduce information on the long-term physical properties of lithospheric plates [*Walcott,* 1970*a, b; Watts and Cochran,* 1974]. The second approach is to use the correlation of relatively long-wavelength gravity anomalies and topography, which exists in some parts of the oceans, to deduce information on the forces which operate on the plates and which may ultimately drive them.

The main features of the Earth's gravity field have now been determined with some confidence up to harmonic degree 16 from combination solutions using both satellite and surface data [*Gaposchkin and Lambeck,* 1971; *Richardson and Lerch,* 1974]. These solutions resolve information of the Earth's gravity field with wavelengths longer than a few thousands of kilometers. There is currently much interest in using long-wavelength gravity anomalies to deduce inhomogeneities deep in the Earth, particularly beneath the lithosphere. However, any attempt to understand contributions to the Earth's gravity field

[1]Lamont-Doherty Geological Observatory Contribution No. 2220.

from heterogeneities beneath the lithosphere is complicated by the contribution to the gravity field of the much larger effects from sources within the lithosphere.

The purpose of this paper is to present new gravity data for parts of the northwest Pacific Ocean basin and its margin. The new data are presented as free-air gravity anomaly maps for parts of the Aleutian island arc–trench system, the Kuril island arc–trench system, the Izu–Bonin island arc–trench system and for the region around Hawaii. In addition, long profiles of free-air anomaly and topography are presented to examine the correlation which exists in parts of the northwest Pacific between long-wavelength gravity anomalies and topography. These maps have outlined a broad belt of positive anomalies seaward of trenches which have been interpreted as caused by an upward flexure of the Pacific lithospheric plate as it approaches a deep-sea trench. The positive anomalies can be easily isolated from the longer wavelength regional gravity field over the northwest Pacific Ocean. The significance of this regional field and its possible relation to sources deep in the Earth, particularly beneath the lithosphere, are discussed. We base this study mainly on data obtained on cruises of Lamont-Doherty vessels in the northwest Pacific.

## GRAVITY MEASUREMENTS

*Previous studies.* The earliest gravity measurements in the Pacific were made by Hecker [*Heiskanen,* 1936] in the region of the Tonga trench. However, reliable measurements were not made in the northwest Pacific until 1926 during the trans-Pacific traverse using the Vening Meinesz pendulum apparatus on board a submarine [*Vening Meinesz,* 1941, 1948]. The traverse was oriented WNW-ESE and crossed the Hawaiian ridge near Oahu. The Vening Meinesz pendulum apparatus continued to be used in the northwest Pacific until about 1954 [*Matuyama,* 1934; *Vening Meinesz,* 1941, 1948; *Worzel,* 1965*b*]. In addition, a Russian pendulum apparatus [*Gainanov,* 1955] obtained early measurements in the region of the Sea of Okhotsk and the Kuril arc.

Since 1959 gravity measurements have mainly been obtained with surface-ship gravimeters aboard U.S., Japanese, Russian, and Canadian research vessels. At least 100,000 surface ship measurements have now been obtained in the northwest Pacific.

Surface-ship and pendulum gravity measurements have been compiled in free-air gravity anomaly maps for part of the northwest Pacific. Maps of the margins of the Pacific have been compiled for the Central Aleutian arc [*Malahoff and Erickson,* 1969], Bering Sea [*Kienle,* 1971; *Pratt et al.,* 1972] and the sea area around Japan [*Tomoda,* 1973*a*; *Segawa and Tomoda,* this volume]. In addition, maps of small parts of the Pacific basin near Hawaii have been compiled [*Rose and Belshé,* 1965; *Dehlinger,* 1969].

Gravity measurements have also been presented as free-air anomaly profiles of the main geological features of the northwest Pacific. These include profiles of the margins of the Pacific: the Aleutian arc [*Hayes and Ewing,* 1970; *Grow,* 1973; *Watts and Talwani,* 1974], the Kuril arc [*Gainanov,* 1955; *Kogan,* 1975], the Izu–Bonin and Mariana arcs [*Segawa,* 1970; *Hayes and Ewing,* 1970; *Watts and Talwani,* 1974] and the Philippine and Ryukyu arcs [*Hayes and Ewing,* 1970; *Watts and Talwani,* 1974]. They also include profiles of the interior of the Pacific basin: Shatsky rise [*Segawa,* 1970; *Tomoda,* 1973*b*], Emperor seamount chain [*Segawa,* 1970; *Watts and Cochran,* 1974] and the Hawaiian ridge [*Worzel,* 1965*b*; *Strange et al.,* 1965; *Watts and Cochran,* 1974]. In addition, *Uspensky* [1972] and *Khan et al.* [1971] have presented long free-air anomaly profiles which cross both the Pacific Ocean basin and its margin.

*Data collection and reduction.* We have used in this study data made available to us up to 1973. A summary of all available surface-ship and pendulum measurements is presented in Figure 1. The data are presented as free-air anomaly maps of the region and as projected profiles along selected ships tracks. Gravity anomalies have been referred to the International Reference Ellipsoid (flattening 1/297).

The principal sources of the pendulum measurements are *Vening Meinesz* [1941, 1948] and *Worzel* [1965*b*]. The principal sources of surface-ship measurements are:

(a) Lamont-Doherty Geological Observatory of Columbia University, during research vessel *Vema* cruises 19, 20, 21, 24 and 28, 1964–1971; research vessel *Robert D. Conrad* cruises 10, 11, 12, 13 and 14, 1965–1971; and USNS *Eltanin* cruise 30, 1967.

(b) University of California, Los Angeles and Scripps Institution of Oceanography during research vessel *Argo* cruise 'Monsoon Expedition', 1960–1961, and 'Lusiad Expedition', 1962.

(c) Bedford Institute of Oceanography during

Canadian survey ship *Hudson* cruise, 1970 [*Von Arx et al.,* 1971].

(d) Ocean Research Institute, University of Tokyo during motor vessel *Hakuhō-Maru* cruise 'Southern Cross Expedition', 1968–1969 [*Horibe,* 1970], and motor vessel *Umitaka-Maru* cruise, 1967–1968 [*Segawa,* 1970].

(e) Environmental Science Services Administration (ESSA, now NOAA) aboard USC&GS ships motor vessels *Pioneer* and *Surveyor,* 1961–1970 [*Chiburis et al., 1972a, b, c; Dowling et al., 1972a, b, c, d*].

(f) Institute of Oceanology, Moscow during research vessel *Vityaz* cruise, 1968 [*Khan et al.,* 1971].

(g) Hawaii Institute of Geophysics during research vessel *Mahi* cruise, 1968 [*Khan et al.,* 1971].

The surface-ship surveys used different measuring instruments and navigation systems. Data collected by Lamont-Doherty and Bedford Institute used the improved versions of the Graf-Askania sea-gravimeter [*Graf and Schulze,* 1961] mounted on gyro-stabilized platforms. Most of the measurements were obtained with an Anschutz platform using an 'electric erection' system. However, since 1971 on research vessel *Vema* (cruise 28) an Aeroflex platform has been used. This platform differs from the Anschutz platform mainly in the vertical reference which utilizes horizontal accelerometers to slave the platform and the fact that torque motors rather than gear trains drive the platform. Data collected by NOAA and Hawaii Institute of Geophysics used the gimbal-mounted LaCoste-Romberg gravimeter [*LaCoste and Harrison,* 1961]. Data collected by the Ocean Research Institute used a vibrating string accelerometer (VSA) mounted on a gyro-stabilized platform [*Tomoda and Kamamori,* 1962].

The accuracy of the surface-ship measurements depends on the measuring instruments used and on the navigation. Measurements obtained with a satellite navigation system are likely to be more accurate than measurements obtained with celestial navigation. A large source of error is in the computation of the Eotvos correction which requires that the E-W component of the ship's speed be accurately known. Measurements which use satellite navigation can determine the Eotvos correction to an accuracy of about ±2 mgal [*Talwani,* 1970]. Measurements obtained with Graf-Askania sea-gravimeters mounted on gyro-stabilized platforms are subject to off-levelling and cross-coupling errors [*LaCoste and Harrison,* 1961]. Cross-coupling errors may be as large as a few tens of milligals for a Graf-Askania gravimeter mounted fore and aft on a surface-ship subject to head or following seas [*Talwani et al.,* 1966]. These errors have been largely reduced on cruises of research vessels *Robert D. Conrad* and *Vema* since 1966 by computing the cross-coupling error in real time and subtracting it from the observed gravity trace. Off-levelling errors are small [*Neuman and Talwani,* 1972] and no correction is applied to the observed trace. The accuracy of measurements obtained with Graf-Askania gravimeters with cross-coupling and satellite navigation is about 2 mgal or less for newer instruments and about 5 mgal for older instruments. Somewhat larger errors would be expected for measurements obtained without cross-coupling and with celestial navigation. Gravity measurements obtained with the gimbal-mounted LaCoste-Romberg gravimeters may be subject to errors due to incorrect computation of the Browne correction. The errors in measurements obtained with these gravimeters in the northwest Pacific have been estimated at about 4 mgal [*Chiburis et al.,* 1972*a, b, c; Dowling et al.,* 1972*a, b, c, d*].

In order to estimate the overall accuracy of the gravity measurements we have analyzed the discrepancy at intersecting ship tracks. We obtained discrepancies in each of the free-air gravity anomaly map areas. A standard deviation of 7.5 mgal (143 discrepancies) was obtained for the Hawaii map area, 6.8 mgal (67 discrepancies) for the Aleutian map area, 14.9 mgal (55 discrepancies) for the Kuril map area, and 7.5 mgal (9 discrepancies) for the Bonin map area; 60% of the total number of discrepancies were less than 10 mgal.

Thus the overall accuracy of the gravity data varies among different map areas. We estimate, however, that the relative accuracy of measurements along individual ship tracks is generally within a few milligals.

## FREE-AIR ANOMALY MAPS

We have compiled all available surface-ship and pendulum gravity measurements in four free-air anomaly maps of parts of the northwest Pacific (Figure 1). These maps are contoured at 50-mgal intervals except in regions of steep gradients where a larger contour interval is used. In general, free-air anomaly contours have not

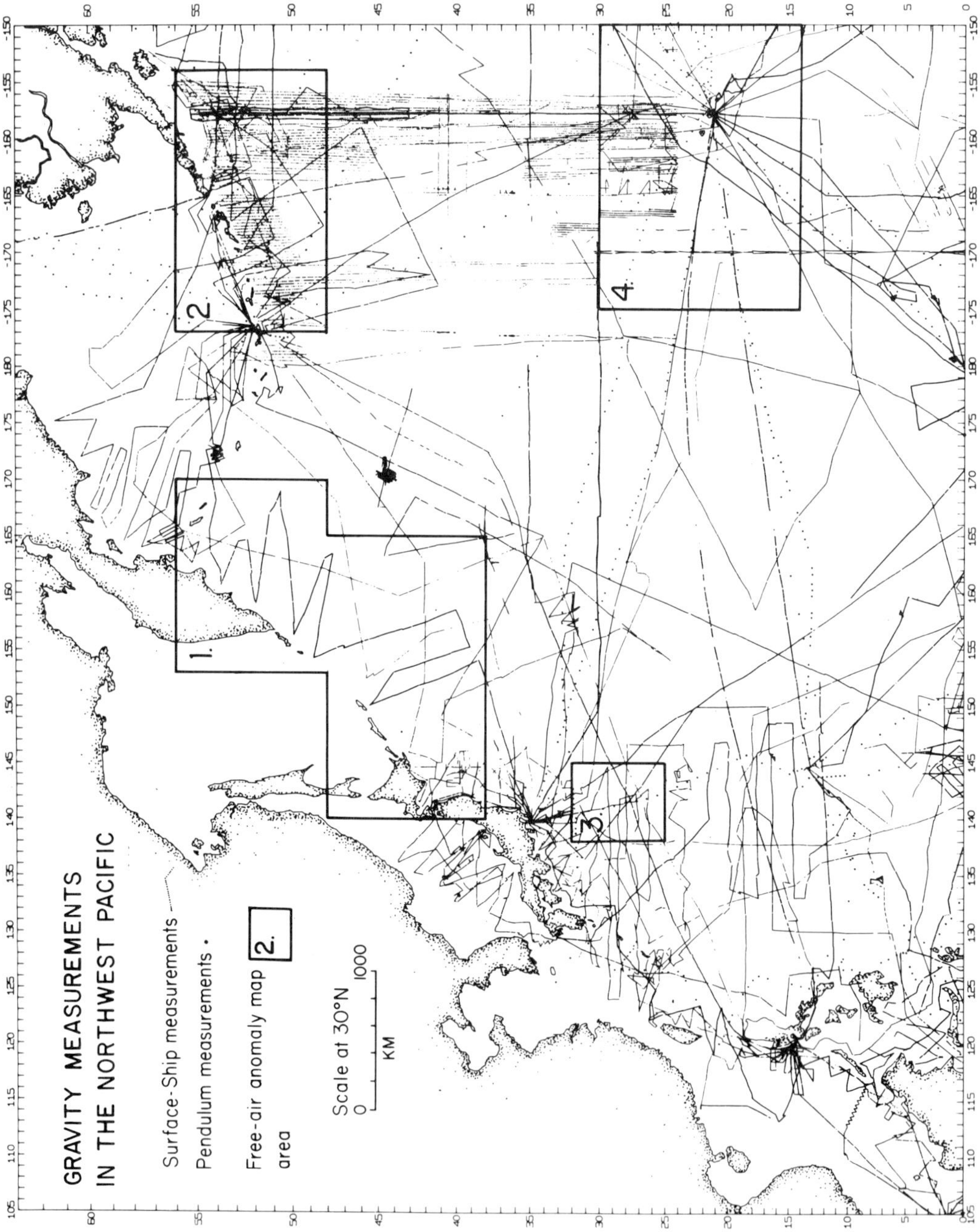


Fig. 1. Gravity measurements obtained in the northwest Pacific Ocean basin and its margin up to 1973. The map areas shown in Figures 2, 3, 4, and 5 are outlined by a heavy line.

been extended into regions of sparse or no gravity data. However, where a correlation has been established between short-wavelength free-air anomalies and topography on individual ship tracks, the bathymetry maps of *Chase et al.* [1970] have been used to extend contours away from the track.

*Kuril trench–Hokkaido rise.* We present a free-air gravity anomaly map of the Kuril trench–Hokkaido rise system in Figure 2. The map has been compiled mainly from data obtained on the recent research vessel *Robert D. Conrad* cruise 14 in the region. The main features shown by the map are positive anomalies over the Kuril Islands ridge, negative anomalies over the Kuril trench, and a broad belt of positive anomalies over the Hokkaido rise, seaward of the trench.

The positive anomalies over the Kuril Islands ridge reach a maximum value of about +300 mgal [Figure 2; *Kogan,* 1975]. Free-air gravity profiles obtained on Russian research vessels show the positive anomalies extend landward across the southeastern part of the Sea of Okhotsk. The positive anomalies appear to reach their maximum values a few tens of kilometers southeast of the active volcanoes of the Kuril Islands. A maximum value of +257 mgal was observed on the research vessel *Robert D. Conrad* cruise southeast of Shiashkotan Island over a shallow topographic ridge. This ridge is a nearly continuous feature of the Kuril Islands ridge and is separated from the active volcanoes of the Kuril Islands by a topo-

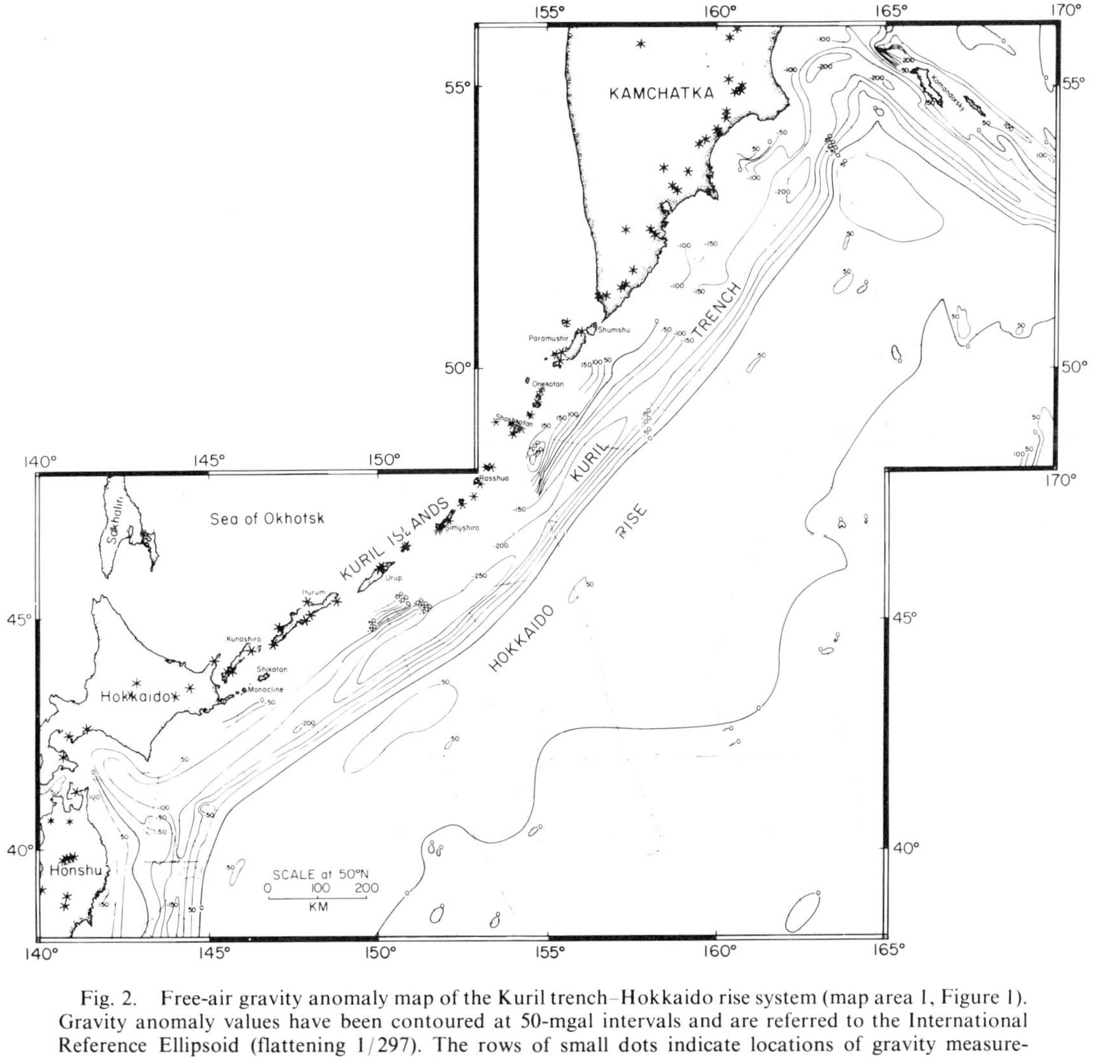


Fig. 2. Free-air gravity anomaly map of the Kuril trench–Hokkaido rise system (map area 1, Figure 1). Gravity anomaly values have been contoured at 50-mgal intervals and are referred to the International Reference Ellipsoid (flattening 1/297). The rows of small dots indicate locations of gravity measurements. Asterisks indicate historically active volcanoes.

graphic basin infilled by substantial thicknesses of sediments.

The negative anomalies over the trench reach minimum values of about −290 mgal and are generally displaced a few kilometers landward of the axis of the trench. The largest displacement occurs off the Kamchatka peninsula. The minimum values over the trench are observed between latitudes 43° and 46°N over a region of the trench where the depth exceeds 9 km.

The positive anomalies over the Hokkaido rise, seaward of the trench, form an uninterrupted belt across the map area. The positive anomalies form a band about 300 km wide (as defined by the 0 mgal contour, Figure 2) and reach values exceeding +50 mgal between about 42° and 43°30′N. The maximum value is reached about 140 km seaward of the trench axis. There is a close correlation between the extent of the positive anomalies and the Hokkaido rise [*Watts and Talwani,* 1974].

The main features of the gravity field of the Kuril island arc–trench system can be extended southwest to the Japan arc although the continuation is complicated by the gravity effect of the Erimo seamount, near the junction of the Kuril and Japan trenches. The positive anomalies of the Kuril Islands ridge continue uninterrupted into the eastern part of Hokkaido. An interesting feature of the gravity field of eastern Hokkaido [*Tomoda,* 1973*a*] is that, as at the Kuril Islands ridge (Figure 2), the maximum gravity anomalies occur not over the area of active volcanoes but some tens of kilometers seaward along the southeast coast of Hokkaido. The negative anomalies over the Kuril trench branch near 42°N, 144°E into a southern extension to the Japan trench and a western extension to the central part of Hokkaido (Figure 2). The negative anomalies of the central part of Hokkaido extend north to seaward of the west coast of Sakhalin [*Tomoda,* 1973*a*]. The positive anomalies seaward of the Kuril trench extend uninterrupted to form a belt seaward of the Japan trench.

The negative anomalies of the Kuril trench and the positive anomalies seaward of the trench extend uninterrupted into the western Aleutian arc. However, the positive anomalies of the Kuril Islands ridge are not continuous into the western Aleutian Islands ridge. A belt of negative anomalies is associated with the southern part of the Kamchatka basin (Figure 2).

*Aleutian island arc–trench system.* We present a free-air anomaly map of part of the Aleutian island arc–trench system in Figure 3. The map area extends between longitudes 154° and 177°W and includes data obtained on recent cruises of NOAA [*Dowling et al.,* 1972*d*] and Lamont-Doherty research vessels. Part of the map area, between 165° and 177°W, has been mapped by *Malahoff and Erickson* [1969]. However, the map in Figure 3 includes more data than were available to Malahoff and Erickson. In addition, the map outlines more clearly gravity anomalies over the Aleutian basin, behind the Aleutian arc, and over the region seaward of the Aleutian trench.

The Aleutian basin is associated with near-0 mgal free-air anomalies (Figure 3). A maximum value of +18 mgal is observed at 55.2°N, 173.7°E over part of the basin which is deeper than 3.4 km. Positive anomalies of up to +40 mgal are associated with the Umnak plateau, a broad relatively flat topographic feature north of Yunaska Island and east of the Aleutian basin.

The positive anomalies over the Aleutian Islands form a nearly continuous feature across the map area. In contrast to the Kuril arc, the maximum values occur over the active volcanoes (Figure 3). The maximum positive anomalies generally decrease from west to east across the map area.

The negative anomalies over the trench form a continuous belt across the map area (Figure 3). The minimum of the negative anomalies is about −275 mgal and is generally displaced a few kilometers landward of the trench axis. This is well seen between 172° and 175°W (Figure 3) where a minimum of −219 mgal is observed southeast of Atka Island even though water depths are 2 km shallower than at the trench axis. The displacement of the negative anomalies are probably caused by substantial thicknesses of low-density sediments underlying the landward wall of the trench [*Hayes and Ewing,* 1970; *Grow,* 1973]. The negative anomalies decrease in size from west to east across the map area. This correlates with a gradual west to east infilling of the trench by sediments.

A similar belt of positive anomalies as identified seaward of the Kuril trench can also be seen seaward of the Aleutian trench (Figure 3). The peak values of the positive anomalies are in the range of +50 to +60 mgal and occur about 140 km seaward of the trench axis. However, the positive anomalies extend some hundreds of kilometers

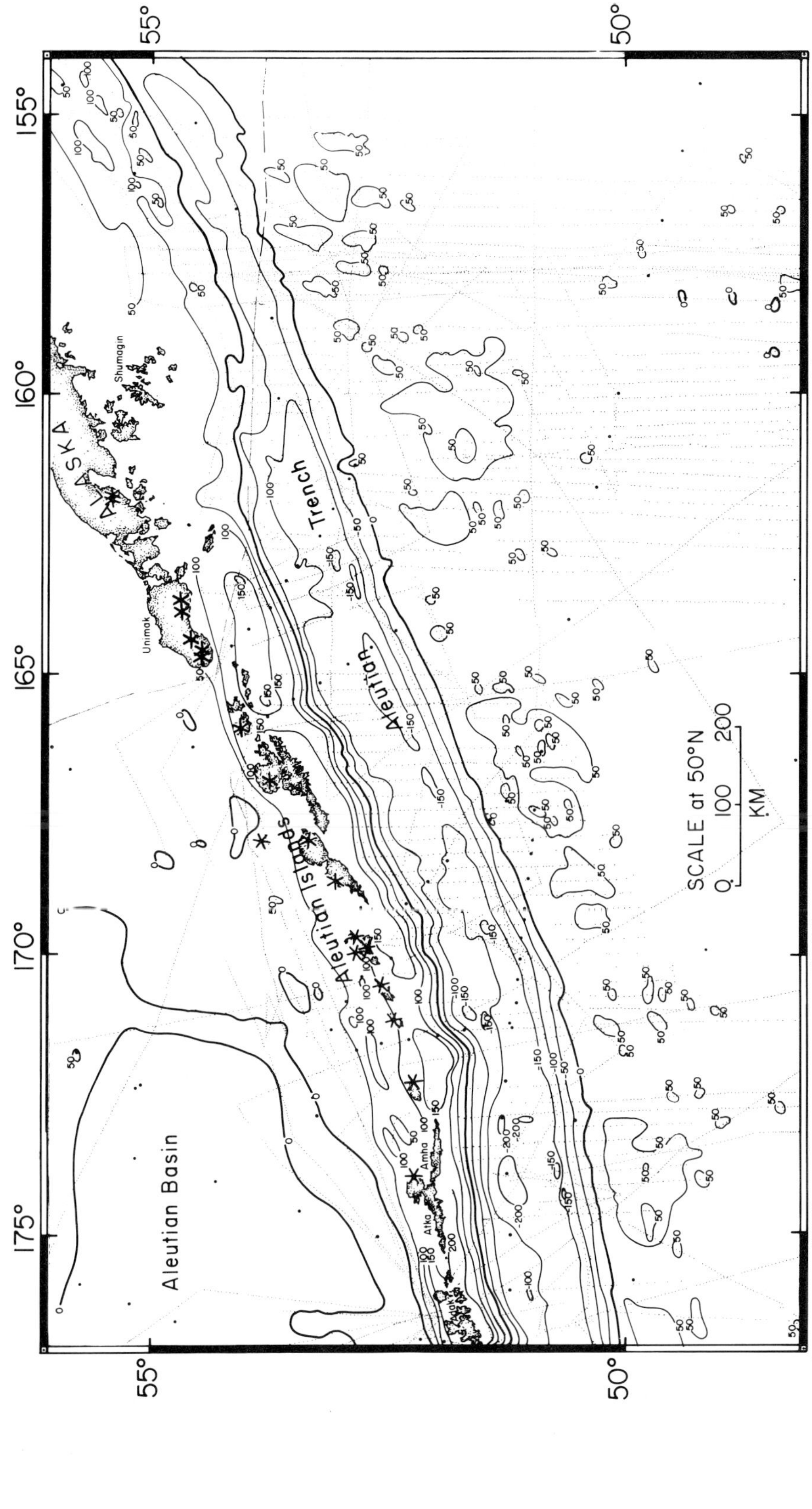


Fig. 3. Free-air gravity anomaly map of the central part of the Aleutian island arc–trench system (map area 2, Figure 1). The map legend for this map and subsequent maps is as given in the caption to Figure 2.

further seaward than at the Kuril trench system. In addition, there is little change in the amplitude of the positive anomalies parallel to the strike of the trench even though the trench decreases gradually in depth from west to east.

*Bonin island arc–trench system.* We present a free-air anomaly map of the southern part of the Izu-Bonin island arc–trench system in Figure 4. The map area includes the region of the arc between latitudes 25° and 32°N. The map has been compiled mainly from data obtained on research vessel *Vema* cruise 28 in the region.

The main features of the gravity field of the Kuril and Aleutian arc systems can also be distinguished in the Bonin arc. Large-amplitude negative anomalies of up to about –250 mgal occur over the Izu–Bonin trench. In addition, large-amplitude positive anomalies of up to about +120 mgal occur over the active volcanoes of the Iwo Jima ridge.

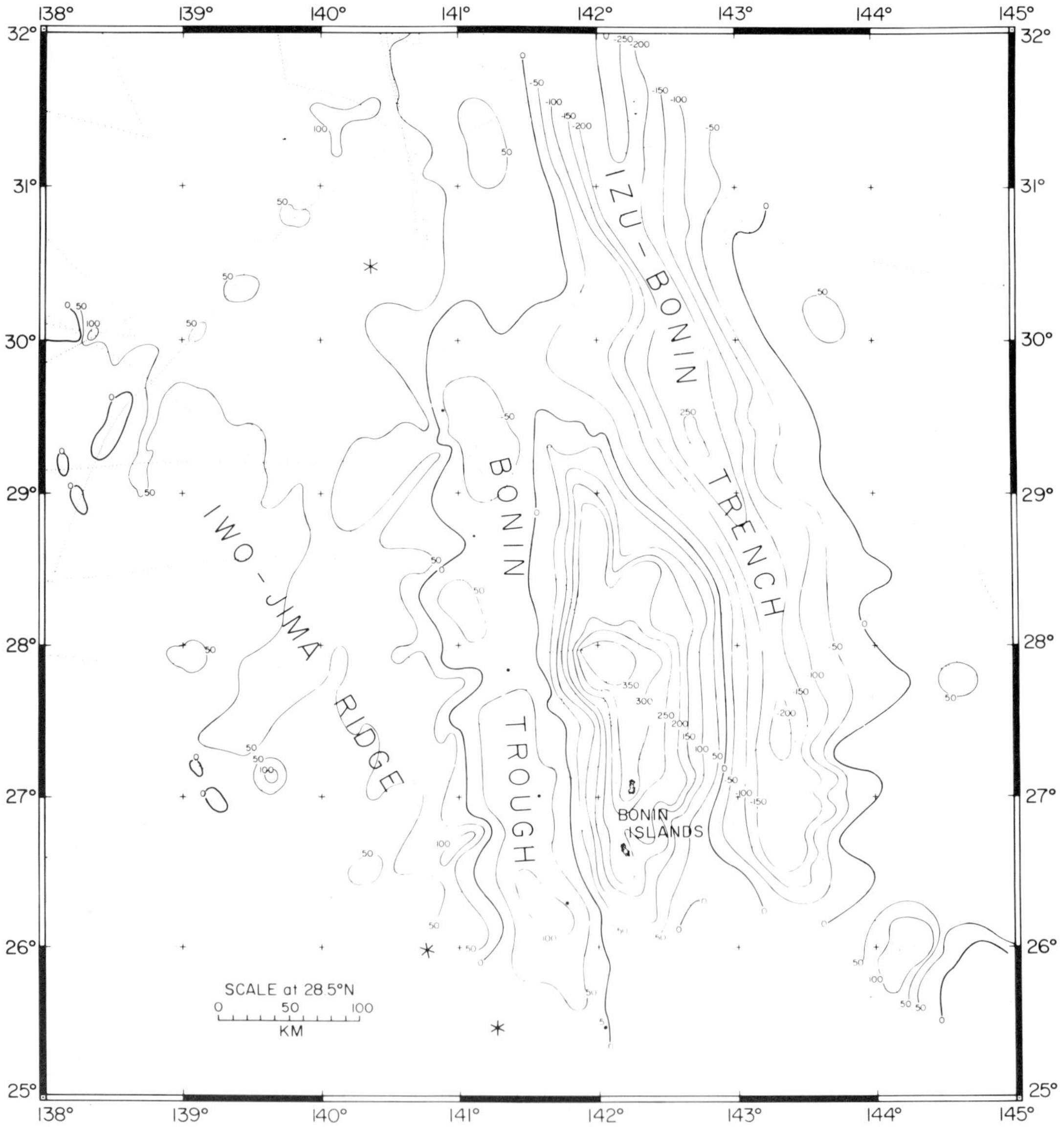


Fig. 4. Free-air gravity anomaly map of the southern part of the Izu–Bonin island arc–trench system (map area 3, Figure 1).

The most striking difference with the Kuril and Aleutian arcs, however, is the presence of belts of large-amplitude positive and negative anomalies located *between* the arc and the trench. These belts of anomalies trend generally N-S and occur between about 25° and 30°N.

The positive anomalies occur over the Bonin Islands ridge where they locally exceed +350 mgal. A maximum value of +397 mgal was observed on research vessel *Vema* cruise 21 near the northern part of the ridge at 27.98°N, 141.95°E in water depths of about 700 m. This value is one of the largest free-air anomalies recorded in oceanic regions.

The negative anomalies occur over the Bonin trough where they locally reach minimum values of -100 mgal. The Bonin trough is a narrow topographic depression about 4 km deep located between the Bonin Islands ridge and the Iwo Jima ridge.

*Hawaii and vicinity.* We present a free-air anomaly map of the eastern part of the Hawaiian ridge in Figure 5. The map has been compiled from shipboard data obtained from several institutions and from land data obtained on the Hawaiian Islands by *Woollard* [1966*a*] and others. A more detailed free-air anomaly map of the region contoured at 25-mgal intervals has been prepared [*Watts and Talwani,* 1975*b*].

The map shows the Hawaiian ridge is associated with large-amplitude positive free-air anomalies with particularly large values over the volcanic centers. The largest anomalies, which reach +700 mgal, are found over Mauna Loa on Hawaii. Subsidiary gravity highs are associated with each of the other major volcanoes and calderas on Hawaii [*Woollard,* 1966*a, b*] and are presumably due to high-density 'plugs' in the volcanic vents [*Adams and Furumoto,* 1965]. In general, the islands on the ridge are associated with anomalies of the order of +200 to +300 mgal, with local highs of up to an additional 115 mgal associated with local volcanic centers [*Strange et al.,* 1965; *Woollard,* 1966*b*].

The Hawaiian ridge is flanked on both sides by a belt of negative anomalies about 100 km wide. The negative anomalies are generally in the range of -25 to -75 mgal and reach a minimum of -136 mgal north of Maui. The negative anomalies appear to be somewhat asymmetric, with values more negative northward of the ridge. The bands of negative anomalies correlate in a general way with a topographic moat which trends parallel to the islands [*Dietz and Menard,* 1953].

The belts of negative gravity anomalies are flanked by belts of positive anomalies which reach maximum values of about +25 mgal. These anomaly belts correlate with a topographic rise known as the Hawaiian swell [*Betz and Hess,* 1942]. The peak of the positive anomalies over curves continuously around the southeastern end of the Hawaiian ridge. The free-air anomalies decrease gradually away from the arch reaching values of about -15 mgal, which are typical of values over the North Pacific basin, about 600 km from the Hawaiian ridge. This decrease correlates with the flanks of a broad topographic rise known as the Hawaiian swell [*Betz and Hess,* 1942]. The peak of the positive anomalies over the arch is about 250 to 300 km from the crest of the ridge and about 150 to 200 km from the axis of the negative anomalies.

The outer belt of positive anomalies is broken on the north side of the Hawaiian ridge from 160° to 162°W by an area of negative anomalies. There is no obvious tectonic or morphologic reason for the interruption of the gravity high, although the area is occupied by the southernmost of the Musicians seamounts. The seamounts themselves would not, however, be expected to disturb the gravity field more than locally, as is the case with a group of similarly sized seamounts WSW of the island of Hawaii. It should be noted, however, that the Murray fracture zone passes just to the north of this area and that several of the southernmost Musicians seamounts are elongated parallel to the Murray fracture zone trend.

The simple pattern of gravity anomalies associated with the Hawaiian ridge may also be disturbed by the Necker ridge, which is a linear topographic feature extending southwestward away from the Hawaiian ridge at Necker Island near 23.5°N, 165°W. Two crossings of the Necker ridge show it is about 25 km wide at the base and rises slightly more than 3 km from the ocean floor to a depth of about 1600 m. The gravity anomaly over the ridge consists of a large positive anomaly, reaching 61 mgal over the peak. The high is flanked by two marked lows which reach -24 mgal or about 30 mgal less than the undisturbed field in that area.

## DISCUSSION OF THE GRAVITY ANOMALY AND TOPOGRAPHY PROFILES

We present two long free-air gravity anomaly and topography profiles across the Pacific Ocean

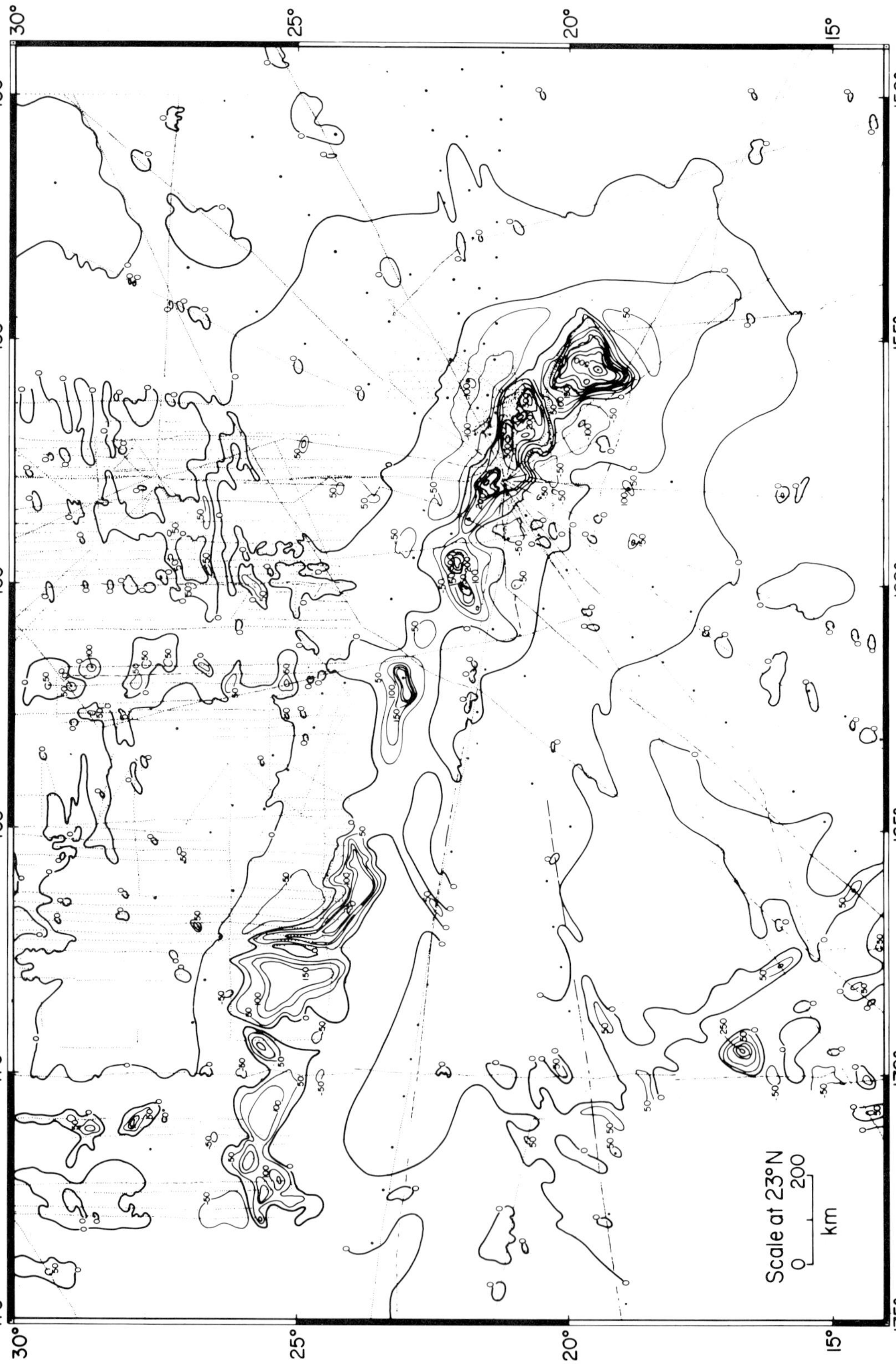


Fig. 5. Free-air gravity anomaly map of the region around Hawaii (map area 4, Figure 1).

basin and its margin in Figures 6 and 7. Figure 6 presents a profile extending from 12.5° N, 158° W across the Hawaiian ridge to the Aleutian trench at 55° N, 158° W; Figure 7, a profile from the Hawaiian ridge at 21° N, 158° W to the Izu–Bonin trench at 31° N, 142° E. This latter profile has been projected E-W along 26° N. We also show on each profile the recent GEM6 [*Richardson and Lerch,* 1974] combination satellite and surface gravity field to harmonic degree 16.

The main features of the gravity field of the Hawaiian ridge shown in Figure 5 can be clearly seen in profile form in Figure 6. This profile shows large-amplitude positive anomalies associated with the Hawaiian ridge, large-amplitude negative anomalies of the flanking moat, and broad positive anomalies over the crest of the arch. The profile shows that beyond the region of the arch, depth and free-air anomalies decrease gradually to typical Pacific basin values of about 6000 m and –10 to –15 mgal, respectively.

The positive anomalies identified seaward of the Aleutian trench (Figure 3) can be clearly seen in Figure 6. This profile shows that the depth decreases to about 5000 m while free-air anomalies increase to about +35 mgal reaching a maximum value about 140 km seaward of the trench axis. The positive anomalies extend about 900 km seaward of the trench—which is significantly wider than seaward of other trenches (Figure 2; *Watts and Talwani* [1974]). Part of the decrease in depth can be explained by an increase in sediment thickness towards the Aleutian trench [*Jones et al.*, 1971].

The most striking feature of the profile (Figure 6) is the close correlation which exists between gravity anomalies and topography. This can be easily seen in the region of the Hawaiian swell flanking the southeastern end of the Hawaiian ridge. Studies of marine magnetic anoma-

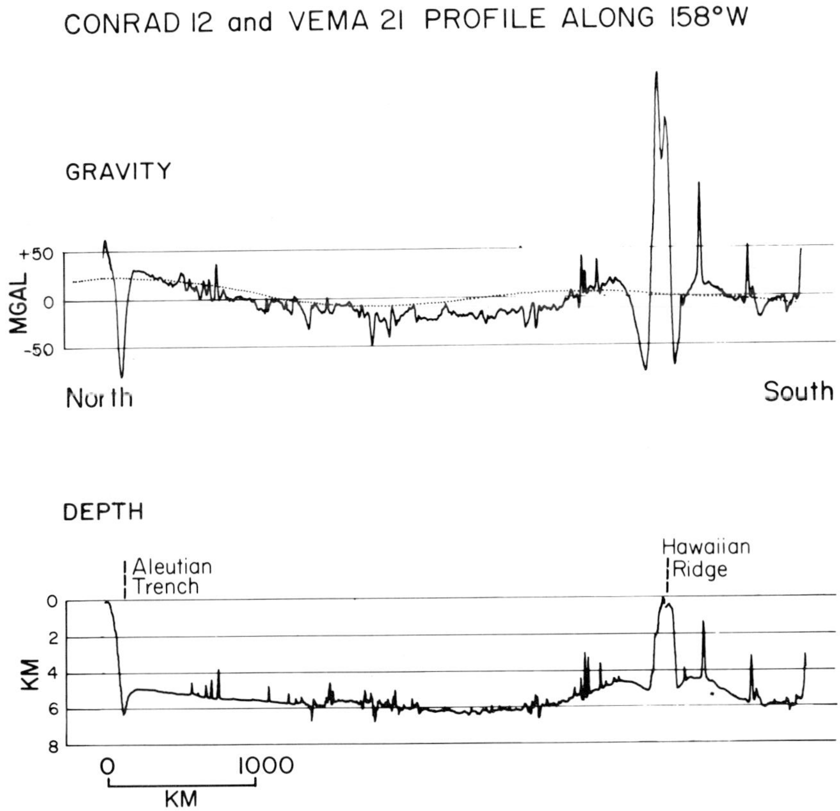


Fig. 6. Free-air gravity anomaly and topography profile between south of the Hawaiian ridge and the Aleutian trench. The faint dots on the free-air anomaly profile indicate the GEM6 combination solution of satellite-derived and surface data to harmonic degree n = 16 [*Richardson and Lerch,* 1974]. Note that in the part of the Pacific Ocean basin crossed by the profile there is a close correlation between gravity anomaly and topography, not only for short wavelengths but also for long wavelengths.

lies [*Atwater and Menard,* 1970; *Larson and Chase,* 1972] show that the age of the sea floor throughout the length of the profile is pre-Tertiary. Thus the regional changes in topography and corresponding changes in gravity anomalies seen in Figure 6 cannot be attributed to normal changes in depth away from mid-ocean ridge systems [*Sclater and Francheteau,* 1970].

The main features of the gravity field west of the Hawaiian ridge can be seen in Figure 7. This profile shows that the Hawaiian swell is generally elongate in an E-W direction extending to about 2000 km west of the Hawaiian islands compared to 600 km to the north. Beyond the region of the Hawaiian swell, the regional depth and gravity anomalies remain nearly constant at about 6000 m and between −10 and −20 mgal, respectively.

The belt of positive anomalies seaward of the Izu–Bonin trench can be seen in Figure 7. The positive anomalies are about 400 km wide and reach maximum values of about +40 mgal. The width and amplitude of the positive anomalies is similar to those observed off the Kuril and Japan trenches.

We have noted that one of the most prominent features seen in the profiles (Figures 6 and 7) is the correlation which exists between topography and gravity anomalies. This is true not only at short wavelengths, but also at long wavelengths. We also note that although the topographic rise seaward of trenches is less than the rise of the Hawaiian swell (about 800 m for the outer rise and 1600 m for the Hawaiian swell) it produces a slightly larger amplitude gravity anomaly (Figures 6 and 7). These differences are significant and probably arise from the different modes of origin of these features.

We present five long free-air anomaly and topography profiles across the margin of the northwest Pacific Ocean basin in Figure 8. These pro-

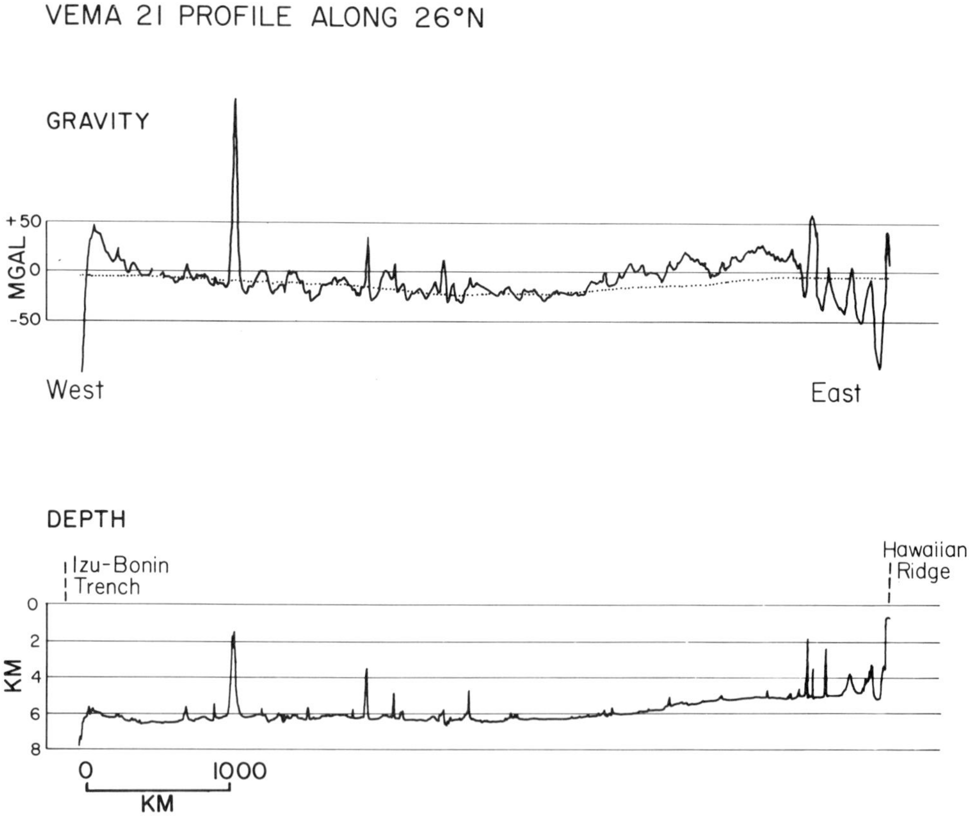


Fig. 7. Free-air gravity anomaly and topography profile between the Hawaiian ridge and the Izu–Bonin trench. Note that on this profile the broad topographic rise and the accompanying increase in free-air anomaly around the Hawaiian ridge is broader (by about 1000 km) than the topographic rise seen around the ridge in Figure 6. Thus the regional rise around Hawaii is elongate in an E–W direction.

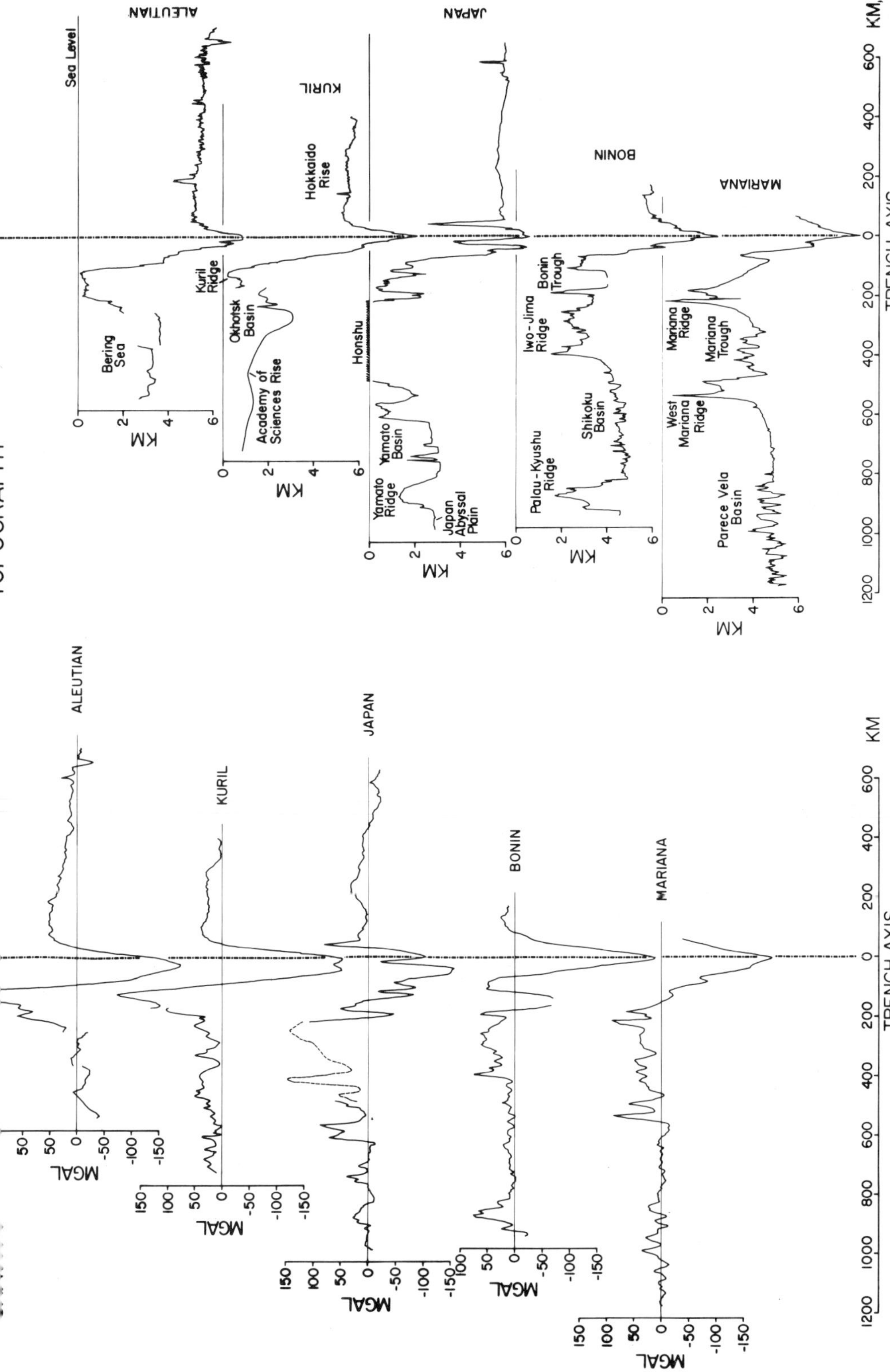


Fig. 8. Selected free-air gravity anomaly and topography profiles across the margins of the northwest Pacific Ocean. Data landward (Sea of Okhotsk side) of the Kuril island arc are from *Kogan* [1975] and across Honshu from *Tomoda* [1973*a*]. Note that the characteristics of the gravity field change for different marginal basins. The Aleutian basin, behind the Aleutian arc, is characterized by nearly 0-mgal free-air anomalies. The Shikoku and Parece Vela basins, behind the Izu–Bonin–Mariana arc, are characterized by slightly positive anomalies in the range of about +5 to +15 mgal. The Mariana trough, behind the Mariana arc, and the Okhotsk basin, behind the Kuril arc, are characterized by positive anomalies of about +40 mgal.

files cross the main features of island arc–trench systems in the northwest Pacific and include the Aleutian basin, Okhotsk basin, Japan Sea, and the Shikoku and Parece Vela basins. These basins, which are oceanic in type, comprise the semi-enclosed seas which occur on the landward side of the island arc–trench systems in the northwest Pacific.

The profiles (Figure 8) show that a great variability occurs in the character of the gravity field across island arcs and marginal basins behind the arcs. The Aleutian basin behind the Aleutian arc is associated with near-0 mgal anomalies (Figures 3 and 8; *Kienle* [1971]). In contrast, the Okhotsk basin behind the Kuril arc, and the Shikoku basin, Mariana trough, and Parece Vela basin behind the Izu–Bonin–Mariana arc are all associated with positive anomalies. The positive anomalies reach maximum values of about +40 mgal over the Mariana trough and somewhat smaller values of +10 to +15 mgal over the Shikoku and Parece Vela basins.

*Karig* [1971*a, b*] and *Packham and Falvey* [1971] proposed that marginal basins formed by crustal extension behind island arcs. Sequential intervals of crustal extension during the Cenozoic are believed to have caused the series of marginal basins and remnant arcs behind the Izu–Bonin–Mariana arc in the northwest Pacific. DSDP drilling [*DSDP Scientific Staff*, 1969*b*, 1973*b*] has confirmed the inferred ages of the Shikoku and Parece Vela basins.

We found that, in general, positive free-air anomalies occur over those basins which are considered to have formed by crustal extension behind island arcs. The Mariana trough, which is characterized by positive anomalies of up to +40 mgal, is considered to be a presently active extensional feature [*Karig*, 1971*a, b*]. The Shikoku and Parece Vela basins, which are associated with positive anomalies of up to +25 mgal, have formed by crustal extension during the latest Oligocene–earliest Miocene [*Karig*, 1971*a, b*].

## INTERPRETATION OF THE MAIN FEATURES OF THE GRAVITY ANOMALY MAPS

We have shown that the margins of the Pacific plate are associated with a broad belt of positive anomalies (Figures 2, 3, 4, 5; *Watts and Talwani* [1974, 1975*b*]). The positive anomalies are well developed seaward of the central Aleutian (Figure 3), Kuril (Figure 2), and Japan trenches where they correlate with a regional rise in topography of a few hundred meters. The positive anomalies reach peak values of +40 to +50 mgal about 140 km seaward of the axis of the trench. The positive anomalies decrease to near-zero values about 400 km seaward of the trench axis.

The positive anomalies and the associated regional rise in topography can be adequately explained by a simple model for the flexure of the oceanic lithospheric plate as the plate approaches a deep-sea trench [*Hanks*, 1971; *Watts and Talwani*, 1974]. The simple model of flexure used treats the oceanic plate as a semi-infinite linearly elastic beam overlying a weak fluid. Similar models have been used to study the deformation of oceanic plates caused by seamounts [*Walcott*, 1970*a*; *Watts and Cochran*, 1974]. *Watts and Talwani* [1974] found that the gravity effect of the simple model of flexure generally explained both the amplitude and wavelength of the observed positive anomalies seaward of trenches.

The most important parameter in deformation studies is the effective flexural rigidity which, in effect, is a measure of the resistance of the plate to deformation. The deflection $y$ of a semi-infinite linearly elastic beam can be expressed by

$$y = \frac{2 P_b \lambda \exp(-\lambda x) \cos \lambda x}{(\rho_m - \rho_w) g} \tag{1}$$

where $\rho_m$ is the density of the substratum, $\rho_w$ is the density of the material above the beam, $P_b$ is the vertical force/unit width of the beam applied along the $x = 0$ axis, $g$ is average gravity, and $\lambda$ is a 'damping factor' related to the effective flexural rigidity $D$ by

$$\lambda = \sqrt[4]{\frac{(\rho_m - \rho_w) g}{4D}} \tag{2}$$

The form of the deflection curve is a damped sinusoid with a large depression beneath the applied load and a bordering peripheral upwarp.

We can estimate a value for the effective flexural rigidity of the oceanic plate from the positive anomalies observed seaward of trenches. For a semi-infinite linearly elastic beam we can express the effective flexural rigidity by

$$D = \frac{64 (\rho_m - \rho_w) g (x_b)^4}{81 (\pi)^4} \tag{3}$$

where $\rho_m$, $\rho_w$, and $g$ have been previously defined and $x_b$ is the distance from the $x = 0$ axis to the crest of the peripheral bulge. The distance to the peak of the positive anomalies seaward of trenches (and the accompanying crest of the regional rise in topography) from the trench axis is about 140 km (Figures 2, 3 and 4). We assume the load required to depress the oceanic plate beneath trenches is applied at the trench axis. With $x_b$= 140 km, $\rho_w$= 1.03 g/cm³, $\rho_m$= 3.3 g/cm³, and $g$ = 981 cm/sec², (3) gives a value of $D = 3 \times 10^{29}$ dyne-cm.

We emphasize that the effective flexural regidity $D$ is not the actual rigidity of the lithosphere. It is, rather, the value the lithosphere would have if it responded to applied loads in the same manner as would a linearly elastic beam overlying a fluid substratum. The lithosphere is, in reality, inhomogeneous and it is unlikely to respond to applied loads as a perfectly elastic material.

We note that the effective flexural rigidity of the Pacific plate deduced from the positive anomalies seaward of trenches is somewhat lower than previously determined values for the Pacific plate. *Watts and Cochran* [1974] deduced values of $5 \times 10^{29}$ dyne-cm and $2 \times 10^{30}$ dyne-cm for the interior parts of the Pacific plate. The age of the interior of the plate is similar (to within a few tens of millions of years) to the age of the edge of the plate. There are therefore two possible explanations for the observed differences. First, the point of application of the load may not be at the trench axis as assumed but farther landward. Second, an external horizontal compressive stress may be transmitted to the Pacific plate as it approaches a deep-sea trench.

That some system of horizontal compressive stress may be transmitted to the approaching plate is supported by the distribution of the positive anomalies in the northwest Pacific. We have shown (Figures 2, 3 and 4; *Watts and Talwani* [1974]), that the positive anomalies are well developed seaward of the Aleutian, Kuril, and Japan trenches but poorly developed seaward of the southern Bonin and Mariana trenches. The island arcs comprising these trenches show a great variability in regional seismicity patterns. The Aleutian and Kuril arcs are characterized by large shallow earthquakes with thrust-fault mechanisms and extensive rupture zones. The Izu-Bonin and Mariana arcs are characterized by an absence of large, shallow earthquakes. *Kanamori* [1971] interpreted the difference in regional seismicity patterns as indicating that a substantially stronger coupling exists between converging plates at the Aleutian and Kuril arcs than at the Izu-Bonin and Mariana arcs. Thus a strong coupling between plates at the Aleutian and Kuril arcs could explain both the regional seismicity pattern of the arcs and the presence of a well-developed belt of positive anomalies seaward of the trench.

We have shown that the interior of the Pacific plate is dominated by the positive and negative anomaly belts associated with the Hawaiian ridge. These belts of anomalies have been clearly defined in map form in Figure 5 and in profile form in Figure 6.

The earliest gravity measurements across the Hawaiian ridge indicated that the ridge was not locally supported by a low-density 'root' such as predicted by the Airy-Heiskanen hypothesis of isostasy but rather was regionally compensated [*Vening Meinesz,* 1941]. *Gunn* [1943] developed a model in which the lithosphere is treated as an elastic beam deformed by imposed loads and showed that *Vening Meinesz's* [1941] gravity data were consistent with the model.

The observation that the Hawaiian ridge is situated on a broad, elongate rise of the sea floor was noted by *Betz and Hess* [1942], who named the rise the Hawaiian swell. *Vening Meinesz* [1941] recognized the presence of a moat east of Hawaii, and *Dietz and Menard* [1953], shortly after the development of accurate recording echo sounders, were able to describe the entire ridge, moat, arch, and swell structure.

*Dietz and Menard* [1953] also proposed an explanation of the origin of the Hawaiian ridge which in some ways is similar to recent explanations in terms of mantle hot spots. It involved broad uparching over the ascending limb of a convection current resulting in fracturing, volcanism, and subsidence due to the load of the volcanic center which develops on the crest of the swell.

Later studies tended to support the concept of elastic bending under a load [*Hamilton,* 1957] as did the original interpretation of seismic refraction results in the area of Hawaii [*Furumoto et al.,* 1965]. The data showed that the depth to the Moho decreases from 22 km under Oahu [*Furumoto et al.,* 1965] to 15.5 km north of Maui to 13 km under the moat and 10 km beneath the arch [*Shor and Pollard,* 1964]. When compared to an average Pacific column, which shows a depth of 12 km to the mantle [*Raitt,* 1956], these data appear to indicate 10 km of subsidence beneath Oahu and 2 km of uplift at the arch.

However, *Woollard* [1966*b*] argued that the greater depths of the Moho beneath Oahu are due to a great thickening of the high-velocity basal crustal layer (layer 3) and that no more than about 2 km of subsidence could have occurred. He did however state that "the shallow Moho depth of 9–10 km on the rise is not due to a subnormal value of crustal thickness but rather to crustal flexure." [*Woollard,* 1966*b,* p. 294].

*Walcott* [1970*a*] and *Watts and Cochran* [1974] explained the main features of the topography and gravity anomalies by a simple model, not unlike that of *Gunn* [1943], which treats the lithosphere as a linearly elastic beam overlying a weak fluid substratum. They found that the effective flexural rigidities which best fit the amplitude and wavelength of the observed gravity anomalies implied that between 8 and 11 km of subsidence have occurred beneath the southeastern end of the Hawaiian ridge, in accordance with the seismic data. The apparent thickening of the basal crustal layer discussed by *Woollard* [1966*b*] might result from the fact that Hawaiian basaltic magma tends to form dense, nonvesicular rocks with densities of 2.9 to 3.0 g/cm$^3$ when erupted at depths of greater than about 1 km [*Strange et al.,* 1965]. Since the majority of the rocks in the Hawaiian ridge were erupted at depth, they could well have seismic velocities similar to those of oceanic layer 3.

*Watts and Cochran* [1974] also found that the gravity anomalies along the entire Hawaiian–Emperor chain could be satisfactorily explained by a single value of the effective flexural rigidity of either $5 \times 10^{29}$ dyne-cm or $2 \times 10^{30}$ dyne-cm depending on the model adopted. The features associated with the flexure thus do not relax with the passage of time and the Pacific plate appears to be rigid enough to support the load of the seamount chain for periods of at least several tens of millions of years. The implication is that the oceanic lithosphere does not respond to sustained

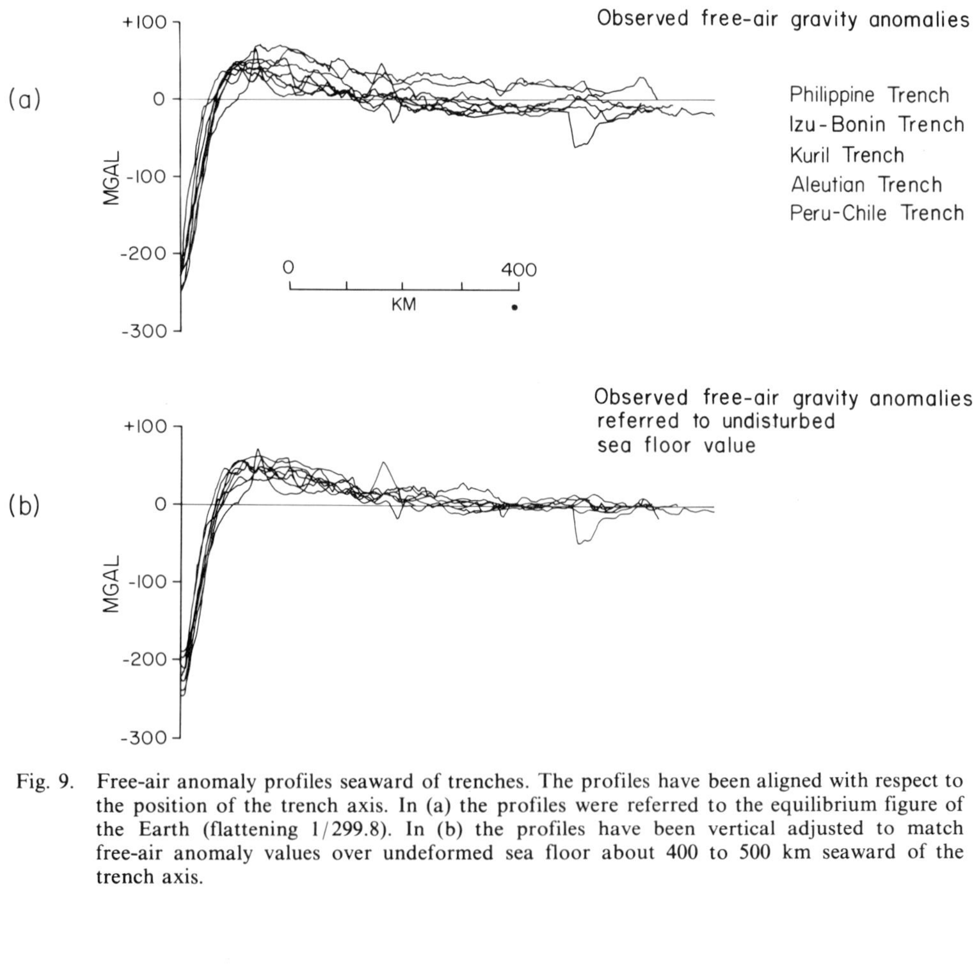


Fig. 9. Free-air anomaly profiles seaward of trenches. The profiles have been aligned with respect to the position of the trench axis. In (a) the profiles were referred to the equilibrium figure of the Earth (flattening 1/299.8). In (b) the profiles have been vertical adjusted to match free-air anomaly values over undeformed sea floor about 400 to 500 km seaward of the trench axis.

loads as a viscoelastic Maxwell body as suggested by *Walcott* [1970*b*], but rather has a finite long-term strength and thus appears to share some of the characteristics of a Kelvin body.

The fact that the effective flexural rigidity remains constant along the chain can also be seen from topographic profiles of the Hawaiian ridge. The distance from the ridge to the crest of the arch, $x_b$, is related to the effective flexural rigidity $D$ by

$$D = \frac{(\rho_m - \rho_w)\ g\ (x_b)^4}{4\ (\pi)^4} \qquad (4)$$

where $\rho_w$, $\rho_m$, and $g$ have been previously defined. The distance to the crest of the arch from the center of the ridge $x_b$ is consistently near 250 km, which implies an effective flexural rigidity of $2.3 \times 10^{30}$ dyne-cm.

Although the arch and moat structure produced by flexure is found along the entire length of the Hawaiian–Emperor chain, the same cannot be said of the broad swell upon which they are superimposed. This feature is limited to the southeastern end of the Hawaiian ridge. This can be clearly seen in Figure 7 which shows a topography and gravity anomaly profile nearly parallel to the ridge. The regional depth increases from 4600 m near Oahu to more than 6000 m 1500 km to the west. The relaxation of the swell can also be clearly seen in long profiles perpendicular to the ridge (Figure 6; *Watts and Cochran* [1974], Figure 2). The swell can be clearly seen in profiles at Oahu and Nihoa, but is absent in profiles at Midway and across the Emperor seamount chain.

We have shown (Figure 6) that the change in gravity anomaly across the Hawaiian swell is similar to that observed across the topographic rises seaward of trenches. This occurs even though the Hawaiian swell is more elevated than the trench outer rise by several hundreds of meters.

## DISCUSSION

The interpretation of many of the prominent regional features of the gravity field of the north-

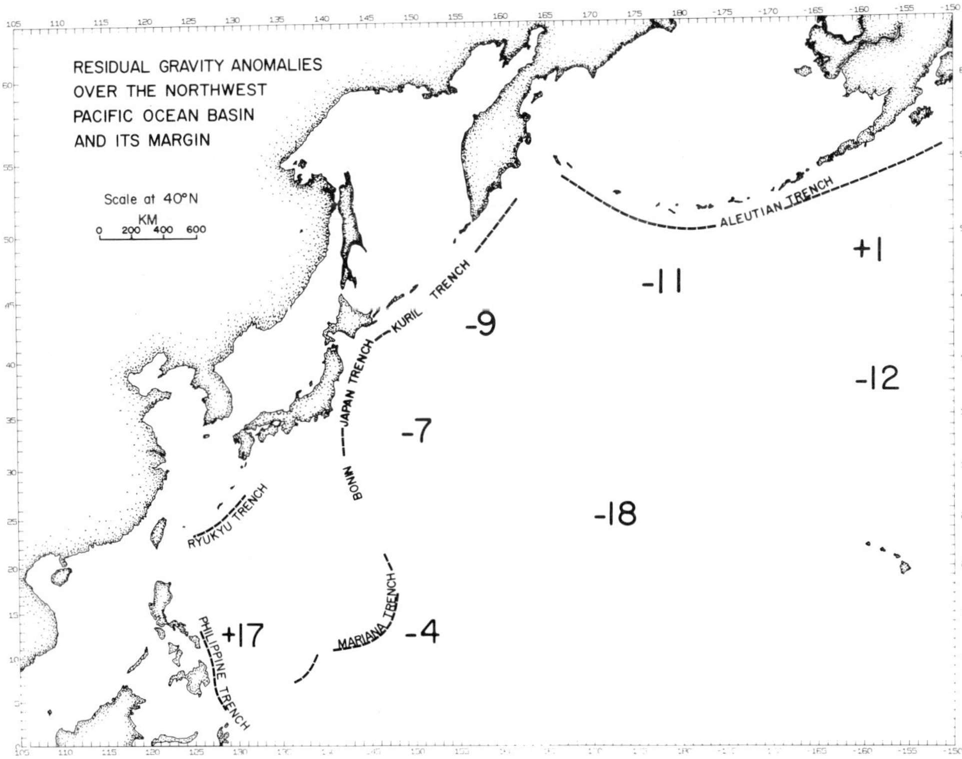


Fig. 10. Residual gravity anomalies in the northwest Pacific. These values are those required to null observed free-air anomalies in the region 400 to 500 km seaward of trenches. Note that residual gravity anomalies over the Pacific ocean are generally negative.

west Pacific as due to causes within the lithosphere has important implications. If these regional gravity anomalies can be successfully identified and removed from the observed gravity anomalies the remaining gravity field can be interpreted as due to deeper sources beneath the lithosphere.

We show in Figure 9 an attempt to isolate the positive anomalies seaward of trenches from the longer wavelength regional gravity field of the northwest Pacific Ocean basin upon which they are superimposed. Figure 9a shows selected profiles seaward of trenches in the northwest and eastern Pacific referred to the hydrostatic figure of the Earth (flattening 1/299.8). The profiles have been aligned with respect to the axis of the trench. In Figure 9b the same profiles have been vertically adjusted to match the observed free-air anomaly value about 400 to 500 km seaward of the trench axis. Studies of topography profiles in the northwest Pacific [*Watts and Talwani,* 1974] show that in the region 400 to 500 km seaward of the trench beyond the outer rise in topography, the sea floor is regionally undisturbed and of normal depth for its age. In Figure 9a the positive anomalies seaward of trenches show a wide scatter of values about the zero line. However, in Figure 9b the scatter is reduced and the characteristics of the positive anomalies can be clearly seen. The positive anomalies are characterized by maximum values of +40 to +50 mgal about 140 km seaward of the trench axis and by widths of about 400 km.

The positive anomalies have been satisfactorily explained by a simple model of flexure of the lithospheric plate as it approaches an island arc [*Watts and Talwani,* 1974]. We therefore assume that the broad belt of positive anomalies isolated in Figure 9*b* is of lithospheric origin.

We present values obtained for the residual gravity field of the northwest Pacific Ocean basin in Figure 10. These values are those required to null observed free-air anomalies in the region 400 to 500 km seaward of trenches. We also include in Figure 10 values obtained from the long free-air anomaly profiles across the Pacific basin (Figures 6 and 7). These values are average observed free-air anomalies which occur over sea floor undisturbed by the regional rise of topography seaward of trenches and the flanks of the Hawaiian swell.

We found that the residual gravity field of the Pacific Ocean basin is generally negative (Figure 10). Slightly positive anomalies of +1 mgal occur seaward of the eastern Aleutian trench. The largest positive anomalies of +17 mgal occur seaward of the Philippine trench. This is over the West Philippine marginal basin however, and not over the Pacific basin.

The mainly negative residual gravity anomalies over the Pacific basin are similar to those found over most ocean basins. *Cochran and Talwani* [1974] have found that with a few notable exceptions (mainly parts of the North Atlantic and southwest Indian oceans) residual gravity anomalies over the world's oceans are generally near-0 or negative and in the range of 0 to −15 mgal. The cause of these generally negative residual gravity anomalies is not at present understood.

*Acknowledgments.* The compilation of the large amount of gravity data obtained by Lamont-Doherty used in this study would not have been possible without the help of the officers, scientists, and gravity observers aboard the research vessels *Vema, Robert D. Conrad,* and *Eltanin.* This work was supported by National Science Foundation grants GV-27472, GA-27281, GA-17761, GA-1434, and GA-1523, and by Office of Naval Research contracts N00014-67-A-0108-0004 and N00014-75-C-0210.

# 3. *Gravity Measurements Near Japan and Study of the Upper Mantle Beneath the Oceanic Trench–Marginal Sea Transition Zones*

JIRO SEGAWA AND YOSHIBUMI TOMODA

*Ocean Research Institute, University of Tokyo, Tokyo 164, Japan*

*Abstract.* A free-air gravity anomaly map of Japan and the surrounding seas is presented, compiled primarily from data obtained by the Tokyo Surface Ship Gravity Meter. By use of gravity and the presently available seismic refraction data obtained for the crustal structures, the density anomalies in the upper mantle near the Japan and Izu–Bonin arcs have been estimated. Plausibilities of the density anomalies thus obtained have been examined by comparing them with the results from anomalous seismic velocities and density–velocity relationships. The residual gravity anomalies near oceanic trenches, which have been considered to be discordant with seismic results, have been estimated as accurately as possible, and then explained in terms of a concentration of mass within a bending lithosphere due to possible elastic compression, phase transformations, or magmatic flows.

Sea-going gravity measurements of Japan were initiated by *Matuyama* [1934] and *Kumagai* [1953], using the Vening Meinesz pendulum, aboard a submarine. Surface ship measurements were initiated by *Tomoda and Kanamori* [1962] in 1961, using a new type of surface ship gravity meter now known as the Tokyo Surface Ship Gravity Meter (TSSG). This meter employs a single vibrating string accelerometer and a gyro-stabilized platform. An earlier model of the single vibrating string accelerometer was subject to nonlinear rectification errors, but recent improvements of the meter have removed such errors, and its performance now appears comparable to other meters such as the LaCoste and Romberg (L & R) meter. A comparison measurement was carried out in 1972, using both a TSSG and an L & R meter, which were installed aboard the research vessel *Hakuho-maru* [*Segawa and Bowin,* 1975]. Figure 1 shows the result of comparison for about 17 hours. Vertical accelerations caused by ship motion during the comparison were about 50 gal. This comparison resulted in a 2.3-mgal averaged difference between results from both meters. Spikes in the curve showing the difference between the TSSG and the L & R meter were caused by rapid variations of gravity, implying that there were differences in the phase lag of measurements which was thought involved in both cases. The TSSG can be made free of this phase lag if it incorporates a digital low-pass filter. In 1973, maps of free-air and Bouguer gravity anomalies in and around Japan were compiled by *Tomoda* [1973*a*], primarily using the data from the TSSG.

Since 1972, we have used both the TSSG and the L & R meter to make extensive surveys in the Japan Sea, the Japan trench, the Philippine Sea, the Izu–Bonin island arc, the Izu–Bonin trench, the Ryukyu island arc, and the Ryukyu trench. During these surveys several gravity profiles were obtained in the areas where previous seismic refraction and reflection surveys had been conducted. In the present paper we discuss the evidence for heterogeneity in the upper mantle inferred from two-dimensional gravity calculations based on seismic data. For such purposes it is necessary to have gravity data at exactly the same location as the seismic profiles. We have chosen two cross sections of the oceanic trench–island arc–marginal sea areas; one extends from the Japan trench to the Japan Sea, and the other from the Izu–Bonin trench to the Shikoku basin. Disagreement between the crustal structures derived from gravity models versus those based on seismic refraction data at oceanic trenches has long been a question [*Watts and Talwani,* 1974; *Grow and Bowin,* 1975]. Later in this paper we derive the residual gravity anomalies at the oceanic trenches, i.e., the anomalies from which the effects of previously known topography, sediments, crust, and upper mantle have been removed. Although interpretations based on gravity anomalies always have a difficulty with non-uniqueness, we have attempted to isolate the residual gravity anomalies which are recognizable in most

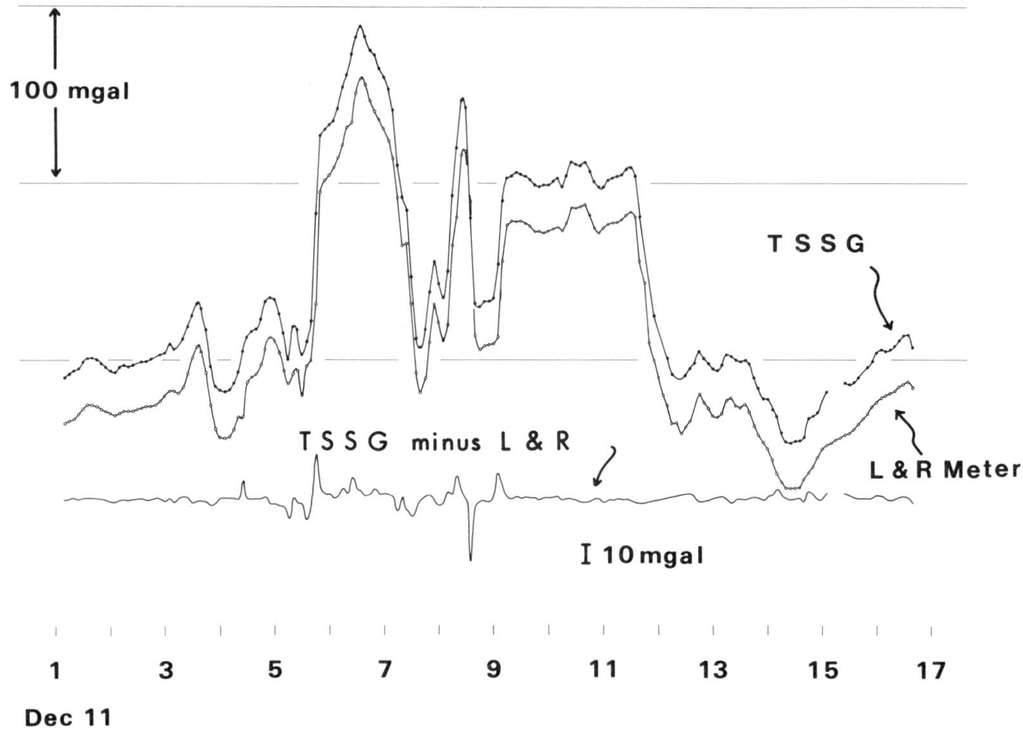


Fig. 1. A comparison of gravity measurements, using a Tokyo Surface Ship Gravity Meter (TSSG) and a LaCoste and Romberg gravity meter (L & R meter), carried out in December 1972. The upper two profiles of the figure indicate the measurements obtained from both meters; the lowest profile shows the difference (TSSG minus L & R meter) between the two.

of the oceanic trenches. These residuals are positive, observed at the outer flanks of the oceanic trenches, and are unaccounted for by the crustal structure alone. Therefore, we have attempted to explain the residual anomalies in terms of a possible concentration of mass associated with the bending of a downgoing lithosphere.

## CHARACTERISTICS OF THE FREE-AIR GRAVITY ANOMALIES IN AND AROUND JAPAN

Figure 2 presents the free-air gravity anomaly map of Japan and its vicinity, compiled by using data obtained prior to 1972. Data sources and coverage are given in the caption to the figure.

Geographic localities relating to the following descriptions are given in Figure 3.

Japan trench: The minimum free-air anomaly at the Japan trench is –200 mgal at the juncture with the Kuril trench. In the middle part of the Japan trench, the minimum is about –180 mgal. At the seaward margin the outer gravity high [*Watts and Talwani,* 1974] can be recognized.

Izu–Bonin trench: The minimum free-air anomaly at the Izu–Bonin trench is –300 mgal, which also occurs at the juncture with the Japan trench. Minimum values across other sections are about –250 mgal. So far as the area north of 30° N is concerned, the free-air minimum increases southward. At the eastern margin, an outer gravity high as large as 50 mgal is well developed.

Nankai trough: This trough, which runs ENE to WSW along the south flank of southwestern Honshu, has a comparatively poor free-air minimum, showing a value of only about –50 mgal. The outer gravity high is not distinct.

Ryukyu trench: The minimum free-air anomaly over this trench is about –100 mgal. A second minimum which occurs between the trench and the

Ryukyu arc is comparable to that of the trench. The outer gravity high shows a value of about 20 mgal, which is not as strong as that of the Izu-Bonin trench.

Japan Sea: In the Japan and Yamato basins the free-air anomalies are approximately zero, or slightly positive. Near the Yamato bank they increase to 80 mgal. A zone of gravity low is found at the northern part of this sea, paralleling coasts of the east Asiatic continent (Primorye, a province of the USSR opposite Japan across the Japan Sea). This low shows a value of -40 mgal.

Shikoku basin: This area shows a comparatively high free-air anomaly, averaging about 15 mgal.

East China Sea: At the southeastern margin of the East China Sea is the Okinawa trough where the free-air anomalies are zero or slightly positive. This means that the narrow topographic low behind the Ryukyu island arc is nearly in isostatic equilibrium, and is therefore considered to be oceanic in crustal structure. Toward the northwest where a shallow continental shelf is developing, the free-air anomalies are almost zero.

Amami plateau and Daito ridges: These ridges are located at the northern part of the western Philippine basin. They show positive free-air anomalies up to 100 mgal or higher at the topographic highs. But, free-air anomalies over the surrounding basin show abnormally low values which range from 0 to -60 mgal.

Kyushu-Palau ridge: This ridge is a linear relief running from Kyushu to the Palau Islands, separating the western Philippine basin from the Shikoku and Parece Vela basins. Free-air anomalies over this ridge are as high as 50 mgal. So far as the northern part is concerned, the eastern side of the ridge shows positive free-air anomalies, while those of the western side are negative. This may imply that the Kyushu-Palau ridge is a type of fault.

## ESTIMATION OF THE UPPER MANTLE STRUCTURE FROM GRAVITY ANOMALIES

In order to estimate inhomogeneity (density anomalies) suspected in the upper mantle, we have selected two areas, an east-west section across the Japan trench, northeastern Honshu, and the Japan Sea, and another across the Izu-Bonin trench, the Izu-Bonin ridge, and the Shikoku basin. These sections align with parallels 39°45′N and 31°47′N, respectively. In Figure 3 bathymetric features related to the present study are indicated together with the locations (profiles *a-b* and *c-d)* where crustal and upper mantle structure is being estimated.

Seismic refraction surveys in Japan have been conducted by the Research Group for Explosion Seismology (RGES) of Japan. In order to study the asthenosphere extending from the Japan trench to the Japan Sea across northeastern Honshu, we have used the results from refraction and reflection surveys obtained by *Hashizume et al.* [1968], *Ludwig et al.* [1966], *Yoshii and Asano* [1972], and *Rikitake et al.* [1968]. For the Izu-Bonin trench, the Izu-Bonin ridge, and the Shikoku basin, we have used results by *Hotta* [1970] and *Murauchi et al.* [1968]. Although the results from seismic refraction and reflection surveys are not necessarily convincing, our policy in the present work is to follow velocity structures obtained from the seismic surveys throughout. Where there are no seismic results, we have adjusted layered structures in order to fit gravity anomalies. Since an attempt to change layer structure is often very arbitrary, we have restricted such modifications to within narrow areas (<50 km), so that such modifications affect only short-wavelength variations of gravity. The relationship between $V_p$ and density of rocks may involve uncertainties on the order of ± 0.1 g/cm³. Such ambiguities make further discussion very difficult. However, for the present, we will assume a systematic relationship

---

Fig. 2. (Next two pages) Free-air anomalies over Japan and vicinity. Solid lines show zero and positive anomalies; dashed lines, negative anomalies. Contour intervals are 20 mgal. Numbers 0, 2, 4, 6, — mean 0, 20, 40, 60, —mgals, and $\bar{2}$, $\bar{4}$, $\bar{6}$, —mean -20, -40, -60 — mgals, respectively. Data sources and coverage are as follows: *Geographical Survey Institute of Japan* [1955, 1957, 1964, 1965, 1968], data for the island of Japan, L & R gravity meters were used; *Segawa* [1970], seas around Japan, TSSG's were used; *Tomoda* [1973*a*], seas around Japan, TSSG's were used; *Segawa and Bowin* [1975], Philippine Sea, Ryukyu island arc, East China Sea, Izu-Bonin island arc, Izu-Bonin trench, and Nankai trough, an L & R meter was used; *Caputo et al.* [1964], Philippine Sea, a gimbal-suspended L & R meter was used; *U.S. Department of Commerce, NOAA* [1969], Philippine Sea and others, a gimbal-suspended L & R meter was used; *Watts and Talwani* [1974], Philippine Sea, Ryukyu trench, and Izu-Bonin trench, a Graf-Askania Gss2 meter was used.

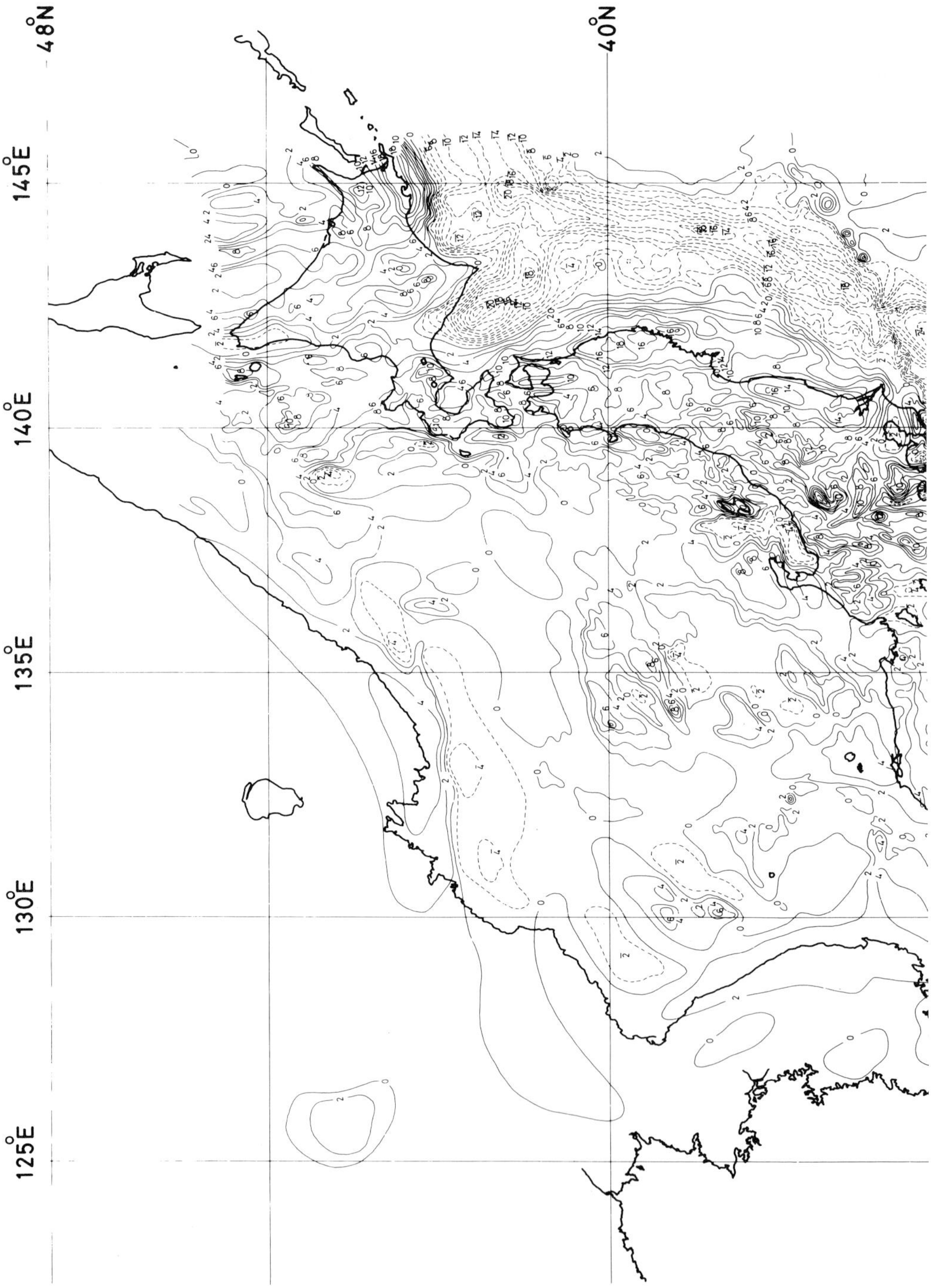


Fig. 2

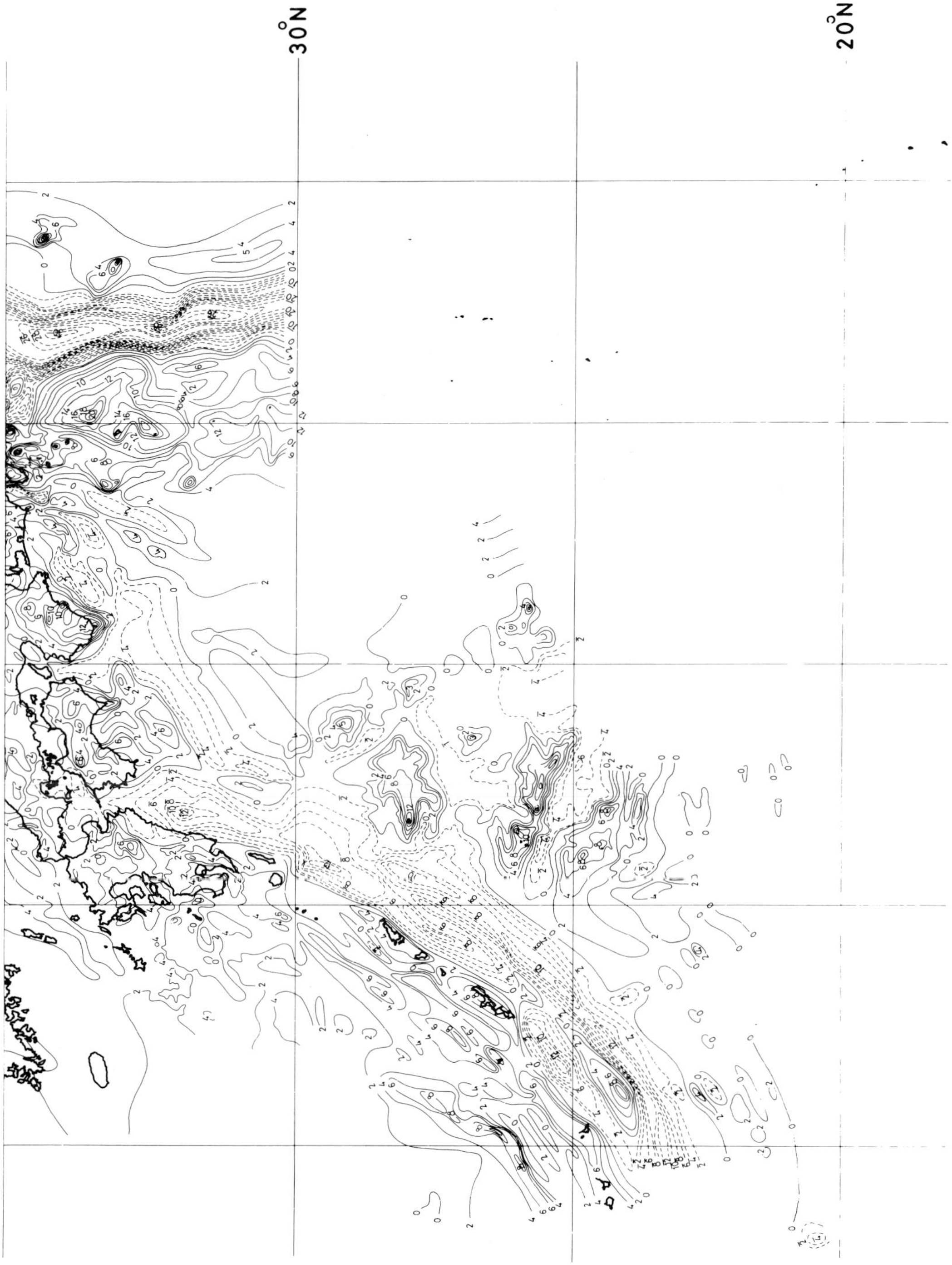


Fig. 2

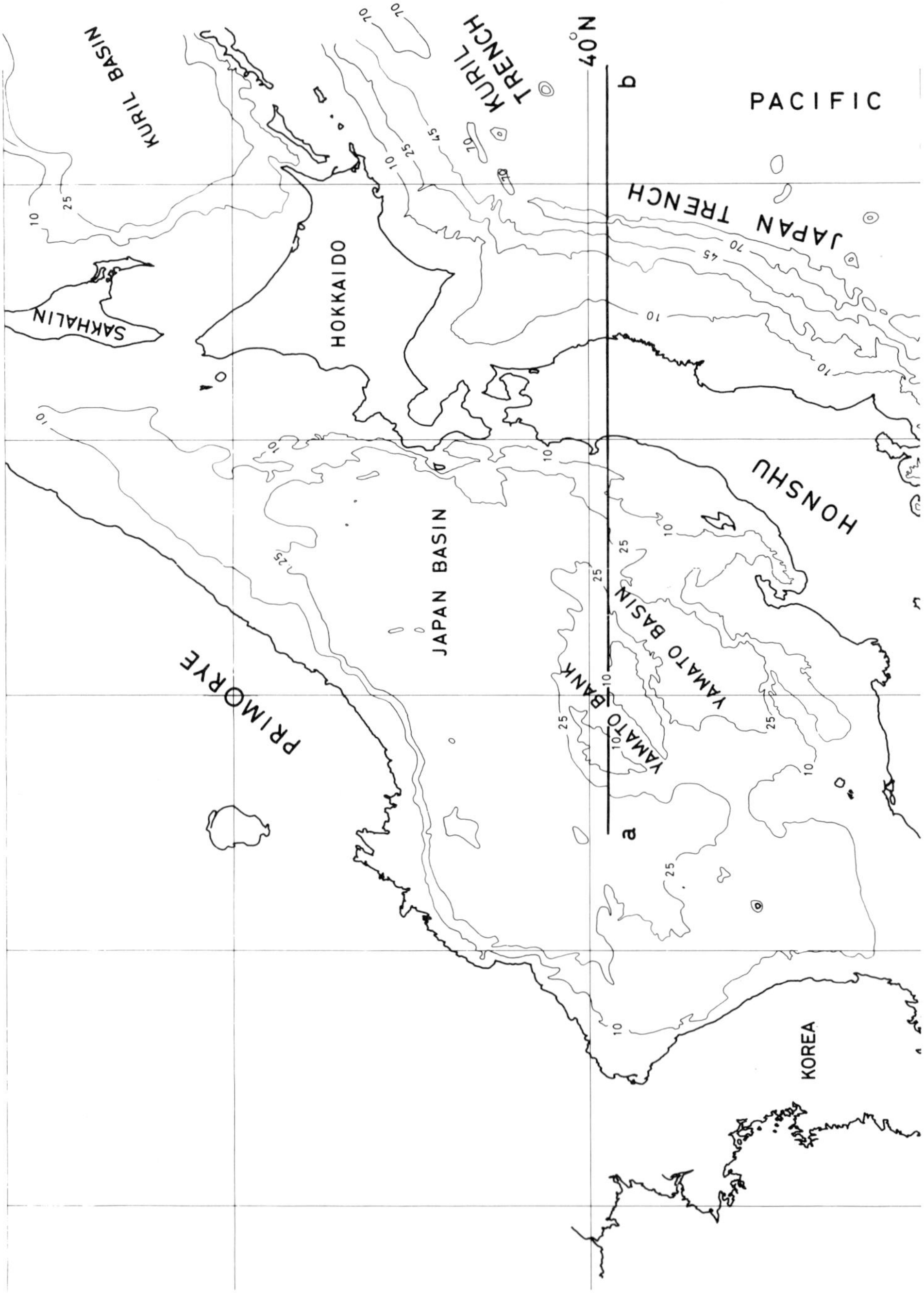


Fig. 3. Geographic and bathymetric map of Japan and vicinity. Numbers attached to bathymetric contours 10, 25, 45,—mean 1000, 2500, 4500,—m, respectively. Thick solid lines *a–b* and *c–d* indicate the location of the crustal and upper mantle structures discussed in the text.

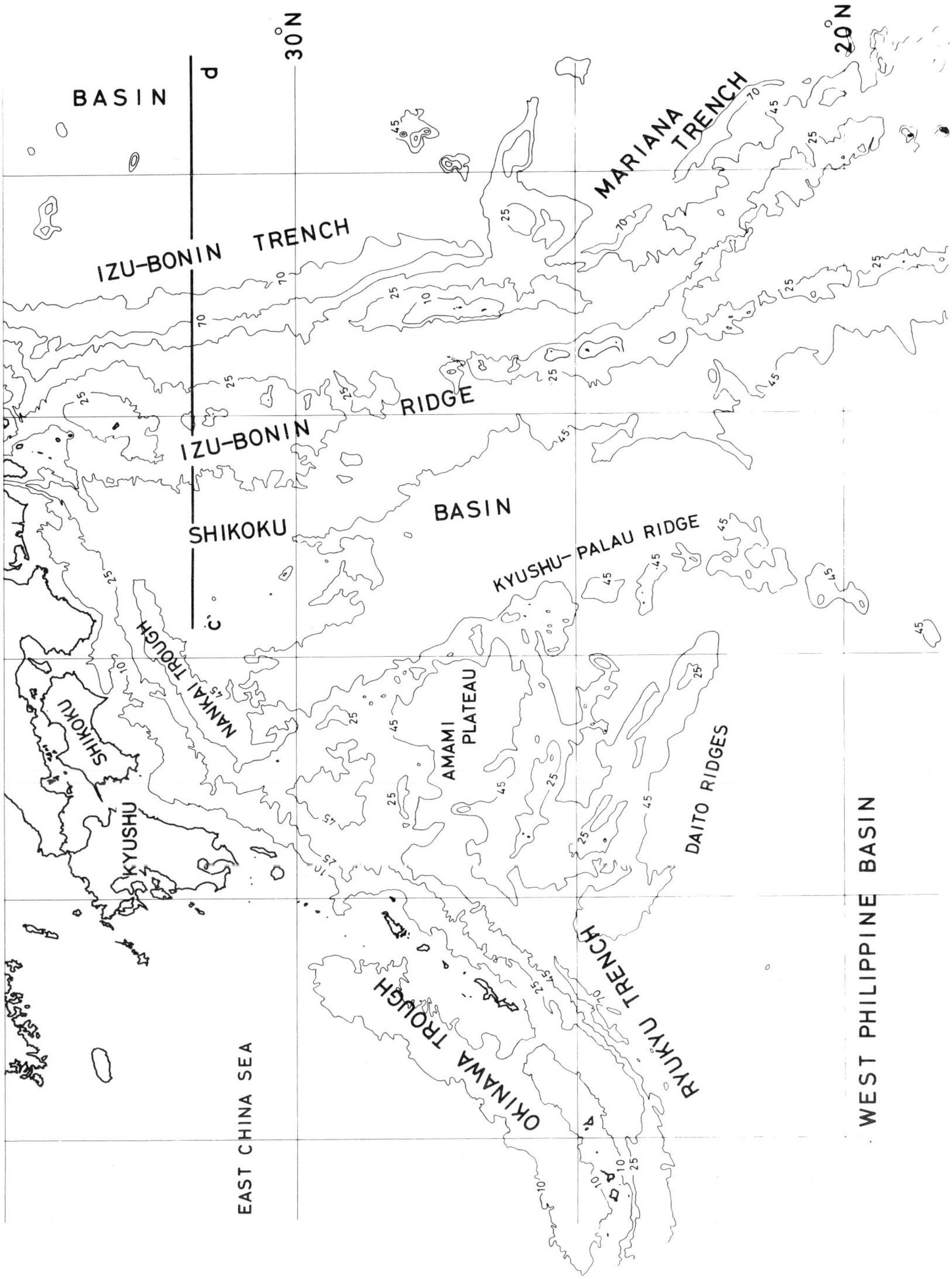
BASIN
d
30°N
20°N
MARIANA TRENCH
IZU-BONIN TRENCH
IZU-BONIN RIDGE
SHIKOKU BASIN
KYUSHU-PALAU RIDGE
c'
NANKAI TROUGH
SHIKOKU
KYUSHU
AMAMI PLATEAU
DAITO RIDGES
WEST PHILIPPINE BASIN
RYUKYU TRENCH
OKINAWA TROUGH
EAST CHINA SEA

Fig. 3

between $V_p$ and density, and then apply it uniformly to the various crustal layers and the uppermost part of the mantle. Thus, we have correlated densities 2.0, 2.3, 2.5, 2.7, 2.9, and 3.3 g/cm³ to *P*-velocities 2.0, 3.5, 4.5, 5.5, 6.5, and 8.0 km/sec, respectively, after *Nafe and Drake* [1957] and *Ludwig et al.* [1970]. Although the observed *P*-velocities of a layer scatter by as much as ±0.5 km/sec in the most extreme case, these variations have been neglected in this study.

*Derivations of crustal structures.* Figure 4 shows the crustal structure over the range from the Japan trench to the Japan Sea across the northeastern part of Honshu (line *a–b* in Figure 3). Where Moho surfaces were not measured by seismic surveys, their depths have been inferred from neighboring Moho surfaces which are known. Crustal structures beneath land areas reported by RGES are continuous profiles of each layer which best fit the observations (a fixed-

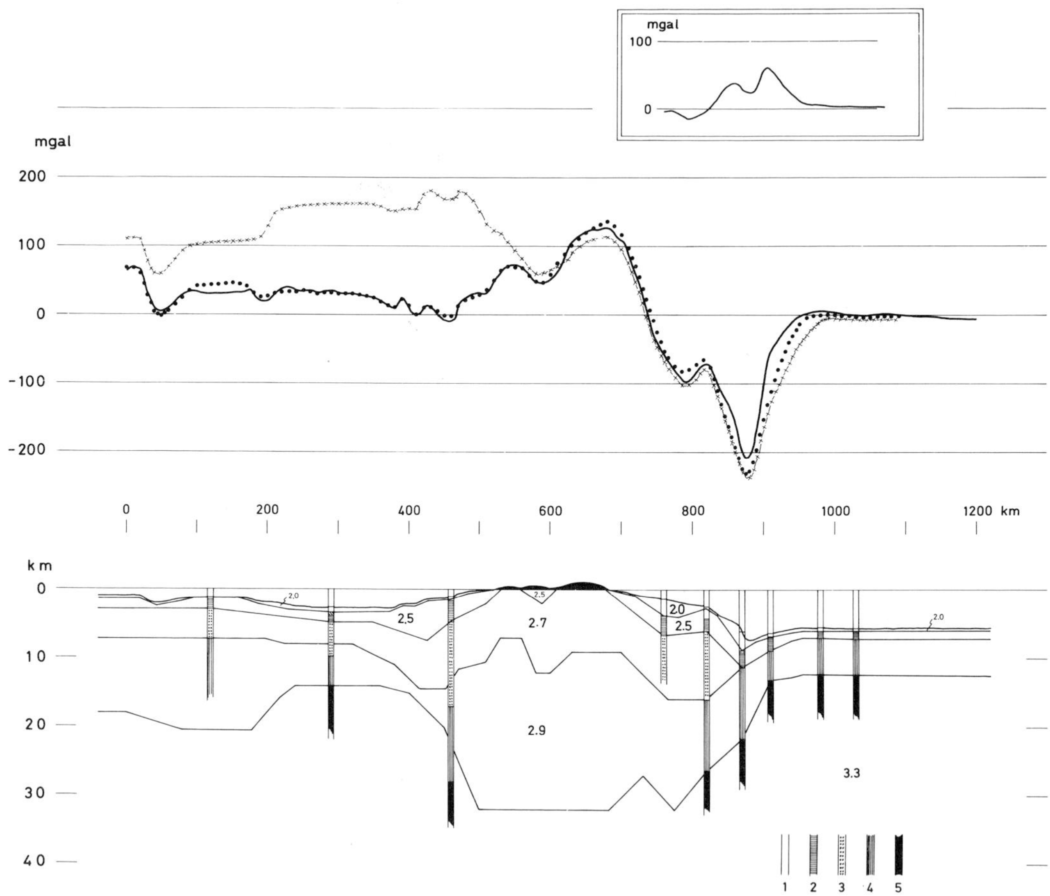


Fig. 4. Free-air anomalies and crustal structures across the Japan Sea, northeastern Honshu, and Japan trench (from left to right) along parallel 39°45′N. The solid line in the middle profile shows observed free-air anomalies. The line with cross marks shows computed free-air anomalies obtained by using only the crustal structures. The dotted line shows computed free-air anomalies obtained by using both the crustal structures and the density anomalies of the upper mantle in Figure 6. The lower part of this figure shows the crustal structures together with columnar descriptions of seismic results. Numbers indicated in each layer are densities in g/cm³. Lateral distances and depths below sea level are indicated in kilometers. Identification of the layers in each column is as follows: (1) sediment layer (2.0 km/sec); (2) volcanic layer (3.5 or 4.5 km/sec); (3) granitic layer (5.5 or 6.5 km/sec); (4) gabbroic layer (6.5 km/sec); and (5) peridotitic layer (8.0 km/sec). The uppermost profile enclosed by a rectangle shows residual anomalies.

array, moving shot point technique was used for this case, using a number of seismograph stations). *Hashizume et al.* [1968] derived an upper layer of 6.0 km/sec $V_p$ with a thickness of about 14 km, and a lower layer $V_p$ of 6.6 km/sec with a thickness of 18 km, resulting in a depth to Moho of 32 km for the middle part of northeastern Honshu. Their results have been re-examined by several later seismic refraction surveys, and *Yoshii and Asano* [1972] presented a different model, obtained from time-term analyses, for the same line as *Hashizume et al.* had previously discussed. Yoshii and Asano's model shows the upper layer of 17-km thickness with $V_p$ of 5.9 km/sec and the lower layer of 12-km thickness with $V_p$ of 6.6 km/sec, resulting in a depth to Moho surface of 29 km. Another seismic explosion survey was carried out recently along the same line as well as its seaward extension by *RGES* [1974]. RGES used large charge-size explosions fired both in the Pacific Ocean and the Japan Sea. One of the shots was located east of the Japan trench, which coincides approximately with the outer rise. From the results of Yoshii and Asano, and later results of the RGES, it is seen that the $P_n$ velocity at Moho surface beneath northeastern Honshu is as low as $7.53 \pm 0.05$ km/sec. Such a low velocity may suggest that a lower density should be assigned to the upper mantle beneath northeastern Honshu. In Figure 4, however, we have not employed their result a priori for the structure, but will wait until we find if such inhomogeneity of the upper mantle can be inferred from gravity anomalies.

The solid curve in the upper part of Figure 4 is the observed free-air anomalies, except for the land area where Bouguer anomalies are used for convenience. Though it is not a rigorous procedure, we have saved the additional labor of making topographic profiles on land by substituting simple Bouguer anomalies which are already given. In order to explain short-wavelength gravity anomalies on land, the thickness of the granitic layer (5.9 km/sec $V_p$) has been slightly modified. The depths to Moho beneath the east and west continental shelves obtained by seismic refraction surveys have also been changed slightly. On the other hand, the seismic controls have been rigorously observed for the structure beneath the Pacific basin, the Japan trench, and the Japan Sea. The curve with cross marks shows the computed free-air anomalies (Bouguer anomalies on land) obtained by assuming that the upper mantle is laterally homogeneous with a density of 3.3 g/cm³. The calculated anomalies were set equal to the observed where they cross zero in the Pacific basin (at 1,110 km in Figure 4). As a standard oceanic crust with zero free-air anomaly for the Pacific basin, we have chosen a crustal section similar to the easternmost column of Figure 4, i.e., of 5.6-km of water of density 1.03 g/cm³, of 0.4-km of sediment of density 2.0 g/cm³, of 1.2 km of volcanic layer of density 2.5 g/cm³, and of 5.2 km gabbroic layer of density 2.9 g/cm³, with a depth to Moho of 12.4 km. The outer topographic rise is recognized in the case of this section, its elevation being about 200 m above the level of the standard Pacific basin, hence it is not very prominent on this profile.

When the curve of computed free-air anomalies is compared with the observed free-air anomalies (computed Bouguer against observed Bouguer on land), we find that the computed anomalies do not agree with the observed anomalies, particularly in the area from west of the western half of northeastern Honshu to as far as the Japan Sea. The computed free-air anomalies are larger by about 130 mgal than those observed in the Yamato basin. Another discrepancy is recognized in the area extending from the Japan trench to the Pacific side of the continental shelf, where the computed anomalies are lower than the observed.

Figure 5 presents the crustal section across the Izu–Bonin trench, the Izu–Bonin ridge, and the Shikoku basin (line *c–d* in Figure 3). The crustal structure looks considerably different from the previous one, showing a Moho surface that is shallower than 20 km, and a $V_p$ layer of 5.4 km/sec (granitic layer?) that is limited to a narrow zone. We lack information on the Moho depth beneath both the western half of the Izu–Bonin ridge and the outer flank of the Izu–Bonin trench. Beneath the latter, we have assumed the ocean bottom and the Moho to run parallel. This condition has been employed from the standpoint that a trench results from bending of horizontal layers. A few more minor adjustments have been applied to the areas where seismic surveys are lacking, so that short-wavelength variations of free-air anomalies are better explained. The computed free-air anomalies shown by the curve with cross marks in Figure 5 have been obtained by assuming a homogeneous mantle with 3.3 g/cm³ density and a crust consisting of 5.7 km of water with 1.03 g/cm³ density, 0.4 km of sediment with 2.0 g/cm³ density, 1.5 km of volcanic layer with 2.5 g/cm³ density, and 5.5 km of gabbroic layer with 2.9 g/cm³ density, with a depth to Moho of 13.1 km, as a standard oceanic crust of zero free-air anomaly for the Pacific basin far beyond the Izu–Bonin trench. The standard crust for this section is slightly thicker than that of the previous section, but the effect is probably not serious. The computed free-air anomalies are much larger than the observed free-air anomalies also in the area from the Izu–Bonin ridge to the Shikoku basin. The difference amounts to 160 mgal. From the Izu–Bonin trench eastward to its outer rise, the computed free-air anomalies are, in turn, lower than the observed, similar to the case with the Japan trench.

*Density anomalies of the upper mantle.* Even if we consider the uncertainty in the density to be on the order of ± 0.1 g/cm³, it is impossible to explain differences as large as 150 mgal between the computed and the observed free-air anomalies simply by adjusting densities of crustal layers. Therefore we have to take into account lateral density variations of the mantle. The upper mantle above roughly 600 km is composed of the lithosphere and asthenosphere. *Kanamori and Press* [1970], *Kanamori* [1968], and *Abe and Kanamori* [1970*a, b*] obtained lithosphere thicknesses from studies of surface seismic waves and shear wave velocities. Their results show that the average thickness of the oceanic lithosphere is about 75 km, but that at marginal seas, such as the Japan basin or the Shikoku basin, the lithosphere is thinner than the average with a thickness of about 35 km. Recently, several authors have proposed that the lithosphere can thicken with age [*Mckenzie,* 1967; *Sclater et al.,* 1971; *Yoshii et al.,* this volume], and that, as the lithosphere grows its mineral composition may change,

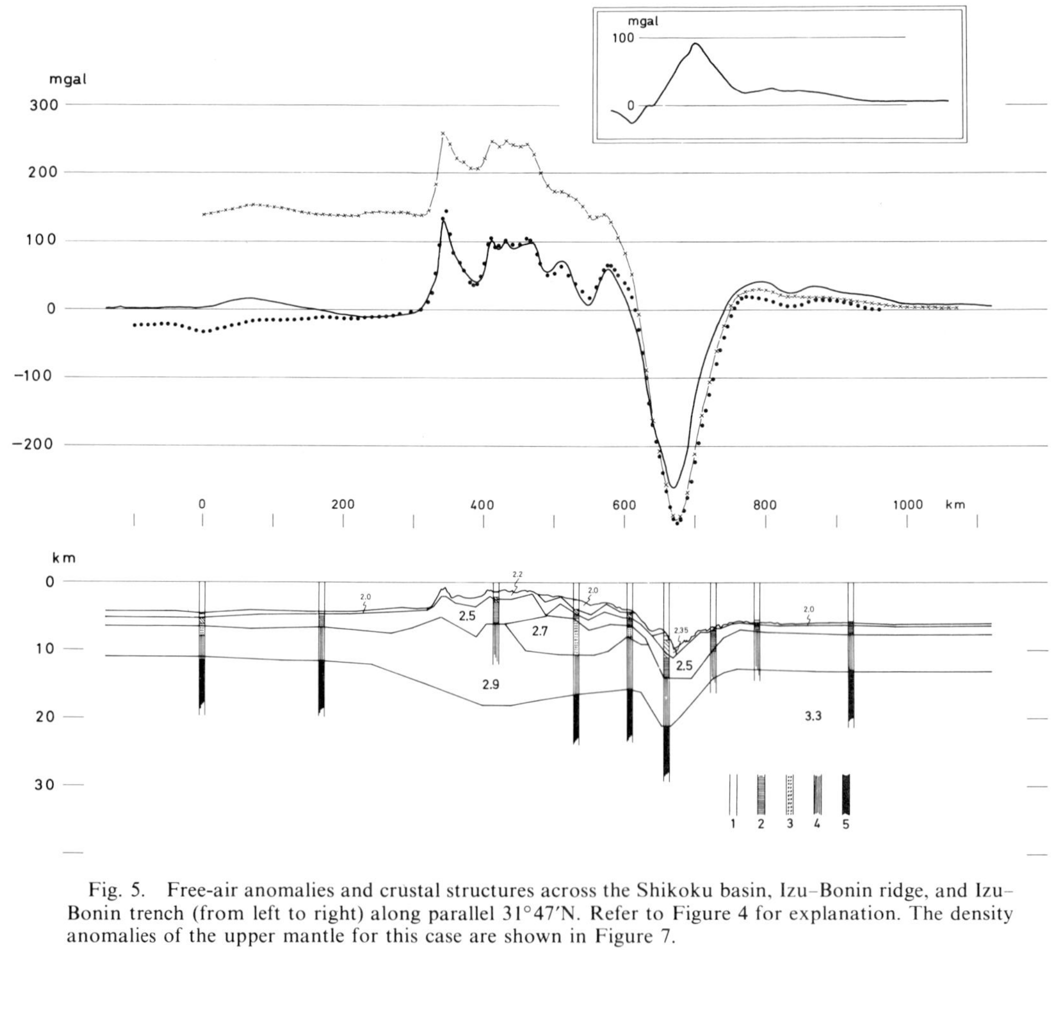


Fig. 5. Free-air anomalies and crustal structures across the Shikoku basin, Izu–Bonin ridge, and Izu–Bonin trench (from left to right) along parallel 31°47′N. Refer to Figure 4 for explanation. The density anomalies of the upper mantle for this case are shown in Figure 7.

so that the average density may change (this means 'increase', if we follow *Yoshii et al.* [this volume]). Therefore there is a strong possibility that a local change of lithosphere may account for the observed gravity anomalies.

On the other hand, the possibility of density anomalies in the asthenosphere may also be important. Since heat flows in marginal seas are often twice as large as those of the western Pacific basin, it is likely that the asthenosphere beneath the marginal seas is hotter than that beneath the Pacific (though some amount of time lag should always be considered when we interpret heat flow data), resulting in a decrease of density due to either thermal expansion or partial melting. Other evidence which supports the decrease of density of the asthenosphere is found in the studies of seismic intensities, amplitudes, wave forms, travel-time anomalies of body waves, and seismic surface waves [*Utsu,* 1967; *Oliver and Isacks,* 1967; *Kanamori,* 1968, 1970; *Wadati et al.*, 1969; *Utsu,* 1971*a*; *Grow and Qamar,* 1973; *Hamada,* 1973]. Most of these studies, however, were concerned with identification of the asthenosphere in contrast to the lithosphere, and less attention was paid to the inhomogeneity of the asthenosphere, except for a few cases. Moreover, since seismic waves had been observed in almost all cases by seismograph stations on islands, the seismic ray paths received were limited to a narrow area which was near the island arc. The previous results can be regarded as strong evidence for the existence of low-*Q*, low-velocity zones between island arcs and descending lithospheres. In this sense a surface wave study of the upper mantle beneath the Japan Sea or the Shikoku basin by *Abe and Kanamori* [1970*a, b*] is considered to be unique. They used seismic waves propagated from the Kuril arc to Seoul in Korea, or from the Flores Sea or the Banda Sea to Japan, crossing the Japan basin or the Shikoku basin, respectively. Their conclusions are that the asthenosphere just beneath the Japan basin shows lower *P*- and *S*-wave velocities than that beneath the islands of Japan. *Hamada* [1971] also discussed lateral heterogeneity of the asthenosphere and derived a lower *P*-velocity of the asthenosphere beneath the islands of Japan than that of the asthenosphere beneath the western Pacific. Thus, if we apply the density–velocity relationship under high pressures and temperatures derived by *Birch* [1969] and *Ringwood* [1969], we get lower densities attributable to layers of lower *P*- or *S*-velocities.

The existence of descending lithosphere beneath island arcs is another source causing heterogeneity of the upper mantle. *Yoshii* [1972*a*], *Grow* [1973], and *Grow and Bowin* [1975] interpreted gravity anomalies in terms of the effect of the descending lithosphere. *Watts and Talwani* [1974] thought that the total gravity effect of the descending lithosphere might be small on the grounds that the outer gravity high can be satisfactorily explained by topography seaward of trenches. *Grow* [1973] and *Grow and Bowin* [1975], on the other hand, pointed out that the approximation procedure used by Watts and Talwani was not accurate near the trench axis. As a counter proposal, Grow and Bowin proposed a density contrast averaging approximately 0.05 g/cm$^3$, but ranging up to 0.28 g/cm$^3$ where the descending oceanic crust transforms into eclogite.

For the sake of estimating density anomalies in the upper mantle, the authors have employed the following constraints:

Since the bases from which density variations of the lithosphere may possibly be derived are still very obscure, only variations of its thickness using a constant density are considered.

The density contrast between the lithosphere and the asthenosphere beneath the western Pacific near Japan is assumed to be 0.05 g/cm$^3$. The density of the descending lithosphere is assumed constant.

The shape of the descending lithosphere is determined from a deep-focused seismic plane, which is taken to be the upper surface of the descending lithosphere. The lower surface of the lithosphere is determined from the temperature distribution of the descending lithosphere, which has been estimated by *Toksöz et al.* [1971].

Using these assumptions, the density anomalies are calculated for the asthenosphere beneath the island arc and that beneath the marginal sea. If the lower boundary of the asthenosphere is taken to be as deep as 600 km, then the density anomaly is expected to be too small, 0.005 g/cm$^3$ or less. This seems incongruent with seismic results, and in this paper we shall assume that the density anomaly is confined within a depth shallower than 150 km.

Figures 6 and 7 show the density structure of the upper mantle thus obtained, for the Japan trench–northeastern Honshu–Japan Sea section and for the Izu–Bonin trench–Izu–Bonin ridge–Shikoku basin section, respectively. The upper surfaces of the descending lithosphere have been determined after *Katsumata* [1956], *Miyamura*

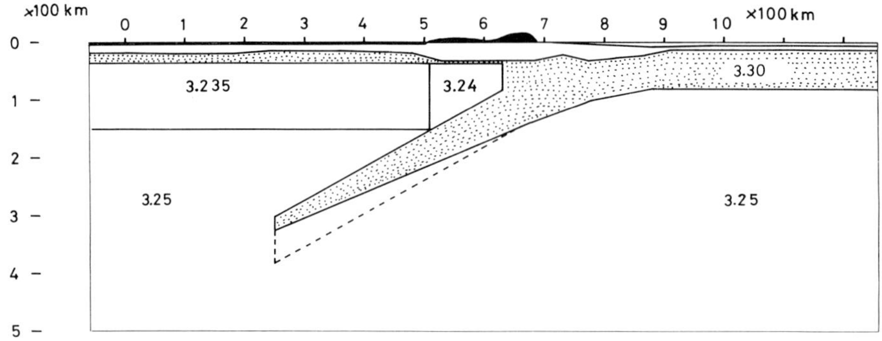


Fig. 6. Density distributions of the upper mantle across the Japan Sea, northeastern Honshu, and Japan trench (from left to right) along parallel 39°45′N. Lateral distances and depths below sea level are indicated in kilometers. The solid area indicates the land of northeastern Honshu. Dotted areas indicate the lithosphere, which is discriminated from the crustal layers above. A dashed line in the lower part of the descending lithosphere indicates an original lower surface of the lithosphere, which is thought to have been softened by heat. Numbers attached to each layer indicate densities in g/cm$^3$.

[1962], and *Katsumata and Sykes* [1969]. The thickness of the lithosphere beneath the western Pacific is assumed to be 75 km [*Kanamori and Press,* 1970]. The thickness of the lithosphere beneath the Japan Sea and the Shikoku basin is assumed to be 30 km [*Abe and Kanamori,* 1970*a*, 1970*b*]. The lower boundaries of the lithosphere beneath island arcs are not clear. Therefore, we have simply extended the lithosphere of marginal seas to beneath the islands. In the case of Honshu, the thickness of the lithosphere thus obtained becomes extremely thin (only a few kilometers), if the crustal part is excluded. The nature of the juncture between the lithosphere beneath the island and the downgoing slab is uncertain. The shapes shown in Figures 6 and 7 are therefore only tentative. The temperature distribution of *Toksöz et al.* [1971] with respect to a downgoing lithosphere sinking at a rate of 8 cm/yr and at an elapsed time of 12.96 m.y. has been used to determine the lithosphere shapes. Although it is difficult to apply their result rigorously to our problem, we have selected a temperature contour within a downgoing lithosphere, which is lower by 500°C than that of the surrounding mantle, and regarded the contour as delineating the downgoing slab. If we follow this method an effective downgoing lithosphere becomes thinner as it goes deeper. According to the olivine mantle model of *Birch* [1969], a 500°C temperature difference yields a density difference of about 0.05 g/cm$^3$ under isobaric conditions. So, the selection of a

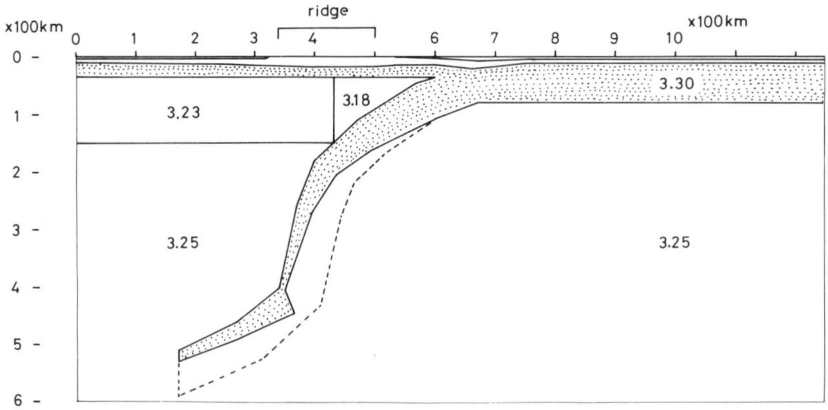


Fig. 7. Density distribution of the upper mantle across the Shikoku basin, Izu-Bonin ridge, and Izu-Bonin trench (from left to right) along parallel 31°47′N. Refer to Figure 6 for explanation.

temperature contrast of 500°C is roughly consistent with the assumed density contrast of 0.05 g/cm³. Since the shapes of the descending slabs indicated in Figures 6 and 7 are very simplified, we include them only to suggest that it might be better to consider the effect of temperature than to not.

If the density of the lithosphere is 3.3 g/cm³, the density of the asthenosphere beneath the western Pacific becomes 3.25 g/cm³. The authors assume that this would hold for the Pacific area—at least not far from the island arcs. Changes in density contrast would cause primarily a biasing of the standard of calculations. Thus, the densities of the asthenosphere in other regions have been determined from gravity in comparison to the asthenosphere beneath the western Pacific. Dotted lines in the upper part of Figures 4 and 5 show computed free-air anomalies (Bouguer anomalies on Honshu) obtained by considering the effect of the upper mantle, while taking the crust and mantle beneath the western Pacific as a standard.

In the area beneath northeastern Honshu the density of the asthenosphere is lower than that on the Pacific side by 0.01 g/cm³, whereas in the area beneath the Japan Sea it is lower by 0.015 g/cm³. The density of the asthenosphere beneath the Izu–Bonin ridge is lower than that on the Pacific side by 0.07 g/cm³, whereas beneath Shikoku basin it is lower by 0.02 g/cm³. The decrease in density beneath the Izu–Bonin ridge is remarkable.

*Residual gravity anomalies observed at oceanic trenches.* Shown in the uppermost part of Figures 4 and 5 are residual gravity anomalies near oceanic trenches which represent the difference between the observed and the computed free-air anomalies. The latter for this case are those obtained when both crustal structure and density anomalies of the asthenosphere are considered, as previously mentioned. Therefore, the residuals are those which have been left unexplained in the previous discussions. In this paper we have intended to extract the residual at the oceanic trenches as genuinely as possible, by removing all other sources which are suspected as influencing gravity near trenches. The characteristics of the residuals can be described as follows:

Japan trench:
- (a) The residual is positive.
- (b) Its maximum amplitude is about 60 mgal.
- (c) The maximum residual is observed at about 20 km outward from the trench axis.
- (d) The width between points of half-amplitude is 70 km.

Izu–Bonin trench:
- (a) The residual is positive.
- (b) Its maximum amplitude is about 95 mgal.
- (c) The maximum residual is observed at about 30 km outward from the trench axis.
- (d) The width between points of half-amplitude is 100 km.

We have not considered the effect of the basalt–eclogite transformation on gravity [*Grow,* 1973; *Grow and Bowin,* 1975] which is expected to occur in the downgoing lithosphere. But the result by Grow shows that such an effect amounts to not more than 5 mgal at the outer flank of the trench. The residual of the Izu–Bonin trench may, to some degree, be changed if the layer with 2.5 g/cm³ density beneath the outer flank of the trench in Figure 5 is assumed otherwise than it is (unfortunately, no seismic result is available here). However, since the amount of the gravity change due to such modifications of layering appears to be less than 20 mgal, the fact that the Izu–Bonin trench is associated with a larger residual than the Japan trench would not be changed. So, if we accept the characteristics described here, they will suggest that the origins of such residual gravity anomalies may exist in shallow zones, presumably, within the lithosphere. Since residual gravity anomalies are common to most of the oceanic trenches, it may not sound unreasonable to say that they are caused by the bending of lithosphere, which is also a feature common to all subduction zones. *Watts and Talwani* [1974] attributed the broad, observed gravity positive seaward of trenches to upwarping of the seafloor over the outer rises, which is in turn caused by bending of the lithosphere. They used a model of lithosphere consisting of 2.7 g/cm³ crust and 3.3 g/cm³ mantle, and calculated gravity caused by a model which best fits the actual bathymetry at trenches and vicinities. They concluded that the edge effect of the downgoing lithosphere is small and limited to the region very near the trench. *Grow and Bowin* [1975] observed that Watts and Talwani's approximation technique was inaccurate near the trench axis and then proposed that important anomalies exist due to phase transforms of rocks which are taking place within a descending lithosphere. Referring to *Ito and Kennedy* [1971] and *Green and Ringwood* [1970], Grow and Bowin proposed that transforms of oceanic basalt (3.0 g/cm³) to eclogite (3.55 g/cm³), and of plagioclase peridotite (3.24 g/cm³) and pyroxene peridotite (3.28 g/cm$^3$) to garnet peridotite (3.38 g/cm$^3$), occur at depths between 30 and 150 km in the downgoing lithosphere, which significantly affects the gravity field near trenches. What Grow and Bowin proposed may be true, but, even after including these phase transforms in their model of the Chile trench, residual gravity anomalies still remain that are similar to those we have observed over the Japan and Izu–Bonin trenches. The remaining residuals observed in the Chile profile have an amplitude of about 30 mgal, and the maximum is found at approximately 30 km outward from the Chile trench axis. This fact as well as the results of our study have made us conclude that the residual anomalies over trenches are caused by unknown excess masses existing within lithospheres just beneath the trenches.

Figure 8 presents a comparison of the residual gravity anomalies and models of excess mass which account for the anomalies over the Japan and Izu–Bonin trenches. The Japan trench residual can be explained by a rectangular block 55 km long and 30 km wide with a top surface at 15-km depth, whose density is higher than the normal lithosphere by 0.1 g/cm$^3$. The Izu–Bonin trench residual can be explained by a rectangular block 60 km long and 35 km wide with a top surface at depth 20 km, whose density is higher by 0.15 g/cm$^3$.

What does the difference in excess mass from one trench to another mean? Although we need to know about more trenches other than the ones under discussion, we can speculate that the residual gravity anomalies become larger as the size of the trench becomes large. We show some characteristic features of the Japan and Izu–Bonin trenches in Table 1 and Figure 9, which suggest that the Izu–Bonin trench is maintained by much greater bending of the lithosphere than the Japan trench. Roughly speaking, the residual gravity anomalies, the excess mass causing the anomalies, and the apparent strain of lithosphere caused by the elastic bending are all 1.5 times larger in the Izu–Bonin trench than in the Japan trench. This implies that, in turn, the residual gravity anomalies may have been caused by the bending of the lithosphere associated with sinking of an oceanic plate.

*Hanks* [1971] and *Watts and Talwani* [1974] showed that the maximum elastic stress due to semi-elastic bending of the lithosphere at trenches takes place at approximately 50 km outward from trench axes in the case of most oceanic trenches. Locations of the maximum bending stress look, as a first approximation, harmonious with the locations of maximum residual gravity anomalies observed in the case of the Japan and Izu–Bonin trenches. This is further evidence to support the hypothesis that excess mass is primarily caused by elastic bending of the lithosphere.

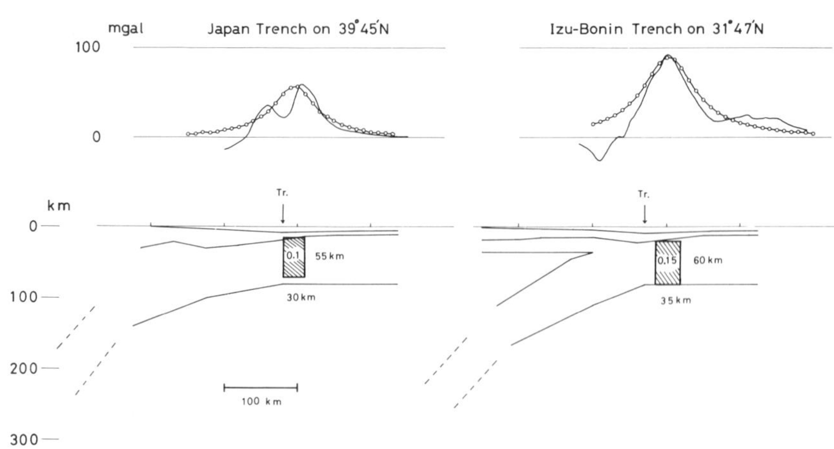


Fig. 8. Comparison of observed residual gravity anomalies (solid lines) and computed residual gravity anomalies (open circles) for the best-fit models of excess masses beneath trenches (hatched areas) between the Japan and Izu–Bonin trenches. In the lower part of the figure are shown the shapes of crust and lithosphere in the vicinity of the trenches. Values within the hatched areas indicate the excess in density over the normal density of the lithosphere (3.3 g/cm$^3$) in g/cm$^3$. Values in kilometers shown close to the hatched areas are width and thickness of the anomalous portions of lithosphere.

TABLE 1. Characteristics of the Japan and Izu–Bonin Trenches

| | Japan Trench, 39°45′N | Izu–Bonin Trench, 31°47′N |
|---|---|---|
| Distance between trench axis and outer margin of trench, $\ell$ | 100 km | 130 km |
| Depth of trench axis below level of the basin, $d$ | 1.5 km | 3.5 km |
| Approximate radius of curvature, $R$ | 3,300 km | 2,400 km |
| Strain of lithosphere | ± 1.1 % | ± 1.6 % |
| Strain-rate | about $10^{-15}$ strain/sec | |
| Density change | ± 0.036 g/cm$^3$ | ± 0.053 g/cm$^3$ |
| Pressure change | ~ ± 30 kbar | ~ ± 50 kbar |

Notes: $R = [d^2 + \ell^2] / 2d$.
Compressibility of a lithosphere assumed to be 0.3 Mbar$^{-1}$.

Correlating the residual gravity anomalies to the elastic bending of the lithosphere is, however, not as straightforward as it appears. Although the existence of outer topographic rises seaward from oceanic trenches seems to support the idea strongly that the bending of the lithosphere at a subduction zone is more or less elastic in nature, there is also some adverse evidence such as the low strain-rate with which the lithosphere is supposed to bend. Assuming the Pacific plate velocity is 10 cm/yr, which is the case with the Japan and Izu–Bonin trenches, the strain-rate may be as low as $10^{-15}$ strain/sec. Therefore, plastic flow is also probably occurring, and simple

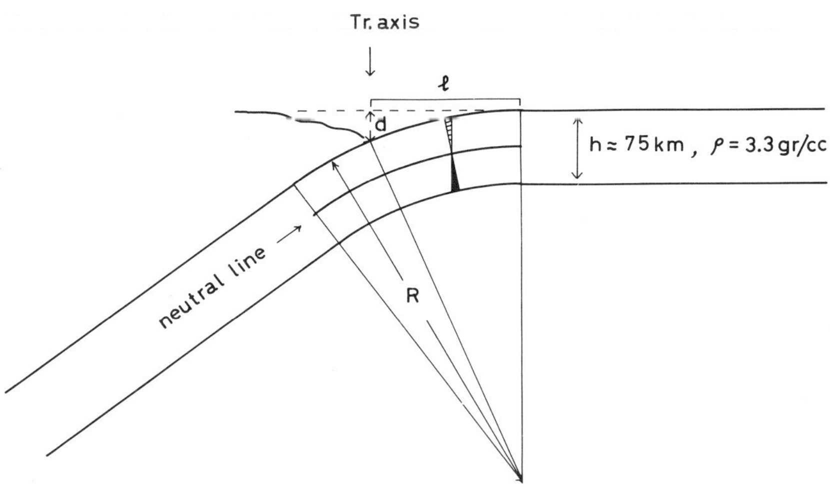


Fig. 9. A schematic view of the elastic bending of a uniform lithosphere assumed at subduction zones. In this case the upper half of the lithosphere is tensioned, while the lower half is compressed. The neutral line is located at the center; $\ell$ indicates distance from the trench axis to the outer margin of the trench; $d$, depth of the trench axis below the level of the neighboring oceanic basin; and $R$, radius of curvature of the descending lithosphere.

elastic bending probably cannot account for all of the facts relating to the downgoing plate. Nevertheless, it is clear from the present knowledge of gravity, bathymetry, and seismicity that a significant amount of elastic stress exists within the lithosphere beneath a trench, and we think the stress or its secondary effects must have caused some excess mass within the bending zone of the lithosphere.

Figure 10 illustrates a possible mechanism by which an increase in density could take place beneath oceanic trenches. If the descending lithosphere is uniform in its physico-chemical condition and behaves as an elastic body, then its upper half is tensioned and its lower half is compressed, leaving its center as a neutral plane. However, since the lower boundary of the lithosphere is considered to be as hot as 1000°–1200°C [*Ringwood,* 1969; *Ito,* 1973], it is reasonable to think that compression in the lower part of the lithosphere proceeds faster than extension in the upper part, resulting in raising the neutral plane and narrowing the zone of extension. This unsymmetric deformation will cause the average density of the lithosphere to increase. In Figure 10, the upper 20 km of the lithosphere is assumed to be in an extensional field, while the lower 60 km is in a compressional field. Bathymetric normal faults [*Ludwig et al.,* 1966; *Hotta,* 1970] as well as shallow earthquakes with tensile focal mechanisms observed at the outer flank of trenches [*Stauder,* 1968*b*; *Shimazaki,* 1972] are regarded as evidence suggesting that the upper portion is in an extensional field. Since the strain of the bending lithosphere is of the order of ± 1–2%, a density change of approximately ±0.03–0.06 g/cm$^3$ can be expected in the bending zone, if the density of the lithosphere is assumed to be 3.3 g/cm$^3$. When an unsymmetric deformation takes place it will cause an increase in the average density on the order of +0.05 g/cm$^3$ or less, because the area of the compressional field exceeds that of the extensional field. But this amount of density increase falls short of what is required to explain the residual gravity anomalies. Therefore, we have to consider other effects in addition to the density increase due to elastic compression.

Just as *Grow and Bowin* [1975] applied phase transforms of rocks to the descending lithosphere, we also apply them to the bending zone of the lithosphere. The phase transformation itself does not cause any change in total mass, hence no appreciable change in gravity. However, the volume change associated with the phase transformation does cause either convergence or divergence of the surrounding mass, which necessarily affects the local gravity. Although the mineral composition of the lithosphere is not known well, we assume that the upper lithosphere (10–30 km deep) is plagioclase peridotite, the middle lithosphere

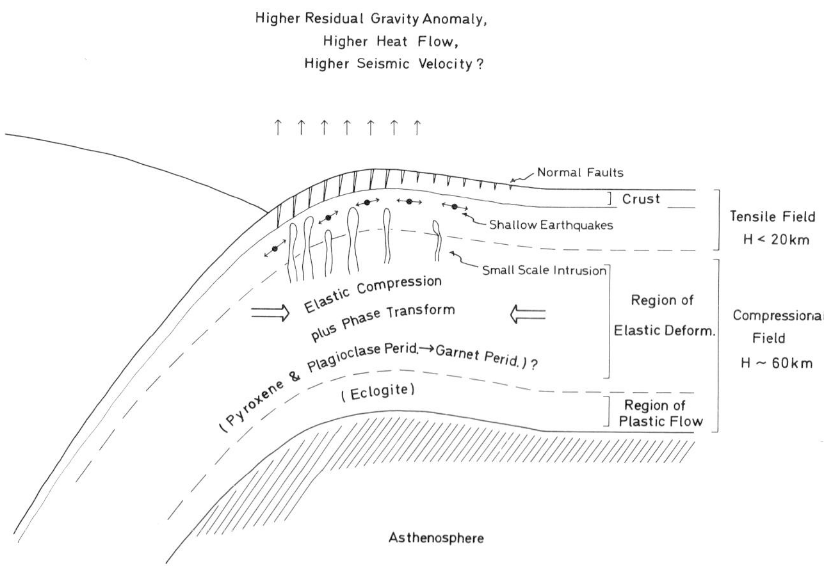


Fig. 10. A speculative schematic showing various phenomena taking place within bending lithosphere. An excess mass beneath oceanic trenches is thought to result from these phenomena.

(30–60 km deep) is pyroxene peridotite, and the lowest lithosphere is eclogite [*Yoshii et al.*, this volume]. Layering of the lithosphere has not been considered in the previous discussions, but it does not cause any significant change in the estimation of the residual gravity anomalies as long as the layers are assumed uniform within the lithosphere. If temperature is kept unchanged, the plagioclase peridotite (3.24 g/cm$^3$) of the upper lithosphere transforms to garnet peridotite (3.38 g/cm$^3$) under pressures higher than 10 kbar, and the pyroxene peridotite (3.28 g/cm$^3$) of the middle lithosphere also transforms to garnet peridotite (3.38 g/cm$^3$) under pressures higher than 26 kbar. If these transforms should occur, a density increase on the order of 0.1 g/cm$^3$ could be expected. An elastic compression of more than 1% in strain would be sufficient to give rise to such a high horizontal pressure, but it is not certain that the strength of the lithosphere would be adequate to support tectonic stresses greater than 1 kbar [*Brace et al.*, 1970]. In this paper, we only note that the increase in density beneath trenches may be partly due to phase transforms of peridotite in the bending zone.

Heat flow values at oceanic trenches have been believed to be subnormal so far. But when previous measurements are examined in detail, those values that range from the outer slope of a trench to its outer topographic rise show slightly higher values than normal, by approximately 0.2–0.4 HFU [see, for example, *Yasui et al.*, 1970]. This higher heat flow may suggest possible magmatic intrusions into the upper part of the bending lithosphere. Although the mechanism of intrusion is not known, we assume here that a kind of magma may flow from the upper surface of the compressed part into the tensioned part. The magma flow may prevent voids from forming in the upper lithosphere by filling the rifts of the upper lithosphere with magmas.

It is also suspected that if this intrusion should reach Moho, some differences in seismic wave velocities must be observed. However, so far as previous seismic results are concerned, no significant differences have been reported. We have found a few instances in which $V_p$ observations higher than those of the surroundings at Moho surfaces beneath the outer slope of oceanic trenches have been reported [*Ludwig et al.*, 1966; *Fisher and Raitt*, 1962], but they do not look convincing.

## DISCUSSION

*Choice of a standard structure and gravity interpretation.* Computed gravity anomalies are based on assuming a gravity at a certain point which we think of as the standard structure. In the present study we have chosen the western Pacific basin and the upper mantle below as a standard. The western Pacific basin near Japan is known from DSDP sample datings [*DSDP Scientific Staff*, 1971*a*] as well as studies of magnetic lineations [*Larson and Pitman*, 1972], as an upper Jurassic or lower Cretaceous basin. On the other hand, recent studies of the evolution of the lithosphere have suggested that not only its thickness but also its mineral composition is subject to considerable change with time while traveling from a ridge axis toward trenches. This means that we should be most careful in choosing a standard point for gravity computation. *Yoshii et al.* [this volume] estimated the thickness of the lithosphere as old as upper Jurassic or lower Cretaceous as about 80 km. This thickness is not so different from the thickness of our standard lithosphere, though it was originally derived from the "standard thickness of the lithosphere" proposed by *Kanamori and Press* [1970]. Our present results can therefore be regarded as an estimation of the density anomalies of island arcs and marginal seas in comparison to the upper Jurassic or lower Cretaceous upper mantle. Since there is still some uncertainty concerning the true thicknesses of the lithosphere, our results may require revision when the lithosphere is better understood.

*Relations between density anomalies and velocity anomalies of the asthenosphere.* *Kanamori* [1968] obtained a 4% $V_p$ contrast in the low-velocity layer beneath the islands of Japan from the travel time anomalies of the LONGSHOT underground explosion of 1965 at Amchitka, Aleutian islands, and explained it in terms of a 2% partial melting coupled with a 500°C temperature difference. *Abe and Kanamori* [1970*b*] obtained an 8% $V_s$ decrease in the asthenosphere beneath the Japan Sea from surface wave studies, and explained it in terms of a 4% partial melting coupled with a 500°C temperature difference. The 8% $V_s$ contrast corresponds to a 6% $V_p$ contrast under the same conditions. From these studies it is apparent that the asthenosphere beneath the Japan Sea may be softer and lower in density than that beneath the islands of Japan. *Hamada* [1973] also discussed compressional velocities in the upper mantle near Japan, from studies of *P*-wave travel time anomalies on the islands of Japan, using the nuclear explosion CANNIKIN in 1971 at Amchitka. His results show $V_p$ decreases of 4% and 7% in the asthenosphere beneath the western Pacific and the islands of Japan, respectively, as compared with the $V_p$ of the lithosphere. Hamada's $V_p$ contrast for the asthenosphere beneath Japan seems a little larger than Kanamori's value. Similar seismic results were obtained for the Shikoku basin [*Abe and Kanamori,* 1970*a*]. In spite of some inconsistencies between seismic results, we conclude that the asthenosphere beneath an island arc and its marginal sea shows densities lower than those beneath the western Pacific basin.

Following *Birch* [1969], the relationship between a $P_v$ change and a density change under isobaric conditions is

$$\left(\frac{\partial V_p}{\partial \rho}\right) \sim 6.0 \text{ km sec}^{-1} \text{ gr}^{-1} \text{ cm}^3$$

which was derived from his laboratory experiments with respect to corundum, forsterite, and olivine. This result applies to the case of temperature change without phase change. If we calculate $P_v$ in the asthenosphere, using densities obtained for the case of the Japan and Izu-Bonin island arcs in the previous section, we obtain the following results:

| | Change in $V_p$, % |
|---|---|
| Asthenosphere beneath the western Pacific | -3.8 |
| Asthenosphere beneath northeastern Honshu | -4.5 |
| Asthenosphere beneath the Japan Sea | -4.9 |
| Asthenosphere beneath the Izu-Bonin ridge | -9.0 |
| Asthenosphere beneath the Shikoku basin | -5.3 |

These estimations almost explain the seismic results, without having to consider the effects of partial melting. The $P_v$ of the asthenosphere beneath the island arcs and marginal seas is lower than that beneath the western Pacific by -0.7 to -5.2%. The differences are consistent with seismic results, though somewhat smaller, and in favor of the assumption of a 150-km depth limit for the anomalous asthenosphere.

If we assume that a density change of 10% occurs when rocks of the upper mantle melt, a 4% partial melting, for example, would result in an average density decrease of 0.4%, i.e., about 0.013 g/cm³ decrease in the case of the mantle with 3.3 g/cm³ density. This implies that the partial melting has less influence on the average density than a temperature change, which is anticipated to be about 500°C. Hence, one possibility is that the lower density of the asthenosphere is primarily determined by a temperature change, and that minor variations of density in the asthenosphere may be correlated to the ratio of the partial meltings.

## CONCLUSION

Two problems have been treated by use of two-dimensional gravity calculations in this paper: one is estimation of lateral density variations within the asthenosphere, the other, a possible density increase caused by bending of the lithosphere at subduction zones.

If the anomalous asthenosphere is assumed to be confined within depths shallower than 150 km, the asthenosphere beneath northeastern Honshu and the Japan Sea along 39°45′N shows density contrasts of -0.01 g/cm³, and -0.015 g/cm³, respectively, as compared with the density (3.25 g/cm³) of the asthenosphere beneath the western Pacific. The density contrasts of the asthenosphere beneath the Izu-Bonin ridge and the Shikoku basin along 31°47′N obtained under the same assumptions are -0.07 and -0.02 g/cm³, respectively. Density reductions of the asthenosphere at island arcs and marginal seas are qualitatively consistent with higher heat flows, which are expected to cause a higher temperature, and a consequent larger thermal expansion and more extensive partial melting of the asthenosphere. Estimations of *P*-velocity contrasts from density anomalies of the asthenosphere, under the assumption that the density contrasts are caused by temperature changes with no melting, are in close agreement with the seismic results. Partial melting of the upper mantle on the order of several percent appears to be less important to density changes than the temperature anomalies.

The existence of positive residual gravity anomalies on the order of 50 to 100 mgal have been identified near the outward slopes of oceanic trenches. We have tried to explain the anomalies in terms of mass concentration within a bending lithosphere beneath a trench, and we have found a positive density contrast on the order of 0.1 to 0.15 g/cm³ assumed within a rectangular block about 30 km wide and 60 km thick. Origins of such an excess mass have been speculated from the point of view of elastic and plastic compression of the lithosphere, phase transforms of peridotite due to compression, and possible flows of magma within the lithosphere.

*Acknowledgments.* The authors thank Carl Bowin, who has been a co-worker in gravity measurements for the last several years. Manik Talwani and Tony Watts kindly provided some of the gravity data used in this paper, which we acknowledge. We express our thanks to George Sutton and Seiya Uyeda, who encouraged us to complete this work.

# 4. Airy–Woollard Isostasy

LUMAN E. WILCOX

*Defense Mapping Agency, Aerospace Center, St. Louis AFS, Missouri 63118*

*Abstract.* It is proposed that the variable crustal density variety of Airy isostatic theory introduced and developed by George P. Woollard be known as Airy–Woollard isostasy. An augmented form of Airy–Woollard isostasy involving a variable mantle density is also derived. Standard Airy–Heiskanen isostasy is compared with Airy–Woollard isostasy using the crustal column concept and basic equations.

The purposes of this paper are (1) to call attention to a significant modification of Airy isostatic theory introduced and developed by George P. Woollard, (2) to define the principal elements of this modified isostatic system, and (3) to propose that this modified system be known as Airy–Woollard isostasy.

The concept of modifying Airy isostatic theory to allow for a crust of variable density as well as variable thickness was suggested and applied by Woollard in a classic document on geophysical interpretation of gravity anomaly data published in 1962 [*Woollard,* 1962]. Since that time, Woollard has authored several additional articles which refine and demonstrate the usefulness of the variable crustal density concept for gravitational analysis of crustal structure. The Airy–Woollard concept differs distinctly from the Pratt isostatic hypothesis which has no provision for a crust of variable thickness and in which the density of the mass above the level of compensation is always related inversely to the topographic elevation.

The present paper includes revised extracts from a previous work of the writer [*Wilcox,* 1974] which briefly but carefully define the principal elements and some of the geophysical aspects of Airy–Woollard isostasy. The concept is recommended for application in studies of crustal and upper mantle structure and the correlation of seismic velocity data with gravity data, but not for computation of isostatic anomalies to be used for geodetic purposes. For comprehensive illustrations of some useful applications of Airy–Woollard concept, the reader is directed to *Woollard* [1959, 1962, 1966*c*, 1968, 1969*a, b,* 1970] and *Woollard and Strange* [1966].

The views expressed herein are those of the writer and do not necessarily reflect the views of the Defense Mapping Agency or the Department of Defense.

## ISOSTASY

Isostasy refers to a state of equilibrium in the outer parts of the earth in which (1) the land masses which extend above sea level are counterbalanced by a compensating mass deficiency beneath sea level, and (2) the ocean basins which contain low-density water are counterbalanced by a compensating mass excess beneath the ocean floor.

The general validity of the isostatic principle has been established conclusively using purely geodetic arguments. For example, the fact that free-air anomalies are largely uncorrelated with regional elevation changes can be cited as evidence of the existence of regional isostatic balance. The reader is referred to *Heiskanen and Vening Meinesz* [1958] for free-air, Bouguer, and isostatic anomaly statistics which demonstrate that, on a regional basis, the mountains and ocean basins are very close to being in complete isostatic equilibrium.

Some departures from regional isostatic balance do exist, for example, recently deglaciated regions. Other examples may include plateaus currently undergoing uplift and mountain areas where isostatic adjustments have not kept pace with erosion of topography [*Woollard,* 1969*a,* p. 325]. In addition, the crust of the earth has sufficient strength to support local mass distribution variations and, consequently, short-period density and topographic variations are largely uncompensated. The strong correlation between free-air anomalies and local topographic variations, for example, proves that the latter are not compensated [*Wolf,* 1973].

The exact physical mechanisms of the isostatic process are unknown. However, there are a number of isostatic theories which probably provide at least a good approximation of the isostatic mechanism. Each of these theories specifies an exact manner in which the compensating mass deficiencies or excesses are distributed within the earth. One such theory must be adopted and used as a basis for determining the isostatic correction in isostatic anomaly computations and, in general, for estimating the gravitational effects of the compensating masses. The theory which is most frequently used as a basis for geophysical analysis of crustal structure is that of G. B. Airy which dates from 1855 [*Airy,* 1855].

## AIRY ISOSTATIC THEORY

Airy, in effect, postulated the existence of a relatively thin crust standing above a denser base material (the mantle). In the Airy system, the crust beneath the continents extends downward into the mantle and, conversely, the mantle under the oceans projects upward into the crust such that the total mass per unit area down to some level just beneath the deepest continental crustal material is everywhere equal. Essentially, Airy's system has a crust of uniform density floating in a denser mantle material, also of uniform density, in accordance with Archimedes' principle.

W. A. Heiskanen developed theory and practical procedures for computing isostatic gravity anomalies using the Airy principle in 1938 [*Heiskanen,* 1938]. Also, using the geophysical knowledge of the day together with geodetic arguments, he proposed mean rock density and crustal thickness values appropriate for these computations. The resulting system is known today as Airy–Heiskanen isostasy.

More recently, George P. Woollard has used modern geophysical and geochemical knowledge and evidence to deduce the most probable mean rock density and crustal thickness parameters for an Airy-type isostatic system. Woollard also introduced and refined the 'crustal column' method of geophysical gravity analysis. All of the foregoing are elements of Airy–Woollard isostasy.

## DERIVATION OF BASIC AIRY SYSTEM FORMULAS

Consider a pair of crustal columns floating in the mantle in accordance with Archimedes' principle, Figure 1. The upper surface of column $S$ is at sea level; the upper surface of column $P$ extends $h$ kilometers above sea level. In order to hydrostatically support its topographic mass, column $P$ has a crustal root which extends a distance $\Delta R$ kilometers deeper into the mantle than column $S$. (Column $S$ is called the standard or reference sea-level column; column $P$ represents a column of any height whose mean density is the same as the standard column.)

Appropriate parameters for these Airy-type crustal columns, as given by *Heiskanen* [*Heiskanen and Vening Meinesz,* 1958] and *Woollard* [1969*a*], are listed in Table 1 where $\sigma_S$ is the

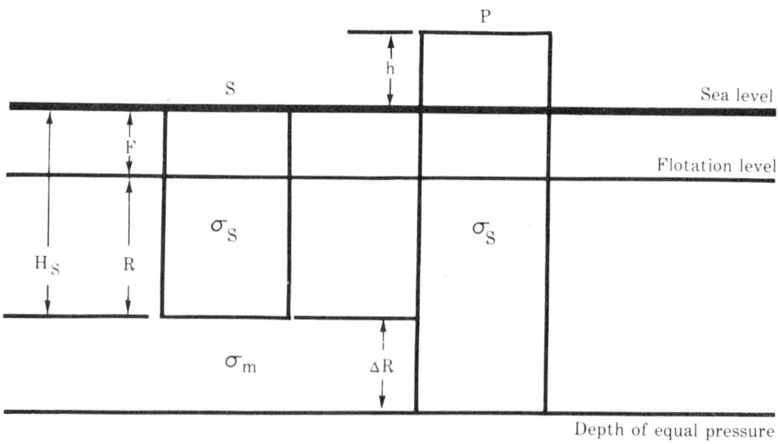


Fig. 1. Crustal columns for Airy isostasy.

TABLE 1. Parameters for Airy Heiskanen and Airy–Woollard Isostatic Models

| Parameter | Heiskanen* | Woollard* |
|---|---|---|
| $\sigma_S$ | 2.67 g/cm³ | 2.93 g/cm³ |
| $\sigma_m$ | 3.27 g/cm³ | 3.32 g/cm³ |
| $H_S$ | 30 km | 33 km |
| $F/R$ | 1/4.45 | 1/7.5 |

*Data from: *Heiskanen and Vening Meinesz* [1958]; *Woollard* [1969*a*].

mean density of the standard crustal column, $\sigma_m$ is the expected mean density at the top of the mantle, $H_S$ is the expected thickness of the reference sea level crustal column, and $F/R$ is the expected ratio of free board to root.

Either set of parameters may be used to compute isostatic anomalies. For geodetic purposes, the isostatic model used need only give agreement between the compensating masses at depth and the surficial masses above sea level for changes in surface elevation. Therefore, the use of more correct values for crustal and upper mantle density in computation of isostatic anomalies according to the Airy hypothesis is not important [*Woollard,* 1969*c*]. The existing procedures for computation of isostatic anomalies using, for example, the Airy–Heiskanen system give values which are good enough for many geodetic and other purposes. In making geophysical interpretations of Airy–Heiskanen isostatic anomalies, on the other hand, the manner in which the Airy–Heiskanen isostatic model differs from physical reality must be considered. Use of more correct density values is also important, for example, in quantitative studies of crustal and upper mantle structure and in correlations of seismic velocity with gravity anomaly data [*Woollard,* 1969*a*]. For the latter purposes, Bouguer anomaly-type gravitational equations (as given in this paper) using the more correct density values of the Airy–Woollard system are adequate. Thus, it is not necessary to actually compute isostatic anomalies using the Airy–Woollard system in order to make geophysical applications of the system.

To develop the basic equations for the Airy isostatic principle note that, according to Archimedes' principle, a floating body displaces its own mass. Therefore, the mass of the standard column, column *S* of Figure 1, is equal to the mass of the mantle material displaced by its root.

Since,

$$m = V\sigma \tag{1}$$

and

$$V = AH \tag{2}$$

then it must be true for the standard column that

$$A\, H_S\, \sigma_S = A\, R\, \sigma_m \tag{3}$$

where

$m$ = mass
$V$ = volume
$A$ = cross sectional area.

With a unit cross-sectional area, the above becomes

$$H_S\, \sigma_S = R\, \sigma_m\ . \tag{4}$$

Equation (4) can be modified to reflect changes in crustal root due to changes in elevation. To convert from the standard sea level column (column *S* of Figure 1) to a general crustal column of elevation *h* (column *P* of Figure 1), it is evident that $H_S$ must be replaced by $h + H_S + \Delta R$ and that *R* must be replaced by $R + \Delta R$. Putting these substitutions into (4) gives

$$(h + H_S + \Delta R)\ \sigma_S = (R + \Delta R)\ \sigma_m \tag{5}$$

or, in another form

$$\Delta R = \frac{h\,\sigma_S}{\sigma_m - \sigma_S}\ . \tag{6}$$

Equations (5) and (6) are basic for Airy–Heiskanen isostasy and show that equilibrium for variations in elevation, *h,* is attained by variations in the length of the crustal root, $\Delta R$, but without variations in density since the values for $\sigma_m$ and

$\sigma_S$ are fixed.

Corresponding gravitational equations are:

$$\Delta g_T = 2\pi k\, \sigma_S\, h$$

for the topography, and

$$\Delta g_C = -2\pi k\, (\sigma_m - \sigma_S)\, \Delta R$$

for the compensation where $k$ is the gravitational constant.

### AIRY-WOOLLARD ISOSTASY

Woollard modifies the basic Airy-type equations to allow for a variation in crustal density as well as in crustal thickness—which is more in keeping with the situation actually found in nature.

Let

$$\sigma_C = \sigma_S + \Delta\sigma_C \tag{7}$$

where

$\sigma_C$ = actual mean crustal density
$\sigma_S$ = expected mean density for the standard sea level crust.

Replace $\sigma_S$ by $\sigma_S + \Delta\sigma_C$ in (5) to obtain

$$(h + H_S + \Delta R)(\sigma_S + \Delta\sigma_C) = (R + \Delta R)\,\sigma_m \tag{8}$$

or,

$$h\,(\sigma_S + \Delta\sigma_C) + \Delta R\,(\sigma_S + \Delta\sigma_C) + H_S\sigma_S + H_S\Delta\sigma_C = R\sigma_m + \Delta R\sigma_m$$

Considering (4) and (7), the above becomes

$$\sigma_C h + \Delta\sigma_C H_S = (\sigma_m - \sigma_C)\,\Delta R \tag{9}$$

or, in another form

$$\Delta R = \frac{\sigma_C h + \Delta\sigma_C H_S}{\sigma_m - \sigma_C}\,. \tag{10}$$

Equations (9) and (10) express Airy-Woollard isostasy in its basic form. Here, compensation of elevation variations is a function of both crustal density variations ($\Delta\sigma_C$) and length of crustal root ($\Delta R$).

The gravitational equations for Airy-Woollard isostasy are

$$\Delta g_T = 2\pi k \sigma_C h$$

for the topography, and

$$\Delta g_C = 2\pi k\,[\Delta\sigma_C H_S - (\sigma_m - \sigma_C)\,\Delta R]$$

for the compensation.

### AUGMENTED AIRY-WOOLLARD ISOSTASY

The writer suggests that one further modification can be made to Airy isostatic theory, namely, inclusion of an anomalous mantle density. Let

$$\sigma_M = \sigma_m + \Delta\sigma_M \tag{11}$$

where

$\sigma_M$ = actual mean density of the upper mantle, and
$\sigma_m$ = expected mean density of the standard upper mantle.

Replace $\sigma_m$ by $\sigma_M + \Delta\sigma_M$ in (8) to obtain

$$(h + H_S + \Delta R)(\sigma_S + \Delta\sigma_C) = (R + \Delta R)(\sigma_m + \Delta\sigma_M)$$

or,

$$h\,(\sigma_S + \Delta\sigma_C) + \Delta R\,(\sigma_S + \Delta\sigma_C) + H_S\sigma_S + H_S\Delta\sigma_C = R\sigma_m + R\Delta\sigma_M + \Delta R\,(\sigma_m + \Delta\sigma_M).$$

Considering (4), (7), and (11), the above reduces to

$$\sigma_C h + \Delta\sigma_C H_S - \Delta\sigma_M R = (\sigma_M - \sigma_C)\,\Delta R \tag{12}$$

or, in another form,

$$\Delta R = \frac{\sigma_C h + \Delta\sigma_C H_S - \Delta\sigma_M R}{\sigma_M - \sigma_C}\,. \tag{13}$$

Since $H_S = F + R$, the two equations above may also be written in the form

$$\sigma_C h + (\Delta\sigma_C - \Delta\sigma_M)\,H_S + \Delta\sigma_M F = (\sigma_M - \sigma_C)\,\Delta R \tag{14}$$

$$\Delta R = \frac{\sigma_C h + (\Delta\sigma_C - \Delta\sigma_M)\,H_S + \Delta\sigma_M F}{\sigma_M - \sigma_C}\,. \tag{15}$$

Equations (12) to (15) express an augmented form of Airy-Woollard isostasy in which compensation of elevation variations is a function of variations in crustal density ($\Delta\sigma_C$), mantle density ($\Delta\sigma_M$), and length of crustal root ($\Delta R$).

The gravitational equations for augmented Airy-Woollard isostasy are

$$\Delta g_T = 2\pi k \sigma_C h$$

for the topography, and

$$\Delta g_C = 2\pi k\,\left[(\Delta\sigma_C - \Delta\sigma_M)\,H_S + \Delta\sigma_M F - (\sigma_M - \sigma_C)\,\Delta R\right]$$

for the compensation.

## SOME IMPLICATIONS OF AIRY-WOOLLARD ISOSTASY

The difference between Airy–Heiskanen isostasy (no density variation) commonly used for isostatic anomaly computations and the geophysically more realistic Airy–Woollard isostasy (density variation possible) is given by the difference between (15) and (6).

$$\delta\Delta R = \frac{\sigma_C h + (\Delta\sigma_C - \Delta\sigma_M) H_S + \Delta\sigma_M F}{\sigma_M - \sigma_C} - \frac{\sigma_S h}{\sigma_m - \sigma_S} \qquad (16)$$

which is approximately equal to

$$\delta\Delta R \approx \frac{\Delta\sigma_C (h + H_S) - \Delta\sigma_M (H_S - F)}{\sigma_M - \sigma_C}. \qquad (17)$$

Equation (17) shows that there is a direct relationship between variations in crustal thickness and variations in crustal density. That is, an increase in crustal thickness is required to maintain isostatic equilibrium when actual mean crustal density exceeds the standard values. Conversely, $\delta\Delta R < 0$ when $\sigma_C < \sigma_S$ in order to maintain isostatic equilibrium. It is also true, from (17), that there is an inverse relationship between variations in crustal thickness and variations in mantle density. Since the velocity of seismic waves shows significantly greater lateral and vertical variations in the crust than in the uppermost mantle, it follows that $|\Delta\sigma_C| > |\Delta\sigma_M|$; also, $|h + H_S| > |H_S - F|$. Therefore, crustal effects are predominant.

Now a greater than normal mean crustal density ($\Delta\sigma_C > 0$) must exert a positive influence on observed gravity and, for this case, an insufficient amount of compensation is predicted by Airy–Heiskanen isostatic theory which ignores the effects of variations in mean crustal density. As a result, the Airy–Heiskanen isostatic anomaly will be positive (greater than zero) in this case even though isostatic compensation is complete. Conversely, a lower than normal mean crustal density can generate a negative Airy–Heiskanen isostatic anomaly even when isostatic compensation is complete.

The reader is directed especially to *Woollard* [1962, 1969*a*] and to *Woollard* [1959, 1966*c*, 1968, 1969*b*, 1970] and *Woollard and Strange* [1966] for comprehensive illustrations of the application of Bouguer-type gravitational equations and the Airy–Woollard isostatic system for studies of crustal structure in conjunction with seismic and other geophysical data.

# 5. *A Simulation Study for Sub-Meter Geodesy in the Pacific Basin*

P. Morgan

*Division of National Mapping, P.O. Box 667, Canberra City, A.C.T., 2601, Australia*

*Abstract.* The advent of Lunar Laser Ranging (LLR), with precisions greater than 10 cm, offers geodesists and geophysicists an unprecedented tool for measuring the dynamic earth and exploring its mysteries. LLR is a fixed-site astronomical technique not readily adaptable to rapid observations in some of the earth's harsher environments nor to the special logistics of those regions. Orbiting artificial satellites, as well as being weather independent, offer techniques that make use of portable equipment. Unfortunately this gain in versatility is at the expense of precision and accuracy. The combination of three Pacific basin LLR observatories collocated with Doppler tracking stations offers the possibility of recovering geodetic information at the sub-meter level. The actual level of the precision of recovery is dependent on the configuration adopted. The Timation III satellite, the only current high precision geodetic satellite capable of spanning the Pacific basin, is particularly suited for this work as offsets and biases of all oscillators taking part in the observing program can now be calibrated, thus considerably tightening up the a priori conditions of the solution. Simulation studies combining Doppler tracking of the Timation satellite in a multi-pass multi-station short arc mode with time transfer information and high precision LLR control show that sub-meter positions can be determined in a unique center of mass reference system over the entire Pacific basin. At the present time the level of precision is 0.3 m per component. This precision should be improved once the satellite is also tracked by laser ranging devices, and transportable LLR becomes operational.

Over the last decade we have seen the birth of satellite geodesy, the fulfillment of its initial goal of a unified global geodetic reference system, and its branching into the exploration of the causative processes that perturb or limit classical geodetic reference systems or datums. The initial goals of the first satellite projects were station coordinates with an error radius of 10 m [*Schmid,* 1964]. Present production accuracies from the U.S. Navy Transit Satellite System are at the 1-m level [*Anderle,* 1973]. This level of precision is not acceptable for most geodynamic investigations where the annual variation of coordinates relative to a fixed or known reference system is of the order of 0.1 m. Thus a number of proposals have been made to obtain higher precision and in many cases greater density of stations useful for geodynamic purposes [*Plotkin,* 1974; *Siry,* 1974; *Silverberg,* 1973]. Unfortunately, many of the new proposals, especially those of *Siry* [1974], are beyond current technology and there is an urgent need to commence monitoring and experimentation now.

Two production systems are currently available to geodesists:

| *System* | *Type of Tracking* | *Precision* |
|---|---|---|
| 1. The U.S. Navy Transit System | Doppler | 1–2 m |
| 2. NASA Goddard Modlas | Satellite laser | 0.5 m |

In addition there are a number of observatory or fixed site systems that determine precise geodetic information:

| *System* | *Type of Tracking* | *Precision* |
|---|---|---|
| 1. SAO Baker Nunn | Optical | 10 m |
| 2. SAO Laser | Satellite laser | 0.5 m |
| 3. LLR Observations | Lunar laser ranging | 0.1 m |

In general, precision is greater at fixed sites and until some of the proposed transportable high precision instruments, e.g., the transportable lunar laser ranging system [*Silverberg,* 1973], become available it is necessary to seek combination solutions where the fixed sites are used to control the scale, orientation, and distortion of a network.

A number of recent advances have been made in satellite hardware, such as Timation III [*Easton,* 1974]. When coupled with increased orbital altitude, these advances render practicable the use of combination methods in covering global areas such as the Pacific basin. Timation III is the third generation satellite in an experimental program to provide precise global real time positioning and navigation. This system, described by *Easton* [1972*a, b*], is the combination of the range rate method of tracking used to generate the Doppler counts and multiplexed frequencies for phase comparison to deduce direct range observations as pioneered by *Wadley* [1957] for the tellurometer. In addition, the satellite carries an optical corner cube array for the retrodirection of laser pulses. This allows us to consider combinatorial techniques using the following:

(1) Precision laser and Doppler tracking from fixed sites coupled with easily moved, weather-independent Doppler observations at the required site.

(2) The combination of satellite observations with baseline information derived from lunar laser ranging, so that orientation and scale of the Doppler network is consistent with the best available reference system.

(3) The introduction of time-scale information into the observations rather than the treatment of time and frequency as unknown parameters or the deduction of mathematical models which are not sensitive to time.

### THE DOPPLER RANGE EQUATION

Doppler tracking observations can be viewed either as ranges or as range differences. The latter approach was used in the initial phases of the TRANET program and is still used in many applications [*Stansell et al.,* 1965]. The more common name for this approach to Doppler tracking is intermittently integrated Doppler (IID). For combined techniques where laser ranges are available, this approach is not as useful as the continuously integrated Doppler technique (CID) since it involves a new zero setting error with each integration, whereas CID has only one such error per pass. The strength and value of the CID approach was first pointed out by *Brown* [*Brown and Trotter,* 1969] and *Krakiwsky* [1970]. However, it was not possible to take full advantage of this approach before the launching of Timation III.

The basic equation for all Doppler work can be found in most physics texts, e.g., *Joos* [1958]. For this application, the following form is derived from the special theory of relativity:

$$\frac{f}{f_r^s} = \left(1 - \frac{\dot{r}}{c}\right) \Big/ \left(1 - \frac{\mathbf{V} \cdot \mathbf{V}}{c^2}\right)^{1/2} . \quad (1)$$

(See the Notation list at the end of this paper.)

In the above equation the rate of change of the scalar quantity range, $\dot{r}$, is the magnitude of the velocity vector $\mathbf{V}$ in the direction of the vector $\mathbf{X}^s - \mathbf{X}^g$, i.e., $\dot{r} = \mathbf{V} \cdot \mathbf{I} = v \cos \theta$. Consequently

$$\dot{r} = c\left[1 - \frac{f}{f_r^s}\left(1 - \frac{v^2}{c^2}\right)^{1/2}\right] . \quad (2)$$

Expand (2) by the binomial theorem, and neglect all terms smaller than $v^2/c^2$, in particular neglect $(1 - f/f_r)(v^2/c^2)$:

$$\dot{r} = c\,\frac{f_r^s - f}{f_r^s} + \frac{v^2}{2c} . \quad (3)$$

Using the wavelength equation, $\lambda = c/f$, and the expression for the frequency difference, $\Delta f$, between the receiver oscillator and the arriving satellite frequency, $\Delta f = f_r^g - f$, (3) can be transformed into:

$$\dot{r} = \lambda_r\left[\Delta f + (f_r^s - f_r^g)\right] + \frac{v^2}{2c} . \quad (4)$$

Under the assumption that $f_r$ and $f_r^g$ are constant over the observing interval, it is possible to determine a range difference by counting the beat frequency cycles as follows:

$$r - r_0 = \int_0^\tau \dot{r}dt = \lambda_r N + \lambda_r(f_r^g - f_r^s)\tau + \int_0^\tau \frac{v^2}{2c}\,dt \quad (5)$$

where

$$N = \int_0^\tau \Delta f dt$$

is the number of beat frequency cycles.

That is, the change in range $(r - r_0)$ from time $t = 0$ to $t = \tau$ is, apart from relativistic corrections

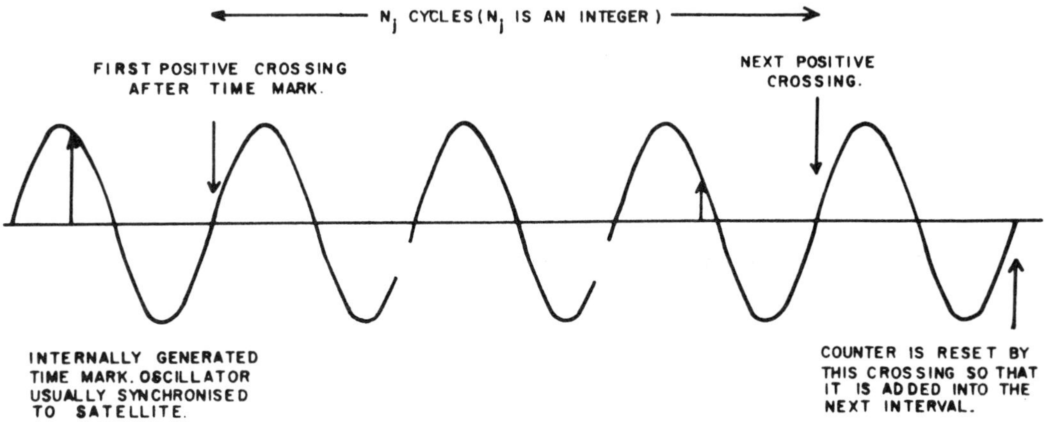


Fig. 1. Schematic for counting the Doppler Beat Frequency using the CID method.

and the linear term involving the frequency offset, $\lambda_r(f_r^g - f_r^s)\tau$, simply the sum of beat frequency cycles; provided that the continuous counting of these cycles is maintained, then the same $r_0$ applies to all points. This places a stability requirement on the oscillators as the frequency offset term must not contain any component resulting from frequency changes or drift.

For the CID method of tracking,

$$N = \int_0^T \Delta f dt = \Delta N_1 + \Delta N_2 + \Delta N_3 + \ldots \Delta N_T . \qquad (6)$$

Thus the error sources in the system are not the integral beats but the start and stop times over which the tracking is accomplished. Figure 1 indicates the CID approach to tracking as used in the receivers.

In an effort to reduce the very rigid requirements on frequency, it is necessary to model the frequency behavior of the oscillators involved so that parameters describing the behavior are either recovered from the solution or determined by external calibration. The most widespread and applicable model for oscillators is the following first order polynomial [*Guinot et al.*, 1971; *Allan et al.*, 1974], although second order models have been used:

$$f_r = f_0 + \delta f_r + \dot{f}_r \tau. \qquad (7)$$

Generalizing the range rate (4) with these frequency models and subsequent simplifications by way of the wavelength equation and the series expansion of $1/f_r$ yields:

$$\dot{r} = \lambda_0 \left(1 - \frac{\delta f_r^s}{f_0^s}\right) (\Delta f - \Delta f_0) + \lambda_0 (\delta f_r^s - \delta f_r^g) + \lambda_0 (\dot{f}_r^s - \dot{f}_r^g)\tau + \frac{v^2}{2c} + \text{high order terms.} \qquad (8)$$

Integration as before yields the following:

$$r - r_0 = \lambda_0 \left(1 - \frac{\delta f_r^s}{f_0^s}\right) (N - \Delta f_0 \tau) + \lambda_0 (\delta f_r^s - \delta f_r^g)\tau + \frac{\lambda_0}{2} (\dot{f}_r^s - \dot{f}_r^g)\tau^2 + \Delta r_{sr} + \Delta r_\epsilon \qquad (9)$$

Simplification results in

$$r^1 + a_0 + a_1\tau + a_2\tau^2 + a_3 r + \Delta r_{sr} + \Delta r_\epsilon = \left[(x^s - x^g)^2 + (y^s - y^g)^2 + (z^s - z^g)^2\right]^{1/2} \qquad (10)$$

where the coefficients are defined as follows:

$a_0 = \left(1 + \frac{\delta f_r^s}{f_0^s}\right) r_0$ The zero set, a constant representing the acquisition range in terms of the adopted frequency.

$a_1 = \lambda_0 (\delta f_r^s - \delta f_r^g)$ A constant representing the differences between true reference frequency and the receiver reference frequency.

$a_2 = \frac{1}{2} \lambda_0 (\dot{f}_r^s - \dot{f}_r^g)$ A constant representing the convergence–divergence of drift rates between the true reference frequency and the receiver reference frequency.

$a_3 = -\frac{\delta f_r^s}{f_0^s}$ This is the fractional frequency bias between the adopted value and the satellite reference value.

$r = \left\| \mathbf{X}^s - \mathbf{X}^g \right\| = \left[ (x^s - x^g)^2 + (y^s - y^g)^2 + (z^s - z^g)^2 \right]^{1/2}$, the range.

$r^1 = \lambda_0 (N - \Delta f_0 \tau)$, the range at an intermediate point of the arc.

Equation (10) is the basic CID range equation. It does not take into consideration the following additional perturbations:

(1) ionospheric refraction;
(2) tropospheric refraction;
(3) propagation delay;
(4) interstation timing offsets and biases;
(5) general relativity and geometrodynamics.

A complete discussion of these perturbations is given by *Nassar* [1972]. Of the perturbations mentioned above, only the interstation timing offsets need be considered for this study. In accordance with the previously adopted linear model for the oscillators, the model for clock offsets is best defined as follows:

$$\delta\tau = \delta\tau_0 + \frac{\delta f_r^s}{f_0^s}\tau = \delta\tau_0 + \lambda_0 \frac{\delta f_r^s}{c}\tau \; . \quad (11)$$

That is, it is necessary to carry as unknown parameters both the clock offsets and the differences in reference frequencies. The usual method of introduction of these terms is as a correction to $r^1$:

$$\Delta r = \dot{r}\delta\tau \; . \quad (12)$$

THE COMBINATION SOLUTIONS

1. An important unknown parameter in (10) is the zero set parameter. Traditionally, this parameter is entered with an a priori estimate of $10^7$ m corresponding to $r_0$. The zero set is coupled with timing biases (see later) and, to a lesser extent, frequency offsets. Hence direct observations of the zero set parameter accomplish the following:

(a) provide the relative Doppler ranges with an absolute starting point;

(b) provide better resolution of coupled effects as mentioned above.

The zero set can be determined by laser ranging systems if the satellite incorporates optical retroflectors and the fixed frequency system necessary for Doppler, with known geometry between the optical and radio centers. Timation III is such a satellite. In such a case, an ephemeris is used to compute the satellite range and hence the zero set, after the laser is discharged. To start the Doppler period, the Doppler timing mark is activated at the center of flight of the photons between the ground stations. Counting is suspended if no return signal is detected and the process is continued until acquisition is obtained. An important reason for considering the direct determination of the zero set parameter rather than software recovery is that there are lunar laser rangers like the Natmap system [*Luck et al.*, 1974] which are not able to slew fast enough to track many of these satellites, but which are capable of maintaining attitude while awaiting the passing of the satellite through their field. Thus the data gathering mode is Doppler with precisions less than 0.10 m supported during the initial acquisition phase by end-point laser data. Despite the similarities of the two systems, no successful simulations are currently available. An indication of possible precisions attainable is given given by *Brown* [1970]. He reports for a pure laser ranging network with a one-sigma range of 0.15 m that the limiting accuracies for a satellite about 10,000 km high like Timation III, are 0.36 m north, 0.31 m east, and 0.30 m vertical over intercontinental distances. *Bender et al.* [1968] obtained similar values. For the natural satellite, the Moon, the values for a 0.15-m range error are 0.53 m north, 0.45 m east, and 0.31 m vertical. *Silverberg* [1974] reports that lunar ranges with a

TABLE 1. The Order of Precision to be Expected from the Combination of LLR-derived Baseline Information and Short-Arc Doppler Tracking over the Pacific Basin

| Station Name | Tracking | Position, deg. | | Sigma, m | | |
|---|---|---|---|---|---|---|
| | | $\phi$ | $\lambda$ | N | E | Vertical |
| Maui (held fixed) | Lunar Laser and Doppler | 20 | 200 | 0.0 | 0.0 | 0.0 |
| McDonald | Lunar Laser and Doppler | 31 | 255 | 0.2 | 0.3 | 0.3 |
| Orroral | Lunar Laser and Doppler | −35 | 150 | 0.2 | 0.3 | 0.3 |
| Wake Island | Doppler with Time | 18 | 170 | 0.4 | 0.5 | 0.3 |
| Samoa | Doppler with Time | −18 | 190 | 0.5 | 0.5 | 0.3 |
| Japan | Fixed Doppler | 35 | 140 | 0.6 | 0.6 | 0.4 |
| Chile | Fixed Doppler | −34 | 290 | 0.8 | 0.8 | 0.5 |
| Papua | Mobile Doppler | −10 | 150 | 0.5 | 0.4 | 0.3 |

precision of 0.06 m have now been obtained with 1-nanosecond-duration pulses.

2. A network of Doppler tracking stations operating in a short arc mode requires the supply of origin and orientation from external observations for a solution to become possible. In this simulation study, the lunar laser observatories at McDonald, Texas; Maui, Hawaii; and Orroral, Australia provided baseline control via ranges at the 0.15-m level. Estimates of azimuth precision were computed by numerical methods, the Rainsford equations [*Rainsford,* 1949], using Bender's Moon standard errors [*Bender et al.,* 1968].

The Timation III satellite was used at an orbital altitude of 14,000 km. A geometric pattern was chosen whose principal points lay at the intersection of the parallels 30°N, 0° and 30°S, and the meridians 150°, 180°, 210°, 240°, and 270°. Six additional mid-Pacific orbits were also selected for additional strength. It was assumed that each lunar laser observatory would operate Doppler and time transfer equipment, and that Wake Island and Samoa would continue to provide the support in Doppler that they currently provide. This distribution of fixed stations showed unfavorable error distributions for South American locations, in excess of 2 m for one-sigma errors in X, Y, and Z. A dramatic reduction of errors in this region was effected by incorporating into the simulation an additional azimuth line from Tokyo, Japan, to Santiago, Chile.

The values given in Table 1 were obtained from a simulation in which Maui was held fixed and the constraining information was only available in one direction across the Pacific basin. There are plans to build LLR's in Japan and Chile but operational sites are some years off. The location of the mobile unit has a considerable effect on the distribution of errors due to the lack of observations in the southwest Pacific.

3. A simulated network of nine stations was established in the Papua New Guinea region to study plate boundary motions. In this case, the standard Navy Navigation Satellites were used in the simulation, augmented with time information derived from portable clock comparisons and Timation values. Figure 2 shows the behavior of various oscillators as a function of time. The Navy Navigation Satellites are never tracked for more than an hour from one site and hence the internal stability of the oscillator for the period of the pass is maximized by the choice of quartz, as internal timing marks at the 1- to 10-sec level are required. It was shown earlier that some advantages do accrue from higher satellites which are visible for much longer periods. Thus it is necessary to look more for long-term stability. This is the case for the Timation satellite in which, from a stability point of view, the minimum data period should not be less than 10 sec since some short-term stability has been sacrificed in order to obtain stability for periods up to $10^4$ sec. This fractional frequency stability does have a small effect on the terms $a_1$ and $a_3$ in (10), but biases and rates of biases have a much greater effect and need to be controlled.

Figure 3 shows the classical methods of comparing oscillators. It is evident from this figure

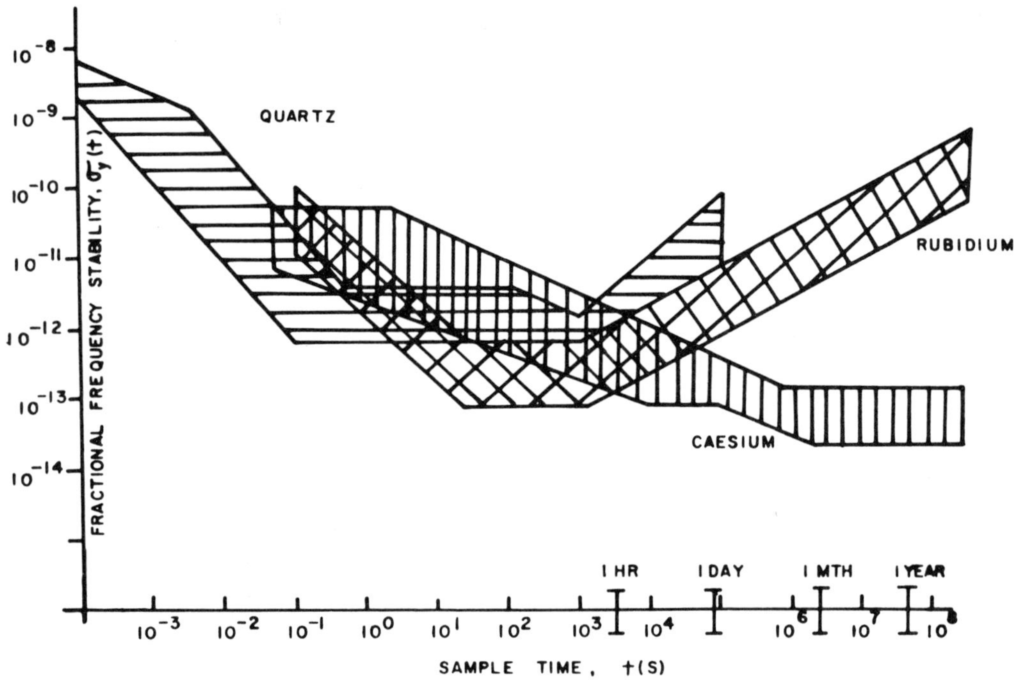


Fig. 2. Fractional frequency stability of frequency standards indicating range of performance available as a function of sample interval.

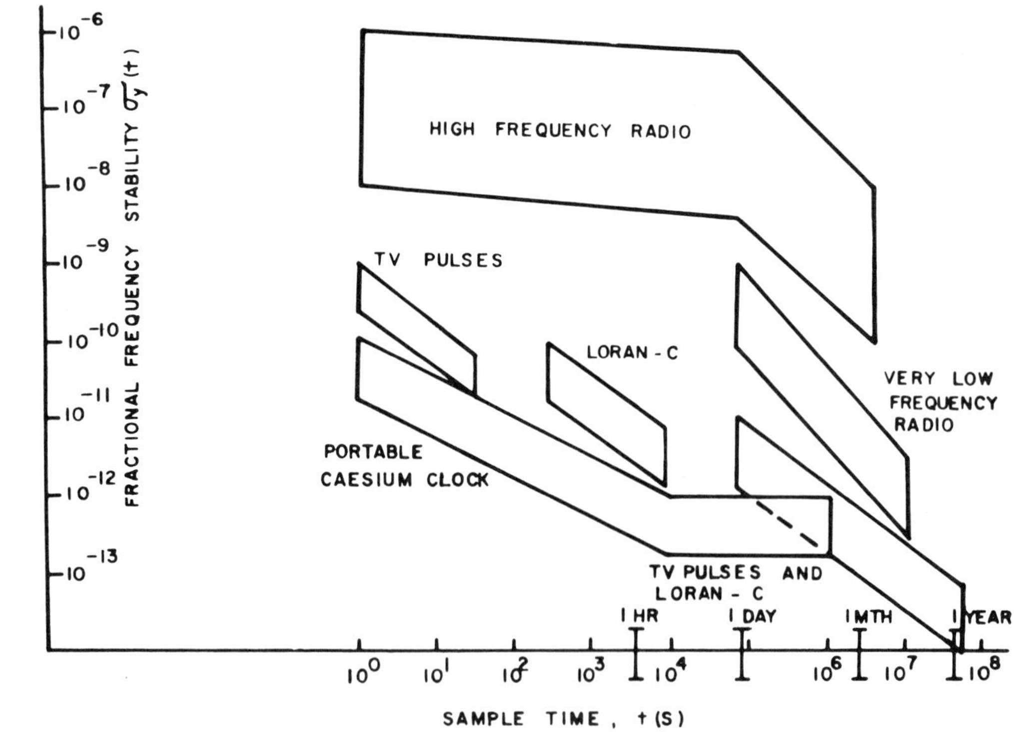


Fig. 3. Fractional frequency stability as a function of sample time for several dissemination techniques.

TABLE 2. Precisions to be Expected when Frequency Biases and Drift Rates are Constrained as Observed Quantities

| Station Name | | Position, deg. | | Sigma, m | | |
|---|---|---|---|---|---|---|
| | | $\phi$ | $\lambda$ | N | E | Vertical |
| Rabaul | (1) | -4.33 | 152.20 | 0.0 | 0.0 | 0.0 |
| Kavieng | (2) | -2.50 | 150.48 | 0.24 | 0.18 | 0.01 |
| Kieta | (3) | -6.20 | 155.50 | 0.56 | 0.47 | 0.10 |
| Alotau | (4) | -10.20 | 150.67 | 0.38 | 1.08 | 0.30 |
| Madang | (5) | -5.20 | 145.90 | 0.87 | 1.12 | 0.98 |

In this table, station (1) is the local origin station while stations (2) and (3) were used to control the orientation of the network. The one-sigma errors can be reduced by increasing the number of passes. The exact nature of this relationship has not yet been determined.

that for speed and accuracy the portable caesium clock is superior; unfortunately, air fares are expensive. *Easton, Smith, Morgan and Luck,* in a series of reports [*Easton et al.,* 1974*a, b; Luck et al.,* 1974], clearly demonstrate the capability of the Timation system to compare oscillators and determine biases. The results so far obtained are superior to LORAN-C but inferior to portable caesium clocks. Improvements are expected with Timation III which will put the method in the upper half of the caesium range and the lower half of the long period LORAN-C range. It will then become possible to determine by observation the constants $a_1$, $a_2$ and $a_3$ of (10) and to constrain the solution to these values.

Simulations using these ideas were performed on twelve passes common to nine observing stations and were later repeated for two groups of five stations. Both gave similar results due to the distribution of stations chosen, but the latter are more logistically feasible. The results of these simulations are presented in Table 2.

## SUMMARY

These simulation studies indicate that the combination of present day technology in time and frequency, laser ranging, and Doppler tracking can support a lightweight portable Doppler system that is capable of producing station values whose one-sigma errors are substantially less than 1 m. The simulations have inherent approximations built into them, but the order of the values is not expected to vary greatly. It must be recognized that the errors in both the Doppler technique and the laser techniques will continue to decrease and that the simulations presented here represent upper limits.

*Acknowledgments.* I am grateful to Mr. George Hadgigeorge for the SAGA program which was adapted for the simulation studies and for his detailed knowledge of such adjustment processes. I also thank Mr. R. James for assistance at the computer in executing the tasks and my colleagues at the Bureau of Mineral Resources who encouraged this work by providing valuable computer time.

## NOTATION

For simplicity, notations used in this paper are defined below.

- $f$ Frequency (Hz), especially the frequency of transmitted signal at the receiver.
- $\lambda$ Wavelength (m); also longitude.
- $c$ Velocity of light (m/s).
- $\mathbf{V}$ Velocity, a vector. $\mathbf{V} = \dot{\mathbf{X}} = (\dot{x}, \dot{y}, \dot{z})^T$.
- $\mathbf{I}$ The unit vector. $\mathbf{I} = (i, j, k)^T$.
- $v$ Magnitude of velocity vector $\mathbf{V}$; speed.
- $\mathbf{X}$ Position vector in center of mass, rectangular coordinate system.
- $r$ Range; the magnitude of the displacement vector. $r = \|\mathbf{X}^s - \mathbf{X}^g\|$ .
- $\dot{r}$ Rate of change of range.
- $\Delta r$ A correction to the range.
- $\tau$ Time elapsed from $t = 0$; the tracking period.
- $N$ The number of Doppler beat frequency counts.
- $\phi$ Latitude.

*Superscripts*

- $g$ Ground station (Geoceiver).
- $s$ Satellite.

*Subscripts*

- $r$ Reference value.
- 0 Nominal value or value at zero time.

# 6. *Regional Gravity of Peru*

L. Ocola and H. Aleman

*Instituto Geofísico del Peru, Lima, Peru*

*Abstract.* A regional gravity anomaly map of Peru and the neighboring ocean reveals the following anomalies of regional extent: (1) a narrow negative anomaly closely following the Peru–Chile trench and reaching values of −200 mgal; (2) a positive anomaly with values up to +50 mgal trending generally along the coastal cordillera; (3) a broad anomaly with extreme values approaching −400 mgal in southern Peru. This anomaly becomes less negative and narrower towards the north. Crustal models constructed to satisfy gravity and available seismic data show an extraordinarily thick crust underlying the Andean mountains.

# 2. SEISMOLOGY

## 7. *Seismological Studies of the Pacific Area: Introduction to Session 2*

JOHN I. EWING[1]

*Lamont-Doherty Geological Observatory, Columbia University, Palisades, New York 10964*

JACK E. OLIVER

*Department of Geological Sciences, Cornell University, Ithica, New York 14850*

Seismological studies have made major contributions to our understanding of the structure and dynamics of the Pacific Ocean basin and its margin. While the following seven papers from the seismology session of the International Woollard Symposium could not possibly cover the entire range of present knowledge and research activity, we feel they do provide excellent examples of the results being obtained through current seismological research. Fairly complete bibliographies associated with the papers extend the range and completeness of topics covered.

Five of the papers are largely concerned with convergent margins, one with a spreading ridge, and one with an active linear island chain. Categorized in another way, three of the papers involve the use of explosive and other controlled seismic sources, three use natural earthquakes, and one uses both types of sources. Crustal and upper mantle structure, physical properties, and dynamics are obtained through studies of both propagation phenomena and seismicity.

The convergent boundary zone between the Nazca and Americas plates is examined in papers 8, 9, and 10 [*Hussong et al.,* this volume; *Rodriguez et al.,* this volume; *Meyer et al.,* this volume]. These papers provide a fairly detailed picture of the complex nature of the zone of interaction between the subducting oceanic plate and the continental plate. However, as might be expected, they also indicate some large gaps in our understanding of the processes involved and some possibly fruitful directions for future research.

Using earthquake hypocenters from a 10-year period, Rodriguez et al. establish the geometry of the leading edge of the Nazca plate, between 2° and 40°S, as it is being subducted beneath South America. They demonstrate that the oceanic plate is broken into discontinuous tongue-like pieces as it underthrusts the continental lithosphere. The authors then try to relate this geometry to the zonation of economic mineral deposits in Bolivia, the gravity field, and near-surface stress patterns obtained both theoretically and from faults and other lineations observed in Bolivia. Indeed, they find intriguing relationships among those parameters, but point out that a direct relationship between the mineralization and subduction is difficult to establish since the age of the mineralization is not known.

Both Hussong et al. and Meyer et al. present models of the subduction zone, between 8° and 12°S and near 4°N, respectively, derived mainly from controlled-source seismic data. These two papers also have in common that they report useful results from refraction observations taken directly across the complex structure of a convergent margin, thus providing structural detail not obtainable from a sequence of lines along strike.

Meyer et al. report on the partial results of a large international sea-to-land seismic refraction experiment, Project Nariño. Shots were fired at sea and recorded by a number of seismographs in the western cordillera of Colombia. The derived model of crust and upper mantle consistent with gravity and supplementary refraction data extends westward from greater than 100 km east of the shoreline, across the Cordillera Occidental to the Malpelo rise about 400 km west of the

[1]Now at Woods Hole Oceanographic Institution, Woods Hole, Massachusetts 02543.

shoreline. It represents the Nazca lithosphere underthrusting a fairly complex continental lithosphere to a depth of about 100 km near the eastern edge of the section and suggests a zone of low upper mantle compressional-wave velocity under the Yaquina graben on the Nazca plate.

Hussong et al. develop a composite model of the subduction zone extending about 400 km from the Nazca plate, across the Peru-Chile trench, onto the continental shelf of Peru. The model is based on a profile of overlapping reversed refraction lines near 9°S; a line of ASPER stations and a multi-channel digital reflection profile near 12°S; and additional standard reversed explosion refraction lines on the oceanic plate and on the shelf. Large-scale compressional deformation predominates in the model, demonstrated by thrust faulting of the entire oceanic crust and imbricate thrusting and uplift of oceanic sediments and upper crustal rocks into the lower continental slope. Nevertheless, in contrast to models of other investigators for different locations, there appears to be nearly complete subduction of oceanic material and erosion and subduction of continental material in this portion of the trench.

The study of surface waves has provided much of our knowledge of the physical properties of the oceanic crust and upper mantle, e.g., the shear velocity distribution in the oceanic lithosphere and asthenosphere. Such studies were represented by one interesting paper at the symposium. Unfortunately only the abstract [*Kausel and Schwab,* this volume] is available for publication here. As indicated in the abstract Kausel and Schwab interpreted the dispersion and attenuation of Rayleigh waves along the East Pacific ridge to indicate low shear velocity and $Q$ values in the upper mantle and a thin, or absent, lid to the channel of low shear velocity often taken to be the asthenosphere.

Papers 12 and 13 [*Asada and Shimamura,* this volume; *Nagumo and Kasahara,* this volume] summarize much of the research conducted by Japanese seismologists in the western Pacific using ocean bottom seismographs (OBS). In addition, Nagumo and Kasahara present a brief review and bibliography of the development of deep OBS techniques and observations in the U.S.A. and U.S.S.R. A large number of OBS observations, mostly conducted since 1966, are discussed in the two papers and they amply demonstrate the importance of OBS's for both seismicity and structural studies of ocean basins and their margins.

Asada and Shimamura discuss results for both natural earthquake and long-range explosion experiments. Among the more important results from their OBS studies of seismicity near western Pacific trenches are (1) that more than 97% of the earthquakes recorded occurred just landward of the trench, and (2) that the seismic $Q$ values in the western Pacific lithosphere exceed 6000 and are generally ten times or more greater than those landward of the trenches. Results from their long-range refraction experiments indicate a stratified oceanic upper mantle with 8.2 km/sec material below Moho, increasing slightly in velocity with depth down to about 60 km where there is an abrupt increase to 8.6 km/sec. The velocity remains at 8.6 km/sec down to about 80 km where it decreases, producing a low-velocity layer less than 50 km thick. The low-velocity layer in turn overlies a region with a positive velocity gradient starting at 8.6 km/sec.

In their paper, Nagumo and Kasahara review observations of micro-earthquake seismicity beneath the ocean floor near the Japan and Mariana trenches and on the Fiji plateau, and discuss some of the tectonic implications of these observations. In contrast with the observations of Asada and Shimamura, Nagumo and Kasahara found high activity both under the marginal escarpment of the Japan trench and in the adjacent ocean basin. They also found high activity in the south Mariana trench and the southern terminus of the Mariana ridge, but no activity in the adjacent East Mariana basin. High activity, including swarm activity, was observed on the Fiji plateau. This activity in conjunction with observations of high and variable heat flow and rough topography suggests a tectonic pattern analogous to that existing on spreading ridges.

In the final paper of the session, paper 14, *Koyanagi et al.* [this volume] describe and interpret the seismic activity on the island of Hawaii from 1970 through 1973 as recorded on the seismograph network of the Hawaiian Volcano Observatory (HVO). The network consists of 33 seismometers connected via radio-telemetry to HVO. Working with a data base of over 15,000 well-located hypocenters, the authors are able to relate the seismicity to eruption activity and to inflation and expansion of Kilauea volcano. An aseismic region between about 4- and 6-km depth beneath Kilauea is believed to outline the principal magma reservoir for the volcano. Feeding conduits for the reservoir are suggested by the distribution of 'long-period' earthquakes extending to about 12-km depth beneath the aseismic region.

# 8. *Crustal Structure of the Peru–Chile Trench: 8°–12° S Latitude*[1]

D. M. HUSSONG

*Hawaii Institute of Geophysics, University of Hawaii, Honolulu, Hawaii 96822*

P. B. EDWARDS[2]

*University of Melbourne, Parkville, Victoria 3052, Australia*

S. H. JOHNSON

*Oregon State University, Corvallis, Oregon 97331*

J. F. CAMPBELL AND G. H. SUTTON

*Hawaii Institute of Geophysics, University of Hawaii, Honolulu, Hawaii 96822*

*Abstract.* Extensive crustal seismic data, including five reversed explosion refraction stations, an eleven-station refraction profile, twelve ASPER stations, and a multi-channel digital reflection profile, were acquired in the vicinity of the Peru–Chile trench between 8° and 12°S latitude in 1972 and 1973. The data are used to develop a complex model of the convergence of the Nazca oceanic plate with the South America continental lithospheric plate. On the Nazca plate, the oceanic crust is thin and composed of rocks of higher seismic velocity than is typical of other oceanic plates. The high-density upper lithosphere may be related to the high spreading rate of the Nazca plate. Near the Peru–Chile trench large-scale compressional deformation predominates, as demonstrated by thrust faulting of the oceanic crust beneath itself prior to and possibly after subduction under the continent. Imbricate thrusting causes uplift of oceanic sediments, and perhaps upper crustal rocks, into the lower continental slope. The upper continental slope, however, appears to be of different origin. This area may be composed of continental rocks that are being fractured by vertical tectonic displacement of the trench inner wall, which is downfaulted and ultimately subducted with resultant erosion of the continental margin. The majority of oceanic rock and sediments, as well as the eroding continental rocks, are subducted and may, therefore, become source material for Andean plutonism and volcanism.

A basic premise of plate tectonic theory is that the oceanic trenches represent zones of convergence and destruction of lithospheric plates, thereby serving as crustal sinks which balance crustal sources at mid-ocean ridge-spreading centers. The plate convergence has been identified as a cause of earthquakes, volcanism, and orogeny. Unfortunately, detailed information concerning crustal models of the ocean–continent interface at a subduction zone has been difficult to obtain, thus the tectonic processes involved are poorly understood.

An excellent example of continental plate–oceanic plate convergence is the Peru–Chile trench along the west coast of South America. Previous studies of the crustal structure [*Aldrich et al.*, 1958; *Fisher and Raitt,* 1962] and marine geological and geophysical relationships [*Hayes,* 1966, 1974; *Scholl et al.*, 1968, 1970; *Herron,* 1972*a*] provide background reconnaissance data on the trench.

In 1971 an extensive geological and geophysical study of the Nazca lithospheric plate and its margins (the Nazca Plate Project) was initiated under the National Science Foundation program for the International Decade of Ocean Exploration (IDOE). Initial surveys were carried out

[1]Hawaii Institute of Geophysics Contribution No. 701.

[2]Now at Exploration Department, Esso Australia, Box 4047, G.P.O. Sydney, Australia 2001.

aboard the research vessels *Kana Keoki* and *Yaquina* in early 1972.

A major objective of the project has been to obtain sufficiently detailed information on the Peru–Chile trench so that a dynamic model of the subduction process can eventually be made. Much work is still in progress. So far, selected geologic studies of the trench near the 8°–12°S region have been published by *Kulm et al.* [1973, 1974], *Prince et al.* [1974], and *Prince and Kulm* [in press].

*Prince and Kulm* [in press] propose that the continental slope of northern Peru is the product of imbricate thrusting and uplift of sheets of upper oceanic crustal material, such as *Seely et al.* [1974] have described. Prince and Kulm suggest that the thrust sheets form along fractures originally developed by normal faulting on the oceanic plate as it bends into the subduction zone.

A substantial portion of the Nazca Plate Project's first field season was devoted to seismic refraction measurements to determine the crustal structure of the Peru–Chile trench subduction zone. This paper reports on these studies as well as the first results from commercially obtained multi-channel digital seismic reflection data that traverse the trench. Expansion and modification is made of the imbricate thrust fault model of the continental slope. The data used to amplify our model of the subduction zone also lend themselves to speculation on the significance of regional variation of oceanic crustal structure, spreading rates, and subduction zone configuration.

OBSERVATIONS

Figure 1 shows the location of the refraction stations described here. Five standard two-ship reversed explosion lines *(1–2, 2–3, A20–A2S, A20–A2E,* and *20–21)* define the crustal velocity structure and serve as end points for line *18–19,* a 375-km-long refraction profile comprised of eleven overlapping split profiles extending from the oceanic plate across the trench and up onto the continental shelf. These refraction data are supplemented by a commercial, multi-channel, common-depth-point (CDP) digital reflection seismic line that transects the trench and slope near 12°S latitude, not far from a line of twelve ASPER (Airgun Sonobuoy Precision Echo Recorder) stations which also extend across the convergence zone and which *Hussong et al.* [1975] have described.

The refraction data were all obtained using free-floating military SSQ-41A sonobuoys modified only by the addition of extra batteries to extend their operational life.

There are several important advantages to using sonobuoys. They are initially inexpensive, and their economy is increased because they allow the ship to move away and perform other work, such as bottom sampling, during refraction recording. No problems are encountered in isolating the sonobuoys from ship noise and motion, as are encountered with hydrophones deployed from a ship. Most importantly, however, sonobuoys can be deployed in large patterns. In the work near the Peru–Chile trench many of the recording stations were linear arrays consisting of

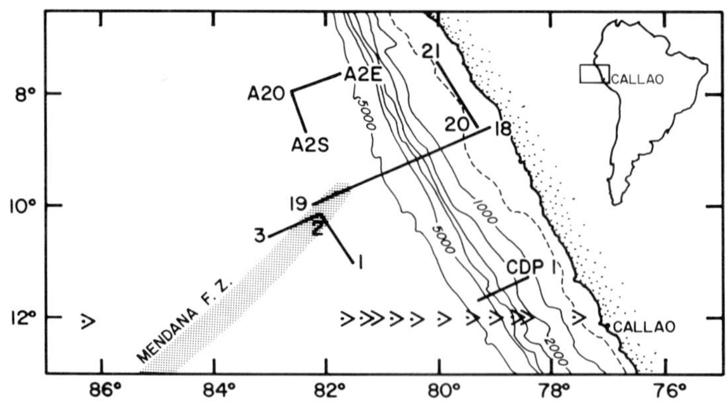


Fig. 1. Location of seismic data described in this paper. The symbols along 12°S indicate ASPER stations; the dots are drop points, the > symbols indicate the direction of shooting away from the sonobuoys. Selected contours, in meters (the dashed line is the 100-m contour), and location of the Mendana fracture zone are from *Mammerickx et al.* [1975].

four receiving sonobuoys spaced along a 1-km line. Such an array permits determination of cross-spread velocities and phase correlation of first and later arrivals having low signal-to-noise ratios. These low-level arrivals, which would probably not be discernible with any confidence from a single data channel, were invaluable in our refraction analysis.

The use of sonobuoys as seismic receivers, however, has two major disadvantages. First, if the sonobuoys drift over sizeable bottom relief or changing crustal structures during the shooting of a line, potentially uncorrectable scatter will be generated in the resultant seismic travel times. Second, response of the expendable military sonobuoys is shaped to suppress both the low-frequency energy in ambient sea noise and the mechanical noise generated by the hydrophone suspension compliance. The buoy response drops below 1 kHz, which reduces the noise (and signal) level at the frequencies of seismic ground arrivals. The first disadvantage was fortunately minimized by favorable sea conditions; the second was overcome by amplifying the suppressed low frequencies and by increasing charge size when needed.

## RESULTS

*Nazca plate oceanic crust.* Four conventional two-ship reversed refraction lines, using explosives for sound sources, were completed on the oceanic plate west of the trench (see Figure 1). The results are listed in Table 1.

The crustal models exhibit velocity structures (Figures 2 and 3) which are similar to those found to be predominant over much of the northern Nazca plate [*Hussong et al.*, 1973, 1975]. The significance of this typical crustal velocity structure for the Nazca plate as compared with typical crustal structure found elsewhere is considered later in the Discussion section. It need only be noted here that four crustal layers are present, with a high-velocity basal crustal layer predominating.

Lines *1-2* and *2-3* (Figure 2) are on 40–50 m.y. old crust (based on the magnetic anomaly identifications of *Herron* [1972*a*]) in the northern Peru basin, and probably cross the Mendana fracture zone (based on the bathymetric map of *Mammerickx et al.* [1975]). No unusual crustal structure could be attributed to the fracture zone.

Possibly the most remarkable result of Line *2-3* is the indication of a high-velocity (8.7 km/sec) layer beneath the Mohorovicic discontinuity. Apparent refracted arrivals from this layer occurred as low-frequency, low-amplitude precursors to the $P_n$ arrival at long shot-to-receiver distances. The arrivals probably would not have been detected except that they could be correlated across an array of four sonobuoys deployed at position 3.

The crust under the more northerly lines *A20–A2E* and *A20–A2S* (Figure 3) has similar velocity layering. The basal crustal layer is thicker, however, with the result that the Moho is depressed to a depth about 2 km greater than is typical elsewhere on the Nazca plate. This crustal thickening may be caused by compression as the plate enters the subduction zone. Whatever the cause, the refraction interpretation indicating deep water and thick crust is supported by the absence of the usual outer-trench gravity high north of 9° S (Tom Getts, personal communication, 1975). South of 9° S, where the Mendana fracture zone intersects the Peru–Chile trench, a gravity high exists along the trench margin of the oceanic plate.

*Refraction profile 18–19.* The reversed refraction technique used to obtain the previously described crustal models is reliable and effective, but has the unfortunate characteristic of smoothing crustal irregularities which might be expected to be present over long sections of oceanic crust (70–100 km) into models consisting of homogeneous velocity layers separated by plane interfaces. For this Nazca plate study, however, a structural model across the complex Peru–Chile trench subduction zone was required. The data generally acquired by standard marine refraction methods were not considered adequate for this purpose. In response, an adaptation of the land seismic method of shooting overlapping refraction lines was run across the trench, from the oceanic plate to the continental shelf. This line, called a refraction profile, generated a sufficiently high density of data points to permit the determination of complex structure across the ocean–continent boundary region.

The data were obtained using two ships underway along the length of the refraction profile (line *18–19* in Figure 1). As sketched in Figure 4, the lead ship steamed about 40–50 km in front of the second ship, deploying sonobuoys at intervals of about 30 km. As the line developed, the lead ship received signals from a given sonobuoy up to at least 30 km astern, by which time the sonobuoy was within telemetry range of the second ship which would then receive signals from the sonobuoy as it steamed up to and past it. The second ship continued to monitor the sonobuoy until

TABLE 1. Locations and Seismic Sections; Two-Ship Refraction Stations

| Position | | | | Velocities, km/sec | | | | | | | | | Thickness, km | | | | | | | | | Summation | | |
|---|---|---|---|---|---|---|---|---|---|---|---|---|---|---|---|---|---|---|---|---|---|---|---|---|
| Line | Posit. | Lat., S. | Long., W. | Sed. | 1 | 2 | 3 | 4 | 5 | 6 | 7 | 8 | Water | Sed. | 1 | 2 | 3 | 4 | 5 | 6 | 7 | Depth to M, km | Crustal Thickness, km | Sub M Depth, km |
| | 1 | 10°59′ | 81°30′ | | | | | | | | | | 4.0 | .5 | | | | 1.9 | 1.1 | 2.7 | | 10.2 | 6.2 | |
| 1–2 | | | | (1.7)[1] | | | | 5.6 | 6.7 | 7.2 | 8.1 | | | | | | | | | | | | | |
| | 2 | 10°14′ | 82°08′ | | | | | | | | | | 4.5 | .3 | | | | 1.1 | 1.9 | 2.0 | | 9.8 | 5.3 | |
| | 2 | 10°14′ | 82°08′ | | | | | | | | | | 4.5 | .3 | | | | .5 | 1.2 | 2.9 | 1.8 | 9.4 | 4.9 | 11.2 |
| 2–3 | | | | (1.7) | | | | 5.6 | 6.6 | 7.1 | 8.1 | 8.7 | | | | | | | | | | | | |
| | 3 | 10°32′ | 83°04′ | | | | | | | | | | 4.5 | .2 | | | | 1.0 | 1.0 | 3.2 | 4.0 | 9.9 | 5.4 | 13.9 |
| A2S– | A2S | 8°44′ | 82°19′ | | | | | | | | | | 4.3 | .4 | | | | .9 | 1.6 | 5.8 | | 13.0 | 8.7 | |
| A20 | | | | (1.7) | | | | 5.9 | 6.4 | 7.0 | 8.5 | | | | | | | | | | | | | |
| | A20 | 7°51′ | 82°37′ | | | | | | | | | | 4.3 | .2 | | | | .4 | 2.2 | 3.8 | | 10.9 | 6.6 | |
| A20– | A20 | 7°51′ | 82°37′ | | | | | | | | | | 4.3 | .2 | | | | .4 | 1.6 | 5.1 | | 11.6 | 7.3 | |
| A2E | | | | (1.7) | | | | 5.8)[2] | 6.3 | 7.3) | 8.5 | | | | | | | | | | | | | |
| | A2E | 7°33′ | 81°41′ | | | | | | | | | | 4.6 | .2 | | | | .9 | 2.2 | 3.8 | | 11.7 | 7.2 | |
| | 20 | 8°38′ | 79°18′ | 1.9 | 2.5 | 4.5 | 5.2 | 5.9 | | | | | .1 | 1.0 | .2 | .7 | 2.0 | 3.4 | 9.4 | 13.1 | | 29.9 | 29.8 | |
| 20–21 | | | | | | | | | 6.8 | 7.0) | 8.2 | | | | | | | | | | | | | |
| | 21 | 7°20′ | 80°03′ | 2.0 | 2.4 | 4.8 | 5.3 | 6.0 | | | | | .1 | .6 | .1 | .5 | 0.8 | 10.4 | 2.8 | 15.2 | | 30.5 | 30.4 | |

[1]Parentheses indicate assumed velocity.
[2]Single parenthesis indicates unreversed velocity.

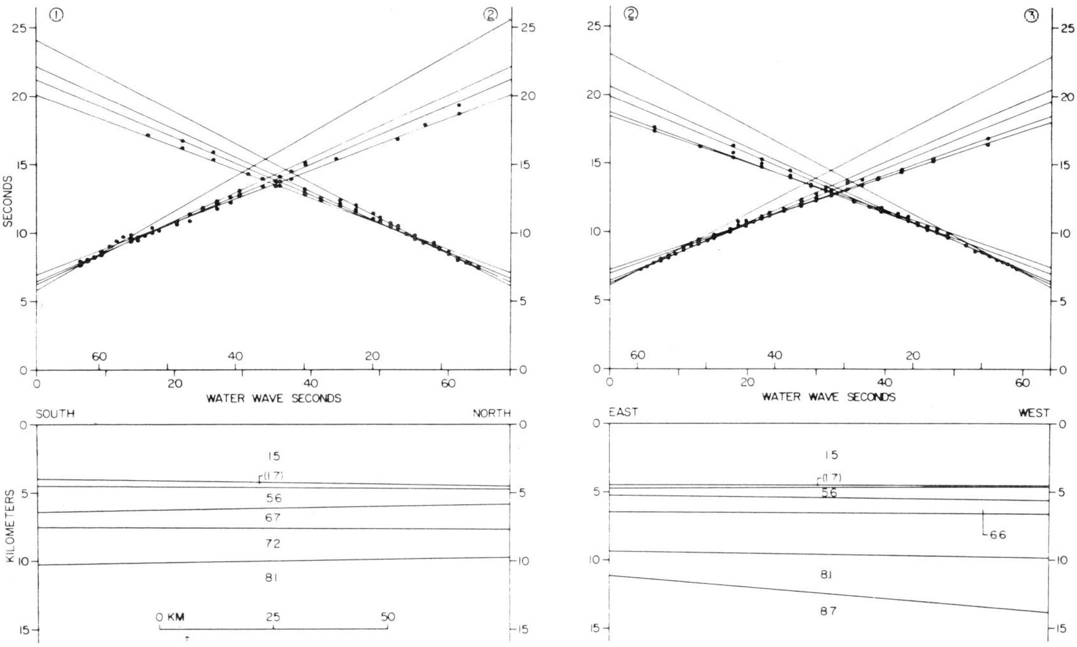


Fig. 2. Crustal velocity structure under lines *1–2* and *2–3*. Velocities are in kilometers per second. The sediment velocity (1.7 km/sec) was not observed.

it was at least 30 km behind.

Since both ships were shooting, the resultant arrivals produced a series of short, reversed refraction lines at ranges of up to the 50 km between the ships. At the same time, large shots from the lead ship were received by the last sonobuoy within telemetry range of the second ship, providing total shot-to-receiver distances of at least 70 km.

On line *18–19,* eleven sonobuoys were successfully deployed over a distance of 375 km. Over four hundred shots were fired along the line. The resultant velocity model, and the travel time plot branches used to prepare the model, are shown

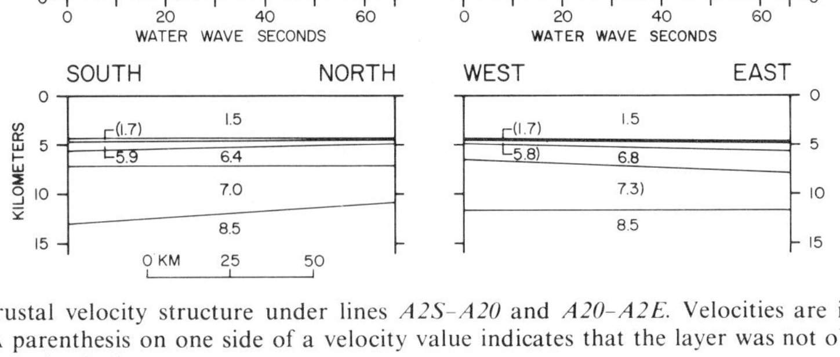


Fig. 3. Crustal velocity structure under lines *A2S–A20* and *A20–A2E*. Velocities are in kilometers per second. A parenthesis on one side of a velocity value indicates that the layer was not observed from that side of the seismic line.

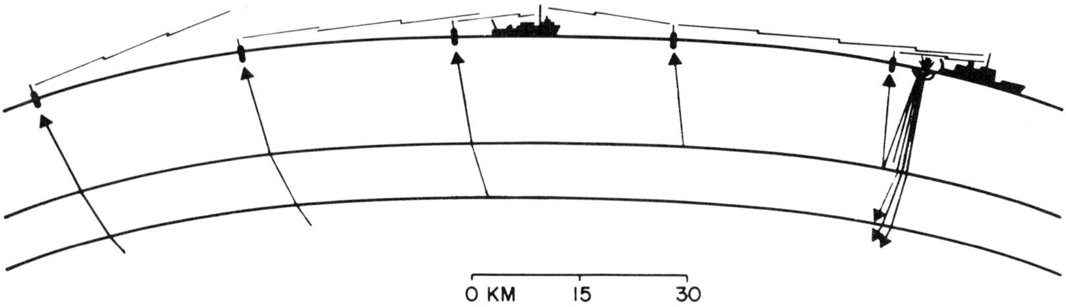


Fig. 4. Schematic of ship and sonobuoy deployment for a refraction profile.

in Figure 5. Details of the data acquisition and interpretation for this line, for which a special technique and computer program were evolved, are described in another paper by Edwards and Hussong (in preparation).

The western end of line *18–19,* on the oceanic plate side of the trench, is near the previously described reversed lines *1–2* and *2–3.* The refraction profile model of crustal velocities on the oceanic plate is similar to that found for lines *1–2* and *2–3,* with the four-layered crust dominated by the basal crustal layer. Suggestions of refractions from a high-velocity sub-Moho layer were also observed on the western ends of the travel-time branches from sonobuoys 13 and 14.

Proceeding east along the refraction profile, some thickening of the oceanic crust can be observed within about 100 km of the trench axis. Most of the increased thickness is caused by enlargement of the 6.2–6.4 km/sec crustal layer to the east between sonobuoys 14 and 13. This irregular layering may be an expression of the Mendana fracture zone which intersects line *18–19* at about this point. The crust on the northwest side of the Mendana fracture zone should be considerably older than the crust on the southeast side [*Herron,* 1972*a*]. As the crust bends into the subduction zone, further offsets in the velocity model are observed, but they are of limited size and are not suggestive of any consistent mode of deformation.

Structure of the axis and the inner wall of the trench is not well defined from the refraction data. A combination of severe topography and frequent explosive charge misfires between about sonobuoys 10 and 6 degraded the data there. The

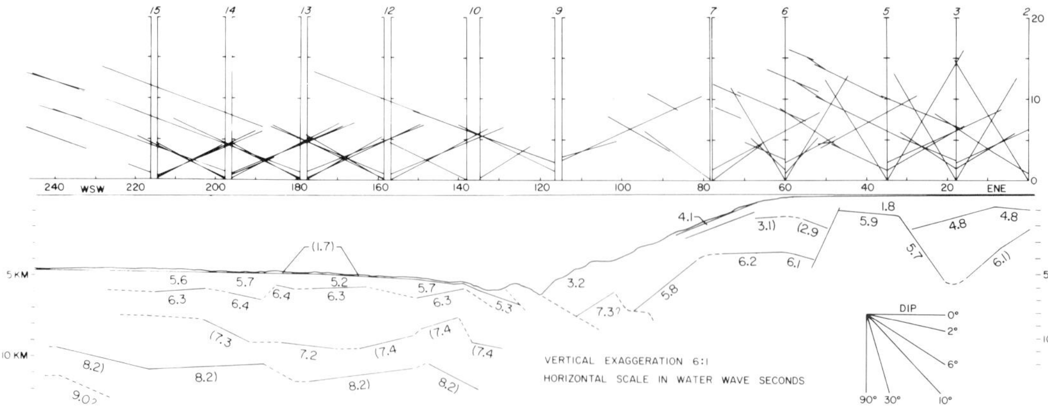


Fig. 5. Refraction profile (line *18–19*) model after P. Edwards and D. Hussong (in preparation). Travel time branches (corrected to the ocean bottom) are shown, without data points, at top. Velocities are in kilometers per second. Numbered positions are sonobuoy locations. Spread positions represent sonobuoy drift during recording interval.

simplified model in Figure 5 shows that about 3 km of perhaps partially indurated sediment (3.2 km/sec) is underlain near the trench axis by rock of about 7.3 km/sec velocity. This high velocity material may be deep oceanic crust uplifted into the inner trench wall by imbricate thrusting.

East of the 7.3 km/sec uplifted block a layer with a velocity of 5.8 km/sec underlies the ~3 km/sec material. This layer increases in velocity toward the continent and rises abruptly under sonobuoy 5 (the rapid change can be seen as offsets in the travel time plots west from sonobuoys 2, 3, and 5). The ~6 km/sec material may correspond to the pre-Mesozoic basement which outcrops [*Cobbing and Pitcher,* 1972] on land north and south of line *18-19.*

A lens of 4.1 km/sec material at the edge of the continental shelf is required for the seismic model beneath sonobuoy 7. The velocity inversion beneath it also satisfies abrupt gaps in arrivals to sonobuoy 6. The lens is not discernible on reflection records, however, and no satisfactory geologic explanation for it is apparent.

*Deep structure under the continental shelf.* Profile *18-19* successfully depicts the shallow crustal structure beneath the continental shelf, but for our study of the subduction zone an idea of deep crustal structure is also required. To satisfy this, a conventional explosion line *(20-21)* using large charges was oriented parallel to the coast. To obtain refractions from the deep crust and upper mantle, the line was reversed at a distance of nearly 190 km, using charge sizes of up to 900 lbs.

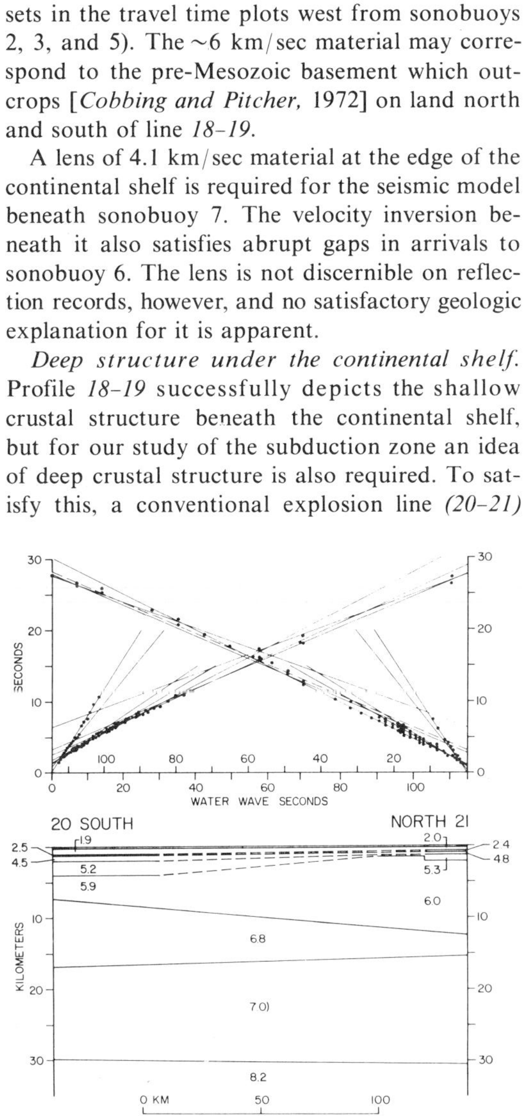


Fig. 6. Crustal velocity model under line *20-21* on the Peru continental shelf. Velocities are in kilometers per second. Upper dashed layers were not reversed because of the length of the line.

The results of line *20-21* are shown in Figure 6. Because of the length of the line the upper layers are unreversed; i.e., for interfaces above the 6.8 km/sec layer, arrivals to each end of the line travel along completely different portions of the interfaces. Refractions from the 6.8 km/sec and lower interfaces did not reverse well, probably because of the approximations used for the shallow layering.

Near the surface, the velocity structure on line *20-21* is similar to that of the eastern end of refraction profile *18-19.* Deeper in model *20-21,* the 7.0 km/sec layer is not well reversed, so the Moho depth beneath station *21* is not as reliable as might be hoped. In general, however, the crustal section is well defined.

The 30-km depth to mantle agrees with earlier estimates by *Woollard* [1960*b*] and *Hayes* [1966]. This depth suggests a somewhat shallow dip, about 7°, of the top of the subducting plate, assuming that the observed Moho is beneath the combined continental and oceanic crust. Analysis of the CDP reflection data, described later in this section, agrees with this model of a shallow interface.

*12°S ASPER profile.* A series of twelve ASPER observations were made along 12°S latitude from 300 km west of the Peru-Chile trench axis, across the trench, and onto the continental shelf near Callao, Peru. The resultant structural model is reported by *Hussong et al.* [1975] and is reproduced here (Figure 7) for use in comparisons with our other data.

The most striking feature of this ASPER profile is the large offset in crustal layering that was interpreted as a major thrust fault seaward of the trench. This apparent crustal shortening suggests that the oceanic crust is undergoing (or has undergone) compression prior to subduction.

*Multi-channel CDP reflection data.* Three 24-channel CDP digital seismic reflection lines were placed across the Peru-Chile trench at roughly 5°40′, 9°10′, and 11°40′S latitude. Each line is about 100 km long. The data were acquired and processed by the Seiscom-Delta Corporation of Houston, Texas, under contract to the Hawaii Institute of Geophysics and Oregon State University for the Nazca Plate Project under the IDOE. Initial processing of the data to produce 1200% stacked time sections has been completed. The

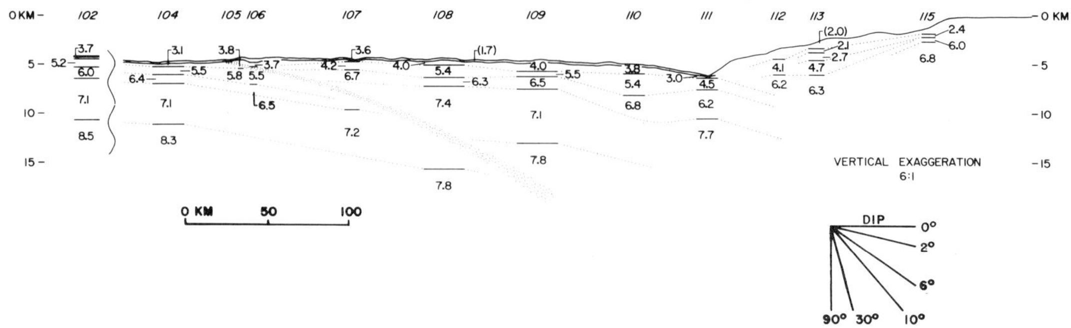


Fig. 7. Crustal sections from ASPER stations along 12°S latitude [from *Hussong et al.,* 1975]. Velocities are in kilometers per second. Note the crustal layering offset, interpreted as a thrust fault, between stations 105 and 109.

southernmost of these lines, labeled *CDP-1* (for location refer to Figure 1), is shown as Figure 8.

Prominent reflectors have been identified and Figure 9 shows the same reflection line with the more important reflectors highlighted. For example, at the western (left) side of the profile, the bottom of the oceanic sediments (the top of oceanic layer 2 using the nomenclature of *Raitt* [1963]) is labeled B and is easily traced from the plate into the trench and beneath the inner wall of the trench. Interface B is an apparent continuous reflector until past kilometer 50, where it fades and rises (in time) before being obscured by the bottom multiple reflection. If the multiple bottom echoes were further repressed, as might occur with additional signal processing, interface B, the top of the oceanic crust, could probably be traced at least to the edge of the continental shelf on the eastern edge of the profile, 100 km landward of the trench axis.

The slope of interface B is misleading because the section is still on a time scale. To orient the reflectors properly, and therefore further facilitate correlation of reflectors, the velocity configuration of the section must be considered.

Detailed analysis by commercial geophysical exploration company methods is now underway to produce depth sections of all the CDP reflection lines. For this paper, however, a simplified approach was employed by the authors to produce a summarized depth section for the southern line *CDP-1.* A variety of velocity data and assumptions were used, generally weighted according to our evaluation of the reliability of the seismic methods involved. In oceanic depths, refraction interpretations from explosion or ASPER stations yield the most reliable velocity models. At shallow depths, the layer interval velocity structure obtained from analysis of wide-angle reflections is generally superior to that which can be obtained from most refraction work. Such velocities are determined with shorter shot-receiver distances, so that lateral variations in interval velocities are also more easily discernible than with refraction determinations.

Seiscom-Delta calculated velocity–depth sections at 27 points (Figure 8) along the line to permit stacking of channels. The locations of the velocity analyses were chosen at points where reflectors were prominent and of potential geologic interest, such as changes in slope, and they were spaced sufficiently close to provide continuity. In many cases, however, the authors felt that alternate (and, for stacking purposes, equally satisfactory) interpretations of the velocity spectra would produce more satisfactory velocity–depth models. Therefore, for this paper we reexamined all the velocity spectra and in many cases calculated revised velocity–depth models for use in preparation of a depth-section of the CDP line. These velocity data were merged with refraction data and smoothed to keep the model shown in Figure 10 as simple as possible.

On the western, oceanic-plate, side of the CDP profile, at a depth of approximately 2.2 (two-way) seconds beneath the ocean bottom, a reflector labeled M (Figure 9) is traced beneath the trench axis and for a distance of 8 km under the continental slope. This interface is at a reflection time which, when converted to depth based on nearby refraction model velocities, agrees with the thickness of typical oceanic crust for that portion of the Nazca plate. Therefore, reflector M is identified as the Mohorovicic discontinuity. The oce-

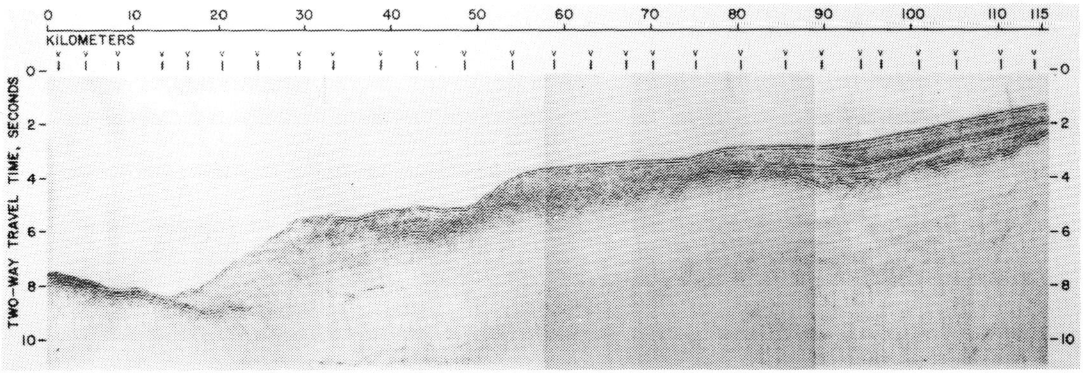


Fig. 8. Twelve-hundred-percent stacked, common-depth-point, seismic reflection line across the Peru–Chile trench near 11°40′S latitude. Data acquired and processed for this project by Seiscom Delta, Inc., Houston, Texas. Data filtered 8–64 Hz; acquired with 5,244-foot, 24-channel cable, using two 300- and two 1,000-cubic-inch airguns. The small v's at the top of the section mark the locations of layer interval velocity analyses from wide-angle reflection data.

anic crust, between interfaces B and M, was assigned a velocity of 6.64 km/sec, based on an average of the nearby refraction data summarized on the left-hand side of Figure 10. Since little variation is observed in the time-separation between the B and M interfaces, it was assumed that the internal layering of the crust was essentially uniform in this portion of the subduction zone.

The oceanic sediment velocity of 1.7 km/sec used for the refraction determination was carried into the sedimentary wedge beneath the trench axis.

Within the continental slope, two reflectors (labeled I in Figure 9) give strong returns although there is little velocity contrast across the interfaces. The lack of a strong velocity contrast, in addition to their general orientation, suggests that these reflectors may be imbricate thrust fault surfaces similar to those described by *Seely et al.* [1974] for the Middle America trench, and by *Kulm and Fowler* [1974] along the coast of Oregon.

The closest refraction data (ASPER 112, Figure 7) are not of good quality [*Hussong et al.*, 1975], but velocity determinations from the multi-channel reflection data are fairly definitive and consistent (four revised velocity analyses were made between the trench axis and the eastern extent of the apparent imbricate thrusts). The generalized velocity structure shows 2.0–2.1 km/sec material overlying a 3.5 km/sec mass which extends down to interface B. The interface between the 2.0–2.1 and the 3.5 km/sec material does not produce a prominent reflection, perhaps

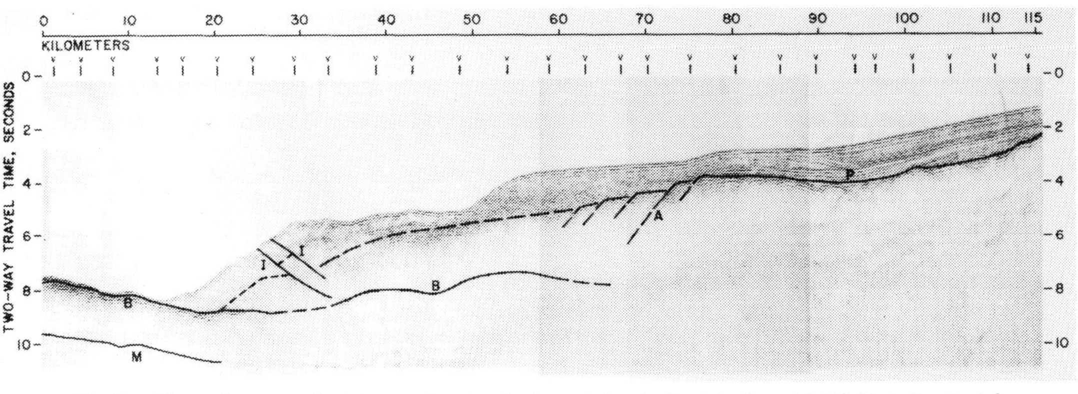


Fig. 9. Figure 8 repeated with prominent reflectors and velocity interfaces highlighted. See text for discussion.

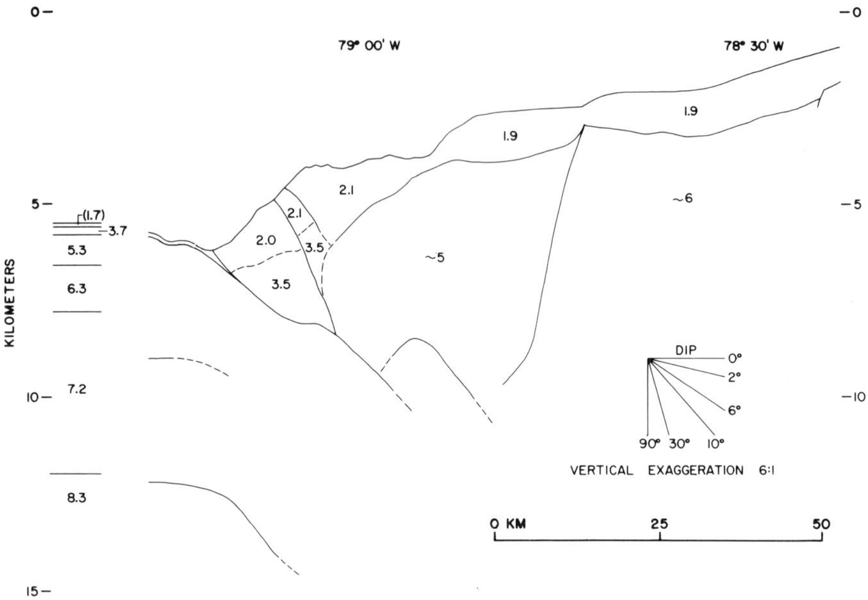


Fig. 10. Prominent reflectors and velocity contrasts in *CDP-1* adjusted from the time section, Figure 8, to true depths based on the calculated velocities as noted. The crustal velocity column at the left is a composite based on the refraction stations nearest the western end of *CDP-1*. Velocities are in kilometers per second.

because it is irregular or gradational. The boundary is, however, sufficiently sharp to be obvious on the velocity spectra.

Interfaces that are identified by velocity contrast but are not good reflectors are highlighted by dashed lines in Figure 9. The velocity spectra at the break in ocean-bottom slope at about kilometer 30 indicate that the low-velocity near-surface sediments thicken markedly. The velocity of the underlying material increases suddenly to about 5 km/sec east of kilometer 30.

The velocity of the surface sediments is somewhat lower as the continental shelf is approached. A distinct change in these sediments occurs near kilometer 80. From this point eastward onto the continental shelf, the sediments are relatively undisturbed and exhibit well-defined, relatively flat-lying interfaces. The base of the sediments on the top of the continental slope (the right-hand side of Figure 9) is also quite distinct, and both the wide-angle reflection and nearby refraction determinations show that the sediments sit on a material with a velocity of ~6 km/sec.

The boundary between the ~6 km/sec and the ~5 km/sec basement material is marked by a very subdued reflector (labeled A in Figure 9), as well as by the change from relatively undisturbed sediments and basement in the east to chaotic structure west of the basement high point near kilometer 76.

A geologic interpretation of the velocity–depth model (Figure 10) is presented later in the paper. It should be noted, however, that the hump on interface B (the top of the oceanic plate) near kilometer 50 could not be removed by any reasonable manipulation of the overlying velocity structure.

The smoothed model shows the oceanic plate to be underthrusting the outer edge of the continent at a dip of about 8°, with the oceanic plate continuing to underthrust itself (the hump at kilometer 50) as it undergoes subduction. From the refraction data discussed earlier, it seems that this shallow dip persists for at least 200 km landward of the trench axis.

## DISCUSSION

*Nazca plate oceanic crust.* It is not apparent what effect the oceanic crustal structure of the Nazca plate has on the subduction processes in the Peru–Chile trench. However, the number of stations reported here is large enough and the results are consistent enough to suggest that the typical crust is shallow (for a given age) and thin and composed of high-velocity material. This crustal character may be related to the very high sea-floor spreading rates on the Nazca plate [*Herron,* 1972*a; Rea et al.,* 1973; *Handschumacher,* this volume], which is the major factor causing the Peru–Chile trench to have the extremely high, calculated plate-convergence rate of 10.1–11.1 cm/yr [*Minster et al.,* 1974].

Figure 11 is a compilation of several composite oceanic crustal velocity columns for the Pacific [*Shor et al.,* 1970; *Hussong,* 1972] as well as the entire world [*Raitt,* 1963; *Worzel,* 1974]. In comparison, the composite Nazca plate section, based on the oceanic crustal models in this paper, is of generally higher velocity at comparable

TABLE 2. Oceanic Crustal Mass Balance

| | Layer Vel., km/sec | Thickness,[1] km | Density, $g/cm^3$ | Mass, $\times 10^5 g/cm^2$ | |
|---|---|---|---|---|---|
| Worldwide | 1.5 | 4.8 | 1.03 | 4.94 | |
| [*Raitt,* 1963] | 2.0 | 0.8 | 1.93 | 1.54 | |
| | 5.07 | 1.71 | 2.56 | 4.38 | |
| | 6.69 | 4.86 | 2.86 | 13.90 | |
| | 8.13 | 2.83 | 3.31 | 9.37 | |
| | | | | | 34.13 |
| Worldwide | 1.5 | 4.8 | 1.03 | 4.94 | |
| [*Worzel,* 1974] | 2.0 | 0.8 | 1.93 | 1.54 | |
| | 5.12 | 1.68 | 2.57 | 4.32 | |
| | 6.70 | 4.77 | 2.86 | 13.64 | |
| | 8.08 | 2.95 | 3.30 | 9.74 | |
| | | | | | 34.18 |
| Pacific | 1.5 | 4.41 | 1.03 | 4.54 | |
| [*Shor et al.,* 1970] | 2.10 | 0.34 | 1.97 | 0.67 | |
| | 5.15 | 1.21 | 2.58 | 3.12 | |
| | 6.82 | 4.57 | 2.88 | 13.16 | |
| | 8.15 | 4.47 | 3.31 | 14.79 | |
| | | | | | 36.28 |
| Pacific | 1.5 | 4.8 | 1.03 | 4.94 | |
| [*Hussong,* 1972] | 2.08 | 0.5 | 1.97 | 0.98 | |
| | 4.46 | 0.95 | 2.46 | 2.34 | |
| | 6.00 | 1.15 | 2.71 | 3.12 | |
| | 6.74 | 1.7 | 2.87 | 4.88 | |
| | 7.47 | 2.9 | 3.08 | 8.93 | |
| | 8.28 | 3.0 | 3.37 | 10.11 | |
| | | | | | 35.30 |
| Nazca Plate | 1.5 | 4.4 | 1.03 | 4.53 | |
| | 1.7 | 0.2 | 1.78 | .36 | |
| | 5.6 | 1.1 | 2.62 | 2.88 | |
| | 6.4 | 1.5 | 2.79 | 4.19 | |
| | 7.2 | 3.6 | 3.00 | 10.80 | |
| | 8.3 | 4.2 | 3.37 | 14.15 | |
| | | | | | 36.91 |

[1]Down to 15 km below sea level.

depths, primarily because of the thick 7.2 km/sec basal crustal layer.

Table 2 summarizes the mass balance between the various models. The results, although a somewhat crude and oversimplified characterization of the structures involved, demonstrate the crustal differences in somewhat quantitative terms. The masses were calculated from sea level to a depth of 15 km, as opposed to similar exercises [*Woollard,* 1962; *Worzel,* 1974] where mass balances are generally calculated to 33–35 km or deeper. The shallower datum was used because only oceanic sections where Moho depths are less than 15 km are compared and we wished to avoid strongly prejudicing the mass relationships by the chosen Moho velocity values. We also feel it unwise to assume that the upper 15–20 km of mantle is homogeneous and of a density that can be determined from the $P_n$ velocity. Our models for line *2–3* and the western end of the refraction profile *18–19,* as well as similar data elsewhere on the Nazca plate [*Hussong et al.,* 1973] and in other parts of the world [*Hales,* 1969; *Odegard,* 1975], suggest substantial velocity variation in the uppermost 20 km of the oceanic mantle.

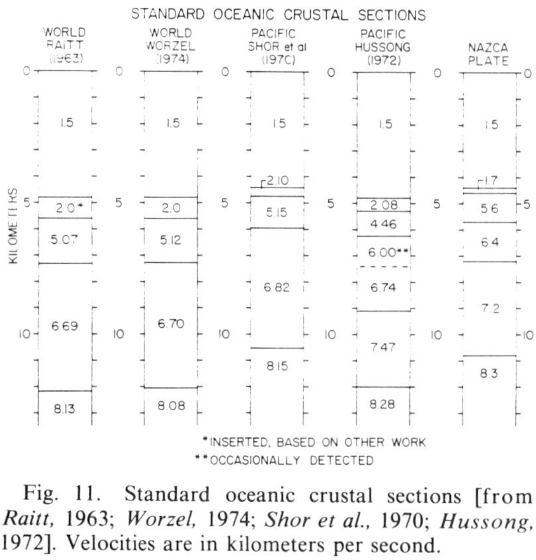


Fig. 11. Standard oceanic crustal sections [from *Raitt,* 1963; *Worzel,* 1974; *Shor et al.,* 1970; *Hussong,* 1972]. Velocities are in kilometers per second.

Referring again to Table 2, note that the *Worzel* [1974] and *Raitt* [1963] worldwide composite sections are very similar and correlate well to 15 km, which is not surprising as the sections are quite likely based on much the same data. The two Pacific columns, one published by *Shor et al.* [1970] and the other calculated by *Hussong* [1972] from data described by *Sutton et al.* [1971], are perhaps more illuminating because they are based on two different sets of data from the same ocean. The model by Shor et al. is based on a much larger data set, and is prepared mostly from conventional two-ship explosion refraction lines. The Sutton et al. model is derived from fewer stations, but shows more detailed crustal structure because it is based largely on models obtained from ASPER stations. The masses are not as close as for the worldwide models (note that in this case if the mass balance had been carried to a depth of 35 km—despite the rationalization presented earlier in this discussion for not doing so—the agreement would be nearly perfect).

The important observation from Table 2 is that the mass of the Nazca plate upper lithosphere is greater than the mass of either set of Pacific data, which are in turn greater than the worldwide averages. Furthermore, since the Nazca plate seafloor spreading rates are the world's fastest and Pacific rates are in general faster than the worldwide average, there is a suggestion that the mass of the upper lithosphere is directly related to seafloor spreading rates. In other words, at a given age, denser upper lithosphere is associated with faster spreading.

Since regional free-air gravity values do not seem to reflect these mass imbalances, the higher upper lithosphere masses must be compensated by mass deficiences at depth, and vice versa. This would accentuate the vertical mass imbalances, and provide additional force to plate-tectonic driving mechanisms such as described by *Hales* [1969], *McKenzie* [1969], *Jacoby* [1970], and *Artyushkov* [1973].

To investigate the relationship of upper lithospheric density and spreading rate, analysis of seismic velocity structure along flow lines on plates moving at a wide range of speeds is required. Particular care has to be made to remove the effect of the expected density redistribution related to the changing water depth with age first described by *Sclater and Francheteau* [1970], so that any pervading difference in structure between plates can be defined.

*Subduction in the Peru-Chile trench.* All the crustal models described in this paper are consolidated in Figure 12, a generalized composite structural section of the Peru-Chile trench region between 8° and 12°S latitude. Our seismic data suggest that the subduction processes within this area may be sufficiently uniform to allow construction of a composite section which is representative. It is likely, however, that very different types or stages of subduction are occurring at other latitudes along the trench.

The oceanic crustal underthrusting (from Figure 7) on the Nazca plate is included in the cross section (Figure 12), although as pointed out by *Hussong et al.* [1975], there is no evidence to show whether the fault is active, or the remanent of compression in the center of the plate or near the spreading center. Regardless of the exact location, the existence of such a thrust fault, supported by the compressional focal mechanism reported by *Mendiguren* [1971] for a central plate earthquake, suggests that the compressional tectonic regime in the subduction zone may also exist far out onto the oceanic plate.

As the lithosphere bends into the trench, surficial normal faulting occurs in the upper crust. These features are observed on reflection seismic records, but large-scale models from refraction data do not show any deep crustal offsets that could be attributed to tensional faulting. Rather, bottom topography and CDP multi-channel reflection records suggest that as the oceanic crust is subducted beneath the continental slope it repeatedly underthrusts itself.

From the deep reflection data in Figures 8 and 9 and the velocity structure described in Figure 10, several speculative interpretations are made and included in the composite structural section. First, it is assumed that the 2.0–2.1 km/sec material near the surface of the toe of the continental slope is oceanic pelagic sediments mixed with terrigenous turbidites and uplifted from the trench floor. Such material has been sampled on the nearby lower continental slope by *Kulm et al.* [1974]. The I reflectors in Figure 9 are interpreted to be particularly coherent examples of the imbricate thrust fault planes that would cause this uplift. The 3.5 km/sec material beneath the uplifted sediments could be consolidated sediments or uplifted slices of the sub-sediment oceanic crustal layer with an average velocity of 3.6 km/sec identified by the ASPER profile in Figure 7. This sub-sediment layer is not observed at the explosion refraction stations on the Nazca plate to the north, but this omission is likely a symptom of the increased sensitivity of the high-data-density ASPER method for detection of thin layers [*Sutton et al.*, 1971], rather than a geologic variation from north to south.

Although we cannot determine its precise composition, the low velocities and interface structure are consistent with the idea that the toe of the continental slope is composed largely of upper oceanic crustal material, as in the model of an accreting continental slope described in this area by *Prince and Kulm* [in press].

Higher in the continental slope, however, our model does not seem consistent with pure imbricate thrusting. East of kilometer 35, where the much thicker sediment cover is of slightly lower velocity and is underlain by consistently high-velocity (~5 km/sec) material, both the velocities and reflector configurations suggest that a geologic interpretation other than the further uplift of the accreted lower slope material is preferable.

The ~5 km/sec material is consistently found directly below less than 1,500 m of low-velocity

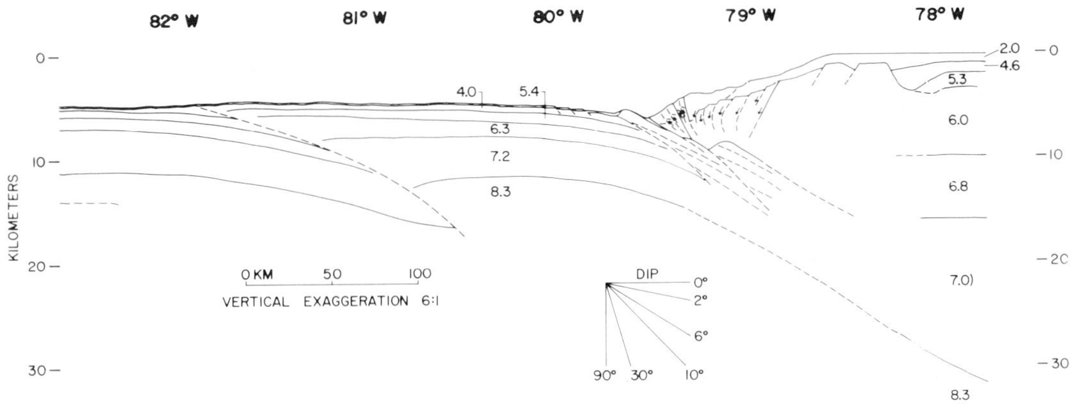


Fig. 12. Composite crustal cross section across the Peru–Chile trench, including the major tectonic features identified by seismic data between 8° and 12°S latitude.

sediments. Although the actual velocities cannot by themselves be diagnostic of rock type, the shallow, abrupt velocity increase makes it difficult to depict the interface as the result of dewatering and low-grade metamorphism of oceanic sediments with slivers of basaltic crust, as would be required by the accreting continental margin model recently generalized by *Karig and Sharman* [1975].

Landward of this mid-slope region, east of about kilometer 75 (Figure 9), the reflectors again change character. On this uppermost slope, adjacent to the continental shelf, relatively undeformed sediments overlie a distinct highly reflective basement (labeled P) with velocities averaging about 6 km/sec. This upper slope region extends onto the continental shelf, and is interpreted as continental rocks, probably of Paleozoic or older age [*Cobbing and Pitcher,* 1972], overlain by terrigenous sediments.

The greatest unknown, then, is the nature of the mid-slope region between kilometers 35 and 75 in Figure 9. Because of the velocity structure, and our interpretation of the many reflection hyperbolas at the top of the ~5 km/sec material in *CDP-1* as suggestive of normal faulting, we think the middle portion of the slope is composed of highly deformed rocks of the same continental origin as the uppermost slope to the east. By this model, chaotic and probably episodic vertical displacement, ultimately resulting in downfaulting, has resulted in the loss of interface coherency required to produce consistent reflectors, as well as the slight lowering of the seismic velocity of the basement rocks. The few faint reflectors, most notably the faint reflector A (which is approximately the boundary between the coherent upper slope and the highly deformed mid-slope continental material), suggest slumping and normal faulting downslope.

In the highlighted reflection section, Figure 9, the thickened sediment at about kilometer 33 is above the center of the crustal sink which, in this model, is the boundary between the oceanic and continental plates.

It should be noted that although our composite section includes crustal features from the entire 8° to 12°S trench region, the continental slope model comes from around 11°40′S latitude. Based on preliminary analysis of CDP data at 9°30′S, it appears that the structure of the slope at that latitude may be considerably different. Indications of normal faulting and high-velocity basement material are not apparent. The nature of the underlying middle slope material is crucial to an understanding of subduction processes. Direct sampling (drilling) is vital to acquire this information.

*Source material for the 'geostill.'* A prime motivation for these seismic experiments was to examine the subduction process as a mechanism for carrying metal-enriched oceanic sediments and upper crust to great depths along the Benioff plane. Partial melting would then remobilize the metals and the enriched felsic magma could rise and be the source of economic ore deposits (notably porphyry copper) in the Andes, according to the 'geostill' concept suggested by *Sillitoe* [1972] and *Sawkins* [1972].

The geostill hypothesis depends, therefore, on the assumption that subduction occurs and that it transports the oceanic sediments and upper crustal rocks deep beneath the continent.

It is, of course, dangerous to speculate on the nature of long-since-consumed crust. However, consideration of a subduction history based on conservative estimates is illuminating. A reasonable assumption based on the time of opening of the South Atlantic [*LePichon and Hayes,* 1971] and the major onset of volcanism and orogenic activity on the west coast of South America [*Jenks,* 1956; *Cobbing and Pitcher,* 1972], is that subduction along the Peru–Chile trench began about 150 m.y. ago. Oceanic sediment thickness cannot be generalized, but we will assume it has been 100 m, somewhat thinner than it is at present, adjacent to the trench. Further, to be conservative, we assume that no terrigenous sediment was supplied to the system (although in fact continental material is probably the major constituent of trench axis sediments). A convergence rate of the plates at the trench is taken as 10 cm/yr [*Minster et al.,* 1974].

If all the oceanic sediments were scraped off the subducting plate, 0.1 km × 10 cm/yr × 150 m.y. = 1,500 $km^2$ of oceanic sediment would have been absorbed in our continental margin cross section. If intense compaction of the sediments is assumed, we might reduce this volume to about 750 $km^2$.

Figure 13 shows the portion of our model which we believe is substantially composed of uplifted oceanic sediments and upper crust. It encompasses only about 25 $km^2$ of the cross section.

Therefore it is likely that almost all the oceanic sediment that entered the trench was subducted.

Furthermore, our model of the continental margin (Figure 12) suggests that not only is practically all the oceanic material being subducted, but also much continental material. In effect, the subduction process is eroding the edge of the continent near 11°40′S latitude. This erosion, by subduction of downfaulted blocks of middle-slope continental rock, and the concurrent vertical displacements of the outer continental shelf, is probably episodic, and may be dependent on the degree of oceanic crustal rupture, the rate of subduction, and the dip of the Benioff plane. The exceptionally high rate of convergence of the Nazca and South America plates may also be a factor giving rise to this continental erosion.

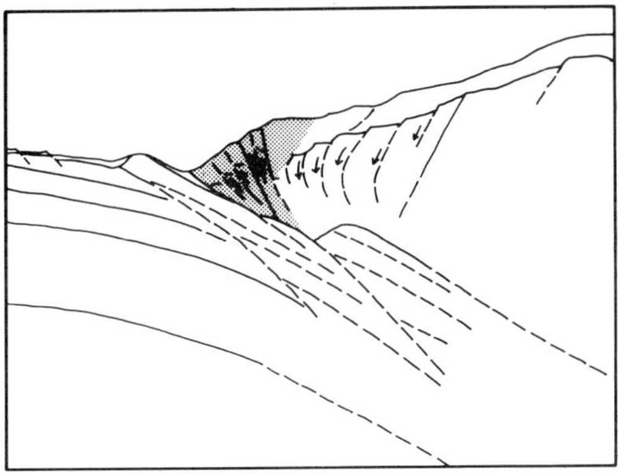

Fig. 13. Continental slope model with possible uplifted oceanic sediments (and upper crustal rocks?) hatchured.

Erosion of the continental margin along the Peru–Chile trench has been suggested to explain the tectonic history [*Rutland,* 1971] and seismicity [*Plafker,* 1972] of northern Chile. In Peru, *Cobbing and Pitcher* [1972] point out that the proximity of Paleozoic and Precambrian gneisses to the trench is inconsistent with the subduction models of *Dewey and Bird* [1970] and *James* [1971*a*]. Subduction accompanied by continental erosion would resolve this disparity.

*Karig* [1974*a*] recognizes the necessity of removing continental crust to explain the land geology, but prefers to rift the margin and slide it obliquely along the subduction zone, resulting in repetitive convergent margins, such as may exist in Colombia and Panama. Although this mechanism probably contributes to the removal of continental material, the location of the huge repetitive margins required to account for translocation of accreted oceanic as well as continental rocks from the west coast of South America is not apparent. For example, *Isaacson* [1975] has recently described evidence of thick sedimentary sequences, principally in Bolivia and southern Peru, which require a large western source of sedimentary detritus. His postulated western land mass occupied an area extending hundreds of kilometers west of the present coast, and probably stretched at least as far north as Lima. Our model of continental erosion by subduction would dispose of the remnants of that huge land mass.

## SUMMARY

Recently collected seismic refraction and reflection data provide a model of crustal structure across the Peru–Chile trench containing sufficient detail to permit discussion of the tectonic processes active along this particular segment of an oceanic–continental plate convergence zone. The fast-spreading Nazca oceanic crust, which is thin and dense relative to other ocean basins, repeatedly underthrusts itself prior to and after subduction beneath the landward wall of the trench, indicating a large-scale compressional tectonic regime. Only the lower toe of the continental slope has a structure and velocity distribution suggestive of uplifted imbricate thrust sheets containing uppermost oceanic sediments and rock. However, this lowermost slope has a volume, above the subducting plate, which is an order of magnitude less than the amount of oceanic sediment that should have been delivered to the trench since the onset of convergence. Above the apparently small accreting toe, at a water depth of 4 km, the slope velocities and structure have changed to what we interpret to be highly disrupted, downfaulted continental rocks. Farther shoreward, at a depth of over 2 km, the continental rocks are less disturbed and are blanketed with more than a kilometer of smoothly stratified sediments. Evidence from nearby land geology indicates that large-scale erosion of the west coast of southern Peru has taken place. In this area, we believe that the intermediate-depth landward trench wall is foundering continental rock that is being eroded by the subducting Nazca plate.

*Acknowledgments.* We thank the scientific parties, officers, and crew of the research vessels *Kana Keoki* and *Yaquina,* whose efforts and enthusiasm permitted the collection of this refraction data. The seismic reflection data were acquired and processed by Seiscom Delta, Inc., of Houston, Texas, under contract to the Nazca Plate Project of the International Decade of Ocean Exploration (IDOE).

We are grateful to Larry Wipperman and Steven Dang for assistance in preparation of the data, and to Daniel Karig for his criticism of the manuscript. We are indebted to George P. Woollard, whose instigation and spearheading of the Nazca Plate Project made this study possible.

This work was supported by grant GX28674 from the National Science Foundation.

# 9. *Geometry of the Nazca Plate and Its Geodynamic Implications*

RENE RODRIGUEZ, E., RAMON CABRE, S. J., AND AUGUSTO MERCADO

*Observatorio San Calixto, La Paz, Bolivia*

*Abstract.* The geometry of the Nazca plate is established on the basis of two-dimensional seismicity sections using NOAA hypocenters from 1963 to 1973. Seismicity as mapped in the region is not a true indicator of the subducted slab due to systematic errors in hypocenter location because there are no seismic stations in the Pacific Ocean. Very narrow seismicity sections, however, are shown to be effective in describing the slab's geometry. Despite the uncertainties, it is clear that the Nazca plate underthrusts the South American continent in individual, discontinuous tongue-like pieces at different dip angles. It is especially true that between latitudes 19° and 21°S, a significantly large change occurs in the direction of underthrusting, from ENE to almost E–W. Accordingly, we should expect to find a transition region somewhere between latitudes 19° and 20°S. Such a region is not detectable from the seismicity alone, however. A simple theoretical model of the collision between the continental and oceanic plates shows two systems of principal stresses: one parallel to the coastline and the other perpendicular to it. The latter system shows a conical concentration of stress lines opening toward the continent, clearly defining a change in underthrusting direction, a conical transition zone near Arica, and the small tongue-like components of the plate coincident with the seismicity data. Thus the slab seems to act as stress guide down to a depth of 250 km. East–west gravity profiles show a minimum in the Bouguer anomaly curve of the order of 300 mgal and agree very closely with the surface topography and the inflexion region of the downgoing slab, but it seems likely that the plate effect is too small to account for the observed Bouguer minimum. The tectonic pattern of Bolivia, studied by means of ERTS satellite images and correlated with the theoretical distribution of stresses, shows that faults, lineations, and other geomorphological features agree well with the two systems of lines of stress derived from the collision of the oceanic plate against the continental lithosphere. Seismicity, gravity, and tectonic data combined with the spatial distribution of mines in Bolivia show very intriguing relationships between the geometry of the slab, gravity minimum, tectonism, and ore deposits. Many questions, however, need to be answered before definite conclusions can be drawn. The ages of most host rocks seem to be too old to have any direct relationship with the present geophysical patterns, but the age of ore mineralization itself is not known. The age of the ore is the critical parameter to be established to see whether or not the mineral deposits are young enough to be attributed to the geometry of the downgoing slab.

The subduction mechanism for absorbing crustal plates is one of the features of geodynamics more clearly understood with the aid of seismology [e.g., *Sykes,* 1966; *McKenzie and Parker,* 1967; *Oliver and Isacks,* 1967; *Isacks et al.,* 1968; *Stauder,* 1968*a, b,* 1973, 1975; *Barazangi and Isacks,* 1971; *Fitch and Scholz,* 1971; *Isacks and Molnar,* 1971; *Johnson and Molnar,* 1972, show precisely this importance]. The distribution of earthquake foci shows very interesting peculiarities of the geometry of the plates when viewed either in two dimensions [*Miyamura,* 1968; *Wojtczak-Gadomska,* 1973] or in three dimensions [*Vesanen et al.,* 1973], and sheds light on mountain building [*James,* 1971*a*] and possibly the metallogenesis of ore deposits.

In the present study we shall attempt to use the seismicity of South America to obtain from the combination of two-dimensional depth sections a three-dimensional picture of the Nazca plate; and to interpret from the geometry, gravity profiles, and surface tectonics, the geodynamic implications of the subducted slab, with special reference to the Andes orogenesis, metallogenesis, and faulting activity in Bolivia.

## SEISMICITY

NOAA catalogs from 1963 to 1973 were used as a primary source of information to compile the seismic data. A computer program was written to select all the earthquakes located within the

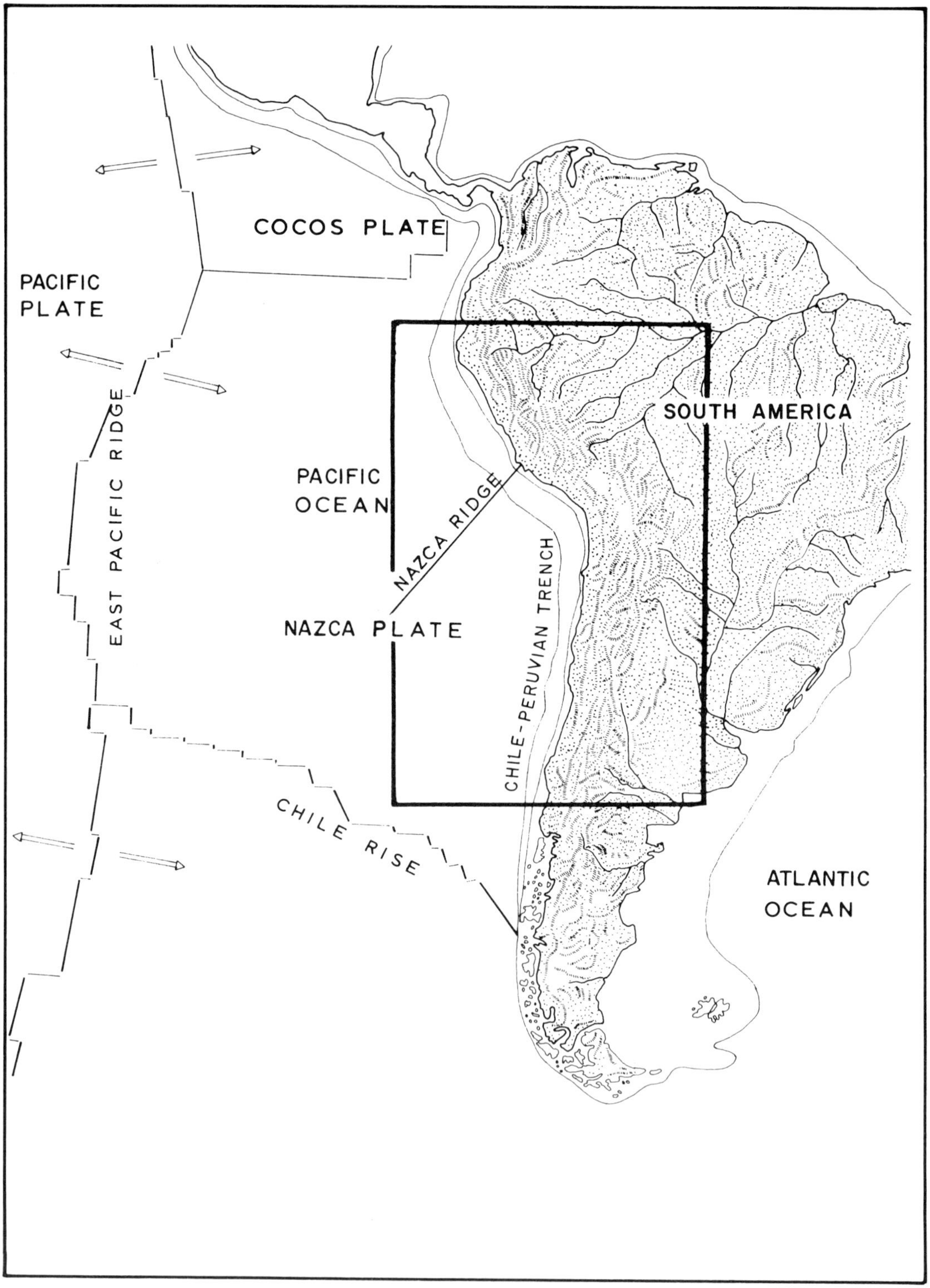


Fig. 1. Rectangle outlines study area between 2° and 40°S latitude and 60° and 85°W longitude.

pre-selected area of study, which is the region between 2° and 40°S latitude, and 60° and 85°W longitude (Figure 1). Two-dimensional, vertical and horizontal sections were made; the vertical sections are along an arbitrary direction with respect to the coastline of South America, and the horizontal sections are at different depth ranges. By combining all these sections, we can delimit the shape of the subducted part of the Nazca plate.

*Vertical sections.* Several vertical sections were made by alternatively changing the direction of the section with respect to the coastline of South America and the width of the section in order to find the section most useful for studying the geometry of the slab. For northern and southern Peru, we found that very narrow sections had to be taken because of the constant change in the direction of underthrusting from one latitude to the next. An example of this is shown in Figure 2, where the section width was changed consecutively from 12° to 6° to 2° at the center latitude of 16°S. It is clear that for the 12° section width (A) the slab is very poorly defined; for the 6°(B), it is more clearly defined; and for the 2°(C), it is very well defined indeed. This example illustrates two important facts: (1) The direction of subduction of the Nazca plate is constantly changing along the coastline; and (2) narrow sections are the only way of defining the geometry of the slab very well, despite the possibility of hypocenter mislocations. For northern and southern Chile, the behavior of the slab is less complicated and the width of the section less critical; however, narrow sections were again found to be best.

Nine (E–W) sections were taken from 6° to 38°S, one degree to each side of the center latitude (Figures 3 and 4). In general, from 7° to 13°S the slab is not well defined, especially near 13°S, where we could hardly define the Benioff zone. The dip angle decreases from 14.1 degrees at 7°S to a minimum of 9.3 degrees at 13°S; from there it increases to a maximum of 28.2 degrees at 22°S and again decreases farther south to 13.3 degrees at 37°S.

The dip angles were obtained by finding the maximum dip from consecutive sections at one point in the trench along different azimuths; this maximum was taken as the dip angle at that particular point, and the direction along which the maximum was obtained was taken as the underthrust direction of the slab. An example of the method is shown in Figure 5. With very few exceptions, the direction of underthrusting or the

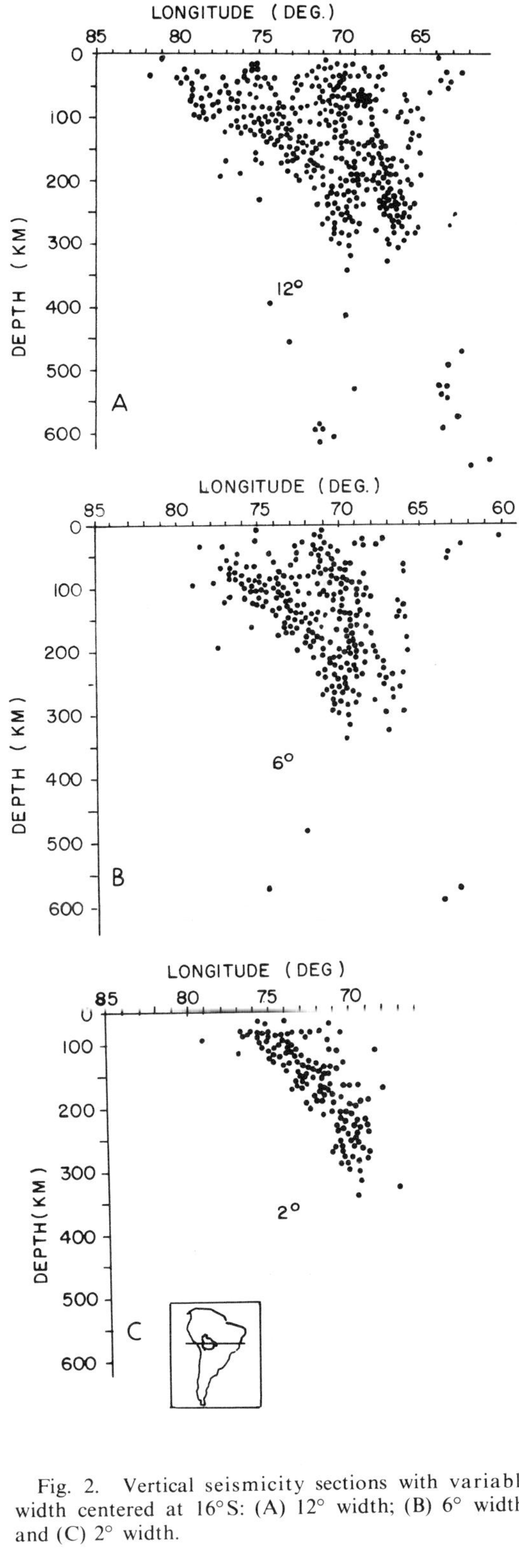


Fig. 2. Vertical seismicity sections with variable width centered at 16°S: (A) 12° width; (B) 6° width; and (C) 2° width.

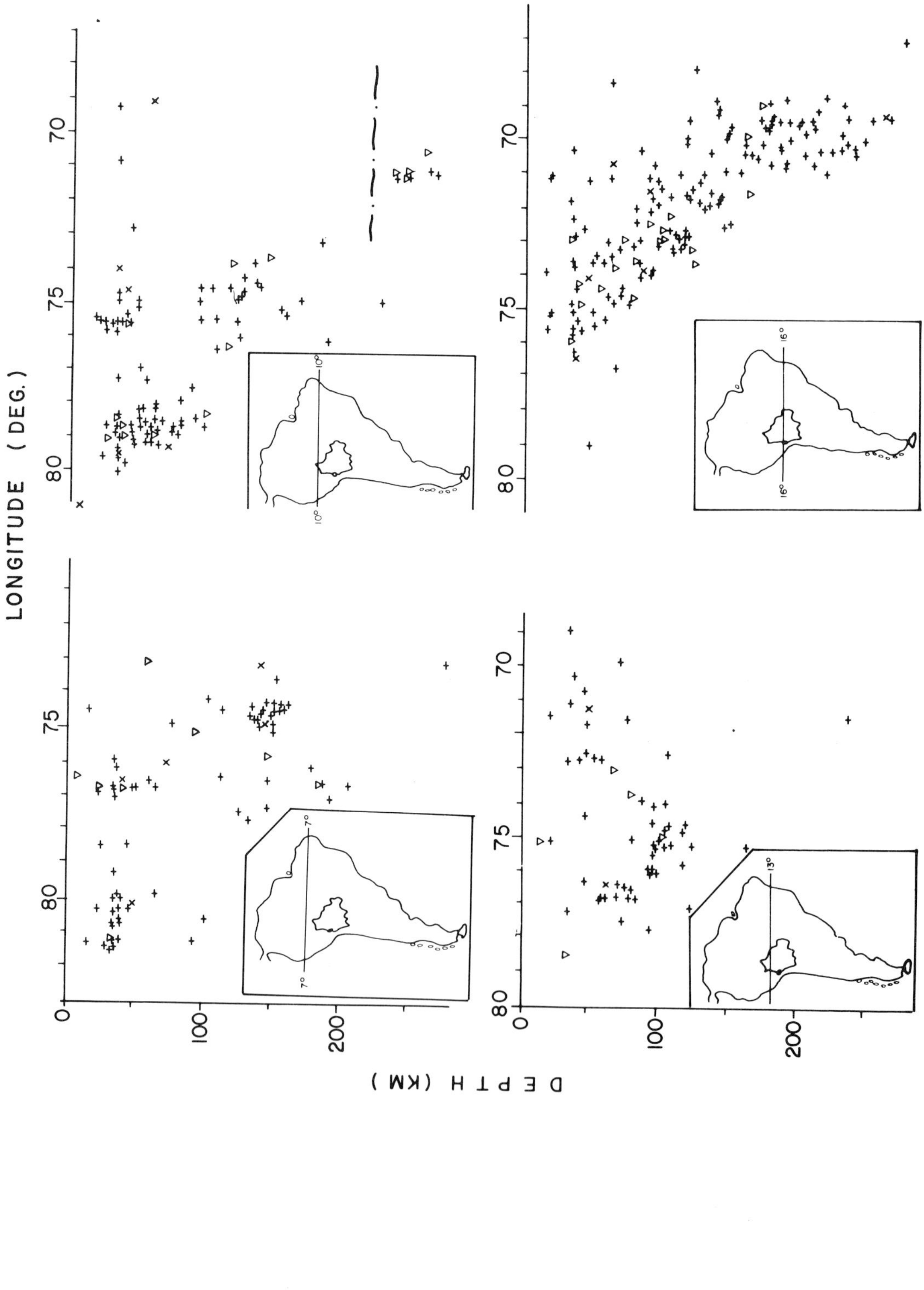


Fig. 3. Transverse (E–W) vertical seismicity sections 2 degrees wide, centered at latitudes 7°, 10°, 13° and 16°S

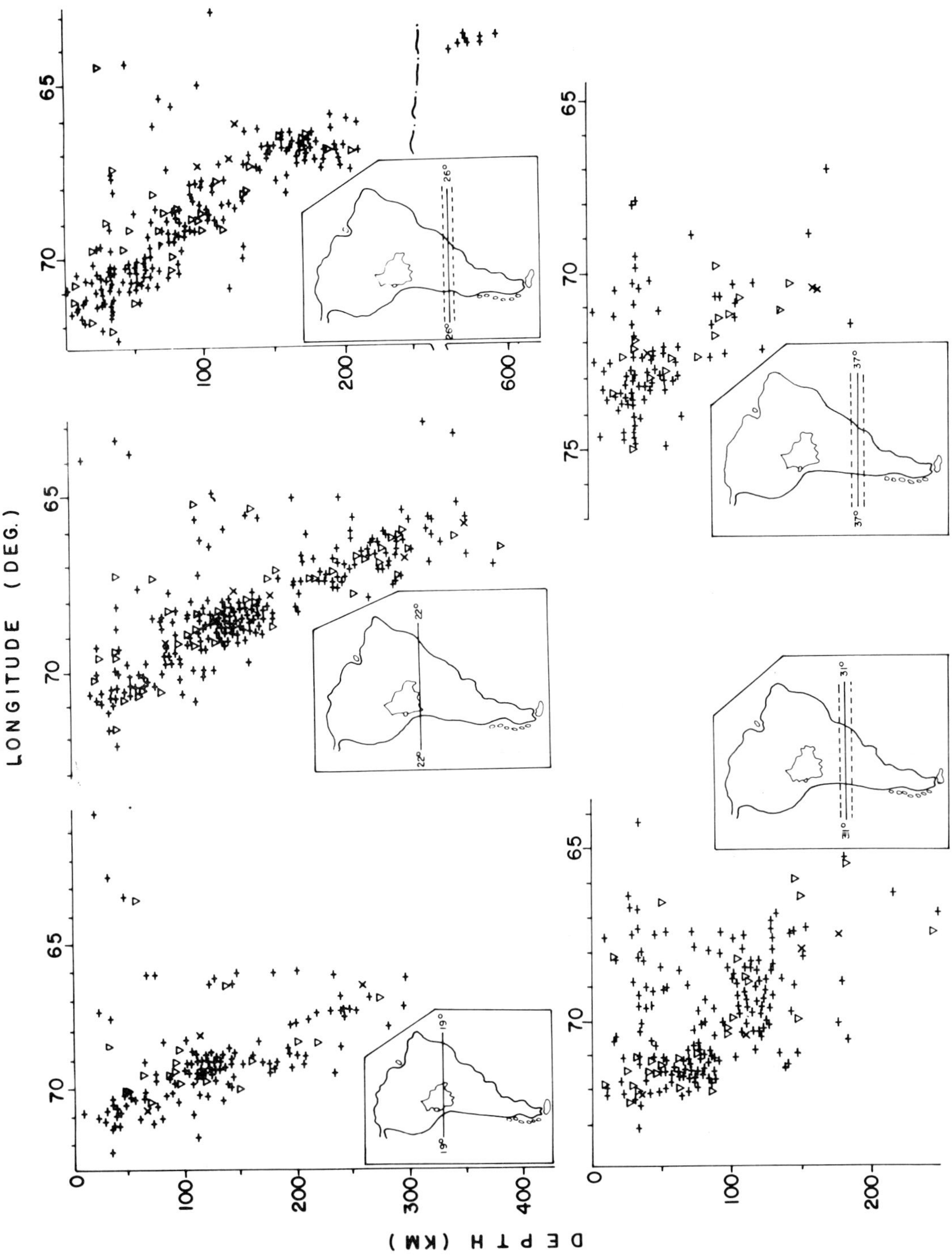


Fig. 4. Transverse (E–W) vertical seismicity sections 2 degrees wide, centered at latitudes 19°, 22°, 26°, 31°, and 37°S.

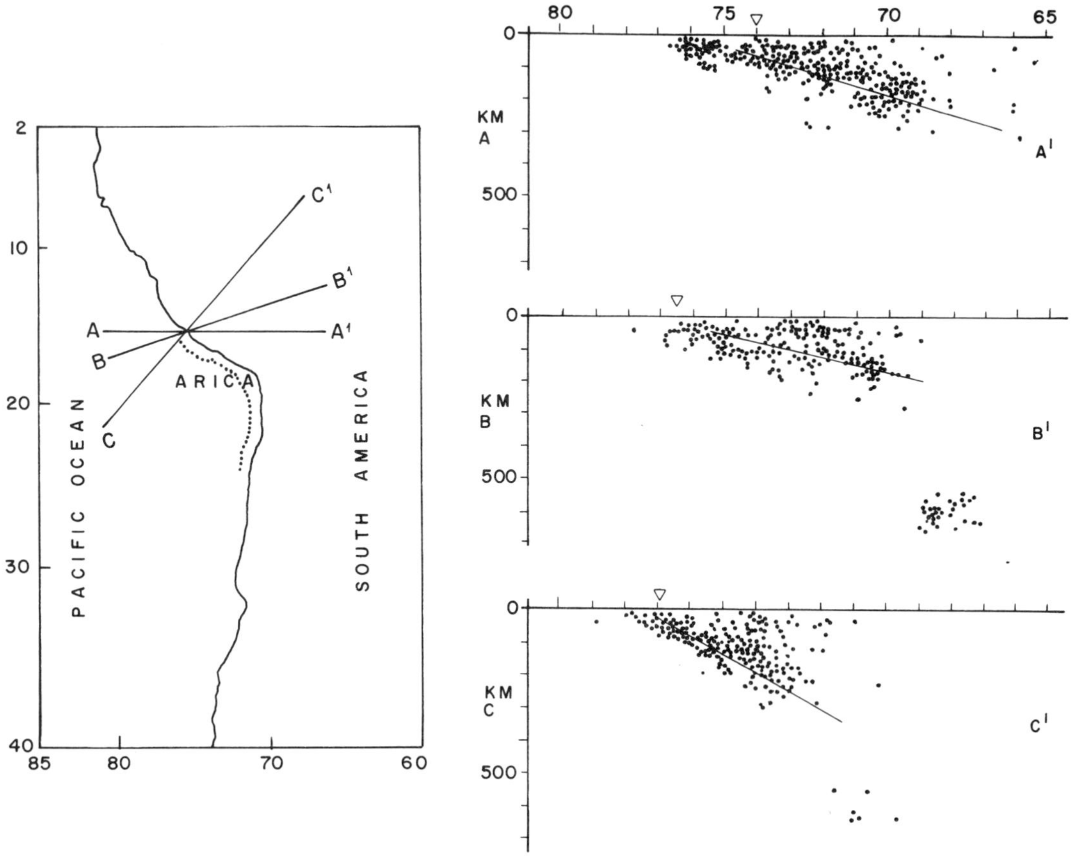


Fig. 5. Vertical sections on the right taken along directions shown on the left to obtain the maximum dip angle and the subduction direction of the slab. Section C-C′ gives the maximum dip; the direction along which it was taken is the underthrusting direction and is nearly perpendicular to the coastline.

TABLE 1. Angles of Underthrusting and Dip Angles

| Latitude at the Trench | Azimuth, deg. | Dip Angle Measured from the Horizontal, deg. | Standard Deviation of the Dip Angle, deg. |
|---|---|---|---|
| 7°S | 58 | 14.1 | 1.0 |
| 10°S | 57.4 | 13.0 | 0.8 |
| 13°S | 51.5 | 9.3 | 1.5 |
| 16°S | 48.5 | 15.0 | 0.8 |
| 19°S | 62.5 | 27.0 | 0.7 |
| 22°S | 94 | 28.2 | 0.5 |
| 26°S | 87 | 17.3 | 0.8 |
| 31°S | 97.6 | 13.7 | 0.9 |
| 37°S | 97.4 | 13.2 | 1.1 |

direction along which the maximum dip was obtained is approximately perpendicular to the coastline of South America. The actual dip values were calculated by a least squares method, considering the denser seismic region in the section of maximum dip. The values so obtained are shown in Table 1.

The observed dip angles agree quite well with the angle of thrusting obtained from focal mechanism solutions of events in the area [*Stauder*, 1973, 1975; *Isacks and Molnar*, 1971; *Fukao*, 1972; *Sykes*, 1972]. They do not agree with those obtained by *Okada* from ScSp forerunners [*Okada*, 1973], with the exception of the maximum dip of 28.2 degrees at 22°S latitude, for which the agreement is fairly close. The discrepancies may be explained by the fact that the Nazca plate does not have a simple plane boundary, as Okada assumed in his calculations, but rather a very complicated geometry that changes from azimuth to azimuth.

From the analysis we can conclude that the Nazca plate underthrusts South America in small tongue-like pieces in different directions and with different dip angles. The subduction direction of the slab is approximately perpendicular to the coastline, and it is especially interesting to note that the direction of underthrusting changes from ENE in southern Peru to almost E in northern Chile. If we assume that the slab is rigid, this sudden change implies the existence of a conical transition region between 19° and 22°S opening toward the continent; actually implying that the slab is split into two pieces in this region. However, the seismicity does not show any special features that would better confirm this phenomenon and therefore either the assumption of a completely rigid slab is not applicable here or the seismicity and the downgoing slab are not interrelated. From attenuation studies *Sacks and Okada* [1973] propose the possibility of a very thick continental lithosphere (300 km) beneath South America, thus the seismicity could be related directly to the continental rather than the oceanic lithosphere; the question remains open for further study.

*Horizontal sections.* Horizontal sections were constructed to see if they would confirm the information obtained by vertical sections and also whether the seismicity at different depths could give more information about the geometry of the slab. We selected five depth intervals for the horizontal sections; namely, from 0 to 50 km, from 50 to 100 km, from 100 to 300 km, from 300 to 500 km, and from 500 to 700 km; the corresponding sections are shown in Figure 6.

In the first section, i.e., between 0- and 50-km depth, the seismicity follows the coastline of South America almost exactly. The shallow activity is concentrated along the trench in the oceanic plate, and in the continental lithosphere. These two regions are well separated in northern Peru between 6° and 12°S. Between 12.5° and 14°S there is a seismic gap in the trench and low activity in the continental lithosphere. From 14° to 40°S the shallow activity in the trench is very narrow along the coastline separated by seismic gaps at 19°S, at 28.5°S, between 35° and 36.5°S, and at 38°S. The activity in the continental lithosphere is greatly dispersed and less well defined. Of interest is the shallow seismicity gap in the area between longitudes 62.5° and 66.5°W and latitudes 17° and 23°S.

In the second section, the seismicity between 50- and 100-km depth again follows the coastline; the principal activity is centered between 19° and 27°S. The seismicity in this depth range is clustered in small groups separated by elongated silent zones oriented approximately perpendicular to the coastline. Here the separation between the activity below the trench and the activity in the continental lithosphere is not so clear.

The seismicity between 100- and 300-km depth, shown in the third section, is again grouped in three regions: the first in northern Peru and Ecuador, the second, the more active one, between 15° and 26°S, and the third between 27.5° and 37°S. These three dense seismic regions are separated by two silent zones, one between 12° and 15°S and the other between 25° and 27°S, both of which have been recognized [*Santo*, 1969; *Stauder*, 1973]. The general shape of the seismicity is again the shape of the coastline, although somewhat distorted.

The fourth horizontal section shows principally the gap in seismicity between 300- and 500-km depth. A few earthquakes define the top of the region of deep seismic activity. The fifth section shows the two regions of deep seismic activity in South America. The first is aligned almost N–S along longitude 71.5°W between latitudes 6.8° and 10°S, and the second is oriented N5°W at 63°W between latitudes 20° and 29.5°S.

The horizontal sections at different depth ranges show that the seismicity is not continuous from one latitude zone to the next or even from one depth to the next. If seismicity defines the geometry of the slab as is assumed in plate tectonics, then we may conclude that the slab is

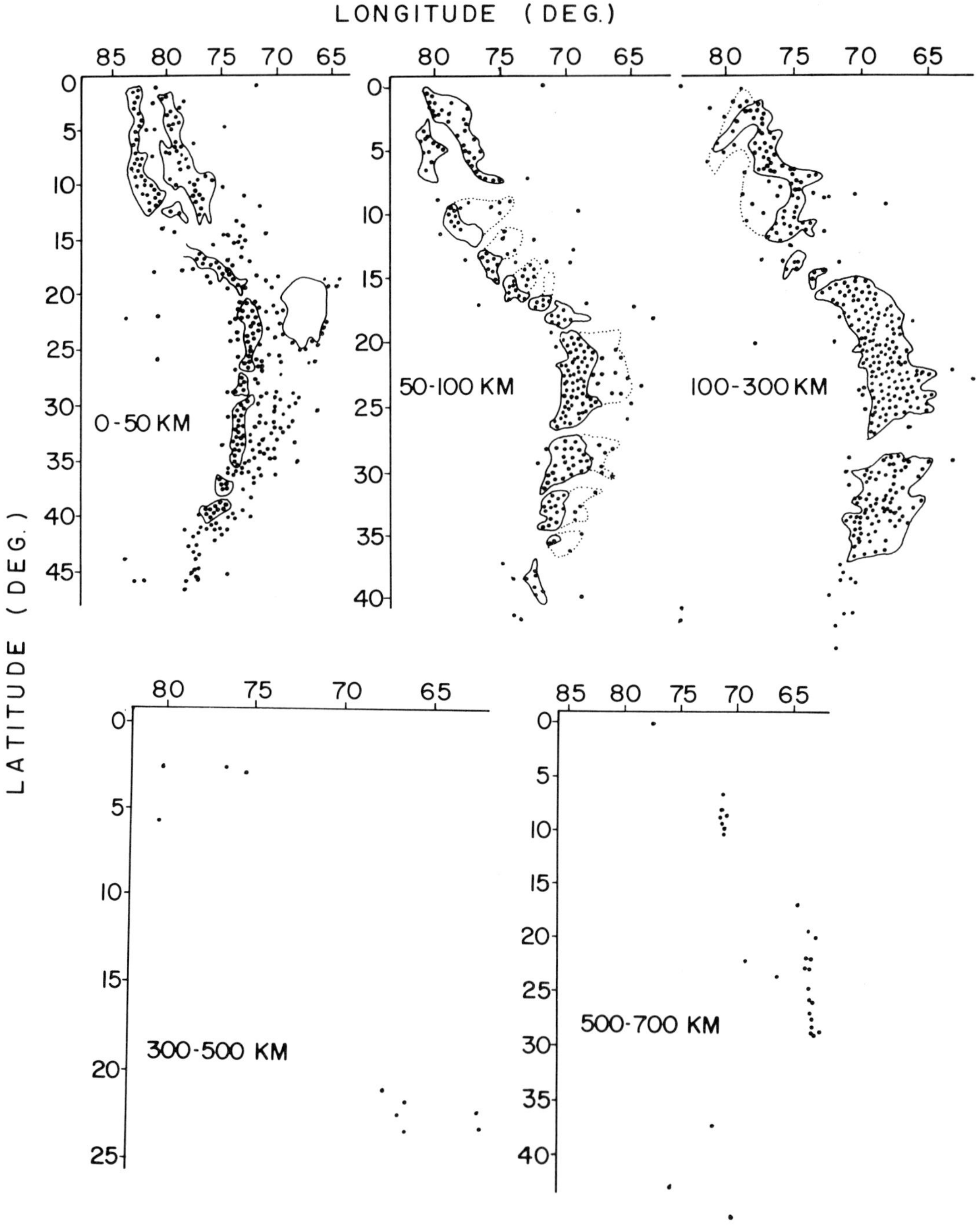


Fig. 6. Horizontal seismicity sections at different depth ranges. The contours outline interpreted areas of high and low seismicity.

broken into a series of tongues which are absorbed independently and quite differently from region to region or even from depth to depth. Only two silent zones as termed by *Santo* [1969] are present at all depths, namely, the elongated zone about 12.5°S oriented N65°E, and the zone about 26°S oriented almost E–W.

Another important conclusion is that the continental coastline significantly controls the shape of the Nazca plate down to a depth of 200–250 km. Farther down there is no clear contact between the continental and oceanic lithospheres.

Finally, we may state that the seismicity is discontinuous from 300- to 500-km depth, and the two regions of deep seismic activity are displaced about 930 km with respect to each other, leaving a seismicity gap near Arica coincident with the transition zone described before.

Combining all these vertical and horizontal sections we have tried to construct a three-dimensional picture of the subducted edge of the Nazca plate. Figure 7 shows the schematics of the slab as seen from seismicity.

In the following discussion, we analyze the rigidity of the Nazca plate and its geodynamic implications as correlated with the seismic data. First we shall attempt to describe the stress field in the plate, and then explain the geometry of the plate due to its curvature, from the viewpoint of the so-called 'rupture-ball-effect' theory.

DISTRIBUTION OF PRINCIPAL LINES OF STRESS

The theory developed by H. Hertz about the pressure between two bodies in contact, and the theory of impact as discussed by *Love* [1963] and *Morton and Close* [1922], were used to obtain a simple model of the stress distribution in the Nazca plate caused by its collision with the continental plate. The fully developed theory is discussed by *Rodriguez* [1975]; here we shall outline only the principal results of the analysis. In general two conjugate systems of principal stresses were found: one nearly perpendicular to the coastline of South America and the other nearly parallel to it. At the contact between the plates, the stresses are compressional; farther away from the contact one set shows tension near the contact and compression in the central portions, the other always shows lines of compression.

Figure 8 shows the set of stress lines perpendicular to the coastline. The lines form conical surfaces opening toward the continent; they seem to be associated with the tongue-like components of the slab and follow the general direction of underthrusting. This is another way of showing that the slab acts as a stress guide. These lines, as will be seen later, define ore provinces in Bolivia and furthermore they coincide with one set of surface faults.

The set of lines parallel to the coast seems to be

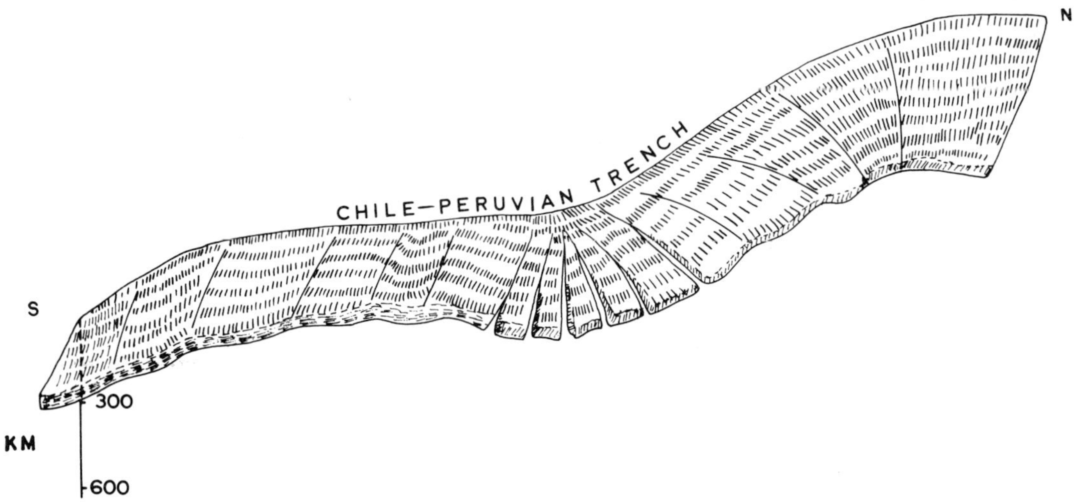


Fig. 7. Three-dimensional schematic of the Nazca plate as viewed from the east. The top edge represents the Chile–Peruvian trench and continental margin where bending begins; below this is represented the downbent upper surface of the Nazca plate. The vertical axis is the depth measured from the trench line.

related to geomorphological features of the continent, such as the Andes Cordillera, and to another set of surface fractures.

## THE CURVATURE OF THE SLAB

The theory developed by *Frank* [1968] and *Van Der Held* [1969] and reviewed by *Strobach* [1973] to explain the circular curvature of trenches and island arcs by means of the so-called 'rupture-ball-effect' is applied to South America to analyze the elastic behavior of the Nazca plate. The effect may be stated as follows: "pushing a dent into the [ping-pong] ball, we notice that the curvature of the inverted surface of the dent is the same as the curvature of the ball, but with negative radius. The rim of the dent is a circle and represents a model of the geographic curvature of a trench or island arc. If the lithosphere is assumed to be rigid and to exhibit considerable strength against lateral deformations, the downgoing slab should behave like the dented part of the ball, i.e., it should transform its shape into a plate with negative (concave) earth's curvature."

Under crucial assumption of plate tectonics, i.e., if a completely rigid lithosphere is assumed, it can be shown [*Strobach,* 1973] that:

$$\beta = \frac{\alpha}{2}$$

where $\beta$ is the radius of the arc or trench in degrees, and $\alpha$ is the dip angle of the downgoing slab. This relation is nearly satisfied in some island arcs (e.g., Aleutian, Kuril), but considerable deviations from the rule do exist (e.g., Tonga, Kermadec). These deviations may be explained by: (1) the lithosphere is not rigid; or (2) the slab undergoes considerable stresses parallel to the strike of the deformation. If the second is true, we have two possibilities, namely: (a) if $\beta > \alpha/2$, the sinking slab would undergo strong compressional stresses parallel to the strike and would be contorted like a curtain; (b) if $\beta < \alpha/2$, the slab would undergo tensional stresses parallel to the strike, presumably tearing the slab into smaller strips.

Deviations from the theory also occur in South America (although the west coast of South America is not really an island arc, the theory is quite general and applicable to any structure on earth where sinking lithosphere is taking place). To see the implications of the 'rupture-ball-effect' on the Nazca plate, we selected four regions along the Peru–Chile trench and measured both the radius in degrees and the dip angle of the slab in each; the results are summarized in Table 2. For regions 1 and 4 (Figure 9), $\beta > \alpha/2$, so in those two regions, the Nazca plate must be undergoing strong compressional stresses in the strike direction. For regions 2 and 3 (Figure 9), $\beta < \alpha/2$, and there the slab must be undergoing tensional stresses parallel to their strike. Region 5 was taken as the whole southern Peru and Northern Chile trench, where the dip angle was averaged,

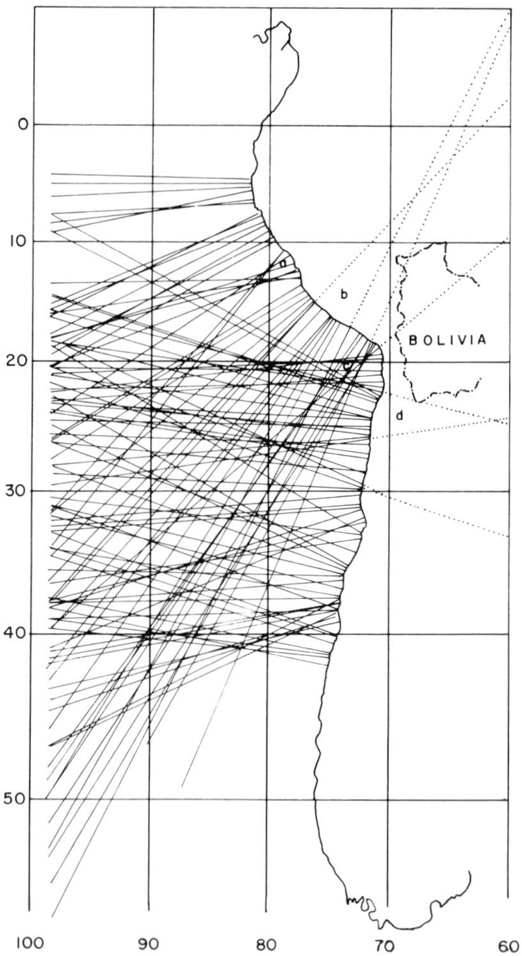


Fig. 8. Theoretical distribution of principal stresses due to the collision of the Nazca plate with the continental plate. This is one set of a two-conjugate system of stresses predicted by the theory; the other system (not shown) is approximately parallel to the coastline.

TABLE 2. Radius of Curvature and Dip Angles

| Region | Radius of Curvature ($\beta$), deg. | Dip Angle ($\alpha$), deg. |
|---|---|---|
| 1 | 8.0 | 10.0 |
| 2 | 2.3 | 15.0 |
| 3 | 1.3 | 27.0 |
| 4 | 18.0 | 17.3 |
| 5 | 7.0 | 26.5 |
| 6 | 25.0 | 17.0 |

and region 6 as the whole central and southern Chile trench. For region 5 there are tensional stresses, while for region 6 there are compressional stresses, parallel to the strike direction of the slab.

We may recall from the seismicity analysis that regions 2 and 3 correspond to the transition zone where the direction of subduction changes from ENE to almost E, and we pointed out that in order to have a *rigid* downgoing slab, it must be broken near Arica. The same kind of conclusion may be drawn from the present discussion, although again it is not conclusive, since the seismicity pattern near Arica does not show any special feature to support this idea unless, as pointed out, the seismicity itself is not interrelated with the sinking slab, but rather is contained within the continental lithosphere.

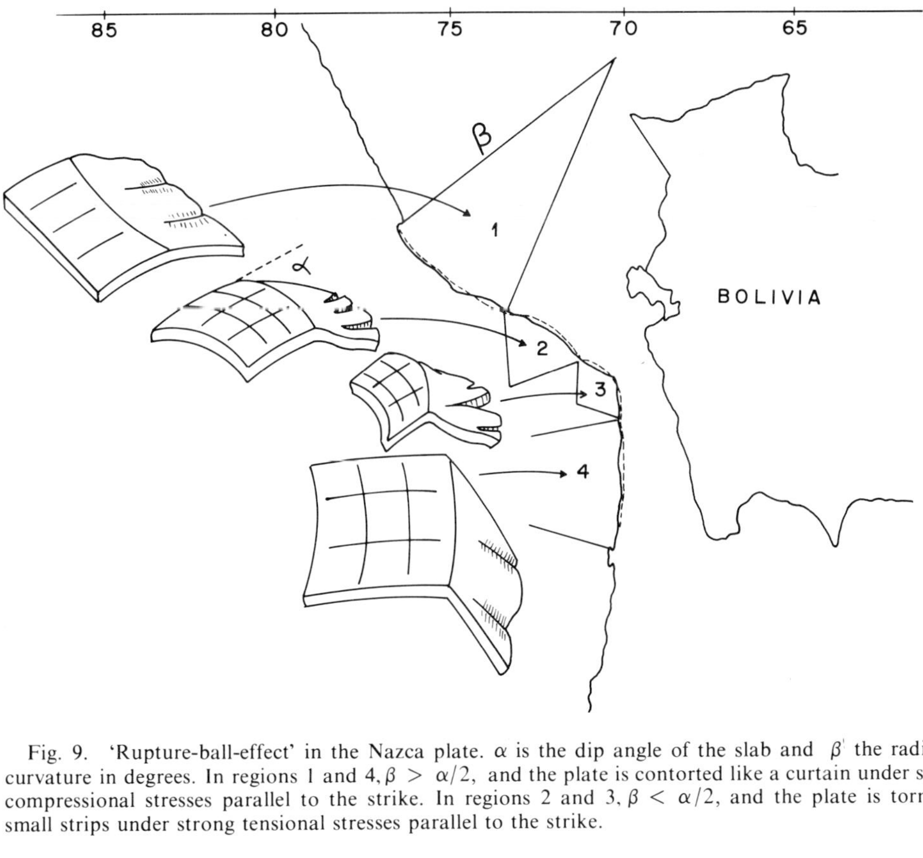


Fig. 9. 'Rupture-ball-effect' in the Nazca plate. $\alpha$ is the dip angle of the slab and $\beta$ the radius of curvature in degrees. In regions 1 and 4, $\beta > \alpha/2$, and the plate is contorted like a curtain under strong compressional stresses parallel to the strike. In regions 2 and 3, $\beta < \alpha/2$, and the plate is torn into small strips under strong tensional stresses parallel to the strike.

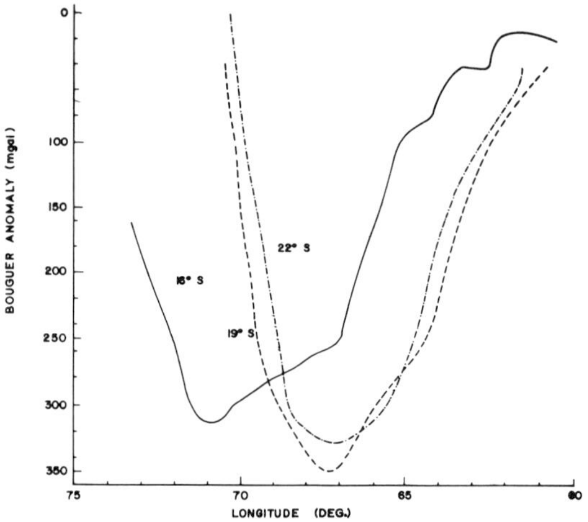


Fig. 10. East–west Bouguer anomaly curves at latitudes 16°, 19°, and 22°S.

GRAVITY DATA

Several E–W gravity profiles have been used in correlation with the seismicity of the region to gain better understanding of the geometric and geophysical features of the Nazca plate. Figure 10 shows three transverse Bouguer anomaly profiles along latitudes 16°, 19°, and 22°S. A prominent minimum of about 300 to 350 mgal near longitudes 70° to 67°W is typical. The profiles reflect mainly the surface topography, as will be shown later.

This minimum does not coincide with the Peru–Chile trench but is within the continent; for instance, the 16°S minimum is at 71°W whereas the trench at this latitude is at 75°W; the 19°S minimum is at 67.2°W, with the trench at 70°W; and the 22°S minimum is at 68°W, with the trench at 71°W.

To correlate these gravity data, we put them together in Figure 11 with the seismicity and surface topography; it is immediately seen that the Bouguer minima closely reflect the surface topography. At 16°S latitude, the western and eastern Andes Cordilleras are well separated by the Altiplano plateau; the 16°S Bouguer anomaly curve reflects the topography, with one minimum corresponding to the western Andes, the other to the eastern Andes, and the flat portion in between to the Altiplano plateau. The minimum at 19°S is narrower, as is the Altiplano plateau. Finally, at 22°S the two cordilleras become one, and this is reflected as a single minimum in the anomaly curve.

Although in all cases the Bouguer minimum (the more prominent) coincides with the inflexion point of the downgoing slab (Figure 11), it is more likely that the cordillera and the plateau are responsible for the minima. Simple calculations show that the slab above has very little effect on the Bouguer anomaly. In contrast to other regions (Aleutian, Puerto Rico, and others), the gravity profiles across the Peru–Chile trench do not show a broad positive anomaly associated with the negative Bouguer anomaly. There are three alternative interpretations: (1) the difference between the density of the mantle through which the slab is dragged and that of the slab itself is very small; (2) the continental lithosphere under the Andes is very thick; or (3) both. *Sacks and Okada* [1973] found high $Q$ values beneath the Andes down to a depth of 300 km between the crust and the sinking slab; this may correspond to the observed gravity data.

Finally we shall interpret the tectonic pattern (faults and lineations) in Bolivia, and correlate it with the geometry of the downgoing slab and mineral deposits in Bolivia.

TECTONIC PATTERN

The tectonic pattern (faults, lineations, volcanoes, and so forth) has been studied by means of Earth Resources Technology Satellite (ERTS) images with a scale of 1 : 1,000,000. The photo interpretations have been made by the ERTS project of the Bolivian Geological Survey (GEOBOL), and the results published by *Fernandez et al.* [1974]. We use their interpretations in correlating the geometric configurations of faults, lineations, and other tectonic features, with the theoretical distribution of stresses where the oceanic Nazca plate collides with the continental lithosphere.

Figure 12 shows the interpreted faults and lineations in the area. From the overall picture we can see three principal sets of faults and lineations: one set is oriented parallel to the coastline of South America, another is nearly perpendicular to the coast, and the third is oriented in a direction different from the two cited. Some examples for the first set of lineations are (from west to east): Coastal 1, at the edge of the continent; Coastal 2, extending from southern Peru to northern Chile; Coipasa, extending from near Lake Titicaca to Uyuni Lake; Viacha; Poopo 1; Poopo 2; and Tunari. For the second set, that is, for lineations oriented approximately perpendicular to the coast, there are (from north to south): Titicaca;

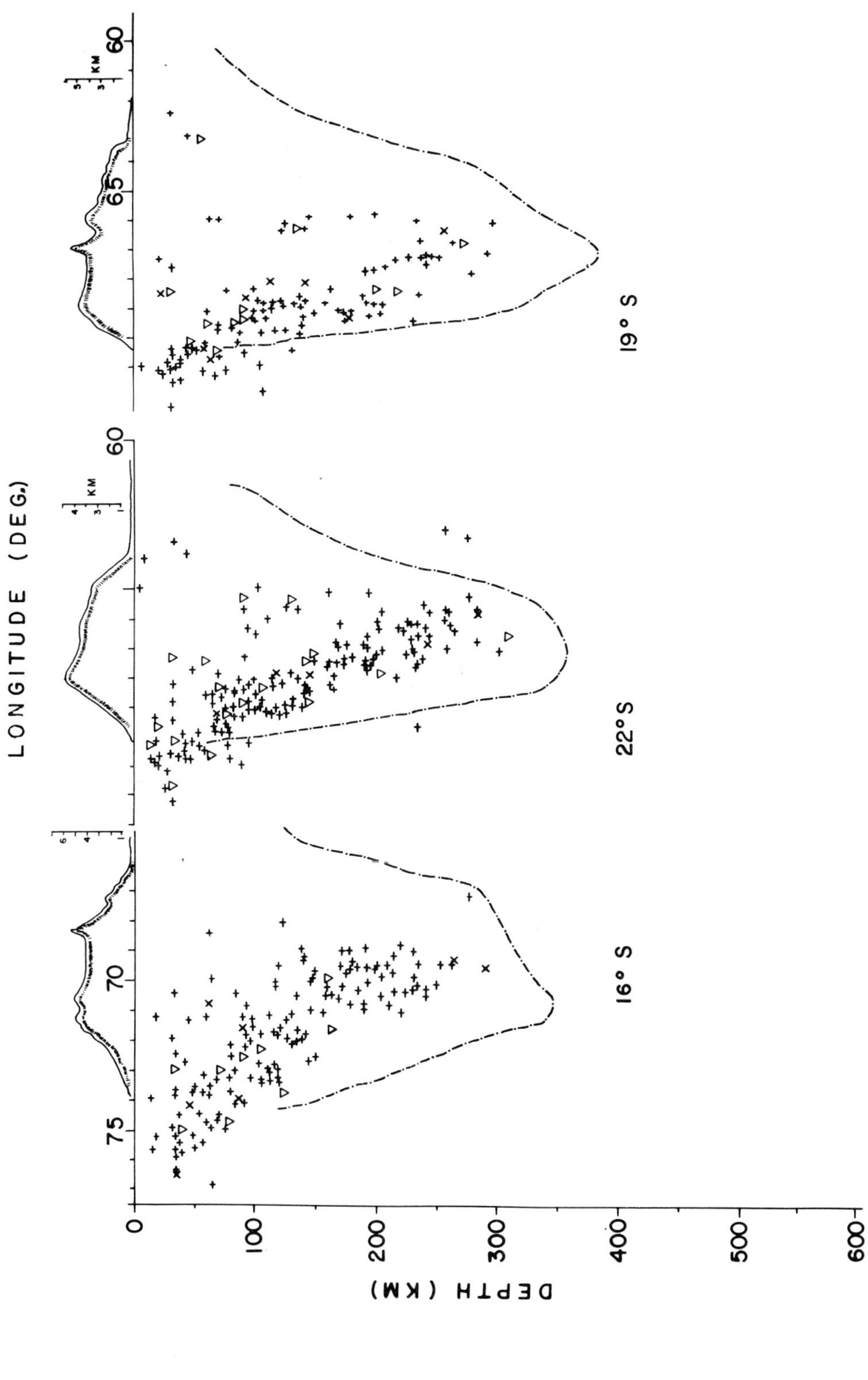


Fig. 11. Seismicity, gravity, and topographic sections along latitudes 16°, 19°, and 22°S. Dashed lines are the Bouguer anomaly at the same scale as on Figure 10. Vertical topographic scale is exaggerated.

La Paz; Julo; Camarones; Machacamarca; Pisagua; Coipasa; and Poopo 3. To the third set belong: Mauri-Desaguardero, oriented E–W; Tapacari, oriented N86° W; etc.

In general the tectonic lineations and the orientation of the Andes Cordillera fit very well with the theoretical stress distribution where the Nazca plate is colliding with the continental lithosphere. Moreover, mineral deposits are concentrated along many of the lineations such as the Machacamarca–Santa Rosa fault system, the Poopo 2 fault, the Illimani lineation, and others. Most of the mineralization associated with these lineations is of hydrothermal origin. The spatial relationship between the Nazca plate, the tectonic pattern, and the mineral deposits is obvious.

### METALLOGENESIS IN BOLIVIA AND THE NAZCA PLATE

Figure 13 shows a schematic plot of the spatial distribution of mines in Bolivia grouped according to thermal origin. High temperature minerals contain Sn, Au, and W; intermediate temperature minerals contain Zn, Pb, and Sb; and the low temperature minerals include Ag and Cu. For the discussion we shall divide the area into three regions: (1) north, between the Titicaca lineation and Machacamarca–Santa Rosa fault system; (2) south, between the Machacamarca–Santa Rosa fault system and the Uyuni lineations; and (3) south of the Uyuni lineations.

In the first region the direction of underthrusting of the slab is approximately ENE and its relationships are clear. The distribution of mines according to thermal origin follows more or less this general ENE trend, with the low temperature ore deposits farthest west, the intermediate in the center, and the high farthest east. This distribution coincides with the depth increase from WSW to ENE of the Nazca plate.

Low-temperature ore deposits are concentrated in the Altiplano plateau along the Bouguer minimum and closely follow lineations paralleling the coastline; this region is called the copper province of the Altiplano. The ore there is present as native copper at the surface and oxides and sulfides at

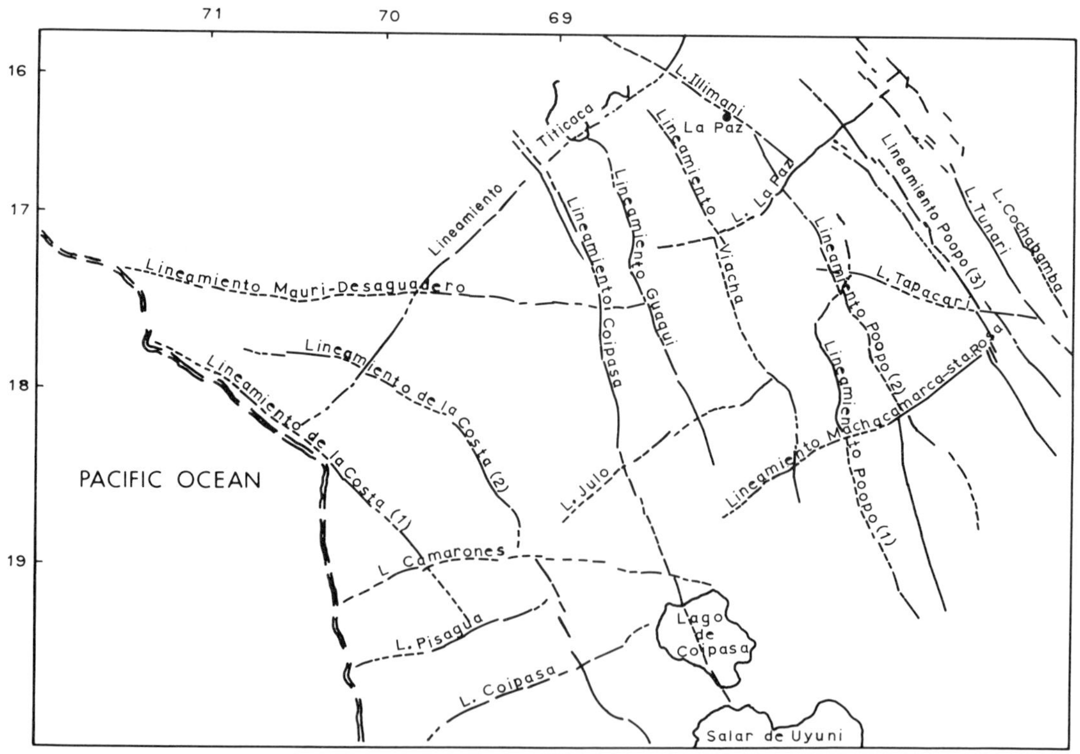


Fig. 12. Tectonic map of lineations and faults photo-interpreted from ERTS images. Heavy lines are the principal lineations and geological faults.

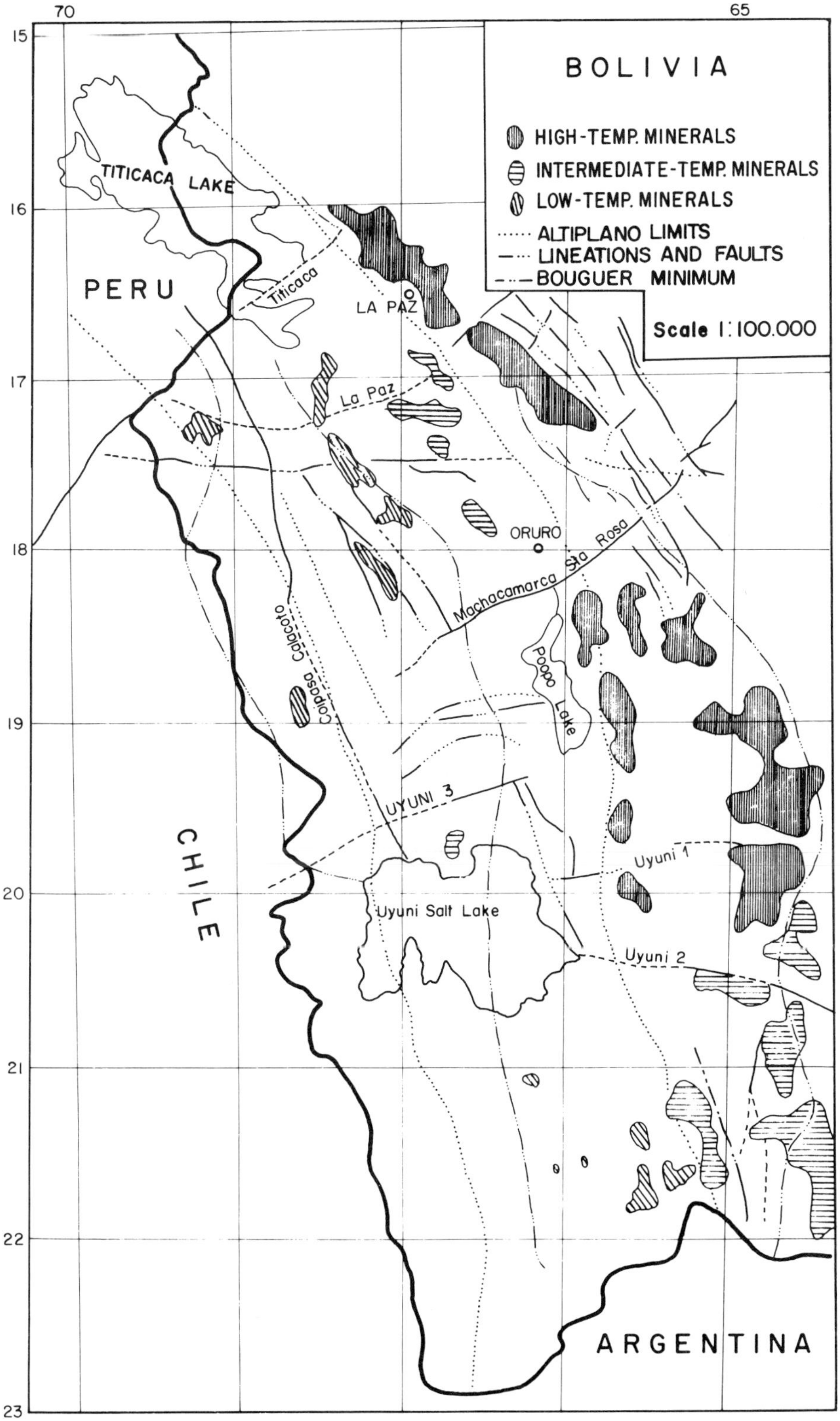


Fig. 13. Map of spatial distribution of mines in Bolivia (see legend for symbols).

some depth. Most of the richest mines are related to the Coniri fault, a long geological fault that extends from Lake Titicaca to the edge of the Machacamarca–Santa Rosa fault system. The age of the host rock is mid- to late Tertiary.

Intermediate-temperature mineral deposits again follow the general trend of lineations parallel to the coast and are located at the eastern edge of the eastern Andes, only indirectly related with intrusives.

High-temperature ore deposits are located within the eastern Andes system and related directly with the intrusive, inasmuch as the host-rock, shales, sandstone, and quartzite have been folded and fractured by the intrusives. These deposits follow the Bouguer minimum related to the eastern Andes. Geochronological studies of the intrusives show two different ages: the intrusive north of La Paz lineation (Huayna Potosi Mountain, Tuni Condoriri Mountain) is 9 m.y. old; the intrusive south of it (Illimani Mountain) is 24 m.y. old. This age difference indicates two intrusive epochs that are separated by the La Paz lineation. If we consider the 9 m.y. host-rock age to also be the age of mineralization of the surrounding zone, then a direct relationship between Nazca plate subduction and the mineral deposits is possible in this ore province of Bolivia. On the other hand, if the 24 m.y. age of the Illimani Mountain host rock is also assigned to its mineralization, there is no apparent temporal relation between the subducted slab and mineralization processes. There remains unexplained, however, the remarkable coincidence between the geometry of the plate and the spatial distribution of mines in the region.

The mineral assemblages are rather similar on both sides of the La Paz lineation, and could indicate that they probably were deposited at the same time, implying that the subducted plate and the mineralization in the whole region could be related to each other.

The second region corresponds to the transition between 19° and 20°S where the subduction direction changes from ENE to E and to the conical surface delimited by stress lines near Arica, where the maximum dip angle and the maximum continuous penetration is achieved by the Nazca plate. There is no normal transition from low-temperature to high-temperature deposits from west to east as in the first region; all the ore deposits are in general high-temperature, mixed with intermediate-temperature deposits of secondary origin. There are no low-temperature mineral deposits and the general trend of the lineations is perpendicular to the coastline. The eastern ore deposits of this region are directly related to intrusives and to the Bouguer minimum corresponding to the eastern Andes, and do not follow any particular tectonic pattern. The mineral deposits to the west define the richest tin belt in Bolivia. They are only indirectly related to the intrusives and are aligned along the eastern flank of the eastern Andes. The host rock of these deposits is the Zapla horizon of Silurian age, in which hydrothermal magma filled old fractures. There is no clear relation with any slab-like geometry and the areal distribution of mineral deposits is chaotic, typical of a transition zone where the slab is presumably broken into small strips. The presence of only high-temperature minerals and the complete absence of low-temperature ore deposits suggests a direct contact with asthenospheric materials of very high temperature, in which magma differentiation originated. However we must point out that this correlation is only tentative, based on geometrical features; it is by no means conclusive since we have neither the precise age of the minerals nor a clear hypothesis to explain how the minerals were formed from that deep region. More definitive data will be necessary for a better understanding of the geochemical process, and, what is more important, we must have reliable geochronological data before any conclusive statements can be made.

In the third region, the slab starts to underthrust eastward but is still within the transition region as defined by the seismicity data and the theoretical model of interaction between continental and oceanic plates. Most of the mineral deposits are of intermediate temperature; there is only one area at the contact with the second region in which there is a high-temperature ore deposit. Again the deposits are aligned along the Bouguer minimum and directly related to intrusives. South of Uyuni 1 lineation there are high-temperature deposits with intermediate- and low-temperature deposits farther south. The thermal gradient appears to run from north to as far south as latitude 21°S. Farther south the gradient again is from west to east.

In the seismicity analysis we found that the maximum dip and the maximum penetration of the slab occur at 22°S and that, southward from there, the dip angle decreases to a minimum in southern Chile. This would imply that the slab is actually contorted sideways somewhere around latitude 21°S; this could be the reason for the north-to-south gradient of hydrothermal mineralization south to 21°S, and the normal west-to-east pattern farther south where the slab underthrusts to the east. We need to say again that this correlation is based purely on geometry and must be considered as preliminary inasmuch as the data used did not involve any geochemical or geochronological parameters. We are in the process of refining the data, collecting field samples, and performing gravity profiles to gain more insight into the physical processes that took place beneath the Andes and are related to the surface mineral deposits.

## SUMMARY

The geometry of the subducted edge of the Nazca plate has been established on the basis of two-dimensional seismicity sections. The features have been discussed with respect to the 'rupture-ball-effect' theory and the distribution of stresses due to the collision of the oceanic plate with the continental lithosphere. It was found that the slab constantly changes its direction of underthrusting from north to south, being in general nearly perpendicular to the coastline. Between 19° and 20° S there is a change in the subduction direction of the plate from ENE to almost E, defining a conical transition region opening toward the continent near Arica, which is not detectable from the shallow or intermediate seismicity but is clearly established from the deep seismicity. The Nazca plate underthrusts the continent in a series of small tongues which are absorbed independently and quite differently from latitude to latitude and even from one depth to the next. Tectonic trends, such as faults and lineations studied by means of ERTS images, correlate very well with the two conjugate systems derived from the collision of plates, and are closely related to ore deposits in Bolivia.

East–West Bouguer anomaly curves show a minimum of about 300 mgal, closely reflecting the surface topography but showing very little effect of the slab, and correlate well with mineral deposits in Bolivia.

Close correlation is seen between the geometry of the plate, gravity data, tectonic lineations, and the spatial distribution of mines in Bolivia. Only geochronological data is lacking to confirm or reject the hypothesis of a direct relationship between mineral deposits in Bolivia and the Nazca plate's geometry.

*Acknowledgments.* Critical discussions with Drs. C. Kisslinger, W. Stauder, S.J., O. Nuttli, I. S. Sacks, A. Linde, A. Snokes, and D. James are gratefully acknowledged. This research was supported by the Air Force Office of Scientific Research Grant AFOSR-72-2397.

# 10. *Project Nariño*[1] *III: Refraction Observation Across a Leading Edge, Malpelo Island to the Colombian Cordillera Occidental*

R. P. Meyer and W. D. Mooney[2]

*Geophysical and Polar Research Center, University of Wisconsin, Madison, Wisconsin 53706*

A. L. Hales[3] and C. E. Helsley

*Institute for Geosciences, University of Texas at Dallas, Richardson, Texas 75080*

G. P. Woollard, D. M. Hussong, and L. W. Kroenke[4]

*Hawaii Institute of Geophysics, University of Hawaii, Honolulu, Hawaii 96822*

J. E. Ramirez

*Instituto Geofísico de los Andes Colombianos, Bogotá, Colombia*

*Abstract.* Energy critically refracted through the continental crust and mantle as well as the oceanic plate including the zone of subduction has been received by seismographs in the western cordillera of Colombia, South America, from shots fired along a 400-km profile from the Pacific coast to the west. With added control provided by sea gravity data, preliminary results from heat flow, reflection and local refraction measurements at sea, and refraction measurements along the western cordillera, all made during Project Nariño, a model crossing the leading edge of the continent at 4°N has been constructed which satisfies the gravity data both at sea and on land with maximum deviation of 22 mgal. The principal results from the study are: (1) the verification of structures expected to be associated with subduction by a controlled seismic source technique in an area of active tectonism; (2) a suggestion of a mechanism for localization of sedimentary basins by continental mantle–oceanic plate interaction; (3) a prediction of the extension of the Bolivar Atrato–San Juan basin to the shelf edge; and (4) a beginning explanation of the source of the unusual positive Bouguer anomaly found in association with the high western cordillera of the Andes of Colombia. In the region from the Yaquina graben to the Malpelo rise, uniform delays in the travel times observed as compared to those computed through the model suggest a soft zone in the oceanic upper mantle in accord with a broad zone of higher than normal heat flow which includes the graben. Multiple observations on land of controlled sources fired at sea appear to be a straightforward and effective method of delineating both oceanic and continental structure.

Direct seismic examination of the shallow portion of subduction zones including the region of downturn may be possible in a previously untried but straightforward way. We have seen refraction arrivals, most logically associated with paths through the subducting slab from shots crossing the oceanic plate, trench, and continental rise, at seismographs placed along the extension of the shot line on the adjacent continent in Colombia. This paper concerns the techniques and evidence, leading to this conclusion, based on a 400-km line

[1]The historical perspectives leading to the undertaking of Project Nariño in 1973, and the objectives and general overview of this, a multinational controlled source seismic investigation in western Colombia and Ecuador, and in the adjacent ocean, have been briefly reviewed by *Aldrich et al.* [1973] and a general report is being prepared by all the participants.

[2] Geophysical and Polar Research Center Contribution Number 320.

[3]Now at the Research School of Earth Sciences, The Australian National University, Canberra, A.C.T., Australia 2600.

[4]Hawaii Institute of Geophysics Contribution No. 702.

of Pacific Ocean shots near 4°N latitude, observed principally in the westernmost Andean range of Colombia, the Cordillera Occidental.

Further, this technique appears to be generally applicable. Its applicability, at least for the eastern Pacific, rests on successful land observations of sea shots, including shots in the region seaward of the trench. As a check, test shots off Peru have been subsequently fired near but on both sides of the trench and successfully recorded in Peru and Bolivia [*Ocola et al.*, 1975]. In 1974, a second full-scale experiment with shots crossing the Mid-America trench and receivers on land in Mexico has been observationally successful. Results appear at the present stage of analysis as supportive of the procedure's general applicability [*Mooney et al.*, 1975; *Helsley et al.*, 1975].

The Colombian study, where this technique was initially applied, had its roots in earlier studies recording mine blasts and special shots in high lakes of the Andes of Peru, Bolivia, and northern Chile with the hope of understanding the structure and tectonic process responsible for the mountains. Here, refracted arrivals through the crust and upper mantle were sought principally along strike, as early attempts across structure were unsuccessful [*Aldrich et al.*, 1958, 1960; *Woollard*, 1960*a, b*; *Ocola et al.*, 1971]. Further attempts along strike were only partially successful in receiving $P_n$ first arrivals. The results nevertheless supported high basal crustal velocities, crustal low-velocity zones, and a very thick crust approximately 65 to 75 km under the mountains [*Woollard*, 1960*b*; *Ocola and Meyer*, 1972], which agrees with estimates by *James* [1971*a*] based on earthquake surface-wave dispersion studies.

## GENERAL SETTING

Western Colombia, the site of Project Nariño (named after a province in southwestern Colombia where much of the work was undertaken) was chosen for a renewed search for an understanding of Andean tectonism for a number of reasons. There, to the north from the Ecuadorian border, the Andes diverge to become three well-separated cordilleras having significantly differing gravity expressions (both positive and strongly negative) and are paralleled at sea by a filled, possibly fossil, trench. These factors, and a more moderate level of earthquake activity, suggested an Andean region undergoing less rapid tectonic evolution and having more separable parts and processes than those found to the south. A major portion of the Nariño experiment was designed to continue the search parallel to structure along the cordilleras using timed explosions in lakes, modeling the previous experiments.

In the light of plate tectonics and subduction processes, profiles perpendicular to structure were planned as well, using the adjacent ocean for shots to be monitored by both land and sea recorders. We attempted this with some trepidation regarding the land observations. Compressional energy transmission might well have been poor across this leading edge, for in addition to the previous experience of poor observations across structural grain, poor transmission had been reported for the $S_n$ phase at land observatories from offshore earthquakes in some regions farther to the south along western South America [*Molnar and Oliver*, 1969]. Fortunately, our experience with compressional wave arrivals was good for stations placed near outcrop even in the easternmost cordillera at ranges of over 800 km from 1800 kg shots (L. C. Ocola, personal communication, 1974). This in part is undoubtedly due to extremely low ambient noise levels, a condition which seems characteristic of large portions of the Andes.

The Nariño experiment (Figure 1) covered a roughly triangular area extending from 4°N to 2°S along its easternmost boundary, the eastern of the three cordilleras of Colombia, and the southern extension of the Andes into Ecuador. The western apex of the triangle, 400 km offshore at Malpelo Island (81.5°W), was the juncture of two lines of sea shots. One line radiated southeastward to terminate near the port of Tumaco in southwestern Colombia; the other radiated eastward toward the port of Buenaventura (77°W) 350 km north of Tumaco. Ten groups of investigators from three continents participated, as did three research vessels.

Other analyses being conducted independently but which we expect will be directly relevant to the observations being reported here are those of the German Group for Explosion Seismology and the University of Hawaii, which have additional perpendicular-to-structure data also within the western cordillera, and those of the Instituto Geofísico del Peru and the Carnegie Institution of Washington, which have data from observations 250 km farther to the east near Bogotá.

## PRELIMINARY RESULTS FROM OTHER STUDIES

For this first discussion of the results from observations of the shot line from Buenaventura to Malpelo Island, roughly along 4°N latitude, we are dependent on initial and preliminary results

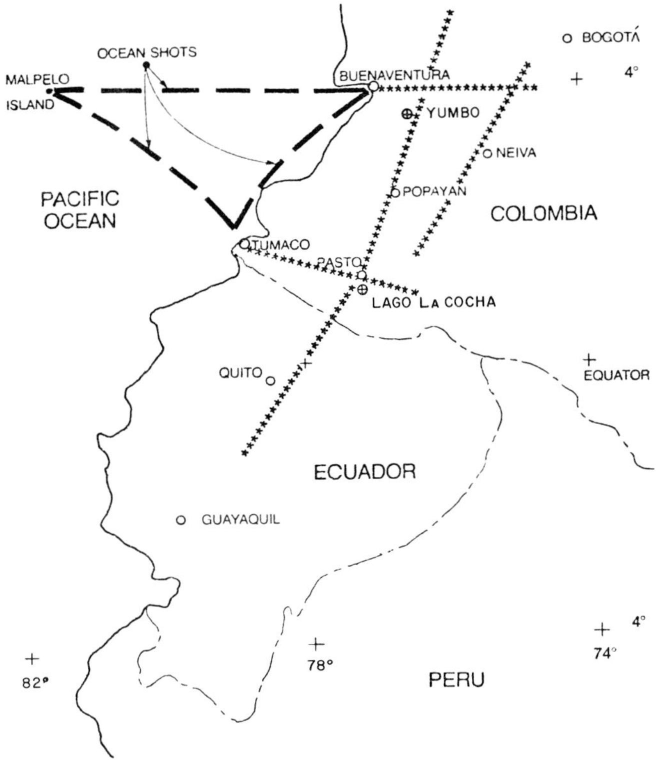


Fig. 1. Sketch map of the Nariño project, Colombia and Ecuador, South America [after *Aldrich et al.*, 1973].

from other observers. These have been generously provided to us in advance of publication (Figure 2).

At the western apex of the triangle in the Malpelo Island region (Figures 1 and 2) free-floating telemetering hydrophone buoys were set from the research vessel *Thomas G. Thompson* by the University of Washington to observe shots fired from the research vessel *Kana Keoki* of the University of Hawaii along the two profiles to the coast. These observations, together with detailed local refraction data taken from the *Thompson* by the University of Washington in the vicinity of Malpelo Island, have provided estimates of the velocity of the oceanic crust in this region (B. T. R. Lewis, personal communication, 1975). A parallel-to-structure profile shot from the Colombian Navy vessel *Gorgona* to a University of Texas ocean-bottom seismometer on the shelf near Buenaventura has provided estimates of velocities and thickness of the sediments filling the basin off Buenaventura, and the velocity of basement rocks underlying it.

Equally important to the interpretation of the sea shots are the results of observations of shots made by the Carnegie Institution of Washington at Laguna La Cocha in southwestern Colombia which were recorded to the north along the western cordillera by the University of Wisconsin and the German Group for Explosion Seismology (R. Meissner, personal communication, 1974). This profile crossed the landward extension of the 4°N profile at the parallel-to-structure profile's approximate end, 400 km to the north of the La Cocha shot point [*Gettrust et al.*, 1973]. Along the western cordillera near the town of Cali and near the intersection of the aforementioned profiles, a partial reverse to the La Cocha profile was provided by three shots from the Yumbo cement quarry of Cementos de Valle, Inc., recorded to the south along the cordillera. These shots, observed by the University of Wisconsin

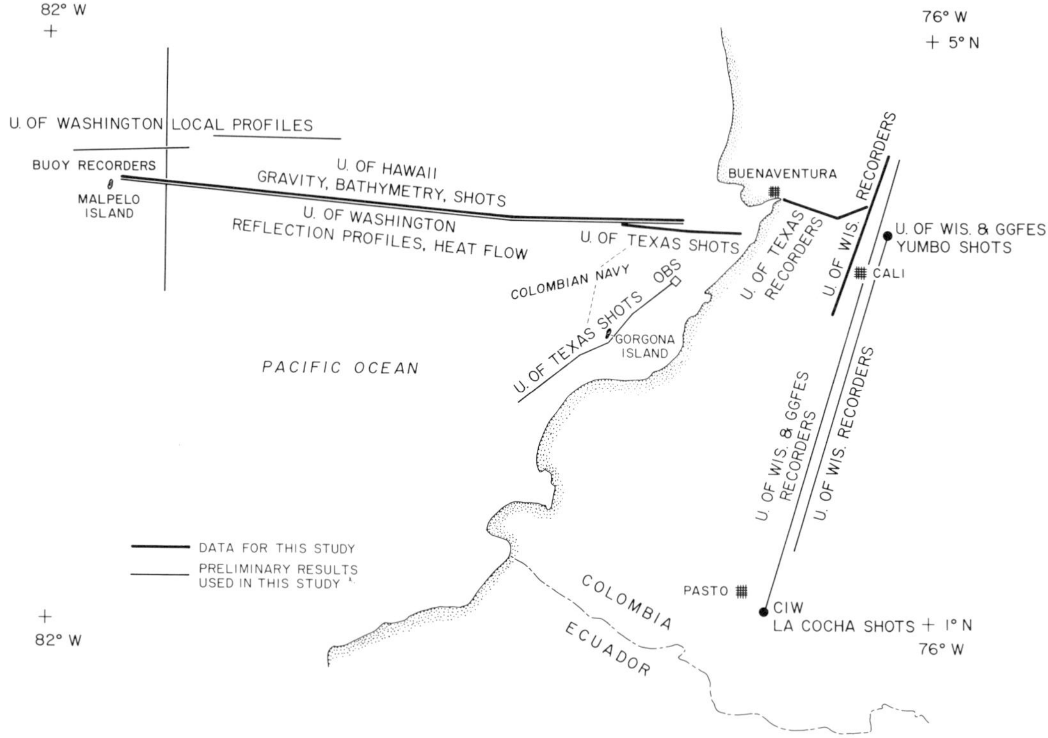


Fig. 2. Summary of the auxiliary data preliminary results used and the data present in this contribution. All of the data shown were taken during Project Nariño, February–March, 1973.

CIW is the Carnegie Institution of Washington, D.C. Department of Terrestrial Magnetism. GGFES is the German Group for Explosion Seismology. The entire study was coordinated by the Instituto Geofísico de Los Andes, Universidad Javeriana, Colombia, and the Carnegie Institution of Washington, D.C. with the cooperation of the following institutions: Bolivia: Comisión de Geofísica; Instituto Pan Americano de Geografía e Historia. Colombia: Instituto Geográfico 'Agustín Codazzi'; La Armada Colombiana; Texas Petroleum Corp.; International Petroleum Corp.; INGEOMINAS; FORGE (Fund of Overseas Research Grants and Education); Ecco Petrol; Espressa Colombianos Petroleros. Ecuador: Instituto Geográfico Militar Ecuatoriano; Dirección National Geológico; Escuela Politecnica; Instituto Hidrográfico de La Armada Ecuatoriana. Germany: German Group for Explosion Seismology. Panama: Inter-American Geodetic Survey. Peru: Instituto Geofísico del Peru; Instituto Geofísico, Universidad Nacional de San Agustín; Centro Regional de Sismología. United States: University of Texas, Dallas; University of Wisconsin, Madison; University of Hawaii; University of Washington; National Science Foundation.

along the eastern flank of the western cordillera, provide an additional measure of the probable velocities and structure under the recording sites used for observing the sea shots [*Meyer and Gettrust,* 1973].

The free-air gravity and bathymetry measurements taken by the University of Hawaii during the shooting of the large-shot profiles have been provided in advance of publication. Reflection time estimates of acoustic 'thickness' within the surficial unconsolidated sediment along the profile line were made by the University of Washington. Two-way times as great as 1 sec were found using match correlator boomer techniques (C. R. B. Lister, personal communication, 1975). There is considerable variability in the times, and these are incorporated in this analysis.

## THE PRINCIPAL DATA SET

Combined here are the results from nine University of Texas and University of Wisconsin observing stations from the six 900-kg and five 1,800-kg shots fired from the *Kana Keoki,* and from the 15 smaller shots made from the *Gorgona,* all roughly along latitude 4°N. Together the shots extend 450 km from Malpelo Island to near the coastal city of Buenaventura and the receiving stations extend from near Buenaventura to the eastern terminus of the Cordillera Occidental (80 km). The equally-spaced shots from the *Kana Keoki* cover the entire sea profile line at about 40-km intervals. The more closely-spaced colinear set of 15 smaller shots covers the shelf and slope region (Figures 2 and 3).

The Texas data taken on line come close to being complete and freestanding, but suffer from signal-to-noise problems attendant in having to put some recorders in regions unfavorable for signal reception, namely, the sediments of the coastal plain and intermountain valley regions (Figure 4). The Wisconsin observing array was placed broadside, within a favorable receiving region, the western cordillera, and crossed the extension of the shot line and the eastern terminus of the in-line Texas array. With the exception of data from Wisconsin station 3, the Wisconsin data are not on the extension of the shot line. The most successful elements of the Texas array were also in the western cordillera, but were near their western limit. Combining the two arrays, we have enough well-observed data for the in-line or colinear analysis.

## REGIONAL SETTING

The general setting of the experiment (Figures 3 and 4) indicates the structural complexity traversed. Overall, from west to east, the profile traverses the Malpelo rise, an oceanic swell culminating in Malpelo Island; the Yaquina graben (believed to be either an analog to or an extension of the Peru–Chile trench [*Hayes,* 1966; *van Andel et al.,* 1971]); an area of rough and rising bathymetric relief; a reasonably flat shelf and coastal plain; and the western range of the Colombian Andes, the Cordillera Occidental.

Near Buenaventura the coastal plain is underlain by Tertiary marine sediments forming the Atrato–San Juan basin, a deep N–S elongated feature [*Case et al.,* 1971] (Figure 4). This basin, considered part of the Bolivar geosyncline [*Nygren,* 1950], is believed to underlie the coastal region of much of western South America. The gravitational effect of the basin is clearly seen in the Bouguer anomaly map as a negative belt which reaches a maximum anomaly of –90 mgal at Buenaventura [*Case et al.,* 1971], as the axis of the basin moves offshore (Figure 5). On geologic evidence, *Nygren* [1950] estimated that 6–11 km of sediment fill the basin, and *Case et al.* [1971], combining geologic and gravity data, have used a value of 10 km in their E–W crustal model of the nearshore and western cordillera near Quibdo, Colombia (Figure 4), 200 km north of Buenaventura.

The topography rises rapidly from the coastal plain into the Cordillera Occidental, reaching a maximum elevation of about 4000 m. This cordillera is unlike the two more eastern cordilleras in Colombia or, for that matter, the bulk of the Andes, for it is characterized by positive Bouguer anomalies [*Case et al.,* 1971] (Figure 5). In the western cordillera, thick Mesozoic eugeosynclinal rocks with lesser amounts of Tertiary sedimentary and volcanic rocks and Mesozoic or Tertiary granitic and dioritic plutons are known, as are common and extensive mafic and ultramafic bodies [*Case et al.,* 1971]. These mafics are in accord with the positive Bouguer anomalies in this portion of the cordillera. The Wisconsin recording stations and some of the Texas stations were located in the southern part of the areas in which greenstone, pillow basalt, and diabase formations were particularly abundant. As will become apparent, because of the regional nature of observations, neither the travel time data nor the gravity data resolve abnormal velocity or high density as being necessarily near the surface; the data, however, do require high velocity and hence high density within the crust under the receiving stations. The southeasternmost station, station 2, although on bedrock, is just east of the Cauca fault, the eastern termination of the Cordillera Occidental; all other stations considered here are west of this fault (Figure 4).

Offshore, positive free-air anomalies characterize the oceanic plate in this region [*Hayes,* 1966]. Along the profile, values measured during the experiment averaged about +25 mgal, reaching as high as +40 mgal seaward of the trench. These values are in contrast with gravity values generally associated with the interior of oceanic plates; for example, free-air values rarely exceed +10 mgal in the Nazca plate to the south.

Regionally considered, the offshore area is a part of the Eastern Panama basin. The N–S Coiba fracture zone at 83°W divides the Panama basin into eastern and western sectors. The western sector is characterized by smooth topography and is the site of N–S spreading along the eastern extension of the Galapagos rift [*van Andel et al.,* 1971]. In contrast, the eastern basin, in which the profile was shot, is characterized by rough topography.

## ON THE QUESTION OF CONSTRAINTS

*Apparent velocities across fixed land array.* The principal data set amounts to a multiply observed single-ended profile crossing a region

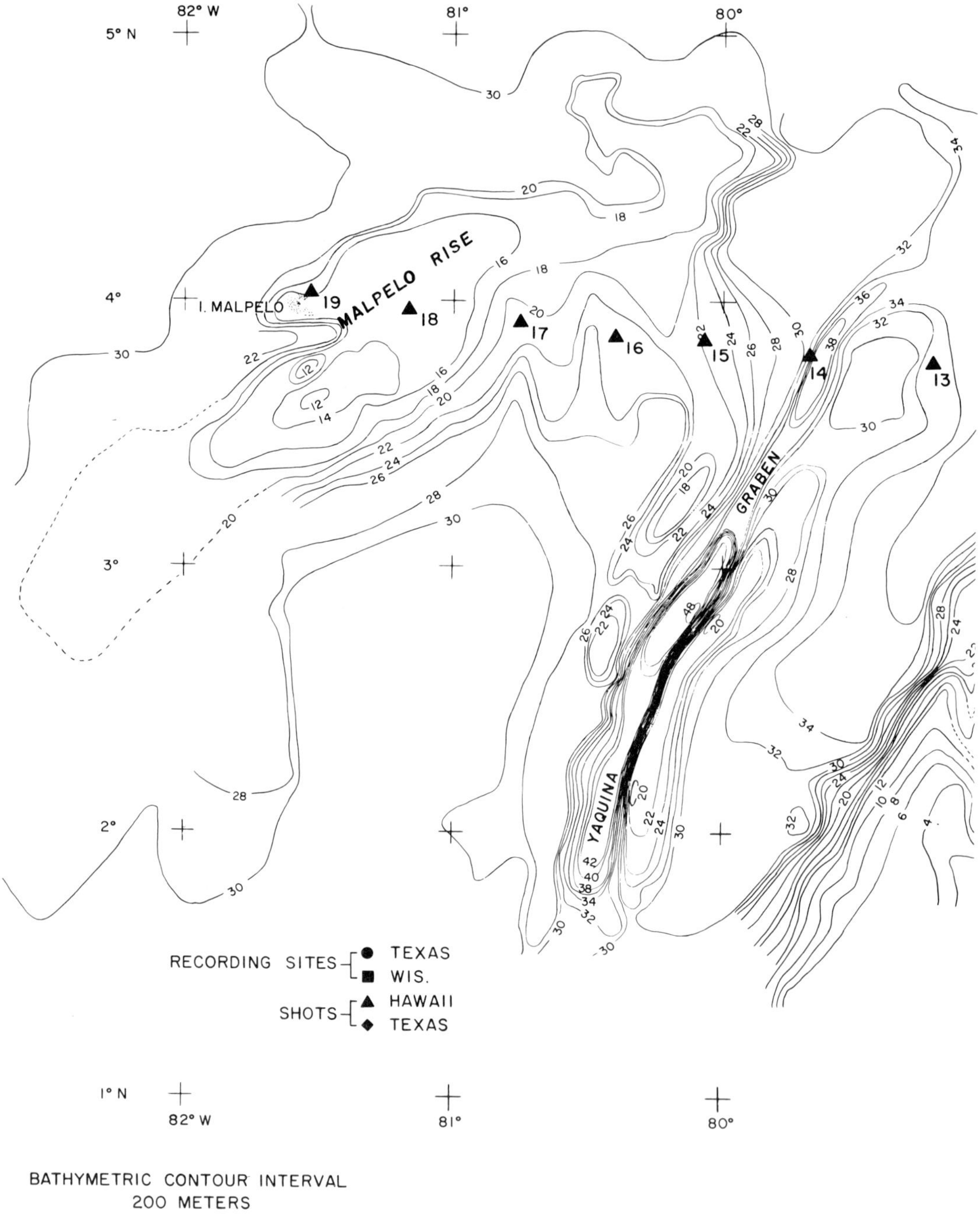


Fig. 3. Bathymetric-topographic map of region examined by Project Nariño. Bathymetric data modified from *van Andel et al.* [1971]; U.S. Naval Oceanographic Office Chart N.O. 22050, 1972 ed., scale 1 : 250,000, in region bounded by north latitudes 3° to 4-1/2°, and west longitudes 77° to 78-1/3°; University of Hawaii *Kana Keoki* tracklines, Malpelo Island to Buenaventura, Buenaventura to Tumaco, and Tumaco to Malpelo Island, February–March, 1973. Topographic data from map, Republica de Colombia, Ministerio de Hacienda y Credito Público, Mapa Fisico–Politico, Elaborado por el Instituto Geográfico de Colombia 'Agustín Codazzi,' 1972, Scale 1 : 500,000.

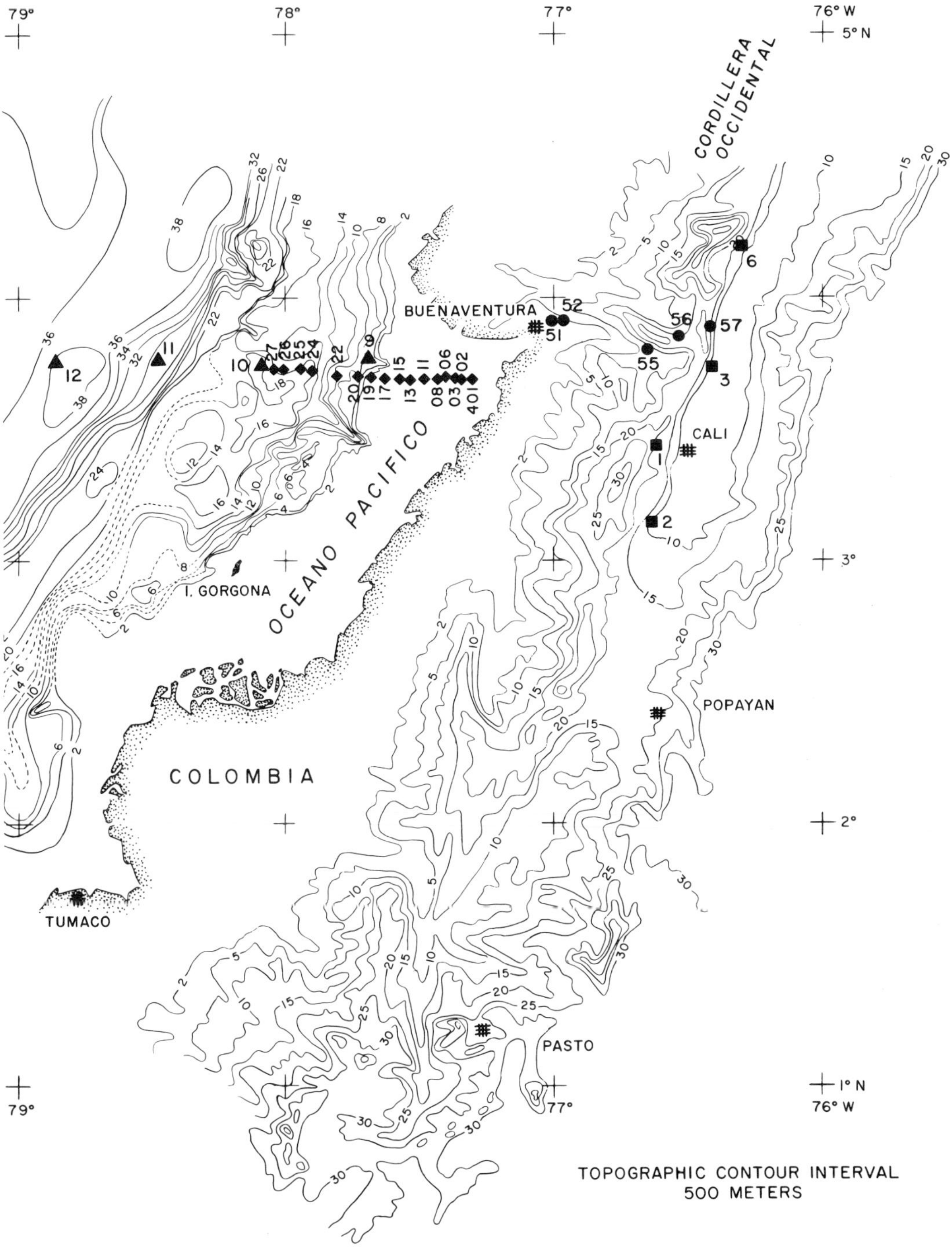
79°
78°
77°
76° W
5° N
3°
2°
1° N
76° W
77°
79°
CORDILLERA OCCIDENTAL
BUENAVENTURA
CALI
POPAYAN
PASTO
TUMACO
I. GORGONA
OCEANO PACIFICO
COLOMBIA
TOPOGRAPHIC CONTOUR INTERVAL
500 METERS

Fig. 3.

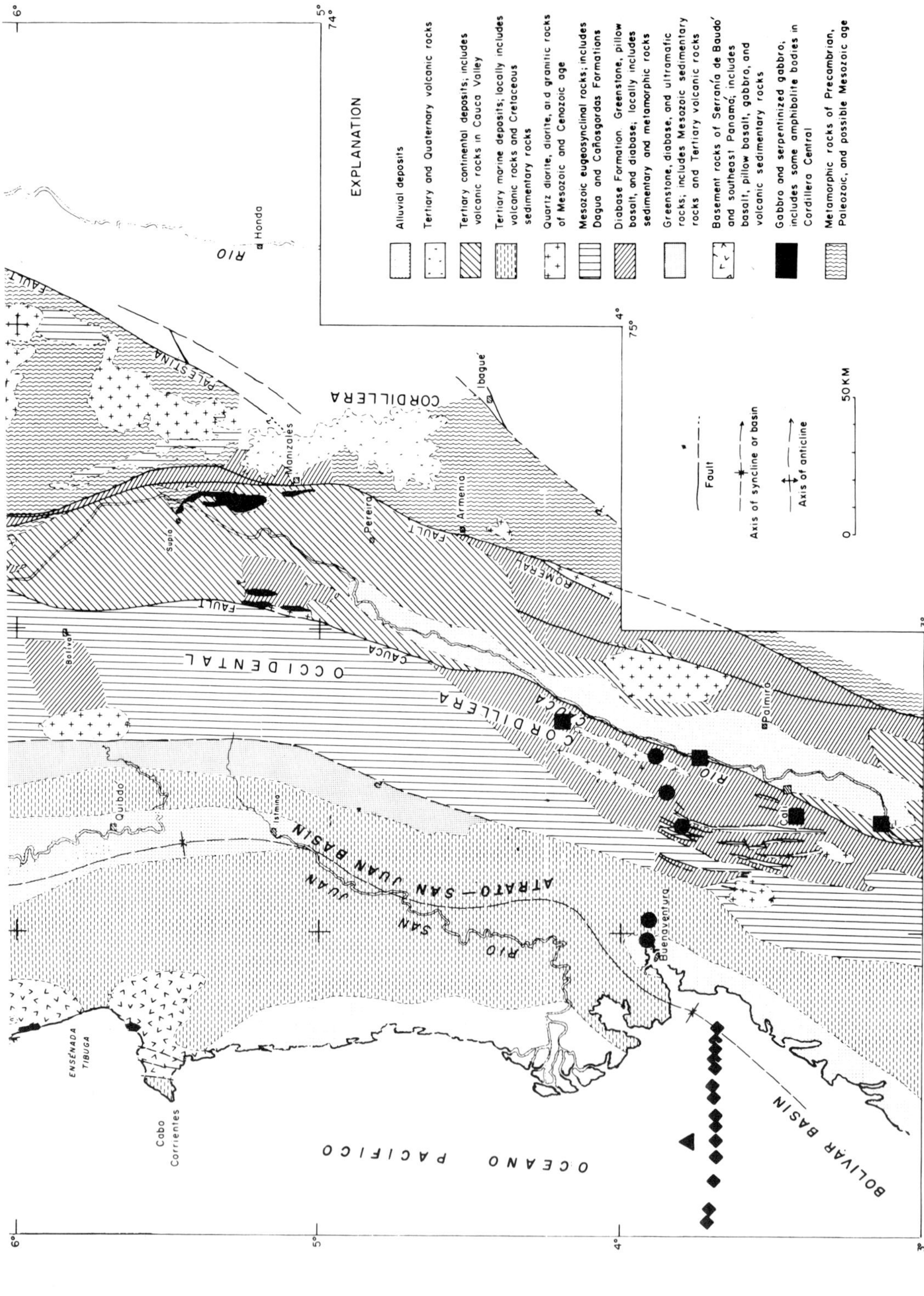
EXPLANATION
Alluvial deposits
Tertiary and Quaternary volcanic rocks
Tertiary continental deposits; includes volcanic rocks in Cauca Valley
Tertiary marine deposits; locally includes volcanic rocks and Cretaceous sedimentary rocks
Quartz diorite, diorite, and granitic rocks of Mesozoic and Cenozoic age
Mesozoic eugeosynclinal rocks; includes Dagua and Cañasgordas Formations
Diabase Formation. Greenstone, pillow basalt, and diabase; locally includes sedimentary and metamorphic rocks
Greenstone, diabase, and ultramafic rocks; includes Mesozoic sedimentary rocks and Tertiary volcanic rocks
Basement rocks of Serranía de Baudó and southeast Panamá; includes basalt, pillow basalt, gabbro, and volcanic sedimentary rocks
Gabbro and serpentinized gabbro; includes some amphibolite bodies in Cordillera Central
Metamorphic rocks of Precambrian, Paleozoic, and possible Mesozoic age
Fault
Axis of syncline or basin
Axis of anticline
0
50 KM
CORDILLERA OCCIDENTAL
CORDILLERA
PALESTINA FAULT
ROMERAL FAULT
CAUCA FAULT
RIO CAUCA
ATRATO–SAN JUAN BASIN
RIO SAN JUAN
BOLIVAR BASIN
OCEANO PACIFICO
RIO
Honda
Manizales
Supia
Pereira
Armenia
Ibagué
Bolivar
Quibdo
Istmina
Palmira
Cali
Buenaventura
ENSENADA TIBUGA
Cabo Corrientes
6°
5°
4°
3°
74°
75°
76°
77°
78°

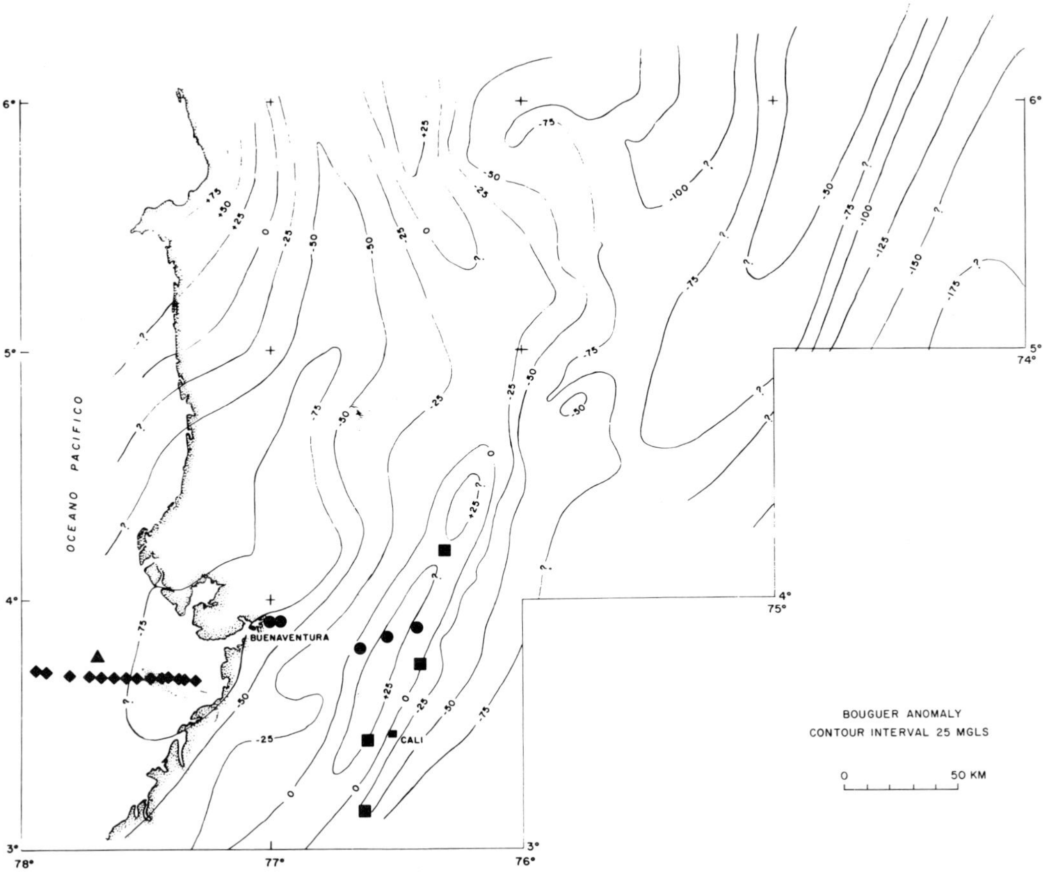


Fig. 5. A Bouguer gravity map covering the area of the recording array and the western cordillera. The maximum positive anomaly near the array is about +35 mgal. The minimum anomaly in the vicinity of the shots on the shelf is estimated to be −90 mgal, and reflects the offshore extension of the Atrato–San Juan sedimentary basin [after *Case et al.,* 1971]. The symbols are the same as those of Figure 3.

in which planar assumptions are unlikely to hold for great distances. We must be prepared to admit that a profile observation at a single station, no matter how perfectly observed, cannot provide enough information to construct a seriously held model for this complex region. There are, however, advantages arising from array observations which make possible the construction of a reasonable model if the true velocities of the principal structural elements are known, e.g., from parallel-to-strike measurements. Besides the obvious advantage of reinforcement of time of arrival estimates and the confidence which array observation engenders, an advantage which is crucial comes from the determination, on a shot-to-shot basis, of the apparent velocity across the receiving array, for these velocities provide an identifier of similar refracting paths and remove the uncertainty in normal travel time analyses caused by time shifts associated with geologic perturbations under the shots.

*Research vessel Kana Keoki shots (KK10–19).* Figure 6, a reduced-time, first arrival plot for stations 1, 2, 3, 6 and 55, includes shots KK10–19. The arrival times clearly are not informative about the critical refractor taken. Viewed on the

Fig. 4. (Opposite) The geologic setting of the land array used in recording the 4° N profile. All recording stations with the exception of those near Buenaventura are on or near basic-rock types. The map is that of *Case et al.* [1971]. The Bolivar basin is described in *Nygren* [1950]. The symbols are the same as those of Figure 3.

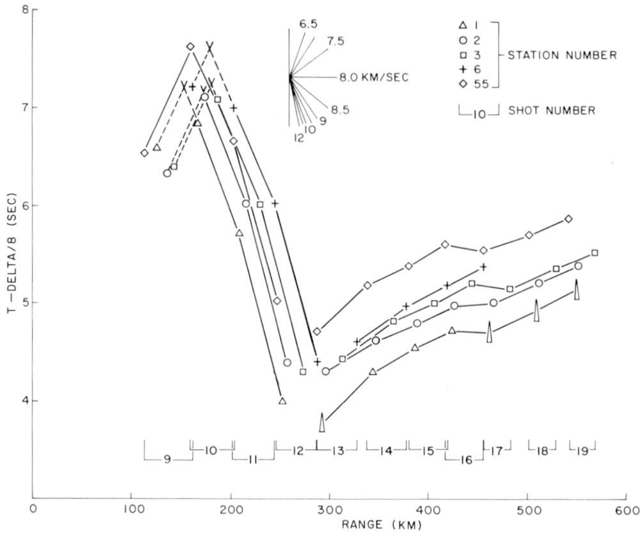


Fig. 6. A reduced first arrival travel time plot for five stations of the land seismograph array (Figure 2) from recordings of the 11-shot sequence fired from the *Kana Keoki* along the 4° N profile, Buenaventura to Malpelo Island. The height of the symbol representing the data from station 1 reflects uncertainty in first arrival times for shots 13, 17, 18, and 19.

most elementary grounds, these data might imply that the two fastest branches represent a layer of about 11 km/s overlying a layer of about 7.8 km/s, a matter hardly expected to be revealed through refracted waves with shots and observations at the surface.

The record sections from these stations (Figures 7, 8, 9, 10, and 11) are more informative.

(Opposite)

Fig. 7. Reduced record section for the vertical component seismometer at station 1. These data, as is true of all others, are from the vertical component of an unattended 3-component seismograph. These, from a location in the western cordillera about 10 km from the industrial city of Cali, are poorest in signal-to-noise ratio of the five stations. The height of the symbol at the beginning of the record trace is proportional to the absolute playback gain for each of the traces shown. The gain of the unattended station was invariant, as was the gain in digitization, and thus the height of the symbol provides an inverse measure and calibration of the signal strength throughout the record. The envelope plot at the top of the diagram is the same data as below, overplotted with the trace inverted. The triangles represent the estimated time of first arrival based on phase matching with the other station's records for the same shot. The data for all stations as displayed in record sections or as travel time plots include first order vertical corrections of the shot to the sea floor using a velocity of 1.5 km/s and a similar correction of receiving stations, a 950-m datum using a velocity of 5.6 km/s.

Fig. 8. Reduced record section for the vertical component seismometer at station 2. The signal-to-noise ratio for this seismograph just east of the Cauca fault is among the best obtained and was a key record in phase-matched picking process. The maximum amplitudes of the first three-fourths second after the first arrival are those used in Figure 19, and of all the stations, the amplitudes at this station are probably the most consistent and smoothly behaved. The behavior of the first arrival from shot KK13 is indicative of a mixed branch 'crossover' first arrival. Ambiguous emergent arrivals from shot KK13 characterize the entire data set from this shot. Corrections applied are identical with those of Figure 7.

Fig. 9. Reduced record section from the vertical component seismometer at station 3. This station in consort with station 55 forms the most critical elements in the analysis. As with stations 2, 3, and 6, essentially no digital processing filter has been applied, as the field analog high-cut filters were set with high-cut filters 3 db points at 18 Hz. The processing high-cut filter was set to start to reject at 20 Hz as indicated. Corrections applied are identical with those of Figure 7.

Fig. 10. Reduced record section for the vertical component seismometer at station 6, the northernmost of the broadside array stations (1, 2, 3, and 6) in the western cordillera. The exceptional signal-to-noise ratio made it, along with station 2 (Figure 8), important in the phase-match process for determination of consistent first arrival times. The unit failed to record shots 17, 18, and 19 because of its removal on suspicion by the local population of the seismometer's being a demolition device.

Fig. 11. Reduced record section from the vertical component seismometer at station 55. This station and station 3 form the most critical elements in the analysis. All of the data presented are from 1-Hz resonant frequency seismometers. As with the other stations, this tape recording instrument was unattended. As it used slow speed AM modulation rather than FM modulation for recording, tape usage was substantially less and field duration substantially greater than at the other stations. This station was located in the western portion of the western cordillera (see Figure 2).

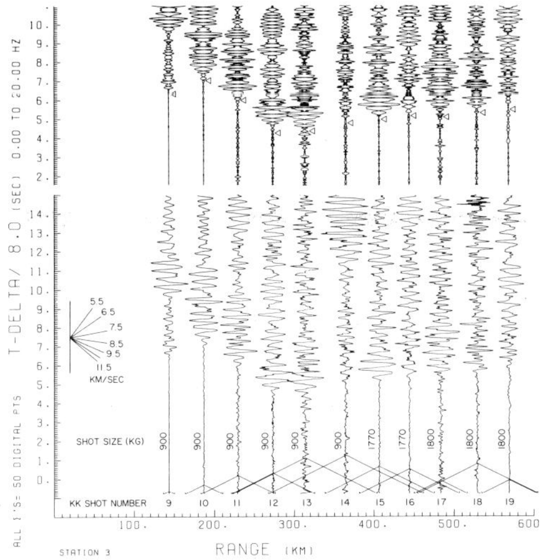

Fig. 9.

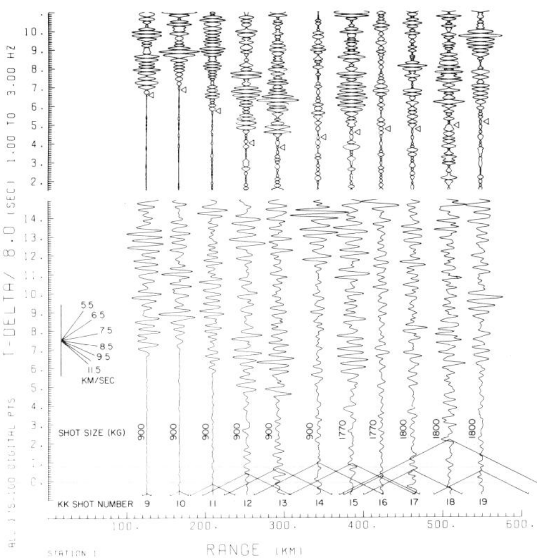

Fig 7.

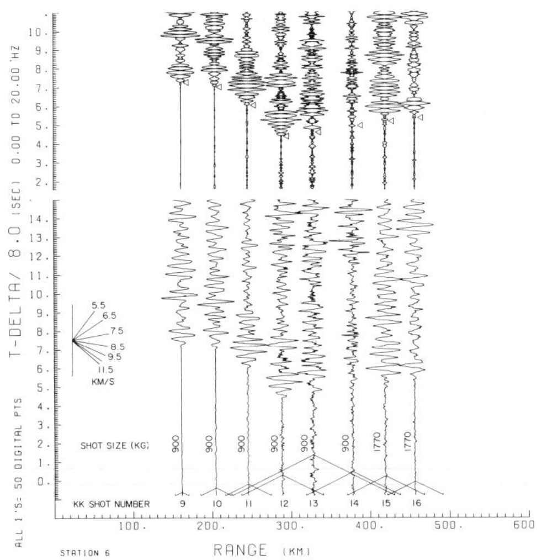

Fig 10.

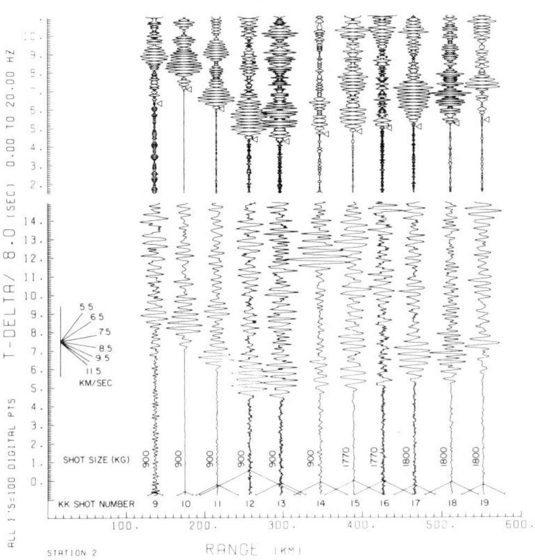

Fig. 8.

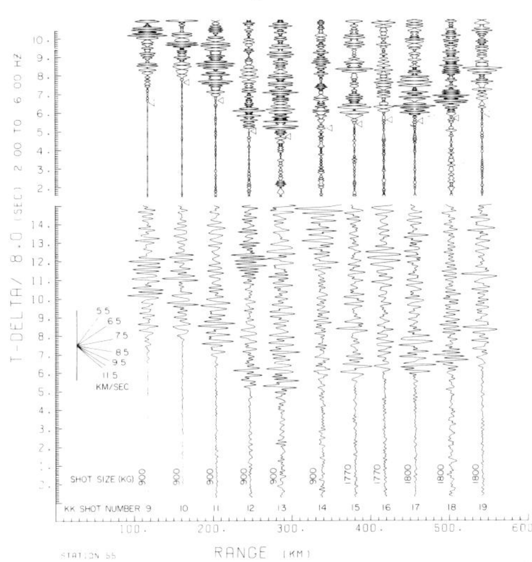

Fig. 11.

Suspicion from the travel times is reinforced regarding the possibility of a different branch bringing first arrival energy to the stations for shot KK9 and perhaps in part for KK10. Observations of shot KK14 consistently show unusually large secondary arrivals or unusually small primary arrivals. Inspection of the playback gain (the symbol at the record start) versus arrival size for the various records from KK14 indicates both effects are present but the prominent secondary arrival is very large. Since it is consistent for all observations, this must be a shot-related effect probably associated with KK14's location at the westernmost edge of the Yaquina graben.

Close inspection on a shot-by-shot basis of the first arrival region for the various seismograms suggested considerable similarity and opened the possibility of obtaining stable, internally consistent first arrival times for the same shot at various stations. Such has been done by overlaying transparencies of record sections and picking all of the records from each shot consistently. This results in the phase-matched first arrival times plotted at their relative range to station 55 for shots KK10–19 in Figure 12. (The same travel times are used in Figure 6.) The similarity in the behavior of first arrival times from shot to shot for particular stations in reference to the least squared lines or in relation to their neighbors suggested strongly that static station corrections would be appropriate for these data. The velocities arising from the application of static station corrections (Figure 13) or the velocities from the least-squared raw data (Figure 12) both indicate that two first arrival branches have been received. The abrupt change in apparent velocity across all the stations between the arrival set associated with shots KK10–12, 9.9 km/s, and

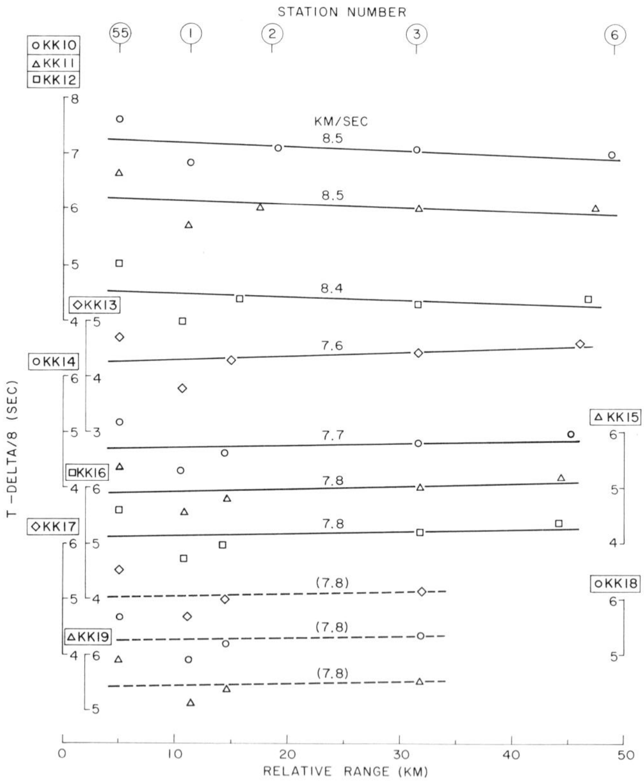


Fig. 12. Phase-matched first arrivals and least squares apparent velocities across the land array, stations 55, 1, 2, 3, and 6. The relative range at which the travel times are plotted is the difference between the actual shot to receiver range and the range to the closest station, 55. The arrivals are at the times indicated on envelope plots of Figures 7 through 11. The velocities shown are least squares estimates except for shots KK17–19, where station 6 data are lacking.

that associated with KK13–19, 9.0 km/s (Figure 13), cannot be due to azimuthal effects because of the abrupt nature of the change.

There is a problem immediately evident, for with normal hypotheses, one does not expect to find an abruptly lower apparent velocity associated with more distant shots. We shall return to this interesting problem and tentatively assign both of these branches to oceanic mantle paths. A byproduct of the phase-matched station-corrected travel time is the ability to make predictions of the times of first arrivals at any particular range or station from the best least squares estimate of that arrival time using the observations of a particular shot at all stations. These times, which are called 'synthetic' times, are the values used in modeling. Comparison between them and the observed (phase-matched) times (Figure 14) shows that deviation—with one exception—is less than .04 sec.

*Shot KK9 and those from the research vessel Gorgona—closer ranges.* Arrivals from shot KK9 were suspected of belonging to a different branch from that of the other KK shots on the

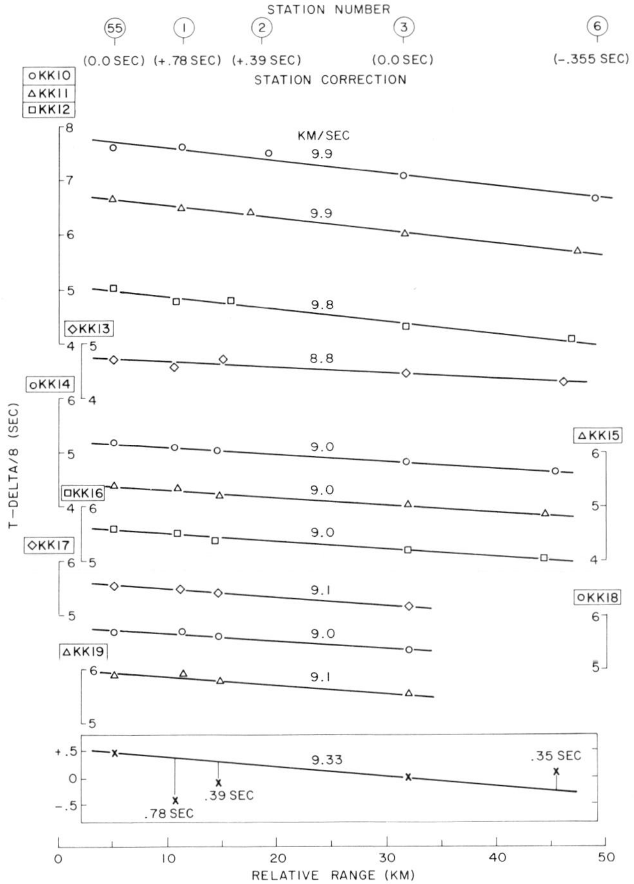


Fig. 13. The data and method of presentation are the same as in Figure 12, except that static station corrections have been applied to travel times for stations 1, 2, and 6. The box at the bottom indicates the static station corrections applied. The crosses represent the mean reduced travel time to each station for the seven shots recorded by all stations less the mean reduced travel time to all stations for the seven shots. The station corrections are the time difference between the above and a line through the values derived for stations 55 and 3, thus forcing the station corrections for in-line stations 55 and 3 to be zero. It is important to note that the application of the static station corrections thus derived biases the least squares apparent velocities, in this case toward higher values, and is not to be construed as an unbiased estimate of the apparent velocity across the array. Rather, this procedure's value lies principally in demonstrating the consistency of the data once the station-related time variations have been removed and in allowing the establishment of a set of consistent synthetic travel times for in-line stations 55 and 3, based on the consensus of arrival times at all stations.

basis of their times and wave forms. The shots of the University of Texas at Dallas from the *Gorgona,* 402 through 427, are more closely spaced, are colinear with the trend of the *Kana Keoki* series, and continue from about the location of KK10 toward shore past KK9 (which is offset to the north) and into shallow water (Figure 3). The *Gorgona* series observations substantially improve the resolution of the observations from KK10 to shore, but there are differences in technique and attendant potential problems which need review before we proceed.

The *Gorgona* used VLF navigation based in part on Colombian maps for a starting reference point (the pier of Buenaventura) while the *Kana Keoki* used satellite navigation to estimate the location of the KK shots. Results obtained by the University of Hawaii using a portable satellite navigator to define positions for certain of the land recording stations suggest that the map and satellite coordinates are in disagreement by about 1 km at the Buenaventura pier reference point [*Woollard and Thompson,* 1974].

Secondly, the water depths at the *Gorgona* shot locations are from estimates, made without the aid of a fathometer, from the bottom reflection times of the shots themselves. Initially, the navigational discordances between the *Kana Keoki* and *Gorgona* shot locations were believed much greater than 1 km, as preliminary depths estimated for the *Gorgona* shots from shooting ship records were inconsistent with the bathymetric estimates from modern charts at the reported shot positions (Figure 3). The charts are consistent with *Kana Keoki* bathymetry along track lines, including the vicinity of the *Gorgona* shots. Reestimates of water depth of *Gorgona* shot bottom-reflection times (as well as the *Gorgona* positions) have substantiated the VLF navigation, have caused revision of the *Gorgona* shot depths, and have pointed to the uncertainty associated with having to estimate water depth, especially in shallow water, from shooting ship shot records.

A third problem concerns station 3, the station most colinear with station 55 and the shot line, for it operated for and received only two of the fifteen most generally received *Gorgona* shots (411 and 422). Times for this station would be most helpful in bridging the results from the *Kana Keoki* shots with those from the *Gorgona* shots, as azimuth effects would be expected to be least when station 3 times are combined with those of station 55. The data sets for the *Kana Keoki* shots (10–19) and the *Gorgona* shots (400 series) would be thus directly comparable through the apparent velocity across these colinear stations.

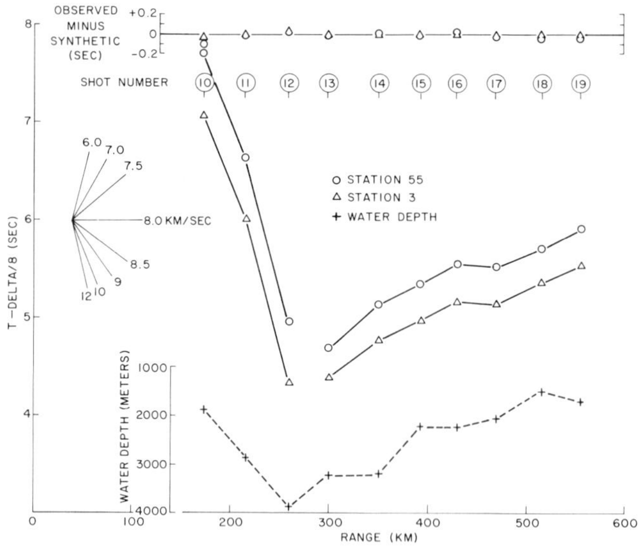


Fig. 14. Least squares travel time plot for stations 55 and 3. Reduced 'synthetic' travel times, as predicted by the least squares (Figure 13) for stations 55 and 3. For direct comparison of the behavior of these synthetics, travel times are plotted at the mean range to stations 55 and 3 from each shot. The plot at the top of the figure shows the time difference between the phase-matched observed and least-squares predicted (synthetic) travel times. The bathymetry at the shot positions is shown below.

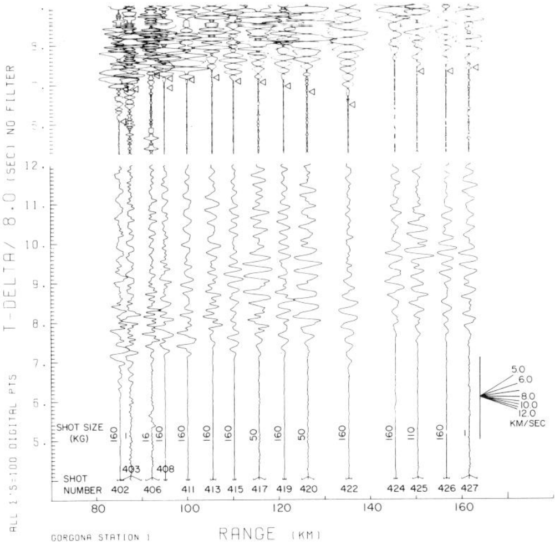


Fig. 15.

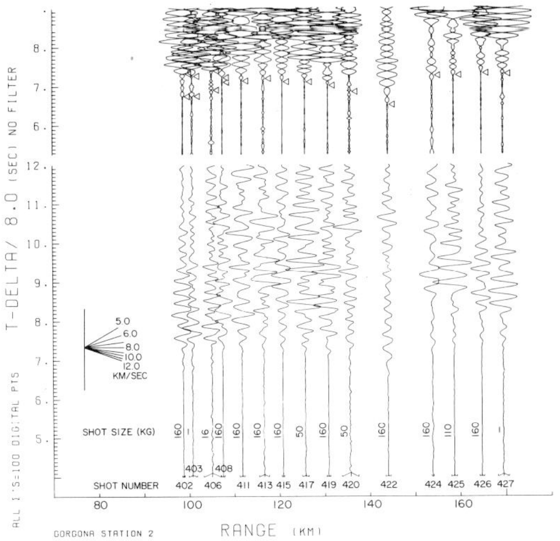


Fig. 16.

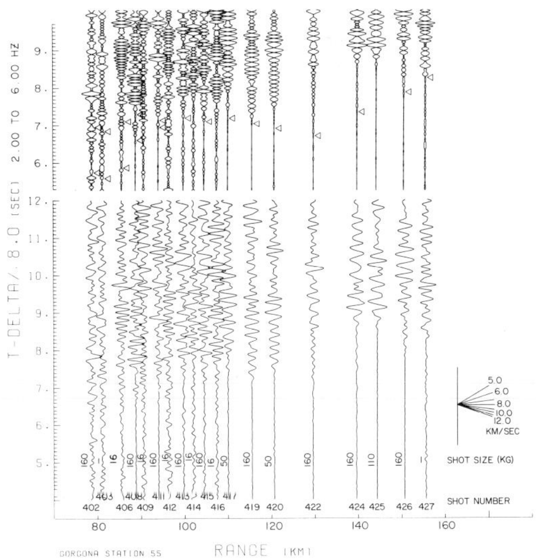


Fig. 17.

Fig. 15. Reduced record section from the vertical component seismometer at station 1 for the nearshore shots from the *Gorgona.* Shots 402–420 are on the continental shelf, shots 422–427 on the continental slope. The playback gain as indicated by the symbol at the beginning of the trace when compared to those of Figure 7 shows that amplitude of the first arrival signals from these small shots is in excess of those of the majority of the larger shots. The continental branches, which we believe these first arrivals entirely represent, are thus superior for transmission of energy to the mixed oceanic and continental paths responsible for first arrival energy for shots KK10–19 at this and the other receiving stations in the western cordillera. Although the first arrival for shot 424 seems clear, no reasonable phase match could be achieved between this record and the records at stations 2 and 55. The methods of reduction and display are identical with those used for the *Kana Keoki* data.

Fig. 16. Reduced record section from the vertical component seismometer at station 2 for the shots from the *Gorgona.* In general, phase-match agreement on the wave form representing the first arrival was achieved between this and station 1 before the arrival times for station 55 data were determined. Stations 1 (Figure 15) and 2 (Figure 16) are the key elements in the determination of missing arrival times at station 3. The methods of reduction and display are identical with those used for the *Kana Keoki* data.

Fig. 17. Reduced record section from the vertical component seismometer at station 55 on the western side of the western cordillera for the shots from the *Gorgona.* Shots 409, 412, and 416 shown are not included in the data set used. The record from shot 425 is not picked as there is no arrival consistent with the data from stations 1 and 2. There is reason for doubt concerning success of phase matching across any of the records from G424–427. The methods of reduction and display are identical with those used for the *Kana Keoki* data.

Stations 1, 2, 55, and, to a somewhat lesser extent, 56 and 57, were successful in receiving most '400' series *Gorgona* shots (16–160 kg) to their maximum range, about 200 km. It is from these successful stations (Figures 15, 16 and 17) that we have synthesized the missing times for station 3. We have determined first arrival times by phase-matching stations 1, 2, and 55, and supplemented them with normal individual picks from stations 56 and 57 when reasonably consistent. In regards to 56 and 57, either the first arrival was recognized and the times found were consistent with the other stations, or the times independently read were very substantially different, usually late, and apparently related to a strong secondary phase. Static corrected travel time data and velocity branches are shown in Figure 18, together with the static correction values. The branches portrayed in Figure 18 are derived through the station correction procedure, since the least squares would be biased by the changes

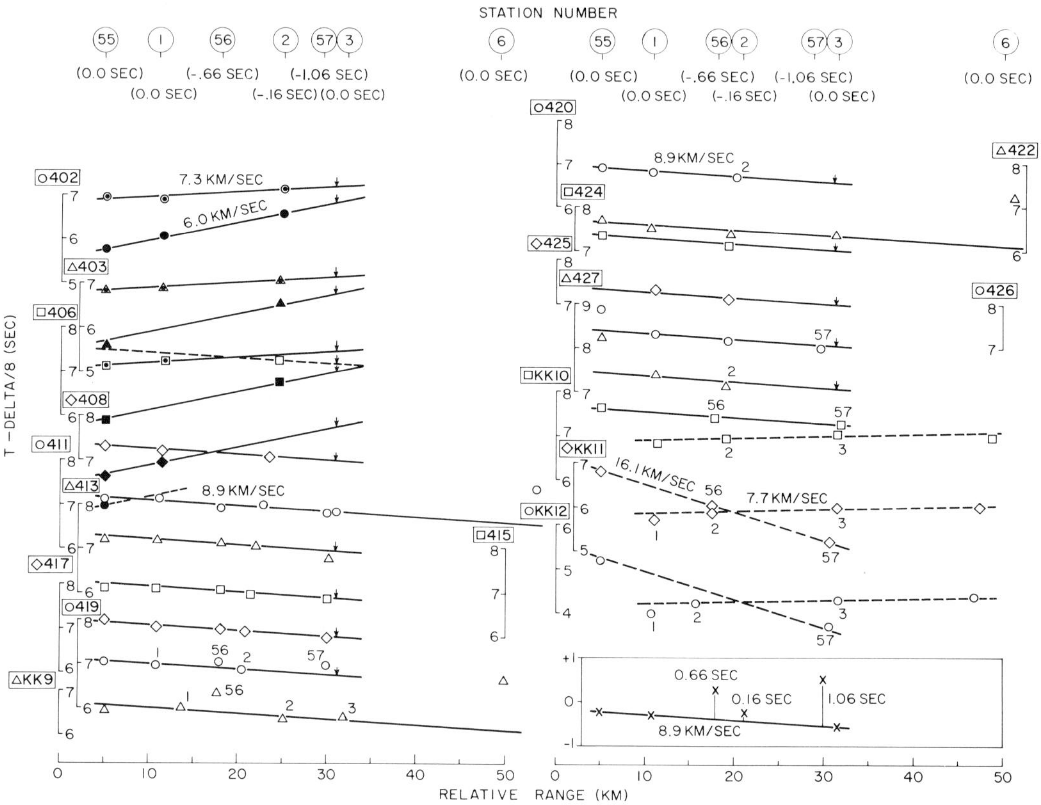


Fig. 18. Travel times to seven land stations for shots of the *Gorgona* 400 series presented in the manner of Figure 13. The solid symbols are the travel times we associate with a 6.0 km/s array apparent velocity refractor, the open symbols with dots with a 7.3 km/s refractor, and the open symbols with higher velocity refractors. The symbols displayed in the box are the differences between the mean reduced travel times for each of the stations for the shot series 411–419, and the mean reduced travel time to all stations for the same shots. As stations 55 and 3 are to be left without correction as was done for the *Kana Keoki* shot data (Figure 13), a best fit between stations 55 and 3 for shots 411, 422, and KK9 (where both stations received arrivals from this branch) forms the correction line, and this has an apparent velocity of 8.9 km/s. The static station corrections shown for the other stations are the differences between the mean reduced travel time for each of these stations, and the correction line. The correction line, apparent velocity 8.9 km/s, was fitted as shown to the data judged to belong to this branch. (To determine static station corrections, a judgment is made from the raw data as to a consistent data set representing a single branch. Individual arrivals are testable after the station corrections are applied; i.e., if the arrival times do not fit through the correction line with other arrivals from the same shot, they either represent poor readings or they belong to another branch.) The arrows indicate the relative range and the predicted (synthetic) arrival time for station 3's missing observations. (No static correction for station 6 was determined and it was assigned a static correction value of zero. It is included for comparison with more distant shots.) Shots KK10–12 are included to establish the transition to the oceanic $P_n$ path. The lack of fit of these data to a single velocity line when applying the corrections derived for shots 411–419 makes it clear that the static station corrections needed for deeper branches (Figure 13) are clearly different from those appropriate here and that new branches are represented by this shorter-range data. Note that the data set for shots 424–427 is less complete and that station 55 arrivals are unreported for shot 425 and late for shots 426 and 427.

in the data set. For shots 411, 422, and KK9, uncorrected times for station 3 fit the predicted synthetic times for the range of station 3 within a few hundredths of a second, providing confidence that the missing times for this station can be and are properly synthesized. More importantly, the principal branch resolved, with an apparent velocity of 8.9 km/s, is different from either of the branches isolated from the KK shots at greater ranges. Synthetic travel time plots for stations 3 and 55 (Figures 19 and 20), provide a summary of the differences between the predictions from all stations, i.e., the synthetic times, and the phase-matched times as well as the bathymetric comparisons between measured and chart values.

The station corrections from the series KK10–19 (Figure 13) are very different from those derived for the 400 series for shots 411 through 420,

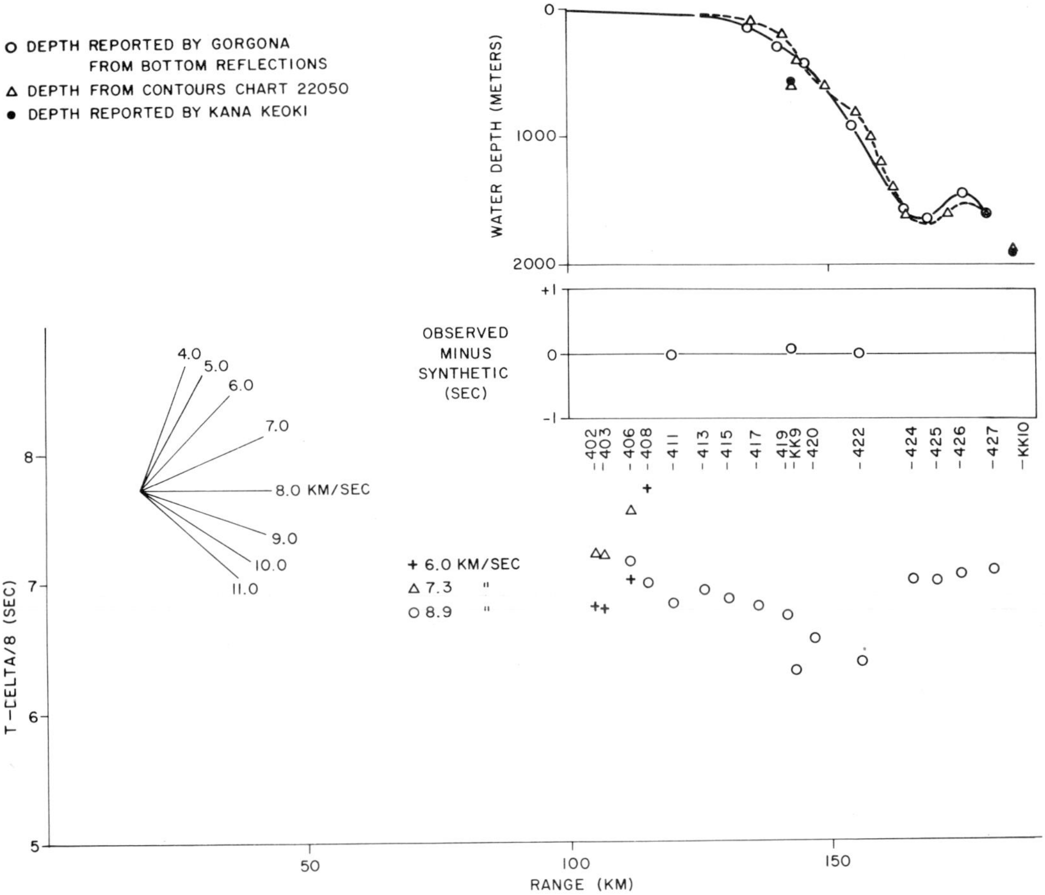


Fig. 19. Bathymetry and travel times to station 3 for shots from the *Gorgona*. Comparison of bathymetry determined from shot to bottom reflections at the *Gorgona* with PDR depths from the *Kana Keoki* and depths from Chart 22050, all at *Gorgona* shot positions. Shown as well are similar data for shots KK9–10, which illustrate the relationship between the bathymetry predicted from Chart 22050 at the satellite navigation-determined positions of the *Kana Keoki* and those measured at these locations with a PDR. These comparisons between the three sources of bathymetric information in this area of great change are taken as partial verification of navigation of the two shooting ships, one of which used satellite navigation and the other VLF navigation. The lower portion of the diagram is a reduced travel time plot for station 3. The times and velocity symbols shown are synthetic array estimates from Figure 18. These are compared above with the observed travel times for shots 411, KK9, and 422, the shots for which station 3 recorded. Shots 411–422 and the data from stations 55, 1, and 2 all show evidence which is similar and reasonably ascribed to landward dip of the continental mantle.

and this is additional proof that new branches have been observed. We have demonstrated this graphically (Figure 18) by plotting the corrected times for KK10–12 using station corrections from the nearshore shots and obtaining a divergent result. The principal branch from the *Gorgona* shots 411–420 and from shot KK9 (8.9 km/s apparent velocity) is attributed to the continental M discontinuity, whereas the branches observed at greater ranges from KK10–19 are to be attributed to the oceanic M discontinuity and deeper oceanic refractions. We will use these branches in construction of a model to fit the seismic and gravity data.

From an examination of the record sections for stations 1, 2, and 55 at short ranges (Figures 15, 16, and 17), an extremely weak earlier arrival can be seen ahead of a more normal-looking later arrival—normal in the sense that it appears like the majority of first arrivals at greater range. The very weak arrival and the stronger secondary arrival have apparent velocities across the receivers of 6.0 and 7.2 km/s respectively (Figure 18). Clearly, the normal-appearing secondary arrival is unrelated to the first arrivals at greater ranges. These are lower apparent velocities than any observed at greater ranges from either *Gorgona* or *Kana Keoki* shots. We have attributed these to continental intracrustal paths.

Thus from the apparent velocities across the

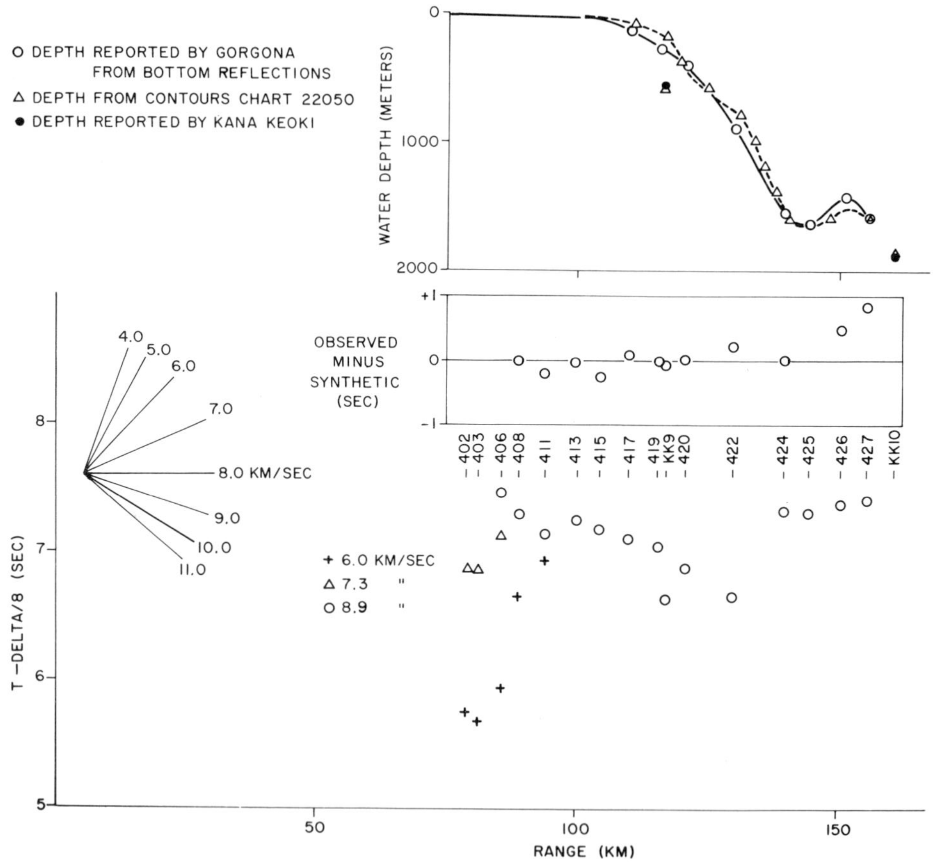


Fig. 20. Travel times to station 55 for shots from the *Gorgona*. This station, which is most colinear with the shot line and station 3, is closer in range and hence shallow refractors are well represented. However, reference to Figure 17 shows a less than favorable signal-to-noise ratio at the time of closer range shots. Phase matching with stations 1 and 2 was essential in estimating first arrival times at station 55. For the data of shots 426 and 427, the deviation of the synthetic times determined from all stations (Figure 18) and the phase-matched times is considered as signalling a new branch. This is reasonable, considering that 55 is the closest station successfully recording these shots. The bathymetry is repeated for reference.

receiving array, we have isolated four first-arrival branches and one set associated with secondary arrivals for the combined *Kana Keoki* and *Gorgona* shot series. In order of decreasing range of occurrence, they are 9.0, 9.9, 8.9, 7.2, and 6.0 km/s.

*Amplitudes.* The relative amplitudes in general substantiate the branches identified on the basis of apparent velocity. Figure 21, showing maximum amplitudes in the first three-fourths second after the first arrival for KK10–19 at station 2 (Figure 8), supports a change of receiver path between shots KK12 and KK13, in accord with the changes in apparent velocity from 9.9 to 9.0 km/s (Figure 13). These data would seem to suggest an additional branch between KK17 and KK19, a possibility not strongly supported by apparent velocity data. Further, as a matter of information, amplitudes and ground motion velocities have been roughly estimated in an absolute sense from separate field amplifier calibration and postmortem geophone sensitivity data so as to provide some indication as to the ground velocity (and amplitude, assuming motion at 4 Hz) seen in the first three-fourths second from these large shots and to make clear the observational requirement for this experiment. Clearly, the first one-half cycle of the first arrivals for the distant shots at station 2 (Figure 8) is about one-fourth of the amplitude seen in the first three-fourths second and only a few Angstrom units. Thus, the noise at the quietest sites is in the few- to sub-Angstrom region during the periods between microearthquake activity and during times of low cultural noise. This is lower by perhaps as much as an order of magnitude than that usually found in North America.

Figure 22 presents similar data for the *Gorgona* shots from amplitude measurements in the first cycle of the first arrival for 160-kg charges. Here again the amplitude data support the conclusions relative to first arrival branches from considerations of apparent velocity, but suggest that the arrivals from shots 424 to 427 represent a new path not recognized on the basis of apparent velocities.

As will become apparent, arrival times from shots 424–427 are not fully satisfied by modeling unless the synthetic predictions are abandoned. The failure to fit the arrivals' transit times through the model by the path appropriate to shots 411 to 422 thus supports the inference of a new path and separate branch from amplitudes.

*Auxiliary profiles parallel to structure.* Preliminary results are available from two areas in which parallel-to-structure measurements have

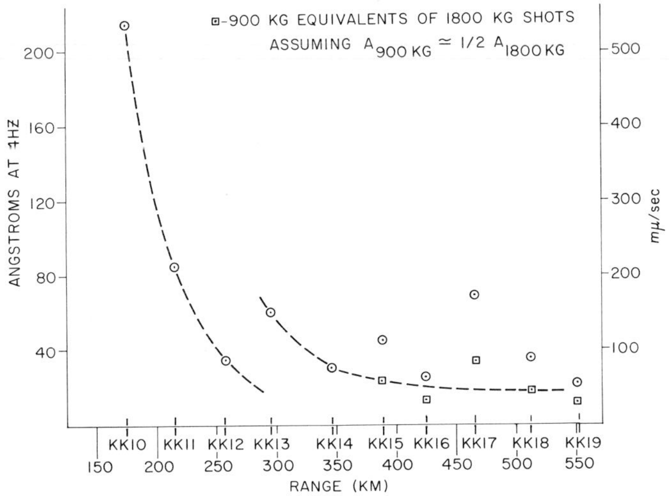


Fig. 21. Maximum velocity in millimicrons per second and displacement at 4 Hz in angstroms for the first three-fourths second of the records at station 2 for shots KK10–19. The offset between shots KK12 and KK13 is attributed to the change in path from head waves at shorter ranges to a diving wave branch from greater ranges (shots KK13–19). There is a suggestion of a possible change in branch between shots KK16 and KK17; however, the array apparent velocity data (Figures 12 and 13) do not support this.

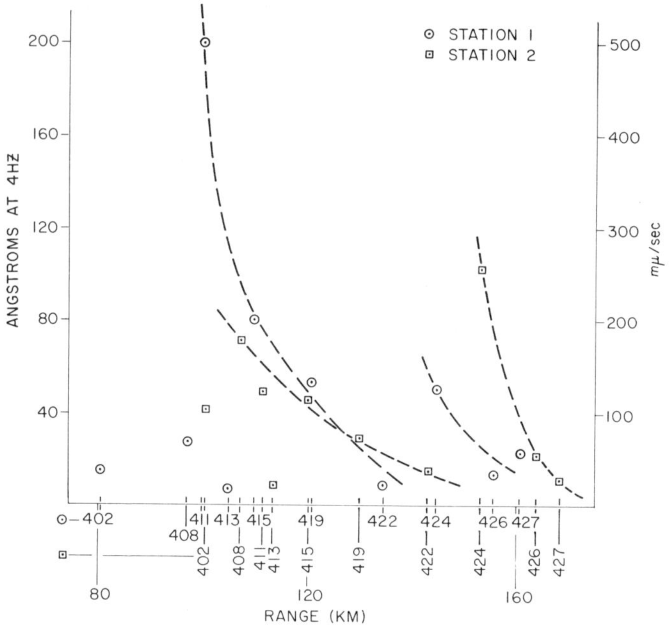


Fig. 22. Maximum velocity in millimicrons per second and displacement at 4 Hz in Angstroms for the first cycle of the first arrivals from 160-kg *Gorgona* shots at stations 1 and 2. The amplitudes fall into three regions according to the range. The closest range region correlates with a 6.0 km/s apparent velocity branch. At about 100 km there is a dramatic rise in first arrival amplitude corresponding to the crossover from the weak 6.0 km/s branch to the 7.8 km/s refractor. At 145 km the diminution of energy signals the return to the path which fits times using the 6.0 km/s branch, a path assigned to critical refraction through the bottom of the Bolivar basin.

been made. The first, at sea, is a single-ended measurement from a line of shots along the continental shelf extending to the southwest from an ocean-bottom seismograph near Buenaventura (Figures 2 and 3). This profile extends successfully to the passage between Gorgona Island and the mainland at a range of about 70 km from the OBS. No shot energy was recorded beyond 70 km. Apparent velocities of the three shortest-range branches (1.7, 2.4, and 3.9 km/s) are appropriate to sediments and a 7.4 km/s branch is logically associated with the basement underlying them. The travel times observed and a solution with the 7.4 km/s branch constrained to a true velocity of 6.0 and 6.1 km/s are shown in Figure 23. The choice of about 6.1 km/s as a true velocity produces a solution consistent with estimates of the depth of the basin on land [*Case et al.,* 1971] and in accord with its horizontal projection to sea and under the OBS. Also, a logical and powerful argument in favor of the assumption of a true velocity of 6.0 or 6.1 km/s is the probable existence of basement rock at the very near surface near Gorgona Island. Furthermore, a structural discontinuity near the island is predicted on the basis of an abrupt cutoff of all energy to the OBS from the shots beyond the passage between Gorgona Island and the mainland.

We shall use the velocity of 3.5 km/s as a reasonable average velocity for materials in the basin and the velocity of 6.1 km/s as typical of the true refractor velocity underlying the basin.

On land, two other auxiliary profiles are relevant to unraveling the continental velocities and depths. They are the profile from La Cocha to the north and the profile from Yumbo to the south (Figure 2). Both are relevant to the velocity–depth structure under the eastern side of the western cordillera. The data set from La Cocha is more complete, and we shall depend primarily on it to constrain the velocity structure under the western cordillera receiving array. The velocity–depth functions derived from these data are shown in Figure 24 [*Gettrust et al.,* 1973; *Meyer and Gettrust,* 1973]. For the purposes of this analysis from first arrival data, we have abandoned the inferences of low-velocity layers shown which are based on ray tracing and synthetic seismograms and have used only the directly observed forward branch observations, essentially first arrival information, to estimate

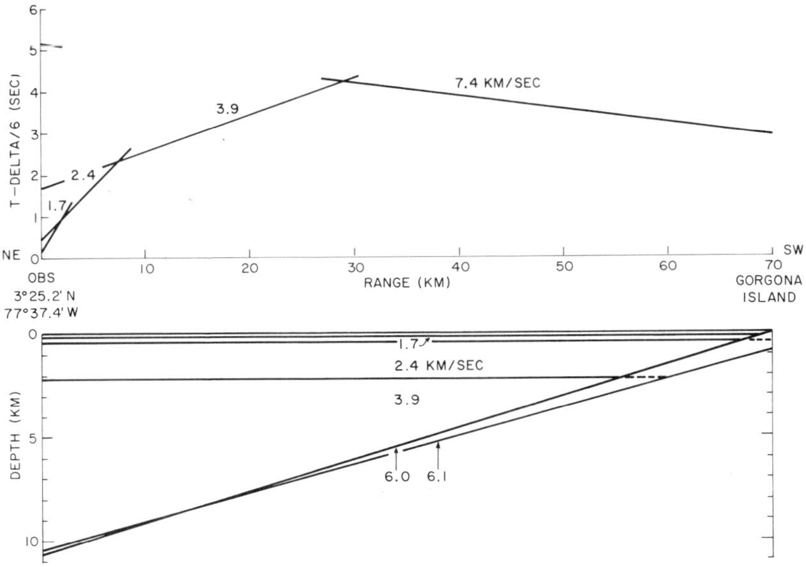


Fig. 23. Travel time diagram for the University of Texas ocean-bottom seismometer (OBS) on the continental shelf (Figure 2) and a solution forcing the 7.4 km/s apparent velocity to a true velocity of 6.0 and 6.1 km/s. Thus, Gorgona Island is taken to signal the outcrop of material flooring the sediments of the Bolivar–Atrato–San Juan basin. No arrivals were seen at the OBS from shots southwest of Gorgona Island. The velocity of the basement underlying the basin is of vital importance to the model to follow.

refractor velocities in these regions. We have allowed for some variation of velocities and depths of interfaces, as these data are single-ended and are derived to a considerable extent from outside the area of the recording array being used here. It is to be noted that there is good agreement between the results from shots at sea and the Yumbo and La Cocha profile data for the velocities of about 6 and 7.3 km/s, but discordance when associating the land-derived parallel-to-structure velocity of 7.8 with 8.9 km/s observed perpendicular to structure and from shots at sea. The apparent velocity of 8.9 km/s across stations 3 and 55 is logically associated with a dip of the 7.8 km/s refractor and/or attendant relative increase in the relative thickness of low-velocity materials above the 7.8 km/s refractor in the crust to the west of station 55 as compared with station 3. The apparent velocity of 8.9 km/s is not an expected crustal or $P_n$ velocity and thus structural complication is certainly to be expected. Station 3 arrivals are likely to be greatly influenced by the structure and high velocities inferred from the gravity to underlie the cordillera, while station 55, nearer the western edge, should be less affected by the high-velocity structure under the cordillera for arrivals from shots to the west, the case for the sea shots.

Finally, regarding the oceanic crust's velocity, we will use a value of 6.8 km/s to model the principal oceanic layer. This has been determined from end-to-end profiles along the Malpelo rise [*Wade et al.,* 1974] to extend as the first arrival branch from short ranges to ranges in excess of 100 km, the maximum range attempted. In accord with the estimates of *Goslin et al.* [1972], a 1.1-km-thick layer of 5 km/s material has been modeled in most regions above the 6.8 km/s layer. For lack of other measurements, and because the entire area has a similar free-air anomaly which may perhaps imply a similar ocean crust, we will use these velocities in modeling the oceanic crust throughout.

## A FIRST MODEL ACROSS THE LEADING EDGE—STARTING ASSUMPTIONS

We have adopted the following assumptions:

*Oceanic velocities.* 2.0 km/s is everywhere appropriate for the sediment measured by reflection profiling; 3.5 km/s is appropriate average velocity for the sediments filling the Bolivar–Atrato–San Juan basin (Figure 4); 1.1 km of 5.1-km/s material is everywhere appropriate for oceanic layer 2; 6.8 km/s is characteristic of the oceanic layer 3 throughout the model; and a velocity of 8.2 km/s is appropriate to the oceanic

$P_n$ regardless of depth or position underlying either continent or ocean. (This somewhat higher-than-normal oceanic velocity perpendicular to structure follows the implications of recent measurements perpendicular to the East Pacific rise (B. T. R. Lewis, in preparation) and is as well a derivative of the modeling process.)

*Continental velocities.* We will assume velocities of 3.5, 5.5, 6.1, 7.1, 7.8, and 8.35 km/s are true velocities. With the exception of 3.5 used for the fill in the Cauca Valley depression for gravity calculation, these are taken to be appropriate single-value estimates from the forward branches of the velocity–depth function inferred from parallel-to-strike measurements (Figure 24).

*Velocity-density relations.* We have assumed that the *Woollard* [1959] velocity–density relations hold in general, but that some deviations are acceptable—either related to microfractures closing (lower density than predicted by velocity) or lowering of velocity due to heat and fluids (higher density than predicted by the velocity).

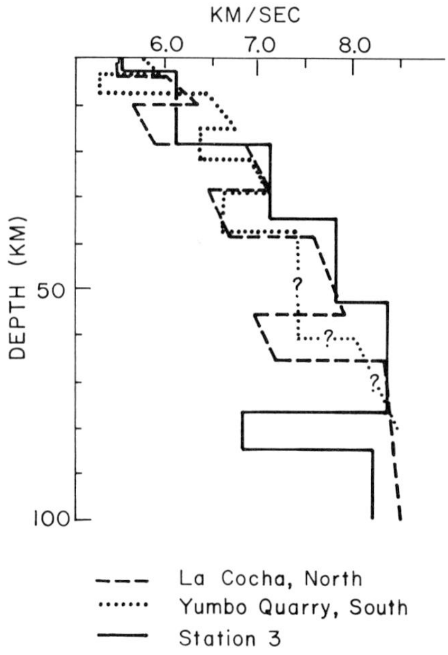


Fig. 24. Velocity-depth functions for the profile north from La Cocha [*Gettrust et al.*, 1973] and south from Yumbo [*Meyer and Gettrust*, 1973], each along the eastern edge of the western cordillera (Figure 2). These functions arise from ray tracing and synthetic seismogram studies. They constitute the only on-strike estimates to date in the western cordillera. From the model (Figure 25), the velocity–depth function underlying station 3 at the eastern edge of the cordillera is shown as well. The Yumbo velocity–depth function data arises from only three quarry shots and twelve records, while the La Cocha result arises from twelve shots and about fifty seismograms. Low-velocity zones have not been considered in this study of first arrivals.

*Gravity calculations.* We have assumed that two-dimensional model gravity calculations [*Talwani et al.*, 1959*b*] are appropriate in all cases.

*First arrival times.* We have assumed in modeling that either head waves or diving waves may be equally appropriate for travel time calculations. Further, head waves are assumed to follow interfaces regardless of their form, and diving waves can be adequately represented by straight line paths. Finally, we assume that positions, elevations, and ranges, as well as estimates of first arrival time onset, are all without significant error.

We have chosen to fit mainly the synthetic travel times appropriate to stations 55 and 3 rather than individually determined times. Synthetic times are little different from those picked except for the data from shots 424–427 (Figures 18, 19, and 20). For these, we have attempted to fit the observed times only at station 55; station 3 did not observe them. The reason for using synthetics where in close accord with observations is that they are internally consistent and that in modeling, relative time perturbations are unable to be accommodated within the same branch, as arrivals from the same branch must follow the same paths under the neighborhood of the stations if there is no change in apparent velocity, as is true in the case at hand.

## THE MODEL AND DISCUSSION

The model conforming to these starting bases, using principally receivers 55 and 3 and checked with the times for the few clear first arrivals at stations 52 and 51 (Figure 3) is shown in Figure 25. This model, which to the extent shown fits the time constraints, the free-air gravity at sea, and the Bouguer gravity on land, convinces us that oceanic lithosphere has been sampled by the compressional waves traveling from shots KK10–19 over a profile distance of 350 km extending oceanward from the continental slope through the Malpelo rise. For shots KK10–12, head waves fit the observed times and the apparent velocities across 55 and 3. For shots KK13–19, diving waves are modeled in a first order sense by straight ray paths to accommodate the apparent velocity and amplitude prediction of a different branch. The apparent dichotomy of a lower apparent velocity across the receiving array for

# *12. Observation of Earthquakes and Explosions at the Bottom of the Western Pacific: Structure of Oceanic Lithosphere Revealed by Longshot Experiment*

TOSHI ASADA

*Geophysical Institute, University of Tokyo, 113, Japan*

HIDEKI SHIMAMURA

*Geophysical Institute, Hokkaido University, Sapporo 060, Japan*

*Abstract.* Results of observations using ocean bottom seismographs (OBS's) for the past five years are reviewed. About seventy operations have been made to date with these instruments. Studies of natural earthquakes near trenches have clearly shown that the frequency content of seismic waves and rates of occurrence of earthquakes are in marked contrast for earthquakes occurring landward as compared with those occurring oceanward of trench axes. $Q$ values in the western Pacific basin exceed 6000, whereas $Q$ values on landward sea floors are generally ten or more times smaller. As for explosion seismology, the OBS has proved to be powerful for detecting refraction signals from small charges; for example, it has detected refraction arrivals from 2 kg charges at distances of more than 100 km in Pacific basins. Two longshot explosion seismology studies were made (August 1973 and September 1974) in the western Pacific basin on which the profile lengths were about 1300 km. Since the ambient noise in deep oceanic basins has been shown to be greater in the optimum frequency range (below 10 Hz) for detecting waves from distant shots, 1.5-ton explosions were found to be inadequate for obtaining clear records, whereas 5- or 7-ton shots (used for the second experiment) proved adequate for the longshot experiments. The experiments have illustrated the stratified oceanic lithosphere. The upper layer, immediately below Moho, is 8.2 km/sec and is underlain by a rather thin, high-velocity layer of 8.6 km/sec. The thickness of the oceanic lithosphere at the site of the second experiment is estimated to be about 80 km. The lithosphere is underlain by a very thin, low-velocity layer of about 8.4 km/sec. Since the thickness of the low-velocity layer is less than 50 km, the layer may not be recognizable by surface wave studies. Beneath that layer is another 8.6 km/sec layer with a positive velocity gradient. Regionality in the structure of the lithosphere is suggested. The depth to the upper boundary and the thickness of the 8.6 km/sec layer are seemingly different between the two experimental sites.

This paper summarizes the results of our studies using ocean bottom seismographs. The ocean bottom seismograph (OBS) was introduced in 1969 in our institutes. It was designed to be self-contained in a small capsule having a volume of 28 liters; its size and weight were intended to be as small as possible. We believe that this is an important feature because if the OBS is small, the mooring system will be small also and so a lot of them can be handled at one time even by a small ship. The pressure case of the OBS is made of high-strength aluminum and can withstand a depth of 9000 m.

The recording period of the OBS is two weeks for the frequency range of 0.2 to 40 Hz. However, we can change such features as frequency range, recording period, or dynamic range, in compliance with the experimental requirements; for instance, if explosion seismology is undertaken, the OBS can be tuned to record higher frequencies compared with those required for the observations of distant earthquakes. In order to gain high sensitivity, various devices have been applied to reduce noises and vibrations which the mechanism of the OBS generates. The weight of the OBS is about 50 kg in air and 20 kg in water.

We have had two objectives with our OBS's: one is to observe natural earthquakes in situ, viz., at the place of interest. The other is explosion seismology studies at the sea floor.

Included in the references are the following papers that detail work with OBS's: *Asada and*

*Shimamura* [1971*a, b*], *Asada and Takano* [1965], *Asada and Matsumoto* [1974], *Bradner* [1964], *Ewing and Ewing* [1961], *Hales et al.* [1970], *Rykunov and Sedov* [1967], *Shimamura and Asada* [1971*a, b,* 1974], and *Shimamura et al.* [1970].

### SOME RESULTS ON OBSERVATIONS OF NATURAL EARTHQUAKES

We have been interested in studying earthquakes which take place at subducting zones of the lithosphere, i.e., in trench–arc systems. Land observations are rather poor for studying such earthquakes, especially for precise location of seismic foci or studies of seismic absorption under oceanic basins.

In these past years we have deployed OBS's mainly at locations along the trenches in the western Pacific. About seventy operations have been made to date; some were deployed seaward of the trenches and some landward (Figure 1). The distances between them and the trench axes were usually from 100 km to 200 km.

When the OBS is positioned seaward of a trench, recorded earthquake frequency is in general two to four an hour. This high level of seismicity was found to be the case at various observation sites at several trenches. We also frequently observed earthquake swarms. More than 97 percent of the earthquakes recorded by the OBS were found to have occurred just landward of the trench axis. If distance from the trench is considered, seismicity just landward of trenches is remarkably high whereas activity seaward is extremely low.

The rates of occurrence and magnitudes of earthquakes occurring landward of trenches are well expressed by the Gutenberg-Richter formula. The number of earthquakes occurring seaward of trenches is very low despite the high sensitivity of the OBS. Although this instrument can record earthquakes with magnitudes of less than −1 at distances of up to 100 km, no earthquakes of magnitudes less than 2 were detected in the areas seaward of trenches. This phenomenon is rather odd for shallow-focus earthquakes; however, it is analogous to the nature of earthquakes occurring in the deep seismic zone.

Some examples of ocean basin seismograms are shown in Figure 2. An outstanding feature of these seismograms is the very short wave train as well as the predominance of high frequency; they are quite different from those obtained from land observations. The appearance of the seismograms implies that the structure of the ocean floor is quite simple and that the absorption of seismic waves is extremely small.

These seismograms are generally rich in high frequencies. In most cases, waves with frequencies greater than 20 Hz travel more than a thousand kilometers. Nevertheless, the amplitudes of the lower frequencies are curiously smaller than expected if we consider that these earthquakes have magnitudes of about 3 to 4. Spectral analyses of seismic waves have been made by the analog method in which the effect of the difference in bandwidth of band-passed channels and the effect of the instrumental response are removed (Figure 3). The results also illustrate the decrease in amplitude at lower frequencies. This phenomenon is frequently found on seismograms recorded in the Pacific Ocean basin for certain ranges of epicentral distance. The lack of lower frequencies in these seismograms is apparently inconsistent with any source spectrum which is generally accepted. This will be discussed later.

On the other hand the gradual decrease of amplitude with increasing frequency in the higher frequency portion of the spectrum is believed to result from absorption in the transmission media. The absorption factor $Q$ is estimated by the inclination of the slope if we assume the source spectrum. Although the value of $Q$ is dependent on the source model assumed, we can obtain the minimum $Q$ value of the media when we assume a source model in which the velocity spectrum is flat, viz., the displacement spectrum is proportional to the frequency [*Asada and Takano,* 1965]. In Figure 3 calculated slopes for various values of $Q$ and various distances to foci are also illustrated for estimation of $Q$ values for the Pacific Ocean basin through comparison with observed data. Hence the $Q$ values are estimated to be 6000 or more, though the resolution of this method becomes poor if $Q$ values exceed 6000. This fact indicates strongly that the oceanic lithosphere is exceedingly transparent for seismic waves with frequencies higher than 10 Hz.

Our observations of earthquakes landward of trenches produced results in marked contrast to those seaward of trenches. One of the differences is that the estimated $Q$ values, which were obtained through the same procedure, were about ten times smaller; generally several hundreds. Another difference is that there is no peak on the spectral curves for any epicentral distance (Figure 4). The implication is that the spectral peaks

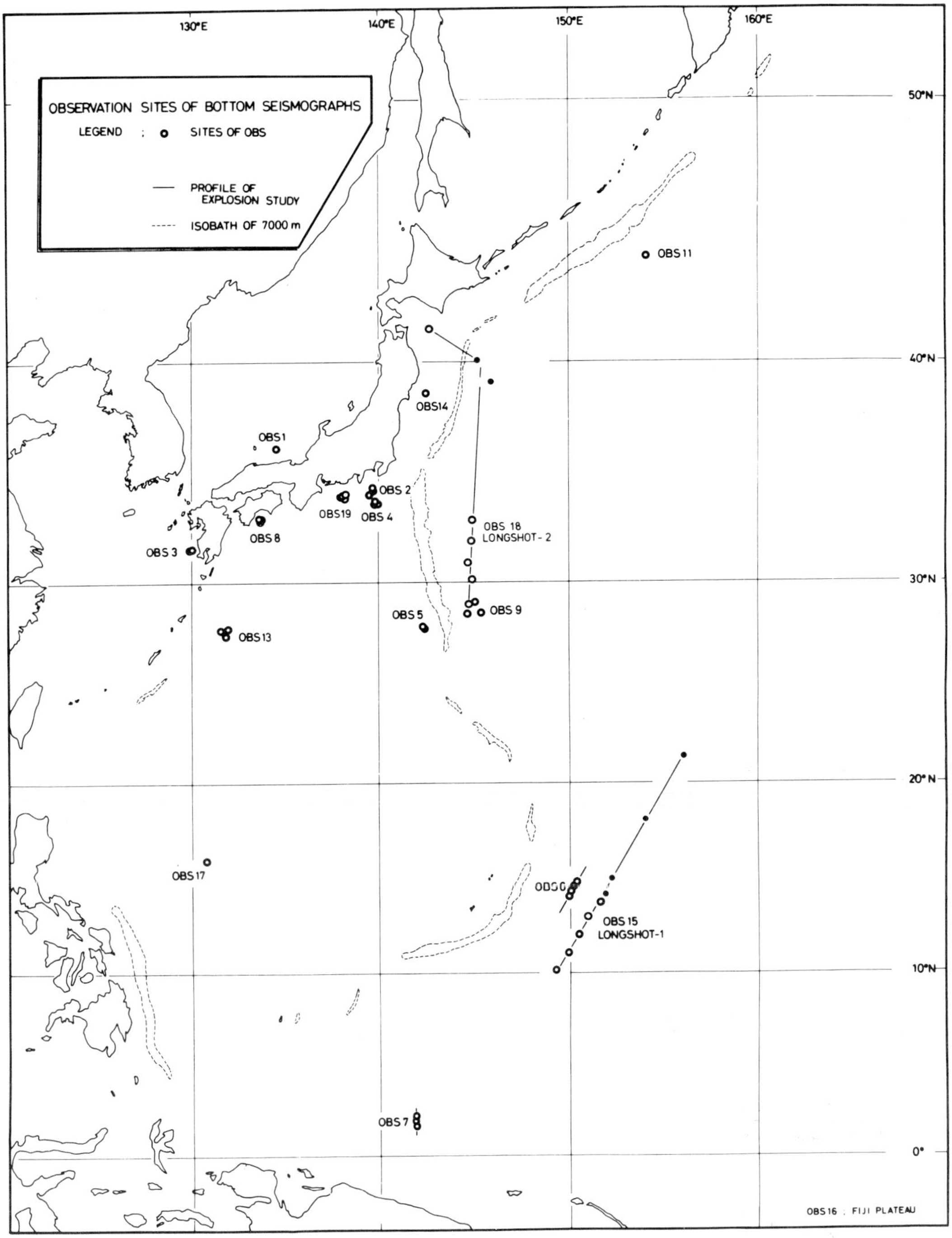


Fig. 1. Observation sites (open circles) of ocean bottom seismographs. Solid circles indicate large explosions for the longshot experiments. Trenches are illustrated by broken lines, showing 7000 m isobath.

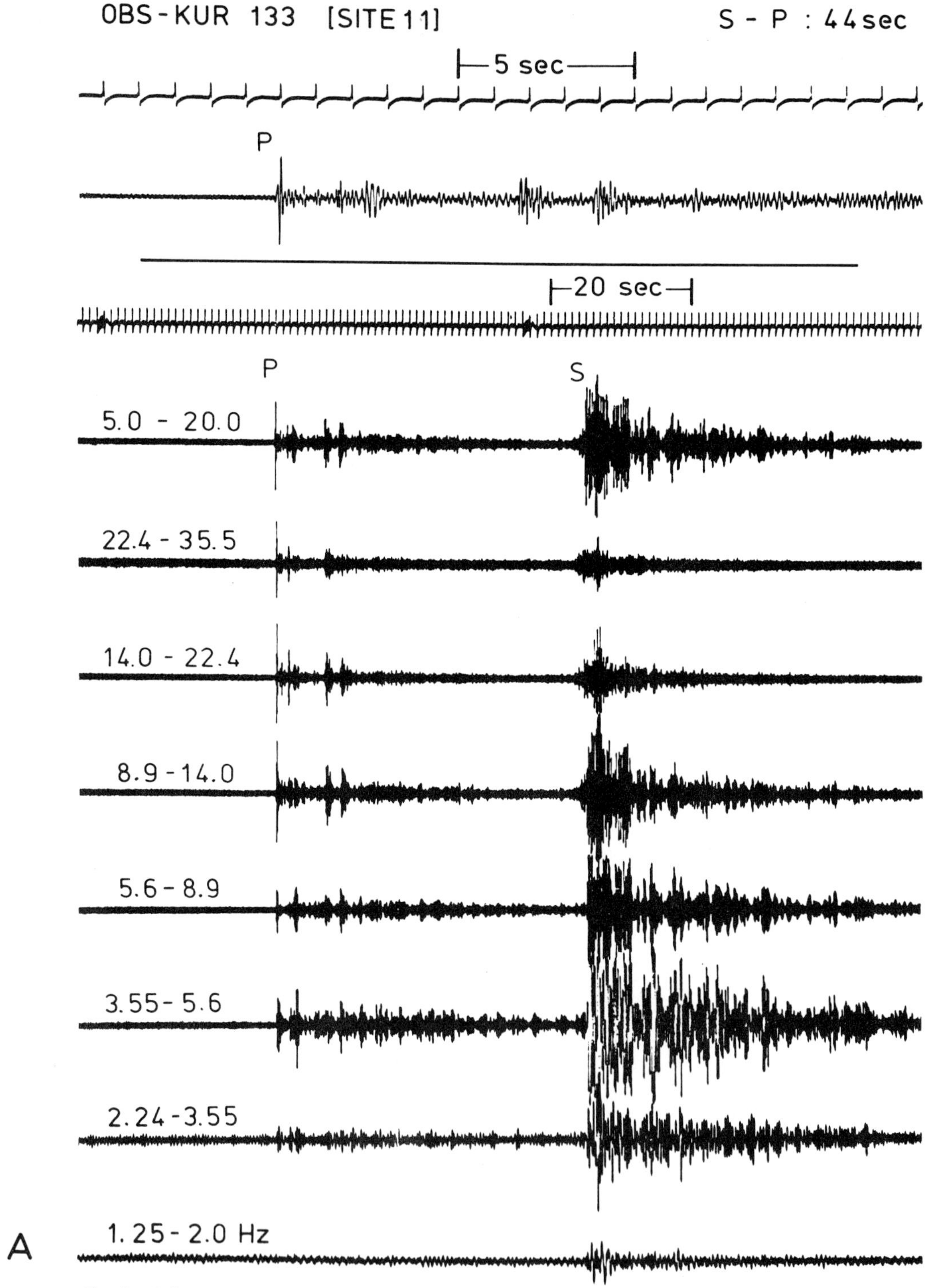


Fig. 2. Seismograms recorded on the Pacific Ocean basin at: (A) site 11 (KUR), and (B) site 9 (OGS) in Figure 1. Seismograms in the figure were obtained from magnetic tapes played through various band-pass filters whose frequencies are indicated on each trace. Note the predominance of high-frequency waves.

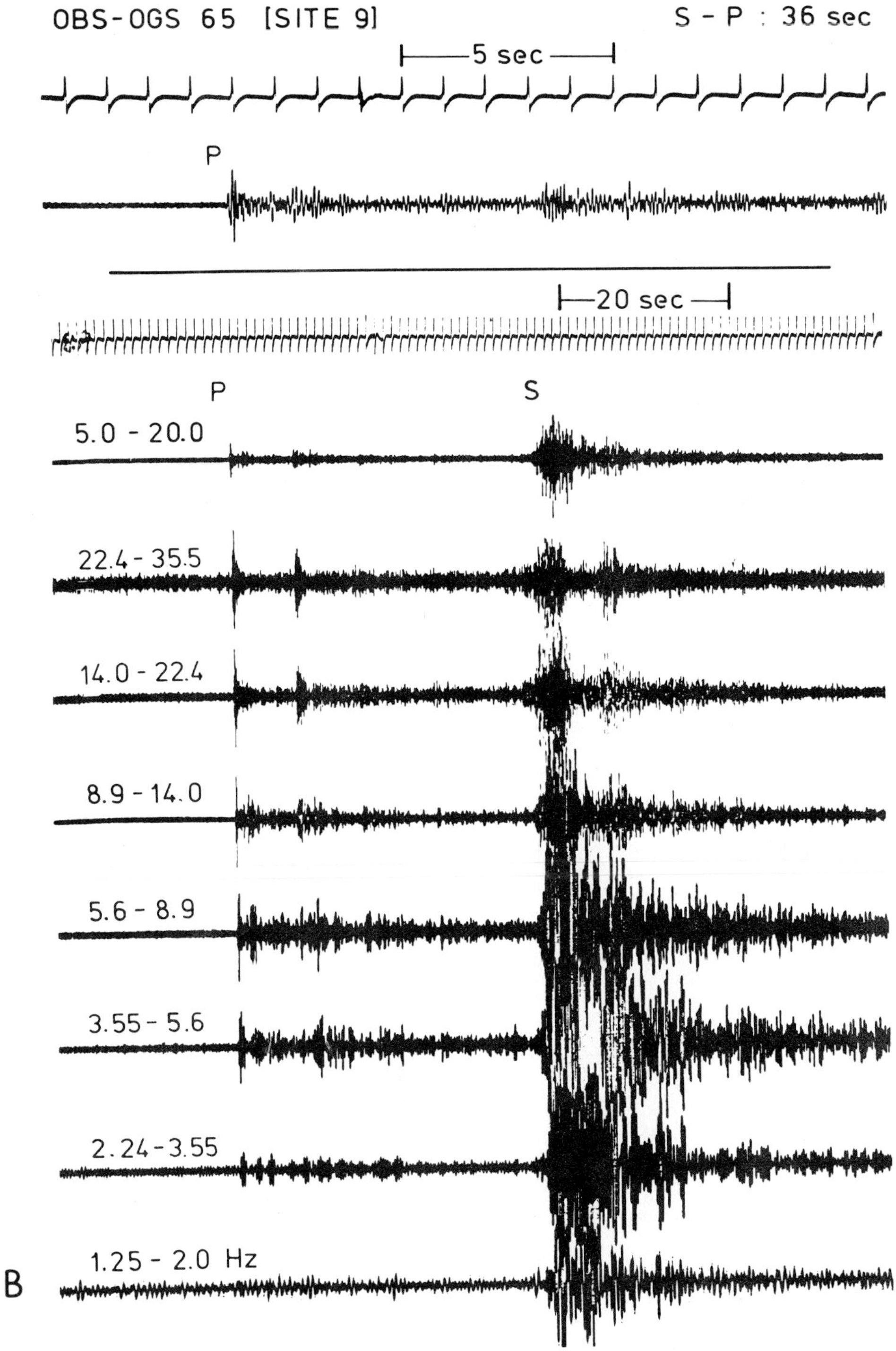
OBS-OGS 65 [SITE 9]
S - P : 36 sec
5 sec
P
20 sec
P
S
5.0 - 20.0
22.4 - 35.5
14.0 - 22.4
8.9 - 14.0
5.6 - 8.9
3.55 - 5.6
2.24 - 3.55
1.25 - 2.0 Hz
B

Fig. 2.

for ocean basins do not come from source spectra but rather from some property of the oceanic basin. The $Q$ values obtained for the spectra at landward sites are quite similar to those obtained on island arcs. It is evident that the structure of the ocean bottom on the seaward side of a trench must be quite different from the structure of the ocean bottom on the landward side as well as from the structure of the upper mantle beneath island arcs.

Another result of the study on $Q$ values is the interesting structure of $Q$ beneath the Central Basin fault of the Philippine Sea, which is thought to be an extinct mid-oceanic ridge. OBS observations showed that $Q$ increased abruptly at a depth of about 80 km [*Shimamura et al.,* 1975].

### EXPLOSION SEISMOLOGY

As stated above, the OBS was designed to observe both natural earthquakes and explosions. For explosion seismology, the intent was to make explosion seismology experiments that would have better signal-to-noise ratios than obtainable with conventional hydrophone methods since the ocean bottom is much quieter than the surface of the sea. Another advantage is that only one ship

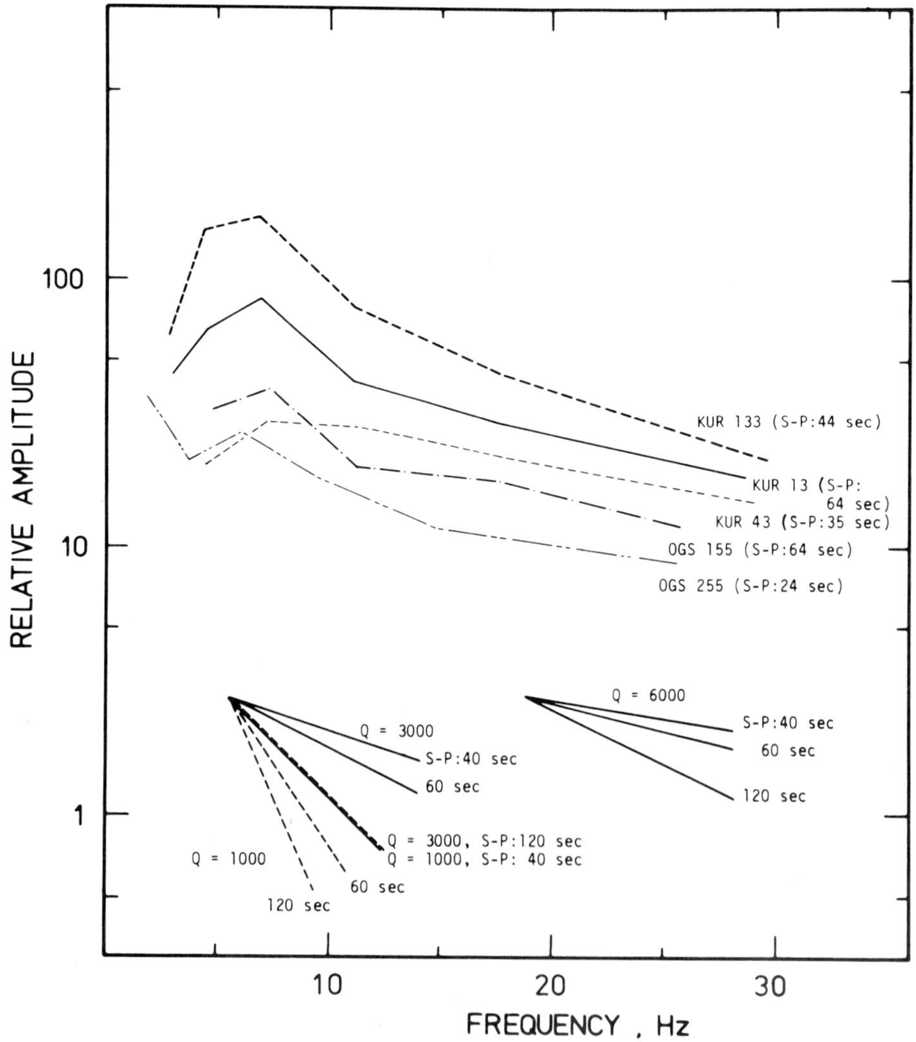


Fig. 3. Examples of unusual frequency spectra obtained from seismograms recorded at site 11 (KUR) and site 9 (OGS). Seismograms with these unusual spectra are generally obtained at the sites on ocean basins for a particular range of epicentral distances. Instrumental response has been taken into account. Calculated inclinations for various $Q$ values and for various $S$–$P$ times are also illustrated.

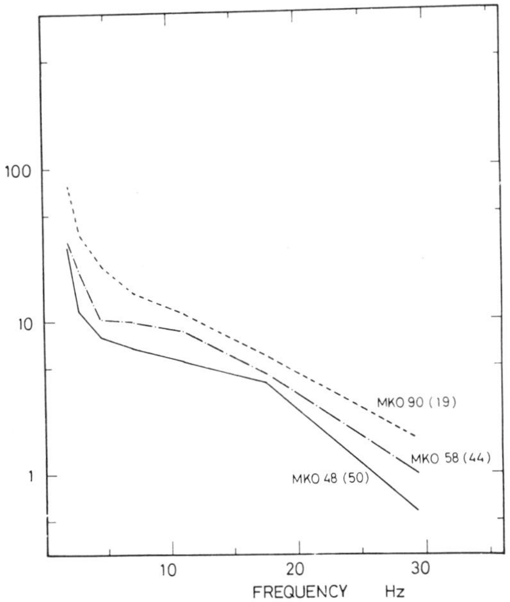


Fig. 4. Example of spectra obtained from records taken on the landward side of trench axes. The examples are from site 5 (MKO) in Figure 1.

is needed in explosion seismology experiments if the OBS is used. These features appreciably enhance the capability as well as mobility of the experiments.

To date, we have conducted four explosion seismology studies in the Pacific Ocean. The first two were small scale as they were intended to study the crustal structure beneath the ocean floor. At the same time, they were to investigate the detection capabilities of the OBS for explosions. Some examples of records taken by the OBS at the bottom of the Pacific Ocean basin are shown in Figure 5. Despite the small amount of explosives, refraction arrivals from the Moho discontinuity were clearly recorded; for instance, shots of 2 or 4 kg were recorded at distances of more than 100 km. We conclude that the OBS is more powerful than the conventional hydrophone, and by its use we can increase detection distances by at least several times, or alternatively reduce the amount of explosives by more than ten times.

We anticipate that the OBS will be useful for crustal studies employing non-explosive sources. Since refraction arrivals from ordinary 0.6 liter air guns are generally recorded at distances of up to 20 km, larger air guns would in all probability be satisfactory for explosion studies of crustal structure down to the Moho discontinuity.

At the same time however, we have found a severe restriction of the detection capabilities of the OBS—the result of the shape of ambient noise spectra at ocean bottoms (Figure 6). The spectra are not the result of instrumental noise, although great care was required in the design of the instruments to be able to detect this level of background noise because we have an active mechanical system within the small pressure case of the OBS.

As seen in Figure 6, the noise level increases greatly at frequencies below 5 or 6 Hz. This phenomenon has been revealed as common for various ocean bottoms of various depths, even if the depth exceeds 6000 m. When ocean-bottom longshot experiments are undertaken, the rapid increase in the noise spectra at lower frequencies constitutes a serious barrier because the waves observed are of lower frequency than in small-scale explosion experiments; the predominant frequency of waves from distant large explosions is around 2 or 3 Hz.

We have carried out two longshot experiments in order to determine the structure of the upper mantle beneath the western Pacific basin. The first experiment was made in August 1973 and the second, in October 1974. Both profiles were designed to be in the vicinity of the region of subducting lithosphere; the directions nearly parallel the trench axes (Figure 1).

The length of the profile was about 1400 km in the first experiment in which eleven OBS's were positioned at five sites. Because of the cost, the large shots were limited to two 1.5-ton explosions. Small shots were also fired on a portion of the profile in order to study the shallow structure.

In the second experiment, the profile extended for about 1250 km on which ten OBS's were positioned at five sites. The explosions fired in this experiment were 7- and 5-ton.

## RESULTS OF THE SECOND LONGSHOT

The two experiments showed that explosions of 1.5 tons are too small to obtain clear first arrivals whereas explosions of 5 tons or more are sufficient.

Examples of seismograms obtained in the second longshot experiment are shown in Figure 7. The difference in charge size between 5 and 7 tons produced no noticeable difference in amplitudes and frequencies on the seismograms throughout the experiment. Nevertheless, exam-

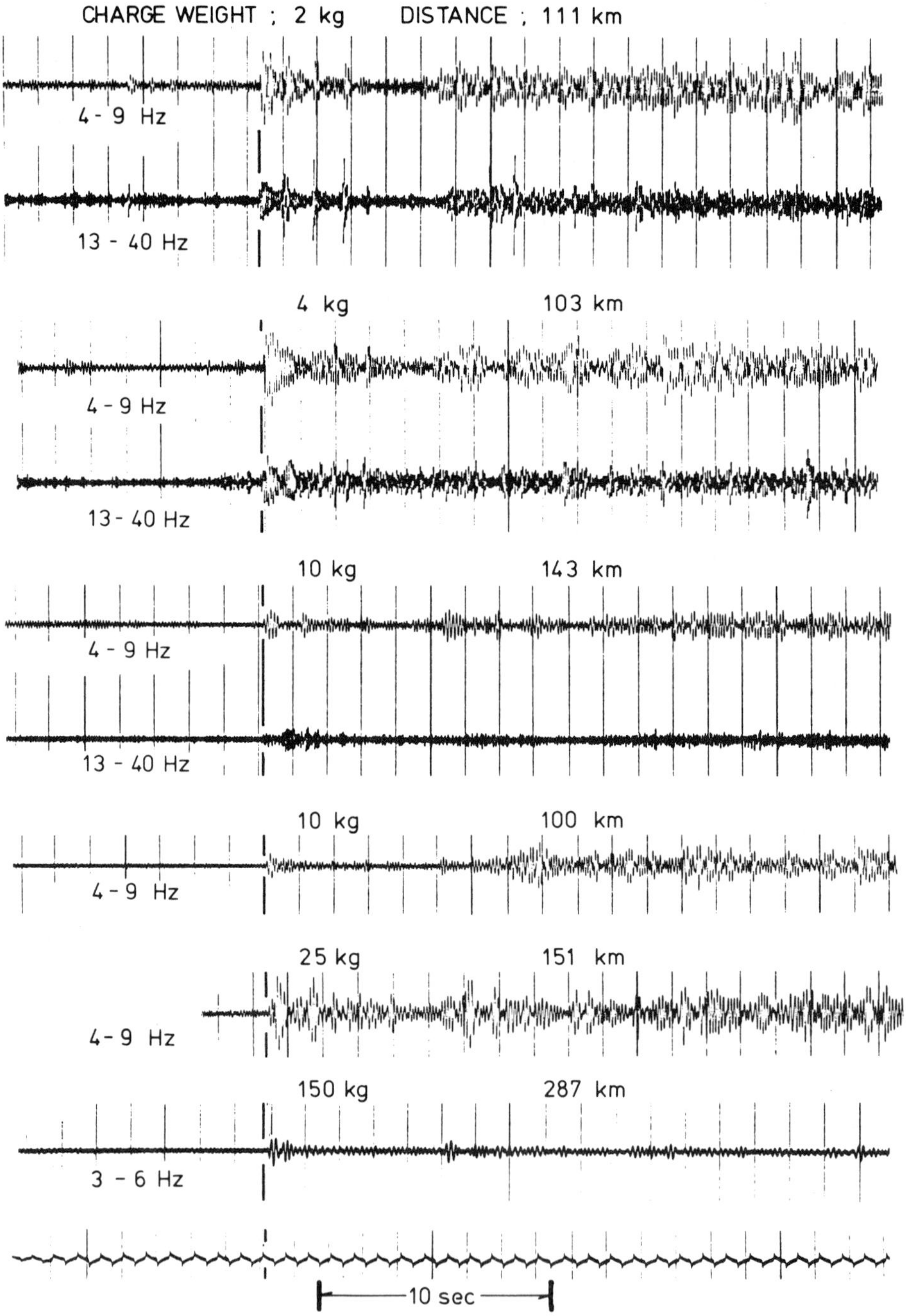


Fig. 5. Seismograms of waves from small explosions taken in the East Mariana basin (site 15). All the onsets are refraction arrivals from the upper mantle.

ples shown in Figure 7 illustrate a striking contrast in apparent frequency content between the two seismograms. At a distance of 1235 km, the signal is rich in low frequency and the first arrival is more easily seen on the lower frequency channels than on the higher frequency channels; but at a distance of 914 km, higher frequencies are abundant whereas lower frequencies are deficient and the first arrival is clearer on the higher frequency channels although the difference in epicentral distances is only about 300 km. This observation can be interpreted as an effect of the low-velocity layer beneath the oceanic lithosphere.

A portion of the record section obtained in the experiment is shown in Figure 8. An offset of the travel times appears clearly on the record section. A large difference in frequency content is shown on the two travel-time branches illustrated in this figure; on the branch up to 1000 km, higher frequencies are predominant whereas on the branch beyond 1000 km, lower frequencies evidently prevail. Moreover amplitudes of lower frequency waves seem to reach their maximum at the near end of the more distant branch in the vicinity of 1000 km. We interpret the rapid increase of the amplitudes as a result of a travel-time cusp which is formed by the waves coming back from a high-velocity layer with a positive velocity gradient that lies beneath the low-velocity layer.

Travel times and amplitudes were calculated for a number of models for inversion. The selected successful models as well as observed travel times are illustrated in Figure 9. Since in the second experiment we concentrated on investigating the deeper structure near the low-velocity layer, we devoted ourselves to obtaining records at distances ranging from 700 to 1200 km. Hence the travel times with distances less than 650 km were obtained not from the second experiment but from the first experiment, and from a nearby but different region.

The branch up to 650 km has an apparent velocity of about 8.2 km/sec. Though not plotted, weak arrivals from the high-velocity layer beneath the 8.2 km/sec layer were seen in the first experiment. By the combination of data

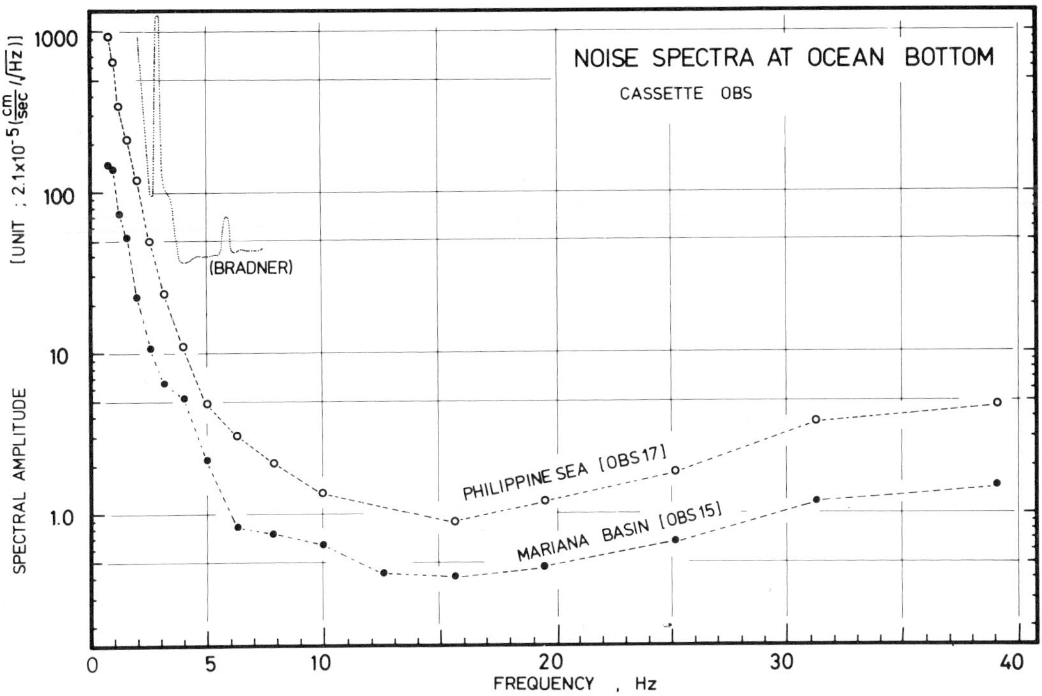


Fig. 6. Typical noise spectra at ocean bottom. Data from site 15 (depth, 5600 m) and site 17 (6500 m). The noise spectra and noise level seem to vary with bottom depth; however when the depth exceeds 5500 m, the noise level is comparable to the quietest places on land, for frequencies more than about 7 Hz. Data from *Bradner* [1964] are inconsistent with our data.

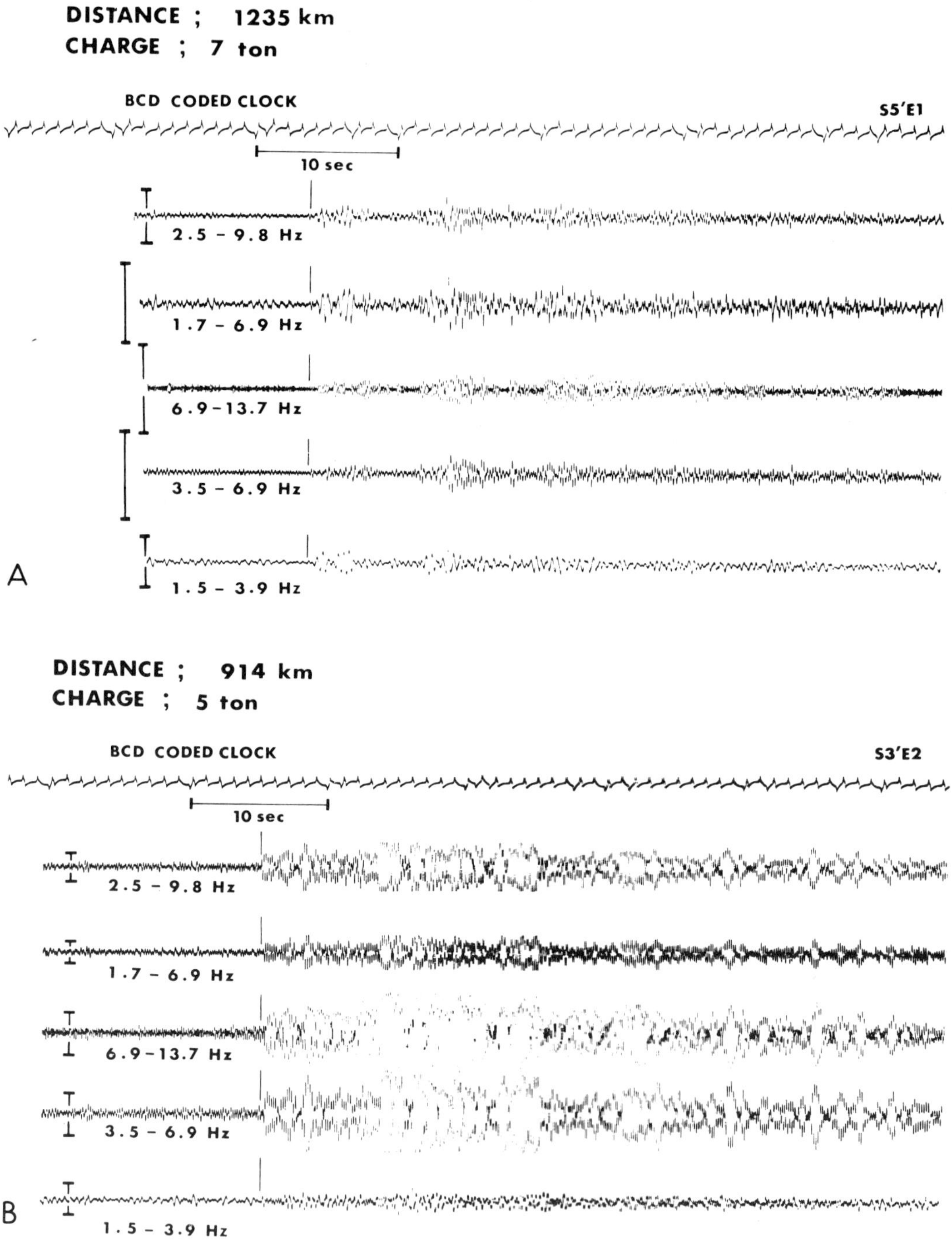


Fig. 7. Seismograms of the longshot observations obtained at site 18. Records were obtained through various band-pass filters whose frequencies are indicated on each trace. Equivalent ground velocity amplitudes of $30 \times 10^{-6}$ cm/s are also shown on traces. Note that amplitudes are much smaller at distances of 1235 km, except for the lowest frequency band, if we consider instrument gain.

mentioned above we have assumed that the uppermost layer of the mantle (8.2 km/sec) is common for these sites; the assumption is likely justified on the basis that this branch of 8.2 km/sec is very commonly observed as an upper mantle velocity in conventional explosion studies made on the western Pacific Ocean basins; however, it was found for the first time by this experiment that the 8.2 km/sec branch extends to more than 650 km distance.

The most successful model is model 115. Model 112 is an example of an unsuccessful model which was ruled out by the travel-time data although it satisfied the amplitude data well.

The crustal structure common to these models is shown in Figure 9. The model is obtained on a portion of the first longshot profile by using hydrophones (N. Den and S. Murauchi, personal communication, 1975). The theoretical amplitudes shown in the figure were obtained by calculating the geometrical spreading factor by the ray theory; the calculation did not include absorption.

## STRUCTURE OF THE UPPER MANTLE

These models are illustrated in Figure 9. The main differences among the models are the velocity gradients in the 8.2 and the 8.6 km/sec layers. Model 115 has a slight velocity gradient in the 8.2 km/sec layer, whereas models 112 and 123 do not.

The velocity gradient in the 8.2 km/sec layer mainly affects the amplitudes on the 8.2 km/sec layer branch, branch *a*, but it does have a slight effect on the travel time data, as shown in Figure 9.

On the other hand the velocity gradients in the 8.6 km/sec layer, which are 0.0, 0.02, 0.05 km/sec for 20 km thickness for models 123, 115, and 112 respectively, affect both the amplitudes on the 8.6 km/sec branch, branch *c*, and the vanishing point of the branch, as seen in Figure 9. Since it is difficult, by refracted wave studies, to determine the bottom of the high-velocity lid upon the low-velocity layer, there is some uncertainty in the thickness and the velocity gradient in the 8.6 km/sec layer; model 115, however, has been chosen through model fitting.

The structure we have obtained in our model exhibits some features of interest; first, the thickness of the oceanic lithosphere in this region is about 80 km; but the lowest, thin stratum of the lithosphere has a remarkably high velocity. The velocity of this high-velocity, thin stratum increases from 8.58 at the top to 8.60 km/sec at the bottom, within a thickness of 20 km, which is shown as model 115 in Figure 9.

The high-velocity layer is underlain by a low-velocity layer. There is a bit of uncertainty in the thickness and the velocity in that lower layer, for example as is shown in Figure 9. Model 112, in which the low-velocity layer is thicker by 10 km than in model 123, is ruled out by the travel time data. In any case the thickness of the low-velocity layer is determined to be very thin in contrast to what it has been thought to be up to this time. The low-velocity layer at this site probably is ineffectual for macroscopic effects such as surface waves or free oscillations of the earth; the existence of the layer is probably beyond the resolution of such macroscopic studies. It is likely that the structure under the site of this experiment, which is near a trench, differs from those obtained from surface wave studies which are inherently an average over a fairly large area.

Beneath the low-velocity layer there is another high-velocity layer; the velocity at the upper edge of this layer together with the velocity of the high-velocity layer which lies at the bottom of the lithosphere is sensitive to the position of the cusp of the travel-time curve. The velocity at the top of this layer is determined to be 8.60 km/sec for model 115, since if the velocity were 8.55 or 8.65 km/sec, the cusp position would either be greater than 1500 km distant or less than 800 km distant, respectively, which would clearly disagree with the observations. The velocity gradient in this high-velocity layer affects the distribution of amplitudes on branch *e*. However it is difficult to compare quantitatively observed amplitudes with theoretical ones since the amplitudes are not very sensitive to the velocity gradient. For instance if the velocity gradient is nearly zero within this layer down to 300 km (hence the velocity is 8.60 km/sec throughout) the amplitudes decrease to about one-third of those for model 115 that has a velocity gradient of 0.20 km/sec for 190 km thickness, viz., $1.1 \times 10^{-3}$ sec$^{-1}$. On the other hand, an increase of the velocity gradient by a factor of 2 results in an increase of amplitude by about a factor of 2. In order to investigate the velocity gradient in this layer more precisely it is necessary to make more extended studies in which the length of the profile should be larger than for the present experiment, since the resolution of amplitude study is surpassed by that of travel time study for this case.

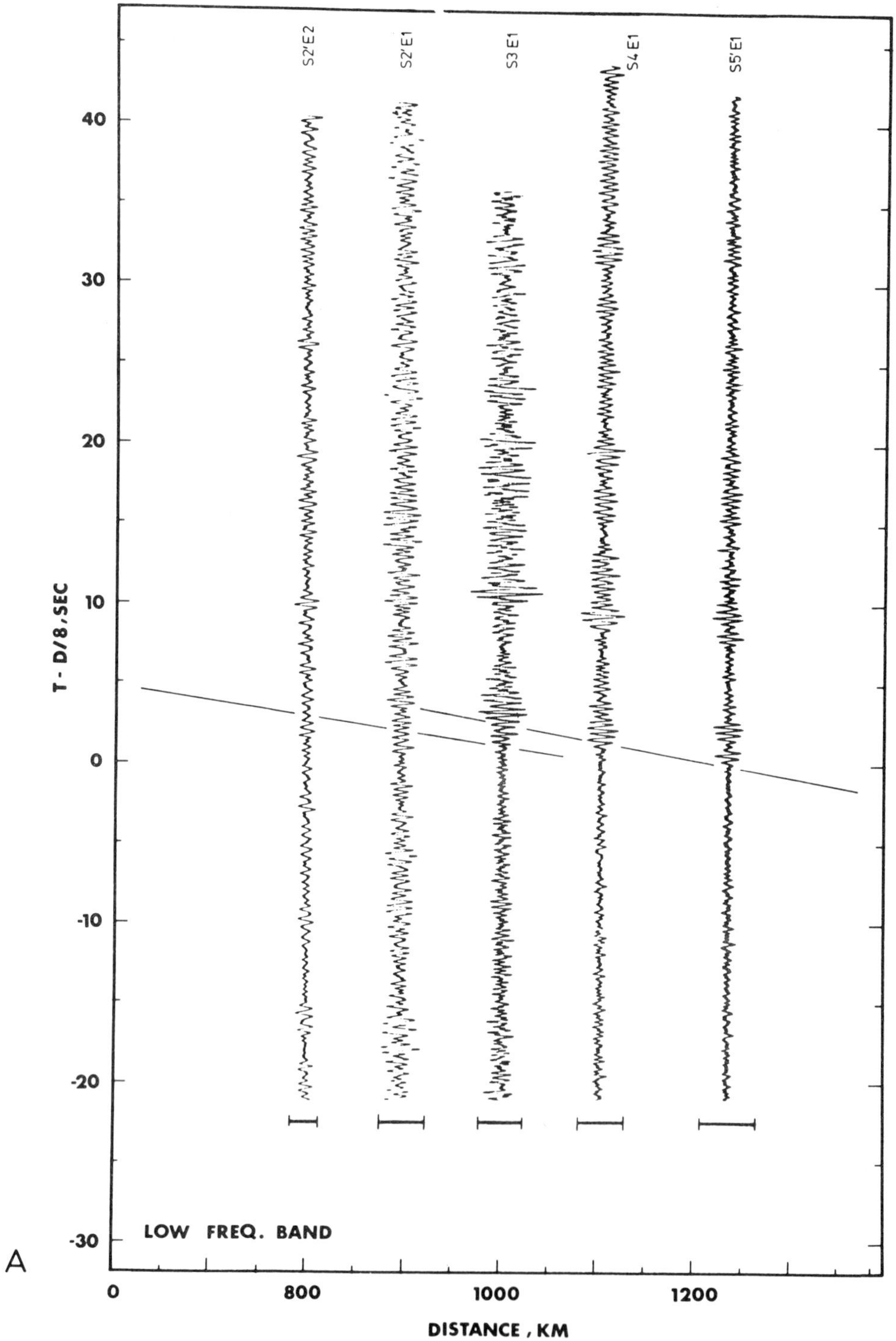


Fig. 8. Record sections obtained in the second longshot experiment. (A) illustrates seismograms through band-pass filter of 1.5–4 Hz; (B) 4–7 Hz. The amplifications are indicated by bars on the traces which denote ground velocity amplitudes of $30 \times 10^{-6}$ cm/sec. Thick lines illustrate calculated travel times for model 115 in Figure 9.

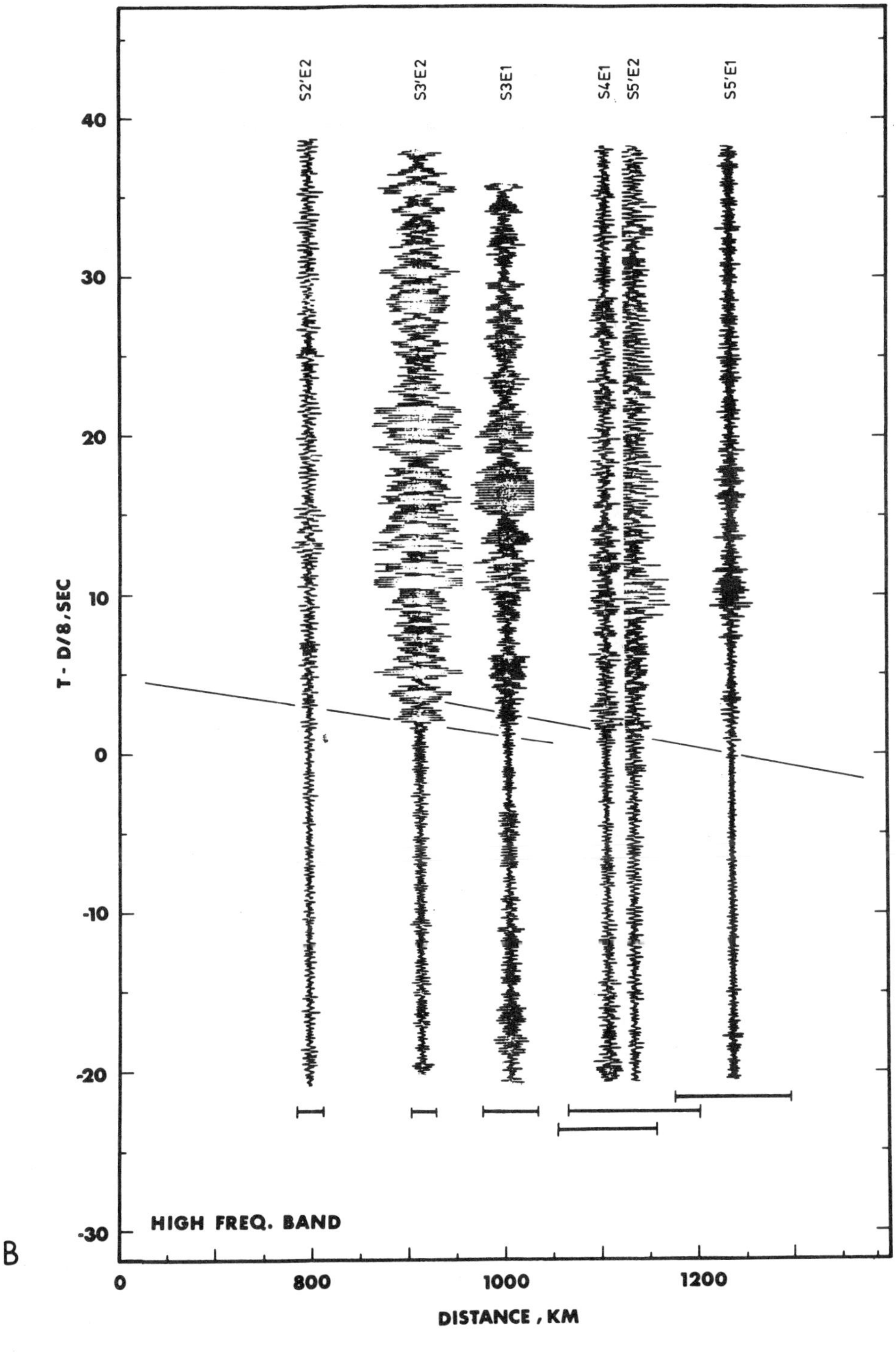
S2'E2
S3'E2
S3E1
S4E1
S5'E2
S5'E1
40
30
20
10
0
-10
-20
-30
T - D/8, SEC
HIGH FREQ. BAND
B
0
800
1000
1200
DISTANCE, KM

Fig. 8.

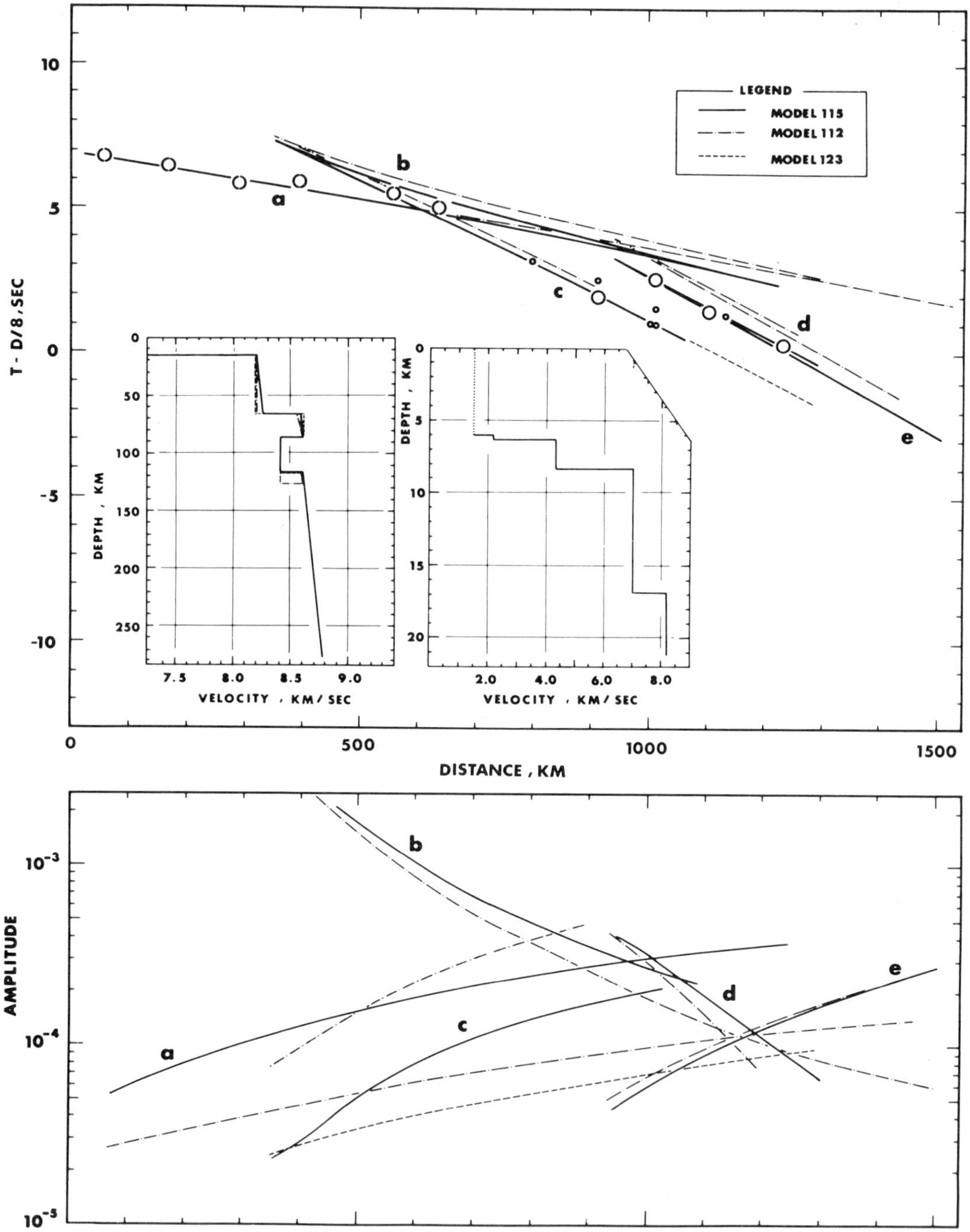


Fig. 9 Reduced travel time curves showing a comparison between observed (open circles) and calculated (lines) values for a series of models. Velocity structure of the models is also shown. The lower half of the figure illustrates calculated amplitudes for the models; as seen, model 115 is the best suited of the three. In the figure, *d* denotes the retrograde branch.

A record section is shown in Figure 10. As previously stated, seismograms with distances of less than 600 km are taken from the first longshot experiment in the Mariana region. Several seismograms were eliminated since their appearances are nearly identical when their epicentral distances are close together. Travel times computed for model 115 are superimposed on this record section. Seismograms farther than 1000 km distant have obviously low predominant frequencies which likely reflect that the waves have twice crossed the low-velocity, and likely low-*Q*, layer. On the contrary, seismograms on the branch of the high-velocity layer at the base of the lithosphere have evidently high prevailing frequencies; the reason has not been clarified yet, but this prevalence can be interpreted as the result of the fact that the low-frequency waves will not transmit well within this layer since the thickness of the layer is less than a few wavelengths, i.e., only about 20 km or less. Hence there seems to be a deficiency of low frequencies on this high-velocity branch.

This contrast between seismograms less than and more than about 1000 km distant has been corroborated on the records of shallow earthquakes. An example of the seismograms recorded by the OBS's are illustrated in Figure 11. The earthquake was located about 100 km north of the site of the northern shot, which was just landward of the trench axis. Depth to focus was very shallow. In Figure 11A low-frequency components are apparently predominant; the seismogram was recorded at 1266 km, which is regarded as beyond the cusp point. Whereas at 929 km, which is likely on the branch of the thin high-velocity layer, the seismogram of the same earthquake differs considerably in appearance (Figure 11B); since the gains of the seismometers and the reproduction instruments are exactly the same, the amplitude from 929 km distance is obviously much larger than that from 1266 km distance,

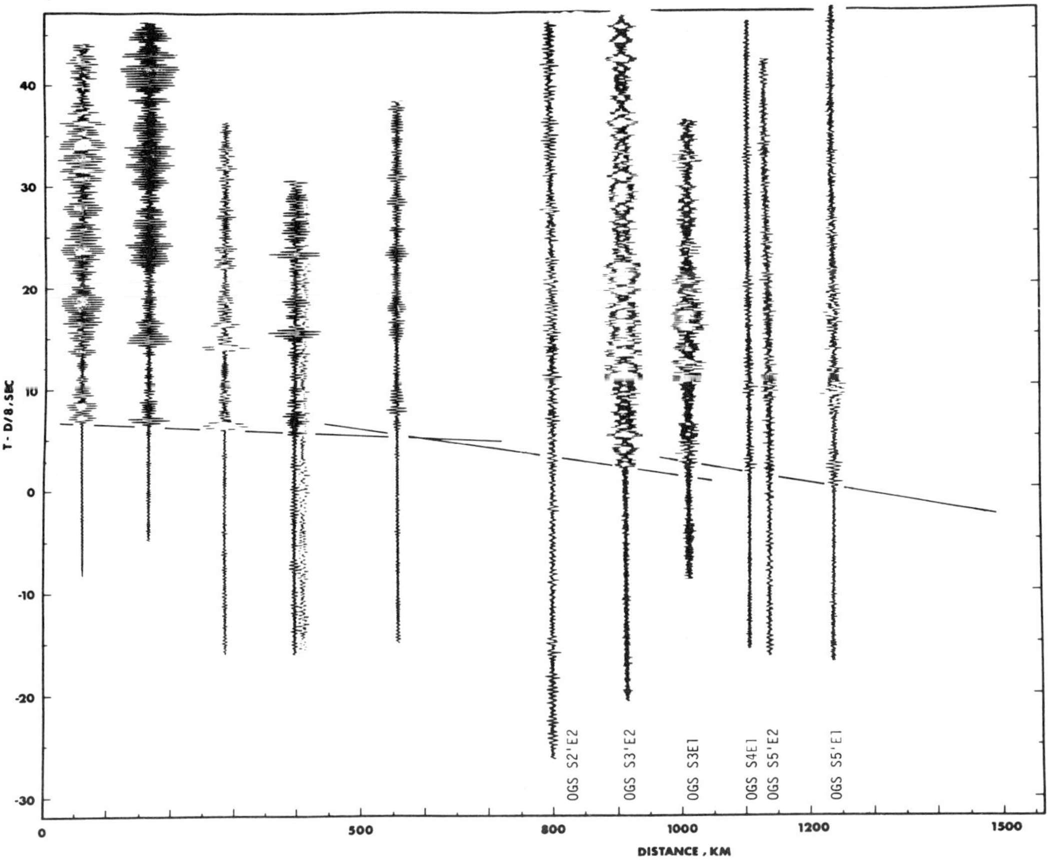


Fig. 10. Record section of the longshot experiment showing travel times for model 115 on Figure 9.

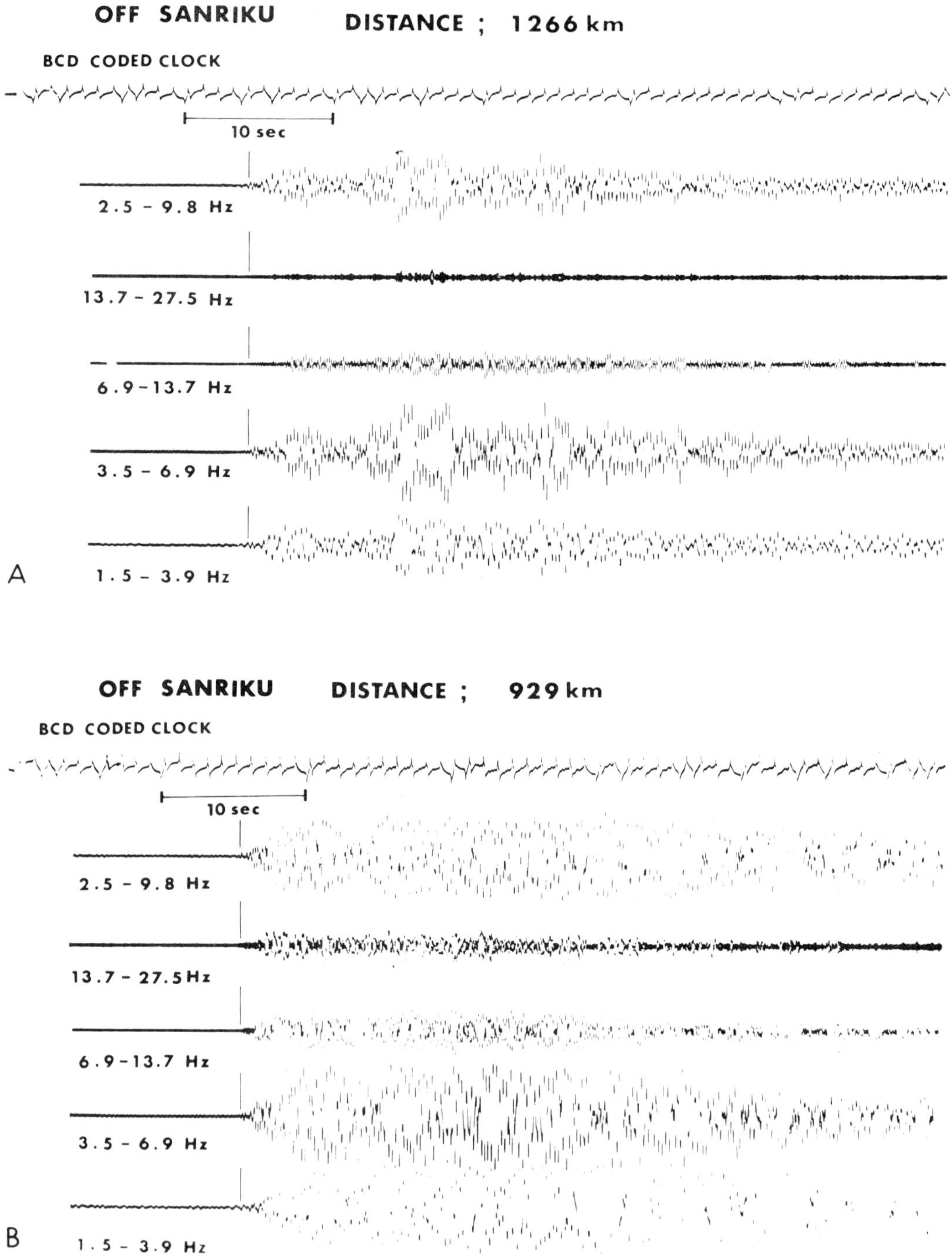


Fig. 11. Seismograms of a natural earthquake (focus: 40.3° N, 143.8° E; depth: shallower than 20 km; magnitude: about 5) recorded by an array of OBS's at site 18. The gains of the instruments are the same for both (A) and (B), except that on the lowermost trace (1.5–3.9 Hz) the amplification of (A) is half that of (B) since the OBS's are slightly different in their frequency responses.

especially in the higher frequencies. Despite the fact that the difference in epicentral distances between the two is only about 300 km, the amplitude decreases by more than three times and high frequency waves of around 20 Hz almost completely disappear.

When a thin high-velocity layer exists as a lid to a low-velocity layer, the spectra of seismic waves recorded on the branch of the thin layer probably suffer from a peculiar distortion in which the lower frequencies decrease markedly. We think that this distortion is a probable cause of the distinctive shape of the spectra obtained from records from ocean basins for a certain range of epicentral distances (Figure 3).

## RESULTS OF THE FIRST LONGSHOT

In the first longshot experiment, four explosions—two 1.5-ton, a 450-kg, and a 150-kg—were set off for the distant shots. Although the ambient noise at ocean basin was the same as that of the second experiment (as is shown in Figure 6), the seismic signals obtained from the shots were in general too small to obtain distinct records. Nevertheless by stacking records from two OBS's placed several hundred meters apart at an observation site, the signal-to-noise ratio was enhanced to an appreciable degree (Figure 12). In Figure 12A rather complex waves having large amplitudes are observed for the first arrival at a distance of 969 km; the distance is regarded as being a little beyond the cusp point. The waves resemble those observed in the second experiment (Figure 8A). In Figure 12B simpler waves are observed at 1319 km; the distance is more likely on the branch beyond the cusp.

The observed travel time data show that the branch with a velocity of 8.2 km/sec is clearly seen from short distances to more than 600 km; the extension of the branch strongly supports the premise that this layer, which is commonly obtained in the upper mantle by explosion seismology studies using hydrophones, extends to considerable depth beneath the ocean basin.

A higher velocity branch, which has a velocity slightly higher than 8.6 km/sec, seems to exist and to intersect the 8.2 km/sec branch at a distance of about 350 km; the amplitudes on this branch however are very small. It means that the thickness of the 8.2 km/sec layer is less than that in the second longshot experiment. Though the branch can not be followed up to 700 km distance or more since the signals are weak, one credible interpretation is that the extreme weakness of signals is caused by a very small or even slightly negative velocity gradient in this high-velocity layer.

An offset of travel times is also seen at around 950 km. Beyond the offset, a branch with a probable cusp can be traced to 1300 km or more. Nevertheless the first arrivals are generally indistinct because the explosive charges were small.

Since the distant signals were too weak in the first longshot experiment, detailed discussions of interpretation will not be attempted here. The third longshot experiment in the Mariana region will be undertaken in early 1976; the forthcoming experimental region will include part of the region of the first longshot experiment.

## DISCUSSION AND CONCLUSIONS

An indication of the regionality in the structure of the oceanic lithosphere is obtained from two longshot experiments. Although the two experimental sites were both on the western margin of the Pacific basin and only about 2000 km apart, the position of the upper edge of the high-velocity layer differs considerably at the two sites. The difference may have some connection with differences in the tectonic patterns; for instance the occurrences of natural earthquakes are evidently different in each area, i.e., at the northern site (the second experimental site) large earthquakes with magnitudes greater than 7.5 or 8 often occur, whereas at the southern site earthquakes with magnitudes exceeding 6 take place very rarely. Moreover, the inclination of the deep seismic zone is larger at the southern site, as is the curvature of the trenches.

As stated before, if the thickness of the high-velocity layer at the bottom of the lithosphere is less than a few wavelengths of seismic waves, the lower frequency waves will have difficulty in transmitting through the layer; this fact is supported by the observational results that the spectral distortion (Figure 3) is clear at the northern site where the high-velocity layer is thinner, whereas it is hardly observed at the southern site. In addition, the spectral distortion is most distinctive at OBS site 11 known from observation of earthquakes (Figure 1) near the Kuril trench. These observations possibly suggest that the high-velocity layer at the bottom of the oceanic lithosphere has developed from north to south, i.e., according to the age of the lithosphere.

The discontinuity between the 8.2 km/sec

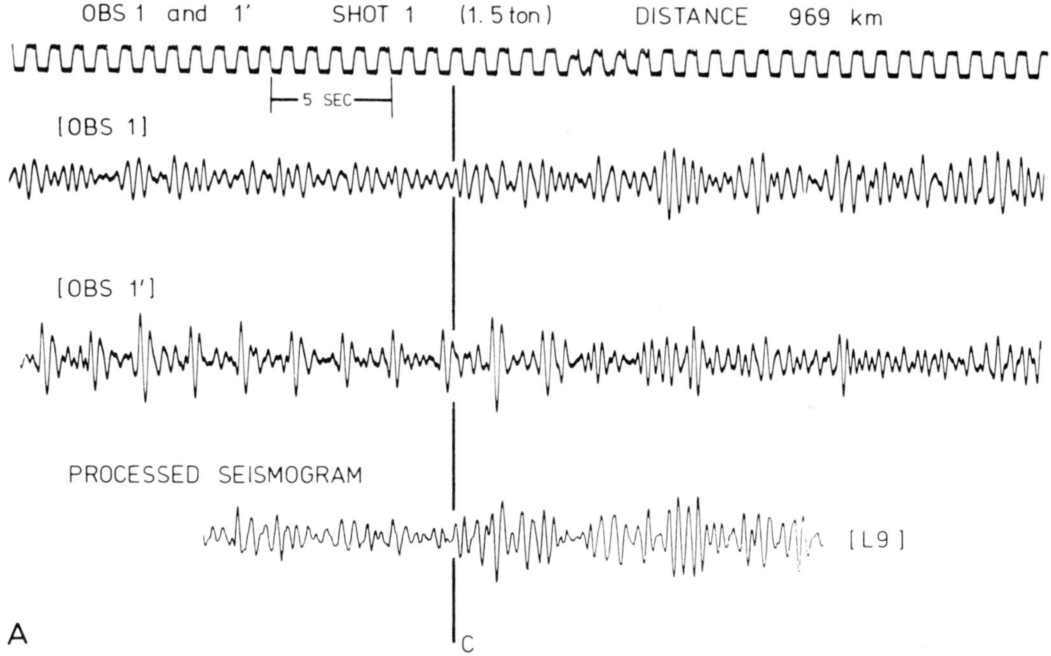


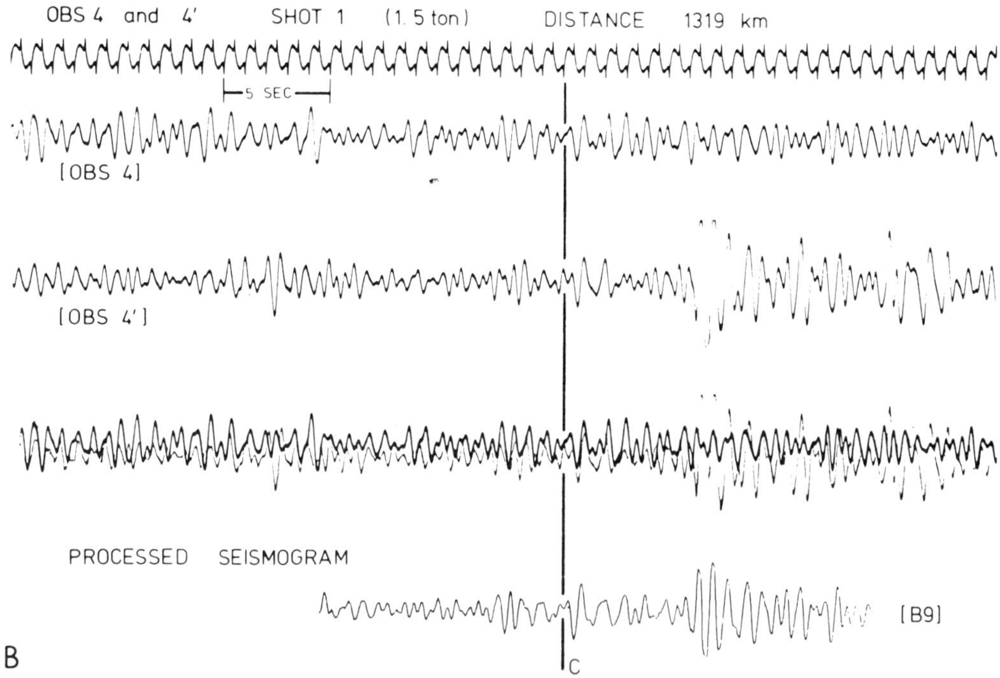


Fig. 12. Examples of seismograms from the first longshot experiment at site 15. The lowermost traces have been obtained by stacking through digital processings. In the figure, *C* denotes interpreted time of the first arrival.

layer and the 8.6 km/sec layer possibly represents a zone of phase transition, since the depth to the discontinuity seems to decrease according to the age of the lithosphere.

The velocity (8.58 to 8.6 km/sec) in the high-velocity layer at the bottom of the oceanic lithosphere, though it possibly differs a little between the experimental sites, is nearly the same as that of the high-velocity layer within the upper mantle obtained for the Gulf of Mexico in the experiment by *Hales et al.* [1970]; the depths to the layers are also nearly the same (66 km for our case and 57 km for their case). However, we lack evidence of whether these high-velocity layers are the same kind or not; we believe that many more longshot experiments on the ocean bottoms should be undertaken in various regions.

*Acknowledgments.* The observations on natural earthquakes were made in cooperation with Professors Y. Tomoda and K. Kobayashi of the Ocean Research Institute of Tokyo University, and the observations on explosion seismology were carried out as a cooperative project with Professors Tomoda and Kobayashi, Dr. S. Murauchi, and Mr. T. Asanuma of the National Museum of Science, Professor N. Den of the Meteorological College, and Professor S. Asano of the Earthquake Research Institute of Tokyo University.

# 13. *Ocean-Bottom Seismograph Study of the Western Margin of the Pacific*

SHOZABURO NAGUMO AND JUNZO KASAHARA

*Earthquake Research Institute, University of Tokyo, Tokyo 113, Japan*

*Abstract.* The locations at which OBS observations have been made since 1966 by the Earthquake Research Institute, University of Tokyo, extend from the Japan trench to the Mariana basin, and to the Fiji plateau along the western margin of the Pacific. The method of observation is the anchored-buoy system, using an OBS containing fully automatic tape recording apparatus. In the Japan trench, we find that very high micro-earthquake activity is taking place both at the marginal escarpment of the Japan trench and in the adjacent ocean basin. This suggests that the oceanic lithosphere is capable of generating earthquakes, and that the marginal escarpment of the Japan trench is an active tectonic element. Micro-earthquake activity is very high in the south Mariana trench. A high concentration of micro-earthquakes is observed along both the midslope basement-high of the trench and the southern terminus of the Mariana ridge. Epicentral distribution suggests uplifting of the Mariana ridge. No micro-earthquakes were observed in the central region of the Mariana basin during a 16-day period. At present the Mariana basin seems to be tectonically very quiet. The ocean floor in the central region of the Fiji plateau is unusually active, with very high micro-earthquake activity including several swarms. The Fiji plateau seems to be tectonically very active. From these observations, it is inferred that mantle diapirism beneath the volcanic islands or ridge is the main cause of the dynamics of the island arc–trench system.

This paper outlines the ocean-bottom seismograph studies of the western margin of the Pacific being conducted by the Earthquake Research Institute, University of Tokyo (Figure 1).

As is well known, deep ocean-bottom seismograph observation was pioneered in 1937 by Professors M. Ewing, G. P. Woollard, A. C. Vine, and J. L. Worzel [*Ewing et al.,* 1946]. At the beginning of the 1960's, the method of OBS observation was intensively developed in connection with nuclear test detection. The method was developed by scientists at Lamont–Doherty Geological Observatory [*Ewing and Ewing,* 1961; *Prentiss and Ewing,* 1963; *Sutton et al.,* 1965; *Thanos and Hubbard,* 1966; *Latham and Sutton,* 1966; *Latham and Nowroozi,* 1968; *Auld et al.,* 1969; *Seeber et al.,* 1970], Professor Bradner of the University of California at La Jolla [*Bradner and Dodds,* 1964; *Bradner et al.,* 1965*a, b*], investigators at Texas Instrument, Inc. [*Thomson and Schneider,* 1962; *Schneider et al.,* 1964; *Schneider and Backus,* 1964; *Arnett and Newhouse,* 1965; *Kimler and Staff,* 1967; *Guidroz et al.,* 1967; *McDermott et al.,* 1967], and, also, Drs. *Monakov* [1961, 1962], and *Ryuknov and Sedov* [1965, 1967] in the USSR.

It was in 1963 that the first OBS observation was conducted in Japan [*Kishinouye et al.,* 1963]. OBS observations have been conducted almost every year since 1966. The locations of OBS stations extend from the Japan trench to the Mariana basin, and to the Fiji plateau along the western margin of the Pacific.

Recently in the 1970's, the progress of OBS observation has become so remarkable that many institutions all over the world have developed various types of instruments and have conducted many cruises in both the Pacific and Atlantic oceans [*Bradner et al.,* 1970; *Whitmarsh,* 1970, 1971; *Shimamura and Asada,* 1971*a*; *Francis and Porter,* 1972, 1973; *Bradner and Brune,* 1973; *Sutton et al.,* 1973; *Reid et al.,* 1973; *Reid and Macdonald,* 1973; *Reichle and Reid,* 1974; *Reid et al.,* 1974; *Orcutt et al.,* 1974; *Prothero,* 1974; *Prothero et al.,* 1974].

This paper presents (1) a brief outline of the instrumentation and the method of observation which we are using; (2) a report on micro-earthquake seismicity under the ocean floor; and (3) some discussion about the tectonic implications of

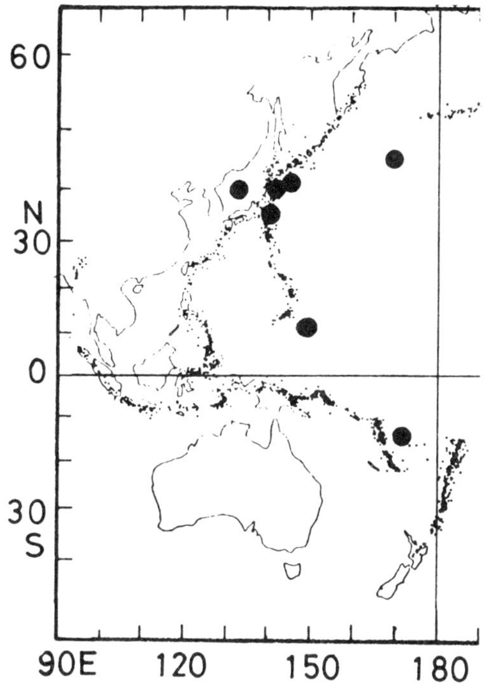


Fig. 1. Locations of OBS stations (solid circles) conducted by the Earthquake Research Institute, University of Tokyo, since 1966.

micro-seismicity.

Figure 1 shows the locations of OBS stations conducted by the Earthquake Research Institute, University of Tokyo, from 1966 to 1974. OBS observations were first performed in Sagami bay, which is the source region of the great Kwanto earthquake [*Nagumo et al.,* 1968*a*]. OBS observations were then conducted for three years, 1968–1970, off Sanriku in the Japan trench [*Nagumo et al.,* 1968*c; Nagumo,* 1971*a, b; Nagumo et al.,* 1970*a, b; Hasegawa and Nagumo,* 1970]. In 1971–1974, OBS observations were again performed in the Sagami trough. These observations were conducted under a project of earthquake prediction research [*Kasahara et al.,* 1973*a, b*]. As regards geodynamics, OBS observations have been performed on one of the Emperor seamounts [*Nagumo,* 1968], North Yamato bank in the Japan Sea [*Nagumo,* 1971*b*], and the Mariana basin and Fiji plateau. Over the past 10 years, the number of OBS stations has increased to about 40.

## INSTRUMENTATION AND METHOD OF OBS

The method employed in our OBS observations is the anchored-buoy system [*Nagumo et al.,* 1970*a*]. A schematic of the system is shown in Figure 2. The system is designed so as to assure reliable recovery even under severe sea conditions, e.g., the strong ocean current of the Kuroshio near Japan, unexpected monsoons, and long observation periods of more than one month. The breaking strength of the rope system is taken as high as about 6 tons. In order to avoid anchor noise, the OBS is placed at least 200 m distant from the anchor.

The whole instrument is installed within a pressure vessel. Completely automatic recording is made within the pressure vessel. The pressure vessel is made of high strength steel, SNC 2; inside diameter is 240 mm, outside diameter, 270 mm; inside length, 1100 mm. Maximum operating water depth is about 8 km.

A block diagram of the OBS recording apparatus is shown in Figure 3. The recording apparatus is composed of geophones, amplifiers, a tape recorder, a crystal clock, and batteries. The vertical geophone is 1.0 Hz, and the horizontal, 4.5 Hz. The recording life of the tape recorder is about 1,000 hours for DR recording by ¼-inch, 1800-foot tape. Power consumption is very low. Batteries of 12 V, 39 AH, operate the instrument for about 500 hours.

Because of very slow tape speed, 0.006 inch/sec, during the recording time several transcriptions are required for changing tape speed to produce the final visible paper record [*Hasegawa and Nagumo,* 1970].

## OBSERVATIONS OFF SANRIKU NEAR THE JAPAN TRENCH

In 1968, we conducted network observations off Sanriku (Figure 4). Accidentally, we recorded the large Tokachi-oki earthquake of M = 7.9. In addition, the OBS recorded abnormally high micro-earthquake activity which is believed to be the forerunner of that large earthquake [*Nagumo,* 1971*a*].

In 1969, another network observation was performed in the same area in order to study the post-seismic activity. At the same time, another station was placed in the oceanic basin near Erimo seamount on the oceanic side of the Japan trench.

*High micro-earthquake activity.* During this observation, we were quite surprised at the intensive micro-earthquake activity that was taking place around the Japan trench, an example of which is shown in Figure 5. As seen, there were three earthquakes in a period of 25 minutes. Their *S–P* times are about 22, 14, and 5 seconds, respectively, showing that they occurred near station

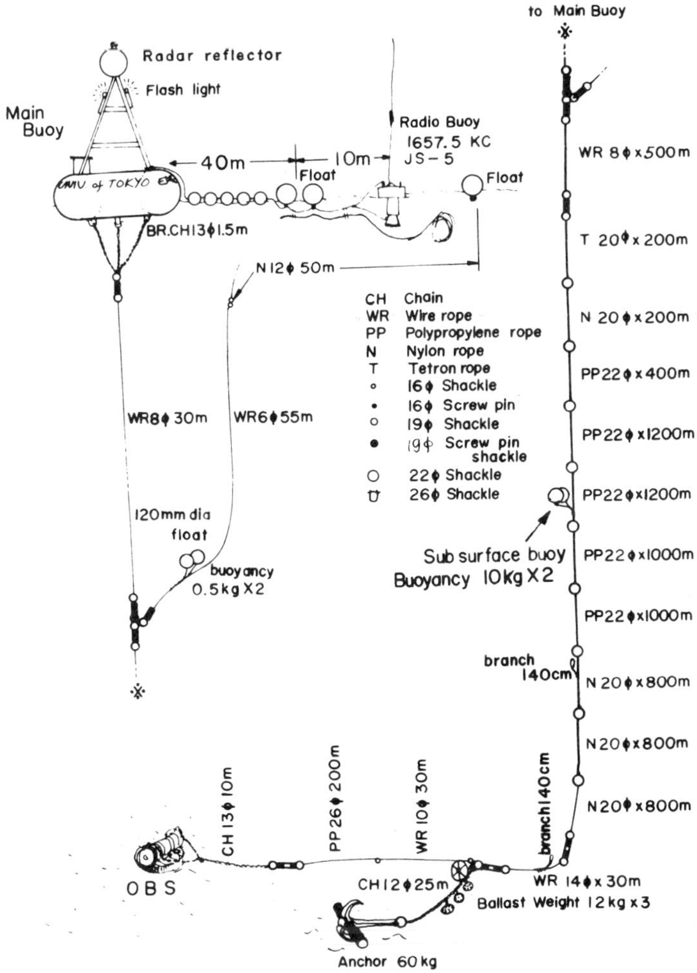


Fig. 2. Anchored-buoy system for mooring OBS, water depth about 6.0 km.

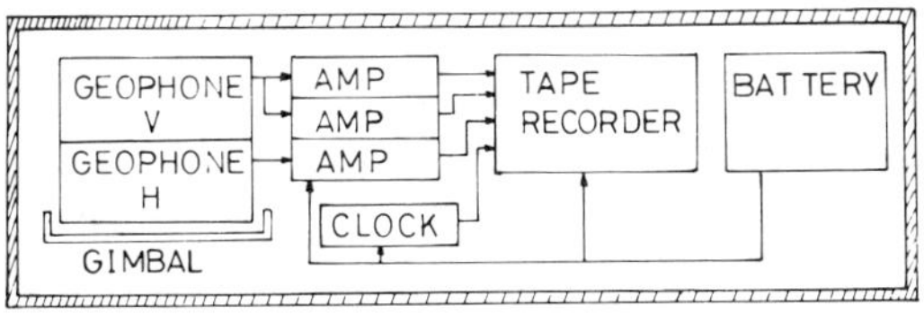


Fig. 3. Block diagram of OBS instruments within the pressure vessel. Geophones: vertical (GeoSpace HS-10-1) $f_0$= 1.0 Hz; horizontal (GeoSpace 8D-L 4B) $f_0$ = 4.5 Hz; gimbal mounted. Amplifiers: 80 dB, 60 dB for vertical component, 80 dB for horizontal component. Magnetic tape recorder: 4-track DR, tape speed 0.006 inch/sec, recording life 1000 hrs by ¼-inch, 1800-foot tape. Frequency response: flat 1-6 Hz. Crystal clock: Seiko Q-952-TF-7. Battery: 12 V, 39 AH (NiCdAl). Pressure vessel: inner diameter 240 mm, outer diameter 270 mm, inner tube length 1100 mm, material SNC2, operating water depth 8.0 km.

Erimo. The first two earthquakes in the figure are accompanied by large *T*-phases. These earthquakes occurred along the steep slope of the inner wall of the Japan trench. The third earthquake took place on the gentle slope of the oceanic side.

There were no particular swarms. The frequency of occurrence was almost stationary. On the average, daily frequency of occurrence was as high as 6 earthquakes per day for earthquakes for which maximum amplitudes are larger than 100 $\mu$kines ($100 \times 10^{-6}$ cm/sec) and *S–P* times are

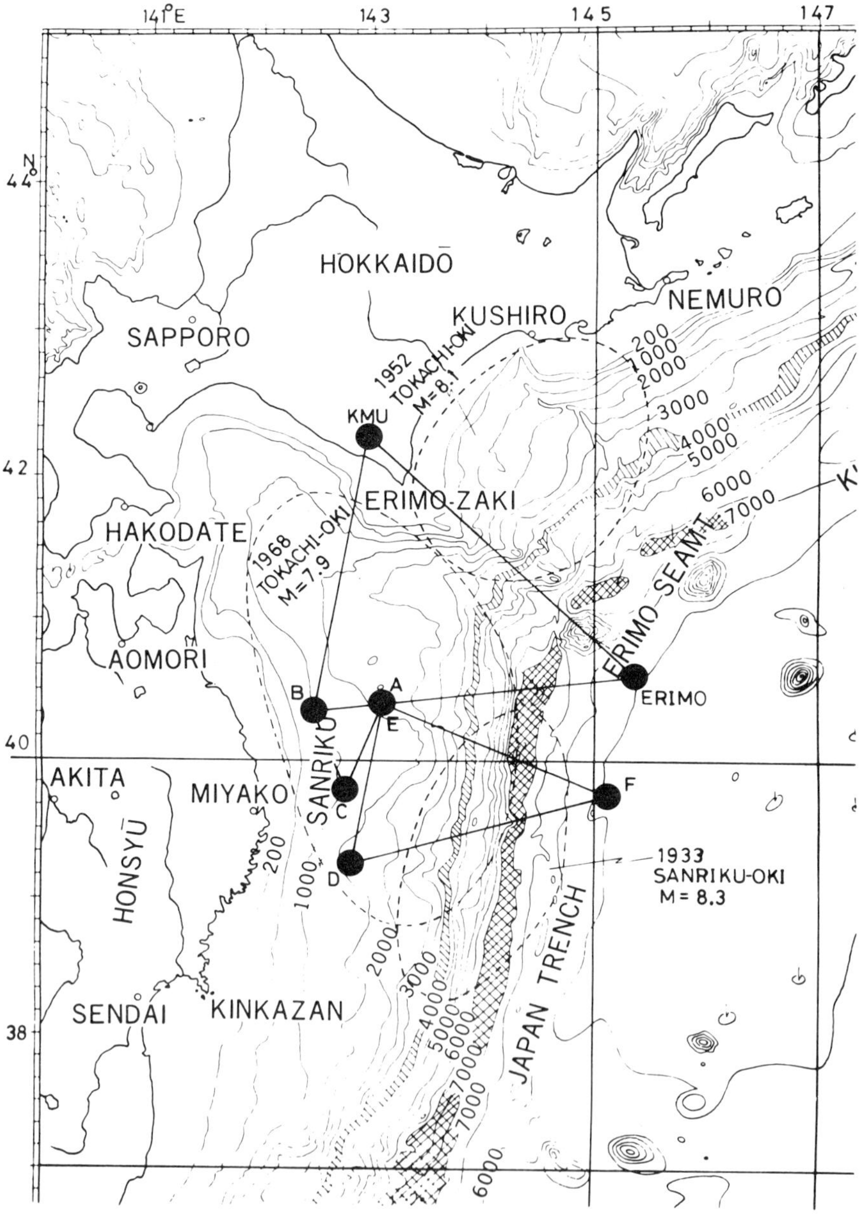


Fig. 4. OBS stations off Sanriku near the Japan trench in 1968–1970. Stations A, B, and C in 1968 (by research vessel *Tansei-Maru*); A, B, C, and ERIMO in 1969 (by research vessel *Hakuho-Maru*); and D, E, and F in 1970 (by research vessel *Tokaidaigaku Maru II*). KMU: Urakawa Micro-earthquake observatory, Hokkaido University. Dotted-line circles indicate aftershock areas of large earthquakes of 1952 Tokachi-oki (M = 8.1), 1968 Tokachi-oki (M = 7.9), and 1933 Sanriku-oki (M = 8.3). Cross-hatched zone: axis of Japan trench and Kuril trench. Hatched zone: marginal escarpment at water depth 3.5–4.0 km.

less than 6 seconds, and 19 earthquakes per day for earthquakes for which *S–P* times are less than 30 seconds.

*Concentration of epicenters.* On making epicentral determinations of the micro-earthquakes we found that the earthquakes were concentrated along the marginal escarpment of the Japan trench [*Nagumo et al.,* 1970*b*]. The location of this concentration corresponds to a structural discontinuity that was discovered by explosion seismology [*Ludwig et al.,* 1966]. It was inferred that the discontinuity is a large active fault between the geophysical ocean basin and the geophysical continent.

In order to confirm this tentative conclusion, we performed another network observation in the same area in 1970. The marginal escarpment was observed from both sides of the trench (Figure 6, stations D, E, and F).

*Micro-earthquakes within the oceanic basin.* *S–P* time distributions (Figure 6) indicate the distribution of epicentral ranges from the observation points. It is seen that micro-earthquakes are taking place both in the ocean basin and on the continental slope. At station F, a large peak is seen for *S–P* times of less than 15 seconds. It should be noted that there are several earthquakes for which *S–P* times are less than 6 seconds. This means that micro-earthquake activity does exist in the ocean basin and that the oceanic lithosphere is capable of producing earthquakes. An example of a micro-earthquake that occurred in the ocean basin is shown in Figure 7. It was observed at station F, off the Japan trench, water depth 5.6 km; *S–P* time is about 3 seconds. It will be noted that high frequency components predominate in both the *P*- and *S*-phases; about 10 to 20 Hz.

*Migration of post-seismic activity.* Many micro-earthquakes are also taking place on the continental slope of the Japanese islands. The earthquakes for which *S–P* times range less than about 18 seconds at stations D and E correspond to those earthquakes.

The fact that the micro-earthquake activity on the oceanic side is about five to six times higher than that on the continental side is very striking.

The average daily frequency of nearby earthquakes for which *S–P* times are 6 seconds or less was about 2 earthquakes per day at station E, and was 12 earthquakes per day at station F in the summer of 1970. This is an unusual phenomenon. It is interpreted that the post-seismic activity of the

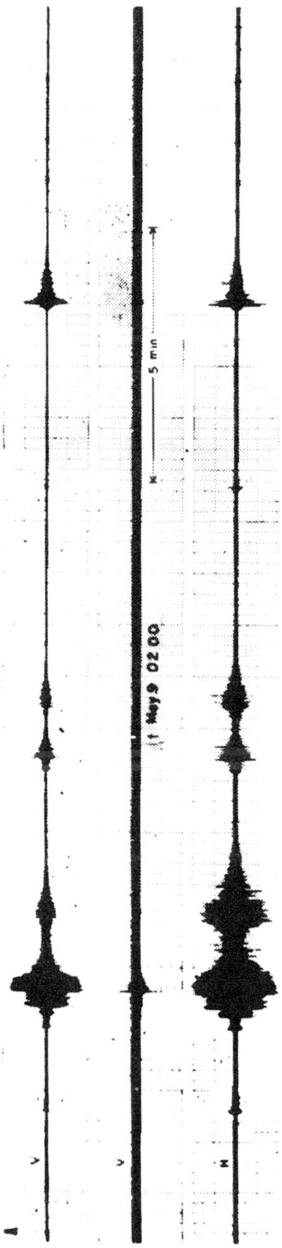

Fig. 5. An example of intense micro-earthquake activity recorded by OBS at station Erimo, off Sanriku, on the oceanic side of the Japan trench, water depth 5.6 km.

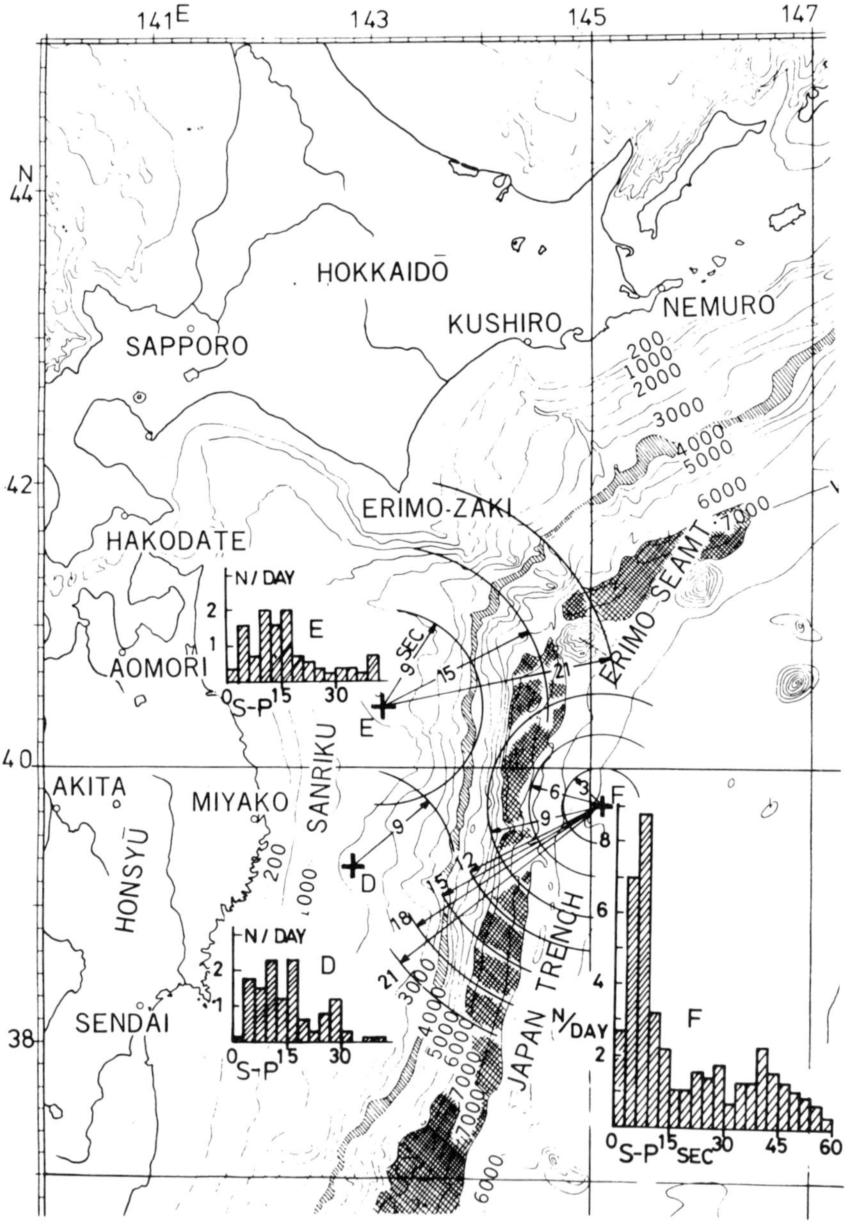


Fig. 6. *S–P* time frequency distribution of micro-earthquakes registered at OBS stations D, E, and F off Sanriku on August 14–19, 1970. The detection level of earthquakes is 100 $\mu$kine (unit $\mu$ kine = $10^{-6}$ cm/sec) at the station. The unit of *S–P* time distribution is the number of earthquakes per day averaged for about five days. The epicentral distances which correspond to the *S–P* times are shown by circles surrounding each station. The Omori coefficients, for computing epicentral distance from *S–P* time, are 8.6 km/sec for station D and E, and 9.4 km/sec for station F [*Nagumo et al.,* 1970*b*].

1968 Tokachi-oki earthquake (M = 7.9) migrated from the continental slope side to the ocean basin side crossing the deepest part of the Japan trench. It also implies that the tectonic deformation of the continental slope caused by the large earthquake gradually migrated into the ocean basin side crossing the deep trench.

*Epicentral distribution in 1970.* The epicentral distribution of micro-earthquakes off Sanriku in the vicinity of the Japan trench observed in August 1970 is shown in Figure 8. There are two groups, one is off Erimo Zaki, Hokkaido, and the other off Sanriku. In the latter group, there is a concentration of epicenters along the steep marginal escarpment of the trench. This observation is in agreement with the result obtained in the former year, 1969. It therefore appears certain that the marginal escarpment is a great active fault lying between the island–arc mass and the oceanic mass. The marginal escarpment corresponds to the mid-slope basement high of *Karig* [1971*b*] and the trench-slope break of *Dickinson* [1973].

Depth distribution of the micro-earthquakes is shown in Figure 9. There are two groups: one, shallow crustal earthquakes, and the other, deeper earthquakes within the lithosphere lying upon an extension of the deep seismic zone. From this cross section, it should be noted that the deep seismic zone, which is usually defined for depths greater than about 100 km, extends into the upper

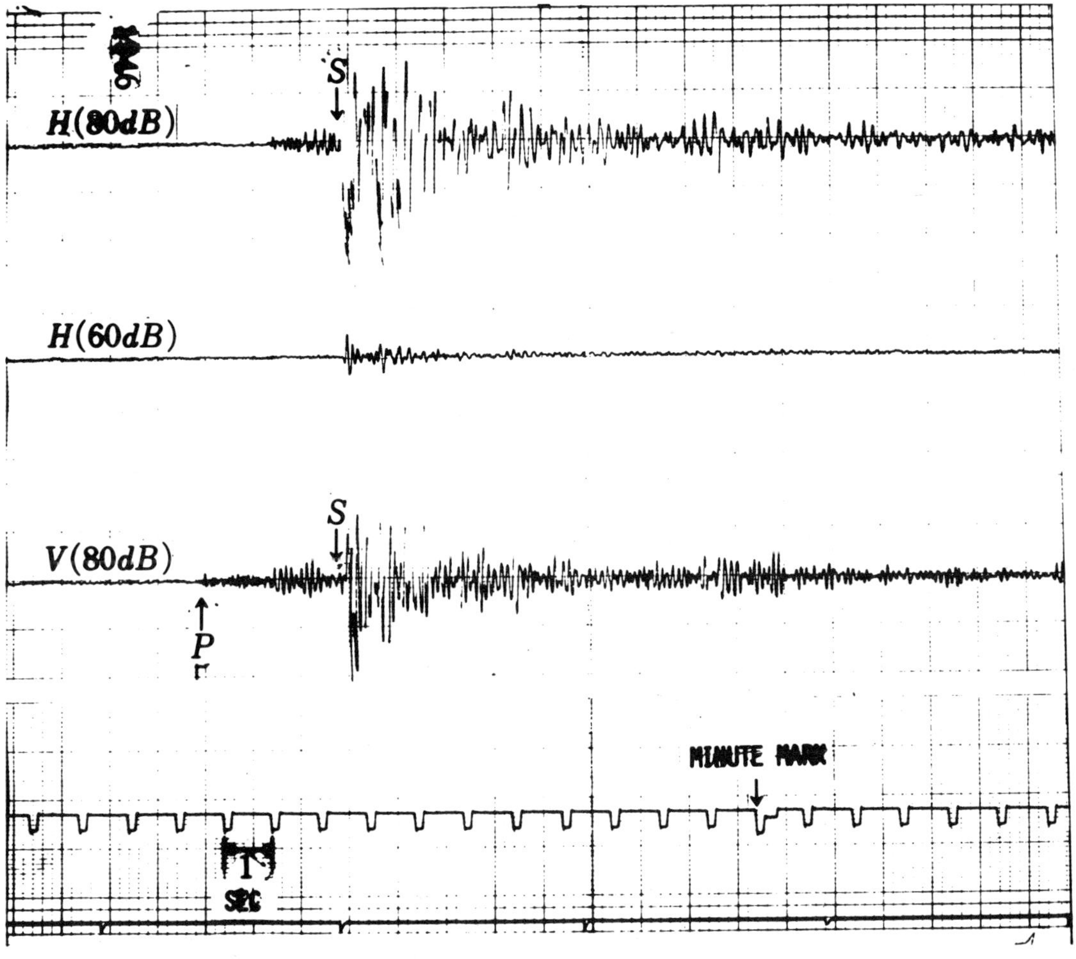


Fig. 7. An example of a micro-earthquake that occurred in the ocean basin, recorded at OBS station F, water depth 5.6 km. The upper two traces are the horizontal component with amplifier gains of 80 dB and 60 dB, respectively. The third trace is the vertical component with amplifier gain 80 dB. The lowest trace is time. Natural frequency of the geophones is 4.5 Hz. *S–P* time is about 3 seconds. High-frequency components predominate.

part of the lithosphere. Heretofore, the existence of the structure of the zone within the upper lithosphere in the range of 0–100 km has been obscured because of the large number of earthquakes. OBS observations over an appropriate period have now revealed this feature.

*Spatial distribution.* The micro-earthquake activity results obtained by OBS will now be generalized using the data of the Japan Meteorological Agency (JMA) to obtain a picture of the seismo-tectonics of the Japan island arc–trench system. The cross-sectional view of the spatial dis-

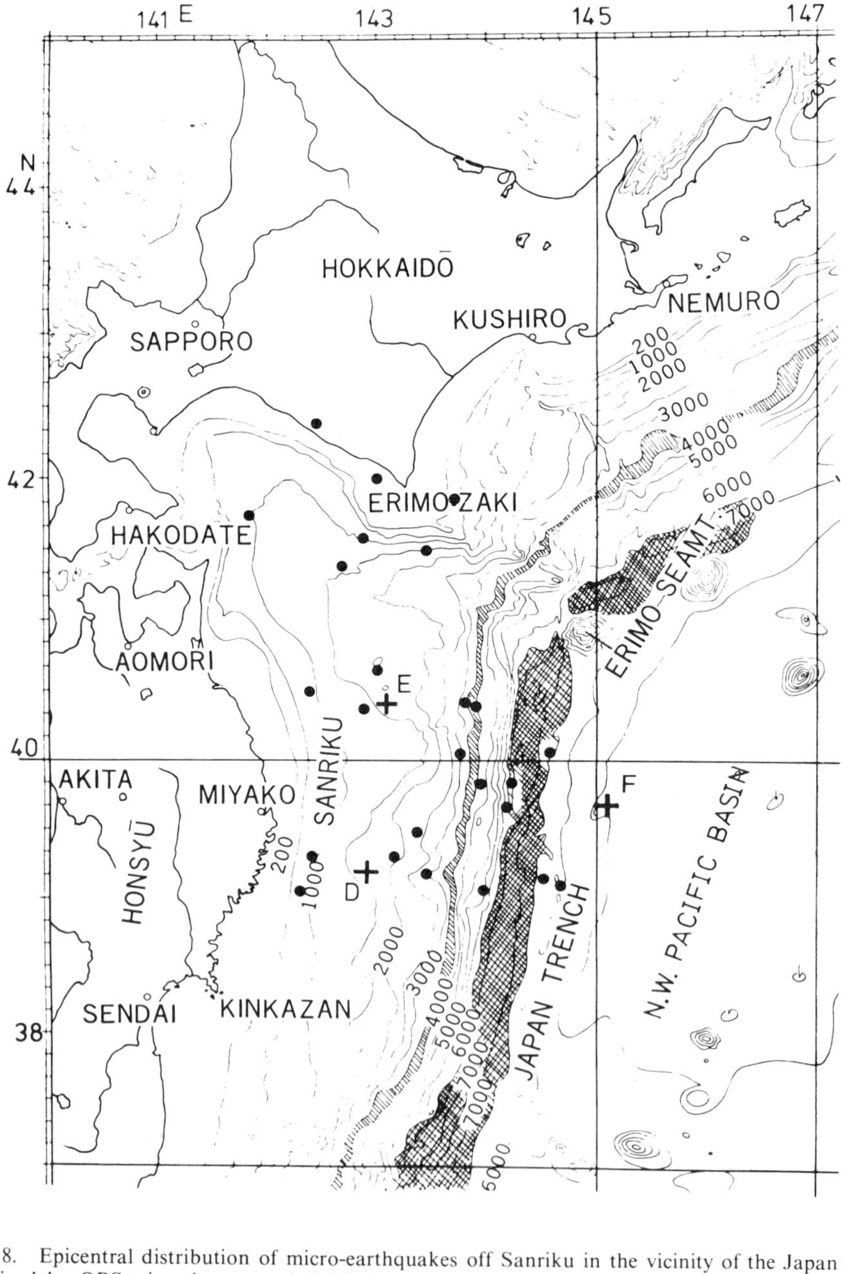


Fig. 8. Epicentral distribution of micro-earthquakes off Sanriku in the vicinity of the Japan trench determined by OBS tripartite network DEF (heavy cross marks) in August 1970. A concentration of hypocenters is seen along the marginal escarpment of the Japan trench. Cross-hatched zone: deepest part of the Japan trench; hatched zone: marginal escarpment at water depths of 3.5–4.0 km.

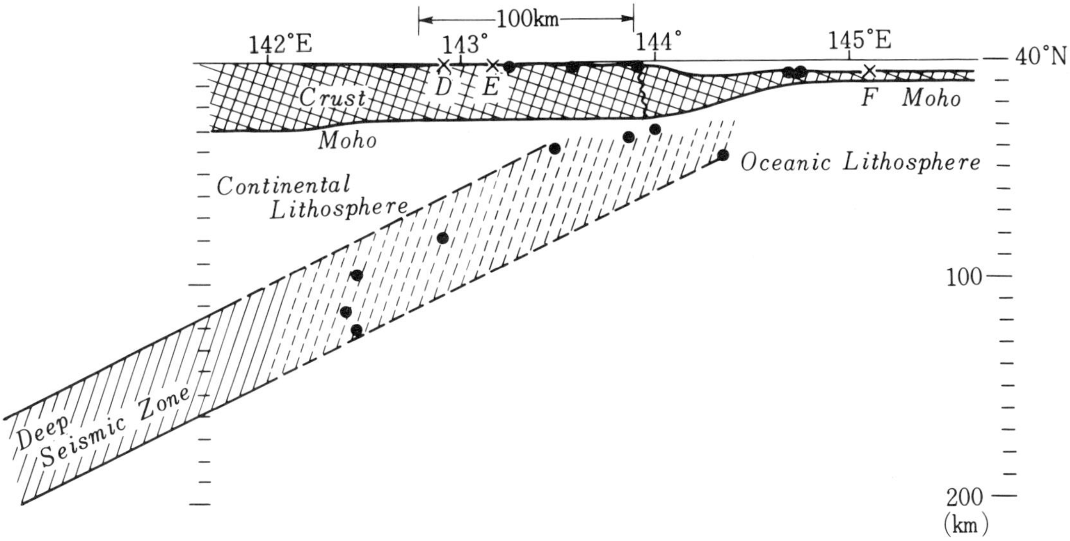


Fig. 9. Depth distribution of micro-earthquakes along latitude 39°–41°N, determined by OBS tripartite network DEF in August 1970.

tribution of seismicity across northeastern Japan along latitude 38°–41°N is shown in Figure 10. The data are taken from JMA bulletins for 1961–1970. Seismic activity is very high in the area between the volcanic island and trench. It should be noted that the seismicity can be segregated into several groups: groups A through $D_4$ denote the shallow crustal earthquakes; group G, those of the deep seismic zone; and group F, those of the upper mantle lithosphere. These seismicity groups correspond to the active elements of the present tectonic activity.

*Seismo-tectonics.* As far as the seismicity distribution is concerned, the seismo-tectonics of the island–arc system of northeastern Japan can be schematically modeled as shown in Figure 11. The volcanic island arc is generated by mantle diapirism. The volcanic island is uplifted. Due to the intrusive mass, the upper mantle lithosphere under the island arc becomes deflected downward. The uplifting motion of the island arc generates shallow earthquakes within the crust at the boundaries of the crustal blocks. The bending of the lithosphere generates upper mantle earthquakes. The thrust motion of island–arc mass against the bending lithosphere generates upper mantle earthquakes along the slip zone.

## OBSERVATIONS IN THE MARIANA REGION

In 1973, research vessel *Hakuho Maru* sailed south to the Mariana Islands and Fiji. Five OBS's were deployed in the Mariana basin, and one back-up station was placed on the island of Guam at the WWSSN station (see Figure 13).

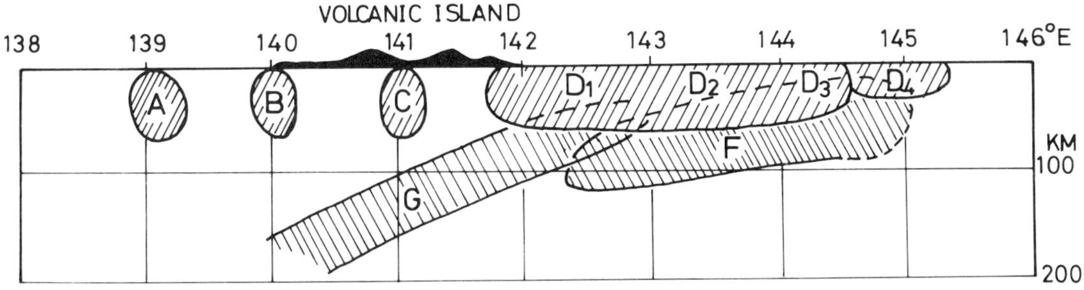


Fig. 10. A schematic cross section of the seismicity distribution in northeastern Japan along latitude 38°–41°N, from the Japan Sea, crossing the Japanese volcanic island and Japan trench, to the Pacific Ocean basin. (Data base: Seismological Bulletin, 1961–1970, Japan Meteorological Agency.)

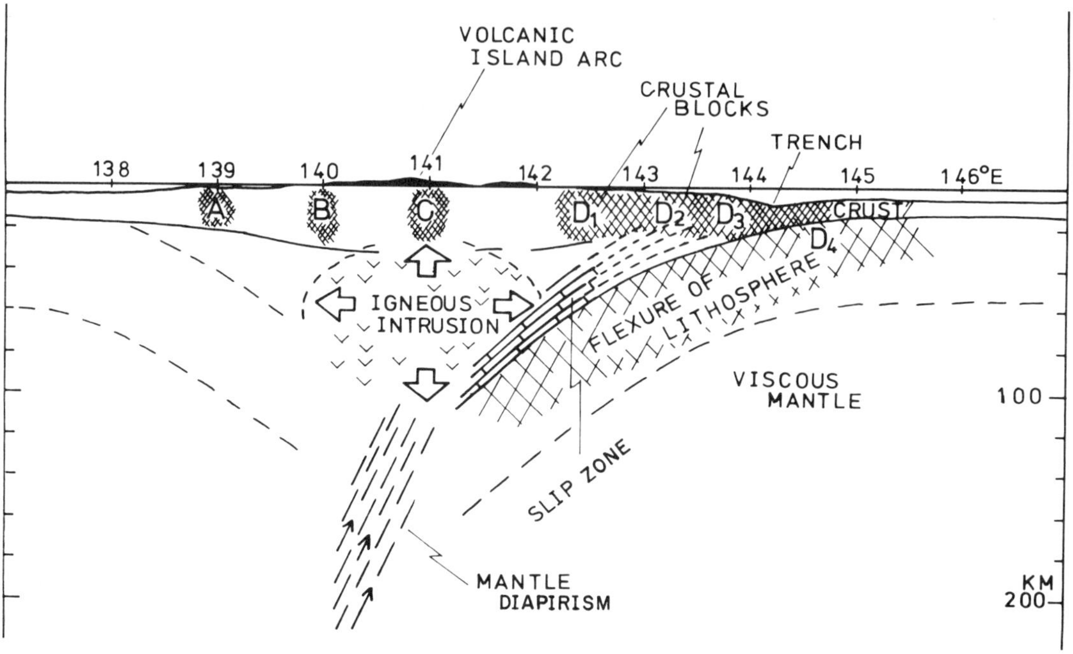


Fig. 11. A schematic of the seismo-tectonics of the island arc–trench system in northeastern Japan.

The Mariana island arc is known as a place where vertical mass mixing has taken place throughout the Cenozoic either on the western margin of the Pacific Ocean [*Karig,* 1971*a, b; Kobayashi,* 1974] or on the western flank of the ancient Darwin rise [*Mogi,* 1970]. It was expected that micro-earthquake activity would reveal the present tectonic structure of the island arc–trench system. The seismicity around the Mariana region was previously studied by *Katsumata and Sykes* [1969].

The *S–P* time distribution obtained at the OBS station in the Mariana basin, St-5, is shown in Figure 12. There is a peak in the 50–70 sec range of *S–P* time. These earthquakes occur in and along the south Mariana trench. The level of the activity is very high; average daily frequency of occurrence is about 5 earthquakes per day. The magnitude of these earthquakes is larger than about ≃ 3.5. It is evident that micro-earthquakes are violently taking place along the Mariana trench. The level of seismicity is as high as in the vicinity of the Japan trench off Sanriku. During the observation period, there was no distinct swarm activity. The high seismic activity in this region seems to be almost steady.

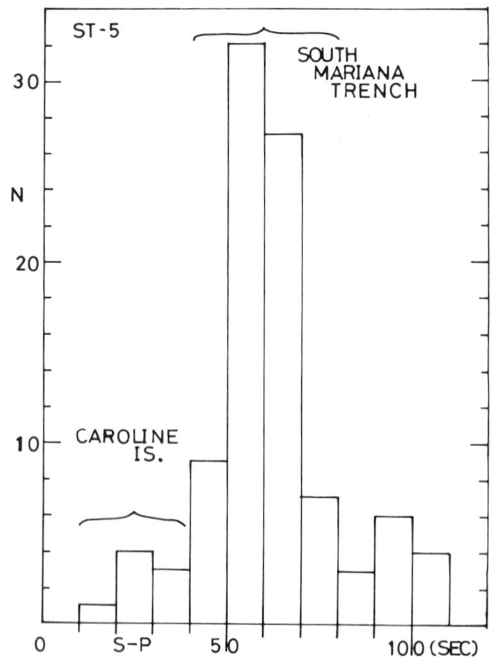


Fig. 12. *S–P* time distribution of micro-earthquakes observed at OBS station St-5 in the Mariana basin, during the period from July 18 to August 2, 1973.

It should be noted also that no micro-earthquakes were detected in the mid-oceanic basin during 16 days of observation in the $S$–$P$ time range of less than 10 seconds at station St-5. Since the sensitivity of the geophone is sufficiently high, 3.0 V/kine, and since the level of the microseisms on the ocean floor is extremely low, less than about 1 $\mu$kine ($1 \times 10^{-6}$ cm/sec) at the frequency range 1–10 Hz, it is clear that the Mariana basin is very quiet. This means that the basin is at present tectonically very stable.

It should also be pointed out that there is some seismic activity in the region of the Caroline Islands. The earthquakes (Figure 12) for which $S$–$P$ time ranges from 10–40 seconds probably correspond to those earthquakes.

The epicentral distribution for this area is shown in Figure 13. There are two distinct groups. One group lies along the mid-slope basement high or the steep slope of the inner wall of the Mariana trench in the range from 13° to 14.5° N. The other lies near the ridge terminus in the range from 143° to 145° E longitude and from 12° to 13° N latitude. There, the activity is very high. We think this is a very important phenomenon. Such high seismicity is in all probability related to the termination of the ridge.

The active seismic area is schematically represented in Figure 14. For group D the dynamics are regarded as two-dimensional motion, which is similar to that in the Japan trench off Sanriku. On the other hand, for groups B and C, the geometry is no longer two-dimensional; the deep trench, which runs parallel to the ridge, curves gradually inward along the ridge terminus.

Similar tectonic patterns are seen in the southern end of the New Hebrides island arc and the northern end of the Tonga island arc.

The tectonics in such regions are very complex. Both vertical and horizontal movements take place. However, the inward inflection of the trench axis and the high seismic activity suggests that the activity is caused by the termination of the uplifting motion of the ridge.

The dynamics of the Mariana island arc–trench system are schematically shown in Figure 15. The uplifting motion of the ridge due to the mantle diapirism and the associated motion of depression at the trench will form a couple of motion. This motion will generate a local convection cell. A strain concentration will be produced at the marginal escarpment along the trench and also around the ridge terminus. Such strain concentration generates the high micro-earthquake ac-

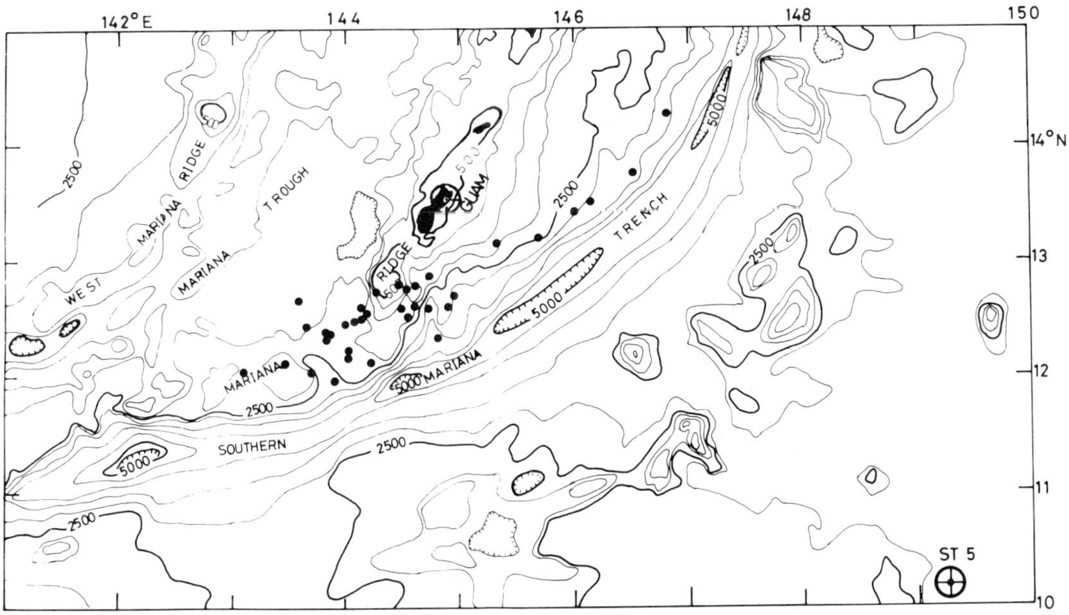


Fig. 13. Epicentral distribution of micro-earthquakes in the southern Mariana region during the period from July 18 to August 2, 1973, determined by OBS station St-5 and the land station at Guam.

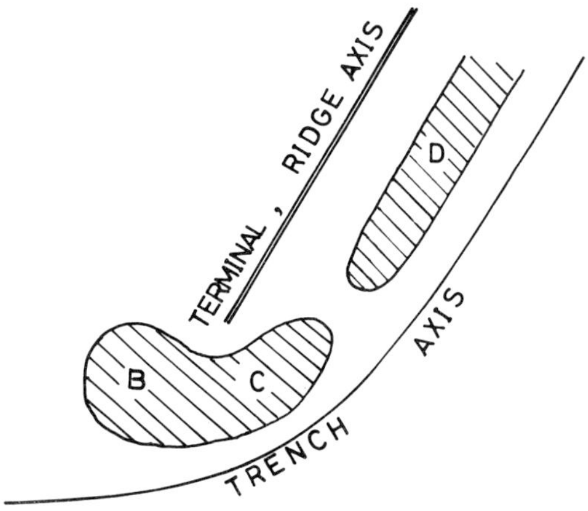


Fig. 14. A schematic of the seismically active area in the southern Mariana region. Area D corresponds to the mid-slope basement high, and areas B and C, to the marginal escarpment around the ridge terminus.

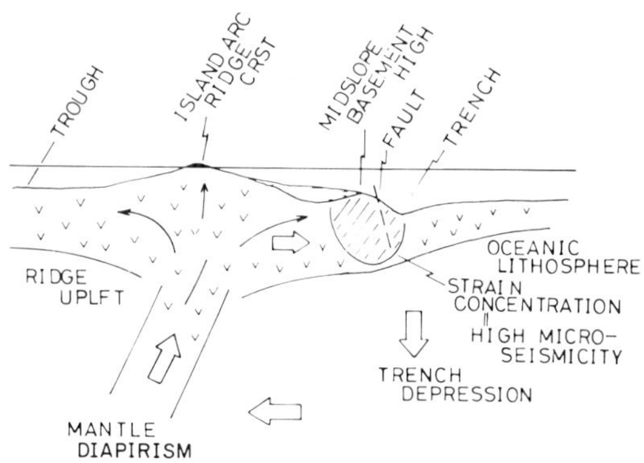


Fig. 15. A schematic of the dynamics of the island arc–trench system in the southern Mariana region.

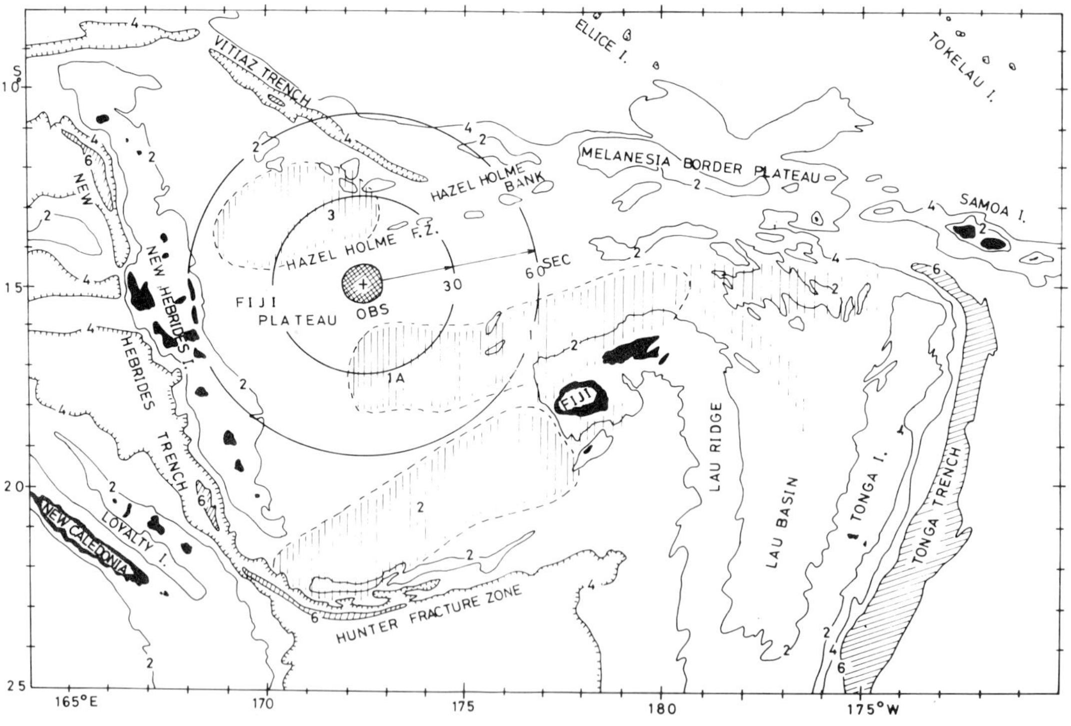


Fig. 16. The location of the OBS station on the Fiji plateau in August 1973. The cross-hatched area around the station represents intense micro-earthquake activity. Circles represent epicentral distances of *S–P* times, 6, 30 and 60 seconds, respectively. Omori coefficient is assumed to be K = 8.0 km/sec. The hatched areas enclosed by dotted lines represent the shallow earthquake groups [*Dubois et al.*, 1973].

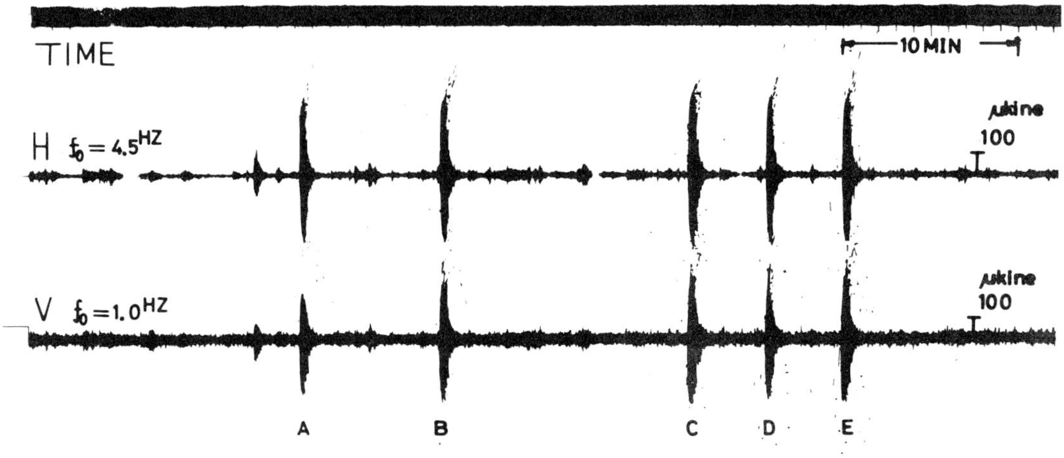


Fig. 17. An example of an earthquake swarm observed by an OBS in the central part of the Fiji plateau in August 1973. First trace: time mark; second trace: horizontal component ($f_0 = 4.5$ Hz); and third trace: vertical component ($f_0 = 1.0$ Hz). Five successive large shocks are seen within an hour.

tivity observed there.

It is concluded that, as far as the local tectonics are concerned, the vertical motion due to mantle diapirism and its associated local convection seems to be adequate to explain the seismicity of the island arc–trench system.

## OBSERVATIONS ON THE FIJI PLATEAU

Research vessel *Hakuho-Maru* sailed farther south to the Fiji plateau in 1973. An OBS was deployed at the center of the plateau; water depth was about 3.3 km (Figure 16). The station was located between the New Hebrides and Fiji islands, south of the Vitiaz trench. The observation was conducted for about 10 days.

The Fiji plateau is known as an area in which intensive tectonics have taken place during the past $10^7$ years since Early Tertiary time [*Menard,* 1967; *Rodda,* 1967]. Many interesting geophysical phenomena have been observed there [*Sclater and Menard,* 1967; *Oliver and Isacks,* 1967; *Solomon and Biehler,* 1969; *Sykes et al.,* 1969; *Karig,* 1970; *Shor et al.,* 1971; *Chase,* 1971; *Dubois,* 1971; *Dubois et al.,* 1973]. From our OBS observation, we found that unusual micro-earthquake activity was occurring in the central part of the plateau. Earthquake swarm activity occurred twice during the 10 days. An example is shown in Figure 17. Almost identical shocks occurred successively. *S–P* times for these shocks were about 2 seconds. They were occurring very near the OBS station. In addition to the shocks, smaller transient noises, probably due to earthquake swarms, took place almost continuously during the entire observation period. Average daily frequency of micro-earthquakes was about 11 earthquakes per day. This is indicative of very high activity.

Such high seismicity implies that the Cenozoic tectonic deformation is still in progress in the lithosphere beneath the plateau. Similar earthquake swarms were recently reported on the mid-Atlantic ridge [*Francis and Porter,* 1972, 1973; *Reid and Macdonald,* 1973] and on the Galapagos spreading center [*Macdonald and Mudie,* 1974; *Northrop,* 1974]. They may be similar phenomena.

*Acknowledgments.* We wish to express our heartfelt thanks to Captain M. Sato of the research vessel *Tokaidaigaku-Maru* II, Tokai University, for designing the rope-buoy system and for conducting shipboard operations. We also thank Captain I. Tamada and the crew of the research vessels *Hakuho-Maru* and *Tansei-Maru,* Ocean Research Institute of the University of Tokyo, for performing shipboard operations. We wish to thank Professors F. Kishinouye, H. Kawasumi, R. Takahashi, and T. Hagiwara of the Earthquake Research Institute for promoting the OBS project, and Professors. Y. Tomoda and K. Kobayashi of the Ocean Research Institute for coordinating the geophysical cruises of the *Hakuho-Maru.* We are very much indebted to Messrs. S. Koresawa, H. Kobayashi, S. Hasegawa, and T. Ouchi for their cooperation in the OBS study.

# 14. *Seismic Activity on the Island of Hawaii, 1970 to 1973*

ROBERT Y. KOYANAGI AND ELLIOT T. ENDO

*U.S. Geological Survey, Hawaii National Park, Hawaii 96718*

PETER L. WARD

*U.S. Geological Survey, Menlo Park, California 94025*

*Abstract.* Seismic activity on the island of Hawaii from 1970 through 1973 is concentrated primarily near Kilauea volcano, with secondary centers of activity about 40 km south of the island, and in an aftershock zone near Hilo. The seismicity near Kilauea is typically associated with eruption activity and inflation and expansion of the volcano. An absence of seismic activity at a depth of 4 to 6 km under Kilauea volcano coupled with deformation data suggests the presence of a small magma chamber.

The Hawaiian Volcano Observatory of the U.S. Geological Survey operates a modern seismograph network on the island of Hawaii. Many hundreds of thousands of events were recorded from 1970 through 1973. Over 15,000 of these events ranging in body-wave magnitude from 0.5 to 6.2 could be located with data from up to 30 stations. About 500 events were felt and had magnitudes greater than 2.5. There were 50 events with magnitudes greater than 4. This paper is a brief summary of these data.

## INSTRUMENTATION AND TIMING

Signals from 33 seismometers were being telemetered by radio to the Hawaiian Volcano Observatory (HVO) and recorded on 16-mm microfilm by late 1973 (Figure 1). Each station consists of a vertical seismometer with a natural frequency of 1 or 2 Hz operated at gains of nearly 100,000 at about 5 Hz. In addition, independent three-component stations, with photographic drum recorders, are operated on the island of Maui, in Hilo, and at the Hawaiian Volcano Observatory.

Arrival times are read from the 16-mm microfilm to within 0.1 sec. The crustal model [*Ward and Gregersen,* 1973] used in computing the locations is an average model based primarily on refraction data reported by *Hill* [1969] and *Ryall and Bennett* [1968]. Variations in travel times for waves passing under the volcano and along the rift zones are observed but are not documented in sufficient detail to be used in the hypocentral calculations. Hypocenters of shallow earthquakes near the summit of Kilauea, where most of the seismic stations are located, are typically accurate to 1 to 2 km. Hypocenters for earthquakes in the upper mantle and all earthquakes near the fringes of the island could be in error by 5 to 10 km.

## SEISMIC ACTIVITY ON THE ISLAND OF HAWAII

The locations of all events in the upper mantle and in the crust under Hawaii with magnitudes greater than 3 are shown in Figures 2 and 3. Some of the significant sequences of earthquakes during the period of time in question are described in Table 1.

Earthquakes in the mantle in Hawaii (Figure 2) are primarily concentrated in three zones that lie roughly along a NNE trending line. The highest and most continuous activity is under Kilauea volcano where the earthquakes outline a roughly cylindrical zone extending from depths of 60 km [*Eaton,* 1962] up into the crust. In recent years, most mantle earthquakes in this area occurred at depths of 30 to 40 km. A second center of activity, which was especially active in late 1971 and early 1972, is located about 40 km south of the island. The third apparent center of activity is on the northeast coast of the island of Hawaii, and this zone was essentially active during an aftershock sequence of a magnitude 6.2 earthquake that occurred at about 45-km depth on

April 26, 1973. This event caused considerable damage in the town of Hilo, 20 km to the southeast [*Unger and Ward,* 1974]. First motions from this earthquake indicate a strike-slip focal mechanism. The data are insufficient to conclusively determine the fault plane but the alternate possibilities are (1) a left-lateral movement along one nodal plane strike about N70° W, and (2) a right-lateral movement along a nodal plane strike about N30° E. Interestingly, the latter nodal plane nearly passes through the other two centers of mantle earthquake activity. Elsewhere in Hawaii, the earthquakes occur scattered under the northern and western parts of the island.

Most crustal earthquakes in Hawaii (Figure 3) occur around Kilauea volcano under the summit area, on the shoreward side of the rift zones, and along the Kaoiki fault zone between Kilauea and Mauna Loa. These events are especially frequent during periods of high volcanic activity. From 1970 through 1973 about 20 of these events measured $4 < M_b < 5$ and were strongly felt; at least

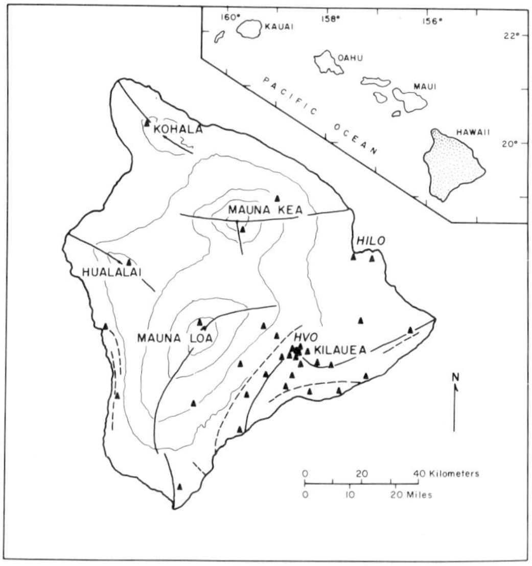


Fig. 1. Map of the island of Hawaii and the five volcanoes that make up the island. The Hawaiian Volcano Observatory (HVO) is situated at the summit of Kilauea, and the town of Hilo is located on the northeast coast of the island. Triangles indicate the locations of seismic stations, thick solid lines indicate the rift zones of the volcanoes, thin lines indicate 1000-m elevation contours, and thick dashed lines indicate the major fault systems. The inset shows the position of Hawaii, the youngest island, with respect to the progressively older islands of the Hawaiian chain to the northwest.

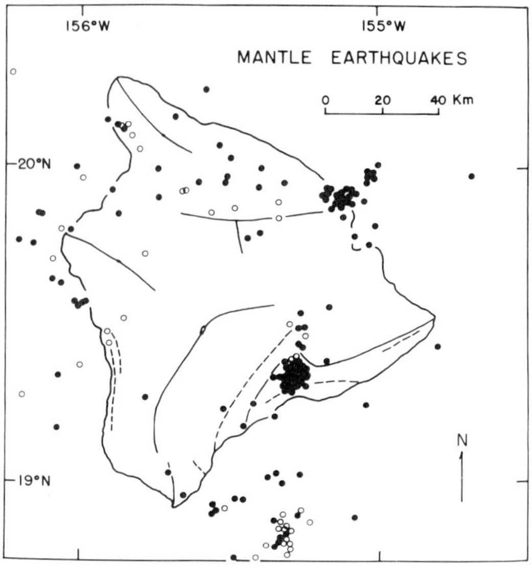


Fig. 2. Epicenters of earthquakes with magnitudes greater than 3 occurring from 1970 through 1973 in the mantle, i.e., at depths of more than 13 km. Open circles represent earthquakes of 14- to 25-km depth; solid circles represent earthquakes of about 26- to 55-km depth. Solid lines outline the island, volcanic centers, and rift zones. Dashed lines show major fault systems.

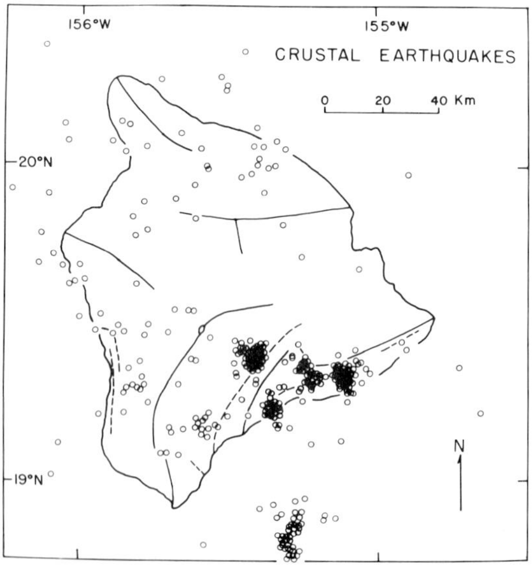


Fig. 3. Epicenters of earthquakes with magnitudes greater than 3 occurring from 1970 through 1973 in the crust, i.e., at depths of less than 13 km. Circles represent the crustal earthquakes. Solid lines outline the island, volcanic centers, and rift zones. Dashed lines show major fault systems.

six of the large events were felt across the entire island. Many of these events occur on the south flank of Kilauea's rift zones and are conspicuously absent from the north flank. When combined with deformation data, these earthquakes seem to accompany seaward movement of the south flank apparently resulting from dilation of the rift zones [*Koyanagi et al.*, 1972]. Most earthquakes located under or near Kilauea caldera repeatedly occur in small areas in response to inflation and eruptive episodes. These events typically have magnitudes less than 3 and consequently are not shown in Figure 3.

SEISMICITY RELATED TO SHALLOW VOLCANISM

The relation of shallow earthquakes to the mechanics of Kilauea's eruptions is shown more clearly in Figures 4 and 5. The time-sequence of the earthquakes occurring in the two clusters south of Halemaumau pit and on the upper east rift together with tilt data, leveling data, and results of electromagnetic studies strongly suggest that these events occurred during intrusion of magma from the Halemaumau area laterally into these two regions [*Koyanagi et al.*, in press]. Detailed first-motion studies of a shallow summit and rift swarm in May 1970 indicate stresses induced by the forceful intrusion of magma from the summit of Kilauea into the upper east rift zone [*Endo*, 1971]. Such swarms seem to clearly mark the movement of magma in the upper few kilometers of the volcano.

A cross section through the epicentral zone shows the depth relations of these summit earthquakes (Figure 5). The microearthquake swarms occur predominantly in the upper few kilometers beneath the Halemaumau area. The long-period earthquakes occur predominantly below 6-km depth; these events seem to delimit a conduit through which magma might be expected to rise

TABLE 1. Some Significant Seismic Sequences Beneath or Near the Island of Hawaii, 1970–1973

| Date | Nature of Event | Earthquake Parameters | | | Associated Volcanic Activity at Kilauea | | |
|---|---|---|---|---|---|---|---|
| | | Magnitude, $M_b$ | Depth, km | Number of Events[1] | Eruption | Harmonic Tremor | Summit Tilt |
| Feb.–May, 1970 | Earthquake swarms in the lower east rift zone of Kilauea | .1 to 3.5 | <10 | 12,000 (Δ= 1 km)[2] | No | No | No |
| May 15–18, 1970 | Earthquake swarm in the upper east rift zone of Kilauea [*Endo*, 1971] | .1 to 3.2 | <7 | 3,000 (Δ= 2 km) | No | Yes | Slight deflation |
| May–Aug., 1971 | Earthquake swarms in the summit area and upper east rift zone of Kilauea | .1 to 3.0 | <7 | 95,000 (Δ= 1 km) | Summit (Aug. 14) | Yes | Moderate deflation |
| Sep.–Oct., 1971 | Earthquake swarm in the summit area and southwest rift zone of Kilauea. Rockfall activity at the summit pit crater Halemaumau | .1 to 3.5 | <7 | 80,000 (Δ= 1 km) | Summit and southwest rift zone (Sept. 24–29) | Yes | Moderate deflation (?) |
| Sep. 1971–Mar. 1972 | Earthquake activity offshore about 70 km SSW of Kilauea | 1.0 to 4.3 | <10 to 50 | 13,000 (Δ= 40 km) | No | No | No |
| Dec. 22–30, 1971 | Earthquake swarm in the southwest flank of Kilauea | .5 to 4.6 | <10 | 6,000 (Δ= 5 km) | No | No | No |
| Aug. 6–12, 1972 | Rockfall activity at Makaopuhi Crater on the upper east rift zone of Kilauea [*Tilling et al.*, in press] | — | — | (Δ= 1 km) | East rift zone | Yes (?) | No (?) |
| Apr. 26, 1973 | Large earthquake north of Hilo and aftershocks [*Unger and Ward*, 1974] | .5 to 6.2 | mostly 40–50 | 1,000 (Δ= 20 km) | No | No | (?) |
| May 5–6, 1973 | Earthquake swarm on the upper east rift zone of Kilauea and adjacent Koae fault system [*Unger*, 1974] | .1 to 3.6 | <7 | 2,000 (Δ= 2 km) | East rift zone and Koae fault system | Yes | Moderate deflation |
| June 8–9, 1973 | Earthquake swarm on the upper east rift zone of Kilauea and adjacent Koae fault system | .1 to 3.2 | <7 | 2,000 (Δ= 2 km) | No | Yes | Slight deflation |
| Nov. 10, 1973 | Earthquake flurry on the upper east rift zone of Kilauea | .1 to 2.5 | <7 | 500 (Δ= 2 km) | East rift zone | Yes | Moderate deflation |

[1]Approximate number of microearthquakes detected on one or more stations situated nearest the seismic zone.
[2]Approximate epicentral distance in kilometers to the nearest seismometer.

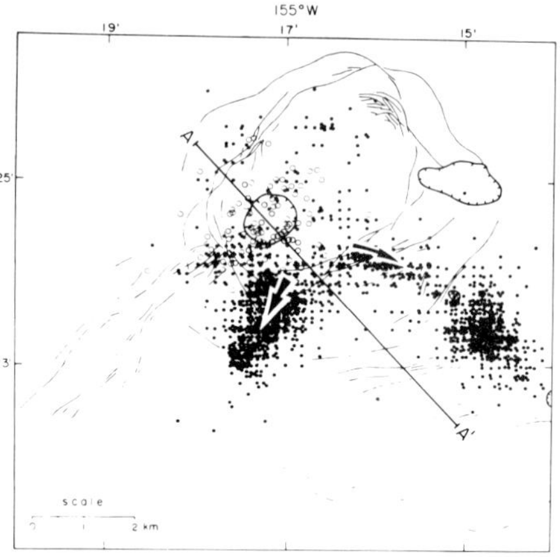


Fig. 4. Epicenter plot of individual earthquakes (0- to 13-km depth) from swarms at the summit region of Kilauea during May–August 1971, September 1971, November 1971–February 1972 (solid dots), and selected L–P earthquakes in 1972–1973 (open circles). The 1-km diameter circular feature located at about 19°24½′N, 155°17′W is the central pit crater Halemaumau, and it is encompassed by an elliptical caldera which is elongated northeast. Although not illustrated here, the east rift zone extends southeastward from the caldera, and the southwest rift zone southwestward from the caldera; the Koae fault system is an ENE-trending structure between the two rift zones located about 5 km south of the caldera. Highly concentrated regions of earthquakes adjacent to the summit caldera suggest lateral intrusive zones, and the arrows show inferred directions of intrusion [from *Koyanagi et al.*, in press]. Line A-A′ is explained in the caption to Figure 5.

from depth. It is not yet known whether the long-period character of these events is caused only by the attenuation of high-frequency waves, or by some special type of seismic source.

There is a volume 2 to 3 km in diameter where no earthquakes occur. This lack of earthquakes coupled with interpretations of the tilt and deformation data imply the presence of a magma chamber at a depth of a few kilometers under Halemaumau and Kilauea caldera. This magma chamber does not appear to be large enough to cause significant attenuation of seismic *P* or *S* waves except possibly for those earthquakes occurring immediately below the magma chamber. The implied upward movement of magma and the location of the shallow magma chamber are stylistically illustrated in Figure 5.

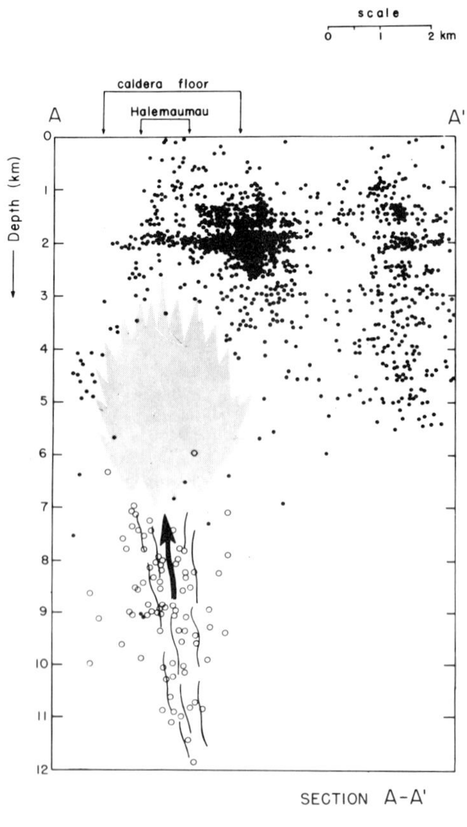


Fig. 5. A cross section along an 8-×5-km zone centered on line A-A′ in Figure 4 showing the depth distribution of the earthquakes. As in Figure 4, the earthquakes were compiled from swarms during May–August 1971, September 1971, November 1971–February 1972 (solid dots), and selected L–P earthquakes in 1972–1973 (open circles). The approximate projections of the locations of Halemaumau, and the caldera floor are indicated. The aseismic part below the caldera (shaded) is interpreted to be the principal magma reservoir zone, and the inferred magmatic feeding conduits from depth suggested by L–P quakes are indicated by sinuous lines and an arrow [from *Koyanagi et al.*, in press].

# 3. MAGNETISM

## *15. Magnetic Studies of the Pacific Area: Introduction to Session 3*

SEIYA UYEDA

*Earthquake Research Institute, University of Tokyo, Tokyo 113, Japan*

Contributions to the main theme of the International Woollard Symposium—the geophysics of the Pacific Ocean basin and its margin—from the investigation of magnetics should, perhaps, be made through three main approaches. They are: (1) surveys and interpretation of marine magnetic anomalies; (2) paleomagnetism of rocks, sediments, seamounts, and the sea floor; and (3) short-period magnetic variation studies. So far, the first approach has made major contributions to the development of the regional Pacific, as well as global, tectonic theories, and the papers presented at this symposium amply covered this approach. A paper from the second approach, i.e., a study based on paleomagnetism of sediment cores [*Hammond and Theyer*, this volume] was presented during Session 4. The third approach needs yet to be technically developed to be of any really useful extent [*Filloux*, 1967]. Because of their potentially high usefulness in assessing the thermal state of the upper mantle, it is hoped that the next similar symposium will carry reports of measurements obtained with ocean-bottom magnetometers.

Now a classic, the Scripps magnetic anomaly chart of the east Pacific [*Mason*, 1958; *Vacquier et al.*, 1961; *Mason and Raff*, 1961; *Raff and Mason*, 1961] revealed the existence of magnetic lineations and their offsets. This map was the first milestone in the dramatic series of studies which established the mobilist view of the earth. That the map was a milestone was not, in fact, realized until the Vine-Matthews [*Vine and Matthews*, 1963]-Morley [*Morley-LaRochelle*, 1964] hypothesis appeared to explain the origin of the lineations, and *Wilson's* [1965*a*] transform fault hypothesis provided elegant explanation of the lineation offsets. Both hypotheses played primary roles in providing convincing support to the more fundamental hypotheses of sea-floor spreading [*Holmes*, 1931; *Hess*, 1962; *Dietz*, 1961] and plate tectonics [*McKenzie and Parker*, 1967; *Morgan*, 1968; *Le Pichon*, 1968]. Quantitative agreement between observed and modelled anomalies for the famous Eltanin-19 profile and other profiles [*Vine and Wilson*, 1965; *Pitman and Heirtzler*, 1966] were so good that even skeptics seemed to find it hard to deny their origin.

Heirtzler and his colleagues at Lamont-Doherty Geological Observatory [*Heirtzler et al.*, 1968] mapped the magnetic anomalies of a considerable portion of the world's ocean basins and assigned ages to the lineations. The northeastern and southeastern parts of the Pacific area were mapped with Cenozoic isochrons ranging from 0 to 80 m.y. old. However, the equatorial part of the eastern Pacific remained a large gap, because of paucity of data and weakness in anomalies, and because of the unusually complicated nature of the anomaly pattern. *Herron* [1972*a*] endeavored to clarify the situation and interpreted the anomaly patterns by a total reorientation of the East Pacific rise in Cenozoic time. The first paper of the present session by *Handschumacher* [this volume] pursues this problem further, with new data obtained during the IDOE Nazca Plate Project and from other sources, by presenting a series of magnetic pattern maps for different Cenozoic periods.

In the northeastern corner of the Pacific, the magnetic lineations change their trend at the Great Magnetic Bight [*Peter et al.*, 1970]. Because of this, the lineations in the area south of the Aleutians are younger toward the Aleutian trench. *Pitman and Hayes* [1968] explained the magnetic bight by assuming the existence of a ridge-ridge-ridge triple junction which migrated northeastward and collided with the American continent. *McKenzie and Morgan* [1969], in their general theory of triple junctions, suggested that a similar collision took place between active mid-ocean ridges and trenches along the western coast of North America, and that the San Andreas fault system was produced as a result of the collision. *Atwater* [1970] examined in great detail the process of the collision of the East Pacific rise and the North American continent in the Cenozoic, providing an excellent explanation for the land geology of the western United States. The philosophy behind the above approach was that mid-ocean ridges or spreading centers migrate and may collide with trenches. This notion demanded a serious reconsideration of the driving mechanism of plate motion. If the mantle convection current is welling up at mid-ocean ridges and sinking at trenches, the migration of ridges and their collision with trenches is hard to envision [*Beloussov,* 1970]. Thus the classical mantle convection hypothesis of the driving mechanism of plate motions was severely questioned and theories putting more emphasis on the various forces inherent to plate boundaries were proposed [e.g., *Artyushkov,* 1973; *Richter,* 1973; *Forsyth and Uyeda,* in press].

Earlier, *Uyeda et al.* [1967] had discussed a set of NE–SW trending magnetic lineations farther to the west of the Emperor trough, which is the westward limit of the Cenozoic anomalies south of the Aleutians [*Hilde et al.,* this volume]. Profiles of these Japanese anomalies were not identifiable with any portion of the Cenozoic reversal models. *Larson and Chase* [1972] made a breakthrough on this problem. They demonstrated that three sets of lineations in the western Pacific, i.e., the Japanese, Hawaiian, and Phoenix lineations, are essentially coeval and represent sea-floor spreading in the Mesozoic (160–110 m.y.B.P.). The time-length of the reversal history of the earth's magnetic field was, in effect, doubled by their work. The magnetic quiet zone in the central Pacific between the Cenozoic and Mesozoic lineations represents sea-floor spreading during 110–85 m.y.B.P., when the earth's magnetic field was maintained in the normal direction. *Larson and Pitman* [1972] examined the quiet zones in the world's oceans and concluded that spreading during the Cretaceous quiet zone was about three times as fast as during other intervals. They attributed this to a pulse of spreading, which they concluded was the cause of the Cretaceous transgression [*Hayes and Pitman,* 1973].

The second paper of this session by *Larson* [this volume] deals with the further examination of the magnetic time scale, using recent data. Only an abstract of this paper is included in this volume as its full account is being published elsewhere by *Larson and Hilde* [1975]. An excellent compilation of magnetic isochrons over the world's oceans as of 1974 is now available in the form of a set of maps [*Pitman et al.,* 1974].

According to these magnetic lineation studies, the Pacific plate has migrated northward by several thousand kilometers since the middle of the Mesozoic, and all of the Kula plate and the Kula–Pacific ridge, as well as a part of the Pacific plate, have been subducted under the Japan, Kuril, and Aleutian trenches. Such large-scale northward movement of the Pacific sea floor had in fact been suggested from the paleomagnetism of seamounts [*Uyeda and Richards,* 1966; *Vacquier and Uyeda,* 1967; *Francheteau et al.,* 1970]. Analysis of the skewness of the anomaly profiles developed by *Shouten and McCamy* [*Shouten,* 1971; *Shouten and McCamy,* 1972] enabled *Larson and Chase* [1972] to interpret the paleomagnetism of the sea floor. Just as paleomagnetism of continental rocks gave evidence for continental drift, that of the ocean floor has shown 'oceanic' drift.

Unlike the west coast of North America, high-rate subduction seems to have continued at the northwestern margin of the Pacific even after the collision of the Kula–Pacific ridge with the trenches, so that key information carried by the lineations younger than the ones now being subducted has been lost. Although this makes the situation harder to see, subduction of the Kula and Pacific plates and collision of the Kula–Pacific ridge with the arcs might have an important bearing on the geologic history of the island arcs of Japan, the Kurils, and the Aleutians, and of the back-arc basins and continents behind them [*Grow and Atwater,* 1970; *Uyeda and Miyashiro,* 1974].

The third paper, by *Hilde et al.* [this volume], also deals with the northwest Pacific. These authors worked on the western Pacific magnetic lineations independently of Larson and his colleagues. With additional critical data, they have mapped the Mesozoic magnetic bight and the lineation pattern for the entire northwest Pacific. Not only have they substantiated the pattern of Mesozoic spreading, but also they have shown the evolution of spreading systems in the North Pacific from earliest time into the Cenozoic. They explain the origin of Shatsky and Hess rises as a result of the evolution of the Kula–Pacific–Farallon ridge triple junction.

In the fourth paper, *Blakely* [this volume] deals with the fidelity of the ocean crust as a magnetic tape-recorder. Observation shows that the width of the transition zone from normal to reverse magnetization increases with the age of the sea floor and Blakely attempts to explain this by a two-layer model. What changes take place in oceanic crust after its formation is one of the basic problems. It seems that only future studies can tell if the model presented here depicts the actual situation.

Another important aspect of marine magnetic anomalies in relation to tectonics is found in the marginal seas. Three main mechanisms have been proposed for the origin of marginal-sea basins behind island arcs: namely, (1) trapping of ocean basin crust by the formation of arcs; (2) extensional opening due to drifting of an arc from the continent; and (3) depression of crust due to oceanization. Although opponents to the mobilist view, such as Beloussov, maintain that formation has been according to the third hypothesis, the second hypothesis has been most popular in recent years [e.g., *Matsuda and Uyeda,* 1971; *Karig,* 1971*b,* 1974*b; Sleep and Toksöz,* 1971]. Magnetic lineations have been studied in many of the marginal basins to see if the Vine–Matthews–Morley mechanism was operative in the basin-forming process [e.g., *van der Linden,* 1969; *Barker,* 1972*a; Luyendyk et al.,* 1973].

The fifth paper by *Kobayashi and Isezaki* [this volume] discusses the possible sea-floor spreading process in the Japan Sea and the Shikoku basin, referring to magnetic data published elsewhere [*Isezaki and Uyeda,* 1973; *Isezaki,* 1975; *Tomoda et al.,* 1975]. Recently, *Watts and Weissel* [1975] also discussed the opening origin of Shikoku basin. The Philippine Sea consists of two major parts, one east of the Kyushu–Palau ridge and the other west of it. The latter, called the West Philippine basin, has a central basin fault. A few years ago, *Ben-Avraham et al.* [1972] proposed that the fault is an extinct spreading center, and subsequently, based on this proposal, *Uyeda and Ben-Avraham* [1972] proposed a trapped-ocean model for the West Philippine basin.

The sixth paper by *Louden* [this volume] compiles all the available magnetic anomaly data around the central basin fault and interprets the lineations as indicating past sea-floor spreading. This kind of study has an important bearing on the problems of origin and development of the arc–trench systems and should be encouraged in the immediate future. Needless to say, all these magnetic anomaly studies rely very heavily on the age data from the DSDP holes, so that many more deep drillings in critical areas must be strongly encouraged.

Obviously, it is not possible to cover all aspects of magnetic studies of the Pacific and its margin in these papers. Still, the author believes that a good overall view is presented from the current investigations reported in this symposium, thanks to the efforts of the participants.

# 16. *Post-Eocene Plate Tectonics of the Eastern Pacific*[1]

DAVID W. HANDSCHUMACHER

*Hawaii Institute of Geophysics, University of Hawaii, Honolulu, Hawaii 96822*

*Abstract.* A Miocene reorientation of spreading patterns in the eastern Pacific is proposed based on post-Eocene tectonic reconstructions using principles of plate theory together with observed magnetic anomaly patterns. The reconstruction shows that the Pacific–Farallon ridge was, prior to 26 m.y.B.P., intact and eastwardly migrating relative to the Americas plate. This ridge, exceeding 11,000 km in overall length, had an equatorial spreading rate of ~13.5 cm/yr. At ~26 m.y.B.P., a section of the ridge collided with the Farallon–Americas trench off the west coast of North America, stopping all spreading and subduction activity at the point of collision and initiating a direct dynamic coupling between the Pacific and the Americas plates. The coupling, presumably unstable to earlier spreading kinematics, required a major reorientation of sea-floor spreading patterns in the eastern Pacific, and included the following correlative events: a clockwise rotation of the southern portion of the ridge; initial N–S spreading along the Galapagos rift zone; break-up of the Farallon plate into the Gorda, Cocos, and Nazca plates; and dual spreading from the Pacific–Antarctic ridge and the Chile ridge. Between ~26 m.y.B.P. and ~4 m.y.B.P., there were further concurrent adjustments to the spreading regime involving the continuing overriding of the ridge off the west coast of North America. Included were numerous sectional jumps of the East Pacific rise, northward migration of the Chile triple junction, westward migration of the Galapagos triple junction, and development of the Gorda–Juan de Fuca–San Andreas–Gulf of California spreading system. It is proposed that a reorientation of sea-floor spreading kinematics on the order of that described for 26 m.y.B.P. would have produced dramatic changes on circum-Pacific plate boundaries and that present circum-Pacific trench–continent and trench–island arc–marginal basin regimes were either created or significantly altered by this early Miocene reorientation of sea-floor spreading kinematics in the eastern Pacific.

Of the numerous geological and geophysical techniques utilized to decipher the tectonic history of ocean basins in terms of plate theory (*McKenzie and Parker,* 1967; *Morgan,* 1968; *LePichon,* 1968), none is more definitive than the analysis of linear marine magnetic anomaly patterns generated by sea-floor spreading (*Dietz,* 1961; *Hess,* 1962; *Vine and Matthews,* 1963). The unique genetic relationship of these patterns to the evolution of new ocean floor along oceanic ridges, when combined with the geomagnetic reversal time scale, provides us with a chronological record of paleo-plate kinematics [*Vine,* 1966; *Pitman and Heirtzler,* 1966; *Pitman et al.,* 1968; *Heirtzler et al.,* 1968]. Extensive magnetic surveys have allowed definitive mapping in the northeastern Pacific [*Raff and Mason,* 1961; *Mason and Raff,* 1961; *Vacquier et al.,* 1961; *Raff,* 1966; *Larson et al.,* 1968; *Bassinger et al.,* 1969; *Naugler and Wageman,* 1973; *Chase et al.,* 1970; *Atwater and Menard,* 1970; *Malahoff and Handschumacher,* 1971] but in the central and southeastern Pacific, sparse data density has severely limited similar studies [*Pitman et al.,* 1968; *Herron,* 1971; *Herron,* 1972*a*; *Molnar et al.,* 1974]. *McKenzie and Morgan* [1969] and *Atwater* [1970] utilized magnetic lineations to reconstruct the Cenozoic history of sea-floor spreading in the northeastern Pacific and to present tectonic models of events related to the partial overriding of the East Pacific rise off the west coast of North America. For the central and southeastern Pacific, *Herron* [1972*a*] interpreted a progressive Cenozoic reorientation of the East Pacific rise, although low data density and technical difficulties inherent in magnetic analysis in the equatorial zone prevented direct tectonic correlation to events in the northeast Pacific.

A new regional magnetic anomaly map is presented in the first section of this paper for the central and southeastern Pacific based on improved data coverage obtained by combining

[1]Hawaii Institute of Geophysics Contribution No. 703.

recent survey results of the I.D.O.E. Nazca Plate Project with data from other sources. Direct mapping of magnetic lineations was augmented by bathymetric analysis to locate topographic features such as oceanic fracture zones and ridges, but regional application of the empirical depth-versus-age curve [*Sclater and Francheteau,* 1970; *Sclater et al.,* 1971] to date oceanic crustal age was utilized only selectively because distinct conflicts with magnetic results were often encountered. In the second section of the paper, a kinematic interpretation is made of the sea-floor spreading history in the eastern Pacific through reconstruction of the paleo-spreading regimes from the observed magnetic patterns.

The study area of new magnetic mapping is shown in Figure 1. The area contains unusually complex structural features and active plate kinematics. Figure 2 identifies the features and shows qualitative vectors of present relative plate motions across spreading centers, which are based largely on the seismic map shown in Figure 3 and magnetic patterns described later.

## DATA PROCESSING AND ANALYSIS

The track lines along which magnetic and bathymetric data were made available for analysis are shown in Figure 4. Regional mapping of magnetic patterns in an area of this size could only be accomplished by incorporating survey data from numerous sources. These sources are acknowledged in the caption to Figure 4.

All magnetic and bathymetric profiles were plotted along tracks on a 1-inch equals 1-degree-longitude scale and the International Geomagnetic Reference Field [*Fabiano and Peddie,* 1969] was subtracted from the measured magnetic total field values. A complete presentation of the magnetic profiles could not be presented here but is available in a project report [*Handschumacher et al.,* 1975]. The analysis of magnetic trends was also supplemented by published

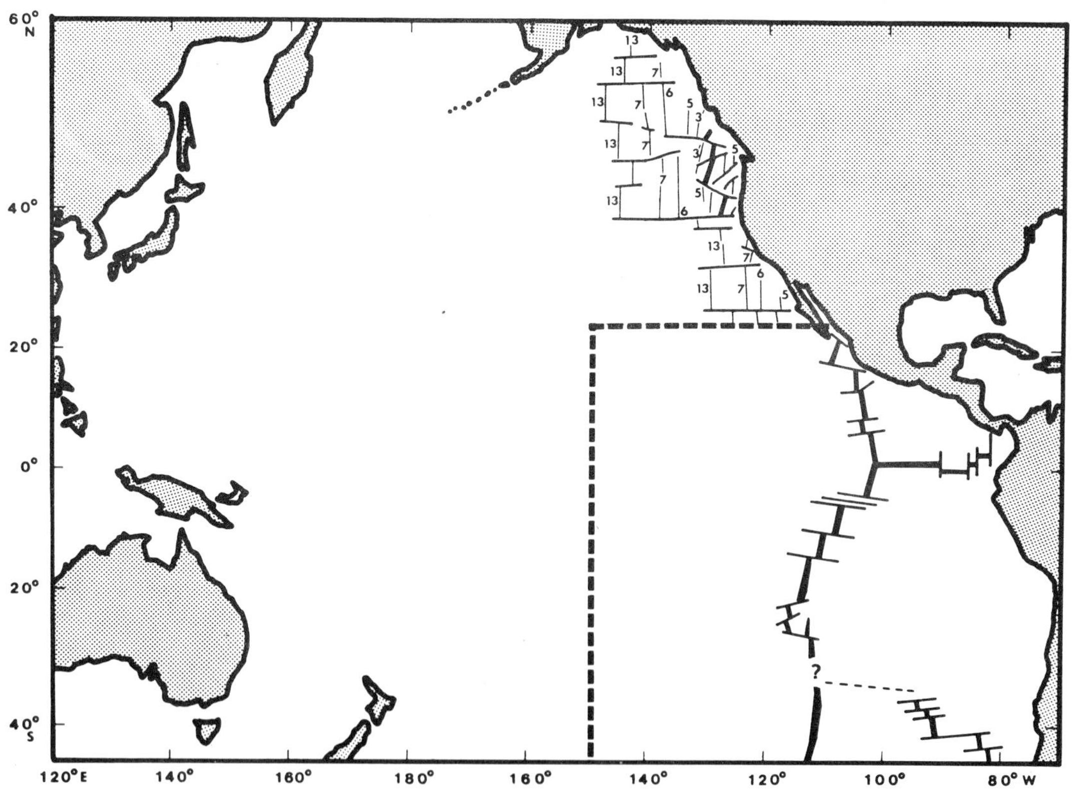


Fig. 1. Area of new magnetic and bathymetric mapping in the central and southeastern Pacific (outlined by heavy dashed line) and active spreading ridges in the eastern Pacific (heavy lines). Also shown are post-Eocene magnetic anomaly patterns in the northeastern Pacific as taken from published sources cited in the text.

magnetic profiles which are cited in the text.

*Magnetic analysis.* The time scale used for 0 to 10 m.y.B.P. is after *Talwani et al.* [1971]; for 10 to 20 m.y.B.P., after *Blakely* [1974]; and for 20 to 36 m.y.B.P., after *Heirtzler et al.* [1968], with adjustment of the older dates based on collective results of the Deep Sea Drilling Project [*Sclater et al.,* 1974]. For convenience in nomenclature and analysis, post-Eocene time has been divided into the following periods: Oligocene, 36–26 m.y.B.P. (anomaly 13 to anomaly 7); Miocene, 26–5.4 m.y.B.P. (anomaly 7 to anomaly 3′); and Plio-Pleistocene, 5.4 m.y.B.P. to present (anomaly 3′ to the axial anomaly).

Because of difficulties encountered in making magnetic identifications in the study area, several methods of establishing reference profiles were attempted. It became apparent during the analysis that computer-generated theoretical profiles [*Vine,* 1966; *Heirtzler et al.,* 1968; *Pitman et al.,* 1968; *Talwani et al.,* 1971; *McKenzie and Sclater,* 1971] were no more successful in identifying anomaly sequences than were direct comparisons with observed profiles which had been correlated

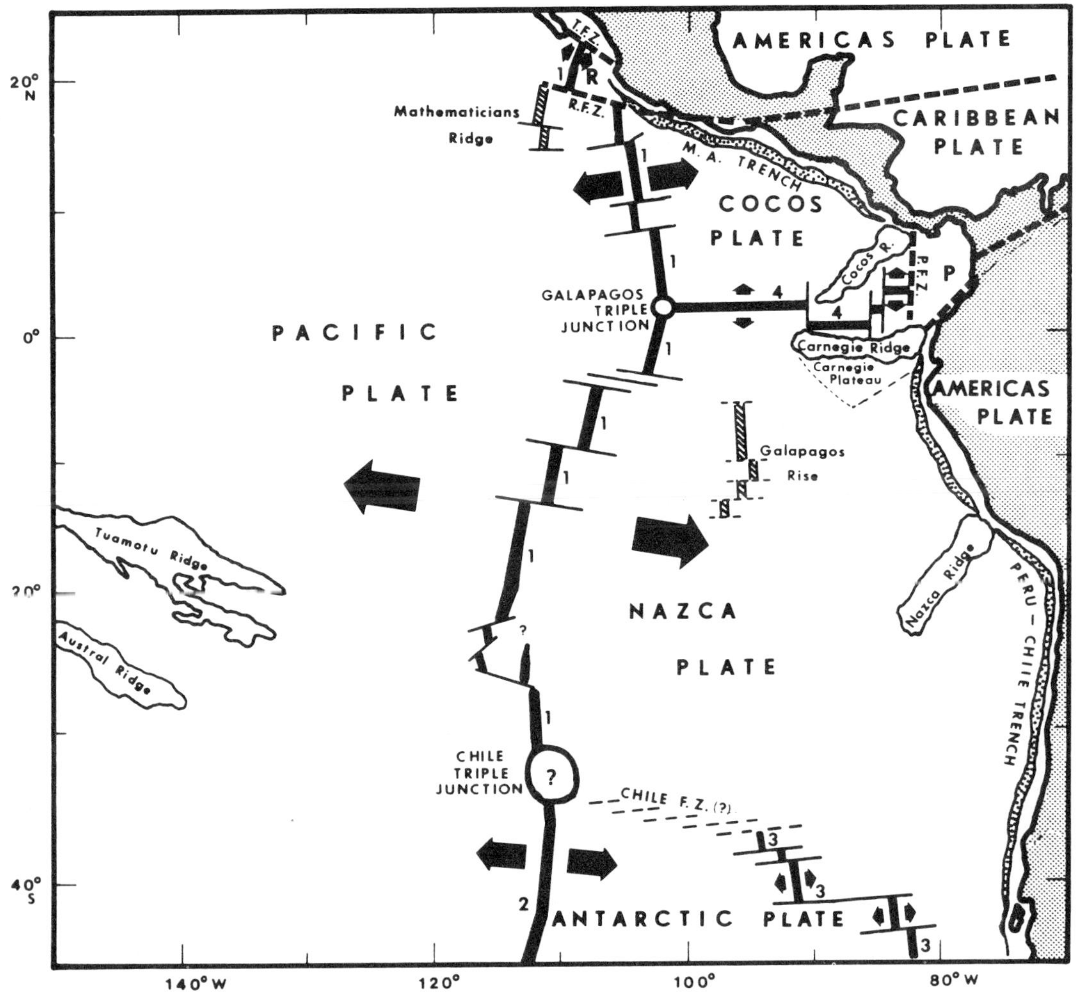


Fig. 2. Components of active plate tectonics and prominent structural features in the area of new mapping as defined by seismic (see Figure 3), magnetic and bathymetric data. Arrows indicate relative plate motions across active spreading centers. R = Rivera plate, P = Panama plate, 1 = East Pacific rise, 2 = Pacific–Antarctic ridge, 3 = Chile rise, 4 = Galapagos rift zone, T.F.Z. = Tamayo fracture zone, R.F.Z. = Rivera fracture zone, P.F.Z. = Panama fracture zone, M.A. trench = Middle America trench.

to the North Pacific Reference Profile. This is demonstrated by close agreement of the composite profile constructed from observed data in the study area (~30°S) with the North Pacific Reference Profile [*Heirtzler et al.,* 1968] and with a more detailed composite profile from a survey at 42°N [*Elvers and Morley,* 1973] as seen in Figure 5. The composite profile from 30°S was therefore used as a data reference profile for anomaly correlations.

*Diurnal variation.* No analysis of magnetic data in equatorial regions should fail to consider the effects of diurnal variation. Diurnal variation, the solar quiet day variations (*Sq*) in the earth's

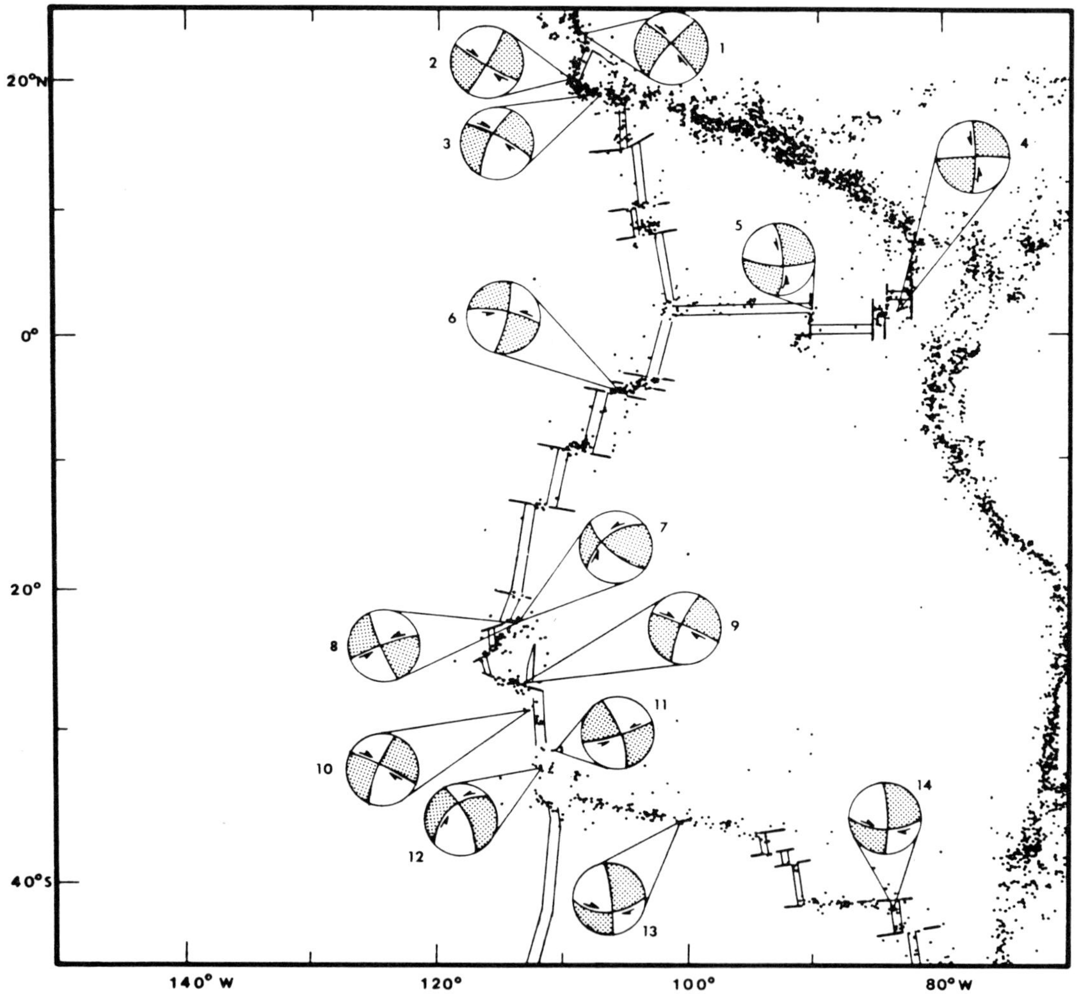


Fig. 3. Earthquake epicenters in the central and southeastern Pacific as plotted from NOAA records for 1961–1973. Also shown are selected focal mechanism determinations from published reports (1–3, *Molnar* [1973]; 4, *Molnar and Sykes* [1969]; 5–14, *Forsyth* [1972]), with relative slip vectors shown on preferred fault planes.

Fig. 4. (Opposite) Tracklines along which magnetic and bathymetric data used in this study were collected. Data sources are: A. Hawaii Institute of Geophysics, University of Hawaii; B. Pacific Oceanographic Laboratories, National Oceanographic and Atmospheric Administration; C. Project Magnet (airborne), U.S. Naval Oceanographic Office; D. Lamont-Doherty Geological Observatory, Columbia University; E. Miscellaneous as shown.

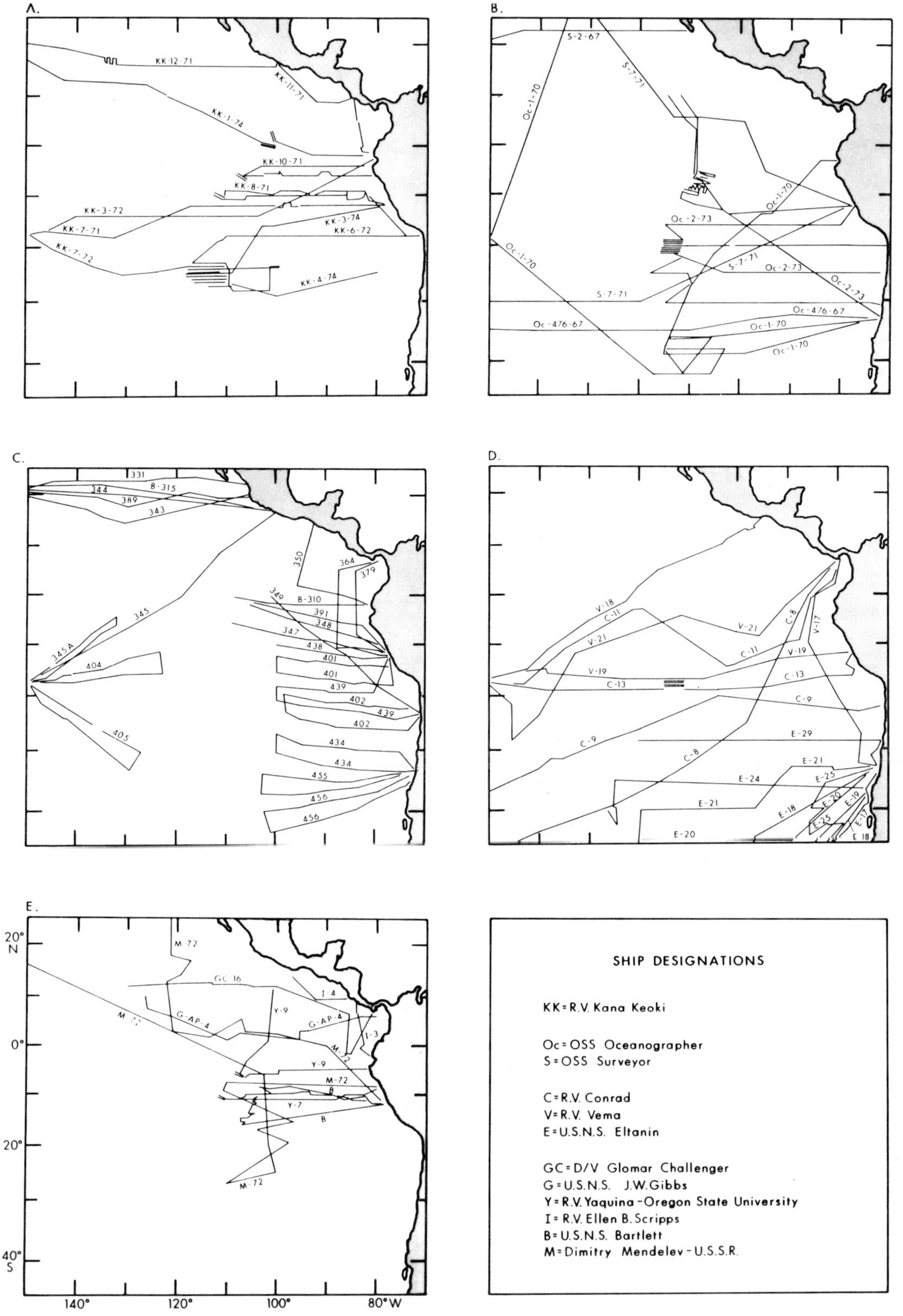
A.
KK-12-71
KK-11-71
KK-1-74
KK-10-71
KK-8-71
KK-3-72
KK-7-71
KK-7-72
KK-3-74
KK-6-72
KK-4-74
B.
S-2-67
S-7-71
Oc-1-70
Oc-2-73
Oc-476-67
C.
331
344
B-315
389
343
350
364
379
345
345A
404
405
349
B-310
391
348
347
438
401
439
402
434
455
456
D.
V-18
C-11
V-21
V-19
C-13
C-9
C-8
V-17
E-29
E-21
E-24
E-25
E-18
E-20
E-19
E-17
E.
M-72
GC 16
I-4
G-AP-4
Y-9
I-3
Y-7
B
20° N
0°
20°
40° S
140°
120°
100°
80°W
SHIP DESIGNATIONS
KK = R.V. Kana Keoki
Oc = OSS Oceanographer
S = OSS Surveyor
C = R.V. Conrad
V = R.V. Vema
E = U.S.N.S. Eltanin
GC = D/V Glomar Challenger
G = U.S.N.S. J.W. Gibbs
Y = R.V. Yaquina - Oregon State University
I = R.V. Ellen B. Scripps
B = U.S.N.S. Bartlett
M = Dimitry Mendelev - U.S.S.R.

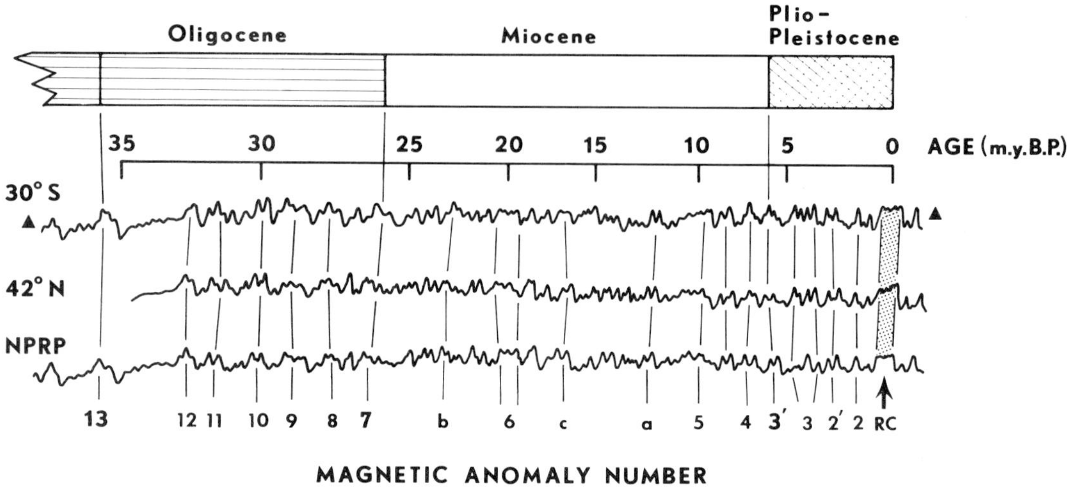


Fig. 5. Comparison of the data reference profile constructed from observed data at ~30°S with the standard North Pacific Reference Profile (NPRP) and a composite profile from 42°N [*Elvers and Morley*, 1973]. The data reference profile from ~30°S is designated in this figure and all subsequent figures by solid triangles. Also shown is the geomagnetic reversal time scale for post-Eocene magnetic anomalies.

geomagnetic field, is magnified near the magnetic dip equator. Figure 6A shows the location of this zone and the horizontal component (*Mx*) variations with latitude. The figure shows that peak variations in excess of 200 gammas occur in the equatorial zone. For a comprehensive discussion of diurnal variation, the reader is referred to *Onwumenchilli* [1967] and *Smith* [1967].

The combined effect of magnified diurnal variations and of diminished amplitudes of N–S trending spreading anomalies in the equatorial zone made the quantitative definition of the diurnal effect on the observed data imperative. To accomplish this, five days of underway shipboard data from ~9°S were compared with simultaneously collected data from the Huancayo Observatory (12°S, 86°W) and with a computed theoretical diurnal curve from 10°S. The excellent agreement of all three profiles when each is plotted against local time can be seen in Figure 6B. Shorter wavelength anomalies superimposed on the diurnal curve are apparent on both the observatory and shipboard data. On the former, they represent daily irregularities in the theoretically computed diurnal curve; on the latter, they may represent either similar irregularities or geological anomalies. Since at present it is impossible to differentiate the source of these anomalies or to correlate diurnal variation over large distances and different environments, identification of N–S trending sea-floor spreading anomalies on shipboard data collected in the equatorial zone is tenuous.

*Bathymetric analysis.* Available bathymetric data for the track lines in Figure 4 were used to locate and define topographic structures, particularly oceanic fracture zones formed by transform faulting [*Wilson*, 1965*a*, *b*]. The bathymetric profiles are not presented here because many of the data have been the subject of a recent bathymetric study [*Mammerickx et al.*, 1975] and are available in a project report [*Handschumacher et al.*, 1975]. Bathymetric contour charts for the central eastern Pacific prepared by *Chase* [1968] were also incorporated in the analysis.

In searching for another interpretive method to date the sea floor where magnetic dating has proved unsuccessful, the use of the empirical relation between crustal age and observed crustal depth [*Menard*, 1969; *Sclater and Francheteau*, 1970; *Sclater et al.*, 1971] was considered. The latter workers suggested that the age of crust formed less than 40 m.y. ago could be determined with better than ±2 m.y. accuracy by comparing present crustal depths below active ridge crest elevation with an established empirical curve. Unfortunately, test correlations of crustal ages determined by the depth-versus-age method with ages determined by well-defined magnetics in the southeastern Pacific have revealed significant errors of up to 20 m.y. in the bathymetrically determined ages [*Handschumacher et al.,* 1975]. It was decided that this method is not generally applicable as a regional indicator of precise age in the study area, but where magnetic data were unavailable or unreliable, it was selectively utilized to interpret Plio-Pleistocene events.

## REGIONAL PATTERNS

Post-Eocene sea-floor spreading patterns as defined by the magnetic and bathymetric data are presented in a composite map in Figure 7. The dissimilarities between the Plio-Pleistocene trends and those generated during the Oligocene are evident, requiring a major Miocene reorientation of spreading dynamics.

*Oligocene patterns (36–26 m.y.B.P., anomalies 13–7).* Oligocene anomalies are relatively easy to identify in geomagnetic reversal profiles because of their unique shape. Two coherent sets of Oligocene anomaly patterns were identified in the study area and are shown in Figure 8. All

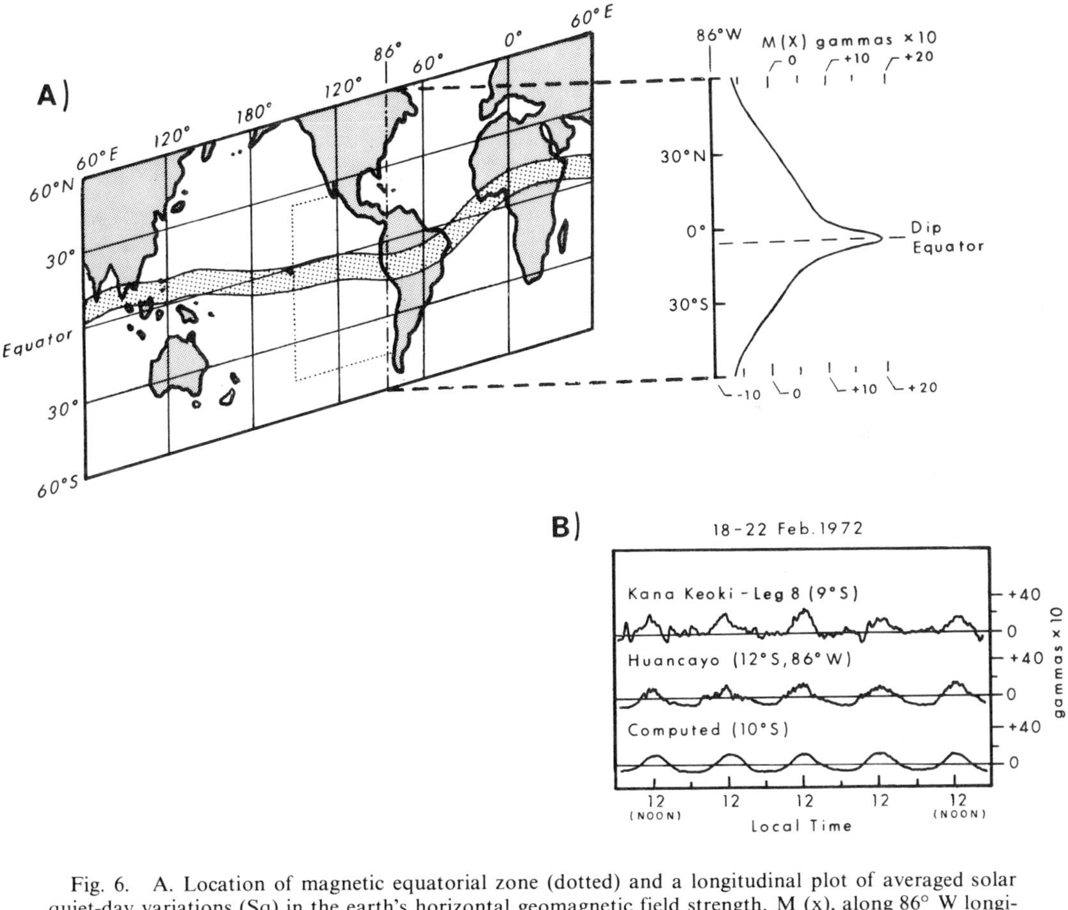


Fig. 6. A. Location of magnetic equatorial zone (dotted) and a longitudinal plot of averaged solar quiet-day variations (Sq) in the earth's horizontal geomagnetic field strength, M (x), along 86° W longitude [adapted from *Smith,* 1967]. Note the +200-gamma daily variation at the magnetic dip equator and the rapid decrease of the diurnal variation away from the dip equator. B. Five days of underway shipboard magnetic data collected aboard the research vessel *Kana Keoki* at ~9°S is compared with stationary data collected simultaneously at the Huancayo Observatory, 12°S, 86°W and a computed diurnal curve for average diurnal variation at 10°S. All are plotted against local time and clearly show 'highs' of ~200 gammas at local noon. The short-wavelength, irregular anomalies observed on the shipboard and station data prevent simple subtraction of the computed curve to correct for the diurnal effect.

fracture zones in both patterns that could be accurately mapped exist perpendicular to the magnetic trends. The uniformity of the Oligocene magnetic trends and the common perpendicularity of the fracture zones indicate that both sets evolved from steady-state spreading ridges. It will be demonstrated later that both sets, in fact, evolved from a common Oligocene spreading ridge. Another important implication of these patterns is that neither has been significantly contorted by subsequent tectonic events.

The selected magnetic profiles which defined the magnetic patterns in Figure 8 are shown in Figure 9. They are correlated to the Oligocene section of the data reference profile and in most cases show an excellent progressive recognition of anomalies from profile to profile despite the fact that correlations near the equatorial zone are tenuous.

The Pacific set begins with the lineations north of the Clarion fracture zone which were mapped from published identifications just south of the Molokai fracture zone [*Atwater and Menard,* 1970] and profile A. Magnetic lineations to the south of the Clarion fracture zone mapped on profiles B, C, and D indicate only slight dextral offset across the Clarion. Because no convincing anomaly identifications could be made between

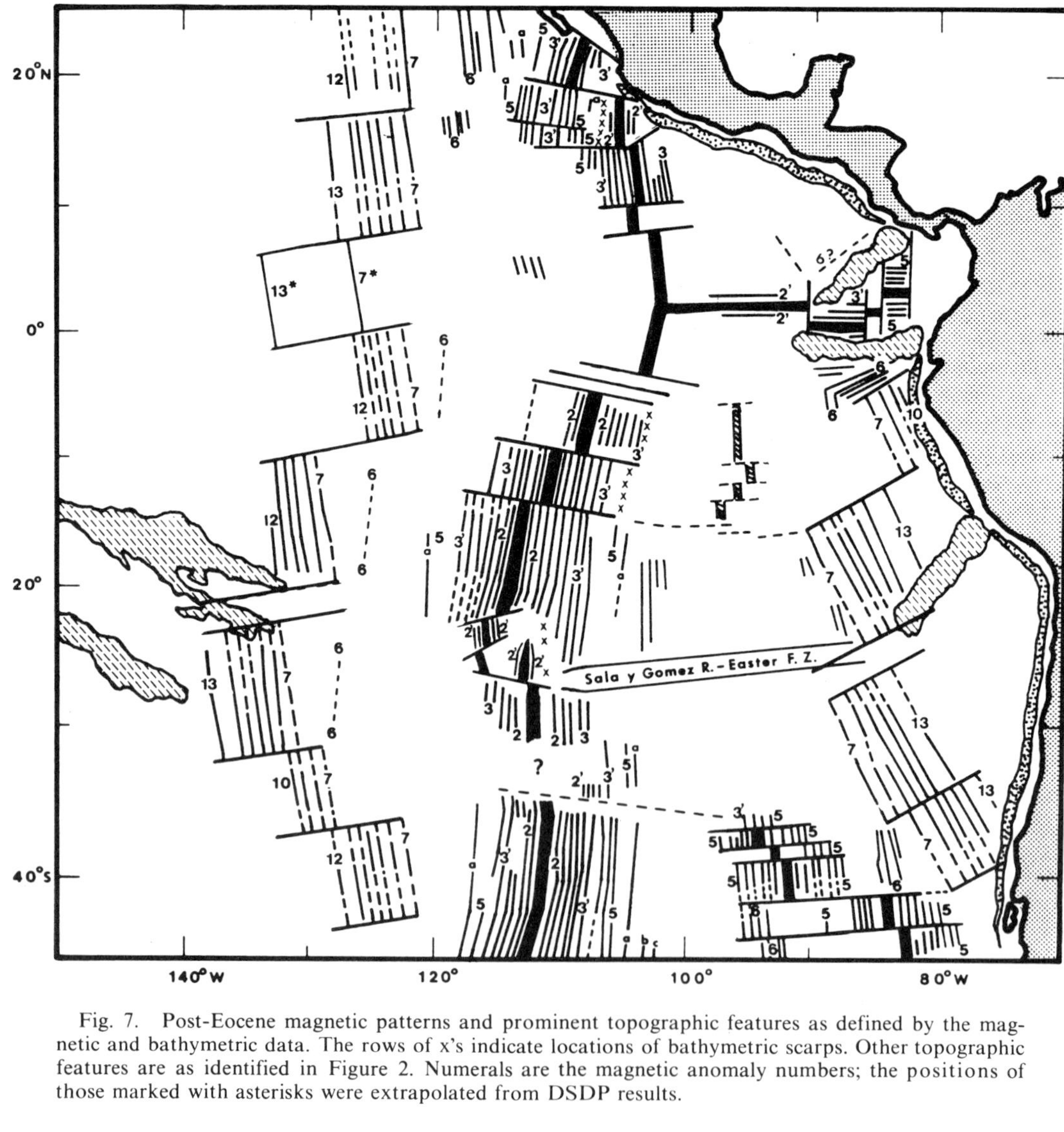


Fig. 7. Post-Eocene magnetic patterns and prominent topographic features as defined by the magnetic and bathymetric data. The rows of x's indicate locations of bathymetric scarps. Other topographic features are as identified in Figure 2. Numerals are the magnetic anomaly numbers; the positions of those marked with asterisks were extrapolated from DSDP results.

the Clipperton and Galapagos fracture zones, the positions of anomalies 7 through 13 were based on the DSDP crustal dates for sites 77 (~38m.y.B.P.) and 79 (~20 m.y.B.P.) [*DSDP Scientific Staff*, 1970]. If the tentative identifications on profile E are correct, their positions define a right-lateral offset on the Clipperton fracture zone and a left-lateral offset on the Galapagos fracture zone. A large (883-km) dextral offset was determined for the Marquesas fracture zone based on the identifications for profile E to the north and profiles F through K to the south. The interpretation of this large offset is also based on its rugged topographic signature and regional shallowing to the south across the Marquesas fracture zone [*Menard*, 1960; *Mammerickx et al.*, 1975]. *Mammerickx et al.* [1975] mapped a major fracture zone, the Tuamotu fracture zone, about 400 km south of the Marquesas fracture zone based on their magnetic and bathymetric interpretation of the data. This fracture zone is eliminated here because of the excellent northward continuation of anomalies 7–11 through profile F (see Figure 9) and the fact that bathymetric expression characteristic of major fracture zones is lacking. The position of the Austral fracture zone is poorly defined by the data but is constrained by magnetic data to the zone indicated. Difficulty in mapping the exact position of this fracture zone arises from the low data density and the proximity of topographic features associated with the Tuamotu ridge. However, the dextral offset on the Austral fracture zone is defined by good magnetic identifications on profiles L through O. The 'A' fracture zone is primarily mapped by the constraining magnetic patterns it separates. Although anomaly correlations on profiles P and Q are considered good, the anomalously rough topography along those tracks adds a certain degree of speculation to these correlations. The Agassiz fracture zone is visible in the bathymetry [*Mammerickx et al.*, 1975] and is also located by the northward termination of magnetic trends delineated on profiles R and S. Correlation on these profiles is excellent and, despite the lack of usable correlations south of profile S, wavelength characteristics on profile E1-24 [*Herron*, 1971] support southward extension of the Oligocene trends to the 'B' fracture zone.

The Nazca set begins with the Grijalva fracture zone whose unusual magnetic [*Raff*, 1968; *Herron and Heirtzler*, 1967] and bathymetric [*Mammerickx et al.*, 1975] signatures suggest a unique evolutionary history (discussed later in this paper). South of this fracture zone good magnetic identifications of anomalies 7 through 10 are made on profiles a and b. The lineations terminate at the Peru–Chile trench where the older anomalies of the Oligocene sequence have been lost due to subduction. A large (889-km) right-lateral offset in the magnetic pattern was determined for the Mendana fracture zone from identifications on profiles a and b to the north and profiles c through g to the south. This fracture zone is also well mapped by bathymetry [*Mammerickx et al.*, 1975], and the large dextral offset is supported both by regional shoaling to the north and by DSDP dates for sites 320 (~30 m.y.) and 321 (~40 m.y.) [*DSDP Scientific Staff*, 1974*b*]. Complex morphology at the junction of the Nazca ridge and Sala y Gomez ridge–Easter fracture zone hinders accurate plotting of the Nazca fracture zone, but magnetic patterns confine its location and offset to the zone indicated. Except for the difficulties encountered near the Nazca fracture zone, anomaly identifications between the Nazca and Challenger fracture zones are excellent (profiles h through l), as are those south of the Challenger fracture zone (profiles m through q). The left-lateral offset on the Challenger fracture zone is defined by these identifications. The Mocha fracture zone was positioned from bathymetry and is suggested to have a sinistral offset in excess of 500 km, although the amount of offset could not be directly determined from magnetic data because the entire Oligocene pattern to the south has been subducted.

*Plio-Pleistocene patterns (5.4 m.y.B.P. to present, anomaly 3' to axis).* Plio-Pleistocene magnetic anomalies are also characteristic and easy to identify, except between 10°N and 5°S on the East Pacific rise. The lack of convincing anomaly correlations in this zone is attributed to its complex tectonics, low data density, and equatorial location. Fortunately, the accentuated topographic relief of youthful spreading components together with the seismic data (Figure 3; also see *Isacks et al.* [1968]) allows general interpretation of Plio-Pleistocene tectonics in the region.

As seen in Figure 10, the Plio-Pleistocene trend configurations are much more complex than those of the Oligocene. The northernmost trends between the Tamayo and Rivera fracture zones were taken from *Larson et al.* [1968] and appear to parallel those found in the Gulf of California [*Moore and Buffington*, 1968]. These trends abruptly terminate at the Rivera fracture zone, the active southern boundary of the Rivera plate

[*Atwater*, 1970; *Molnar*, 1973]. Two anomaly patterns are mapped south of that fracture zone, the western pattern is symmetric about the axis of the Mathematicians ridge and the eastern pattern corresponds to the East Pacific rise between the Rivera and Orozco fracture zones [*Sclater et al.*, 1971]. New identification of the symmetric anomalies about Mathematicians ridge presented in the next section supports *Sclater et al.*'s [1971] contention that this ridge is a fossil segment of the East Pacific rise that was active until about 4 m.y.B.P. At that time, spreading activity jumped 4 degrees eastward between the Rivera and Orozco fracture zones, accounting for the failure to identify consecutive anomalies beyond anomaly 2′ (3 m.y.B.P.) about the East Pacific rise. The lineations shown south of the Orozco fracture zone were mapped by *Larson and Chase* [1970] and new data have extended them to the Siquieros fracture zone. Bathymetric trends were used to interpret ridge positions for the East Pacific rise from the Siquieros fracture zone to 5°S because no convincing magnetic sequences can be identified.

The Galapagos ridge–ridge–ridge triple junction [*McKenzie and Morgan*, 1969] was mapped at 2°11′N, 102°10′W by *Hey et al.* [1972]. Magnetic anomalies generated by N–S spreading along the Galapagos rift zone are easily correlated to the geomagnetic reversal time scale because

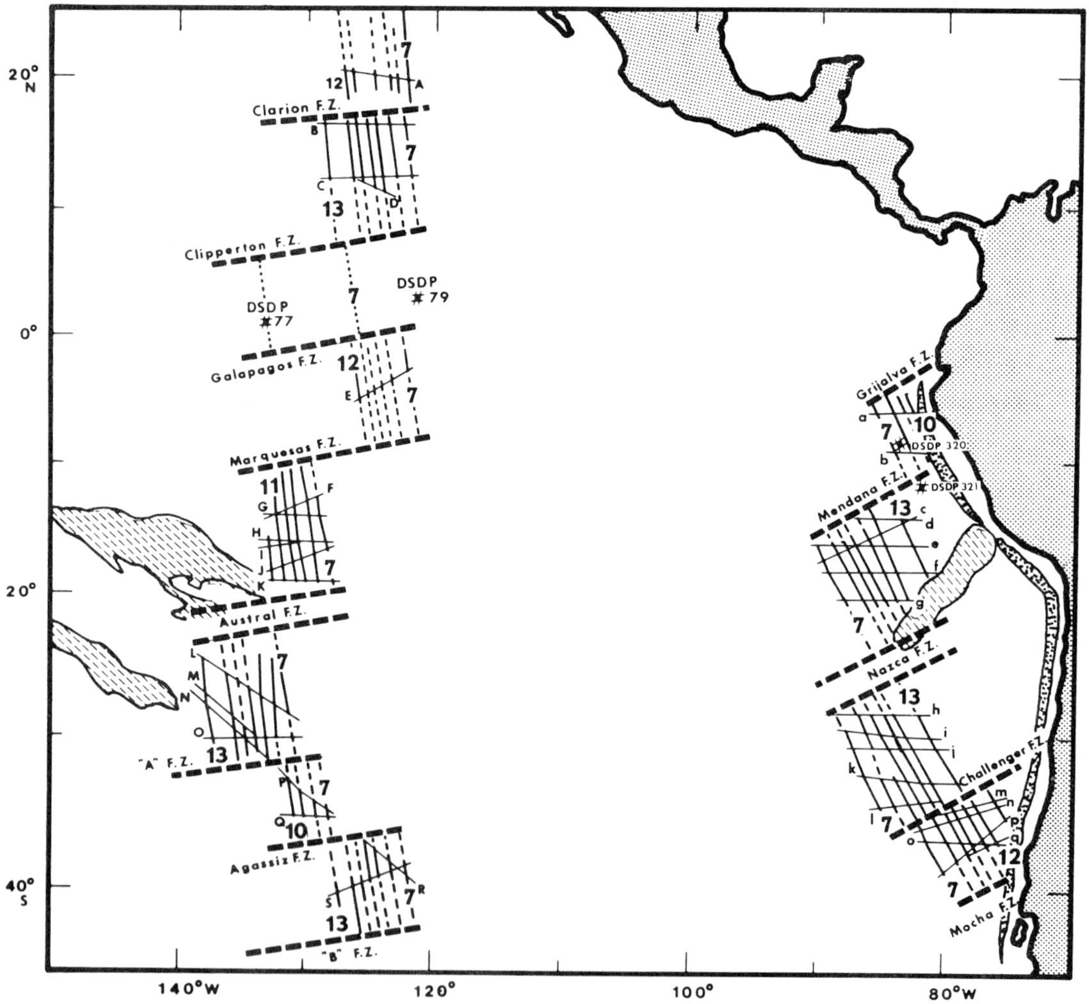


Fig. 8. Oligocene magnetic patterns and track lines for profiles shown in Figure 9; Pacific set is on the left, Nazca set on the right. DSDP site locations and numbers are also shown.

the zone's E–W trending anomalies have amplitudes of 500 to 1000 gammas [*Herron and Heirtzler*, 1967; *Raff*, 1968]. Lineation patterns shown in Figure 10 for the Galapagos rift zone are taken from published sources [*Herron and Heirtzler*, 1967; *Raff*, 1968; *Grim*, 1970; *Sclater and Klitgord*, 1973].

A consistent trend is determined for the East Pacific rise between the Galapagos triple junction and the northern boundary of an anomalous area of spreading at 24°S. Seismic and bathymetric data were used to interpret the 427-km dextral offset on a series of closely spaced dextral transforms around 5°S [*Herron*, 1972*a*; *Mammerickx*

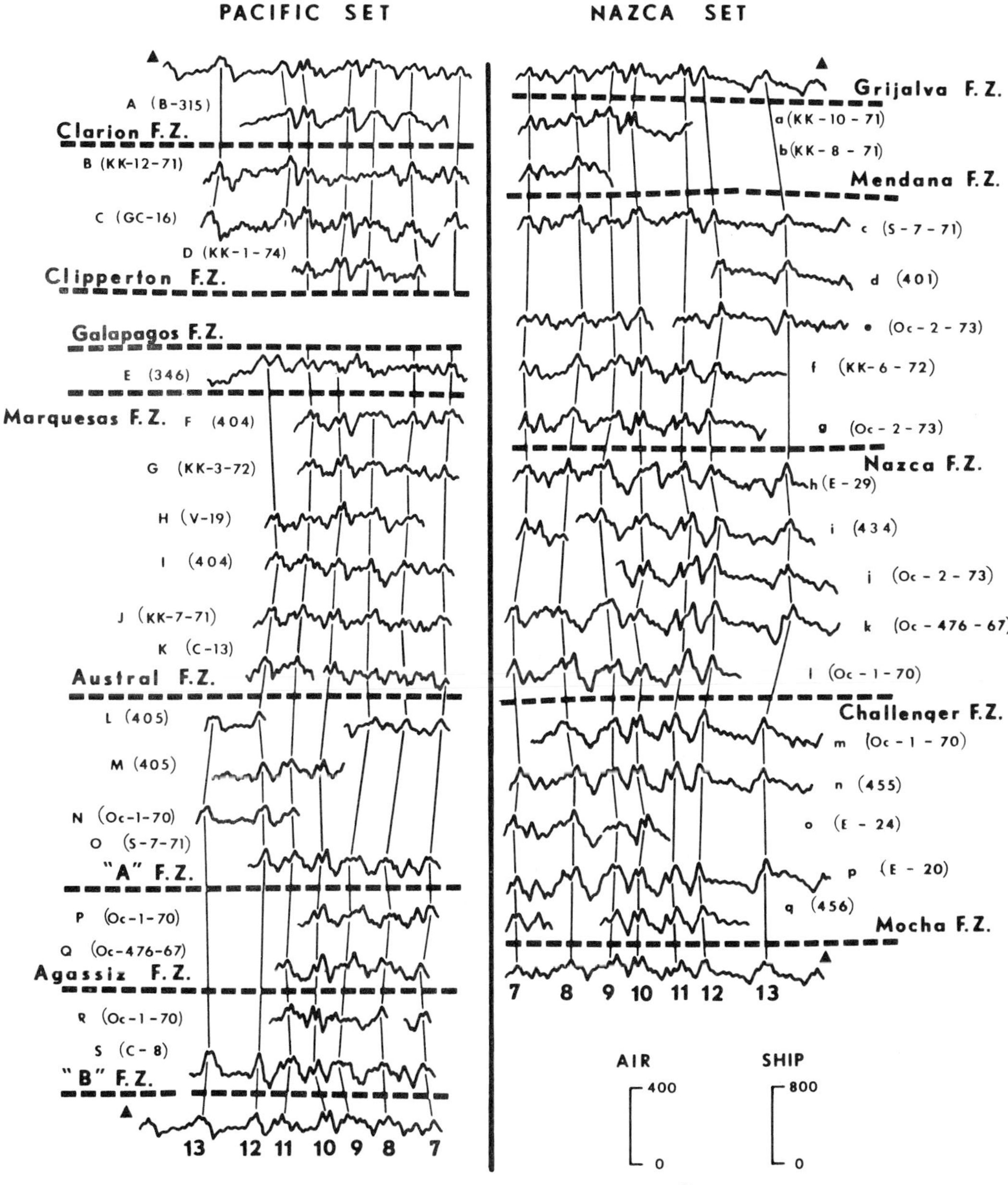


Fig. 9. Magnetic profiles for the track lines in Figure 8 correlated to the Oligocene section of the data reference profile (solid triangles). Specific track line identifications are included in parentheses for reference to Figure 4.

*et al.,* 1975]. South of these transforms, tentative identifications on profile A (Figure 11) show consecutive anomalies to anomaly 3 or 3′ on the eastern ridge flank, and to anomaly 2 (not shown) on the western flank. A 195-km dextral offset is interpreted for the fracture zone at 9°S based on these identifications and the mapping of symmetric anomalies to anomaly 3′ about the ridge crest to the south of this fracture zone (profiles B through E). The fracture zone at about 14°S corresponds roughly to fracture zone VII of *Mammerickx et al.* [1975] and its 184-km right-lateral offset is well defined by the magnetic data. However, a N85°W trend is interpreted here for this fracture zone rather than the N70°W trend described in that study. Between this fracture zone and 24°S, more than thirty survey lines across the ridge province clearly define the magnetic lineations; the selected profiles shown in Figure 11 (F through M) are representative of this data coverage.

*Herron's* [1972*b*] mapping of the area outlined by the seismic ring near 24°S (see Figure 3) contained two active ridge segments subparallel to each other. New data collected during the Nazca Plate project have allowed a much improved description of this anomalous area (to be presented in a separate publication, Handschumacher,

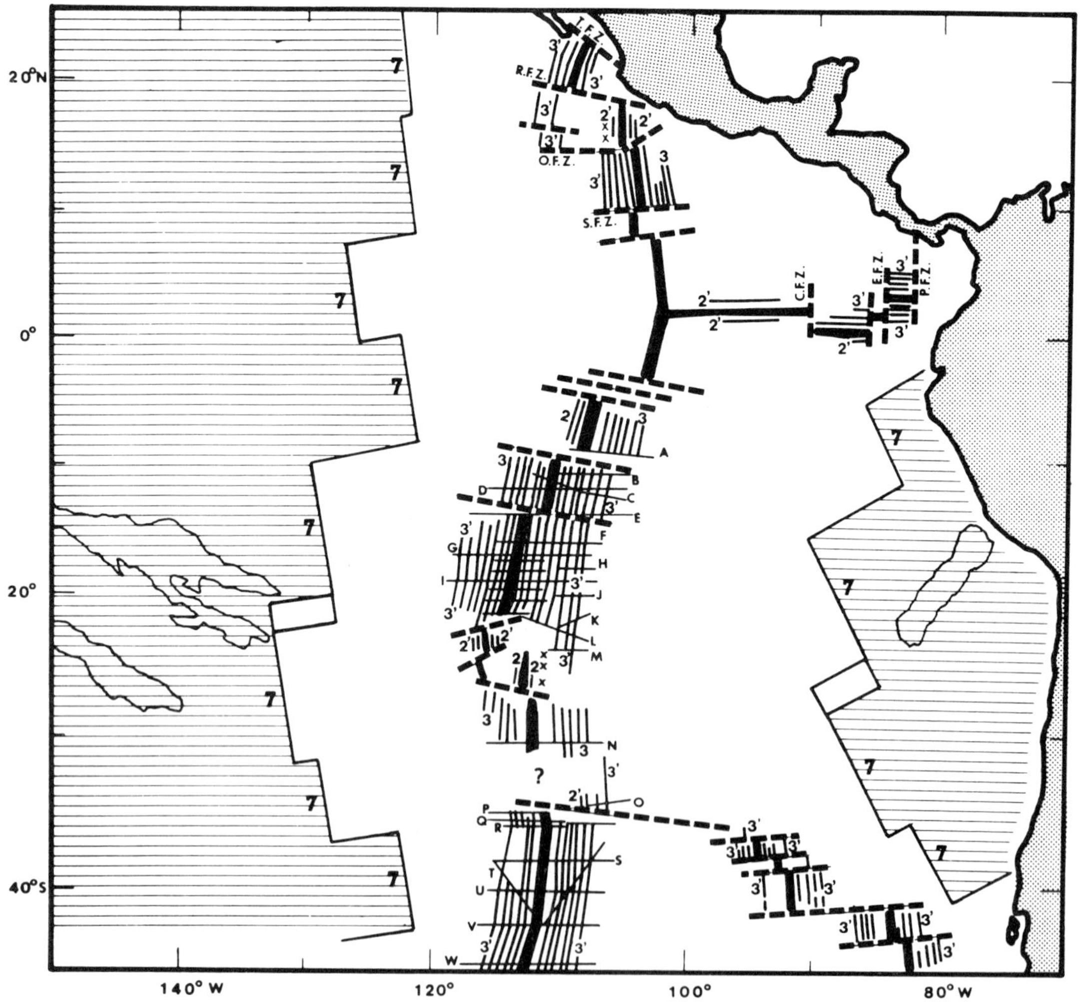


Fig. 10. Plio-Pleistocene magnetic patterns and track lines for profiles shown in Figure 11. O.F.Z. = Orozco fracture zone, S.F.Z. = Siquieros fracture zone, C.F.Z. = Carnegie fracture zone, E.F.Z. = Ecuador fracture zone, others as identified in Figure 2 and text. Note asymmetry of Plio-Pleistocene patterns between the Pacific and Nazca Oligocene patterns outlined by anomaly 7.

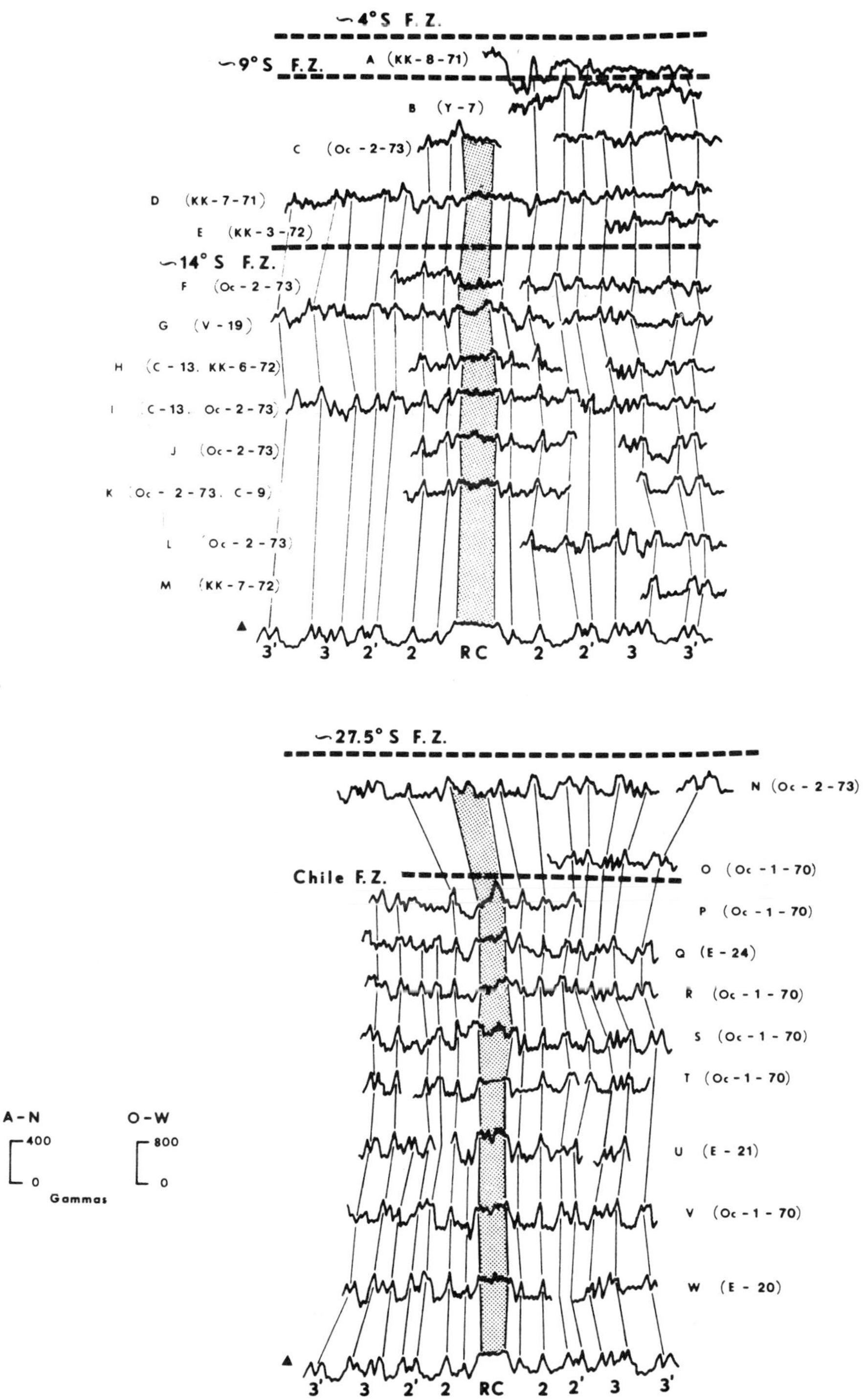


Fig. 11. Magnetic profiles for the track lines in Figure 10 correlated to the Plio-Pleistocene section of the data reference profile (solid triangles). Specific track line identifications are included in parentheses for reference to Figure 4.

in preparation). The trends shown in Figure 10 (taken from that study) reveal three active ridge segments whose spreading activity must be consistent with the spreading geometry of the bordering sections of the East Pacific rise since neither the Pacific nor the Nazca plate shows any seismic activity outside the anomalous region. The southern boundary of the anomalous area is defined by seismic and magnetic data to be at 27.5°S [*Herron,* 1972*b*]. Profiles N and O show that the lineations between 27.5°S and another anomalous region of seismic activity near 33°S are consistent out to anomaly 3′. The southern anomalous region is also outlined by seismicity (see Figure 3), and has been interpreted as a small plate between two subparallel ridge crests [*Herron,* 1972*b*]. It is apparent that the Chile rise or fracture zone intercepts the East Pacific rise in this area, forming the Chile triple junction [*Herron,* 1972*b; Mammerickx et al.,* 1975], but because of low data density, the nature of the triple junction is not clear.

Magnetic anomalies across the Pacific–Antarctic ridge were first mapped by *Pitman and Heirtzler* [1966] and more recently by *Herron* [1971] and *Molnar et al.* [1974]. Lineations shown in Figure 10 were constructed from profiles P through W. Although the western portion has less bathymetric and magnetic continuity than the eastern portion, a consecutive sequence out to anomaly 3′ is easily identified along both sides

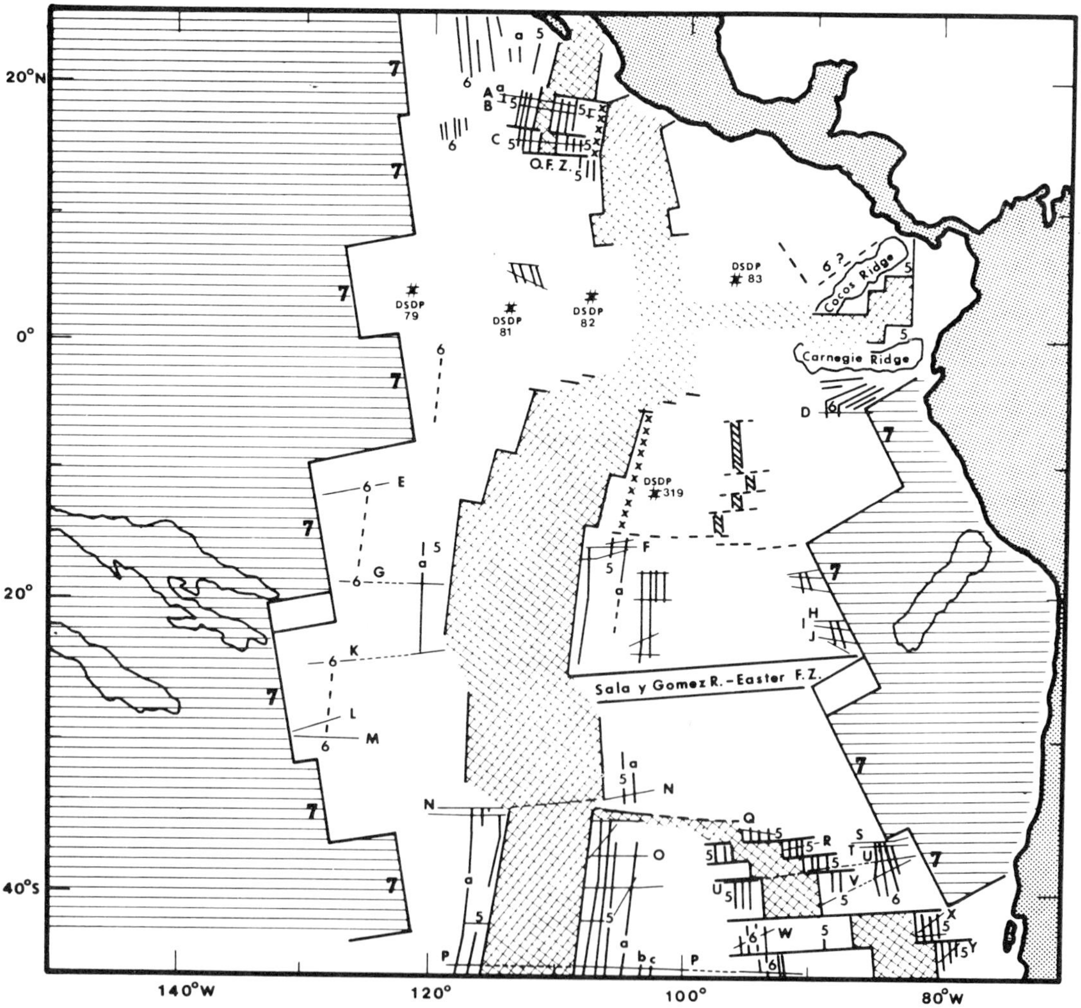


Fig. 12. Miocene magnetic patterns and track lines for profiles shown in Figure 13. The Oligocene and Plio-Pleistocene patterns are outlined by anomalies 7 and 3′, respectively. Note sparsity of Miocene identifications.

of this ridge segment.

Interpretations of the Plio-Pleistocene trends associated with the Chile rise vary considerably in the literature [*Herron and Hayes,* 1969; *Morgan et al.,* 1969; *Klitgord et al.,* 1973; *Molnar et al.,* 1974]. Although very little new data were available for this study, an analysis of magnetic and bathymetric data has led to the interpretation shown in Figure 11, which is similar to those of *Klitgord et al.* [1973] and *Molnar et al.* [1974]. The interpretation maintains the mutual perpendicularity of magnetic lineations to fracture zones, eliminating the need for oblique spreading during formation. The implication of this inter-

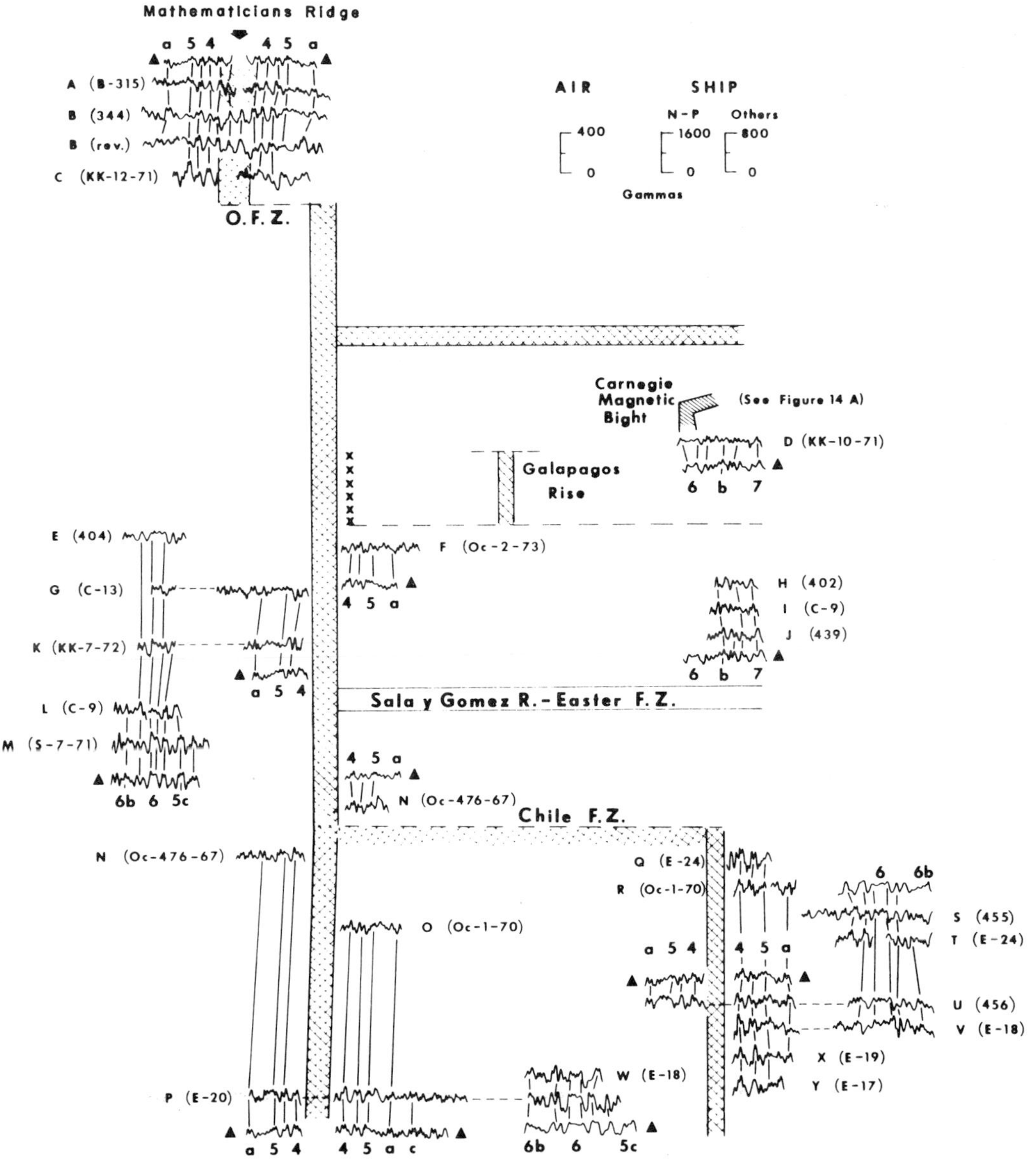


Fig. 13. Magnetic profiles for the track lines in Figure 12 correlated to Miocene sections of the data reference profile (solid triangles). Profiles are schematically positioned to their relative positions in Figure 12 and prominent topographic features. Specific track line identifications are included in parentheses for reference to Figure 4.

pretation is that the Chile ridge and associated Plio-Pleistocene trends are not continuous with the Oligocene trend to the east.

*Miocene patterns (26–5.4 m.y.B.P. anomalies 7–3′).* Mapping of Miocene anomalies in the central and southeastern Pacific was extremely difficult and resulted in only a few homogeneous patterns as seen in Figure 12. The failure to clearly define complete patterns is due partly to the relatively indistinguishable nature of the short wavelength anomalies observed between anomalies 5a and 7 (Figure 13), and to the substantial deformation of the magnetic patterns which accompanied the major reorientation of spreading patterns that occurred during the Miocene. Such deformation has been observed in the northeastern Pacific [*Menard and Atwater,* 1968], the North Atlantic [*Vogt et al.,* 1969, 1971*a*] and the Indian Ocean [*Fisher et al.,* 1971; *McKenzie and Sclater,* 1971] where complete magnetic patterns were accurately mapped from detailed surveys. Miocene spreading history must therefore be interpreted by combining existing identifications with geometric constraints of plate theory, as will be seen in the interpretive phase of this study.

Miocene lineations shown north of the Clarion–Rivera fracture zone, taken from *Chase et al.* [1970] and *Atwater and Menard* [1970], appear to indicate a continuous sequence from anomaly 7 to anomaly 3′. South of this fracture zone, a continuous sequence is also indicated on the western flank of the Mathematicians ridge, but the eastern sequence is disrupted just east of anomaly 5A by Plio-Pleistocene crust formed after the 4-degree eastward jump of the East Pacific rise about 4 m.y.B.P. [*Sclater et al.,* 1971]. The symmetric magnetic anomaly identifications about the Mathematicians ridge are shown in Figure 13 (profiles A, B, and C). Identification of consecutive anomalies out to anomaly 5 on the western flank of the East Pacific rise south of the Siqueiros fracture zone negate a jump of spreading activity in that region subsequent to 10 m.y.B.P.

Few magnetic identifications were made between 10°N and 5°S and tectonic events are not directly interpretable from magnetic data. An exception is the E–W anomalies associated with spreading from the Galapagos rift zone. The identification of Miocene anomalies (Carnegie sequence) on the Carnegie plateau and the presence of a Miocene magnetic bight (Carnegie bight) joining these anomalies with N–S trending anomalies on the Nazca plate (Handschumacher, in preparation) are critical to the tectonic reconstructions that will be presented in the next section. For this reason, a more detailed map of the region is presented in Figure 14A. The lineations are from Handschumacher (in preparation), and a selected Profile (V-17) has been included here for comparison with the data reference profile.

No convincing anomaly identifications or symmetry such as *Herron* [1972*a*] has reported were found on the fossil Galapagos rise. A linear bathymetric scarp (designated by x's in Figure 12) lies between the East Pacific and Galapagos rises. This feature was first identified by *Herron* [1972*a*] who interpreted it as a boundary between the fossil Galapagos rise spreading system and the presently active East Pacific rise spreading system. *Anderson and Sclater* [1972] redefined the morphology of this region and interpreted the scarp as a feature formed by a westward jump of spreading activity at ~6 m.y.B.P. Identification here of Plio-Pleistocene anomalies just west of this scarp support the interpretation of a westward ridge jump ~6 m.y.B.P. However, the morphology is not fully understood and suitable magnetics are lacking on the Galapagos rise at this time.

Magnetic correlations between about 15°S and about 25°S, though only fair, indicate extension of the Plio-Pleistocene trend out to at least anomaly 5a on both sides of the East Pacific rise (profiles G, K, and F). Since Plio-Pleistocene patterns are asymmetrically positioned between the Oligocene patterns on the Pacific and Nazca plates in this region (as they are to the north), a westward-ridge jump here has also been proposed

---

Fig. 14. (Opposite) A. Miocene magnetic lineations on the Carnegie plateau (Carnegie sequence) and interpretation of the Grijalva scarp as a tectonic boundary. Also shown is the Carnegie magnetic bight which joins N–S trending anomalies on the Nazca plate to E–W trending anomalies in the Carnegie sequence. Tenuous identification of anomaly 6 north of the Cocos ridge is also shown. The magnetic lineations in this figure are copied from Handschumacher (in preparation). A selected profile, V-17, is correlated to the data reference profile to demonstrate the excellent correlation of the Carnegie sequence. B. The clockwise rotation of anomaly lineations from anomaly 7 to anomaly 5e south of the Challenger fracture zone. Magnetic profiles on which this diagram is based are shown in Figures 12 and 13.

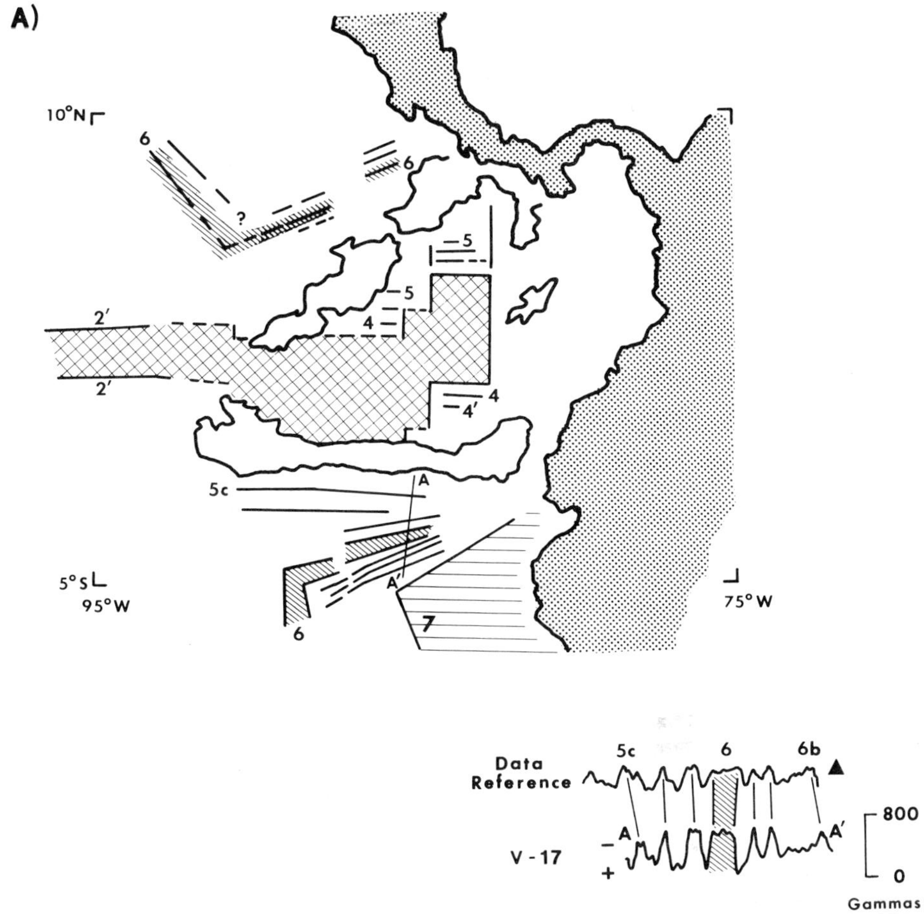
A)
10°N
6
6
?
5
5
4
2′
2′
4
4′
5c
A
A′
6
7
5°S
95°W
75°W
Data
Reference
5c
6
6b
A
A′
V-17
800
0
Gammas

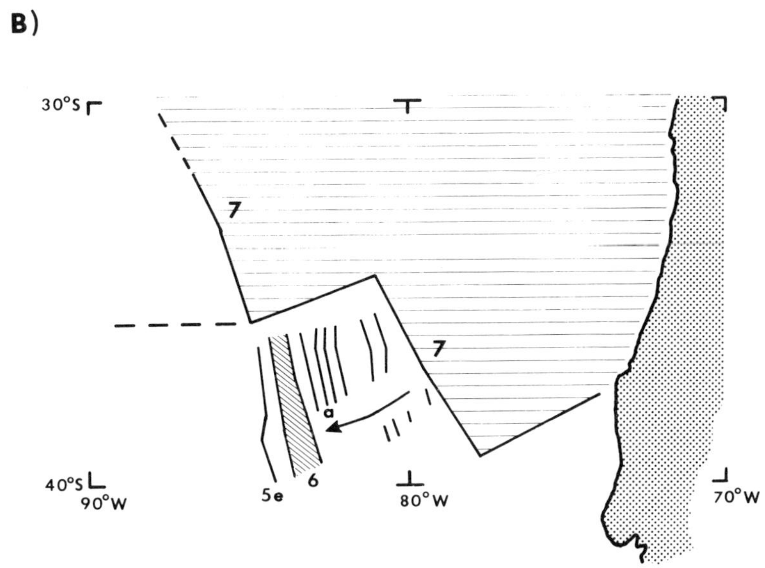
B)
30°S
7
7
a
5e
6
40°S
90°W
80°W
70°W

at about 10 m.y.B.P. based on magnetics [*Herron*, 1972*a*] and bathymetry [*Anderson and Sclater*, 1972]. The author does not feel that the data in either of these studies are conclusive as to the nature or time of the jump, although the geometry requires a westward ridge jump here confined to a position between the Plio-Pleistocene and Oligocene patterns on the Nazca plate. The fossil spreading center appears to have become dormant between 23 and 12 m.y.B.P. The older age limit of 23 m.y.B.P. is defined by two lineation sets just west of the Oligocene-Miocene boundary on the Nazca plate (profiles H, I, and J) which suggest a clockwise rotation of spreading patterns between the formation of anomaly 7 and anomaly 6b (23 m.y.B.P.). The younger age limit of 12 m.y.B.P. is based on the identification of anomaly 5a on the flanks of the East Pacific rise. The absence of a prominent topographic high over the predicted fossil ridge axis indicates sufficient time has elapsed since spreading activity ceased to allow crustal subsidence of the fossil ridge in this region. Identifications of anomaly 6 (profiles E and G) on the Pacific plate would lower the older age limit to 20 m.y.B.P., but these identifications are considered speculative because no convincing sequence of anomalies was found to support them. The N-S trending anomalies just east of anomaly 5a on the Nazca plate appear to be well defined by the data, but are not uniquely correlatable to the reversal chronology. They therefore could have been generated by the proposed fossil ridge to the east or by the East Pacific rise to the west.

The Sala y Gomez ridge–Easter fracture zone, a prominent topographic feature between Easter Island and the southern tip of the Nazca ridge, is structurally not typical of either an aseismic ridge or an oceanic fracture zone. Instead, it is characterized by irregular seamounts and is a significant regional bathymetric boundary, separating normal oceanic crustal depth to the north and anomalously shallow crustal depths to the south [*Handschumacher et al.*, 1975]. There is bathymetric evidence that this zone of rugged topography extends west of Easter Island to the Tuamotu ridge, although this western extension is not well defined [*Menard*, 1966; *Mammerickx et al.*, 1975].

South of this topographically complex zone to 35°S, identifications are made only of anomaly 6 (profiles K, L, and M) on the Pacific plate and anomalies 3′ to 5a on the Nazca plate (33°S, 106°W; profile N). Between 35°S and 45°S, Plio-Pleistocene anomaly trends are continued to at least anomaly 5a across both the Pacific–Antarctic ridge (profiles N, O, and P) and segments of the Chile rise (profiles Q through Y). The identifications require that both of these spreading centers have been active since 12 m.y.B.P., although the magnetic and bathymetric data indicate a slowdown in spreading rates of the Chile rise occurred at about 5 m.y.B.P. Anomaly sets containing anomaly 6 are identified on both flanks of the Chile rise and require that the southern portion of the Chile rise has been active since 20 m.y.B.P. On the western flank, anomaly 6 is mapped on the Antarctic plate near 43°S, 94°W (profiles W and P). This identification differs from that of *Herron* [1972*a*] and *Molnar et al.* [1974] who both identified the same anomaly as anomaly 9. On the eastern flank the mapping of anomalies 6 to 7 (profiles S through V) shows a clockwise rotation occurring after the formation of anomaly 7. Again, a large-scale draft of these trends (Figure 14B) is included for later reference.

## REGIONAL KINEMATIC ANALYSIS

In the sections that follow, paleo-spreading kinematics are reconstructed utilizing the observed magnetic patterns and geometric constraints of plate theory. Steady-state conditions [*McKenzie and Parker*, 1967; *Morgan*, 1968] are assumed in the regional dynamics, i.e., symmetric spreading with pure transcurrent motion on transform faults, unless this assumption is violated by the magnetic and bathymetric data. Where the data do violate the steady-state assumption by indicating a reorientation of the sea-floor spreading dynamics, the simplest geometric solutions are applied to the regional reconstruction. Several of the events described are not defined by existing data and are presented only as possible solutions which conform to the geometric constraints. They are not unique. For example, asymmetry in older magnetic patterns about an active spreading center may be accounted for by asymmetric spreading, ridge jumping, or former dual spreading from two spreading ridges where one has subsequently ceased spreading. Without adequate magnetic or bathymetric data to resolve the mechanism, any of these mechanisms could be offered as a solution for the observed asymmetry. It is important for the reader to keep this in mind, especially in the discussion of Miocene events which are not well defined by the data.

*Oligocene kinematics.* Two coherent Oligocene magnetic anomaly patterns, the Pacific and the Nazca sets, are mapped in Figure 8, each having an internal consistency that is indicative of formation from a steady-state spreading center. If both patterns were generated by a common Oligocene spreading ridge and if neither has been significantly contorted by subsequent tectonic events, it should be possible to fit them together and, in so doing, reconstruct the Oligocene spreading system. As seen in Figure 15, an excellent fit of anomaly 7 is obtained by rotation of the Nazca set into the fixed Pacific set along the path shown by arrows. The quantitative aspects of this fit, and fracture zone offsets and spacing, are summarized in Table 1.

Another test of the symmetry of this reconstruction was to calculate Oligocene half-spreading rates from the opposing anomaly sets. These spreading-rate calculations are shown in Figure 16. Wherever possible, the rates were determined between anomalies 7 and 12 to avoid apparent discrepancies resulting from inaccuracies in the geomagnetic reversal time scale [*Watkins,* 1972]. As seen in the figure, opposing rates are nearly symmetrical where they could be calculated on both sides of the reconstructed ridge. The maximum asymmetry (<7%) present is not considered significant.

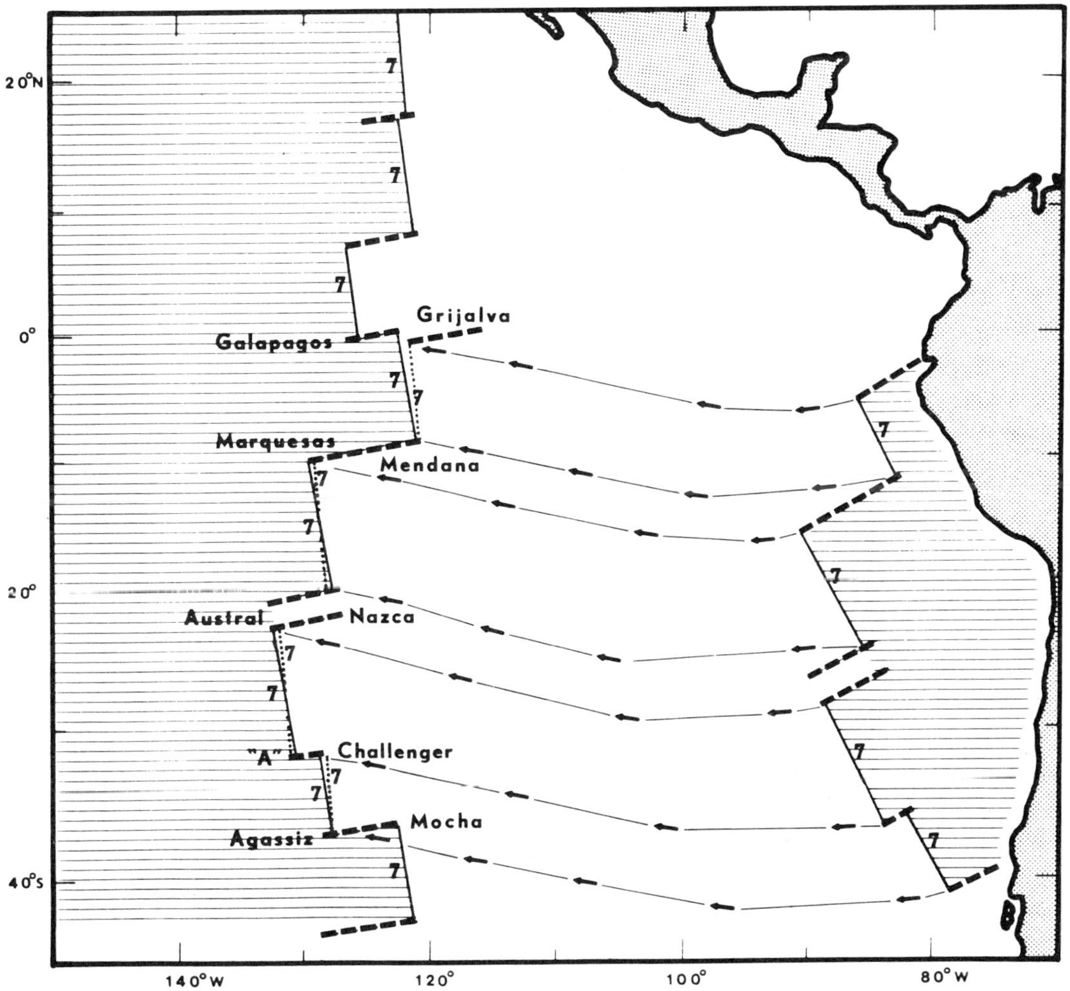


Fig. 15. Suggested fit of the Pacific and Nazca Oligocene magnetic patterns about anomaly 7. With the Pacific plate fixed, the Nazca set is rotated about the path shown by the arrows. Dotted lines represent anomaly 7 of the Nazca set after rotation. The quality of the 'mirror image' fit is described in the text and in Table 1.

TABLE 1. Oligocene Fit of the Pacific–Nazca Sets

| Pacific Set | | Nazca Set | |
|---|---|---|---|
| | *A. Fracture Zone Offsets, km** | | |
| Marquesas | 883RL | Mendana | 889RL |
| Austral | 425RL | Nazca | 485RL |
| "A" | 155LL | Challenger | 200LL |
| Agassiz | 518LL | Mocha | 500 + LL |
| | *B. Fracture Zone Spacing, km* | | |
| Galapagos to Marquesas | 430 | Grijalva to Mendana | 390 |
| Marquesas to Austral | 550 | Mendana to Nazca | 550 |
| Austral to "A" | 510 | Nazca to Challenger | 480 |
| "A" to Agassiz | 280 | Challenger to Mocha | 330 |
| | *C. Spreading Rates Between Fracture Zones, cm/yr* | | |
| Galapagos to Marquesas | 6.5 | Grijalva to Mendana | 6.8 |
| Marquesas to Austral | 6.8 | Mendana to Nazca | 7.2 |
| Austral to "A" | 6.1 | Nazca to Challenger | 6.2 |
| "A" to Agassiz | 5.7 | Challenger to Mocha | 5.5 |

*RL = right lateral (dextral); LL = left lateral (sinistral).

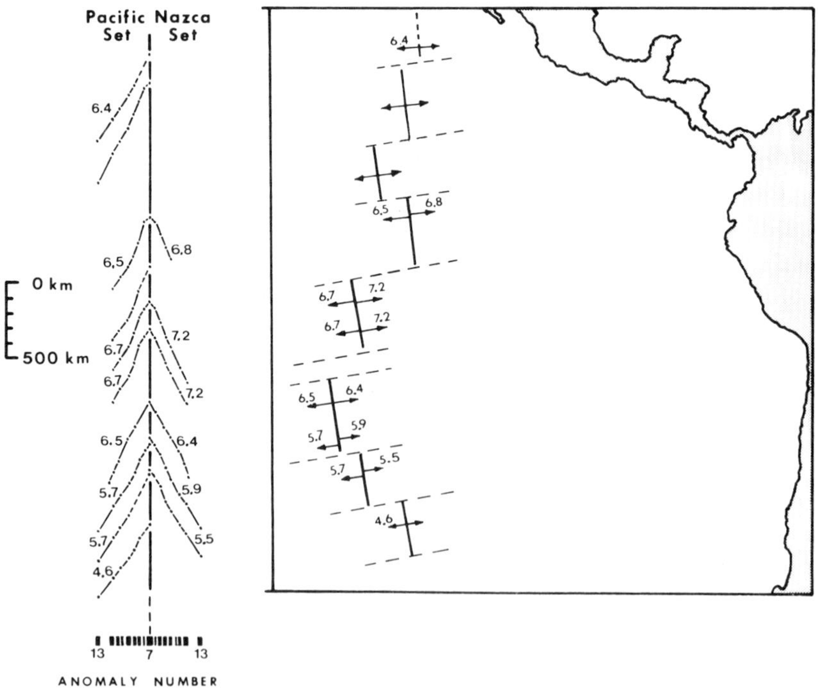


Fig. 16. Averaged Oligocene half-spreading rates (cm/yr) as calculated about the reconstructed ridge at anomaly 7 and lineations in Figure 8. On the left are the time–distance plots; on the right, the results plotted on the reconstructed ridge.

The Oligocene reconstruction is expanded in Figure 17 to include the northeastern Pacific by including the magnetic anomaly patterns in Figure 1 and assuming a fixed Pacific plate. Total rates were obtained by taking the sum of the opposing half-rates across ridge segments when data permitted (Figure 16) and doubling the western half-rate, assuming symmetrical spreading, where only the western rate could be calculated from observed data. The various components of this reconstructed spreading pattern, ridge strike, fracture zone azimuths, and spreading rates, were then used to calculate the appropriate pole of opening. A pole at 70° N, 145° W was determined and the theoretical lines of latitude and longitude for this pole [*Wong et al.*, 1975] were then superimposed on the Mercator grid plot of Figure 17. As seen, the steady-state dynamics of the Oligocene ridge system between 55° N and 43° S (>11,000 km) is supported by the near-perfect fit of the observed data to the theoretical coordinate system. This fit includes the parallelism of fracture zone strikes to the theoretical latitude lines and of ridge strikes to theoretical longitude lines, as well as the progressive decrease of observed spreading rates to the theoretical decrease of values from a maximum at the 'equator' of 13.5 cm/yr. A lone break in the continuity of the ridge occurs off the west coast of North America where anomaly 7 is missing just south of the Mendocino fracture zone and some distortion of the Oligocene anomaly patterns is noted between the Mendocino and Murray fracture zones. The disturbed patterns in this region may reflect the

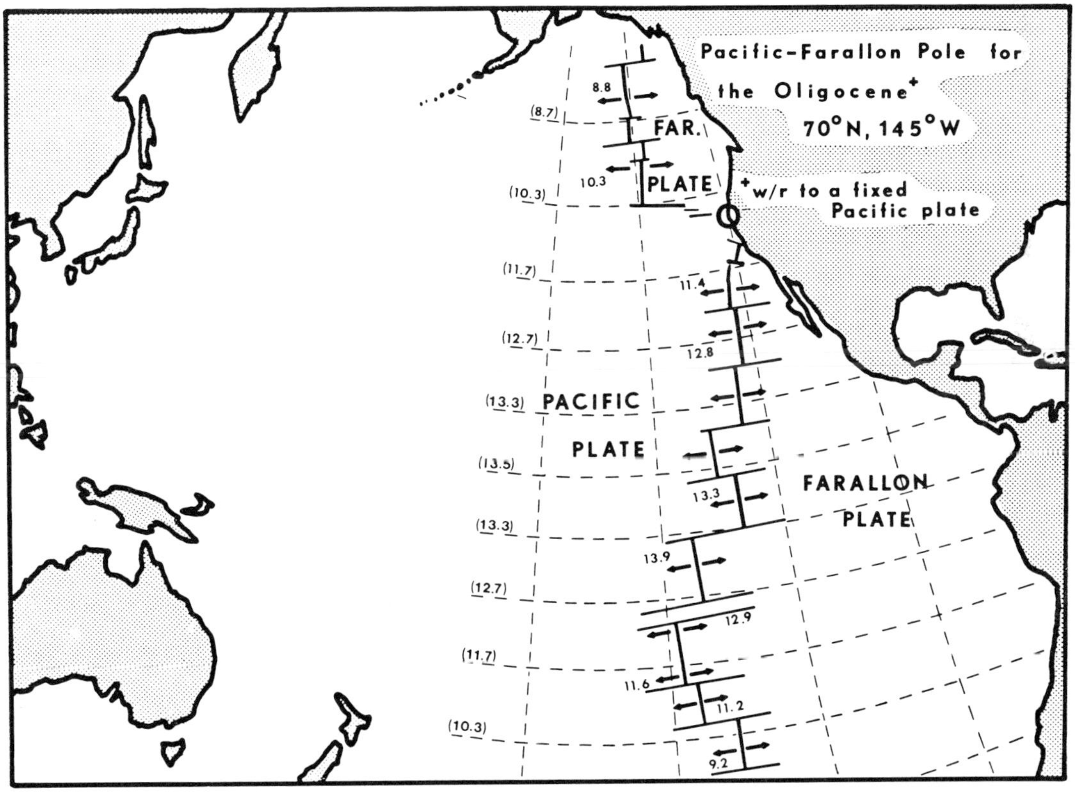


Fig. 17. Total Oligocene spreading rates (cm/yr) for the reconstructed ridge at anomaly 7 in the eastern Pacific. The Pacific plate is held fixed and a total rate calculated by doubling the half-rate where only the western half of the Oligocene pattern can be observed. Theoretical longitude and latitude (dashed lines) projected about a fictitious pole of spreading at 70° N, 145° W are shown in parentheses. Note the excellent parallelism of ridge segments to longitude lines and that of fracture zones to latitude lines. Note also the close agreement of spreading rates calculated from magnetic patterns with theoretical spreading rates (in parentheses).

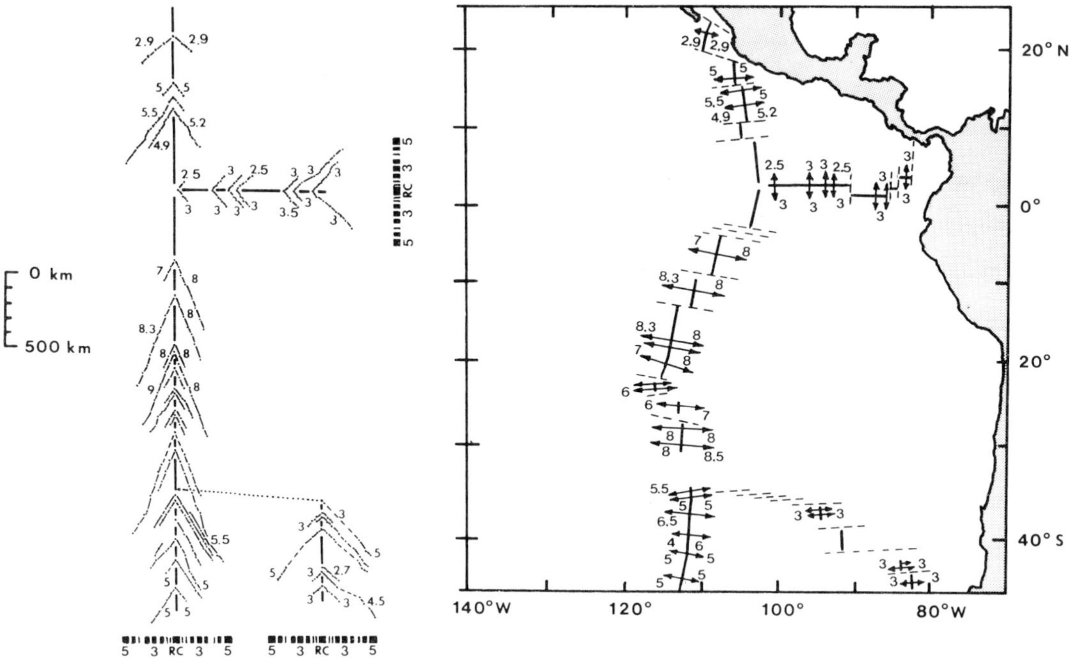


Fig. 18. Averaged half-spreading rates (cm/yr) calculated for the Plio-Pleistocene period as calculated from lineations in Figure 10.

collision of the eastwardly migrating Pacific–Farallon ridge with the formerly active Farallon–Americas trench system off the west coast of North America about 29 m.y.B.P. [*Atwater*, 1970]. However, the continuity of the reconstructed ridge dynamics north and south of the region negates major disruption of the Pacific–Farallon ridge prior to ~26 m.y.B.P. An important corollary of this reconstruction is that the Galapagos spreading center evolved after 26 m.y.B.P. since the Farallon plate is required to be homogeneous by the reconstruction in Figure 17.

*Plio-Pleistocene kinematics.* Since Plio-Pleistocene magnetic anomalies are symmetrically identified on almost all active ridge segments, half-spreading rate calculations could be made on opposing flanks with accuracy dependent only on the quality of individual anomaly identifications. The averaged Plio-Pleistocene spreading rates to anomaly 3′ are shown in Figure 18. The rate variations seen on opposite sides of ridge segments are not consistent and may again be derived from data errors or reflect random small-scale irregularities of sea-floor spreading. For example, the asymmetry in rates seen on the East Pacific rise (4°S–24°S) is neither large nor consistent in direction.

The anomaly pattern in the northeastern Pacific was included in constructing Figure 19, and the total spreading rates were again obtained by summing the opposing half-rates. The rate of 5.8 cm/yr on both the Juan de Fuca ridge and the East Pacific rise segment between the Tamayo and Rivera fracture zones compares well with those determined by *Atwater* [1970] and *Larson et al.* [1968]. The trends across the Mathematicians ridge (see Figures 10 and 12) suggest that this fossil spreading ridge was, prior to the 4-degree eastward jump at about 4 m.y.B.P., an extension of the ridge north of the Rivera fracture zone, and that the Orozco fracture zone was the former plate boundary between the Rivera and Cocos plates. Rate determinations about the presently active East Pacific rise reveal a uniform spreading rate of 10 cm/yr (Figure 18) between the Rivera and Orozco fracture zones to anomaly 2′, and the Orozco and Siquieros fracture zones to anomaly 3′. However, the author does not believe that these results were sufficiently controlled by the data used for this study to contradict the finding of *Larson and Chase* [1970] that the rates systematically increase from 9.6 cm/yr just south of the Orozco fracture zone to 13.8 cm/yr just south of the Siquieros fracture zone.

Along the Galapagos rift zone, calculated rates are in close agreement with the results of previous studies [*Herron and Heirtzler,* 1967; *Raff,* 1968; *Grim,* 1970; *van Andel et al.,* 1971; *Sclater and Klitgord,* 1973] which cite a total rate of about 6 cm/yr for the western portion of the ridge (83° to 90° W). This 6 cm/yr rate does not appear to vary significantly about the Galapagos rift zone except in the immediate vicinity of the Galapagos triple junction where there is an apparent wedging of the bathymetry and magnetics associated with the rift zone [*Hey et al.,* 1972]. East-west trending anomaly lineations, terminating at anomaly 2′ or anomaly 3 on both sides of the Galapagos rift zone between the triple junction and the Carnegie fracture zone (90° W) indicate an age limit of about 4 m.y. activity on this segment of the ridge. This would imply that, prior to 4 m.y.B.P., a former triple junction was located near the junction of the Cocos and Carnegie ridges, accounting for earlier N–S spreading in the Panama basin and on the Carnegie plateau (Handschumacher, in preparation).

The kinematics on the East Pacific rise extending from the Galapagos triple junction at 2° N to the Chile triple junction at about 35° S appears

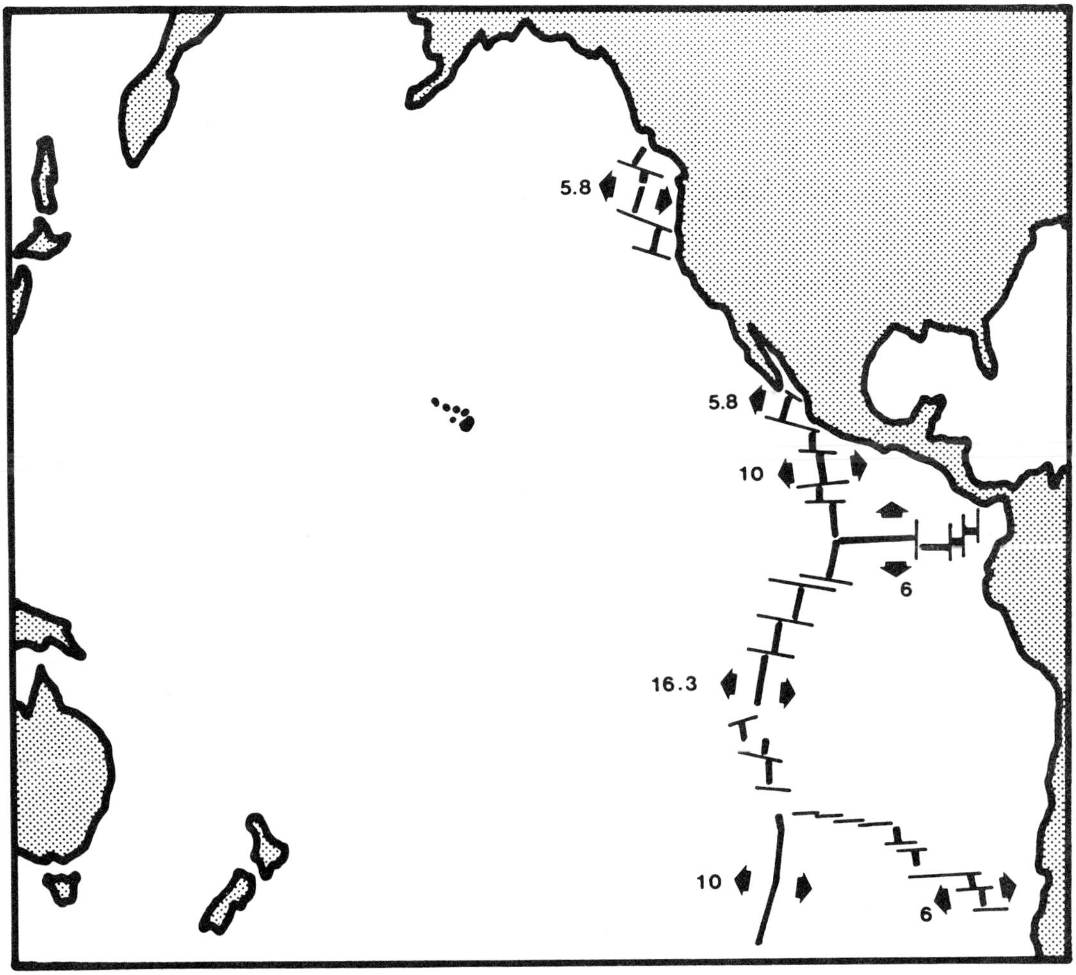


Fig. 19. Total spreading rates (cm/yr) for Plio-Pleistocene spreading are shown on ridge segments in the eastern Pacific. Comparison of the kinematic regime in this diagram with that of the Oligocene in Figure 17 demonstrates the radical reorientation of sea-floor spreading patterns that must have occurred during the Miocene.

to be consistent with near steady-state spreading between the Pacific and Nazca plates since 4 or 5 m.y.B.P. Since the spreading rate calculations for the anomalous spreading activity at ~24°S do not affect the regional kinematics and are the subject of a separate study (Handschumacher, in preparation), they are not considered here. Spreading rates in Figure 18 show little change from the en echelon fractures at 4°S to the Chile triple junction, indicative of a distant pole of spreading. The interpretation differs from that of *Herron* [1972*a*] in that a nearly consistent trend of ~N5°E is given for the ridge axis from the Galapagos triple junction to 24°S. Herron's determination of the present trend of the East Pacific rise north of 14°S as 25° to 40° east of north requires a change exceeding 20 degrees in the ridge strike crossing the fracture zone at 14°S. Her determination was partly based on the high anomaly amplitudes observed on two data profiles crossing the ridge north of 14°S. The observed amplitudes were larger than those which could be theoretically justified from nearly N–S trending source bodies. In this study, these higher amplitudes are attributed to E–W trending source bodies introduced along the fracture zones at 4°S and 9°S, eliminating the 'kink' in the East Pacific rise at 14°S and the geometric requirement for oblique spreading between 14°S and the Galapagos triple junction. This interpretation is supported by numerous additional profile crossings of the ridge crest between 2°N and 14°S [*Handschumacher et al.,* 1975] which have not revealed the high amplitude anomalies cited by Herron.

South of the Chile triple junction, plate theory requires that spreading from the Chile rise and Pacific–Antarctic ridge must be collectively compatible with spreading north of the triple junction both in direction and in rate if the Nazca plate is to remain rigid to 45°S [*Herron,* 1972*a*]. These conditions are met by the Plio-Pleistocene trends shown in Figure 10 and the kinematics indicated in Figures 18 and 19. As noted earlier, there was a slowdown in total spreading rate on the Chile rise from ~9 cm/yr to ~6 cm/yr at about ~5 m.y.B.P., as indicated in Figure 18.

*Miocene kinematics.* The paleo-spreading ridge configurations at ~26, ~20, ~10, and ~5.4 m.y.B.P. are shown in Figure 20. Magnetic patterns from the northeastern Pacific (Figure 1) have been included, but spreading rate calculations for specific periods are not included because comprehensive measurements cannot be made from the limited number of Miocene identifications.

A comparison of Figure 20A with Figure 20B shows the radical change in spreading kinematics that resulted from the direct coupling of the Americas and Pacific plates at ~26 m.y.B.P. By 20 m.y.B.P., the southern portion of the ridge had rotated in a clockwise direction, resulting in a divergent stress condition on the Farallon plate and the initiation of N–S spreading from the E–W trending Galapagos rift zone. Evidence for the clockwise rotation of the southern ridge segments comes from the interpreted reorientation of magnetic trends between anomalies 7 and 6 as shown in Figure 12. The mapping of these trends is sparse and a continuous anomaly sequence has only been defined south of the Challenger fracture zone (Figure 14B). Evidence for the development of N–S spreading from an E–W trending ridge comes from the identification of (1) E–W trending, early Miocene anomalies north of the Grijalva fracture zone, and (2) the Carnegie magnetic bight (Figure 14A). The final feature of note in this reconstruction is the development of dual spreading in the southernmost region. This dual spreading is defined by the identifications of

Fig. 20. (Opposite) Reconstructions of spreading patterns and related tectonic events during the Miocene. Although these diagrams are schematic and contain scale distortions where necessary for clarity, they are consistent with the observed magnetic and bathymetric data. In each, the active spreading ridge is indicated by the heavy line; and magnetic and bathymetric trends generated by preceding stage or stages, by light solid or dashed lines. The Pacific plate is held fixed; and the relative motion of other plates is shown by arrows. A. ~26 m.y.B.P., Pacific–Farallon ridge intact but colliding with the Americas plate (Farallon–Americas trench system) off the west coast of North America. B. ~20 m.y.B.P., a clockwise rotation of the southern portion of the Pacific–Farallon ridge, development of the Galapagos rift zone, and development of dual spreading from the Pacific–Antarctic ridge and Chile rise. C. ~10 m.y.B.P., breakup of the ridge south of Baja California, northward extension of dual spreading of the Pacific–Antarctic ridge and Chile rise, and indicated ridge jumps (small arrows). D. ~5 m.y.B.P., development of the Gorda–Juan de Fuca–San Andreas–Gulf of California spreading system, eastward jump of the East Pacific rise from its former position at the crest of the Mathematicians ridge, and westward jumps of the Galapagos triple junction and of the East Pacific rise from its former position on the Galapagos rise.

anomaly 6 on both flanks of the Chile rise (Figure 12) and on both flanks of the Pacific–Antarctic ridge at about 45°S [*Herron,* 1971; *Molnar et al.,* 1974].

Events occurring between 20 and 10 m.y.B.P. (Figure 20C), have been interpreted almost entirely from bathymetric trends and geometric constraints due to the extreme sparsity of magnetic identifications. During this period, the ridge was still operative in a consistent orientation

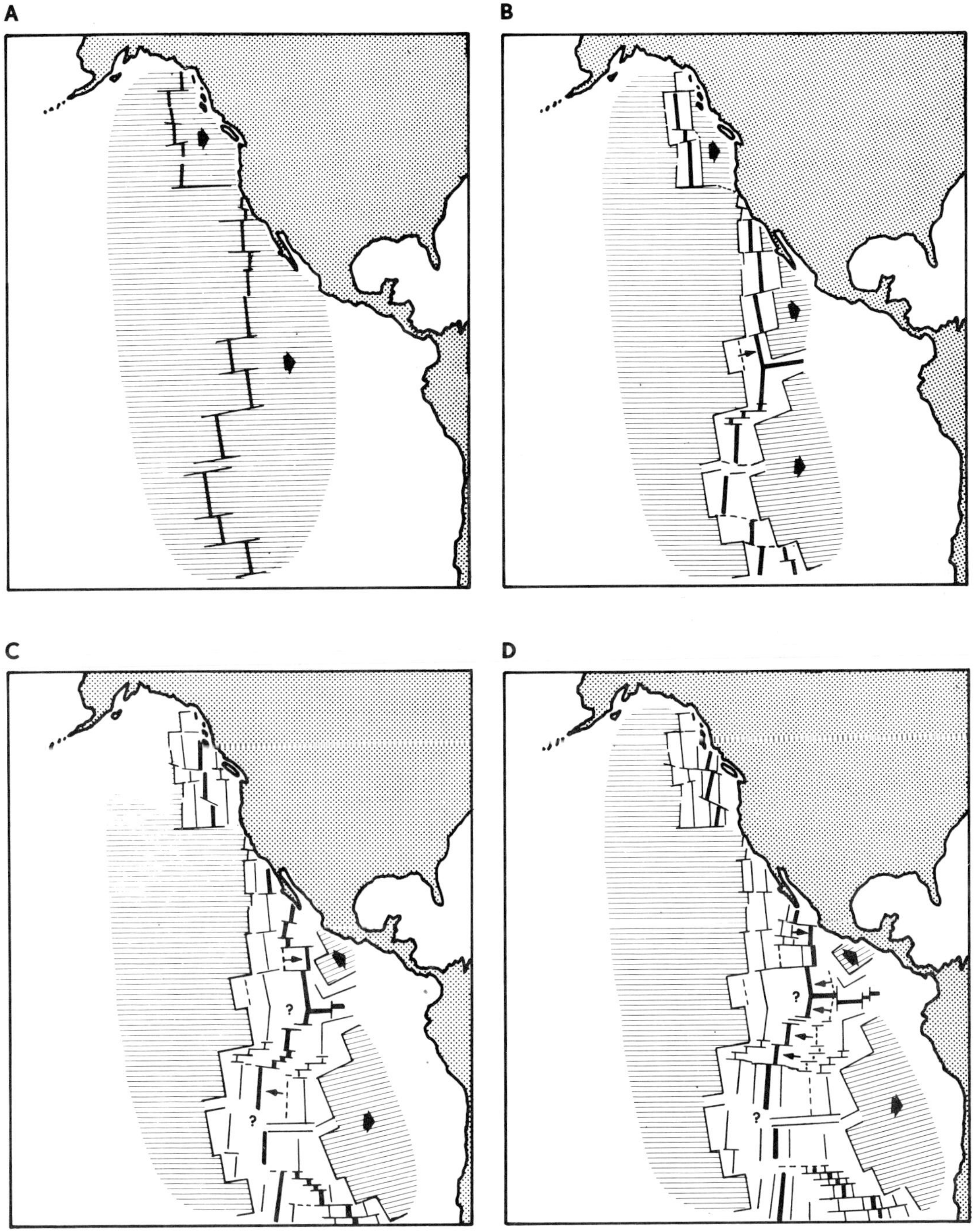

north of the Mendocino fracture zone, although the progressive collision of the ridge with the Americas plate had significantly distorted spreading patterns south of Baja California [*Chase et al.,* 1970]. There was an eastward jump of the ridge south of the Orozco fracture zone and a westward jump between ~14°S and the Sala y Gomez ridge–Easter fracture zone during this period. The dual spreading had moved farther northward with the development of all the Chile rise spreading segments that are actively spreading today. No data were found to define the activity in the equatorial region. However, it is suggested that the geometric complications in the continuing development of the ridge–ridge–ridge triple junction between 20 and 10 m.y.B.P. led to the formation of the Cocos and Carnegie ridges.

Figure 20D was constructed by bringing opposing trends of anomaly 3′ (Figure 10) together across appropriate ridge segments. However, some overlap of the Plio-Pleistocene period has been allowed to show some of the borderline events. The spreading patterns off the west coast of North America had undergone a major reorientation with the development of the Gorda–Juan de Fuca–San Andreas–Gulf of California spreading system. South of this system, there was an eastward jump of the ridge between the Rivera, and Orozco fracture zones which left the Mathematicians ridge as a topographic high. There also appears to have been a westward jump of the Galapagos triple junction and its southward ridge component. The westward jump of the ridge between 4° and 14°S left the Galapagos rise as a topographic high. Specific evidence for these events was discussed in previous sections to which the reader is referred.

### CIRCUM-PACIFIC TECTONIC IMPLICATIONS

One consequence of the radical reorientation of sea-floor spreading kinematics ~26 m.y. ago would have been simultaneous and catastrophic changes in the circum-Pacific tectonic regimes. Could evidence be found to support such changes? One approach would be to define absolute plate motions across the circum-Pacific boundaries. Since neither current analytical methods nor data as yet provide accurate determinations for absolute motion of purely oceanic plates—such as the Pacific, Cocos, Rivera, and Nazca—studies of this nature have been confined to the continental margin of western North America. A determination of relative motion for the Americas continental plate and Pacific oceanic plate was made directly across a spreading ridge [*Atwater,* 1970; *Silver,* 1971]. These studies were based on magnetic anomalies that were less than 5 m.y. old and therefore Atwater and Silver's construction of previous kinematic events required inclusion of kinematic assumptions.

An alternative approach is to examine circum-Pacific geology for an orogeny coeval with the reorientation. In this regard, *Dott* [1969] reported on a major mid-Cenozoic (30–25 m.y.B.P.) discontinuity in the tectonic history of circum-Pacific land areas. Noting the coeval nature of the mid-Cenozoic tectonic events, he suggested that they may be genetically related to Pacific sea-floor spreading dynamics. It is proposed that the circum-Pacific mid-Cenozoic orogeny and the major reorientation of sea-floor spreading dynamics at ~26 m.y.B.P. are both tectonic consequences of the coupling of the Americas and Pacific plates off the west coast of North America. It is further proposed that the present trench–continent boundaries in the eastern Pacific and the trench–island arc-marginal basin boundaries in the western Pacific were either created or significantly altered by the early Miocene reorientation of sea-floor spreading patterns in the eastern Pacific.

*Acknowledgments.* I would like to give special thanks to R. H. Higgs and the staff of the U.S. Naval Oceanographic Office for inspiring and encouraging my interest in marine magnetic studies. I also thank Dr. G. P. Woollard for creating an atmosphere in which I, as a student, was allowed to pursue those studies freely. For advice and critical reviews on the manuscript, I thank J. E. Andrews, S. Margolis, and G. H. Sutton. P. K. Wong and S. T. Okumura helped significantly in the data processing and analysis that are so essential for a regional study of this size. Data contributors included B. Erickson (Pacific Oceanographic Laboratories), D. Heinrichs (Oregon State University), H. Eppert (U.S. Naval Oceanographic Office), Prof. A. P. Lisitzin (U.S.S.R.), and H. Stockard and R. Lorentzen (Project Magnet, U.S. Naval Oceanographic Office).

This study was supported by the National Science Foundation (I.D.O.E. Nazca Plate Project) and the Office of Naval Research (contract N00014-70-A-0016-0001). This paper is part of work done toward a Ph.D. degree at the University of Hawaii, Honolulu, Hawaii.

# *17. The Early Cretaceous–Late Jurassic Magnetic Reversal Time Scale, and the Phoenix Magnetic Lineations Revisited*

R. L. Larson

*Lamont-Doherty Geological Observatory of Columbia University, Palisades, New York 10964*

*Abstract.* The Hawaiian and Phoenix lineation patterns of the western Pacific are the type localities for a time scale of magnetic reversals for the period 100–160 m.y.B.P. By analyzing the anomaly patterns, a relative stratigraphy of normal and reversed intervals has been derived for the M-reversal sequence. This stratigraphy has been placed in a geochronologic framework by analyzing the magnetic anomalies and basement ages in the vicinities of five DSDP holes. The analysis shows the M-reversal sequence to range in age from Oxfordian (~153 m.y.) to Aptian (~106 m.y.) and to be bounded both before and after by long intervals of normal magnetic polarity. Since the initial charting of the Phoenix lineations, additional anomalies have been identified as M11 to M19 just east of the Gilbert Islands, and M11 to M25 have been mapped between the Marshall–Gilbert islands and the Ontong Java plateau. The older anomalies (M20–M25) have a skewness (cross-sectional shape) indicating that these lineations were formed closer to the paleo-equator than were those of the eastern and younger portion of the Phoenix anomaly pattern.

# 18. *Mesozoic Sea-Floor Spreading in the North Pacific*

THOMAS W. C. HILDE

*Committee for Co-ordination of Joint Prospecting for Mineral Resources in Asian Offshore Areas (CCOP), United Nations, Bangkok, Thailand*

NOBUHIRO ISEZAKI

*Meteorological College, Kashiwa 277, Japan*

JOHN M. WAGEMAN

*Gulf Research and Development Company, Pittsburgh, Pennsylvania 15230*

*Abstract.* Sea-floor spreading magnetic anomalies of Mesozoic age have been mapped for the entire northwest Pacific, from the mid-Mesozoic quiet zone to the Cretaceous quiet zone. There are two general lineation directions: NE–SW between Shatsky rise and Japan, and NW–SE to the east and southeast of Shatsky rise. These two patterns are joined in a magnetic bight southwest of Shatsky rise. The bight is assumed to have been produced by spreading from ridges of a former ridge–ridge–ridge (R–R–R) triple junction between the Pacific, Kula, and Farallon plates which existed at the beginning of the late Mesozoic geomagnetic reversal period (anomaly M-26, 155 m.y.B.P.). Evolution of the triple junction resulted in the formation of Shatsky rise during 142 to 116 m.y.B.P. and Hess rise during 116 to 95 m.y.B.P. Identification of the anomalies is based on a late Mesozoic reversal model constructed from four closely spaced, parallel profiles recorded across the complete late Mesozoic sequence of reversal anomalies near the Hawaiian ridge. Anomalies which are repeated in all four profiles are considered due to reversals, while those not repeated are shown to be associated with seamounts or other topographic features. It is concluded that at 155 m.y.B.P. (M-26) the Kula–Pacific, Pacific–Farallon, Pacific–Phoenix ridge system described a comparatively small triangle about the Pacific plate. Assuming that spreading existed at the same rates prior to the period of late Mesozoic reversals, and projecting it back in time, the system would have started essentially at a point in the vicinity of 15° N and 155° E at 185 to 190 m.y.B.P. Spreading was faster on the Kula–Pacific and the Pacific–Phoenix ridges than on the Pacific–Farallon ridge during the late Mesozoic. Spreading direction of the Kula–Pacific and northern Pacific–Farallon ridges changed clockwise with time (40° from M-26 to Cenozoic anomaly 32), while spreading direction changed counterclockwise on the Pacific–Phoenix ridge. This evolution of the spreading system has persisted to the present time; thus, the Pacific plate has apparently formed from a small cell by the unfolding of the spreading ridges bordering it.

The discovery that linear magnetic anomalies parallel to mid-ocean ridges are formed by alternate normal and reversed field magnetization of the oceanic crust as it is produced at spreading ridges [*Vine and Matthews*, 1963; *Vine*, 1966] has brought about a revolution in our knowledge of the tectonic history of the oceans. Age determinations of magnetic reversals recorded in basalt flows [*Cox et al.*, 1964] and in deep sea sediments [*Opdyke et al.*, 1966] allowed the development of the Cenozoic sea-floor spreading, geomagnetic field reversal model by *Heirtzler et al.* [1968]. Most oceanic magnetic anomalies were then identified as Late Cretaceous to Recent in age and their distribution and pattern mapped [*Pitman et al.*, 1968; *Dickson et al.*, 1968; *Le Pichon and Heirtzler*, 1968], providing confirmation of the theory of sea-floor spreading [*Dietz*, 1961; *Hess*, 1962]. Subsequent work with oceanic magnetic anomalies related nearly all of the North Atlantic Ocean [*Pitman et al.*, 1971; *Pitman and Talwani*, 1972], Indian Ocean [*McKenzie and Sclater*,

1971], and eastern Pacific Ocean [*Atwater and Menard,* 1970; *Herron,* 1972*a*], to origin by Cenozoic sea-floor spreading within the framework of plate tectonics [*Le Pichon,* 1968; *Morgan,* 1968].

Linear magnetic anomalies that can be correlated between widely separated regions of the western Pacific Ocean have been found not to match the Cenozoic model [*Hilde and Raff,* 1970; *Hayes and Pitman,* 1970; *Larson et al.,* 1972; *Isezaki,* 1973]. These anomalies were first mapped east of Japan by *Uyeda et al.* [1962] and it was recognized that they were similar in character to those found in the eastern Pacific by the Pioneer surveys [*Mason and Raff,* 1961]. Further investigations [*Uyeda et al.,* 1967; *Uyeda and Vacquier,* 1968] showed the greater distribution and prominent NE–SW linearity of these anomalies, but no spreading ridge exists in this region to which they could be attributed. *Hayes and Pitman* [1970] found another group of anomalies west of the Hawaiian ridge and, on the basis of Mesozoic geologic samples from the region, suggested that these and the lineations east of Japan were formed between Cretaceous and Triassic or Jurassic time. Although they could not demonstrate that the two sets of anomalies were connected, they speculated that they might form a 'magnetic bight' in the vicinity of Shatsky rise similar to that mapped by *Grim and Erickson* [1969] for the Cenozoic anomalies in the northeast Pacific.

Based on an advance copy of the Hayes and Pitman manuscript, a survey of the northwest Pacific was planned with a major objective of determining details of the distribution of these anomalies and whether or not the Hawaiian and Japanese lineations connect in a magnetic bight (Figure 1). This investigation employed two ships, the *USNS Silas Bent* and *F. V. Hunt,* for a period of four months ending in December of 1969. Using these and extensive other data, *Isezaki* [1973] mapped an extended distribution of the lineations. He demonstrated by model studies that the latitude of formation was between the equator and 20°S, in good agreement with the paleolatitudes determined from magnetic surveys of western Pacific seamounts by *Uyeda and Richards* [1966], *Richards et al.* [1967], and *Vacquier and Uyeda* [1967].

*Larson et al.* [1972] mapped a third group of western Pacific magnetic anomalies in the vicinity of the Phoenix Islands. *Larson and Chase* [1972] correlated all three groups of lineations (Japanese, Hawaiian, and Phoenix sets) and *Larson and Pitman* [1972] derived a reversal model for the Pacific anomalies. On the basis of Deep-Sea Drilling Project results and previously determined geologic ages, they dated the anomalies M-1 to M-22 as 112 to 148 m.y.B.P. *Larson and Chase* [1972] proposed a history of evolution for the western Pacific from Mesozoic to Cenozoic time. While their proposed history infers that a magnetic bight existed between the Japanese and Hawaiian lineations, they did not map this connection.

The subject of this paper is the evolution of the sea-floor spreading system that formed the present northwest Pacific Ocean crust, from the very earliest stages of the spreading system to its Cenozoic continuation, as deduced from the magnetic lineation pattern. The research presented here is an extension of *Hilde* [1973].

The late Mesozoic magnetic reversal sequence is revealed from four closely spaced, parallel magnetic profiles recorded across the Hawaiian lineations. Because these profiles clearly extend into the quiet zones on each side of the Hawaiian lineation pattern and because the anomalies are repeated in detail from profile to profile (Figure 2), it is assumed that these data have recorded the complete series of reversals. These data are the basis for further anomaly identifications and construction of the late Mesozoic reversal model (Figures 3 and 4).

The magnetic anomaly lineations are mapped for the northwest Pacific region west of the Emperor seamount–Hawaiian ridge and north of the Mid-Pacific mountains (Figure 5). Fracture zones are mapped and relative spreading rates and directions are determined (all motions are considered relative to a hypothetically fixed Pacific plate unless otherwise stated). Explanations for the origin of major ocean floor features including Shatsky rise, Hess rise, the Emperor seamount chain, and the Hawaiian ridge are offered. Finally, the evolution of the spreading system is considered in the context of plate tectonics theory.

## GEOLOGY AND REGIONAL DESCRIPTION

The northwest Pacific basin is generally flat at about 5500–6000 m depth. It is bordered on the north and west by the active Kuril, Japan, and Bonin trenches. To the south are the Mid-Pacific mountains, a broad E–W zone of seamounts extending from the central Pacific to the Bonin and Mariana trenches. On the east is the Emperor seamount chain, a narrow N–S line of large seamounts, and the Hawaiian ridge which extends southeast from the southern end of the Emperor seamount chain. Clusters of seamounts are located near the trench east of Japan and along the southern portion of the basin. (For bathymetry of the North Pacific see *Chase et al.* [1971].)

In the center of the northwest Pacific basin is a large shoal, the Shatsky rise, which rises to a minimum depth of about 2000 m. Slopes are gentle and it is linear in a NNE direction. As will be shown later this is believed to result from the evolution of the spreading system in the northwest Pacific during the Mesozoic.

Other significant topographic features whose origin will be related to the evolution of the seafloor spreading system are Hess rise, the Mendocino fracture zone, Emperor trough, and the Surveyor fracture zone.

Hess rise is similar in nearly all respects to Shatsky rise, except it is in mirror image with respect to its location and shape. Depth, slope, size, and general shape are otherwise very similar.

The North Pacific is now known to extend in age from at least Jurassic in the west to the present along the East Pacific rise in the east. *Sclater and Francheteau* [1970] divided the North Pacific into age provinces on the basis of Deep Sea Drilling Project results [*McManus and Burns,* 1969; *DSDP Scientific Staff,* 1969*a*] and the distribution of magnetic anomalies in the northeastern Pacific [*Atwater and Menard,* 1970]. *Larson and Chase* [1972] compiled the later Deep Sea Drilling Project dates for the northwest Pacific and by correlating a section of the magnetic anomalies (M-1 to M-10) between Shatsky

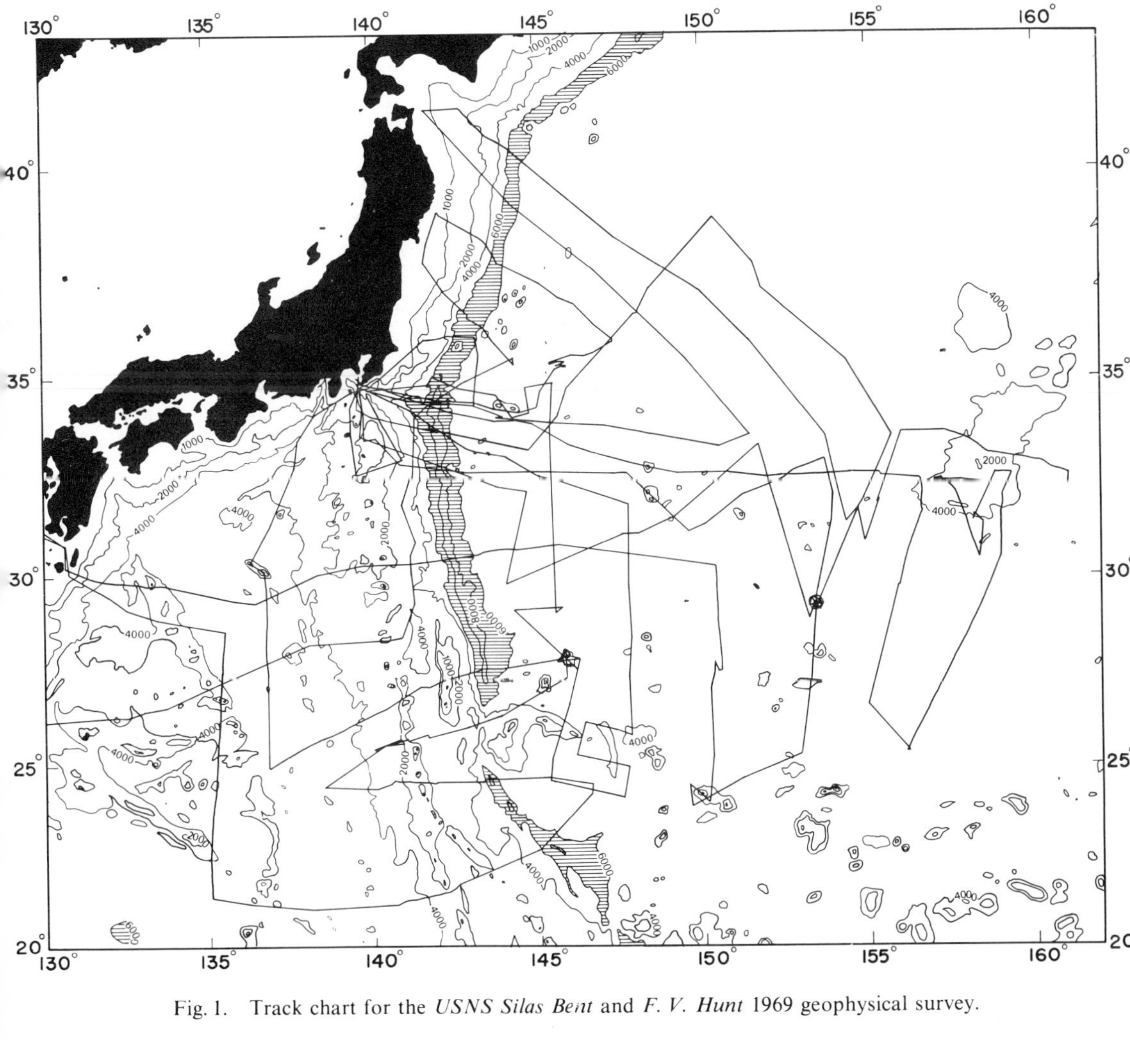


Fig. 1. Track chart for the *USNS Silas Bent* and *F. V. Hunt* 1969 geophysical survey.

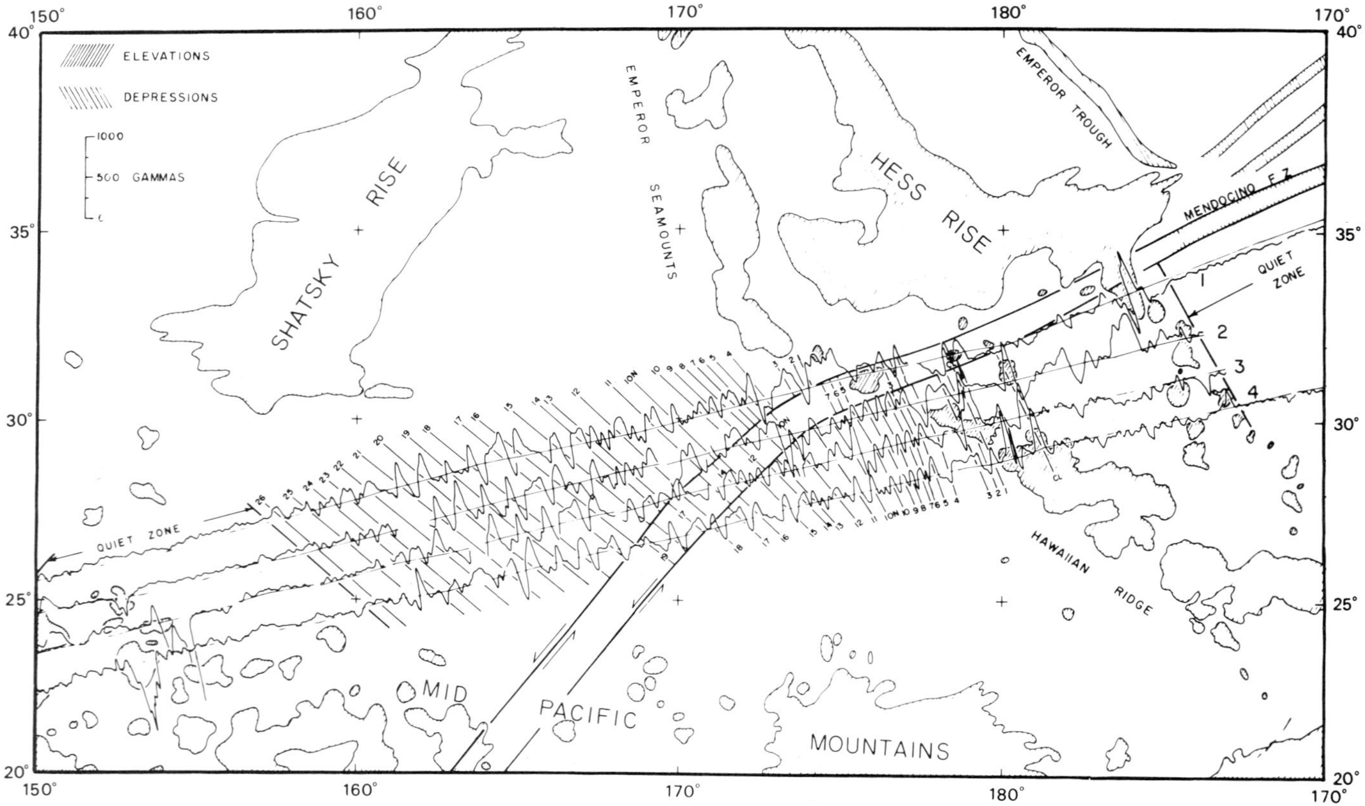


Fig. 2. Late Mesozoic magnetic anomalies and correlations. The data were collected by the ocean survey ship *Oceanographer* during the NOAA Western Pacific Geotraverse Project, 1972. Numbering is after *Larson and Pitman* [1972] with the additions of CL (Cretaceous lineation), M-10N (after Naugler), M-23, -24, -25, and -26. The complete late Mesozoic reversal epoch is recorded from the mid-Mesozoic quiet period to the Cretaceous quiet period. Arrows shown on the Mendocino fracture zone indicate offset of the lineations, not fault motion.

rise and Japan with dated anomalies from the Hawaiian and Phoenix sets, and Atlantic Keathley lineations [*Vogt et al.*, 1971*b*], they concluded that this region of ocean crust was 110 to 130 m.y. old.

*J. Ewing et al.* [1968] mapped the distribution of an opaque layer in the lower sediment section throughout the northwest Pacific by seismic reflection profiling. They concluded that its upper surface was as old as Cretaceous on the basis of a Cretaceous sediment sample from this layer on Shatsky rise, dated as Albian [*M. Ewing et al.*, 1966]. The distribution and general Mesozoic age of the opaque layer has since been confirmed by Deep Sea Drilling Project results. The thickness of this layer is greatest at Shatsky rise, however, the opaque layer uniformly thickens toward about 15°N and 155°E. As will be shown later the magnetic data suggest that this region is very likely the oldest ocean crust of the Pacific.

While Cretaceous fossils have been sampled from the Mid-Pacific mountains [*Hamilton*, 1956], the crust of this region is most likely older. It is unclear whether the Mid-Pacific mountains are contemporaneous with the origin of the surrounding ocean crust or were formed shortly afterwards. Cretaceous fossils have also been sampled from Erimo seamount, located at the

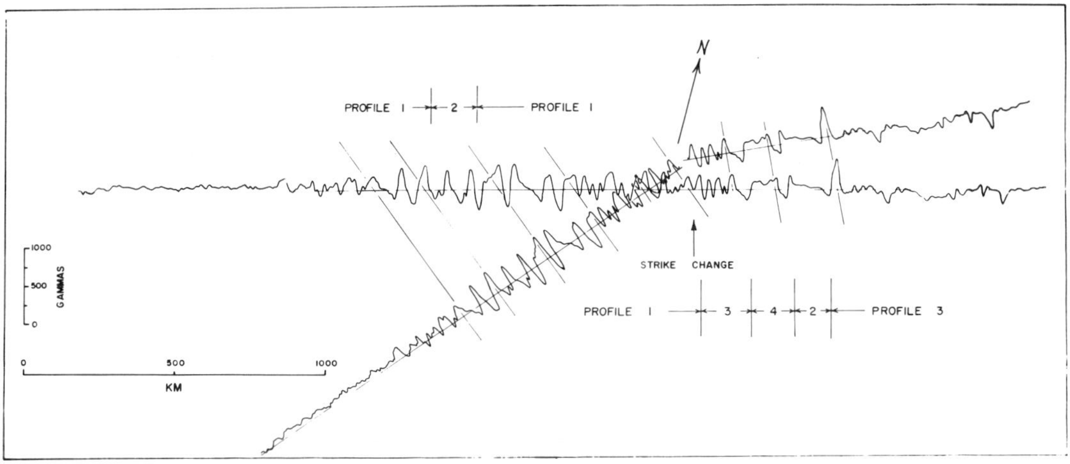


Fig. 3. Construction of the composite profile for the Hawaiian anomalies from the indicated sections of the four profiles in Figure 2.

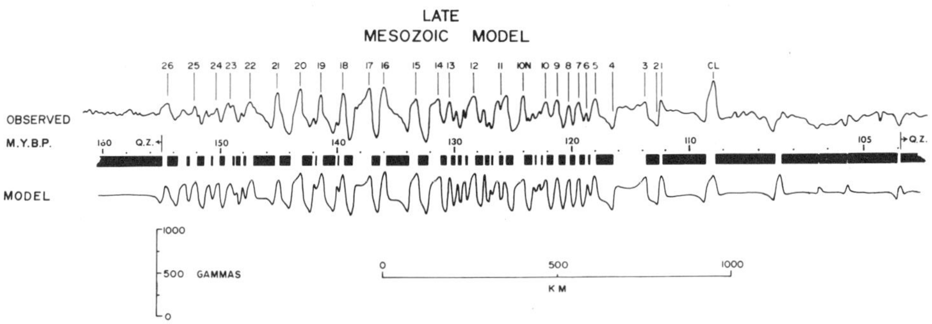


Fig. 4. The late Mesozoic model. The model profile was computed at −15° magnetization inclination and varied emu/cc; emu/cc equals $1.0 \times 10^{-3}$ (M-26 to M-21), $1.5 \times 10^{-3}$ (M-20 to M-1), and $1.0 \times 10^{-3}$ (CL to Cretaceous quiet zone). The time scale is taken from *Larson and Pitman* [1972] and *Hilde* [1973] with the beginning of the reversal period dated as 155 m.y.B.P. based on basement ages for DSDP sites 100 and 105 adjacent to M-26 in the Keathley lineations, and between M-7 and M-8 dated as 120 m.y.B.P. based on that basement age for DSDP Site 166. The time scale younger than 110 m.y.B.P. is expanded based on the 10 cm/yr spreading for this time and place proposed by *Larson and Pitman* [1972]. Other parameters are given in Table 1.

intersection of the Japan and Kuril trenches [*Kagami et al.*, 1964].

Ages of seamounts in the Emperor seamount chain have been shown to be progressively younger from the north, where a Cretaceous sample has been recovered from Meiji seamount [*DSDP Scientific Staff*, 1971*b*] to the south [*Ozima et al.*, 1970; *Clague and Dalrymple*, 1973]. *Clague and Jarrard* [1973] reviewed the dates of seamounts in the Hawaiian–Emperor seamount chain and concluded that the Emperor chain is 42–44 m.y. old at the connection it makes with the Hawaiian ridge. The ages of seamounts of the Hawaiian ridge likewise are progressively younger from the connection to Hawaii which is volcanically active at the present time. *Handschumacher* [1973] suggested that the Emperor and Hawaiian ridges are a continuous boundary between the eastern and western portions of the North Pacific, formed by volcanic activity along faults or zones of weakness, and that origin of the basins on either side of this boundary was by unrelated tectonic activity. Neither of these suggestions seems likely in view of this investigation and

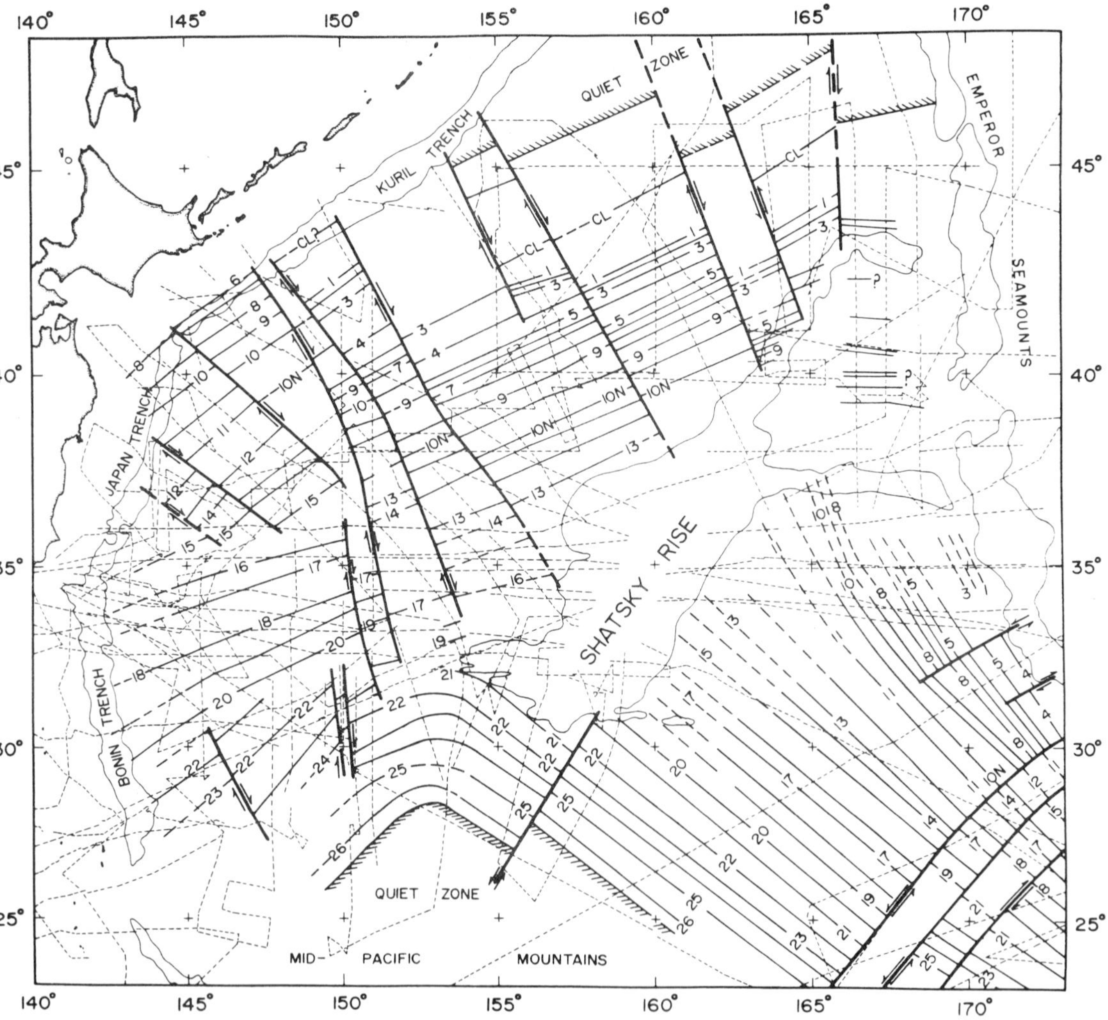


Fig. 5. Mesozoic magnetic anomaly lineations in the northwest Pacific Ocean. Solid lines are the anomaly correlations. Heavy solid lines are fracture zones. Dashed portions of these lines mean either a projection of the correlation or less than positive identification. Light dashed lines are ship tracks for the magnetic data used in mapping the pattern. Arrows on the fracture zones indicate offset of the lineations, not fault motion.

previous studies. The Hawaiian–Emperor seamount chain is apparently of later origin than the surrounding crust; it may have been created by a 'hot spot' as proposed by *Wilson* [1963*a, b*] and later expanded on by *Morgan* [1971, 1972*a, b*].

## MAGNETIC ANOMALIES

In addition to the data collected by the authors (Figure 1), the magnetic data collections for this region from Scripps Institution of Oceanography and the Earthquake Research Institute of the University of Tokyo were used in mapping of the magnetic lineations (Figure 5). Several institutions have contributed to these collections including the Meteorological Agency of Japan, the Ocean Research Institute of the University of Tokyo, Lamont-Doherty Geological Observatory, the U.S. Naval Oceanographic Office, and the Pacific Oceanographic Laboratory of the National Oceanographic and Atmospheric Administration (NOAA).

In nearly all cases the total intensity of the earth's magnetic field was measured with a proton-precession magnetometer. Precision of these measurements is generally within one gamma. Navigation was by various means, including satellite navigation system positioning, LORAN A and C, and star fixes. The accuracy of positioning of the magnetic data is generally within ± 1 km and probably only rarely worse than ± 3 km.

The magnetic anomalies were derived by subtracting the International Geomagnetic Reference Field [*IAGA, Commission 2 Working Group 4,* 1969] from the total field measurements. They were then plotted as profiles along the ship tracks. Initial correlations were made from compilations in this form. After lineation directions were established, several profiles (Figure 6) were projected normal to the strike of the anomalies for further evaluation of the correlations (Figures 7 through 11, 15, and 16). Identification of the anomalies is based on the anomaly sequence of the Hawaiian set (Figure 2) and the late Mesozoic reversal model (Figure 4) derived from these anomalies [*Hilde,* 1973]. Sea-floor topography was examined where unusual anomalies were observed and in nearly every case these anomalies were found to be associated with seamounts or other topographic features. Fracture zones are mapped along discontinuities in the anomaly lineations.

The Hawaiian anomalies (Figure 2) are discussed first because it is from these that the late

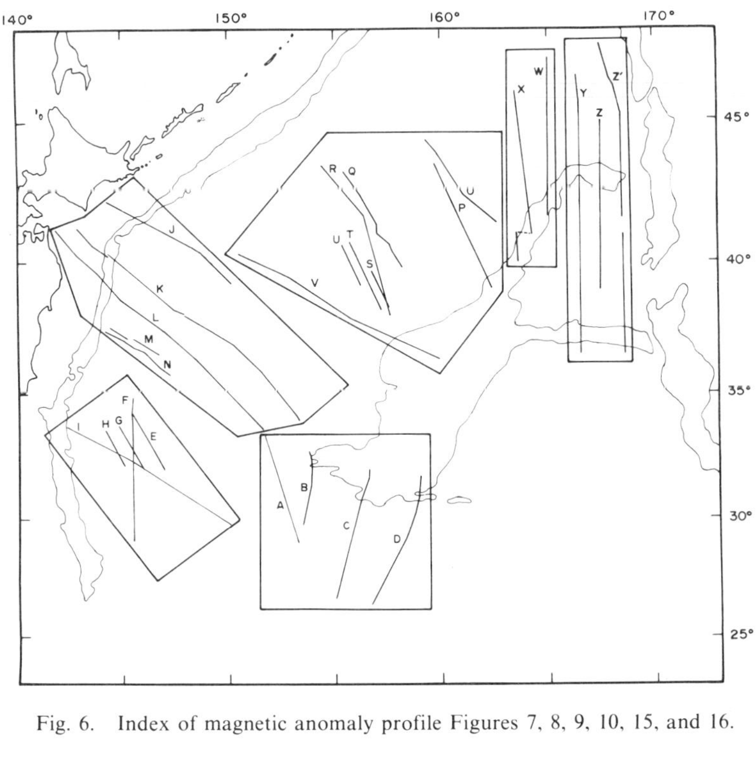


Fig. 6. Index of magnetic anomaly profile Figures 7, 8, 9, 10, 15, and 16.

Mesozoic sequence is identified and the reversal model constructed. The connection of the Hawaiian and Japanese lineations is then established followed by anomaly identifications and a discussion of the lineation pattern for the northwest Pacific. The spreading system that formed this pattern will be described from its oldest stage to the Cenozoic continuation.

*The Hawaiian lineations.* Four parallel profiles of magnetic data were recorded by the ocean survey ship *Oceanographer* during NOAA's Western Pacific Geotraverse Project in a WSW direction across the western Pacific from 800 km east of the Hawaiian ridge. These lines, spaced approximately 110 km apart, clearly extend from the Cretaceous quiet zone in the east, across the complete sequence of late Mesozoic reversals to the mid-Mesozoic quiet zone in the west (Figure 2). The identification attached to the anomalies follows the numbering system of *Larson and Chase* [1972] with the major additions being anomaly M-10N, anomaly CL (younger than anomaly M-1), and anomalies M-23 through M-26 [*Hilde,* 1973]. The N designation for anomaly M-10N is after the late Fred Naugler, co-chief scientist of the NOAA Western Pacific Geotraverse Project. CL means Cretaceous Lineation. If it is later discovered that individual reversals can be clearly identified in this time period, they may be labeled as CL-1, CL-2, etc., without confusion.

Inspection of the four profiles in Figure 2 shows that the anomalies are recorded in good detail in each profile. Because of the near-perfect repetition of the anomalies from one profile to another, and the clear definition of the quiet zones on either side of the sequence, the complete late Mesozoic reversal epoch is believed represented in these data. It appears that the anomaly amplitude of the profiles systematically increases from the beginning of the reversal epoch to a high near the middle of the epoch (M-16 and -17) and then decreases, although less systematically, towards the Cretaceous quiet period (Figure 2). While lack of reversals appears the most likely explanation for the quiet periods [*McElhinny and Burek,* 1971], another explanation is required for the different amplitudes. Differences in thickness or magnetization of the blocks, or a change in the intensity of the Earth's field, which seems more likely, could produce the different amplitudes. Decay of the magnetized layer in such a systematic manner seems unreasonable.

There are numerous seamounts of the Mapmaker seamount group in this region [*Chase et al.,* 1971] and the few peculiarities that are found in the anomalies are associated with seamount crossings. Anomalies M-4 to M-8 on profile 1, M-14 and M-16 on profile 2, M-10N on profile 3, and anomalies M-23 through M-26 on profile 4, are disturbed due to seamount crossings. Anomaly CL on profile 1 is likewise disturbed between the

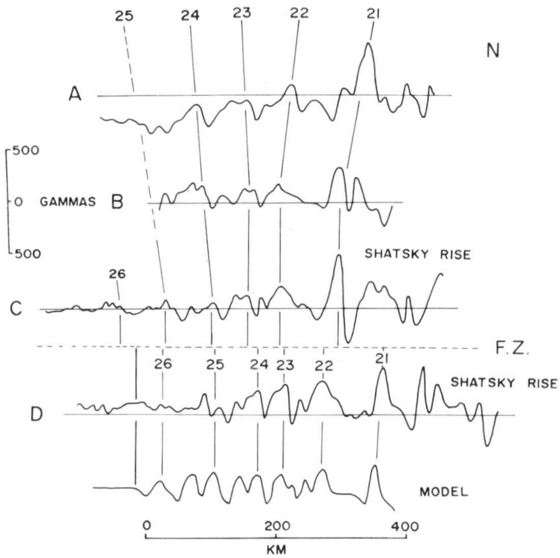


Fig. 7. Magnetic anomaly profiles and correlations south of Shatsky rise through the magnetic bight. Profiles are projected normal to the strike of the lineations.

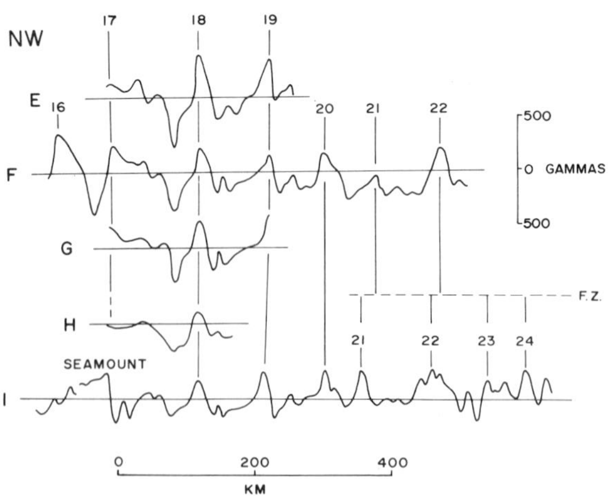


Fig. 8. Magnetic anomaly profiles east of the Bonin trench. The profiles have been projected normal to the strike of the lineations. Spreading was 1.5×that for the Hawaiian lineations.

fracture zones due to a seamount. Except for these cases and at the Hawaiian ridge, the anomalies are repeated. The Mapmaker seamounts are not shown in Figure 2 for the sake of clarity.

Two parallel fracture zones [*Hayes and Pitman,* 1970] offset the lineations by a total of 310–410 km, but the anomalies are disturbed very little at the offsets and can easily be identified between them. These fracture zones are traced east through the Hawaiian ridge, directly into the Mendocino fracture zone at the southeast edge of Hess rise. We conclude that they are the westward extension of the Mendocino. They are projected southwest into the Mid-Pacific mountains region by assuming their trend is maintained normal to the magnetic lineations.

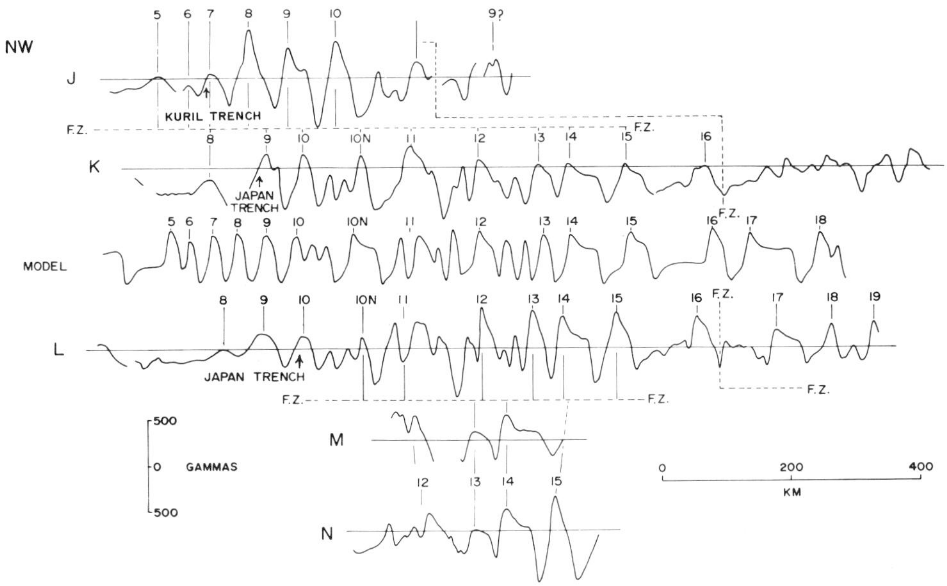


Fig. 9. Magnetic anomaly profiles east of the Japan trench. The profiles have been projected normal to the strike of the lineations. Spreading was 1.35X that for the Hawaiian lineations.

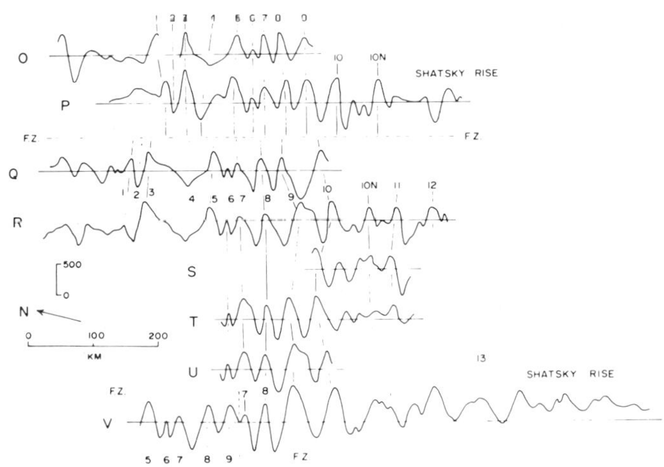


Fig. 10. Magnetic anomaly profiles northwest of Shatsky rise, projected normal to the lineation direction. Spreading was 1.1X that for the Hawaiian lineations.

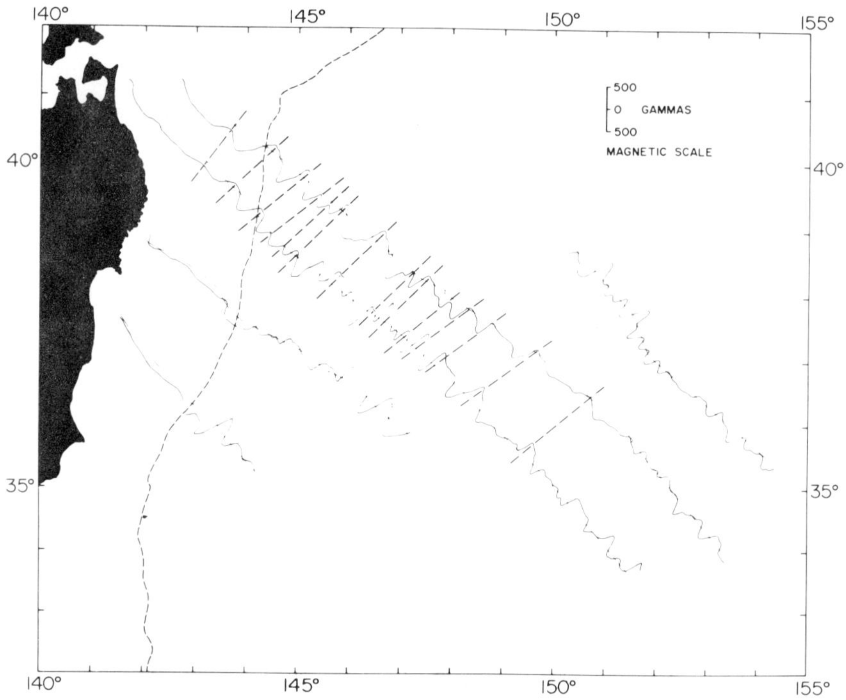


Fig. 11. Correlations of profiles K and L (Figure 9) along the ship tracks. The anomalies clearly strike across the trench near northern Honshu.

Strike of the anomalies north of the Mendocino was shown by *Hilde* [1973] to be N50° W until about anomaly M-13. At this time the spreading direction started to change clockwise and continued to do so until M-4 to M-1 (Figures 2 and 5). South of the Mendocino the change in strike of the anomalies, and therefore spreading direction, was accomplished mostly between M-13 and M-10 [*Hilde,* 1973]. By the time of M-1, both south and north of the Mendocino, the strike of the anomalies is N25° W.

While offset of the Mesozoic lineations across the Mendocino is 310 to 410 km, there is 1,200 km of offset across the fracture zone for Cenozoic lineation 21. *Larson and Chase* [1972] noted this increased offset and concluded that it must be the result of faster spreading on the Pacific–Farallon ridge between the Mendocino and Murray fracture zones from 110 to 85 m.y.B.P. An alternate explanation is for the spreading center south of the Mendocino having jumped several hundred kilometers to the east. If this was the case then a large section of the Farallon plate must be welded into the Pacific plate south of the Mendocino fracture zone and east of the Hawaiian lineations for which reversed distribution and age of the sediments older than the time of the jump should exist. There is no evidence of jumps in the Hawaiian lineations (Figure 2), which would mean that the jump would have had to take place during the Cretaceous quiet period when no magnetic lineations were formed that could have been useful in examining this possibility. If the increased offset is due to more rapid spreading for this period south of the Mendocino, then inversely proportional asymmetric spreading must have existed each side of the fracture zone or the two parts of the Farallon plate north and south of the Mendocino would have had to move as individual plates during this time.

It should be noted that, as observed by *Hayes and Pitman* [1970], the magnetic lineations strike through the Hawaiian ridge with no offset. This is well established by these data and means that no lateral movement has occurred along the Hawaiian ridge. Because anomalies M-3 through CL show no disruption of strike through the ridge, any horizontal separation of the plate that may have occurred, as suggested by *Molnar and Atwater* [1973*a*] in attributing the Hawaiian ridge origin to that of a propagating crack, must have been very little indeed. In addition, there is no observable offset of the Hawaiian ridge by the Mendocino fracture zone which means that no movement has occurred along the fracture since this part of the Hawaiian ridge was formed; at least for the last 25 m.y. [*Clague and Jarrard,* 1973]. These determinations support transform fault origin of the Mendocino. The observed relationship of the magnetic anomalies and Mendocino fracture zone to the Hawaiian ridge support 'hot spot' origin of the Hawaiian ridge [*Wilson* 1963*a, b; Morgan,* 1971].

*The late Mesozoic reversal model. Hilde* [1973] constructed a late Mesozoic reversal model from a composite profile of the four profiles in Figure 2 which contained nearly twice the number of reversals as the original model by *Larson and Pitman* [1972]. Major additions in the *Hilde* [1973] model were reversals for anomalies M-23 through M-26, M-10N, and for anomaly CL and unlabeled reversals on the younger side of M-1. *DSDP Scientific Staff* [1973*c*] also found M-23 through M-25 during *Glomar Challenger* Leg 32 of the Deep Sea Drilling Project.

In this paper (Figure 4) we present the block model of *Hilde* [1973]. M-26 and the reversals younger than CL are less obvious in the profiles but are retained because we have been able to use them in correlating the corresponding portions of the Hawaiian and Japanese lineation patterns (Figures 2, 5, and 19). While these anomalies are not recognized everywhere that the Mesozoic lineations have been found, those younger than CL are believed to be recognized in the Phoenix Islands region (Bruce Rosendahl personal communication, 1973), and we think we can see M-26 in profiles of the Keathley lineations [*Vogt et al.,* 1971*b*]. The Cretaceous quiet zone boundary for the model is primarily based on the difference in character of the anomalies on either side of the boundary as shown on the profiles in Figure 2. The strike of the edge of the Cretaceous quiet zone, east of the Hawaiian ridge, may be slightly different than shown in Figure 2 because seamounts extend along this line NNW from the Gardner Pinnacles on the Hawaiian ridge, making precise representation of this border difficult. However, the difference in amplitude of the magnetic anomalies on profiles either side of this location strongly suggests that

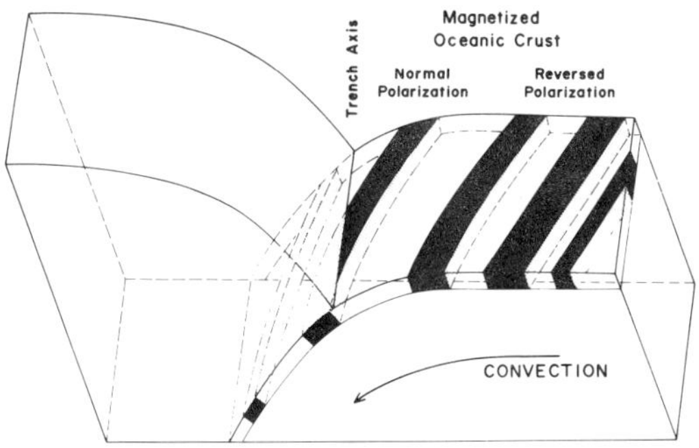


Fig. 12. Three-dimensional view of the magnetized ocean crust in relation to the Japan trench.

this is the start of the Cretaceous quiet zone. The line of seamounts is believed to be coincident.

The time scale shown with our model (Figure 4) is from *Larson and Pitman* [1972] and *Hilde* [1973]. *Larson and Pitman* [1972] relied primarily on basement ages from DSDP sites 100, 105, and 166 for fixing a time scale to their model. We have done the same, differing only in that our expanded model was available for comparison to the Keathley lineations and adjacent DSDP sites 100 and 105. DSDP sites 100 and 105, which have basement ages of about 155 m.y.B.P., correspond closely to what we believe is M-26 in the Keathley lineations. The 155 m.y.B.P. age for the beginning of the reversal period [*Hilde,* 1973] is therefore applied to our model. The 120 m.y.B.P. basement age for DSDP Site 166 is applied to our model between M-7 and M-8 according to *Larson and Pitman* [1972]. For the portion of our model younger than 110 m.y.B.P. we have shown a time scale based on the 10 cm/yr spreading rate suggested by *Larson and Pitman* [1972] for the Hawaiian lineations south of the Mendocino fracture zone. So that the model can be easily compared with the four profiles in Figure 2, we have kept a constant distance scale, which means the time scale calibration has different spacing either side of 110 m.y.B.P. According to our model the late Mesozoic reversal period was from 155 to 104 m.y.B.P.

The variables used in computing our model profile and the reversal boundaries of the model are given in Table 1. The observed profile shown in Figure 4 is a composite of the four profiles

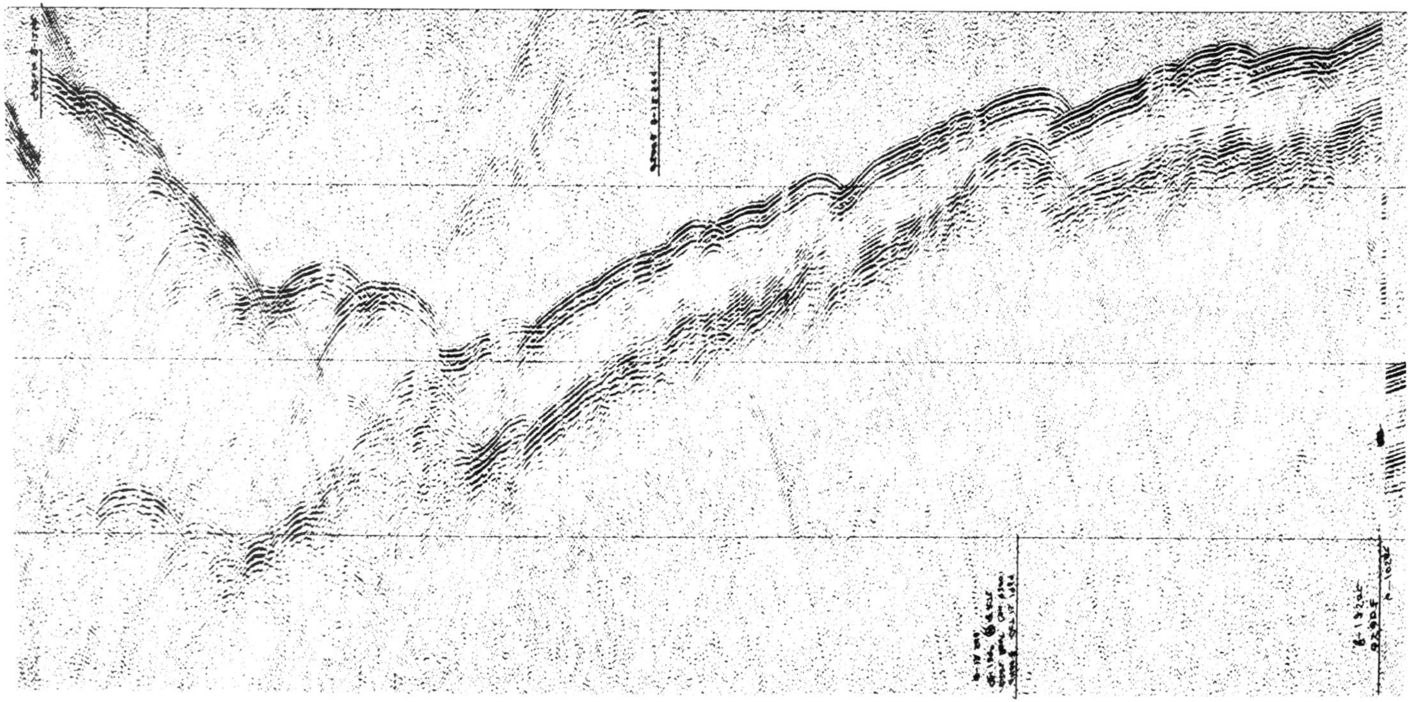

Fig. 13. Seismic reflection profile of the Japan trench recorded along profile *K*. Horizontal length equals approximately 80 km and the vertical scale is 4 sec. Sediment is folded against the shoreward slope at the trench axis, and a faulted oceanic crust dips beneath the shoreward slope.

across the Hawaiian lineations shown in Figure 2. The construction of the observed profile is shown in Figure 3. In the remaining discussion this is referred to as the observed profile.

In deriving the model, a combination of block thickness and magnetization was selected consistent with existing knowledge of the thickness of the basaltic crust for the western Pacific and measurements of the remanent magnetization of the oceanic basalts that gave an amplitude that matches the observed data. It is not intended to examine crustal thickness–magnetization relationships, and the thickness of the magnetized blocks could well be other than what we selected. However, only amplitude of the computed anomalies is affected by changes in either the intensity of remanent magnetization or block thickness in the model and identical anomalies can be produced by decreasing one and increasing the other. So, for the purpose of building the reversal model it is only important that amplitude be matched by some combination of these two variables.

For our model computations, assumptions were made that the present declination is zero and that there has been no change since time of origin as suggested by the seamount magnetic studies of *Uyeda and Richards* [1966]. Any rotation of the Pacific plate, other than northward relative to the magnetic pole [*Vacquier and Uyeda,* 1967; *Larson and Chase,* 1972] has probably been relatively small and would not significantly affect the paleolatitude determinations or the primary objective of constructing an accurate

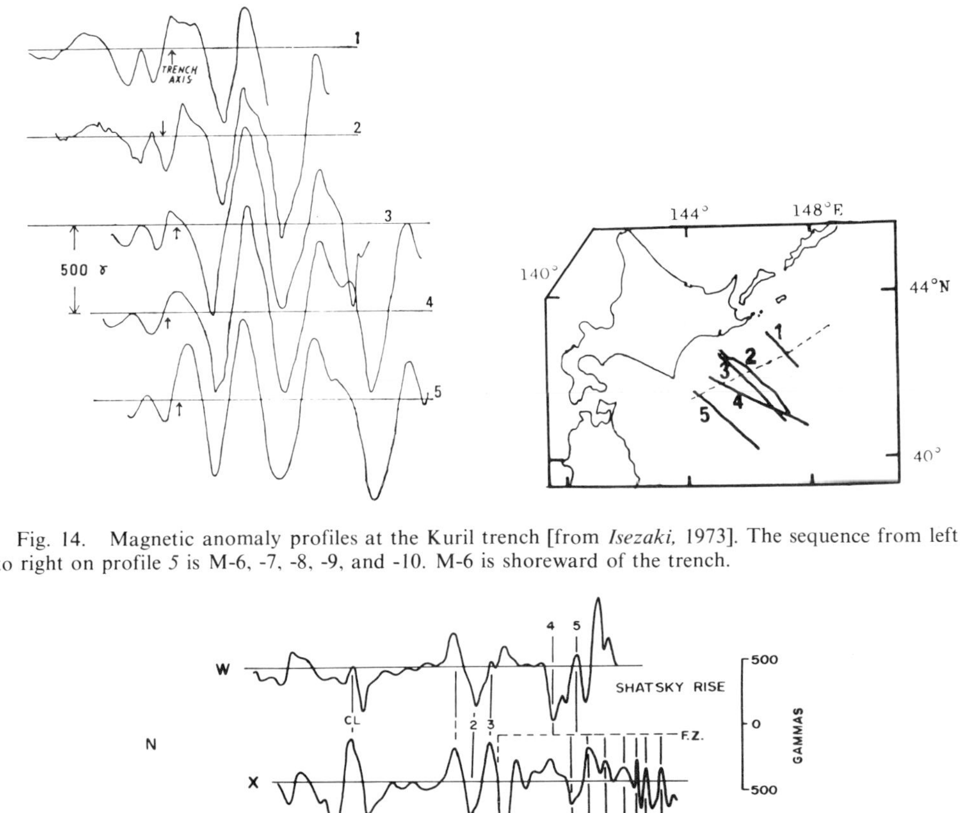


Fig. 14. Magnetic anomaly profiles at the Kuril trench [from *Isezaki,* 1973]. The sequence from left to right on profile *5* is M-6, -7, -8, -9, and -10. M-6 is shoreward of the trench.

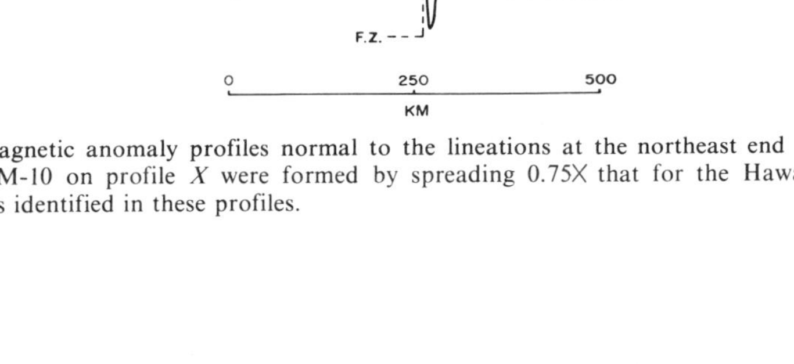


Fig. 15. Magnetic anomaly profiles normal to the lineations at the northeast end of Shatsky rise. M-4 through M-10 on profile *X* were formed by spreading 0.75X that for the Hawaiian lineations. Anomaly CL is identified in these profiles.

model of the late Mesozoic reversal epoch. The other variables are: present magnetic inclination, taken from the geomagnetic field for 1965.0 [*IAGA, Commission 2 Working Group 4,* 1969] for the latitude at which the anomalies are observed, the apparent spreading direction as measured, the magnetic inclination at latitude of formation, and the distribution of the blocks. As in other model studies the dike model has been used which assumes the blocks to be infinitely long, and normal periods of polarity are used for the quiet zones based on rock paleomagnetic studies [*McElhinny and Burek,* 1971]. Model profiles were computed for various inclinations of magnetization to find the best match with the observed profile. The best-fitting model thus provides an estimate of latitude of formation.

The best match of the computed profiles to the observed profile is for an inclination of −15° which indicates this section of ocean crust has moved 30° to 35° north since its time of origin. This is in general agreement with the results of *Uyeda and Richards* [1966], *Vacquier and Uyeda* [1967], *Francheteau et al.* [1970], and *Larson and Chase* [1972], and particularly the results of *Richards et al.* [1967] for seamount studies just west of the Hawaiian ridge.

The shape and distribution of the computed anomalies closely match the observed profile, but the amplitude for anomalies M-22 through M-26 and for the unlabeled anomalies younger than CL is clearly too great if constant thickness and intensity of magnetization are used in computation of the model profile. We have used a lesser intensity of magnetization for these portions of the model (Figure 4).

Because the late Mesozoic reversals are clearly repeated in each of the four profiles across the Hawaiian lineations and because the bordering quiet zones are well defined, we propose that this model be referred to as the late Mesozoic model, rather than the Pacific model as was done by *Larson and Pitman* [1972]. It is likely that all of the reversals for this period are represented in these data. Inclusion in the model of the short-period reversals has proved extremely useful in identifying widely separated magnetic anomalies.

*The magnetic bight.* The magnetic lineations of the Hawaiian and Japanese sets are assumed to have been produced by two ridges of a triple junction, the Pacific–Farallon ridge and Kula–Pacific ridge, respectively. A reconstruction of this assumed junction and the rest of the spreading system on the basis of the paleomagnetic poles for the Pacific and North American plates at 110 m.y.B.P. [*Larson and Pitman,* 1972] provides a good approximation of the Kula–Pacific, Pacific–Farallon, Kula–Farallon triple junction. Existence of the triple junction is confirmed by mapping the connection of the Hawaiian and Japanese lineations for anomalies M-26 through M-21 southwest of Shatsky rise (Figure 5). Here, they form a magnetic bight and define the location of the R–R–R triple junction at the margin of the Pacific plate during the earliest part of the late Mesozoic reversal epoch, 144–155 m.y.B.P. (M-21 to M-26). Correlation of these anomalies is shown in Figure 7 (the locations of all profiles are shown in Figures 5 and 6). The match between these profiles and anomalies M-21 through M-26 of the Hawaiian lineations is clear.

A small fracture zone is mapped south of Shatsky rise on the basis of the well-established lineations on either side (Figure 5). North of anomaly M-21 on profiles B, C, and D (Figure 7), the lineations are obliterated by the effects of Shatsky rise. West of the bight (Figure 5) spreading during this time was faster and more irregular than to the east. Between the bight and the Bonin trench, spreading was in a S50°E direction until M-20, when it changed to about S30°E. As can be seen in Figure 5 the spreading direction during this early period of reversals fluctuated, but generally changed clockwise as spreading continued on the Kula–Pacific ridge.

Failure to map anomalies M-23 through M-26 and the anomalies older than M-15 west and southwest of Shatsky rise [*Larson and Chase,* 1972] led *Larson and Pitman* [1972] to conclude that spreading started on the Kula–Pacific ridge at about 130 m.y.B.P. With the mapping of M-26 just west of the bight, we can see that spreading occurred at least 155 m.y.B.P. on both ridges and probably long before this time. The quiet zone before M-26 makes it difficult to determine when spreading started. If spreading was continuous prior to this time, it would have started at essentially a point described by the intersection of the orthogonals to the Kula–Pacific, Pacific–Farallon, and Pacific–Phoenix Mesozoic lineations, in the vicinity of 15°N and 155°E. The spreading rates for the Pacific Mesozoic anomalies determined by *Larson and Pitman* [1972] of 3.0 to 5.0 cm/yr would indicate an age of between 185 and 190 m.y. for this point.

*The Japanese lineations.* These are the magnetic anomaly lineations of a general NE strike west of Shatsky rise and the magnetic bight (Figure 5). Correlations of individual anomalies and fracture zone offsets are shown in Figures 8, 9, 10, and 15. Origin of these anomalies is presumed to have been by spreading from the Kula–Pacific ridge.

In the earliest stage of spreading observable from the anomalies, spreading was faster west of the bight, for the period M-26 to M-16. Correlations of these anomalies, just east of the Bonin trench are mostly from data collected at small angles to the strike of the lineations, which required lengthy analysis to establish their identity. However, the identification can be convincingly made from some of the more normal profiles (Figure 8). Identification of this sequence demonstrates the importance of the short period reversals in correlation of the profiles, particularly

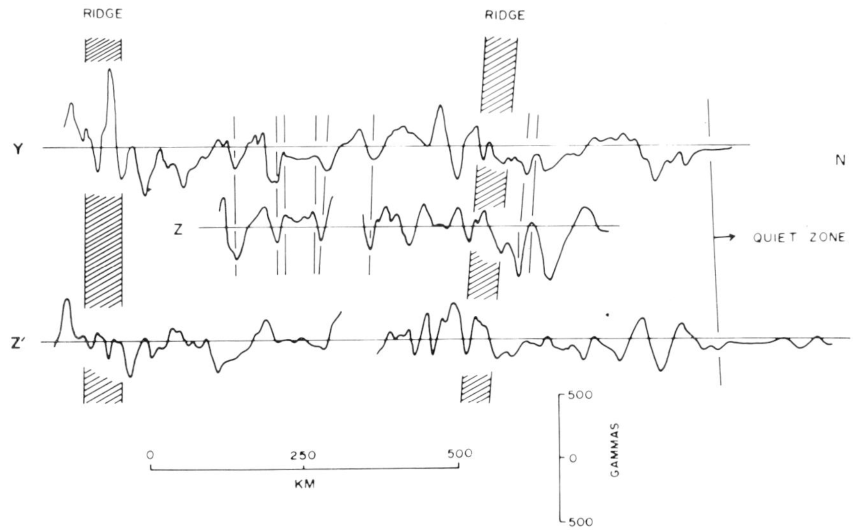


Fig. 16. Magnetic anomaly profiles east of northern Shatsky rise. Correlations between profiles $Y$ and $Z$ are east–west. There is a suggestion of symmetry about the anomaly halfway between the two ridges. Correlation with the late Mesozoic model is uncertain.

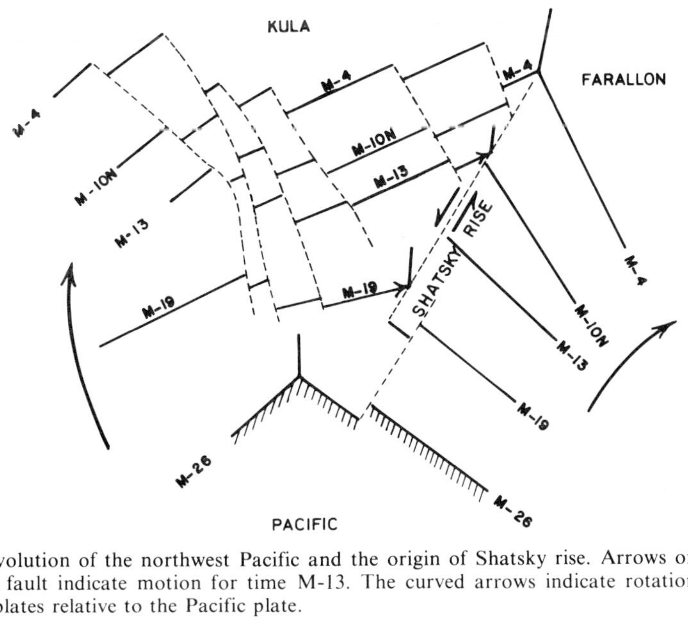


Fig. 17. Evolution of the northwest Pacific and the origin of Shatsky rise. Arrows on the Shatsky rise transform fault indicate motion for time M-13. The curved arrows indicate rotation of the Kula and Farallon plates relative to the Pacific plate.

those between M-18, -19, and -20.

Spreading between M-26 and M-16 is 1.5X that of the Pacific–Farallon ridge (Hawaiian lineations) for the same period. This difference in spreading rates between the two ridges may have initiated new fracture zones on the Kula–Pacific ridge in the vicinity of the bight at the time of M-21 (Figure 5). These fracture zones, extending NNW from between 150° and 155°E to the Kuril trench between 145° and 150°E, indicate that offset of the spreading ridge increased as spreading continued. Comparison of distances between anomalies either side of these fracture zones (Figures 9 and 10) shows spreading was slower on the northeast side.

Several model profiles were computed for comparison with the recorded profiles east of the Japan trench and the best match was obtained for between –10° and –20° magnetization inclination, indicating about 40° of northward drift since formation. The –20° magnetization inclination model profile is shown in Figure 9. Spreading that produced these anomalies was 1.35X that for the Hawaiian lineations. It is obvious by the difference in amplitudes and shapes between the recorded profiles that making an exact match with a model for this group is difficult. These differences are particularly evident for anomalies M-11 through M-13 on profiles K and L (Figure 9). However, the match is still convincing. The southeast end of profiles K and L are near the southwest edge of Shatsky rise and for profile K there may be more fractures crossing the profile to the right of M-16 than shown.

At the Japan trench the anomalies can be observed to strike across the trench and anomaly M-8 is recognized on profile L on Figure 9, 120 km shoreward of the trench axis. The strike of these anomalies across the trench axis is more easily seen in Figure 11. The correlations are proof that the oceanic crust dips west beneath the trench and continental slope. The counterclockwise-rotated strike of the anomalies at and shoreward of the trench may indicate the horizontal shortening of the anomaly spacing due to bending of the crust. The relationship of the magnetized ocean crust to the trench is shown in the diagrammatic model of Figure 12. Seismic reflection records across the trench axis for profile K of Figure 9 also show the ocean crust dipping west beneath the trench (Figure 13). The oceanic sediments are apparently folded against the shoreward trench slope and the faults in the ocean crust have progressively greater vertical throw toward the trench. These were probably originally normal faults near the apex of the bend in the crust east of the trench that later evolved into thrust faults as the ocean crust neared the trench axis. This interpretation suggests thrusting of the oceanic crust beneath the trench. The magnetic

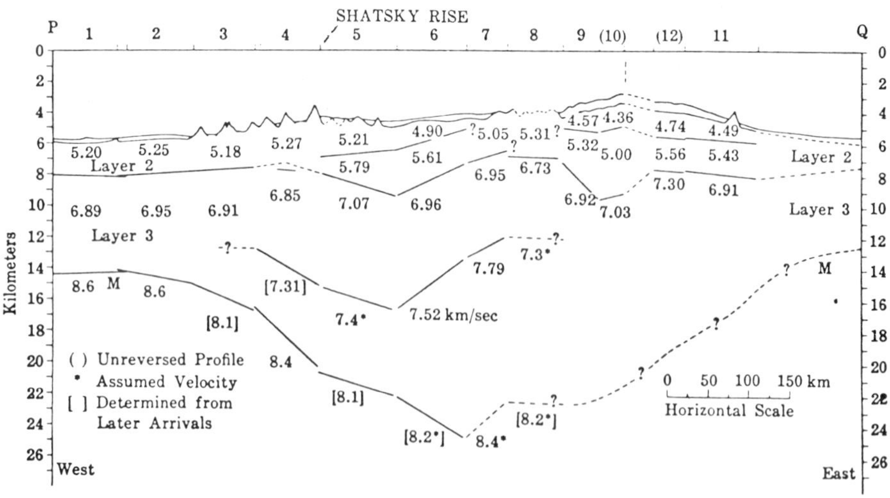


Fig. 18. East–west seismic refraction section along southern Shatsky rise [from *Den et al.*, 1969].

anomalies can also be observed beneath the Kuril trench as indicated by profile J (Figure 9) and as shown in Figure 14 [*Isezaki,* 1973].

The spreading direction for this portion of the plate (west of 150°E between 35°N and 47°N, Figure 5) changed sharply in a counterclockwise direction from M-16. This change seems in conflict with the general clockwise change of spreading direction throughout the sea-floor spreading history of the North Pacific (Figure 19). *Atwater and Menard* [1970] and *Atwater* [1970] showed that similar abrupt spreading changes for the Cenozoic Pacific–Farallon ridge were related to the ridge colliding with a trench along North America. Possibly this similar change on the Kula–Pacific ridge at about 136 m.y.B.P. likewise indicates collision of the ridge with a former trench along the Pacific margin of Asia[1]. If so, geologic evidence at the continental margin may be difficult to find because of subsequent geologic structures. Any evidence that does exist should be far to the south because of the approximately 40° of northward drift of the Pacific plate since that time. *Uyeda and Ben-Avraham* [1972] have suggested that collision of the Kula–Pacific ridge with Asia at 80–90 m.y.B.P. initiated the formation of the Japan Sea, and *Larson and Chase* [1972] and *Larson and Pitman* [1972] suggested the Kula–Pacific ridge was subducted beneath the Ryukyu–Japan–Kuril trenches about 100 m.y.B.P. Initial subduction was probably earlier and farther south.

Northwest of Shatsky rise (Figure 5) spreading was 1.10X that for the Hawaiian lineations. Anomalies M-1 to M-13 are identified in profiles *O* through *V* (Figure 10). While anomaly CL and other specific anomalies were not identified younger than M-1, it is still apparent that there were reversals in this time range. The character of these anomalies is very similar to those younger than CL in the Hawaiian set (Figure 2). Anomaly M-14 is not observed on profile *V* (Figure 10) because of the effect of Shatsky rise and M-10N is the last anomaly identified farther to the northeast along Shatsky rise on profile *P*. The fracture zones here show the same right lateral offset of the lineations with progressively slower spreading to the northeast side of each, as was seen to be the case for the lineation pattern nearer Japan (Figures 5 and 9). Between profile *O* (Figure 10) and profile *X* (Figure 15) there is another fracture zone (Figure 5) northeast of which spreading from anomaly M-10 became much slower. This coincides in time with the increase in spreading rate and change of spreading direction just southeast of Shatsky rise on the Pacific–Farallon ridge (Figures 2 and 5). Anomaly CL is clearly observed in profiles *W* and *X*.

*East-West lineations.* Between 165° and 170°E, east of the northern portion of Shatsky rise are two E–W topographic swells or ridges with associated seamounts [*Chase et al.,* 1971]. These are shown schematically on Figures 5 and 19 at the northeast end of Shatsky rise along 37° and 43°N. Between these ridges are E–W striking anomalies (Figure 16). Attempts to match them with the late Mesozoic model proved unconvincing. The only possibility of a match seems to be for reversals from M-4 to the Cretaceous quiet period. There is a strong suggestion of symmetry about the anomaly located mid-way between the two ridges, and if they are of late Mesozoic age, they may identify a brief period of interplate spreading at about 110 m.y.B.P. This time is consistent with projection of identified anomalies into this region. These anomalies are apparently associated with readjustments in the triple junction at the north end of Shatsky rise from the time of M-4 (116 m.y.B.P.).

The Cretaceous quiet zone can be seen on profiles *Y* and *Z'* (Figure 16), north of Shatsky rise. The large negative anomaly, about 80 km south of the quiet zone, is probably the same as the negative anomalies observed farther west near the north ends of profiles *O, Q,* and *R* (Figure 10). While the Emperor seamount chain forms an eastern topographic boundary to the northwest Pacific basin (Figure 5), it is not believed to be related to the origin of the surrounding ocean crust, because of its younger age.

[1]Footnote added in press: It has been learned from M. Steckler, T. Atwater, and N. Isezaki that new data collected by research vessel *Vema* during the summer of 1975 suggest that the magnetic lineations just east of Japan may have a trend of about N70°E, similar to the rest of the Japanese lineations. If so, the need for a special explanation of this region is obviated. See also discussion of this point on page 225.

## EVOLUTION OF THE NORTH PACIFIC

The oldest identifiable Japanese, Hawaiian, and Phoenix lineations describe a triangular Pacific plate boundary during the late Mesozoic. Spreading existed on both the Kula–Pacific and Pacific–Farallon ridges at M-26, and it is likely that spreading also existed on the Pacific–Phoenix ridge at that time. M-26 defines the ridge configuration at the beginning of the reversal epoch and from that time the details of the spreading system evolution can be observed in the lineations. If the assumption that spreading was continuous prior to M-26 (155 m.y.B.P.) is correct, then the Pacific plate originated at a point that is now located at 15° N, 155° E, and is 185 to 190 m.y. old. Prior to this time the Kula, Farallon, and Phoenix plates may have been in relative position to each other about an R–R–R triple junction.

An R–R–R triple junction between the Kula, Pacific, and Farallon plates existed between M-26 and M-20 (155 to 142 m.y.B.P.). Spreading on the Kula–Pacific ridge was 1.5× that of the Pacific–Farallon ridge. Whether or not this difference in spreading rates for the two ridges is the cause or the effect, the two became separated by a transform fault where they had previously joined at the R–R–R triple junction. This occurred at the time of M-20 at what is now the southwest edge of Shatsky rise and the transform fault increased in length along the long axis of the rise

Fig. 19. (Below and opposite) Magnetic anomaly lineations in the North Pacific. Arrows on transform faults indicate offset of lineations, not fault motion. The lineations west of the Emperor trough are of the late Mesozoic reversal epoch mapped in this investigation. East of Emperor trough, the lineations are all of the Cenozoic reversal epoch [after *Atwater and Menard,* 1970; *Peter et al.,* 1970; *Naugler and Wageman, 1973*].

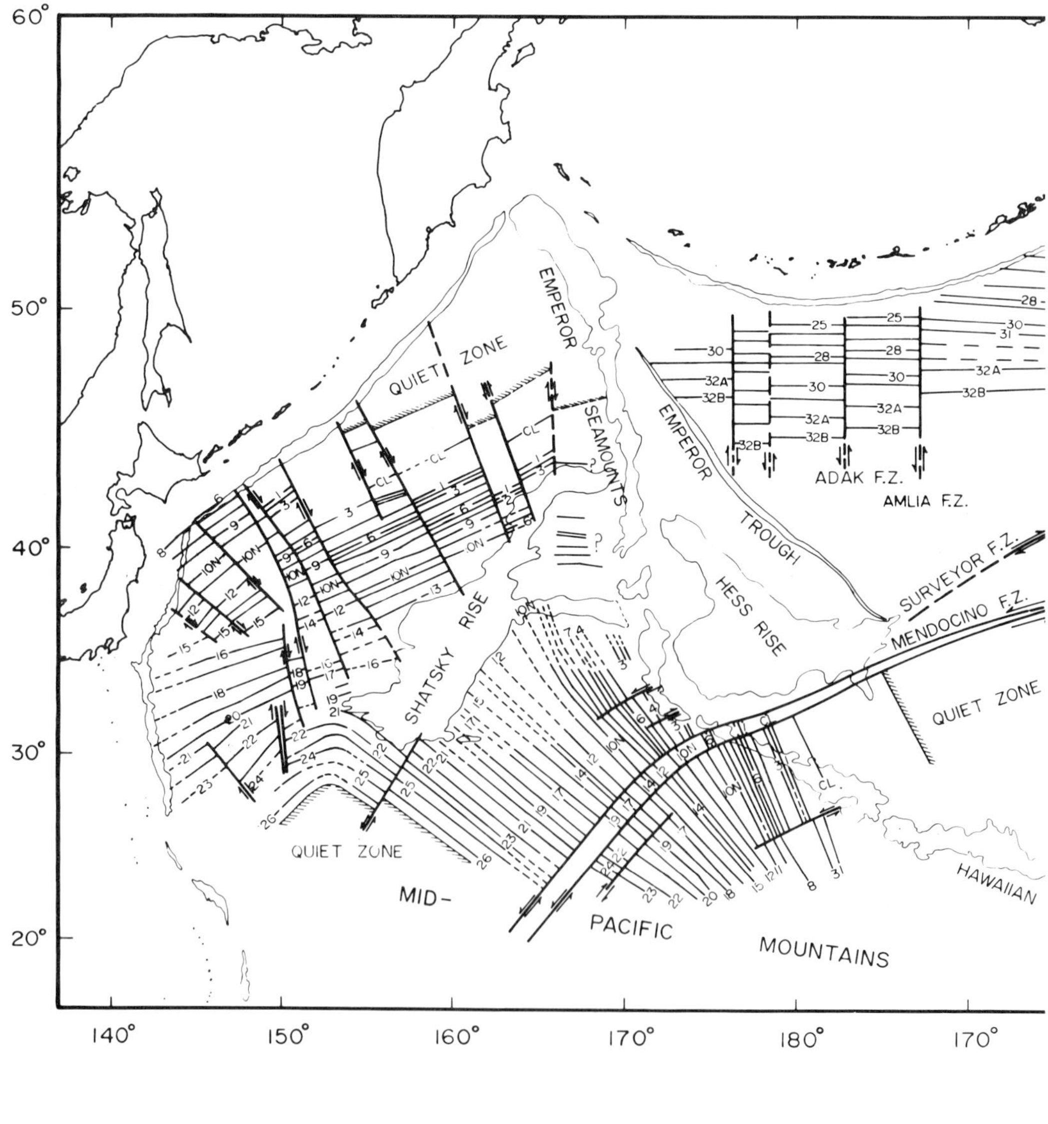

as spreading continued; that is, the two ridges became farther separated as the Pacific plate accreted faster along its northwest side (Figure 17). This is best indicated by lineation M-13 which does not connect when projected beneath Shatsky rise (Figures 5 and 17). There is no direct evidence of how the Kula–Farallon ridge related to these two at this time.

The Kula–Pacific ridge essentially grew from its northeast end as spreading continued after the time of M-20. As it grew in length, spreading became progressively slower along its newly formed length, which resulted in the formation of several NW striking transform faults (Figure 17). The faults divided the ridge into sections, each of which spread slower, from southwest to northeast, as evidenced by the closer spacing of lineations northeast of each fracture zone in the Japanese lineations.

This condition continued until about M-13 to M-10, when the Pacific–Farallon ridge changed spreading direction by 25° clockwise. The ridge then migrated northeast along the Shatsky rise transform fault more rapidly until its northwest end caught up with the northeast end of the Kula–Pacific ridge and an R–R–R triple junction formed again at about M-4 (116 m.y.B.P.) at the northeast end of Shatsky rise (Figure 17).

Most transform faults have not formed large features like Shatsky rise, but as we have already seen, Shatsky rise was not a simple transform fault. In this case the initial development probably involved a break in one or more of the plates near the triple junction which resulted in large-

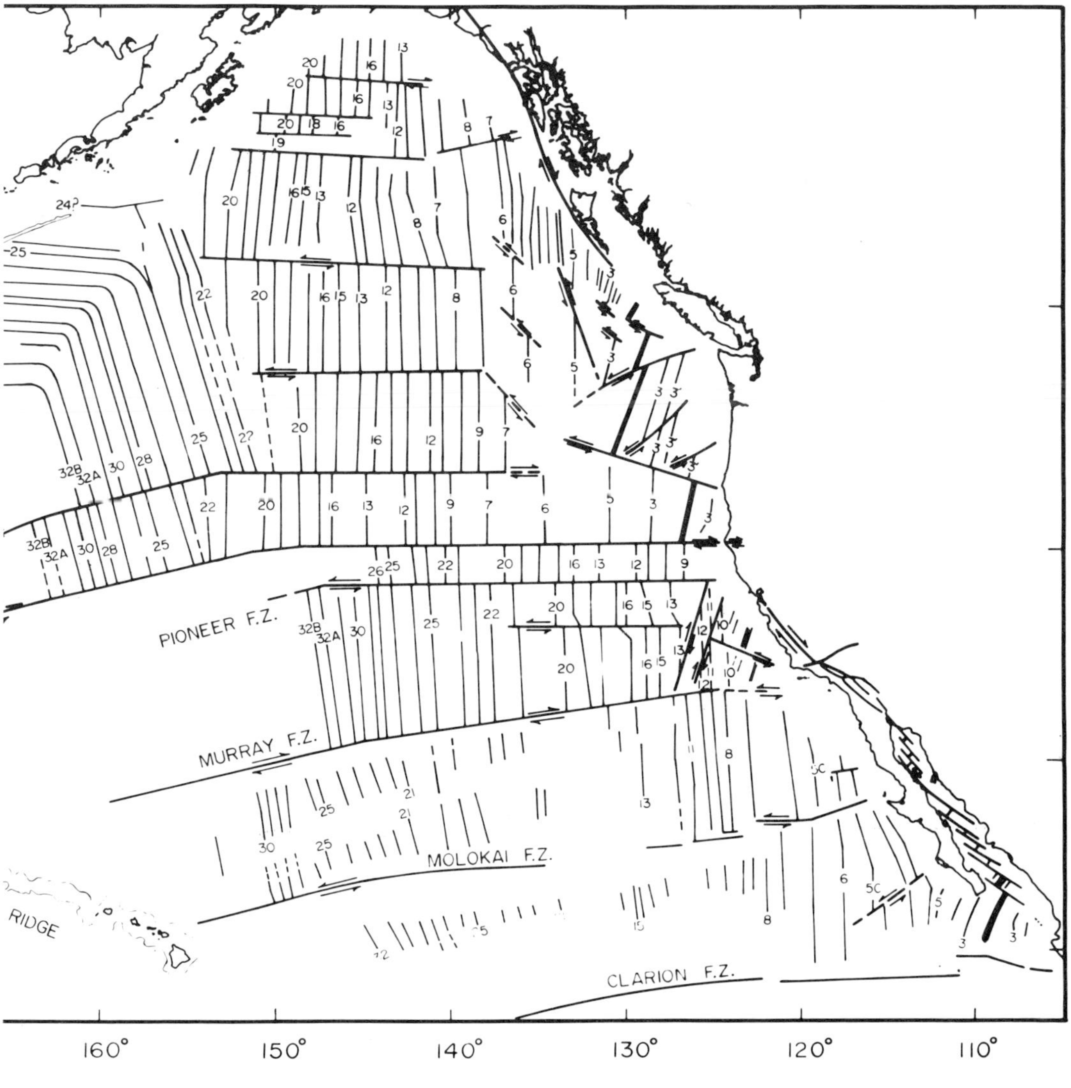

TABLE 1. Late Mesozoic Model

| | |
|---|---|
| Geomagnetic Field | |
| Inclination | 38° |
| Declination | 0° |
| Magnetization | |
| Inclination | −15° |
| Declination | 0° |
| Intensity | $1.5 \times 10^{-3}$ emu/cc (M-20 to M-1) |
| | $1.0 \times 10^{-3}$ emu/cc (Mid-Mes. Q.Z. to M-21, CL to Cret. Q.Z.) |
| Spreading Direction | |
| Mid-Mesozoic Quiet Zone to M-10N | 40° |
| M-10 to Cretaceous Quiet Zone | 65° |
| Depth Below Sea Level | |
| Upper Surface of Block | 5.7 km |
| Lower Surface of Block | 8.6 km |

Block Boundaries from Mid-Mesozoic Quiet Zone to Cretaceous Quiet Zone in km

0, 15, 37, 63, 71, 94, 114, 133, 139, 159, 173, 195, 201, 207, 218, 229, 235, 258, 320, 334, 365, 399, 427, 438, 441, 461, 495, 503, 506, 523, 546, 602, 625, 645, 710, 738, 766, 803, 820, 834, 845, 854, 862, 876, 882, 905, 924, 933, 941, 950, 953, 972, 981, 986, 1012, 1046, 1066, 1077, 1080, 1094, 1097, 1111, 1128, 1145, 1162, 1176, 1190, 1204, 1218, 1229, 1235, 1252, 1300, 1396, 1433, 1444, 1568, 1597, 1769, 1789, 1896, 1899, 1978, 1981, 2121, 2130.

scale volcanic extrusions.

*Francheteau et al.* [1970] suggested on the basis of seamount paleomagnetic studies that Shatsky rise may have been an active boundary between rigid oceanic plates. *Ozima et al.* [1971] concluded that higher than normal oceanic basalt $(^{87}Sr/^{86}Sr)_0$ ratios for samples from Shatsky rise could be due to mixing of crustal and mantle material. A seismic refraction section (Figure 18) shows that the crust does dip into the mantle beneath this rise from the west [*Den et al.*, 1969]. These data suggest that convergence as well as lateral fault motion occurred along the rise. It may be that a portion of the former Farallon plate was thrust and trapped beneath the Pacific plate along the southeast side of the Shatsky rise transform fault. This activity would have been maintained only during the 142–116 m.y.B.P. period. The elevated layer 2 (Figure 18) can be explained by overlap and piling up of the crust and by post-116 m.y.B.P. isostatic adjustment of a trench-arc that may have developed in response to the crustal convergence.

At the north end of Shatsky rise from M-4 (116 m.y.B.P.) the R–R–R triple junction may have rotated further clockwise. Clockwise rotation in spreading is indicated in the Japanese lineations M-26 to M-16 and 25° clockwise rotation is observed in the Hawaiian lineations at about M-10. It is suggested that the R–R–R triple junction was again disrupted; this time with the Pacific–Farallon ridge being separated from the Kula–Pacific ridge by a SE-trending transform fault (Figures 19 and 20). This SE migration continued until about 95–100 m.y.B.P. and it is during this time, from ~116 to 95 m.y.B.P., that Hess rise was formed, probably in the same manner that Shatsky rise formed earlier.

While the north end of the Pacific–Farallon ridge was migrating southeast, the Kula–Pacific ridge continued to migrate northwest as the Pacific plate accreted along its northwest edge. The latter ridge was probably subducted in a trench to the northwest at this time, which left the NE-trending Kula-Farallon ridge and the SSE-1 trending Pacific–Farallon ridge either joined directly or connected by a transform fault in the vicinity of what is now the southeast edge of Hess rise (Figure 20). A third ridge, the new Kula–Pacific ridge, developed extending west from the junction of the other two ridges, again forming an R–R–R triple junction. This is shown by the imaginary lineations $QZ_2$ at the southeast edge of Hess rise in Figure 20.

This junction then migrated NE as an R–R–R triple junction during the Cretaceous quiet period and later, as is observed in the Cenozoic lineations of the northeast Pacific. Meanwhile, the western end of the new Kula-Pacific ridge moved NNW forming the Emperor trough as it moved (Figures 19 and 20). The Emperor trough was a left-lateral transform fault between the Kula plate and the Pacific plate to the west. The Surveyor fracture zone apparently originated on the Pacific–Farallon ridge near the new triple junction at the southeast end of Hess rise, and its increased separation from the Mendocino fracture zone from west to east reveals growth in the length of the Pacific–Farallon ridge north of the Mendocino fracture zone as spreading continued.

All of these activities remained confined to the area north of the Mendocino fracture zone.

The major factors in the evolution of the North Pacific appear to have been a progressive clockwise change in sea-floor spreading direction on the Kula–Pacific and Pacific–Farallon ridges (a total of 40° from M-26 to Cenozoic anomaly 32), and faster spreading on the Kula–Pacific ridge from middle Mesozoic time. This suggests that the Kula and Farallon plates rotated clockwise relative to the Pacific plate.

The disruption of the R–R–R triple junction which existed between the Kula, Pacific, and Farallon plates prior to M-20 is not easily explained by plate tectonics since R–R–R triple junctions are always supposed to be stable. With faster spreading on the Kula–Pacific ridge it would be expected that the triple junction would have migrated to the northwest relative to the Pacific plate and that the R–R–R junction would have been preserved. This would have been the simplest evolution, but this clearly was not the case.

Regarding the Japanese lineations, if the Pacific plate is considered to be rigid throughout its growth then the progressively greater separation of the Kula–Pacific ridge segments between fracture zones, as indicated by the lineations, suggests that the Kula plate either rotated rapidly clockwise relative to the Pacific plate, or that the Kula plate was actually several plates separated by transform faults which extended to a trench

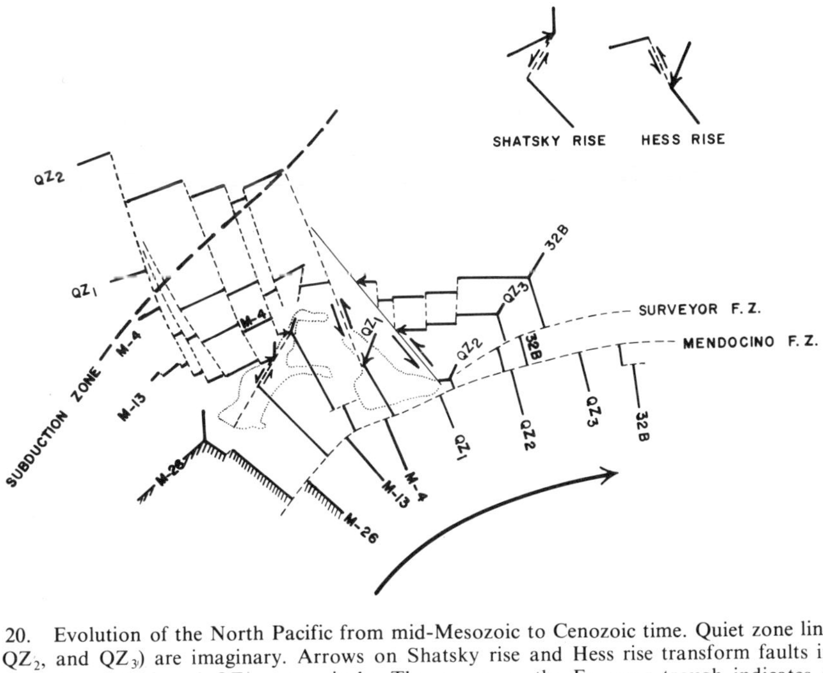


Fig. 20. Evolution of the North Pacific from mid-Mesozoic to Cenozoic time. Quiet zone lineations ($QZ_1$, $QZ_2$, and $QZ_3$) are imaginary. Arrows on Shatsky rise and Hess rise transform faults indicate motion for times M-13 and $QZ_1$, respectively. The arrow on the Emperor trough indicates motion of the Kula plate on the east relative to the Pacific plate on the west after time $QZ_2$. The curved arrow indicates rotation of spreading directions relative to the northern Pacific plate from mid-Mesozoic to Cenozoic time.

on the northwest. If the Kula plate was also a single rigid plate which rotated clockwise relative to the Pacific plate, then spreading between the two plates must have been asymmetric, varying inversely along the length of the ridge, in order to have produced the observed lineation pattern. We recognize that the lineation and fracture zone pattern between 150°E and Japan, from M-15 to the north (Figure 19), is not consistent with this conclusion. The data for this region are believed to be accurate and we could not have shown trends different than we show without having invoked fracture zones not evident in the data. Possibly more detailed surveys can resolve this problem.

We have seen that spreading on the Kula–Pacific ridge was faster than spreading on the Pacific–Farallon ridge from middle Mesozoic time, and that the spreading system rotated clockwise. Spreading on the Pacific–Phoenix ridge was also faster than spreading on the Pacific–Farallon ridge and apparently the spreading direction for this ridge rotated in a counterclockwise direction relative to the Pacific plate [*Larson and Chase,* 1972]. These factors appear to have resulted in an unfolding of the spreading system about the Pacific plate.

Magellan rise and Manihiki plateau in the south-central Pacific are located near the east end of the Phoenix lineations. Formation of these features has been attributed to former spreading ridges [*Larson and Chase,* 1972; *Winterer et al.,* 1974]. It may be that they were formed in the same manner as Shatsky and Hess rises as spreading took place about the Pacific–Farallon–Phoenix triple junction during late Mesozoic time.

*Acknowledgments.* The authors are indebted to the U.S. Naval Oceanographic Office for the 1969 Northwest Pacific Survey; to the National Oceanographic and Atmospheric Administration for the data from the 1972 Western Pacific Geotraverse Project; and to Scripps Institution of Oceanography of the University of California and the Earthquake Research Institute of the University of Tokyo, for use of their magnetic data collections; and to the many institutions and individuals who contributed data to those collections.

We are grateful to S. Uyeda, M. Yasui, T. Watanabe, R. L. Larson, G. Sharman, M. A. Hobart, and J. G. Sclater for their many helpful suggestions, and the senior author expresses special gratitude to the University of Tokyo for support of his research on this subject for his D.Sc. degree.

# *19. An Age-Dependent, Two-Layer Model for Marine Magnetic Anomalies*

RICHARD J. BLAKELY

*School of Oceanography, Oregon State University, Corvallis, Oregon 97331*

*Abstract.* Marine magnetic anomalies of the North Pacific Ocean are analyzed to determine the recording fidelity of the oceanic crust as a function of crustal age. The parameter describing fidelity is the width $W$ of the transition from crust of predominantly normal polarity to crust of predominantly reversed polarity. The width of the transition zone is related to the distance $\Delta X$ between the zeros of the first derivative of the total-field anomaly. Average values of $\Delta X$ are found for specific crustal ages, using profiles from the North Pacific, and then converted to transition widths by comparing with computer-generated profiles. Transition width increases monotonically with age throughout the Cenozoic, starting at 2 km and increasing to 20 km, assuming a layer thickness of 0.5 km. Increasing transition width is considered to be a manifestation of a more complicated crustal source consisting of two discrete layers. In this model, an upper layer corresponds to pillow basalt with a high initial magnetization and a narrow transition zone. A thicker, underlying layer corresponds to basaltic dikes with lower magnetization and wider transition zone. It is suggested that the wide transition zone of the second layer is the result of a Curie-point isotherm which slopes gently away from the spreading center. As the crust ages, the upper layer loses its high intensity of magnetization because of progressive, low-temperature oxidation of titanomagnetite in the pillow basalts. Consequently, surface anomalies over old crust are more representative of deeper magnetic sources. Such a model predicts a general deterioration of the recording fidelity of the ocean crust and may partially explain the general lack of very short polarity intervals retained by older crust.

The boundary between crust with opposite polarity, formed at the time of a geomagnetic reversal, is commonly depicted as both vertical and discontinuous; i.e., a block of constant normal polarity juxtaposed with a block of constant reversed polarity [e.g., *Vine and Matthews,* 1963]. Several attempts have been made to determine a more realistic representation for polarity contrasts. *Matthews and Bath* [1967] considered the injection of vertical dikes as the mechanism by which new material is introduced into the crust, the injection zone being normally distributed about the center of spreading. *Harrison* [1968] used a Monte Carlo technique to estimate an upper limit of 12 km for the width of this dike-injection zone. As a consequence of a finite-width injection zone, the width of the transition zone is also non-zero. 'Transition zone' is defined here as the distance through which the crust switches from predominantly normal polarity to predominantly reversed polarity as the result of a reversal of the earth's magnetic field.

This paper investigates the possibility that, not only is the transition zone finite in width, but that it effectively increases with the age of the crust. There is some evidence in the literature that suggests the existence of such an increase. For example, data from a near-bottom magnetometer has been used to determine the transition width over crust young in age (less than 4 m.y. in each case) and estimates all are less than 2 km [*Larson and Spiess,* 1969; *Atwater and Mudie,* 1973; *Klitgord et al.,* 1972; *Klitgord,* 1974]. Using anomalies measured at the surface, *Blakely* [1974] found an upper limit for the transition width of 6 km for Miocene crust (8 to 22 m.y. in age) and *Blakely and Cox* [1972*b*] found a width of 12 km for early Cenozoic crust (54 to 70 m.y. in age). While the data and the methods are different, there appears to be a significant increase in the three determinations of the width of the transition zone.

A method is developed here to analyze marine magnetic profiles rapidly in order to determine the width of the transition zone. The method is then applied to a large volume of data from the North Pacific. It is found that, if the magnetic layer is assumed to be 0.5 km thick, the transition zone apparently increases in width from about 2 km for recently created crust to about 20 km for crust with an age of 85 m.y. An alternative and preferred interpretation is that the source of marine anomalies is located in two distinct layers rather than one and, because of a redistribution of the relative magnetizations of the two layers as they age, the effective thickness and the effective transition width increase.

## METHOD

Assume for the present discussion that the time spent during a reversal of the geomagnetic field is negligible (Figure 1). This is not strictly true since the reversal process apparently takes between $10^3$ and $10^4$ years [*Ninkovich et al.*, 1966; *Harrison and Somayajulu*, 1966; *Cox and Dalrymple*, 1967], but is a reasonable assumption for the present discussion. With this assumption, the field intensity $H(t)$ is discontinuous at the time of a reversal. The recording process is modeled as a convolution of $H(t)$ with a function $g(x)$

$$J(x) = g(x) * CH(vt) \qquad (1)$$

where $J(x)$ is the average magnetization of the crustal layer, $C$ is a factor which converts field

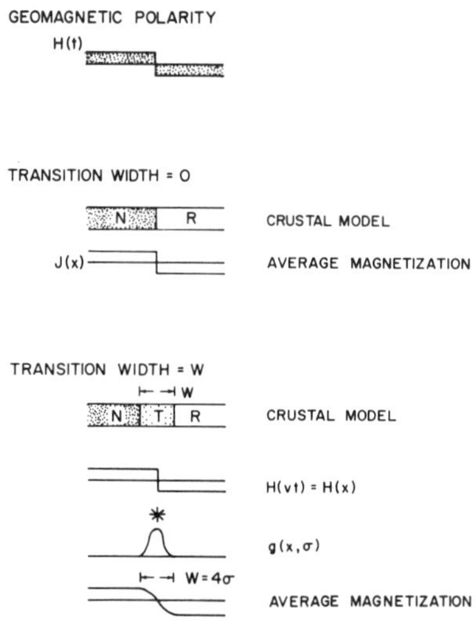


Fig. 1. Mathematical model for the reversal transition zone in oceanic crust. The earth's field $H(t)$ during a reversal is assumed discontinuous. If the crust perfectly records $H(t)$, then the average magnetization of the layer $J(x)$ is also discontinuous. The case of a nonperfect recorder is modeled by convolving $C \cdot H(vt)$ with normal distribution $g(x,\sigma)$ where $C$ is a scale factor, $v$ is the spreading rate, and $\sigma$ is the standard deviation of $g(x)$. N, T, and R refer to crust with normal, transitional, and reverse polarities, respectively. $W = 4\sigma$ is the width of the transition zone.

intensity to magnetization, $v$ is the half-spreading rate, and $g(x)$ is a normal distribution.

$$g(x) = \exp(-x^2/2\sigma^2)/\sigma\sqrt{2\pi} \qquad (2)$$

with the standard deviation $\sigma$. It can be shown that 95.4% of the change from crust with normal polarity to crust with reversed polarity takes place in a region with width $4\sigma$. The transition width is therefore defined as

$$W = 4\sigma \qquad (3)$$

If the crustal 'tape recorder' were perfect, then $W = \sigma = 0$, so that

$$J(x) = CH(vt)$$

and $J(x)$ is discontinuous at the polarity boundary. If $W \neq 0$, $J(x)$ will not be an accurate reproduction of $H(vt)$, so that $W$ represents the fidelity of the crustal recording process. There is no a priori reason to select a normal distribution for $g(x)$; it is merely an easy distribution with which to work and produces a smooth transition in $J(x)$ at any reversal boundary.

Assuming the magnetization and regional field are vertically directed, the peak-to-trough distance $\Delta X$ of the total-field anomaly increases as $\sigma$ increases. Figure 2 shows the dependence of $\Delta X$ on $\sigma$ for the case of a layer with 0.5 km thickness. Notice that not only does $\Delta X$ increase monotonically with increasing $\sigma$, but that it also increases with the depth to the magnetic layer. One could therefore measure the peak-to-trough distance of an anomaly produced by a single reversal, assume a layer thickness and a depth to the top of the layer, and determine the transition width $W$ of the reversal boundary. This in essence is the method employed by *Atwater and Mudie* [1973] to analyze near-bottom measurements. In the present study, $\Delta X$ was determined by first differentiating the anomaly, with a technique employing a second-order Lagrangian polynomial, and then searching for the zeros of the first derivative. Of course, a zero in the first derivative corresponds to either a peak or a trough of the anomaly, but is computationally simple to work with. The various steps in the analysis are: (1) transform the anomaly $h(x)$ to the pole [*Blakely and Cox*, 1972*a*] in order to bring the vectors representing the magnetization and the regional field to a vertical direction, (2) determine the first derivative $h'(x)$ of the transformed anomaly, (3) locate and compute the distances between the various zeros of the first derivative, (4) assume

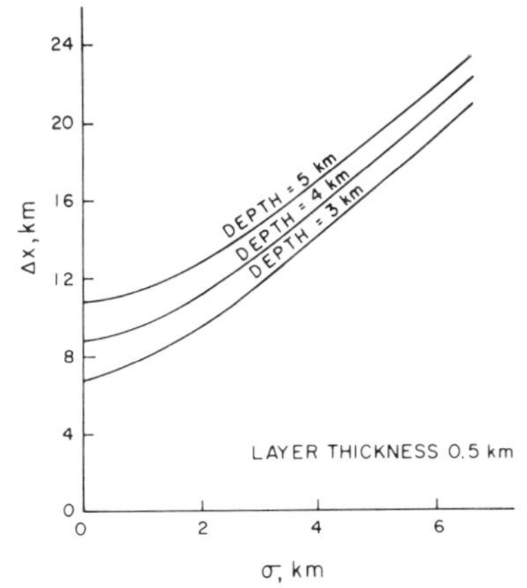


Fig. 2. The distance $\Delta X$ between the peak and the trough of the anomaly produced by a single reversal as a function of $\sigma$. For each of the three depths, the layer thickness is 0.5 km and the regional field and magnetization are assumed vertically oriented.

both a thickness and a depth for the layer, and (5) convert $\Delta X$ to transition width $W$ using functions similar to those graphed in Figure 2.

## DATA ANALYSIS

The locations of data analyzed by this method are shown in Figure 3. The thin E–W lines are tracklines from two major surveys done in 1971 by the National Oceanic and Atmospheric Administration. The thick dashed lines indicate the specific anomalies used in the analysis:

| Anomaly | Central Age (m.y.) | Reference |
|---|---|---|
| 1 | 0 | *Klitgord* [1974] |
| 2 | 1.71 | *Klitgord* [1974] |
| 5C | 17.05 | *Blakely* [1974] |
| 6 | 20.71 | *Blakely* [1974] |
| 10 | 31.84 | *Heirtzler et al.* [1968] |
| 12 | 34.76 | *Heirtzler et al.* [1968] |
| 13 | 37.94 | *Heirtzler et al.* [1968] |
| 31 | 72.11 | *Heirtzler et al.* [1968] |
| 33 | 74.7 | *Larson and Pitman* [1972] |
| Q | 84.7 | *Larson and Pitman* [1972] |

The abbreviation Q refers to the anomaly representing the younger boundary of the quiet zone.

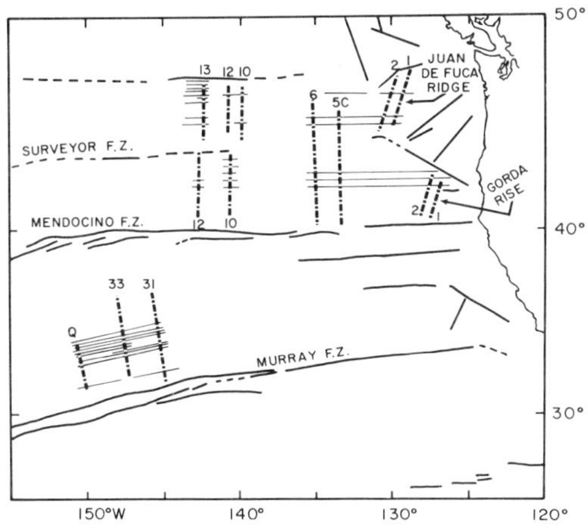


Fig. 3. Location map of the data. Thin E-W lines are ship tracklines. Thick dashed lines are locations of the anomalies used in the analysis. Heavy lines indicate fracture zones.

Nine specific examples are shown in Figure 4: three profiles each for anomalies 6, 13, and 31. In each case, the anomaly $h(x)$ is shown below its first derivative $h'(x)$. Peaks and troughs of the anomaly are connected by vertical lines to the zeros of its derivative. The average $\Delta X$, using only the three profiles in Figure 4, is 12.8 km for anomaly 6, 14.7 km for anomaly 13, and 24.2 km for anomaly 31. In general, $\Delta X$ is increasing with age for the examples shown in Figure 4, but in each of the three cases, there is a good deal of scatter about the average. This is probably the result of irregularities of the crustal magnetization, such as topographic distortion or inhomogeneities in the magnetic petrology of the crust, or caused by temporal variations of the main field during the survey. Because of this large amount of deviation, at least ten reversal boundaries were used for each of the ten points in time so as to obtain a reasonable estimate for the average $\Delta X$.

Values of $\Delta X$ were found for both the young and old boundaries of each of the ten specific anomalies (except anomaly Q, where only the younger boundary was available) for all of the profiles shown in Figure 3. These were averaged together to provide representative values of $\Delta X$ for each of the ten points in time. Figure 5 illustrates the compiled results. The open circles are values of $\Delta X$ and the error bars are two standard deviations in length. According to the analysis summarized by Figure 5, the distance between peak and trough above polarity reversals recorded in the crust generally increases with crustal age.

As indicated by Figure 2, the peak-to-trough distance also increases with increasing depth to the magnetic layer. The increase in $\Delta X$ with age as shown in Figure 5 may, therefore, only be a manifestation of increasing depth, since the depth to basement increases with the age of the ocean floor [*Sclater et al.*, 1971]. To remove basement depth as a variable, model profiles were computed for various values of $\sigma$, using proper values for the depth to the magnetic layer. Specific anomalies were generated by using the geomagnetic time scale rather than a single, isolated reversal. Curves of $\Delta X$ as a function of $\sigma$ were thereby created for each of the ten specific anomalies. As an example, anomaly 6 and its derivative are shown in Figure 6 for various values of $\sigma$ and the relationship between $\Delta X$ and $\sigma$ is plotted in Figure 7. $\Delta X$ increases monotonically with increasing $\sigma$ for both the younger and the older reversal boundary of the anomaly, as expected, but not in a simple way as for a single, isolated reversal (Figure 2). Herein lies the major limitation to the method: The relationship between $\Delta X$ and $\sigma$ depends on the assumed configuration of the geomagnetic time scale, the details of which are still under substantial scrutiny [*Klitgord*, 1974; *Blakely*, 1974; *Blakely and Cox*,

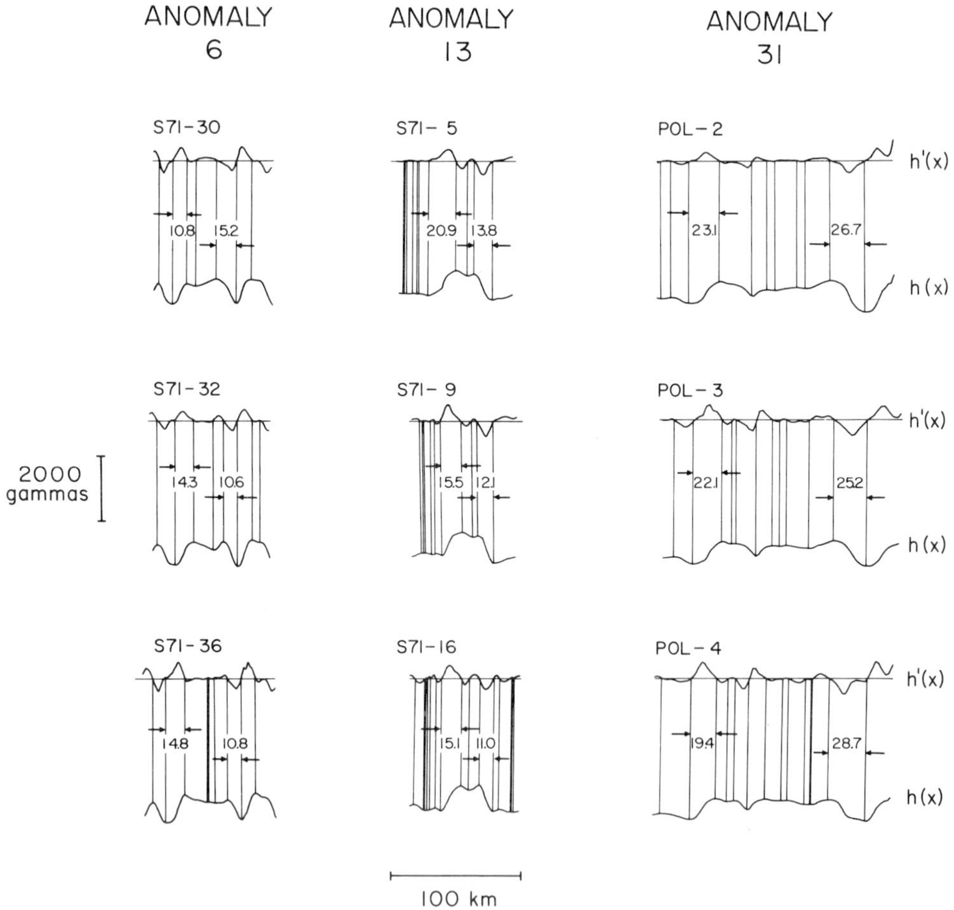


Fig. 4. Specific examples of the determination of $\Delta X$. Three cases are shown for each of three anomalies: 6, 13, and 31. The measured anomaly is $h(x)$, after transformation to the pole, and $h'(x)$ is the first derivative of the anomaly. The vertical lines connect zeros of the first derivative with the corresponding peak or trough of the anomaly. Values of $\Delta X$ are shown for both the young and old reversal boundaries of each of the anomalies.

1972*b*; *Theyer and Hammond,* 1974; *Talwani et al.,* 1971]. It was with this understanding that the ten specific anomalies were selected. First, anomalies 1 and 2 are considered well defined from paleomagnetic and potassium–argon measurements of continental rocks [e.g., *Cox,* 1969] and from detailed marine anomaly studies [*Klitgord,* 1974]. Second, anomalies 5C, 6, 10, 12, 13, 31, 33, and the quiet zone were selected since they are widely separated from other reversals, according to the most recently published scales. The geomagnetic time scale used is a composite of four efforts: *Klitgord* [1974] for anomalies 1 and 2; *Blakely* [1974] for anomalies 5C and 6; *Heirtzler et al.* [1968] for anomalies 10, 12, 13, and 31; and *Larson and Pitman* [1972] for anomalies 33 and the quiet zone.

Curves similar to Figure 7 were constructed for each of the ten anomalies using correct depths and assuming a layer thickness of 0.5 km. Then

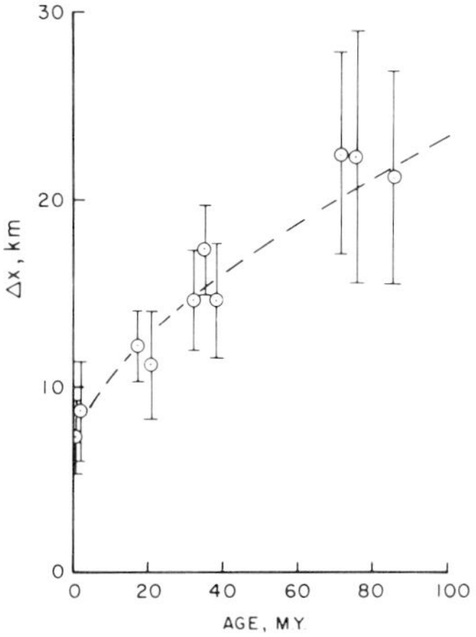


Fig. 5. The peak-to-trough distance as a function of crustal age for the North Pacific. Open circles are average values of $\Delta X$ and the error bars are two standard deviations in length.

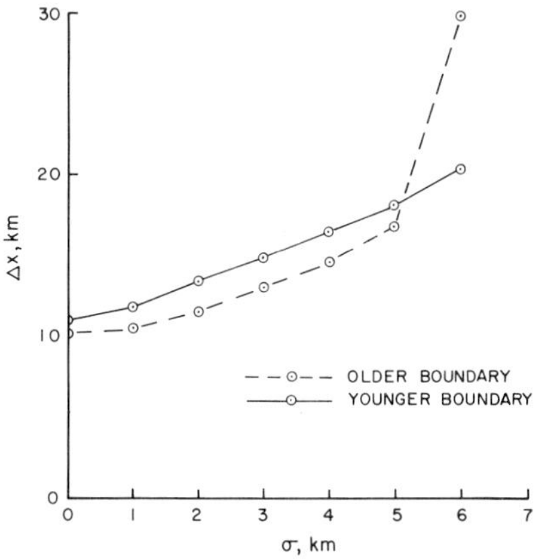


Fig. 7. Peak-to-trough distance of computer-generated anomaly 6 plotted against various values of $\sigma$. With graphs such as this, values of $\Delta X$ from measured anomalies can be converted to values of $\sigma$.

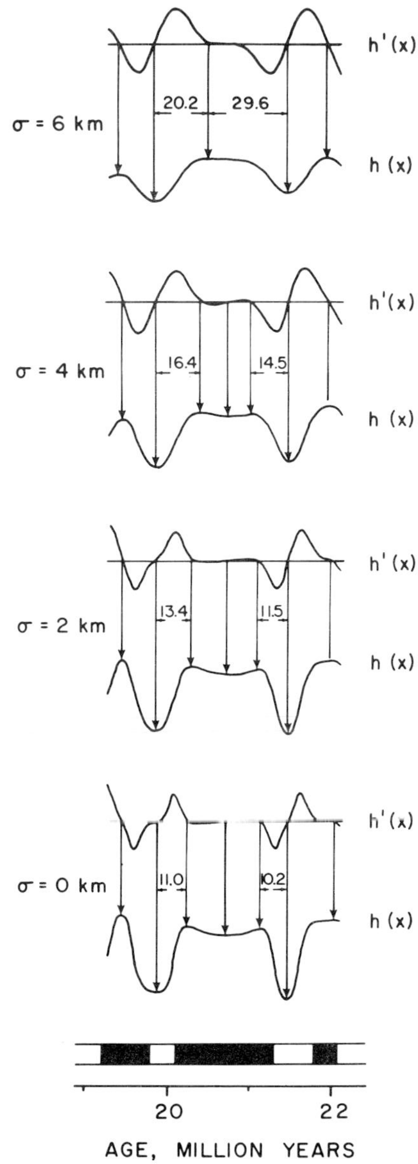


Fig. 6. Determination of peak-to-trough distance of anomaly 6 for various values of $\sigma$. The anomaly $h(x)$, computationally derived, and its first derivative $h'(x)$ are shown for four values of $\sigma$ with vertical lines connecting the zeroes of $h'(x)$ with corresponding peaks and troughs of $h(x)$. Values of $\Delta X$ are shown for both the younger and older reverasal boundaries of the anomalies.

given the average $\Delta X$ and the error bars of Figure 5, these curves were used to obtain $\sigma$ (and consequently $W$) as a function of age. The compiled results are shown in Figure 8 where the transition width $W$ ($W = 4\sigma$) increases in a general sense with age. For very young crust, the transition width is approximately 2 km, an encouraging result since it agrees with recent determinations by *Atwater and Mudie* [1973], *Klitgord et al.* [1972], and *Klitgord* [1974]. At 85 m.y., the transition width is approximately 20 km, an impressive increase. The error bars in Figure 8 are large enough that no significance should be attached to the *shape* of the dashed line; it is only included to emphasize the general increase in the width of the transition zone as a product of the aging process of the oceanic crust.

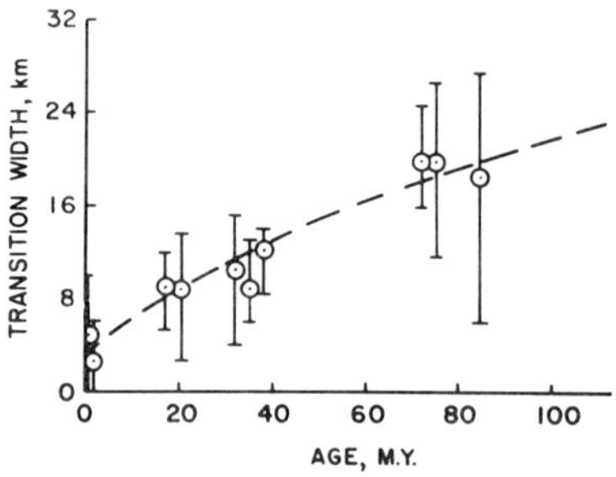


Fig. 8. Increase in transition width as a function of age. Open circles are average values of $W$ and error bars are two standard deviations in length. Because of the large error bars, no special significance should be attached to the *shape* of the dashed line. The dashed line is only included to indicate the general increase of $W$ with age.

DISCUSSION

An explanation for this apparent deterioration of the recording fidelity of the ocean crust is not straightforward, if the layer thickness is assumed constant. Perhaps the region about the spreading center into which new crustal material is introduced has narrowed with time, or perhaps the length of time spent by the geomagnetic field during a reversal is shorter now than earlier in the Cenozoic. While either of these interpretations would explain the above results, they violate the principle of uniformitarianism.

A third possibility is that the transition width has not changed, but that the effective thickness of the magnetic layer increases with age. Figure 9 shows the relationship between peak-to-trough distance $\Delta X$ and the layer thickness for an anomaly produced by an isolated reversal, assuming that the transition width is zero. Notice that $\Delta X$ increases with increasing thickness. One might therefore assume a negligible width for the transition zone for all time and develop a curve of layer thickness as a function of age. This requires, however, that the magnetic layer be some 12 km in thickness by Late Cretaceous. Since this depth extends well below seismic layers II and III [*Fox et al.,* 1973] and into the upper mantle, it is not entirely acceptable [*Vine and Moores,* 1972].

We therefore face a dilemma. If a constant thickness is assumed for the layer, a mechanism is required to increase the transition width as the crust ages. If, on the other hand, a negligible transition width is assumed for all time, then the layer thickness for old crust is unreasonably large. A solution to the problem involves a combination of the two models: an apparent increase in *both* the thickness of the layer and the width of the reversal transition zone. The mechanism for such a phenomenon is a two-layer model for the crustal magnetization as proposed by *Cox et al.* [1972] and illustrated in Figure 10. The model consists of ocean crust composed of two layers: (1) a thin upper layer with a narrow transition zone and (2) a thicker lower layer with a larger transition zone. When the crust is young, the upper layer has high magnetization relative to the lower layer. As the crust ages, the magnetization of the

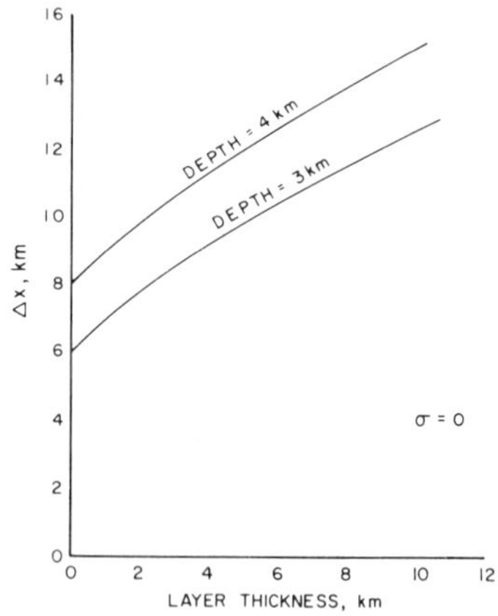


Fig. 9. Increase in the peak-to-trough distance $\Delta X$ of an anomaly produced by a single reversal boundary as a function of layer thickness. Transition width is zero and the regional field and magnetization are assumed vertically oriented.

upper layer decreases, presumably due to low-temperature oxidation of titanomagnetite to titanomaghemite [e.g., *Ozima and Ozima,* 1971] so that the two layers become more equally magnetized. If in an analysis, such as the preceding one, a single magnetic layer is assumed, but in fact there are two layers whose magnetizations equilibrate with time, the assumed single-layer will appear to thicken and the transition width to widen with age. Figure 10 provides one set of parameters which satisfies the data. It is senseless at this point to speculate further about the exact values for parameters since there now are six unknowns (i.e., magnetization, thickness, and width of the transition zone for both layers) in a problem which to begin with was non-unique.

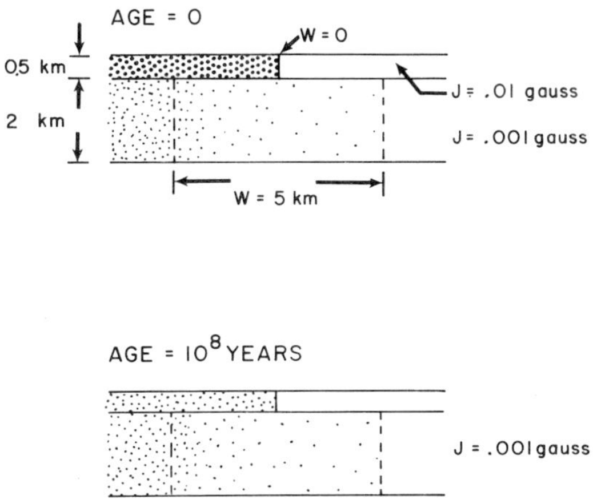


Fig. 10. Schematic of the two-layer crustal model. The upper layer corresponds to pillow basalt and is 0.5 km thick, with a high initial magnetization of 0.01 gauss and a transition zone of zero. The lower layer corresponds to basaltic dikes and is 2 km thick, with a magnetization of .001 gauss and a transition zone of 5 km in width. At $10^8$ years, the model is the same except the magnetization of the upper layer has decreased to .001 gauss.

Adding variables to a model always permits a better fit to the observed data and, in general, one is not justified in arbitrarily adding parameters unless there is some physical evidence to do so. An age-dependent, two-layer crustal model does have a good deal of the required physical basis. Two layers which are magnetically important have been observed in exposed oceanic crust of the Troodos massif of Cypress [*Vine and Moores,* 1972] and at Macquarie Island [*Butler and Banerjee,* 1973; *Banerjee et al.*, 1974]: an upper layer of pillow basalt and a lower layer of doleritic and basaltic dikes. Even a third layer of gabbroic intrusives is not totally insignificant in both of these examples. *Banerjee et al.* [1974] report the upper pillow layer of Macquarie Island, a 30 m.y. old section of the Indian plate, to have a low magnetic stability and low natural remanent magnetization (NRM) ranging from $8 \times 10^{-4}$ to $6 \times 10^{-3}$ gauss. These low values of NRM and the low stability are apparently the result of low-temperature oxidation or slight hydrothermal alteration of the magnetic minerals. The basaltic dikes, in contrast, have high magnetic stability and significant NRM ($6 \times 10^{-3}$ gauss) and although high temperature deuteric oxidation is present, no alteration has occurred since initial cooling. *Vine and Moores* [1972], working with substantially more data from the Troodos complex, find values of NRM intensity ranging from $10^{-3}$ to $10^{-1}$ gauss for pillow basalts and from $10^{-5}$ to $10^{-2}$ for underlying dikes. Butler et al. conclude, therefore, that the underlying dike layer, and perhaps even a third layer of intrusives, has substantial importance to the production of marine anomalies, especially for older crust where low-temperature oxidation of the top pillow layer has proceeded to a large degree. Hopefully, the very confusing, preliminary results from recent cruises of the Deep Sea Drilling Project [e.g., *DSDP Scientific Staff,* 1974*c*], when thoroughly analyzed, will provide additional information on the relative importance of underlying layers to the generation of magnetic anamalies. For example, samples from hole 319 A of Leg 34 which are apparently from unoxidized massive basalts are substantially more magnetic than either oxidized pillow basalt or oxidized massive basalt. (P. Johnson, personal communication, 1975). This preliminary result supports the idea of the importance of the lower layers of oceanic basalt to the production of marine anomalies.

It has been amply demonstrated that the uppermost part of the pillow layer loses its intensity of NRM with age. Certainly other properties of the crust change with time, presumably also due to weathering of the crust. *Christensen and Salisbury* [1973], for example, find that both density and velocity of layer II basalts decrease with age. *Hart* [1973] interprets these and other refraction results in terms of a low-velocity cap of weathered basalt which increases in thickness due to progressive alteration of the basalt. *Marshall and Cox* [1971*a*, *b*] studied the magnetic properties of dredged pillow fragments and found that weathering proceeded from the outside to the inside of the pillow as time progresses. *Irving* [1970] showed an extremely rapid decrease in the NRM of dredged pillow basalt as a function of the distance from the spreading center at the mid-Atlantic ridge at 45°N, decreasing from about .1 gauss to .01 gauss in only about 20 km. While this is much more rapid than required by the two-layer model presented here, it is reasonable to assume that the reduction continues to even lower values of NRM as the crust ages and that penetration of oxidation to greater depths in the pillow layer is also age-dependent.

A narrow transition zone for the upper layer has reasonable physical basis. A narrow ridge, approximately 1 km wide and broken into en echelon segments, has been observed in the center of the inner rift valley of the mid-Atlantic ridge by detailed mapping [*Moore et al.,* 1974] and by manned, deep-submersible dives to the region [*Bellaiche et al.,* 1974]. This narrow topographic feature is interpreted as the actual fissure separating the two lithospheric plates. Extrusive flows which spread large distances from such narrow fissures would tend to 'smear out' polarity boundaries creating a large transition width [*Atwater and Mudie,* 1973]. Extrusives do not, however, flow for more than about 1 km from the fissure [*Gass,* 1958; *Moore et al.,* 1974]. The zone into which new material is injected is apparently very narrow, certainly less than 2 km.

A more speculative feature of the two-layer model is the requirement of a wide transition zone for the lower layer. An injection zone which is wider for the second layer than for the first is not realistic since it requires an extension of the second layer which is not compensated in the upper layer. This would imply slippage between the two layers. One possibility, however, to achieve a wider transition zone for the second layer is with a depth-dependent time of magnetization. *Parker and Oldenburg* [1973] have calculated theoretical isotherms in the region of active spreading centers. These isotherms increase in depth only very slowly as the crust moves away from the spreading center. Consider a 'test dike' which extends from 0.5 to 2 km in depth directly after it enters the crust (Figure 11) and assume the Curie temperature for the magnetic materials of the dike to be 200°C. According to *Parker and Oldenburg* [1973], the top of the dike will be magnetized at

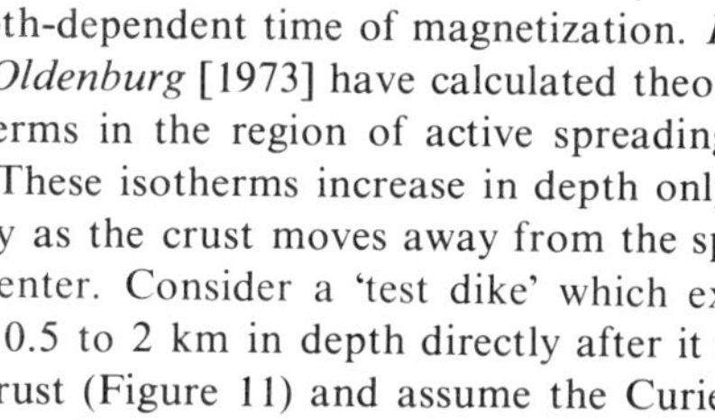


Fig. 11. Mechanism to produce a wide transition zone in the lower layer. If the Curie-point isotherm slopes gently away from the spreading center, then reversal boundaries will also be sloped in the dike layer. The pillow flows of the first layer, on the other hand, are magnetized in a much shorter time since they are quenched.

a shorter distance from the center of spreading than the bottom of the dike. An entire slab moving through this gently sloping Curie-point isotherm will obtain reversal boundaries with the same gentle slope. A sloped, reversal boundary will appear at the surface as a wide transition zone. As a test for such a model, one might investigate marine anomalies from older sections of the crust (where the top layer is reduced in its intensity of NRM). If the reversal boundaries are indeed sloped, then the surface anomalies should possess a phase shifting to an extent dependent on both the slope of the Curie-point isotherm and the thickness of the layer.

## CONCLUSIONS

The model of two layers, a thin layer of pillow basalts with an initially high but decaying NRM overlying a thicker layer of dikes, satisfies the data presented here if the upper layer has a narrow transition zone relative to the lower layer. This model predicts an apparent increase in (1) the thickness of the layer, (2) the width of the transition zone, or (3) a combination, should a single layer be incorrectly assumed.

One possible explanation for a wide transition zone in the lower layer is a gently sloping reversal boundary due to a sloping Curie-point isotherm at the spreading center. This mechanism produces a transition zone which, at the ocean surface, appears to be wide at reversal boundaries without requiring new crustal material, either flows or dikes, to be introduced over a wide region about the spreading center. The actual zone of material injection is probably quite narrow, which is consistent with recent concepts explaining ridge crest topography by the relationship between spreading rate and injection rate [*Deffeyes,* 1970; *Anderson and Noltimier,* 1973].

The aging phenomenon described here produces a deterioration of the fidelity of the magnetic tape recorder. This deterioration may partially explain the general lack of short polarity intervals observed in older ocean crust [*Blakely and Cox,* 1972*b;* *Blakely,* 1974; *Naidu,* 1971, 1974].

*Acknowledgments.* Many of the ideas discussed in this paper, such as the two-layer model and the sloping reversal boundary, are the product of group discussions at Stanford University among Allan Cox, Jeffrey Phillips, myself, and others.

This research was supported by the Office of Naval Research through contract N00014-67A-0369-0007 under project NR 083-102, and by the U.S. National Science Foundation through the IDOE Nazca Plate Project.

# 20. Magnetic Anomalies in the Sea of Japan and the Shikoku Basin: Possible Tectonic Implications

KAZUO KOBAYASHI

*Ocean Research Institute, University of Tokyo, Tokyo 164, Japan*

NOBUHIRO ISEZAKI

*Meteorological College, Kashiwa 277, Japan*

*Abstract.* Recent magnetic surveys in the Sea of Japan and the Shikoku basin indicate the existence of linear magnetic anomalies and their offsets by transform faulting. Age identification of these anomalies has not yet been made, but there are indications that the ages of the anomalies in the Sea of Japan may be of pre-Cenozoic age, and those in the basin may be of Miocene age. Possible spreading centers were identified in both basins through analysis of magnetic anomaly profiles reduced to meridian. Based on these and other geophysical observations, a hypothesis is proposed to explain the evolution of marginal seas by sea-floor spreading.

The northwestern border of the Pacific Ocean is characterized by a chain of island arc–trench systems. On the continental side of these island arc–trench systems are marginal seas, i.e., the Okhotsk Sea behind the Kuril–Kamchatka arc, the Sea of Japan (hereinafter called the Japan Sea) behind the Japanese Islands, and the East China Sea behind the Ryukyu Islands (Figure 1).

The Philippine Sea is somewhat exceptional in its geographical setting. The Izu–Bonin–Mariana island arc–trench system that separates the Philippine Sea from the Pacific Ocean is situated at an extraordinarily great distance from the Asiatic continent. Moreover, the Ryukyu Islands and the Philippine systems exist behind the Izu–Bonin–Mariana arc.

Detailed bathymetric studies have shown that these marginal seas consist of two or more smaller basins. In the Japan Sea a submarine topographic high in its south-central part (the Yamato rise) defines three separate deep basins: the Japan, Yamato and Tsushima basins. The Philippine Sea is divided into west and east basins by a N–S trending chain of submarine seamounts called the Kyushu–Palau ridge. The west basin is called the Philippine basin. The elongated east basin is subdivided into the Shikoku basin to the north and the Parece Vela (West Mariana) basin to the south, although no distinct physiographic boundary has yet been found between them.

Recent explosion seismological studies have convincingly shown that the crust beneath these marginal basins is similar to that of normal oceanic basins, with a few exceptions such as the Yamato rise and the Kyushu–Palau ridge that have a crust of continental structure [*Murauchi et al.*, 1968; *Ludwig et al.*, 1966; *Den*, 1972]. It therefore seems very likely that these basins were formed by a process of sea-floor spreading from a linear center of crustal upwelling in a manner similar to that at the crest of the mid-oceanic ridge. If magnetic lineations are identified in these marginal basins, they could provide valuable information on the evolutionary history of the basins.

The present article is a review of recent marine geomagnetic investigations of the Japan Sea [*Yasui et al.*, 1967; *Isezaki*, 1973, 1975; *Isezaki and Uyeda*, 1973] and the Shikoku basin [*Tomoda et al.*, 1968, 1973, 1975]. Also, a hypothesis on the evolution of the marginal basins

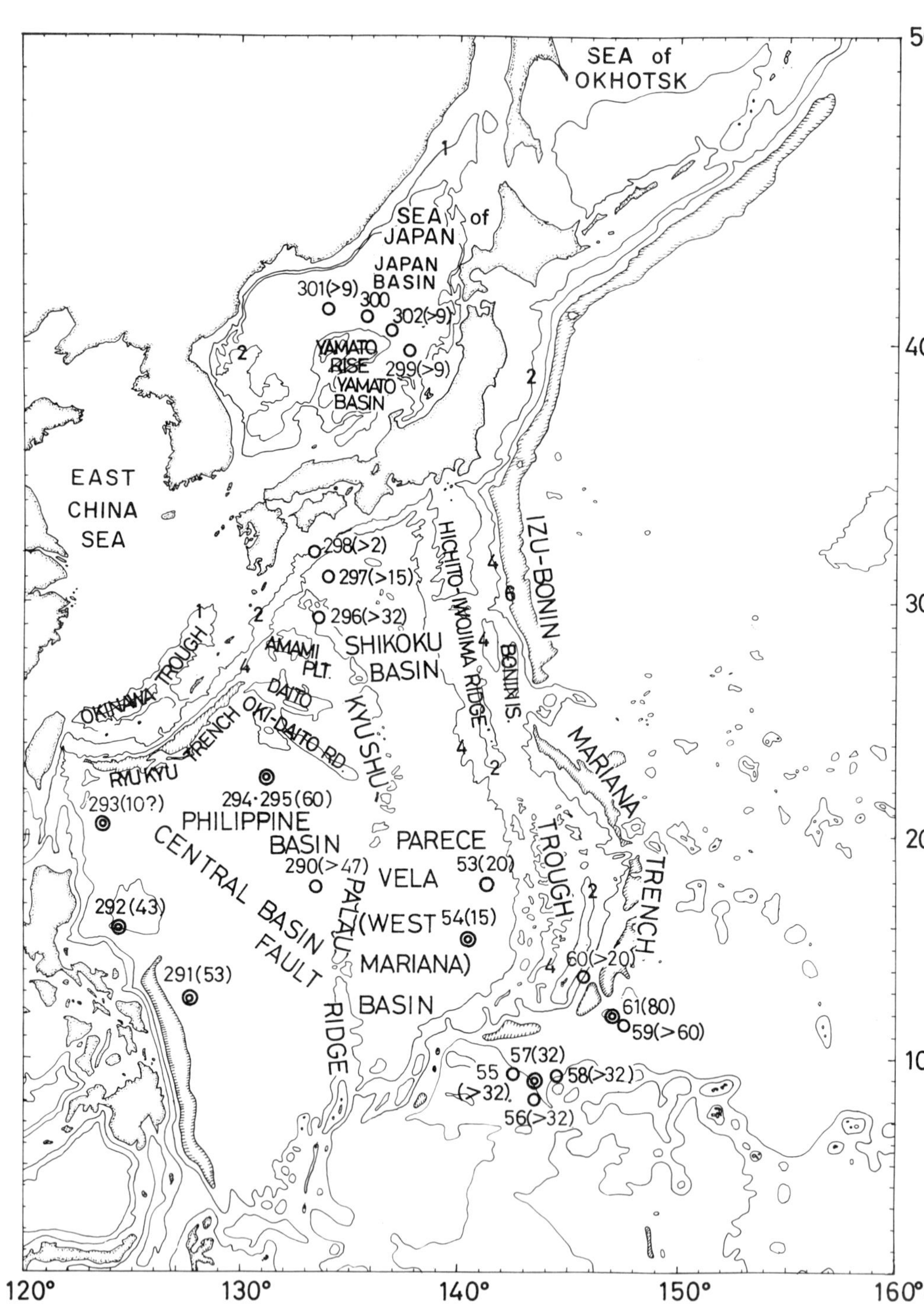


Fig. 1. Sketch of bathymetric chart of the western Pacific margins (depth in km, after Hydrog. Dept. Japan Chart No. 6302) and location of DSDP holes with hole number and bottom age (in m.y.B.P.). Double circles indicate holes penetrating basement rock. DSDP data are from *Fischer, Heezen et al.* [1971] and *DSDP Scientific Staff* [1973*b*].

around Japan and the tectonic implications on some geologic events occurring in the Japanese Islands is presented as one of the possible interpretations of the magnetic data in the framework of plate tectonics.

## MAGNETIC ANOMALIES IN THE JAPAN SEA

Geomagnetic total intensity in the Japan Sea has been extensively measured by several Japanese ships: *Seifu-maru* (Maizuru Marine Meteorological Observatory, *Yasui et al.* [1967]),

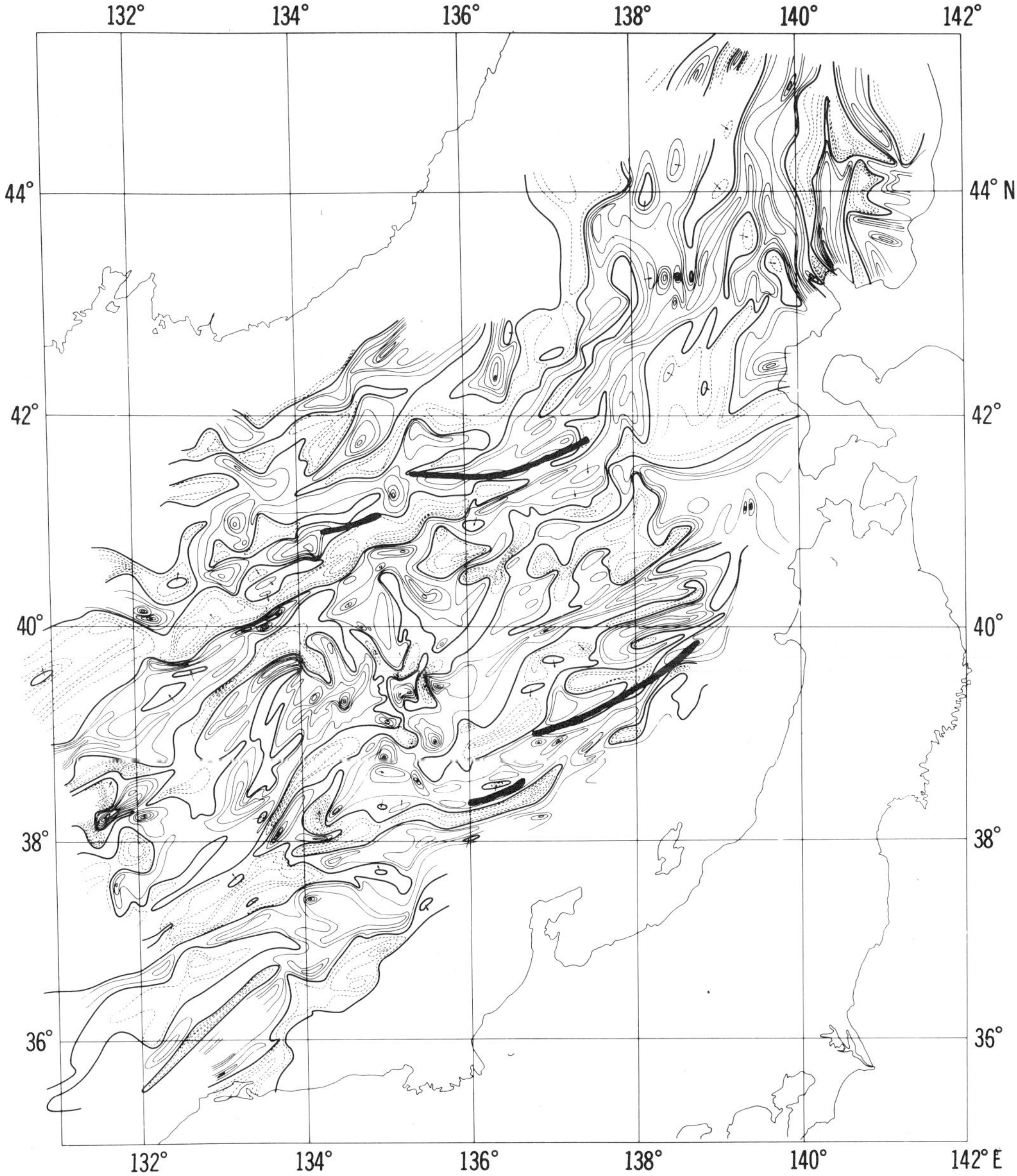


Fig. 2. Contours of total force magnetic anomalies with proposed centers of spreading in the Japan Sea. Solid and broken lines indicate positive and negative anomalies. Contour interval: 50 gammas. Heavy solid lines indicate possible spreading centers [from *Isezaki,* 1975].

*Kofu-maru* (Hakodate Marine Meteorological Observatory, *Isezaki et al.* [1971]), *Hakuho-maru* (Ocean Research Institute, *Tomoda and Nasu* [1971]), *Takuyo* (Hydrographic Department, *Matsuzaki* [1966]), and *Daiei-maru No. 83* (chartered fishing boat, *Isezaki and Uyeda* [1973]). Spacing of these tracks is close enough (less than 10 km) to draw a reliable contour map of geomagnetic total intensity anomalies as shown in Figure 2 [*Isezaki and Uyeda,* 1973].

General characteristics of the magnetic anomaly distribution in the Japan Sea are as follows:

1. Amplitudes of anomalies (peak-to-peak) are usually smaller than 300 gammas, and their wavelengths range from 20 to 40 km.

2. Linearity of anomalies is fairly distinct in the central region of the Japan basin. Linearity is less distinct but recognizable in the Yamato basin and the western part of the Japan basin. Anomalies are irregular in the Tsushima basin and over the Yamato rise.

3. Trend of lineations in the basins is roughly ENE to WSW, except for the northeastern section of the Japan basin in which anomalies trend NE–SW.

## MAGNETIC ANOMALIES IN THE SHIKOKU BASIN

*Tomoda et al.* [1975] compiled results of their marine geomagnetic measurements made on board the research vessels *Hakuho-maru* and *Tansei-maru* (Ocean Research Institute), *Umitaka-maru* (Tokyo University of Fisheries), and *Bosei-maru* (Tokai University) together with the aeromagnetic survey carried out by the Hydrographic Department, Maritime Safety Agency of Japan. They showed the existence of magnetic lineations in the Shikoku basin as illustrated in Figure 3. Correlation of anomalies may not be unique for some of the profiles. However, the following conclusions may be drawn from the data.

1. Peak-to-peak amplitudes of the magnetic anomalies in the basin are smaller than 300 gammas. Anomalies related to changing bottom topography are negligibly small or only of secondary importance when compared to the linear anomalies. Exceptions are the anomalies that are associated with a few seamounts in the Kinan seamount chain situated in the central part of the basin.

2. The magnetic lineations nearly parallel the trend of the Kyushu–Palau ridge.

3. The magnetic lineations are offset by at least four fracture zones trending NE–SW.

4. Wavelengths of magnetic anomalies in southern Shikoku basin are nearly one-half those in the northern basin.

5. The same magnetic lineations seem to exist north of the Nankai trough, although anomaly amplitudes gradually decrease (see the upper left corner of Figure 3).

## NUMERICAL TEST OF SYMMETRY AND POSSIBLE SPREADING CENTERS

In both the Japan Sea and the Shikoku basin it is very difficult to compare the magnetic anomalies with the chronological standard of magnetic polarity history [*Heirtzler et al.,* 1968; *Larson and Pitman,* 1972; *Blakely,* 1974]. First, it is not easy to discern whether the magnetic anomalies are symmetric. Second, even when the magnetic anomalies prove to be symmetric, a possible axis of symmetry is very difficult to locate. This is mainly due to the deformed wave form of the magnetic anomalies which possibly results from the irregular magnetization of the basement rocks, but more likely is due to the large angles between the trend of magnetic lineations and the geomagnetic meridian. The dominant trend of magnetic lineations in the Japan Sea is ENE–WSW, or 60° east of the present-day geomagnetic meridian. Likewise, magnetic lineations in the Shikoku basin trend about 35° west of the meridian.

In order to reproduce a de-skewed shape of the anomalies, *Isezaki* [1975] has introduced a new method according to which the observed anomalies are transformed to those caused by the alternately magnetized two-dimensional source aligned parallel to the geomagnetic meridian. This method is similar to the methods developed by *Bott* [1967], *Schouten* [1971], and *Blakely and Cox* [1972*a*] in the sense that it also utilizes the technique of Fourier transform. However, the new method may be more general and possibly of greater use for the symmetry test within one profile and for the comparison of two or more anomaly profiles with different azimuth in nearly the same latitudes, since in this method the anomalies are reduced to the geomagnetic meridian. In contrast, Blakely and Cox reduced the anomalies to the pole at which the regional field and magnetization are vertical. Their method is more appropriate for comparison and stacking of many profiles measured in various latitudes [*Blakely,* 1974].

After the Fourier transform was carried out, the correlation coefficients were calculated from the de-skewed wave forms to test symmetry. The correlation coefficient $R_f$ at a point $f$ is defined

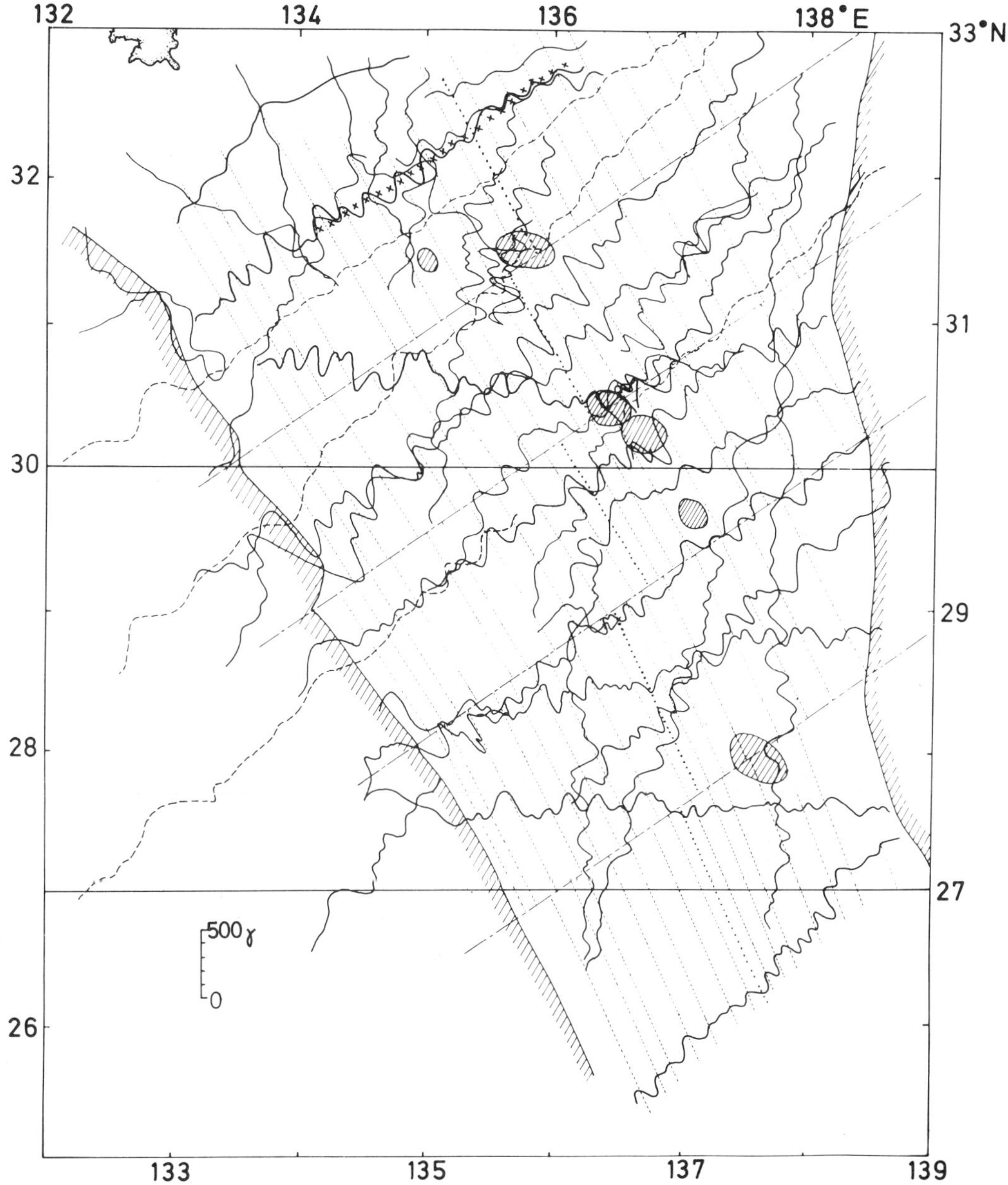


Fig. 3. Magnetic anomaly profiles along ship and aircraft tracks with proposed magnetic lineations in the Shikoku basin. Solid curves represent surface marine magnetic anomalies; broken curves, aeromagnetic anomalies. Light, dashed lines connect corresponding anomalies, and dotted lines show the proposed centers of symmetry. The dash-dot lines trending roughly NE–SW are postulated fracture zones. The row of X marks denotes the axis of the Nankai trough. Shaded areas are the margins of topographic highs [modified from *Tomoda et al.*, 1975].

as the correlation between the profiles of magnetic anomalies extending to the east and the west of point $f$:

$$R_f = \frac{S_1}{\sqrt{S_2 \cdot S_3}}$$

where

$$S_1 = \sum_i a_s(f-i) \cdot a_s(f+i),$$

$$S_2 = \sum_i a_s^2(f-i)$$

and

$$S_3 = \sum_i a_s^2(f+i)$$

with $a_s(x)$ denoting the amplitude of the deskewed anomaly at the point $x$. If a center of symmetry exists, $|R_f|$ will be maximum there. Note that values of $R_f$ at points near the edges of the profile are unreliable because the amount of data that can be used for the analysis is small.

The coefficients were calculated for six long profiles traversing the Japan and Yamato basins (Figures 4a, 4b). As seen in Figure 4b, two centers of symmetry seem to exist in the Japan Sea: one in the central zone of the Japan basin and the other in the Yamato basin. It must be noted that the centers of symmetry can be clearly defined for each of the profiles. Trajectories of the symmetry centers thus determined are indicated in Figure 2. It is remarkable that the two trajectories almost parallel each other, and are in turn parallel to the trend of magnetic lineations. It is inferred that the two axes of symmetry represent the centers of spreading of the two basins.

In the Shikoku basin most of the profiles are either truncated by the fracture zones or incomplete so that significant correlation coefficients were obtained for only one profile KH72-2 (Figures 4a, 4c). A center of symmetry determined on these profiles is located near the center of the Shikoku basin. If the center of symmetry is projected along the trend of magnetic lineations on other profiles, an axis of symmetry is obtained as shown in Figure 3. The axis of symmetry thus obtained shows good agreement with that postulated by *Tomoda et al.* [1975] based on other geophysical evidence.

## OTHER GEOPHYSICAL EVIDENCE

*Morphology and sediment distribution.* It has been recognized that the Japan basin differs from the Yamato basin in morphology and geologic structure [*Den,* 1972; *Kovylin,* 1966]. Water depth in the Japan basin is 3000 to 3500 m, and the bottom is primarily covered by undeformed, flat-lying sediments approximately 2 km thick on the average. The Yamato basin, by contrast, is about 2500 m deep, and contains thinner sediment overlying an uneven basement of apparent igneous origin. *Hilde and Wageman* [1973] pointed out, from their seismic reflection profiler records, that the sediments in both basins are thickest on the basin margins and thin along the central axis of the basin. Interpretation of Hilde and Wageman's results led to the inference that initial rifting and opening of the Japan Sea had started in the axial zone of the Japan and Tsushima basins, and later spreading centers migrated to the axial zone of the Yamato basin. Although their profiler records are not complete enough to deduce the exact location of the centers, the general patterns of sediment and basement topography in the Japan Sea seem to be consistent with the proposed origin of the Japan Sea by spreading. Jigsaw-fitting of the shelf lines of the Asiatic continent, the Japanese Islands, and the Yamato rise [*Murauchi,* 1971] first provided a plausible configuration for these continental masses before spreading of the basins began.

The average water depth of the northern Shikoku basin is approximately 4600 m; the floor south of 30°N is slightly deeper. In both basins, the water deepens toward the west (Figure 5). The sediment thickness gradually increases toward the north (landward) [*Ludwig et al.,* 1973]. *Murauchi and Asanuma* [1974] have shown by seismic reflection studies that sediment is thin along the central zone of the basin, thickening toward the eastern and western margins. A boundary of thin and thick sediments seems to exist nearly halfway between the margins and the central axis. The sediment cover is much thinner toward the central axis. This may imply that the central zone is younger than the margins. The basement relief of the central zone is similar to that of the spreading ridge of the ocean. This observation is consistent with our proposition that the center of symmetry of magnetic anomaly is also a center of spreading. Asymmetric topography of the basin is probably due to later volcanic activity of the Izu–Bonin ridge.

*Heat flow.* Heat flow data in the Japan Sea and Shikoku basin have been summarized by *Uyeda* [1972]. Figure 6 is the most recent contour map of heat flow by Watanabe (personal communication, 1974) supplemented with some new values. It may be stated from this figure that the extraordinarily high values of heat flow are located near the proposed spreading centers. This does not necessarily mean that these spreading centers are still active. Spots of high heat flow may be remnants of ridge spreading activities that ceased long ago.

*Seismicity.* Seismic studies have shown that no detectable shallow earthquakes occur under either the floor of the Japan Sea [*Utsu,* 1971*b*] or the axial zone of the Shikoku basin [*Katsumata and Sykes,* 1969]. Inasmuch as all active spreading ridges are accompanied by shallow earthquakes, the observation would imply that the spreading centers in the Japan Sea and the Shikoku basin are now extinct. It must be noted that shallow to intermediate earthquakes occur

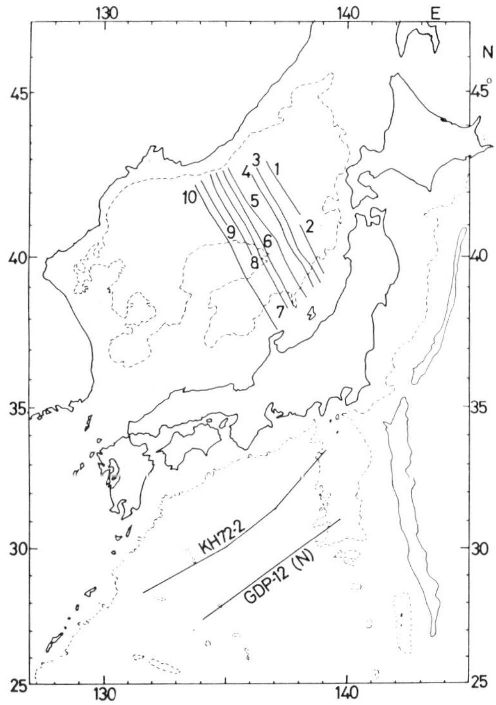


Fig. 4a.

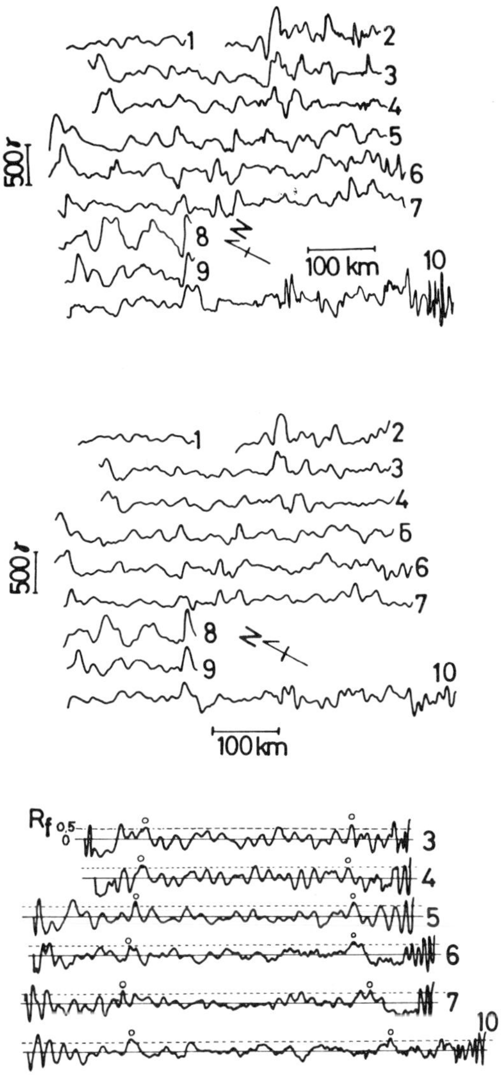


Fig. 4b.

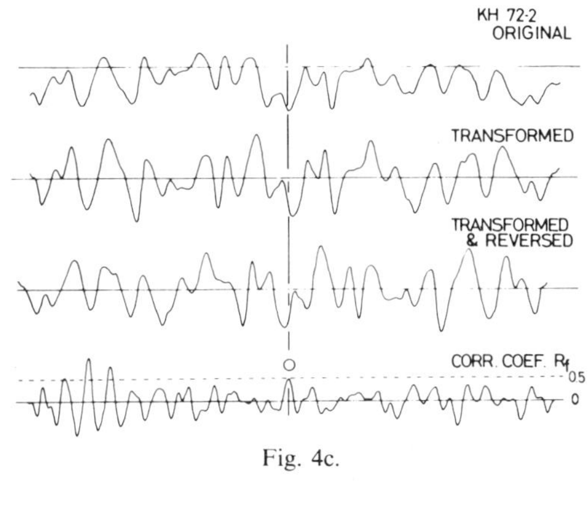


Fig. 4c.

Fig. 4. Observed and transformed anomaly profiles and correlation coefficients of symmetry, $R_f$. (a) Index map for quoted magnetic anomaly profiles. (b) Japan Sea: (upper) Observed anomalies arranged from NE to SW; (middle) transformed anomalies; (lower) correlation coefficients, $R_f$. (c) Shikoku basin. The vertical line marked by an open circle denotes a possible center of symmetry.

in the eastern part of the Nankai trough, which very likely is one of the active subduction zones [*Kanamori,* 1971].

*Gravity.* Marine gravity anomaly data around Japan were compiled by *Tomoda* [1973*a*]. *Segawa and Tomoda* [this volume] summarized recent results and discussed the pattern of anomalies. Positive free-air gravity anomalies, +20 to +30 mgal in the Japan and Yamato basins, and a nearly null anomaly in the Shikoku basin indicate that these basins are isostatically compensated. This observation also supports the idea that the spreading centers in those basins are inactive.

## AGE CONSTRAINTS ON SPREADING PERIODS

Comparison of geologic formations in the Japanese islands and Korea showed that the opening of the Japan Sea occurred between 60 and 20 m.y.B.P. [*Ichikawa,* 1972]. The isotope age of an olivine basalt collected by dredge haul from the northeastern flank of the Yamato rise is 21 m.y.B.P. [*Ozima et al.,* 1970]. Much younger ages (8 and 4 m.y.B.P.) have been obtained for andesite samples collected at the eastern margin of the Yamato basin. However, these samples were possibly a product of the more recent island-arc volcanism, hence not representative of the age of the spreading basin. DSDP hole 299 at the western margin of the Yamato basin did not reach the basement rock. The data therefore provide only a minimum age for the basin, about 9 m.y.B.P. [*DSDP Scientific Staff,* 1973*b*].

Age data for the Shikoku basin were provided by DSDP holes 296 and 297 (Figures 1 and 7). In particular, the data from hole 297 are useful because the hole is situated in the northwestern quarter of the basin in which the magnetic lineation is well defined. The age of the sediment in hole 297—extrapolated to the basement that lies about 100 m below the bottom of the hole with the aid of the sedimentation rate determined in the adjacent hole 296—was 20 m.y.B.P. The age data from holes 53 and 54 on the eastern side of

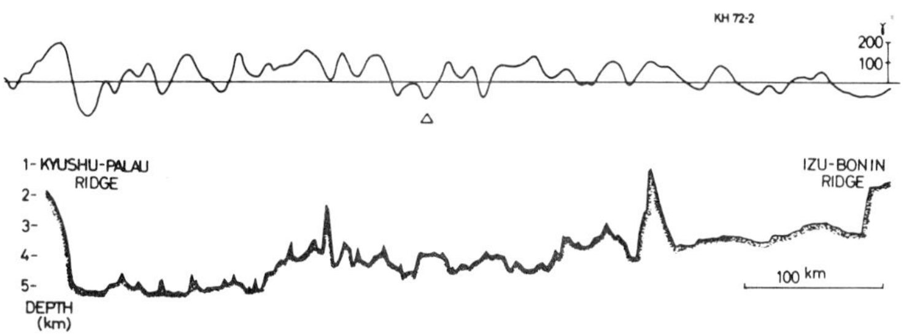


Fig. 5a.

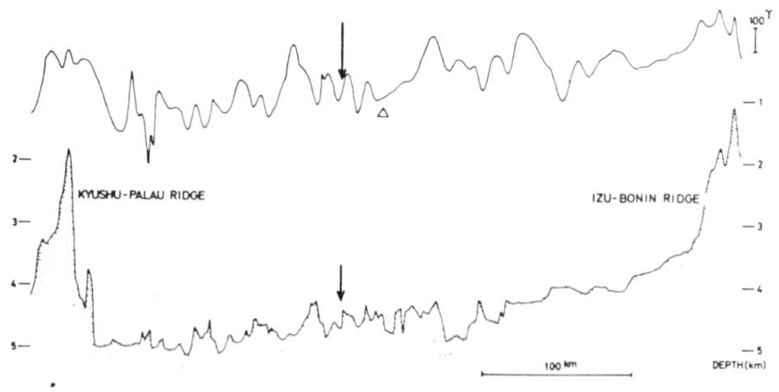


Fig. 5b.

Fig. 5. Magnetic anomaly (upper) and topographic (lower) profiles across the Shikoku basin. (a) KH72-2 cruise. (b) GDP-12 cruise (*N*-profile). The triangle indicates the proposed center of spreading and the arrow (in (b) only) denotes position of fracture zone. Note that the magnetic anomalies are almost independent of the bottom topography.

Parece Vela basin [*Fischer, Heezen et al.*, 1970, 1971] will also provide an age limit for the southern Shikoku basin assuming that the two basins opened simultaneously. In hole 54, altered high-alumina olivine tholeiite that closely resembles ocean ridge tholeiite was cored [*Ridley et al.*, 1974]. The tholeiite is overlain by middle Miocene ash deposits and, based on combined stratigraphic data from holes 53 and 54, is considered to be early Miocene or late Oligocene in age (20 to 25 m.y.B.P.) [*Fischer, Heezen et al.*, 1971].

Occurrence of turbidite in a limited range of depth in DSDP hole 297 indicates that the Nankai trough, which traps seaward-moving bottom materials, was non-existent between about 5 and 3 m.y. B.P. and was rejuvenated about 3 m.y. B.P.

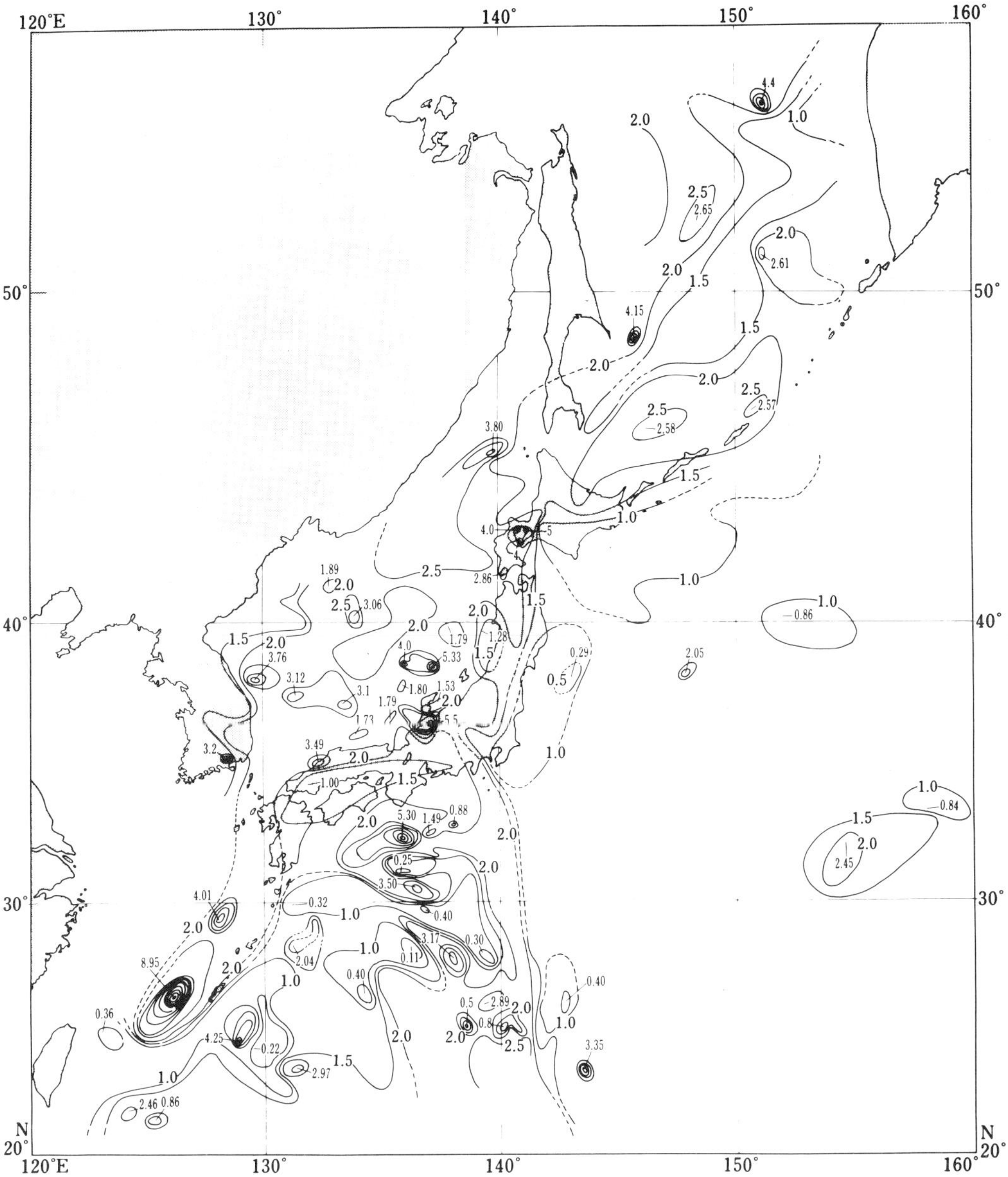


Fig. 6. Contoured distribution of heat flow values around Japan (after Watanabe, personal communication, 1974) (modified from *Uyeda* [1972], with some additional data).

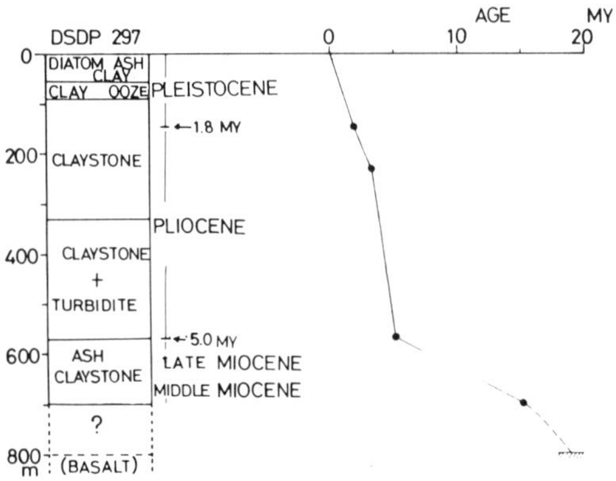


Fig. 7a.

Lithology of hole 296 drilled at a saddle point on the Kyushu–Palau ridge indicates that the volcanic activity of this part of the ridge ceased about 28 m.y.B.P. This age may limit the maximum age for the initiation of spreading, because it is widely accepted that the remnant ridge, like the Kyushu–Palau ridge, became inactive after the inter-arc basin started opening. During the GDP-8 cruise

Fig. 7. Diagrams showing age and lithology of sediments as a function of four DSDP holes in the Shikoku and Parece Vela basins [compiled from *DSDP Scientific Staff*, 1973*b*; and *Fischer, Heezen et al.*, 1971]. Absolute ages are according to *Berggren* [1972]. (a) DSDP hole 297, Shikoku basin. (b) DSDP hole 296, Shikoku basin (c) DSDP hole 53, Parece Vela basin. (d) DSDP hole 54, Parece Vela basin.

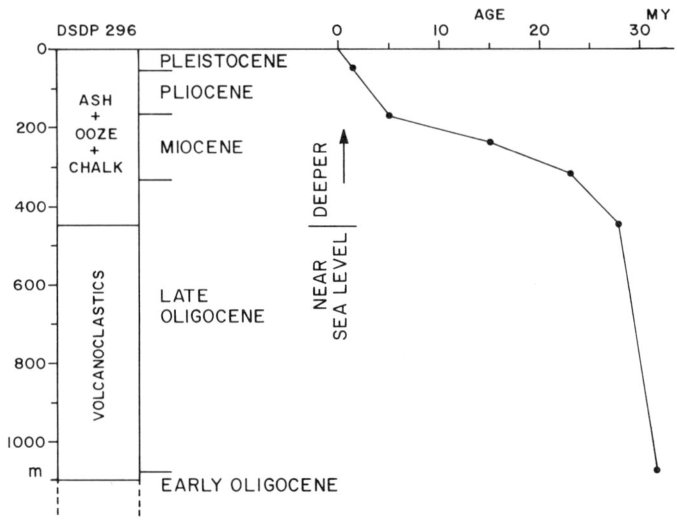


Fig. 7b.

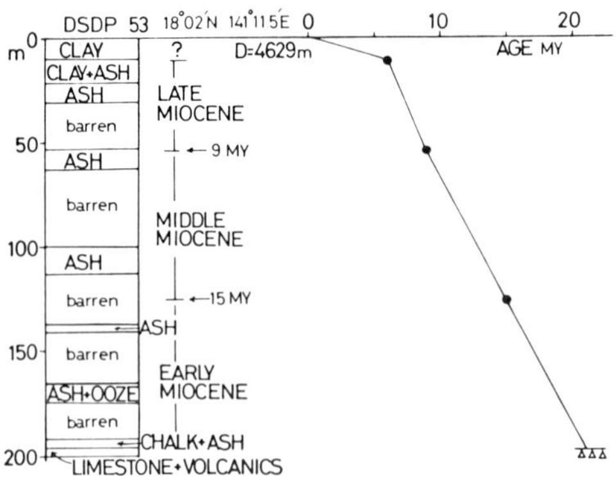


Fig. 7c.

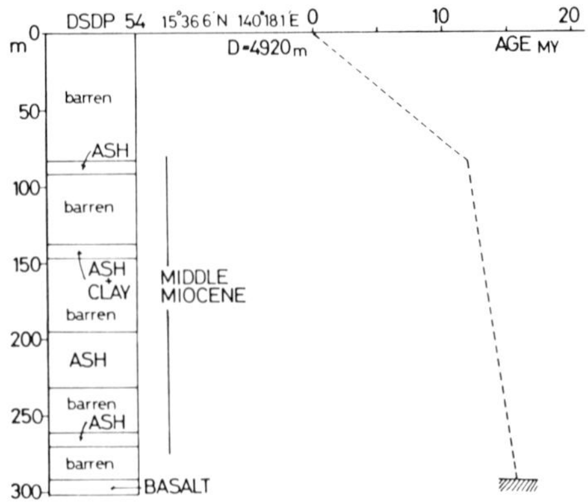


Fig. 7d.

of the *Bosei-maru,* a large number of angular pebbles coated with ferromanganese oxides were collected by dredge hauls at the northern end of the Kyushu–Palau ridge (Komahashi II seamount). The dredged samples consist of a variety of rocks including quartz diorite, diorite porphyry, biotite granite, granite porphyry, and coarse-grained leucocratic granite, which are supposed to occur in island arcs [*Shiki et al.,* 1974]. Isotope age determined for one of the granite samples is 37.8 m.y.B.P. [*Shibata and Okuda,* 1975], which seems to date an event that occurred prior to the opening of the Shikoku basin for the same reason as previously stated.

The oldest ages of rocks in the Bonin Islands are 40 m.y.B.P. at Hahajima and 26 m.y.B.P. at Chichijima [*Kaneoka et al.,* 1970]. The latter age is nearly concordant with the bottom age of DSDP hole 60 which is located east of Guam, Mariana Islands. The ages of the Bonin Islands volcanism can be either contemporaneous with or prior to the spreading of the Shikoku basin. Note that the volcanoes in these islands are all inactive. Active volcanoes now exist farther west on the Hichito–Iwojima ridge.

Considering all these constraints on ages, *Tomoda et al.* [1975] have dated the linear magnetic anomalies in the Shikoku basin to be early to middle Miocene. According to their interpretation, the Shikoku basin began opening about 22 m.y.B.P. and ceased spreading about 13 m.y.B.P. Recently, *Watts and Weissel* [1975] postulated a slightly different age of magnetic anomalies by analyzing a number of the U.S. data in addition to *Tomoda et al.'s* [1975] results. They inferred that the Shikoku basin began spreading about 27 m.y.B.P., although they have not proposed the exact position of the spreading center and the date when the opening ceased. Both models indicate that the separation of the Izu–Bonin island arc from the Kyushu–Palau ridge started after the late Oligocene.

## A HYPOTHESIS FOR THE EVOLUTION OF MARGINAL BASINS AROUND THE JAPANESE ISLANDS

*Background information.* Evolution of the marginal basins including the Japan and Shikoku basins has already been discussed by several authors [*cf., Sugimura and Uyeda,* 1972; *Matsuda and Uyeda,* 1971; *Packham and Falvey,* 1971]. The basic concept common to most of the recent models is the formation of marginal basins by a seaward drift of island arcs from the continental margin, and intrusion of mantle materials filling up the void between. Upward injection of mantle materials is either by thermal upwelling of magma from the partially molten upper surface of the sinking lithosphere [*Hasebe et al.,* 1970; *Karig,* 1971*a, b; Bracey and Ogden,* 1972] or by a countercurrent in the asthenosphere beneath the overriding continental lithosphere [*Sleep and Toksöz,* 1971].

*Karig* [1971*a, b*] further suggested, from the available data of sediment structure and basement ages in Tonga and the Mariana Islands, that the opening of marginal basins is caused by an extensional force and may be discontinuous in time. He visualized a typical 'pulse' of tectonic activity in the trench–arc–marginal sea system as beginning with trench formation, followed by volcanism, and later by an extensional development of the basin. The extensional direction of the basin seems to be roughly normal to the trend of arc–trench systems, although there often exist 'fixed points' at both ends of the system which usually give a concave shape to the final trench–arc unit. According to Karig's postulation, the Mariana trough, the basin on the concave side of the Mariana volcanic chain, is of the most recent extensional origin, the Parece Vela basin in the west was opened by an earlier pulse, and the Philippine basin was formed in the earliest epoch in this sequence of events.

The six DSDP holes (290 to 295) in the Philippine basin indicate that the basin was formed during the Eocene as proposed by Karig [*DSDP Scientific Staff,* 1973*b*; *Karig et al.,* 1973]. Magnetic anomaly profiles and bottom relief across the Central Basin fault suggest that the origin of the Philippine basin is due to spreading, with a center along the Central Basin fault trending WNW–ESE [*Ben-Avraham et al.,* 1972]. Patterns of magnetic lineations obtained by *Chao-Shing and Hilde* [1971], *Louden* [this volume] and Isezaki (in preparation) are more complicated but essentially the same as obtained by *Ben-Avraham et al.* [1972]. From comparison of observed magnetic anomaly profiles and simulated models, the latter suggested that one of the following two movements has occurred since the basin was formed: (1) the basin has rotated anticlockwise by about 40°; (2) the basin has drifted northward across the equator. No evidence has been found to justify either of these postulates.

*Uyeda and Ben-Avraham* [1972] and *Uyeda and Miyashiro* [1974] postulated that the Central Basin fault was a part of the Kula–Pacific ridge which has been lost under the circum-Pacific trenches. They assumed that the Kyushu–Palau ridge was a ridge–ridge transform fault connecting the two parts of the Kula–Pacific ridge and was later converted to an island arc. However, an objection to this hypothesis arises from the depth–age relationship of the sea floor proposed by *Sclater et al.* [1971]. As *Yoshii* (1973*a, b*) pointed out, the water depths of marginal seas are usually inconsistent with those predicted by the empirical relation obtained in the northeast Pacific basin. The Shikoku and Parece Vela basins are about 1000 m deeper than the depth predicted by *Sclater et al.'s* [1971] depth–age relation. Water depth of the central Japan basin is about 3500 m, and apparently shallower than the depth which the 40 to 50 m.y.B.P. age suggests. However, if the thickness of sediment cover amounting to 2000 m is taken into account, the basement depth (about 5500 m) is deeper than the predicted depth. Water depths of the Philippine basin are clearly much deeper than water depths of the Pacific basin of the same age.

*Yoshii* [1973*a, b*] estimated the 'residual gravity anomalies' (RGA) using the crustal structure

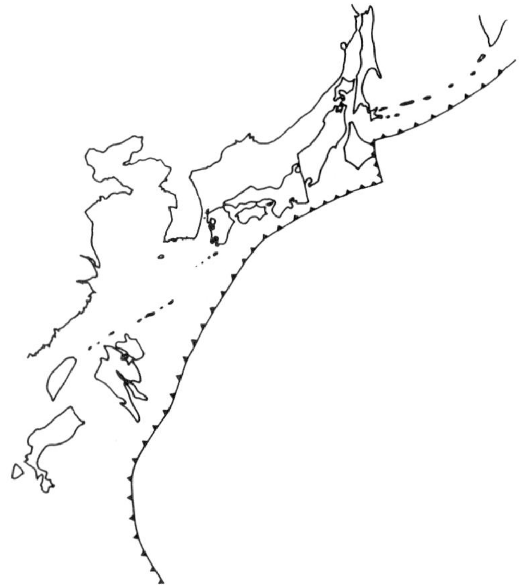

Fig. 8a.

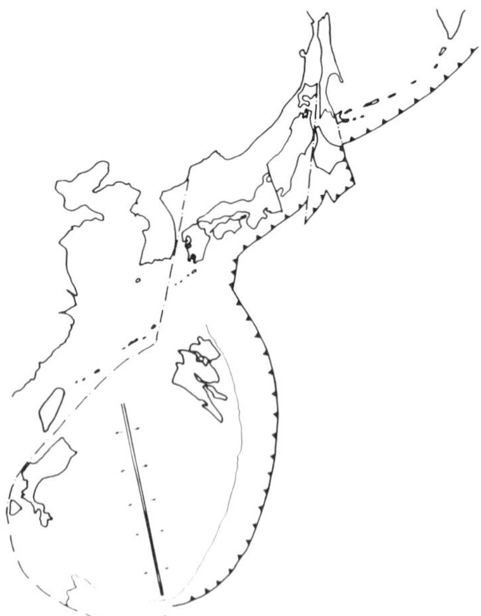

Fig. 8b.

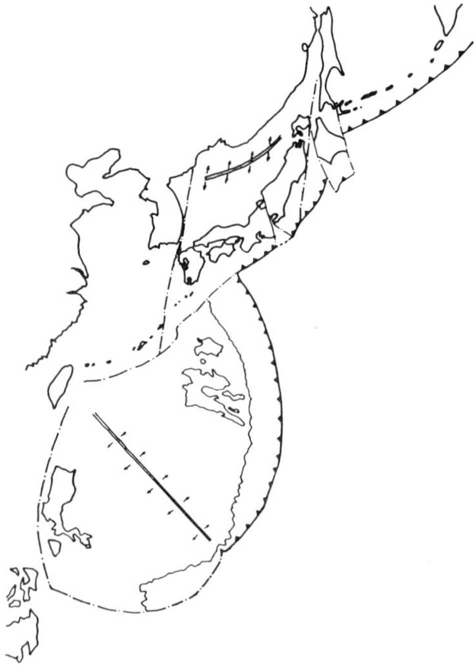

Fig. 8c

Fig. 8. A reconstruction of the trench–arc–marginal sea systems in the northwestern Pacific. (a) For a period older than about 60 m.y.B.P. (b) For a period about 50 to 40 m.y.B.P. (c) For a period about 40 to 30 m.y.B.P. (d) Present configuration of the trench–arc–marginal sea systems with postulated centers of the past spreading and fracture zones (double lines show the spreading centers determined from magnetic data). Dashed lines are spreading centers speculated only from high heat flow and morphological evidences. Lines marked with a query are highly speculative.

revealed by the seismic studies. Systematic relationship between the age of basement and the RGA was obtained for the north Pacific basin. However, the RGAs of marginal basins were discordant with that of the north Pacific of the same age. The estimated value of RGA in both the Parece Vela and Philippine basins is nearly 100 mgal higher than that of the normal Pacific basin, implying that the upper mantle density is slightly higher in those basins. This result seems

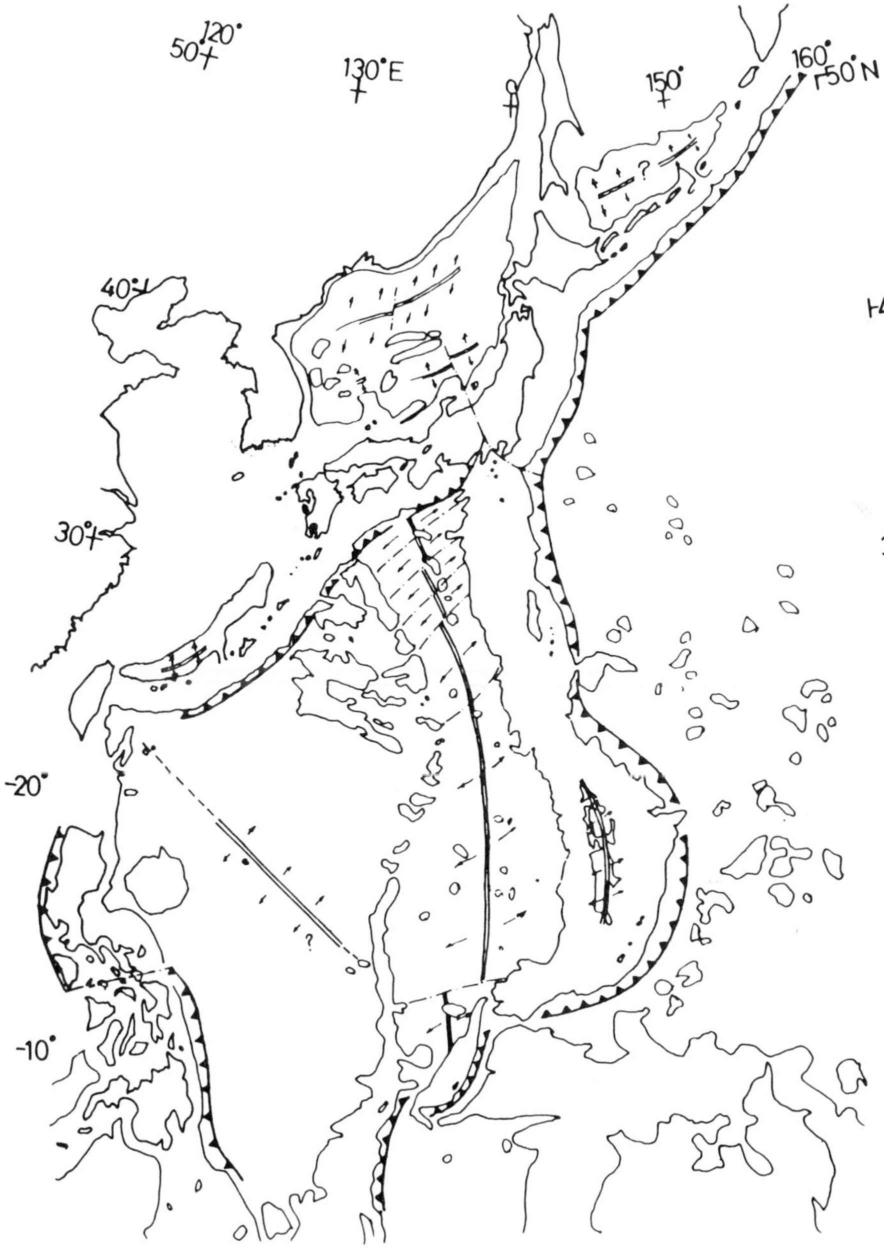


Fig. 8d.

to indicate that the Philippine basin was not a relic of the Kula–Pacific plates, but possibly originated in the same manner as other marginal basins, unless the ages given for the basins so far are all erroneous.

*A proposed hypothesis.* We propose a model which is similar to that of *Karig* [1971*a, b*] but is specifically oriented to explain the evolutionary history of marginal basins and island arcs around Japan. We begin with an epoch in which there was no marginal sea along the northeastern margin of the Asiatic continent. At about 60 m.y.B.P., a trench was formed along the margin, at which occurrence volcanism soon became active under the influence of the sinking lithosphere (Figure 8a). The Philippine basin then began to open, but the direction of spreading was, we propose, roughly normal to the trend of trench. The arc–trench system shifted eastward and eventually migrated to the present location of the Kyushu–Palau ridge (Figure 8b).

Shortly after the extinction of spreading of the Philippine basin (possibly 45 to 40 m.y.B.P.), the Japan basin began to open. The southward drift of the main islands of Japan, the Ryukyu Islands, and possibly Taiwan rotated the Philippine basin anticlockwise about 40° in accordance with one of the interpretations of the magnetic anomaly across the Central Basin fault [*Ben-Avraham et al.,* 1972]. A high tensional force induced by the rotation of the Philippine basin tore the northern part of the 'Kyushu–Palau Island' into pieces to form the Oki-daito and Daito ridges and the Amami plateau (Figure 8c). Kobayashi (unpublished data) has found a rifting of possibly extensional origin in the Amami plateau. Free-air gravity anomalies in the narrow, deep basins between these topographic highs amount to –30 to –40 mgal [*Segawa and Tomoda,* this volume], implying mass deficiency beneath these basins.

Subsequently (possibly 40 to 30 m.y.B.P.), subduction of the oceanic lithosphere commenced again along the Japan and Kyushu–Palau trenches. It seems likely that this rejuvenation of subduction was associated with the change in direction of motion of the Pacific plate. After development of the subduction zone, the Shikoku and Yamato basins began to spread. As a consequence of this event, the Izu–Bonin–Mariana island arc drifted to its present position. If such has been the case, the Izu Peninsula may also have been transported from the west to its present position and may have collided with the main island (Honshu) of Japan drifting southeastward. It is also inferred that spreading of the Yamato basin caused the drift of northeastern Japan southeastward by about 100 km relative to southwestern Japan. The Nankai trough that had been a subduction zone until the Shikoku basin began to spread was then transformed into a trench–trench type transform fault.

The opening of the Shikoku basin ceased by middle Miocene. By middle Pliocene the basin again began to underthrust the Nankai trough, but the duration of subduction was too short to result in active volcanism in southwestern Japan and spreading of the Tsushima basin. Small magnetic anomalies recognized on the landward side of the trough (*Tomoda et al.* [1975]; see also Figure 3) may possibly be caused by the subducted lithosphere. The Mariana trough may have been opening since early Quaternary, although seismic and other data are still insufficient to support the idea. The Bonin trough, located between the extinct Bonin island arc and the Hichito–Iwojima ridge, now active as is evidenced by the active Iwojima volcano, may also be of spreading origin. However, the low heat flow values measured in the trough [*Saki,* 1975] provide negative evidence for recent opening of the trough.

*A plate tectonic model for pulsive evolution of marginal basins.* Repeated pulses in the history of the Japan and Philippine seas may become more understandable if the disruption and detachment of the sinking lithosphere are taken into account. The model that will be discussed below is nearly the same as proposed by *Kanamori* [1971] to explain the focal mechanism of large earthquakes occurring around Japan, but is modified to fit the new data. Figure 9 illustrates the simplified cycle of evolution of a marginal sea. The top of the figure (stage 1) is a state in which no subduction is occurring (a passive margin). (The Japan basin adjoining the Asian continent is in this stage at present.) Stage 2 is, by contrast, an initial stage of subduction in which denser oceanic lithosphere is descending beneath the lighter continental lithosphere. Interaction of the two lithospheres is detected as shallow to intermediate earthquakes with focal mechanisms of reverse fault type. Materials near the upper surface of the sinking lithosphere begin to melt partially either from the heat generated by friction between the two lithospheres [*Hasebe et al.,* 1970] or from the influence of $H_2O$ and $CO_2$ [*Ringwood,* 1974; *Kushiro,* 1974]. High heat flow and volcanic activity at the margin of the continent are probably due to diapiric injection of the molten material. (Recent Alaskan volcanism

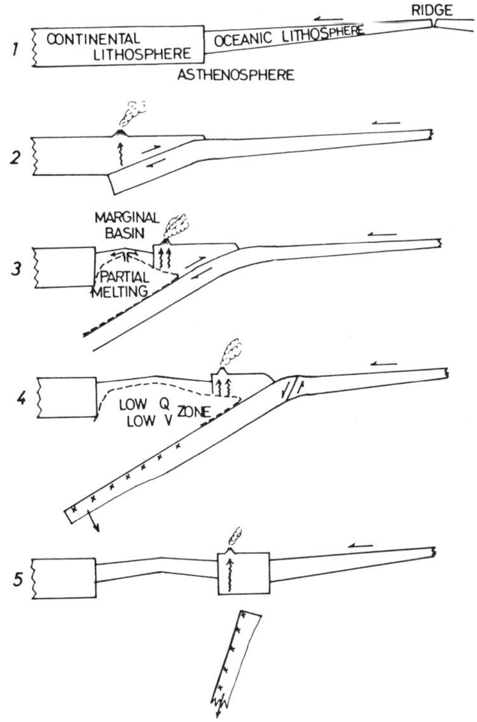


Fig. 9. Evolution of marginal sea and island arc systems [largely modified from *Kanamori,* 1971]. See text for further details.

seems to originate from this stage of subduction.)

As the oceanic lithosphere descends further, a large volume of molten materials forms beneath the continent. Then the margin of continental lithosphere rifts and the molten material intrudes to form the marginal basin (stage 3). If the subduction continues and partial melting near the upper surface of the oceanic lithosphere proceeds, coupling between the continental and oceanic lithospheres is greatly reduced. It is very likely at this stage that the descending lithosphere is dislocated and disrupted from the horizontal oceanic lithosphere along a plane near the migrating edge of the continental lithosphere due to the gravitational instability (stage 4). Gigantic earthquakes of the normal fault type may occur at the location of disruption. *Kanamori* [1971] postulated this mechanism to explain the origin of a strong earthquake of the normal fault type occurring at a site off the Sanriku coast, northeastern Japan.

When this process, which is accompanied by a repetition of several large earthquakes, is completed, the detached slab of oceanic lithosphere sinks down to the deeper asthenosphere (stage 5). (The nearly vertical dip of the zone of deep earthquakes at the present Izu–Bonin arc [*Katsumata and Sykes,* 1969] seems to be such a detached lithosphere.) In *Kanamori's* model [1971], linear opening of the marginal basin and seaward drift of the island arc were disregarded and a catastrophic subduction of the oceanic lithosphere at an arbitrary site in the ocean was assumed. The present revised model based upon the assumption that subduction always starts at the boundary between the continental and oceanic lithospheres seems to be more consistent with geophysical observations such as magnetic anomalies, thickness of sediment cover, and age of the sea floor.

The aforementioned cycle of events will repeat itself. There are two possible sites at which a new subduction may occur; either at the continental margin or at the trench side of the island arc where the underthrusting of the oceanic lithosphere had been taking place. The choice of site may simply depend upon the thermal structure of the two oceanic lithospheres. (Underthrusting now proceeding in the Nankai trough [*Fitch and Scholz,* 1971] seems to be the initial stage of this second cycle.) As there the length of the deep focus earthquake zone is only 100 to 150 km, subduction must have started 3 to 5 m.y.B.P. if the horizontal motion of the lithosphere is about 3 cm/yr. Thus the estimated age of the rejuvenation of the Nankai trough agrees with that suggested by the lithology of DSDP hole 297 and also with the age postulated by *Hilde et al.* [1969]. Rejuvenation of the Ryukyu trench may be older, because the length of the descending slab there is about 300 km.

## SOME IMPLICATIONS ON THE CENOZOIC GEOLOGY OF JAPAN

The evolution of marginal seas around Japan as depicted above is in harmony with the tectonic evolution of the Japanese Islands. The correspondence between the igneous activities and the spreading of the Shikoku basin is particularly good. Figures 10a, b, and c show the distribution of volcanic rocks in Japan in three different periods; late Mesozoic and early Tertiary, early Neogene, and Quaternary. The transition of volcanic activity in southwestern Japan is most remarkable. Extensive volcanism occurred there to form the Late Cretaceous Nohi Group and other masses in earlier periods (Figure 10a). In the early Neogene appreciable amounts of volcanic materials erupted only in the northern zone of southwestern Japan in marked contrast to the thick piles of volcanic layers found in the western zone of northeastern Japan (Figure 10b). On the other hand, Quaternary volcanic activities are limited to northeastern Japan except for a very few locations in southwestern Japan (Figure 10c).

Figure 11 presents a quantitative estimate of the volcanic rocks that have formed in Japan since 25 m.y.B.P. The diagram clearly shows that volcanic activity began 23 m.y.B.P. Figure 10b shows that this activity was restricted mainly to northeastern Japan. Tectonic activities in southwestern Japan for the period are shown in Figure 12. The activities were minor compared with those in northeastern Japan.

All of these geologic events can be explained satisfactorily in terms of the evolution of the marginal basin system. Extensive rhyolitic volcanism throughout the Japanese Islands in pre-Cenozoic time was very likely caused by a trench system that developed along the oceanic side of the islands. After a pause in volcanic activity in middle to late Oligocene, which corresponds to the initial stage of a new subduction cycle (stages 1 or 2 in Figure 9), there came a period of violent volcanism in the islands that was accompanied by the opening of the Shikoku and Yamato basins in the middle to late Miocene. It is still uncertain whether the maximum stage of island volcanism and the spreading of basins are contemporaneous. However, it seems likely that volcanic activity in the frontal arc continued for more than 10 m.y. after the opening of the basins ceased. Late Miocene volcanism in northeastern Japan may possibly have been of this type.

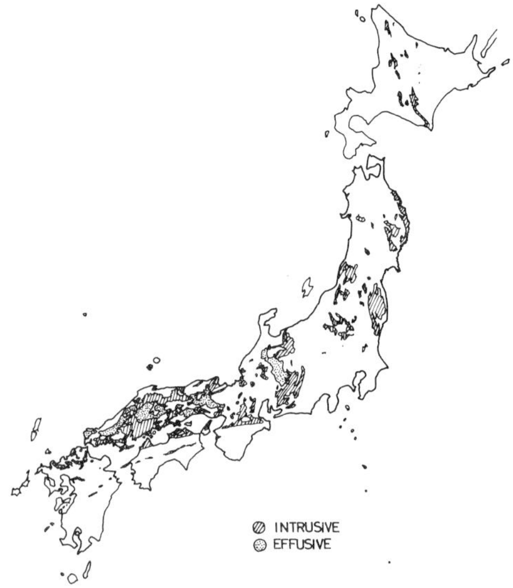


Fig. 10a.

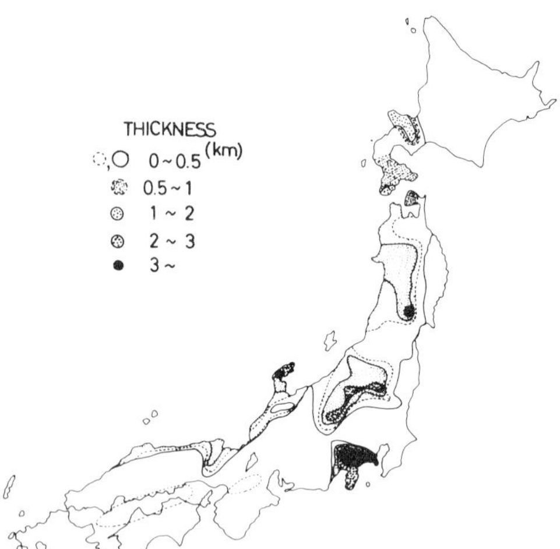


Fig. 10b.

Fig. 10c.

Fig. 10. Distribution of volcanic and intrusive rocks (a) Late Mesozoic to early Tertiary [after Geological Map of Japan, *Geological Survey of Japan,* 1971]. (b) Early Neogene [modified from *Sugimura et al.,* 1963]. (*c*) Quaternary, contour interval: 20 m thick; dashed lines denote a thickness of 10 m plus the adjacent contour [modified from *Sugimura et al.,* 1963].

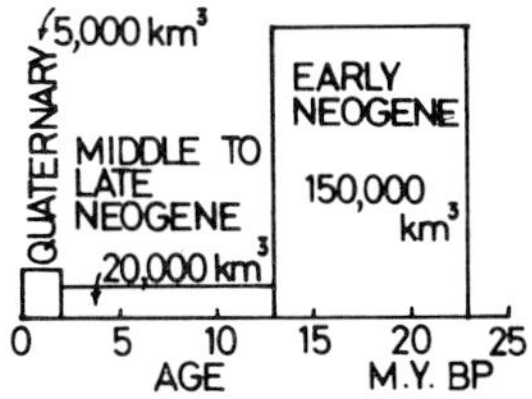


Fig. 11. Histogram showing time variation of total amount of volcanic rocks in the main islands of Japan [after *Sugimura et al.*, 1963]. Note that the absolute ages are largely revised according to the new chronological scale proposed by *Ikebe et al.* [1972].

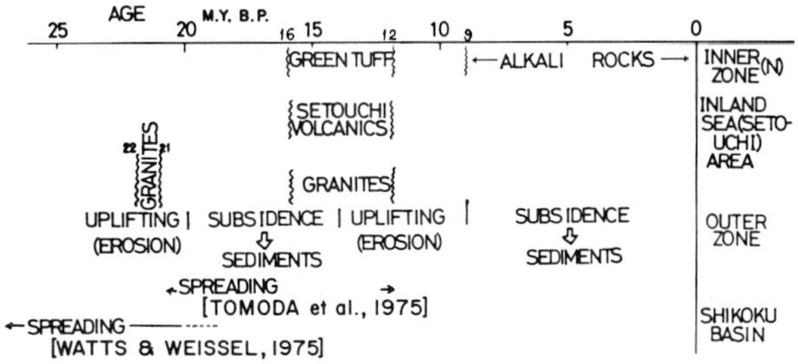


Fig. 12. Tectonic events in southwestern Japan and in the Shikoku basin since 27 m.y.B.P.

It seems plausible from the distribution of volcanic materials shown in Figure 10b that northeastern Japan drifted southeastward relative to southwestern Japan by a distance of about 100 km, possibly along a line paralleling the Itoigawa–Shizuoka and Kashiwazaki–Chyoshi fault lines. The drift may have been caused by the opening of the Yamato basin and probably was completed by the time the Izu Peninsula reached its present position.

Further investigation of local geology of the Japanese and Izu–Bonin islands as well as marine geological and geophysical studies of the surrounding seas will provide data to further elaborate the present model of tectonic history of these areas.

# 21. *Magnetic Anomalies in the West Philippine Basin*[1]

KEITH E. LOUDEN

*Department of Earth and Planetary Sciences, Massachusetts Institute of Technology, Cambridge, Massachusetts 02139*

*and*

*Department of Geology and Geophysics, Woods Hole Oceanographic Institution, Woods Hole, Massachusetts 02543*

*Abstract.* We present a comprehensive collection of magnetic anomalies in the West Philippine basin together with several bathymetric profiles and nannofossil age dates for DSDP sites 290 and 291. From these data we locate symmetric magnetic anomalies that have strikes roughly parallel to the Central Basin fault. They can be identified as anomalies 17–21 if the basin originated south of the equator and opened at a rate of 41–44 mm/yr. This adds justification but not total proof for the theory that the Central Basin fault may be an extinct spreading center which slowed and then ceased spreading 40 m.y. ago. It is still possible that island-arc spreading created these anomalies, but if so, the origin and evolution of such an arc cannot be simply connected to the later development of the Parece Vela and Shikoku basins to the east. Other results show that the elevation of this region is 1 km lower than would be observed for similar-aged crust in the North Pacific. If the DSDP dates and our anomaly interpretations are correct, then the lack of a correspondingly large negative gravity anomaly suggests major differences between this extinct ridge and those known to be currently spreading.

One of the most significant difficulties in the original formulation of plate tectonics was its failure to account for tensional features on the concave side of island arcs. *Karig's* later analysis of these features [*Karig,* 1970, 1971*a*] which he synthesized into his theory of inter-arc spreading [*Karig,* 1971*b*, 1974*b*] has therefore been a major contribution to our understanding of these marginal basins. Unfortunately, there is as yet little geophysical proof that older, and now no longer active, sections of marginal basins were formed by the same processes that we observe in active troughs. This is important since only a small number of these basins are presently active.

The studies of magnetic anomalies which could presumably yield new insights into the origins of inactive marginal basins have so far had mixed success. Results fall into two general categories:

(1) Characteristic anomalies are not apparent, but there are low-amplitude lineations parallel to the trench.

(2) Characteristic anomalies that can be correlated to worldwide events are observed together with a well-defined bathymetric ridge.

The Japan Sea typifies the first case [*Isezaki and Uyeda,* 1973] and is similar to what is observed in the actively spreading basins of the Tonga (L. A. Lawver, personal communication, 1974) and Mariana [*Karig,* 1971*a*] island arcs. The apparent lack of characteristic magnetic signatures is explained by the diffuse nature of inter-arc as opposed to mid-ocean ridge spreading, and is one of the differences that distinguish these two spreading modes [*Karig,* 1970; *Matsuda and Uyeda,* 1971]. The second type is observed in the Tasman [*Hayes and Ringis,* 1973] and Scotia [*Barker,* 1972*a*] seas and is typical of mid-ocean ridges.

The possibility of ridge spreading in marginal basins raises several questions. If ridge spreading exists, is it related to inter-arc spreading or are there two separate processes that together determine the history of these basins? Can such ridge spreading be related to mid-ocean ridge systems of the past or does it have a separate origin?

A good place to seek the answer to these questions lies in the Western Philippine Sea. This region (Figure 1) is separated from the eastern sections of the Philippine Sea by the Oki–Daito and Palau–Kyushu ridges. Its main bathymetric feature is the Central Basin fault [*Hess,* 1948],

[1]Woods Hole Oceanographic Institution Contribution No. 3551.

which runs NW–SE and splits the area into nearly equal halves. This trend is quite different in direction from that observed in the ridges to the east, which run N–S and parallel to the Mariana and Bonin trenches. What makes the West Philippine basin of interest is that it is the only region where ideas favoring alternative modes of spreading have been in direct dispute. Karig proposed that this region was formed in an Eocene pulse of inter-arc spreading—an older manifestation of what is presently occurring in the Mariana trough [*Karig,* 1971*a*]. He treated the Central Basin fault as a strike-slip fault with several hundred kilometers of left-lateral offset [*Karig et al.,* 1973]. Counter to this was evidence presented by *Ben-Avraham et al.* [1972] of NW–SE trending magnetic anomalies symmetric about the Central Basin fault. *Uyeda and Ben-Avraham* [1972] therefore proposed that the Central Basin fault is an extinct spreading center that is related to the Kula–Pacific ridge system of *Larson and Chase* [1972].

This paper is a study of this region as a particularly interesting example of an inactive marginal basin. By such a study we hope to clarify the processes that have shaped its history. Specifically, we synthesize magnetic, bathymetric, and Deep Sea Drilling data in an attempt to:

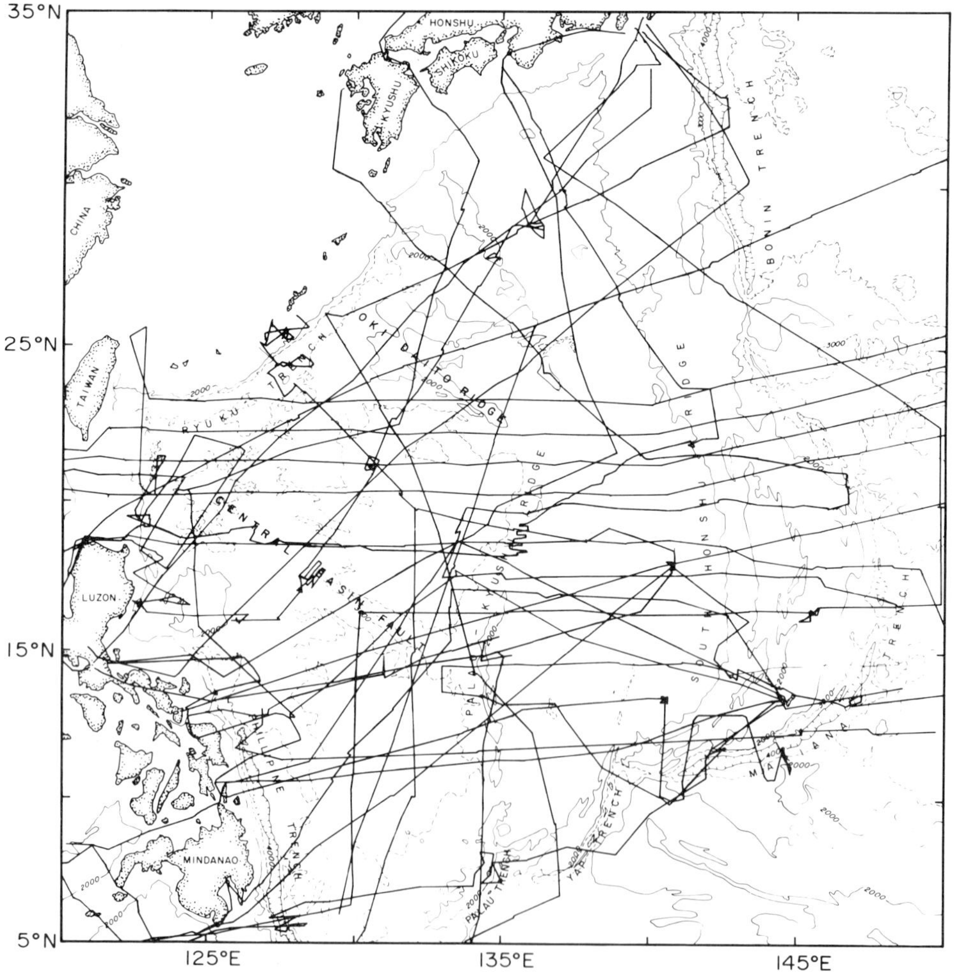


Fig. 1. Ship tracks superimposed on a 2000-, 3000-, and 4000-m contour map of the Philippine Sea [*Chase and Menard,* 1969]. Major bathymetric features are identified.

(1) locate any magnetic lineations and confirm or reject those published by *Ben-Avraham et al.* [1972];

(2) correlate specific anomalies with worldwide reversals;

(3) determine the nature of the Central Basin fault; and

(4) relate our findings to possible theories of the basin's origin.

## ANALYSIS OF THE MAGNETICS

*Lineations.* Our first step was to collect all the known magnetic data in the Philippine Sea. Those lines which are utilized for this work are shown in Figure 1. While processing these data we were able to run our own line during Leg 7 of the Tasaday Expedition of Scripps Institution of Oceanography (Nov.–Dec., 1973). We ran this line at right angles to the Central Basin fault and close to the lines which *Ben-Avraham et al.* [1972] had published earlier. We hoped that if lineations existed, they would at least be traceable in this limited area of denser coverage.

Figure 2 shows the magnetic anomalies in the western basin plotted at right angles to each ship's track. The scale is smaller than usual because of the low amplitudes of the anomalies. There are two difficulties that such low-amplitude anomalies pose. One is the considerable effect that diurnal and semi-diurnal variations

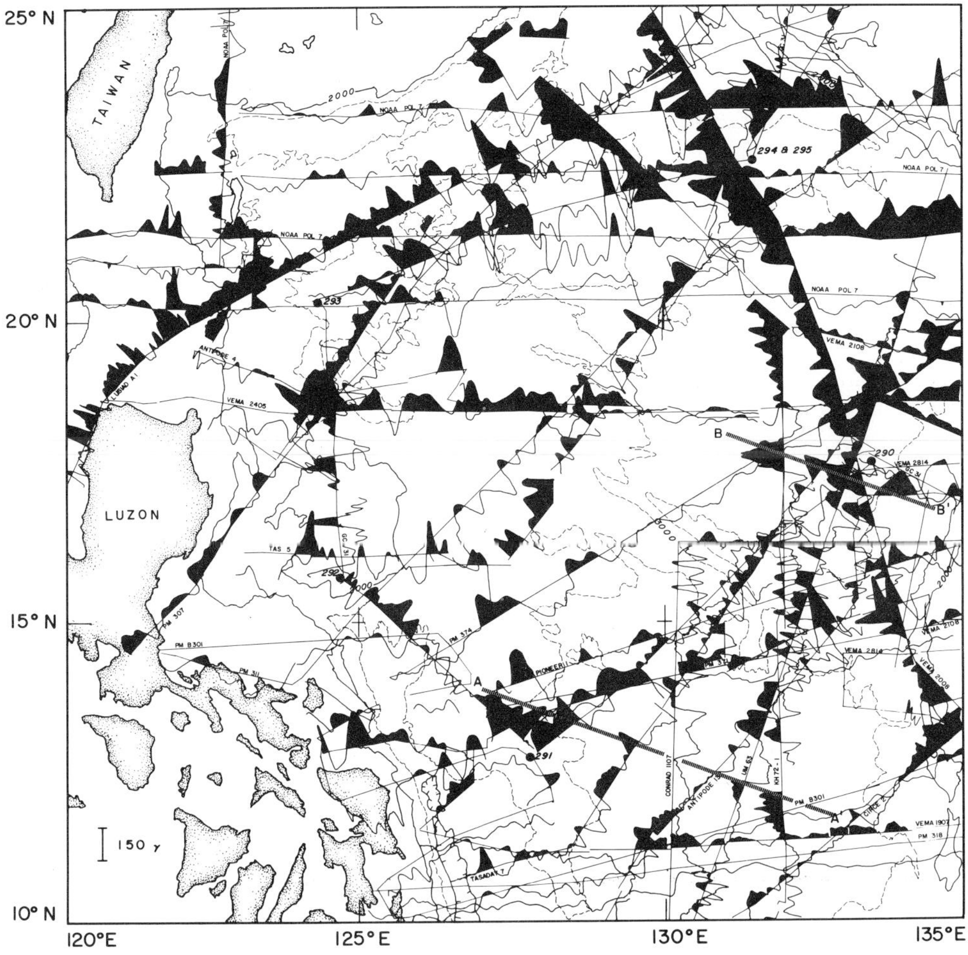


Fig. 2. Magnetic anomalies in the west basin of the Philippine Sea plotted at right angles to track together with an identification of each ship track, 2000- and 3000-m contours, and the location of DSDP sites 290 through 295 [*DSDP Scientific Staff,* 1973*b*].

in the Earth's present field have upon the amplitude. This might be on the order of 10 to 30 gammas for each component, so that the total effect could be as large as 50 gammas [*Chapman and Bartels,* 1940]. The other factor that affects the zero level of the anomalies but leaves the shape unchanged is the secular variation, which in this region is as great as tens of gammas per year [*Vestine,* 1967]. Both of these effects can be seen in the large offsets in certain profiles in Figure 2, but can be neglected in this study since our anomalies have a wavelength of 50 nm or less and we are only matching shapes and not contouring equal values.

We have concentrated our study on the anomalies located between lines *A-A'* and *B-B'* in Figure 2. This is where our coverage is greatest and where we can use DSDP data taken in the West Philippine basin during Leg 31 of the Glomar Challenger [*DSDP Scientific Staff,* 1973*b*] most effectively. The long wavelength anomalies close to the lines are particularly noticeable and distinctive. It is possible that there are more lineations to the north of this area; if so, however, they are certainly more confused than those in the south and we feel that the orientation and density of our tracks is not sufficient to unravel the confusion into an indisputable pattern. In Figure 3 we project the anomalies between these two lines along a direction N21°E. This direction is determined primarily by the strike of the large wavelength anomaly near line *A-A'.* Notice also in Figure 2 that the anomalies for tracks in a direction perpendicular to the projection show low amplitudes and long wavelengths, as expected when running parallel to a mid-ocean ridge. But caution must still be taken in accepting this result since offsets between tracks can produce a spurious trend. At the present we have no evidence from either magnetics or topography that such offsets occur in this restricted area, and have therefore positioned our projected tracks only by the crossings between them in Figure 2.

Even with consideration for these precautionary words, we feel that these lineations are real, and that our measured trend of N69°W (at right angles to the direction of projection) is valid. This confirms that the anomalies of *Ben-Avraham et al.* [1972] form traceable lineations (but in a direction 20° counterclockwise to their original prediction of N50°W). The extent of these lineations is rather limited, but one may still conclude from this evidence that at least the southern part of the West Philippine Sea opened in a nearly N–S direction. This is contradictory to the early inter-arc spreading hypotheses, which would predict a west-to-east direction of opening—the same as occurred in the Parece Vela basin to the east [*Karig,* 1971*a*]. Another objection is the lack of a uniform progression in the dates for DSDP sites 290 through 295 that were drilled during Leg 31 of the Deep Sea Drilling Project. We agree with Karig's more recent statement [*Karig et al.,* 1973] that this basin must have had a more complex history than was originally thought. If inter-arc spreading occurred at all, it must have been in a direction significantly different from that observed to the east.

The absence of recorded seismic events within the West Philippine basin [*Barazangi and Dorman,* 1969; *Fitch,* 1972; *Shimamura et al.,* 1975] probably indicates that there is no current generation of crust. It is most likely older than the Parece Vela basin to the east [*Karig,* 1971*a*] as indicated by its thicker sediment cover [*Ingle et al.,* 1975]. We conclude that the West Philippine basin opened before the start of inter-arc spreading in the Parece Vela basin by one of the following two processes:

(1) symmetric spreading by the Central Basin fault; or

(2) spreading from a ridge that no longer exists, such as in a piece of trapped oceanic crust.

In the first case we would see symmetry about the Central Basin fault in our magnetic anomalies. This is the contention of *Ben-Avraham et al.* [1972], which we will refer to as the symmetric model. In the second case there would be no symmetry and our anomalies would be matched by a model of spreading from only one side of a ridge. This we will refer to as the non-symmetric model. Notice that each model could have originated either in a mid-ocean ridge system or as spreading behind an island arc.

*Phase shifting.* Relative poles of rotation for the Pacific–Philippine plate system have been

Fig. 3. (Opposite) Stacked magnetic anomalies which occur between lines *A-A'* and *B-B'* of Figure 2 projected along a direction N21°E. The relative position of each profile is indicated by the location of points *A, A', B,* and *B'* as well as by relative numbers to the right and left of each profile. These numbers increase in an eastward direction. DSDP sites 290 and 291 are located on profile *GC 31.*

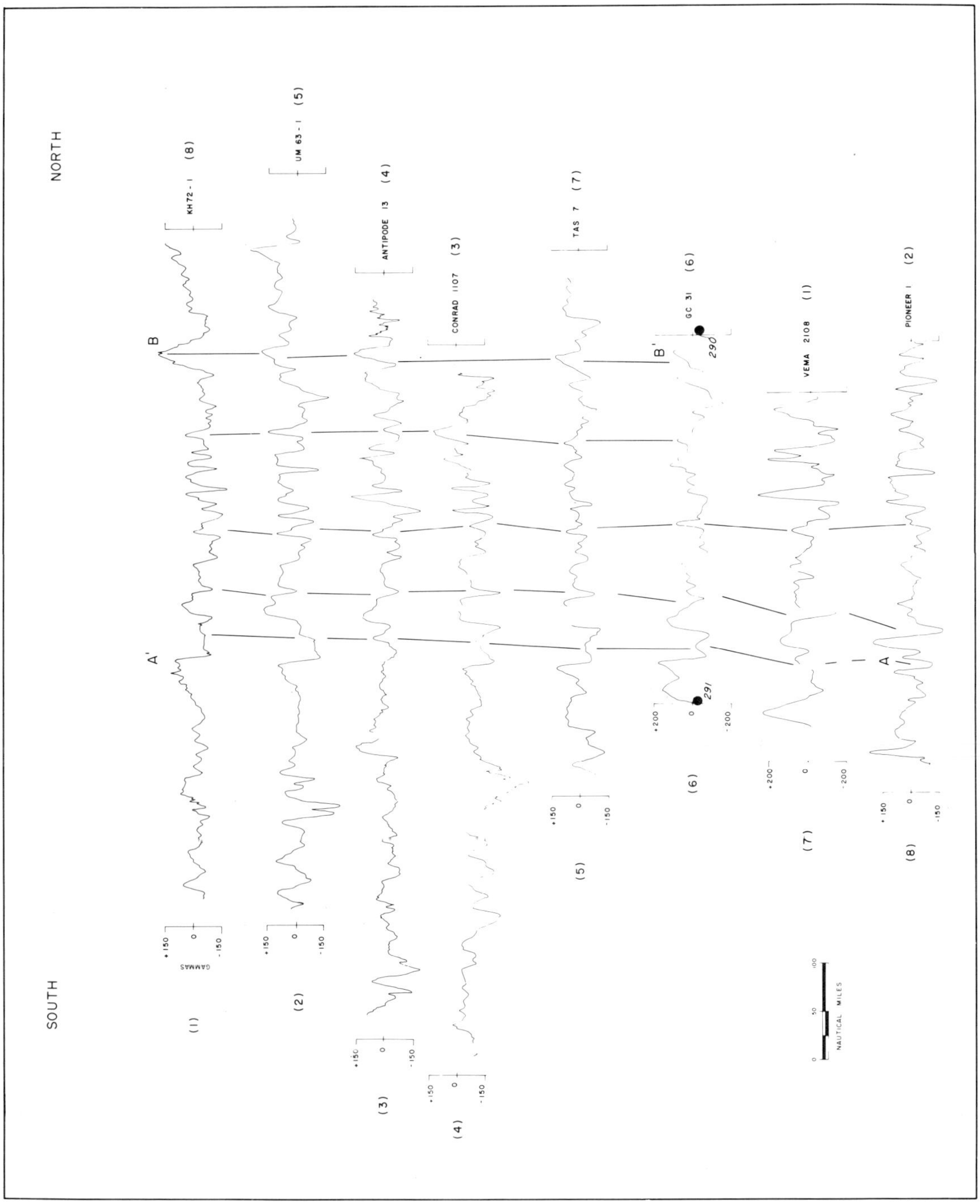
NORTH
SOUTH
KH72-1 (8)
UM 63-1 (5)
ANTIPODE 13 (4)
CONRAD 1107 (3)
TAS 7 (7)
GC 31 (6)
VEMA 2108 (1)
PIONEER 1 (2)
B
B'
A
A'
290
291
GAMMAS
+150 0 -150
+200 0 -200
(1)
(2)
(3)
(4)
(5)
(6)
(7)
(8)
0 50 100
NAUTICAL MILES

determined by fault plane solutions [*Fitch*, 1972], but they are poorly constrained and still a matter of debate (D. E. Karig, personal communication, 1974). Otherwise very little is known about the plate motions of the Philippine Sea. Such knowledge is important if we wish to compare our magnetic anomalies to those representative of other oceans or to a theoretical model based on a standard time scale. Changes in paleolatitude and paleostrike produce changes in the inclination of the remanent magnetization which can cause large changes in the amplitude and shape of the anomalies, making valid correlations difficult or impossible. This is especially true when close to the equator, as we are in this case.

A very useful approach to the problem of matching anomalies when little is known about their original orientation is a method developed by Schouten and other workers [*Cande and Schouten*, 1972; *Cande et al.*, 1973] (Cande and Schouten, in preparation). In this method the anomaly is formulated in terms of an angle $\theta$ called the skewness parameter:

$$\theta = I'_r + I' - 180° \quad \text{(Cande and Schouten, in prep.)} \quad (1)$$

where

$I'_r$ = effective inclination of the remanent field

$I'$ = effective inclination of the ambient field

The influence of both the magnetic inclination and the strike of the anomaly structure are combined in the angle $\theta$ since the effective inclinations are defined by,

$$\tan I' = \tan I_0 / \sin \alpha \quad [\textit{Gay}, 1963] \quad (2)$$

where

$I_0$ = inclination of the magnetic field

$\alpha$ = strike of the lineations.

The Fourier transform of an anomaly formulated in this manner can be represented as,

$$M(k) = C \cdot J(k) \cdot E(k) \cdot e^{j\theta}. \quad [\textit{Schouten}, 1971] \quad (3)$$

where

$$C = (\sin I \sin I_r) / (\sin I' \sin I'_r) \quad (4)$$

is a real constant. $J(k)$ is the Fourier transform of the function that represents the distribution of magnetization in the anomaly-producing layer.

$$E(k) = 2\pi (e^{-ak} - e^{-bk}) \quad (5)$$

where $a$ and $b$ are the depths to the top and bottom of the magnetized layer. If two sets of anomalies have the same origin but are now separated from each other, $E(k)$ and $J(k)$ would be identical. Changes in amplitude and shape from one set to the other would be determined solely by the term $C \cdot e^{j\theta}$. Multiplication of this term by $e^{-j\theta}$ would make $M(k)$ real and independent of $\theta$. That would leave us with de-skewed (symmetric) anomalies which have different amplitudes but identical shapes.

The method involves an input profile of magnetic anomalies for which the value of $\theta$ is changed until no skewness is observed in the output profile. It is therefore the actual shape of the anomalies that determines a value for $\theta$. Using (1) and (2) one can then calculate possible sets of values for the paleolatitude and strike of the anomalies. In the most general case one can determine two possible values for $\theta$. These would be 180° apart and represent two de-skewed solutions, one of which is the inverse of the other and occurs when inverted positive anomalies become negative anomalies. Recognition of the possibility that anomalies can be inverted was the key to the *Larson and Pitman* [1972] correlation of the Keathley lineations to three other sets of anomalies in the Pacific. This established a worldwide correlation of Mesozoic anomalies and led to the first attempt at a Mesozoic time scale. In their case the Pacific anomalies were inverted with respect to the Keathley lineations, and this was the main reason why they had been invalidly correlated in an earlier attempt by *Vogt et al.* [1971]. We believe that a similar situation led to earlier unsuccessful attempts by Ben-Avraham and the author to correlate the Philippine anomalies of *Ben-Avraham et al.* [1972].

Results of our application of Schouten's method to the anomalies of Figure 3 are shown in Figure 4. These anomalies represent the de-skewed inverse of the original anomalies. The other, non-inverted solution has been eliminated because it would lead to values of $I'_r$ greater than 90° and therefore is not a physically meaningful solution. Thus, from the inversion of our original anomalies, we are left with only one possible $\theta$ value, which ranges from −160° to −180°. In our case $I_0 = 28°$ and $\alpha = 110°$, which means that $I' = 30°$ from (2). Thus $I'_r$ can range from −10°

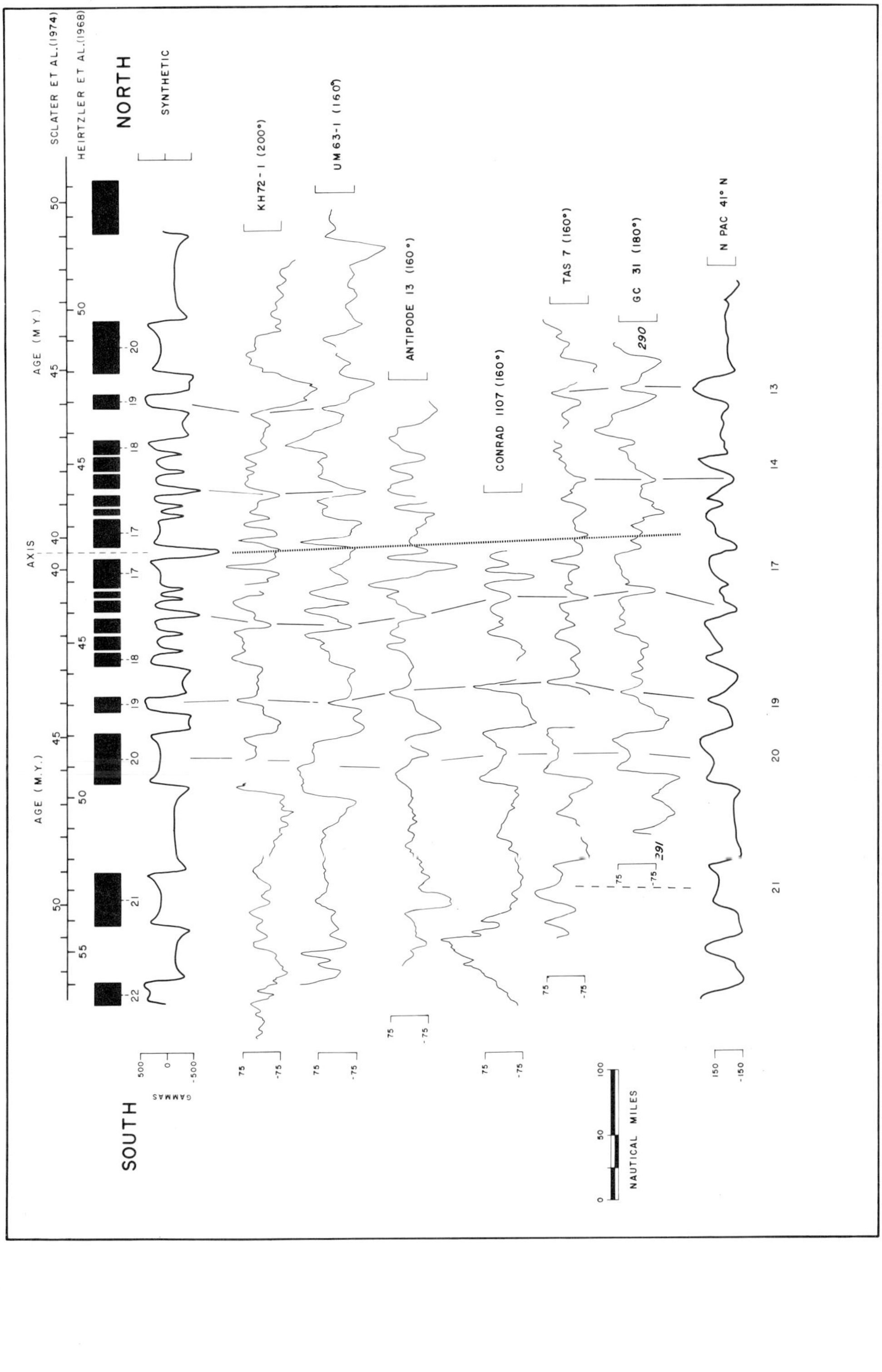


Fig. 4. Phase-shifted magnetic profiles compared with a symmetric spreading model at top and a non-symmetric model (spreading from only one side of a ridge) at the bottom. The degree of phase shifting is indicated in parentheses next to the identification of each profile. The dotted line locates the central valley of the Central Basin fault. Anomaly numbers are based on the system of *Heirtzler et al.* [1968].

to −30° according to (1). If $\alpha$ has remained constant, then these anomalies were formed between −4.5° and −15° latitude; if $\alpha$ has changed, the value of the paleolatitude could range from 0° to −16°. These results disagree with earlier solutions given by *Ben-Avraham et al.* [1972] which predict formation north of the equator. It is unfortunate that we cannot determine if $\alpha$ has changed since these anomalies were formed. Changes in $\alpha$ near the equator affect anomaly amplitude (the value of $C$ in (3)) rather than skewness [*Schouten,* 1971], and too little is known regarding the predicted amplitudes of magnetic anomalies for us to use them as a criterion (Cande and Schouten, in preparation).

*Identification of anomalies.* The nannofossil ages reported from Leg 31 of the Glomar Challenger [*DSDP Scientific Staff,* 1973*b*] were all in the Eocene or younger, except for holes 294/295 which were reported to be Paleocene. This last date has been recently revised and now seems to be much the same as the others (D. E. Karig, personal communication, 1974). At first these dates were thought to rule out the possibility of symmetric spreading from the Central Basin fault [*DSDP Scientific Staff,* 1973*b; Karig et al.,* 1973]. We do not subscribe to such a conclusion. These dates do rule out a speculation by *Ben-Avraham et al.* [1972] that his Philippine anomalies might be part of the Mesozoic. But beyond telling us in what age province to look for matching our anomalies to standard reversals, they tell us little. Spreading rates must have been large, but the direction of spreading is not clearly indicated. We must therefore assume that the anomalies get younger as they approach the Central Basin fault from the south, so as not to rule out the symmetric model from the outset. We cannot say a priori whether this is solid ground upon which to build further analysis. If we assume this nonetheless and then look at sea-floor spreading anomalies of 50 m.y.B.P., we find that a progression from shorter to longer wavelengths exists for anomalies 22 to 17. This fits the general trend of long to short to long wavelengths observed in the anomalies of Figure 3 as one crosses the Central Basin fault. By using the standard North Pacific profile at 41°N [*Pitman et al.,* 1968], shifting it to remove its slight skewness and changing its horizontal scale to form a best fit, we produce the model at the bottom of Figure 4. This is a non-symmetric model starting at anomaly 21 and ending at anomaly 13, where only the direction of spreading has been assumed. To contrast it with the symmetric model, we have constructed the top model of Figure 4 from a standard magnetics program [*McKenzie and Sclater,* 1971]. It adheres to the same fit in the region south of the Central Basin fault as does the non-symmetric model. Notice, however, that they are not identical reproductions. The synthetic anomalies include some additional short reversals not found in the Pacific data. This is to be expected since the reversal history for that age is not known with precision.

Except for these few short reversals, both models are the same south of the Central Basin fault. If we now look at the north side where the models should differ, it appears that the symmetric model forms a better match to our lineations. The primary basis for this is the presence of some positive events that do not appear in the non-symmetric model. It should be noted, however, that both models show the same general trend from long to short to long wavelengths already noted in the real data. We therefore cannot be absolutely certain which model to discard on the basis of anomaly matching alone.

*Testing the models.* The locations of DSDP sites 290 and 291 occur at either end of the anomalies presented in Figure 4. This allows us to use the dates of these sites for confirmation of our identifications to the south of the Central Basin fault, and as a test for the age that each model would predict to the north. In order to do this we must: (1) analyze the fossil record for sites 290 and 291 to determine their range of possible ages; and (2) date our anomalies by the use of a standard time scale.

Table 1 presents the DSDP data for sites 290, 291, 19, and 32. The latter two sites are of similar age to the former two but taken from other ocean basins. They were used by *Sclater et al.* [1974] to revise the *Heirtzler et al.* [1968] time scale, which was found to predict progressively older ages than the fossil record. We will refer to their revision as the modified Heirtzler scale. Notice that this revision still does not account for all of the discrepancies between fossil and magnetic ages. Site 19 has a magnetic anomaly age based on the modified Heirtzler scale of 1–6 m.y. older than its nannofossil age as determined according to the *Berggren* [1972] Cenozoic time scale. A similar discrepancy appears in the ages for site 291, where we predict a magnetic anomaly age 3 to 10 m.y. older than the nannofossils allow. Since site 19 has well-defined magnetics but still shows a similar variance, we believe that this justifies

TABLE 1. Stratigraphic Ages of Selected Magnetic Anomalies

| DSDP Site | Lat. | Long. | Contact Core | Diagnostic Calcareous Zone | Nannofossil Age, m.y.* | Magnetic Anomaly | *Heirtzler et al.* [1968] Age, m.y. | *Sclater et al.* [1974] Age, m.y. |
|---|---|---|---|---|---|---|---|---|
| 19 | 28°32.1′S | 23°40.6′W | 12 | *Chiphragmolithus quadratus* | 46–48 | Just before 21 | 53–56 | 49–52 |
| 291 | 12°48.4′N | 127°49.9′E | 5 | *Discoaster barbadiensis* *Cyclococcolithus formosus* | 38–43 | End of 21 | 50–52 | 46–48 |
| 32 | 37°07.6′N | 127°33.4′W | 14 | *Coccolithus bisectus* | 34–37.5 | 13 | 37–39 | 35–37 |
| 290 | 17°44.9′N | 133°28.1′E | 9 | *Reticulofenestra hillae* | 36–38 | Just before 19 | 47–49 | 44–46 |
| | | | | *R. umbilica* | | Just after 13 | 36–38 | 34–36 |

From *Berggren* [1972] and *Bukry* [1974].

our magnetic identifications to the south of the Central Basin fault.

Let us now analyze what our two spreading models predict for Site 290. The magnetic versus fossil ages should exhibit a similar pattern as do those of Site 32. This means that they should agree to within 3 m.y. At first glance this would preclude the symmetric model, but actually this evidence is not very convincing. In Figure 5, which is a graph of measurements taken from Figure 4 and tabulated in Table 2, neither the symmetric nor the nonsymmetric models uniquely fit the DSDP data. The primary reason for this is the difficulty in dating the DSDP holes due to a lack of fossils close to basement for Site 291 and failure to hit basement for Site 290. If sedimentation rates are extrapolated to the basement of Site 291, one obtains an age 5 m.y. older than the lowermost nannofossils indicate [*Ingle et al.*, 1975]. If one does the same for Site 290, an age in excess of 43 m.y.B.P. results.

These paleontological ages, once corrected to represent true basement (dotted arrows in Figure 5) better, have now become much more definitive. Site 291 becomes an even better fit to our magnetic identifications. Site 290 now adds further evidence in favor of magnetic symmetry across the Central Basin fault, since it is unreasonable to suggest that magnetic basement could be younger than the age of the earliest sediments.

## BATHYMETRIC DATA

A summary diagram (Figure 6) shows our magnetic lineations and their possible interpretation. The lineations of Figure 4 are mapped onto the anomalies of Figure 2 for several specific reversals. We also include the bathymetric trace of the Central Basin fault indicated in Figures 4 and 7. Notice that in this limited area it runs at a 20° angle to its overall trend, and is nearly parallel to our magnetic lineations. It seems highly unlikely that this would occur unless both magnetics and bathymetry are genetically related. This could only mean that this section of the Central Basin fault was a former spreading center. Further confirmation is seen in Figure 7, which includes profiles of bathymetry projected along the same direction as the magnetics of Figures 3 and 4. We observe that:

(1) In all but the *TAS 7* profile, the Central Basin fault is a sharp trough bounded by two peaks, which bottoms at a depth equal to that of the outer flanks. The *TAS 7* profile has this same sharp trough but also has an adjacent broad trough, which somewhat distorts the pattern.

(2) If a smooth average is drawn through the profiles, the area immediately adjacent to the Central Basin fault is elevated relative to the flanks. This can be observed even though the flanks of several profiles are distorted by the

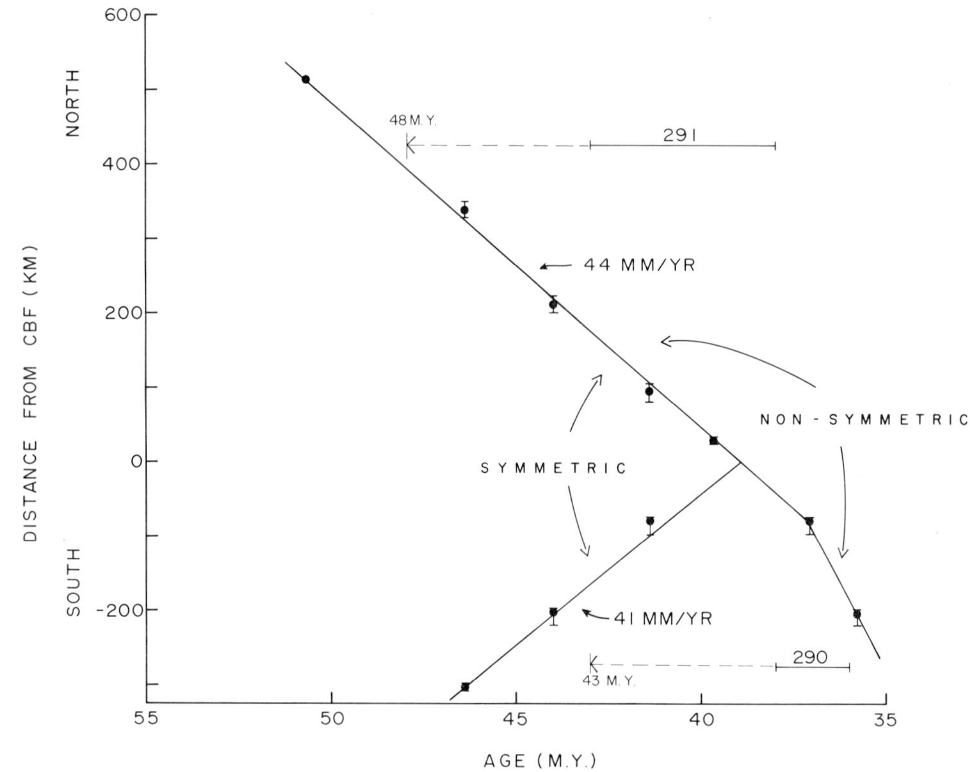


Fig. 5. Graph of distance from the Central Basin fault versus age, taken from values measured in Figure 4 and listed in Table 2. Symmetric versus non-symmetric models for ages given by the revised Heirtzler time scale [*Sclater et al.,* 1974] are compared to nannofossil ages for DSDP sites 290 and 291. Arrows indicate ages adjusted to better represent the basement contact.

outer rise of the Philippine trench [*Watts and Talwani,* 1974] (southern sections of profiles *ANT 13, TAS 7, GC 31,* and *PIONEER 1*) and the Palau–Kyushu ridge (northern sections of profiles *ANT 13* and *PIONEER 1*). However, the rate of drop in elevation differs for each profile. In the *GC 31* profile the drop is somewhat obscured by large-amplitude, short-wavelength fluctuations. The *TAS 5* line, located to the north of the magnetic lineations (Figure 2), shows a slight drop that is not well determined. The Central Basin fault has an anti-symmetric shape in this profile which differs from those to the south.

A detailed survey during Leg 5 of cruise Tasaday indicates that at that point the Central Basin fault is a N–S en echelon feature which appears highly fractured (D. E. Karig, personal communication, 1974). This could have occurred during a later tectonic activity postdating its spreading history and thereby changing its original character. If such activity was strongest to the north, it might explain why the bathymetry and magnetic patterns become more confused there. The remaining question is whether our region of study to the south represents a small remainder of the original configuration of the entire Central Basin fault or whether the West Philippine basin was always a collection of distinct tectonic provinces. Unfortunately the answer is at present beyond the limits of our data to decide.

The profiles south of the *TAS 5* track suggest two further complications. First, the deep trough at the axis of the Central Basin fault is morphologically similar to slow spreading ridges ($v_{1/2}$ = 20 mm/yr) which contradicts our magnetic anomaly evidence (Figure 5) of a faster spreading rate ($v_{1/2}$ = 41–44 mm/yr). The faster spreading rate is also supported by the sediment pattern which shows a thin (<200 m), uniform layer. Perhaps we can explain this conflict by suggesting that the spreading decreased just before it stopped, thereby producing a limited extent of greatly

TABLE 2. Anomalies of the West Philippine Basin, and Their Distances from the Central Basin Fault

| | Distance from Central Basin Fault, nm | | | | | | | | Age, m.y. | |
|---|---|---|---|---|---|---|---|---|---|---|
| Anomaly | *KH 72-1* | *UM 63-1* | *ANT 13* | *C 1107* | *TAS 7* | *GC 31* | Average, nm | Average, km | *Heirtzler et al.* [1968] | *Sclater et al.* [1974] |
| Beginning of 21 | | | | | 277 | | 277 | 513 | 54.2 | 50.7 |
| Beginning of 20 | 183 | 189 | 177 | 180 | 180 | 183 | 183±6 | 339±11 | 49.6 | 46.4 |
| 19 | 114 | 117 | 114 | 109 | 109 | 120 | 114±6 | 211±11 | 47 | 44 |
| Between 17 and 18 | 54 | 57 | 54 | 26* | 46 | 43 | $51^{+6}_{-8}$ | $95^{+11}_{-15}$ | 44.1 | 41.4 |
| End of 17 | 14 | 14 | 31* | | 17 | 14 | $15^{+2}_{-1}$ | $28^{+4}_{-2}$ | 42.3 | 39.7 |
| Between 15 and 14<br>or<br>Between 17 and 18 | -43 | -46 | -34 | | -46 | -43 | $-43^{+3}_{-9}$ | $-80^{+5}_{-17}$ | 39.5<br><br>44.1 | 37.1<br><br>41.4 |
| 19<br>or<br>13 | -100 | -109 | | | -112 | -112 | $-109^{+3}_{-9}$ | $-202^{+5}_{-17}$ | 47<br><br>38 | 44<br><br>35.8 |
| Beginning of 20 | -166 | -160 | | | | | -163±3 | -302±5 | 49.6 | 46.4 |
| DSDP Site 290 | | | | | -149 | | -149 | -276 | | |
| DSDP Site 291 | | | | | 229 | | 229 | 424 | | |

*Not included in average.

elevated and rifted topography. This would also help explain why the magnetic pattern changes so drastically from one profile to the next in the vicinity of the Central Basin fault. Second, if we plot depth versus age, we see from Table 3 and Figure 8 that all the depths, both for DSDP ages and for ages based on our magnetic anomalies, are universally greater than for the North Pacific [*Sclater et al.,* 1971]. Thus according to these plots, the age of this basin should be greater than 80 m.y.B.P.

We do not believe that the depth of this basin gives a true indication of age for the following reasons:

(1) Nowhere in DSDP cores 290–295 are there sediments older than the Paleocene [*Ingle et al.,* 1975].

(2) The correspondence of the younger sediment ages for DSDP sites 290 and 291 to a plausible magnetic anomaly identification seems too coincidental to be an artifact. In our view this rules out the older 'depth' age even if possible magnetic anomaly identifications could be found in the Mesozoic.

(3) There are recent seismic studies which indicate that the crust in the West Philippine basin is not similar to North Pacific crust [*Shimamura et al.,* 1975; *Henry et al.,* 1975]. In particular, the thin 'oceanic' crust suggested by the refraction studies of *Henry et al.* [1975] can help to explain the occurrence of a 1-km negative anomaly without a correspondingly large negative gravity anomaly [*Sclater et al.,* in press].

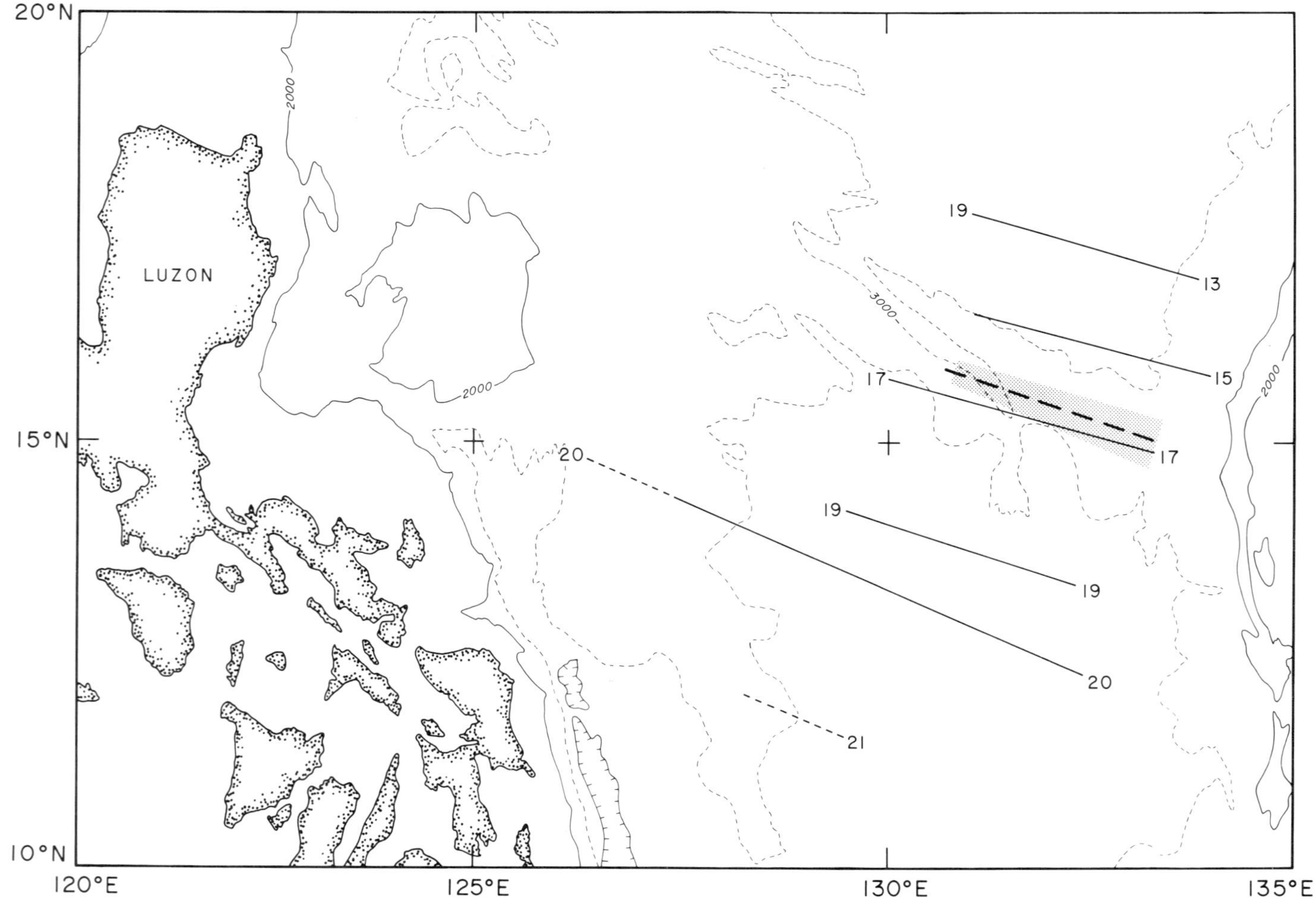


Fig. 6. Map of magnetic lineations (heavy lines) and their possible interpretation according to both symmetric and non-symmetric spreading models. Less certain identifications are dotted. Heavy dashed line indicates the central valley of the Central Basin fault which occurs between peaks on either side. Total extent from peak to peak is indicated by shaded region.

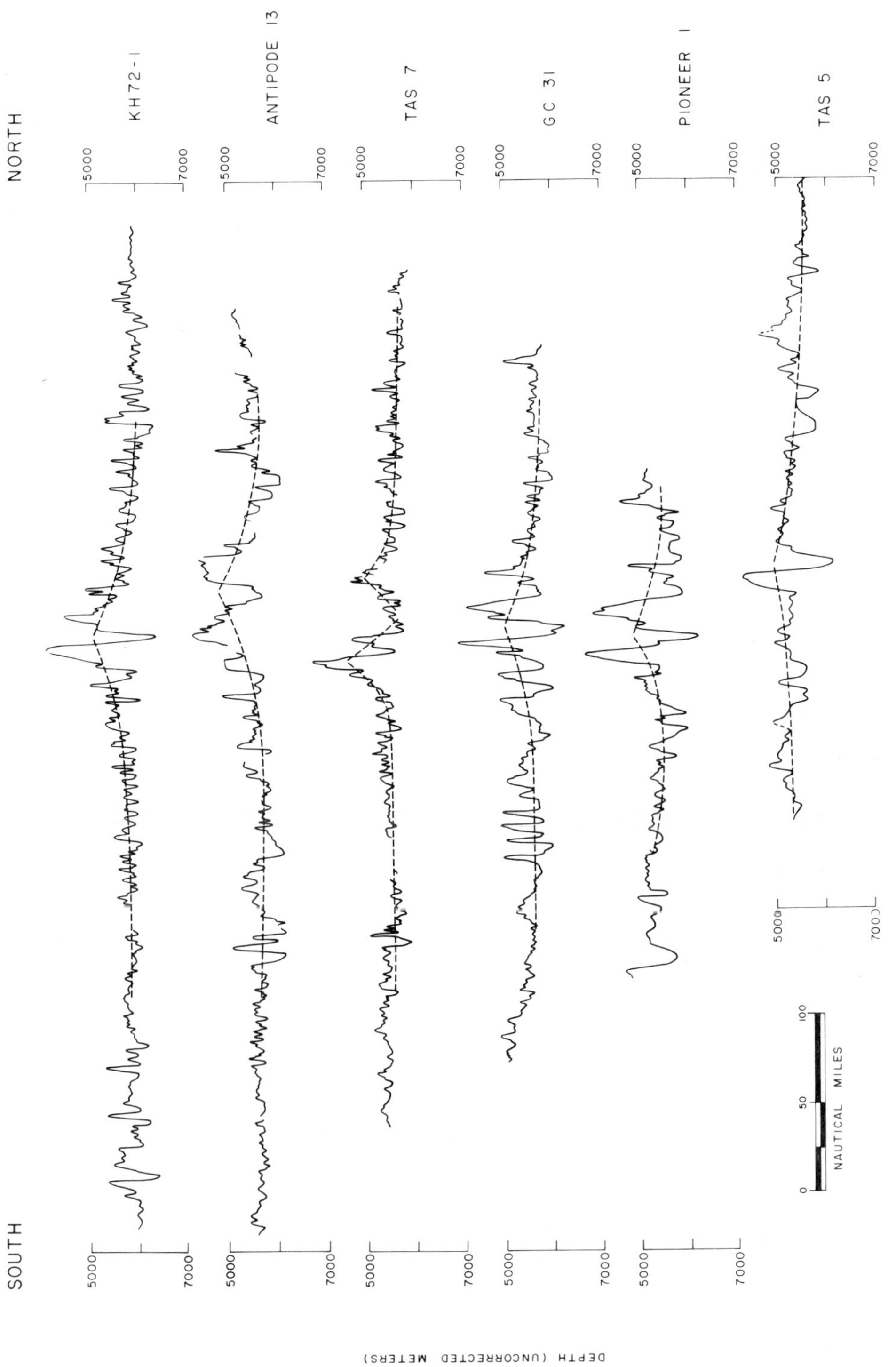


Fig. 7. Bathymetric profiles across the Central Basin fault projected along a direction N21°E. Dashed line represents visual smoothing of overall topographic trends.

TABLE 3. Depth of the West Philippine Basin

| Cruise | Smoothed Bathymetric Depths, uncorrected km | | |
|---|---|---|---|
| | *South Flank* | *Ridge Crest* | *North Flank* |
| *KH 72-1* | 5.8 | 5.1 | 6.0 |
| *ANT 13* | 5.7 | 4.8 | 5.7 |
| *TAS 7* | 5.6 | 4.7 | 5.7 |
| *GC 31* | 5.6 | 5.0 | 5.8 |
| *PIONEER 1* | — | 4.9 | — |
| *TAS 5* | 5.3 | 4.9 | 5.6 |
| Average | 5.6±0.3 | 4.9±0.2 | 5.8±0.2 |

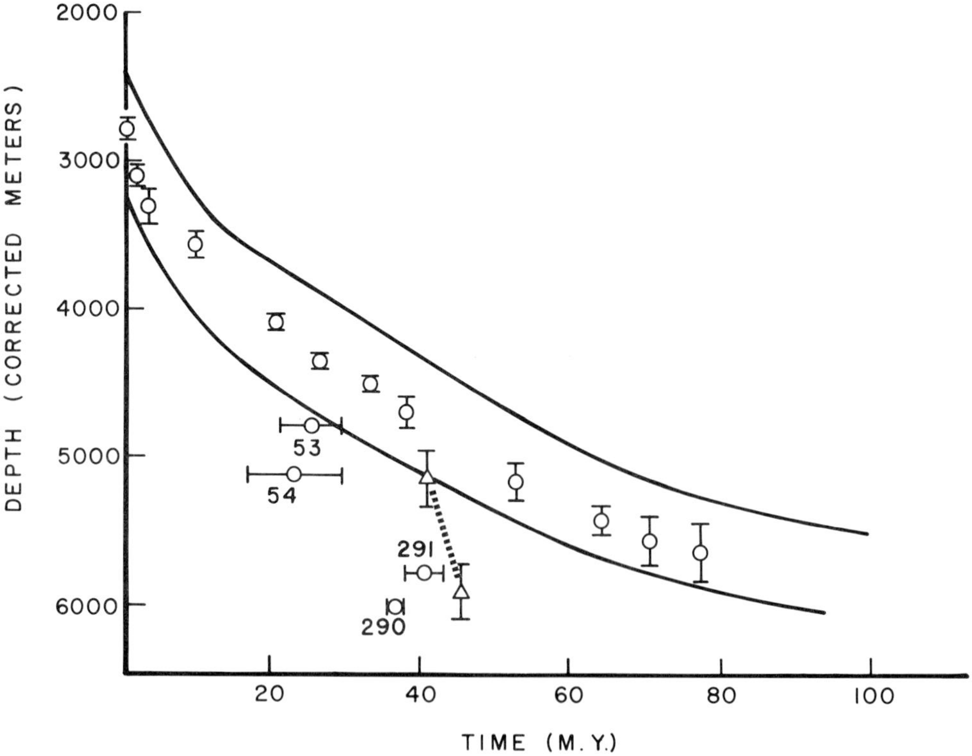


Fig. 8. Depth of smoothed topography versus age. Data for the North Pacific (circles between heavy lines) is from *Sclater et al.* [1971] and is compared with values for DSDP sites 53, 54, 290 and 291 [*DSDP Scientific Staff*, 1969*a*, 1973*b*]. We also plot values treating the Central Basin fault as a symmetric spreading ridge (triangles) taken from Table 3 and Figure 5. Dashed line connects ridge crest to flank.

## CONCLUSIONS

All of the data presented in this paper can be accounted for by treating the Central Basin fault as a spreading center that slowed and stopped 40 m.y. ago. The ridge-like bathymetry and symmetry in the magnetic lineations are the strongest support for such a conclusion. The ages of DSDP sites 290 and 291 also can be explained with this model, but because of the incomplete agreement between fossil ages and magnetic time scales for this period, coupled with the limited extent of our lineations and the fast spreading rates needed to match the anomalies, they cannot remove all uncertainty in our identifications.

It is very tempting to go beyond the limit of such a conclusion and propose a comprehensive tectonic model that connects the formation of this basin with a proto-Pacific reconstruction. Such an attempt has already been made by *Uyeda and Ben-Avraham* [1972] and *Uyeda and Miyashiro* [1974]. They propose that the Philippine ridge (Central Basin fault) was connected by means of a long transform fault to the Kula–Pacific ridge system. When the Pacific plate changed its direction of motion from N to NW 42–44 m.y. ago [*Clague and Jarrard,* 1973], the transform fault became a subduction zone, leading to the start of inter-arc spreading and the formation of the Parece Vela basin. There are several specific problems with such a model:

(1) The Palau–Kyushu ridge is asserted to be the current expression of the transform fault between the Kula–Pacific and the Philippine ridges. In that case, the fault would have extended for several thousand kilometers, a much greater length than for most currently active transform faults. The only purely oceanic transform fault of such a large extent that has been magnetically documented is the no longer active Ninety-east ridge [*Sclater and Fisher,* 1974]. There are also other interpretations of the origin of this feature [*Bowin,* 1973]. Other long transform faults are either on continental crust (San Andreas fault) or are not yet supported by unequivocal data (the connecting link between the South Atlantic ridge and the West Indian Ocean ridge). It is possible that the mechanics of ocean-floor spreading make unlikely the possibility of such a long transform fault lasting over an extended period of time [*Lachenbruch,* 1973].

(2) The magnetic anomalies identified in this paper appear to have originated near to or just south of the equator. This invalidates the location of the ridge in its present latitude when spreading ceased, which is central to the *Uyeda and Ben-Avraham* [1972] model. Their model also predicts very different subsequent motions between the Philippine and Pacific plates. It is interesting to note, however, that our results (formation of anomalies from 0° to 16°S) can be explained by using the same pole of 'absolute' rotation for the Philippine plate as has been determined for the Pacific plate by the hot-spot theory [*Clague and Jarrard,* 1973]. A pole at 72°N, 83°W and a rate of $13.10^{-7}$ deg/yr yield a paleolatitude of 10°S for 50 m.y.B.P.

(3) The elevation of the West Philippine basin is depressed by 1 km from standard oceanic depths of corresponding ages taken from the North Pacific [*Sclater et al.,* 1971]. Such a negative residual depth anomaly should produce a negative gravity anomaly of $30 \pm 10$ mgals if it follows the relationship of gravity to bathymetry found in the North Atlantic [*Sclater et al.,* in press]. This is not observed. It indicates that the crustal structure of the basin is different from that of the North Atlantic and possibly from other ocean basins. It follows that it could not have originated in an identical process.

*Acknowledgments.* This research was supported by contract N00014-67-A-0204-0048 of the Office of Naval Research. John Sclater originally suggested looking at the magnetics of the Philippine Sea and later continued to give invaluable support and advice. Zvi Ben-Avraham helped in the initial stages of locating data. Peter Molnar suggested using the North Pacific profile in Figure 4. Computer programs for data compilation and presentation were written by Linda Meinke. Pam Thompson and Linda Meinke drafted the final diagrams. Yolanta Geissler helped in the digitization of the Project Magnet data. Hugo Bezdek, chief scientist on Leg 7 of cruise Tasaday, kindly offered us the ship time for our track across the Central Basin fault. I would like to thank Roger Larson and Walter Pitman for allowing me to use some of L.D.G.O.'s magnetic data, and Jiro Segawa for obtaining the original *Umitake-maru* data. I would also like to thank Dan Karig, John Sclater, and Barry Parsons for reading early versions of this paper and offering helpful suggestions for its improvement.

# 4. MARINE GEOLOGY AND TECTONICS

## 22. *Anomalies in the Tectonic Evolution of the Pacific*

EDWARD L. WINTERER

*Scripps Institution of Oceanography, La Jolla, California 92093*

*Abstract.* Patterns of plate evolution over the past 130 m.y. show an anomalous band of sea floor trending northwest across the central Pacific where large changes in the location and orientation of spreading ridges have been more frequent than elsewhere. In this band regional depths are shallower than expected, and there is a concentration of plateaus, ridges, seamount chains, guyots, atolls, and high islands, some of which contain mildly to strongly alkaline rocks. In this same region spreading rates have been generally higher than elsewhere, commonly exceeding 8 cm/yr, and motions of this part of the Pacific plate relative to the earth's spin axis have been even faster. It is suggested that the high velocities may lead to instabilities in normal plate evolution, expressed as jumping of spreading ridges, and to voluminous mid-plate volcanism.

The papers grouped under the title Marine Geology and Tectonics in this symposium are so diverse in their subject matter that I will not attempt in this introductory paper to find the common thread that stitches them together. But since at least three of the papers—those by van Andel, Hammond and Theyer, and Dott—deal in one way or another with the kinematics of plate motions in the Pacific, I have elected to talk about some features associated with these plate motions that I think are a little odd.

The working out of the detailed history of plate motions in the Pacific, in the several different frames of reference—plate versus plate, plate versus melting anomalies, plate versus magnetic field, and plate versus spin axis—is still a matter of considerable importance. The history of motion recorded is a long one that extends back into the Jurassic, and is a very complex one, with evidence of major reorientation of motions in the several frames of reference. During much of the last 150 m.y., the major lithospheric plates in the Pacific have had mainly convergent boundaries with adjacent continental plates, and thus any quantitative estimates of the rates and direction of convergence prevailing during the development of circum-Pacific orogenic belts of the type from southernmost South America described by Dott in his paper depend heavily on our working out the detailed kinematic history of the oceanic plates, and particularly of the Pacific plate, which preserves the oldest records of plate motions. Furthermore, until middle Cenozoic times, the Pacific plate was not in direct contact with the Americas plates, but was separated from them by other plates now either wholly or partly consumed at trenches. Interpretations of the Mesozoic and early Cenozoic history of the western American orogenic belts are thus greatly dependent on working out the kinematics of now-vanished plates by way of the kinematics of the remaining parts of the Pacific plate, with which the other plates shared a common spreading-ridge boundary.

Turning to the paper by van Andel, we see that a detailed four-dimensional knowledge of the stratigraphy of pelagic sediments, which have facies patterns controlled grossly by depth of accumulation and by fertility patterns in the surface waters, has led in itself to a history of plate motions, because water depth is a function largely of crustal age, and because oceanic fertility is so closely linked to latitudinal belts. Van Andel's paper takes us quite beyond this kinematic puzzle-solving stage and into the second-order effects on pelagic sedimentation patterns of the changing intensities of surface- and bottom-water circulation over the past 40 m.y. But if we try to extend this method of analysis to areas beyond the fairly close control from deep-sea drilling holes, predictive models of sediment facies distribution must lean very heavily on having a reliable crustal-age map, and a knowledge also of the history of latitudinal motions of the sea floor. The crustal age map is now taking shape for the areas where sea-floor spreading magnetic anomalies can be mapped; that is, where crustal ages range from 0 to about 80 m.y., and from about 115 to 150 m.y. The big gap is in the time represented by the magnetic quiet zone, from about 80 to 115 m.y. ago. These are, in a sense, the dark ages, and crust produced during those times occupies a wide band in the Pacific.

The orientation of major spreading ridges changed significantly during this time interval, particularly in the south central Pacific [*Winterer et al.,* 1974], and there may have been substantial changes in spreading rates during this time interval [*Larson and Pitman,* 1972]. In trying to reconstruct the kinematic history of these times, detailed bathymetry may be our best hope, old-fashioned as this may appear.

One bright part of the history of quiet zone kinematics is the chapter on seamounts of this age, because they tend to give excellent virtual magnetic pole positions [*Francheteau et al.,* 1970], whereas seamounts built during times of frequent magnetic reversals tend to yield poor results, probably because the seamounts incorporate lavas of both normal and reversed polarity. The other ray of hope in the darkness is the paleomagnetic information that can be extracted from the set of deep-sea drilling cores of quiet-zone age, especially since the advent of the cryogenic apparatus that permits rapid analyses and that can be used on materials of very low intensity of magnetization [*Cockerham and Jarrard,* in press]. The coherent paleomagnetic results on piston cores of Cenozoic sediments, presented by Hammond and Theyer in this symposium, augur well for the future of this kind of work in older sediments. But their results also should serve as a warning that things may not be simple. The discrepancy between their results and that obtained by facies analysis of equatorial sediments [*Winterer,* 1973; *Berger,* 1973*a; van Andel and Moore,* 1974] suggests that some of the assumptions of either the magneticians or the sedimentologists or both are wrong.

*Special features of the central Pacific.* The region I want to focus on is the broad band extending northwest across the central Pacific from the Nazca plate area through Hawaii to Japan—the very region where George Woollard has sent the ships of his Institute over the past decade. In contrast to the northeast and South Pacific, where there is a more or less regular increase in depth from young to old crust, in a manner like most of the Atlantic, there are innumerable special features of the central region.

First, there has been especially abundant mid-plate volcanism to create volcanic chains like the Hawaiian, Samoan, Society, and Marquesas islands. Older chains in the same region include the Emperor and the Marshall–Gilbert–Ellice groups of early Cenozoic age, and the Marcus–Necker chains, of Cretaceous age. Some of the young chains, like the Hawaiian, surmount broad arches shallower than the expected depth for the age of crust. Mid-plate volcanism has left fewer traces on the Pacific plate in the southwest and northeast Pacific.

Next, there are large areas in the band that do not fit at all the normal age-versus-depth curve [*Sclater et al.,* 1971]. Foremost among these areas are the plateaus—Ontong Java, Shatsky, Manihiki, Magellan, and possibly the Mid-Pacific mountains. But there are also regions such as that within the Phoenix magnetic lineation pattern near the equator south of Magellan rise where depths are as much as 1000 m shallower than predicted by their crustal age [*Winterer, Ewing et al.,* 1973]. In the region of the East Pacific rise, the misfit is even asymmetric: The west side is too shallow and the east side too deep, as compared to the South Pacific curve [*Mammerickx et al.,* 1975].

Then, there are ridges of yet undetermined origin, like the Line Islands and Tuamotu ridges. Age data from Line Islands area [*DSDP Scientific Staff,* 1974*a*] suggest that the ridge may be of nearly uniform age along much of its length. Age data from the Tuamotus are too sparse yet to know whether the ridge is progressive or not. I will return to the Line Islands chain later.

Finally, in this band there have been rather complicated patterns of sea-floor spreading, extending right from the present East Pacific rise to the oldest crust yet identified in the Pacific; i.e., late Jurassic. These complications are manifested in several ways, but especially by sudden shifts of position of actively spreading ridges—i.e., by ridge-crest jumping. These jumps have been best documented in Cenozoic crust along the Nazca–Pacific plate boundary by *Herron* [1972*a*], *Sclater* [1971], and *Mammerickx et al.* [1975]. Some of the jumps leave behind distinct ridges like the Mathematicians seamounts, while others are recognizable only as broad arches in the bathymetry. Herron's picture of the ridge-crest jumping and reorientation is that it was essentially simultaneous over most of the area about 20 m.y. ago, while Mammerickx et al. interpret magnetic and bathymetric patterns as indicating a difference in timing of the jumps from one fracture zone block to another. Some of the abandoned ridge crest areas on both the Nazca and Pacific plates show anomalously high heat flow values [*Herron,* 1972*a*; *Langseth,* 1968]. It may be fair to ask how long it takes for an abandoned spreading center to die. Very slow spreading after abandonment may account for some of the sharp relief of certain fossil ridges.

There is evidence on the Pacific plate in the eastern equatorial Pacific of other Cenozoic episodes of ridge-crest jumping, or perhaps asymmetric spreading. If we look at data compiled by *van Andel and Bukry* [1973] on the ages of basement in drill holes in the eastern equatorial Pacific, where magnetic data are very difficult to interpret, we are led to the likelihood that some still undetected pieces of Cocos plate have been captured by the Pacific plate, even farther west than the fossil ridge shown on *Herron's* [1972*a*] map. On Figure 1, average spreading rates for a part of the Cenozoic (anomalies 8–21, = ~29–53 m.y.) are plotted against latitude from the contemporaneous pole of rotation for Pacific plate versus Farallon plate motion. If we accept the northeast Pacific rates, where magnetic anomalies are relatively easy to identify, then the average equatorial Pacific rates are somewhat high, as compared to the predicted rate, which should follow a sine curve through the northeast Pacific points. Somewhere in the region between drill sites 159 and 161, and sites 77 and 79, one may suspect the presence of pieces of captured Cocos or Farallon plate. There may have been many small jumps rather than one big jump that would leave its mark as a prominent topographic feature. Very close spacing of data points is required to address the problem of jumping versus asymmetric spreading. As shown on Figure 1, the departures from the predicted curve seem to begin where spreading rates exceed 8 or 10 cm/yr.

The response to changes in spreading direction at a ridge crest may be different in the fast-spreading parts of the Pacific from what it is in the slow regions. For example, the reorientation of spreading directions along the East Pacific rise which began sometime between 10 and 20 m.y. ago was not uniformly absorbed along the crest of the rise. Slight differences in spreading directions along the rise north and south of the equator gave birth to a triple junction and the creation of the Galapagos spreading ridge system. Where spreading rates are fast, such slight differences in direction can give resultant vectors that may not easily be accommodated by oblique spreading such as one sees farther north off the coast of Oregon, near the Blanco fracture zone [*Elvers et al.,* 1973]. Perhaps the slower spreading rates in the north permit adjustments without recourse to new triple junctions.

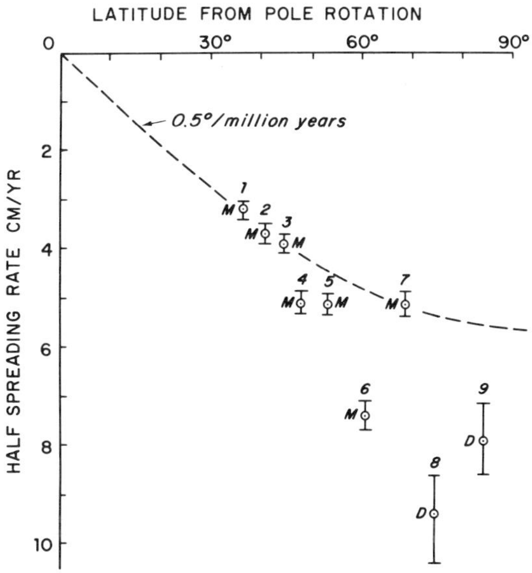


Fig. 1. Average spreading rates in the northeast Pacific between 29 and 53 m.y. ago (anomalies 8–21); M indicates magnetic data; D, drill data. Data from *Atwater and Menard* [1970] and *van Andel and Bukry* [1973]. The dashed curve is the rate for rotation at 0.5°/m.y. around the pole of rotation for contemporaneous Pacific versus Farallon plate motion.

*Cretaceous plate tectonic evolution of central Pacific.* The evolution of the central Pacific during Cretaceous times was complicated not only by ridge crest jumping, changes in ridge orientation, and the development of plateaus, but also by extraordinary mid-plate volcanism, which not only built chains of volcanoes and flooded large areas of deep-sea floor, but also strongly overprinted certain older structures, leaving them as palimpsest features mainly buried beneath younger volcanic rocks which tended to be guided by the older structural trends. In the following paragraphs, a cinematic glimpse of the main features of the Cretaceous history of the central Pacific is presented. A more detailed analysis, including new bathymetric charts, is given elsewhere [*Winterer*, in press].

The early Cretaceous plate evolution of the region can be inferred from the magnetic anomaly pattern worked out by *Hayes and Pitman* [1970], *Larson et al.* [1972], *Larson and Chase* [1972], and *Hilde* [1973]; supplemented by DSDP drill data [*Winterer, Riedel et al.*, 1971; *van Andel, Heath et al.*, 1973; *Winterer, Ewing et al.*, 1973; *Heezen, MacGregor et al.*, 1973; *DSDP Scientific Staff*, 1973*a*, 1973*c*, 1974*a*]; by dredging [*Hamilton*, 1956; *Heezen, Matthews et al.*, 1973]; and bathymetry [*Chase et al.*, 1971; *Winterer et al.*, 1974; *Mammerickx et al.*, 1974; *Winterer*, in press]. As shown in Figure 2, the triple junction among the Pacific, Farallon, and Phoenix plates evolved from the beginning of the Cretaceous to about 110 m.y. ago in a fairly simple way, with a bight between the Hawaiian and Phoenix lineations in the central basin, between the present-day Line Islands and Magellan plateau. Recent detailed surveys by the Scripps Institution of Oceanography (Eurydice Expedition, Leg 2, Sept.–Oct., 1974) show that the Phoenix linea-

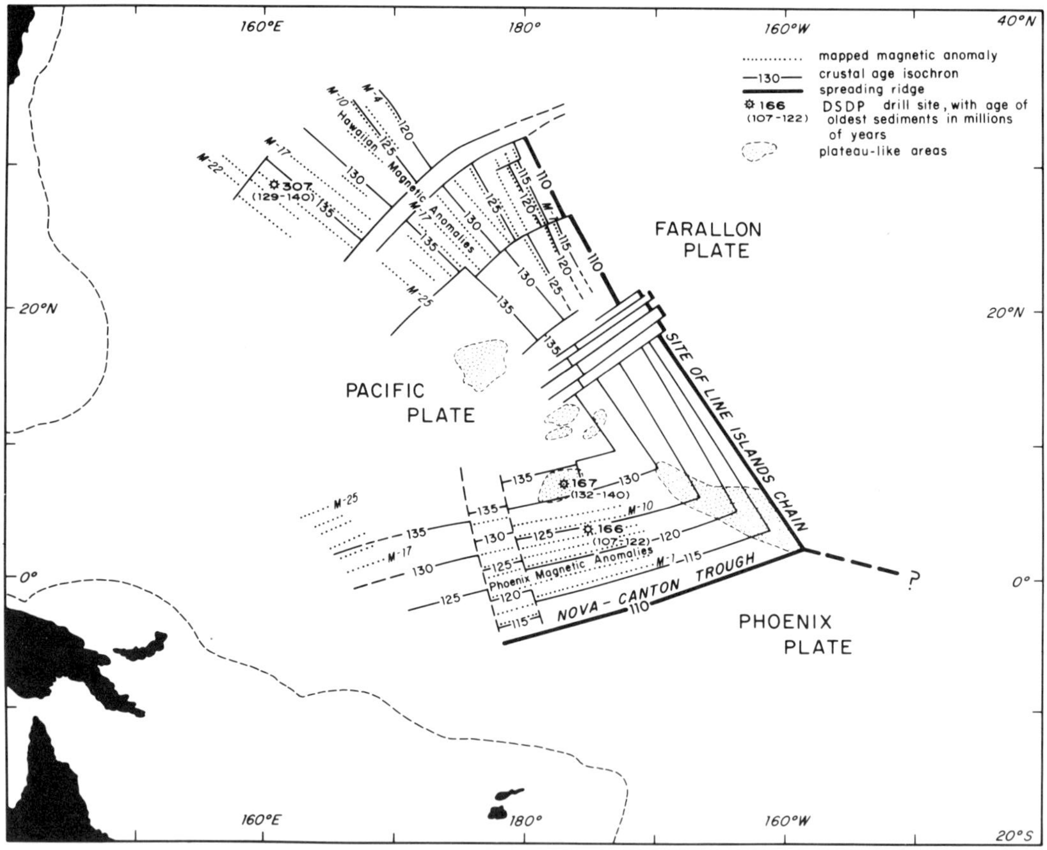


Fig. 2. Configuration of plates in the central Pacific about 110 m.y. ago. The Pacific and Phoenix plates were separated by a spreading ridge at the present-day location of Nova–Canton trough, and the Pacific and Farallon plates were separated by a spreading ridge at the present-day location of the Line Islands chain. Magellan plateau and other anomalously shallow areas formed during late Jurassic and early Cretaceous times in the vicinity of the bight between the two spreading ridges. Magnetic anomalies [from *Pitman et al.*, 1974] are shown as dotted lines. Ages at DSDP drill sites 166 and 167 from *Winterer, Ewing et al.* [1973]; at Site 307, from *DSDP Scientific Staff* [1973*c*]. Assignment of ages in the magnetic anomaly regions adjusted for an even spreading rate from Site 167 to the Nova–Canton trough. Isochrons and fracture zones in areas beyond mapped magnetic anomalies were extrapolated largely on the basis of bathymetric trends.

tions can be traced as a band of parallel bathymetric trends nearly as far as the Line Islands, in the vicinity of Fanning and Christmas islands [*Winterer*, in press]. Nova-Canton trough [*Rosendahl*, 1972] parallels these lineations and can be traced to the Line Islands, but no farther. The parallelism of the Line Islands chain with the Hawaiian magnetic lineations farther to the northwest suggests that this structural trend had its origin as a spreading ridge, coeval with a spreading ridge along Nova-Canton trough, about 110 m.y. ago. Magellan plateau developed near the triple junction in earliest Cretaceous times, and other smaller plateaus and broad arches likewise grew along the bight as it developed. Whether the Mid-Pacific mountains were recognizable as a set of plateau-like features during this time is not established; later volcanism obscures their foundations.

About 110 m.y. ago, the Pacific-Phoenix spreading ridge suddenly changed its orientation, and the ridge jumped southward, leaving pieces of Phoenix plate trapped on the Pacific plate (Figure 3). The abandoned spreading ridge may have continued to spread very slowly for a little while, resulting in the deep (~ 7500 m) cleft in Nova-Canton trough. Spreading on the new ridge was complex, at least in the vicinity of Manihiki plateau, which was built up nearly to sea level by massive outpourings of abyssal tholeiites [*Jack-*

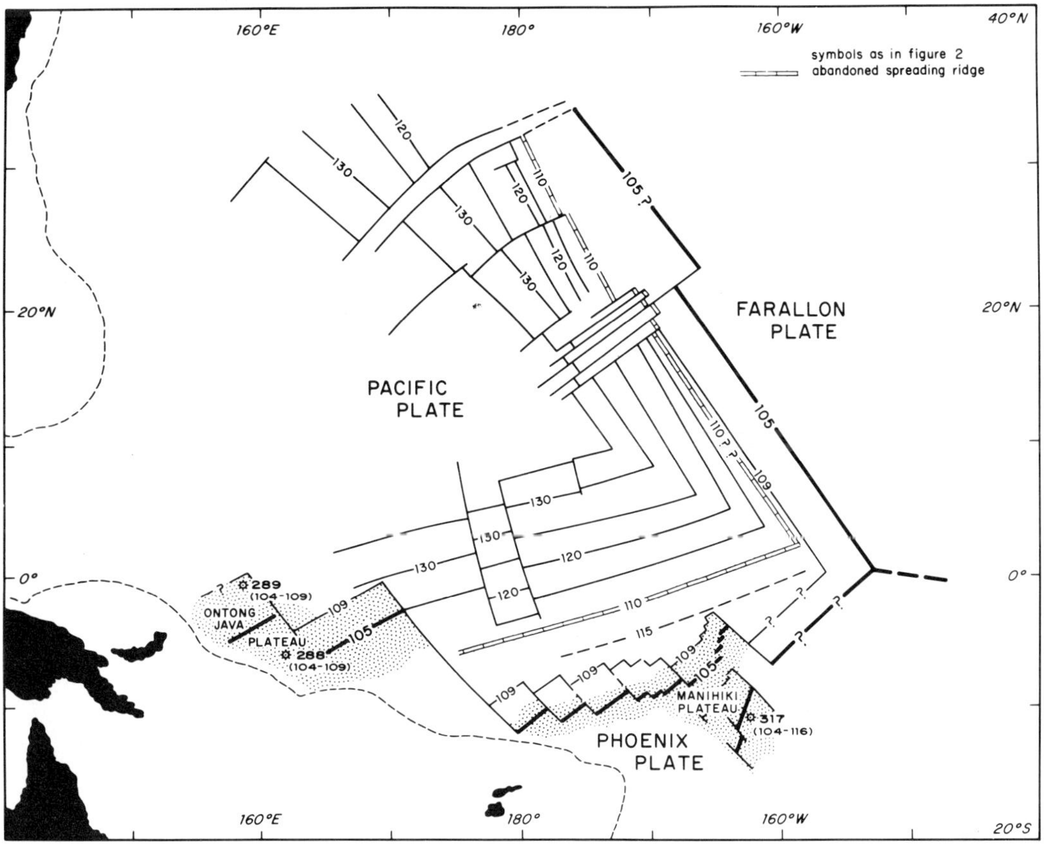


Fig. 3. Configuration of plates in the central Pacific about 105 m.y. ago. The former spreading ridge at Nova-Canton trough was abandoned during a change in orientation of the Pacific-Phoenix spreading ridge. The new ridge was located on the Manihiki plateau, but its connections to the Pacific-Farallon ridge are not known. There may have been a small extra plate on part of the Manihiki plateau. The history along the Pacific-Farallon ridge during this time is not known. In the version shown here spreading continued, with perhaps a change in rate at about 110 m.y. ago, but the alternative of a jump in the Pacific-Farallon ridge at this time is also possible. Ages at DSDP sites from *DSDP Scientific Staff* [1973*a*, 1974*a*].

*son et al.*, in press], and which may have included a small separate plate on the High plateau [*Winterer et al.*, 1974] analogous to the modern small plate on the East Pacific rise, near Easter Island [*Herron*, 1972*b*]. Through a system of transform faults, the Pacific–Phoenix ridge was probably connected to the Ontong Java plateau [*Winterer et al.*, 1974], where the age of basement is essentially the same as on Manihiki plateau [*DSDP Scientific Staff*, 1973*a*].

On the Pacific–Farallon spreading ridge, the story during this time is nearly undecipherable, in large part because of later volcanism which has obscured bathymetric trends and probably buried much of the older crust. No reorientation of this ridge is required at the time of the change on the Pacific–Phoenix ridge, since fracture zones farther northwest; e.g., the Mendocino, do not show any marked changes in direction in the region east from magnetic anomaly M-1. It is possible that this ridge was abandoned by a jump of the ridge to the east, although this leaves the short segment of the main Line Islands lineation south of Christmas Island as an anomalous element. In the interpretation shown on Figure 3, this portion of the ridge is produced by a southward shift of the triple junction which requires a southward extension of the Pacific–Farallon ridge. The alternate hypothesis of a jumped ridge, on the other hand, is attractive because it makes the Line Islands lineation a special bathymetric and crustal feature, which may be recognizable in spite of—or even because of—younger volcanism in the same area. With no jump, the unusual character of the lineation would have to derive from some marked change in spreading rate, and it is this hypothesis that is shown in Figure 3.

The era of spreading that created the Manihiki plateau was short-lived: after only about 5 m.y., the Pacific–Farallon ridge changed its orientation, and the triple junction shifted to a location south of Manihiki plateau, possibly in the vicinity of present-day Palmerston Island (Figure 4). The Pacific–Phoenix ridge was abandoned on Manihiki plateau, and pieces of old Phoenix plate were trapped on the Pacific plate between Manihiki and the new spreading ridge [*Winterer et al.*, 1974]. Judging by the orientation of bathymetric features in the southwest part of the Pacific plate, east of the Tonga trench, the new ridge is logically identified as the Pacific–Antarctic ridge, and this reorientation and jump about 100–105 m.y. ago would plausibly be associated with the beginning of opening of the oceanic wedge that later opened the gap between Antarctica and New Zealand [*Molnar et al.*, 1974].

The directions of spreading on the Pacific–Farallon ridge between this time and the imprinting of magnetic anomaly 32, about 75 m.y. ago, can be read over much of the region by using the trends of the great fracture zones that are so clearly marked out in the bathymetry—the Mendocino, Murray, Molokai, Clarion, and Clipperton, but the poorly surveyed region east of Manihiki plateau cannot yet be fitted into this scheme satisfactorily. The rates of spreading on this ridge system, on the other hand, are not easy to establish, because the distance eastward to anomaly 32 is significantly different for each fracture-zone-bounded block (Figure 4). This geometry requires there to have been ridge-crest jumping, but it does not tell us exactly when, where, or in which direction.

At about the same time as this reorientation of spreading ridges, there was a great outburst of mid-plate volcanism on the Pacific plate that was especially voluminous along the trace of the early Cretaceous bight between the Phoenix and Hawaiian lineations. Evidence for this comes from DSDP sites 169 and 170 in the deep (>5400 m) central basin (Figure 4), where Upper Albian (~100–102 m.y.) nannofossil limestones rest on basalt [*Winterer, Ewing et al.*, 1973]. These basalts probably flooded the deeper parts of the old central basin floor, burying older Cretaceous and Jurassic sediments, while higher-standing plateaus like Magellan plateau, located along and near the old bight between former spreading ridges, escaped inundation. At the same time, volcanism was probably building some of the nearby seamounts. The massive volcanism that built up the Mid-Pacific mountains may have taken place mainly at this time; there the only available dates are minimum ages for the cessation of volcanism, and these come from the shallow-water and reef fossils that cap the volcanic edifices. None of the sunken reefs is younger than Cenomanian (92–100 m.y.) or older than Aptian (104–110 m.y.) [*Heezen, MacGregor et al.*, 1973]. The Mid-Pacific mountains comprise two topographic levels, a general plateau level at about 3500 m and guyots which have their summits at about 1500 m. The plateau is blocky in plan view and clearly follows patterns laid down by the old Jurassic and early Cretaceous fracture zone and spreading ridge system. The guyots, on the other hand, tend to be roughly aligned in chains trending a little south of east.

The mid-plate volcanism continued to spread eastward toward the Line Islands lineation, progressing along several parallel chains in the central basin (Figure 4). The shapes of individual volcanoes along the chains tend strongly to reflect underlying old crustal structural trends; e.g., some volcanoes show shapes parallel to old spreading ridges, and others are parallel to old fracture zones. The ESE trend of the chains, on the other hand, is most likely an indication of the direction of motion of the Pacific plate with respect to melting anomalies in the asthenosphere.

Several of the chains intersect the Line Islands trend, and one of these chains continues eastward from the central part of the Line Islands as a series of parallel seamounts and ridges (Figure 4). Detailed surveys show that the ESE trends are a prominent feature *within* the Line Islands chain (Figure 5), and this pervasive en echelon structure, when taken together with the continuity of

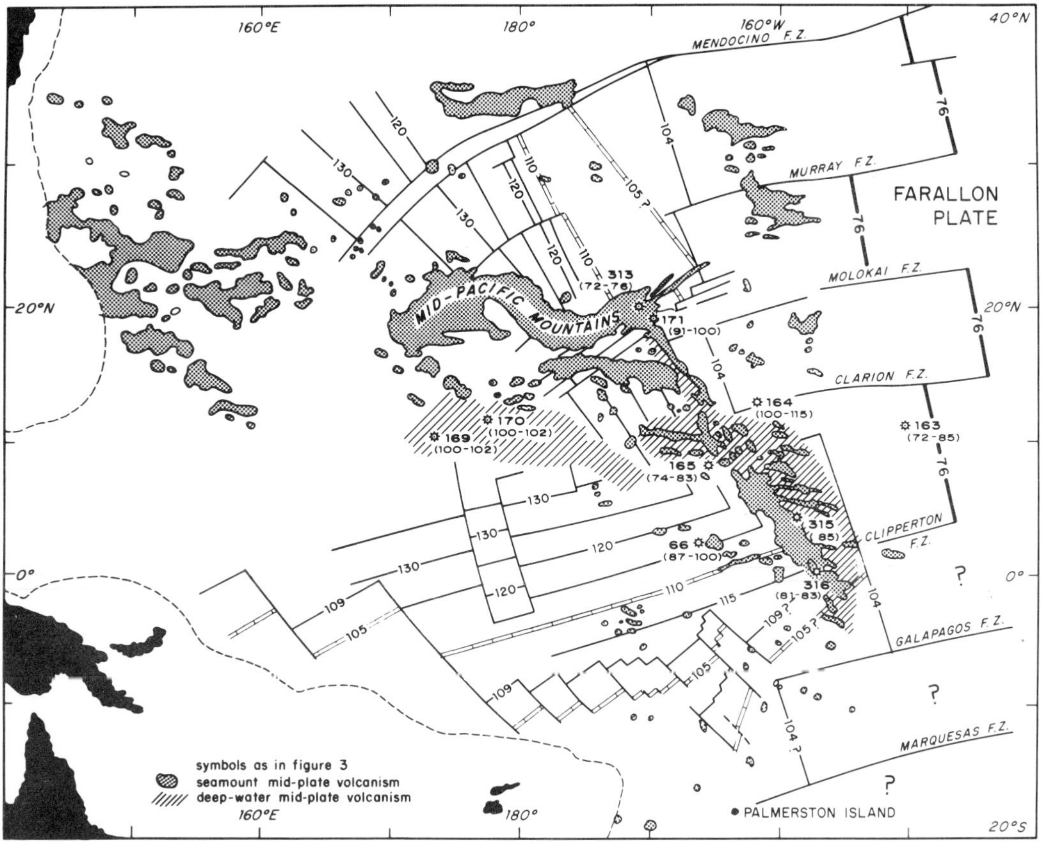


Fig. 4. Configuration of plates in the central Pacific about 76 m.y. ago (anomaly 32). A reorientation of spreading direction on the Pacific–Farallon and Pacific–Phoenix (Antarctic) ridges about 105 m.y. ago stranded the old ridges on Manihiki plateau. The new triple junction moved south, probably to the vicinity of Palmerston Island. Spreading on the Pacific–Farallon ridge between 105 and 75 m.y. ago probably included ridge-crest jumping, but the location and timing of these events is not yet known. Between about 105 and 85 m.y. ago mid-plate volcanism built chains of cross trend seamounts in a broad band trending WNW from the Line Islands to the Bonin trench. Lavas of this age were extruded in deep water over large regions in the central basin and around the Line Islands. The volcanism was probably progressive, propagating toward the ESE along parallel lines. Where Pacific plate motion carried the old Line Islands spreading ridge over the cross-trend melting anomalies, the lavas tended to be guided by older structural trends. The older crust in the Line Islands area is thus mainly buried beneath younger volcanics. Ages at DSDP Site 66 from *Winterer, Reidel et al.* [1971]; at Site 163, from *van Andel, Heath et al.* [1973]; at sites 164, 165, 169, 170, and 171, from *Winterer, Ewing et al.* [1973]; at Site 313, from *DSDP Scientific Staff* [1973*c*]; and at sites 315 and 316, from *DSDP Scientific Staff* [1974*a*].

cross trend chains across and through the Line Islands chain, is interpreted as indicating that the cross trend is younger than the Line Islands trend. It furthermore suggests that cross trend volcanism was strongly influenced by the pre-existing Line Islands lineation; that is, old structural trends were followed during the emplacement of younger lavas. In this interpretation, the present-day Line Islands chain consists mainly of 80–90 m.y.-old mid-plate volcanic rocks emplaced as an overprint on a lineation produced about 110 m.y. ago at a ridge crest. As to just why this particular old lineation would be the locus of so much younger volcanism, one can only speculate. A plausible suggestion is that the lineation is the site of an abandoned ridge crest, where continued

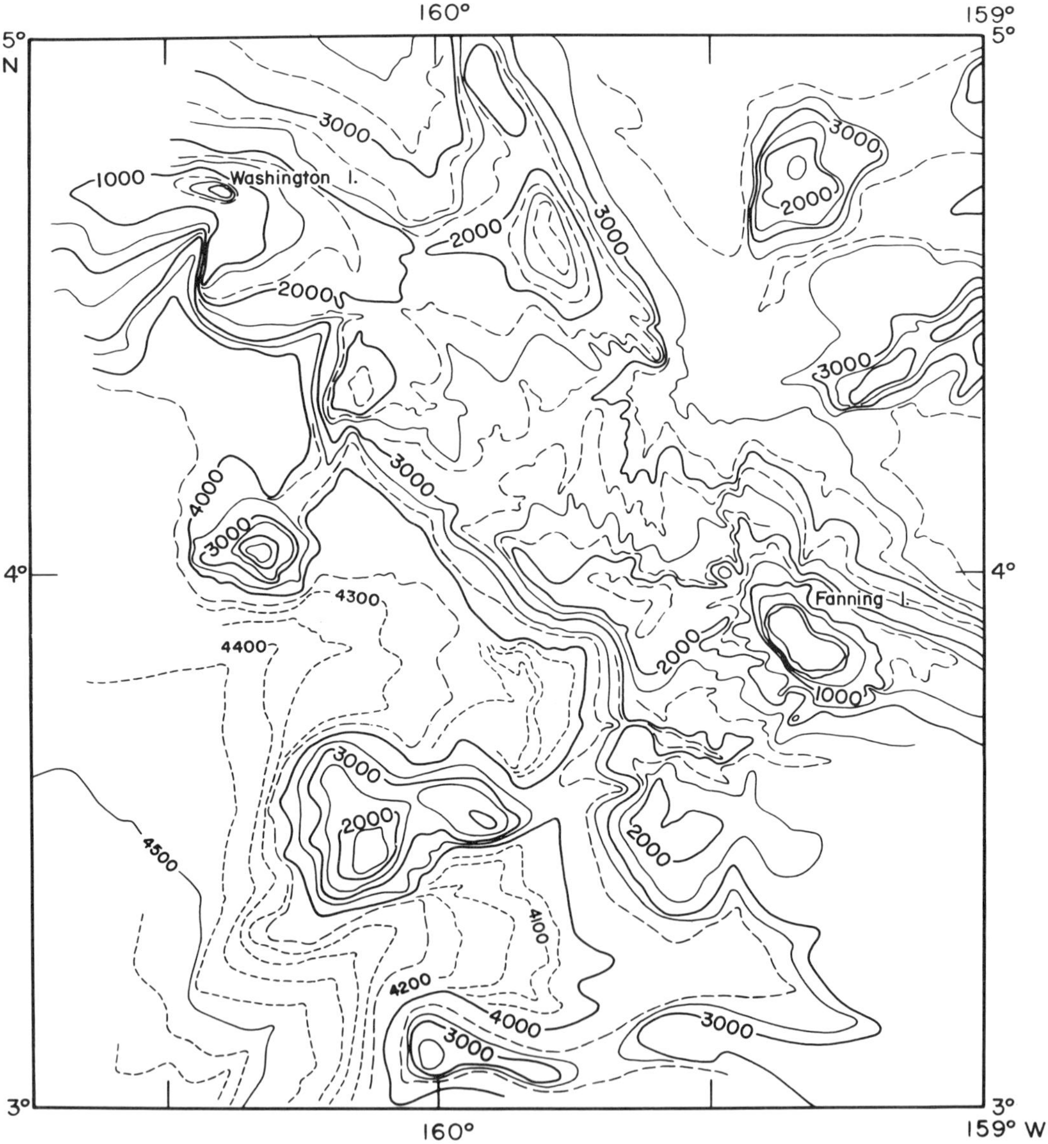


Fig. 5. Bathymetric chart of a part of the Line Islands ridge near Fanning Island. Contour interval 500 m (solid line), and 50-m contours (short dashed lines). Precision sounding data from Scripps Institution of Oceanography, Hawaii Institute of Geophysics, Lamont-Doherty Geological Observatory, supplemented by U.S. Navy and British Admiralty soundings. For track chart see *Winterer* [in press].

very slow spreading might have cracked the lithosphere so as to provide easy upward movement of lavas during later volcanism. The continued high heat flow on old abandoned ridges on the Nazca and Pacific plates [*Herron,* 1972*a*] suggests that the intersection of a mid-plate melting anomaly or 'hot spot' with a still-hot jumped ridge might result in unusually voluminous volcanism.

The middle and late Cretaceous mid-plate volcanism in the central basin and Line Islands region is peculiar in several ways. First, lavas of mildly alkalic to ocean-island tholeiitic character were extruded onto the deep-sea floor as well as on seamounts. These abyssal alkalic rocks were encountered at each of the three drill sites (165, 315, 316) close to the Line Islands, and at one site (170) in deep water in the central basin. The basement rocks at the other deep-water central basin site (169), on the other hand, have the chemistry of abyssal tholeiites [*Bass et al.,* 1973; *Jackson et al.,* in press]. Second, the lavas on several of the dredged seamounts are not only alkaline, but are potassic: potash exceeds soda in some of the rocks. According to *Natland and Jackson* [in press], these rocks are chemically distinct from known oceanic rocks, but are similar to certain rocks known only from the East African rifts. The potash-rich rocks have also been dredged from a seamount in the cross trend chain that extends WNW through Marcus Island, west of the Mid-Pacific mountains. Finally, at both sites 169 and 170 in the central basin, sills occur in late Cretaceous strata, giving evidence of repeated episodes of mid-plate volcanism in deep water. The considerable difference in age reported for the oldest volcanic rocks drilled at Site 171 (91–100 m.y.) [*Winterer, Ewing et al.,* 1973] and at Site 313 (72–76 m.y.) [*DSDP Scientific Staff,* 1973*c*] located only about 150 km apart in the Mid-Pacific mountains, also suggests repeated mid-plate volcanism.

If we take the difference between the youngest ages for volcanism in the Line Islands ( ~82 m.y. at Site 316) [*DSDP Scientific Staff,* 1974*a*], and the oldest ages reported for volcanism in the central basin 'upstream' from the Line Islands ( ~102 m.y. at Site 170) [*Winterer, Ewing et al.,* 1973], the average rate of propagation of volcanism during this time (Pacific plate versus melting anomaly) would be about 15 cm/yr. The age of the basal sediments at Site 66, on a low hill in the southeast part of the central basin (Figure 4) along the same flow line, is consistent with this estimate. The absence of limestone in the basal sediments at this site [*Winterer, Riedel et al.,* 1971], suggests accumulation on mid-plate sea floor, below the compensation depth for calcite.

The direction of volcanic propagation during this time was ESE, parallel to the cross trend, but the several advancing chains may not have been exactly abreast of one another. Thus, the exact timing of volcanism along various parts of the Line Islands chain would depend on the timing of the arrival of each cross trend chain at the old ridge axis, and, of course, on the duration of volcanism along the segments of cross trend chains at the intersection. The cross trend chains far to the west, between the Mid-Pacific mountains and the present-day Mariana–Japan trench system (Figure 4) are most likely about the same age as the central basin chains, judging from the ages of dredged fossils [*Heezen, MacGregor et al.,* 1973] and their geometric relation to the younger Marshall chain [*Chase et al.,* 1971].

The cross trend direction of volcanic propagation probably persisted until about 70 m.y. ago, if we regard Meiji seamount at the north end of the Emperor chain to be close to the point marking a change in direction of plate versus melting spot motion. The age of the basal sediments on Meiji guyot (DSDP Site 192) is Early Maestrichtian (70–72 m.y.) [*Creager, Scholl et al.,* 1973]. This same bend would be reflected farther south in the change from the ESE cross trend to the SSE Marshall Islands trend. No obvious change in direction can be seen for the cross trend chains that intersect the Line Islands, and the melting anomalies may have died out in the region before the change in direction of plate motion occurred.

## CONCLUSIONS

The tentative version of the history of Cretaceous plate evolution in the west central Pacific, sketched out briefly here, is comparable in many features to the Cenozoic history of the east central Pacific that has been much more firmly documented by study of magnetic, bathymetric, and drill hole data: (1) moderately fast spreading (>8 cm/yr) was common in the Cretaceous along both the Pacific–Farallon and Pacific–Phoenix ridges. Very fast spreading ( ~20 cm/yr) may have prevailed during middle and late Cretaceous times (anomaly M-1 to 32) on the Pacific–Farallon ridge, but the possibility of important amounts of ridge crest jumping makes these fast rates uncertain. (2) At least two major mid-Cretaceous reorientations of ridge crests took place.

The first of these, about 110 m.y. ago, resulted in the abandonment of the old Pacific–Phoenix ridge at Nova–Canton trough, and the second, about 105 m.y. ago, resulted in the abandonment of the newer Pacific–Phoenix ridge at Manihiki plateau. A small extra plate, like the modern-day small extra plate near Easter Island, may have functioned during the fast spreading that produced Manihiki plateau. (3) Middle and late Cretaceous mid-plate volcanism, recording plate versus melting anomaly motions, was directed ESE (in the present-day coordinate system), very nearly parallel to the Neogene motion direction recorded by the Hawaiian chain. The two big changes in direction of volcanic propagation, first near the end of the Cretaceous (cross trend–Marshall bend) and then in the Eocene (Emperor–Hawaiian bend), are not reflected by conspicuous widespread changes in spreading rate or direction on the Pacific–Farallon or Pacific–Antarctic ridges, as judged by the trends of fracture zones and the spacing of magnetic anomalies [*Pitman et al.,* 1974]. (4) The pattern of Cretaceous mid-plate volcanism suggests that old structural trends—fracture zones and ridge crests—influenced the local shapes of seamounts, and in some places—e.g., in the Line Islands—the old pattern may have strongly directed the younger volcanism. The sigmoidal en echelon trends in the Neogene Hawaiian chain [*Jackson et al.,* 1972] are likewise nearly parallel to isochrons in the adjacent Cretaceous oceanic crust, as are the en echelon trends in the southern part of the Paleogene Emperor chain [*Clague,* 1975].

The repeated occurrences of ridge-crest jumping, commonly, but not necessarily, associated with sudden reorientations of spreading directions in crust from early Cretaceous to late Cenozoic age in a wide band across the central Pacific may be somehow associated with relatively fast spreading rates. This is not to say that slow spreading prohibits jumping. Jumps in the slow-spreading north Atlantic are well documented [*Pitman and Talwani,* 1972], and small jumps are known in the slower spreading parts of the Pacific [*Molnar et al.,* 1974]. It may be that in fast spreading the position of the ridge crest is not as stable as in slow spreading, because of differences in crustal or thermal structure. Fast-spreading ridges may cool down more slowly than slow-spreading ridges, and may thus be more susceptible to re-breaking. If there is general applicability of the observation by *Shaw* [1973] that in mid-plate volcanism in the Hawaiian chain the volume rate of volcanism is correlated with the local rate of volcanic propagation along the chain, then it might be inferred that the massive outpourings of mid-plate lavas in the central Pacific in mid- and late Cretaceous times were the consequence of shear melting during relatively fast motions of the Pacific plate over a region of melting anomalies. These anomalies may have interacted with a region of residual high heat flow along the old Line Islands lineation to produce even greater-than-normal volumes of melt.

*Acknowledgments.* I am grateful for the contributions of many Scripps colleagues and graduate students who participated with me in expeditions to the central Pacific over the past few years. I would especially like to acknowledge the help of the late E. C. Allison and that of D. Clague, R. Jarrard, J. B. Keene, P. F. Lonsdale, J. L. Matthews, J. Natland, and B. R. Rosendahl.

This work was supported by the National Science Foundation (Grant GA-33227) and the Office of Naval Research (Contract N00014-69-A-0200-6006). Duplication for any purpose of the U.S. Government is permitted.

# 23. *Cenozoic Plate Motion and the Magnetic Time Scale, Equatorial Pacific*

S. R. HAMMOND AND F. THEYER

*Hawaii Institute of Geophysics, University of Hawaii, Honolulu, Hawaii 96822*

*Abstract.* This discussion summarizes (1) present results of a continuing paleomagnetic investigation of Pacific plate kinematics using chronologically overlapping sediment (piston) cores, and (2) concurrent efforts to refine and extend a proposed sediment-based magnetostratigraphic time scale. Most of the 16 cores used were recovered within a 15-degree-square region centered slightly north of the equator in the Central Pacific basin. The average length of the cores is about 13 m and they consist mainly of brown radiolarian-rich clays. The average water depth at the coring sites exceeds 5 km. Submarine erosion throughout the general area has exposed older Tertiary sediments and the cores together span an age interval exceeding 30 m.y. Of the cores, eight are fully oriented by special instrumentation which recorded both the vertical and azimuthal orientations of the corer. Paleomagnetic results based on these cores indicate that the Central Pacific basin lithosphere has been displaced northward by about 16 degrees during the past 30 m.y. The average rate of motion was about 6 cm/yr, but the data suggest that there may have been one or more changes in either the rate of motion or the direction of plate motion during that interval. The most obvious paleomagnetic evidence in the cores for northward plate motion consists of the following: (1) Within individual cores which span several million years, angles of inclination steepen toward the tops of the cores (i.e., with decreasing age), whereas the opposite is true for those cores taken in the southern hemisphere. The only exceptions proved to be relatively old cores, recovered north of the equator, in each of which measured inclination angles became shallower toward the tops of the cores. Most of the latter cores, however, are oriented and the directions of their paleomagnetic vectors, relative to true north and horizontal, show that they contain sediment which accumulated in the southern hemisphere prior to the time their coring sites crossed the equator and moved into the northern hemisphere. (2) The older the mean age of a core, the greater is the difference between its mean inclination value and the dipole-predicted inclination, or dip, of the field at the latitude of the coring site. Abundant evidence supports a net northward displacement of the Pacific plate (or plates) since the Late Cretaceous. The paleomagnetically determined rate of motion reported here, however, is significantly more rapid than the average northward velocities predicted by plate motion models, based either on movement of the lithosphere over a stationary Hawaiian hotspot or on the migration of equatorial sedimentary facies, for the latest Paleogene and the Neogene in the Central Pacific basin. The reason for this disagreement is unclear at present. Assuming, nevertheless, that the paleomagnetism of equatorial Pacific sediments does preserve an essentially reliable record of plate motion, the present results, together with the initial paleomagnetic results from DSDP Site 289 (located approximately 25° due west of the Central Pacific basin on the crest of the Ontong Java plateau), suggest that relative motion occurred between the western and central parts of the Pacific lithosphere sometime during the past 30 m.y. because the Site 289 data indicate that the plateau had already reached its present latitude by about 30 m.y.B.P. Paralleling the paleomagnetic-tectonic investigations, research toward extending and confirming a proposed, sediment-based, magnetic time scale has continued using magneto- and biostratigraphic data from these piston cores, which collectively span from latest Pleistocene to latest Eocene. Although stratigraphic continuity is still lacking in mid-Oligocene time and at the Oligocene/Eocene boundary, it can be concluded that 26 to 27 polarity epochs may have occurred during the time from earliest Late Oligocene (about 32 m.y. ago) to Holocene. Available geochronological data for the first 24 of these Cenozoic epochs indicate that a typical polarity epoch lasted about 1.0 m.y., the shortest about 0.6 m.y. (E 8 and 13), and the longest as long as 2.8 m.y. (E 16). Correlation between polarity epochs and magnetic anomaly patterns is complete in the interval between anomalies 1 and 5, nearly impossible between anomalies 5 and 6, and somewhat clearer between anomalies 6 and 9. Correlation between the sediment- and anomaly-based time scales may be aided by paleomagnetic, biostratigraphic, and possibly radiometric analyses of sediments and basalts recovered by drilling at well-defined Cenozoic magnetic anomalies.

# 24. *Cenozoic History of the Central Equatorial Pacific: A Synthesis Based on Deep Sea Drilling Project Data*

TJEERD H. VAN ANDEL, G. ROSS HEATH[1], AND T. C. MOORE, JR.[1]

*School of Oceanography, Oregon State University, Corvallis, Oregon 97331*

Cenozoic sediments of the equatorial Pacific show large spatial and temporal changes resulting from the tectonic and paleoceanographic evolution of the region. The tectonic evolution involves rotation of the Pacific plate across the equator, a gradual subsidence with increasing age of the flanks of the ancestral East Pacific rise, and changes in height, shape, and position of the rise. The principal paleoceanographic changes include variations in biological productivity at the sea surface, and temporal changes in the characteristics of the bottom water and its circulation pattern which are related to the history of Antarctic glaciation.

Between 45 and 38 m.y.B.P., extensive carbonate dissolution resulted in low carbonate accumulation rates, extensive erosion, and the formation of only a narrow equatorial carbonate zone. Around 38 m.y.B.P., carbonate input increased and carbonate dissolution decreased abruptly, the equatorial calcareous zone widened, and accumulation rates greatly increased. This was caused by a change in bottom water characteristics resulting from the first significant development of sea ice around Antarctica. About 33 m.y. ago, carbonate dissolution at depth began to increase again, but initially this increase was compensated either by a large increase in the carbonate supply or by a depression of the lysocline. As a result, a very broad equatorial carbonate zone with maximal accumulation rates persisted until about 26 m.y. ago. The steepening of the dissolution gradient may have resulted from a decreasing influx of Antarctic bottom water, while the change in supply may have been related to the final closure of the Tethys seaway. Around 15 m.y.B.P., major growth of the Antarctic ice took place, leading to a further increase in the dissolution rate. At that time, development of the configuration of sources and mechanisms characteristic of the present circulation of intermediate and deep water may have begun. As a result, erosion increased and the equatorial zone of carbonate deposition narrowed. It is also possible that the fertility of the equatorial region increased somewhat. Finally, around 3–4 m.y.B.P., the onset of the Arctic glaciation marked the beginning of a period of large and rapid changes in depositional conditions which cannot be resolved with Deep Sea Drilling Project data.

The cruises of the drilling vessel *Glomar Challenger* have gathered a large collection of samples and geologic data which contains much information on the Cenozoic and Mesozoic history of the ocean basins. In this paper we summarize the results of a synthesis of the data for the central equatorial Pacific with emphasis on the tectonic and paleo-oceanographic history of the region. This summary does not include data listings or details of data processing, which are described in *van Andel et al.* [1975]. In addition, the background material on today's ocean, including a full review of previous work, has been slighted in order to present, in the space available, as complete a picture as possible of the Cenozoic history of the area. We apologize for any superficiality or dogmatic aspect of portions of the presentation that results from this decision. Geographically, the area of study is limited by the 150° and 95° meridians (Figure 1), while beyond 15° N and 15° S drill data are too sparse and biostratigraphic control too inadequate. Since early Cenozoic and late Mesozoic deposits are found only in the extreme western part of the region, we have limited our study to the last 45–50 m.y.

[1]Present address: Graduate School of Oceanography, University of Rhode Island, Kingston, R.I. 02881.

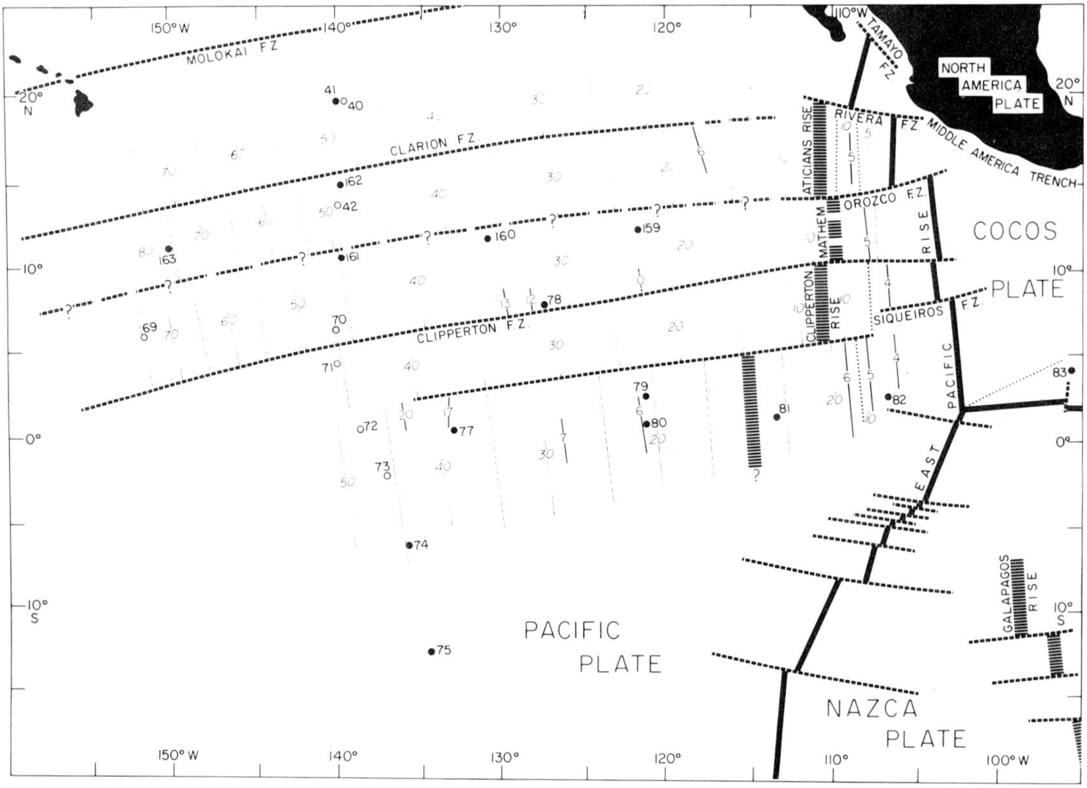


Fig. 1. Tectonic map of the central equatorial Pacific with basement-age isochrons. Heavy line is present rise crest; broad, striped zone is extinct rise crest (Mathematicians–Clipperton ridge). Very thin lines numbered 10–70 are basement-age isochrons in m.y. obtained from drill site basement ages. Heavier lines numbered 4–20 are magnetic anomalies. Dotted line indicates boundary between crust generated from East Pacific rise and that of Mathematicians–Clipperton ridge. Black dots: drill sites reaching basement; open circles: drill sites terminating in sediment.

The depositional history of the central equatorial Pacific has been controlled by four major environmental variables: the northward migration of the Pacific plate across the equator; subsidence of the oceanic basement with age; temporal variations of the rate of calcium carbonate dissolution in deep water; and probably also variations with time of the biological productivity in equatorial surface waters. The sediments in this region are entirely pelagic. Sedimentation rates decrease rapidly with increasing depth (and thus increasing age) of the basement as a result of carbonate dissolution. Moreover, the production of biogenic material, which is the principal sediment source, decreases sharply with increasing distance from the equatorial zone. Consequently, accumulation rates, subsidence histories, and paleopositions of the drill sites are important components in the data set needed to unravel the depositional history of the region. The primary data come from twenty drill sites (Figure 1) described in volumes 5, 8, 9, and 16 of the Initial Reports of the Deep Sea Drilling Project, and include mainly the biostratigraphic zonations, lithologic descriptions, and quantitative measurements of calcium carbonate content, wet bulk density, and porosity of the sediments. With the aid of a minor revision of the biostratigraphic zonation and with the absolute chronology of *Berggren* [1972], each drill record was subdivided into one-million-year increments. For each increment, accumulation rates of carbonate and carbonate-free sediment in g/cm$^2$/1000 yrs were computed. By a procedure analogous to that of *Berger* [1972, 1973*a*], subsidence histories were determined for each drill site. A plate rotation scheme based on tectonic data and on the northward migration with time of the axis of maximum equatorial deposition [*van Andel*, 1974] was used to determine the paleolatitude and paleolongitude of each drill site for each time interval of 1 m.y. The subsequent interpretation is based on an absolute time scale with confidence limits ranging from ±1 m.y. for the youngest to ±3 m.y. for the oldest intervals [*van Andel et al.,* 1975, Appendix 1].

REGIONAL LITHOLOGY AND TECTONIC HISTORY

All but two of the drill sites used in this study are located on the Pacific plate (Figure 1). Of the remaining two, site 83 is located on the Cocos plate while site 81 began its history on that plate but shifted to the Pacific plate about 15 m.y. ago when the rise crest jumped from the Mathematicians–Clipperton ridge system to the present East Pacific rise. Thus, the history of the Pacific plate, in particular its gradual subsidence with increasing age and distance from the rise crest and its rotation with respect to the equator, are of prime importance.

With increasing water depth, dissolution of calcium carbonate increases sharply, sedimentation rates decrease and deposits change from calcareous to siliceous. The large decrease in sedimentation rate produces diachronous facies boundaries and the isochrons converge upward and away from the rise crest. These features are illustrated by a profile through drill sites north of the equator and perpendicular to the East Pacific rise (Figure 2). The thin deposits at the western end of this profile are particularly susceptible to erosion and a large unconformity here in actuality represents a relatively limited amount of sediment. In some ways, the profile appears to be anomalous. At site 162, the oldest deposit on basement is siliceous rather than calcareous and is overlain by Oligocene calcareous sediments. Since the basement has subsided continuously since its formation, this unexpected sequence suggests either a change in the depth dependence of carbonate dissolution, a change in the width of the equatorial carbonate zone, or passage of this site across the equator in the early Oligocene.

A N–S profile through the region (Figure 3) is quite different. It is dominated by the large change in sedimentation rates across the equator that results from the equatorial zone of high biological productivity. As a result, each stratigraphic interval is lens-shaped and thins rapidly to the

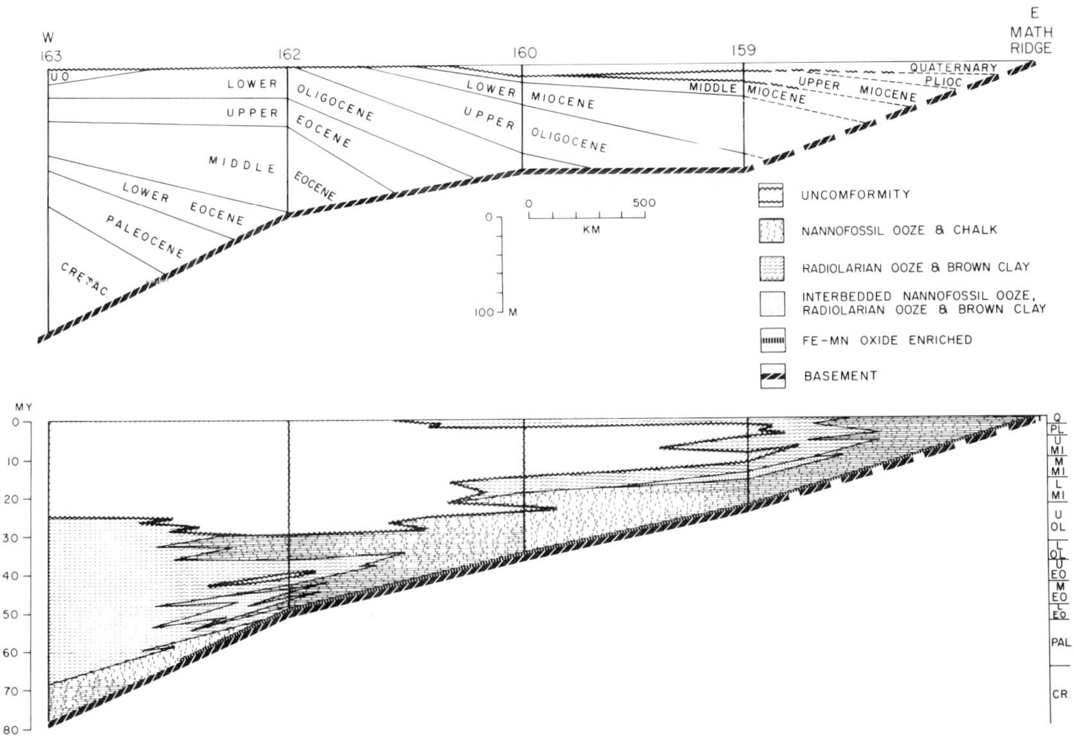


Fig. 2. E–W biostratigraphic and lithologic profile north of the equator. Upper section based on depth in hole, lower section on age in m.y. Locations of drill sites given on Figure 1.

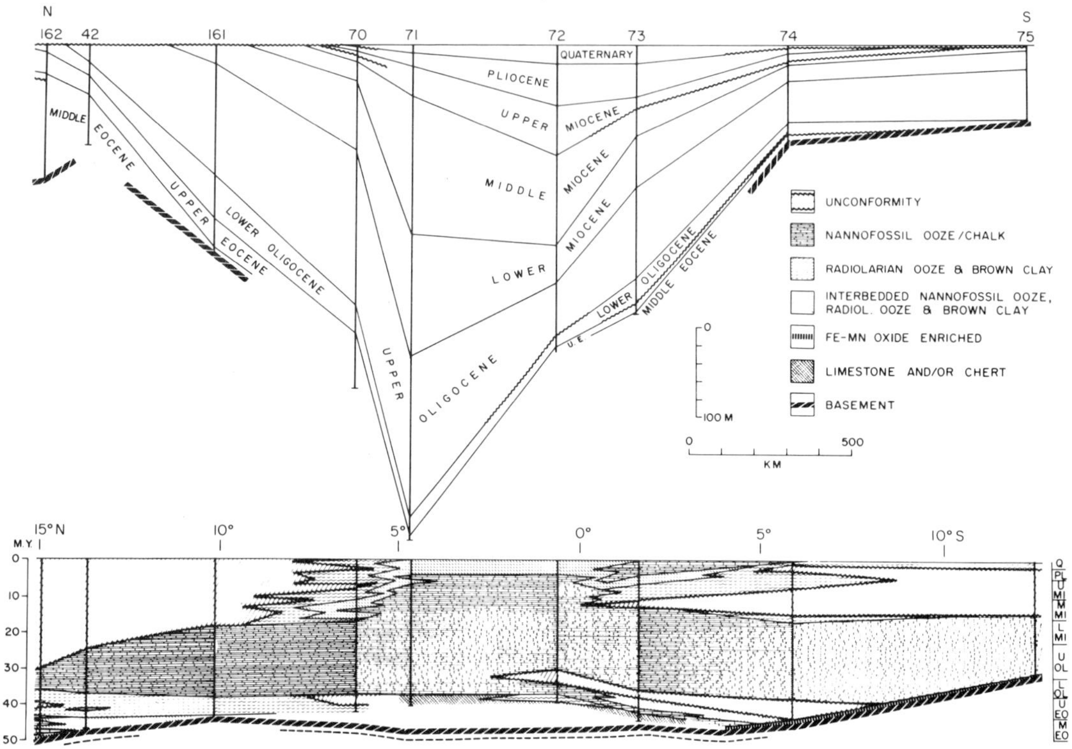


Fig. 3. N–S biostratigraphic and lithologic profile across equator along 140°W. Upper section based on depth in hole, lower section on age in m.y. Locations of drill sites given on Figure 1.

north and south with an accompanying change from calcareous in the center to siliceous at each end. Major unconformities which include much or all of the middle and late Cenozoic occur at both ends. Toward the equator, they break up into smaller units separated by zones of deposition. This suggests that the erosion was not restricted to the Quaternary but may have begun much earlier. An earlier major unconformity is restricted to the southern end of the profile.

The point of maximum thickness of individual stratigraphic units is progressively displaced northward with increasing age of the unit, with a total displacement of about 15° for the middle Eocene. Moreover, the calcareous center was very narrow during the Eocene, widened enormously in the Oligocene and earlier Miocene, and narrowed again in the later Cenozoic. Plate migration, changes in the depth and intensity of carbonate dissolution, and possibly changes in the equatorial zone of biological productivity are implied.

Northward displacement of the Pacific plate with time was first described by *Francheteau et al.* [1970]. Subsequently, numerous authors have attempted to determine the pole and rate of rotation of the plate using both hot spot trails and magnetic anomalies and transform faults. The evidence has been summarized by *van Andel* [1974]. The northward shift of the axis of equatorial sedimentation noted above can be used as an independent means to describe the history of plate rotation for the past 45 m.y. Axes of maximum deposition can be determined from isopach maps for various time intervals. The best fit of these axes to the equator is obtained by rotation around a pole at 67°N, 59°W at a rate of 0.25°/m.y. for the interval 45–25 m.y.B.P., and of 0.8°/m.y. since that time [*van Andel,* 1974]. The fit is shown in Figure 4, and the drill site tracks when rotated back to their positions according to this scheme, in Figure 5.

*Berger* [1973*a*] and *van Andel and Bukry* [1973] have shown that the relation between basement age and basement depth in the equatorial Pacific parallels the curve postulated by

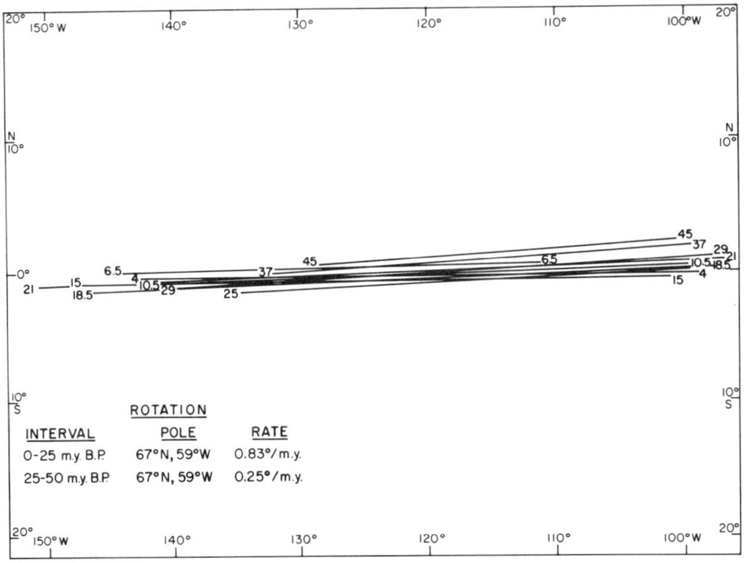


Fig. 4. Axes of maximum equatorial deposition, rotated according to scheme shown on diagram. Position of axes from *van Andel et al.* [1975]. Numbers are ages in m.y. Some idea of the uncertainty in the position of these axes can be gained by examination of the accumulation-rate isopleths of Figure 7.

*Sclater et al.* [1971]. With this relation the subsidence of the drill sites with time can be determined with an accuracy of about ± 200 m. Combining the paleodepth values with paleolatitudes and paleolongitudes obtained from Figure 5, paleobathymetric maps can be constructed. The initial positions of the drill sites that reached basement indicate positions of the axis of the ancestral East Pacific rise (Figure 6). *Sclater et al.* [1971] and *Herron* [1972*a*] have shown that the present East Pacific rise is a young feature and that the eastern boundary of the Pacific plate south of 4.5°S was formerly located at the Galapagos rise and, north of that latitude, at the Mathematicians–Clipperton ridge. Readjustment to the present position began about 15 m.y. ago at the Galapagos ridge and ended 5–7 m.y. ago at the Mathematicians ridge (references cited in *van Andel et al.* [1975]). Figure 6 shows that, notwithstanding this jump, the position of the eastern plate edge has been essentially stationary for the last 25 m.y., but that prior to that time the edge migrated eastward from 115°W at 50 m.y.B.P. to 105°W at 25 m.y.B.P. The early rapid migration was a consequence of a mismatch between the rotation rate of the plate and the spreading rate which began to balance when the rotation rate increased threefold about 25 m.y. ago.

In general the past topography is rather similar to the present one and mainly reflects a gradual change in orientation and longitudinal position of the ancestral East Pacific rise as a result of the rotation. There are, however, a number of interesting second order deviations from the present morphology. One of these is the gradual southward shift of a broad shallow area which at present lies west of the East Pacific rise and south of 10°S. Its existence during the past 50 m.y. is confirmed by the persistence through time of an anomalously calcareous zone extending well south and west of its normal position on the equator and on the crest of the East Pacific rise. During the middle Cenozoic this shoal zone was located considerably farther north.

Figure 6 also shows some significant changes in the shape of the ancestral East Pacific rise. The crestal depth may have been as much as 3000 m in the Late Cretaceous and early Cenozoic, shoaling to 2600–2700 m in the late Eocene, and gradually deepening again to its present depth near 3000 m. Parallel with this ran a change in the profile of the ridge, which was narrower and rather steep on the west flank during the period 45–25 m.y.B.P., and acquired its present rather broad and gentle western slope only during the last 20 m.y. At the present time, the rise is asymmetric with a steep and narrow eastern slope. Apparently, this asymmetry is of recent origin and resulted from a broadening of the west flank.

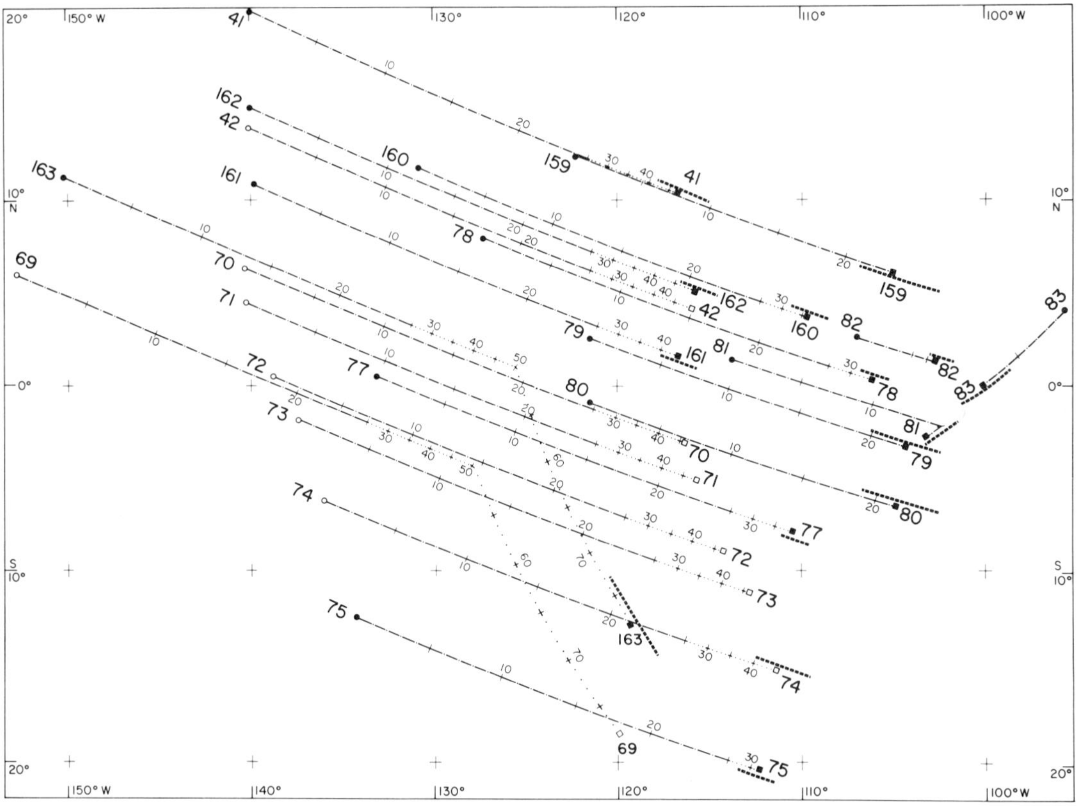

Fig. 5. Migration tracks of drill sites rotated according to scheme of Figure 4. Sites 83 and first 2.5 m.y. of site 81 rotated around Cocos plate pole at 23°N, 119°W at 1.5°/m.y. Black dots are present positions, black squares indicate initial positions of drill sites that reached basement; open circles and squares similarly refer to drill sites that terminated in sediment. Migration tracks are marked in 1-m.y. steps beginning with present. Dashed bars indicate age uncertainty of initial position. Rotation of sites 69 and 163 prior to 50 m.y. is around Emperor pole position at 11°N, 84°W at 0.8°/m.y., obtained by migration of present Emperor pole around later Pacific pole.

## SEDIMENT ACCUMULATION AND CARBONATE DISSOLUTION

The sediments of the central equatorial Pacific consist mainly of calcareous and siliceous (opaline) biogenic material. Other components, such as terrigenous clay and volcanic ash, play a minor role. Along the equator, upwelling associated with the equatorial current system provides nutrient-rich surface water that supports a large standing crop of plankton. As a result, accumulation rates of biogenic material are at a maximum along the equator. This is reflected in a well-defined narrow equatorial zone of rapid deposition of mainly calcareous sediments. Northward and southward, these sediments grade first into dominantly biogenic siliceous deposits and then into pelagic, brown, fossil-free clays of the central Pacific gyres. This pattern has persisted for at least 50 m.y., but during this time the width, westward extent, and maximum accumulation rates in the equatorial zone have varied considerably (Figure 7). Most of the variation was due to changes in the rates of accumulation of carbonate, although smaller changes in the rates of accumulation of non-calcareous sediment have also occurred, for example about 10 and 25 m.y. ago.

The accumulation rate of calcium carbonate reflects the balance between carbonate production in surface waters and carbonate dissolution at depth. Below the upper few hundred meters, equatorial Pacific waters are undersaturated with respect to calcium carbonate and dissolution takes place. Near 3700 m (at the lysocline) the dissolution rate increases sharply, leading to a large change in the faunal composition of the sediments. Below the lysocline, the dissolution rate increases with depth until the rates of supply and dissolution are equal. Below that level, called the calcite compensation depth (CCD), the sediments are carbonate-free. At present in the equatorial Pacific the CCD lies below 5000 m, rising southward and northward to about 4500 m under the central gyres [*Berger and Winterer,* 1974]. Thus the carbonate content of the sediments in the central Pacific is a function of the depth of the ocean floor relative to the local levels of the lysocline and CCD as well as of the rate of supply. We will show below that the changes in carbonate accumulation rates over the last 50 m.y. can be accounted for mainly in terms of changes in the rates of dissolution rather than changes in carbonate production.

The CCD has varied significantly in time [*Berger,* 1973*a; van Andel and Moore,* 1974]. In Figure 8 this change is shown separately for the equatorial zone ($CCD_{eq}$; between 3°N and 3°S) and the Pacific north of 4°N and south of 4°S ($CCD_{pac}$), together with changes in the width and maximum accumulation rate of the equatorial carbonate zone. The most striking event is a large and abrupt change of the CCD at the Eocene/Oligocene boundary which is

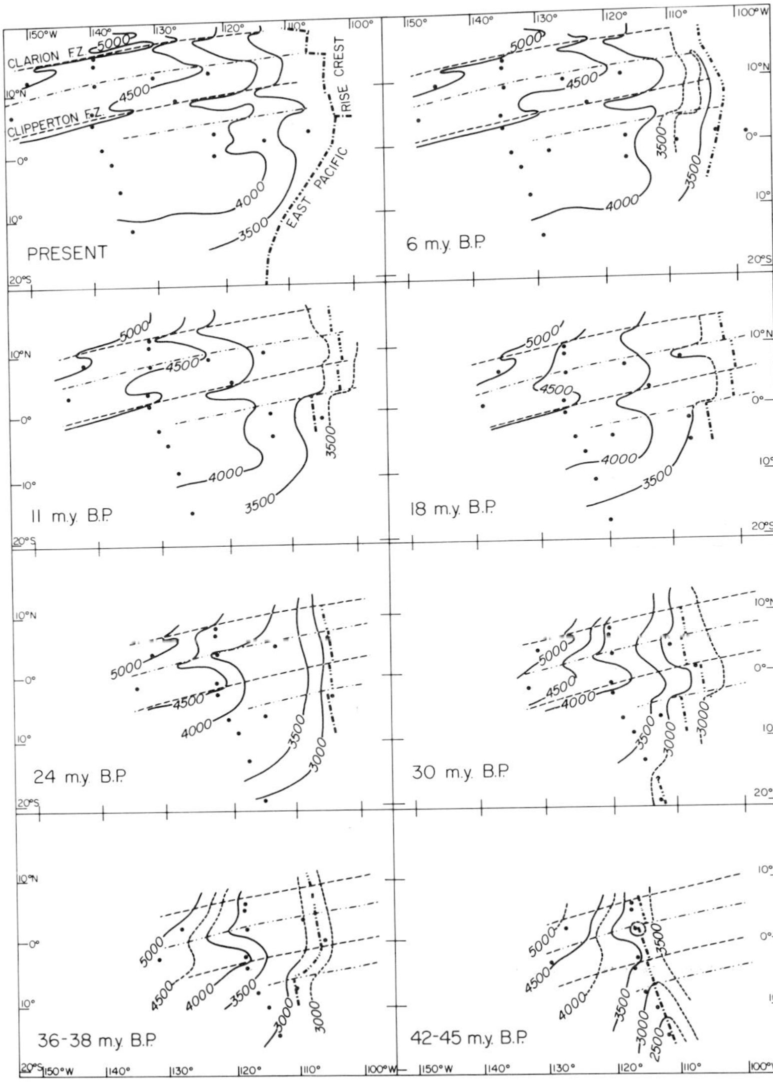


Fig. 6. Paleobathymetry of central equatorial Pacific. Positions of fracture zones obtained by rotation, plate edge by interpolation from initial positions of drill sites. Depth contours in meters. Black dots are drill site locations.

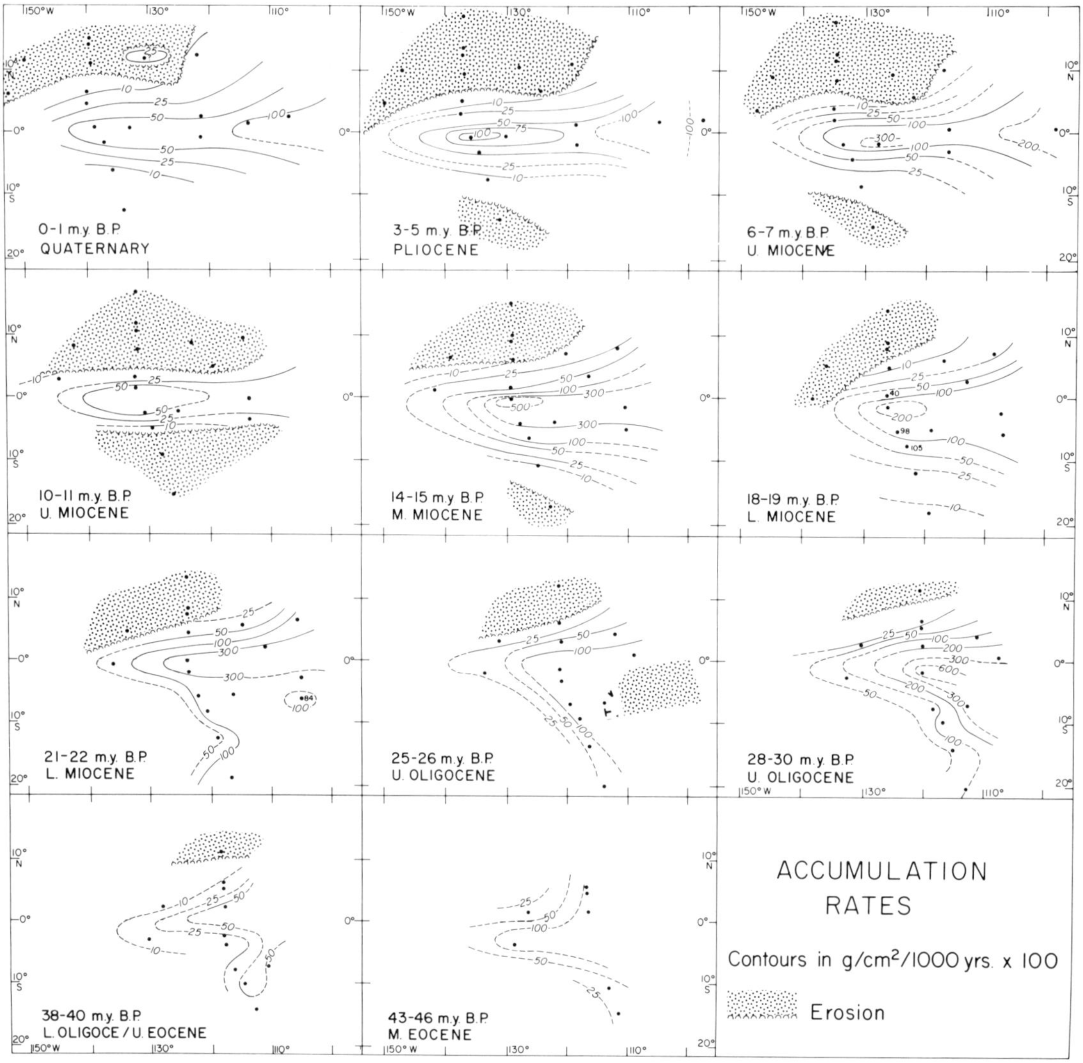


Fig. 7. Isopleths of accumulation rates for various times during the last 50 m.y. Black dots are drill site positions rotated according to Figure 5. Rate data (small numbers near dots) shown only where they do not fit isopleths.

closely followed by a large increase in the width and accumulation rate of the equatorial calcareous zone. About 20 m.y. ago, the $CCD_{eq}$ began to rise again in parallel with a decrease in the width and accumulation rate of the equatorial calcareous zone. About 10 m.y. ago, the maximum accumulation rate at the equator decreased drastically and since then has remained constant. The $CCD_{pac}$ also dropped about 38 m.y. ago but to a much shallower level than the $CCD_{eq}$, and did not begin to rise again until about 10–12 m.y.B.P. It reached its shallowest level about 7 m.y. ago, then dropped to its present position accompanied by an increase in the width of the equatorial calcareous zone (but not in the carbonate accumulation rate).

These relations suggest that changes in the dissolution rate rather than changes in productivity played a principal role. The CCD itself, however, is a function of three more fundamental variables: the position of the lysocline, the gradient of the dissolution rate below the lysocline, and the supply of carbonate from the surface. In the following discussion, we will assume a linear increase in the solution rate from the lysocline to the CCD. This assumption is consistent with Peterson's buoy experiment data [*Peterson,* 1966; *Heath and Culberson,* 1970], but is not required by these data [*Berger,* in press]. As we shall show, however, the linear approximation fits the DSDP data extremely well. An attempt can be made to distinguish variations in supply from changes in dissolution because they have different effects on the rate of carbonate accumulation as a function of depth, as Figure 9 shows. If at time A (Figure 9a) the supply rate is S and the dissolution rate increases below the lysocline (L) according to curve D, the accumulation rate will decrease with depth according to curve A(1). The CCD is located by definition where A(1) reaches zero. A move of the CCD at time B to another depth CCD(B) can be due to a change in any of the three parameters. The three extreme cases are shown in Figure 9b (change due entirely to change in supply), Figure 9c (change in dissolution gradient), and Figure 9d (change in depth of the lysocline). Therefore, by themselves, changes in the CCD will not allow a segregation of the effects of variations in input and dissolution.

It is clear that knowledge of the variation of the rate of carbonate accumulation as a function of depth will define all three parameters provided the paleodepth range for the time interval is adequate. The available data have been used to prepare plots of the depth variation of the carbonate accumulation rate for many time intervals separately for the equatorial zone between 3° N and 3° S and for the Pacific between 4° and 10° latitude [*van Andel et al.,* 1975]. Two examples of the depth dependence of carbonate accumulation are shown in Figure 10. In each case, the data are segregated by paleolatitude and longitude (all within a 20° span) to minimize the effects of N–S and E–W variations in carbonate supply. All four data sets fit the assumption of a linear increase in solution rate with depth extremely well, and further indicate that the rate of solution is independent of the rate of carbonate deposition at a given depth—strong evidence for control of dissolution by a chemical rather than a physical mechanism. The plots can be described by the relation:

accumulation rate = S – D X water depth

where S is the 'extrapolated surface production' (the rate of supply at the surface if the lysocline did not exist) and D the vertical gradient of the rate of carbonate dissolution. Unfortunately, the paleodepth range of the drill data is insufficient to define the position or even the existence of the

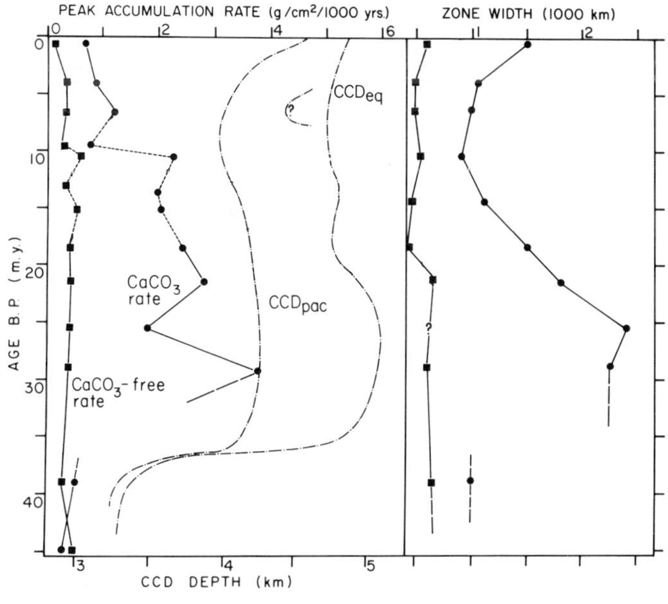


Fig. 8. Variation with time of CCD between 3° N and 3° S ($CCD_{eq}$), CCD north of 4° N and south of 4° S ($CCD_{pac}$), with width and maximum accumulation rate of equatorial zone of maximum deposition. Black dots: carbonate zone; squares: carbonate-free zone. Dot-dash curve with question mark: possible high position of $CCD_{eq}$.

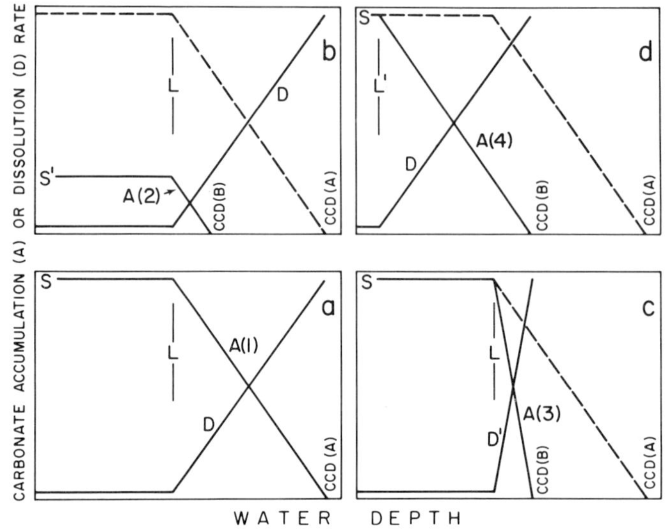


Fig. 9. Simplified model of carbonate dissolution (D) and rate of accumulation (A). CCD: calcite compensation depth; L: lysocline; S: rate of supply of carbonate. A shift of the CCD from CCD(A) to CCD(B) may result from: a decrease in S (Figure 9b); an increase in the depth gradient of the solution rate (Figure 9c); or a rise of the lysocline (Figure 9d).

lysocline. Thus, although the variation of the dissolution rate gradient with time can be quantitatively estimated, the effects of changes in lysocline depth and surface input cannot be separated.

In Figure 11, the D and S values of numerous plots of this type for the equatorial zone (3°N to 3°S) have been plotted as a function of age. The $CCD_{eq}$ curve is the depth where the accumulation rate extrapolates to zero. It is quite similar to the $CCD_{eq}$ curve of Figure 8, although different in detail (mainly because the curve of Figure 8 is based on a value of 20% $CaCO_3$ as the limit). Isopleths for the depths of 2 and 6 $g/cm^2/1000$ yrs accumulation rates were calculated from the equation; 6 $g/cm^2/1000$ yrs is a value close to the present lysocline, while 2 $g/cm^2/1000$ yrs is representative for the present equatorial zone of maximum deposition.

Figure 11 clearly shows that the dissolution rate has varied greatly (by almost a factor of 30) during the last 35 m.y., notwithstanding the impression of stability given by the $CCD_{eq}$. The rapid drop of the CCD at the end of the Eocene was due to a threefold drop in the vertical gradient of the dissolution rate even though the rate of supply (curve at bottom) also decreased somewhat. Of course, the lysocline could have risen almost to the surface to produce the same effect—but this seems unlikely. During the early Oligocene, the supply rate increased somewhat (or the lysocline dropped) to a maximum near 30 m.y., but the effect was largely offset by the increased vertical gradient of the dissolution rate. It is likely that the supply rate was the

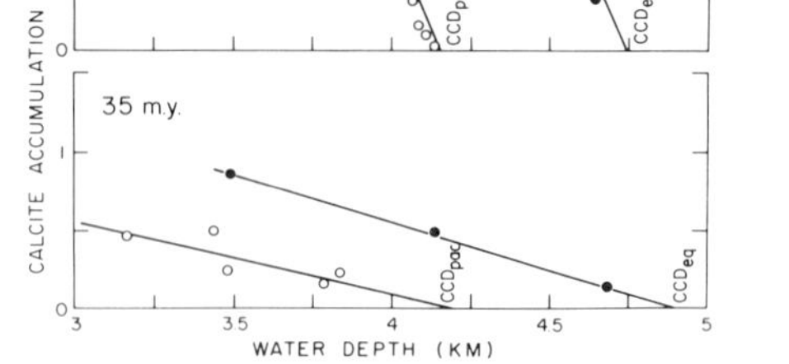


Fig. 10. Variation in the accumulation rate of calcium carbonate as a function of paleodepth for samples within 3° of the equator (solid circles) and 4–10° from the equator (open circles) during the early Oligocene (35 m.y.) and early Miocene (18–20 m.y.). The variations of the slope of the equatorial best-fit line ('dissolution rate') and of the accumulation rate extrapolated to the sea surface ('extrapolated surface supply') during the Cenozoic are shown in Figure 11.

# 25. *Sea Water, Sea-Floor Spreading, Subduction, and Ore Deposits*

J. B. Corliss

*School of Oceanography, Oregon State University, Corvallis, Oregon 97331*

*Abstract.* There is growing evidence that sea water hydrothermal systems serve to extract and transport metals initially disseminated in newly erupted submarine volcanic rocks, and that this is a fundamental geochemical process which influences the composition of sea water and has led to the formation of a variety of economically important ore bodies now preserved on the continents. This evidence includes: (1) observations of the effects of hydrothermal circulation in both dredged and cored oceanic crustal rocks; (2) the nature of the nearly ubiquitous basal metalliferous sediment layer in the oceans; (3) the pattern of heat flow over spreading centers; (4) anomalies in vertical profiles of suspended matter and dissolved gases in the deep oceans; (5) observations of submarine hot springs in Indonesia and the Mediterranean and on the crest of the Galapagos rift; (6) experimental investigations of sea water–basalt interaction; and (7) observations of ore deposits which can be interpreted as resulting from submarine hydrothermal systems. Some examples of such ore deposits are: massive sulfides in Precambrian greenschist belts such as the Noranda deposits of the Abitibi belt in Canada; gold ores in possible Precambrian 'metalliferous sediment' of the Homestake mine of South Dakota; the Precambrian Jerome, Arizona copper ores; the massive sulfides of the Troodos ophiolite complex on Cyprus, deposited in Cretaceous seas; and the Miocene Kuroko ore bodies in Japan. The hydration of oceanic crust during this process leads to the recycling of sea water first into the calc-alkaline magmas which form plutonic and volcanic rocks overlying subduction zones, and ultimately into the residual chloride-rich aqueous fluids which form during crystallization of these magmas. These fluids contribute to the formation of ore deposits accompanying the magmatic activity. These include the porphyry copper ore bodies and related vein-type deposits of western North and South America and the southwest Pacific.

Fig. 4. Hypothetical restorations of the southeastern Pacific showing opening of the Scotia Sea based upon sea-floor magnetic anomaly patterns [from *Pitman et al.*, 1974], interpretations of *Herron* [1972*a*, 1974] and DSDP Leg 35 results (courtesy C. Craddock), together with evidence from onshore geology. Dashed lines in Scotia Sea are undated anomaly trends after *Barker* [1972*b*]. S.G. is South Georgia microcontinent; S.S., South Shetland arc; S.O., South Orkney Islands. Presently active East Pacific spreading ridge lies west of area of A, having jumped about 25–30 m.y. ago.

20 to 40 m.y. to move South Georgia 2,000 km. A faster *average* rate is unrealistic; therefore, it is probable that South Georgia separated *at least* 20 m.y. ago.

*Antarctandes.* The Antarctic Peninsula south of the Scotia Sea shows many similarities with the southern Andes, and the two were certainly more closely connected in the past. *Dalziel and Elliot* [1973] have provided a recent detailed summary of the evidence for former connection, which lies chiefly in the strong similarity of Jurassic volcanic rocks already noted, widespread dioritic batholiths virtually identical both petrologically and chronologically (Figure 3), similarity of tectono-stratigraphic zones, and Cenozoic fossil plants and marine invertebrates. Very *unlike* Tierra del Fuego is surprisingly open folding *and* the lack of prominent ophiolites (thus apparently the absence of a late Mesozoic marginal ocean basin here unless it be the Weddell Sea). Also different is the presence along the northwest side of the Peninsula of a sizable Cenozoic volcanic archipelago, the South Shetland Islands (Figure 3). Volcanic rocks with associated fossil floras ranging from Eocene to Miocene and sea-floor magnetic anomalies to the northwest point to subduction of lithosphere beneath the Shetlands since at least the early Cenozoic (Figure 4C).

*How old is the Scotia Sea?* At a 1967 UNESCO meeting at Montevideo on continental drift, I argued that, by analogy with the rest of the circum-Pacific rim, the breakup of the formerly continuous Andean orogenic belt to form the Scotia arc most probably began in mid-Cenozoic time [*Dott,* 1972]. While new evidence seems to support that conclusion in a general way, it is now clear that the East Pacific rim is more complex and heterogeneous than it seemed in 1967. The oldest dated volcanic rock from the South Sandwich arc is only 4 m.y. old [*Baker,* 1968], and the oldest discernible north-south-trending magnetic anomalies in the eastern Scotia Sea just west of the arc (Figure 4A) are correlated with better-dated ones elsewhere that are 8 m.y. old. Anomalies in the Drake Passage between Cape Horn and the South Shetlands seem to date back at least 20 m.y. according to *Barker* [1972*b*] and *Pitman et al.* [1974]. Therefore, the western Scotia Sea floor seems to be at least 20 m.y. old.

The late Cenozoic sea floor in Drake Passage was generated along a northeast-trending spreading ridge recognized by *Griffiths and Barker* [1972]. This ridge extends diagonally across the western Scotia Sea and terminates near the middle of the North Scotia ridge at the western end of the South Georgia microcontinent (Figure 4A). It appears most probable that the South Georgia block was moved eastward along a major transform by spreading from this ridge at a moderately high rate in the past. At present the low level of seismicity in the Scotia Sea region and the patterns of magnetic anomalies, fragmentary as they are, suggest that this ridge is virtually inactive. The meager sea-floor evidence indicates that Antarctica and South America separated no later than 20 m.y. ago (early Miocene); *Herron* [1974] concluded that the split probably occurred between 30 and 50 m.y. ago. Considering the structural and stratigraphic evidence from the land on each side of the Scotia Sea, the major break would seem to have begun in early Oligocene time (about 35 m.y. ago) because it appears that Andean tectonic and stratigraphic zones remained intact at least through the Eocene. Moreover, around the north end of the Antarctic Peninsula, just as in Tierra del Fuego, late Cenozoic sediments are totally unalike, and are unconformable upon, older ones, while Pliocene–Recent basalts (e.g., James Ross Island Group) also are widely scattered and discordant with older terranes [summarized in *Dalziel and Elliot,* 1973].

*Synthesis.* We see then that the Scotia Sea region represents a situation still different in important ways from the other portions of the eastern Pacific rim discussed herein. Perhaps part of its uniqueness lies in the fact that the area lay near the junction of a region that was to become three separate continents—South America, Antarctica, and Africa. The complex geometry of Cenozoic plates here must be better defined, however, before the history of the Scotia Sea region can be satisfactorily deduced; we are forced to speculate until new sea-floor data become available.

Figure 4 offers a hypothetical sequential restoration of the Scotia Sea region suggesting a manner in which South Georgia was separated from Tierra del Fuego and carried eastward along a transform postulated by *Wilson* [1966]. The transform seems to have resulted by spreading from the northeast-trending ridge described above, which was the third arm of a triple junction that must have lain slightly west of the present site of the Scotia Sea (Figure 4C; see also *Herron* [1974]). That triple junction seems to have been destroyed by shearing and/or subduction in late Cenozoic time. If the separation of

South Georgia and general breakup of the Andean belt occurred around 20–35 m.y. ago as postulated above, then this event would seem to correlate roughly with the major jump of the East Pacific ridge farther north [*Herron,* 1972*a*; *Handschumacher,* this volume] and the initial collision of southern South America with the Chile rise (Figure 4B) as well as the collision of North America with the East Pacific ridge. Indeed, it may have been the encounter of the triple junction shown in Figure 4C that initiated the opening of the Scotia Sea. Seemingly these collisions had profound effects upon the entire plate-motion regime of the Pacific. For reasons that are not clear, seismicity and spreading were greatly reduced subsequently in the Scotia Sea region. Based upon fault-plane solutions of recent earthquakes, *Forsyth* [in press] postulates that *both* the North and South Scotia ridge transforms show left-lateral motion and that there is not a single, clear boundary between the South American and Antarctic plates at the present time. Rather, he believes that the entire region is in a transitional stage following the ridge–continent collision. Forsyth envisions a separate "slowly-moving Scotia plate with uncertain boundaries," and which seems to suffer non-rigid internal deformation.

## CONCLUSIONS

When plate theory was formulated around 1970, the eastern Pacific rim appeared rather homogeneous. Superficially, structural trends and facies belts seemed more or less continuous from the Aleutians to Antarctica. Moreover, certain events were at least broadly synchronous. For example, all around the Pacific the greatest bulk of granitic batholiths date from the latest Mesozoic, and the mid-Cenozoic tectonic reorganization seemed almost universally evident. Nonetheless, seismicity and volcanism show compartmentalization today, and, when examined in more detail, the ancient record of the Pacific rim is also seen to be subdivided into segments that differ markedly in detail even while showing certain broad similarities.

Unusually large exposed areas of ophiolites and melanges of the Franciscan complex of California and Oregon may simply be due to the nearly unique San Andreas fault system. Perhaps that faulting has raised a large volume of trench rocks by oblique slip that normally are not exposed (as in Peru and northern Chile). The locus of ophiolites inland in Tierra del Fuego reflects rifting of a late Mesozoic marginal ocean basin, but why such rifting occurred here and not more generally is unclear. The different positions of batholiths relative to coastlines noted at the outset may be due to large transform or transcurrent motions as on the central California coast (Figure 2). The extreme uplift of the central Andes may be the result of complications in the subducting plate beneath the sharp bend of the Pacific rim in this region. *Rodriguez et al.* [this volume] suggest that subducting plates may be wrinkled like a curtain or torn apart at points of sharp bends in arc–trench systems. It may be noteworthy that the very high central Andes, Mt. McKinley in Alaska, and the highest Himalaya all occur at or near sharp concavities of plate margins.

Lack of deep seismicity beneath the Cascade volcanic arc corresponds with present dormancy of the volcanoes. This may only represent a temporary lull, but it may also indicate something unique about the spreading rate or coupling characteristics between plates there. The absence today of a companion trench seems to be the result of high rates of sediment input from the Columbia and other rivers draining humid, glaciated hinterlands. The paucity of seismicity between the Chile rise and Cape Horn may be related in part to the clockwise shift of spreading direction around 25–30 m.y. ago in the southeastern Pacific [*Herron,* 1972*a; Handschumacher,* this volume] or to consumption of the southern Chile rise [*Forsyth,* in press], producing a nearly tangential and much slower motion between the Antarctic and Americas plates in southern Chile (Figure 4).

A direct linkage between sea-floor spreading at ridges and diastrophism at convergent plate margins is debatable. It is an open question whether major changes on one plate are felt at great enough distances to cause wholesale reorganization of several plates. And if so, then how fast are such changes propagated? Much empirical evidence suggests that long-distance (perhaps even global) interactions do occur and that propagation time may be of the order of at least 10 or 20 m.y. While the evidence is all circumstantial, to be sure, it is intriguing to wonder, for example, if the generation of circum-Pacific batholiths was somehow causally related to the speedup of Pacific spreading in Late Cretaceous time. One also wonders if major late Cenozoic tectonic changes around the Pacific do not reflect a combination of ridge jumping and continent–ridge collisions, which happened to coincide very roughly in time. Furthermore, changes of rate, direction, or vertical angle of subduction may account for some of the segmentation of orogenic belts, and surely the consumption of a triple junction would affect a convergent margin.

If no preserved ocean-basin spreading chronology is present, as for basins older than 200 m.y., then we naturally look to the old convergent margins (orogenic belts) for clues about the ancient oceans. But only if there is a valid basis for assuming direct causal interactions between oceanic spreading and tectonic events on those convergent margins can there be any hope of deducing ancient oceanic plate patterns. Testing the reality of long-distance plate interactions for the past 200 m.y., therefore, seems to be a matter of first importance.

*Acknowledgments.* Work in Oregon, California, and Nevada on which this paper is partly based was supported by the Graduate School of the University of Wisconsin from funds granted by the Wisconsin Alumni Research Foundation. The research in South America and Antarctica was made possible by a series of grants from the Office of Polar Programs of the National Science Foundation (current grant number is OPP72-05799). I am, of course, greatly indebted to my current co-workers in the Scotia Sea region: I. W. D. Dalziel, R. D. Winn, Jr., R. L. Bruhn, and M. J. de Wit, for constant stimulation. R. D. Winn, Jr. read an early draft, and I. W. D. Dalziel, Tj. H. van Andel and W. R. Dickinson reviewed the manuscript. Discussions with E. M. Herron, D. W. Handschumacher, D. W. Forsyth, and others at the Woollard Symposium were especially helpful in developing the interpretations presented here, as were subsequent discussions with Campbell Craddock. D. W. Forsyth provided a copy of an important manuscript, which is in press. Last, it is a great pleasure to acknowledge my debt to George P. Woollard for first encouraging me about 1960 to undertake research in the Scotia Sea region.

# 5. VOLCANOLOGY AND PETROLOGY

## *27. Some Highlights in the Development of Volcanology in the Pacific Area*[1]

GORDON A. MACDONALD

*Hawaii Institute of Geophysics, University of Hawaii, Honolulu, Hawaii 96822*

Originally, it was intended to review in this paper the entire development of volcanology in the Pacific area, but it quickly became apparent that it is not possible to do so within the limit of length set for the papers in this volume. Considering that about 80% of the world's recent and active volcanoes lie in and around the Pacific, it is hardly surprising that the amount of volcanological material is enormous. A review much longer than that possible here could easily be written on any one of a dozen parts of the Pacific area, and a complete review could constitute a very lengthy volume. The review presented here is limited to some of the early work in the mid-Pacific, and to a few of the major problems of volcanology.

The early explorers, such as *Cook, Vancouver,* and *La Perouse,* contributed little to volcanology. Their interests were largely geographical. But it is sad to pass over with bare mention the work of some of the truly great investigators: *Charles Darwin* [1844], who recognized gravitative crystal differentiation in the lavas of the Galapagos and the alteration of glassy basaltic ejecta to what *Sartorius von Waltershausen* later called palagonite (according to *E. D. Jackson* [1971], crystal settling had previously been described in the Canary Islands by *Leopold von Buch* [1825]); *Alexander von Humboldt* [1824, 1845], whose work on the volcanoes of Mexico, Central America, and South America helped disprove *von Buch's* [1818] craters-of-elevation hypothesis; *Ferdinand von Hochstetter's* [1864] superb work on the volcanic geology of New Zealand; the descriptions of volcanoes and volcanic rocks by *Ferdinand, Baron von Richthofen* [1868]; the work of *Alphons Stübel* [1897, 1906] on the Andean volcanoes and *Karl Sapper* [e.g., 1926, 1927; *Sapper and Termer,* 1930] on those of Central America, of great importance in demonstrating the structure and mode of eruption of continental volcanoes; the work of *C. N. Fenner* [1923] in Alaska and *Patrick Marshall* [1935] in New Zealand on ash flows; the work of *Fusakichi Omori* [1911–1920] and *Bundjiro Koto* [1916] on ground deformation associated with volcanic eruptions; *Hisashi Kuno's* [1959] studies on the distribution of types of volcanic rocks in island arcs and the origin of magmas; and many others. It is particularly sad to pass over the superb work of the Volcanological Service of the Netherlands Indies. Truly amazing is the tremendous contribution to volcanology of this very small group of dedicated men, which at no time numbered more than three college-trained scientists, but over the years included at one time or another such famous volcanologists as *G. L. L. Kemmerling, N. J. M. Taverne, C. E. Stehn, R. W. van Bemmelen,* and *M. Neumann van Padang.*

### EARLY WORK IN HAWAII

In Hawaii, as elsewhere, we must pass over many of the early workers with bare mention, or none at all. *Archibald Menzies,* the botanist of the Vancouver expedition, and a small group of seamen were the first Europeans to climb Hualalai and Mauna Loa volcanoes, on the island of Hawaii. *Menzies* recognized them to be volcanoes, and described them in his journal [*Menzies,* 1907, 1909]. Although the journal was not published for more than a century, at least some of the information in it apparently was available to other writers. *G. Vancouver* [1798] and *J. Turnbull* [1805] recorded the existence of volcanic activity on Hawaii, but the first published descriptions of the volcanoes were by *William Ellis* [1825, 1826] and *Joseph Goodrich* [1826]. Both were Protestant missionaries. *The Rev. William Ellis* had been loaned to the Hawaiian Mission by the London

[1]Hawaii Institute of Geophysics Contribution No. 705.

Missionary Society to aid in establishing an alphabet and grammar for the Hawaiian language. He accompanied a group of American missionaries on a tour around the island of Hawaii in 1823, and his broad interests led him to record incidents and facts bearing on many facets of the ethnology and natural history of Hawaii. *The Rev. Joseph Goodrich* had studied mineralogy and geology at Yale University, and his account of the volcano was in the form of a letter to his former professor, *Benjamin Silliman,* who published it, along with excerpts from *Ellis'* account, in "Silliman's Journal." *Goodrich* was the first to describe Mauna Kea, and to note the inclusions of coarse-grained 'granite' in its pyroclastic materials.

*George Anson, Lord Byron,* visited Hawaii in 1825 as commander of HMS *Blonde.* He described Kilauea caldera and the volcanic activity in it and published the first map of the caldera, which had been made by Lt. Malden, one of his officers [*Byron,* 1826].

*The Rev. Titus Coan* observed the results of the Kilauea eruption of 1840 in the course of riding his circuit from Hilo around the south end of the island. Later in the year he made the friendship of *J. D. Dana,* and for the next four decades he maintained a continuous chronicle of Hawaiian volcanic activity in letters written to *Dana* and published in the American Journal of Science, and other notes sent to other publications [*Coan,* 1841, 1843, 1852, 1869, etc.]. (For a more complete listing of publications by *Coan* and other early workers on Hawaiian volcanoes see *G. A. Macdonald,* 1947.)

The United States Exploring Expedition visited Hawaii in 1840–41. *Lt. Charles Wilkes,* the commander of the expedition, was much interested in the volcanoes. His narrative contains accounts of the activity he observed in Kilauea caldera, and of the eruption on the east flank of the volcano earlier in the year, as well as a new map of Kilauea caldera, and a description and the first map of the summit of Mauna Loa [*Wilkes,* 1845]. He proposed the name 'pit crater' for those craters that were formed by collapse and, in his words, "never throw out lava."

The geologist of the United States Exploring Expedition was *James Dwight Dana.* His report, published in 1849, contains many contributions to Pacific volcanic geology, such as his description of the volcanic terrane around Bay of Islands in New Zealand, and of the Samoan Islands, particularly Upolu.

*Dana's* most important contributions to volcanology were in Hawaii. Among other things, he described [1849] the lava lake activity and the formation of driblet spires and Pele's hair, and recognized that many eruptions take place from points on the walls of the caldera tens to hundreds of feet above other open vents on the caldera floor, and also that some cinder and ash cones are formed by disruption of lava flows by steam explosions where the flows enter water. (These were later named littoral cones by *C. K. Wentworth* [1938].) *Dana* described the principal types of Hawaiian lava flows, although he did not use the Hawaiian names for them, and he called attention to the quiet character of Hawaiian eruptions. He remarked on the independence of action of Kilauea and Mauna Loa, and pointed out that the open vent of Kilauea did not act as a 'safety valve' to deter eruptions of Mauna Loa.

*Dana* [1849] recognized that the island of Oahu consists of the eroded remains of two big contiguous volcanoes like those of the island of Hawaii. He concluded that the seaward sides of both volcanoes had been destroyed by subsidence, and believed the windward Pali to be a fault scarp with displacement of more than 3,100 feet—a view only disproved by detailed mapping by *Harold Stearns* and his associates nearly a century later [*Stearns and Vaksvik,* 1935]. He also recognized the tuff cones of Diamond Head, Punchbowl, the Koko Craters, and Ulupau Head. Moving on, he recognized Kauai to consist of a single big volcano, and described the low-dipping lava beds that built it, the presence of many phenocrysts of olivine, and the many dikes that cut the lavas.

*Dana* [1849] concluded, from the degree of erosion of the successive volcanic mountains, that volcanic activity had progressed southeastward along the Hawaiian chain. The progression was concurred with by later workers [e.g., *Stearns,* 1946], and eventually confirmed by potassium–argon dating [*McDougall,* 1964]. *Dana* pointed out that, from Oahu southeastward to Hawaii, the major volcanoes form two subparallel curved lines, which he called the Loa and Kea ranges. The Loa (southern) range can be extended northwestward through Kauai, with a sigmoidal form [*Jackson et al.,* 1972]. Farther northwest a series of similar sigmoidal lines can be drawn through the major volcanic centers. The lines overlap and are offset en echelon [*Jackson et al.,* 1972]. The en echelon arrangement was recognized by *Dana* [1863], although the hydrographic data available to him did not reveal the sigmoidal form.

Early in its existence, the U.S. Geological Survey decided that a systematic study should be made of the volcanoes of the Cascade Range. In preparation for this, *C. E. Dutton* was sent, in 1882, to study the Hawaiian volcanoes. Unfortunately, the Cascade investigation did not materialize, but *Dutton's* time in Hawaii was fruitful. His report [*Dutton,* 1884], which appeared in the *Fourth Annual Report of the Geological Survey,* is a beautifully written description of all the volcanoes on the island of Hawaii. In it, he gave a partial summary of the historic eruptions, and pointed out that flank eruptions of Mauna Loa usually occur shortly after summit eruptions. Thus, he probably was the first to attempt to find a pattern in the eruptive history of the active volcanoes. He dealt with the problem of maintenance of heat in the long-continued activity of the Halemaumau lava lake, and concluded that it was the result of convection currents and the rise of gas bubbles, thus anticipating parts of the "two-phase convection" idea later advanced by *R. A. Daly* [1911]. The spasmodic character of eruptions in general he attributed to the activation of successively deeper levels of magma in the reservoir by, in his words, "the acquisition of elastic expansive force." He probably was the first to call attention to this important mechanism, by which progressive unloading results in vesiculation of successively deeper portions of the magma body, with resultant eruption of huge volumes of pumice and ash, as in the great eruption of Krakatau (which took place while he was preparing his report). *Dutton* examined and rejected the hypothesis, then very popular, that volcanic activity results from the absorption of vadose water into underlying magma. He also introduced into geologic literature the Hawaiian names for types of lava flows—pahoehoe and aa. Some geologists objected to the names, calling them "barbarous" [*Bonney,* 1899, p. 79]. *Thomas A. Jaggar* [1917*b*] proposed to improve the cultural quality of the nomenclature by substituting the Greek-rooted words, dermolith and aphrolith, but we still use pahoehoe and aa.

*Dutton's* report inspired the British consul in Honolulu, *William Lothian Green* (author of the *Tetrahedral Theory of the Earth*) to publish his own extensive observations on Hawaiian volcanoes. In volume 2 of *Vestiges of the Molten Globe* [1887] *Green* argued strongly for the anhydrous character of the gases given off in Hawaiian eruptions, an idea advanced again a quarter of a century later by the Swiss geochemist, *Albert Brun* [1911]. It was disproved for Hawaiian gases when *Day and Shepherd* [1913] published the results of their analyses of gases collected on the Halemaumau lava lake. It is noteworthy that, although most analyses of volcanic gases since then have contained large proportions of water, some recent analyses, such as those by *H. Tazieff* and his co-workers at Stromboli and Etna, have contained very little water [*Tazieff,* 1971], and the problem of the origin of the water in all collections is still very much alive.

*Green* [1887] pointed correctly to the small total amount of gas in the erupting magma as the reason for the non-violence of Hawaiian eruptions, and he described the gravitative sinking of olivine phenocrysts in some flows as evidence of the high fluidity of the lava. He believed that the principal cause of the rise of magma is the weight of the surrounding crust on an underlying fluid basalt layer. He also gave us our first description of what almost certainly was the first stage in the formation of pillow lava, as the liquid basalt drooled in great blebs over the low sea cliff during the 1859 eruption of Mauna Loa. The same sort of behavior has been observed in recent years in flows on the south shore of Kilauea.

Shortly afterward, *J. D. Dana* appeared again on the Hawaiian scene. In a series of papers published in the American Journal of Science in 1887–1888 [see *Macdonald,* 1947], he recounted his observations during a visit in 1887, and summarized the information on all Hawaiian historical eruptions up to that time. He described the cycle of Kilauea as: (1) a rise in level of the liquid lava and the floor of the crater; (2) a discharge of the accumulated lava down to some level in the conduit determined by a [flank] outbreak; and (3) a down-plunge of more or less of the region undermined by the drainage [notably the caldera floor]. In a general way, our picture remains much the same today. He discussed the volcanic mechanism, and the influence of volatiles, concluding that the lava fountains of Mauna Loa that had been observed up until then were larger than those of Kilauea because they contained more gas. He believed the gas cloud to be "at least ninety-nine percent steam," but that the steam is largely derived from water of meteoric origin.

On Maui, *Dana* [1889] suggested that Haleakala Crater and the Koolau and Kaupo valleys leading out of it are the result of rifting of the mountain, with the two major segments moving apart in a northeast-southwest direction. A decade later *Whitman Cross* visited the crater and, in a report not published until 1915, suggested that it was formed by erosion—a hypothesis later proved by *Harold Stearns* [1942]. The volcano has indeed been rifted, with great distension, but the distension has been in a northwest-southeast direction, roughly at right angles to the direction suggested by *Dana.*

*Dana's* well-known *Characteristics of Volcanoes* [1890] is largely a reprinting of his articles in the *American Journal of Science.*

Comprehensive records of historic eruptions in Hawaii were compiled by *William T. Brigham* [1909] and *Charles H. Hitchcock* [1909]. Both have been of great value to later workers.

### THE HALEMAUMAU LAVA LAKE

From the time of the first visit of foreigners to Kilauea, in 1823, to the turn of the century, there was nearly always an active lava lake in the caldera, sometimes as many as six. The lakes were described in general terms by several writers, including *Ellis* [1825], *Goodrich* [1826], *Count Strzeleki* [1838], *Wilkes* [1845], *Dana* [1849], and *C. S. Lyman* [1851]. However, most descriptions were based on visits of only a day or two, and such statements as that by *Ellis,* that the southwestern and northern parts of the crater floor were "one vast flood of burning matter, in a state of terrific ebullition, rolling to and fro its 'fiery surge' and flaming billows" [1917, p. 177], while picturesque, tell us little except that the lakes were very active. The first good account of the lakes, based on repeated and extended visits, is by *Green* [1887]. He describes the streaming of lava in the lake, the stationary and moving fountains, the islands, the blowing cones, the foundering of crust fragments at sinkholes, and the occasional periods of overturn when the gray crust over the lake abruptly disintegrates and sinks, transforming the surface for a brief time into orange-hot liquid—all features that are very familiar to recent observers.

During his visit to Kilauea in 1909, *R. A. Daly* became acutely aware of the problem of how the high temperature of the lava lake is maintained, in the case of Halemaumau for years on end. Examining various methods of heat transfer from depth, including thermal convection in the viscous magma, he concluded they were inadequate. In their place *Daly* [1911] suggested "two-phase convection." At depth, gas comes out of solution in the magma, and vesicles form; the vesiculated magma, lighter than the non-vesiculated magma, rises. At the surface the magma loses its gas and cools, and becoming heavy, sinks.

With the establishment of the Hawaiian Volcano Observatory, in 1912, *T. A. Jaggar* immediately turned his attention to the Halemaumau lava lake, and the origin of its heat. Since the days of *Ellis,* most observers had believed the lakes to be shallow, largely on the basis of the behavior of the currents in them, but real proof was lacking. *Daly* [1911] diagrammed the Halemaumau lake as occupying a shallow saucer-like basin, fed by a central pipe with persistent lava fountains on the lake surface above it, and enclosed in solid older lava. *Jaggar's* [1917*a*] observations soon confirmed the shallowness of the lake, but also established its variability and complexity. At times the bottom of the lake was upheaved to form islands; at times of rapid lowering of lake level, landslides from the collapsing crater wall caused the liquid to surge far up the opposite bank of the lake, momentarily revealing the shallow bottom; and during larger scale rapid subsidences the liquid drained completely away, leaving the bottom exposed. Moreover, it soon became apparent that the lake bottom and sides were not solid, but were mobile, though their mobility was far less than that of the lake itself. Loading of one portion of the lake margin by lava overflows caused it to sink, while the opposite side rose in compensation. Islands of former lake bottom rose and sank, and at times moved majestically across the lake as part of what appeared to be a slow convectional circulation. Earlier observers, including *Brigham* [1909] and *F. A. Perret* [1913], had believed these islands to be floating, like icebergs, in the liquid lake, perhaps buoyed up in part by entrapped gas; but the independence of their motion from that of the liquid streaming around them, their rise and fall, their mergings with the bank and separations from it, all indicated to *Jaggar* [1917*a*] that they were actually projecting parts of a mobile lake bottom.

To crown the demonstration, occasional rapid subsidences resulted in the entire ensemble of liquid lake and semi-solid bottom sinking hundreds of feet within a few hours (e.g., ~ 120 m in June 1916), leaving a great fuming pit into which the collapsing walls sent landslides that consisted partly of black solid older rock, but partly also of red-hot still-plastic fragments. Because the semi-solid material formed a sort of bench around the lake within the older-rock walls of the crater, *Jaggar* [1917*a*] called it "bench magma," and the liquid of the lake, "lake magma." Later [1920] he changed the names to epimagma and pyromagma. Thus the picture became one of "hypomagma" at depth, in which the gases were still in solution, grading upward into vesiculating pyromagma, which rose through a series of small shafts leading to a shallow lake on the surface of a mass of semi-solid incandescent epimagma, which was constantly being added to by overflows from the lake onto the bench and by accumulation of sunken crusts on the lake bottom.

But *Jaggar* was still not satisfied. He wanted direct measurements of the depth of the pyromagma lake and determinations of the physical character of its bottom. In January 1917 conditions made it possible to go out onto the bench of epimagma and thrust a long one-inch galvanized steel pipe over the spatter rampart at the edge of the lake into the pyromagma. The experiment encountered difficulties: the heat was intense, the crust on the lake was too tough to thrust the pipe through; the pipe could be thrust through cracks in the crust, but the cracks tended to close and make it impossible to withdraw the pipe (two pipes were lost); edge fountains spattered the nearby bench from time to time, and occasional overflows of lava came over the rampart. But *Jaggar* enlisted the aid of loyal friends, themselves volcano buffs, and the job was accomplished. The lake was found to be about 14 m deep, with a highly viscous plastic bottom [*Jaggar,* 1917*a*].

The pipe experiment also made it possible to measure the temperature at various depths in the lake. A temperature of about 1000°C had been measured with a thermo-element lowered into the surface lava of the lake near the source fountains by *E. S. Shepherd and Perret* in 1911 [*Shepherd,* 1914], and in 1912 *Day and Shepherd* [1913], using an optical pyrometer, had obtained readings up to 1185°C in grottoes where gases were burning. *Jaggar* packed series of Seger cones into his pipe at intervals, then thrust the pipe into the lake. The temperature was found to be about 1000°C at the surface, decreasing to 860°C at a depth of 1 m, and then gradually increasing again to 1170°C at 13 m. Temperatures of 1100° to 1200°C were obtained in fountains over sinkholes and in grottoes [*Jaggar,* 1917*a*].

The high surface temperature of the lake was attributed to surficial heating by oxidation of gases. Flames had been observed over the lake by *Brigham* [1887], and were abundantly confirmed by *Jaggar* [1917*a*]. Analyses of gases collected from the lake in 1912 by *Day and Shepherd* [1913] had shown the presence of several percent of combustible gases (S, CO, H), and it is to the oxidation of these gases, by air entrapped in sunken crusts or drawn into the magma in the turbulent down-draft of sinkholes, that *Jaggar* [1917*a*] attributed the high temperature of the bottom layer and much of the maintenance of the heat of the lake.

Our picture of the processes in lava lakes remained essentially the same until the recent series of superb determinations of physical properties and crystallization sequence in these 'natural laboratories' by *Jaggar's* lineal descendents, the U.S. Geological Survey scientists at the Hawaiian Volcano Observatory.

### CALDERAS AND VOLCANO-TECTONIC DEPRESSIONS

The origin of the over-size volcanic craters known as calderas was long the subject of debate. The term was first used by *von Buch,* in 1818, for the Taburiente depression on the island of La Palma (Canary Islands). He believed it to have been formed by splitting open of the summit of an up-domed cone, but his hypothesis of craters-of-elevation was soon discredited by *G. Poulet Scrope* [1825] and *Sir Charles Lyell* [1830]. Many geologists now believe that the Taburiente basin was formed by erosion. But it is with the depressions directly related to volcanic action that we are concerned here.

Most calderas are considered to have originated either by explosive action, or by sinking in of the summit of a volcanic mountain. According to *Darwin* [1844], about 1800 the French explorer, *Bailly,* described the caldera of Mauritius, and attributed it to sinking of the summit of the volcano. *Darwin* agreed. In 1884, *Dutton* described the summit depressions of the Hawaiian volcanoes, and concluded that they formed by collapse. Thus there was good agreement on the origin of the calderas of oceanic shield volcanoes, where it was obvious that no large amount of debris had been thrown out by explosion. Furthermore, several episodes of abrupt sinking of the floor of Kilauea caldera had been actually observed. But around the calderas of some continental volcanoes there are enormous volumes of material thrown out by explosion.

The results of the great eruption of Krakatau, in 1883, were studied by an expedition dispatched by the Royal Society of London. Huge amounts of pyroclastic material were found, and a caldera 7 km across on the sea floor where part of the island of Krakatau had been. *J. W. Judd* [1888] believed that the caldera had been blasted out by the explosions that produced the tephra. But a more thorough study by *R. D. M. Verbeek*, a member of the Geological Service of the Netherlands Indies, demonstrated that only a small fraction of the tephra was debris from the old cones. *Verbeek* [1886] concluded that the eruption of a huge volume of magma, represented by some 15 $km^3$ of vitric ash and pumice, had removed support and allowed the overlying rocks to subside. *Dana* [1890] accepted *Verbeek's* explanation; and in later years it has been confirmed repeatedly by studies by *Charles Stehn* [1929], *R. W. van Bemmelen* [1929*b*], *C. A. van den Bosch* [1930], *M. Neumann van Padang* [1933*a*], and *Howel Williams* [1941].

*Von Richthofen* studied the southern end of Kyushu Island, Japan, in the early 1860's, and concluded that Kagoshima Bay is a subsidence basin related to volcanic action [*Richthofen*, 1903]. After the great eruption of Sakurajima, in the southern part of the bay, in 1914, releveling by the Japanese army enabled both *Koto* [1916] and *Omori* [1916] to demonstrate that the volcano and much of the bay had sunk more than 8 m, with the point of maximum subsidence nearly 10 km north of the summit of the mountain. Both believed that this indicated that the magma chamber feeding the volcano lay under the northern part of the bay. Later work by *T. Matsumoto* [1937, 1943] has identified two big calderas, respectively at the north and south ends of the bay. The formation of the calderas accompanied or followed the eruption of widespread blankets of vitric ash, pumice, and ash flows. The maximum sinking in 1914 was approximately at the center of the northern (Aira) caldera, and Sakurajima lies near its southern boundary. Between the Aira Caldera and the southerly Ibusuki Caldera, Kagoshima Bay is a graben-like structure, known as a volcano-tectonic depression. *Von Richthofen's* sharp geological intuition has been confirmed. It is noteworthy that he [1903] compared the bay with the Taal Lake basin in the Philippines, also a volcano-tectonic depression.

Farther north in Kyushu lies the well-known caldera of Aso, the subject of repeated studies. Again, one finds large deposits of 'mud lava' or 'Aso lava', now known to have been formed by ash flows, and along with them some layers of air-laid tuff and probably some lahar deposits. The enormousness of the deposits is emphasized by the fact that, even though they have been much eroded, their remnants still cover almost half of the island of Kyushu [*Williams*, 1941, p. 278]. *Williams*, who examined the area in company with *Matsumoto*, writes, "At the close of this eruptive cycle, and just before the collapse which produced the caldera, Aso must have been an unusually broad but low cone or complex of cones." It was an ignimbrite shield, like those recently described in the Tibesti region of north Africa by *Pierre Vincent* [1963]. *Williams* [1941] followed *Matsumoto* in his conclusion that the eruption of this great volume of magma resulted in the collapse that formed the caldera.

In his study of Crater Lake, in Oregon, about the turn of the century, *Josiah Diller* still adhered to the old idea of origin of the caldera by explosive decapitation of Mt. Mazama [*Diller and Patton*, 1902]. However, detailed studies by *Williams* [1942] have shown the sparsity of fragments of the old cone in the deposits of 42 $km^3$ of vitric ash and pumice. In his superb summary of the caldera problem, *Williams* [1941, p. 324] writes, "Clearly these volcanoes did not 'blow their heads off'; on the contrary, they were disemboweled and their heads fell in."

Still greater depressions also have been attributed to collapse of segments of the earth's crust owing to eruption of great volumes of magma from underlying chambers. They are commonly referred to as 'volcano-tectonic depressions.' As early as 1855, *Richard Taylor* [1855] had suggested that the volcanic zone of the Taupo area, on the North Island of New Zealand, had subsided because of the removal of magma from beneath it. *Von Hochstetter* [1864] recognized the area as a graben, and subscribed to the idea of its subsidence due to the eruption of large amounts of rhyolite—the rocks that *Marshall* [1935] later named ignimbrite. The concept has been disputed in some details, but as a generality it is accepted by most geologists today. *Van Bemmelen* [1929*a*, 1949] has drawn a similar conclusion as to the origin of the Toba, Ranau, and other basins that lie along the backbone of Sumatra. They are associated with great volumes of rhyolitic ignimbrite, but *J. Westerveldt* [1952] and others believe that the basins were already in existence before the eruption of the ash flows.

Recent workers have suggested that the Mono Lake basin, in California, also is a volcano-tectonic depression.

Many of the world's major volcanoes are aligned within major grabens, or are closely associated with them. But whether the grabens are the result of withdrawal of underlying magma, or whether both grabens and volcanoes are the result of crustal tension caused by some other force (such as an upwelling and diverging convection current in the mantle), is still a moot question. The latter seems the more probable.

### DOMES

The steep-sided accumulations of viscous lava now known as volcanic domes were first recognized in 1789 in the Auvergne region of France by *Count Reynaud de Montlosier* [cited by *von Zittel,* 1901], and their general character was correctly deduced by *Scrope* [1825]. The actual growth of domes was observed at Santorini in 1866, and described by *F. Fouqué* [1879]. During the eruption of Mont Pelée in Martinique in 1902, the growth of the dome and spine was studied and described by several geologists. *Angelo Heilprin* [1903] and *Jaggar* [1904] considered the dome to be a cone of pyroclastic material, but *Alfred Lacroix* [1904], whose study was by far the most comprehensive, concluded that the dome grew largely by expansion of a newly extruded mass of viscous lava enclosed in a solidified shell, through which many semi-solid protrusions were squeezed to form spines, which soon crumbled down to add to the volume of the dome. Although at first he agreed with *Heilprin* and *Jaggar, E. O. Hovey* [1903] soon came around to *Lacroix's* view, and *Jaggar* also later modified his opinion to agree with that of *Lacroix.*

The eruption of Tarumai volcano, on Japan's northern island of Hokkaido, in 1909, was described in Japanese by *Y. Oinouye* [1909] and *D. Sato* [1909], and in German by *Hidazo Simotomai* [1912]. *Jaggar* also studied the eruption about a month after its climax. *Simotomai,* who later wrote under his title of *Prince Tanakadate,* was to become the leading Japanese volcanologist of the early part of the 20th century. From 1914 to 1936 he recorded the activity of Japanese volcanoes, and reported many special studies of volcanic phenomena [e.g., *Tanakadate,* 1917, 1931, 1934, 1936], and late in life he prepared a report on the volcanoes of the northern Mariana Islands, then mandated to Japan [1940]. The most interesting feature of the 1909 eruption was the growth of a dome 200 m high in the crater of the volcano. Flow lines in the lava were nearly parallel to the summit of the dome, indicating lateral expansion of the growing mass [*Simotomai,* 1912; *Tanakadate,* 1917.]

In 1910, the eruption of Usu volcano, also in northern Hokkaido, contrasted strongly with that of Tarumai. Ground at the northern base of the mountain was upheaved to form a "Roof Mountain" 155 m high—probably the first actual observation of the formation of a laccolith. The eruption was described by *Simotomai* [1911], *Omori* [1911, 1920], and *Immanuel Friedlaender* [1912]. The new activity stimulated the study of older domes at the summit of Usu, which were shown to have formed not entirely by inflation and gradual expansion of a mass of very viscous lava, like that at Tarumai, but by essentially vertical protrusion of a nearly solid cylinder, carrying upward on its top gravel and sand that had formerly been the old crater bottom. They are true 'plug domes.' The contrast between the two types of domes was emphasized by *Tanakadate* [1930]. The growth of the "Roof Mountain" was duplicated in 1944 by an eruption on the east flank of Usu, except that during the formation of the "New Roof Mountain" (Showa Sin-zan) the roof of the laccolith eventually was ruptured and magma was extruded to form a surficial dome [*Minakami et al.,* 1951].

Many domes have been recognized in Central and South America. *Stübel* [1897] separated volcanoes into "monogenous" and "polygenous," the latter built by a series of eruptions, and the former essentially by a single eruption. Among monogenous volcanoes he included a group of "summit pyramids" which he considered to be domes. He described several of them among the high peaks of Ecuador, and later in Colombia. In Mexico, domes have been described on Nevado de Toluca by *A. Bergeat* [1907], on Colima and Popocatépetl by *P. Waitz* [1906, 1921], and on Orizaba by *Friedlaender* [1930]. The most famous of Central American domes is that of Santa Maria, in Guatemala, the growth of which has been closely followed since its inception in 1922 to the present [*Sapper,* 1926; *Sapper and Termer,* 1930; *Stoiber and Rose,* 1969].

The Bogoslof Islands, which lie in the Bering Sea just north of the Aleutian Islands, are domes at the summit of a submarine volcano. Part of Old Bogoslof originated in 1796, but an older island at the site was recorded by *Krenitzin and Levashef* in 1768 and by *Cook* in 1778 [*Byers,* 1959]. Another dome appeared above water in 1883 [*Hilgard,* 1884], and two more in 1906. The latter were studied by *Jaggar* [1908]. Still another dome, which grew in 1927, also was studied by *Jaggar* [1930, 1931]. Knowledge of domes in the Pacific area up to 1915 was summarized by *Sidney Powers* [1916].

The Katmai eruption of 1912 produced the famous ash flow of the Valley of Ten Thousand Smokes, and also a dome of glassy rhyolite, named Novarupta by *R. F. Griggs* [1918]. The dome was studied by *C. N. Fenner* [1923], who described a fan-like orientation of flow banding that indicated the dome to have grown by expansion upward and outward.

Exploration of the Indonesian volcanoes also revealed the presence of many domes. Some of them, such as those of Merapi, Galunggung, and Ruang, have been active in historic times. They have been described by *H. A. Brouwer* [1921]. The domes of Merapi were studied especially by *G. L. L. Kemmerling* [1921] and *Neumann van Padang* [1931, 1933*b*]. Starting in 1883, a new dome began growing in the crater of Merapi, and by 1909 the crater was filled and blocks from the crumbling dome were rolling down the flanks of the mountain. Shortly afterward a new dome started to grow to the west of the earlier one, and by 1920 it had spread onto the flank of the volcano. Collapse of the side of the growing dome repeatedly sent avalanches of red-hot rock fragments down the mountainside, taking many lives in villages near its foot. Thus there was recognized a third mode of formation of glowing avalanches (nuées ardentes) to be added to those already recognized at Mt. Pelée and St. Vincent.

In the mid-1920's *Williams* began his study of Lassen Volcanic National Park, in California. Lassen Peak itself is one of the largest domes on record, and other domes lie nearby. *Williams* [1932*a*] showed one of them, at Raker Peak, to have the onion-like structure found in artificial viscous protrusions [*Reyer,* 1888], but seldom found in nature. The Lassen dome is nearly devoid of flow lines, but the domes of Chaos Crags, at the north base of Lassen, show the fan structure noted at Novarupta and found in many other domes. The Lassen studies led *Williams* [1932*b*] to review the knowledge of similar structures worldwide. His paper is the best existing summary of the character and origin of domes.

## GLOBAL TECTONICS AND VOLCANISM

The spacial relationships of belts of recent mountain-building and volcanoes is obvious, and attempts to link them causally started early. Impressed by the upheaval of the sea bottom on the top of a growing volcanic dome near Santorini, *Antonio Lazaro Moro* [1740] suggested that all mountains, and indeed the continents themselves, have been elevated from beneath the ocean by volcanic forces. The primary relationship between mountain-building and volcanism was elaborated by *Elie de Beaumont* [1829] in the early 19th century, and emphasized by *Green* [1887] in the later part of the century. *Dana* [1863] distinguished between volcanic island arcs related to the circum-Pacific zone of mountain building, and the linear volcanic chains of the mid-Pacific.

The last two decades have seen a steady procession of papers treating the generation of magmas in the zones of andesitic volcanism around the Pacific and relating it in one way or another to the Benioff zone, at first simply as a zone of thrust faulting (usually considered overthrust, but occasionally underthrust), and later as the zone of subduction at the edge of plates of the spreading sea floor [e.g., *Kuno,* 1959; *Coats,* 1962; *Hatherton and Dickinson,* 1969]. Particularly in recent years agreement has been growing that the formation of andesitic magma does not depend on assimilation of sialic continental crust [e.g., *Macdonald,* 1960; *Gorshkov,* 1962, 1970], but that it is more likely the result of partial melting, at mantle depths and under appropriate pressures, of subducted oceanic crust and/or upper mantle, perhaps dependent on the presence of water [*Green and Ringwood,* 1969; *McBirney,* 1969]. From this vantage point it is interesting to look back at some earlier ideas.

*Arthur Holmes'* [1931] pioneering suggestion of the general concept of sea-floor spreading, later elaborated and more specifically related to mid-ocean rises by *R. S. Dietz* [1961] and *H. H. Hess* [1962], is generally recognized, but another very noteworthy hypothesis seems to be fading from memory.

About 1906, *Daly* started to put together his once-famous "eclectic theory" of igneous action, set forth at length in a paper on the nature of volcanic action [1911], and in his book, *Igneous Rocks and their Origin* [1914]. The theory was eclectic in the sense that he had adopted what he thought were the best ideas of others, but much of it was his own original contribution. The fundamental concept was characterized by *Daly,* himself, as the "substratum-injection hypothesis." The earth's crust is underlain, he believed, by a universal layer of eruptible basalt. To start with, he considered the basalt layer to be liquid, but later the demonstration of the passage of earthquake shear waves through that part of the earth led him to believe it to be vitreous. From the substratum, basalt magma rises into fissures, presumably opened by tension in the lower part of the earth's crust, and works its way to the surface. The diversity of volcanic products was attributed to differentiation of the original basalt, assimilation of foreign rock in the basalt magma, and differentiation of the resulting syntectic magmas.

To summarize the tectonic part of *Daly's* hypothesis [*Igneous Rocks and the Depths of the Earth,* 1933]: Beneath a rigid outer shell the underlying part of the earth spheroid becomes deleveled, probably as a result of differential heating, by radioactivity or other means. The high parts tend to be beneath the continents, and the continents tend to split and slide off the highs. The movement may be aided by convection currents beneath. Thus, segments of the outer rigid shell shift laterally over a more easily deformed substratum, fissures form behind the 'upstream' edge of the shifting rigid segment and basalt magma rises into them from the substratum, to form dikes, or pour out onto the surface to form voluminous accumulations of lava flows. Repeated dike filling forms new crust, and the lava flows form a scab over the rent. Thus the shifting continental segment leaves behind it a new crust that differs from the continental segment in its lack of sial. At the 'downstream' edge of the shifting mass, where it impinges against another rigid mass, folding and thrust faulting occur, and the front of the moving segment plunges downward, partly driven by the moving material behind it, but partly dragged down because it is denser than the substratum. Assimilation of crustal material in the substratum forms intermediate magmas, which rise to form batholiths or belts of andesitic volcanoes in or close to the folded zones.

Particularly if we do some word substitution: plates for rigid segments, spreading centers for the areas upstream from the moving masses, subduction zones for the downstream edges, and *J. Barrell's* [1914] term "asthenosphere" for the more easily deformed substratum, the similarity to modern ideas of the formation of the Atlantic basin is striking. Though the basic idea is still applicable, the similarity is less for the Pacific. The information then available forced *Daly* to consider the Pacific floor a passive element, being actively overthrust by the continents, but he did believe that it was the denser sub-Pacific crust that was being subducted around its edges [*Daly,* 1934].

It is as interesting to speculate on the evolution of scientific ideas as it is on the evolution of ocean basins!

*Acknowledgments.* I am grateful to my friends, Donald W. Peterson and E. Dale Jackson, of the U.S. Geological Survey, for critical reading of this manuscript and for useful suggestions. Jackson's extensive knowledge of the history of petrological and structural thought has been especially helpful.

# 28. *Linear Volcanic Chains on the Pacific Plate*

EVERETT D. JACKSON

*U.S. Geological Survey, 345 Middlefield Road, Menlo Park, California 94025*

*Abstract.* It has been proposed that linear volcanic chains on the Pacific plate, of which the Hawaiian chain is the best known example, formed by the steady passage of older Pacific lithosphere over fixed melting anomalies in the asthenosphere or the deep mantle, at least during the last 70–80 m.y. It has further been proposed that the active, shield-building, southeastern ends of these chains form part of a fixed reference frame on Earth from which 'absolute' plate motions can be derived. Age data on volcanoes of chains in the Pacific are not linear when plotted against distance along the chains. This scatter appears to be a result, in some cases, of the use of inappropriate age data; in others, of dating lavas of volcanoes whose tholeiitic and alkalic suite rocks have extended life spans; and in still others, of dating volcanic rocks that represent a period of rejuvenated volcanism on older shields. However, in some cases, the scatter is demonstrably the result of real variable rates of emplacement of volcanic edifices along chains. The time scale of apparent irregular progression, where documented or suspected, ranges from 1 or 2 to 30 or 40 m.y., and the detailed rate of volcanic progression in the Hawaiian chain ranges from as little as −7 to as much as +24 cm/yr. On the other hand, students of magnetic anomaly time scales maintain that spreading rates along oceanic rift systems in the major oceans of the world, including the Pacific, have been relatively constant over the last 70–80 m.y. At least three major possibilities may explain this apparent lack of kinematic agreement: (1) the rate of volcanic propagation of linear volcanic chains is not directly proportional to the rate of Pacific plate motion; (2) the East Pacific rise, while maintaining a steady spreading rate, has jumped or migrated in an irregular manner with time; or (3) the magnetic anomaly time scale is not linear. Of these alternatives, the weight of evidence at present favors the first, although all three mechanisms may be involved.

*Gilluly* [1973] pointed out that the correlated pattern of magnetic stripes on the sea floor and the implication that rates of sea-floor spreading have been relatively uniform over at least the last 80 m.y. were seemingly inconsistent with the sporadic and episodic phenomena of orogeny, mountain uplift, and magmatism on the continents. His discussion largely concerned the interaction between the Kula, Farallon, and Pacific plates acting along the boundary of western North America. I wish to raise the same issue with regard to the linear island chains within the Pacific plate itself; i.e., despite the implication that the rate of ocean-floor spreading in the Pacific has been reasonably uniform about relative rotation poles to the plate over the last 80 m.y., the rate of volcanic propagation along linear island chains has been variable, irregular, and, in the best-documented area, accelerating, during the same period of time. The rates of volcanic propagation of these chains, therefore, cannot be directly correlated with relative rates of plate motion as determined by magnetic lineations, and models that equate the currently active ends of these chains with fixed melting anomaly sites in the deep mantle from which absolute plate motion can be derived are open to further examination.

## RELATIVE SPREADING RATES IN THE PACIFIC

The relatively parallel magnetic anomalies in the Pacific plate, although many of its neighbors have been subducted, are generally believed to conform to the symmetrical spreading hypothesis of *Vine and Matthews* [1963]. *Vine* [1966] proposed a magnetic reversal time scale based on a steady rate somewhat adjusted from known magnetic reversals [*Cox et al.*, 1964] and extrapolated this scale backward in time some 11 m.y.B.P. *Heirtzler et al.* [1968] assembled all magnetic and age data available at that time and proposed a linear geomagnetic time scale based on a profile in the south Atlantic believed to have resulted from constant spreading, and extended the scale backward in time nearly 80 m.y. from a 3.35 m.y.B.P. age for the end of the Gilbert reversed epoch. At the time, these authors were aware of the possible error in this extrapolation. They were also aware that the comparisons of anomalies assuming a steady spreading rate in the South Atlantic did not lead to perfectly linear relations in other oceanic plates (Figure 1). Both *Heirtzler et al.* [1968] and *Pitman et al.* [1968] were particularly concerned about relative variations in spreading rate in the South Pacific, near anomalies 5 and 24, when compared with the Atlantic profile. *Larson and Pitman* [1972] further recognized apparent changes in spreading rate in the Pacific compared to *Heirtzler et al.'s* [1968] time scale, and *Rona* [1973] suggested similar changes in the North Atlantic, although these were of minor importance compared with proposed variations in mid-Cretaceous and older spreading rates. Subsequently, *Atwater and Mudie* [1973] reported that a decrease in spreading rate from 3.7 cm/yr to 1.2 cm/yr occurred at the Gorda rise about 2 m.y.B.P. They considered this change in rate to be a local feature related to rotation of spreading directions in this complex ridge system. *Blakely* [1974] made a detailed study of fourteen Miocene magnetic anomalies in the Pacific between latitudes 42° and 44°N and longitudes 130° and 136°W, which were produced by spreading from what are now the Gorda and Juan de Fuca ridges. These were transformed to the pole and stacked. *Blakely* [1974] constructed a time scale for the Miocene from these profiles, and the differences between his time scale and *Heirtzler et al.'s* [1968] time scale for the Miocene were expressed as changes in spreading rate for his northeast Pacific profiles (Figure 2). *Blakely* [1974] followed the same procedure for the Eltanin 20 (Pacific–Antarctic ridge) and the Eltanin 41 (Indian–Antarctic ridge), and identical changes in spreading rate were found for the latter profiles. *Blakely* [1974] reasoned that if *Heirtzler et al.'s* [1968] South Atlantic time scale was correct, synchronous accelerations of at least four plates—the Farallon, Pacific, Antarctic, and Indian—would be required. He therefore found it reasonable to replace *Heirtzler et al.'s* [1968] Miocene time scale. Of course, if the latters' original scale for the South Atlantic were correct, one

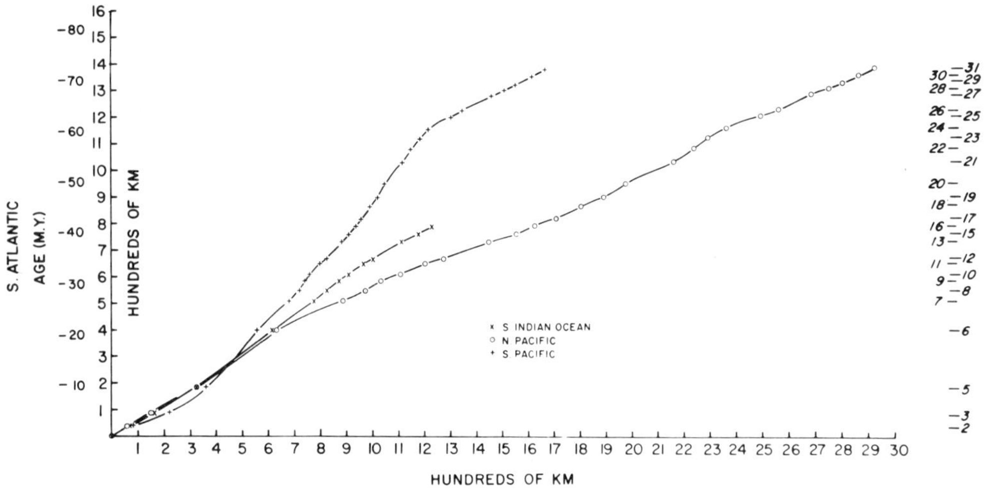


Fig. 1. The distance to a given magnetic anomaly in the South Atlantic Ocean versus distance to the same anomaly in the South Indian, North Pacific, and South Pacific oceans. Numbers on right refer to anomaly numbers. [After *Heirtzler et al.*, 1968.]

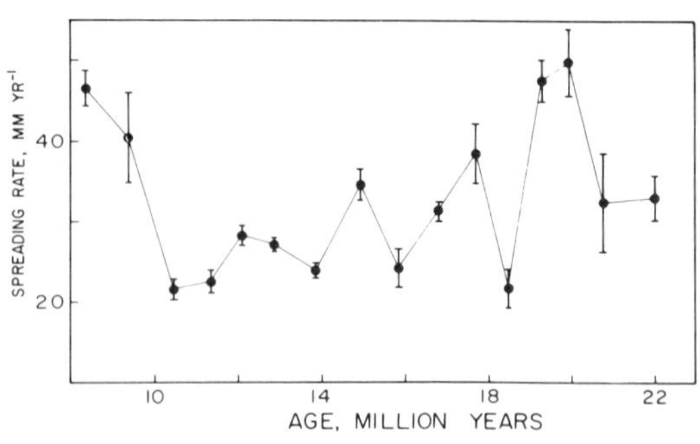


Fig. 2. Apparent spreading rates for the northeast Pacific during the Miocene. The solid dots are spreading rates averaged over 14 profiles. The error bars are four standard deviations in length. [After *Blakely*, 1974.]

could assign the changes in spreading rate to only two plates—the South American and African—spreading from a common ridge. In the absence of synchronous variations in either topography or magnetic anomalies on other plates, Blakely contended that Pacific spreading rates were constant over the period 10–20 m.y.B.P., adjusted the Heirtzler scale for that segment of time, and calculated an average spreading rate of 3.12 cm/yr for the Gorda and Juan de Fuca ridges. Although the literature is replete with examples of local changes in East Pacific ridge spreading rates [*Menard and Atwater,* 1968; *Herron,* 1972*a*; *Elvers et al.,* 1973; *Hein,* 1973], *Blakely* [1974] made the point that in the absence of evidence of synchronous spreading rate changes on several plates, the assumption of minimal acceleration of most plates was the simplest.

Deep Sea Drilling in the Atlantic [*Maxwell et al.,* 1970; *Sclater et al.* 1974] has tended to confirm *Heirtzler et al.'s* [1968] prediction of a steady spreading rate for the western part of that plate. *Van Andel and Bukry* [1973], on the other hand, have used drilling data to confirm a roughly uniform spreading rate in the southern part of the Pacific plate. These latter authors point out, however, that the resolution of Deep Sea Drilling data is poor and inferences as to detailed changes in spreading rates based on that data are arbitrary.

It seems safe to say that most students of Pacific magnetic anomalies or ridge bathymetry would be reluctant to accept the idea of highly variable or irregular changes in the spreading rate at the Pacific ridges over the last 80 m.y. For spreading rates prior to that time, there is little consensus of opinion [see *Baldwin et al.,* 1974].

### THE FIXED HOT SPOT HYPOTHESIS

*Wilson* [1963*a*] proposed that Hawaiian volcanic edifices were formed and carried northwest as the Pacific lithosphere passed northwestward over a fixed 'hot spot' in the asthenosphere or deep mantle beneath. *Schminke and Bufe* [1968] extended Wilson's idea, calculated a Hawaiian rotation pole, and noted, as did *Christofferson* [1968], that the Emperor seamount chain bent sharply north from the Hawaiian chain, and lay about a different pole. *Morgan* [1972*a, b*] further proposed that the Marshall–Ellice–Gilbert and Austral chains, and the Line Islands and Tuamotu chains were related in a similar way, and lay about Emperor and Hawaiian poles, respectively. (Figure 3 shows the generalized locations of these chains.) *Morgan* [1972*a, b*] further proposed that the currently active ends of these chains sit over convective thermal plumes located in the deep mantle and, along with a number of other plume localities on Earth, formed a fixed reference frame for plate motion. A number of objections have been raised about plumes themselves as physical entities [*Shaw and Jackson,* 1973; *Khan,* 1973; *Tozer,* 1973], but the point I wish to raise here regards whether they have been fixed through time. In view, however, of the controversy regarding the existence of thermal plumes, it seems preferable to refer to whatever mechanisms produce linear island chains as melting anomalies.

Given fixed rates of plate motion, the Wilson–Morgan hypothesis of fixed melting anomalies requires fixed rates of volcanic propagation along linear island chains. *Shaw* [1973] objected to this hypothesis on the grounds that motion vectors in the Pacific plate could not be directly related to rates of progression of volcanism in the Hawaiian chain due to time–distance–volume relations among the volcanoes that form the Hawaiian chain. *Clague and Jarrard* [1973] found the fixed melting anomaly hypothesis tenable if the rate of Pacific plate motion was very slow in the period from about 20 to 44 m.y.B.P. (between anomalies 6 and 17 in *Heirtzler et al.'s* [1968] time scale). They also pointed out that the Line Islands did not fit the same rotation pole as the Marshall–Gilbert–Ellice and Emperor chains. *Molnar and Atwater* [1973*b*] showed by rotating plates back to their 13 and 38 m.y. positions that most melting anomalies had moved with respect to one another, some at rates of as much as 1 to 2 cm/yr. *Burke et al.* [1973] came to the same conclusion using a different framework of rotation.

In spite of the controversy with regard to whether melting anomalies are indeed fixed with respect to the spin axis of the earth, two models of 'instantaneous' plate motion indicated that *Morgan's* [1972*a, b*] ideas were consistent with the kinematics of plate motion and could form a fixed reference frame for plate tectonics. *Minster et al.* [1974] systematically inverted worldwide spreading rates, fracture zone trends, and earthquake slip vectors to obtain a consistent model of 'instantaneous' relative motions for the major plates over the last 10 m.y. They then inverted the trends of twenty linear island chains and aseismic ridges (including three in the Pacific) under the assumption that they were fixed with respect to the deep mantle. They concluded that

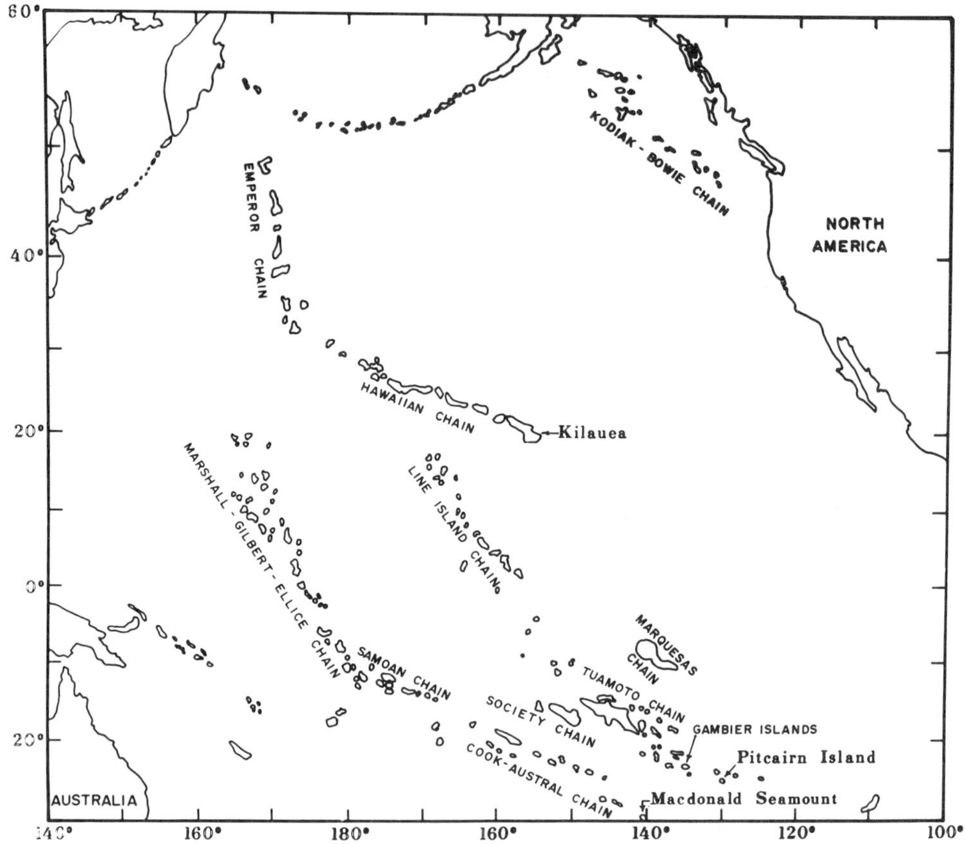


Fig. 3. Generalized Mercator projection of linear island chains in the Pacific. Kilauea volcano and Macdonald seamount are currently active; Pitcairn Island has been dated as less than 1 m.y. old [*Duncan et al.*, 1974.]

the ends of these chains have had no significant relative motions to one another over the last 10 m.y., and they derived a rate of plate motion at the Hawaiian chain of 8.9 cm/yr, at the Austral chain of 8.5 cm/yr, and at the Tuamotu chain of 9.1 cm/yr over that period of time. *Solomon and Sleep* [1974] calculated several models of 'absolute' plate motion velocity fields using assumptions involving viscous drag to derive the forces driving plate motion. These models gave 'absolute' plate velocities that were very similar to those they generated using the assumption that a set of fixed melting anomalies provided a fixed reference frame for plate motion. These models yield 'instant' Pacific plate velocities seen at Hawaii of from 7.16 to 9.66 cm/yr. *Solomon and Sleep* [1974] felt their model to be consistent with that of *Minster et al.* [1974] and that no significant movement of melting anomalies with respect to each other has occurred over the last 10 m.y., although they admitted the possibility of relative movement prior to that time. Bearing in mind that *Minster et al.'s* [1974] rates as seen at the three Pacific chains were not inconsistent with fixed melting anomalies, it should be pointed out that other models that require no direct relation between rates of plate motion and rates of volcanic propagation of linear island chains have been proposed, as discussed below.

Hypotheses of origin of linear island chains by propagating fractures [*Betz and Hess,* 1942; *Jackson and Wright,* 1970; *Green,* 1971; *Jarrard and Clague,* 1974] require no direct relation between the rate of plate motion and the rate of volcanic propagation. *McDougall* [1971] suggested equal and opposite movement of melting anomalies in the asthenosphere to plate motion of the lithosphere to account for the greater rate of volcanic propagation than the rate of spreading at the East Pacific rise. If the lithosphere and melting

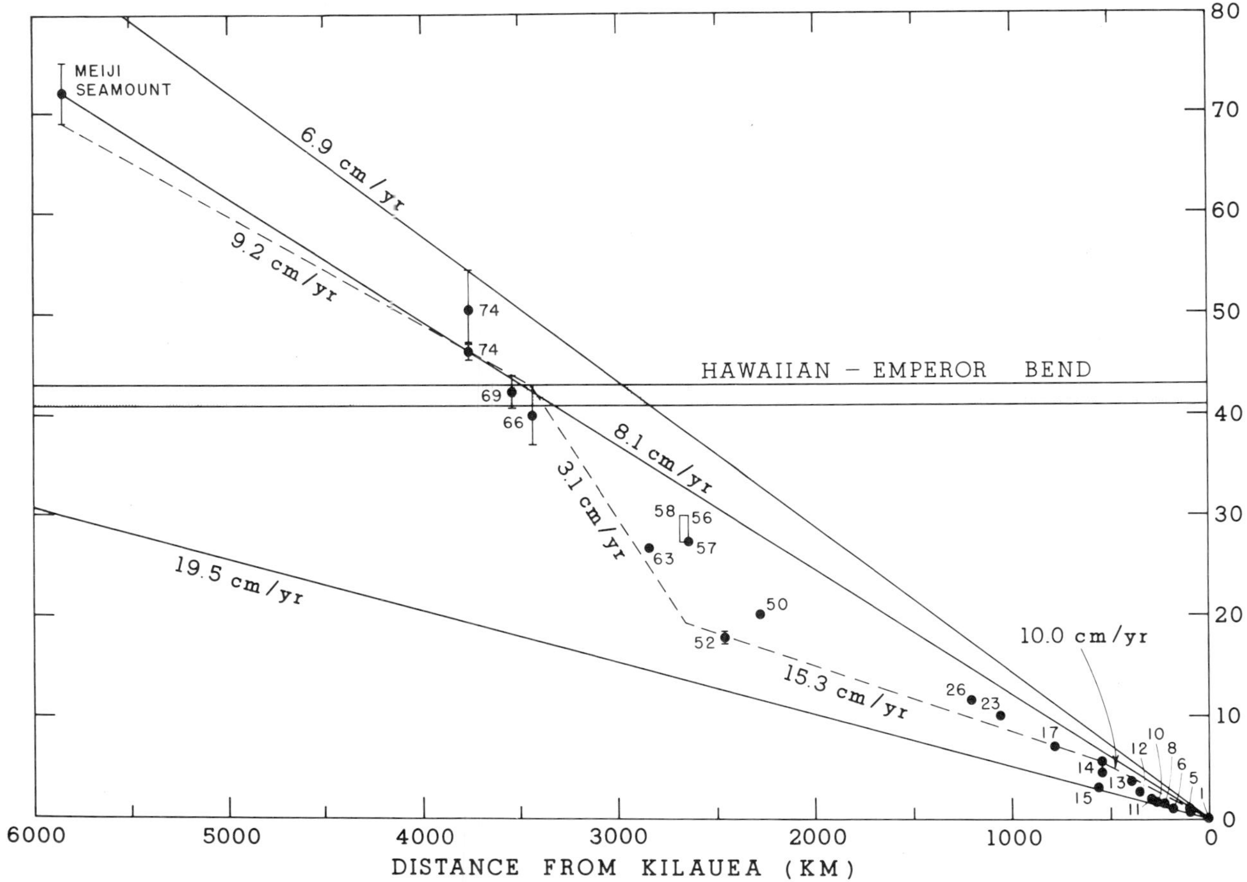


Fig. 4. Ages of rocks from islands and seamounts of the Hawaiian–Emperor chain as a function of distance from Kilauea. Circles represent dated edifices; bars represent limits of uncertainty in age. Numbers beside circles are edifice numbers in the system of *Bargar and Jackson* [1974]. Box represents a fossil age and its uncertainty. (Fossils were recovered from sediment between two edifices.) Age and distance data and sources, in Table 1.

asthenosphere are of the same thickness, the model requires the rate of volcanic propagation to be twice that of the rate of plate motion. The gravitational anchor model of *Shaw and Jackson* [1973], in which dense refractory residua are postulated to sink as volcanoes grow, requires only a relatively fixed melting anomaly, inasmuch as the dense anchor in the asthenosphere could advance or retreat depending on the vagaries of flow and shear melting in the asthenosphere. The rate of volcanic progression in *Shaw and Jackson's* [1973] model may accelerate or decelerate with respect to the rate of plate motion.

## GEOCHRONOLOGICAL DATA ON LINEAR ISLAND CHAINS IN THE PACIFIC

The age data given below are, with few exceptions, whole-rock potassium–argon ages of basaltic rocks collected directly from linear island chains or dredged from seamount chains. In a few cases paleontological ages are given where no other data are available. The ages are selected only on the basis that they represent the best age data now available. Distances between dated volcanic edifices were corrected for Mercator projections from bathymetric data of *Chase et al.* [1973] and *Mammerickx et al.* [1973] by the method outlined in *Bargar and Jackson* [1974]. Not all chains noted on Figure 3 are included; age data for some are non-existent or thought to be patently unreliable.

*Hawaiian–Emperor chain.* The Hawaiian chain is by far the best dated in the Pacific. A plot of all reliable ages as a function of distance from Kilauea is given in Figure 4. The lack of any sort of linearity is painfully obvious. The limiting slopes of straight lines that enclose all the data are 6.9 and 19.5 cm/yr. A straight line connecting the best age of Meiji seamount (assuming it to be a part of the Emperor chain) and Kilauea is 8.1 cm/yr. The dashed line of Figure 4 is that of *Clague and Jarrard* [1973], based on ages of volcanoes known at that time, and contains a break in slope in the period 20–44 m.y.B.P., although no change in spreading rate at Pacific ridges has been noted between anomalies 6 and 17 based on the magnetic time scale. Nevertheless, *van Andel* [*1974*] lends support to *Clague and Jarrard's* [1973] break in slope, finding a slower rate during that period of time to provide the best fit for the positions of thick equatorial sedimentation through time. It is apparent from Figure 4 that, taken literally, the rate of volcanic propagation along the chain has been far from constant. This is emphasized in Table 1, where rates between individual volcanic edifices in the chain have been calculated. These rates range from –7 to +24 cm/yr. The rates between Kauai and Kilauea in Table 1 are especially worthy of attention, because they represent the most completely dated part of the chain. The oldest reliable age obtained from a reported tholeiite in these islands is plotted as a function of distance from Kilauea in Figure 5. This figure suggests that, where best known, the rate of volcanic progression has been smoothly accelerating and that, if the ages of all edifices in the chain were known, other accelerating portions would be found. The age of the tholeiitic lavas of Niihau have been reported to be younger than those of Kauai, although the former island is farther from Kilauea. This suggests that there may be other areas of overlapping volcanism in older parts of the chain, and is further supported by negative rates of volcanic progression in Table 1.

*Line Islands chain.* This chain, which *Morgan* [1972*a, b*] considered to have formed over the same time interval as the Emperor chain, has been drilled at three Deep Sea Drilling sites [*DSDP Scientific Staff,* 1973*c*]. It appears to be older than the Emperor chain, and was considered to be coeval along at least 1300 km of its length (Figure 6). Subsequent to the publication of the figure by *DSDP Scientific Staff* [1973*c*] (see Figure 6), *Ozima and Saito* [1974] gave two new ages for Line Islands edifices. These data

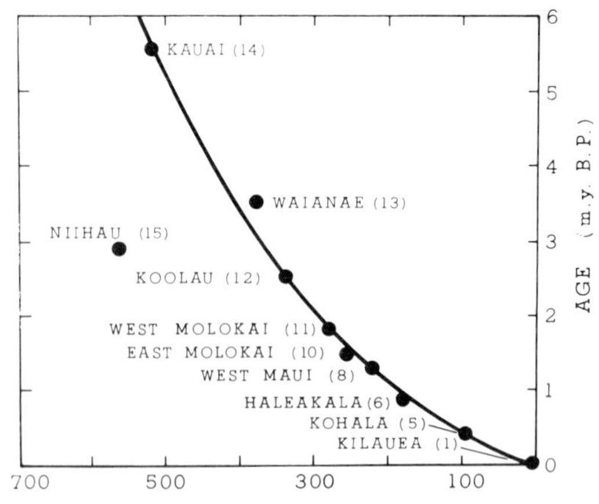


Fig. 5. Ages of oldest tholeiitic basalts from islands at the southeastern part of the Hawaiian chain plotted as a function of distance from Kilauea. Numbers from the system of *Bargar and Jackson* [1974]. [After *Jackson et al.,* 1972.]

TABLE 1. Rate of Volcanic Propagation: Hawaiian–Emperor Chain

| Edifice | Distance to Next Dated Edifice, km | Age, m.y. | Source | Rate, cm/yr |
|---|---|---|---|---|
| Meiji seamount | | 72 ± 3 | *Scholl and Creager* [1973] | |
| Koko seamount | 2056 | 46.4[1] ± 1.1<br>50.5 ± 3.5 | *Clague and Dalrymple* [1973]: *DSDP Scientific Staff* [1973*c*]; Bukry (Pers. Comm.) | 8.0 |
| Yuryaku seamount | 264 | 42.3 ± 1.6 | *Clague et al.* [1975] | 6.4 |
| Kanmu seamount | 106 | 40 ± 3.0 | Quoted in *Clague and Jarrard* [1973] | 4.6 |
| Unnamed seamount #63[2] | 589 | 26.7 ± 0.5 | *Clague et al.* [1975] | 4.4 |
| Unnamed seamounts #56 and 58[2] | 153 | 27–30[1] | *DSDP Scientific Staff* [1973*c*] | -4.6 |
| Unnamed seamount #57[2] | 72 | 27.3 ± 0.4 | *Clague et al.* [1975] | 2.7 |
| Midway Island | 168 | 17.9 ± 0.6 | *Dalrymple et al.* [1974] | 1.8 |
| Pearl and Hermes Reef | 151 | 20.0 ± 0.5 | *Clague et al.* [1975] | -7.2 |
| La Perouse Pinnacles (French Frigate Shoals) | 1072 | 11.7 ± 0.4 | *Dalrymple et al.* [1974] | 12.9 |
| Necker Island | 151 | 10.0 ± 0.4 | *Dalrymple et al.* [1974] | 8.9 |
| Nihoa Island | 278 | 7.0 ± 0.3 | *Dalrymple et al.* [1974] | 9.3 |
| Niihau Island | 215 | 3.0 ± 0.2 | *Jackson et al.* [1972] | 5.4 |
| Kauai Island | 46 | 4.50 – 5.62[1] | *McDougall* [1964] | -1.8 |
| Waianae volcano | 145 | 3.60 ± 0.4 | *Jackson et al.* [1972] | 7.2 |
| Koolau volcano | 35 | 2.55 ± 0.10 | *McDougall* [1964]; *Jackson et al.* [1972]; *Doell and Dalrymple* [1973] | 3.3 |
| West Molokai volcano | 59 | 1.84 ± 0.06 | *McDougall* [1964] | 8.3 |
| East Molokai volcano | 24 | 1.48 ± 0.04 | *McDougall* [1964] | 6.7 |
| West Maui volcano | 35 | 1.30 ± 0.04 | *McDougall* [1964] | 19.4 |
| Haleakala volcano | 39 | 0.84 ± 0.03 | *McDougall* [1964] | 8.5 |
| Kohala volcano | 82 | 0.42 ± 0.02 | *McDougall and Swanson* [1972] | 19.5 |
| Kilauea volcano | 100 | 0 | | 23.8 |

[1]Where more than one edifice age is given, the underlined age was the one used in the rate calculation, either because it was considered more reliable or because the older of a range of dates is more likely to present the true edifice age.

[2]Numbered volcanoes refer to the numbering system of *Bargar and Jackson* [1974].

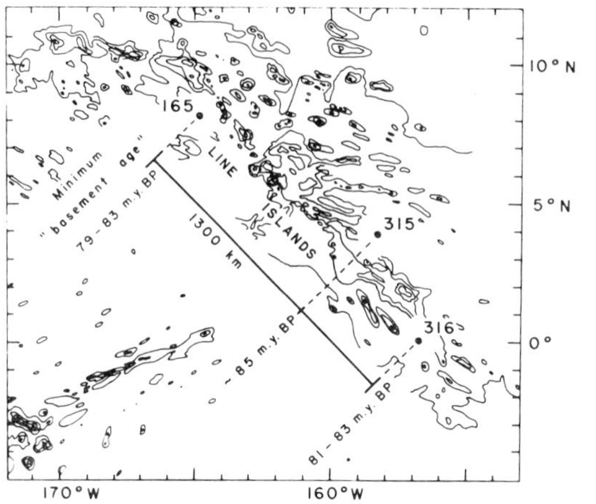


Fig. 6. Map showing location of DSDP sites 165, 315, and 316 along the Line Islands chain. Bracket shows minimum basement age at the three sites. [Modified after *DSDP Scientific Staff,* 1974*a*.]

are summarized in Table 2. It would seem apparent that the Line Islands show no uniform rate of volcanic propagation, and that, in contradiction to the *Morgan* [1972*a, b*] hypothesis, all of the Line Islands edifices are older than those of the Emperor chain.

*Marshall-Gilbert-Ellice chain.* Although *Clague and Jarrard* [1973] gave some age data for the Marshall-Gilbert-Ellice chain, insufficient data exist to make any meaningful statement about the rate of volcanic progression along the third northerly trending chain of *Morgan* [1972*a, b*].

*Kodiak-Bowie (Pratt-Welker) chain.* This chain lies on a concentric small circle to the same pole as the Hawaiian chain [*Morgan,* 1972*a, b; Clague and Jarrard,* 1973] but, in lying closer to that pole, should have slower rates of volcanic propagation. Four ages from edifices of the chain have been reported, and these are given, along with rates of progression between edifices, in Table 3. While slower than 'average' Hawaiian rates, as predicted, they do appear to indicate changes in rate of volcanic progression.

*Marquesas chain.* While this chain does not lie on a small circle about the Hawaiian pole but trends somewhat northwest of it [*Jarrard and Clague,* 1974], it does appear to be a progressive chain formed during part of the same time interval. *Brousse and Bellon* [1974] have reason to suspect excess argon in the volcanic rocks of Eiao, and report both conventional K/Ar and isochron ages. Rates are calculated for both in Table 4. *Duncan and McDougall* [1974] give a range of ages from each island, and in each case the oldest was used in the calculations of Table 4 as being most likely to represent the age of the edifice. *Duncan and McDougall* [1974], however, draw a best-fit slope through the center of their range of ages and report a linear rate of 9.9 cm/yr. The scatter of rates in using an island-by-island approach is evident in Table 4.

TABLE 2. Rate of Volcanic Propagation: Line Islands Chain

| Edifice | Distance to Next Dated Edifice, km | Age, m.y. | Source | Rate, cm/yr |
|---|---|---|---|---|
| 18°05'N, 169°04'W | | 128 | *Ozima and Saito* [1974] | |
| 12°04'N, 165°50'W | 756 | 84.4 | *Ozima and Saito* [1974] | 1.7 |
| 08°11'N, 164°51.5'W | 446 | 79–83[1] | *Winterer, Ewing et al.* [1973] | 31.9 |
| 04°10'N, 158°31.5'W | 832 | 85 | *DSDP Scientific Staff* [1974*a*] | –41.6 |
| 00°05.5'N, 157°08'W | 473 | 81–83[1] | *DSDP Scientific Staff* [1974*a*] | 23.7 |

[1]Same statement as on Table 1.

TABLE 3. Rate of Volcanic Propagation: Kodiak–Bowie (Pratt–Welker) Chain

| Edifice | Distance to Next Dated Edifice, km | Age, m.y. | Source | Rate, cm/yr |
|---|---|---|---|---|
| Kodiak seamount | | 22.6 ± 1.1 | *Turner et al.* [1973] | |
| Giacomini seamount | 164 | 19.9 ± 1.0 | *Turner et al.* [1973] | 6.1 |
| Dickens seamount | 587 | 3.7 ± 0.2 | *Turner et al.* [1974] | 3.6 |
| Hodgkins seamount | 89 | 2.67 ± 0.2 | *Turner et al.* [1974] | 8.6 |

*Society Islands.* These islands follow the Hawaiian pole and appear to be progressively younger in age to the southeast, albeit with a great deal of scatter in the data. *Krummenacher and Noetzlin* [1966] do not give petrologic descriptions of the rocks they dated, but it seems probable that the 147 and 156 m.y. dates reported for Tahiti (see Table 5) were from the gabbroic rocks that characterize the calderas of Tahiti and, in fact, all of the other islands in the Society chain. *Krummenacher et al.* [1972] reported an anomalous age of 833 m.y. for one such gabbro, and, inasmuch as this age is considerably older than the oceanic crust upon which Tahiti rests, concluded that some gabbros contain excess argon. *Denoufbourg's* [1965] excellent geologic maps of Tahiti show, as do his maps of all the other islands of the chain, that the Tahitian gabbros are the youngest rocks on that volcano. The older ankaramites have been independently dated on only two islands [*Dymond,* 1974]. The practice of using the older ages to calculate the rates

TABLE 4. Rate of Volcanic Propagation: Marquesas Chain

| Edifice | Distance to Next Dated Edifice, km | Age, m.y. | Source | Rate, cm/yr |
|---|---|---|---|---|
| Eiao Island | | 4.92–8.72[1] (4.8 ± 0.2–5.18 ± 0.2[1])[2] | *Brousse and Bellon* [1974] | |
| Nuku Hiva Island | 115 | 2.99–4.23[1] | *Duncan and McDougall* [1974] | 2.6<br>12.1 |
| Ua Huku Island | 62 | 2.71–2.78[1] | *Duncan and McDougall* [1974] | 4.3 |
| Hiva Oa Island | 109 | 1.59–2.48[1] | *Duncan and McDougall* [1974] | 36.3 |
| Tahuata Island | 19 | 1.78–2.04[1] | *Duncan and McDougall* [1974] | 5.6 |
| Fatu Hiva Island | 78 | 1.30–1.39[1] | *Duncan and McDougall* [1974] | 12.0 |

[1]Same statement as on Table 1.

[2]Parentheses indicate isochron ages; both raw K/Ar and isochron ages were used in calculation of rates.

TABLE 5. Rate of Volcanic Propagation: Society Chain

| Edifice | Distance to Next Dated Edifice, km | Age, m.y. | Source | Rate, cm/yr |
|---|---|---|---|---|
| Bora Bora Island | | 0.25–4.0[1] | *Krummenacher and Noetzlin* [1966] | |
| Tahaa Island | 37 | 2.8–6.7[1] | *Krummenacher and Noetzlin* [1966] | –1.4 |
| Raiatea Island | 25 | 2.5–2.8[1] | *Krummenacher and Noetzlin* [1966] | .6 |
| Huahine Island | 50 | 1.5–5.4[1] | *Krummenacher and Noetzlin* [1966] | –1.9 |
| Moorea Island | 149 | 1.2–2.6[1] | *Krummenacher and Noetzlin* [1966] | |
| | | 1.5–2.0[1] | *Dymond* [1974] | 4.4 |
| Tahiti volcano | 43 | 0.2–1.4[1] | *Dymond* [1974] | 7.2 |
| | | 1.4–156. | *Krummenacher and Noetzlin* [1966] | |
| | | <4.0–30. | *Krummenacher et al.* [1972] | |
| Tahiti Iti volcano | 40 | 0.2 ± 0.3 | *Krummenacher and Noetzlin* [1966] | 3.6 |

[1]Same statement as on Table 1.

of volcanic progression on the older islands, which have presently been dated only by *Krummenacher and Noetzlin* [1966], leave open the possibility that gabbros with excess argon have in some cases been dated there as well. The scatter in rates in Table 5 may therefore have more than one cause.

*Austral Islands.* Nineteen new ages in the Austral Islands have recently been reported by *Dalrymple et al.* [in press]; their Figure 1 is reproduced here as Figure 7. These authors noted that the ages do not follow the predicted pattern of increasingly younger ages to the southeast, and that the relation of the dated rocks to the shield-building stages of these volcanoes is in doubt. It is interesting to note that the young ages on Rurutu, Rarotonga, and Aitutaki were, with one exception, obtained from strongly alkalic rocks, whereas the ages on Mangaia were obtained from ankaramites.

*Easter Island–Gambier–Tuamotu chain. Duncan et al.* [1974] have dated Pitcairn Island and have suggested that a progressive linear chain extends from Pitcairn through the Gambier Islands, and through Mururoa Atoll into the Duke of Gloucester Islands. Using age data postdated by that in Table 6, they arrived at an average rate of volcanic propagation of 11 cm/yr. The Duke of Gloucester Islands have not, to my knowledge, been dated. Ages in the northern end of the Tuamoto chain have been included in Table 6 to indicate that the islands are related to Mururoa by a rate of volcanic propagation of only 2.4 cm/yr, and that the northeasterly edifices of the chain were built considerably prior to the Hawaiian–Emperor bend, dated by *Clague et al.* [1975] at 42 m.y.B.P.

*Summary.* It is quite apparent that age data on all Pacific linear island chains show rates of volcanic propagation that are irregular to accelerating when examined in detail. On the other hand, most Pacific chains younger than 50 m.y. appear to fall near small circles about *Morgan's* [1972*a*, *b*] Hawaiian pole, or one near it, and most become younger to the southeast in a very general way [*Clague and Jarrard,* 1973; *Jarrard*

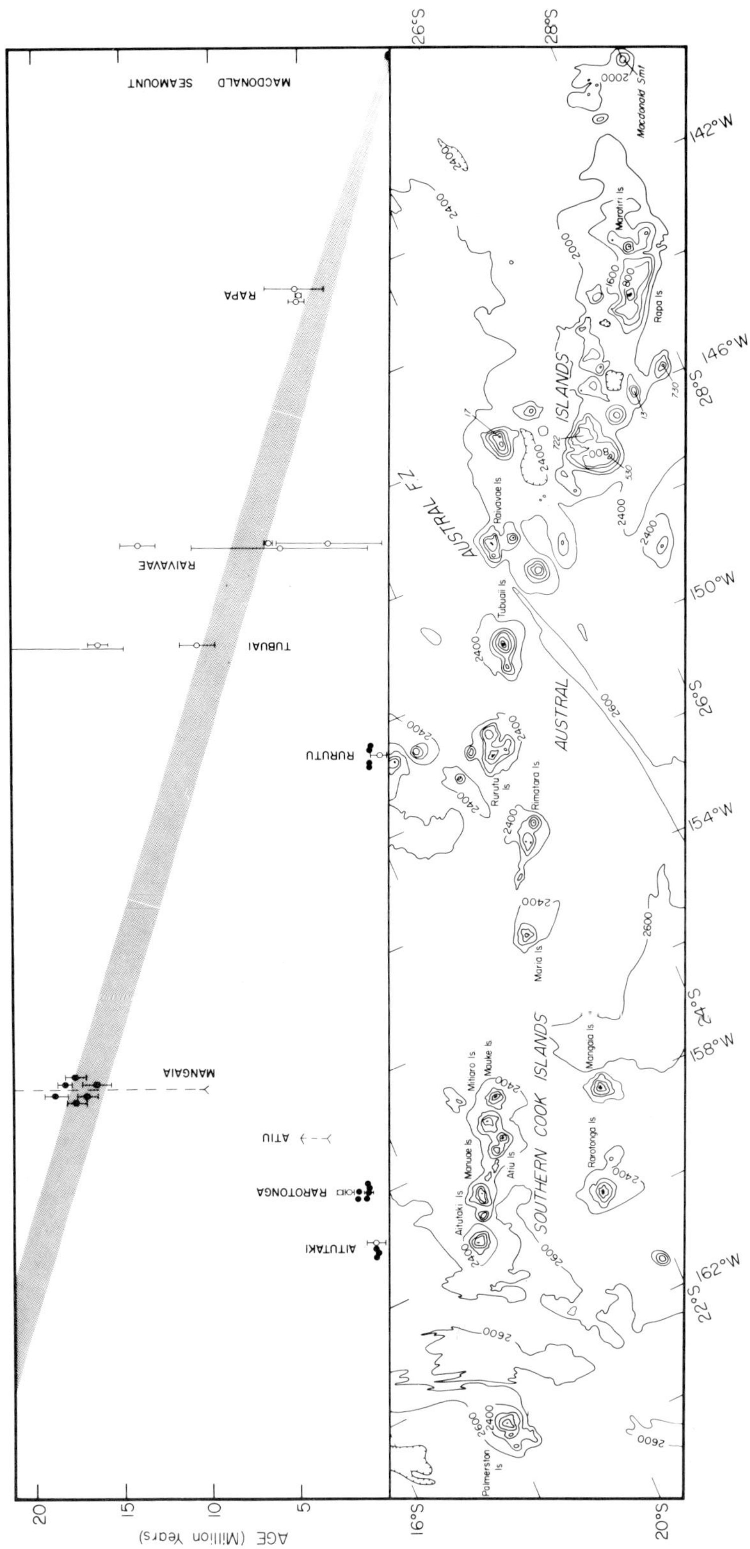


Fig. 7. Bathymetry of the southern Cook–Austral chain as a function of distance from Macdonald seamount. Open circles: data from *Krummenacher and Noetzlin* [1966]; solid circles, *Dalrymple et al.* [in press]. Bars indicate analytical uncertainty at the 2σ level. Arrows show minimum fossil ages. Shaded band represents predicted age pattern based on Hawaiian data. [After *Dalrymple et al.*, in press.]

TABLE 6. Rate of Volcanic Propagation: Pitcairn–Gambier–Tuamotu Chain

| Edifice | Distance to Next Dated Edifice, km | Age, m.y. | Source | Rate, cm/yr |
|---|---|---|---|---|
| 14°50′S, 146°51.5′W | | 49–53.5[1] | *DSDP Scientific Staff* [1974*a*] | |
| 14°06′S, 145°40′W | 155 | 49–53.5[1] | *Hays, Cook et al.* [1972] | |
| Mururoa Island | 1110 | 7.0 ± 1.0 | *Chevallier* [1973] | 2.4 |
| Mangareva Island | 423 | 4.44–6.31[1]<br>(4.77 ± 0.2–5.98[1] ± 0.2[2]) | *Bellon* [in press] | 61.3<br>41.5 |
| Aukena Island | 7 | 5.91–5.62[1]<br>(4.77 ± 0.2–5.98[1] ± 0.2[2]) | *Bellon* [in press] | 1.0<br>2.1 |
| Makapu Island | 7 | 4.72–5.62[1]<br>(4.77 ± 0.2–5.98[1] ± 0.2[2]) | *Bellon* [in press] | ∞<br>–1.9 |
| Pitcairn Island | 537 | 0.45 ± 0.2–0.95 ± 0.2[1] | *Duncan et al.* [1974] | 11.5<br>10.7 |

[1]Same statement as on Table 1.
[2]Same statement as on Table 4.

TABLE 7. Average Rates of Volcanic Propagation over Long Chains[1]

| Chain | Length, km | Age Span, m.y. | Rate, cm/yr |
|---|---|---|---|
| Emperor chain | 2320 | 72–42.3 | 7.8 |
| Line chain | 2507 | 128–83 | 5.6 |
| Kodiak–Bowie (Pratt–Welker) chain | 840 | 22.6–2.7 | 4.2 |
| Hawaiian chain | 3520 | 42.3–0 | 8.3 |
| (Necker Island–Kilauea volcano)[2] | (1058) | (10.0) | (10.6) |
| Marquesas chain | 383 | 5.2–1.4 | 10.1 |
| Society chain | 344 | 4.0–0.3 | 9.3 |
| Australs chain | 1932 | 17.8–0 | 10.9 |
| Pitcairn–Gambier–Mururoa chain | 974 | 7.0–0.95 | 16.1 |

[1]Data from Tables 1–6. In each case only the oldest and youngest dated edifices were considered; all intervening dated edifices were omitted.

[2]The Necker Island–Kilauea volcano segment is for comparison with *Minster et al.'s* [1974] calculated rate of plate motion as seen at Hawaii over the last 10 m.y.

*and Clague,* 1974]. When these chains are viewed as very long segments, and when ages between only the oldest and youngest volcano of each chain are considered, the rate of volcanic progression of these chains is considerably smoothed (Table 7). The average rate of volcanic progression from Necker Island to Kilauea is included for comparison with the 10 m.y. average plate motion of 8.9 cm/yr calculated by *Minster et al.* [1974]. Their calculated average rates of plate motion as seen at the Austral and Tuamotu chains are also considerably slower than the values of Table 7. Present data do not appear to support the Wilson-Morgan fixed melting anomaly hypothesis, unless melting anomalies are many hundreds to thousands of kilometers in length, and overlapping volcanism is more common than previously suspected. It appears appropriate to consider the sources of scatter in the rate of volcanic propagation observed in all dated chains.

## CAUSES CONTRIBUTING TO IRREGULAR VOLCANIC PROPAGATION

*Are some of the age data incorrect?* It is certainly possible that some of the age data on which Figures 4-7 and Tables 1-6 were constructed are not representative of the real cooling age of the rocks upon which they were measured, in which case irregular propagation would be apparent, not real. Reasons for this could include excess argon, alteration, or faulty techniques. Certainly the Tahitian gabbros contain excess argon [*Krummenacher et al.,* 1972], and Brousse and his co-workers [*Brousse et al.,* 1972; *Brousse and Bellon,* 1974; *Bellon,* in press] now routinely reduce their raw K/Ar data to $^{40}K/^{36}Ar$–$^{40}Ar/^{36}Ar$ isochrons, which involve assumptions that make their data difficult to compare with other age data, but which may be correct. *Dalrymple* and his co-workers [1973, 1974, in press] check selected analyses by incremental heating or $^{39}Ar/^{40}Ar$ techniques, and *Turner* and his co-workers [1973, 1974] date rocks by two methods —conventional K/Ar and fission track methods. *Ozima and Saito* [1974] now use differential heating and $^{39}Ar/^{40}Ar$ techniques. Certainly all dated rocks should be carefully examined petrographically for alteration and, insofar as possible, dated by more than one method.

*Are the wrong rocks being dated?* The life span of a shield edifice in a linear volcanic chain will probably never be known until one is drilled from top to sea floor, and even then evidence in Hawaii suggests that the length of time of construction of volcanic shields will not be a constant. *McDougall* [1964], for example, gives evidence that the time span of presently exposed tholeiitic volcanism on Kauai is at least 1 m.y. long, and that the entire span of volcanism on the island is 4.3 m.y. On the other hand, *McDougall and Swanson* [1972] give an entire time span of exposed volcanism at Kohala volcano of only 0.39 m.y. and consider the entire five volcanoes of the island of Hawaii to have been built in the last 0.45 m.y., the oldest age they obtained on Kohala tholeiites. *Stearns* [1973] has objected to this conclusion on geomorphologic grounds, but *Dalrymple* [1971] had previously obtained an age of less than 0.7 m.y.B.P. on the Kohala tholeiites. In Hawaii, tholeiites are thought to form, in general, very rapidly, and to constitute the major parts of the Hawaiian shields [*Macdonald,* 1949; *McDougall,* 1964; *Stearns,* 1966]. Tholeiites occur abundantly throughout most of the Hawaiian chain [*Macdonald and Katsura,* 1964; *Macdonald,* 1969; *Dalrymple et al.,* 1974], and at least as far into the Emperor chain as Kōko seamount [*Clague and Dalrymple,* 1973]. In the better known southeastern Hawaiian Islands, the volcanic edifices pass through an alkalic, and finally a nephelinitic suite of activity prior to complete extinction. Table 8 shows the generalized Hawaiian sequence, and is largely taken from *Macdonald and Katsura* [1964] and *Macdonald* [1968]. The diagram is modified to show the presence of pre-erosional alkalic basalts among the rocks of the alkalic suite immediately following tholeiite eruption, and again among the rocks of the post-erosional nephelinitic suite. These latter alkalic basalts occur on Oahu, where they postdate tholeiitic ages by 2 m.y. [*Gramlich et al.,* 1971], (G. B. Dalrymple, personal communication, 1974); on Kauai, where they postdate tholeiitic eruption by 3-4 m.y. [*McDougall,* 1964], and on Niihau, where they postdate tholeiitic eruption by at least 2 m.y. (G. B. Dalrymple, personal communication, 1974). These basalts are indistinguishable in major element chemistry from those of the alkalic suite (see Figures 8, 9, and 10). *Jackson and Wright* [1970] have postulated that nephelinitic suite lavas were formed by the passage of the Hawaiian arch [*Hamilton,* 1957] over the position of Oahu some 2-4 m.y. after the formation of the main edifices on that island. Age dates on dredged alkalic basalts in the Hawaiian-Emperor chain may, therefore, be 2-4 m.y. younger than the tholeiites of seamounts from which they are dredged. It seems reason-

TABLE 8. Hawaiian Lava Types
[Modified from *Macdonald and Katsura*, 1964, and *Macdonald*, 1968.]

| | Lava Suite | Lava Type | Estimated Abundance | |
|---|---|---|---|---|
| General Sequence of Eruption ↑ | Nephelinitic | Nepheline–melilite basalt | | |
| | | Nepheline basalt | <<1% | Post-erosional |
| | | Basanite | | |
| | | Basanatoid | | |
| | | Alkalic basalt | | |
| | Alkalic | Trachyte | | |
| | | Mugearite | | |
| | | Hawaiite | <1% | |
| | | Ankaramite | | |
| | | Alkalic basalt | | Pre-erosional |
| | Tholeiitic | Tholeiite | 99% | |

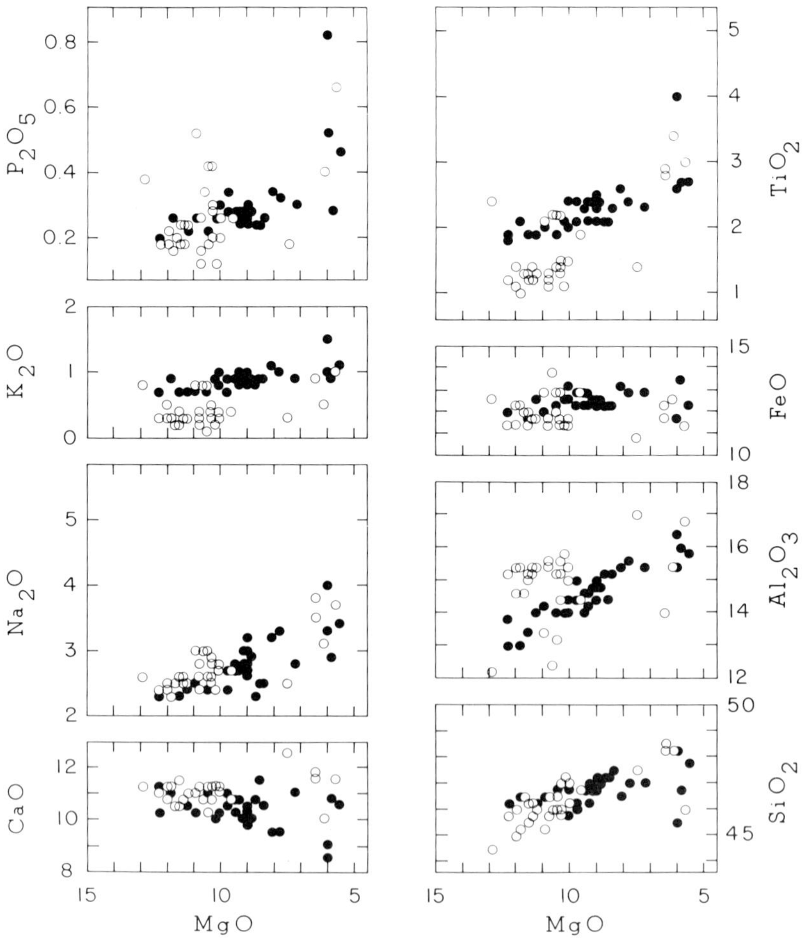

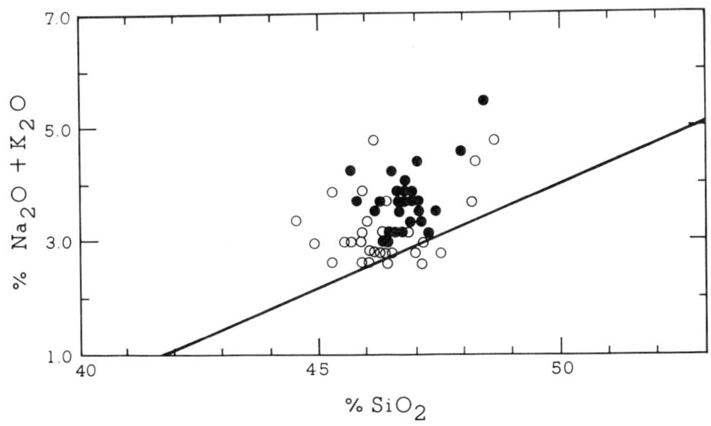


Fig. 9. Total alkali–silica plot of the same samples with the same symbols as those of Figure 8.

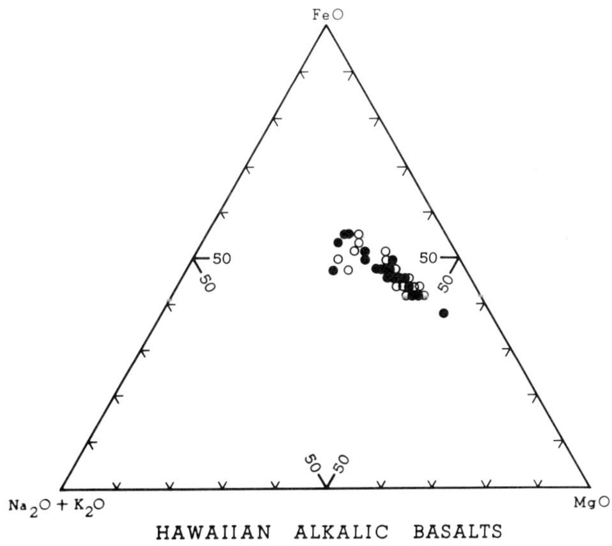


Fig. 10. FeO–MgO–total alkali plot of the same samples with the same symbols as those of Figure 8.

---

Fig. 8. (Opposite) MgO variation diagram of the dry, reduced, chemical composition of pre-erosional alkalic basalt lavas from Hualalai volcano (open circles) and of post-erosional basalts from Oahu, Kauai, and Niihau (solid circles). Data sources: Hualalai basalts, *Macdonald* [1968], E. D. Jackson (unpublished data), and T. L. Wright (unpublished data); post-erosional basalts, *Macdonald* [1968], E. D. Jackson and M. H. Beeson (unpublished data).

able, in the Hawaiian–Emperor chain, to establish a datum of edifice formation based on tholeiitic rocks, where possible. Where tholeiites are not present in exposures or dredge hauls, then ankaramites, hawaiites, and mugearites appear to be more certain of being closer to that datum than are alkalic basalts.

Of course, all linear island chains in the Pacific plate may not conform to the Hawaiian pattern—tholeiites have been reported only in the Gambier Islands [*Brousse et al.,* 1972] and in western Samoa [*Hedge et al.,* 1972]. *McBirney* [1967] and *McBirney and Aoki* [1968] argue that among Pacific linear island chains, there are two separate mantle sources for oceanic islands, and imply that tholeiites need not underpin all islands, particularly Tahiti. But it is interesting that in the case of Samoa, where the older tholeiites remain undated, *Hawkins and Natland* [in press] report a 1200-km-long lineament of recent post-erosional basanite and olivine nephelinite. They attribute the formation of the Samoan chain to its position along the hinge line of flexure of the Pacific plate into the Tonga trench. In the absence of age data on the pre-erosional lavas, I would agree that the post-erosional volcanic rocks are due to the mechanism they suggest. In fact, in view of *Menard's* [1973] summary of epeirogeny and plate tectonics, it seems possible that any release of pressure beneath any chain of pre-existing volcanic edifices could initiate post-erosional volcanism of any age younger than the formation of the major shields. In view of the composition of rocks from the Hawaiian post-erosional nephelitic suite and the compositions of post-erosional rocks from Samoa, one would expect such lavas to have a strongly alkaline character but to include alkalic basalts comparable in major element composition to alkalic basalts of the Hawaiian alkalic suite. Post-erosional volcanism might account, for example, for the apparently erratic rate of volcanic propagation in the Austral chain [*Jarrard and Clague,* 1974; *Dalrymple et al.,* in press], and in the Line Islands chain [*DSDP Scientific Staff,* 1973*c*].

In summary it seems quite likely that some of the scatter in age–distance relations along linear volcanic chains is caused by dating volcanic rocks that were formed at ages considerably younger than those of the main edifices upon which they rest. Dated rocks of strongly alkalic character are especially suspect in this regard.

*Is there extensive overlap in the ages of shields along linear volcanic chains? Jackson et al.* [1972] reported that the tholeiites of Niihau Island were about the same age as those of Waianae volcano, located nearly 200 km southeast along the Hawaiian chain, and suggested that overlapping shield-building volcanism could indeed occur. *Shaw* [1973], in calculating time–distance–volume relations along the Hawaiian chain, considered that overlapping volcanism must have occurred repeatedly along the chain. Yet the idea remains untested, although *Grommé and Vine* [1972] measured paleolatitudes in the Midway drill core and suggested that Midway volcano formed at about the present latitude of Kilauea. If overlapping volcanism is widespread, one would expect no direct proportionality between the rate of volcanic progression and the rate of plate motion.

*Is there a real periodic acceleration in the rate of volcanic progression? Jackson et al.* [1972] pointed out that an acceleration appeared to exist in the best-dated part of the Hawaiian chain—from Kauai to Kilauea—where the oldest-dated tholeiites on each volcano were plotted against distance from Kilauea (see Figure 5). *Shaw* [1973] calculated the volumes of volcanoes along this part of the chain and plotted the change in volume with distance as a function of rate of volcanic propagation along this segment of the chain (Figure 11). He also calculated the volume

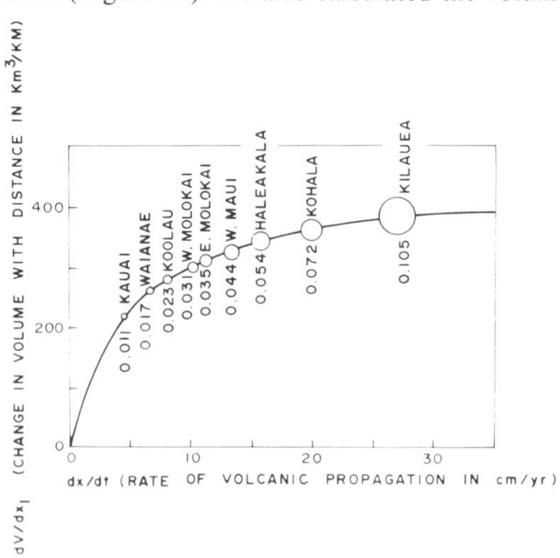


Fig. 11. Volumes per unit length between volcanic edifices of the southeastern Hawaiian chain plotted against propagation rates between edifices. Numbers on the curve are *Shaw's* [1973] calculated volume rates of eruption in $km^3/yr$. Values of $dV/dx_{\parallel}$ given by *Shaw* [1973] were normalized to his idealized model of behavior along a single locus. Diameters of circles on curve are approximately proportional to volume rates of eruption. [After *Shaw,* 1973; *Dalrymple et al.,* 1973.]

rates of eruption from the identity $dV/dt = (dV/dx)(dx/dt)$. These volume rates of eruption are plotted on Figure 11 at the relative positions of the volcanoes involved, and range from about 0.01 cubic km/yr at Kauai to nearly 0.11 cubic km/yr at Kilauea. His calculated volume rate of eruption compares with a measured volume rate of eruption of Kilauea of 0.11 cubic km/yr based on direct measurement by *Swanson* [1972]. *Shaw's* [1973] calculations were consistent with those of a shear melting model using independently derived parameters. Essentially, Shaw proposed that the lithosphere–asthenosphere interface heats when sheared by plate motion, and viscosity decreases, which allows a further increase in the rate of shearing. As ascending magma carries off excess heat, temperatures drop, viscosity increases, and a new cycle begins. If *Shaw* [1973] is correct, then again the rate of volcanic propagation bears only an indirect relation to the rate of plate movement.

## CONCLUSIONS

In spite of the fact that some radiometric ages have not yielded the true age of volcanic material dated, and that some rocks that are probably younger than true ages of the volcanic edifices on which they rest have been dated, it is apparent from the data assembled here that linear island chains on the Pacific plate, with few exceptions, do tend to lie on small circles about common poles, and to become younger toward the southeast or south. It is also apparent that the rate of volcanic propagation of these chains is irregular if not periodically accelerating.

It now seems clear that *Morgan's* [1972*a*, *b*] model was considerably over-simplified, in places wrong (e.g., the Line Islands), in places incomplete (in failing to note the random beginning in time and space of short chains such as the Societies and the Marquesas), and in part still speculative (the very existence of thermal plumes, their fixity to the Earth's spin axis, and their ability to drive plates have been questioned). Nevertheless, his idea stimulated thinking and research into a megatectonic problem largely ignored since the time of *Dana's* [1849; 1876] work.

At the present state of knowledge there appear to be three major possibilities to explain the irregular rates of volcanic propagation along linear island chains on the Pacific plate:

(1) The rate of volcanic propagation of linear island chains along older parts of the Pacific lithosphere is not directly proportional to the rate of Pacific plate motion, and has not been over the last 70–80 m.y. If so, the positions of melting anomalies may deviate from fixed positions relative to a geocentric coordinate system.

(2) The East Pacific rise, while maintaining a steady spreading rate as evidenced by a linear magnetic time scale, is linked to the melting anomaly in the deep mantle, and has migrated irregularly eastward over the last 70–80 m.y. at the same rates as the rates of volcanic progression. The most recent magnetic anomaly map [*Pitman et al.,* 1974] shows what appears to be a series of fossil ridges west of the present East Pacific rise.

(3) The magnetic time scale in the Pacific plate, as based on *Heirtzler et al.'s* [1968] assumption and *Blakely's* [1974] modification, is wrong, and the spreading rate has varied irregularly with time. This possibility would receive little support from kinematicists at this time.

In view of data given by *Jackson et al.* [1972], in view of *Shaw's* [1973] impressive arguments for a thermal feedback, shear-melting origin for the chains, in view of recurrent hypotheses of origin of the chains by propagating fractures, of which *Jarrard and Clague's* [1974] is the most recent, and, in view of the generally faster rates of volcanic propagation over long segments of chains (Table 7) compared with kinematically derived models of the rate of Pacific plate movement, the first of these possibilities seems most probable; but, in the light of present data, all three alternatives may have been operative during the growth of the chains.

It seems apparent that attempts to determine absolute plate motion by assuming that melting anomalies in the Pacific are fixed [*Minster et al.,* 1974] are premature. So are attempts to calculate asthenospheric thickness based on existing age data [*Forristal,* 1974]. What is needed is more good data on composition, ages, and paleomagnetic positions [*Grommé and Vine,* 1972] of edifices of the chains. Ultimately the only way to obtain this information is by mapping and intelligently collecting from exposed edifices, and drilling exposed edifices, atolls, and seamounts.

*Acknowledgments.* I thank my colleagues G. B. Dalrymple, C. S. Grommé, and D. A. Clague of the U.S. Geological Survey, and G. A. Macdonald of the University of Hawaii, for critically reviewing the manuscript.

# 29. *Mechanics of Benioff Zone Magmatism*

BRUCE D. MARSH

*Department of Earth and Planetary Sciences,*
*The Johns Hopkins University, Baltimore, Maryland 21218*

*Abstract.* By deducing that andesitic magmas are strongly undersaturated with regard to water in the near-surface environment, the hypothesis that andesitic magmas arise from partial melting of peridotitic upper mantle material under hydrous conditions is questioned. Andesitic lavas are more likely to have formed by partial melting of subducted oceanic crust, quartz eclogite, under nearly anhydrous conditions. The distinctive $K_2O$, $SiO_2$, $TiO_2$, and REE concentrations displayed by andesitic lavas can be explained by the mineralogy of the eclogite source. In particular, the characteristic low $TiO_2$ content of andesites is probably a result of the highly refractory nature of rutile ($TiO_2$) in quartz eclogite. The zone of melting along the slab–mantle interface probably extends from a depth of about 100 to 150 km downdip, although in some instances it may extend to a depth of 200 km. The temperature of the magma at a depth of about 110 km is ~ 1400°C. The low density layer of andesitic magma is gravitationally unstable and as it strives to stabilize itself blobs of magma ascend to the surface at regularly spaced distances along the layer, parallel to the island arc. The surface manifestation of this instability is the regular spacing of island arc volcanic centers, which is about 70 km in Tonga and the Aleutian Islands. Cooling curves have been calculated for an ascending body of magma by assuming a uniform ascent velocity and a heat conduction approximation. These curves, along with boundary conditions on the initial depth of crystallization deduced from lavas, suggest that andesitic magmas travel to within 25 km, or less, of the surface in a superheated state. Thus differentiation through crystal settling is probably ruled out as an agent in chemically modifying the ascending magma. The shear stress needed along the slab–mantle interface to cause melting is shown to be about 1 kb. The thickness of the shear zone is also calculated by modelling it as a Couette flow; the thickness is inversely proportional to the shear stress. For a Newtonian material the shear zone thickness is less than 10 km with a 1-kb shear stress. For a non-Newtonian material, the shear zone thickness is about 1 m for a 1-kb shear stress. Thus, at any place and time, a layer of subducted oceanic crust perhaps only a few tens of meters thick is involved in partial melting and andesitic magma production.

Andesitic volcanism occurs in long narrow chains of volcanoes in regions of lithospheric subduction. This association of volcanism and plate consumption appears intimate and its explanation therefore might be helpful in understanding the mechanical and thermal nature of regions of plate convergence. Volcanism can be viewed as a thermal probe into the earth's interior which, if properly understood, is a fundamental and direct source of information pertaining to the earth's thermal regime, and the mechanics of magma generation and transfer within the earth.

It is generally concluded that andesitic magma is produced either from partial melting of the subducted oceanic crust or from hydrous melting of peridotitic upper mantle material (PUMM) a short distance above the subducting plate (see *Marsh and Carmichael* [1974] for references.) For the PUMM source, water is envisioned to diffuse upward as a result of dehydration of the downgoing slab. This water lowers the solidus of the PUMM, initiating melting and magma production. It is unclear whether siliceous melts of the andesitic type can be produced from PUMM at the prevailing pressures near the Benioff zone [*Mysen et al.,* 1974]. Some authors would rather envision the diffusing water to travel a great distance upward so that initial melting takes place at much lower pressures where siliceous melt production from PUMM may be less difficult. Dry partial melting of PUMM at, say, 30 kb produces a melt highly deficient in silica. A large amount of water—probably a minimum of about 20 wt. %—is probably needed to produce a siliceous melt [*Mysen et al.,* 1974], and water is also necessary to lower the PUMM solidus to allow melting under the ambient T-P conditions. To produce a siliceous magma (which may not necessarily turn out to be an andesite) by partial melting of PUMM, then, water is the critical necessary agent and water diffusion is the critical rate-controlling step.

Partial melting of the subducted oceanic crust, on the other hand, of predominantly tholeiitic composition and quartz eclogite mineralogy, needs no water to produce an andesitic magma [*Marsh and Carmichael,* 1974]. Petrologists have been hesitant to ascribe andesites to evolve from such a simple process primarily because of the thermal models calculated for the downgoing plate [e.g., *Toksöz et al.,* 1973]. These thermal models show the mantle isotherms being strongly depressed along the top of the downgoing plate implying that melting of the uppermost portion of the subducted oceanic crust might be unlikely. Thermal model calculations, however, are in dire need of accurate boundary conditions (a *known* temperature at a *known* pressure) and if petrologists could decide on the T-P conditions of andesite genesis these could be used to calibrate thermal models [*Turcotte and Schubert,* 1973].

Through the presence of a substantial quantity of water as a necessary condition for only the PUMM source, the two possible andesitic magma sources can be contrasted. Several lines of evidence strongly indicate that andesitic magma is undersaturated with respect to water at a total pressure of a few kilobars, where initial crystallization takes place. For instance, a Tongan basaltic-andesite melted under $P_{H_2O} = P_{total}$ laboratory conditions showed olivine as the stable liquidus phase at low pressures [*Nicholls and Ringwood,* 1973] while indigenous olivine has never been observed in any Tongan lava under natural conditions (A. Ewart, personal communication, 1974). The phenocryst sequence of crystallization displayed by andesites, plagioclase as the liquidus phase, similarly reflects a paucity of water in the magma. At a total pressure of a few kilobars, addition of water to an initially dry silicate melt causes a great reduction in the stability field of plagioclase; with addition of enough water, plagioclase crystallizes last in the sequence of crystallization. In Aleutian and Tongan andesitic lavas, however, plagioclase seems to be one of the earliest (if not the earliest) phase to begin crystallizing [*Marsh,* in press, *a*]. Thus andesitic magmas are probably highly undersaturated with water in the near-surface environment and are therefore extremely undersaturated at depth in their source region. The solubility of water in a silicate melt is directly proportional to the confining pressure at constant temperature [*Hamilton et al.,* 1964]. *Hydrous melting of a peridotitic source therefore seems excluded as a source of andesitic magma.*

This conclusion is further substantiated by noting that it is probably exceedingly difficult to transport water from the slab into the overlying mantle, the force of buoyancy is negligible and diffusion is very slow. An appropriate cation bulk diffusivity for this situation at, say, 1300°C may be about $10^{-3}$ to $10^{-4}$ $cm^2$ $sec^{-1}$ or less [*Fisher and Elliot,* 1974], implying that for water to diffuse a distance of 10 km takes on the order of between 30 and 300 m.y.; island arc magmatism seems to occur on a much shorter time scale than this. These chemical diffusivities are about 10 to 100 times smaller than probable mantle thermal diffusivities, which implies that the downgoing slab will retain its chemical identity about 10 to 100 times longer than it retains its thermal identity. Since the time constant for thermal absorption of the downgoing plate seems to be about 10 m.y. [e.g., *Isacks et al.,* 1968, p. 5885], the downgoing plate may retain its chemical identity for 100 m.y. or more. Were diffusion of water from the downgoing plate into the overlying mantle an important process in island arcs, one would expect invariably to find very large regions of partial melting in the mantle above the slab. A comprehensive comparison of the characteristics of *P* and $_pP$ waves produced by earthquakes within the downgoing plates has shown that many island arcs display no zone of high attenuation in the wedge above the plate [*Barazangi et al.,* 1975], although small, localized zones of magma production, as in the case of oceanic crust melting, are not precluded. In sum, andesitic magma production in island arcs appears to be the result of partial melting of the subducted oceanic (quartz eclogite) crust.

Andesitic lavas reflect their novel source by exhibiting mineralogical and bulk chemical properties distinctly different from that of ocean-ridge tholeiites (ORT) and nodule-bearing alkali basalts (NAB); ORT and NAB form a convenient chemical reference frame by which to compare other lavas. ORT and NAB (Table 1) arise from partial melting of PUMM. ORT are produced at low pressures, about 10 kb, and NAB are produced much deeper, probably at pressures of 30 to 50 kb. The degree of partial melting is of the same order in each case, or NAB is a result of a slightly smaller fraction of melting [*Green,* 1971]. In both cases the magmas are formed *below* the lithosphere. Their respective compositions reflect the variation in depth of magma formation. When PUMM is melted at increasingly higher pressures, the resulting magmas are progressively depleted in $SiO_2$ and enriched in $K_2O$ and $P_2O_5$ (which reflects the abundance of the rare earth elements). Calculations concerning $SiO_2$ [*Carmichael et al.,* 1970] and the experimentally determined change in distribution coefficients for the rare earth elements (REE) with increasing pressure [*Masuda and Kushiro,* 1970] bear out these deductions.

TABLE 1. Compositions of Ocean-Ridge Tholeiites (ORT), Nodule-bearing Alkali Basalts (NAB), and an Andesite (AND)

| | ORT[a] | AND[b] | NAB[c] |
|---|---|---|---|
| $SiO_2$ | 49.34 | 57.30 | 47.72 |
| $TiO_2$ | 1.49 | 0.77 | 2.29 |
| $Al_2O_3$ | 17.04 | 18.21 | 15.62 |
| $Fe_2O_3$ | 1.99 | 2.98 | 2.61 |
| FeO | 6.82 | 4.54 | 7.92 |
| MnO | 0.17 | 0.16 | 0.19 |
| MgO | 7.19 | 3.06 | 8.02 |
| CaO | 11.72 | 8.01 | 8.92 |
| $Na_2O$ | 2.73 | 3.04 | 3.81 |
| $K_2O$ | 0.16 | 1.24 | 1.58 |
| $P_2O_5$ | 0.16 | 0.19 | 0.53 |
| $H_2O^+$ | 0.58 | 0.18 | 0.09 |
| $H_2O^-$ | | 0.05 | |
| $CO_2$ | | <.05 | |
| Total | 100.08 | 99.73 | 99.66 |

[a]From *Engel et al.* [1965].
[b]Sample CB-12 from *Marsh* [in press, *a*].
[c]Average of eleven analyses from *Smith and Carmichael* [1969] and *Bacon and Carmichael* [1973].

Since andesites contain more $SiO_2$ than either ORT or NAB one would expect, all else being equal, were a PUMM source responsible for them, that they would be produced at shallower depths than ORT. Andesites, however, contain about five to ten times more $K_2O$ than ORT does, but usually only slightly less than NAB. From the $K_2O$ content, then, the PUMM source would be expected to render an andesite at great depth, similar to that yielding NAB. The $P_2O_5$ content of andesites, on the other hand, is low like that of ORT, and would reflect a shallow PUMM source. This chemical incongruency between andesites and magmas obviously derived from PUMM also may indicate that andesites are from a distinctly different source.

The uniformity within a string of volcanoes producing andesitic lavas is amazing. Such a tendency surely points to a common, universal source without complicated schemes of crystal fractionation wholeheartedly changing some never-seen precursor magma to the same, monotonous, end product. The Benioff zone magma source thus offers a unique opportunity to relate magma production and ascent to the thermal and mechanical processes taking place within the earth. In the remainder of this paper the physics and chemistry of the generation and evolution of andesitic magmas is discussed.

## PETROLOGY OF ANDESITE PRODUCTION

When the oceanic crust, of predominantly tholeiitic composition, is thrust deep into the earth it undergoes a phase transition induced by increased pressure. At 1000°C this transition would begin at about 15 kb and is spread over several kilobars. The high pressure mineral assemblage is eclogite and consists predominantly of quartz, garnet, pyroxene(s), rutile, and probably sanidine and apatite as minor constituents [*Marsh and Carmichael,* 1974, *T. Green,* 1970; *Green and Ringwood,* 1967]. The major phases and apatite are commonly observed in quartz eclogites; however, the presence of sanidine is largely conjectural. It is the densest, most common K–Al silicate and is probably the only likely phase to hold a significant quantity of potassium under such conditions. *Green and Ringwood* [1967*b*] alluded to its presence from their experimental work. *Green and Lambert* [1965] showed very clearly that during such a basalt–eclogite transition the original plagioclase first lost its calcium component, rendering an alkali–feldspar, and then lost its sodium component through a reaction to form jadeite and quartz. A virtually pure sanidine phase was left. *T. Green* [1970] has also outlined these steps and identified the emergence of sanidine in gabbro and diorite. *Eggler and McCallum* [1974] have recently confirmed the presence of orthoclase (sanidine) in a natural quartz eclogite. There is some uncertainty as to the presence of mica or hornblende under the conditions of andesite formation, ~35 kb and 1350°–1450°C, but these conditions seem well beyond their stability regions [*Marsh and Carmichael,* 1974; *Holloway and Ford,* 1975].

Quartz eclogite has a solidus which is virtually coincident with the andesite liquidus, a necessary condition for the direct generation of andesite from partial fusion of quartz eclogite. It begins to melt approximately 200 degrees lower than for PUMM [*Lambert and Wyllie,* 1972] under similar conditions. The following will attempt to interpret the major chemical features of andesitic lavas in light of the eclogite mineral assemblage. The principal features are (1) the curious correlation between $K_2O$ content and depth to the Benioff zone, (2) the high $SiO_2$ content, and (3) the low $TiO_2$ and $P_2O_5$ content.

*$K_2O$:* Perhaps the best known property of andesites is the distinctive positive correlation between their mean $K_2O$ content and depth to the Benioff zone [*Hatherton and Dickinson,* 1969]. *Marsh* [1973] and *Marsh and Carmichael* [1974] have suggested an explanation of this feature by

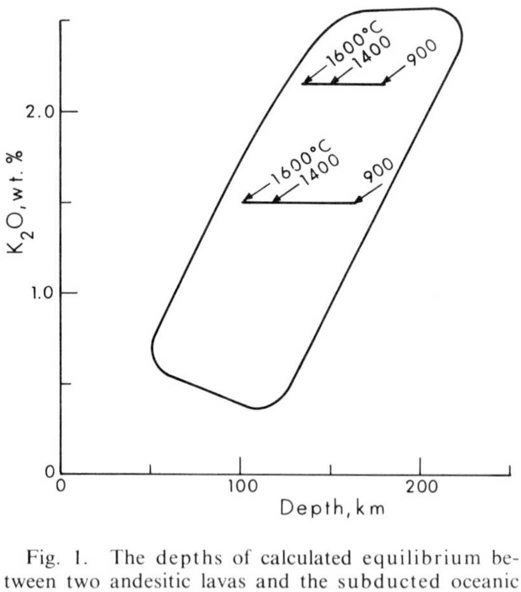


Fig. 1. The depths of calculated equilibrium between two andesitic lavas and the subducted oceanic crust, assumed to be a quartz eclogite, are shown as horizontal bars over a range of temperatures. These temperatures do not pertain to the field as a whole. The circle enclosing these calculated points is the range of observed $K_2O$ values associated with a certain depth to the Benioff zone found in present day island arcs [*Dickinson,* 1970]. [After *Marsh and Carmichael,* 1974.]

showing that the chemical potential of $K_2O$, defined by the above eclogite mineralogy, will, beyond a certain depth, increase with increasing pressure. Hence, when the slab undergoes partial melting the chemical potential of $K_2O$ in the magma will reflect the pressure at which the melting takes place. Given a lava of known composition at the earth's surface, it is possible theoretically to reverse the process and estimate the temperature and pressure at which the magma—fully molten lava—was in equilibrium with the eclogite [*Marsh and Carmichael,* 1974]. By using lavas of different $K_2O$ content, these results produced a theoretical plot of lava $K_2O$ content versus depth to the Benioff zone which matched the observed plots (Figure 1).

The $K_2O$ content of the incipient magma is buffered by the presence of sanidine in the eclogite. Sanidine and quartz probably define the solidus, although this has not been shown under dry conditions for sanidine, and once sanidine disappears it will no longer buffer the chemical potential of $K_2O$ in the magma. For similar degrees of melting, the final value of this chemical potential, for a group of lavas, will only differ from the estimated value by a common constant. This constant will be small relative to the other inherent uncertainties (e.g., exact mineralogy of eclogite). Previously presented calculations [*Marsh and Carmichael,* 1974] hence seem sufficiently accurate to describe the behavior of $K_2O$ during fusion.

*Dickinson* [1975] has suggested that there may be discernible, systematic changes in $K_2O$ versus depth to the Benioff zone as a function of an island arc's tectonic position. Andesitic volcanoes associated with continental areas seem to show higher $K_2O$ content than andesitic lavas from intra-oceanic island arcs. This difference in $K_2O$ content probably reflects a difference in the mantle and, hence, slab thermal regime between these areas. For instance, with a cooler mantle the slab will probably have to penetrate deeper before melting begins; once the slab melts, the magma $K_2O$ content, which is largely controlled by pressure, will be higher than for the case of a hotter mantle [*Marsh and Carmichael,* 1974].

*$SiO_2$:* With the presence of quartz in the source rock as a solidus phase [*Green and Ringwood,* 1967*a*], it is easy to understand why the derived magmas exhibit a relatively large activity of silica.

*$TiO_2$ and REE:* Quartz eclogite is one of the few rocks that contain rutile, a very refractory

phase (i.e., having a large entropy of fusion per gram atom) which will remain in the eclogite until the last stages of melting [*Green and Ringwood,* 1967*a*]. Thus, any ensuing partial melt will be very depleted in $TiO_2$, a universal characteristic of andesitic lavas. This is in contrast to the PUMM source where $TiO_2$ is controlled by clinopyroxene, the solidus phase.

The oceanic crust has low, unfractionated rare earth element (REE) abundances similar to those of andesitic lavas [*Kay et al.,* 1970; *Ewart and Bryan,* 1973]. REE partition coefficients ($X^{REE}$ magma/$X^{REE}$ source) for garnet and clinopyroxene probably increase greatly with pressure [*Masuda and Kushiro,* 1970], consequently, melting these phases at high pressure would produce a magma richer in REE than the source. Rutile could also possibly accommodate large amounts of REE as well as other transition elements. The riddle why andesites have such low abundances of REE could possibly be explained by rutile; and apatite, commonly seen in California eclogites (F. J. Turner, personal communication, 1974), may also be an important phase in determining the abundance of the light REE in andesitic magmas.

## TEMPERATURE OF THE SOURCE REGION

There is reason to believe that, to a good approximation, magmas from the Benioff zone ascend isochemically. One infers this both from the calculated equilibria between lavas and eclogite and from the possibility that the magma is superheated over the majority of its ascent (see below). Thus, experimentally determined P–T phase diagrams for the quartz–eclogite solidus [*Green and Ringwood,* 1967*a*] and the basaltic–andesite or andesite liquidus [*T. Green,* 1972] may give a good estimate of temperature during magma formation. These results indicate temperatures of about 1350° to 1450°C at depths of 100–130 km, which tally with the results of chemical equilibrium calculations [*Marsh and Carmichael,* 1974]. These experimentally determined solidus temperatures, however, may be too high as they represent a 'clean' laboratory situation whereas in the earth they could be 50–100 degrees less. This is the case for Hawaiian tholeiites which have an observed solidus, carefully measured in a lava lake, significantly less than that experimentally determined [*Peck et al.,* 1966]. This temperature determination at a known depth within the earth has important implications.

Not only does this temperature serve as a boundary condition on thermal models of subducting slabs, but also it may place a constraint on the amount of viscous heating necessary to produce melting. A bound is thereby placed on the magnitude of the deviatoric shear stress in this region through the relation $q = \tau u$, where $q$ is heat flow, $\tau$ is shear stress, and $u$ is velocity. This derivation will be discussed in a later section.

## MAGMA DYNAMICS

*Volcano spacing and Rayleigh-Taylor fluid instability.* When a lighter fluid lies beneath a denser one, the system is gravitationally unstable. This type of fluid instability, commonly called Rayleigh-Taylor instability, may be important in the Benioff zone magmatic environment; andesitic magma is much less dense than the overlying peridotitic mantle. The manner in which such a system stabilizes itself and transmits andesitic magma to the surface gives insight into the basic dimensions and physical properties of the magmatic environment.

When perturbed from equilibrium, the two-fluid interface takes on a sinusoidal pattern which in time tends to a characteristic or dominant wavelength. This particular wavelength exhibits the fastest rate of growth, relative to all others, and is determined by the viscosity and relative thickness of the two layers. Should the amplitude of the perturbation continue to increase and blobs of magma migrate to the surface to form volcanic centers, the spacing between these centers will correspond to the dominant wavelength. Volcanic centers in island arcs indeed display an amazing regularity in spacing which can be interpreted to arise from Rayleigh-Taylor instability [*Marsh,* 1973; *Marsh and Carmichael,* 1974].

In most previous studies the problem of determining the dominant wavelength has been assumed to be linear, two-dimensional, isothermal, incompressible, and Newtonian; for incipient instability these assumptions, for the most part, seem justifiable. The two-dimensional solution assumes horizontally infinite layers, while the long strings of island arc volcanoes may perhaps be better represented by a line or strip source of buoyant fluid. Nevertheless, to anticipate a bit, the magma source is probably extremely thin (meters) compared to its width (kilometers), implying that the two-dimensional results should be applicable here (see also below). *Biot and Odé*

[1965] and *Selig* [1965] showed that when the upper layer is about ten or more times thicker than the lower layer, the dominant wavelength is independent of the upper layer thickness. Surely, this is the case for initial instability along the Benioff zone. The dominant wavelength is then given by [*Selig*, 1965]:

$$\lambda = \left(\frac{2\pi h_2}{2.15}\right) \left(\frac{\eta_1}{\eta_2}\right)^{1/3}$$

where

$\lambda$ = dominant wavelength or spacing of volcanic centers
$h_2$ = thickness of the lower density layer
$\eta_1$ = viscosity of denser layer (mantle)
$\eta_2$ = viscosity of lighter layer (andesitic magma)

The volcanic center spacing is about 70 km in Tonga and the Aleutians; a characteristic order of magnitude mantle viscosity is fairly well known, but the viscosity of the magma is highly uncertain due to the unknown initial crystal content and pressure effect. A pressure of 30–40 kb probably increases the viscosity of a silicate liquid by 3 or 4 orders of magnitude [*Marsh*, in press, *b*]. Since it is expected that the magma layer thickness ($h_2$) will be initially small, but finite, a reasonable value for $\eta_1/\eta_2$ might be about $10^9$; $h_2$ would then be about 25 m. This pertains to the magma layer thickness at the time of initial perturbation, or, one could say, the instability which gave rise to the 70-km spacing took place when the layer was about 25 m thick.

Once formed, these zones of magma ascent, or diapiric conduits, will remain as the path of least resistance for a magmatic blob to travel to the surface. This may result from the thermal, and hence viscosity and density, anomaly created in the overlying medium above the initial instability due to the upward passage of magma, either as a continuous seep or as discrete blobs. Thus, the layer thickness might increase with time but the pattern of volcanic centers may remain fixed. This seems to be the case in the Aleutians where older, Tertiary islands are sites for the recent volcanic activity which began about 5 m.y. ago [*Marsh*, in press, *a*].

The above two-dimensional solution also holds quite closely in three dimensions [*Selig*, 1965; *Biot*, 1966]. That is, given a wide horizontal layer of fluid beneath a denser fluid the spacing of the domes produced is similar to that for a thin strip of fluid. One might say that the spacing of volcanic centers produced by such a process is quantized. Indeed, this does seem to be the case for the Aleutian Islands. There, two very young volcanic centers (Bogoslof and Amak) have appeared about 50 km north of the major arc. If the source grows with time downward along the Benioff zone, the low density fluid layer would be dipping northward at an angle of about 45 degrees. The next instability, perpendicular to the arc and along the Benioff zone, might be about 70 km downdip which is about 50 km on a horizontal surface (i.e., $70 \cdot \cos 45° \simeq 50$ km). The delay in this three-dimensional instability could be due to the competition between vertical diapiric instability or fluid movement upward along the dipping layer to the first diapiric zone.

*Segregation and Ascent.* When diapiric instability commences the magma may contain a substantial fraction of dense crystals (e.g., garnet, pyroxene, and rutile) which can be expected to remain in place as the liquid buoyantly rises away. Whether the body remains attached to the supply region is unclear. If the supply of magma keeps up with the rate of ascent, the diapir will remain attached and may grow into a ball of magma followed by a long, thin continuous column [*Whitehead and Luther*, 1975]. If the supply does not keep up, the magma will probably detach itself and rise as globules. Both cases may occur; however, such small quantities of magma erupt from island arc volcanoes so briefly and infrequently that the latter is probably more common. Although in the following treatment the ascending magma will be treated as a detached globule [*Marsh*, in press, *b*], the results can be easily amended to the other case as well.

Upon ascension the mantle will behave like a fluid and the magma will assume a geometrical form of least resistance or drag. Different body shapes of course experience different degrees of drag; for slow velocities in very viscous fluids (low Reynolds-number flows) *Pironneau* [1973] and *Bourot* [1974] have determined the optimum-shaped prolate spheroid to have a drag of 0.95 that of a sphere of equal volume. Thus the globule can be assumed to be spherical with little loss of generality.

The ascent velocity of a viscous sphere (viscosity $\eta_2$) in a medium with viscosity $\eta_1$ is [*Lamb*, 1932]

$$V = \frac{1}{3}\,\frac{a^2 g \Delta\rho}{\eta_1} \left(\frac{\eta_1 + \eta_2}{\eta_1 + \frac{3}{2}\eta_2}\right) \qquad (1)$$

where $a$ is the body radius, $g$ is the gravitational acceleration, and $\Delta\rho$ is the density contrast. Since $\eta_2 << \eta_1$, (1) can be simplified to,

$$V = \frac{1}{3}\,\frac{a^2 g \Delta\rho}{\eta_1} \tag{2}$$

This relation (2) can be further restricted by allowing the body only a specific amount of time to rise to the surface. This time will be less than the time necessary for the body to solidify before reaching the surface. From heat conduction theory this time is proportional to $a^2/K$ where $K$ is the thermal diffusivity and $a$ is the body radius. Noting that $V = Z/t$ or $ZK/a^2C$, where $Z$ is the ascent distance and $C$ is a constant dependent on the degree of cooling, and conveniently chosen as .3 for about 25 percent cooling [*Jaegar*, 1968],

$$\frac{Z}{a} = \frac{a^3 g \Delta\rho}{10\eta_1 K} \tag{3}$$

The nondimensional quantity on the right-hand side of (3) is similar to the Rayleigh number for convective instability.

It is instructive to use (3) and construct a graph of mantle viscosity ($\eta_1$) versus body radius ($a$), all other parameters being known. This is shown in Figure 2 and indicates that unless the body is quite large, 20 km or more in diameter, the local mantle viscosity must be very low. Figure 2 may be somewhat misleading, however, since (1) is found by assuming an isothermal situation ($\eta$ constant) whereas in this case the body is heating up the surrounding mantle and forcing its way through. The viscosity of the wall rock is very sensitive to temperature($\eta \sim \exp 1/T$). In the thermal boundary layer surrounding the body, the wall rock viscosity will thus be anomalously low, allowing the body to pass. The body in effect 'burns' its way to the surface. This might be called a hot-Stokes problem, and, to my knowledge, has not been solved. Nevertheless, Figure 2 is instructive in pointing out some of the necessary interactions involved in magma ascent. A typical body radius is difficult to choose. A characteristic ascent time, consequently, is also uncertain; however, for $a = 10^6$ cm and $10^5$ cm the respective ascent times using $t = .3\,a^2/K$ ($K = 10^{-2}$ cm$^2$ sec$^{-1}$) are about $10^6$ and $10^4$ years, or, over a distance of 100 km, the respective velocities are 10 cm/yr and 1000 cm/yr. There does seem to be a close time correspondence between increased

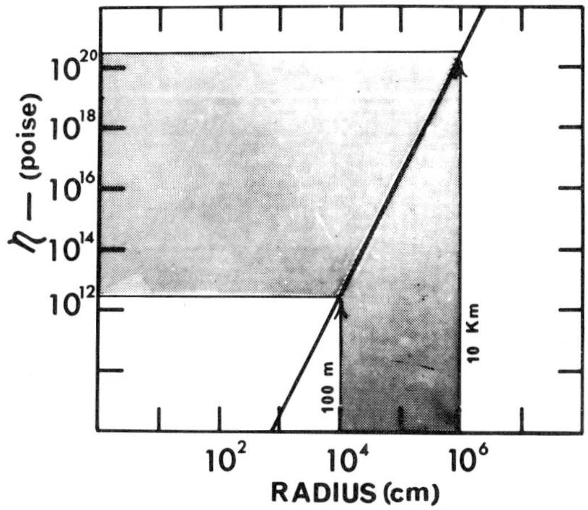


Fig. 2. Plot of (3) with $Z = 120$ km, $\Delta\rho = 0.5$g cm$^{-3}$, $g = 980$ cm sec$^{-2}$, $K = 10^{-2}$ cm$^2$sec$^{-1}$, or $\eta = (10^{-3.4})a^4$ poise. The apparent mantle viscosity as felt by the ascending magma is given by the vertical axis and the associated magma body radius ($a$) is given by the horizontal axis.

rate of plate convergence and volcanic activity which may imply that the smaller ascent time is more characteristic and, hence, a characteristic body radius is closer to $10^5$ cm.

*Ascent cooling curves.* In the conduction approximation, a set of $T$ versus depth magma cooling curves can be calculated by assuming that the heat transfer is similar to that of a spherical body when moving through a bath at some lower temperature, namely that of the mean mantle temperature encountered over an arbitrarily small ascent distance. Other thermal effects such as adiabatic cooling and viscous heating can also be taken into account. Details are given in *Marsh* [in press, *b*]. A set of curves calculated in this manner is given in Figure 3 for a uniform ascent velocity. An appropriate andesite liquidus and solidus and a peridotite solidus are also shown.

Physically these curves seem reasonable and fit the conditions of crystallization observed in andesite lavas, if not in most lavas. In the early ascent stages the temperature difference between the magma and the mantle wall rock will be small and magma cooling will be minimal; because the viscosity of the mantle may increase upward toward the surface, the ascent velocity will probably be fastest in this region. The magma may rise nearly isothermally for much of its early ascent, and internal convection will be forced by shear on the boundary. When the wall rock temperature diminishes rapidly upon approaching the surface, cooling becomes increasingly rapid

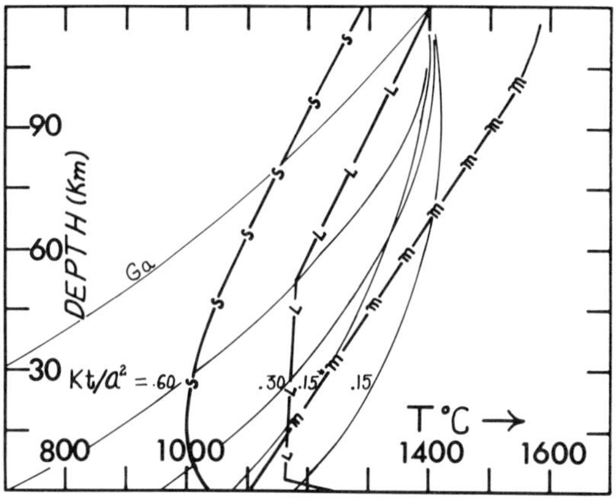


Fig. 3. Cooling curves for a spherical body of magma ascending at constant velocity. The curve Ga is the assumed mantle geotherm. The ascent velocity and body radius are characterized by the dimensionless parameter $Kt/a^2$, where $K$ is thermal diffusivity, $t$ is time and $a$ is body radius. Total viscous heating equal to the loss of potential energy upon ascent is assumed in all cases except for those values of $Kt/a^2$ marked with an asterisk where viscous heating cancels adiabatic cooling [see *Marsh,* in press, *b*]. The curves labeled S, L, and m are, respectively, the andesite solidus, liquidus (with 1.0% (wt.) $H_2O$), and a mantle peridotite solidus.

and the magma may undergo free convection. If for any reason the magma should tarry as it nears the surface, its chances of solidifying before erupting increase rapidly.

The above argument has assumed only conductive cooling; convective heat transfer within the magma is also surely important but is difficult to represent analytically. In a study of the importance of heat transfer by circulation (forced convection) in liquid drops moving through various fluids, *Calderbank and Korchinski* [1956] found that internal circulation increased heat transfer by a factor of 2.25. Free internal convection may also contribute to the total heat transfer, although this effect is difficult to evaluate; it is probably only important when the body is near the surface. Nevertheless, the rate-controlling step is the ability of heat to escape from the magma into the wall rock, and conduction calculations should therefore be reasonable to first order.

The general shape of these curves indicates that the magma is likely to be superheated over a portion of its ascent path. Which of the family of cooling curves is most applicable cannot be decided without additional information. An estimate of the point where the desired cooling curve intersects the liquidus can be found by determining the depth at which initial crystallization began. This depth can be found by studying the phenocrysts found in the lava. Once it is known, an appropriate cooling curve, and value of $Kt/a^2$, can be found.

Phenocrysts in andesitic lavas seldom, if ever, indicate crystallization under a pressure greater than about 5 kb [*Marsh,* in press, *a*]. This is shown for instance, by the $Na_2O$ content of clinopyroxene, which is directly proportional to the pressure of crystallization [*Marsh,* in press, *b*]; clinopyroxene phenocrysts in andesitic lavas show $Na_2O$ contents identical with that of the ground-mass clinopyroxenes which crystallized at about 1 bar. It might be reasoned that all the high-pressure phenocrysts settled out before eruption. Nevertheless, occasionally some should show up at the surface, but none have been observed. *Bartlett* [1969] has shown that, due to magma convection, crystal settling in andesitic magmas may be unlikely in the near-surface environment.

Mantle-like olivine xenocrysts, however, are occasionally found which are large, slightly strained, and typically forsteritic $Fo_{92}$ [*Ewart et al.,* 1973; *Marsh,* in press, *a*]. These xenocrysts are strikingly different from any olivine indigenous of the magma. Oddly enough, no coeval ortho- or clinopyroxene is found, in contrast to the case of the ultramafic nodules so common in alkali basalts. This singular presence of olivine suggests that when the magma acquired the mantle fragment the magma temperature was hot enough to melt all the mantle constituents except olivine, which is the peridotite liquidus phase. Even if a small amount of water is present in the mantle, a minimum magma temperature for this to occur is given by the dry peridotite solidus shown in Figure 3. An appropriate value of the parameter $Kt/a^2$ is likely to be less than or equal to about 0.3. The mineralogical constraints thus lead to the result that the magma is probably superheated over the majority of its ascent. Crystallization probably begins within about 15 km of the surface.

Since no indigenous crystals can exist in the superheated magma for the greater part of its ascent, one is compelled to conclude that crystal fractionation is precluded as an agent of chemical differentiation. Even in the near-surface environment fractionation seems unlikely because of convection [*Bartlett,* 1969]. Convection will cease with increasing viscosity when crystallization begins, but the magma may then be so viscous that gravitational settling may still not occur. The partial melting process which generates andesitic magma and its interaction with wall rock during its ascent, which surely will be much less important than the former, seem to dominate its chemical evolution.

*Source geometry and secondary phenomena.* It was indicated earlier, when discussing Rayleigh-Taylor instability, that two very young volcanic centers, Bogoslof and Amak, have appeared about 50 km north of the main Aleutian arc. Their positions seem consistent with a three-dimensional pattern of fluid instability. If this interpretation is correct, a lower bound on the downdip extent (width) of the magma source region is about 70 km. There is, as usual, an alternate interpretation. As an increasing amount of lithosphere is thrust into the mantle the thermal regime along the slab–mantle interface may slowly change to lower temperatures. If melting, initially restricted to a short distance downdip, say, about 30 km, is to be maintained, the source region will migrate along the interface to greater depth according to the ambient temperature. Magma will continue to ascend the initial diapiric conduit until the source has moved so far as to render it more economical, in terms of energy, for the magma to burn a new diapiric conduit to the surface at this greater depth. The original volcanic center, fed by the first diapiric conduit, would then become extinct and the new one would take over. During a certain transition period, however, both conduits would be drawing magma from the same source. Thus, the chemistry of coeval lavas from each center erupted during this transition period provides a check on which interpretation is the more correct. If the lavas exhibit similar $K_2O$ contents, the migrating, more localized source is favored; distinctly different lavas, on the average, with the younger group showing more $K_2O$, favor the long inclined source with each center drawing from regions of different pressure and temperature.

Table 2 gives several chemical analyses of lavas of Pleistocene age from the volcanic center pair represented by Frosty volcano, near Cold Bay, Alaska, and Amak Island, about 50 km north of Frosty. The lavas from Amak exhibit a distinct enrichment in $K_2O$ over any yet found at Frosty volcano. This favors the long inclined, continuous magma layer. For the other available Aleutian pair, consisting of Okmok caldera and Bogoslof Island, the test is more difficult to perform.

TABLE 2. Composition of Pleistocene Lavas from Frosty and Amak Volcanic Centers, Alaska

| | Frosty Volcanic Center | | | Amak Volcanic Center | | |
|---|---|---|---|---|---|---|
| $SiO_2$ | 57.30 | 57.83 | 59.21 | 56.64 | 59.09 | 59.33 |
| $TiO_2$ | 0.77 | 0.64 | 0.62 | 0.71 | 0.54 | 0.70 |
| $Al_2O_3$ | 18.21 | 18.17 | 18.02 | 18.82 | 18.97 | 18.52 |
| $Fe_2O_3$ | 2.98 | 3.37 | 3.57 | 4.12 | 5.24 | 2.32 |
| FeO | 4.54 | 3.83 | 3.22 | 2.84 | 0.30 | 3.58 |
| MnO | 0.16 | 0.17 | 0.16 | 0.21 | 0.20 | 0.13 |
| MgO | 3.06 | 3.46 | 2.96 | 2.36 | 1.46 | 2.39 |
| CaO | 8.01 | 7.81 | 7.26 | 7.52 | 5.95 | 6.82 |
| $Na_2O$ | 3.04 | 3.19 | 3.47 | 4.65 | 5.14 | 4.18 |
| $K_2O$ | 1.24 | 1.05 | 1.25 | 1.88 | 2.28 | 1.86 |
| $P_2O_5$ | 0.19 | 0.16 | 0.23 | 0.23 | 0.28 | 0.20 |
| $H_2O^+$ | 0.18 | 0.29 | 0.21 | 0.28 | 0.38 | 0.25 |
| $H_2O^-$ | 0.05 | 0.07 | 0.09 | 0.07 | 0.11 | 0.07 |
| $CO_2$ | <.05 | | <.05 | | | |
| Total | 99.73 | 100.04 | 100.27 | 100.33 | 99.94 | 100.35 |

Data from *Marsh* [in press, *a* and unpublished].

Bogoslof appeared only about 200 years ago, and most analyses to date have been of highly oxidized, altered rocks. These analyses are, nevertheless, similar to those of Amak [*Byers,* 1961]. The great caldera, Okmok, on Umnak Island, shows some recent lavas not typically andesitic (>1.% $TiO_2$). It may be that, owing to the rise of superheated andesitic magma, the mantle may be partially fused at shallow depths producing, ironically (i.e., the andesite was produced from a subducted tholeiite and now it produces a tholeiite), a more tholeiitic lava; that is, the casing of the pathway may be ascending and erupting. Although such behavior could be a consequence of the superheated model, Okmok seems the only area in the Aleutians where this may be happening. It, however, probably is one of the oldest and most voluminous centers in the Aleutians and therefore it is probable that such a process would be initiated there. Aside from these difficulties, the inclined layer model seems more probable. No island arcs today have a well-developed secondary arc 'downdip' from the original arc, from which one might infer that the source region re-

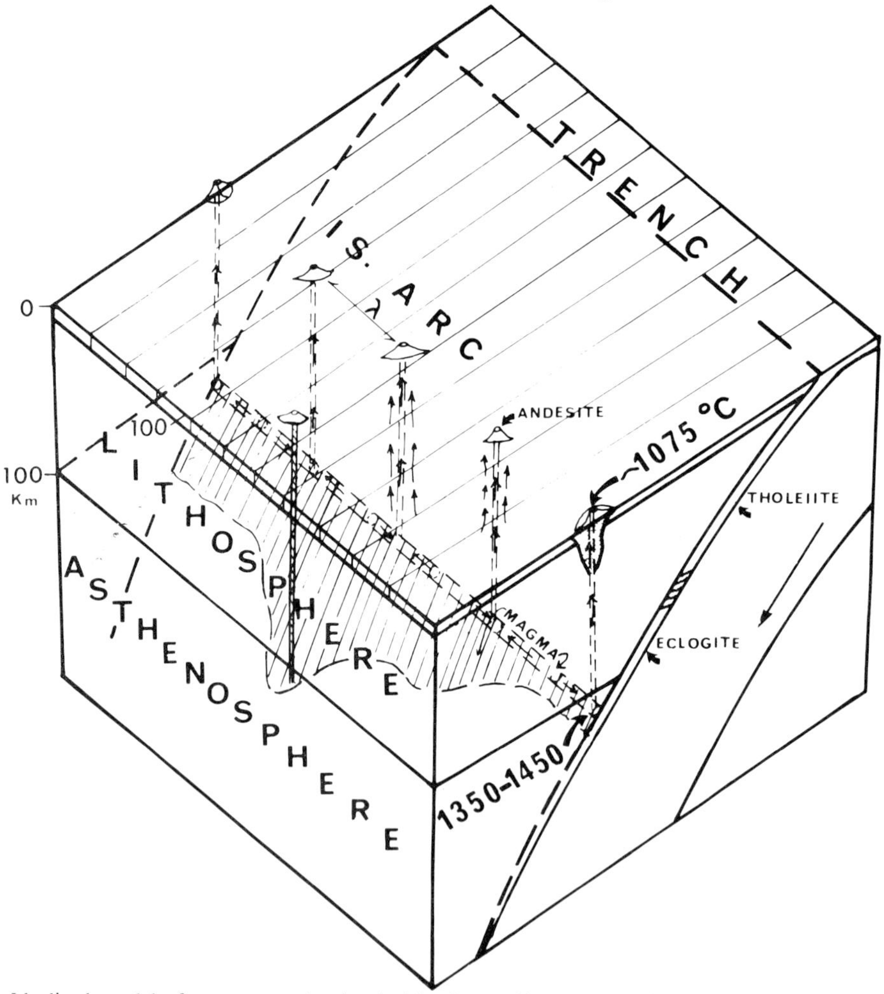


Fig. 4. Idealized model of magma production in island arcs. Upon subduction the oceanic crust undergoes an isochemical phase transition to quartz eclogite. At a depth of about 110 km the quartz eclogite is partially melted to produce an andesitic magma. Initially a very localized strip of magma is produced on the Benioff zone along the entire length of the island arc. This layer of low-density magma is gravitationally unstable, and as it strives to stabilize itself it tends to form volcanic centers at equally spaced distances along the island arc. The Tongan island arc may be an example of this type of youthful island arc magmatism. With time the zone of magma production may grow downdip along the Benioff zone, and eventually undergo instability producing a volcanic center about 50 km behind the initial island arc. Bogoslof and Amak islands in the Aleutian island arc are probably examples of this more mature type of island arc magmatism.

mains relatively fixed in time. In conclusion, the maximum downdip extent of the zone of magma production is perhaps 100 km, but in general it is probably only several tens of kilometers. This interpretation of the mechanics of island arc development in later stages can be combined with that given previously [*Marsh,* 1973; *Marsh and Carmichael,* 1974] to generate a more general model shown in Figure 4.

There is, however, another way in which the width of the zone of magma production can be estimated (J. Verhoogen, personal communication, 1974). The oceanic crust undergoes a temperature increase of about 1400°C in descending about 110 km vertically. It has gained roughly 12.7 degrees, or 3.8 calories per vertical kilometer. To melt, say, 50% of the eclogite an additional heat of fusion must be added; this is about 50 cal per gram. To acquire this energy the slab must move an additional 13 km vertically or about 18 km downdip, assuming a dip of 45 degrees. This must be considered a minimum distance as the energy gain is surely nonlinear, being less with increasing depth. The initial zone of magma production, then, may extend 30–40 km downdip. Similarly, if the slab descends at 6 cm $yr^{-1}$, its uppermost portion gains energy at a mean rate of about $6.5 \times 10^{-4}$ erg $cm^{-3}sec^{-1}$ ($1.55 \times 10^{-11}$ cal $cm^{-3}sec^{-1}$) over the first, say, 150 km of descent. Under what conditions can this be supplied? To answer this question the thermal history of the slab must be considered.

## THERMAL MODELS, SHEAR STRESS AND SHEAR ZONE THICKNESS

A sharp depression of mantle isotherms along the upper edge of the subducting lithosphere, as suggested by some thermal models [e.g., *Toksöz et al.,* 1973] is incompatible with the generation of andesitic magma from partial melting of the subducted oceanic crust. The calculated temperatures are much too low to produce melting at the proper depths. Although the 'normal' thermal regime of the upper mantle is poorly known, the previously suggested temperatures for andesite generation (1350°–1450°C, at depths of 100–150 km) seem to imply that the mantle isotherms, to first order, are not depressed and may even locally be deflected slightly upward along the slab–mantle interface. From the great variety of thermal models calculated by Toksöz and his associates [e.g., *Minear and Toksöz,* 1970; *Toksöz et al.,* 1973] the magnitude of the shear strain heating along the slab–mantle interface appears to be a critical quantity in deciding the isotherm geometry in this region [*Oxburgh and Turcotte,* 1970].

A probable, first order, necessary condition for slab melting at the appropriate depth, thus, is that the mantle isotherms continue approximately horizontally until the slab (eclogite) is encountered. To cause this, the total heat produced in the shear zone must balance that lost upward to the mantle and downward into the cooler slab plus that absorbed by the shear zone. (This is usually the case with analytical thermal models [e.g., *Griggs,* 1972] which choose the temperature gradient along the slab–mantle interface to be the 'normal' mantle temperature gradient, and thus implicitly assume a good deal of shear strain heating.) Once the total necessary heat flow is found, this can be balanced through the product of velocity and shear stress to find the magnitude of the required shear stress. The upward component must be as large as the 'normal' vertical component of heat flow in the mantle at this depth were the slab not present; estimating from the mantle thermal models used by *Toksöz et al.* [1973], at 100–150 km this component is about $.45 \times 10^{-6}$ cal $cm^{-2}sec^{-1}$. The component into the slab, as an extreme, can be estimated by assuming a thermal gradient of 500 degrees over 10 km which gives about $3.8 \times 10^{-6}$ cal $cm^{-2}sec^{-1}$. The amount of heat needed to raise the temperature of the subducted oceanic crust to fusion temperature was previously found to be about $1.55 \times 10^{-11}$ cal $cm^{-3}sec^{-1}$, and to anticipate later results, for a maximum fusion zone thickness of, say, 100 m the necessary heat flow 'into' this region is then $1.55 \times 10^{-7}$ cal $cm^{-2}$sec. The total heat produced in the zone needed to balance these flows is thus $1.84 \times 10^{2}$ ergs $cm^{-2}sec^{-1}$.

This value is now equated to the heat flux produced in simple plane shearing flow at constant shear stress (e.g., Couette flow at vanishing Reynolds number [see *Turcotte and Oxburgh,* 1968]), that is:

$$q_T = \tau V_0 = 1.84 \times 10^2 \text{ erg/cm}^2 \text{ sec} \quad (4)$$

$$\tau = q_T / V_0 \quad (5)$$

where $\tau$ is shear stress and $V_0$ is velocity. It is emphasized that (4) is valid for any type of creeping process; it holds equally well for Newtonian or non-Newtonian materials. Assuming a spreading velocity of 6 cm $yr^{-1}$ ($1.9 \times 10^{-7}$ $cmsec^{-1}$) gives a shear stress of about 970 bars. Taking a much higher mantle heat flow of $.75 \times 10^{-6}$ cal $cm^{-2}sec^{-1}$ gives a stress of about 1036 bars. Similar calculations by *Turcotte and Schubert* [1973] and *Oxburgh and Turcotte* [1970] give a maximum shear stress of about 1–2 kb (when corrected for a 1400°C fusion temperature).

In order to determine the total heat production in the shear zone and also to gain a feeling for how much of the oceanic crust is involved in the fusion process, the shear zone thickness ($h$) must be known. In numerical thermal models $h$ is constrained by the computational grid size used, to be about 10 km or more [e.g., *Toksöz et al.*, 1973]. In such a case, the total heat produced in the shear zone may be larger than the amount of energy available from loss of gravitational potential energy from the descending slab [*Hanks and Whitcomb,* 1971]. This difficulty is overcome, however, if the shear zone is much thinner than 10 km [*Hanks and Whitcomb,* 1971]. *Turcotte and Oxburgh* [1968], cleverly using a model of Newtonian fluid flow between parallel plates, showed that for a shear stress of 1 kb the shear zone thickness would be about 20 km. It is now apparent, however, that a non-Newtonian or power law stress–strain rate constitutive relation is probably a more accurate rheological description of earth material (e.g., see AGU Geophys. Monogr., vol. 16), especially in regions of large velocity gradients such as the shear zone considered here. The method of analysis of *Turcotte and Oxburgh* [1968] can be extended to the non-Newtonian case; however, *Gavis and Laurence* [1968] have given a somewhat more general result which can be used to calculate the shear zone thickness.

In Couette flow between parallel isothermal plates it has been shown [*Gavis and Laurence,* 1968] that, associated with the flow region of thickness $h$, there is a maximum shear stress which can be exerted on the fluid. That is, as the shear stress is increased beyond a certain point the fluid 'viscosity' decreases, due to viscous heating, faster than the strain rate increases, thus a maximum is reached in the shear stress which can be applied to the fluid. Conversely, by posing the shear stress, and various other known parameters, the thickness of the shear zone can be estimated. The maximum shear stress for such fluid flow is given by

$$\tau \leqslant \left( \frac{3.5knT_0 m_0^{\frac{1}{n}}}{\beta h^2} \right)^{\frac{n}{n+1}} \tag{6}$$

A similar group of parameters is given by Turcotte and Oxburgh for the Newtonian case; here $k$ is thermal conductivity ($4.186 \times 10^5$ egs $deg^{-1}$ $cm^{-1}sec^{-1}$), $n$ is the exponent from the constitutive relation given below, $T_0$ is the characteristic temperature of the shear zone ($1.5 \times 10^3$ deg K) and $\beta$ and $m_0$ are parameters described below. The fluid is described by the constitutive relation

$$\tau = -m \left| \dot{e} \right|^{n-1} \dot{e} \tag{7}$$

where $\tau$ is the shear stress, $\dot{e}$ is the strain rate, and $m$ is represented by

$$m = m_0 \exp\left(-\beta (T - T_0)/T_0\right). \tag{8}$$

For Newtonian fluids $n = 1$ and $m$ is the viscosity. Both Newtonian ($m_0 = 10^{22}$ poise, $n = 1$, $\beta \simeq 25$) and non-Newtonian cases can be considered by putting the data of *Post and Griggs* [1973] into the form of (7) and (8) with $m_0 = 2.62 \times 10^{10}$ poise, $n = 1/3.18$, and $\beta$ about 13.66. The shear zone thickness ($h$) for each case is given in Figure 5 as a function of shear stress; these are upper bounds for $h$. In the Newtonian case the shear zone thickness for a 1-kb shear stress is, at most, 10 km, while in the other case it is extremely thin ($\simeq$60 cm). Assuming the shear stress is as previously calculated (1–2 kb), the maximum slip zone thickness is 5–10 km; while if the non-Newtonian model holds, it is of knife-edge thickness as in continental thrust faults. The plug-flow exhibited by salt domes and the thin shear zone at the base of glaciers are probably other examples of this type of non-Newtonian or power-law fluid flow.

Since the rheology of the earth in this region is probably accurately described as a power-law fluid, and the shear stress necessary to produce melting is probably about 1–2 kb, the shear zone must be very thin if it is a continuum; its thickness is perhaps on the order of a kilometer, maybe even a few meters. The suspicions of *Minear and Toksöz* [1970] and *Hanks and Whitcomb* [1971] regarding a thin shear zone seem correct; this feature has not been properly portrayed by the previous numerical models. Using a shear zone thickness of, say, 100 m and combining this with the previous shear zone heat flux calculation ($q_T = 1.84 \times 10^2$ erg $cm^{-2}$ $sec^{-1}$) implies a heat production rate of $1.84 \times 10^{-2}$ erg $cm^{-3}$ $sec^{-1}$, or about a hundred times larger than that used by *Toksöz et al.* [1973] for a shear zone 14 km thick. Thus, since thin shear zones are difficult to model

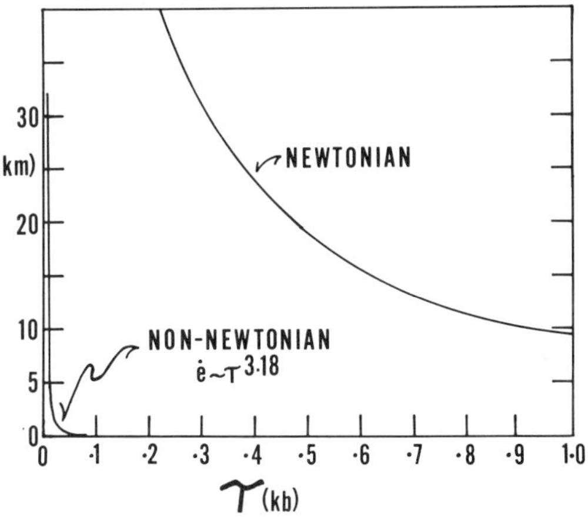


Fig. 5. Thickness of the slip zone in the region of the Benioff zone as a function of shear stress. The upper curve ($n = 1$) is for a viscous ($10^{22}$p) Newtonian earth while the lower curve is for a non-Newtonian constitutive law as suggested by *Post and Griggs* [1973] where $\dot{e} = A\tau^x$ and $x = 3.18$ (see text).

numerically, it may be easiest to increase the heat production in the numerical layer by about a hundred; even though the associated shear stress will be unrealistic, the results may be more reasonable.

*Wyss* [1970], among others, has pointed out that estimated stress drops, associated with intraplate earthquakes, when corrected for inefficiency, pass through a maximum ( ~ 3 kb) at a depth of about 100 km. It is interesting that this is also the characteristic depth of magma production. For near-surface and very deep earthquakes, the actual stress drops are probably much lower (50–100 bars). Since the zone of intraplate earthquakes in the Aleutian Islands has been determined to be about 10–20 km below the upper plate edge [*Engdahl,* 1973], this may imply that along the lower edge of the shear zone the plate is able to withstand a 1–2 kb shear stress. This is also suggested by the calculations of *Griggs* [1972, Figure 12] showing that the strength of the downgoing plate at a depth of about 100 km may be as large as 5 kb within 5 km of its upper edge.

In sum, the shear stress needed along the upper edge of the subducting lithosphere to produce melting of the subducted oceanic crust is probably about 1–2 kb, and the width of the shear zone is perhaps much less than a kilometer. Consequently, at any time a layer of subducted oceanic crust perhaps only a few tens of meters thick is involved in the generation of andesitic magma at the slab-mantle interface.

## DIRECT SOURCE DETECTION

*Grow and Qamar* [1973] have examined the near-source arrivals of rays emerging from intraplate earthquakes beneath the western Aleutian Islands near Amchitka. They found strong attenuation in *P* and *S* phases; in some instances the *S* phase was absent. Good instrument coverage transverse to the arc is possible in this region due to the divergence between the islands representing the older, Tertiary, arc (e.g., Amchitka Island) and the Recent volcanic islands (e.g., Semisopochnoi Island). The arrivals recorded on the active volcanic islands showed strong attenuation for earthquakes deeper than 25 km; some arrivals recorded at Semisopochnoi showed no *S* phase. Grow and Qamar's interpretation of the geometry of the highly attenuating region places its lower boundary coincident with the slip zone, extending from about 90 to 200 km downdip and, at the upper end, protruding to within about 50 km of the surface. Laterally the zone is about 40 km wide in this region. Since linear ray paths were assumed, these dimensions are probably an upper limit. Ray tracing may show the zone of high attenuation to be more restricted; however, due to the thermal effects of the rising magma and probable complexity of the ascent region, a certain seismic fuzziness is to be expected.

*Grow* [1973] has also presented a detailed interpretation of a gravity profile through the central Aleutian Islands, nearly midway between the active volcanic centers of Atka and Seguam. Although his most plausible interpretation shows a low-density column (20 km wide) extending from the surface to the Benioff zone, there are no volcanoes at the surface nor any sign of submarine volcanic activity where the gravity profile pertains. Thus, the mass deficiency should probably be placed at a greater depth with a larger density contrast distributed over a smaller volume. Nevertheless, Grow's results may indicate the presence of a much larger negative anomaly associated with the nearest active volcanic center, Atka, of which the interpreted profile only shows the anomaly edge. A gravity profile transverse to the arc and very close to one of the large active volcanic centers may give a more complete picture of the diapiric conduit geometry. These results imply the existence of a relatively small and restricted zone of magma immediately beneath the island arc. There is no sign of widespread melting in the upper mantle above and behind the island arc [*Barazangi et al.,* 1975]. These results are in good agreement with the preceding results reported in this paper.

*Acknowledgments.* It is a pleasure to acknowledge John Verhoogen for his many helpful comments on this work. Tony Ewart, Dave Elliott, Ian Carmichael, W. Glen Zinn, F. D. Bennett, and R. Zimmerman also kindly helped in a number of ways. This work was supported in part by NSF Grant DES-75-18451. This support is gratefully appreciated. Some of this work was supported by NSF Grant GA-38086 (Carmichael).

# 31. $^{40}Ar/^{39}Ar$ Ages of Submarine Rocks from the Line Islands: Implications on the Origin of the Line Islands

K. SAITO and M. OZIMA

*Geophysical Institute, University of Tokyo, Tokyo 113, Japan*

*Abstract.* $^{40}Ar/^{39}Ar$ ages of two submarine basalts dredged from the north to central part of the Line Islands volcanic chain and of two DSDP basalts (DSDP-66 and DSDP-167) drilled from the adjacent ocean floor are presented. The ages of the two dredged basalts are inconsistent with a hot spot origin of the Line Islands chain. Sample DSDP-66 gives a $^{40}Ar/^{39}Ar$ age of 52 m.y., which is younger than the fossil age in the overlying sediments. It is suggested that basalt DSDP-66 is a later intrusive.

*Wilson* [1963*a*] first proposed that the linear Hawaiian volcanic chain formed as oceanic crust moved over a magma source in the mantle. *Morgan* [1972*a*] extended this hypothesis to explain other volcanic chains such as the Austral–Marshall–Gilbert, and the Tuamotu–Line as well as the Hawaiian–Emperor, on the basis of a 'hot spot' fixed to the mesosphere. *Clague and Jarrard* [1973] showed that the movement of the Pacific plate deduced from analyses of the Hawaiian-Emperor chain on the basis of the hot-spot hypothesis agrees with the movements expected from paleomagnetism of seamounts as well as from the sedimentation pattern in the equatorial Pacific. Radiometric ages of volcanic rocks [*McDougall,* 1964; *Clague and Dalrymple,* 1973] and ages of fossils [*DSDP Scientific Staff,* 1971*b*] on seamounts in the Hawaiian–Emperor chain are consistent with a hot-spot origin of this chain. However, age data by which to test such an origin for the Tuamotu–Line and the Austral–Marshall–Gilbert chains are still limited. Recently, data from DSDP legs 17 [*Winterer, Ewing et al.,* 1973] and 33 [*DSDP Scientific Staff,* 1974*a*] in the central to south part of the Line chain have shown that the basement fossil ages in the holes are almost the same, ranging from 79 m.y. to 85 m.y. These simultaneous ages appear to argue against a hot-spot origin for that chain [*DSDP Scientific Staff,* 1974*a*]. However, as the three DSDP samples are from the south to central part of the chain and no radiometric ages of basement rocks from the chain have been reported, we applied $^{40}Ar/^{39}Ar$ stepwise dating [*Merrihue and Turner,* 1966] to rocks dredged from seamounts in the north to central part of the chain and to rocks drilled from the adjacent ocean floor.

In contrast to conventional K/Ar dating, whose validity is often obscured by Ar loss or Ar excess commonly encountered in submarine rocks, the $^{40}Ar/^{39}Ar$ stepwise dating method has the following advantage. $^{40}Ar/^{39}Ar$ age is determined from concordant ages obtained for several gas fractions released at different temperatures. Gas fractions released at the same temperature must come from certain lattice sites or interstitial positions which are likely to be different from those released at other temperatures. If rocks have suffered Ar loss or contain excess Ar, it would be too fortuitous to expect that exactly the same degree of Ar loss or Ar excess had affected all the gas fractions giving concordant ages. Since the $^{40}Ar/^{39}Ar$ stepwise degassing age is determined from concordant ages on different temperature fractions, the reliability of the age thus obtained is self-assuring. An age determined by the $^{40}Ar/^{39}Ar$ method is therefore analogous to concordant K/Ar ages determined on separate minerals. The purpose of the present paper is to offer reliable absolute ages determined by the $^{40}Ar/^{39}Ar$ stepwise dating method of some submarine rocks from the Line chain and adjacent area. Such information should be vital for further understanding of the origin of this island chain.

## EXPERIMENTAL METHOD AND SAMPLES

The $^{40}Ar/^{39}Ar$ stepwise-dating method is based on the same principle as the conventional K/Ar dating method, i.e., age is calculated from the ratio of radiogenic $^{40}Ar$ to $^{40}K$ in a sample. In the $^{40}Ar/^{39}Ar$ dating method, $^{40}K$ is determined by measuring $^{39}Ar$ which is produced from $^{39}K$ in a sample by fast neutron irradiation, i.e., $^{39}K(n,p)^{39}Ar$. $^{40}Ar/^{39}Ar$ stepwise degassing data can be treated in two ways. One is to plot an apparent age for each temperature fraction against the percentage fraction of $^{39}Ar$ released [*Turner et al.,* 1966]. The calculation of the apparent age is essentially the same as in a conventional K/Ar dating apart from the use of $^{39}K$-induced $(^{39}Ar)_K$ instead of $^{40}K$ content. (Such age spectrum plots are shown on the left in Figures 2 to 5.) An age spectrum that is flat then indicates concordance in the ages determined for different temperature fractions, assuring the reliability of the age. However, it is to be noted that in order to construct the age spectrum one must assume an atmospheric argon isotopic ratio ( $(^{40}Ar/^{36}Ar) = 295.5$) for non-radiogenic trapped Ar in the sample. Since there are several reports that the isotopic ratio of contaminated Ar is different from the atmospheric one [*Brereton,* 1972; *Dalrymple and Lanphere,* 1974], one must be careful in interpreting the age spectrum.

The other way to treat the experimental data is to plot $(^{40}Ar^*/^{36}Ar^*)$ ratio against $(^{39}Ar^*/^{36}Ar^*)$ ratio, in which the asterisk indicates Ar isotopes corrected for interfering Ar isotopes produced by neutron irradiation [*Merrihue and Turner,* 1966]. (Such correlation plots are shown on the right in Figures 2 to 5.) If the experimental data fall on a straight line in the plot, we can calculate an age for the sample by comparing the gradient of the correlation line with that of a standard sample of known age. If some data points lie off the straight line, this indicates that the sample has not remained as a closed system with respect to K and Ar or contains non-uniform trapped Ar. Even in the latter case, however, it is still possible in some instances to resolve the primary solidification age by ignoring the data points lying off the straight line [*Ozima and Saito,* 1973]. In the correlation plot it is not necessary to make an assumption about the isotopic ratio of the contaminating Ar, for the isotopic ratio can be determined from the $y$-intercept of the correlation line. In addition, since the statistical error evaluation is more easily made for the correlation age than for ages where ratios are assumed, it would be advisable to discuss the age data in terms of the correlation age so that an inter-laboratory comparison can be made on a more objective basis.

Except for sample DSDP-66, interfering Ar isotopes produced from K and Ca in each sample were corrected with correction factors $(^{37}Ar/^{36}Ar)_{Ca}=0.00026$, $(^{39}Ar/^{37}Ar)_{Ca}= 0.0017$, and $(^{40}Ar/^{39}Ar)_K= 0.043$. These correction factors were determined for each experiment using $K_2SO_4$ and $CaCO_3$. In the case of sample DSDP-66, on which the work was done at an early stage in this study, the correction factor $(^{37}Ar/^{36}Ar)_{Ca}$ was determined by neutron irradiation of potassium-poor pyroxene, and other factors were adopted from the literature [*Ozima and Saito,* 1973]. The uncertainty due to the correction factors is indicated by a bar attached to each experimental point in Figure 4. Corrections for K-induced Ar isotopes are insignificant except for a standard sample. Only the corrections for Ca-induced $^{37}Ar$ are significant for higher temperature fractions. Since the higher temperature fractions are in all cases insensitive in defining the correlation lines, the corrections due to interfering Ar isotopes did not affect the resultant $^{40}Ar/^{39}Ar$ ages. $^{40}Ar/^{39}Ar$ ages were calculated from the comparison of the slopes of the correlation lines in the $(^{40}Ar^*/^{36}Ar^*)$–$(^{39}Ar^*/^{36}Ar^*)$ diagrams between the sample and a standard muscovite (Bern 4M) prepared by Professor E. Jager.

Samples were first wrapped in aluminum foil and then sealed in a quartz tube at a pressure of about $10^{-3}$ torr. Several vacuum-sealed quartz tubes were then irradiated by a neutron flux of about $10^{18}$ ntv. Ar gases were extracted in a high-vacuum line from samples for 90 min (45 min for sample DSDP-66) at each temperature. After purifying the extracted gases with a hot Ti sponge, isotopic measurements were made by a Reynolds-type mass spectrometer.

Samples 133D and 142D were collected on Expedition Seven-Tow of the Scripps Institution of Oceanography from the north to the central part of the Line chain. DSDP-66 and -167 were drilled from the sea floor adjacent to the Line chain. These sampling sites are shown in Figure 1.

## $^{40}Ar/^{39}Ar$ AGES

In Figures 2 and 3, the results of $^{40}Ar/^{39}Ar$ stepwise dating for the seamount samples are represented both on an age spectrum and on a

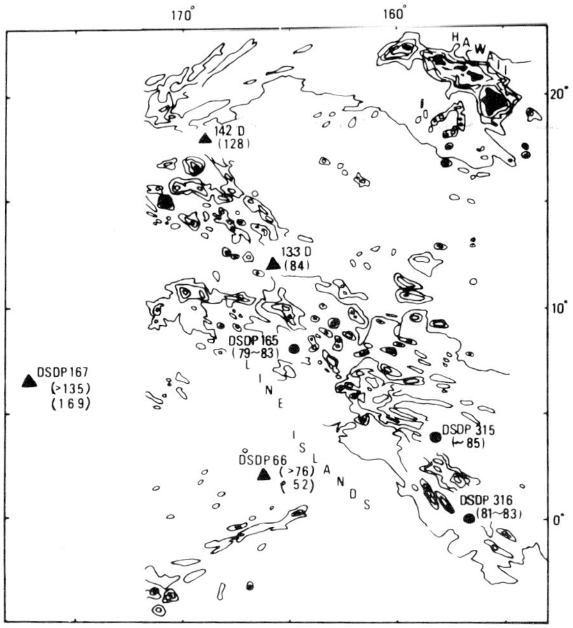


Fig. 1. Map showing the sampling sites. Triangles indicate rock samples for which $^{40}Ar/^{39}Ar$ ages have been obtained; lower bracket indicates $^{40}Ar/^{39}Ar$ age, upper bracket, fossil age. Solid circles indicate basement fossil samples. Fossil age given in parentheses.

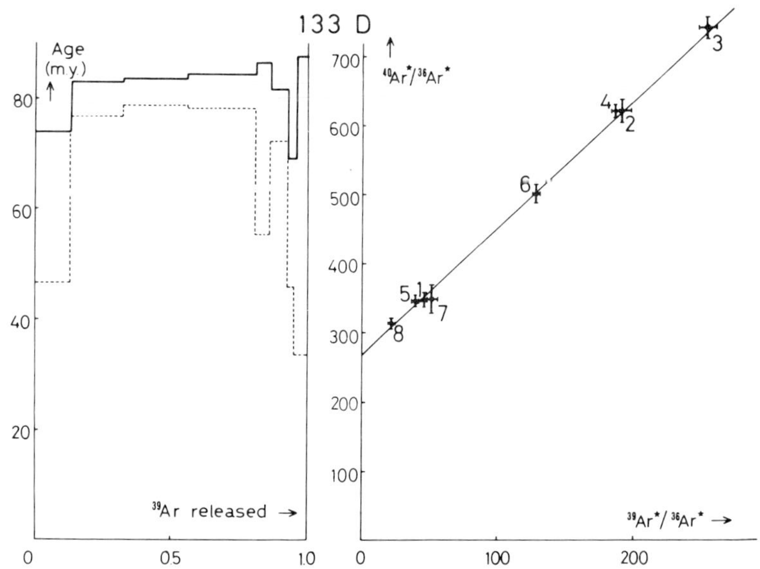


Fig. 2. Sample: 7-TOW-VI-133D-1. Apparent K/Ar age plot (on left) and $^{40}Ar^*/^{36}Ar^*$–$^{39}Ar^*/^{36}Ar^*$ correlation plot (on right) for the stepwise degassing data. The correlation line was calculated by *York's* [1969] method of least-squares fits with r (correlation coefficient) = 1. Error bars indicate 1 standard deviation. Numerals 1, 2, . . . in the correlation plot indicate the lowest temperature fraction, the second lowest, and so on.

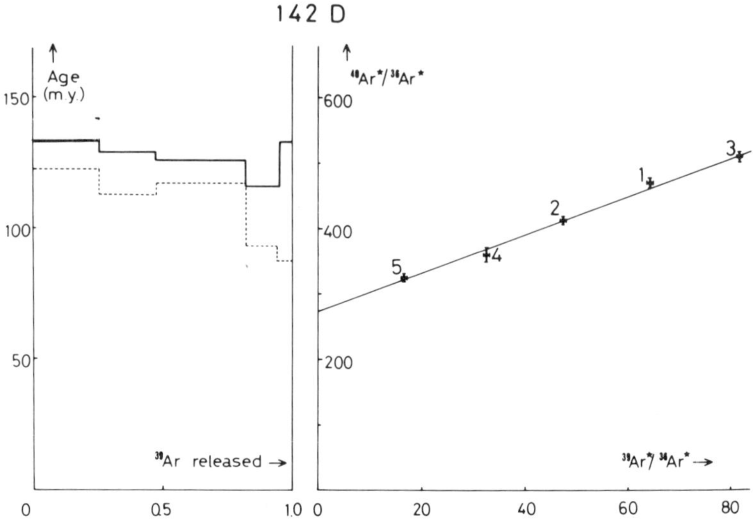


Fig. 3. Sample: 7-TOW-VI-142D-1. Notations are the same as in Figure 2.

correlation plot. In the age spectrum apparent ages were calculated by assuming that (1) nonradiogenic trapped Ar was atmospheric (shown by broken lines in Figures 2 and 3), and additionally by assuming that (2) the trapped Ar has an isotopic ratio ($^{40}Ar/^{36}Ar$) which was determined from the *y*-intercept in the correlation diagram (solid lines in Figures 2 and 3). It is seen that the latter assumption gives a flatter age spectrum. In both cases about 90% of the released gas fractions give concordant ages, suggesting the reliability of the ages. Also in the correlation plots for both samples, all the temperature fractions fall on a straight line. The gradients of the correlation lines give a $^{40}Ar/^{39}Ar$ age of 84.4±0.9 m.y. for sample 133D and 128 ± 5.0 m.y. for sample 142D. An error in the final age is essentially due to errors in the isotopic ratios. The standard deviations in the ages were calculated by means of York's method with correlation coefficient r = 1.0 [*York*, 1969].

Figures 4 and 5 present the experimental data for the two DSDP samples from the sea floor adjacent to the Line chain. Sample DSDP-66 gives a fairly flat age spectrum as well as a straight correlation line. The $^{40}Ar/^{39}Ar$ age is determined to be 52.0 ± 5.0 m.y. However, this age is much younger than the basement fossil age indicated by "poorly preserved Cretaceous radiolarians of probable pre-Campanian age (76 m.y.)" [*DSDP Scientific Staff*, 1969*b*]. Although there is no evidence of a baked contact between the sediment and the basalt, we may consider that the basalt is a later intrusive, probably a sill. The age spectrum is not flat for sample DSDP-167, but it is still possible to define a correlation line, the slope of which gives an age of 169 m.y. The age can be compared with the basement fossil age which is possibly Jurassic [*Winterer, Ewing et al.*, 1973]. Experimental data are shown in Table 1, including pertinent fossil ages.

## DISCUSSION

An older age to the northwest and younger age to the southeast for the two dredged samples is consistent with what would be expected from a hot-spot origin of the Line Islands, though obviously only two radiometric ages are not enough to prove the hot-spot origin. If we assume that the Line chain was formed by a mantle hot spot, the present $^{40}Ar/^{39}Ar$ ages would imply 0.2°/m.y. for the rotation rate of the Pacific plate with respect to the Emperor seamount pole of rotation [*Morgan*, 1972*a*] during a period from 84 m.y. to 130 m.y. The rotation rate is about one-fourth of the 0.8° per m.y. during the period from 42 m.y. to 70 m.y. estimated by *Clague and Jarrard* [1973]. On the contrary, the DSDP Leg 33 expedition found that basement fossil ages in the south to central part of the Line Islands are almost simultaneous, which is at variance with a hot-spot origin of the chain.

Although it is impossible to derive a unique interpretation of these age data, there are several possibilities. First, we may argue that the Line

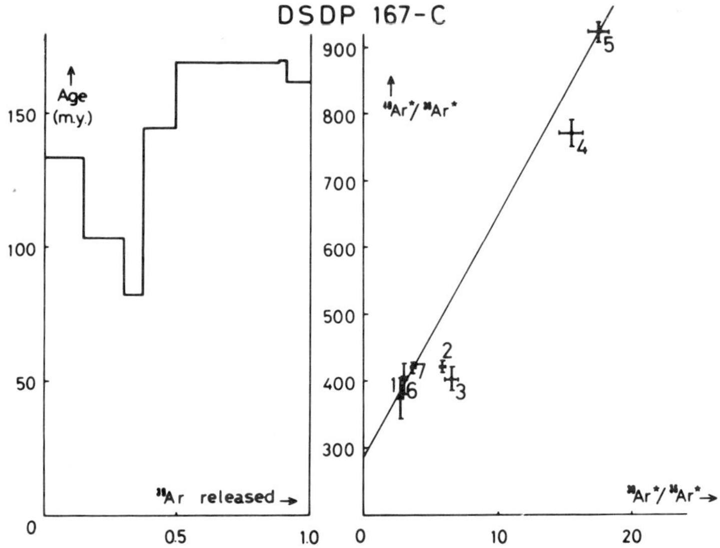


Fig. 4. Sample: DSDP-66-0-11-cc. Notations are the same as in Figure 2. See text for explanation of error bars attached to each experimental point in the correlation plot.

chain was not of hot-spot origin. *Clague and Jarrard* [1973] pointed out that the Tuamotu chain shows better parallelism to the Emperor chain than does the Line chain. Hence, they question Morgan's hypothesis that the Line chain was formed by a rotation of the Pacific plate around the Emperor pole, while the Tuamotu chain was formed by a rotation around the Hawaiian pole. In addition, the topography of the Line chain is not as simple as that of the Emperor chain. The northern end of the chain merges into the Mid-Pacific Mountains and the center of the chain is cut by the Line cross-chain. These observations suggest that the Line chain may consist of several independent chains which may or may not be of hot-spot origin.

Second, the fossil ages above basement at DSDP sites 165, 315, and 316, and the radiometric age of dredged basalt 133D indicate that the Line chain formed almost simultaneously at about 80 to 85 m.y. ago. If the entire chain formed simultaneously, perhaps the most easily understood mechanism would be a spreading-ridge origin of the Line chain. However, the absence of magnetic anomaly lineations parallel to the chain precludes the possibility that it was a spreading ridge.

As other possibilities, we may suggest that the

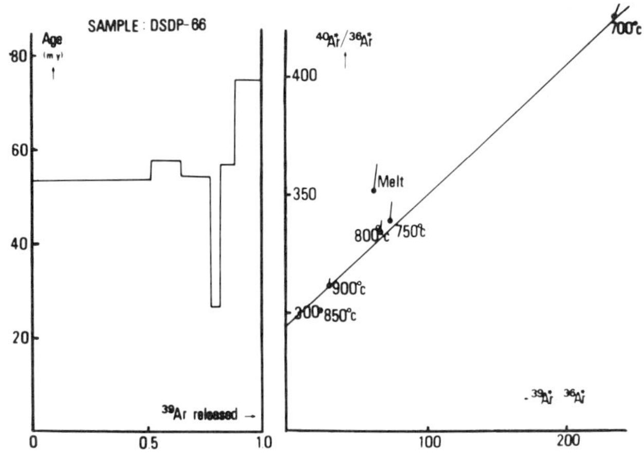


Fig. 5. Sample: DSDP-167-95-2. Notations are the same as in Figure 2.

TABLE 1. Radiometric and Fossil Ages of Samples

| Sample | Water Depth, m | Rock Type | J[1] | $^{40}Ar$–$^{39}Ar$ Age[2] | Fossil Age |
|---|---|---|---|---|---|
| 7-TOW-VI-133D-1 (Kapsitotwa S.M.) 165°50′W, 12°04′N | 1550 | Altered olivine basalt with intergranular groundmass plagioclase: relatively fresh | 0.0241 | 84.4 ± 0.9 | Late Cretaceous (Senonian) planktonic foraminifera (87 ~ 72 m.y.)[4] |
| 7-TOW-VI-142D-1 169°04′W, 18°05′N | 1570 | Fine grained trachyte plagioclase: relatively fresh | 0.0241 | 128 ± 5.0 | Quaternary planktonic foraminifera[4] |
| DSDP-66-0-11-cc 166°07.3′W, 2°23.6′N | 5293 | Vesicular basalt | 0.0493 | 52 ± 5.0 | Late Cretaceous fossils (100 ~ 87 m.y.)[5] |
| DSDP-167-95-2 176°49.5′W, 7°04.1′N | 3176 | Altered basalt | 0.00260 | (169)[3] | Tithonian–Berriasian–Turonian fossils (146 ~ 136 m.y.)[6] |

[1] $J = (^{39}Ar^*/^{40}K)\ (\lambda_e + \lambda_\beta/\lambda_e)$.

[2] $\lambda_e = 0.585 \times 10^{-10} y^{-1}$, $\lambda_\beta = 4.72 \times 10^{-10} y^{-1}$.

[3] Tentative age.

[4] E.L. Winterer (personal communication, 1973).

[5] *DSDP Scientific Staff* [1969*b*].

[6] *Winterer, Ewing et al.* [1973].

basement fossil ages (DSDP-165, -315, and -316) do not represent the final cessation of the volcanic activity in this area, or there may be a considerable time gap between the fossil age of the basement of the seamount and the age of the top of the seamount, the latter representing the final stage of the volcanic activity. Although there are few data to indicate the time span for the growth of a seamount, the long time span for the growth of St. Helena Island is very suggestive; i.e., *Abdel-Momen and Gast* [1967] found that about 5 m.y. elapsed during the formation of the part of St. Helena Island now above sea level. A simple linear extrapolation would then indicate a time span of a few tens of millions of years to permit the growth of St. Helena Island from the ocean bottom to the present elevation. It is then not entirely unreasonable to speculate that rocks dredged from the top of a seamount in the central to south part in the Line chain may give ages much younger than the fossil ages. In this respect it is interesting to recall the $^{40}Ar/^{39}Ar$ age of DSDP-66 basalt, which was drilled about 500 km to the west of the southern part of the Line chain. The age seems to indicate that volcanic activity still existed in that area long after formation of the surrounding ocean floor.

At any rate the problem seems to be still open to further investigation. $^{40}Ar/^{39}Ar$ stepwise dating of seamount rocks in the central to southern part of the Line chain should help to resolve the question. The attempt is currently being made in the authors' laboratory.

*Acknowledgments.* We thank Dr. R. D. Jarrard for many helpful suggestions on the manuscript, and Dr. D. York for a Fortran Program for least-squares fit. We wish to express our appreciation to the Scripps Institution of Oceanography and to Drs. E. L. Winterer, R. D. Jarrard, and J. N. Natland for providing samples (133D and 142D), and to the Deep Sea Drilling Project for Samples DSDP-66 and DSDP-167. We are indebted to Professor S. Yajima and Mr. M. Narui for the neutron irradiation.

# 32. *Kilauea Lava Lakes: Natural Laboratories for Study of Cooling, Crystallization, and Differentiation of Basaltic Magma*

THOMAS L. WRIGHT, DALLAS L. PECK, AND HERBERT R. SHAW

*U.S. Geological Survey, Reston, Virginia 22092*

*Abstract.* Three Kilauea eruptions have produced accessible lava ponds in the pit craters Kilauea Iki (1959), Alae (1963), and Makaopuhi west pit (1965). These have provided a unique laboratory in which to study the cooling and crystallization of basaltic magma. The results of field and laboratory studies conducted at Kilauea place constraints on processes that take place during cooling of shallow magma chambers and of thick basaltic lava flows. Field methods of study include repeated core drilling to determine the thickness of upper crust, collection of gas and measurement of temperature and oxygen fugacity in uncased drill holes, sampling and measurement of viscosity of molten basalt, and periodic measurement of changes in surface altitude and crack configuration. Laboratory studies include measurement of density, chemical composition, and conductivity of drill core, megascopic and microscopic study of crystal and glass content as a function of depth, time, and temperature, and finite-element modeling of the thermal history. Liquidus temperatures of each magma are at or above 1200°C depending on MgO content. Solidus temperatures are all near 980°C. The interface between rigid 'crust' and fluid 'melt' occurs at 1070 ± 5°C. The thickness of upper crust in each lake during the first several months increased at a nearly linear rate of 40 cm per unit change in $\sqrt{\text{time}}$ measured in days. Finite-element modeling of the cooling of Alae lava lake has established that cooling is controlled principally by conductive heat transfer. Most measured temperatures there can be computed within 50°C with a model assuming a latent heat of 80 ± 5 cal/gm and a constant diffusivity of .006 $cm^2$/sec and taking into account the heating and vaporization of the measured rainfall of approximately 250 cm/yr. The rainfall considerably hastened post-solidification cooling of the lake. Measured temperatures can be more closely duplicated using a diffusivity based on the measured density and calculated heat content of the lava and on a conductivity that decreases with porosity and increases with temperature. The cooling rate of Makaopuhi and Kilauea Iki may have been increased by convective heat transfer in the melt, as indicated by measured temperature fluctuations and melt differentiation in Makaopuhi. Three kinds of crystal–liquid differentiation are observed in the lava lakes as inferred from chemical analysis of drill core: (1) gravitative settling of olivine phenocrysts, (2) filter-pressing of low-temperature (1030–1070°C) liquids into open fractures, and (3) downward concentration of augite and plagioclase during inferred convective flow. Similar chemical variation observed in Kilauea rift eruptions suggests that these differentiation processes occur on a larger scale within the Kilauea conduits. Some features are unique to each lake. One example is the pattern of vesiculation represented by changes in crust and melt density with depth and by surface altitude changes. Vesiculation is important in determining the density stratification in each lake, and hence the convective history. The distribution of vesicles depends in part on the initial volatile content and on differences in the mode of filling of each lake. These observations made for any single lava lake cannot, therefore, be directly applied to other basaltic lava flows without consideration of evidence for initial gas content and vesiculation history.

One of the most fundamental problems of igneous petrology concerns the mechanisms by which lavas cool, crystallize, and differentiate. Studies of these mechanisms have, in the past, been confined either to laboratory investigations of idealized magma systems, or to reconstructions based on observations of completely solidified lavas which, in many cases, are modified by post-crystallization alteration processes. For study of small bodies of lava, Kilauea volcano affords a unique combination of frequent eruption and numerous pit craters in which lava can pond to form 'lava lakes', which become natural laboratories for the study of physical and chemical processes of magma cooling and crystallization through time. The lava lakes described here

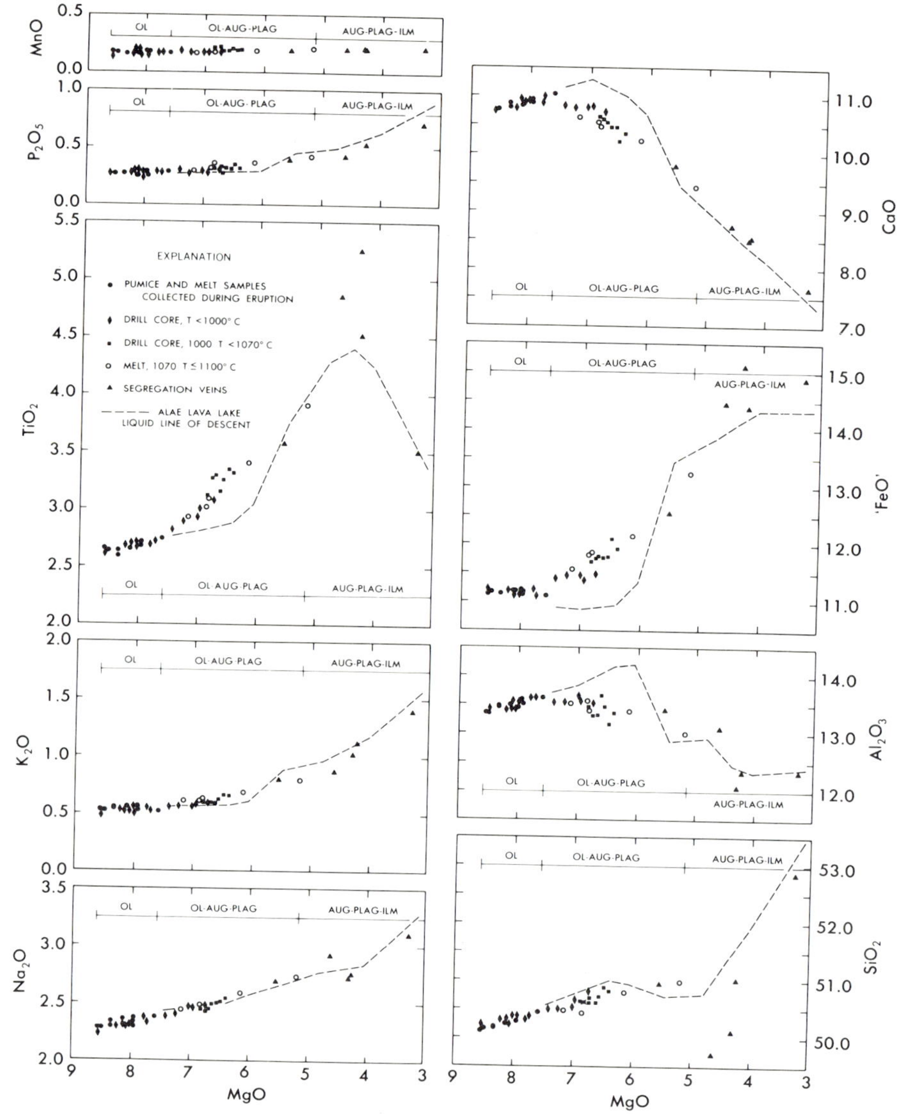


Fig. 6. MgO variation diagram for pumice and drill core, Makaopuhi lava lake. Dashed line is liquid line of descent for Alae lava lake, given for reference. 'FeO' is sum of analyzed FeO and .9 $Fe_2O_3$.

Subsequent studies have shown *no* evidence of differentiation attributable to gaseous transfer, nor can this process be detected in any Kilauea lava lakes or flows [*Wright and Fiske,* 1971]. Thus the Kilauea lava lakes offer no evidence to support this process, often cited to explain differentiation of other types of lava.

## INTERPRETATION OF SURFACE ALTITUDE CHANGES: DENSITY CHANGE ON SOLIDIFICATION AND DISTRIBUTION OF VESICLES

The surface of each lake underwent differential changes of altitude as the lake cooled. The surface of Kilauea Iki, for example, subsided as much as 2 m during the fifteen years following its formation, and the surface of Alae was uplifted as much as 25 cm during its year of solidification. The changes in altitude are a one-dimensional product of a number of processes beneath the surface of the lake, such as thermal contraction on cooling of both molten and solidified lava and changes in lava density during crystallization and vesiculation.

The contoured data for Alae form a consistent

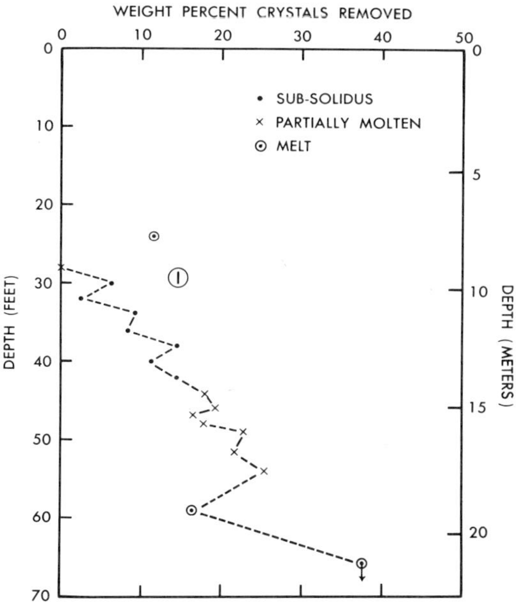


Fig. 7. Amount of differentiation as a function of depth, Makaopuhi lava lake. Samples, connected by dashed lines, to 16.5 m were collected by core drilling. Samples off the dashed line apparently lost crystals (became more differentiated) during flow into drill casings. Crystals removed are olivine, augite, and plagioclase in a ratio of 2:3:3.

pattern amenable to interpretation [*Peck et al.*, 1966, p. 16–17]. During the year of solidification, the central part of the lake overlying the lens of fluid and partly molten lava was uplifted (Figure 8). The surface over the solidified margin, in contrast, subsided due to thermal contraction, as did the entire lake surface after complete solidification. The line between uplift and subsidence corresponded to the outer edge of the 1000°C isothermal surface, i.e., very nearly to the solidus (980°C). The central part of the lake was uplifted because solidification of this thin lake was accom-

TABLE 4. Corrected Analyses, Kilauea Iki Drill Core

| Sample No.[1] | Original Analysis[1] | Corrected Analysis[2] and Comments |
|---|---|---|
| KI-112 (melt) | | |
| $K_2O$ | 0.65 | Analysis is good. Sample is |
| $Na_2O$ | 2.25 | contaminated by differentiated ooze |
| KI-165 | | |
| $K_2O$ | 0.83 | 0.49[3]; 0.52[4] |
| $Na_2O$ | 1.73 | 1.97[3]; 2.03[4] |
| | | Similar to other analyses of olivine-controlled samples |
| KI-168 | | |
| $K_2O$ | 0.79 | 0.48[3]; 0.50[4] |
| $Na_2O$ | 1.67 | 1.91[3]; 1.97[4] |
| | | Similar to other analyses of olivine-controlled samples |
| KI-184 | | |
| $K_2O$ | 0.55 | 0.56[3]; original analysis is good |
| $Na_2O$ | 2.24 | 2.22[3]; original analysis is good |
| KI-186 | | |
| $K_2O$ | 0.48 | 0.49[3]; original analysis is good |
| $Na_2O$ | 2.02 | 2.02[3]; original analysis is good |

[1]From *Richter and Moore* [1966, Tables 2 and 3].
[2]Analyses done in the U.S. Geological Survey, Denver rock analysis laboratory under the supervision of L. C. Peck.
[3]Flame photometer; analyst, Violet Merritt.
[4]Wet chemical methods; analyst, Elaine Munson.

panied by vesiculation and the formation of porous basalt less dense than the parent melt. Analysis of the data yields a coefficient of linear thermal expansion of $3 \times 10^{-6}$, and indicates that fluid lava at 1130°C (melt plus 15% crystals) had a density of 2.78 gm/cc [*Peck,* 1969]. Analysis of the pattern and timing of uplift suggests that the rate of vesiculation varied markedly with temperature and that most of the vesicles were formed between 1090° and 1040°C.

The simple model for Alae does not hold for either Makaopuhi or Kilauea Iki. The central part of Makaopuhi lava lake subsided for 27 months after the eruption and thereafter was uplifted. The contrast with Alae is probably caused by the greater thickness and different density distribution in Makaopuhi lake (Figure 9) which, in turn, reflects difference in gas content, history of degassing, and mechanical effects related to pressure in the respective melts.

The surface of Kilauea Iki lava lake has subsided throughout its entire history; since 1968 the rate of subsidence has accelerated from an average of 3 cm/yr to 30 cm/yr, which may indicate an increased rate of solidification of the lake.

SUMMARY

Studies of Kilauea lava lakes provide the only detailed and quantitative 'real time' observations of the cooling, crystallization, and differentiation of tholeiitic basalt. The information gained on thermal history, mechanical properties, oxygen fugacity, viscosity, liquidus and solidus temperatures, mineral paragenesis, thermal conductivity, gas content, and vesiculation history cannot in general be inferred from study of solidified basaltic rocks. Although the lakes show many similarities, significant differences are apparent when cooling rates and the chemistry and petrology of the lavas are studied in detail. Knowledge of the history of each eruption is essential in specifying the initial temperature distribution in each lake, and the history may also affect the later pattern of vesiculation in ways as yet poorly defined. Thermal coefficients (diffusivity and heat of crystallization) have been derived by finite-element modeling of measured temperature distributions and are in agreement with laboratory determined values; in general, the diffusivities vary through different regimes of cooling, and average thermal properties are similar, but not identical, for lava in the different lakes. Evaporation of rainfall is important in cooling the lakes, particularly at temperatures below the solidus. Cooling by convection of the melt is suggested for the larger

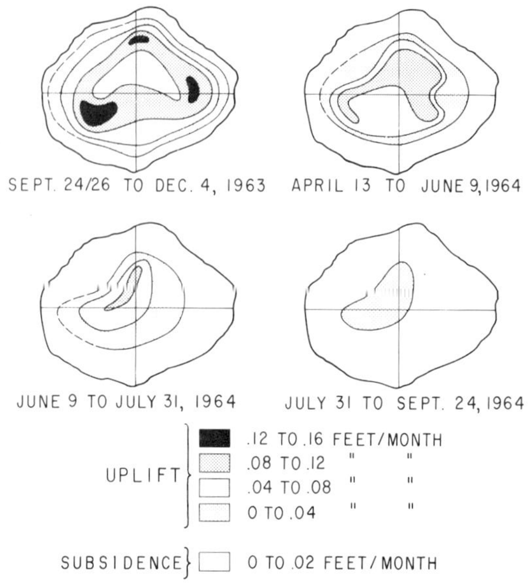


Fig. 8. Alae lava lake: elevation changes during solidification. The line separating subsidence (outside) from uplift (inside) corresponds approximately to the edge of the 1000°C isothermal surface and moved inward as the lake cooled.

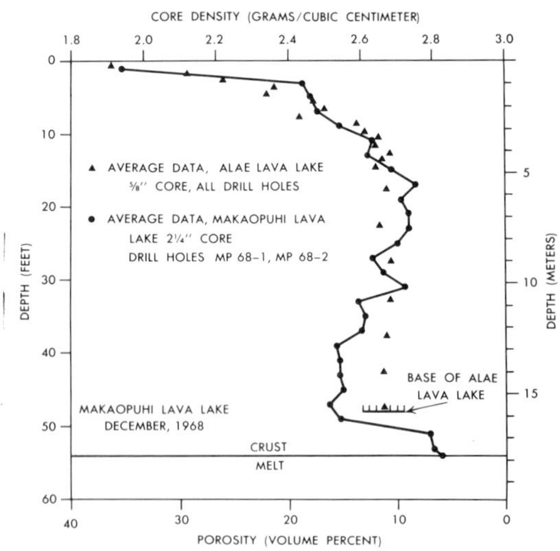


Fig. 9. Average core density and porosity plotted against depth for Alae (▲) and Makaopuhi (●-●) lava lakes. Each point represents two to ten measurements. Core densities are similar in the upper part of the lakes but the Makaopuhi profile shows decreasing density in the zone of partially molten crust and very high density (2.82) in nonvesiculated melt collected at 16.5–18.3 m depth. Alae densities are constant (2.66–2.68) at depths greater than 4.5 m.

lakes. Convection depends critically on the suppression of significant vesiculation to temperatures below 1070°C and is manifested by flow differentiation of lava within the convective cooling regime.

Crystal–liquid fractionation occurs in at least three different ways—gravitational settling of olivine, flow differentiation of crystals and melt, and segregation of differentiated melt in partially crystallized crust. Each of these processes takes place within a fixed temperature interval. The chemistries of natural differentiates tend to group into specific compositional ranges that reflect the different differentiation processes. Thus, separation of liquid and crystals is governed by fluid–mechanical considerations and is a discontinuous process, even though the liquid composition changes in a continuous fashion during crystallization.

The patterns of vesiculation in each lake are shown by the observed distribution of core density and are reflected in altitude changes of the lava lake surface. Vesiculation, the least well-understood process in the lava lakes, has important effects on both the cooling and differentiation regimes.

These studies provide guidelines for interpretation of the cooling history of other ponded lava flows and shallow intrusive masses. However, the Kilauea studies cannot be applied quantitatively to other ponded flow without consideration of the factors mentioned above.

## APPENDIX: AN ANNOTATED BIBLIOGRAPHY OF KILAUEA LAVA LAKE STUDIES

### Kilauea Iki lava lake

*Rawson,* 1960.
Account of the first drill hole to reach the crust–melt interface with a temperature profile measured at that time (July 1960).

*Ault et al.,* 1961.
Early temperature profiles, through June 1960.

*Macdonald and Katsura,* 1961.
Chemical analyses of crust and melt from the hole drilled in July 1960.

*Ault et al.,* 1962.
Temperature profiles through October 1961.

*Decker,* 1963.
Results of ground magnetic surveys made in 1961.

*Frischknecht,* 1967 (p. 18).
Electrical resistivity of Kilauea Iki melt found to be 2.07 ohm-m.

U.S. Geological Survey Professional Paper 537—A, B, C, D, and E, The 1959–60 eruption of Kilauea volcano, Hawaii. (See individual entries for *Murata and Richter* [1966*a*]; *Richter and Moore* [1966]; *Murata* [1966]; *Richter and Murata* [1966*b*]; and *Richter et al.* [1970].)

A. Chemical composition of the erupted lavas.
B. Summary of the chronology of filling of Kilauea Iki pit crater. Temperature profiles and both modal and bulk chemical analyses of drill core are given from drilling up to December 1962.
C. Analyses of fumarolic gas collected from the pumice cone following the eruption.
D. Petrographic data on the erupted lavas.
E. Chronological account of the eruption.

Kilauea Iki was re-drilled in 1967 and 1975 and each time a 6-cm core was recovered from three drill holes. Petrochemical study is being made by R. L. Helz and R. S. Fiske.

### Alae lava lake

*Peck et al.,* 1964.
Description of the eruption and temperature measurements to February 1964.

*Peck,* 1966.
Description of surface features that indicate shear sense during deformation of a lava lake surface.

*Peck et al.,* 1966.
Preliminary summary of the cooling and crystallization history through August 1965. Data on rate of crustal growth, chemistry of crust, melt, differentiates, and deuterically altered crust, and petrography of core samples.

*Decker and Peck,* 1967.
Measurement of infrared radiation and its contribution to cooling of the lake.

*Peck and Minakami,* 1968.
Data and interpretation regarding the temperature and mode of formation of columnar joints in the lava lakes. Data principally from Alae and Makaopuhi.

*Peck,* 1969.
Calculation of density of fluid lava in the lake.

*Skinner and Peck,* 1969.
Mineralogy and composition of immiscible sulfide blebs from Alae formed at 1070–1100°C.

*Peck,* 1974.
Summary of thermal properties of basalt in all three lakes based on field and laboratory measurements and preliminary modeling of the Alae temperature data by J. C. Jaeger.

*Robertson and Peck,* 1974.
Results of the thermal conductivity measurements on Hawaiian lavas including samples from all three lava lakes.

U.S. Geological Survey Professional Paper [in press], Solidification of Alae lava lake, Hawaii.
A. The eruption of August 1963 and the formation of Alae lava lake, Hawaii [*Dallas L. Peck and W. T. Kinoshita,* in press]. Describes the eruption chronology and surface features of the lake, including joint cracks.

For preparation of the present paper unpublished data were used relating to temperature measurements, surface altitude measurements, and related analysis of cooling and vesiculation (D. L. Peck and W. T. Kinoshita); chemical and petrographic data related to crystallization of the Alae magma (T. L. Wright and D. L. Peck); and finite element modeling of Alae lava lake (H. R. Shaw, M. Hamilton, and D. L. Peck). (Papers involving this material are in preparation.)

**Makaopuhi lava lake**

*Sato and Wright,* 1966.
Preliminary measurements of $FO_2$—temperature distribution, including interpretation of oxidation in the lava lake.

*Wright and Weiblen,* 1967.
Composition and temperature of first appearance of minerals.

*Finlayson et al.,* 1968.
Analyses of gas collected from drill holes, principally in Makaopuhi lava lake. Description of change of gas composition with time and discussion of reactions within the gas and between gas and wall-rocks.

*Shaw et al.,* 1968.
Data and interpretation of the direct measurement of melt viscosity.

*Wright et al.,* 1968.
An account of the March 1965 eruption and the chronology of formation of Makaopuhi lava lake.

*Grommé et al.,* 1969.
Data on composition and curie point temperatures of Fe-Ti oxide minerals in Alae and Makaopuhi lava lakes.

*Shaw,* 1969.
Experimentally determined viscosities of basalt from Makaopuhi lava lake. Analysis of the variation of viscosity with temperature during crystallization of Makaopuhi lava lake.

*Evans and Wright,* 1972.
Chemical study of chromite precipitation in Makaopuhi and Kilauea Iki lava lakes.

For preparation of the present paper extensive use was made of unpublished data on temperature and surface altitude measurements and chemical and petrographic analyses, interpreted in terms of possible cooling regimes, vesiculation, and crystal–liquid differentiation in Makaopuhi lava lake (T. L. Wright and R. T. Okamura).

**Prehistoric Makaopuhi lava lake**

*Moore and Evans,* 1967.
Bulk chemical analyses, modes, and electron probe analyses of olivine, plotted according to depth in the lake. Analyses of a cooling model and comments on differentiation processes.

*Evans and Moore,* 1968.
Electron probe analyses of augite, pigeonite, plagioclase, ilmenite, magnetite, chromite, and residual glass. Discussion of mineral paragenesis as a function of depth in the lake.

*Acknowledgments.* The lava lake studies involved participation by most of the staff of the Hawaiian Volcano Observatory from 1959 to 1969. Shop and electronic work in support of the studies was ably provided by B. J. Loucks, William Francis, John Forbes, and George Kojima. R. T. Okamura was the principal driller in the studies of Makaopuhi lava lake and the later study of Kilauea Iki. Other participants were: S. Aramaki, W. U. Ault, I. L. Barnes, R. W. Decker, J. P. Eaton, E. Endo, J. B. Finlayson, R. S. Fiske, S. Hiraga, J. Judd, W. T. Kinoshita, T. Minakami, J. G. Moore, D. L. Peck, D. H. Richter, H. R. Shaw, D. B. Stewart, D. A. Swanson, T. L. Wright, K. Yamashita, and members of the Kilauea Job Corps Camp. The paper has been improved after review by E. D. Jackson, D. A. Swanson, and R. T. Helz.

# 6. TECTONOPHYSICS

## 33. *Phase Transformations in Descending Plates: Implications for Mantle Dynamics and Differentiation*

A. E. RINGWOOD

*Research School of Earth Sciences, Australian National University, Canberra 2600, Australia*

*Abstract.* The role of phase transformations in a mantle of pyrolite composition is reviewed in the light of recent experimental data. The pyroxene component of pyrolite transforms to the garnet structure at 300–350 km while olivine transforms to beta-$Mg_2SiO_4$ near 400 km. Between about 500 and 550 km, beta-$Mg_2SiO_4'$ probably transforms to a partially inverse spinel structure while the $CaSiO_3$ component of the complex garnet solid solution exsolves and transforms to the perovskite structure. The major seismic discontinuity near 650–700 km is probably caused by disproportionation of $Mg_2SiO_4$ spinel into periclase plus stishovite. At a slightly greater depth, the remaining magnesian garnet transforms to the corundum or ilmenite structure. Finally, at a depth probably in the vicinity of 800–1000 km, the (Mg,Fe)$SiO_3$ component of the ilmenite phase transforms to a perovskite structure while stishovite and some of the periclase recombine to form perovskite also. The mineral assemblage so formed is about 4% denser than mixed oxides (MgO + FeO + $Al_2O_3$ + CaO + stishovite) isochemical with pyrolite. The above sequence of phase transformations in pyrolite provides a satisfactory general explanation of the elastic properties and density distribution in the mantle. In particular, there is no evidence requiring an increase of FeO/FeO+MgO ratio with depth. The depths at which major phase transformations occur in subducted lithosphere differ from those in 'normal' mantle. These differences are caused by two factors: (1) Temperatures within sinking plates are much lower than in surrounding mantle to depths of 700 km or more; and (2) irreversible chemical differentiation of pyrolite occurs at oceanic ridges. Lithosphere plates so formed consist of a layer of basaltic rocks underlain successively by layers of harzburgite, lherzolite, and pyrolite slightly depleted in highly incompatible elements (e.g., La, Ba, Rb, U). The phase transformation behavior of the first three of these layers differs from that of pyrolite. The effects of these and other factors connected with phase transformations on the dynamics of plate subsidence are discussed. It appears quite likely that plates penetrate the 650–700 km discontinuity, largely because the slope of the spinel disproportionation is probably positive, *not* negative as generally supposed. The former basaltic oceanic crust probably sinks deeply into the lower mantle, while the former harzburgite component of the plate may collect above the perovskite transition boundary. Phase transformations may thus serve as a kind of filter, leading to increased and irreversible mantle heterogeneity with time. The observed depletions of highly incompatible elements such as Ba, Rb, La, and U in oceanic tholeiites are also discussed in terms of this model.

A wide range of petrological and geochemical evidence leads to the conclusion that the upper mantle is chemically zoned, consisting of an uppermost layer of peridotite which passes downward into a more primitive rock type, termed pyrolite. When subjected to varying degrees of partial melting, pyrolite has the property of yielding basaltic magmas, leaving behind residual, refractory peridotite. The composition of pyrolite is derived from experimental petrological studies of the chemical and phase relationships between basaltic magmas and their complementary peridotites [*Green and Ringwood,* 1967*b*; *Green,* 1970*a*]. A detailed review of these topics is given in *Ringwood* [1975].

Several seismic discontinuities or regions of anomalously high velocity gradient, or both, are believed to exist in the mantle between depths of approximately 300–900 km [*Anderson and Toksöz,* 1963; *Johnson,* 1967; *Green and Hales,* 1968; *Whitcomb and Anderson,* 1970; *Wiggins and Helmberger,* 1973] (see Figure 1). It is now known that these are caused, at least to a considerable extent, by phase transitions. The principal upper mantle minerals—olivine, pyroxenes, garnet—become unstable at very high pressures and transform to denser, closer packed structures. Considerable progress has been made during the last ten years in studying the nature of the phase transformations which may occur in this region of the mantle. It is useful to explore the extent to which presently known transformations in a mantle of pyrolite composition are capable of explaining the observed variation of seismic velocities and density with depth throughout the mantle. An attempt in this direction is made in the next section. It is emphasized that we are dealing with a *model* subject to changes in the light of further experimental data on phase transformations which are likely to be obtained during the next few years.

## PHASE TRANSITIONS IN A MANTLE OF PYROLITE COMPOSITION

The most probable sequence of phase transitions in a pyrolite mantle based upon currently available experimental data is given in Figure 2. The following discussion is based upon reviews

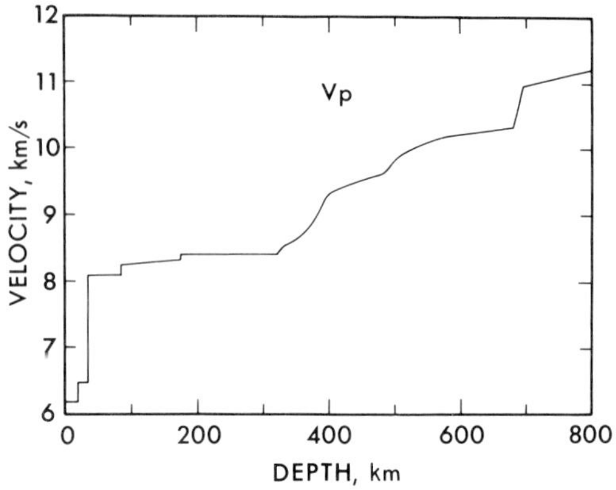


Fig. 1. *P*-wave distribution for the upper mantle and transition zone beneath northern Australia based on data obtained by the Warramunga array, Tennant Creek [from *Simpson*, 1973].

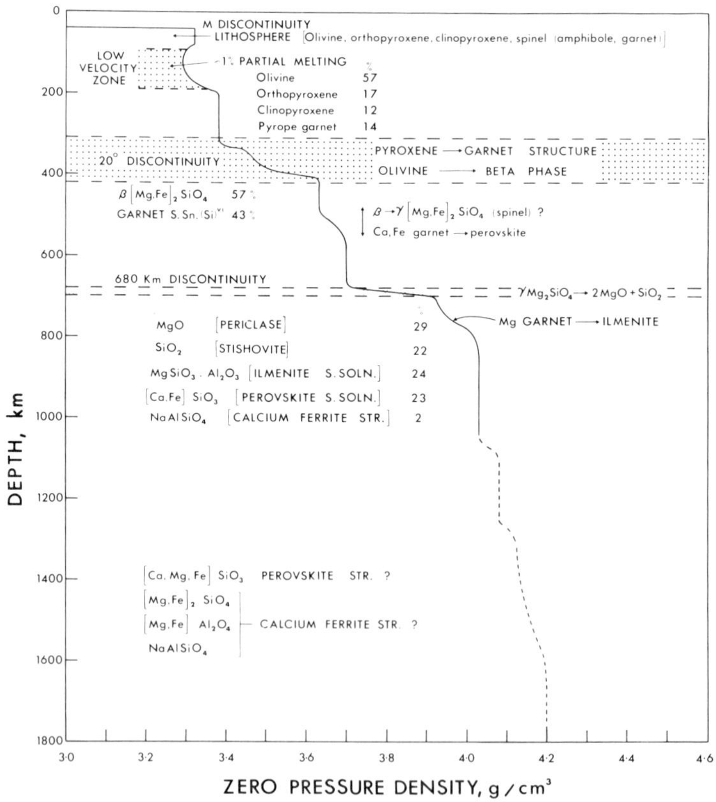


Fig. 2. Probable mineral assemblages and corresponding zero pressure densities for a model mantle of pyrolite composition [from *Ringwood*, 1975].

given by *Ringwood* [1970, 1975] plus other detailed references subsequently cited. Down to depths of 300–350 km, pyrolite would crystallize to a mineral assemblage consisting of olivine + orthopyroxene + clinopyroxene + garnet. At 300–350 km, pyroxenes transform to a new garnet structure characterized by partial octahedral coordination of silicon, while near 400 km, olivine transforms to a polymorph possessing the beta-$Mg_2SiO_4$ structure. These transformations involve density increases of 8–10% and are responsible for the major seismic discontinuity(s) in the vicinity of 350–400 km (Figure 1).

At a depth in the vicinity of 500–550 km, the beta phase transforms to a true spinel structure accompanied by a 2% increase in density. It seems likely that around this depth, the $CaSiO_3$ component of the garnet solid solution may transform to the perovskite structure. These two transformations may be responsible for high velocity gradients in the vicinity of 500–550 km such as reported by *Whitcomb and Anderson* [1970].

A major seismic discontinuity occurs near 680 km. It appears most likely that this is caused by the disproportionation of $(Mg, Fe)_2SiO_4$ into the isochemical mixed oxides (Mg,Fe)O and $SiO_2$ (stishovite) [*Kumazawa et al.*, 1974; *Ming and Bassett*, 1974]. A further major transformation involving the magnesian garnet component must also occur in this region, but the corresponding discontinuity has not yet been resolved by seismologists. This transformation most probably involves the change of garnet into an $MgSiO_3$-$Al_2O_3$ ilmenite-type solid solution [*Kawai et al.*, 1974]. By a depth of 800 km, pyrolite would occur in a mineral assemblage consisting of rocksalt, rutile, ilmenite, and perovskite-type structures. The density of this mineral assemblage would be very similar to that of mixed oxide components ($MgO + FeO + CaO + Al_2O_3 + SiO_2$) isochemical with pyrolite.

There are indications based on shock-wave data [*McQueen et al.*, 1967; *Trunin et al.*, 1965] that the density of the lower mantle (below 800 km) might be a few percent greater than pyrolite in a 'mixed-oxide' assemblage. In principle, this could be caused by additional phase transformations to denser assemblages, or by an increase in iron content with depth. There has been some controversy on this subject but it has been largely resolved by a discovery by *Liu* [1974] (predicted by *Ringwood* [1962] and *Reid and Ringwood* [1974]) that $MgSiO_3$ transforms to the perovskite structure at about 350 kbar. $MgSiO_3$ perovskite is 4% denser than an isochemical mixture of periclase plus stishovite. This result implies that some of the MgO and $SiO_2$ occurring below the 680-km discontinuity recombine at greater depth to form $MgSiO_3$ perovskite which may well be the most abundant mineral in the mantle.

The sequence of mineral assemblages shown in Figure 2 occurring in a mantle of pyrolite composition appears capable of explaining the gross variation of seismic velocities and density throughout the mantle. In particular, there is no requirement for an increase in Fe/Mg ratio near the 680 km discontinuity as previously argued by, for example, *Anderson* [1970]. The principal evidence which was once held to support this interpretation was an offset in the linear bulk sound velocity versus density plot between the upper and lower mantle [*Press*, 1970*b*] in the direction of higher mean atomic weight below 680 km. It is now known that this offset is to be expected when phase transformations occur involving an increase of cation-oxygen coordination [*Liebermann*, 1974; *Davies*, in press].

The occurrence of an increase of iron content at the 680-km discontinuity would effectively inhibit the sinking of lithosphere plates to greater depths. We conclude that, *at least on these grounds*, this discontinuity no longer can be regarded as providing a barrier to plate subsidence.

On the other hand, the lowermost 200 km of the mantle appears to be inhomogeneous and may well have an increased iron content [*Bullen*, 1963]. A speculative interpretation of this feature is as follows: Segregation of the earth's core occurred very early in its history, before the earth had finally swept up all the residual planetesimals in its neighborhood of the solar nebula. The population of late planetesimals falling on the earth would be similar to low temperature condensate to be expected in a cool solar nebula—i.e., compositions resembling the Type 1 carbonaceous chondrites. Such material would accumulate at the surface to form cool, iron-rich, dense plates, perhaps 10–20 km thick. Ultimately, because of gravitional unstability, they would subside into the mantle analogously to present lithosphere plates. Because such plates would be intrinsically denser than surrounding mantle *at all depths*, they would probably sink directly to the base of the mantle where they would accumulate to form an iron-rich inhomogeneous layer surrounding the core.

When the structure and detailed physical properties of the lowermost mantle are better understood, it might be possible to reverse this hypothesis and estimate the mass of planetesimals which fell on the earth subsequent to core-formation.

PHASE TRANSITIONS IN SINKING SLABS—THERMAL EFFECTS

The depths at which the phase transformations discussed in the previous section occur within sinking slabs of subducted lithosphere differ from those in 'normal' mantle because of two factors: (1) temperature, and (2) chemical heterogeneity within the lithosphere [*Toksöz et al.*, 1971; *Minear and Toksöz*, 1970; *McKenzie*, 1969; *Turcotte and Schubert*, 1971, 1973]. We will consider these in turn.

Because of its large thermal inertia in relation to the velocity of sinking, the interior of the slab remains much cooler than surrounding mantle until great depths [e.g., *Griggs*, 1972]. Thus, at 400 km, the slab interior may be 1000°C cooler than surrounding mantle, while at 700 km, this difference may amount to 500°C (Figure 3). The depths at which phase transitions occur in the mantle are temperature dependent. The transitions occurring near 400 km are believed to possess a positive slope with a mean $dP/dT$ value of about 30 bars/°C [*Ringwood*, 1975]. With a temperature differential of 1000°C at this depth, the 400-km discontinuity will be elevated within the slab by about 100 km (Figure 4). Formation of this dense upward-extending lobe of high-pressure phases within the slab has many important geophysical effects which have been discussed by several authors. It creates a large body force tending to pull the slab deeper into the mantle, thereby playing an important role as one of the driving forces of plate motions. It also contributes to the formation of positive gravity anomalies over island arcs and to the high mean seismic velocities within the slab. Finally, the lobe strongly influences the stress distribution within the slab, tending to place it in compression below 300 km and in tension above 300 km.

Because of uncertainties regarding the nature of the phase transitions occurring around 680 km, there has been some debate as to the overall effect on plate motions. If the mean gradient $(dP/dT)$ were positive, then the 500°C temperature differential would cause an elevation of the phase boundaries within the slab analogous to the situation at 400 km. This would provide an additional body force tending to drive the slab through the 680-km discontinuity into the lower mantle. On the other hand, a mean negative $dP/dT$ value for the transitions at this level would cause a lobe of low pressure–low density phases to extend downward into the slab, relative to surrounding mantle. This would inhibit sinking of the slab through the discontinuity.

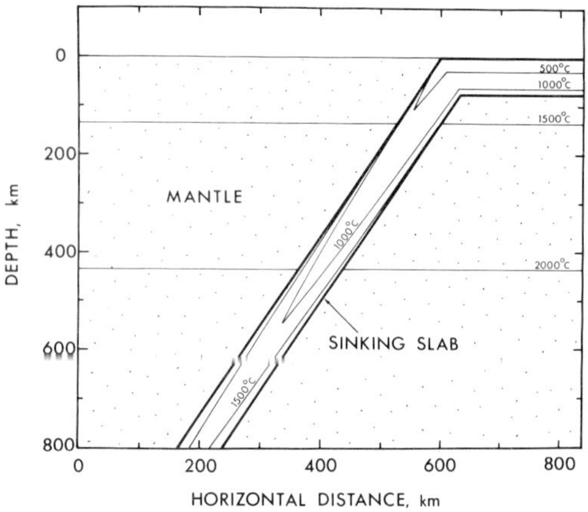


Fig. 3. Temperature distribution in the sinking lithosphere compared to surrounding mantle. Thickness of lithosphere, 80 km; velocity of sinking, 8 cm/yr; thermal conductivity of slab = 0.005 C.G.S. [after *Griggs*, 1972].

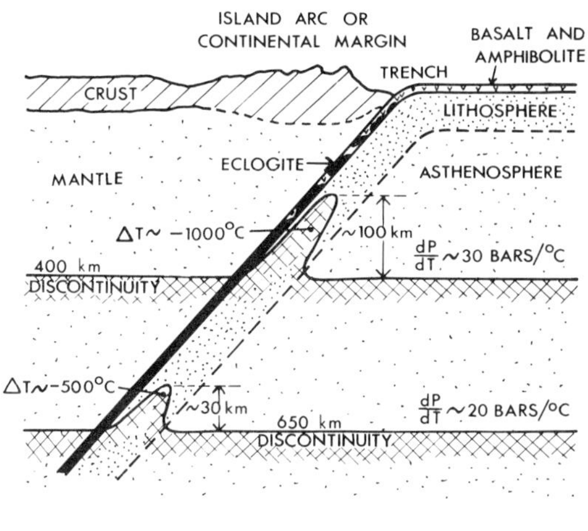


Fig. 4. Simplified sketch showing possible relationships between sinking lithosphere plate and phase transformations. ΔT is the maximum temperature difference between interior of sinking slab and normal mantle at given depth (not to scale) [after *Ringwood*, 1972].

The principal transformation occurring near 680 km is now believed to be disproportionation of $(Mg,Fe)_2SiO_4$ spinel into (Mg,Fe)O plus stishovite. Earlier thermodynamic calculations [*Ahrens and Syono*, 1967] yielded a negative gradient for this transition of −13 bars/°C, and on this basis it was concluded by *Kumazawa et al.* [1974] that the slab would not penetrate this discontinuity.

However, these conclusions have been questioned by *Jackson et al.* [1974]. On the basis of an improved estimate of the entropy of $Mg_2SiO_4$ (normal) spinel, they recalculated the gradient for disproportionation into oxides and found $dP/dT$ to be indistinguishable from zero. They also pointed out the importance of another factor—namely, the degree of disorder or inverse character in the spinel, which has an important effect upon its entropy.

The spinel, $Mg_2SnO_4$, is known to disproportionate into a mixture of $MgO + SnO_2$ (rutile structure at high pressure [*Ringwood and Reid*, 1968]). In this sense it behaves as a model for $Mg_2SiO_4$. *Jackson et al.* [1974] determined the slope of the disproportionation reaction for $Mg_2SnO_4$ and found that it was *positive*, amounting to 40 bars/°C (Figure 5). They showed that this was a consequence of the inverse nature of $Mg_2SnO_4$, leading to increased configurational entropy. Turning to $Mg_2SiO_4$ spinel, they noted that if it displayed any significant degree of disorder or inverse character in the mantle, then the slope of its disproportionation would be substantially positive. Evidence suggesting that $Mg_2SiO_4$ spinel in the mantle would indeed be likely to display significant disorder was cited. *Jackson et al.* [1974] concluded that the slope of $Mg_2SiO_4$ spinel disproportionation was more likely to be positive than negative, in which case, providing the slab were cooler than the surrounding mantle, it would be likely to penetrate the 680-km discontinuity. That the slab is indeed likely to be substantially cooler than surrounding mantle at this depth is strongly indicated by studies of its thermal development [*Griggs*, 1972; *Schubert et al.*, 1975] and by the observations that slab seismic velocities are several percent higher than those in surrounding mantle at depths near 600 km [*Sondergeld et al.*, 1974; *Fitch*, 1975]. It should also be noted that another major transformation (garnet → ilmenite) also probably occurs near 680 km (Figure 2). In most known phase transitions not involving disproportionation, the denser phase has a lower entropy and hence the slopes *(dP/dT)* are positive. There are no reasons to suspect that this will not be the case for the garnet–ilmenite transition, in which case, this would also contribute to the body force driving the slab through the discontinuity.

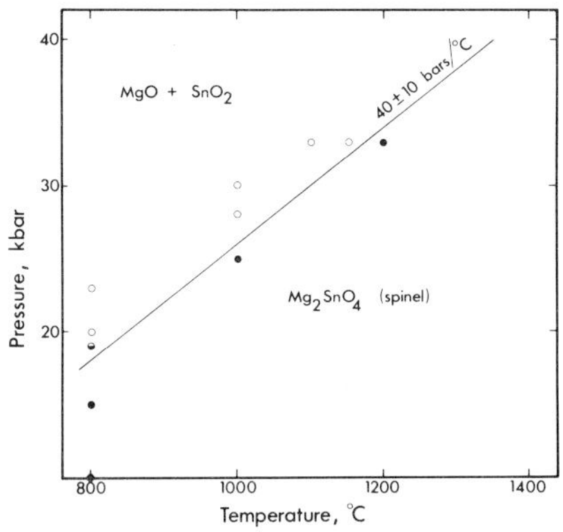


Fig. 5. Equilibrium boundary showing disproportionation of $Mg_2SnO_4$ (spinel) into mixture of $MgO + SnO_2$ [after *Jackson et al.*, 1974].

If lithospheric slabs indeed descend into the deep mantle, the question of the absence of deep focus earthquakes below 700 km is raised. At least three factors may be involved. *Griggs* [1972] has pointed out that the shear-failure mechanism responsible for earthquakes may be possible only below a certain critical temperature which may well be much lower than the temperature in normal surrounding mantle. Above this critical temperature, deformation in response to stresses occurs by ductile flow and catastrophic failure is not possible. It may well be that by a depth of 700 km, the interiors of the slabs have reached this critical temperature through thermal conduction from surrounding mantle, but are still far from thermal equilibration. A second factor concerns the possibility that stresses ultimately caused by phase transitions within the slab are responsible for deep-focus earthquakes [e.g., *Ringwood*, 1975]. Absence of earthquakes below 700 km would then be a consequence of the absence of phase transitions in the deep mantle characterized by *large* density changes. A third possibility is that the rheological properties of the deep mantle are very different from those in the mantle shallower than 680 km. One of the phases which is important below 680 km is periclase. At relatively low pressures and temperatures, periclase is much weaker and more ductile than structurally more complex silicates such as olivine. If this differential behavior is also exhibited at very high pressures and temperatures (and also at very low strain rates) it would be possible for the mantle below 700 km to exhibit smaller ultimate strength and viscosity than in the region above. In that case, the resistance to sinking slabs would also be smaller below 700 km and this might conceivably affect the stress distribution within the slabs, which is ultimately responsible for earthquakes. It would certainly lead to an increased rate of sinking of the slabs into the deep mantle.

## PHASE TRANSITIONS IN SINKING SLABS: EFFECTS OF CHEMICAL HETEROGENEITY

The formation of new lithosphere plates at oceanic ridges proceeds through the uprise of primitive pyrolite from the low-velocity zone accompanied by partial melting and differentiation. This results in the formation of a mafic oceanic crust overlying a complementary residual peridotitic lithosphere. It is likely moreover that the lithosphere is chemically and petrologically zoned [*Ringwood,* 1975]. The highest degree of partial melting occurs in the axial regions where pyrolite ascends closest to the surface before segregation of magma. Thus, the uppermost layer of the mantle probably consists of a harzburgite (e.g., ol 80%, opx 20%) which is complementary to oceanic tholeiite. Upwelling pyrolite displaced from the axis does not rise so high before turning over and hence suffers a lesser degree of partial melting. The liquid produced has the characteristics of an alkali basalt which is complementary to a layer of lherzolite underlying the harzburgite [*Oxburgh and Turcotte,* 1968; *Gast,* 1968; *Ringwood,* 1975]. The lherzolite is strongly depleted in incompatible elements and other low-melting components and is incapable of yielding a basaltic magma if subjected to a further stage of partial melting. It is suggested, moreover, that the pyrolite underlying the lherzolite layer may have been subjected to a *very small* degree of partial melting ($< 1\%$) resulting in the loss of a highly alkalic liquid very strongly enriched in the most incompatible elements of Rb, Ba, La, and U. This layer of pyrolite is insufficiently depleted in other incompatible elements so that it could not serve as a source of oceanic tholeiite magma if subjected to a further extensive degree of partial melting. A sketch of the proposed zoned lithosphere is given in Figure 6.

Consider now the behavior of the differentiated lithosphere when it sinks into the mantle in a subduction zone. Because of its chemically heterogeneous nature, the differentiated region's response to phase transformations will be different from that of the surrounding pyrolite. As they sink into the mantle, phase transformations in the respective mafic, harzburgite, and lherzolite layers will occur at different depths because of the differences in chemical composition. The depths to which the lithospheric layers sink may be controlled by the sequence and properties of the phase transformations which occur.

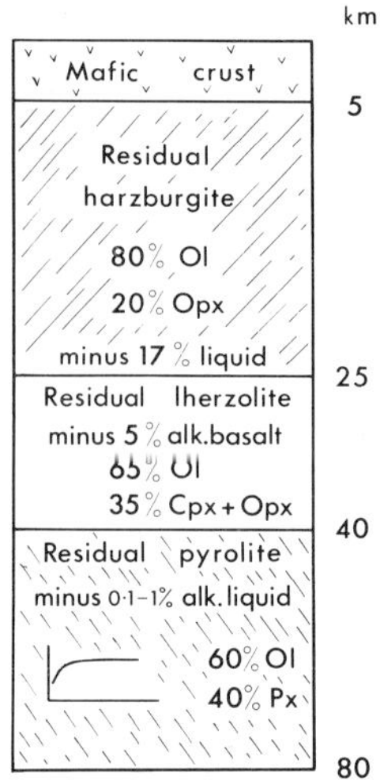


Fig. 6. Proposed structure of oceanic lithosphere showing chemical and mineralogical zoning.

For example, consider the sinking mafic crust which transforms successively to eclogite ($\rho \sim 3.5$ g/cm$^3$) at depths less than 100 km and then to garnetite ($\rho \sim 3.8$ g/cm$^3$) at about 300 km. At greater depths, further transformations occur. It appears likely that the former mafic crust will attain states which always remain denser than surrounding pyrolite [*Ringwood,* 1975]. This is largely a consequence of the relatively important role of the garnet–perovskite transition and the higher Fe/Mg ratio. Thus, the former mafic crust is likely to sink very deeply into the lower mantle, providing this is permitted by rheologic properties [*Ringwood,* 1975].

The situation is different in the case of the residual peridotite layer underlying the oceanic crust. This may consist partly of harzburgite having a higher ratio of normative olivine to pyroxene (80/20) than underlying lherzolite and pyrolite (55/45). If $Mg_2SiO_4$ indeed transforms to the 'post-spinel' state at a depth which is less than that at which Mg garnet transforms to denser phases, then the differentiated part of the sinking lithosphere will become substantially denser than surrounding material of pyrolite composition below 680 km and will therefore penetrate this discontinuity. But, when the garnet component of the undifferentiated mantle transforms at greater depths, the differentiated former harzburgitic component of the sinking slab may or may not be able to break through, depending upon the sequence of phase transformations in the slab and in the surrounding mantle and upon the density changes associated with these transformations. If unable to penetrate a particular transformation, the differentiated material would spread out to form a distinct layer. Thus, phase transformations may act as a kind of filter, leading to significant degrees of lateral and vertical chemical heterogeneity in the mantle. Perhaps some of the differences between seismic velocity distributions observed by workers in different parts of the world can be attributed to this cause. These heterogeneities may also be responsible for some of the seismic reflections which have been observed from deep-mantle horizons [*Whitcomb and Anderson,* 1970].

## DEPLETION OF LIGHT REE, Ba, U, AND Rb IN OCEANIC THOLEIITES

*Gast* [1968] drew attention to a widespread characteristic of the rare earth element (REE) distributions in oceanic tholeiites as shown diagramatically in Figure 7. It is seen that whereas most of the REE pattern is 'flat', displaying chondritic relative abundances, marked depletions occur of the extremely light REE, La, and Ce. The magnitudes of these depletions average about twofold, but range from essentially zero to fourfold. Gast also showed that Ba is similarly depleted. Moreover, it had long been known that the amount of Ru in oceanic tholeiites was insufficient to account for the radiogenic Sr present, assuming nearly total partition of Rb and Sr into the melt during partial melting. This implied that Rb also was a member of this depleted group. Finally, *Gast* [1968] noted that significant fractionations of uranium relative to lead had occurred in the source regions of oceanic tholeiites, possibly in a series of episodes extending back two or three billion years.

Gast explained these observations by postulating that the source region of oceanic tholeiites

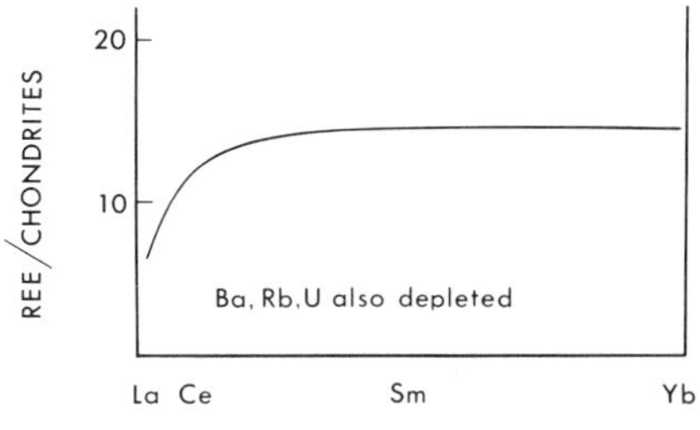


Fig. 7. Characteristic rare earth abundance pattern (normalized to chondritic abundances) displayed by oceanic tholeiites.

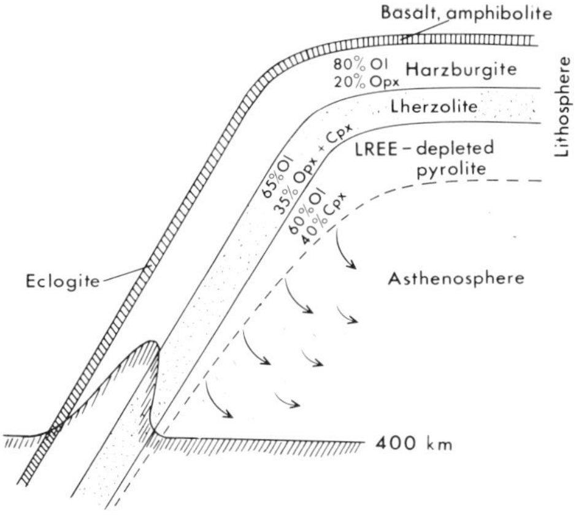


Fig. 8. Sketch showing possible two-stage evolution of source regions of oceanic tholeiites. See text for further details.

had previously been subjected to very small degrees of partial melting, resulting in the extraction of a highly alkalic liquid extremely enriched in the most incompatible elements, La, Ce, Ba, Rb, and U. The isotopic evolution of lead in the source regions showed, however, that the loss of this alkalic liquid did not occur near the time of generation of oceanic tholeiites at ridge crests, nor indeed, had it occurred previously on the time scale of $10^8$ years associated with the creation, migration, and consumption of oceanic lithosphere. As noted above, the previous melting episodes occurred over a much longer time scale, at least two billion years, implying the existence of processes not directly connected with the present regime of lithosphere plate evolution.

A possible cause of this phenomenon is illustrated in Figure 8. Previously we postulated that during the creation of plates at oceanic ridges, a chemically zoned lithosphere was created, and that the lowermost layer consisted essentially of pyrolite which had lost the very minute amount of alkalic liquid proposed by Gast (Figure 6). As the lithosphere subsides into the mantle beneath trenches, the effect of thermal conduction is such as to cause the isotherm marking the lower boundary of the lithosphere to migrate upward toward the more differentiated harzburgite zone, so that the thickness of the cool region of the descending plate decreases as the lithosphere descends. A corollary is that the lowermost depleted pyrolite zone of the lithosphere becomes thermally and rheologically resorbed in the upper mantle and is available for a subsequent final episode of partial melting which would be capable of yielding oceanic tholeiites with their characteristic depletion patterns. Clearly, the time scale for this two-stage evolution of oceanic upper mantle would be very long. It may well be on the order of billions of years, thereby providing a possible explanation of the isotopic development of mantle lead.

# 34. *Mineralogic Distribution of Iron in the Upper Half of the Transition Zone in the Earth's Mantle*

SYUN-ITI AKIMOTO, MASAKI AKAOGI, KAORU KAWADA, AND OSAMU NISHIZAWA[1]

*Institute for Solid State Physics, University of Tokyo, Roppongi, Minato-ku, Tokyo 106, Japan*

*Abstract.* Chemical equilibrium partitioning of $Mg^{2+}$ and $Fe^{2+}$ between $(Mg, Fe)_2SiO_4$ olivine or spinel and $(Mg,Fe)_3Al_2Si_3O_{12}$ garnet was investigated at pressures of 40 to 100 kbar and at temperatures of 1000° to 1200°C, using tetrahedral- and cubic-anvil types of high pressure apparatus. Relative enrichment of $Fe^{2+}$ in garnet compared to olivine was observed at 40 kbar and 1050°C. When olivine transforms to spinel at the higher pressures, however, the partition relation reverses; relative concentration of $Fe^{2+}$ in spinel is remarkably high. Apparent partition coefficient $K' = \{X_{Fe}^{Sp}/(1-X_{Fe}^{Sp})\}/\{X_{Fe}^{Ga}/(1-X_{Fe}^{Ga})\}$, where $X_{Fe}^{Sp}$ and $X_{Fe}^{Ga}$ denote the mole fraction of $Fe_2SiO_4$ in $(Mg,Fe)_2SiO_4$ spinel and that of $Fe_3Al_2Si_3O_{12}$ in $(Mg,Fe)_3Al_2Si_3O_{12}$ garnet, ranges from 2.7 to 3.1 at 80–100 kbar and 1000°–1200°C. Based on these partition data and on Ringwood's pyrolite model, chemical composition of $(Mg,Fe)_2SiO_4$ modified spinel or true spinel in equilibrium with complex garnet solid solution in the mantle transition zone was estimated to be $(Mg_{0.87}Fe_{0.13})_2SiO_4$. It is concluded that 2% increase in $Fe_2SiO_4$ component in $(Mg,Fe)_2SiO_4$ solid solution is expected in the transition zone as a coupling effect of the olivine–modified spinel–true spinel transformation and the pyroxene–garnet transformation. By combining the present mineralogical model with the olivine–spinel transformation diagram of the $Mg_2SiO_4$–$Fe_2SiO_4$ system and with seismic profiles in the mantle, temperatures at depths of 380 and 500 km were provisionally estimated as 1400° and 1620°C, respectively.

Olivine, pyroxene, and garnet are generally accepted to be the major component minerals in the earth's upper mantle. Accordingly, laboratory experiments on the high-pressure phase transformations in these silicate minerals have been carried out extensively in the past decade by many investigators for the purpose of clarifying the mineralogical and chemical composition of the mantle. Since ferromagnesian olivine $(Mg,Fe)_2SiO_4$ is believed to be the most abundant mineral in the upper mantle, special attention has been directed to the comprehensive investigation of the high-pressure transformation diagram of the $Mg_2SiO_4$–$Fe_2SiO_4$ system from olivine to the denser structures. The phase diagram for the olivine($\alpha$)—modified spinel($\beta$)—true spinel ($\gamma$) transformation of the $Mg_2SiO_4$–$Fe_2SiO_4$ system has been approximately completed through a series of investigations using modern, very high pressure apparatus [*Ringwood and Major,* 1966*a*, 1970; *Akimoto and Fujisawa,* 1968; *Akimoto,* 1972; *Kawai et al.,* 1970; *K. Ito et al.,* 1971; *Suito,* 1972; *E. Ito et al.,* 1974]. The comparison of these results with seismologically determined profiles in the mantle constitutes a powerful method for deducing the chemical, mineralogical, and physical character of the mantle.

Analysis of the mantle transition zone using a single-component olivine model has already been made by a number of investigators [*Anderson,* 1967; *Akimoto and Fujisawa,* 1968; *Fujisawa,* 1968; *Ringwood and Major,* 1970]. However, experimental uncertainty of the phase diagram used by them produces considerable ambiguities in their conclusions. In particular, the lack of precise information on the gradient dP/dT for the olivine–modified spinel and the modified spinel–true spinel transformation is detrimental to the detailed discussion of the transition zone. The single-component olivine model is too simple to be applied to the actual earth's mantle; other major component minerals, such as orthopyroxene, clinopyroxene, and garnet, cannot be ignored. Recent laboratory experiments demonstrate that phase transformations take place in those component minerals, too, possibly in the transition zone. *Ringwood and Major* [1966*b*] and *Ringwood*

[1]Now at Geological Survey of Japan, Hisamoto-cho, Kawasaki 213, Japan.

perovskite-like structure. These recent experimental results imply that the complex garnet solid solution in the pyrolite mantle transforms to denser phases such as perovskite at pressures above 300 kbar. However, phase transformations accompanying the large density increase would not occur in the complex garnet solid solution at depths shallower than 650 km (≃230 kbar).

## $Mg^{2+}$–$Fe^{2+}$ PARTITION IN THE UPPER HALF OF THE TRANSITION ZONE

In a previous paper, *Nishizawa and Akimoto* [1973] estimated the chemical composition of $(Mg,Fe)_2SiO_4$ spinel in equilibrium with $(Mg,Fe)SiO_3$ pyroxene in the transition zone. They concluded that the partition reaction results in an increase in the iron content of the $(Mg,Fe)_2SiO_4$ solid solution of 1–2% through the olivine–spinel transformation. If we recall the pyroxene–garnet transformation described in a previous section, however, we know that the chemical equilibrium between $(Mg,Fe)_2SiO_4$ spinel and $(Mg,Fe)SiO_3$ pyroxene is meaningless in the actual transition zone.

In the present study we have obtained the thermochemical data necessary to calculate the distribution of $Fe^{2+}$ and $Mg^{2+}$ for the most probable minerals in the upper half of the transition zone. The following calculation is generally based on the mineralogic model just described for the mantle.

First we assume that the Fe: (Fe+Mg) ratio of total mantle material is 0.106 according to *Ringwood* [1966]. We also assume that the upper half of the mantle transition zone between 430 and 650 km consists of 60 wt % $\beta$ and/or $\gamma(Mg,Fe)_2SiO_4$ and 40 wt % complex garnet solid solution, ignoring the small quantity of jadeite component. This model corresponds to the mineral assembly of 80 mole % $\beta$ and/or $\gamma(Mg,Fe)_2SiO_4$ and 20 mole % complex garnet solid solution.

Further, it is assumed that the $Mg^{2+}$–$Fe^{2+}$ partition relation between the $\beta$ or $\gamma$ phases and the complex garnet can be approximated by the simple $(Mg,Fe)_2SiO_4$ spinel–$(Mg,Fe)_3Al_2Si_3O_{12}$ garnet system. Then, we can calculate the chemical composition of the modified spinel or the true spinel solid solution in the mantle using the following two equations:

$$\{X_{Fe}^{Sp}/(1 - X_{Fe}^{Sp})\}/\{X_{Fe}^{Ga}/(1 - X_{Fe}^{Ga})\} = K'$$

$$(16/22)X_{Fe}^{Sp} + (6/22)X_{Fe}^{Ga} = X_{Fe}^{Total}$$

where $X_{Fe}^{Sp}$ and $X_{Fe}^{Ga}$ represent the mole fraction of $Fe_2SiO_4$ in $(Mg,Fe)_2SiO_4$ spinel and the mole fraction of $Fe_3Al_2Si_3O_{12}$ in $(Mg,Fe)_3Al_2Si_3O_{12}$ garnet, respectively. Conservation of Fe and Mg is involved in the second equation, where $X_{Fe}^{Total}$ is the Fe:(Fe+Mg) ratio of total mantle material. In the present calculation we considered no discrimination in the partition behavior between the $\beta$ and $\gamma$ phases based on the similarity of divalent cation sites in both structures.

Figure 1 indicates that the apparent partition coefficient $K'$ varies within a rather narrow range of 2.7 to 3.1 in the present experimental condition. This permits us to estimate tentatively the value of $K'$ for the magnesium-rich side as 3. Substituting the value of 0.106 for $X_{Fe}^{Total}$, we finally obtained the value of $X_{Fe}^{Sp} = 0.13$. Thus the chemical composition of the modified spinel or true spinel phase in the transition zone was estimated to be $(Mg_{0.87}Fe_{0.13})_2SiO_4$.

Chemical composition of the upper mantle olivine was taken as $(Mg_{0.89}Fe_{0.11})_2SiO_4$ in Ringwood's pyrolite model. Accordingly, it is concluded that a 2% increase in $Fe_2SiO_4$ component in $(Mg,Fe)_2SiO_4$ solid solution is expected as a coupling effect of the olivine–modified spinel–true spinel transformation and the pyroxene–garnet transformation.

Stability relations of the $Mg_2SiO_4$–$Fe_2SiO_4$ system (Figure 2) indicate that the transformation pressure from ($\alpha + \gamma$) to ($\alpha + \beta$) is scarcely affected by such a small compositional change in $(Mg,Fe)_2SiO_4$. Figure 2 also suggests, however, that compositional effect on the transformation pressure from $\beta$ to ($\beta + \gamma$) and ($\beta + \gamma$) to $\gamma$ is extremely large. About 2% increase in $Fe_2SiO_4$ component results in the decrease of the transformation pressure from $\beta$ to ($\beta + \gamma$) by approximately 10 kbar at around 1500°C. These conclusions derived from the partition reaction are of great importance for refining the structure model of the mantle transition zone.

## TEMPERATURE GRADIENT IN THE TRANSITION ZONE

Previously, *Akimoto and Fujisawa* [1968] and *Fujisawa* [1968] obtained a temperature estimate of 1150°–1530°C at a depth of 370 km based on their preliminary phase diagram for the olivine–spinel transformation in the $Mg_2SiO_4$–$Fe_2SiO_4$ system. In the present paper, we assigned seismic discontinuities at 380–430 km and 500–550 km to the olivine–modified spinel and modified spinel–true spinel transformations, respectively. We now

TABLE 2. Chemical Analyses of Selected Papuan Ophiolites, wt %

| | Gabbros | | Harzburgite | Pyroxenites | | |
|---|---|---|---|---|---|---|
| | (2612) | (2613) | (2608) | (2602) | (2606) | (2611) |
| $SiO_2$ | 50.30 | 49.35 | 52.54 | 48.42 | 53.40 | 51.30 |
| $TiO_2$ | 0.12 | 0.12 | 0.00 | 0.03 | 0.03 | 0.05 |
| $Al_2O_3$ | 15.22 | 16.55 | 0.32 | 1.32 | 5.38 | 2.10 |
| $Fe_2O_3$ | 0.87 | 0.94 | 0.50 | 2.21 | 0.73 | 0.71 |
| FeO | 4.98 | 4.66 | 5.01 | 9.82 | 6.77 | 3.63 |
| CaO | 14.61 | 15.02 | 0.02 | 1.56 | 4.94 | 18.85 |
| MgO | 12.65 | 12.13 | 38.97 | 35.41 | 27.01 | 21.41 |
| MnO | 0.01 | 0.09 | 0.10 | 0.18 | 0.15 | 0.09 |
| $Na_2O$ | 0.59 | 0.60 | 0.15 | 0.10 | 0.11 | 0.15 |
| $K_2O$ | 0.01 | 0.01 | 0.03 | 0.02 | 0.02 | 0.02 |
| $H_2O(+)$ | 1.07 | 1.05 | 1.72 | 1.29 | 1.09 | 1.19 |
| $H_2O(-)$ | 0.13 | 0.16 | 0.33 | 0.27 | 0.20 | 0.24 |
| $P_2O_5$ | 0.00 | 0.01 | 0.00 | 0.00 | 0.00 | 0.00 |
| Total | 100.65 | 100.69 | 99.69 | 100.63 | 99.83 | 99.74 |

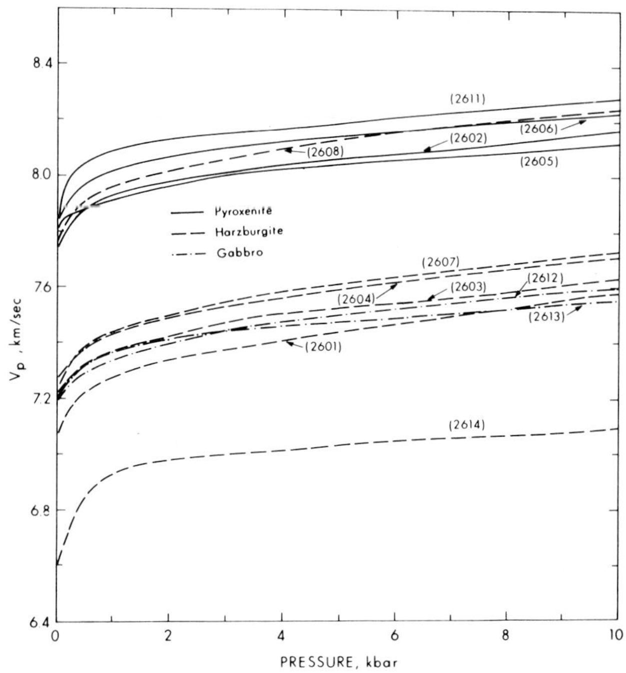


Fig. 3. Mean curves representing $V_p$ versus pressure for Papuan ophiolites.

TABLE 3. Compressional ($V_p$) and Shear Wave ($V_s$) Velocities in Papuan Ophiolites to 10 kbar

| Rock Type and Sample No. | Core Designation | Ambient Density, g/cm³ | Velocity, km/sec; Pressure, kbar: .05 | | .50 | | 1 | | 2 | |
|---|---|---|---|---|---|---|---|---|---|---|
| | | | $V_p$ | $V_s$ | $V_p$ | $V_s$ | $V_p$ | $V_s$ | $V_p$ | $V_s$ |
| Granular gabbro (2612) | a | 3.04 | 7.10 | 3.84 | 7.21 | 3.87 | 7.27 | 3.88 | 7.33 | 3.89 |
| | b | 3.03 | 6.96 | 3.96 | 7.20 | 3.76 | 7.30 | 3.78 | 7.40 | 3.80 |
| | c | 3.03 | 7.15 | 3.75 | 7.34 | 3.77 | 7.42 | 3.78 | 7.50 | 3.79 |
| | Mean | 3.03 | 7.07 | 3.76 | 7.25 | 3.80 | 7.33 | 3.81 | 7.41 | 3.83 |
| Granular gabbro (2613) | a | 3.03 | 7.35 | 3.95 | 7.50 | 4.00 | 7.57 | 4.03 | 7.62 | 4.04 |
| | b | 2.92 | 7.12 | 3.83 | 7.20 | 3.87 | 7.23 | 3.89 | 7.29 | 3.90 |
| | c | 3.03 | 7.16 | 3.85 | 7.24 | 3.87 | 7.28 | 3.88 | 7.33 | 3.89 |
| | Mean | 2.99 | 7.21 | 3.88 | 7.31 | 3.91 | 7.36 | 3.93 | 7.41 | 3.94 |
| Gabbro pegmatite (2614) | a | 2.78 | 6.74 | 3.59 | 6.89 | 3.64 | 6.95 | 3.67 | 7.01 | 3.69 |
| | b | 2.79 | 6.42 | 3.35 | 6.84 | 3.47 | 6.93 | 3.49 | 6.98 | 3.50 |
| | c | 2.77 | 6.64 | 3.71 | 6.86 | 3.78 | 6.91 | 3.81 | 6.95 | 3.82 |
| | Mean | 2.78 | 6.60 | 3.55 | 6.86 | 3.63 | 6.93 | 3.66 | 6.98 | 3.67 |
| Harzburgite (2601) | a | 3.19 | 7.76 | 3.42 | 7.88 | 3.44 | 7.94 | 3.45 | 7.98 | 3.46 |
| | b | 3.19 | 6.61 | 3.38 | 6.82 | 3.44 | 6.89 | 3.48 | 6.97 | 3.50 |
| | c | 3.13 | 6.87 | 3.60 | 6.96 | 3.63 | 7.00 | 3.65 | 7.03 | 3.68 |
| | Mean | 3.17 | 7.08 | 3.47 | 7.22 | 3.50 | 7.28 | 3.53 | 7.30 | 3.55 |
| Harzburgite (2607) | a | 3.19 | 7.30 | 3.79 | 7.49 | 3.88 | 7.57 | 3.91 | 7.65 | 3.94 |
| | b | 3.13 | 7.30 | 3.72 | 7.38 | 3.74 | 7.41 | 3.75 | 7.46 | 3.76 |
| | c | 3.16 | 7.11 | 3.78 | 7.29 | 3.83 | 7.33 | 3.86 | 7.40 | 3.87 |
| | Mean | 3.16 | 7.24 | 3.76 | 7.39 | 3.82 | 7.44 | 3.84 | 7.50 | 3.86 |
| Harzburgite (2603) | a | 3.20 | 7.01 | 3.55 | 7.09 | 3.59 | 7.13 | 3.61 | 7.17 | 3.62 |
| | b | 3.23 | 6.90 | 3.50 | 7.09 | 3.54 | 7.45 | 3.57 | 7.20 | 3.59 |
| | c | 3.23 | 7.68 | 3.58 | 7.77 | 3.62 | 7.81 | 3.63 | 7.78 | 3.64 |
| | Mean | 3.22 | 7.20 | 3.54 | 7.32 | 3.58 | 7.46 | 3.60 | 7.41 | 3.62 |
| Harzburgite (2604) | a | 3.27 | 7.72 | 4.01 | 7.83 | 4.07 | 7.88 | 4.10 | 7.95 | 4.12 |
| | b | 3.24 | 6.86 | 3.44 | 6.97 | 3.48 | 7.01 | 3.50 | 7.07 | 3.52 |
| | c | 3.21 | 7.26 | 3.71 | 7.34 | 3.75 | 7.39 | 3.78 | 7.46 | 3.80 |
| | Mean | 3.24 | 7.28 | 3.72 | 7.38 | 3.77 | 7.43 | 3.79 | 7.49 | 3.81 |
| Harzburgite-pyroxenite (2608) | a | 3.22 | 8.06 | 4.26 | 8.17 | 4.31 | 8.24 | 4.33 | 8.31 | 4.35 |
| | b | 3.22 | 7.56 | 4.25 | 7.68 | 4.29 | 7.73 | 4.30 | 7.78 | 4.32 |
| | c | 3.26 | 7.68 | 3.97 | 7.83 | 4.09 | 7.90 | 4.11 | 7.95 | 4.12 |
| | Mean | 3.23 | 7.77 | 4.16 | 7.89 | 4.23 | 7.96 | 4.25 | 8.01 | 4.26 |
| Pyroxenite-harzburgite (2602) | a | 3.17 | 7.96 | 4.18 | 8.03 | 4.23 | 8.07 | 4.25 | 8.11 | 4.27 |
| | b | 3.29 | 7.67 | 4.00 | 7.78 | 4.08 | 7.85 | 4.12 | 7.92 | 4.16 |
| | c | 3.30 | 7.62 | 4.18 | 7.81 | 4.22 | 7.86 | 4.24 | 7.91 | 4.26 |
| | Mean | 3.25 | 7.75 | 4.12 | 7.87 | 4.18 | 7.93 | 4.20 | 7.98 | 4.23 |
| Pyroxenite (2606) | a | 3.37 | 7.82 | 4.22 | 7.94 | 4.25 | 7.99 | 4.26 | 8.05 | 4.28 |
| | b | 3.38 | 7.89 | 4.33 | 8.00 | 4.38 | 8.05 | 4.40 | 8.09 | 4.41 |
| | c | 3.28 | 7.83 | 4.31 | 7.96 | 4.37 | 8.01 | 4.37 | 8.07 | 4.39 |
| | Mean | 3.34 | 7.85 | 4.29 | 7.97 | 4.32 | 8.02 | 4.34 | 8.07 | 4.36 |
| Clino-pyroxenite (2611) | a | 3.28 | 8.11 | 4.31 | 8.24 | 4.36 | 8.28 | 4.38 | 8.32 | 4.39 |
| | b | 3.29 | 7.59 | 4.22 | 7.81 | 4.26 | 7.86 | 4.28 | 7.89 | 4.29 |
| | c | 3.28 | 7.90 | 4.31 | 8.06 | 4.36 | 8.12 | 4.39 | 8.17 | 4.42 |
| | Mean | 3.28 | 7.87 | 4.28 | 8.04 | 4.33 | 8.09 | 4.35 | 8.13 | 4.37 |
| Websterite (2605) | a | 3.27 | 7.68 | 4.17 | 7.76 | 4.21 | 7.80 | 4.22 | 7.85 | 4.23 |
| | b | 3.28 | 7.80 | 4.11 | 7.86 | 4.14 | 7.90 | 4.14 | 7.94 | 4.16 |
| | c | 3.26 | 7.98 | 4.26 | 8.03 | 4.29 | 8.06 | 4.31 | 8.10 | 4.32 |
| | Mean | 3.27 | 7.82 | 4.18 | 7.88 | 4.21 | 7.92 | 4.22 | 7.96 | 4.24 |

| 3 | | 5 | | 10 | | $(\partial V_p/\partial P)_T$, km/sec · kbar × 100 | $(\partial V_s/\partial P)_T$, km/sec · kbar × 100 |
|---|---|---|---|---|---|---|---|
| $V_p$ | $V_s$ | $V_p$ | $V_s$ | $V_p$ | $V_s$ | | |
| 7.36 | 3.90 | 7.39 | 3.91 | 7.47 | 3.93 | 1.81 | .38 |
| 7.45 | 3.81 | 7.52 | 3.82 | 7.63 | 3.84 | 2.93 | .53 |
| 7.54 | 3.78 | 7.60 | 3.79 | 7.66 | 3.80 | 1.37 | .30 |
| 7.45 | 3.83 | 7.50 | 3.84 | 7.59 | 3.86 | 2.04 | .40 |
| 7.65 | 4.06 | 7.69 | 4.08 | 7.79 | 4.11 | 2.21 | .46 |
| 7.30 | 3.90 | 7.34 | 3.91 | 7.41 | 3.92 | 1.83 | .33 |
| 7.36 | 3.89 | 7.40 | 3.90 | 7.49 | 3.91 | 2.04 | .36 |
| 7.44 | 3.95 | 7.48 | 3.96 | 7.56 | 3.98 | 2.03 | .38 |
| 7.03 | 3.71 | 7.06 | 3.73 | 7.12 | 3.75 | 1.57 | .28 |
| 7.00 | 3.50 | 7.05 | 3.51 | 7.11 | 3.51 | 1.47 | .30 |
| 6.97 | 3.83 | 6.98 | 3.85 | 7.03 | 3.87 | 1.32 | .63 |
| 7.00 | 3.68 | 7.03 | 3.70 | 7.09 | 3.71 | 1.45 | .40 |
| 8.02 | 3.47 | 8.08 | 3.48 | 8.19 | 3.52 | 2.10 | .74 |
| 7.02 | 3.51 | 7.10 | 3.52 | 7.26 | 3.56 | 2.46 | .75 |
| 7.06 | 3.69 | 7.13 | 3.70 | 7.28 | 3.72 | 2.29 | .43 |
| 7.37 | 3.56 | 7.43 | 3.57 | 7.58 | 3.60 | 2.28 | .64 |
| 7.69 | 3.97 | 7.74 | 4.00 | 7.86 | 4.05 | 2.09 | .73 |
| 7.49 | 3.78 | 7.57 | 3.80 | 7.70 | 3.83 | 2.20 | .53 |
| 7.46 | 3.90 | 7.52 | 3.93 | 7.64 | 3.97 | 1.87 | .73 |
| 7.55 | 3.88 | 7.61 | 3.91 | 7.73 | 3.95 | 2.05 | .66 |
| 7.20 | 3.63 | 7.25 | 3.65 | 7.38 | 3.67 | 1.64 | .58 |
| 7.32 | 3.60 | 7.30 | 3.62 | 7.42 | 3.65 | 2.54 | .64 |
| 7.91 | 3.65 | 7.98 | 3.66 | 8.08 | 3.67 | 1.61 | .35 |
| 7.48 | 3.63 | 7.51 | 3.63 | 7.63 | 3.66 | 1.93 | .52 |
| 7.99 | 4.14 | 8.05 | 4.16 | 8.15 | 4.19 | 2.07 | .56 |
| 7.11 | 3.53 | 7.20 | 3.55 | 7.31 | 3.58 | 2.19 | .46 |
| 7.50 | 3.81 | 7.57 | 3.89 | 7.68 | 3.87 | 1.92 | .64 |
| 7.53 | 3.83 | 7.61 | 3.87 | 7.71 | 3.88 | 2.06 | .55 |
| 8.36 | 4.37 | 8.43 | 4.40 | 8.57 | 4.44 | 2.86 | .87 |
| 7.83 | 4.34 | 7.90 | 4.37 | 8.03 | 4.42 | 2.62 | .92 |
| 8.00 | 4.13 | 8.05 | 4.14 | 8.16 | 4.17 | 2.31 | .65 |
| 8.06 | 4.28 | 8.13 | 4.30 | 8.25 | 4.34 | 2.60 | .81 |
| 8.14 | 4.28 | 8.18 | 4.30 | 8.28 | 4.32 | 2.15 | .50 |
| 7.96 | 4.17 | 8.02 | 4.18 | 8.14 | 4.21 | 2.47 | .44 |
| 7.94 | 4.26 | 7.99 | 4.28 | 8.10 | 4.29 | 2.56 | .40 |
| 8.01 | 4.24 | 8.06 | 4.25 | 8.17 | 4.27 | 2.39 | .45 |
| 8.08 | 4.29 | 8.13 | 4.35 | 8.23 | 4.33 | 2.03 | .65 |
| 8.12 | 4.42 | 8.16 | 4.43 | 8.27 | 4.45 | 2.43 | .51 |
| 8.10 | 4.41 | 8.15 | 4.42 | 8.29 | 4.45 | 2.15 | .68 |
| 8.10 | 4.37 | 8.14 | 4.40 | 8.24 | 4.41 | 2.20 | .61 |
| 8.34 | 4.40 | 8.38 | 4.41 | 8.48 | 4.43 | 2.24 | .59 |
| 7.92 | 4.30 | 7.95 | 4.31 | 8.03 | 4.34 | 1.89 | .62 |
| 8.19 | 4.45 | 8.23 | 4.49 | 8.35 | 4.54 | 2.36 | 1.00 |
| 8.15 | 4.38 | 8.19 | 4.40 | 8.29 | 4.44 | 2.16 | .74 |
| 7.87 | 4.25 | 7.90 | 4.26 | 8.01 | 4.29 | 1.88 | .72 |
| 7.97 | 4.17 | 8.02 | 4.18 | 8.09 | 4.20 | 1.54 | .57 |
| 8.13 | 4.33 | 8.18 | 4.35 | 8.27 | 4.38 | 1.30 | .64 |
| 7.99 | 4.25 | 8.03 | 4.26 | 8.12 | 4.29 | 1.57 | .64 |

and mean velocities at 10 kbar, calculated elastic moduli, and mean atomic weight ($\overline{M}$) values are listed in Table 4. Most of the rocks studied have $\overline{M}$ values of ~21 and are characterized by high values (~0.31) of Poisson's ratio ($\sigma$). The $V_p$ and $V_s$ versus density data fit well with Birch relationships $V_p = -2.55 + 3.31\rho$ and $V_s = -0.23 + 1.38\rho$ for $\overline{M} \sim 21$ (Figures 5 and 6), except for the three harzburgites (2601, 2603, and 2604) which fall below the $\overline{M} \sim 21$ line by about 0.5, 0.6, and 0.7 km/sec, respectively. The subnormal velocity values can be explained by serpentinization effects [*Birch,* 1961; *Christensen,* 1966]. Although harzburgites 2601 and 2607 contain the same amounts of serpentine, the $V_p$ and $V_s$ values for 2607 are higher because of higher olivine:pyroxene ratio (Table 1). We calculated $V_p$ and $V_s$ values for the four harzburgites from the velocity–composition triangular diagrams involving olivine, enstatite, and serpentine (*Christensen* [1966]; Figures 5 to 7) and found good agreement between calculated and experimental $V_p$ values (to within 3%). The agreement between the $V_s$ values was rather poor (the experimental values were 10–16% too low).

At 10 kbar, the mean $V_p$ and $\rho$ values for the two granular gabbros (2612 and 2613) range from 7.56 to 7.59 km/sec and from 3.02 to 3.06 g/cm³, respectively. The gabbros do not exhibit significant velocity anisotropy; the difference between maximum and minimum velocities among the three cores, $\Delta V$, expressed as a percentage of the mean velocity value, is less than 5%. On the other hand, both $\Delta V_p$ and $\Delta V_s$ values for harzburgites 2601 and 2604, which have cumulate, layered texture, are significantly higher (ranging from ~6 to almost 16%) than those found in gabbros. In harzburgites, $\Delta V_s$ values are not well defined but $\Delta V_p$ values are more or less affected by texture and composition. In harzburgite 2608 (noncumulate?), whose composition trends toward that of pyroxenite, $\Delta V_p$ is intermediate (~7%), whereas in harzburgite 2607, with noncumulate texture, $\Delta V$ values are lowest ($\Delta V_p \sim 2.9\%$; $\Delta V_s \sim 1.9\%$).

Three of the pyroxenites (2602, 2606, 2611) contain significant amounts (14–66 wt %) of orthopyroxene (enstatite). Although the four pyroxenites, in general, are less anisotropic than the harzburgites ($\Delta V_p$ and $\Delta V_s$ less than 5%), the clinopyroxenite (2611) exhibits a higher velocity and higher anisotropy than the orthopyroxenites (2602, 2605, 2606). The mean 10-kbar $V_p$ and $\rho$ values for the pyroxenites range from 8.12 to 8.29 km/sec and from 3.28 to 3.36 g/cm³, respectively.

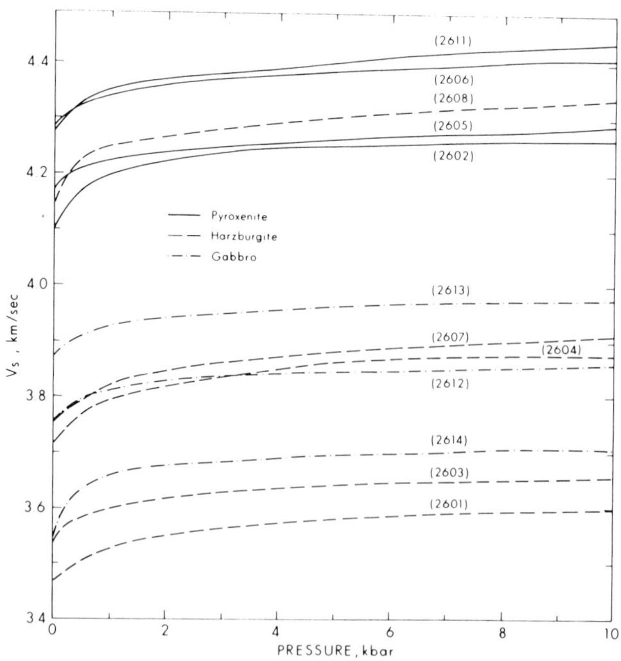


Fig. 4. Mean curves representing $V_s$ versus pressure for Papuan ophiolites.

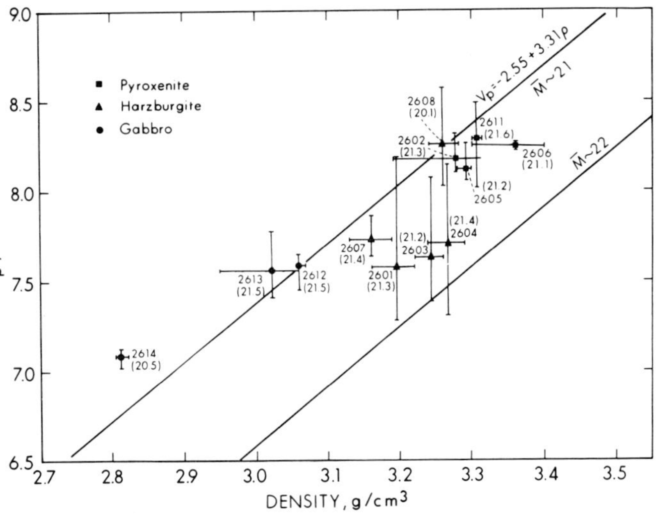


Fig. 5. $V_p$ versus density for Papuan ophiolites at 10 kbar. Values in parentheses are $\overline{M}$ values in g/mole.

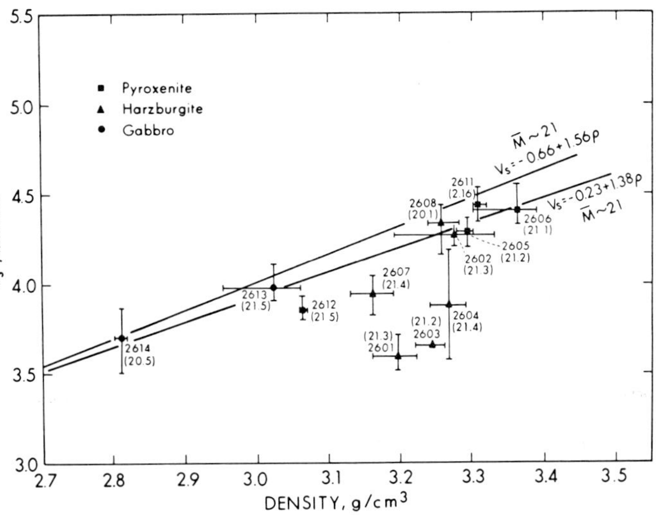


Fig. 6. $V_s$ versus density for Papuan ophiolites at 10 kbar. Values in parentheses are $\overline{M}$ values in g/mole.

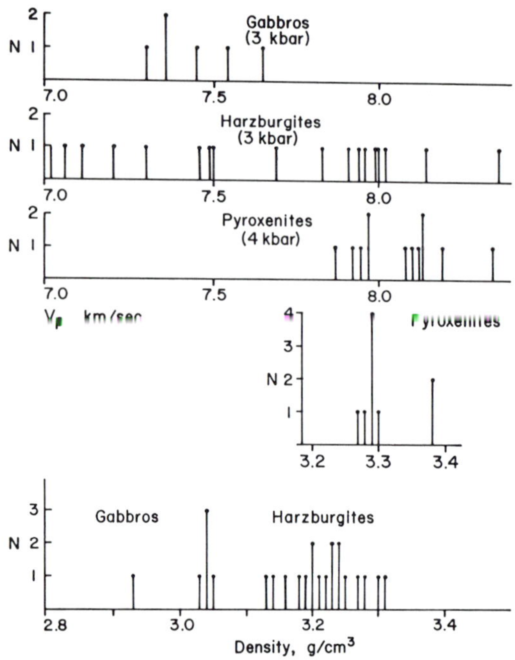


Fig. 7. $V_p$ and density values for individual core specimens of gabbros and harzburgites (at 3 kbar), and pyroxenites (at 4 kbar). N is the number of cases.

The $V_p$ values for these pyroxenites, in general, are significantly higher than the range of values reported for bronzitites and bronzite: 7.89 km/sec, reported by *Babuška* [1972] for a bronzitite from the Stillwater complex; 7.80 to 8.05 km/sec, reported by *Wang* [1974] for a bronzitite also from the Stillwater complex; 7.97 km/sec, reported by *Chung* [1971] for a hot-pressed polycrystalline specimen of bronzite; and a Voigt-Reuss-Hill value of 8.07 km/sec, reported by *Frisillo and Barsch* [1972], for a single crystal bronzite. The especially high range of $V_p$ values (8.03 to 8.48 km/sec) observed for clinopyroxenite 2611 may be explained by the significantly high CaO content (~18.9wt %) of the sample (see Table 2). The effect of high CaO content in increasing $V_p$ is also observed in pyroxenite 2606, which contains 4.9% CaO (28% diopsidic augite)

TABLE 4. Mean Density and Velocity, Velocity Anisotropy, and Calculated Elastic Moduli for Papuan Ophiolitic Rocks at 10 kbar

| Specimen No. | Rock Type | $\overline{M}$, g/mole | $\rho$, g/cm³ | $V_p$, km/sec | $\Delta V_p$,[a] % | $V_s$, km/sec | $\Delta V_s$,[a] % | $K_s$, kbar | $\mu$, kbar | $\sigma$ |
|---|---|---|---|---|---|---|---|---|---|---|
| 2612 | Granular gabbro | 21.5 | 3.063 | 7.59 | 2.5 | 3.86 | 2.3 | 1155 | 455 | .33 |
| 2613 | Granular gabbro | 21.5 | 3.023 | 7.56 | 5.0 | 3.98 | 5.0 | 1091 | 479 | .31 |
| 2614 | Gabbro pegmatite | 20.5 | 2.810 | 7.09 | 1.3 | 3.71 | 9.7 | 895 | 388 | .31 |
| 2601 | Harzburgite | 21.3[b] | 3.197 | 7.58 | 12.3 | 3.60 | 5.6 | 1289 | 415 | .35 |
| 2603 | Harzburgite | 21.2[b] | 3.243 | 7.63 | 9.2 | 3.66 | 0.6 | 1310 | 436 | .35 |
| 2604 | Harzburgite | 21.4[b] | 3.267 | 7.71 | 10.9 | 3.88 | 15.7 | 1288 | 494 | .33 |
| 2607 | Harzburgite | 21.4[b] | 3.187 | 7.74 | 2.9 | 3.95 | 1.9 | 1244 | 497 | .32 |
| 2608 | Harzburgite-pyroxenite | 20.5 | 3.257 | 8.25 | 6.5 | 4.34 | 6.5 | 1401 | 615 | .31 |
| 2602 | Pyroxenite-harzburgite | 21.3 | 3.277 | 8.17 | 2.2 | 4.27 | 2.6 | 1390 | 599 | .31 |
| 2606 | Pyroxenite | 21.1 | 3.363 | 8.24 | 0.5 | 4.41 | 2.7 | 1419 | 654 | .30 |
| 2611 | Pyroxenite | 21.6 | 3.310 | 8.29 | 5.4 | 4.44 | 2.9 | 1404 | 651 | .30 |
| 2605 | Websterite | 21.2[b] | 3.293 | 8.12 | 3.2 | 4.29 | 4.0 | 1366 | 606 | .31 |

Explanation of symbols: $\overline{M}$, mean atomic weight; $\rho$, density; $V_p$, compressional wave velocity; $V_s$, shear wave velocity; $K_s$, adiabatic bulk modulus (incompressibility); $\mu$, shear modulus; $\sigma$, Poisson's ratio.

[a] $\Delta V_p$ and $\Delta V_s$ are maximum differences in $V_p$ and $V_s$, respectively, expressed as percentages of the mean velocity values.

[b] Estimated from modal analysis.

and has a mean $V_p$ value of 8.24 km/sec, slightly higher than that of bronzite.

The relations defining variations in velocity with rock type also hold for $V_s$. As seen in Figure 4, $V_s$ values for pyroxenites are higher than those for harzburgites and gabbros. Similarly, the $V_s$ versus pressure curve for the freshest harzburgite falls in the range of pyroxenite curves. However, the $V_s$ curves for partially serpentinized harzburgites (2606 and 2603) fall below the gabbro curves (2612 and 2613). At 10 kbar the range of mean $V_s$ for pyroxenites is 4.27 to 4.44 km/sec; for harzburgites, 3.60 to 3.95 km/sec; and for gabbros (excluding pegmatite gabbro 2614), 3.86 to 3.98 km/sec.

We determined the $V_p$ values for the Papuan rock under in situ conditions at realistic pressures, that is, for gabbros and harzburgites under confining pressures of 3 kbar (~9 km overburden), and for pyroxenites under confining pressures of 4 kbar (~12 km overburden). Figure 7 shows velocity and density value ranges for all the cores of three rock types under the specified confining pressures.

At 4 kbar, the range of mean $V_p$ values for the four pyroxenites is distinctly larger than for the other two rock types (harzburgites and gabbros) as brought out in Figure 7 (see also Figure 3). For the pyroxenites, the mean values for $V_p$ range from 8.01 to 8.17 km/sec, corresponding to density values of 3.28 and 3.29 g/cm$^3$, respectively; for the harzburgites, the values (at 3 kbar) fall over a wide range: 7.37 to 8.06 km/sec, corresponding to a density range of 3.22 to 3.26 g/cm$^3$, reflecting the variation in chemical composition of the zone. Among the harzburgites, sample 2608 (which is intermediate between harzburgite and pyroxenite) has the highest mean $V_p$ (8.06 km/sec) and density (3.24 g/cm$^3$) values, both of which fall in the range of the pyroxenites (Figure 7). The other harzburgites are partially altered (serpentinized) and their anomalously low $V_p$ values fall in the range of the gabbro velocities (see Figures 3 and 7). There is continuous lowering of $V_p$ values from pyroxenite to harzburgite, the lowering being dependent upon the type of pyroxene present and the degree of alteration of the rock. $V_p$ values for the two gabbros at 3 kbar are about the same: 7.44 to 7.45 km/sec, corresponding to density values of 3.00 to 3.04 g/cm$^3$. This range of $V_p$ values is comparable with the peak distribution of 1.5-kbar $V_p$ values (7.4 to 7.5 km/sec) reported for cumulate gabbros from the Mings Bight complex [*Peterson et al.*, 1974], but higher than the range of values (6.8 to 6.9 km/sec) for the two fresh gabbros from the Troodos massif [*Poster,* 1973], and appreciably higher than the in situ $V_p$ value (5.7 km/sec) for a massive gabbro (slightly altered to greenstone) dredged from the Mid-Atlantic ridge (45°N) [*Barrett and Aumento,* 1970].

*Effect of temperature.* The temperature dependence of $V_p$ was studied to 350°C at 3 kbar for a gabbro, a pyroxenite-harzburgite, and a pyroxenite (see Figure 8). The temperature coefficient $(\partial V_p/\partial T)_p$, for the three rock types in the 150–250°C range is about the same: −15.0, −13.9, and −14.8 × 10$^{-4}$ km/sec deg, respectively. These values are quite high (by a factor of ~3) as compared with that for a polycrystalline specimen of MgO (−4.9 × 10$^{-4}$ km/sec deg, at 1 bar) reported by *Spetzler* [1970], and may be accounted for by the development of microcracks, even though the samples were under confining pressures of 3 kbar. Because the jackets were ascertained to be gastight and the velocity versus temperature curves were reproducible, we consider the values to be real and have used an average value of $(\Delta V_p/\Delta T) = -14 \times 10^{-4}$ km/sec deg for estimating the ranges in velocity and density values at depth in the following test of a crustal and upper mantle model

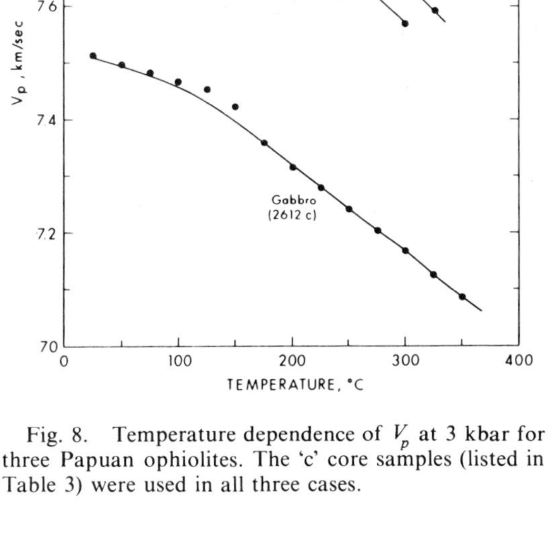


Fig. 8. Temperature dependence of $V_p$ at 3 kbar for three Papuan ophiolites. The 'c' core samples (listed in Table 3) were used in all three cases.

based on *Davies'* [1971] east Papua geologic section.

TEST OF DAVIES' CRUSTAL MODEL FOR COASTAL EASTERN PAPUA

Figure 2, showing the geologic section of the Papua ultramafic belt, represents an oceanic crust-upper mantle sequence [*Davies,* 1971]. The velocity data for the Papuan rocks were corrected for lithostatic pressure and temperature to derive the velocity–density structure for an oceanic crustal model in the following manner. The pressure at a given depth in the section was calculated assuming that the offshore portion of the section is under a water column 1–2 km deep ($\rho$ = 1.03 g/cm$^3$), a sediment layer 1–2 km thick ($\rho$ = 2.2 g/cm$^3$), a basalt layer 4 km thick ($\rho$ = 2.70 g/cm$^3$), and a gabbro layer 4–6 km thick ($\rho$ = 2.98 g/cm$^3$), as indicated in Davies' section (Figure 2). Temperature at a given depth was calculated assuming temperature gradient of 20°C/km throughout the entire column below the sediments zone. Knowing $V_p$ at a given pressure (see Figure 3), the temperature correction was then applied assuming a value of $(\Delta V_p/\Delta T) = -14 \times 10^{-4}$ km/sec deg for all the rock types. We realize that the temperature correction would give the lower limits of velocity and density values for the model. Knowing the in situ $V_p$, the density value was determined from *Birch's* [1961] relation, $V_p = -2.55 + 3.31\rho$, which holds for $\bar{M} \sim 21$. The in situ $V_p$ and density value ranges thus calculated for gabbroic, and ultramafic noncumulate and cumulate, rocks in the geologic section are listed in Table 5. The present laboratory results, together with preliminary seismic data in the Solomon Sea [*Finlayson et al.*, in press *a, b*; *Broyles and Furumoto,* 1974] and previous laboratory data for oceanic tholeiitic basalts from Hawaii [*Manghnani and Woollard,* 1968] and the Mid-Atlantic ridge [*Christensen and Shaw,* 1970] were used to derive a probable velocity–density structure of the oceanic crust extending into the Solomon Sea.

We used the density–velocity data for the Hawaiian and Mid-Atlantic ridge tholeiitic basalts because the basalts found in the Papuan complex are of similar tholeiitic composition [*Davies,* 1971].

*Composition of lower crust (layer 3) and upper mantle.* There has been some speculation [*Peterson et al.,* 1974] on the probable composition of the widely occurring 7.1 to 7.7 km/sec (average 7.4 km/sec) layer in the Pacific reported by *Sutton et al.* [1971] and *Hussong* [1972]. In terms of all the velocity values reported from recent studies of oceanic crust in the Pacific basin using air guns in combination with explosives [*Sutton et al.,* 1971; *Hussong,* 1972; *Hussong et al.*, this volume; *Woollard,* 1975], the gabbro velocity values (7.2 to 7.3 km/sec, Table 5) correspond well with those of the basal layer (layer 3) of the oceanic crust. Such an interpretation of the composition of the basal crustal layer was also made in two other areas (Newfoundland and Cyprus) by other investigators [*Peterson et al.,* 1974; *Poster,* 1973].

The in situ $V_p$ values of the cumulate ultramafics (harzburgites and a harzburgite-pyroxenite), representing a thin transition zone (0.5 km thick in Davies' section (Figure 2)), range widely, from about 7.1 to 7.9 km/sec (Table 5); values for the pyroxenites are generally higher. However, it is difficult to distinguish the cumulate from the noncumulate ultramafics on the basis of $V_p$ values (Table 5).

Among the cumulate ultramafics, the clinopyroxenite 2611 has the highest mean in situ $V_p$ (7.9 km/sec) and density (3.15 g/cm$^3$) values. Thus the upper range of $V_p$ values for the cumulate ultramafics would correspond to upper mantle velocities of 8.0 and 8.1 km/sec. In view of higher observed $V_p$ (8.1 to 8.4 km/sec) in the Solomon Sea [*Finlayson et al.,* in press *b*], the composition of the upper mantle is more likely to be peridotitic. On the other hand, based on the geologic data [*Davies,* 1971], pyroxenite could partly constitute the uppermost ultramafic zone in the upper mantle and still not violate the velocity structure based on the laboratory measurements. More such measurements on a wider selection of rocks from the region are needed to resolve this problem.

*The tectonics and the gravity field.* The tectonic model of the Papua New Guinea region based on the geologic section (section *A–B* in Figures 1 and 9) proposed by *Davies* [1971] emphasizes that a slice of oceanic crust and upper mantle has been 'obducted' [*Coleman,* 1971] onto Mesozoic low-grade metamorphics which form the upper part of the continental (sialic) crust.

The large gravity anomalies and steep gravity gradients across the Papuan ultramafic belt reported earlier by *St. John* [1970] have been recently studied in greater detail [*Milsom,* 1973*a, b*]. In spite of similarities between the gravity profiles for the sections across the Papuan belt (such as section *A–B,* Figure 1) and those for the

TABLE 5. Assigned Ranges of Pressure and Temperature Values for Various Zones in *Davies'* [1971] Geologic Section, and the Derived in situ $V_p$ and $\rho$ Values from Laboratory Measurements

| Zone | Range of In Situ Conditions | | | In Situ Values | | |
|---|---|---|---|---|---|---|
| | Thickness, km | Lithostatic pressure, kbar | Temperature, °C | $V_p$,[a] km/sec | $V_p$, km/sec | $\rho$, g/cm³ |
| Oceanic Crust | | | | | | |
| Water | 0–2 | 0–0.25 | 22–0 | | 1.50 | 1.03 |
| Sediments | 1–2 | 0.25–0.65 | 0–25 | | 2.75[b]–2.80[c] | 2.2[d] |
| Basalt | 4–6 | 0.65–2.0 | 25–125 | | 5.3–5.8[c] | 2.65–2.75[e] |
| Gabbro | 4 | 2.0–3.0 | 125–200 | 7.44–7.45 | 7.2–7.3 | 2.95–2.98 |
| Upper Mantle | | | | | | |
| Cumulate ultramafics | | | | | | |
| Harzburgites | 0.5 | 3.0 | 200 | 7.37–8.12 | 7.09–7.84 | 2.91–3.14 |
| Pyroxenites | | | | 8.10–8.15 | 7.84–7.89 | 3.14–3.15 |
| Noncumulate ultramafics | | | | | | |
| Harzburgite | 4–8 | 3.0–4.0 | 200–280 | 7.58 | 7.19 | 2.94 |
| Pyroxenite | | | | 8.10 | 7.71 | 3.10 |
| Peridotite | | >4 | >280 | | 8.1–8.4[c] | 3.3–3.4 |

[a] Not corrected for temperature.
[b] Average value for the Pacific Ocean basin [*Woollard,* 1975].
[c] Observed seismic values from *Broyles and Furumoto* [1974] and Furumoto and Broyles (personal communication, 1975), based on preliminary studies in the Solomon Sea, near section *A–B*.
[d] From velocity–density relationship for sediments [*Nafe and Drake,* 1963*b*].
[e] Computed from pressure–velocity–density relationships for oceanic basalts [*Manghnani and Woollard,* 1968; *Christensen and Shaw,* 1970].

sections across the Semail ophiolitic belt in Oman and the Troodos massif in Cyprus, both of which represent overthrusting of a slab of oceanic crust and upper mantle, there seems to be some uncertainty as regards the uniqueness of the solution for the 'obduction' hypothesis as applied to the Papua New Guinea region [*Milsom,* 1973*a, b*]. *Milsom* [1973*b*] has suggested an alternate possibility for the emplacement of the Papuan ophiolite belt that involves a large-scale splitting of the lithosphere.

*Rod* [1974], arguing against Davies' overthrust model, has proposed yet another interpretation. His model involves the cessation of subduction of an oceanic plate westward beneath the Australian continental plate, which resulted in thrust faulting at the leading edges of the continental plate and upward intrusions from the lower subduction zone [*Rod,* 1974] (see lower section of Figure 9). More recently, however, Davies' model has been defended by *Milsom and Smith* [1975] on the basis of detailed geological and geophysical (gravity and magnetic) data from eastern Papua. We consider Davies' model to be correct in principle for the purposes of this paper.

Figure 9 (middle) shows a simplified model of a three-layer oceanic crust, 12–15 km thick, to the east of Owen Stanley fault, based on Davies' section *A–B* (modified without changing the basic concept of overthrusting). Also shown in the figure is Rod's original model. The Bouguer gravity profiles for the two models were calculated by two-dimensional gravity analysis [*Talwani et al.,* 1959*b*], using the density values of 1.03, 2.2, 2.7, 2.98, and 3.35 g/cm³ for the water column, sediments, basaltic and gabbroic layers, and upper

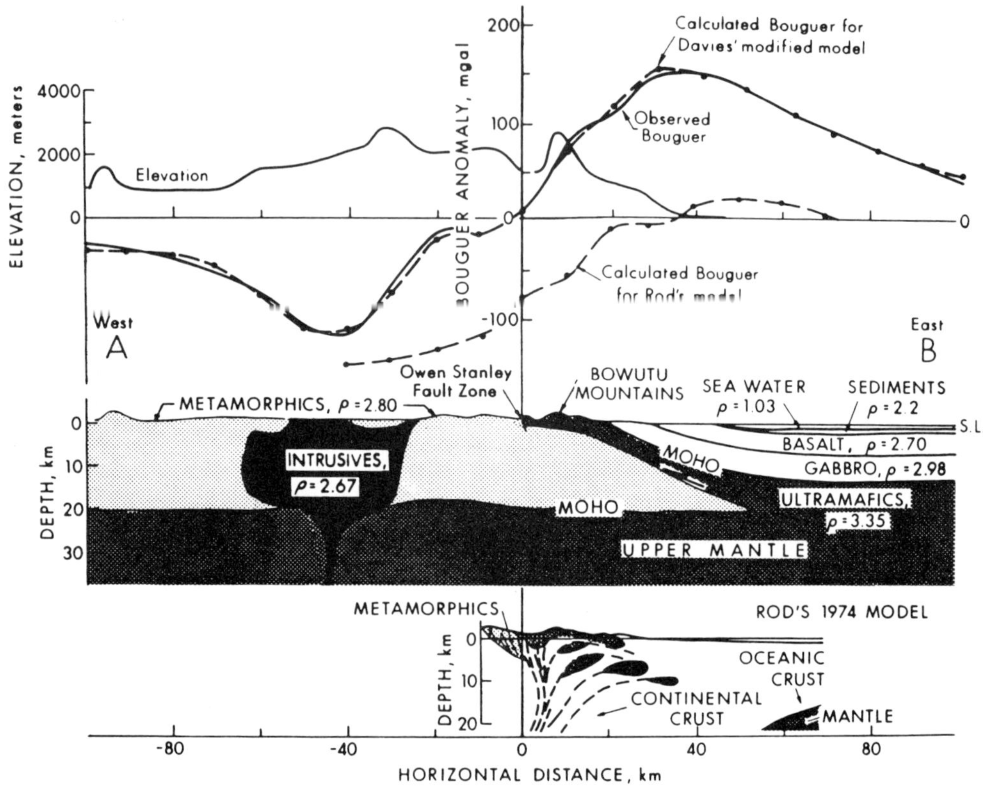


Fig. 9. Comparison between (top) observed and calculated Bouguer anomaly profiles based on (middle) *Davies'* [1971] model, which has been modified, and on (bottom) *Rod's* [1974] model. Cross section *A-B* is that shown in Figure 1.

mantle, respectively (Table 5). The observed Bouguer anomaly profile is based on G. P. Woollard's (personal communication, 1975) unpublished Bouguer anomaly map of Papua New Guinea, and is terrain-corrected, following the method of *Voss* [1974]. As noted by *Davies* [1971] and *Milsom* [1973*b*], the Bouguer gravity-high is offset to the east with respect to the surface exposure of high-density ultramafics. As seen in Figure 9, the computed Bouguer anomaly profile of Rod's model is completely discordant with the observed profile, whereas the modified profile based on Davies' section *A–B* is in good agreement. One finds it difficult to explain the observed gravity profile by any other than the 'overthrust' model. Davies' model is further strengthened by good agreement between the observed and the calculated Bouguer anomalies 100 km offshore based on this model.

There is some uncertainty about the geologic interpretation of the gravity low (40–50 km wide) over the metamorphics ($\rho = 2.80$ g/cm$^3$) to the west of the Owen Stanley fault. This low may reflect either a change in the thickness of the crust or the presence of a near-surface intrusive. The crustal thickness at Port Moresby, slightly southeast of section *A–B*, has been found to be ~25 km [*Broyles and Furumoto*, 1974; *Finlayson et al.*, in press *b*]. The depth to the Moho below the metamorphics has therefore been assumed to be 25 km in the present model. An increase of about 12 km in crustal thickness would be required to account for the observed gravity low. This view is less tenable in light of the recent seismic studies in the area which show that the depth of the Moho decreases eastward in going from offshore (in the Coral Sea, south of section *A–B*) toward the interior of East Papua (A. S. Furumoto, personal communication, 1975). Based on the second hypothesis, the gravity low would be caused by some low-density plutonic intrusions ($\rho \sim 2.67$ g/cm$^3$), similar to the tonalite stocks found in the area. Such a possibility fits well with the observed gravity (Figure 9).

## CONCLUSIONS

The ultramafic and mafic rocks from the Papuan ophiolite complex, representing oceanic upper mantle and basal crust, with $\overline{M} \sim 21$, are characterized by high Poisson's ratio ($\sigma \sim 0.31$), and exhibit velocity-density systematics at 10 kbar that fit well with Birch's law.

Parity between the in situ gabbroic $V_p$ values (7.2 to 7.3 km/sec) and the reported 7.1 to 7.7 km/sec values for the oceanic basal crustal layer further strengthens the proposal that the lower crust in the region is gabbroic.

Among the cumulate ultramafics, harzburgites are the most anisotropic; their velocities are, as a result of serpentinization, generally lower than those for pyroxenites. The upper range of in situ $V_p$ values (7.8 to 7.9 km/sec) for unaltered harzburgites and pyroxenites are subnormal as compared with the average upper mantle values (8.1 to 8.4 km/sec) reported for the region. Although the higher observed values would indicate that the upper mantle composition is peridotitic, the presence of lower velocity harzburgite-pyroxenite material in the uppermost part of the upper mantle (4–8 km thick ultramafic zone of Davies) cannot be precluded.

On the basis of Bouguer gravity anomaly analysis, Davies' 'overthrust' model is found to be more acceptable.

*Acknowledgments.* The authors wish to dedicate this paper to Professor G. P. Woollard, with whom it has been their privilege to be associated as students and as colleagues. The authors are most grateful for his personal interest and always present guidance and help in pilot studies such as this in the Hawaii Institute of Geophysics.

Thanks are due Dr. H. L. Davies for collecting the rock specimens for this study and for constructive criticism of the manuscript; Drs. Finlayson, J. C. Dooley, and N. G. Chamberlain of the Australian Bureau of Mineral Resources (BMR) for their review of the paper and for providing the preprints of their forthcoming papers on the Papuan crustal structure; Dr. G. A. Macdonald for review and helpful comments; Dr. A. S. Furumoto and Mr. M. Broyles for helpful discussion on the preliminary seismic data obtained under a co-operative project organized by the BMR; Dr. K. A. Pankiwskyj for help in petrographic study; and Dr. J. C. Rose and Mr. T. Getts for help in programming.

The authors also acknowledge the incorporation of some of the BMR gravity data from G. P. Woollard's unpublished map.

Financial support for the research was provided by National Science Foundation grant GA-37195X1.

# 36. Thickening of the Oceanic Lithosphere

TOSHIKATSU YOSHII

*Earthquake Research Institute, University of Tokyo, Tokyo 113, Japan*

YOSHITERU KONO

*Faculty of Science, Kanazawa University, Kanazawa, Japan*

KEISUKE ITO

*Faculty of Science, Kobe University, Kobe, Japan*

*Abstract.* A concept of thickening of the oceanic lithosphere is presented. From the 'residual gravity anomaly' *(RGA)*, defined as a mantle gravity anomaly, the thickness of the lithosphere $\ell$ (km) is inferred to increase as the sea-floor age $t$(m.y.) gets older. The relation is given by $\ell = 7.49\sqrt{t}$. A thermal model of the thickening lithosphere is derived so that the lithosphere grows from a molten asthenosphere and the latent heat is released at the lithosphere-asthenosphere boundary. The thermal model is concordant with the thickness of the lithosphere inferred from the *RGA*. The possible presence of a highly molten asthenosphere just beneath the lithosphere is implied from the thermal model. Surface wave and other geophysical studies strongly support the concept of the thickening of the lithosphere. Petrological considerations suggest a distinct layered structure within the lithosphere; namely, dunite or harzburgite, garnet pyroxenite, and eclogite, from the top to bottom layers.

The elevation of the ocean-floor topography near mid-ocean ridges is a remarkable feature that has been noticed even before the beginning of the sea-floor spreading hypothesis [e.g. *Heezen et al.*, 1953]. The elevation has been accounted for by an abnormal upper mantle structure beneath the ridges [e.g., *Talwani et al.*, 1965]. Although the characteristic features near the ridges have often been considered to indicate such deep-seated phenomena as upward currents of the mantle convection, recent studies show the existence of a gradual but remarkable topographic change in older basins where rising mantle currents are not suspected [e.g., *Sclater et al.*, 1971].

*McKenzie* [1967], examining the heat flow and gravity near mid-ocean ridges, proposed a model of the cooling lithosphere with a uniform thickness. The same model has been used to explain the topographic change of the ocean floor [*Sclater et al.*, 1971; *Forsyth and Press*, 1971; *Sclater and Francheteau*, 1970; *Sleep*, 1969]. It should be remembered, however, that McKenzie's model with uniform thickness of the lithosphere lacks a physical basis.

In the present paper, we present an idea of the 'thickening lithosphere model.' Thickening of the lithosphere has been suspected by many authors [e.g., *Atwater*, 1970; *Dewey and Bird*, 1970; *Jacoby*, 1970; *Deffeyes*, 1972; *Vogt*, 1974*a*, *b*] and, recently, *Parker and Oldenburg* [1973] and *Yoshii* [1973] independently proposed physical bases for the thickening process. They stressed an important role of release of latent heat at the base of the lithosphere where the asthenosphere solidifies to become the lithosphere. In this respect, the thickening of the lithosphere strongly resembles that of floating ice sheets or of surface crusts in lava lakes [*Peck et al.*, 1966].

We believe that the concept of thickening of the lithosphere provides an important insight into various kinds of discussions of plate tectonics, because the evolution and the present structure of plates can be deduced from this concept.

## MANTLE GRAVITY ANOMALY AND THE UPPER MANTLE STRUCTURE

*Yoshii* [1972*a*, *b*; 1973*a*, *b*], examining the

upper mantle structure beneath the Pacific and marginal seas, introduced the 'residual gravity anomaly' *(RGA)*, or the mantle gravity anomaly, in the following form:

$$RGA = FA - 2\pi k^2 \ \Sigma \ H_i \Delta \rho_i$$

where $FA$ and $k^2$ indicate the free-air anomaly and the gravity constant, respectively; $H_i$ and $\Delta \rho_i$ are the thickness of the $i$th crustal layer and the density difference between the $i$th layer and the upper mantle, 3.4 g cm$^{-3}$, respectively. The $RGA$ is thus defined as the difference between the observed gravity anomaly and the theoretical one calculated from the crustal structure, and can be regarded as a kind of Bouguer anomaly implying the mass anomaly in the upper mantle.

The $RGA$ has a clear correlation with heat flow as shown in Figure 1. *Yoshii* [1972*b*] considered two thermal models to explain the correlation (cases I and II in Figure 1; see Figures 5 and 7 in *Yoshii* [1972*b*] for explanation). In both models, the heat flow in the oceanic region is controlled essentially by the thickness of the lithosphere. According to Yoshii's models, the thickness of the lithosphere $\ell$ (km) is given by the following simple relation:

$$\ell = RGA \, / \, 4.5 + \text{const}, \qquad (1)$$

where $RGA$ is given in mgal. This relation corresponds to assuming a constant density contrast of 0.107 g cm$^{-3}$ between the lithosphere (averaged vertically) and the asthenosphere.

The $RGA$ shows a good correlation also with the age of the sea-floor inferred from geomagnetic lineations [*Larson and Chase,* 1972; *Atwater and Menard,* 1970]. This relation, shown in Figure 2, is well approximated by

$$RGA = 33.7 \sqrt{t} + 398, \qquad (2)$$

where $t$ is the age in m.y. [*Kono and Yoshii,* 1975]. Consequently, the thickness of the lithosphere is given from (1) and (2) as follows:

$$\ell = 7.49 \sqrt{t} + \ell_0 , \qquad (3)$$

where $\ell_0$ indicates the initial thickness and is assumed to be zero in the following discussion. The thickness given by (3) is, of course, only an approximate one, critically depending on assumptions made. However, it is very interesting that an average thickness for 0–150 m.y. from (3) is about 61 km and is very close to that obtained by studies of surface waves [e.g., *Dorman et al.,* 1960].

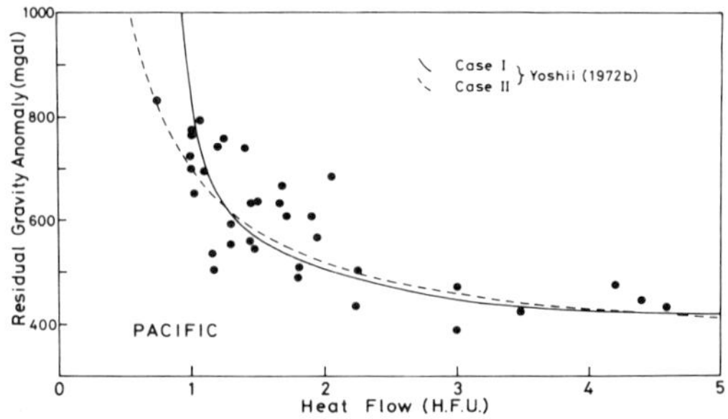


Fig. 1. $RGA$/heat-flow relation in the Pacific, reproduced from *Yoshii* [1972*b*]. Theoretical curves for two thermal models are also given.

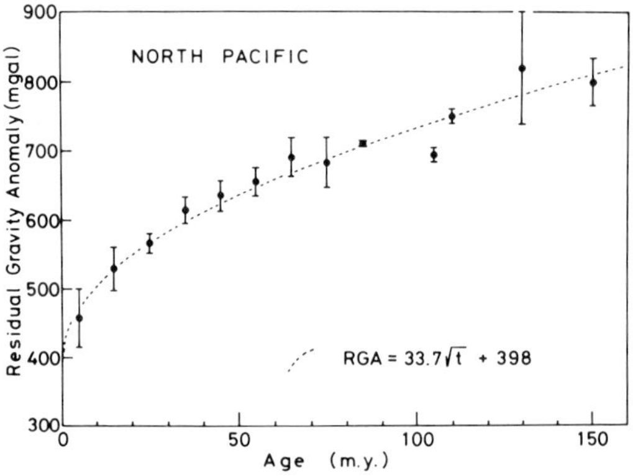


Fig. 2. $RGA$/age relation in the north Pacific, reproduced from *Yoshii* [1973*a, b*]. A dotted curve $33.7 \sqrt{t} + 398$ was determined by the least-squares method.

## THERMAL MODEL OF THE THICKENING LITHOSPHERE

The changes in ocean-floor topography and heat flow correlate well with ocean-floor age, but are essentially independent of the spreading velocity of the plate [*Sclater et al.,* 1971; *Sclater and Francheteau,* 1970]. Ignoring horizontal conduction of heat, we will consider our problem to be one-dimensional in which the coordinate system is fixed on the moving plate. The temperature distribution in a thickening lithosphere is given by an equation of thermal conduction,

$$\rho C_p \frac{\partial T}{\partial t} = K \frac{\partial^2 T}{\partial z^2} + H , \qquad (4)$$

with a boundary condition at the base of the lithosphere.

$$\rho L \frac{d\ell}{dt} = K \left(\frac{dT}{dz}\right)_{\ell}, \qquad (5)$$

where $\rho$, $C_p$, $T$, $K$, $H$, and $L$ indicate density, specific heat, temperature, thermal conductivity, heat generation, and latent heat of fusion, respectively.

The thermal equation in steady state is considered for a first approximation. When the temperature at the base of the lithosphere is assumed to be $T_1$ and the initial thickness of the lithosphere to be zero, the following solution is obtained [*Yoshii,* 1973*a*]:

$$\ell^2 = \frac{2KT_1}{H} \left\{1 - \exp\left(-\frac{H}{\rho L} t\right)\right\}. \qquad (6)$$

In Figure 3, two theoretical curves from (6) are given by solid curves together with estimated thicknesses of the lithosphere calculated from (1) with data plotted in Figure 2. Physical constants used are: $K = 0.007$ cal cm$^{-1}$ sec$^{-1}$ °C$^{-1}$, $T_1 = 1100$°C, $H = 10^{-14}$ cal cm$^{-3}$ sec$^{-1}$, and $\rho = 3.4$ g cm$^{-3}$. Although a theoretical curve for $L = 250$ cal g$^{-1}$ fits well with the estimated values, the latent heat is too large, because the latent heat of total fusion for the upper mantle material is believed to be about 100 cal g$^{-1}$.

*Kono and Yoshii* [1975] improved the thermal model, modifying it in the following two ways: (1) a thermal equation in unsteady state was used; and (2) upward heat flow $Q_0$ in the asthenosphere was introduced. In the new model, (5) is rewritten as

$$\rho L \frac{d\ell}{dt} = K \left(\frac{dT}{dz}\right)_{\ell} - Q_0. \qquad (5')$$

Solutions for (4) and (5′) are obtained by the use of a numerical integration method. Two examples of the theoretical curves are given in Figure 3 by dotted curves. The specific heat $C_p$ is taken to be 0.25 cal g$^{-1}$ °C$^{-1}$, and the heat generation $H$ is ignored because it only insignificantly affects the result.

A solution for $L = 80$ cal g$^{-1}$ and $Q_0 = 0.6$ HFU has a good fit with the estimated thicknesses and is comparable with the steady-state solution for $L = 250$ cal g$^{-1}$ and $Q_0 = 0$. Although solutions for various combinations of $L$ and $Q_0$ show satisfactory fits with the estimated thicknesses, reducing the latent heat to less than 10 cal g$^{-1}$ is very difficult. The high values of $L$ that are required imply that the asthenosphere just beneath the lithosphere may be molten to a high degree. The thin molten layer may behave as a lubricant to make the moving lithosphere slippery on the asthenosphere.

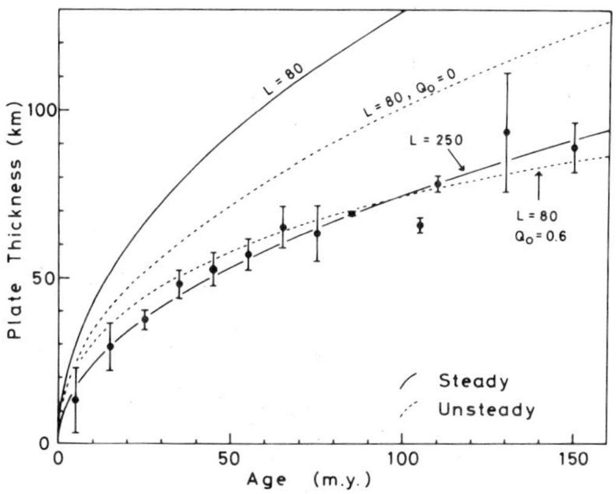


Fig. 3. Theoretical plate thicknesses calculated from the thermal models and estimated ones from the *RGA*.

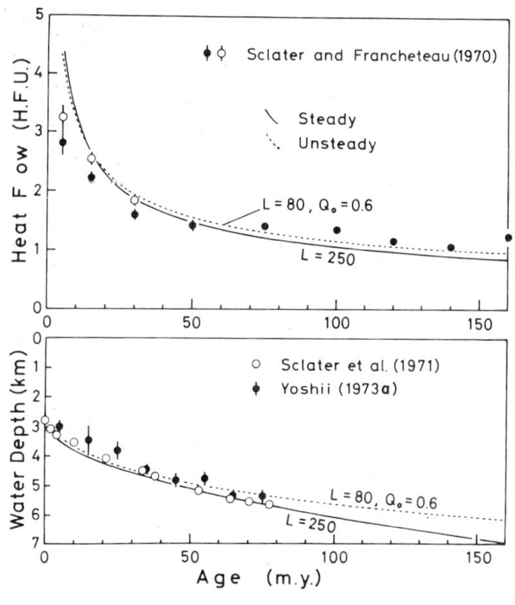


Fig. 4. Comparisons between the observed and theoretical values of heat flow (upper) and topography (lower).

Our thickening plate model also predicts heat flow and topographic changes. In calculation of the topographic changes, isostasy is assumed and densities of the lithosphere and the asthenosphere are taken to be 3.4 g cm$^{-3}$ and 3.3 g cm$^{-3}$, respectively. The predicted changes are compared with observed ones in Figure 4. They show good agreement.

## REGIONAL VARIATIONS OF RAYLEIGH WAVE DISPERSION IN THE PACIFIC

The thickness of the lithosphere in the oceanic region has been inferred to be 60–70 km from studies of surface-wave velocities [e.g., *Kanamori and Press,* 1970; *Press,* 1970*b*; *Dorman et al., 1960*]. However, it has been an open question whether the lithosphere has a uniform thickness in the whole oceanic region or whether it thickens away from the ridges. In the later case, the thickness estimated from surface-wave studies implies an averaged one over the investigated paths. Recently, *Leeds et al.* [1974] and *Yoshii* [1975] estimated regional changes in the thickness of the lithosphere beneath the Pacific on the basis of variations in Rayleigh wave dispersion.

*Santo* [1963] noted regional variations of Rayleigh wave dispersion, but his data are limited to Rayleigh waves of rather short periods (less

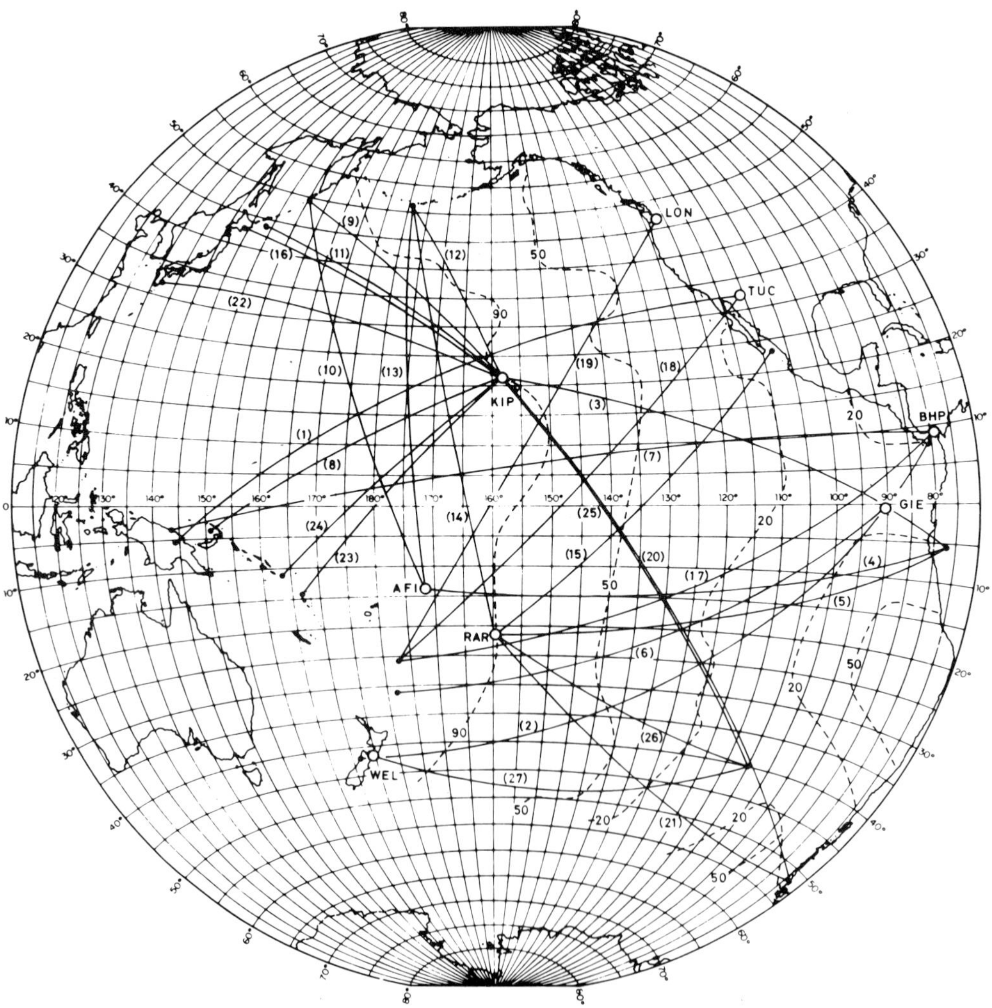


Fig. 5. Map of stations, epicenters, and great circle paths used in the analysis of Rayleigh wave dispersion.

than 40 sec), and therefore it is difficult to interpret a detailed structure of the upper mantle from his data. *Yoshii* [1975] analyzed group velocities of Rayleigh waves in the period range of 40–90 sec using long-period records of W.W.S.S.N.

Figure 5 shows stations (open circles), epicenters (dots), and 27 great-circle paths (solid curves) used in the analysis. The Pacific basin was divided by isochrons of 20, 50, and 90 m.y. based on the geomagnetic lineations [*Larson and Chase,* 1972; *Atwater and Menard,* 1970]; namely, region 1: 0–20 m.y.; region 2: 20–50 m.y.; region 3: 50–90 m.y.; and region 4: older than 90 m.y. Path numbers are given in parentheses.

A 'key' station in the analysis may be KIP (Kipapa, Hawaii) located in the midst of the Pacific (see Figure 5). The Rayleigh wave velocities observed at this station depend greatly on the direction of propagation as shown in Figure 6. Group velocities of Rayleigh waves traveling from the east are very low and those from the west are very high as compared with a theoretical curve for the famous 8099 model [*Dorman et al.,* 1960].

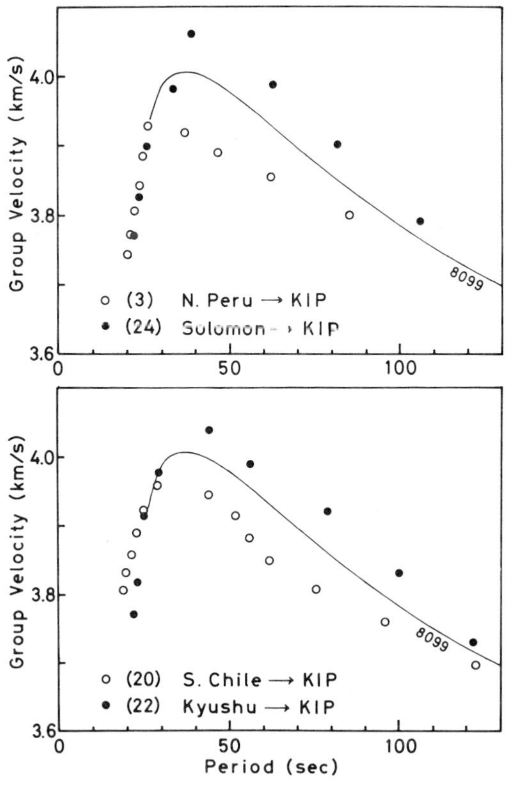


Fig. 6. Some examples of group velocities of Rayleigh waves recorded at station KIP (Kipapa, Hawaii).

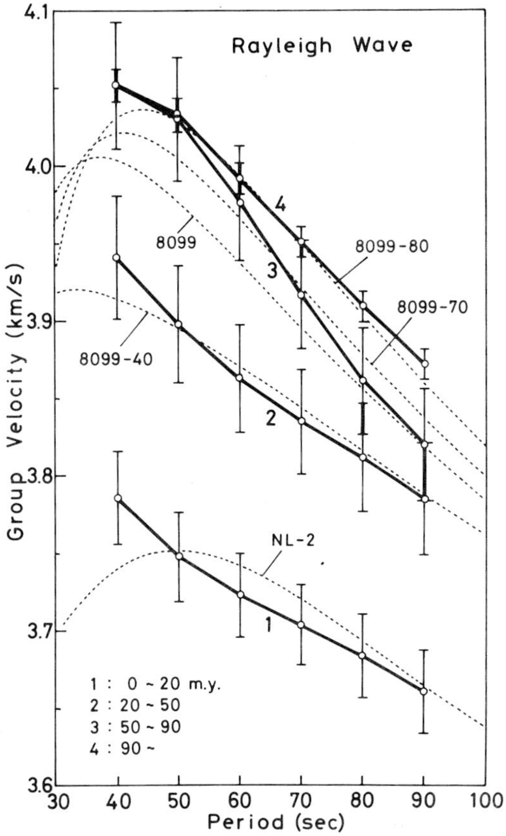


Fig. 7. Regionalized group velocities of Rayleigh waves. Theoretical curves are those for 'modified 8099 models' and model NL-2. Plate thicknesses in the models 8099–40, 8099–70, and 8099–80 are 40, 70, and 80 km, respectively.

Regional group velocities in the four regions are separated by the use of the usual least-squares method [e.g., *Dziewonski,* 1971*b; Hamada,* 1972]. The result is given in Figure 7. The error bar indicates twice the standard deviation. A significant regional change in the group velocity is found; namely, the group velocity increases as the age of the ocean floor becomes older.

The observed group velocities in regions 2, 3, and 4 are well explained by modified 8099 models in which only the thickness of the lithosphere is changed to 40, 70, and 80 km, respectively (see Figure 7). The observed velocity in region 1 is very low and cannot be explained by an 8099 model modified in this manner. A model NL-2 with a pronounced low-velocity layer having a shear velocity of 4.22 km sec$^{-1}$ and without a high-velocity lid satisfies the observed data. Although the present NL-2 model does not have a high-velocity lid, a thin lid may exist because the effect of a thin lid on the theoretical group velocity is rather small. Despite large uncertainties in the inversion of data for region 1, the thickness of the lithosphere is required to be less than about 25 km.

Regional thicknesses of the oceanic lithosphere have been estimated by various methods, such as seismological analysis, electrical conductivity analysis, and others; they are compiled in Figure 8.

As seen in Figure 8, all the data are very close to the dotted curve given by $\ell = 7.49\sqrt{t}$, and strongly suggest the thickening of the lithosphere.

*Kausel et al.* [1974] and *Leeds et al.* [1974] also confirmed the thickening of the lithosphere from studies of phase velocities of Rayleigh waves. However, the thickness obtained by them is somewhat large as compared with the data in Figure 8, especially in older regions.

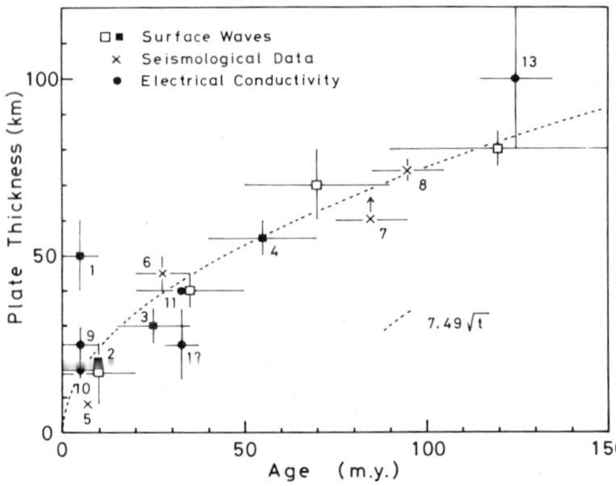


Fig. 8. Regional plate thicknesses estimated by various methods. The dotted curve denotes the thickness inferred from the *RGA* using (3). The open squares represent the present estimate from surface-wave velocities. Other estimates are from (1) *Knopoff et al.* [1970], Rayleigh waves, East Pacific rise; (2) *Weidner* [1974], Rayleigh waves, Mid-Atlantic ridge; (3) *Kanamori and Abe* [1968], surface waves, Parece Vela basin; (4) *James* [1971*b*], Rayleigh waves, Peru–Chile trench region; (5) *Yoshii* [1973*b*], $P_n$ velocities, Pacific; (6) *Nuttli and Bolt* [1969], P wave travel times, west coast of the United States; (7) *Koyanagi and Endo* [1971], earthquake distribution, Hawaii; (8) *Shimamura* [1973], seismic waves reflected at the plate bottom, Kurile–Kamchatka trench region; (9) *Hermance and Garland* [1968], electrical conductivity, Iceland; (10) *Hermance and Grillot* [1970], electrical conductivity, Iceland; (11) *Cox et al.* [1970], electrical conductivity, off the California coast; (12) *Launay* [1974], electrical conductivity, off the California coast; and (13) *Rikitake* [1969], electrical conductivity, off northeastern Japan. In all cases, ocean-floor ages are inferred from geomagnetic lineations or DSDP drillings.

## PETROLOGICAL CONSIDERATION

When the lithosphere grows from the asthenosphere, chemical differentiation may take place. Such differentiation depends on the mechanism of growth. We have considered the following three cases for the thickening mechanism: (1) non-fractional accretion model; (2) fractional accretion model; and (3) liquid injection model.

*Non-fractional accretion model.* The liquid in the asthenosphere is assumed to exist interstitially between mineral grains. The asthenosphere solidifies to become the lithosphere when the liquid crystallizes in situ. No fractionation takes place when the lithosphere is created, thus the chemical composition of the lithosphere is that of the asthenosphere. The apparent latent heat released when the lithosphere grows should be less than 3 cal g$^{-1}$, because the degree of partial melting in the asthenosphere is generally considered to be only 1–3%. Because it is difficult to reconcile the low latent heat with our thermal model, the non-fractional accretion model is not preferred. A peridotitic, non-fractional accretion model in which the latent heat is taken to be 20 cal g$^{-1}$ is shown in Figure 9 for the purpose of comparison.

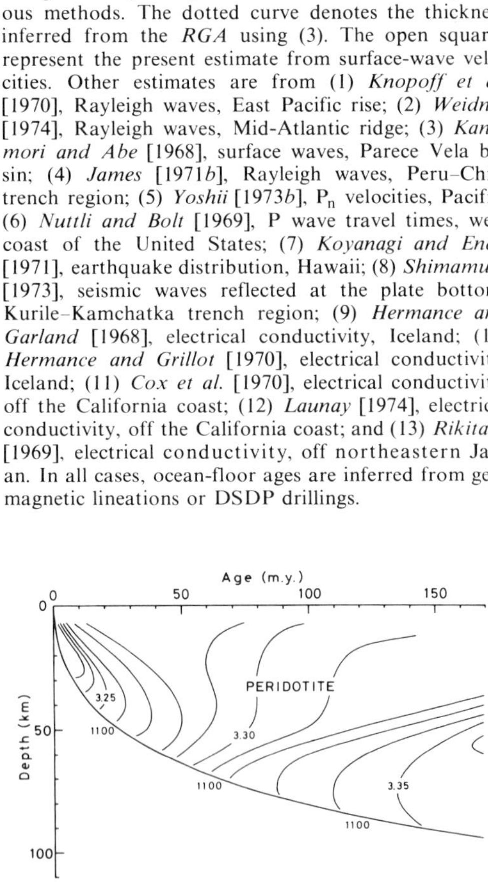


Fig. 9. A contour map of densities in a peridotitic lithosphere based on the non-fractional accretion model. The temperature at the base of the lithosphere is the wet solidus, 1100°C. $L = 20$ cal g$^{-1}$, $Q_0 = 0.4$ HFU, $K = 0.006$ cal cm$^{-1}$ sec$^{-1}$ °C$^{-1}$, $C_p = 0.25$ cal g$^{-1}$ °C$^{-1}$, $\alpha = 3 \times 10^{-5}$ °C$^{-1}$ (thermal expansion), and $K_T = 1.26$ mb (bulk modulus).

*Fractional accretion model.* The asthenosphere is assumed to be chemically differentiated, liquid being concentrated in the top of the asthenosphere. The lithosphere thickens as minerals crystallize fractionally from the liquid and accrete to the base of the lithosphere. Because crystallizing minerals are chemically different from the liquid in which they are forming, the composition of the lithosphere is different from that of the asthenosphere.

The type and composition of the minerals crystallizing from the liquid depend on the pressure and also on the degree of partial melting, but depend little on the chemistry of the asthenosphere as long as the mineral assemblages are the same. At low pressures, a well-known reaction of olivine with liquid to form orthopyroxene [*Bowen*, 1928] plays an important role. The sequence of rocks in layered ultramafic and mafic complexes shows that mafic magma derived by partial melting of peridotite, upon cooling and crystallization, yields the assemblages in the following sequence: dunite (olivine), then harzburgite (olivine + orthopyroxene), and finally gabbro (orthopyroxene + clinopyroxene + plagioclase ± olivine). Therefore, minerals crystallizing from the liquid in the asthenosphere at shallow depths are either olivine or olivine + orthopyroxene if the degree of partial melting is great, but the mineral assemblage is pyroxene + plagioclase if partial melting is very limited.

Since ocean-ridge basalt magmas are considered to be formed by more than 30% partial melting of the asthenosphere [*Green*, 1970*b*], the asthenosphere at shallow depths near mid-ocean ridges may be molten to a high degree, although the asthenosphere at greater depths is only weakly molten. Therefore, the upper part of the lithosphere is considered to consist of either dunite or harzburgite.

Although the reaction of olivine with liquid is suppressed by pressures of 5 kb [*Boyd et al.*, 1964], other reactions appear under other pressures. At intermediate pressures, the reaction olivine + orthopyroxene + liquid = clinopyroxene + spinel, has been inferred from nodule assemblages in alkali basalts [*O'Hara*, 1968] and is confirmed by experiments at about 15 kb [*Ito and Kennedy*, 1967].

Under high pressures, the presence of eclogite nodules as well as peridotite nodules in Hawaiian alkali basalts and kimberlites suggests that a liquid formed from peridotite precipitates eclogite composed of clinopyroxene and garnet. An inferred reaction, olivine + orthopyroxene + liquid = clinopyroxene + garnet, has been demonstrated in experiments at 20–30 kb [*O'Hara and Yoder*, 1967; *Tilley and Yoder*, 1964; *Ito and Kennedy*, 1968; *Green and Ringwood*, 1967*c*].

The reactions mentioned above suggest that upon crystallization, the small fraction of melt in the deep asthenosphere forms clinopyroxenite at intermediate pressures (10–20 kb) and eclogite at high pressures ( $>$ 20 kb). Thus, clinopyroxenite forms the intermediate part of the lithosphere. As the lithosphere further thickens and cools, the clinopyroxenite exsolves garnet and transforms into garnet pyroxenite. The boundary between the upper part and the intermediate part is tentatively assumed to coincide with the phase boundary between the plagioclase-bearing mineral assemblage and the plagioclase-free assemblage on the solidus of the asthenosphere material (see phase diagrams in *Green and Ringwood* [1967*c*] and *Ito and Kennedy* [1974]).

The lowest part of the lithosphere is composed of eclogite. The boundary between the intermediate part and the lower one is tentatively assumed to coincide with the phase boundary between spinel peridotite and garnet peridotite in a peridotite composition (see a phase diagram in *Ito and Kennedy* [1967]).

Although mineral assemblages crystallizing from liquid as stated above are known, ratios of minerals and densities of rocks are rather uncertain. Densities of rocks are inferred from those of nodules in basalts and kimberlites.

In this fractional accretion model, the latent heat of the asthenosphere-lithosphere transition is that of crystallization of a silicate melt, thus this mechanism is consistent with the present thermal model.

*Liquid injection model.* If a liquid layer exists beneath the lithosphere, the lithosphere may thicken as liquid injects into the basal part of the lithosphere and freezes. The composition of liquid formed by partial melting of peridotite is basaltic or gabbroic at low pressures, but becomes picritic, more MgO-rich and $SiO_2$-poor as the pressure increases [*Ito*, 1973; *Kushiro*, 1968]. Therefore, the lithosphere in the liquid injection model consists of gabbro or garnet granulite and of eclogite at great depths. We do not prefer this model because (1) the observed seismic wave velocities at the upper part of the sub-oceanic lithosphere (about 8.1 km $sec^{-1}$) cannot be explained if the lithosphere is gabbroic or picritic in chemistry, and (2) it is doubtful that injection of liquid which needs fracturing of the lithosphere is of much importance at great depths

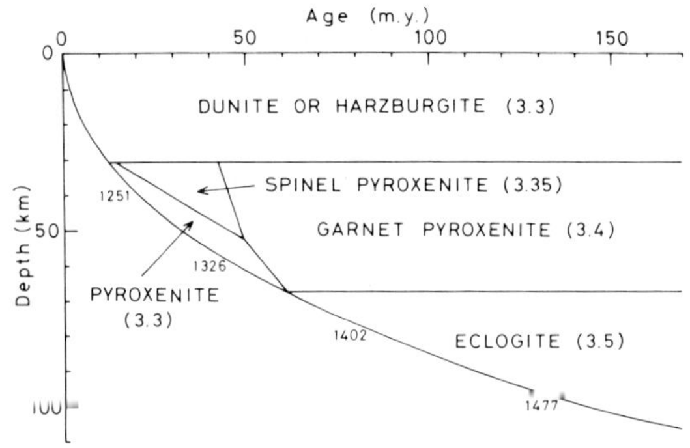


Fig. 10. Rocks and phases in a lithosphere based on the fractional accretion model. The temperature at the base of the lithosphere is the dry solidus [*Ito,* 1974]. Densities in parentheses are those at the room temperature and pressure. $L = 100$ cal g$^{-1}$ and $Q_0 = 0$.

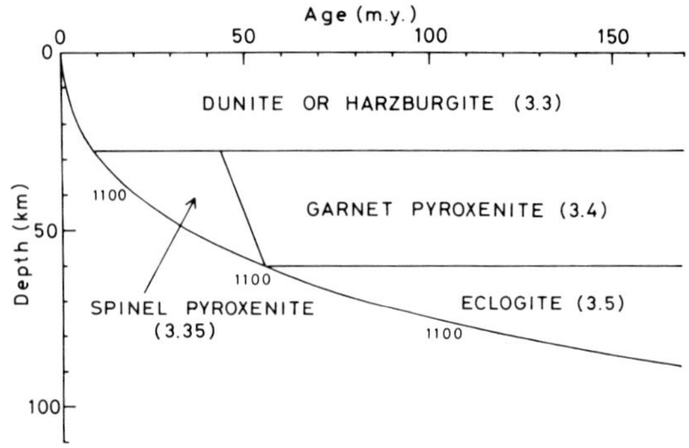


Fig. 11. Rocks and phases in a lithosphere based on the fractional accretion model. The temperature at the base of the lithosphere is the wet solidus, 1100°C. $L =$ 80 cal g$^{-1}$ and $Q_0 = 0.6$ HFU.

where the confining pressures are high.

In conclusion, we prefer the fractional accretion model for the mechanism of thickening of the lithosphere. Petrological models of the lithosphere formed by this mechanism are shown in Figure 10 (a dry solidus model) and Figure 11 (a wet solidus model). Dry solidus models require temperatures higher than 1400°C at the base of the lithosphere where the thickness exceeds 80 km. Because of very high temperatures required at the base, dry solidus models are not preferred. The preferred model shown in Figure 11 gives basal temperatures determined from the wet solidus.

## CONCLUSION

Studies of gravity, heat flow, surface waves, seismology, and electrical conductivity strongly support the concept of thickening of the lithosphere. We are able to obtain a very simple and physically plausible thermal model for the thickening lithosphere. There is no doubt that understanding the fine structure of the lithosphere is one of the most essential bases for discussion of plate tectonics. In this respect, the concept of the thickening lithosphere seems to be very important.

In our model, the mid-ocean ridge is nothing more than a 'rift' in the lithosphere, and the role of large-scale mantle convection is less important. Although the present model can be regarded as a kind of thermal convection in a broad sense, it is a very special one because a dense, rigid 'crustal plate' is formed at the surface. The plate has a density that is higher than that of the underlying, partially molten asthenosphere. We suggest that this density inversion and the gravity instability caused by the thickening of the lithosphere are the most important components of the driving forces of the plate movement. Moreover, the possible presence of a highly molten layer just beneath the lithosphere should be noticed in discussion of the driving mechanism of plates, because it makes the movement of plates very easy.

Petrological consideration suggests a distinct layered structure of the thickening lithosphere. Long-range experiments of explosion seismology in the oceanic region are needed in order to confirm this layered structure. Seismic velocity jumps at depths of about 30 and 55 km, found in the eastern Atlantic [*Steinmetz and Hirn,* 1974; Y. Bottinga, personal communication, 1974], may be pertinent to the layered structure.

*Acknowledgments.* We wish to thank S. Uyeda, K. Nakamura, S. Asano, T. Watanabe, and A. R. Leeds for their discussions and comments. We also wish to thank Y. Bottinga, A. Hirn, and L. Steinmetz for supplying the results of an explosion seismic experiment in the Atlantic before publication.

# 37. Lithospheric Flexure, Analysis of Gravity Anomalies, and the Propagation of Seamount Chains

R. I. WALCOTT

*Geophysics Division, Department of Scientific and Industrial Research, Wellington, New Zealand*

*Abstract.* Topography on the earth's surface loads the lithospheric plates producing vertical displacements, gravity anomalies, and plate stresses. The displacements and gravity anomalies of the deformation produced by individual loads where so distinguished give estimates of the thickness and viscosity of the lithosphere. The statistical analysis of gravity anomalies over large areas is also possible and can lead to the identification of areas of different plate thicknesses. The stresses generated within a plate by the loads of the great seamount chains like the Hawaiian–Emperor chain are sufficiently large to produce elastic failure and even complete rupture of the lithosphere. Depending on the availability of magma, the rupture could result in further growth of seamounts and a self-propagating mechanism for the chain. Such a mechanism is inadequate to explain all features believed to be observed in the Pacific seamount chains, but it is a mechanism that is likely to be superimposed on others such as the drift of the lithosphere over an asthenospheric hot spot.

The mechanical model of plate tectonics proposes that strong lithospheric plates float on and in places plunge down into the underlying weak shell of the asthenosphere. The existence of a lithosphere and an asthenosphere is based on some simple observational evidence known since the turn of the century [*Barrell,* 1914]. The existence of an asthenosphere, a weak layer in which flow of the mantle takes place, is required because of gravitational compensation for surface loads; all surface loads such as accumulations of sediment in large deltaic bodies, or ice caps like Greenland and Antarctica, are gravitationally compensated to the extent that free-air anomalies measured over them are very much smaller than the gravity effect of the load itself. As these loads involve the lateral transfer of mass at the surface, then to achieve compensation, mass must be able to move within the asthenosphere. The existence of a lithosphere, an outer shell capable of supporting large stress differences near the crushing strength of rock for many millions of years, is required because of the persistence of such load differences on the surface of the earth as mountain ranges and valleys or large bodies of different densities within the crust.

Although their existence is demonstrated by these observations, knowledge of the structure, thickness, and rheology of the lithosphere and asthenosphere is principally based on the geometry and time response of the deformation produced by surface loads. The structure and rheology of the asthenosphere is known to a limited extent from the variations in velocity and attenuation of seismic waves. However, it is relaxation of vertical movements that occur in such regions as shield areas unloaded by Late Quaternary ice sheets that provides the most relevant information for processes on a geological time scale. The relaxation time of vertical movements in these areas is of the order of $10^4$ years so we deduce that for loading times exceeding $10^6$ years, as in the case of most geological processes, the asthenosphere can be regarded as a dense inviscid fluid within which the state of stress is hydrostatic. The structure of the lithosphere, on the other hand, is investigated through its deformation by surface loads. In many situations the lithosphere appears to behave like an elastic plate where the wavelength and amplitude of flexure depends on the flexural rigidity of the plate [*Walcott,* 1970*b*]. The flexural rigidity of an elastic plate is a measure of its resistance to bending and is principally determined by the plate thickness. If the plate exhibits a non-elastic (e.g., a viscous) component of deformation, then we can ascribe an apparent flexural rigidity to the plate from the geometry of the deformation upon the supposition that the plate is purely elastic. In this case the non-elastic property is determined by the variation in apparent flexural rigidity with time [*Walcott,* 1970*b; Watts and Cochran,* 1974].

In this paper I examine the structure and rheology of the lithosphere in terms of its deformation by topographic loads, modelling it as an elastic and visco-elastic body floating on a fluid. Two separate aspects are examined: in Part A, the interpretation of gravity anomalies over large areas from the perspective of plate theory; and in Part B, the stresses produced in a plate loaded by chains of seamounts like those of the Pacific.

## GENERAL THEORY OF PLATE LOADING

A particularly simple and general development of the theory of bending a visco-elastic plate is given by *Foucher* [1974]. A surface load $P(r, \kappa)$ of wave number $\kappa$ on a plate floating on an inviscid fluid of density contrast (the difference in density between the asthenosphere and any fluid overlying the plate, water, sediments, air, etc.) $\Delta\rho$ will produce a deflection at point $r$ such that

$$w(r, \alpha) = \frac{P(r, \kappa)}{\Delta\rho \cdot g} \cdot \phi(\kappa, \alpha) \qquad (1)$$

where $g$ is gravity. $\alpha$ represents the physical parameters of the lithosphere, and its form varies as the model of the lithosphere varies. If the symmetry of the problem is two-dimensional then $P$ is a simple harmonic $P_o \cdot \cos \kappa r$ of wavelength $\lambda = 2\pi/\kappa$, where $P_o$ is amplitude, and $r$ is distance from an arbitrary origin. The problem can be generalized for an arbitrary two-dimensional load by the summation of Fourier components. If the symmetry is axisymmetric, the $w$ and $P$ in (1) are the Hankel transforms of the displacement and load respectively, and $r$ is the radial distance from the axis of symmetry.

Irrespective of the symmetry, $\phi(\kappa, \alpha)$ is the same and is called the flexural response function of the model. For a thin elastic plate (the plate thickness $H$ being much less than wavelength and amplitude of bending) the deformation is given by the differential equation [e.g., *Nadai,* 1963]

$$N\left(\frac{\partial^2}{\partial x^2} + \frac{\partial^2}{\partial y^2}\right)^2 w + \Delta\rho \cdot g \cdot w = P \qquad (2)$$

where $N$ is the flexural rigidity $= EH^3/12(1-\sigma^2)$ with $E$ as Young's modulus and $\sigma$ as Poisson's ratio. Then $\phi(\kappa, \alpha)$ has the form

$$\phi(\kappa, \alpha) = \left(1 + \left(\frac{\alpha\kappa}{\sqrt{2}}\right)^4\right)^{-1} \qquad (3)$$

where, in the thin plate model, $\alpha$ = flexural parameter

$$= \left(\frac{4N}{\Delta\rho \cdot g}\right)^{1/4}. \qquad (4)$$

For a thin visco-elastic plate of viscosity $\eta$ the appropriate differential equation is given by *Walcott* [1970*b*], and the form for the flexural response function is [*Foucher,* 1974]

$$\phi(\kappa, \alpha, t) = 1 - \frac{(\alpha\kappa/\sqrt{2})^4}{1 + (\alpha\kappa/\sqrt{2})^4} \cdot \exp\left(\frac{-t}{(1 + (\alpha\kappa/\sqrt{2})^4) \cdot T_o}\right) \qquad (5)$$

with

$$T_o = 3\eta/E.$$

A plot of the visco-elastic thin plate flexural response function $\phi(\kappa, \alpha, t)$ is given in Figure 1 as solid lines. The function varies between 0 for short wavelength and 1 for long wavelength, large $\kappa$ and small $\kappa$, respectively. Note that for different times $t$ the shape of the function is very similar and a curve for a specific time $t$ can be closely matched by the curve for $t = 0$, equivalent to (3), by simply changing the horizontal scale $\alpha$. In other words, rather than have three independent parameters $\kappa$, $\alpha$, and $t$, we can take the flexural parameter $\alpha$ to be time dependent with its instantaneous value ($t = 0$) giving $N$ and its long-term response being determined by the viscosity $\eta$. This procedure has some advantages principally in requiring just one parameter to be determined which is called the apparent flexural parameter ($\alpha'$). As an equivalent, we can use the concept of apparent flexural rigidity as outlined in the introduction.

Foucher (personal communication, 1974) has also calculated the value for the flexural response function rigorously using thick plate theory, where the thin plate assumption is not made. The calculation is numerical and no analytical form for the function is possible. Some curves given by him are shown in Figure 1 as dotted lines. As long as time is comparatively short or wavelengths of loading are long, thin-plate theory is a good approximation, but for short wavelengths of deformation and long times of loading the theory is inadequate. For the analysis given in this paper thin-plate theory only is used and will be adequate if the value of $T_o$ exceeds $10^5$ years, i.e., if

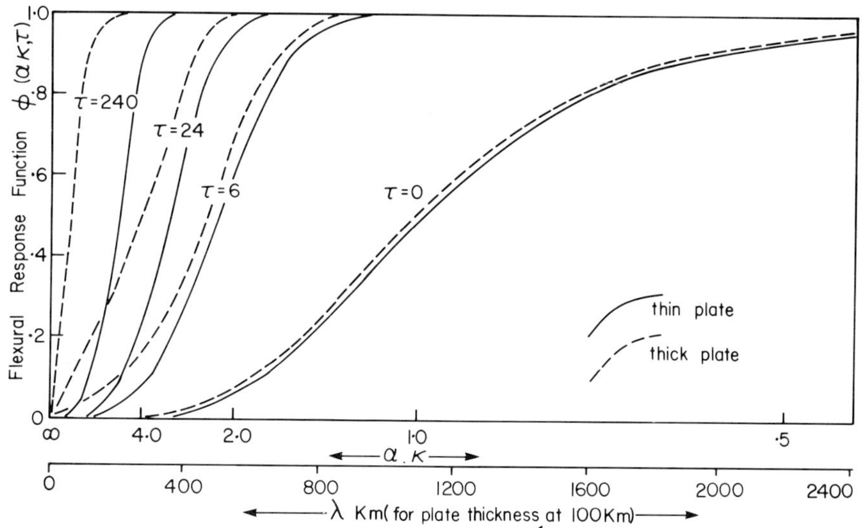


Fig. 1. Flexural response function plotted against the product of wave number $\kappa$ and flexural parameter $\alpha$. The solid lines are those given by (5) in the text where $\tau = t/T_o$. The dotted lines are calculated for a thick plate. Note that for smaller values of $\alpha \cdot \kappa$ or $\tau$ the thin plate approximation is close to the thick plate. Note also that the shape of the curves $\tau = 6$ and 24 is similar to that for $\tau = 0$ and that a close match with $\tau = 0$ can be obtained by treating $\alpha$ as a scaling parameter.

the viscosity of the lithosphere exceeds $10^{24}$ poise ($10^{23}$ $Nsm^{-2}$).

## PART A: STATISTICAL ANALYSIS OF GRAVITY ANOMALIES

The gravity effect produced by the flexural deformation by topographic loads results from the displacement of mass in the asthenosphere of density contrast $\Delta\rho$. But density varies little from the asthenosphere to the Mohorovicic discontinuity and most of the density contrast occurs in the top 30 to 50 km. For long wavelengths of deformation exceeding a hundred kilometers or so, therefore, the gravity effect ($g_w$) at the surface of the earth by the flexure of the lithosphere is given to a good approximation by the infinite slab equation

$$g_w = 2\pi \cdot G \cdot \Delta\rho \cdot w \qquad (6)$$

where $G$ is the gravitational constant and $w$ is, as before, the displacement.
By (1)

$$w(\alpha') = \frac{P(\kappa)}{\Delta\rho \cdot g} \cdot \phi(\kappa, \alpha')$$

where $P(\kappa)$ is the amplitude of the topographic load as a function of wave number $\kappa$. The load itself can be written in terms of topographic height $H(\kappa)$ and density $\rho_T$ as $H(\kappa) \cdot \rho_T \cdot g$. The gravity effect of the topography similarly is $g_t(\kappa) = 2\pi G\rho_T H(\kappa)$, so that the total gravity effect expressed in terms of wave number is

$$g_{t+w}(\kappa) = 2\pi G\rho_T H(\kappa) \left[1 - \phi(\kappa, \alpha')\right] \qquad (7)$$

We normalize the gravity effect by the elevation and obtain

$$g_{t+w}(\kappa)/H(\kappa) = 2\pi G\rho_T \left[1 - \phi(\kappa, \alpha')\right] \qquad (8)$$

The first expression on the right is the Bouguer correction, about 0.112 mgal/m for a load density of 2.67 g/cm³. The free-air anomaly over a large area will contain the gravity effects measured by (7) and those arising from crustal and deeper mass anomalies unrelated to topography and its compensation. If we choose a large area, these anomalies presumably can be regarded as uncorrelated, and values of free-air anomaly as a function of wave number normalized with respect to elevation give the flexural response function.

As an example of the analysis, I use the values for normalized anomalies computed from data in the continental United States (and part of Canada) by *Dorman and Lewis* [1972]. These data are derived from a spectral analysis over a large part of continental North America and include information from areas of probably very different lithospheric thickness and hence flexural parameter.

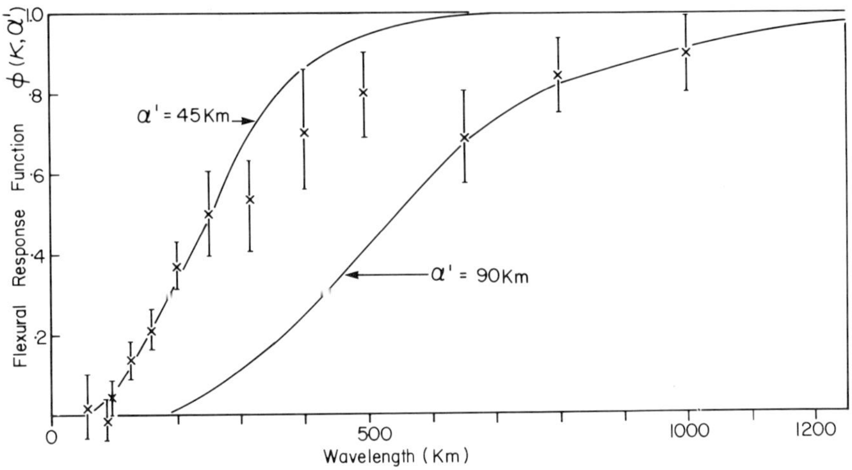


Fig. 2. The crosses and standard deviation bars are from *Dorman and Lewis* [1972], and are the normalized free-air gravity anomalies at specific wavelengths over North America. They lie within the envelope given by $\alpha' = 45$ km fitting short wavelength observations and $\alpha' = 90$ km fitting longer wavelength observations.

The curve (Figure 2) given by the data does not define a single flexural response function value but lies in an envelope bounded by function values of $\alpha' = 45$ km and $\alpha' = 90$ km. I interpret this to mean that the short wavelength data which fit the curve for $\alpha' = 45$ km are dominated by signals arising from a region of comparatively small apparent flexural parameter and, conversely, the longer wavelength data are strongly affected by signals from a region of larger apparent flexural parameter. This is reasonable in the light of values for the apparent flexural parameter estimated independently. The short wavelengths of the spectrum presumably arise largely from the topography of the Canadian Cordillera and Basin and Range Province as the amplitudes in these areas will be considerably larger than in the craton in this part of the spectrum. Estimates of the apparent flexural parameter in the Lake Bonneville area of the Basin and Range Province [*Walcott,* 1970*b*] and in the Canadian Cordillera [*Fulton and Walcott,* in press] give values of about 50 km. The long-wavelength data presumably arise largely from the craton, for a greater total signal will be contributed by its large area to this part of the spectrum. An estimate of the apparent flexural parameter for topographic loading in the Interior Plains gives a value of 110 km [*Walcott,* 1970*c*] and values as high as 180 km for very short term loading by late Quaternary ice sheets [*Walcott,* 1970*b*].

## DISCUSSION

The extension of the technique to other areas, in particular the Pacific basin, should be possible. The calculation of the power spectrum of free-air gravity anomalies over the ocean can be expected to show variation in the flexural rigidity of the oceanic lithosphere. If the lithosphere does thicken away from the spreading ridge as suggested for instance by *Vogt* [1974*b*], then comparison of the spectrum of one-dimensional profiles parallel to the ridges or of areas sorted according to age of the lithosphere should show a systematic increase in the ability of the lithosphere to support longer wavelength loads with age. The analytical problem is, however, not easy because of several complicating factors. The wavelength of the load itself, quite independently of any flexural considerations, may be correlated with age of the lithosphere; because of anelastic effects within the plate, the apparent flexural rigidity may decrease with age and so produce a pattern like that for short-term loading on a thinner plate; simple elastic theory is probably inadequate for certain large loads like seamount chains which may rupture the lithosphere by their weight and would require either the exclusion of gravity information from these areas or development of a more complicated theory than that given above. This last effect is discussed in the following part of the paper.

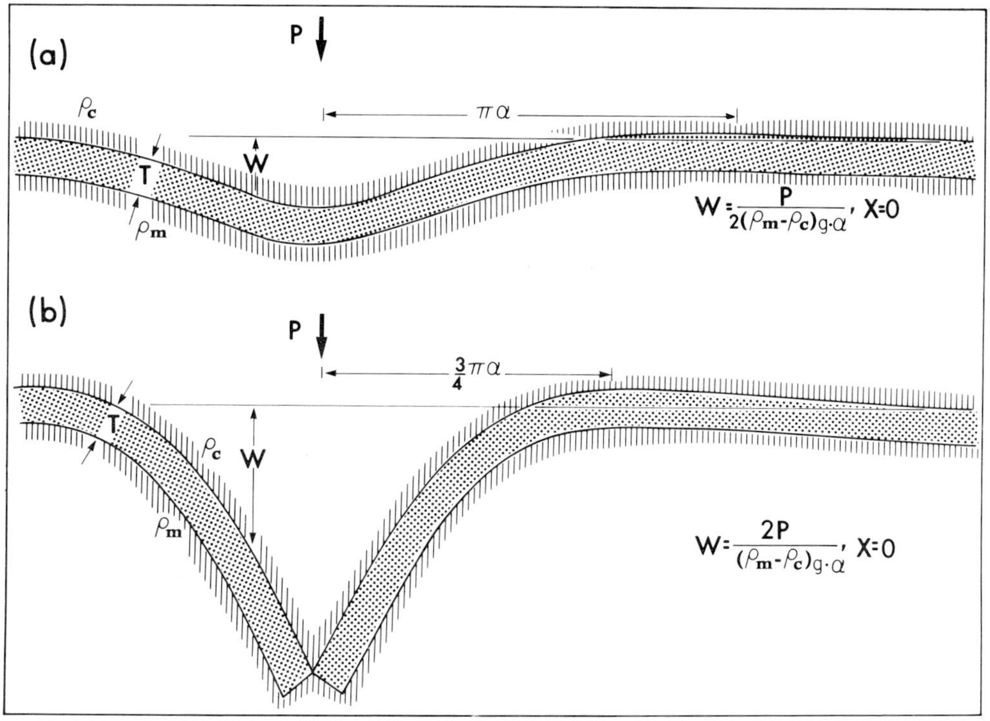


Fig. 3. (a) The flexure by a line load $P$ of an unbroken elastic plate floating on a fluid of density $\rho_m$ and covered by a fluid of density $\rho_c$. The expression for $\alpha$ is given in the text in (4). (b) The flexure of a broken plate, ruptured under the line load $P$ which acts equally on both sides of the plate. The total load supported, therefore is $2P$.

## PART B: PROPAGATION OF SEAMOUNT CHAINS

For narrow and intense loads approximating a point or line load such as an individual or a line of seamounts, the flexural deformation of the lithosphere extends far beyond the load itself. If identifiable surfaces deformed by the flexure can be found, then the flexural parameter and consequently the flexural rigidity can be deduced. Such surfaces could be the Mohorovicic discontinuity or the sea floor if it is flat and not too irregular. In addition, the displacement of these surfaces will produce a gravity anomaly of 0.1 mgal/m displacement and this will be superimposed on the gravity effects of the load and any other mass anomalies about. The deformation of a horizontal surface by a line load on a plate is shown in Figure 3a. A horizontal distance such as that from the load to the positive bulge is a function only of the flexural parameter and this distance can be used for its calculation. The flexural parameter in the region of Oahu was calculated this way on the assumption that the arch of the Hawaiian swell was the positive bulge [*Walcott,* 1970*a*]; the value obtained was 85 km[1]. Alternatively, the amplitude of deflection is a function of both the flexural parameter and the load so that if the load is known, say from bathymetry, then the flexural parameter can be calculated. For the Hawaiian load and a flexural parameter of 85 km, the deflection under Oahu is only about 7 km. Yet according to *Woollard* [1966*b*], the depth of the Moho under Oahu is 21 km and under undisturbed ocean it is 9 km, implying a deflection of the Moho of about 12 km, and a much smaller value of flexural parameter. Similarly the observed 1.6-km displacement under the island of Aitutaki, one of the Cook Islands [*Hochstein,*

[1] Because of an error in the formulation of the Hertz integral solution for the flexural parameter of axisymmetric loading in papers [*Walcott,* 1970*a, b*], the values cited for the flexural parameters are in error by a factor of root two, thus 60 km given in *Walcott* [1970*a*] should be 85 km. The error was pointed out by X. LePichon (personal communication, 1973).

1967], requires a low value for the flexural parameter of about 30 km. There is, therefore, a conflict in the flexural hypothesis between values for the flexural parameter estimated from observations of wavelength or horizontal distance on the one hand and amplitude or vertical distance on the other.

The conflict can be resolved if instead of a continuous lithospheric sheet the sheet is fractured immediately under the load so that it behaves as independent one-sided beams, as in Figure 3b. If this is the case then the value of the flexural parameter required to match both wavelength and amplitude is 125 km, giving a flexural rigidity of $5 \times 10^{31}$ dyne cm ($5 \times 10^{24}$ Nm) [*Walcott,* 1970*c*].

This is the value appropriate for the profile northeast of Oahu. A problem with the flexural hypothesis is that when considering a larger area, the Hawaiian Chain and its associated deformation is non-symmetrical, with the arch to the south being both of smaller amplitude and closer to the ridge than that to the north. *Watts and Cochran* [1974] give a number of profiles of gravity anomalies over the Hawaiian–Emperor structure and fit calculated curves to the data, using amplitude of anomaly as the variable. They obtained a value for the flexural parameter of only 54 km on average. However, their fit north of Oahu is very poor. We can improve the fit by assuming that the lithosphere is of different thicknesses and hence different flexural rigidities on either side of the ridge. However, until some independent evidence of the required variation in structure is available, such ad hoc modifications of the model are suspect. The flexural hypothesis does explain a number of the gross features of the structure, but it is clear that considerable modifications of the assumed uniform plate structure are required to match the observations precisely along the length and on either side of the ridge. Taking the amplitude as a variable in least-squares solution is probably not very reliable as it is strongly affected by surface mass heterogeneities; horizontal distances, i.e., wavelength, however, should be little affected by them and would provide a better variable for analysis.

*Flexural stresses.* Another consideration in flexural theory is the stresses generated by the flexure. In the bending of a thin elastic plate the horizontal stress produced by displacement $w$ is

$$\sigma_x = -Ezd^2w/dx^2 \qquad (9)$$

for two-dimensional bending with $x$ horizontal and normal to the symmetry axis. $E$ is Young's modulus and $z$ is the distance from the mid plane positive downwards, and, as usual, tensile stresses are positive.

For a line load $P$ and an unfractured plate [*Walcott,* 1970*a*]

$$w = \frac{P}{2\Delta\rho_m g\alpha} \exp(-x/\alpha) \cdot (\cos x/\alpha + \sin x/\alpha).$$

Therefore

$$\frac{d^2w}{dx^2} = \frac{-P}{\Delta\rho_m g\alpha^3} \exp(-x/\alpha) \cdot (\cos x/\alpha - \sin x/\alpha),$$

so that at $x = 0$, directly under the load

$$\left.\frac{d^2w}{dx^2}\right|_{x=0} = \frac{-P}{\Delta\rho_m g\alpha^3} = -\frac{2w_o}{\alpha^2}. \qquad (10)$$

Therefore a plate of thickness 60 km, where $\alpha$ will be 160 km, $E = 8 \times 10^{10}$ Nm$^{-2}$ and $z = 30$ km, i.e., the bottom of the plate

$$\sigma_x \simeq 2 \times 10^4 \cdot w_o \text{ dynes/cm}^2$$

i.e., ($2 \times 10^3 \cdot w_o$ Nm$^{-2}$).

A displacement of 1 km, therefore, produces a horizontal stress difference of $2 \times 10^9$ dyne/cm$^2$ ($2 \times 10^8$ Nm$^{-2}$) or about the crushing strength of rock. So there is a very real probability that elastic failure will occur and simple flexure of an elastic sheet is inadequate for loads large enough to cause displacement to the order of 1 km.

We can estimate the height of a seamount chain that will give a displacement of 1 km by considering a plate of the same thickness and flexural parameter and a load with a triangular cross section of height $H$, density contrast with water $\Delta\rho_c$ and half width $d$

$$\text{Load } P = H \cdot d \cdot \Delta\rho_c \cdot g = 2 \cdot \Delta\rho_m \cdot \alpha \cdot w_o \cdot g$$

Therefore,

$$H = \frac{2 \cdot \Delta\rho_m}{\Delta\rho_c} \cdot \frac{\alpha}{d} w_o,$$

so that for

$\Delta\rho_m = 0.5$ g/cm$^3$, $\Delta\rho_c = 1.8$ g/cm$^3$,
$d = 30$ km
$H = 3$ km.

This calculation is for a line load. For a load

standing by itself or at the end of the line, $H$ can be greater to produce the same flexure and the same stresses.

On these arguments I deduce that while the smaller seamounts can be supported by simple elastic flexure of the lithosphere, the larger seamounts will produce stresses during their growth large enough to cause a disruption of the plate, its tearing apart under the line of the seamounts, and a downward collapse under the load. This appears to have happened in the Hawaiian-Emperor chain, as I argued earlier, to reconcile the differences in amplitude and wavelength of the flexure.

*Self-propagation.* The question of the direction of propagation of the rupture is an interesting one in the light of some of the proposed mechanisms of plate kinematics. Consider first a point load (Figure 4). The tensile stresses at the base of the lithosphere under the load are radially symmetric and the rupture could occur in any direction. If, however, the lithosphere is under a normal state of horizontal compression—as *Hanks* [1971] has inferred from study of island arc earthquakes and structure, *Artyushkov* [1973] requires from the topographic irregularities of the earth's surface, and *Hales* [1969] implies in his mechanism for plate movement—then failure will occur preferentially parallel to the direction of ambient compressive stress. If the ambient stress direction is maintained then the tear will be linear, but if the direction changes so will the direction of rupture.

The speed of propagation will be limited by the speed of formation of the load, i.e., by the rate at which magma can collect and move upward to produce the volcano. This rate can be very rapid, as the individual shield volcanoes of the Hawaiian chain appear to form in half a million years or so [*Jackson et al.,* 1972]. If this is correct, and if magma were universally and abundantly available, the propagation of the chain by rupture should be very rapid. The fact that the rate of propagation of chains of seamounts is only about that of the rate of movement of a plate over the asthenosphere indicates that availability of magma is the real limiting factor.

Whereas lines of seamounts could originate by a self-propagating rupture, this model is inadequate to explain why the seamounts in general decrease in age in one direction or why the average velocity of propagation is about that of plate movement. We still seem to require a hot spot moving relative to the lithosphere to explain these observations. However, the self-propagation of the rupture could still be important on the small scale, over say a few hundred kilometers. Once a new volcano is formed to a sufficient size to cause rupture, the crack can propagate rapidly in both directions to the limits of the availability of magma. The volcanoes obtain a height sufficient to stifle the magma flow and volcanic activity ceases and remains dormant until, by plate movement, a new region of virgin lithosphere is moved over the hot spot. At that time a new volcano could break through producing an en echelon fracture or a continuation of the rupture depending on whether the ambient stress field is at an angle or parallel to the direction of relative movement between the hot spot and lithosphere.

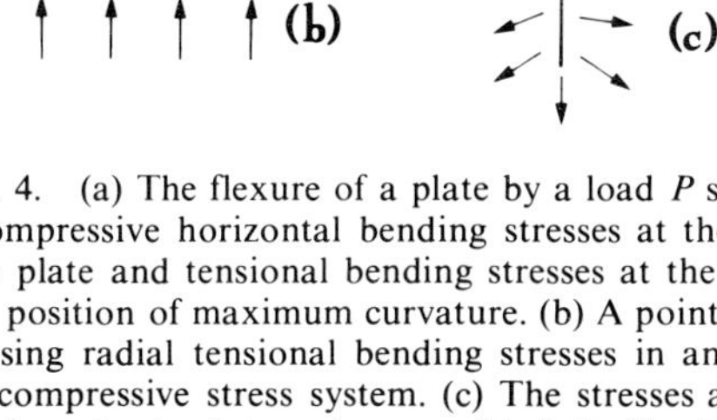


Fig. 4. (a) The flexure of a plate by a load $P$ showing compressive horizontal bending stresses at the top of the plate and tensional bending stresses at the base in the position of maximum curvature. (b) A point load $P$ causing radial tensional bending stresses in an ambient compressive stress system. (c) The stresses at the base of an elastic plate under a semi-infinite line load.

## SUMMARY

The mechanism of rupture by a surface load appears to be important in considering the stresses generated by a large shield volcano and by the type of model required to explain the amplitude and wavelength of associated deformation of the lithosphere. How important the mechanism is in controlling the growth of chains is not clear. Certainly the present inferences of pattern of growth as a simultaneous procession along parallel sigmoidal arcs [*Jackson et al.,* 1972] do not appear to fit at all well. But it is possible to interpret the order of tholeiitic eruption in a different manner from the perspective of flexural theory. *Jackson et al.* [1972] give the order as West Molokai-East Molokai, Lanai-West Maui, Kahoolawe-Haleakala, Kohala, Hualalai-Mauna Kea, Mauna Loa-Kilauea. (I have replaced some of their commas with dashes.) The pairs joined by dashes are interpreted here as nearly simultaneous eruptions produced along an E–W trending rupture, the suggested trend of the ambient stress field. The movement of the plate over the hot spot, however, is suggested to be from SE to NW, giving a series of parallel en echelon ruptures whose individual lengths are limited by the diameter of the hot spot.

This suggestion hypothesizes a direction of movement, an areally restricted hot spot, and a specific direction for the stress field, and no independent information is available for any of them. However, the interpretation of structure depends to a large extent on the intuitively held notions of earth processes, and the suggestion is given here more as an example of a possible process that should be considered in the analysis of field observations than as a preferred mechanism in itself. Clearly the direction of ambient stress is important information which we hope can be determined someday.

elastic moduli, and weathering-age relations for Pacific layer 2 basalts, *Earth Planet. Sci. Lett., 19,* 461–470, 1973. **(19)**

Christensen, N. I., and G. H. Shaw, Elasticity of mafic rocks from the Mid-Atlantic ridge, *Geophys. J. Roy. Astron. Soc., 20,* 271–284, 1970. **(35)**

Christiansen, R. L., and P. W. Lipman, Cenozoic volcanism and plate-tectonic evolution of the western United States, II, Late Cenozoic, *Phil. Trans. Roy. Soc. London A., 271,* 249–284, 1972. **(26)**

Christofferson, E., The relationship of sea-floor spreading in the Pacific to the origin of the Emperor seamounts and the Hawaiian Island chain, (abstract), *Trans. AGU, 49,* 214, 1968. **(28)**

Chung, D. H., Equation of state of pyroxenes in the $(Mg,Fe)SiO_3$ system (abstract), *Eos Trans. AGU, 52,* 919, 1971. **(35)**

Churkin, M., Jr., Paleozoic marginal ocean basin–volcanic arc systems in the Cordilleran foldbelt, *Soc. Econ. Paleo. Mineral. Spec. Publ., 19,* 174–192, 1974. **(26)**

Clague, D., The Hawaiian–Emperor seamount chain: Its petrology, origin, and implications for plate tectonics, Ph.D. dissertation, Univ. Calif., San Diego, 1975. **(22)**

Clague, D. A., and G. B. Dalrymple, Age of Kōko seamount, Emperor seamount chain, *Earth Planet. Sci. Lett., 17,* 411–415, 1973. **(18, 28, 31)**

Clague, D. A., and R. D. Jarrard, Tertiary Pacific plate motion deduced from the Hawaiian–Emperor chain, *Geol. Soc. Amer. Bull., 84,* 1135–1154, 1973. **(18, 21, 28, 31)**

Clague, D. A., G. B. Dalrymple, and R. Moberly, Petrography and K/Ar ages of dredged volcanic rocks from the western Hawaiian ridge and the southern Emperor seamount chain, *Geol. Soc. Amer. Bull., 86,* 991–998, 1975. **(28)**

Clarke, S. H., Jr., and T. H. Nilsen, Displacement of Eocene strata and implications for the history of offset along the San Andreas fault, central and northern California, *Proc. Conf. Tectonic Problems of the San Andreas Fault System,* pp. 358–367, Stanford Univ., 1973. **(26)**

Coan, T., Great eruption of the volcano of Kilauea, *Missionary Herald, 37,* 283–285, 1841. **(27)**

Coan, T., Volcano of Mauna Loa, *Missionary Herald, 39,* 463–464, 1843. **(27)**

Coan, T., On the eruption of Mauna Loa in 1851, *Amer. J. Sci., Ser. 2, 13,* 395–397, 1852. **(27)**

Coan, T., Notes on the recent volcanic disturbances of Hawaii, *Amer. J. Sci., Ser. 2, 47,* 89–98, 1869. **(27)**

Coats, R. R., Magma type and crustal structure in the Aleutian arc, in *The Crust of the Pacific Basin, Geophys. Monogr. Ser.,* vol. 6, edited by G. A. Macdonald and H. Kuno, pp. 92–109, AGU, Washington, D.C., 1962. **(27)**

Cobbing, E. J., and W. S. Pitcher, Plate tectonics and the Peruvian Andes, *Nature Phys. Sci., 240,* 51–53, 1972. **(8)**

Cochran, J. R., and M. Talwani, Gravity anomalies over mid-ocean ridges (abstract), *Abstr. Program Geol. Soc. Amer., 6,* 691, 1974. **(2)**

Cockerham, R. S., and R. D. Jarrard, Paleomagnetic studies, in *Initial Reports of the Deep Sea Drilling Project,* vol. 33, S. O. Schlanger, E. D. Jackson et al., U.S. Govt. Printing Office, Washington, D.C., in press. **(22)**

Coleman, R. G., Low-temperature reaction zones and alpine mafic rocks of California, Oregon, and Washington, *U.S. Geol. Surv. Bull., 1247,* 46 pp., 1967. **(35)**

Coleman, R. G., Plate tectonic emplacement of upper mantle peridotites along continental edges, *J. Geophys. Res., 74,* 1413–1420, 1971. **(35)**

Coleman, R. G., and W. P. Irwin, Ophiolites and ancient continental margins, in *The Geology of Continental Margins,* edited by C. A. Burk and C. L. Drake, pp. 921–931, Springer-Verlag, New York, 1974. **(35)**

Cox, A., Geomagnetic reversals, *Science, 163,* 237–245, 1969. **(19)**

Cox, A., and G. B. Dalrymple, Statistical analysis of geomagnetic reversal data and the precision of potassium–argon dating, *J. Geophys. Res., 72,* 2603–2614, 1967. **(19)**

Cox, A., R. J. Blakely, and J. D. Phillips, A two-layer model for marine magnetic anomalies (abstract), *Eos Trans. AGU, 53,* 974, 1972. **(19)**

Cox, A., R. R. Doell, and G. B. Dalrymple, Reversals of the Earth's magnetic field, *Science, 144,* 1537–1543, 1964. **(18, 28)**

Cox, C. S., J. H. Filloux, and J. C. Larson, Electromagnetic studies of ocean currents and electrical conductivity below the ocean-floor, in *The Sea,* vol. 4, pt. 1, edited by A. E. Maxwell, pp. 637–693, Wiley-Interscience, New York, 1970. **(36)**

Creager, J. S., D. W. Scholl et al., *Initial Reports of the Deep Sea Drilling Project,* vol. 19, 913 pp., U.S. Govt. Printing Office, Washington, D.C., 1973. **(22)**

Cross, W., Lavas of Hawaii and their relations, *U.S. Geol. Surv. Prof. Pap., 88,* 97 pp., 1915. **(27)**

Crowell, J. C., Displacement along the San Andreas fault, California, *Geol. Soc. Amer. Spec. Pap., 71,* 61 pp., 1962. **(26)**

Dalrymple, G. B., Potassium–argon ages from the Pololu Volcanic Series, Kohala volcano, Hawaii, *Geol. Soc. Amer. Bull., 82,* 1997–2000, 1971. **(28)**

Dalrymple, G. B., and M. A. Lanphere, $^{40}Ar/^{39}Ar$ age spectra of some undisturbed terrestrial samples, *Geochim. Cosmochim. Acta, 38,* 715–738, 1974. **(31)**

Dalrymple, G. B., R. D. Jarrard, and D. A. Clague, Potassium–argon ages of some volcanic rocks from the Cook and Austral islands, *Geol. Soc. Amer. Bull.,* in press. **(28)**

Dalrymple, G. B., M. A. Lanphere, and E. D. Jackson, Contributions to the petrography and geochronology of volcanic rocks from the leeward Hawaiian Islands, *Geol. Soc. Amer. Bull., 85,* 727–738, 1974. **(28)**

Dalrymple, G. B., E. A. Silver, and E. D. Jackson, Origin of the Hawaiian Islands, *Amer. Scient., 61,* 294–308, 1973. **(28)**

Daly, R. A., The nature of volcanic action, *Proc. Amer. Acad. Arts Sci., 47*(3), 45–122, 1911. **(27)**

Daly, R. A., *Igneous Rocks and Their Origin,* 563 pp., McGraw-Hill, New York, 1914. **(27)**

Daly, R. A., *Igneous Rocks and the Depths of the Earth,* 598 pp., McGraw-Hill, New York, 1933. **(27)**

Daly, R. A., The sub-Pacific crust, *Proc. Fifth Pacific Sci. Cong.,* 2503–2510, 1934. **(27)**

Dalziel, I. W. D., The evolution of the margins of the

Scotia Sea, in *The Geology of Continental Margins,* edited by C. A. Burk and C. L. Drake, pp. 567–579, Springer-Verlag, New York, 1974. **(26)**

Dalziel, I. W. D., and D. H. Elliot, The Scotia arc and Antarctic margin, in *The Ocean Basins and Margins,* vol. 1, edited by A. E. M. Nairn and F. G. Stehli, pp. 171–246, Plenum Press, New York, 1973. **(26)**

Dalziel, I. W. D., M. J. de Wit, and K. F. Palmer, A fossil marginal basin in the southern Andes, *Nature, 250,* 291–294, 1974*a.* **(26)**

Dalziel, I. W. D., R. Caminos, K. F. Palmer, F. Nullo, and R. Casanova, South extremity of Andes: Geology of Isla de los Estados, Argentine Tierra del Fuego, *Amer. Assoc. Petrol. Geol. Bull., 58,* 2502–2512, 1974*b.* **(26)**

Dalziel, I. W. D., R. H. Dott, Jr., R. D. Winn, Jr., and R. L. Bruhn, Tectonic relations of South Georgia Island to the southernmost Andes, *Geol. Soc. Amer. Bull., 86,* in press. **(26)**

Dana, J. D., *Geology, Volume 10 of the United States exploring expedition during the years 1838–1839, 1840, 1841, 1842,* 756 pp., Sherman and Co., Philadelphia, Pa., 1849. **(27, 28)**

Dana, J. D., *Manual of Geology,* 1st ed., 798 pp., Philadelphia, 1863. **(27)**

Dana J. D., *Manual of Geology,* 2nd ed., 828 pp., Iveson, Blakeman, Taylor and Company, New York, 1876. **(28)**

Dana, J. D., History of the changes in the Mt. Loa craters, on Hawaii, *Amer. J. Sci. Ser. 3, 33,* 433–451; *34,* 81–97, 349–364, 1887; *35,* 15–34, 213–228, 282–289; *36,* 14–32, 81–112, 167–175, 1888. **(27)**

Dana, J. D., Points in the geological history of the islands of Maui and Oahu, *Amer. J. Sci. Ser. 3, 37,* 81–103, 1889. **(27)**

Dana, J. D., *Characteristics of Volcanoes, with Contributions of Facts and Principles from the Hawaiian Islands,* 399 pp., Dodd, Mead & Co., New York, 1890. **(27)**

Darwin, C., *Geological Observations on Volcanic Islands, with Brief Notices on the Geology of Australia and the Cape of Good Hope,* 95 pp., London, 1844. **(27)**

Davies, G. F., The effects of polymorphic phase transitions on elasticity, *Earth Planet. Sci. Lett.,* in press. **(33)**

Davies, H. L., Papuan ultramafic belt, *XIII Int. Geol. Cong., Sess. 23, 1,* 209–220, 1968. **(35)**

Davies, H. L., Peridotite–gabbro–basalt complex in eastern Papua: An overthrust plate of oceanic mantle and crust, *Australia Bur. Min. Res. Bull., 128,* 48 pp., 1971. **(35)**

Davies, H. L., and I. E. Smith, Geology of eastern Papua, *Geol. Soc. Amer. Bull., 82,* 3299–3312, 1971. **(35)**

Day, A. L., and E. S. Shepherd, Water and volcanic activity, *Geol. Soc. Amer. Bull., 24,* 573–606, 1913. **(27)**

Decker, R. W., Magnetic studies on Kilauea Iki lava lake, Hawaii, *Bull. Volcanol., 26,* 23–35, 1963. **(32)**

Decker, R. W., and D. L. Peck, Infrared radiation from Alae lava lake, Hawaii, in *Geological Survey Research 1967, U.S. Geol. Surv. Prof. Pap., 575-D.* D169–D175, 1967. **(32)**

Deffeyes, K. S., The axial valley: A steady-state feature of the terrain, in *Megatectonics of Continents and Oceans,* edited by H. Johnson and B. L. Smith, pp. 194–222, Rutgers Univ. Press, New Jersey, 1970. **(19)**

Deffeyes, K. S., Plume convection with an upper-mantle temperature inversion, *Nature, 240,* 539–544, 1972. **(36)**

Dehlinger, P., Gravity and its relation to topography and geology in the Pacific Ocean, in *The Earth's Crust and Upper Mantle, Geophys. Monogr. Ser.,* vol. 13, edited by P. J. Hart, pp. 352–364, AGU, Washington, D.C., 1969. **(2)**

Den, N., Crustal structures in the western Pacific Ocean, in *The Crust and Upper Mantle of the Japanese Area, part I, Geophys.* Jap. Com. UMP, pp. 57–68, 1972. **(20)**

Den, N., W. J. Ludwig, S. Murauchi, J. I. Ewing, H. Hotta, N. T. Edgar, T. Yoshii, T. Asanuma, K. Hagiwara, T. Sato, and S. Ando, Seismic refraction measurements in the northwest Pacific Basin, *J. Geophys. Res., 74,* 1421–1434, 1969. **(18)**

Denoufbourg, G., *Notice explicative sur la feuille Tahiti,* 24 pp., Bur. de Recherches géolog. et minières, Paris, 1965. **(28)**

de Roever, W. P., Sind die Alpinotypen Peridotitmassen vielleicht tectonisch verfrachette Bruchstücke der Peridotitschale? *Geol. Rundsch., 46,* 137–146, 1957. **(35)**

Dewey, J. F., and J. M. Bird, Mountain belts and the new global tectonics, *J. Geophys. Res., 75,* 2625–2647, 1970. **(8, 26, 36)**

Dewey, J. F., and K. M. Burke, Hot spots and continental break-up: Implications for collisional orogeny, *Geology, 2,* 57–60, 1974. **(26)**

Dickinson, W. R., Relation of andesites, granites, and derivative sandstones to arc–trench tectonics, *Rev. Geophys. Space Phys., 8,* 813–860, 1970. **(26, 29)**

Dickinson, W. R., Plate tectonic models of geosynclines, *Earth Planet. Sci. Lett., 10,* 165–174, 1971. **(26)**

Dickinson, W. R., Width of modern arc–trench gaps proportional to past duration of igneous activity in associated magnetic arcs, *J. Geophys. Res., 78,* 3376–3389, 1973. **(13)**

Dickinson, W. R., Potash-depth (K–h) relations in continental margin and intra-oceanic magmatic arcs, *Geology, 3,* 53–56, 1975. **(29)**

Dickinson, W. R., and E. I. Rich, Petrologic intervals and petrofacies in the Great Valley sequence, Sacramento Valley, California, *Geol. Soc. Amer. Bull., 83,* 3007–3024, 1972. **(26)**

Dickinson, W. R., D. S. Cowan, and R. A. Schweickert, Test of new global tectonics: Discussion, *Amer. Assoc. Petrol. Geol. Bull., 56,* 375–384, 1972. **(26)**

Dickson, G. O., W. C. Pitman, III, and J. R. Heirtzler, Magnetic anomalies in the South Atlantic and ocean-floor spreading, *J. Geophys. Res., 73,* 2087–2100, 1968. **(18)**

Dietz, R. S., Continent and ocean basin evolution by spreading of the sea floor, *Nature, 190,* 854–857, 1961. **(15, 16, 18, 27)**

Dietz, R. S., and H. W. Menard, Hawaiian swell, deep, and arch, and subsidence of the Hawaiian Islands, *J. Geol., 61,* 99–113, 1953. **(2)**

Diller, J. S., and H. B. Patton, The geology and petrography of Crater Lake National Park, *U.S. Geol. Surv. Prof. Pap., 3,* 167 pp., 1902. **(27)**

Doell, R. R., and G. B. Dalrymple, Potassium–argon ages and paleomagnetism of the Waianae and Koolau Volcanic Series, Oahu, Hawaii, *Geol. Soc. Amer. Bull., 84,* 1217–1242, 1973. **(28)**

Dorman, J., Seismic surface wave data on the upper mantle, in *The Earth's Crust and Upper Mantle, Geophys. Monogr. Ser.,* vol. 13, edited by P. J. Hart, pp. 257–265, AGU, Washington, D.C., 1969. **(1)**

Dorman, J., M. Ewing, and J. Oliver, Study of shear-velocity distribution in the upper mantle by mantle Rayleigh waves, *Bull. Seism. Soc. Amer., 50,* 87–115. 1960. **(36)**

Dorman, L. M., and B. T. R. Lewis, Experimental isostasy: 3, inversion of the isostatic Green's function and lateral density changes, *J. Geophys. Res., 77,* 3068–3077, 1972. **(37)**

Dott, R. H., Jr., Superimposed rhythmic stratigraphic patterns in mobile belts, *Geol. Surv. Kans. Bull., 169*(1), 69–86, 1964. **(26)**

Dott, R. H., Jr., Mesozoic–Cenozoic tectonic history of the southwestern Oregon coast in relation to cordilleran orogenesis, *J. Geophys. Res., 70,* 4687–4707, 1965. **(26)**

Dott, R. H., Jr., Circum-Pacific Late Cenozoic structural rejuvenation: Implications for sea-floor spreading, *Science, 166,* 874–876, 1969. **(16, 26)**

Dott, R. H., Jr., Geology of the southwestern Oregon coast west of the 124th meridian, *Oreg. Dept. Geol. Min. Ind. Bull., 69,* 63 pp., 1971. **(26)**

Dott, R. H., Jr., The antiquity of the Scotia arc (abstract), *Eos Trans. AGU, 53,* 178–179, 1972. **(26)**

Dott, R. H., Jr., The geosynclinal concept, in *Soc. Econ. Paleo. Mineral. Spec. Publ., 19,* 1–13, 1974. **(26)**

Douglas, R. G., and S. M. Savin, Isotopic analyses of planktonic foraminifera from the Cenozoic of the northwest Pacific: Leg 6, in *Initial Reports of the Deep Sea Drilling Project,* vol. 6, A. G. Fischer, B. C. Heezen et al., pp. 1123–1127, U.S. Govt. Printing Office, Washington, D.C., 1971. **(24)**

Dowling, J. J., E. F. Chiburis, P. Dehlinger, and M. J. Yellin, Pacific SEAMAP 1961–70 data for area 15524-10, *NOAA Technical Rep., NOS 45,* U.S. Govt. Printing Office, Washington, D.C., 1972*a*. **(2)**

Dowling, J. J., E. F. Chiburis, P. Dehlinger, and M. J. Yellin, Pacific SEAMAP 1961–70 data for area 15530-10, *NOAA Technical Rep., NOS 46,* U.S. Govt. Printing Office, Washington, D.C., 1972*b*. **(2)**

Dowling, J.J., E.F. Chiburis, P. Dehlinger, and M.J. Yellin, Pacific SEAMAP 1961–70 data for area 15248-14, *NOAA Technical Rep., NOS 47,* U.S. Govt. Printing Office, Washington, D.C., 1972*c*. **(2)**

Dowling, J. J., E. F. Chiburis, P. Dehlinger, and M. J. Yellin, Pacific SEAMAP 1961–70 data for area 16648-14, *NOAA Technical Rep., NOS 48,* U.S. Govt. Printing Office, Washington, D.C. 1972*d*. **(2)**

DSDP Scientific Staff, Deep Sea Drilling Project: Leg 6, *Geotimes, 14*(8), 13–17, 1969*a*. **(18, 21)**

DSDP Scientific Staff, Deep Sea Drilling Project: Leg 7, *Geotimes, 14*(10), 12–13, 1969*b*. **(2, 31)**

DSDP Scientific Staff, Deep Sea Drilling Project: Leg 9, *Geotimes, 15*(4), 11–13, 1970. **(16)**

DSDP Scientific Staff, Deep Sea Drilling Project: Leg 17, *Geotimes, 16*(9), 12–14, 1971*a*. **(3)**

DSDP Scientific Staff, Deep Sea Drilling Project: Leg 19, *Geotimes, 16*(11), 12–15, 1971*b*. **(18, 31)**

DSDP Scientific Staff, Deep Sea Drilling Project: Leg 30, Southwest Pacific structures, *Geotimes, 18*(9), 18–21, 1973*a*. **(22)**

DSDP Scientific Staff, Deep Sea Drilling Project: Leg 31, Western Pacific floor, *Geotimes, 18*(10), 22–25, 1973*b*. **(2, 20, 21)**

DSDP Scientific Staff, Deep Sea Drilling Project: Leg 32, *Geotimes, 18*(12), 14–17, 1973*c*. **(18, 22, 28)**

DSDP Scientific Staff, Deep Sea Drilling Project: Leg 33, Testing a hot spot theory, *Geotimes, 19*(3), 16–20, 1974*a*. **(22, 28, 31)**

DSDP Scientific Staff, Deep Sea Drilling Project: Leg 34, Oceanic basalt and the Nazca plate, *Geotimes, 19*(4), 20–24, 1974*b*. **(16)**

DSDP Scientific Staff, Deep Sea Drilling Project: Leg 37, The volcanic layer, *Geotimes, 19*(12), 16–18, 1974*c*. **(19)**

Dubois, J., Propagation of *P* waves and Rayleigh waves in Melanesia: Structural implications, *J. Geophys. Res., 76,* 7217, 1971. **(13)**

Dubois, J., G. Pascal, M. Barazangi, B. L. Isacks, and J. Oliver, Travel times of seismic waves between the New Hebrides and Fiji islands: A zone of low velocity beneath the Fiji plateau, *J. Geophys. Res., 78,* 3431–3436, 1973. **(13)**

Duffield, W. A., A naturally occurring model of global plate tectonics, *J. Geophys. Res., 77,* 2543–2555, 1972. **(32)**

Duncan, R. A., and I. McDougall, Migration of volcanism with time in the Marquesas Islands, French Polynesia, *Earth Planet. Sci. Lett., 21,* 414–420, 1974. **(28)**

Duncan, R. A., I. McDougall, R. M. Carter, and D. S. Coombs, Pitcairn Island—another Pacific hot spot? *Nature, 251*(5477), 679–682, 1974. **(28)**

Dutton, C. E., Hawaiian volcanoes, *U.S. Geol. Survey Fourth Ann. Rep.,* 75–219, 1884. **(27)**

Dymond, J., Geochronology of Tahiti and Moorea, Society Islands (abstract), *Abstr. Program Geol. Soc. Amer. 6*(7), 1036, 1974. **(28)**

Dziewonski, A. M., Upper mantle models from 'pure-path' dispersion data, *J. Geophys. Res., 76,* 2587–2601, 1971*a*. **(1)**

Dziewonski, A. M., On the regional differences in dispersion of mantle Rayleigh waves, *Geophys. J. Roy. Astron. Soc., 22,* 289–325, 1971*b*. **(36)**

Easton, R. L., The role of time/frequency in Navy Navigation Satellites, *Proc. IEEE, 60*(5), 557–563, 1972*a*. **(5)**

Easton, R. L., Timing receiver for Timation Satellite, *Proc. Third Precise Time and Time Interval Planning Meeting,* U.S. Naval Observatory, Washington, D.C., 1972*b*. **(5)**

Easton, R. L., Timation III clocks and sidetone ranging subsystem description, draft report, Naval Research Laboratories, Washington, D.C., 1974. **(5)**

Easton, R. L., P. Morgan, and H. M. Smith, Satellite time transfer from the U.S. Naval Observatory to the Royal Greenwich Observatory and to Australia, presented at the Conference on Precision Electromagnetic Measurements, London, July 1974*a*. **(5)**

Easton, R. L., H. M. Smith, and P. Morgan, Submicrosecond time transfer between the United States, United Kingdom and Australia via satellite, *Proc. Fifth Precise Time and Time Interval Planning Meeting, NASA X-814-74-225,* Goddard Space Flight Center, Greenbelt, Md., 1974*b*. **(5)**

Eaton, J. P., Crustal structure and volcanism in Hawaii, in *The Crust of the Pacific Basin, Geophys. Monogr. Ser.*, vol. 6, edited by E. T. Endo, pp. 13–29, AGU, Washington, D.C., 1962. (**14**)

Eggler, D. H., and M. E. McCallum, Xenoliths in diatremes of the western United States, *Carnegie Inst. Washington, Annual Report of the Director of the Geophysical Laboratory, 73–74,* 294–300, 1974. (**29**)

Ellis, W., *A Journal of a Tour Around Hawaii, the Largest of the Sandwich Islands,* 264 pp., Crocker and Brewster, Boston, 1825. (**27**)

Ellis, W., *A Narrative of a Tour Through Hawaii, or Owhyhee,* 442 pp., H. Fisher, Son, and P. Jackson, London, 1826; reprinted, Honolulu, 1917. (**27**)

Elvers, D., and J. Morley, Northwest Pacific IDOE (2), *Geotimes, 18*(6), 25–26, 1973. (**16**)

Elvers, D., S. P. Srivastava, K. Potter, J. Morley, and D. Spiedel, Asymmetric spreading across the Juan de Fuca and Gorda rises as obtained from a detailed magnetic survey, *Earth Planet. Sci. Lett., 20,* 211–219, 1973. (**22, 28**)

Endo, E. T., Focal mechanisms for the May 15–18, 1970 shallow Kilauea earthquake swarm, M.S. thesis, San Jose State College, Calif., 1971. (**14**)

Engdahl, E. R., Depth of Aleutian earthquakes located by ray tracing, *Nature Phys. Sci., 245,* 23–25, 1973. (**29**)

Engel, A. E. J., C. G. Engel, and R. G. Havens, Chemical characteristics of oceanic basalts and the upper mantle, *Geol. Soc. Amer. Bull., 76,* 719–734, 1965. (**29, 30**)

Evans, B. W., and J. G. Moore, Mineralogy as a function of depth in the prehistoric Makaopuhi tholeiitic lava lake, Hawaii, *Contr. Mineral. Petrol., 17,* 85–115, 1968. (**32**)

Evans, B. W., and T. L. Wright, Composition of liquidus chromite from the 1959 (Kilauea Iki) and 1965 (Makaopuhi) eruptions of Kilauea volcano, Hawaii, *Amer. Mineral., 57,* 217–230, 1972. (**32**)

Ewart, A., and W. B. Bryan, The petrology and geochemistry of the Tongan Islands, in *The Western Pacific: Island Arcs, Marginal Seas, Geochemistry,* edited by P. J. Coleman, pp. 503–522, Univ. Western Australia Press, 1973. (**29**)

Ewart, A., W. B. Bryan, and J. B. Gill, Mineralogy and geochemistry of the younger islands of Tonga, S.W. Pacific, *J. Petrol., 14,* 429–465, 1973. (**29**)

Ewing, J., and M. Ewing, A telemetering ocean-bottom seismograph, *J. Geophys. Res., 66,* 3863–3878, 1961. (**12, 13**)

Ewing, J., M. Ewing, T. Aitkin and W. J. Ludwig, North Pacific sediment layers measured by seismic profiling, in *The Crust and Upper Mantle of the Pacific Area, Geophys. Monogr. Ser.,* vol. 12, edited by L. Knopoff, C. L. Drake, and P. J. Hart, pp. 147–173, AGU, Washington, D.C., 1968. (**18**)

Ewing, M., L. V. Hawkins, and W. J. Ludwig, Crustal structure of the Coral Sea, *J. Geophys. Res., 75,* 1953–1962, 1970. (**35**)

Ewing, M., T. Saito, J. I. Ewing, and L. H. Burckle, Lower Cretaceous sediments from the northwest Pacific, *Science, 152,* 751–755, 1966. (**18**)

Ewing, M., G. P. Woollard, A. C. Vine, and J. L. Worzel, Recent results in submarine geophysics, *Bull. Geol. Soc. Amer., 57,* 909–934, 1946. (**13**)

Fabiano, E. B., and N. W. Peddie, Grid values of total magnetic intensity IGRF-1965, *ESSA Tech. Rep. C and GS 38,* U.S. Dept. Commerce, 55 pp., 1969. (**16**)

Fenner, C. N., The origin and mode of emplacement of the great tuff deposit in the Valley of Ten Thousand Smokes, *National Geogr. Soc. Contrib. Tech. Papers, Katmai Series, 1,* 74 pp., 1923. (**27**)

Fernandez, A., H. Claure, and F. Alcazar, Lineamientos de imagenes ERTS, su significado geologico y su aplicacion en exploracion de minerales, *Boletin GEOBOL,* 1974. (**9**)

Filloux, J. H., Oceanic electric currents, geomagnetic variations and deep conductivity structure of the ocean–continent transition of central California, thesis, Univ. Calif., San Diego, 1967. (**15**)

Finlayson, D. M., B. J. Drummond, C. D. N. Collins, and J. B. Connelly, Crustal structure under the Mt. Lamington region of Papua New Guinea, in *Volcanism in Oceania,* edited by R. W. Johnson, Elsevier Scientific Publishing Company, Amsterdam, in press *a.* (**35**)

Finlayson, D. M., K. E. Muirhead, J. P. Webb, G. Gibson, A. S. Furumoto, R. J. S. Cooke, and A. J. Russell, Seismic investigation of the Papuan ultramafic belt, *Geophys. J. Roy. Astron. Soc.,* in press *b.* (**35**)

Finlayson, J. B., I. L. Barnes, and J. J. Naughton, Developments in volcanic gas research in Hawaii, in *The Crust and Upper Mantle of the Pacific Area, Geophys. Monogr. Ser.,* vol. 12, edited by L. Knopoff, C. L. Drake, and P. J. Hart, pp. 428–438, AGU, Washington, D.C., 1968. (**32**)

Fischer, A. G., B. C. Heezen, R. E. Boyce, D. Bukry, R. G. Douglas, R. E. Garrison, S. A. Kling, V. Krasheninnikov, A. P. Lisitzin, and A. C. Pimm, Geological history of the western North Pacific, *Science, 168,* 1210–1214, 1970. (**20**)

Fischer, A. G., B. C. Heezen et al., *Initial Reports of the Deep Sea Drilling Project,* vol. 6, 1329 pp., U.S. Govt. Printing Office, Washington, D.C., 1971. (**20**)

Fisher, G. W., and D. Elliott, Criteria for quasi-steady diffusion and local equilibrium in metamorphism, in *Geochemical Transport and Kinetics,* edited by A. W. Hofmann, B. J. Giletti, H. S. Yoder, Jr., and R. A. Yand, *Pub. 634,* pp. 231–241, Carnegie Inst., Washington, D.C., 1974. (**29**)

Fisher, R. L., and R. W. Raitt, Topography and structure of the Peru–Chile trench, *Deep-Sea Res., 9,* 423–443, 1962. (**3, 8**)

Fisher, R. L., J. G. Sclater, and D. McKenzie, Evolution of the central Indian ridge, western Indian Ocean, *Geol. Soc. Amer. Bull., 82,* 553–562, 1971. (**16**)

Fisher, R. L., E. T. Bunce, P. J. Cernok, D. C. Clegg, D. S. Cronan, V. Damiani, L. V. Dmitriev, D. J. J. Kinsman, P. H. Roth, J. Thiede, E. Vincent, Site 236, in *Initial Reports of the Deep Sea Drilling Project,* vol. 24, R. L. Fisher, E. T. Bunce et al., pp. 327–389, U.S. Govt. Printing Office, Washington, D.C., 1974. (**30**)

Fitch, T. J., Plate convergence, transcurrent faults, and internal deformation adjacent to southeast Asia and the western Pacific, *J. Geophys. Res., 77,* 4432–4460, 1972. (**21**)

Fitch, T. J., Compressional velocity and velocity contrast in the source region of deep earthquakes, *Earth Planet. Sci. Lett., 26,* 156–166, 1975. (**33**)

Fitch, T. J., and C. H. Scholz, Mechanism of underthrusting in southwest Japan: A model of convergent plate interactions, *J. Geophys. Res., 76,* 7260–7292, 1971. **(9, 20)**

Forristal, G. Z., The thickness of the asthenosphere deduced from the thickness of the Hawaiian melting spot, *Geophys. Res. Lett., 1,* 131–133, 1974. **(28)**

Forsyth, D. W., Mechanisms of earthquakes and plate motions in the East Pacific, *Earth Planet. Sci. Lett., 17,* 189–193, 1972. **(16)**

Forsyth, D. W., Fault plane solutions and tectonics of the South Atlantic and Scotia seas, *J. Geophys. Res.,* in press. **(26)**

Forsyth, D. W., and F. Press, Geophysical tests of petrological models of the spreading lithosphere, *J. Geophys. Res., 76,* 7963–7979, 1971. **(36)**

Forsyth, D., and S. Uyeda, On the relative importance of driving forces of plate motion, *Geophys. J. Roy. Astron. Soc., 41,* in press. **(15)**

Foucher, J. P., Thin or thick viscoelastic lithospheric plate, *Eos Trans. AGU, 55,* 438, 1974. **(37)**

Fouqué, F., *Santorin et ses Eruptions,* 440 pp., G. Masson, Paris, 1879. **(27)**

Fox, P. J., E. Schreiber, and J. J. Peterson, The geology of the oceanic crust: Compressional wave velocities of oceanic rocks, *J. Geophys. Res., 78,* 5155–5172, 1973. **(19)**

Francheteau, J., C. G. A. Harrison, J. G. Sclater, and M. L. Richards, Magnetization of Pacific seamounts: A preliminary polar curve for the northeastern Pacific, *J. Geophys. Res., 75,* 2035–2061, 1970. **(15, 18, 22, 24)**

Francis, T. J. G., and I. T. Porter, A microearthquake survey of the Mid-Atlantic ridge, *Nature, 240,* 547–549, 1972. **(13)**

Francis, T. J. G., and I. T. Porter, Median valley seismology: The Mid-Atlantic ridge near 45° N, *Geophys. J. Roy.Astron. Soc., 34*(3), 279–311, 1973. **(13)**

Frank, F. C., Curvature of island arcs, *Nature, 220,* 363, 1968. **(9)**

Frey, F. A., W. B. Bryan, and G. Thompson, Atlantic Ocean floor: Geochemistry and petrology of basalts from Legs 2 and 3 of the Deep Sea Drilling Project, *J. Geophys. Res., 79*(35), 5507–5528, 1974. **(30)**

Friedlaender, I., Über den Usu in Hokkaido und über einige andere Vulkane mit Quellkuppenbildung, *Petermann's Geogr. Mitt., 58,* 309–312, 1912. **(27)**

Friedlaender, I., Über die mexicanischen Vulkane, Pico de Orizaba, Cerro de Tequila, und Colima, *Z. f. Vulkanol., 13,* 154–164, 1930. **(27)**

Frischknecht, F. C., Fields about an oscillating magnetic dipole over a two-layer earth, and application to ground and airborne electromagnetic surveys, *Quart. Colorado School of Mines, 62*(1), 1–326, 1967. **(32)**

Frisillo, A. L., and G. R. Barsch, Measurement of single-crystal elastic constants of bronzite as a function of pressure and temperature, *J. Geophys. Res., 77,* 6360–6384, 1972. **(35)**

Fujisawa, H., Temperature and discontinuities in the transition layer within the earth's mantle: Geophysical application of the olivine–spinel transition in the $Mg_2SiO_4$–$Fe_2SiO_4$ system, *J. Geophys. Res., 73,* 3281–3294, 1968. **(34)**

Fukao, Y., Source process of large deep-focus earthquake and its tectonic implications—the western Brazil earthquake of 1963, *Phys. Earth Planet. Interiors, 5,* 61–76, 1972. **(9)**

Fulton, R., and R. I. Walcott, Lithospheric flexure as shown by deformation of glacier lake shore lines in southern British Columbia, in *Krumbein Volume, Geol. Soc. Amer. Mem.,* in press. **(37)**

Furumoto, A. S., N. J. Thompson, and G. P. Woollard, Seismic refraction studies of the crustal structure of the Hawaiian archipelago, *Pacific Sci., 19,* 315–319, 1965. **(2)**

Furumoto, A. S., W. A. Wiebenga, J. P. Webb, and G. H. Sutton, Crustal structure of the Hawaiian archipelago, northern Melanesia and the central Pacific basin by seismic refraction methods, *Tectonophysics, 20,* 153–164, 1973. **(35)**

Gainanov, A. G., Pendulum determinations of gravity in the northwest Pacific Ocean, *Trans. Int. Okeanol., AN SSR,* in Russian, 145–154, 1955. **(2)**

Galli-Olivier, C., Pediplain in northern Chile and the Andean uplift, *Science, 158,* 653–655, 1967. **(26)**

Ganser, A., Ausseralpine ophiolith problem, *Econ. Geol. Helv., 52,* 659–680, 1959. **(35)**

Gaposchkin, E. M., and K. Lambeck, Earth's gravity field to the sixteenth degree and station coordinates from satellite and terrestrial data, *J. Geophys. Res., 76,* 4855–4883, 1971. **(2)**

Gass, I. G., Ultrabasic pillow lavas from Cypress, *Geol. Mag., 95,* 241–251, 1958. **(19)**

Gast, P. W., Trace element fractionation and the origin of tholeiitic and alkaline magma types, *Geochim. Cosmochim. Acta, 32,* 1057–1086, 1968. **(33)**

Gavis, J., and R. L. Laurence, Viscous heating of a power-law liquid in plane flow, *Ind. Eng. Chem. Fundamentals, 7,* 525–527, 1968. **(29)**

Gay, S. P., Standard curves for interpretation of magnetic anomalies over long tabular bodies, *Geophysics, 28,* 161–200, 1963. **(21)**

Geographical Survey Institute of Japan, Gravity survey in Japan: Parts 1, 2, 3, 4, 5, *Bull. Geogr. Surv. Inst.,* 1955, 1957, 1964, 1965, 1968. **(3)**

Geological Survey of Japan, Geological map of Japan, 1971. **(20)**

Gettrust, J. F., R. P. Meyer, and L. C. Ocola, Crustal structure in the western-central cordillera region of southern Colombia (abstract), *XVII Mtg. Int'l Assoc. Seismol. Phys. Earth's Interior and Int. Union Geod. Geophys. Proc.,* p. 247, Lima, Peru, 1973. **(10)**

Gilluly, J., Steady plate motion and episodic orogeny and magmatism, *Geol. Soc. Amer. Bull., 84,* 499–514, 1973. **(28)**

Goodrich, J., Notice of the volcanic character of the island of Hawaii, *Amer. J. Sci. Ser. 1, 11,* 2–7, 1826. **(27)**

Gorshkov, G. S., Petrochemical features of volcanism in relation to the types of the earth's crust, in *The Crust of the Pacific Basin, Geophys. Monogr. Ser.,* vol. 6, edited by G. A. Macdonald and H. Kuno, pp. 158–170, AGU, Washington, D.C., 1962. **(27)**

Gorshkov, G. S., *Volcanism and the Upper Mantle: Investigations in the Kurile Island Arc,* 385 pp., Plenum, New York, 1970. **(27)**

Goslin, J., P. Beuzart, J. Francheteau, and X. Le Pichon, Thickening of the oceanic layer in the Pacific Ocean, *Marine Geophys. Res., 1,* 418–427, 1972. **(10)**

Graf, A., and R. Schulze, Improvements on the sea gravimeter Gss2, *J. Geophys. Res., 66,* 1813–1821, 1961. (**2**)
Gramlich, J. W., V. A. Lewis, and J. J. Naughton, Potassium–argon dating of Holocene basalts of the Honolulu Volcanic Series, *Geol. Soc. Amer. Bull., 82,* 1399–1404, 1971. (**28**)
Green, D. H., The origin of the eclogites from Salt Lake Crater, Hawaii, *Earth Planet. Sci. Lett., 1,* 414–420, 1966. (**36**)
Green, D. H., The origin of basaltic and nephelinitic magmas, *Trans. Leiscester Lit. Philos. Soc., 64,* 28–54, 1970*a*. (**33**)
Green, D. H., A review of experimental evidence on the origin of basaltic and nephelinitic magmas, *Phys. Earth Planet. Interiors, 3,* 221–235, 1970*b*. (**36**)
Green, D. H., Composition of basaltic magmas as indicators of conditions of origin: Application to oceanic volcanism, *Phil. Trans. Roy. Soc. London, Ser. A, 268,* 707–725, 1971. (**28, 29**)
Green, D. H., and I. B. Lambert, Experimental crystallization of anhydrous granite at high pressures and temperatures, *J. Geophys. Res., 70,* 5259–5268, 1965. (**29**)
Green, D. H., and A. E. Ringwood, An experimental investigation of the gabbro-eclogite transformation and its petrological applications, *Geochim. Cosmochim. Acta, 31,* 767–833, 1967*a*. (**29**)
Green, D. H., and A. E. Ringwood, The genesis of basaltic magmas, *Contr. Mineral. Petrol., 15,* 103–190, 1967*b*. (**33**)
Green, D. H., and A. E. Ringwood, The stability fields of aluminous pyroxene peridotite and garnet peridotite and their relevance in upper mantle structure, *Earth Planet. Sci. Lett., 3,* 151–160, 1967*c*. (**36**)
Green, D. H., and A. E. Ringwood, Mineralogy of peridotitic compositions under upper mantle conditions, *Phys. Earth Planet. Interiors, 3,* 359–371, 1970. (3)
Green, R. W. E., and A. L. Hales, The travel times of *P* waves to 30° in the central United States and upper mantle structure, *Bull. Seism. Soc. Amer., 58,* 267–289, 1968. (**33**)
Green, T. H., High pressure experimental studies on the mineralogical constitution of the lower crust, *Phys. Earth Planet. Interiors, 3,* 441–450, 1970. (**29**)
Green, T. H., Crystallization of calc–alkaline andesite under controlled high pressure hydrous conditions, *Contr. Mineral. Petrol., 34,* 150–166, 1972. (**29**)
Green, T. H., and A. E. Ringwood, High pressure experimental studies on the origin of andesites, *Andesite Conference Proc., Oregon Dept. Geol. Min. Ind. Bull., 65,* 21–32, 1969. (**27**)
Green, W. L., *Vestiges of the Molten Globe, II, The Earth's Surface Features and Volcanic Phenomena,* 337 pp., Honolulu, 1887. (**27**)
Griffiths, D. H., and P. F. Barker, Review of marine geophysical investigations in the Scotia Sea, in *Antarctic Geology and Geophysics,* pp. 3–12, Universitetsforlaget, Oslo, 1972. (**26**)
Griggs, D. T., The sinking lithosphere and the focal mechanism of deep earthquakes, in *The Nature of the Solid Earth,* edited by E. C. Robertson, pp. 361–384, McGraw-Hill, New York, 1972. (**29, 33**)
Griggs, R. F., The Valley of Ten Thousand Smokes, *Natl. Geogr. Mag., 33,* 115–169, 1918. (**27**)
Grim, P. J., Connection of the Panama fracture zone with the Galapagos rift zone, eastern tropical Pacific, *Marine Geophys. Res., 1,* 85–90, 1970. (**16**)
Grim, P. J., and B. H. Erickson, Fracture zones and magnetic anomalies south of the Aleutian trench, *J. Geophys. Res., 74,* 1488–1499, 1969. (**18**)
Grommé, C. S., and F. J. Vine, Paleomagnetism of Midway atoll lavas and northward motion of the Pacific plate, *Earth Planet. Sci. Lett., 17,* 159–168, 1972. (**28**)
Grommé, C. S., T. L. Wright, and D. L. Peck, Magnetic properties and oxidation of iron–titanium oxide minerals in Alae and Makaopuhi lava lakes, Hawaii, *J. Geophys. Res., 74,* 5277–5293, 1969. (**32**)
Grow, J. A., Crustal and upper mantle structure of the central Aleutian arc, *Geol. Soc. Amer. Bull., 84,* 2169–2192, 1973. (**1, 2, 3, 29**)
Grow, J. A., and T. Atwater, Mid-tertiary tectonic transition in the Aleutian arc, *Geol. Soc. Amer. Bull., 81,* 3715–3722, 1970. (**13**)
Grow, J. A., and C. Bowin, Evidence for high density crust and mantle beneath the Chile trench due to the descending lithosphere, *J. Geophys. Res., 80,* 1449–1458, 1975. (3)
Grow, J. A., and A. Qamar, Seismic wave attenuation beneath the central Aleutian arc, *Bull. Seism. Soc. Amer. 63,* 2155–2166, 1973. (**3, 29**)
Guidroz, R., F. Howard, and J. Frederic, Explosion program Kurile islands experiment, *Spec. Rep. 2,* Texas Instrument Inc., Houston, Texas, 1967. (**13**)
Guinot, B., M. Feissel, and M. Granveaud, *Bureau International de l'Heure, Annual Report for 1970,* Bureau International de l'Heure, Paris, 1971. (**5**)
Gunn, R., A quantitative evaluation of the influence of the lithosphere on the anomalies of gravity, *J. Franklin Inst., 236,* 373–396, 1943. (**2**)

Hales, A. L., Gravitational sliding and continental drift, *Earth Planet. Sci. Lett., 6,* 31–34, 1969. (**8, 37**)
Hales, A. L., C. E. Helsley, and J. B. Nation, *P* travel times for an oceanic path, *J. Geophys. Res., 75,* 7362–7381, 1970. (**12**)
Halpern, M., Regional geochronology of Chile south of 50° latitude, *Geol. Soc. Amer. Bull., 84,* 2407–2422, 1973. (**26**)
Halpern, M., M. U. Urquhart, and E. Linares, Radiometric ages of crystalline rocks from southern South America as related to Gondwana and Andean geologic provinces, *Upper Mantle Symps., Buenos Aires, 2,* 345–356, 1972. (**26**)
Hamada, K., Regionalized shear-velocity models for the upper mantle inferred from surface-wave dispersion data, *J. Phys. Earth, 20,* 301–326, 1972. (**36**)
Hamada, K., The upper mantle beneath Japan inferred from *P* travel time anomalies by means of three dimensional ray tracing, Part 1, *J. Phys. Earth, 21,* 463–474, 1974. (**3**)
Hamilton, D. L., C. W. Burnham, and E. F. Osborn, The solubility of water and effects of oxygen fugacity on crystallization in mafic magmas, *J. Petrol., 5,* pt. 1, 21–39, 1964. (**29**)
Hamilton, E. L., Sunken islands of the Mid-Pacific mountains, *Geol. Soc. Amer. Mem., 64,* 97 pp., 1956. (**18, 22**)
Hamilton, E. L., Marine geology of the southern Hawaiian ridge, *Bull. Geol. Soc. Amer., 68,* 1011–1026, 1957. (**2, 28**)

Hamilton, W., Mesozoic California and the underflow of Pacific mantle, *Geol. Soc. Amer. Bull., 80,* 2409–2430, 1969. **(26)**

Hamilton, W., Tectonics of the Indonesian region, *Geol. Soc. Malaysia Bull., 6,* 3–10, 1973. **(26)**

Hammond, S. R., and F. Theyer, Cenozoic plate motion and the magnetic time scale, equatorial Pacific (abstract), this volume. **(15)**

Handschumacher, D. W., Formation of the Emperor seamount chain, *Nature, 244,* 150–152, 1973. **(18)**

Handschumacher, D. W., Post-Eocene plate tectonics of the eastern Pacific, this volume. **(8, 15, 26)**

Handschumacher, D. W., S. T. Okumura, and P. K. Wong, Magnetic and bathymetric profiles from the central and southeastern Pacific: 10°N–45°S, 70°W–150°W, *Hawaii Inst. Geophys. Rep., HIG-75-18,* 1975. **(16)**

Hanks, T. C., The Kuril trench–Hokkaido rise system: Large shallow earthquakes and simple models of deformation, *Geophys. J. Roy. Astron. Soc., 23,* 173–189, 1971. **(2, 3, 37)**

Hanks, T. C., and J. H. Whitcomb, Comments on paper by J. W. Minear and N. N. Toksöz, *J. Geophys. Res., 76,* 613–616, 1971. **(29)**

Harrington, H. J., Paleogeographic development of South America, *Amer. Assoc. Petrol. Geol. Bull., 46,* 1773–1814, 1962. **(26)**

Harrison, C. G. A., Formation of magnetic anomaly patterns by dyke injection, *J. Geophys. Res., 73,* 2137–2142, 1968. **(19)**

Harrison, C. G. A., and B. L. K. Somayajulu, Behaviour of the earth's magnetic field during a reversal, *Nature, 212,* 1193–1195, 1966. **(19)**

Hart, R. A., A model for chemical exchange in the basalt–seawater system of oceanic layer II, *Can. J. Earth. Sci., 10,* 799–816, 1973. **(19)**

Hasebe, K., N. Fujii, and S. Uyeda, Thermal processes under island arcs, *Tectonophysics, 10,* 335–355, 1970. **(20)**

Hasegawa, S., and S. Nagumo, Construction of a long life magnetic tape recorder and some features of ocean-bottom seismograms, *Bull. Earthq. Res. Inst., 48,* 967–981, 1970. **(13)**

Hashizume, M., K. Oike, S. Asano, H. Hamaguchi, A. Okada, S. Murauchi, E. Shima, and M. Nogoshi, Crustal structure in the profile across the northeastern part of Honshu, Japan, derived from explosion seismic observations: Part 2, crustal structure, *Bull. Earthq. Res. Inst., 46,* 607–630, 1968. (3)

Hatherton, T., and W. R. Dickinson, The relationship between andesitic volcanism and seismicity in Indonesia, the Lesser Antilles, and other island arcs, *J. Geophys. Res., 74,* 5301–5310, 1969. **(27, 29)**

Hawkins, J. W., Jr., and J. H. Natland, Nephelinites and basanites of the Samoan linear volcanic chain, *Earth Planet. Sci. Lett.,* in press. **(28)**

Hayes, D. E., A Geophysical Investigation of the Peru–Chile trench, *Mar. Geol., 4,* 309–351, 1966. **(8, 10)**

Hayes, D. E., Continental margin of western South America, in *The Geology of Continental Margins,* edited by C. A. Burk and C. L. Drake, pp. 587–590, Springer-Verlag, New York, 1974. **(8)**

Hayes, D. E., and M. Ewing, Pacific boundary structure, in *The Sea,* vol. 4, pt. 2, edited by A. Maxwell, pp. 29–72, John Wiley & Sons, N.Y., 1970. **(2)**

Hayes, D. E., and W. C. Pitman, III, Magnetic lineations in the North Pacific, in *Geological Investigations of the North Pacific,* edited by J. D. Hayes, *Geol. Soc. Amer. Mem. 126,* 291–314, 1970. **(18, 22)**

Hayes, D. E., and J. Ringis, Sea-floor spreading in the Tasman Sea, *Nature, 243,* 454–458, 1973. **(21)**

Hays, J. D., and W. C. Pitman, III, Lithospheric plate motion, sea level changes and climatic and ecological consequences, *Nature, 246,* 18–22, 1973. **(15)**

Hays, J. D., T. Saito, N. D. Opdyke, and L. H. Burckle, Pliocene–Pleistocene sediments of the eastern equatorial Pacific: Their paleomagnetic, biostratigraphic and climatic record, *Geol. Soc. Amer. Bull., 80,* 1481–1495, 1969. **(24)**

Hays, J. D., H. E. Cook et al., *Initial Reports of the Deep Sea Drilling Project,* vol. 9, 1205 pp., U.S. Govt. Printing Office, Washington, D.C., 1972. **(28)**

Heath, G. R., Dissolved silica and deep sea sediments, in *Studies in Paleo-oceanography,* edited by W. W. Hay, *Soc. Econ. Paleon. Mineral. Spec. Pub. 20,* p. 77, 1974. **(24)**

Heath, G. R., and C. Culberson, Calcite: Degree of saturation, rate of dissolution, and the compensation depth in the deep ocean, *Geol. Soc. Amer. Bull., 81,* 3157–3160, 1970. **(24)**

Hedge, C. E., Z. E. Peterman, and W. R. Dickenson, Petrogenesis of lavas from western Samoa, *Geol. Soc. Amer. Bull., 83,* 2709–2714, 1972. **(28)**

Heezen, B. C., M. Ewing, and E. T. Miller, Trans-Atlantic profile of total magnetic intensity and topography, Dakar to Barbados, *Deep-Sea Res., 1,* 25–33, 1953. **(36)**

Heezen, B. C., J. L. Matthews, R. Catalano, J. Natland, A. Coogan, M. Tharp, and M. Rawson, Western Pacific guyots, in *Initial Reports of the Deep Sea Drilling Project,* vol. 20, B. C. Heezen, I. D. MacGregor et al., pp. 653–723, U.S. Govt. Printing Office, Washington, D.C. 1973. **(22)**

Heezen, B. C., I. D. MacGregor et al., *Initial Reports of the Deep Sea Drilling Project,* vol. 20, 958 pp., U.S. Govt. Printing Office, Washington, D.C., 1973. **(22)**

Heilprin, A., *Mont Pelée and the Tragedy of Martinique,* 335 pp., New York, 1903. **(27)**

Hein, J. R., Increasing rate of movement with time between California and the Pacific plate, from Delgada submarine fan source areas, *J. Geophys. Res., 78,* 7752–7762, 1973. **(28)**

Heirtzler, J. R., G. O. Dickson, E. M. Herron, W. C. Pitman, III, and X. Le Pichon, Marine magnetic anomalies, geomagnetic field reversals and motions of ocean floor and continents, *J. Geophys. Res., 73,* 2119–2136, 1968. **(15, 16, 18, 19, 20, 21, 28)**

Heiskanen, W., Das problem der isostasie, *Handbuch der Geophysik, 1,* sect. 12, 878–951, Berlin, 1936. **(2)**

Heiskanen, W. A., New isostatic tables for the reduction of the gravity values calculated on the basis of Airy's hypothesis, *Publ. Isostatic Inst., 2,* 1938. **(4)**

Heiskanen, W. A., and F. A. Vening-Meinesz, *The Earth and Its Gravity Field,* McGraw-Hill, New York, 1958. **(4)**

Hekinian, R., Petrology and geochemistry of basalts and related rocks from sites 214, 215, 217, DSDP Leg 22, Indian Ocean, in *Initial Reports of the Deep Sea Drilling Project,* vol. 22, C. C. von de Borch, J. G. Sclater et al., pp. 459–468, U.S. Govt. Printing Office, Washington, D.C., 1974. **(30)**

Helmberger, D., and R. A. Wiggins, Upper mantle structure of midwestern United States, *J. Geophys. Res., 76,* 3229–3245, 1971. **(34)**

Helsley, C. E., J. B. Nation, and R. P. Meyer, Seismic refraction observations in southern Mexico, *Eos Trans. AGU, 56,* 452, 1975. **(10)**

Helwig, J., Plate tectonic model for the evolution of the central Andes: Discussion, *Geol. Soc. Amer. Bull., 84,* 1493–1496, 1973. **(26)**

Helwig, J., Eugeosynclinal basement and a collage concept of orogenic belts, *Soc. Econ. Paleon. Mineral. Spec. Pub., 19,* 359–376, 1974. **(26)**

Henry, M., D. E. Karig, and G. G. Shor, Two seismic refraction profiles in the West Philippine Sea, in *Initial Reports of Deep Sea Drilling Project,* vol. 31, D. E. Karig, J. C. Ingle, Jr. et al., pp. 611–614, U.S. Govt. Printing Office, Washington, D.C., 1975. **(21)**

Hermance, J. F., and G. D. Garland, Deep electrical structure under Iceland, *J. Geophys. Res., 73,* 3797–3800, 1968. **(36)**

Hermance, J. F., and L. R. Grillot, Correlation of magnetotelluric, seismic, and temperature data from southwest Iceland, *J. Geophys. Res., 75,* 6582–6591, 1970. **(36)**

Herron, E. M., Crustal plates and sea-floor spreading in the southeastern Pacific, *AGU, Antarctic Oceanol. I, Antarctic Res. Ser., 15,* 229–238, 1971. **(16)**

Herron, E. M., Sea-floor spreading and the Cenozoic history of the East Central Pacific, *Geol. Soc. Amer. Bull., 83,* 1671–1692, 1972*a*. **(8, 15, 16, 18, 22, 24, 26, 28)**

Herron, E. M., Two small crustal plates in the South Pacific near Easter Island, *Nature, 240,* 35–37, 1972*b*. **(16, 22)**

Herron, E. M., Magnetic lineations in the extreme southeast Pacific and their bearing on the evolution of the Scotia arc and Patagonia (abstract), *Abstr. Program Geol. Soc. Amer.,* 791, 1974. **(26)**

Herron, E. M., and D. E. Hayes, A geophysical study of the Chile ridge, *Earth Planet. Sci. Lett., 6,* 77–83, 1969. **(16)**

Herron, E. M., and J. R. Heirtzler, Sea-floor spreading near the Galapagos, *Science, 158,* 775–780, 1967. **(16)**

Hess, H. H., Major structural features of the western and North Pacific, *Geol. Soc. Amer. Bull., 59,* 417–446, 1948. **(21)**

Hess, H. H., History of ocean basins, in *Petrologic Studies: A. F. Buddington volume,* edited by A. E. J. Engel, H. L. James, and B. F. Leonard, *Geol. Soc. Amer. Mem.,* pp. 559–620, New York, 1962. **(15, 16, 18, 27)**

Hey, R. N., K. S. Deffeys, G. L. Johnson, and A. Lowrie, The Galapagos triple junction and plate motions in the East Pacific, *Nature, 237,* 20–22, 1972. **(16)**

Hilde, T. W. C., Mesozoic sea-floor spreading in the North Pacific, D.Sc. dissertation, 84 pp., Univ. Tokyo, Tokyo, 1973. **(18, 22)**

Hilde, T. W. C., and A. D. Raff, Evidence of oceanic crust beneath the shoreward slope of the Japan trench from magnetic and seismic reflection data (abstract), *EOS Trans. AGU, 51,* 330, 1970. **(18)**

Hilde, T., and J. M. Wageman, Structure and origin of the Japan Sea, in *The Western Pacific: Island Arcs, Marginal Seas, Geochemistry,* edited by P. J. Coleman, pp. 415–434, Univ. Western Australia Press, 1973. **(20)**

Hilde, T. W. C., Nobuhiro Isezaki, and J. M. Wageman, Mesozoic sea-floor spreading in the North Pacific, this volume. **(15)**

Hilde, T. W. C., J. M. Wageman, and W. T. Hammond, The structure of Tosa terrace and Nankai trough off southeastern Japan, *Deep-Sea Res., 16,* 67–75, 1969. **(20)**

Hilgard, J. E., The new Bogoslof volcano in Bering Sea, *Science, 3,* 282–286, 1884. **(27)**

Hill, D. P., Crustal structure of the Island of Hawaii from seismic refraction measurements, *Bull. Seism. Soc. Amer., 59,* 101–130, 1969. **(14)**

Hitchcock, C. H., *Hawaii and Its Volcanoes,* 314 pp., Honolulu, 1909. **(27)**

Hochstein, M. P., Seismic measurements in the Cook Islands, southwest Pacific Ocean, *N.Z.J. Geol. Geophys., 10,* 1499–1526, 1967. **(37)**

Hochstetter, F. von, *Geologie von Neu-Seeland, Beiträge zur Geologie der Provinzen Auckland und Nelson, Novara Expedition, Geol. Theil, 1*(1), 273 pp. Vienna, 1864; reprinted in English transl. by C. A. Fleming, Govt. Printer, Wellington, N.Z., 1959. **(27)**

Holloway, J. R., and C. E. Ford, Fluid-absent melting of fluorohydroxy amphibole pargasite to 35 kilobars, *Earth Planet. Sci. Lett., 25*(1), 44–48, 1975. **(29)**

Holmes, A., Radioactivity and earth movements, *Geol. Soc. Glasgow Trans., 18,* 559–606, 1931. **(15, 27)**

Horibe, Y., Preliminary report of the Hakuho Maru cruise KH-68-4 (Southern Cross Cruise), 170 pp., Ocean Research Inst., Univ. Tokyo, 1970. **(2)**

Hotta, H., A crustal section across the Izu–Ogasawara (Bonin) arc and trench, *J. Phys. Earth, 18,* 125–141, 1970. **(3)**

Hovey, E. O., The new cone of Mont Pelée and the gorge of the Rivière Blanche, Martinique, *Amer. J. Sci. Ser. 4, 16,* 269–281, 1903. **(27)**

Humboldt, A. von, Über den Bau und die Wirksamkeit der Vulkane in vershiedenen Erdstrichen, *Leonhards Taschenbuch für Mineralogie, 18,* 1824; (cited by K. Sapper, 1927, p. 382). **(27)**

Humboldt, A. von, *Des Kosmos,* 1, Stuttgart, 1845. **(27)**

Hussong, D. M., Detailed structural interpretations of the Pacific oceanic crust using ASPER and ocean-bottom seismometer methods, Ph.D. dissertation, 165 pp., Univ. Hawaii, 1972. **(8, 35)**

Hussong, D. M., L. K. Wipperman, and M. E. Odegard, Compressional faulting of the oceanic crust prior to subduction in the Peru–Chile trench, *Geology, 3,* 601–604, 1975. **(8)**

Hussong, D. M., P. Edwards, S. H. Johnson, J. F. Campbell, and M. E. Odegard, Crustal and upper mantle structure of the Nazca plate (abstract), *Int. Assoc. Seismol. Phys. Earth's Interior, XVII, Lima Program,* p. 221, 1973. **(8)**

Hussong, D. M., P. B. Edwards, S. H. Johnson, J. F. Campbell, and G. H. Sutton, Crustal structure of the Peru–Chile trench: 8°–12°S latitude, this volume. **(7, 26)**

IAGA, Commission 2 Working Group 4, Analysis of the geomagnetic field: The international geomagnetic field 1965.0, *J. Geophys. Res., 74,* 4407–4408, 1969. **(18)**

Ichikawa, K., Correlation of geological formations between Japan and Korea in the Mesozoic and Paleo-

zoic, *Japan Sea*, in Japanese, *8*, 9–13, 1972. **(20)**
Ikebe, N., Y. Takayanagi, M. Chiji, and K. Chinzei, Neogene biostratigraphy and radiometric time scale of Japan—an attempt at intercontinental correlation, *Pacific Geol., 4*, 39–78, 1972. **(20)**
Ingle, J. C., D. E., Karig et al., *Initial Reports of the Deep Sea Drilling Project*, vol. 31, 927 pp., U.S. Govt. Printing Office, Washington, D.C., 1975. **(21)**
Irving, E., The Mid-Atlantic ridge at 45°N, XIV, Oxidation and magnetic properties of basalt: Review and discussion, *Can. J. Earth Sci., 7*, 1528–1538, 1970. **(19)**
Isaacson, P. E., Evidence for a western extracontinental land source during the Devonian in the central Andes, *Geol. Soc. Amer. Bull., 86*, 39–46, 1975. **(8)**
Isacks, B., and P. Molnar, Distribution of stresses in the descending lithosphere from a global survey of focal mechanism solutions of mantle earthquakes, *Rev. Geophys. Space Phys., 9*, 103–174, 1971. **(9)**
Isacks, B., J. Oliver, and L. R. Sykes, Seismology and the new global tectonics, *J. Geophys. Res., 73*, 5855–5899, 1968. **(9, 16, 29)**
Isezaki, N., Geomagnetic anomalies and tectonics around the Japanese Islands, *Oceanogr. Mag., 24*(2), 107–158, 1973. **(18, 20)**
Isezaki, N., Possible spreading centers in the Japan Sea, *J. Mar. Geophys. Res., 4*, 53–65, 1975. **(15, 20)**
Isezaki, N., and S. Uyeda, Geomagnetic anomaly pattern of the Japan Sea, *J. Mar. Geophys. Res., 2*, 51–59, 1973. **(15, 20, 21)**
Isezaki, N., K. Hata, and S. Uyeda, Magnetic survey of the Japan Sea (part 1), *Bull. Earthq. Res. Inst., 49*, 77–83, 1971. **(20)**
Ito, E., Y. Matsui, K. Suito, and N. Kawai, Synthesis of $\gamma$-$Mg_2SiO_4$, *Phys. Earth Planet. Interiors, 8*, 342–344, 1974. **(34)**
Ito, E., T. Matsumoto, K. Suito, and N. Kawai, High pressure breakdown of enstatite, *Proc. Japan Acad., 48*, 412–415, 1972. **(34)**
Ito, K., Analytical approach to estimating the source rock of basaltic magmas: Major elements, *J. Geophys. Res., 78*, 412–431, 1973. **(3, 36)**
Ito, K., Petrological models of the oceanic lithosphere, geophysical and geochemical tests, *Earth Planet. Sci. Lett., 21*, 169–180, 1974. **(36)**
Ito, K., and G. C. Kennedy, Melting and phase relations in a natural peridotite to 40 kilobars, *Amer. J. Sci., 265*, 519–539, 1967. **(36)**
Ito, K., and G. C. Kennedy, Melting and phase relations in the plane tholeiite–lherzolite–nepheline basanite to 40 kilobars with geological implications, *Contrib. Mineral. Petrol., 19*, 177–211, 1968. **(36)**
Ito, K., and G. C. Kennedy, An experimental study of the basalt–garnet granulite–eclogite transition, in *The Structure and Physical Properties of the Earth's Crust, Geophys. Monogr. Ser.*, vol. 14, edited by J. G. Heacock, pp. 303–314, AGU, Washington, D.C., 1971. **(3)**
Ito, K., and G. C. Kennedy, The composition of liquids formed by partial melting of eclogites at high temperatures and pressures, *J. Geol., 82*, 383–392, 1974. **(36)**
Ito, K., S. Endo, and N. Kawai, Olivine–spinel transformation in a natural forsterite, *Phys. Earth Planet. Interiors, 4*, 425–428, 1971. **(34)**

Jackson, E. D., The origin of ultramafic rocks by cumulus processes, *Fortschr. Mineral., 48*, 128–174, 1971. **(27)**
Jackson, E. D., and T. P. Thayer, Some criteria for distinguishing between stratiform, concentric and alpine peridotite-gabbro complexes, *Int. Geol. Cong., 25th Session, 2*, 289–296, 1972. **(35)**
Jackson, E. D., and T. L. Wright, Xenoliths in the Honolulu Volcanic Series, *J. Petrol, 11*, 405–430, 1970. **(28)**
Jackson, E. D., E. A. Silver, and G. B. Dalrymple, Hawaiian–Emperor chain and its relation to Cenozoic circum-Pacific tectonics, *Geol. Soc. Amer. Bull., 83*, 601–618, 1972. **(22, 26, 27, 28, 37)**
Jackson, E. D., K. E. Bargar, B. P. Fabbi, and C. Heroponlos, Petrology of the basaltic rocks drilled on Deep Sea Drilling Project 33, in *Initial Reports of the Deep Sea Drilling Project*, vol. 33, S. O. Schlanger, E. D. Jackson et al., U.S. Govt. Printing Office, Washington, D.C., in press. **(22)**
Jackson, I., R. Liebermann, and A. E. Ringwood, Disproportionation of spinels to mixed oxides: Significance of cation coordination and implications for the mantle, *Earth Planet. Sci. Lett., 24*, 203–208, 1974. **(33)**
Jacoby, W. R., Instability in the upper mantle and global plate movement, *J. Geophys. Res., 75*, 5671–5680, 1970. **(8, 36)**
Jaegar, J. C., Cooling and solidification of igneous rocks, in *Basalts*, vol. 2, edited by H. H. Hess and A. Poldervaart, pp. 503–536, Interscience Pub., New York, 1968. **(29, 32)**
Jaggar, T. A., The initial stages of the spine on Pelée, *Amer. J. Sci. Ser. 4, 17*, 34–40, 1904. **(27)**
Jaggar, T. A., The evolution of Bogoslof volcano, *Amer. Geogr. Soc. Bull., 40*, 385–400, 1908. **(27)**
Jaggar, T. A., Volcanological investigations at Kilauea, *Amer. J. Sci. Ser. 4, 44*, 161–220, 1917*a*. **(27)**
Jaggar, T. A., On the terms aphrolith and dermolith, *Washington Acad. Sci. J., 7*, 277–281, 1917*b*. **(27)**
Jaggar, T. A., Seismometric investigation of the Hawaiian lava column, *Seism. Soc. Amer. Bull., 10*, 155–275, 1920. **(27)**
Jaggar, T. A., Recent activity of Bogoslof volcano, *Volcano Letter, 275*, 1–3, 1930. **(27)**
Jaggar, T. A., Evolution of Bogoslof volcano, *Volcano Letter, 322*, 1–3, 1931. **(27)**
James, D. E., Plate tectonic model for the evolution of the central Andes, *Geol. Soc. Amer. Bull., 82*, 3325–3346, 1971*a*. **(8, 9, 10, 26)**
James, D. E., Andean crustal and upper mantle structure, *J. Geophys. Res., 76*, 3246–3271, 1971*b*. **(36)**
Jarrard, R. D., and D. A. Clague, Implications of Pacific island and seamount ages for the origin of volcanic chains, in The Hawaiian–Emperor seamount chain, its origin, and implications for plate tectonics, D. A. Clague, Ph.D. dissertation, pp. 103–153, Univ. Calif., San Diego, 1974. **(28)**
Jenks, W. F., Peru, in *Handbook of South American Geology*, edited by W. F. Jenks, *Geol. Soc. Amer. Mem., 65*, 215–247, 1956. **(8)**
Johnson, L. R., Array measurements of *P* velocities in the upper mantle, *J. Geophys. Res., 72*, 6309–6325, 1967. **(33)**
Johnson, T., and P. Molnar, Focal mechanism and plate tectonics of southwest Pacific, *J. Geophys. Res., 77*, 5000–5032, 1972. **(9)**

Jones, E. W. J., J. Ewing, and M. Truchan, Aleutian plain sediments and lithospheric plate motions, *J. Geophys. Res., 76,* 8121–8127, 1971. **(2)**

Joos, G., *Theoretical Physics,* 3rd ed., Blackie and Son, London, United Kingdom, 1958. **(5)**

Judd, J. W., The eruption of Krakatoa, and subsequent phenomena; on the volcanic phenomena of the eruption, and on the nature and distribution of the ejected materials, *Royal Soc. London, Krakatoa Comm. Rept.,* 1–56, 1888. **(27)**

Kagami, H., T. Saito, K. Oshite, J. Ishii, R. Tsuchi, H. Mitsui, K. Koizumi, H. Shiozawa, and M. Nogami, Survey of the Japan trench (abstract in Japanese), *J. Geol. Soc. Japan, 70,* 414, 1964. **(18)**

Kanamori, H., Attenuation of *P* waves in the upper and lower mantle, *Bull. Earthquake Res. Inst., 45,* 299–312, 1967*a*. **(33)**

Kanamori, H., Upper mantle structure from apparent velocities of *P* waves recorded at Wakayama microearthquake observatory, *Bull. Earthquake Res. Inst., 45,* 657–678, 1967*b*. **(34)**

Kanamori, H., Travel times to Japanese stations from LONGSHOT and their geophysical implications, *Bull. Earthquake Res. Inst., 46,* 841–859, 1968. **(3)**

Kanamori, H., Mantle beneath the Japanese arc, *Phys. Earth Planet. Interiors, 3,* 475–483, 1970. **(3)**

Kanamori, H., Great earthquakes at island arcs and the lithosphere, *Tectonophysics, 12,* 187–198, 1971. **(2, 20)**

Kanamori, H., and K. Abe, Deep structure of island arcs as revealed by surface waves, *Bull. Earthquake Res. Inst., 46,* 1001–1025, 1968. **(36)**

Kanamori, H., and F. Press, How thick is the lithosphere?, *Nature, 226,* 330–331, 1970. **(3, 36)**

Kaneoka, I., N. Isshiki, and S. Zashu, K–Ar ages of the Izu–Bonin Islands, *Geochem. J., 4,* 53–60, 1970. **(20)**

Karig, D. E., Ridges and basins of the Tonga–Kermadec island arc system, *J. Geophys. Res., 75,* 239–255, 1970. **(13, 21)**

Karig, D. E., Structural history of the Mariana island arc system, *Geol. Soc. Amer. Bull., 82,* 323–344, 1971*a*. **(2, 13, 20, 21)**

Karig, D. E., Origin and development of marginal basins in the western Pacific, *J. Geophys. Res., 76,* 2542–2561, 1971*b*. **(2, 13, 15, 20, 21)**

Karig, D. E., Tectonic erosion at trenches, *Earth Planet. Sci. Lett., 21,* 209–212, 1974*a*. **(8)**

Karig, D. E., Evolution of arc systems in the western Pacific, *Ann. Rev. Earth Planet. Sci., 2,* 51–75, 1974*b*. **(15, 21)**

Karig, D. E., and G. F. Sharman, Subduction and accretion in trenches, *Geol. Soc. Amer. Bull., 86,* 377–389, 1975. **(8)**

Karig, D. E., J. C. Ingle, A. H. Bouma, H. Ellis, N. Haile, I. Koizumi, I. D. MacGregor, C. Moore, H. Ujiie, T. Watanabe, S. M. White, M. Yasui, and H. Yi Ling, Origin of the West Philippine basin, *Nature, 246,* 458–461, 1973. **(20, 21)**

Kasahara, J., S. Nagumo, and S. Koresawa, Ocean-bottom seismographic observation east of Hachijojima and Sagami Bay, *Report of the Coordinating Committee for Earthquake Prediction,* in Japanese, *10,* 33–36, 1973*a*. **(13)**

Kasahara, J., S. Nagumo, S. Koresawa, and K. Kobayashi, Ocean-bottom seismographic observation at Suruga Bay utilizing oil filled battery, *Zisin, J. Seism. Soc. Japan, 2nd Ser., 26,* 171–177, 1973*b*. **(13)**

Katsumata, M., Vertical distribution of earthquake foci in and around Japan, *Geophys. Mag., 27,* 483–486, 1956. **(3)**

Katsumata, M., and L. R. Sykes, Seismicity and tectonics of the western Pacific: Izu–Mariana–Caroline and Ryukyu–Taiwan region, *J. Geophys. Res., 74,* 5923–5948, 1969. **(3, 13, 20)**

Katz, H. R., Fracture patterns and structural history in the sub-Andean belt of southern Chile, *J. Geol., 70,* 595–603, 1962. **(26)**

Katz, H. R., Continental margin in Chile—is tectonic style compressional or extensional?, *Amer. Assoc. Petrol. Geol. Bull., 55,* 1753–1758, 1971. **(26)**

Katz, H. R., and W. A. Watters, Geological investigation of the Yahgan Formation (upper Mesozoic) and associated igneous rocks of Navarino Island, southern Chile, *New Zealand J. Geol. Geophys., 9,* 323–359, 1966. **(26)**

Kausel, E., and F. Schwab, Attenuation of Rayleigh waves along the East Pacific ridge (abstract), this volume. **(7)**

Kausel, E. G., A. R. Leeds, and L. Knopoff, Variations of Rayleigh wave phase velocities across the Pacific Ocean, *Science, 186,* 139–141, 1974. **(36)**

Kawai, N., S. Endo, and K. Ito, Split sphere high pressure vessel and phase equilibrium relation in the system $Mg_2SiO_4$–$Fe_2SiO_4$, *Phys. Earth Planet. Interiors, 3,* 182–185, 1970. **(34)**

Kawai, N., M. Tachimori, and E. Ito, A high pressure hexagonal form of $MgSiO_3$, *Proc. Japan Acad., 50,* 378–380, 1974. **(33, 34)**

Kay R., N. J. Hubbard, and P. W. Gast, Chemical characteristics and origin of ocean ridge volcanic rocks, *J. Geophys. Res., 75,* 1585–1613, 1970. **(29)**

Kemmerling, G. L. L., De hernieuwde werking van den Vulkaan G. Merapi (midden Java) van begin Augustus 1920 tot en met einde Februari 1921, *Dienst v.h. Mijnwezen in Nederlandsch-Oost-Indie, Vulkanologische Meded., 3,* 30 pp., 1921. **(27)**

Kempe, D. R. C., Petrology of basalts, Leg 26, in *Initial Reports of the Deep Sea Drilling Project,* vol. 26, T. A. Davies, B. P. Luyendyk et al., pp. 465–504, U.S. Govt. Printing Office, Washington, D.C., 1974. **(30)**

Kennett, J. P., R. E. Burns, J. E. Andrews, M. Churkin, Jr., T. A. Davies, P. Dumitrica, A. R. Edwards, J. S. Galehouse, G. H. Packham, and G. J. van der Lingen, Australian–Antarctic continental drift, paleocirculation changes and Oligocene deep-sea erosion, *Nature Phys. Sci., 239,* 51–55, 1972. **(24)**

Kennett, J. P., R. E. Houtz, P. V. Andrews, A. R. Edwards, V. A. Gostin, N. Hajós, M. Hampton, D. G. Jenkins, S. V. Margolis, A. T. Ovenshine, and K. Perch-Nielsen, Cenozoic paleo-oceanography in the southwest Pacific Ocean and the development of the circum-Antarctic current, in *Initial Reports of the Deep Sea Drilling Project,* vol. 29, J. P. Kennett, R. E. Houtz et al., pp. 1155–1169, U.S. Govt. Printing Office, Washington, D.C., 1975. **(24)**

Khan, M. A., Plumes in the mantle, *Natl. Aeronautics and Space Admin. (Goddard) Rep., X-592-73-334,* 27 pp., 1973. **(28)**

Khan, M. A., G. P. Woollard, and K. I. Daugherty, Statistical analysis of the relation of marine gravity anomalies to bathymetry, *Hawaii Inst. Geophys.*

*Rep., HIG-71-20,* 141 pp., 1971. **(2)**

Kienle, J., Gravity and magnetic measurements over Bowers ridge and Shirshov ridge, Bering Sea, *J. Geophys. Res., 76,* 7138–7153, 1971. **(2)**

Kimler, B. F., and Staff, Field operation and ocean-bottom seismograph performance and evaluation Kurile Island experiment, *Spec. Rep. 1,* Texas Instrument Inc., Houston, Texas, 1967. **(13)**

Kishinouye, F., Y. Yamazaki, H. Kobayashi, and S. Koresawa, A submarine seismograph, *Bull. Earthquake Res. Inst., 41,* 819–824, 1963. **(13)**

Klitgord, K. D., Near bottom geophysical surveys and their implication on the crustal generation process, sea-floor spreading history of the Pacific, and the geomagnetic time scale of 0–6 million years B.P., Ph.D. dissertation, Scripps Inst. Oceanog., La Jolla, Calif., 1974. **(19)**

Klitgord, K. D., J. D. Mudie, and W. R. Normark, Magnetic lineations observed near the ocean floor and possible implications on the geomagnetic chronology of the Gilbert epoch, *Geophys. J., 28,* 35–48, 1972. **(19)**

Klitgord, K. D., J. D. Mudie, P. A. Larson, and J. A. Grow, Fast sea-floor spreading on the Chile ridge, *Earth Planet. Sci. Lett., 20,* 93–99, 1973. **(16)**

Knopoff, L., J. W. Schlue, and A. A. Schwab, Phase velocity of Rayleigh waves across the East Pacific rise, *Tectonophysics, 10,* 321–324, 1970. **(36)**

Kobayashi, K., Sea-floor spreading of the Shikoku basin and the Japanese island arc, *Marine Sciences, 6,* 27–34, 1974. **(13)**

Kobayashi, K., and N. Isezaki, Magnetic anomalies in the Sea of Japan and the Shikoku basin: Possible tectonic implications, this volume. **(15)**

Kogan, M. G., Gravity field of the Kuril–Kamchatka arc and its relation to the thermal regime of the lithosphere, *J. Geophys. Res., 80,* 1381–1390, 1975. **(2)**

Kono, Y., and T. Yoshii, Numerical experiments on the thickening plate model, *J. Phys. Earth, 23,* 63–75, 1975. **(36)**

Koto, B., The great eruption of Sakura-jima in 1914, *Tokyo Imp. Univ., J. Coll. Sci., 38*(3), 237 pp., 1916. **(27)**

Kovylin, V. M., Results of seismic studies in the southwest of the Sea of Japan abyssal basin, *Oceanology, 6*(2), 238–249, 1966. **(20)**

Koyanagi, R. Y., and E. T. Endo, Hawaiian seismic events during 1969, *U.S. Geol. Surv. Prof. Pap., 750-C,* C158–C164, 1971. **(36)**

Koyanagi, R. Y., E. T. Endo, and P. L. Ward, Seismic activity on the island of Hawaii, 1970 to 1973, this volume. **(7)**

Koyanagi, R. Y., D. A. Swanson, and E. T. Endo, Distribution of earthquakes related to mobility of the south flank of Kilauea Volcano, Hawaii, *U.S. Geol. Surv. Prof. Pap., 800-D,* D89–D97, 1972. **(14)**

Koyanagi, R. Y., J. D. Unger, E. T. Endo, and A. T. Okamura, Shallow earthquakes associated with inflation episodes at the summit of Kilauea volcano, Hawaii, *Bull. Volcanol.,* in press. **(14)**

Krakiwsky, E. J., Satellite geodesy, class notes, *S.E. 6250,* Dept. Surveying Engineering, Univ. New Brunswick, Fredericton, New Brunswick, 1970. **(5)**

Kranck, E. H., Geological investigations in the cordillera of Tierra del Fuego, *Acta Geographica, 4,* 1–231, 1932. **(26)**

Krummenacher, D., and J. Noetzlin, Ages isotopiques K/A de roches prélevées dans les possessions françaises du Pacifique, *Bull. Soc. Geol. France, 8,* 173–175, 1966. **(28)**

Krummenacher, D., D. H. Dowd, V. F. Duda, W. B. Cunningham, F. L. Kingery, and W. F. Speidel, Potassium–argon ages from xenoliths and differentiates in coarse-grained rocks from the center of the island of Tahiti (French Polynesia) (abstract), *Abstr. Program Geol. Soc. Amer., 4*(7), 186, 1972. **(28)**

Kuhn, T. S., *The Structure of Scientific Revolutions,* 2nd. ed., 210 pp., Univ. Chicago Press, Chicago, Ill., 1970. **(26)**

Kulm, L. D., and G. A. Fowler, Oregon continental margin structure and stratigraphy: A test of the imbricate thrust model, in *The Geology of Continental Margins,* edited by C. A. Burk and C. L. Drake, pp. 261–283, Springer-Verlag, New York, 1974. **(8)**

Kulm, L. D., J. M. Resig, T. C. Moore, Jr., and V. J. Rosato, Transfer of Nazca ridge pelagic sediments to the Peru continental margin, *Geol. Soc. Amer. Bull., 85,* 769–780, 1974. **(8)**

Kulm, L. D., K. F. Scheidigger, R. F. Prince, R. A. Dymond, T. C. Moore, Jr., and D. M. Hussong, Tholeiitic basalt ridge in the Peru trench, *Geology, 1,* 11–14, 1973. **(8)**

Kumagai, N., Results of measurements of gravity in Japan and her vicinity, spec. ed., *Geolog. Mineralog. Inst.,* Fac. Science, Kyoto Univ., 1953. **(3)**

Kumazawa, M., H. Sawamoto, E. Ohtani, and K. Masaki, Post spinel phase of forsterite and evolution of the earth's mantle, *Nature, 247,* 356–358, 1974. **(33)**

Kuno, H., Origin of Cenozoic petrographic provinces of Japan and surrounding areas, *Bull. Volcanol., Ser. 2, 20,* 37–76, 1959. **(27)**

Kushiro, I., Compositions of magmas formed by partial zone melting of the earth's upper mantle, *J. Geophys. Res., 73,* 619–634, 1968. **(36)**

Kushiro, I., Melting of hydrous upper mantle and possible generation of andesitic magma: An approach from synthetic systems, *Earth Planet. Sci. Lett., 22,* 294–299, 1974. **(20)**

Lachenbruch, A. H., A simple mechanical model for oceanic spreading centers, *J. Geophys. Res., 78,* 3395–3417, 1973. **(21)**

LaCoste, L. J. B., and J. C. Harrison, Some theoretical considerations in the measurement of gravity at sea, *Geophys. J. Roy. Astron. Soc., 5,* 89, 1961. **(2)**

Lacroix, A., *La Montagne Pelée et ses Éruptions,* 662 pp., Paris, 1904. **(27)**

Lamb, H., *Hydrodynamics,* p. 600, Dover Pub., New York, 1932. **(29)**

Lambert, I. B., and P. J. Wyllie, Melting of gabbro (quartz eclogite) with excess water to 35 kilobars with geological applications, *J. Geol., 80,* 693–708, 1972. **(29)**

Langseth, M. G., Jr., The flow of heat from the Earth and its global distribution at the surface, *AIAA Pap.,* 4th Thermophysics Conf., pp. 96–589, 1968. **(22)**

Larson, R. L., The Early Cretaceous–Late Jurassic magnetic reversal time scale, and the Phoenix magnetic lineations revisited (abstract), this volume. **(15)**

Larson, R. L., and C. G. Chase, Relative velocities of the Pacific, North America and Cocos plates in the Middle American region, *Earth Planet. Sci. Lett., 7,* 425–428, 1970. **(16)**

Larson, R. L., and C. G. Chase, Late Mesozoic evolution of the western Pacific Ocean, *Geol. Soc.*

*Amer. Bull., 83,* 3627–3644, 1972. **(2, 15, 18, 21, 22, 36)**

Larson, R. L., and T. W. C. Hilde, A revised time scale of magnetic reversals for the Early Cretaceous and Late Jurassic, *J. Geophys. Res., 80,* 2586–2594, 1975. **(15)**

Larson, R. L., and W. C. Pitman, III, World-wide correlation of Mesozoic magnetic anomalies and its implications, *Geol. Soc. Amer. Bull., 83,* 3645–3662, 1972. **(3, 15, 18, 19, 20, 21, 22, 28)**

Larson, R. L., and F. N. Spiess, East Pacific rise crest: A near-bottom geophysical profile, *Science, 163,* 68–71, 1969. **(19)**

Larson, R. L., H. W. Menard, and S. M. Smith, Gulf of California: A result of ocean-floor spreading and transform faulting, *Science, 161,* 781–782, 1968. **(16)**

Larson, R. L., S. M. Smith, and C. G. Chase, Magnetic lineations of Early Cretaceous age in the western equatorial Pacific Ocean, *Earth Planet. Sci. Lett., 15,* 315–319, 1972. **(18, 22)**

Latham, G., and A. A. Nowroozi, Waves, weather, and ocean-bottom microseisms, *J. Geophys. Res., 73,* 3945–3956, 1968. **(13)**

Latham, G. V., and G. H. Sutton, Seismic measurements on the ocean floor, 1, Bermuda area, *J. Geophys. Res., 71,* 2545–2573, 1966. **(13)**

Launay, L., Conductivity under the oceans: Interpretation of a magneto-telluric sounding 630 km off the California coast, *Phys. Earth Planet. Interiors, 8,* 83–86, 1974. **(36)**

Leeds, A. R., L. Knopoff, and E. G. Kausel, Variations of upper mantle structure under the Pacific Ocean, *Science, 186,* 141–143, 1974. **(36)**

Le Pichon, X., Sea-floor spreading and continental drift, *J. Geophys. Res., 73,* 3661–3697, 1968. **(15, 16, 18)**

Le Pichon, X., and D. E. Hayes, Marginal offset, fracture zones, and the early opening of the South Atlantic, *J. Geophys. Res., 76,* 6283–6293, 1971. **(8)**

Le Pichon, X., and J. R. Heirtzler, Magnetic anomalies in the Indian Ocean and sea-floor spreading, *J. Geophys. Res., 73,* 2101–2117, 1968. **(18, 26)**

Liebermann, R. C., Elasticity of the olivine–spinel and olivine–$\beta$ phase transformations and the 400-km discontinuity of the mantle, *J. Geophys. Res., 78,* 7015–7017, 1973*a.* **(34)**

Liebermann, R. C., Elastic properties of germanate analogues of olivine, spinel and $\beta$ polymorphs of $(Mg, Fe)_2SiO_4$, *Nature Phys. Sci., 244,* 105–107, 1973*b.* **(34)**

Liebermann, R. C., Elasticity of pyroxene–garnet and pyroxene–ilmenite phase transformations in germanates, *Phys. Earth Planet. Interiors, 8,* 361–374, 1974. **(33)**

Lisitzin, A. P., and Associates, Geochemical, mineralogical, and paleontological studies, in *Initial Reports of the Deep Sea Drilling Project,* vol. 6, A. L. Fischer, B. C. Heezen et al., pp. 829–960, U.S. Govt. Printing Office, Washington, D.C., 1971. **(30)**

Liu, L., Silicate perovskite from phase transformations of pyrope garnet at high pressure and temperature, *Geophys. Res. Lett., 1,* 277–280, 1974. **(33, 34)**

Lloyd, J. J., Tectonic history of the South Central American orogen, in *Backbone of the Americas, Amer. Assoc. Petrol. Geol. Mem., 2,* 88–100, 1963. **(26)**

Lohmann, H. H., Un nuevo metodo para la determinacion de la direccion del movimiento glacial, aplicado al horizonte Zapla en el Gotlandico de Bolivia, *Petroleo Boliviano, 3,* 32–41, 1961. **(26)**

Louden, K. E., Magnetic anomalies in the West Philippine basin, this volume. **(15, 20)**

Love, A. E. H., *The Mathematical Theory of Elasticity,* 4th rev. ed., Cambridge Univ. Press, 1963. **(9)**

Luck, J. McK., and P. Morgan, Statistical analysis of time transfer data from Timation III, presented at the Sixth Precise Time and Time Interval Planning Meeting, Goddard Space Flight Center, Greenbelt, Md., 1974. **(5)**

Luck, J. McK., M. J. Miller, and P. J. Morgan, The national mapping lunar laser program, *Proc. Symposium Earth's Gravitational Field and Secular Variations in Position, School of Surveying,* Univ. New South Wales, Sydney, Australia, 1974. **(5)**

Ludwig, W. J., N. Den, and S. Murauchi, Seismic reflection measurements of southwest Japan margin, *J. Geophys. Res., 78,* 2508–2515, 1973. **(20)**

Ludwig, W. J., J. E. Nafe, and C. L. Drake, Seismic refraction, in *The Sea,* vol. 4, part 1, edited by E. C. Bullard and A. E. Maxwell, pp. 53–84, Wiley-Interscience, New York, 1970. **(3)**

Ludwig, W. J., J. I. Ewing, M. Ewing, S. Murauchi, N. Den, S. Asano, H. Hotta, M. Hayakawa, T. Asanuma, K. Ichikawa, and I. Noguchi, Sediments and structure of the Japan trench, *J. Geophys. Res., 71*(8), 2121–2137, 1966. **(3, 13, 20)**

Luyendyk, B. P., D. Forsyth, and J. D. Phillips, Experimental approach to the paleocirculation of the oceanic surface waters, *Geol. Soc. Amer. Bull., 83,* 2649–2663, 1972. **(24)**

Luyendyk, B. P., K. C. McDonald, and W. B. Bryan, Rifting history of the Woodlark basin in the southwest Pacific, *Geol. Soc. Amer. Bull., 84,* 1125–1134, 1973. **(15)**

Lyell, C., *Principles of Geology,* 1st ed., London, 1830. **(27)**

Lyman, C. S., On the recent condition of Kilauea, *Amer. J. Sci., Ser. 2, 12,* 75–80, 1851. **(27)**

Macdonald, G. A., Bibliography of the geology and water resources of the Island of Hawaii, *Hawaii Div. Hydrog. Bull., 10,* 191 pp., 1947. **(27)**

Macdonald, G. A., Petrology of the Island of Hawaii, *U.S. Geol. Surv. Prof. Pap., 214-D,* 51–96, 1949. **(28)**

Macdonald, G. A., Dissimilarity of continental and oceanic rock types, *J. Petrol., 1,* 172–177, 1960. **(27)**

Macdonald, G. A., Composition and origin of Hawaiian lavas, in *Studies in Volcanology,* edited by R. R. Coats, R. L. Hay, and C. A. Anderson, *Geol. Soc. Amer. Mem., 116,* 477–522, 1968. **(28)**

Macdonald, G. A., Petrology of the basalt cores from Midway Atoll, *U.S. Geol. Surv. Prof. Pap., 680-B,* 10 pp., 1969. **(28)**

Macdonald, G. A., and T. Katsura, Variations in the lava of the 1959–60 eruption in Kilauea Iki, *Pacific Sci., 15,* 358–369, 1961. **(32)**

Macdonald, G. A., and T. Katsura, Chemical composition of Hawaiian lavas, *J. Petrol., 5,* 82–133, 1964. **(28)**

Macdonald, K. C., and J. D. Mudie, Microearthquakes on the Galapagos spreading center and the seismicity of fast spreading ridges, *Geophys. J. Roy. Astron. Soc., 36,* 245–257, 1974. **(13)**

Macdonald, K. C., B. P. Luyendyk, and R. P. Von Herzen, Heat flow and plate boundaries in Melanesia, *J. Geophys. Res., 78,* 2537–2546, 1973. **(13)**

MacLeod, N. S., and R. M. Pratt, Petrology of volcanic rocks recovered on Leg 18, in *Initial Reports of the Deep Sea Drilling Project,* vol. 18, L. D. Kulm, R. von Heune et al., pp. 935–945, U.S. Govt. Printing Office, Washington, D.C., 1973. **(30)**

Malahoff, A., and D. W. Handschumacher, Magnetic anomalies south of the Murray fracture zone: New evidence for a secondary sea-floor spreading center and strike-slip movement, *J. Geophys. Res., 76,* 6265–6275, 1971. **(16)**

Mammerickx, J., R. N. Anderson, H. W. Menard, and S. M. Smith, Morphology and tectonic evolution of the East Central Pacific, *Geol. Soc. Amer. Bull., 86,* 111–118, 1975. **(8, 16, 22)**

Mammerickx, J., S. M. Smith, I. L. Taylor, and T. E. Chase, Bathymetry of the South Pacific, charts 11–20, Inst. Mar. Resour., Scripps Inst. Oceanogr., San Diego, Calif., 1973. **(28)**

Mammerickx, J., S. M. Smith, I. L. Taylor, and T. E. Chase, Bathymetry of the South Pacific, charts 12 and 13, Inst. Mar. Resour., Scripps Inst. Oceanogr., Univ. Calif., San Diego, 1974. **(22)**

Manghnani, M. H., and G. P. Woollard, Elastic wave velocities in Hawaiian rocks at pressures to ten kilobars, in *The Crust and Upper Mantle of the Pacific Area, Geophys. Monogr. Ser.,* vol. 12, edited by L. Knopoff, C. L. Drake, and P. J. Hart, pp. 501–516, AGU, Washington, D.C., 1968. **(35)**

Marsh, B. D., Island arc magmatism, (abstract) *Eos Trans. AGU, 54,* 1205, 1973. **(29)**

Marsh, B. D., Some Aleutian andesites: Their nature and source, *J. Geol.,* in press *a.* **(29)**

Marsh, B. D., Magma ascent in island arcs, *Proc. Int. Symp. Geothermometry and Geobarometry,* Penn. State Univ., in press *b.* **(29)**

Marsh, B. D., and I. S. E. Carmichael, Benioff zone magmatism, *J. Geophys. Res., 79*(8), 1196–1206, 1974. **(29)**

Marshall, M., and A. Cox, Effect of oxidation on the natural remanent magnetization of titanomagnetite in suboceanic basalt, *Nature, 230,* 28–31, 1971*a.* **(19)**

Marshall, M., and A. Cox, Magnetism of pillow basalts and their petrology, *Geol. Soc. Amer. Bull., 82,* 537–552, 1971*b.* **(19)**

Marshall, P., Acid rocks of the Taupo–Rotorua District, *Royal Soc. New Zealand Trans., 64,* 323–366, 1935. **(27)**

Mason, R. G., A magnetic survey off the west coast of the United States between latitudes 32° and 26°N, longitudes 121° and 128°W, *Geophys. J., 1,* 320–329, 1958. **(15)**

Mason, R. G., and A. D. Raff, A magnetic survey off the west coast of North America, 32°N latitude to 42°N latitude, *Geol. Soc. Amer. Bull., 72,* 1259–1270, 1961. **(15, 16, 18)**

Masuda, A., and I. Kushiro, Experimental determination of partition coefficients of ten rare earth elements and barium between clinopyroxene and liquid in the synthetic silicate system at 20 kilobar pressure, *Contr. Mineral. Petrol., 26,* 42–49, 1970. **(29)**

Matsuda, T., and S. Uyeda, On the Pacific-type orogeny and its model—extension of the paired belts concept and possible origin of marginal seas, *Tectonophysics, 11,* 5–27, 1971. **(15, 20, 21)**

Matsuda, T., K. Nakamura, and A. Sugimura, Late Cenozoic orogeny in Japan, *Tectonophysics, 4,* 349–366, 1967. **(26)**

Matsumoto, T., Four gigantic calderas in Kyushu (abstract), *Proc. 17th Int. Geol. Cong., Moscow,* 138, 1937. **(27)**

Matsumoto, T., The four gigantic caldera volcanoes of Kyushu, *Japanese J. Geol. Geophys., 19, spec. number,* 57 pp., 1943. **(27)**

Matsuzaki, T., Magnetic anomalies over and around Japan trench off Sanriku and the Yamato bank in the Japan Sea, *Rep. Hydrographic Res., 1,* 1–10, 1966. **(20)**

Matthews, D. H., and J. Bath, Formation of magnetic anomaly pattern of Mid-Atlantic ridge, *Geophys. J. Roy. Astron. Soc., 13,* 349–357, 1967. **(19)**

Matuyama, M., Measurements of gravity over Nippon trench on board H.I.J.M. submarine Ro-57, *Preliminary Report Proc. Imp. Acad. Japan, 10,* 625–628, 1934. **(2, 3)**

Maxwell, A. E., R. P. Von Herzen, K. J. Hsü, J. E. Andrews, T. Saito, S. F. Percival, Jr., E. D. Milow, and R. E. Boyce, Deep-sea drilling in the South Atlantic, *Science, 168,* 1047–1059, 1970. **(28)**

McBirney, A. R., Genetic relations of volcanic rocks of the Pacific Ocean, *Geol. Rundsch., 57,* 21–33, 1967. **(28)**

McBirney, A. R., Compositional variations in Cenozoic calc–alkaline suites of Central America, *Oregon Dept. Geol. Min. Ind. Bull., 65,* 185–189, 1969. **(27)**

McBirney, A. R., and K. Aoki, Petrology of the island of Tahiti, *Geol. Soc. Amer. Mem, 116,* 523–556, 1968. **(28)**

McDermott, J. G., R. J. Labhart, and V. O. Marshall, Preliminary bulletin Kurile islands experiment ocean-bottom seismographic experiment, Texas Instrument, Inc., Houston, Texas, 1967. **(13)**

McDougall, I., Potassium–argon ages from lavas of the Hawaiian Islands, *Geol. Soc. Amer. Bull., 75,* 107–127, 1964. **(27, 28, 31)**

McDougall, I., Volcanic island chains and sea-floor spreading, *Nature, 231,* 141–144, 1971. **(28)**

McDougall, I., and D. A. Swanson, Potassium–argon ages of lavas from the Hawi and Polulu volcanic series, Kohala volcano, Hawaii, *Geol. Soc. Amer. Bull., 83,* 3731–3738, 1972. **(28)**

McElhinny, M. W., and P. J. Burek, Mesozoic paleomagnetic stratigraphy, *Nature, 232,* 98–102, 1971. **(18)**

McKenzie, D. P., Some remarks on heat flow and gravity anomalies, *J. Geophys. Res., 72,* 6261–6273, 1967. **(3, 36)**

McKenzie, D. P., Speculations on the consequences and causes of plate motions, *Geophys. J. Roy. Astron. Soc., 18,* 1–32, 1969. **(8, 33)**

McKenzie, D. P., and W. J. Morgan, Evolution of triple junctions, *Nature, 224,* 125–133, 1969. **(15, 16)**

McKenzie, D., and R. L. Parker, The North Pacific: An example of tectonics on a sphere, *Nature, 216,* 1276–1280, 1967. **(9, 15, 16)**

McKenzie, D., and J. G. Sclater, The evolution of the Indian Ocean since the Late Cretaceous, *Geophys. J. Roy. Astron. Soc., 25,* 437–528, 1971. **(16, 18, 21, 30)**

McLaughlin, D. H., Geology and physiography of the Peruvian Cordillera, Departments of Junin and Lima, *Bull. Geol. Soc. Amer., 35,* 591–632, 1924. **(26)**

McManus, D. A., and R. E. Burns, Scientific report on

Deep Sea Drilling Project: Leg 5, *Ocean Ind., 4*(8), 40–42, 1969. **(18)**

McQueen, R. G., S. P. Marsh, and J. N. Fritz, Hugoniot equation of state of twelve rocks, *J. Geophys. Res., 72,* 4999–5036, 1967. **(33)**

Melson, W. G., Basaltic glasses from the Deep Sea Drilling Project, chemical characteristics, compositions of alteration products and fission track 'ages', *Eos Trans. AGU, 54*(11), 1011–1014, 1973. **(30)**

Menard, H. W., The East Pacific rise, *Science, 132,* 1737–1766, 1960. **(16)**

Menard, H. W., Fracture zones on offsets of the East Pacific rise, *J. Geophys. Res., 71,* 682–685, 1966. **(16)**

Menard, H. W., Transitional types of crust and small ocean basins, *J. Geophys. Res., 72,* 3061–3073, 1967. **(13)**

Menard, H. W., Elevation and subsidence of oceanic crust, *Earth Planet. Sci. Lett., 6,* 275–284, 1969. **(16)**

Menard, H. W., Epeirogeny and plate tectonics, *Eos Trans. AGU, 54*(12), 1244–1355, 1973. **(28)**

Menard, H. W., and T. Atwater, Changes in the direction of sea-floor spreading, *Nature, 219,* 463–467, 1968. **(16, 28)**

Mendiguren, J. S., Focal mechanism of a shock in the middle of the Nazca plate, *J. Geophys. Res., 76,* 3861–3879, 1971. **(8)**

Menzies, A., An early ascent of Mauna Loa, *Hawaiian Annual for 1908,* 99–112, 1907. **(27)**

Menzies, A., Ascent of Mount Hualalai, *Hawaiian Annual for 1910,* 72–89, 1909. **(27)**

Merrihue, C., and G. Turner, Potassium–argon dating by activation with fast neutrons, *J. Geophys. Res., 71,* 2852–2857, 1966. **(31)**

Meyer, R. P., and J. F. Gettrust, Project Nariño: Vp/Vs ratios along the western cordillera from the Yumbo quarry (Cali, Colombia) (abstract), *XVII Mtg. Int. Assoc. Seismol. Phys. Earth's Interior and Int. Union Geod. Geophys.,* Lima, Peru, 1973. **(10)**

Meyer, R. P., W. D. Mooney, A. L. Hales, C. E. Helsley, G. P. Woollard, D. M. Hussong, L. W. Kroenke, and J. E. Ramirez, Project Nariño III: Refraction observation across a leading edge, Malpelo Island to the Colombian Cordillera Occidental, this volume. **(7)**

Milsom, J., Gravity field of the Papuan peninsula, *Geol. Mijnbouw, 52,* 13–20, 1973*a*. **(35)**

Milsom, J., Papuan ultramafic belt: Gravity anomalies and the emplacement of ophiolites, *Geol. Soc. Amer. Bull., 84,* 2243–2258, 1973*b*. **(35)**

Milsom, J., and I. E. Smith, Southeastern Papua: Generation of thick crust in a tensional environment?, *Geology, 3,* 117–120, 1975. **(35)**

Minakami, T., T. Ishikawa, and K. Yagi, The 1944 eruption of volcano Usu in Hokkaido, Japan, *Bull. Volcanol., Ser. 2, 11,* 45–157, 1951. **(27)**

Minato, M., M. Gorai, M. Hunahashi, editors, *The Geologic Development of the Japanese Islands,* 442 pp., T. Shokan Co., Tokyo, 1965. **(26)**

Minear, J. W., and M. N. Toksöz, Thermal regime of a downgoing slab and new global tectonics, *J. Geophys. Res., 75,* 1397–1419, 1970. **(1, 29, 33)**

Ming, L., and W. Bassett, Post-spinel phases in $Mg_2SiO_4$–$Fe_2SiO_4$ system up to 80% of $Mg_2SiO_4$ (abstract) *Eos Trans. AGU, 55,* 416–417, 1974. **(33)**

Minster, J. B., T. H. Jordan, P. Molnar, and E. Haines, Numerical modeling of instantaneous plate tectonics, *Geophys. J. Roy. Astron. Soc., 36,* 541–576, 1974. **(8, 28)**

Miyamura, S., Seismicity and structures, *Zisin, J. Seism. Soc. Japan,* in Japanese, 15, 23–52, 1962. **(3)**

Miyamura, S., Seismicity of island arcs and other arc tectonic regions of the circum-Pacific zone, in *The Crust and Upper Mantle of the Pacific Area, Geophys. Monogr. Ser.,* vol. 12, edited by L. Knopoff, C. L. Drake, and P. J. Hart, pp. 60–69, AGU, Washington, D.C., 1968. **(9)**

Miyashiro, A., Classification, characteristics and origin of ophiolites, *J. Geol., 83,* 249–281, 1975. **(35)**

Mogi, A., *Southwest Japan Trench Related to the Philippine Sea, Island Arc and Ocean,* pp. 137–148, Tokai University Press, Japan, 1970. **(13)**

Molnar, P., Fault plane solutions of earthquakes and direction of motion in the Gulf of California and on the Rivera fracture zone, *Geol. Soc. Amer. Bull., 84,* 1651–1658, 1973. **(16)**

Molnar, P., and T. Atwater, Relative motion of hot spots in the mantle, *Nature, 246,* 288–291, 1973*a*. **(18)**

Molnar, P., and T. Atwater, Magnetic anomalies and evolution of the South Pacific: Implications for fixed hot spots (abstract), *Eos Trans. AGU, 54*(4), 240, 1973*b*. **(28)**

Molnar, P., and J. Oliver, Lateral variations of attenuation in the upper mantle and discontinuities in the lithosphere, *J. Geophys. Res., 74,* 2648–2682, 1969. **(10)**

Molnar, P., and L. R. Sykes, Tectonics of the Caribbean and Middle American region from focal mechanisms and seismicity, *Geol. Soc. Amer. Bull., 80,* 1639–1684, 1969. **(16)**

Molnar, P., T. Atwater, J. Mammerickx, and S. M. Smith, Magnetic anomalies, bathymetry and the tectonic evolution of the South Pacific since the Late Cretaceous, *Geophys. J. Roy. Astron. Soc., 40,* 383–420, 1974. **(16, 22)**

Monakov, F. I., Microseisms on the bed of the Black Sea, *Izv. Geophys. Ser.,* English transl. *no. 5,* 461–462, 1961. **(13)**

Monakov, F. I., Microseisms at the bottom of the Baltic Sea and in the northern part of the Atlantic Ocean, *Izv. Geophys. Ser.,* English transl., *no. 7,* 573–580, 1962. **(13)**

Monger, J. W., and C. A. Ross, Distribution of fusulinaceans in the western Canadian Cordillera, *Can. J. Earth Sci., 8,* 259–278, 1971. **(26)**

Montlosier, R., Count de, *Essai des volcans d'Auvergne,* Paris, 1789. **(27)**

Mooney, W. D., R. P. Meyer, C. E. Helsley, C. Lomnitz, and B. T. R. Lewis, Refracted waves across a leading edge: Observations of Pacific shots in southern Mexico (abstract), *Eos Trans. AGU, 56,* 452, 1975. **(10)**

Moore, D. G., and E. C. Buffington, Transform faulting and growth of the Gulf of California since the Late Pliocene, *Science, 161,* 1238–1241, 1968. **(16)**

Moore, J. G., and B. W. Evans, The role of olivine in the crystallization of the prehistoric Makaopuhi tholeiitic lava lake, Hawaii, *Contr. Mineral. Petrol., 15,* 202–223, 1967. **(32)**

Moore, J. G., H. S. Fleming, and J. D. Phillips, Preliminary model for extrusion and rifting at the axis of the Mid-Atlantic ridge, 36°48′N, *Geology, 2,* 437–440, 1974. **(19)**

Moore, T. C., Jr., DSDP: Successes, failures, proposals, *Geotimes, 17*(7), 27–31, 1972. **(24)**

Moore, T. C., Jr., Tj. H. van Andel, C. Sancetta, and N.

Pisias, Cenozoic hiatuses in marine sediments, in *Marine Plankton and Sediments,* edited by W. R. Riedel and T. Saito, Micropaleontology Press (Amer. Mus. Natural History), in press. **(24)**
Moores, E. M., Tectonic significance of ultramafic rocks, *Earth-Science Rev., 9,* 241–258, 1973. **(35)**
Moores, E. M., and E. D. Jackson, A comparison of selected ophiolites and oceanic crust, *Nature,* in press. **(35)**
Moores, E. M., and F. J. Vine, The Troodos massif, Cyprus, and other ophiolites as oceanic crust: Evaluations and implications, *Phil. Trans. Roy. Soc. London, Ser. A., 268,* 443–466, 1971. **(35)**
Morgan, W. J., Rises, trenches, great faults and crustal blocks, *J. Geophys. Res., 73,* 1959–1982, 1968. **(15, 16, 18)**
Morgan, W. J., Convection plumes in the lower mantle, *Nature, 230,* 42–43, 1971. **(18)**
Morgan, W. J. Deep mantle convection plumes and plate motions, *Amer. Assoc. Petrol. Geol. Bull., 56* (2), 203–213, 1972*a.* **(18, 28, 31)**
Morgan, W. J., Plate motions and deep mantle convection, in *Studies in Earth and Space Sciences: Harry H. Hess Volume,* edited by R. Shagam et al., *Geol. Soc. Amer. Mem., 132,* 7–22, 1972*b.* **(18, 28)**
Morgan, W. J., P. R. Vogt, and D. F. Falls, Magnetic anomalies and sea-floor spreading on the Chile rise, *Nature, 222,* 137–142, 1969. **(16)**
Morley, L. W., and A. Larochelle, Paleomagnetism as a means of dating geological events, *Roy. Soc. Can. Spec. Publ., 8,* 39, 1964. **(15)**
Moro, A. L., *Dei crostacei e degli altri marini corpi che si truovano su'monti,* Venice, 1740. [Cited by *Sapper,* 1927, p. 368.] **(27)**
Morris, R. C., Cretaceous sedimentary and tectonic history in northwest Peru, *IXth Int. Sed. Cong., Nice,* in press. **(26)**
Morton, W. R., and L. J. Close, Notes on Hertz's theory of the contact of elastic bodies, *Phil. Mag., 43,* 320, 1922. **(9)**
Murata, K. J., An acid fumarolic gas from Kilauea Iki, Hawaii, *U.S. Geol. Surv. Prof. Pap., 537-C,* 61–66, 1966. **(32)**
Murata, K. J., and D. H. Richter, Chemistry of the lavas of the 1959–60 eruption of Kilauea volcano, Hawaii, *U.S. Geol. Surv. Prof. Pap., 537-A,* 26 pp., 1966*a.* **(32)**
Murata, K. J., and D. H. Richter, The settling of olivine in Kilauean magma as shown by lavas of the 1959 eruption, *Amer. J. Sci., Ser. 4, 264,* 194–203, 1966*b.* **(32)**
Murauchi, S., The renewal of island arcs and the tectonics of marginal sea (in Japanese with English abstract) in *Island Arc and Marginal Sea,* edited by S. Asano and G. B. Udintsev, pp. 39–56, Tokai Univ. Press, 1971. **(20)**
Murauchi, S., and T. Asanuma, Seismic profiler and sonobuoy refraction measurements during GDP-6 to -8 voyages (in Japanese with English abstract), *Mar. Sci., 6*(8), 23–27, 1974. **(20)**
Murauchi, S., N. Den, S. Asano, H. Hotta, T. Yoshii, T. Asanuma, K. Hagiwara, K. Ichikawa, T. Sato, W. J. Ludwig, J. Ewing, N. T. Edgar, and R. E. Houtz, Crustal structure of the Philippine Sea, *J. Geophys. Res., 73,* 3143–3171, 1968. **(3, 20)**
Myers, J. S., Cretaceous stratigraphy and structure, western Andes of Peru, between latitudes 10°–10°30′, *Amer. Assoc. Petrol. Geol. Bull., 58,* 474–487, 1974. **(26)**
Mysen, B. O., I. Kushiro, I. Nicholls, and A. E. Ringwood, A possible mantle origin for andesitic magmas; discussion of a paper by Nicholls and Ringwood, *Earth Planet. Sci. Lett., 21,* 221–229, 1974. **(29)**

Nadai, A., *Theory of Flow and Fracture of Solids,* vol. 2, 705 pp., McGraw-Hill, New York, 1963. **(37)**
Nafe, J. E., and C. L. Drake, Variation with depth in shallow and deep-water marine sediments of porosity, density and the velocities of compressional and shear waves, *Geophysics,* 22, 523–552, 1957. **(3)**
Nafe, J. E., and C. L. Drake, Physical properties of marine sediments, in *The Sea,* vol. 3, edited by M. N. Hill, pp. 794–815, Interscience Publishers, New York, 1963*a.* **(1)**
Nafe, J. E., and C. L. Drake, Physical properties of marine sediments, in *The Sea,* vol. 3, edited by M. N. Hill, pp. 85–102, John Wiley and Sons, New York, 1963*b.* **(35)**
Nagumo, S., Ocean-bottom seismographic observation and bottom-current observation, in *Preliminary Cruise Results of R/V Hakuho Maru, No. 5, Cruise No. KH-68-3, Northwest Pacific Ocean,* pp. 48–59, Ocean Research Institute, Univ. Tokyo, Japan, 1968. **(13)**
Nagumo, S., Ocean-bottom seismographic observation of the Tokachi-oki earthquake of 1968, in *General Report of the Tokachi-oki Earthquake of 1968,* pp. 225–240, Keigaku-sha Co., Tokyo, 1971*a.* **(13)**
Nagumo, S., Ocean-bottom seismographic observation, *Preliminary Report of the Hakuho Maru Cruise KH-69-2,* pp. 173–196, Ocean Research Institute, Univ. Tokyo, Japan, 1971*b.* **(13)**
Nagumo, S., and J. Kasahara, Ocean-bottom seismographic study of the western margin of the Pacific, this volume. **(7)**
Nagumo, S., H. Kobayashi, and S. Koresawa, Sea-bottom seismic observation at Sagami Bay, (1) seismic activity, *Bull. Earthquake Res. Inst., 46,* 877–888, 1968*a.* **(13)**
Nagumo, S., H. Kobayashi, and S. Koresawa, Improvements of ocean-bottom seismograph-construction of a long-life magnetic tape recorder, *Bull. Earthquake Res. Inst., 46,* 861–875, 1968*b.* **(13)**
Nagumo, S., H. Kobayashi, and S. Koresawa, Foreshock phenomena of the 1968 Tokachi-oki earthquake observed by ocean-bottom seismographs off Sanriku, *Bull. Earthquake Res. Inst., 46,* 1355–1368, 1968*c.* **(13)**
Nagumo, S., S. Hasegawa, S. Koresawa, and H. Kobayashi, Ocean-bottom seismographic observation at the off-side of Japan trench near the Erimo seamount —seismic activity of the oceanic lithosphere and velocity structure around the geophysical ocean–continent boundary, *Bull. Earthquake Res. Inst., 48,* 769–792, 1970*a.* **(13)**
Nagumo, S., S. Hasegawa, S. Koresawa, and H. Kobayashi, Ocean-bottom seismographic observation off Sanriku aftershock activity of the 1968 Tokachi-oki earthquake and its relation to the ocean–continent boundary fault, *Bull. Earthquake Res. Inst.,* 48, 793–809, 1970*b.* **(13)**
Naidu, P. S., Statistical structure of geomagnetic field reversals, *J. Geophys. Res., 76,* 2649–2662, 1971. **(19)**
Naidu, P. S., Are geomagnetic field reversals indepen-

dent?, *J. Geomagn. Geoelec., 26,* 101–104, 1974. **(19)**

Nassar, M. M., Least squares adjustment of Doppler satellite observations, *Tech. Rep. 21,* Dept. Surveying Engineering, Univ. New Brunswick, Fredericton, New Brunswick, 1972. **(5)**

Natland, J. H., and E. D. Jackson, Volcanic rocks from the Line Islands, in *Initial Reports of the Deep Sea Drilling Project,* vol. 33, S. O. Schlanger, E. D. Jackson et al., U.S. Govt. Printing Office, Washington, D.C., in press. **(22)**

Naugler, F. P., and J. M. Wageman, Gulf of Alaska: Magnetic anomalies, fracture zones, and plate interaction, *Geol. Soc. Amer. Bull., 84,* 1575–1584, 1973. **(16, 18)**

Neuman, L. D., and M. Talwani, Accelerations and errors in gravity measurements on surface ships, *J. Geophys. Res., 77,* 4330–4338, 1972. **(2)**

Neumann van Padang, M., Der Ausbruch des Merapi (Mittel Java) im Jahre 1930, *Z. Vulkanol., 14,* 135–148, 1931. **(27)**

Neumann van Padang, M., De Krakatau voorheen en thans, *De Tropische Natuur., 22,* 137–150, 1933*a.* **(27)**

Neumann van Padang, M., De uitbarsting van den Merapi (Midden Java) in de jaren 1930–1931, *Ned. Indies, Dienst Mijnbouwk. Vulkan. Seism. Meded., 12,* 135 pp., 1933*b.* **(27)**

Newell, N. D., J. Chronic, and T. G. Roberts, Upper Paleozoic of Peru, *Geol. Soc. Amer. Mem., 58,* 276 pp., 1953. **(26)**

Nicholls. I. A., and A. E. Ringwood, Effect of water on olivine stability in tholeiites and production of silica-saturated magmas in the island arc environment, *J. Geol., 81,* 285–300, 1973. **(29)**

Nicolas, A., and E. D. Jackson, Repartition en deux provinces des peridotites des châines alpines longeant La Mediterranée: Géotectoniques, *Bull. Suisse Mineral. Petrog., 53,* 385–401, 1972. **(35)**

Ninkovich, D., N. Opdyke, B. C. Heezen, and J. H. Foster, Paleomagnetic stratigraphy, rates of deposition and tephrachronology in North Pacific deep-sea sediments, *Earth Planet. Sci. Lett., 1,* 476–492, 1966. **(19)**

Nishizawa, O., and S. Akimoto, Partition of magnesium and iron between olivine and spinel and between pyroxene and spinel, *Contr. Mineral. Petrol., 41,* 217–230, 1973. **(34)**

Nixon, P. H., and F. R. Boyd, The discrete nodule association in kimberlites from northern Lesotho, in *Lesotho Kimberlites,* edited by P. H. Nixon, pp. 67–75, Lesotho National Development Corporation, Cape Town, 1973. **(34)**

Northrop, J., Seaquakes on the Nazca–Cocos plate boundary, *Bull. Seism. Soc. Amer., 64,* 595–606, 1974. **(13)**

Nuttli, O. W., and B. A. Bolt, *P* wave residuals as a function of azimuth, 2, Undulations of the mantle low-velocity layer as an explanation, *J. Geophys. Res., 74,* 6594–6602, 1969. **(36)**

Nygren, W. E., Bolivar geosyncline of northwestern South America, *Amer. Assoc. Petrol. Geol. Bull., 34,* 1998–2005, 1950. **(10)**

Ocola, L. C., and R. P. Meyer, Crustal low-velocity zones under the Peru–Bolivia altiplano, *Geophys. J. Roy. Astron. Soc., 30,* 199–209, 1972. **(10)**

Ocola, L. C., R. P. Meyer, and L. T. Aldrich, Gross crustal structure under Peru-Bolivia altiplano, *Earthquake Notes, 42,* 33–48, 1971. **(10)**

Ocola, L. C., J. F. Gettrust, R. P. Meyer, and G. P. Woollard, Seismic energy transmission from ocean to continent in western South America, *Geophys. Res. Lett., 2*(8), pp. 361–364, 1975. **(10)**

Odegard, M. E., Upper mantle structure of the North Pacific, Ph.D. dissertation, Univ. Hawaii, 271 pp., 1975. **(8)**

O'Hara, M. J., The bearing of phase equilibria studies in synthetic and natural systems on the origin and evolution of basic and ultra-basic rocks, *Earth Sci. Rev., 4,* 69–133, 1968. **(36)**

O'Hara, M. J., and Y. S. Yoder, Jr., Formation and fractionation of basic magmas at high pressures, *Scot. J. Geol., 3,* 67–117, 1967. **(36)**

Oinouye, Y., Report on the eruption of Tarumai-dake, in Japanese, *Imp. Earthquake Invest. Comm., Pub. 64,* 23 pp., 1909. **(27)**

Ojakangas, R. W., Cretaceous sedimentation, Sacramento Valley, California, *Geol. Soc. Amer. Bull., 79,* 973–1008, 1968. **(26)**

Okada, H., Converted *P* phase from the ScS phase at the inclined deep seismic zone, *Carnegie Inst. Washington Yearb., 72,* 238, 1973. **(9)**

Oliver, J., and B. Isacks, Deep earthquake zones, anomalous structures in the upper mantle, and the lithosphere, *J. Geophys. Res., 72,* 4259–4275, 1967. **(3, 9, 13)**

Omori, F., The Usu-san eruption and the earthquake and elevation phenomena, *Imp. Earthquake Invest. Comm. Bull., 5*(1), 101–107, 1911. **(27)**

Omori, F., The Sakurajima eruptions and earthquakes, I and II, *Imp. Earthquake Invest. Comm. Bull., 8*(2 and 3), 1916. **(27)**

Omori, F., The Usu-san eruption and the earthquake and elevation phenomena, *Imp. Earthquake Invest. Comm. Bull., 9,* 41–76, 1920. **(27)**

Onwumechelli, A., Geomagnetic variations in the equatorial zone, in *Physics of Geomagnetic Phenomena,* edited by S. Matsushita and W. H. Campbell, pp. 426–507, Academic Press, New York, 1967. **(16)**

Opdyke, N. D., B. Glass, J. D. Hays, and J. Foster, Paleomagnetic study of Antarctic deep-sea cores, *Science, 154,* 349–357, 1966. **(18)**

Orcutt, J., L. M. Dorman, P. Spudich, and W. A. Prother, A seismic refraction survey of the East Pacific rise using ocean-bottom seismographs (abstract), *Eos Trans. AGU, 56,* 1146, 1974. **(13)**

Oxburgh, E. R., and D. L. Turcotte, Mid-ocean ridges and geotherm distribution during mantle convection, *J. Geophys. Res., 73,* 2643–2661, 1968. **(33)**

Oxburgh, E. R., and D. L. Turcotte, Thermal structure of island arcs, *Geol. Soc. Amer. Bull., 81,* 1665–1688, 1970. **(29)**

Ozima, M., and M. Ozima, Characteristic thermomagnetic curve in submarine basalts, *J. Geophys. Res., 76,* 2051–2056, 1971. **(19)**

Ozima, M., and K. Saito, $^{40}Ar$–$^{39}Ar$ stepwise degassing experiments on some submarine rocks, *Earth Planet. Sci. Lett., 20,* 77–87, 1973. **(31)**

Ozima, M., and K. Saito, $^{40}Ar$–$^{39}Ar$ ages of some submarine rocks and their geophysical implications (abstract), *Abstr. Program Int. Woollard Symposium,* p. 33, Hawaii Inst. Geophys., Univ. Hawaii, 1974. **(28)**

Ozima, M., I. Kaneoka, and S. Aramaki, K–Ar ages of submarine basalts dredged from seamounts in the western Pacific area and discussion of ocean crust, *Earth Planet. Sci. Lett., 8,* 237–245, 1970. **(18, 20)**

Ozima, M., S. Zashu, and N. Ueno, I/Rb and $(^{87}Sr/^{86}Sr)_0$ ratios of dredged submarine basalts, *Earth Planet. Sci. Lett., 10,* 23–24, 1971. **(18)**

Packham, G. H., and D. A. Falvey, An hypothesis for the formation of marginal seas in the western Pacific, *Tectonophysics, 11,* 79–109, 1971. **(2, 20)**

Parker, R. L., and D. W. Oldenburg, Thermal model of ocean ridges, *Nature Phys. Sci., 242,* 137–139, 1973. **(19, 36)**

Peck, D. L., Lava coils of some recent historic flows, Hawaii, in Geological Survey Research 1966, *U.S. Geol. Surv. Prof. Pap., 550-B,* B148–B151, 1966. **(32)**

Peck, D. L., Density of molten lava in Alae lava lake, Hawaii (abstract), *Trans. AGU, 50,* 339, 1969. **(32)**

Peck, D. L., Thermal properties of basaltic magma; results and practical experience from study of Hawaiian lava lakes, in *The Utilization of Volcano Energy,* edited by J. L. Colp and A. S. Furumoto, pp. 287–298, Sandia Laboratories, Albuquerque, New Mexico, 1974. **(32)**

Peck, D. L., and W. T. Kinoshita, The eruption of August, 1963 and the formation of Alae lava lake, Hawaii, *U.S. Geol. Surv. Prof. Pap.,* in press. **(32)**

Peck, D. L., and T. Minakami, The formation of columnar joints in the upper part of Kilauean lava lakes, Hawaii, *Geol. Soc. Amer. Bull.,* 79, 1151–1166, 1968. **(32)**

Peck, D. L., J. G. Moore, and G. Kojima, Temperatures in the crust and melt of Alae lava lake, Hawaii, after the August 1963 eruption of Kilauea volcano—a preliminary report, in Geological Survey Research 1964, *U.S. Geol. Surv. Prof. Pap., 501-D,* D1–D7, 1964. **(32)**

Peck, D. L., T. L. Wright, and J. G. Moore, Crystallization of tholeiitic basalt lava in Alae lava lake, Hawaii, *Bull. Volcanol., 29,* 629–656, 1966. **(29, 32, 36)**

Perret, F. A., The floating islands of Halemaumau, *Amer. J. Sci., Ser. 4*(35), 273–282, 1913. **(27)**

Peter, G., B. H. Erickson, and P. J. Grim, Magnetic structure of the Aleutian trench and northeast Pacific basin, in *The Sea,* vol. 4, edited by A. E. Maxwell, 191–222, Wiley-Interscience, New York, 1970. **(15, 18)**

Peterson, M. N. A., Calcite: Rates of dissolution in a vertical profile in the central Pacific, *Science, 154,* 1542–1544, 1966. **(24)**

Peterson, J. J., P. J. Fox, and E. Schreiber, Newfoundland ophiolites and the geology of the oceanic layer, *Nature, 247,* 194–196, 1974. **(35)**

Pironneau, O., On optimum profiles in Stokes flow, *Fluid Mech., 59,* 117–128, 1973. **(29)**

Pitcher, W. S., The coastal batholith of Peru: Some structural aspects, *24th Int. Geol. Cong., sec. 2,* 156–163, 1972. **(26)**

Pitman, W. C., III, and D. E. Hayes, Sea-floor spreading in the Gulf of Alaska, *J. Geophys. Res., 73,* 6571–6580, 1968. **(15)**

Pitman, W. C., III, and J. R. Heirtzler, Magnetic anomalies over the Pacific–Antarctic ridge, *Science, 154,* 1164–1171, 1966. **(15, 16)**

Pitman, W. C., III, and M. Talwani, Sea-floor spreading in the North Atlantic, *Geol. Soc. Amer. Bull., 83,* 619–646, 1972. **(18, 22)**

Pitman, W. C., III, E. M. Herron, and J. R. Heirtzler, Magnetic anomalies in the Pacific and sea-floor spreading, *J. Geophys. Res., 73,* 2069–2085, 1968. **(16, 18, 21, 28)**

Pitman, W. C., III, R. L. Larson, and E. M. Herron, Age of the ocean basins determined from magnetic anomaly lineations, *Geol. Soc. Amer. Charts,* 1974. **(15, 22, 26, 28, 30)**

Pitman, W. C., III, M. Talwani, and J. R. Heirtzler, Age of the North Atlantic Ocean from magnetic anomalies, *Earth Planet. Sci. Lett., 11,* 195–200, 1971. **(18)**

Plafker, G., Alaskan earthquake of 1964 and Chilean earthquake of 1960: Implications for arc tectonics, *J. Geophys. Res., 77,* 901–925, 1972. **(8)**

Plotkin, H. H., Laser technology for high precision satellite tracking, *Proc. Symposium Earth's Gravitational Field and Secular Variations in Position,* School of Surveying, Univ. New South Wales, Sydney, Australia, 1974. **(5)**

Post, R. L., and D. T. Griggs, The earth's mantle: Evidence of non-Newtonian flow, *Science, 181,* 1242–1244, 1973. **(29)**

Poster, C. K., Ultrasonic velocities in rocks from the Troodos massif, Cyprus, *Nature Phys. Sci., 243,* 2–3, 1973. **(35)**

Powers, H. A., Composition and origin of basaltic magma of the Hawaiian Islands, *Geochim. Cosmochim. Acta, 7,* 77–107, 1955. **(32)**

Powers, S., Volcanic domes in the Pacific, *Amer. J. Sci., Ser. 4, 42,* 261–274, 1916. **(27)**

Pratt, R. M., M. S. Rutstein, F. A. Walton, and J. A. Buschur, Extension of Alaskan structural trends beneath Bristol Bay, Bering shelf, Alaska, *J. Geophys. Res., 77,* 4994–4999, 1972. **(2)**

Pratt, W. P., Local evidence of Pleistocene to Recent orogeny in the Argentine Andes, *Geol. Soc. Amer. Bull., 72,* 1539–1550, 1961. **(26)**

Prentiss, D. D., and J. Ewing, The seismic motion of the deep ocean floor, *Bull. Seism. Soc. Amer., 53,* 765–781, 1963. **(13)**

Press, F., Earth models obtained by Monte Carlo inversion, *J. Geophys. Res., 73,* 5223–5234, 1968. **(1)**

Press, F., Regionalized earth models, *J. Geophys. Res., 75,* 6575–6581, 1970*a.* **(1)**

Press, F., Earth models consistent with geophysical data, *Phys. Earth Planet. Int., 3,* 3–22, 1970*b.* **(33, 36)**

Prince, R. A., and L. D. Kulm, Crustal rupture and the initiation of imbricate thrusting in the Peru trench, *Geol. Soc. Amer. Bull.,* in press. **(8)**

Prince, R. A., J. M. Resig, L. D. Kulm, and T. C. Moore, Jr., Uplifted turbidite basins on the seaward wall of the Peru trench, *Geology, 2,* 607–611, 1974. **(8)**

Prothero, W., I. Reid, and M. Reichle, Deployment of OBS capsules on the East Pacific rise (abstract), *Eos Trans. AGU, 56,* 1151, 1974. **(13)**

Prothero, W. A., Jr., An ocean-bottom seismometer capsule, *Bull. Seism. Soc. Amer., 64,* 1251–1262, 1974. **(13)**

Raff, A. D., Boundaries of an area of very long magnetic anomalies in the northeast Pacific, *J. Geophys. Res., 71,* 2631–2636, 1966. **(16)**

Raff, A. D., Sea-floor spreading—another rift, *J. Geophys. Res., 73,* 3699–3705, 1968. **(16)**

Raff, A. D., and R. G. Mason, Magnetic survey off the

west coast of North America 40°N latitude to 52°N latitude, *Geol. Soc. Amer. Bull., 72,* 1267–1270, 1961. **(15, 16)**

Rainsford, H. F., Long lines on the earth, in *Empire Survey Review, 10*(71 and 72), London, 1949. **(5)**

Raitt, R. W., Seismic refraction studies of the Pacific Ocean Basin, Part I: Crustal thickness of the central equatorial Pacific, *Bull. Geol. Soc. Amer., 67,* 1623–1640, 1956. **(2)**

Raitt, R. W., The Crustal Rocks, in *The Sea,* vol. 3, edited by M. N. Hill, pp. 85–102, Wiley-Interscience, New York, 1963. **(8, 35)**

Raitt, R. W., R. L. Fisher, and R. G. Mason, Tonga trench, *Geol. Soc. Amer. Spec. Pap., 62,* 237–254. 1955. **(1)**

Rawson, D. E., Drilling into molten lava in the Kilauea Iki volcanic crater, Hawaii, *Nature, 188,* 930–931, 1960. **(32)**

Rea, D. K., J. Dymond, G. R. Heath, D. F. Heinrichs, S. H. Johnson, and D. M. Hussong, New estimates of rapid sea-floor spreading rates and the identification of young magnetic anomalies on the East Pacific rise, 6° and 11°S, *Earth Planet. Sci. Lett., 19,* 225–229, 1973. **(8)**

Reichle, M., and I. Reid, Sonobuoy studies of two earthquake sequences in the Gulf of California (abstract), *Eos Trans. AGU, 56,* 1149, 1974. **(13)**

Reid, A. F., and A. E. Ringwood, New dense phases of geophysical significance, *Nature, 252,* 681–682, 1974. **(33)**

Reid, I., and K. Macdonald, A microearthquake study of the Mid-Atlantic ridge near 37°N, using sonobuoys, *Nature, 246,* 88–90, 1973. **(13)**

Reid, I., W. Protero, M. Reichle, and J. Bruno, Ocean-floor microearthquake studies of the East Pacific rise and Rivera fracture zone (abstract), *Eos Trans. AGU, 56,* 1149, 1974. **(13)**

Reid, I., M. Reichle, J. Brune, and H. Bradner, Microearthquake studies using sonobuoys: Preliminary results from Gulf of California, *Geophys. J. Roy. Astron. Soc., 34,* 365–379, 1973. **(13)**

Research Group for Explosion Seismology of Japan, Local changes of *Pn* velocity of the upper mantle across the northeastern Honshu and the Japan trench as derived from large charge size explosion seismic surveys, paper read at the Fall Meeting of the Japan Seismological Society, Nagoya Univ., Nagoya City, October, 1974. **(3)**

Reyer, E., *Theoretische Geologie,* Stuttgart, 1888. **(27)**

Richards, M. L., V. Vacquier, and G. D. Van Voorhis, Calculation of magnetization of uplifts from combining topographic and magnetic surveys, *Geophysics, 32,* 678–701, 1967. **(18)**

Richardson, J. A., and F. J. Lerch, Gravitational field models, GEM 5 and GEM 6, *Eos Trans. AGU, 55,* 218, 1974. **(2)**

Richter, D. H., and J. G. Moore, Petrology of the Kilauea Iki lava lake, Hawaii, *U.S. Geol. Surv. Prof. Pap., 537-B,* 26 pp., 1966, **(32)**

Richter, D. H., and K. J. Murata, Petrography of the lavas of the 1959–60 eruption of Kilauea volcano, Hawaii, *U.S. Geol. Surv. Prof. Pap., 537-D,* 12 pp., 1966. **(32)**

Richter, D. H., J. P. Eaton, K. J. Murata, W. U. Ault, and H. L. Krivoy, Chronological narrative of the 1959–60 eruption of Kilauea volcano, Hawaii, *U.S. Geol. Surv. Prof. Pap., 537-E,* E1–E73, 1970. **(32)**

Richter, F., Dynamical models for sea-floor spreading, *Rev. Geophys. Space Phys., 11,* 223–287, 1973. **(15)**

Richthofen, F. von, The natural system of volcanic rocks, *California Acad. Sci. Mem., 1,* part 2, 98 pp., 1868. **(27)**

Richthofen, F. von, Morphologische Studien aus Ostasien, III, *Kgl. Preuss. Akad. Wiss. Berlin, Sitzungsber., pt. 40,* 1–52, 1903. **(27)**

Ridley, W. I., J. M. Rhodes, A. M. Reid, P. Jakes, C. Shim, and M. N. Bass, Basalts from Leg 6 of the Deep Sea Drilling Project, *J. Petrol., 15,* 140–159, 1974. **(20)**

Rikitake, T., The undulation of an electrically conducting layer beneath the islands of Japan, *Tectonophysics, 7,* 257–264, 1969. **(36)**

Rikitake, T., S. Miyamura, I. Tsubokawa, S. Murauchi, S. Uyeda, H. Kuno, and M. Gorai, Geophysical and geological data in and around the Japan arc, *Can. J. Earth Sci., 5,* 1101–1118, 1968. **(3)**

Ringwood, A. E., Mineralogical constitution of the deep mantle, *J. Geophys. Res., 67,* 4005–4010, 1962. **(33)**

Ringwood, A. E., Mineralogy of the mantle, in *Advances in Earth Science,* edited by P. Hurley, pp. 357–398, MIT Press, Boston, 1966. **(34)**

Ringwood, A. E., The pyroxene–garnet transformation in the earth's mantle, *Earth Planet. Sci. Lett., 2,* 255–263, 1967. **(34)**

Ringwood, A. E., Composition and evolution of the upper mantle, in *The Earth's Crust and Upper Mantle, Geophys. Monogr. Ser.,* vol. 13, edited by P. J. Hart, pp. 1–17, AGU, Washington, D.C., 1969. **(3)**

Ringwood, A. E., Phase transformations and the constitution of the mantle, *Phys. Earth Planet. Interiors, 3,* 109–155, 1970. **(33, 34)**

Ringwood, A. E., Phase transformations and mantle dynamics, *Earth Planet. Sci. Lett., 14,* 233–241, 1972. **(33)**

Ringwood, A. E., The petrological evolution of island arc systems, *J. Geol. Soc., 130,* 183–204, 1974. **(20)**

Ringwood, A. E., *Composition and Petrology of the Earth's Mantle,* McGraw-Hill, New York, 1975 **(33)**

Ringwood, A. E., and A. Major, Synthesis of $Mg_2SiO_4$–$Fe_2SiO_4$ spinel solid solutions, *Earth Planet. Sci. Lett., 1,* 241–245, 1966*a.* **(34)**

Ringwood, A. E., and A. Major, High pressure transformations in pyroxenes, *Earth Planet. Sci. Lett., 1,* 351–357, 1966*b.* **(34)**

Ringwood, A. E., and A. Major, High pressure transformation of $FeSiO_3$ pyroxene to spinel plus stishovite, *Earth Planet. Sci. Lett., 1,* 135–136, 1966c. **(34)**

Ringwood, A. E., and A. Major, High pressure transformations in pyroxenes II, *Earth Planet. Sci. Lett., 5,* 76–78, 1968. **(34)**

Ringwood, A. E., and A. Major, The system $Mg_2SiO_4$–$Fe_2SiO_4$ at high pressures and temperatures, *Phys. Earth Planet. Interiors., 3,* 89–108, 1970. **(34)**

Ringwood, A. E., and A. F. Reid, High pressure transformations of spinels, *Earth Planet. Sci. Lett., 5,* 245–250, 1968. **(33)**

Robertson, E. C., and D. L. Peck, Thermal conductivity of vesicular basalt from Hawaii, *J. Geophys. Res., 79,* 4875–4888, 1974. **(32)**

Rod, E., Geology of eastern Papua: Discussion, *Geol. Soc. Amer. Bull., 85,* 653–658, 1974. **(35)**

Rodda, P., Outline of the geology of Viti Levu, *New Zealand J. Geol. Geophys., 10,* 1260–1268, 1967. **(13)**

Rodriguez, R., Lineas de esfuerzo debidas al choque de

la placa de Nazca y el continente Americano, *Publicacion No. 24 del Observatorio San Calixto,* 1975. **(9)**

Rodriguez, R., E., Ramon Cabre, S. J., and A. Mercado, Geometry of the Nazca plate and its geodynamic implications, this volume. **(7, 26)**

Rona, P. A., Relations between rates of sediment accumulation on continental shelves, sea-floor spreading, and eustacy inferred from the central North Atlantic, *Geol. Soc. Amer. Bull., 84,* 2851–2872, 1973. **(28)**

Rose, J. C., and J. C. Belshe, Gravity and magnetic fields over the proposed Moho Hole site north of Maui, *Pacific Sci., 19,* 374–380, 1965. **(2)**

Rose, J. C., G. P. Woollard, and A. Malahoff, Marine gravity studies of the Solomon Islands, in *The Crust and Upper Mantle of the Pacific Area, Geophys. Monogr. Ser.,* vol. 12, edited by L. Knopoff, C. L. Drake, and J. H. Pembroke, pp. 379–410, AGU, Washington, D.C., 1968. **(35)**

Rosendahl, B. R., Geological and geophysical studies of the Canton trough and surrounding regions, M.S. thesis, 110 pp., Univ. Hawaii, 1972. **(22)**

Ruiz, C., L. Aguirre, J. Corvalan, H. J. Rose, Jr., K. Segerstrom, and T. W. Stern, Ages of batholithic intrusions of northern and central Chile, *Geol. Soc. Amer. Bull., 72,* 1551–1560, 1961. **(26)**

Rutland, R. W. R., Andean orogeny and ocean-floor spreading, *Nature, 233,* 252–255, 1971. **(8)**

Ryall, A., and D. L. Bennett, Crustal structure of southern Hawaii related to volcanic processes in the upper mantle, *J. Geophys. Res., 73,* 4561–4582, 1968. **(14)**

Rykunov, L. N., and V. V. Sedov, Seismic noise in the 2–15 cps frequency range on the bottom of the Black Sea, *Izv. Phys. Solid Earth,* English transl., *7,* 443–448, 1965. **(13)**

Rykunov, L. N., and V. V. Sedov, An ocean-bottom seismograph, *Izv. Phys. Solid Earth,* English transl., *8,* 537–541, 1967. **(12, 13)**

Sacks, I. S., and H. Okada, A comparison of the anelasticity structure between western South America and Japan, *Carnegie Inst. Washington Yearb., 72,* 226, 1973. **(9)**

Saki, T., Investigation of the Philippine Sea floor by the ship Bosei Maru, paper presented at the Japanese Geodynamics Symposium, Tokyo, 1975. **(20)**

Santo, T., Division of the Pacific area into seven regions in each of which Rayleigh waves have the same group velocities, *Bull. Earthquake Res. Inst., 41,* 719–741, 1963. **(36)**

Santo, T., Characteristics of seismicity in South America, *Bull. Earthquake Res. Inst., 47,* 635–672, 1969. **(9)**

Sapper, K., Die vulkanische Tätigkeit in Mittelamerika im 20 Jahrhundert, *Z. Vulkanol., 9,* 156–203, 1926. **(27)**

Sapper, K., *Vulkankunde,* 424 pp., Stuttgart, 1927. **(27)**

Sapper, K., and F. Termer, Der Ausbruch des Vulkans Santa Maria in Guatemala vom 2–4 Nov., 1929, *Z. Vulkanol., 13,* 73–101, 1930. **(27)**

Sato, D., Report on the Tarumai eruption, in Japanese, *Imp. Geol. Survey Japan, Pub. 14,* 1–33, 1909. **(27)**

Sato, M., and T. L. Wright, Oxygen fugacities directly measured in magmatic gases, *Science, 153,* 1103–1105, 1966. **(32)**

Sawkins, F. J., Sulfide ore deposits in relation to plate tectonics, *J. Geol., 80,* 377–397, 1972. **(8)**

Schmid, H. H., Accuracy aspects of a world-wide passive satellite triangulation system, U.S. Dept. Commerce, Coast and Geodetic Survey, Washington, D.C., 1964. **(5)**

Schminke, G. R., and C. G. Bufe, Relationship of Chataqua seamount to sea-floor spreading in the North Pacific Ocean, *Nature, 219,* 1351–1352, 1968. **(28)**

Schneider, E. D., and P. R. Vogt, Discontinuities in the history of sea-floor spreading, *Nature, 217,* 1212–1222, 1968. **(26)**

Schneider, W. A., and M. M. Backus, Ocean-bottom seismic measurements off the California coast, *J. Geophys. Res., 69,* 1135–1143, 1964. **(13)**

Schneider, W. A., P. J. Farrell, and R. E. Brannian, Collection and analysis of Pacific ocean-bottom seismic data, *Geophysics, 29,* 745–771, 1964. **(13)**

Scholl, D. W., and J. S. Creager, Geologic synthesis of Leg 19 (DSDP) results: Far North Pacific, Aleutian ridge, and Bering Sea, in *Initial Reports of the Deep Sea Drilling Project,* vol. 19, J. S. Creager, D. W. Scholl et al., pp. 897–913, U.S. Govt. Printing Office, Washington, D.C., 1973. **(28)**

Scholl, D. W., R. von Huene, and J. B. Ridlon, Spreading of the ocean floor: Undeformed sediments in the Peru–Chile trench, *Science, 159,* 869–871, 1968. **(8)**

Scholl, D. W., M. N. Christensen, R. von Huene, and M. S. Marlow, Peru–Chile trench sediments and sea-floor spreading, *Geol. Soc. Amer. Bull., 81,* 1339–1360, 1970. **(8)**

Schouten, H., and K. McCamy, Filtering marine magnetic anomalies, *J. Geophys. Res., 77,* 7089–7099, 1972. **(15)**

Schouten, J. A., A fundamental analysis of magnetic anomalies over oceanic ridges, *Mar. Geophys. Res., 1,* 111–144, 1971. **(15, 20, 21)**

Schubert, G., D. Yuen, and D. Turcotte, Role of phase transitions in a dynamic mantle, *Geophys. J. Roy. Astron. Soc., 42,* 705–735, 1975. **(33)**

Sclater, J. G., and R. L. Fisher, Evolution of the East Central Indian Ocean with emphasis on the tectonic setting of the ninety-east ridge, *Geol. Soc. Amer. Bull., 85,* 683–702, 1974. **(21)**

Sclater, J. G., and J. Francheteau, The implications of terrestrial heat flow observations on current tectonic and geochemical models of the crust and upper mantle of the earth, *Geophys. J. Roy. Astron. Soc., 20,* 509–542, 1970. **(2, 8, 16, 18, 36)**

Sclater, J. G., and K. D. Klitgord, A detailed heat flow, topographic, and magnetic survey across the Galapagos spreading center at 86°W, *J. Geophys. Res., 78,* 6951–6975, 1973. **(16)**

Sclater, J. G., and H. W. Menard, Topography and heat flow of the Fiji Plateau, *Nature, 216,* 991–993, 1967. **(13)**

Sclater, J. G., R. N. Anderson, and M. L. Bell, Elevation of ridges and evolution of the central eastern Pacific, *J. Geophys. Res., 76,* 7888–7915, 1971. **(3, 16, 19, 20, 21, 22, 24, 36)**

Sclater, J. G., R. D. Jarrard, and S. Gertner, Jr., Comparison of the magnetic and biostratigraphic time scales since the Late Cretaceous, in *Initial Reports of the Deep Sea Drilling Project,* vol. 22, C. C. von der Borch, J. G. Sclater et al., pp. 381–386, U.S. Govt. Printing Office, Washington, D.C., 1974. **(16, 21, 28)**

Sclater, J. G., L. A. Lawver, and B. Parson, Comparison of long wavelength residual elevation and free-air gravity anomalies in the North Atlantic and possible implications for the thickness of the lithospheric plate, *J. Geophys. Res., 80,* 1031–1052, 1975. **(21)**

Scott, K. M., Sedimentology and dispersal pattern of a Cretaceous flysch sequence, Patagonian Andes, southern Chile, *Amer. Assoc. Petrol. Geol. Bull., 50,* 72–107, 1966. **(26)**

Scrope, G. P., *Considerations on Volcanoes . . . ,* London, 1825; 2nd ed., 490 pp., 1862. **(27)**

Seeber, L., M. Barazangi, and A. Nowroozi, Microearthquake seismicity and tectonics of coastal northern California, *Bull. Seism. Soc. Amer., 60,* 1669–1699, 1970. **(13)**

Seely, D. R., P. R. Vail, and G. G. Walton, Trench slope model, in *The Geology of Continental Margins,* edited by C. A. Burk and C. L. Drake, pp. 243–260, Springer-Verlag, New York, 1974. **(8)**

Segawa, J., Gravity measurements at sea by use of the TSSG, Part 2, *J. Phys. Earth, 18,* 203–284, 1970. **(2, 3)**

Segawa, J., and C. Bowin, The Tokyo surface ship gravity meter: Recent developments and results of comparison measurements, *Geophysics, 40,* 246–255, 1975. **(3)**

Segawa, J., and Y. Tomoda, Gravity measurements near Japan and study of the upper mantle beneath the oceanic trench-marginal sea transition zones, this volume. **(2, 20)**

Segerstrom, K., Structural geology of an area east of Copiapo, Atacama Province, Chile, *21st Int. Geol. Cong., part XVIII,* 14–20, 1960. **(26)**

Selig, F., A theoretical prediction of salt dome pattern, *Geophysics, 30,* 633–643, 1965. **(29)**

Shackleton, N. J., and J. P. Kennett, Paleotemperature history of the Cenozoic and the initiation of Antarctic glaciations: Oxygen and carbon isotope analyses at DSDP sites 277, 279, and 281, in *Initial Reports of the Deep Sea Drilling Project,* vol. 29, J. P. Kennett, R. E. Houtz et al., pp. 743–755, U.S. Govt. Printing Office, Washington, D.C., 1975. **(24)**

Shaw, H. R., Rheology of basalt in the melting range, *J. Petrol., 10,* 510–535, 1969. **(32)**

Shaw, H. R., Mantle convection and volcanic periodicity in the Pacific: Evidence from Hawaii, *Geol. Soc. Amer. Bull., 84,* 1505–1526, 1973. **(22, 28)**

Shaw, H. R., and E. D. Jackson, Linear island chains in the Pacific: Result of thermal plumes or gravitational anchors?, *J. Geophys. Res., 78,* 8634–8652, 1973. **(28)**

Shaw, H. R., T. L. Wright, D. L. Peck, and R. Okamura, The viscosity of basaltic magma: An analysis of field measurements in Makaopuhi lava lake, Hawaii, *Amer. J. Sci., 266,* 225–264, 1968. **(32)**

Shepherd, E. S., Temperature of the fluid lava of Halemaumau, July 1911, *Hawaiian Volcano Observatory, 1st Rep.,* 47–51, 1914. **(27)**

Shibata, Y., and Y. Okuda, K–Ar ages of a granitic fragment dredged from the 2nd Komahashi seamount, *Bull. Geol. Sur. Japan, 26,* 71–72, 1975. **(20)**

Shido, F., and A. Miyashiro, Compositional difference between abyssal tholeiites from north and south of the Azores on the Mid-Atlantic ridge, *Nature Phys. Sci., 245,* 59–60, 1973. **(30)**

Shiki, T., H. Aoki, H. Suzuki, M. Musashino, and Y. Okuda, Geological and petrographical results of the GDP-8 cruise in the Philippine Sea (in Japanese with English abstract), *Mar. Sci., 6*(8), 51–55, 1974. **(20)**

Shimamura, H., Plate thickness in the Kuril–Kamchatka region and earthquake distribution within the plate (in Japanese), *Abstracts for 1973 Meeting of the Seismological Society of Japan, No. 1,* 118, 1973. **(36)**

Shimamura, H., and T. Asada, Ocean-bottom seismograph as a tool to study the arc–trench system (in Japanese), in *Island Arc and Marginal Sea, Proceedings of the 1st Japan–USSR Symposium,* edited by S. Asano and G. B. Udintsev, pp. 187–199, Tokai University Press, 1971*a*. **(12, 13)**

Shimamura, H., and T. Asada, Explosion seismological observation by ocean bottom seismographs, *Preliminary report of the Hakuho Maru cruise KH 71-1,* pp. 11–39, Ocean Res. Inst., Tokyo Univ., Tokyo, 1971*b*. **(12)**

Shimamura, H., and T. Asada, A cassette recorder for an ocean-bottom seismometer, *Geophys. Bull. Hokkaido Univ., 32,* 17–24, 1974. **(12)**

Shimamura, H., T. Asada, and K. Takano, Measurements of micro-earthquakes at sea bottom, *La Mer, 8,* 6–12, 1970. **(12)**

Shimamura, H., Y. Tomada, and T. Asada, Seismographic observation at the bottom of the central basin fault of the Philippine Sea, *Nature, 253,* 177–179, 1975. **(12, 21)**

Shimazaki, K., Focal mechanism of a shock at the northeastern boundary of the Pacific plate: Extensional feature of the oceanic lithosphere and compressional feature of the continental lithosphere, *Phys. Earth Planet. Interiors, 6,* 397–404, 1972. **(3)**

Shor, G. G., and D. D. Pollard, Mohole site selection studies north of Maui, *J. Geophys. Res., 69,* 1627–1637, 1964. **(2)**

Shor, G. G., Jr., H. K. Kirk, and H. W. Menard, Crustal structure of the Melanesian area, *J. Geophys. Res., 76,* 2562–2586, 1971. **(13)**

Shor, G. G., Jr., H. W. Menard, and R. W. Raitt, Structure of the Pacific basin, in *The Sea,* vol. 4, part 2, edited by A. E. Maxwell, pp. 3–28, Wiley-Interscience, New York, 1970. **(8)**

Shortley, G., and D. Williams, *Elements of Physics,* 2nd ed., Prentice-Hall, Englewood Cliffs, N.J., 1955. **(5)**

Silberling, N. J., Geologic events during Permian–Triassic time along the Pacific margin of the United States, in *The Permian and Triassic Systems and Their Mutual Boundary,* edited by B. W. Logan and L. V. Hills, *Alberta Soc. Petrol. Geol. Mem., 2,* pp. 345–362, 1973. **(26)**

Sillitoe, R. H., Relation of metal provinces in western America to subduction of oceanic lithosphere, *Geol. Soc. Amer. Bull., 83,* 813–818, 1972. **(8)**

Silver, E. A., Small plate tectonics in the northeastern Pacific, *Geol. Soc. Amer. Bull., 82,* 3491–3496, 1971. **(16)**

Silverberg, E. C., Proposal to National Aeronautics and Space Administration for construction of a transportable lunar laser ranging facility, McDonald Observatory, Univ. Texas, 1973. **(5)**

Silverberg, E. C., Report on the lunar ranging at McDonald Observatory for the period June 20, 1974 to October 15, 1974, *Res. Memorandum Astron. 74-007,* Univ. Texas at Austin, Texas, 1974. **(5)**

Simotomai, H., Nachrichten über den Ausbruch des Vulkans Usu in Japan im Jahre 1910, *Gesell. Erdkunde Berlin Z., 12,* 705–710, 1911. **(27)**

Simotomai, H., Der Tarumai–Ausbruch in Japan, 1909, *Gesell. Erdkunde Berlin Z., 13,* 433–444, 1912. **(27)**

Simpson, D. W., *P* wave velocity structure of the upper mantle in the Australian region, Ph.D. dissertation, Australian National Univ., 1973. **(33)**

Siry, J. W., Satellite techniques for the study of secular variations in position, *Proc. Symp. Earth's Gravitational Field and Secular Variations in Position,* the School of Surveying, Univ. New South Wales, Sydney, Australia, 1974. **(5)**

Skinner, B. J., and D. L. Peck, An immiscible sulfide melt from Hawaii, *Economic Geol. Monogr. 4,* 310–322, 1969. **(32)**

Sleep, N. H., Sensitivity of heat flow and gravity to the mechanism of sea-floor spreading, *J. Geophys. Res., 74,* 542–549, 1969. **(36)**

Sleep, N. H., and M. N. Toksöz, Evolution of marginal basins, *Nature, 233,* 548–550, 1971. **(15, 20)**

Smith, A. L., and I. S. E. Carmichael, Quaternary trachybasalts from southeastern California, *Amer. Mineral., 54,* 909–923, 1969. **(29)**

Smith, E. K., Temperate-latitude sporadic E, in *Physics of Geomagnetic Phenomena,* edited by S. Matsushita and W. H. Campbell, pp. 615–623, Academic Press, New York, 1967. **(16)**

Smith, J. V., and B. Mason, Pyroxene–garnet transformation in Coorara meteorite, *Science, 168,* 832–833, 1970. **(34)**

Solomon, S., and S. Biehler, Crustal structure from gravity anomalies in the southwest Pacific, *J. Geophys. Res., 74,* 6696–6701, 1969. **(13)**

Solomon, S. C., and N. H. Sleep, Some simple physical models for absolute plate motions, *J. Geophys. Res., 79,* 2557–2567, 1974. **(28)**

Sondergeld, C., B. Isacks, and M. Barazangi, Velocities of *P* and *S* waves at depths of 450 to 650 km inside the lithospheric slab descending beneath the Tonga island arc (abstract), *Eos Trans. AGU, 55,* 349, 1974. **(33)**

Spetzler, H., Equation of state of polycrystalline and single-crystal MgO to 8 kilobars and 800°K, *J. Geophys. Res., 75,* 2073–2087, 1970. **(35)**

Stanley, K. O., Tectonic and sedimentologic history of Lower Jurassic Sunrise and Dunlap formations, West-Central Nevada, *Amer. Assoc. Petrol. Geol. Bull., 55,* 454–477, 1971. **(26)**

Stansell, T. A., W. A. Bourguignon, C. R. Karl, and P. C. Marth, Geoceiver: An integrated Doppler geodetic receiver, *TG-710,* Applied Physics Laboratory, Johns Hopkins Univ., Silver Spring, Md, 1965. **(5)**

Stauder, W., Mechanism of the Rat Island earthquake sequence of February 4, 1965, with relation to island arcs and sea-floor spreading, *J. Geophys. Res., 73,* 3847–3858, 1968*a*. **(9)**

Stauder, W., Tensional character of earthquake foci beneath the Aleutian trench with relation to sea-floor spreading, *J. Geophys. Res., 73,* 7693–7701, 1968*b*. **(3, 9)**

Stauder, W., Mechanism and spatial distribution of Chilean earthquakes with relation to subduction of the oceanic plate, *J. Geophys. Res., 78,* 5033–5061, 1973. **(9)**

Stauder, W., Subduction of Nazca plate under Peru as evidenced by focal mechanism and by seismicity, *J. Geophys. Res., 80,* 1053–1064, 1975. **(9)**

Stearns, H. T., Origin of Haleakala Crater, island of Maui, Hawaii, *Bull. Geol. Soc. Amer., 53,* 1–14, 1942. **(27)**

Stearns, H. T., Geology of the Hawaiian Islands, *Hawaii Div. Hydrog. Bull., 8,* 105 pp., 1946. **(27)**

Stearns, H. T., *Geology of the State of Hawaii,* p. 266, Pacific Books, Palo Alto, Calif., 1966. **(28)**

Stearns, H. T., Potassium–argon ages of lavas from the Hawi and Polulu volcanic series, Kohala volcano, Hawaii: Discussion, *Geol. Soc. Amer. Bull., 84,* 3483–3484, 1973. **(28)**

Stearns, H. T., and K. N. Vaksvik, Geology and groundwater resources of the island of Oahu, Hawaii, *Hawaii Div. Hydrog. Bull., 1,* 479 pp., 1935. **(27)**

Stehn, C., The geology and volcanism of the Krakatau group, *Guidebook, Fourth Pacific Sci. Cong., Batavia,* 1–55, 1929. **(27)**

Steinmetz, L., and A. Hirn, Sub-Moho stratification from an oceanic long-range profile, *Eos Trans. AGU, 55,* 360, 1974. **(36)**

Stewart, J. H., Initial deposits in the cordilleran geosyncline: Evidence of a Late Precambrian (<850 m.y.) continental separation, *Geol. Soc. Amer. Bull., 83,* 1345–1360, 1972. **(26)**

Stewart, R. J., J. H. Natland, and W. R. Glassby, Petrology of volcanic rocks from the North Pacific Ocean and the Bering Sea, in *Initial Reports of the Deep Sea Drilling Project,* vol. 19, J. S. Creager, J. S. Scholl et al., pp. 615–628, U.S. Govt. Printing Office, Washington, D.C., 1973. **(30)**

St. John, V. P., The gravity field and structure of Papua and New Guinea, *APEA J., 10,* 41–55, 1970. **(35)**

Stoiber, R. E., and W. I. Rose, Jr., Recent volcanic and fumarolic activity at Santiaguito volcano, Guatemala, *Bull. Volcanol., Ser. 2, 33,* 475–502, 1969. **(27)**

Strange, W. E., G. P. Woollard, and J. C. Rose, An analysis of the gravity field over the Hawaiian Islands in terms of crustal structure, *Pacific Sci., 19,* 381–389, 1965. **(2)**

Strobach, K., Curvature of island arcs and plate tectonics, *Z.f. Geophysik, 39,* 819, 1973. **(9)**

Strzeleki, P. E., Count de, Sandwich Islands—Crater of Kiraueа, Hawaii, *Hawaiian Spectator, 1,* 435, 1838. **(27)**

Stübel, A., *Die Vulkanberge von Ecuador,* 557 pp., Berlin, 1897. **(27)**

Stübel, A., *Die Vulkanberge von Colombia,* Dresden, 1906. **(27)**

Sugimura, A., and S. Uyeda, *Island Arcs, Japan and Its Environs,* Elsevier Scientific Publishing Company, Amsterdam, 1972. **(20)**

Sugimura, A., T. Matsuda, K. Chinzei, and K. Nakamura, Quantitative distribution of late Cenozoic volcanic material in Japan, *Bull. Volcanol., 26,* 125–140, 1963. **(20)**

Suito, K., Phase transformations of pure $Mg_2SiO_4$ into a spinel structure under high pressures and temperatures, *J. Phys. Earth, 20,* 225–243, 1972. **(34)**

Suppe, J., Offset of late Mesozoic basement terrains by the San Andreas fault system, *Geol. Soc. Amer. Bull., 81,* 3253–3258, 1970. **(26)**

Sutton, G. H., G. L. Maynard, and D. M. Hussong, Wide-spread occurrence of a high-velocity basal crustal layer in the Pacific crust found with repetitive sources and sonobuoys, in *The Structure and Physical Properties of the Earth's Crust, Geophys. Monogr. Ser.,* vol. 14, edited by J. G. Heacock, pp. 193–209, AGU, Washington, D.C., 1971. **(8, 35)**

Sutton, G. H., M. E. Odegard, and D. M. Hussong, Telemetering ocean-bottom seismographs (abstract), *Earthquake Notes, 44,* 12, 1973. **(13)**

Sutton, G. H., W. G. McDonald, D. D. Prentiss, and

S. N. Thanos, Ocean-bottom seismic observatories, *Proc. IEEE, 53,* 1909–1921, 1965. **(13)**

Swanson, D. A., Magma supply rate at Kilauea volcano, 1952–1971, *Science, 175,* 169–170, 1972. **(28)**

Swanson, D. A., W. A. Duffield, D. B. Jackson, and D. W. Peterson, The complex filling of Alae Crater, Kilauea volcano, Hawaii, *Bull. Volcanol., 36,* 105–126, 1972. **(32)**

Swe, W., and W. R. Dickinson, Sedimentation and thrusting of late Mesozoic rocks in the Coast Ranges near Clear Lake, California, *Geol. Soc. Amer. Bull., 81,* 165–188, 1970. **(26)**

Sykes, L. R., The seismicity of deep structures of island arcs, *J. Geophys. Res., 71,* 2981–3006, 1966. **(9)**

Sykes, L. R., Seismicity as a guide to global tectonics and earthquake prediction, *Tectonophysics, 13,* 393, 1972. **(9)**

Sykes, L. R., B. L. Isacks, and J. Oliver, Spatial distribution of deep and shallow earthquakes of small magnitudes in the Tonga–Fiji region, *Bull. Seism. Soc. Amer., 59,* 1093–1112, 1969. **(13)**

Talwani, M., Developments in navigation and measurement of gravity at sea, *Geoexploration, 8,* 151–183 1970. **(2)**

Talwani, M., W. P. Early, and D. E. Hayes, Continuous analog computation and recording of cross-coupling and off-levelling errors, *J. Geophys. Res., 71,* 2079–2090, 1966. **(2)**

Talwani, M., X. Le Pichon, and M. Ewing, Crustal structure of the mid-ocean ridges, 2, computed model from gravity and seismic refraction data, *J. Geophys. Res., 70,* 341–352, 1965. **(36)**

Talwani, M., W. C. Pitman, III, and J. R. Heirtzler, Magnetic anomalies in the North Atlantic (abstract), *Trans. AGU, 50,* 189, 1969. **(26)**

Talwani, M., G. H. Sutton, and J. L. Worzel, Crustal section across the Puerto Rico trench, *J. Geophys. Res., 64,* 1545–1555, 1959*a*. **(1)**

Talwani, M., J. L. Worzel, and M. Landisman, Rapid gravity computation for two-dimensional bodies with application to the Mendocino submarine fracture zone, *J. Geophys. Res., 64,* 49–59, 1959*b*. **(10, 35)**

Talwani, M., C. C. Winsich, and M. G. Langseth, Reykjanes ridge crest: A detailed geophysical study, *J. Geophys. Res., 76,* 473–517, 1971. **(16, 19)**

Talwani, M., J. L. Worzel, and M. Ewing, Gravity anomalies and crustal section across the Tonga trench, *J. Geophys. Res., 66,* 1265–1278, 1961. **(1, 2)**

Tanakadate, H., Physiographic development of the Tarumai dome in Japan, *Amer. J. Sci., Ser. 4, 44,* 87–97, 1917. **(27)**

Tanakadate, H., Two types of volcanic dome in Japan, *Proc. Fourth Pacific Sci. Cong., Batavia, 2b,* 695–704, 1930. **(27)**

Tanakadate, H., Volcanic activity in Japan and vicinity during the period between 1924 and 1931, *Japanese J. Astron. Geophys., 9,* 47–61, 1931. **(27)**

Tanakadate, H., Volcanic activity in Japan during the period between June 1931 and June 1934, *Japanese J. Astron. Geophys., 12,* 89–108, 1934. **(27)**

Tanakadate, H., Volcanic activity in Japan during the period between July 1934 and Ocotber 1935, *Japanese J. Astron. Geophys., 13,* 121–139, 1936. **(27)**

Tanakadate, H., Volcanoes in the Mariana Islands in the Japanese mandated south seas, *Bull. Volcanol., Ser. 2, 6,* 199–223, 1940. **(27)**

Taylor, R., *Te Ika a Maui, or New Zealand and its Inhabitants,* London, 1855. **(27)**

Tazieff, H., New investigations on eruptive gases, *Bull. Volcanol., Ser. 2, 34,* 421–438, 1971. **(27)**

Thanos, S. N., and A. C. Hubbard, Two-way hydroacoustic communications link for an ocean-bottom seismograph, *IEEE, GE4,* 17–24, 1966. **(13)**

Theyer, F., and S. R. Hammond, Cenozoic magnetic time scale in deep-sea cores: Completion of the Neogene, *Geology, 2,* 487–492, 1974. **(19)**

Thomson, J. T., and W. A. Schneider, An automatic marine seismic monitoring and recording device, *Proc. IRE, 50,* 2209–2216, 1962. **(13)**

Tilley, C. E., and Y. S. Yoder, Jr., Pyroxene fractionation in mafic magma at high pressures and its bearings on basalt genesis, *Carnegie Inst. Washington Yearb., 63,* 114–121, 1964. **(36)**

Tilley, C. E., H. S. Yoder, Jr., and J. F. Schairer, Melting relations of volcanic tholeiite and alkali rock series, *Ann. Rept. Dir. Geophys. Lab., Carnegie Inst. Washington Yearb., 64,* 69–82, 1965. **(32)**

Tilley, C. E., H. S. Yoder, Jr., and J. F. Schairer, Melting relations of volcanic rock series, *Ann. Rept. Dir. Geophys. Lab., Carnegie Inst. Washington Yearb., 65,* 260–269, 1966. **(32)**

Tilling, R. I., R. Y. Koyanagi, and R. T. Holcomb, Rockfall seismicity-correlation with field observations, Makaopuhi Crater, Kilauea volcano, Hawaii, *U.S. Geol. Surv., J. Res., 3,* in press. **(14)**

Toksöz, M. N., J. W. Minear, and B. R. Julian, Temperature field and geophysical effects of a downgoing slab, *J. Geophys. Res., 76,* 1113–1138, 1971. **(3, 33)**

Toksöz, M. N., N. H. Sleep, and T. A. Smith, Evolution of the downgoing lithosphere and mechanisms of deep focus earthquakes, *Geophys. J., 35,* 285–310, 1973. **(29)**

Tomoda, Y., *Maps of Free Air and Bouguer Gravity Anomalies in and Around Japan,* Univ. Tokyo Press, 1973*a*. **(2, 3, 20)**

Tomoda, Y., Gravity anomalies in the Pacific Ocean, in *The Western Pacific,* edited by P. J. Coleman, pp. 5–20, Univ. Western Australia Press, 1973*b*. **(2)**

Tomoda, Y., and H. Kanamori, Tokyo surface ship gravity meter α-1, *J. Geodetic Soc. Japan, 7,* 116–145, 1962. **(2, 3)**

Tomoda, Y., and N. Nasu (editors), *Preliminary Report R. V. Hakuho Maru Cruise KH-69-2,* Ocean Res. Inst., Univ. Tokyo, 1971. **(20)**

Tomoda, Y., K. Ozawa, and J. Segawa, Measurement of gravity and magnetic force on board a cruising vessel, *Bull. Ocean Res. Inst., Univ. Tokyo, 3,* 1–170, 1968. **(20)**

Tomoda, Y., K. Kobayashi, J. Segawa, and M. Nomura, Magnetic anomalies in the Shikoku basin, northern Philippine Sea, paper presented at the IAGA general scientific assembly at Kyoto, Sept. 1973; abstract in *IAGA Bull., 34,* 1973. **(20)**

Tomoda, Y., K. Kobayashi, J. Segawa, M. Nomura, K. Kimura, and T. Saki, Linear magnetic anomalies in the Shikoku basin, northeastern Philippine Sea, *J. Geomag. Geoelect., 28,* 47–56, 1975. **(15, 20)**

Tozer, D. C., Thermal plumes in the Earth's mantle, *Nature, 244,* 398–400, 1973. **(28)**

Trunin, R. F., V. I. Gon'Shakova, G. U. Simakov, and N. E. Galdin, A study of rocks under the high pressures and temperatures created by shock compression, *Izv. Phys. Solid Earth,* in Russian, *9,* 1–12, 1965; English transl., *9,* 579–586, 1966. **(33)**

Tsuboi, C., Crustal structure along a certain profile across the East Indies as deduced by a new calculation method, in *Verhandelingen van het Koninklijk Nederlandsch Geologisch-Mijnbouwkundig Genootschap,* N. V. Boek-EN Kunstdrukkerij V/H Mouton & Co., 'S-Gravenhage, 1957. **(1)**

Turcotte, D. L., and E. R. Oxburgh, A fluid theory for the deep structure of dip-slip fault zones, *Phys. Earth Planet. Interiors, 1,* 381–386, 1968. **(29)**

Turcotte, D. L., and G. Schubert, Structure of the olivine–spinel phase boundary in the descending lithosphere, *J. Geophys. Res., 76,* 7980–7987, 1971. **(33)**

Turcotte, D. L., and G. Schubert, Frictional heating of the descending lithosphere, *J. Geophys. Res., 78,* 5876–5886, 1973. **(29, 33)**

Turnbull, J., *A Voyage Around the World,* vol. 2, pp. 88–89, London, 1805. **(27)**

Turner, D. L., R. B. Forbes, and C. W. Naeser, Radiometric ages of Kodiak seamount and Giacomini seamount in the Gulf of Alaska: Implications for circum-Pacific tectonics, *Science, 182,* 589–591, 1973. **(28)**

Turner, D. L., R. B. Forbes, and C. W. Naeser, Geochronology and genesis of the Kodiak–Bowie seamount chain in the Gulf of Alaska (abstract), *Eos Trans. AGU, 56*(12), 1186, 1974. **(28)**

Turner, G., J. A. Miller, and R. L. Grasty, The thermal history of the Bruderheim meteorite, *Earth Planet. Sci. Lett., 1,* 155–157, 1966. **(31)**

U.S. Department of Commerce, NOAA, Bathymetry, geomagnetic, and gravity data, *International Indian Ocean Expedition USC&GS Ship Pioneer 1964, 3,* 1969. **(3)**

U.S. Geological Survey, Set of 7 colored slides illustrating crystallization of tholeiitic basalt in Makaopuhi lava lake, Hawaii, U.S. Geol. Surv. Photographic Library, Denver Federal Center, Denver, Colorado, 1967. **(32)**

Unger, J. D., Seismic evidence for magma intrusion in the eastern Koae fault system, Kilauea volcano, Hawaii (abstract), *Eos Trans. AGU, 54,* 1216, 1974.

Unger, J. D., and P. L. Ward, Traveltime delays and tectonic stress from a large, subcrustal Hawaiian earthquake (abstract), *Eos Trans. AGU, 56,* 1150, 1974. **(14)**

Uspensky, D. G., Geological interpretation of gravity data for a profile across the Pacific Ocean, *J. Geophys. Res., 77,* 6316–6326, 1972. **(26)**

Utsu, T., Anomalies in seismic wave velocity and attenuation associated with a deep earthquake zone (1), *J. Fac. Sci. Hokkaido Univ., Ser. 7, 3,* 1–25, 1967. **(3)**

Utsu, T., Anomalous structure of the upper mantle beneath the Japanese Islands, *J. Fac. Sci. Hokkaido Univ., Ser. 7, 7,* 99–127, 1971*a.* **(3)**

Utsu, T., Seismological evidence for the anomalous structure of island arcs with special reference to the Japanese region, *Rev. Geophys. Space Phys., 9,* 839–890, 1971*b.* **(20)**

Uyeda, S., Heat flow, in *The Crust and Upper Mantle of the Japanese Area,* part I, Jap. Comm. UMP, pp. 97–105, 1972. **(20)**

Uyeda, S., and Z. Ben-Avraham, Origin and development of the Philippine Sea, *Nature, 240,* 176–178, 1972. **(15, 18, 20, 21)**

Uyeda, S., and A. Miyashiro, Plate tectonics and the Japanese Islands: A synthesis, *Geol. Soc. Amer. Bull., 85,* 1159–1170, 1974. **(15, 20, 21)**

Uyeda, S., and M. Richards, Magnetization of four Pacific seamounts near the Japanese Islands, *Bull. Earthquake Res. Inst., 44,* 179–213, 1966. **(15, 18)**

Uyeda, S., and V. Vacquier, Geothermal and geomagnetic data in and around the island of Japan, in *The Crust and Upper Mantle of the Pacific Area, Geophys. Monogr. Ser.,* vol. 12, edited by L. Knopoff, C. L. Drake, and P. J. Hart, pp. 349–366, AGU, Washington, D.C., 1968. **(18)**

Uyeda, S., M. Yasui, K. Horai, and T. Yabe, Report on geomagnetic survey during JEDS-4 Cruise, *Oceanogr. Mag., 13,* 167–183, 1962. **(18)**

Uyeda, S., V. Vacquier, M. Yasui, J. Sclater, T. Sato, J. Lawson, T. Watanabe, F. Dixon, E. Silver, Y. Fukao, K. Sudo, N. Nishikawa, and T. Tanaka, Results of geomagnetic survey during the cruise of R/V ARGO in western Pacific 1966 and the compilation of magnetic charts of the same area, *Bull. Earthquake Res. Inst., 45,* 799–814, 1967. **(15, 18)**

Vacquier, V., and S. Uyeda, Paleomagnetism of nine seamounts in the western Pacific and of three volcanoes in Japan, *Bull. Earthquake Res. Inst., 45,* 815–848, 1967. **(15, 18)**

Vacquier, V., A. D. Raff, and R. E. Warren, Horizontal displacements in the floor of the northeastern Pacific Ocean, *Geol. Soc. Amer. Bull., 72,* 1251–1258, 1961. **(15, 16)**

van Andel, Tj. H., Cenozoic migration of the Pacific plate, northward shift of the axis of deposition and paleobathymetry of the central equatorial Pacific, *Geology, 2,* 507–510, 1974. **(24, 28)**

van Andel, Tj. H., and D. Bukry, Basement ages and basement depths in the eastern equatorial Pacific from Deep Sea Drilling Project legs 5, 8, 9, and 16, *Geol. Soc. Amer. Bull., 84,* 2361–2369, 1973. **(22, 24, 28)**

van Andel, Tj. H., and G. R. Heath, Geological results of Leg 16: The central equatorial Pacific west of the East Pacific rise, in *Initial Reports of the Deep Sea Drilling Project,* vol. *16,* Tj. H. van Andel, G. R. Heath et al., pp. 937–949, U.S. Govt. Printing Office, Washington, D.C., 1973. **(24)**

van Andel, Tj. H., and T. C. Moore, Jr., Cenozoic calcium carbonate distribution and calcite compensation depth in the central equatorial Pacific Ocean, *Geology, 2,* 87–92, 1974. **(22, 24)**

van Andel, Tj. H., G. R. Heath, and T. C. Moore, Jr., Cenozoic history and palaeoceanography of the central equatorial Pacific Ocean, *Geol. Soc. Amer. Mem., 143,* 134 pp., 1975. **(24)**

van Andel, Tj. H., J. D. Phillips, and R. P. Von Herzen, Rifting origin for the Vema fracture in the North Atlantic, *Earth Planet. Sci. Lett., 5,* 269–300, 1969. **(26)**

van Andel, Tj. H., G. R. Heath, B. T. Malfait, D. F. Heinrichs, and J. I. Ewing, Tectonics of the Panama basin, eastern equatorial Pacific, *Geol. Soc. Amer. Bull., 82,* 1489–1508, 1971. **(10, 16)**

van Andel, Tj. H., G. R. Heath et al., *Initial Reports of the Deep Sea Drilling Project,* vol. 16, 949 pp., U.S. Govt. Printing Office, Washington, D.C., 1973. **(22)**

Vancouver, G., *A Voyage of Discovery to the North Pa-*

*cific Ocean, 3,* 8, 1798. **(27)**

Van Der Held, E. F. M., Currents inside the earth, *Proc. Kon. Ned. Akad. van westenschappen-Amsterdam, Ser. B, 72,* 256, 1969. **(9)**

Van Der Linden, W. J. M., Extinct mid-ocean ridges in the Tasman Sea and in the western Pacific, *Earth Planet. Sci. Lett., 6,* 483–490, 1969. **(15)**

Vening Meinesz, F. A., Gravity over the Hawaiian Archipelago and over the Madeira area; conclusions about the earth's crust, *Proc. Kon. Ned. Akad. V. Wetens., 44,* 1, 1941. **(2)**

Vening Meinesz, F. A., Gravity expeditions at sea 1923–1938, vol. IV, *Publ. Netherlands Geod. Comm.,* 233 pp., 1948. **(2)**

Verbeek, R. D. M., *Krakatau,* 495 pp., Batavia, Java, 1886. **(27)**

Vesanen, E., H. V. Tuominen, U. Luosto, and M. L. Maki, The spatial distribution of earthquakes in Kurile–Okhotsk region, *Bull. Int. Inst. Seism. Eng., 10,* 1973. **(9)**

Vestine, E. H., Main geomagnetic field, in *Physics of Geomagnetic Phenomena,* edited by S. Matsushita and W. H. Campbell, Academic Press, New York, 1967. **(21)**

Vincent, P., Le volcanisme ignimbritique du Tibesti (Sahara tchadien), Essai d'interpretation dynamique, *Bull. Volcanol., Ser. 2, 26,* 259–272, 1963. **(27)**

Vine, F. J., Spreading of the ocean floor: New evidence, *Science, 154,* 1405–1415, 1966. **(16, 18, 28)**

Vine, F. J., and D. H. Matthews, Magnetic anomalies over oceanic ridges, *Nature, 199,* 947–949, 1963. **(15, 16, 18, 19, 28)**

Vine, F. J., and E. M. Moores, A model for the gross structure, petrology, and magnetic properties of oceanic crust, in *Studies in Earth and Space Sciences, Geol. Soc. Amer. Mem. 132,* edited by R. Shagam, R. B. Hargraves et al., pp. 195–205, 1972. **(19, 35)**

Vine, F. J., and J. T. Wilson, Magnetic anomalies over a young oceanic ridge off Vancouver Island, *Science, 150,* 485–489, 1965. **(15)**

Vogt, P. R., Volcano spacing, fractures, and thickness of the lithosphere, *Earth Planet. Sci. Lett., 21,* 235–252, 1974*a*. **(36)**

Vogt, P. R., Volcano height and plate thickness, *Earth Planet. Sci. Lett., 23,* 337–348, 1974*b*. **(36, 37)**

Vogt, P. R., and G. L. Johnson, Magnetic telechemistry of oceanic crust, *Nature, 254,* 373–375, 1973. **(30)**

Vogt, P. R., O. E. Avery, E. D. Schneider, C. N. Anderson, and D. R. Bracey, Discontinuities in sea-floor spreading, *Tectonophysics, 8,* 285–317, 1969. **(16)**

Vogt, P. R., G. L. Johnson, T. L. Holcombe, J. G. Gilg, and O. E. Avery, Episodes of sea-floor spreading recorded by the North Atlantic basement, *Tectonophysics, 12,* 211–234, 1971*a*. **(16)**

Vogt, P. R., C. N. Anderson, and D. R. Bracey, Mesozoic magnetic anomalies, sea-floor spreading, and geomagnetic reversals in the southwestern North Atlantic, *J. Geophys. Res., 76,* 4796–4823, 1971*b*. **(18, 21)**

Von Arx, W., D. C. Bowen, J. P. Dean, and R. T. Haworth, Hudson 70 gravity observations 62.9°S to 57.5°N along 150°W, *Woods Hole Report 71-68,* 1971. **(2)**

von der Borch, C. C., J. G. Sclater, S. Gartner, Jr., R. Hekinian, D. A. Johnson, B. McGowran, A. C. Pimm, R. W. Thompson, J. J. Veevers, and L. S. Waterman, Site 212, in *Initial Reports of the Deep Sea Drilling Project,* vol. 22, C. C. von der Borch, J. G. Sclater et al., pp. 37–83, U.S. Govt. Printing Office, Washington, D.C., 1974*a*. **(30)**

von der Borch, C. C., J. G. Sclater, S. Gartner, Jr., R. Hekinian, D. A. Johnson, B. McGowran, A. C. Pimm, R. W. Thompson, J. J. Veevers, and L. S. Waterman, Site 215, in *Initial Reports of the Deep Sea Drilling Project,* vol. 22, C. C. von der Borch, J. G. Sclater et al., pp. 193–211, U.S. Govt. Printing Office, Washington, D.C., 1974*b*. **(30)**

Voss, J. T., An empirical method for the determination of gravity terrain corrections, M.S. thesis, Univ. Hawaii, 1974. **(35)**

Wadati, K., T. Hirono, and T. Yumura, On the attenuation of *S* waves and the structure of the upper mantle in the region of Japanese Islands, *Papers Meteor. Geophys., 20,* 49–78, 1969. **(3)**

Wade, U. S., Jr., B. T. R. Lewis, and C. R. B. Lister, Structure near Malpelo ridge from seismic refraction results (abstract), *Eos Trans. AGU, 56,* 1146, 1974. **(10)**

Wadley, T. L., The Tellurometer system of distance measurement, *Empire Survey Review, 14*(105 and 106), 1957. **(5)**

Waitz, P., Le volcan de Colima, *Tenth Int. Geol. Cong., Mexico, Guide to Exc. 13,* 28 pp., 1906. **(27)**

Waitz, P., Popocatepetl again in activity, *Amer. J. Sci., Ser. 5, 1,* 81–87, 1921. **(27)**

Walcott, R. I., Flexure of the lithosphere at Hawaii, *Tectonophysics, 9,* 435–446, 1970*a*. **(2, 37)**

Walcott, R. I., Flexural rigidity, thickness and viscosity of the lithosphere, *J. Geophys. Res., 75,* 3941–3954, 1970*b*. **(2, 37)**

Walcott, R. I., Isostatic response to crustal loading in Canada, *Can. J. Earth Sci., 7,* 716–727, 1970*c*. **(37)**

Wang, Chi-Yuen, Density and constitution of the mantle, *J. Geophys. Res., 75,* 3264–3284, 1970. **(1)**

Wang, C. Y., Pressure coefficients of compressional wave velocity for a bronzitite, *J. Geophys. Res., 79,* 689–776, 1974. **(35)**

Ward, P. L., and S. Gregersen, Comparison of earthquake locations determined with data from a network of stations and small tripartite arrays on Kilauea volcano, Hawaii, *Bull. Seism. Soc. Amer., 63,* 679–711, 1973. **(14)**

Watkins, N. D., Review of the development of the geomagnetic polarity time scale and discussion of prospects for its finer definition, *Geol. Soc. Amer. Bull., 83,* 551–574, 1972. **(16)**

Watts, A. B., and J. R. Cochran, Gravity anomalies and flexure of the lithosphere along the Hawaiian–Emperor seamount chain, *Geophys. J. Roy. Astron. Soc., 38,* 119–141, 1974. **(2, 37)**

Watts, A. B., and M. Talwani, Gravity anomalies seaward of deep-sea trenches and their tectonic implications, *Geophys. J. Roy. Astron. Soc., 36,* 57–90, 1974. **(2, 3, 21)**

Watts, A. B., and M. Talwani, Gravity effect of downgoing lithospheric slabs beneath island arcs, *Geol. Soc. Amer. Bull., 86,* 1–4, 1975*a*. **(2)**

Watts, A. B., and M. Talwani, Gravity field of the northwest Pacific Ocean basin and its margin: Hawaii and vicinity, *Geol. Soc. Amer. Spec. Map Ser.,* MC-9, 1975*b*. **(2)**

Watts, A. B., and J. K. Weissel, Tectonic history of the Shikoku marginal basin, *Earth Planet. Sci. Lett., 25,*

239–250, 1975. **(15, 20)**

Weidner, D. J., Rayleigh wave phase velocities in the Atlantic Ocean, *Geophys. J., 36,* 105–139, 1974. **(36)**

Wentworth, C. K., Ash formations of the island Hawaii, *Hawaiian Volc. Res. Assoc., Hawaiian Volc. Obs. Spec. Rep. 3,* 183 pp., 1938. **(27)**

Westerveldt, J., Quaternary volcanism on Sumatra, *Bull. Geol. Soc. Amer., 63,* 561–594, 1952. **(27)**

Whitcomb, J. H., and D. L. Anderson, Reflection of *P'P'* seismic waves from discontinuities in the mantle, *J. Geophys. Res., 75,* 5713–5728, 1970. **(33, 34)**

Whitehead, J. A., Jr., and D. S. Luther, Dynamics of laboratory diapir and plume models, *J. Geophys. Res., 80,* 705–717, 1975. **(29)**

Whitmarsh, R. B., An ocean-bottom pop-up seismic recorder, *Marine Geophys. Res., 1,* 91–98, 1970. **(13)**

Whitmarsh, R. B., Seismic anisotropy of the uppermost mantle absent beneath the east flank of the Reykjanes ridge, *Bull. Seism. Soc. Amer., 61,* 1351–1369, 1971. **(13)**

Whitmarsh, R. B., Some aspects of plate tectonics in the Arabian Sea, in *Initial Reports of the Deep Sea Drilling Project,* vol. 23, R. B. Whitmarsh, O. E. Weser et al., pp. 527–536, U.S. Govt. Printing Office, Washington, D.C., 1974. **(30)**

Whitmarsh, R. B., O. E. Weser, S. Ali, J. E. Boudreaux, R. L. Fleisher, D. Jipa, R. B. Kidd, T. K. Malik, A. Matter, C. Nigrini, H. N. Siddiquie and P. Stoffers, Site 220, in *Initial Reports of the Deep Sea Drilling Project,* vol. 23, R. B. Whitmarsh, O. E. Weser et al., pp. 117–166, U.S. Govt. Printing Office, Washington, D.C., 1974*a*. **(30)**

Whitmarsh, R. B., O. E. Weser, S. Ali, J. E. Boudreaux, R. L. Fleisher, D. Jipa, R. B. Kidd, T. K. Malik, A. Matter, C. Nigrini, H. N. Siddiquie and P. Stoffers, Site 221, in *Initial Reports of the Deep Sea Drilling Project,* vol. 23, R. B. Whitmarsh, O. E. Weser et al., pp. 167–210, U.S. Govt. Printing Office, Washington, D.C., 1974*b*. **(30)**

Wiggins, R. A., and D. V. Helmberger, Upper mantle structure of western United States, *J. Geophys. Res., 78,* 1870–1880, 1973. **(33)**

Wilcox, L. E., An analysis of gravity prediction in continental areas, *Aerospace Center Ref. Pub. 74-001,* Defense Mapping Agency, St. Louis, Mo., 1974. **(4)**

Wilkes, C., *Narrative of the United States Exploring Expedition, 1838–1842, 4,* 539 pp., Philadelphia, Pa., 1845. **(27)**

Williams, H., Geology of the Lassen Volcanic National Park, California, *Univ. Calif., Bull. Dept. Geol. Sci., 21,* 195–385, 1932*a*. **(27)**

Williams, H., The history and character of volcanic domes, *Univ. Calif., Bull. Dept. Geol. Sci., 21,* 51–146, 1932*b*. **(27)**

Williams, H., Calderas and their origin, *Univ. Calif. Publ. Geol. Sci., 25,* 239–346, 1941. **(27)**

Williams, H., The geology of Crater Lake National Park, Oregon, with a reconnaissance of the Cascade Range southward to Mount Shasta, *Carnegie Inst. Washington Pub. 540,* 162 pp., 1942. **(27)**

Wilson, J. T., A possible origin of the Hawaiian Islands, *Can. J. Phys., 41,* 863–870, 1963*a*. **(18, 28, 31)**

Wilson, J. T., Evidence from islands on the spreading of the ocean floors, *Nature, 197,* 536–538, 1963*b*. **(18)**

Wilson, J. T., A new class of faults and their bearing on continental drift, *Nature, 207,* 343, 1965*a*. **(15, 16)**

Wilson, J. T., Transform faults, oceanic ridges, and magnetic anomalies southwest of Vancouver Island, *Science, 150,* 482–485, 1965*b*. **(16)**

Wilson, J. T., Are the structures of the Caribbean and Scotia arc regions analogous to ice rafting?, *Earth Planet. Sci. Lett., 1,* 335–338, 1966. **(26)**

Winterer, E. L., Sedimentary facies and plate tectonics of the equatorial Pacific, *Amer. Assoc. Petrol. Geol. Bull., 57,* 265–282, 1973. **(22)**

Winterer, E. L., Line Islands: Bathymetry and tectonic setting, in *Initial Reports of the Deep Sea Drilling Project,* vol. 33, S. O. Schlanger, E. D. Jackson et al., U.S. Govt. Printing Office, Washington, D.C., in press. **(22)**

Winterer, E. L., P. F. Lonsdale, J. L. Matthews, and B. R. Rosendahl, Structure and acoustic stratigraphy of the Manihiki Plateau, *Deep-Sea Res., 21,* 793–814, 1974. **(18, 22)**

Winterer, E. L., J. Ewing et al., *Initial Reports of the Deep Sea Drilling Project,* vol. 17, 930 pp., U.S. Govt. Printing Office, Washington, D.C., 1973. **(22)**

Winterer, E. L., W. R. Riedel et al., *Initial Reports of the Deep Sea Drilling Project,* vol. 7, 1757 pp., U.S. Govt. Printing Office, Washington, D.C., 1971. **(22)**

Wojtczak-Gadomsha, B., Distribution of seismic energy released in the Aleutian–Alaska region and the relation with the dislocation flow, *Inst. Geophys. Pol. Ac. Sci., 60,* 1973. **(9)**

Wolf, H., Die Hohenregression der Schwereanomalien und ihre Deutungen, *Acta Geodet. Geophys. Montan. Academ. Scient. Hungaricae, 8*(1–2), 143–157, 1973. **(4)**

Wong, P., K. C. Gregory, and D. W. Handschumacher, Computer program projecting fictitious coordinate system on standard Mercator projections, *Hawaii Inst. Geophys. Rep., HIG-75-19,* 1975. **(16)**

Woollard, G. P., Crustal structure from gravity and seismic measurements, *J. Geophys. Res., 64*(10), 1521–1544, 1959. **(4, 10)**

Woollard, G. P., Seismic crustal studies during the IGY, part I: Marine program, IGY Bulletin No. 33, *Trans. AGU, 41,* 107–113, 1960*a*. **(10)**

Woollard, G. P., Seismic crustal studies during the IGY, part II: Continental program, IGY Bulletin No. 34, *Trans. AGU, 41,* 351–355, 1960*b*. **(8, 10)**

Woollard, G. P., The relation of gravity anomalies to surface elevation, crustal structure, and geology, *Res. Rep. 6-2a,* Geophysical and Polar Research Center, Univ. Wisconsin, 1962. **(4, 8)**

Woollard, G. P., Principal facts for gravity observations in the Hawaiian archipelago, Johnston Island, American Samoa, and Society Islands, *Hawaii Inst. Geophys. Rep., HIG-66-20,* 1966*a*. **(2)**

Woollard, G. P., Crust and mantle relations in the Hawaiian area, in *Continental Margins and Island Arcs,* edited by W. H. Poole, *Geol. Surv. Can. Paper, 66-15,* 294–310, 1966*b*. **(2, 37)**

Woollard, G. P., Regional isostatic relations in the United States, in *The Earth Beneath the Continents, Geophys. Monogr. Ser.,* vol. 10, edited by J. S. Steinhart and T. J. Smith, pp. 557–594, AGU, Washington, D.C., 1966*c*. **(4)**

Woollard, G. P., The interrelationship of the crust, the upper mantle, and isostatic gravity anomalies in the United States, in *The Crust and Upper Mantle of the Pacific Area, Geophys. Monogr. Ser.,* vol. 12, edited by L. Knopoff, C. L. Drake, and P. J. Hart, pp. 312–341, AGU, Washington, D.C., 1968. **(4)**

Woollard, G. P., Regional variations in gravity, in *The Earth's Crust and Upper Mantle, Geophys. Monogr. Ser.*, vol. 13, edited by P. J. Hart, pp. 320–341, AGU, Washington, D.C., 1969*a*. (**4**)

Woollard, G. P., A regional analysis of crustal structure in North America and a study of problems associated with the prediction of gravity in Europe, AF f23(601)-68-C-0186, *Hawaii Inst. Geophys. Rep., HIG-69-12*, 1969*b*. (**4**)

Woollard, G. P., Standardization of gravity measurements, in *The Earth's Crust and Upper Mantle, Geophys. Monogr. Ser.*, vol. 13, edited by P. J. Hart, pp. 283–293, AGU, Washington, D.C., 1969*c*. (**4**)

Woollard, G. P., Evaluation of the isostatic mechanism and the role of mineralogic transformations from seismic and gravity data, *Phys. Earth Planet. Interiors, 3,* 484–498, 1970. (**4**)

Woollard, G. P., The interrelation of crustal and upper mantle parameter values in the Pacific, *Rev. Geophys. Space Phys., 13,* 87–137, 1975. (**35**)

Woollard, G. P., and W. E. Strange, The prediction of gravity, in *Gravity Anomalies: Unsurveyed Areas, Geophys. Monogr. Ser.*, vol. 13, edited by P. J. Hart, pp. 96–113, AGU, Washington, D.C., 1966. (**4**)

Woollard, G. P., and N. J. Thompson, Problems in the use of a Magnavox 706 marine-type satellite navigator for land station position control, Part II, in Reliability of satellite determined positions of latitude and longitude, K. L. Daugherty, G. P. Woollard, and N. J. Thompson, *Hawaii Inst. Geophys. Rep., HIG-74-1,* 1974. (**10**)

Worzel, J. L., Deep structure of coastal margins and mid-oceanic ridges, in *Submarine Geology and Geophysics,* edited by W. F. Whittard and R. Bradshaw, pp. 335–359, Butterworthes, London, 1965*a*. (**1**)

Worzel, J. L., *Pendulum Gravity Measurements at Sea, 1936–1959,* 422 pp., John Wiley and Sons, New York, 1965*b*. (**1, 2**)

Worzel, J. L., Standard oceanic and continental structure, in *The Geology of Continental Margins,* edited by C. A. Burk and C. L. Drake, pp. 59–66, Springer-Verlag, New York, 1974. (**1, 8**)

Wright, T. L., Descriptive material to accompany set of 7 colored slides illustrating crystallization of basalt in Makaopuhi lava lake, Hawaii, U.S. Geol. Surv., Denver Federal Center, Denver, Colorado, 1967. (**32**)

Wright, T. L., Composition of Kilauea and Mauna Loa lava in space and time, *U.S. Geol. Surv. Prof. Pap., 735,* 1–40, 1971. (**32**)

Wright, T. L., and R. S. Fiske, Origin of the differentiated and hybrid lavas of Kilauea volcano, Hawaii, *J. Petrol., 12,* 1–65, 1971. (**32**)

Wright, T. L., and D. L. Peck, Crystallization and vesiculation of the Alae magma, *U.S. Geol. Surv. Prof. Pap.*, in press. (**32**)

Wright, T. L., and P. W. Weiblen, Mineral composition and paragenesis in tholeiitic basalt from Makaopuhi lava lake, Hawaii (abstract), *Program 1967 Annual Meeting, Geol. Soc. Amer.*, 242–243, 1967. (**32**)

Wright, T. L., W. T. Kinoshita, and D. L. Peck, March 1965 eruption of Kilauea volcano and the formation of Makaopuhi lava lake, *J. Geophys. Res., 73,* 3181–3205, 1968. (**32**)

Wyss, M., Stress estimates for South American shallow and deep earthquakes, *J. Geophys. Res., 75,* 1529–1544, 1970. (**29**)

Yasui, M., Y. Hashimoto, and S. Uyeda, Geomagnetic studies of the Japan Sea—anomaly pattern in the Japan Sea, *Oceanogr. Mag., 19,* 221–231, 1967. (**20**)

Yasui, M., D. Epp, K. Nagasaka, and T. Kishii, Terrestrial heat flow in the seas around the Nansei shoto (Ryukyu Islands), *Tectonophysics, 10,* 225–234, 1970. (**3**)

Yeats, R. S., W. C. Forbes, G. Heath, D. A. Ross, and K. Scheidegger, Petrology and geochemistry of DSDP Leg 16 basalts, in *Initial Reports of the Deep Sea Drilling Project,* vol. 16, Tj. H. van Andel, G. R. Heath et al., pp. 617–640, U.S. Govt. Printing Office, Washington, D.C., 1973. (**30**)

York, D., Least square fitting of a straight line with correlated errors, *Earth Planet. Sci. Lett., 5,* 359–408, 1969. (**31**)

Yoshii, T., Features of the upper mantle around Japan as inferred from gravity anomalies, *J. Phys. Earth, 20,* 23–34, 1972*a*. (**3, 36**)

Yoshii, T., Terrestrial heat flow and features of the upper mantle beneath the Pacific and the Sea of Japan, *J. Phys. Earth, 20,* 271–285, 1972*b*. (**36**)

Yoshii, T., Normal ocean, marginal seas and hot spots, *Nature Phys. Sci., 244,* 92–93, 1973*a*. (**20, 36**)

Yoshii, T., Upper mantle structure beneath the North Pacific and the marginal seas, *J. Phys. Earth, 21,* 313–328, 1973*b*. (**20, 36**)

Yoshii, T., Regionality of group velocities of Rayleigh waves in the Pacific and thickening of the plate, *Earth Planet Sci. Lett., 25,* 305–312, 1975. (**36**)

Yoshii, T., and S. Asano, Time term analyses of explosion seismic data, *J. Phys. Earth, 20,* 47–57, 1972. (**3**)

Yoshii, T., Y. Kono, and K. Ito, Thickening of the oceanic lithosphere, this volume. (**3**)

Zittel, K. A. von, *History of Geology and Paleontology,* transl. by M. M. Ogilvie-Gordon, 562 pp., London, 1901. (**27**)

# Subject Index

# Genuine Use of Trademarks

**AIPPI Law Series**

VOLUME 4

**Series Editors**

AIPPI

**Introduction & Contents/Subjects**

Books in this series are developed within the framework of the International Association for the Protection of Intellectual Property (AIPPI), a non-affiliated non-profit organization dedicated to the development and improvement of legal regimes for the protection of intellectual property at both national and international levels.

**Objective & Readership**

The aim is to publish innovative work appealing to practitioners, other users of IP systems and academics.

*The titles in this series are listed at the back of this volume.*

# Genuine Use of Trademarks

*Edited by*

**Eléonore Gaspar**

Wolters Kluwer

*Published by:*
Kluwer Law International B.V.
PO Box 316
2400 AH Alphen aan den Rijn
The Netherlands
Website: www.wolterskluwerlr.com

*Sold and distributed in North, Central and South America by:*
Wolters Kluwer Legal & Regulatory U.S.
7201 McKinney Circle
Frederick, MD 21704
United States of America
Email: customer.service@wolterskluwer.com

*Sold and distributed in all other countries by:*
Quadrant
Rockwood House
Haywards Heath
West Sussex
RH16 3DH
United Kingdom
Email: international-customerservice@wolterskluwer.com

*Printed on acid-free paper.*

ISBN: 978-94-035-0563-3
Web PDF ISBN: 978-94-035-0565-7
e-ISBN: 978-94-035-0564-0

AIPPI Law Series ISBN 98-888-8020-8



Printed in the United Kingdom.

FSC
MIX
FSC® C103993

# List of Contributors

**ARGENTINA**

*Graciela Claudia Pérez De Inzaurraga*
Graciela Pérez de Inzaurraga is an Argentine lawyer with over 25 years of experience in IP. She is Head of the Foreign Clients Department at Hausheer Belgrano & Fernandez, where she counsels clients on trademark, design and patent protection and enforcement strategies and takes an active role in settlement negotiations and litigation planning. She has lectured on a variety of IP related topics at local universities and local and international seminars and congresses, and conducted training seminars in Argentina and foreign countries for in-house staff at companies and for government officials at Trademark Offices. She has also authored and published several articles on IP issues. She actively participates in the Argentine IP Association (Chair of the Trademarks Committee and Secretary to the Academic Committee), AIPPI (Vice-President of the Argentine Group and Secretary to the Standing Committee on Trademarks) and INTA (Chair of the Legislation and Regulation Committee). gcperez@hbf.com.ar

**AUSTRALIA**

*Mark Metzeling*
Mark is a qualified solicitor and registered trade marks attorney, he is a special counsel at the law firm Macpherson Kelley, and also has a Masters of Intellectual Property Laws, a Master of Laws and a Bachelor of Applied Science (Biochemistry). He has never lost a contested matter. He is an expert in brand protection and securing Trade Mark rights. Mark has received numerous awards for his work as a Trade Mark Lawyer. Mark is ranked as a leading lawyer by the World Trade Mark Review 2018. In January 2017, Mark was appointed to the Trade Marks Standing Committee; and in January 2018, Mark was appointed to the Standing Committee on Commercialisation of IP, of the International Association for the Protection of Intellectual Property, due to his knowledge and experience across all areas of Trade Mark law. Mark.Metzeling@mk.com.au

*Sarah Matheson*
Sarah Matheson BA LLB is an attorney-at-law and a partner of Allens. She advises in contentious and non-contentious IP matters. She is an experienced trade mark litigator and is also a registered Trade Marks Attorney. Sarah is the Reporter General of the International Association for the Protection of Intellectual Property (AIPPI). sarah.matheson@allens.com.au

**BENELUX**

*Emmanuel Cornu*
Emmanuel Cornu is member of the Brussels Bar and partner at the law firm Simont Braun. He is a specialist in intellectual property law (trademarks, patents, designs and copyrights). He assists clients in matters related to intellectual property law, litigation and dispute resolution, before both the Belgian courts and the Benelux court, the European Court of Justice, the General Court and the EUIPO. Emmanuel Cornu is the editor-in-chief of the *Revue de Droit Intellectuel-L'Ingénieur-conseil* (founded in 1911) and is the author of numerous publications on intellectual property law, including the fifth edition of *Précis des Marques* with Antoine Braun (Larcier, 2009, 1070 pages). emmanuel.cornu@simontbraun.eu

**BRAZIL**

*Elisabeth Siemsen do Amaral*
Elisabeth Siemsen do Amaral is a partner of Dannemann Siemsen Bigler and Ipanema Moreira, where she leads the firm's trademark department. She is also Vice-President of Asociación Interamericana de la Propiedad Intelectual (ASIPI), member of the Board of Directors of the Brazilian Association of Intellectual Property (ABPI) and member of the Trademark Committee of the Association Internationale pour la Protection de la Propriété Intellectuelle (AIPPI). esiemsen@dannemann.com.br

*Rafael Atab*
Rafael Atab is a partner of Dannemann Siemsen Bigler and Ipanema Moreira. He is also Chair of the Standing Committee on ADR of Association Internationale pour la Protection de la Propriété Intellectuelle (AIPPI) and Coordinator of the Standing Committee on Trademarks of Associação Brasileira da Propriedade Intelectual (ABPI). He has a master degree on International Law from the University of the State of Rio de Janeiro and is a Professor at the Post-Graduation Course on Intellectual Property Law of Pontífice Universidade Católica do Rio de Janeiro (PUC-Rio). atab@dannemann.com.br

**CANADA**

*Paula Clancy*
Paula Clancy is a lawyer and trademark agent based in Ottawa, Canada. After working at one of Canada's largest firms for nearly 15 years, Paula founded Clancy P.C. in 2008, and later affiliated with the patent agency, Brion Raffoul. Clancy P.C. + Brion Raffoul has been consistently recognized as a leading Canadian IP boutique. Ms. Clancy has won numerous awards including Top Forty Under 40 (2008), Entrepreneur of the Year (2017), and was a finalist for the Businesswoman of the Year award

(2014). She has been recognized by the WTR 1000 as one of the world's leading trademark practitioners for 7 consecutive years. Ms. Clancy currently serves as Chair of INTA's Public Information Committee, as well as IPIC's International Trademarks Issues Committee, and is a member of AIPPI's Canadian Trademarks Committee. She has also served as Co-Chair of INTA's North America Global Advisory Council, as the Canadian National Delegate for ASIPI, and as Chair of ASIPI's Social Action Committee. pclancy@cbrcanada.com

**CHINA**

*Lena (Lanying) Shen*
Lena started her career in IP in 2000 and is now a partner in a top 10 Chinese IP law firm, Beijing Sanyou Intellectual Property Agency Ltd., in charge of the Trademark Department. She is an attorney-at-law in China and has rich experience in trademark prosecution and IP enforcement. Lena and her team provides various services relating to trademark matters in China including but not limited to search, watch, registration, opposition, invalidation, licensing, assignment, customs recordal, litigation, administrative enforcement, etc. They also help client register and enforce copyright in China. Lena received education in both China and UK and is Chair of Trademark Committee of AIPPI China and Chair of CET 9 (Asian Group) of FICPI. lenashen@sanyouip.com

**EUROPEAN UNION**

*Alexander Von Mühlendahl*
Alexander von Mühlendahl is a lawyer (Rechtsanwalt) with BARDEHLE PAGENBERG in Munich, Germany. Previously he was Vice-President of the Office for Harmonization in the Internal Market (OHIM), now the European Union Intellectual Property Office (EUIPO). He has a doctorate degree from the University of Munich, J.D. and LL.M. from Northwestern University. He teaches at various universities and is Visiting Professor at Queen Mary University of London, Centre for Commercial Law Studies. vonmuhlendahl@bardehle.de

**FRANCE**

*Eléonore Gaspar*
Eléonore Gaspar is a partner of the law firm Duclos, Thorne, Mollet-Viéville & Associés (DTMV) in Paris. Eléonore is an attorney-at-law specialized in IP (trademarks, designs, copyright and patents) and has been litigating and prosecuting trademarks for more than 20 years. She represents her clients before the French Courts, the trademark offices (INPI, EUIPO, WIPO) and before the General Court of the EU and the Court of Justice of the EU. Eléonore's clients are mainly involved in the film, entertainment, fashion, e-commerce, energy and travel industries. Eléonore is the President of AIPPI France and the chair of AIPPI's Standing Committee on Trademarks. She is a frequent lecturer on trademark and copyright topics. She was voted Best Lawyer in Media for France by Best Lawyers in 2017. GASPAR@dtmvparis.com

*Laurine Janin-Reynaud*
Laurine Janin-Reynaud is an Attorney-at-Law, a member of the Paris Bar since 2006, and an Of Counsel at the law firm Duclos Thorne Mollet-Viéville & Associés. She is a specialist in Intellectual Property Law (trademarks, patents, designs and copyrights). She assists clients in matters related to Intellectual Property Law, litigation and dispute resolution, before the French courts and before the French and European offices. Her practice and experience also focus on legal counseling and on prosecuting and defending trademark and design rights in France and abroad. She is an active member of the AIPPI and APRAM associations.JANINREYNAUD@dtmvparis.com

**GERMANY**

*Rainer Böhm*
Rainer Böhm is one of the two founding partners of the Hamburg office of Eisenführ Speiser. He is an IP litigator with a wide range of expertise which includes trademark law and the law of marks and signs, IT law, Internet/multimedia law, outsourcing, patent law/know-how law, advertising law and competition law as well as licensing law (including anti-trust law). Rainer has long-term experience in managing and coordinating multinational litigations, i.a. in the USA and India. He lectures regularly at the Hamburg School of Business Administration (HSBA) and for the American Bar Association. The magazine World Trademark Review recommends Rainer as being praised for his 'eloquent reasoning' which is 'absolutely first class'. rbohm@ eisenfuhr.com

*Sandra Müller*
Sandra Müller is an experienced IP litigator at the Hamburg office of Eisenführ Speiser. She has represented clients from the luxury brand and consumer goods industries in various infringement proceedings before the German courts and in cancellation and opposition proceedings before the GPTO, the EUIPO and the ECJ. Sandra holds a German PhD (Dr. jur) in comparative law on publishing agreements and copyright. She lectures on IP and unfair competition law at University of Hanover law school and is a member of the MARQUES Education Team. sbmuller@eisenfuhr.com

**ISRAEL**

*Avi Ordo*
Avi Ordo is a partner at S. Horowitz & Co. He has more than 25 years of extensive IP experience, including advising and representing global and local clients with respect to all IP rights, on contentious and non-contentious matters. Avi also heads the firm's Trademarks Prosecution Department. Avi has been ranked by prominent international legal directories as one of Israel's leading intellectual property lawyers. Avi has been involved in many precedent-setting and high-profile cases, and has been frequently retained to assist clients in co-ordinating their cross-border IP claims. avio@s-horowitz.com

*Moran Katz*
Moran Katz is a partner at S. Horowitz & Co., where her IP practice encompass a wide spectrum of contentious and non-contentious IP matters, including the creation, commercial exploitation, protection and enforcement of patents, trademarks, brand

names, copyright, designs and trade secrets. Moran represents clients from a variety of sectors, ranging from manufacturers of pharmaceutical, cosmetics, beverages, appliances, machinery and vehicles to communication, software and more. moran-k@s-horowitz.com

**ITALY**

*Gualtiero Dragotti*
Gualtiero Dragotti is a partner of the law firm DLA Piper in Italy and has extensive experience in IP matters, with a focus on patents and the protection of innovation. He advises leading Italian and international clients on various contentious IP cases, including some of the key patent litigation issues in Italy. He has acquired a deep knowledge of the Life Sciences sector and has dealt with numerous pharmaceutical patent cases. In addition, Gualtiero has experience in dealing with a variety of corporate matters related to protection and management of IP rights, both for domestic and international clients. He's President of AIPPI Group in Italy, Vice-president of the Anti-Counterfeiting Study Center and he served as Juridical Expert by CNAC (Anti-Counterfeiting National Agency). Gualtiero.Dragotti@dlapiper.com

*Elena Varese*
Elena Varese is an attorney-at-law at the law firm DLA Piper in Italy has gained extensive experience in contentious and non-contentious Intellectual Property, Fashion Law, unfair competition and advertising matters. She has a strong focus in advising clients active especially in the fashion, food & beverages, publishing and entertainment sectors. She is an occasional guest-speaker in Universities, specialized courses and seminars on Intellectual Property matters and Fashion Law. Elena.Varese@dlapiper.com

**JAPAN**

*Masako Nishimura*
Masako is partner of Oshima, Nishimura & Miyanaga PPC and has more than 25 years of experience in handling trademarks and designs for domestic and overseas clients. In May 2007, she founded a boutique trademark firm, in Tokyo. Masako focuses on providing expert opinions on trademark and design matters requested by major brands from both legal and practical standpoints. Masako is also qualified as an IP Litigation attorney, handling disputes and litigation in the areas of trademarks and designs, and unfair competitions relating to both. Masako is active in the educational field, with many students both domestic and international, especially as a professor of Tokyo University of Science Graduate School of Master of Intellectual Property, during 2007-2018. She has also written a large volume of articles, chapters and textbooks on law and practice relating to trademarks and unfair competition.office@onm-tm.jp

**REPUBLIC OF KOREA**

*Seung Yong Lee*
Seung Yong Lee is a patent attorney and a partner of Y.P. Lee, Mock & Partners in Korea. His practice primarily focuses on counseling and representing clients in the area of filing and prosecution of trademark applications in Korea and abroad. He

is a member of AIPPI's Standing Committee on Trademarks and APAA's Copyright Committee. He also acts as a panel of the Korean Internet Address Disputes Resolution Committee and a Vise-President of the Korean Patent Attorney Association. leesy@leemock.com

*Doug Jay Lee*
Doug Jay Lee is a partner of Yoon & Yang (IP) LLC and has been a trademark attorney for more than 25 years. Mr. Lee is a vice president of AIPPI Korea and a key member of the Korea Trademark & Design Association who contributed to the establishment of the organization. He is a well-reputed trademark specialist and his ability to successfully navigate challenges is evidenced in his strong prosecution skills, which has earned him referrals from all over the country. He has been named as one of the leading trademark/design practitioners in Korea by World Trademark Review 1000 for several years, including recognition in 2018. His contributions to the local community include speaking at forums, guest lecturing, writing and contributing to articles, and serving on various national boards. djlee@hwawoo.com

**MALAYSIA**

*Karen Abraham*
Karen Abraham is partner of Shearn Delamore & Co and has broad experience in contentious IP and technology matters, IP litigation, enforcement and licensing programmes, anti-counterfeiting, exploitation of IP rights, competition law and broadcasting. She negotiates and advises on the exploitation and enforcement of IP rights for leading multinational companies around the world. Karen has also designed and crafted anti-piracy and anti-counterfeiting programmes, as well as brand management schemes for small to leading local and global companies. She further provides legal counsel on all allied IP rights relating to matters such as ambush marketing, food and drugs, domain name disputes, technology & communications, licensing, agency franchising, merchandising, commercial sales contracts, sponsorship, advertising and entertainment, and media broadcasting laws. Karen is also the Past Director of INTA and Past Assistant Secretary General of the International Association for the Protection of Intellectual Property (AIPPI) as well as the Past President of the Malaysian Chapter of the International Association for the Protection of Intellectual Property (AIPPI). karen@shearndelamore.com

*Janet Toh Yoong San*
Janet Toh is partner of Shearn Delamore & Co and and focuses on intellectual property and technology issues in a wide variety of transactions, and has worked on a wide range of agreements, including distributorship, licensing, outsourcing, service and consultancy agreements. She has advised clients on intellectual property protection and ownership issues, advertising issues, consumer protection, copyright, domain names, e-commerce, franchise, gaming, regulatory approvals for food and drugs, and telecommunications issues. Janet has also conducted due diligence for various acquisition projects, IT-related contracts, e-commerce and telecommunications, and has advised leading multinational companies in the pharmaceutical, tobacco and Internet-related services industries on IP protection. janet.toh@shearndelamore.com

**MEXICO**

*Eduardo Kleinberg*

Eduardo Kleinberg is Partner since 2003 and Managing Partner since 2014 of Basham, Ringe y Correa, S.C. He has been practicing Intellectual Property since 1993. He is the Head of the Trademark, Licensing and Franchising practices with vast experience in domestic and international matters. He was designated by CONCAMIN (Confederación de Cámaras Industriales de los Estados Unidos Mexicanos) as the representative of the Mexican private sector with respect to the Intellectual Property chapter of the Trans-Pacific Partnership Agreement (TPP) ), Past-President of the Mexican chapter of the International Association for the Protection of Intellectual Property (AIPPI), Member of the Nominations Committee of AIPPI, Past-Member of the Permanent Committee on Trademarks of AIPPI, Past-President of Licensing Executives Society – Mexico (LES-Mexico), Member of the Licensing Executives Society International (LESI), Member of the International Trademark Association (INTA), Member of the Inter-American Association for the Protection of Intellectual Property (ASIPI), Member of the Association of European Trade Mark Owners (MARQUES), Member of the Mexican Franchise Association. Member of the International Franchise Association (IFA). kleinberg@basham.com.mx

**PHILIPPINES**

*Aleli Angela G. Quirino*

Aleli Angela G. Quirino is an Of Counsel of Angara Abello Concepcion Regala & Cruz Law (ACCRALAW) Offices. The main focus of Ms. Quirino's practice is in the areas of intellectual property (IP) law and business law. She headed the Firm's Intellectual Property Department from 1992 to 2007. She has helped place the Firm at the forefront in the fight against counterfeiting and piracy of intellectual property. Ms. Quirino has written several articles on intellectual property law and actively participates in various international and local seminars and conferences on intellectual property law as speaker or panellist. She is currently the President of the ASEAN Intellectual Property Association (ASEAN IPA). agquirino@accralaw.com

*Victor Basilio n. De Leon*

Victor Basilio n. De Leon is a Partner in the Intellectual Property Department of Angara Abello Concepcion Regala & Cruz Law Offices and was its Monitor from 2008-2014. He assists the Firm's clients in prosecuting trademark applications, maintaining trademark portfolios, conducting IP due diligence, and filing/prosecuting enforcement actions with administrative tribunals and regular courts. Mr. De Leon has represented the Firm in foreign IP conferences such as the Association Internationale pour la Protection de la Propriete Intellectuelle (AIPPI), International Trademark Association (INTA) and the Asian Patent Attorneys Association (APAA). Mr. De Leon has been a speaker and a participant in local and international intellectual property conferences and seminars. vndeleon@accralaw.com

**POLAND**

*Bartosz Krakowiak*

Bartosz Krakowiak is a Polish patent and trademark attorney, European patent attorney and professional representative before the EUIPO. He is Co-Head of Trademark Department at POLSERVICE Patent and Trademark Attorneys Office in Warsaw, Poland, additionally leading the firm's legal team. Bartosz specializes in trademark and design prosecution, maintenance, licensing and protection, as well as in unfair competition matters, including litigation and anticounterfeiting actions. Since 2010, Bartosz has been President of AIPPI Poland; he is also a seasoned member of the AIPPI Standing Committee – Trademarks. He is, for the second term, Chair of the Europe and Central Asia Sub-Committee of the INTA Legislation and Regulation Committee; previously, he chaired the INTA Anticounterfeiting Committee; in 2016 he was honored with the INTA Volunteer Service Award for the Advancement of Trademark Law. Bartosz is also Vice President of the Disciplinary Court of Appeal of the Polish Chamber of Patent Attorneys and Vice Chair of the Chamber's Legislation Committee. Bartosz.Krakowiak@polservice.com.pl

*Klaudia Błach-Morysińska*

Klaudia is an Advocate, Patent Attorney and European Trademark and Design Attorney in ZM Legal, Warsaw, Poland. Klaudia specializes in intellectual and industrial property law, copyright and unfair competition law. She provides expert-level legal assistance in courtroom litigations and in administrative disputes involving IP. Klaudia has a vast expertise before Polish courts and authorities and international authorities. She authored numerous publications in the field of IP and gives lectures in IP rights. Klaudia is Disciplinary Proceedings Representative of the Polish Patent Attorney Chamber, AIPPI (Board Member of AIPPI Poland, Design Committee), PTMG, INTA, ECTA (Law Committee). blach@zmlegal.pl

**RUSSIA**

*Denis Voevodin*

Denis Voevodin is a partner in Dentons' Moscow office and the Head of the European Intellectual Property and Technology practice. Denis has more than 30 years of experience in a broad spectrum of areas of contentious and non-contentious IP matters and is recognised as one of the top IP practitioners in Russia and the CIS. He has considerable experience of acting in trademark, patent and copyright litigation proceedings as well as advising on anti-counterfeit activity. His work actively concerns acquisition and disposal of companies with valuable intellectual property rights and complex technology as well as copyright, trademark and patent rights. denis.voevodin@dentons.com

*Valeria Larionova*

Valeria Larionova is an Intellectual Property Consultant in Dentons' Moscow office. Valeria has graduated from the National Research University Higher School of Economics, Faculty of Law. Valeria has a 3 years' experience of acting in trademark and copyright litigation proceedings and advising on anti-counterfeit activity. Valeria has a broad experience in representing clients in various trademark infringement on the Internet matters. valeria.larionova@dentons.com

**SINGAPORE**

*Stanley Lai*
Dr Stanley Lai, SC leads the Intellectual Property practice at Allen & Gledhill. Stanley specialises in all forms of IP litigation and information technology disputes, and is also a commercial litigator. Stanley has strong experience in brand management and trade mark portfolio protection. He has acted in many trade mark disputes before the Singapore Courts. In the biomedical and pharmaceutical sectors, he has substantial experience in advising on healthcare and medical IP and regulatory issues concerning medicines, generics, patent linkage, bio-similars, medical devices, clinical trials, product recalls and product liability. Stanley also maintains a strong advisory practice for IP management and strategy, serving a broad spectrum of clients. stanley.lai@allenandgledhill.com

**SPAIN**

*David Gómez*
David Gómez holds a Doctor Degree from the Max Planck Institute (Munich) and the Alcalá de Henares University, as well as a Master Degree in Private Law from the ICAM. Currently, he is a Lecturer of Procedural Law at the Rey Juan Carlos University of Madrid, as well as a regular speaker at conferences and post-graduate courses, both in Spain and abroad, and the author of many articles published in specialised magazines. His professional career is focused on the fields of civil and criminal litigation, as well as in the areas of intellectual and unfair competition, representing companies across all industrial sectors. He is a partner at BAYLOS Abogados. d.gomez@baylos.com

*Pedro Merino Baylos*
Pedro Merino Baylos is a graduate in Law from the CEU San Pablo University in Madrid. He has a broad professional experience and is specialised in Intellectual Property law and unfair competition, mainly in the civil sphere. He is usually a speaker in seminars and an author of several IP publications. In February 2017, he was appointed President of the Spanish Group of AIPPI. He is a partner at BAYLOS Abogados. p.merino@baylos.com

**SWEDEN**

*Karin Cederlund*
Karin Cederlund (LL.M. Stockholm University, MBA Stockholm School of Economic), Attorney-at-law and partner of Sandart & Partners, specializes in IP and marketing law. She has extensive experience from litigation and transactional work within these fields of law. Since several years, Karin is a member of the board of the Swedish Association for the Protection of Intellectual Property (SFIR) and of the board of the Institute of Intellectual Property and Marketing Law at the Stockholm University (IFIM). Karin is top ranked in the intellectual property field by Chambers & Partners and other guides. karin.cederlund@sandart.se

*Anders Kylhammar*
Anders Kylhammar (LL.M. Stockholm, LL.M. Amsterdam), Attorney-at-law and partner of Sandart & Partners, specializes in IP law, marketing law and protection of

trade secrets. He has extensive experience from court litigation in these fields of law. Anders is regularly top ranked in several guides, e.g. WTR1000, Chambers and Partners and Legal 500.Before joining Sandart & Partners, Anders was an Associate Judge at Svea Court of Appeal in Stockholm. Further, he was a Legal Adviser at the Swedish Ministry of Justice, the Division for IP Law. The Swedish Government appointed Anders as an expert in The Trade Mark Committees of 1998-2001 and 2015-2016. He was an expert in the Committee on Criminal Sanctions of IP infringement of 2017. Anders has published several articles and books on IP law.anders.kylhammar@sandart.se

*Dan Eklöf*
Dan Eklöf (LL.D., Stockholm), Attorney-at-law at Sandart & Partners, specializes in IP and competition law. He has extensive experience from litigation and also regularly provides strategic advice in these fields. Dan Eklöf has academic background and has published a book and several articles on IP and competition law. dan.eklof@sandart.se

**SWITZERLAND**

*Thierry Calame*
Thierry Calame is a partner at Lenz Staehelin. He has over 20 years of experience in intellectual property litigation and IP-related arbitration. He also sits as an arbitrator and advises on advertising, licensing and the IP aspects of other transactions. Thierry is a permanent lecturer at the University of St. Gallen (HSG) and at the Swiss Federal Institute of Intellectual Property (FIIP). He served as Reporter General of AIPPI from 2010-2014. He is currently Vice-President of the European Patent Lawyers Association (EPLAW) and member of the board of Innosuisse, the federal agency for science-based innovation. thierry.calame@lenzstaehelin.com

**THAILAND**

*Say Sujintaya*
Say Sujintaya is the Regional Chair of Baker McKenzie's Intellectual Property Group in Asia Pacific and a partner in the firm's Bangkok office, where she is co-head of the Intellectual Property Practice Group. Say's clients include many well-known consumer goods, food and beverage, pharmaceutical, telecommunications, electronics, and media industry leaders. She manages the patent and trademark portfolios of some of the world's largest companies and, in this capacity, regularly advises on prosecution, licensing, and enforcement of IP rights, including coordination of anti-counterfeiting and anti-piracy campaigns, as well as IP-related litigation. Say is a longstanding member of the Thai AIPPI Group, which she currently serves as Secretary.Say.Sujintaya@bakermckenzie.com

**TURKEY**

*E. Kerim Yardimci*
Kerim is a founding Partner and Executive Board Member at Deris and Kerim is a highly experienced and respected intellectual property attorney locally and internationally. Fluent in English and French, His international expertise is complemented

by his extensive network. Kerim's innovative approach to intellectual property litigation, complex court proceedings, and portfolio management has helped drive Deris' status as a top, forward-thinking IP law firm. He is also an active panellist at the World Intellectual Property Organization's Uniform Domain Name Dispute Resolution (UDRP). Since his graduation from the University of Geneva, Kerim has honed his knowledge of the establishment, licensing, acquisition and use of IP assets in Turkey and globally.Kerim.Yardimci@deris.com

*Sarper Sever*
Sarper is an intellectual property lawyer at the law firm Deris who counsels local and international clients for intellectual property prosecution in trademark, copyright, design, and geographical indications matters; including registrations, oppositions, counter-statements, objections and any other types of filing. He also provides legal advice and opinion on disputes involving litigation or non-contentious matters. He has been working with clients active in fashion, food and beverage, automobiles and parts, pharmaceuticals, software and computer services, media and entertainment sectors. He graduated from Maltepe University and received his master's degree from Coventry University. For the time being, he continues his PhD. at Istanbul University. Sarper.Sever@deris.com

**UK**

*Paul A. Harris*
Paul Harris is the senior lawyer at the IP firm, Venner Shipley LLP, with over 35 years of practising IP law in the UK. He combines a mixed practice of litigation and non-contentious (mainly licensing and strategy) work, in all aspects of IP, but particularly patents and trade marks. In relation to trade marks, Paul conducted the first case under the 1994 Trade Marks Act (Wagamama Ltd v. City Centre Restaurants plc [1995] FSR 713 (Ch.D)) which also settled the issue that 'association' was a sub-set of likelihood of confusion and not vice versa. Paul lectures on the Oxford University IP Diploma Post-Graduate course on the topics of Revocation and Invalidity, and Enforcement and has done since 1996. Paul has been a UK Council Member of AIPPI since 2000. pharris@vennershipley.co.uk

*Jan M. Vleck*
Jan Vleck is an UK Chartered Trade Mark Attorney, an UK Chartered Patent Attorney, European Patent Attorney and qualified UK attorney Higher Courts litigator. Jan is partner in the UK firm Reddie & Grose and has an extensive patent and trade mark practice involving the creation of trade mark portfolios around the world and advising on disputes in the UK and EU. Jan is a member of the AIPPI Trade Marks committee and the Trade Marks committee of the UK Chartered Institute of Patent Attorneys and lectures regularly on trade marks and other aspects of intellectual property. He has lead the UK response on a number of AIPPI trade mark questions and chaired the discussions in Buenos Aires on the protection of major sports events. Jan is a longstanding member of the Council of the UK Group of AIPPI. Jan.Vleck@reddie.co.uk

**USA**

*Dennis S. Prahl*
Dennis S. Prahl is a Partner of Ladas & Parry in the firm's New York office. Mr. Prahl has been consistently recognized as one of the leading trademark practitioners in the world by the World Trademark Review and Euromoney/MIP. He specializes in the areas of international trademark, copyright, domain name, right of publicity and unfair competition law and has represented and advised a wide range of clients in diverse fields, from multinational corporations to individuals. Mr. Prahl received his B.A. from Northwestern College and his J.D. from New York Law School. In addition, he holds a diploma in Advanced International Legal Studies from the McGeorge School of Law. Dennis also serves as president of Principium Strategies – Ladas & Parry's ICANN-accredited domain name registrar and brand monitoring subsidiary.
DPrahl@ladas.com

# Summary of Contents

# Table of Contents

# Table of Abbreviations

| | |
|---|---|
| AAA | American Arbitration Association |
| AAA Rules | International Arbitration Rules of the American Arbitration Association |
| ACICA | Australian Centre for International Commercial Arbitration |
| ACTAPS | Association of Contentious Trust and Probate Specialists |
| ADR | Alternative Dispute Resolution |
| ADRLJ | The Arbitration and Dispute Resolution Law Journal, London |
| AJCL | American Journal of Comparative Law |
| AJIL | American Journal of International Law, Washington |
| AJP | Aktuelle juristische Praxis, Lachen |
| All E.R. | All England Law Reports |
| Am.Rev.Int'lArb. | American Review of International Arbitration, New York |
| APP.L.R. | Arbitration, Practice & Procedure Law Reports |
| ArbInt. | Arbitration International |
| ARIAS | AIDA Reinsurance and Insurance Arbitration Society |
| Art. / Arts. | Article / Articles |
| ASA | Swiss Arbitration Association (Association suisse de l'arbitrage) |
| ASA Bull. | ASA Bulletin, The Hague |
| ASDI | Annuaire Suisse de droit international |
| AVEG | Federal Act on the General Application of Collective Employment Agreements (Bundesgesetz vom 28. September 1956 über die Allgemeinverbindlicherklärung von Gesamtarbeitsverträgen, SR 221.215.311) |
| BBl | Records of the Swiss Parliament (Bundesblatt / Feuille Fédérale / Foglio Federale) |

| | |
|---|---|
| BGE | Decision of the Swiss Federal Supreme Court, published in the Court's Official Compilation (Entscheidungen des Schweizerischen Bundesgerichts), available online at the Court's website (www.bger.ch) |
| BGer. | Decision of the Swiss Federal Supreme Court, not published in the Court's Official Compilation, available online at the Court's website (www.bger.ch) |
| BGFA | Federal Act on Attorneys-at-Law (Bundesgesetz über die Freizügigkeit der Anwältinnen und Anwälte, SR 935.61) |
| BGG | Federal Statute on the Swiss Federal Supreme Court (Bundesgesetz vom 17. Juni 2005 über das Bundesgericht, SR 173.110) |
| BIT | Bilateral Investment Treaty |
| BJM | Basler Juristische Miteilungen, Basel |
| BLI | Business Law International |
| BLR | Business Law Review, London |
| Brussels Convention | Convention on Jurisdiction and the Enforcement of Judgments in Civil and Commercial Matters (1968) |
| B.Y.I.L. | British Yearbook of International Law |
| BYU Law Review | Brigham Young University Law Review |
| CAFTA | Central American Fee Trade Agreement |
| CAS | Court of Arbitration for Sport (Tribunal Arbitral du Sport, TAS) |
| CaS | Causa Sport (CaS), Die Sport-Zeitschrift für nationales und internationales Recht sowie für Wirtschaft |
| CAS Bull. | CAS Bulletin |
| CAS Digest | Digest of CAS Awards (Recueil des sentences du TAS) |
| CC | Swiss Civil Code (Schweizerisches Zivilgesetzbuch vom 10. Dezember 1907, ZGB, SR 210) |
| CEDR | Centre for Effective Dispute Resolution, London |
| CEPANI Rules | Rules of the Belgian Centre for Mediation and Arbitration (CEPANI), Brussels |
| cf. | confer (compare) |
| Ch. | Chapter |
| Chambers | Chambers of Commerce and Industry of Basel, Berne, Geneva, Ticino, Neuchatel, Vaud and Zurich |
| CHF | Swiss francs |
| CIArb | Chartered Institute of Arbitrators, London |
| CISG | United Nations Convention on Contracts for the International Sale of Goods |
| CLOUT | Case Law on UNICITRAL Texts |

| | |
|---|---|
| CMR | Convention on the Contract for the International Carriage of Goods by Road (Convention Relative au Contrat de Transport International de Marchandises par Route) |
| CO | Swiss Code of Obligations (Bundesgesetz betreffend die Ergänzung des Schweizerischen Zivilgesetzbuches, Fünfer Teil: Obligationenrecht, vom 30. März 1911, SR 220) |
| Collection | Collection of ICC Arbitral Awards (I: 1974-1985; II: 1986-1990; III: 1991-1995; IV: 1996-2000, etc.) |
| Comp. L. Yb. Int'l Bus. | The Comparative Law Yearbook of International Business, The Hague |
| Concordat | Concordat on Arbitration (Konkordat über die Schiedsgerichtsbarkeit vom 27 März 1969, SR 279), repealed as from 1 January 2011 by the Swiss Code of Civil Procedure (ZPO) |
| CONI | Comitato Olimpico Nazionale Italiano |
| ConLR | Construction Law Reports |
| CPR | International Institute for Conflict Prevention and Resolution |
| CPR Rules | CPR Rules for Non-Administered Arbitration of International Disputes |
| CTR | (Iran-U.S.) Claims Tribunal Reports, Cambridge |
| DAC | Departemental Advisory Committee on Arbitration Law |
| DEBS | Swiss Debt Enforcement and Bankruptcy Statute (Bundesgesetz über Schuldbetreibung und Konkurs vom 11. April 1889, SchKG, SR 281.1) |
| DIS | Deutsche Institution für Schiedsgerichtsbarkeit (German Institution of Arbitration) |
| DIS Rules | Arbitration Rules of the DIS (German Institution of Arbitration) |
| Disp.Res.J. | Dispute Resolution Journal, AAA |
| DOCDEX | ICC Rules for Documentary Credit Dispute Resolution Expertise |
| DPCI | Droit et Pratique du Commerce International, Paris |
| DRJ | Dispute Resolution Journal, London |
| EBOLR | European Business Organization Law Review, The Hague |
| ECHR | European Convention on Human Rights of 4 November 1950 |
| ECJ | European Court of Justice |
| ECR | European Court Reports |
| ed./eds. | editor/editors |
| e.g. | exempli gratia (for example) |
| EIJL | European Journal of International Law |
| ELJ | European Law Journal |

| | |
|---|---|
| ERA | Academy of European Law (Europäische Rechtsakademie) |
| ERA Forum | Journal of the Academy of European Law (ERA) |
| et al. | et alii (and others) |
| etc. | et cetera (and so forth) |
| et seq./et seqq. | et sequens (and the following one)/et sequentes (and the following ones) |
| EU | European Union |
| EUR | Euro |
| Eur.Rev.Priv.Law | European Review of Private Law |
| European Convention 1961 | The European Convention of 21 April 1961 on International Commercial Arbitration |
| EuZW | Europäische Zeitschrift für Wirtschafsrecht |
| EWCA | England and Wales Court of Appeal |
| FAA | Federal Arbitration Act (USA) |
| FC | Federal Constitution of the Swiss Confederation (Bundesverfassung der Schweizerischen Eidgenossenschaft vom 18. April 1999, SR 101) |
| Fed.R.Civ.P. | Federal Rules of Civil Procedure (USA) |
| FIDIC | Fédération Internationale des Ingénieurs-Conseils (International Federation of Consulting Engineers), Geneva |
| FIFA | Fédération Internationale de Football Association |
| GA J.Int'l & Comp.L. | Georgia (University) Journal of International and Comparative Law |
| GAFTA | Grain and Feed Trade Association |
| Gaz.Pal. | Gazette du Palais, Paris |
| GesKR | Schweizerische Zeitschrift für Gesellschafts- und Kapitalmarktrecht sowie Umstrukturierungen, Zurich |
| GestG | Swiss Federal Statute on the Jurisdiction of Courts in Civil Matters (Bundesgesetz vom 24. März 2000 über den Gerichtsstand in Zivilsachen, GestG, SR 272), repealed as from 1 January 2011 by the Swiss Code of Civil Procedure (ZPO) |
| Giust.civ. | Giustizia civile, Milano |
| Harv.L.Rev. | Harvard Law Review |
| HAVE | Haftung und Versicherung (Responsabilité et Assurance, REAS), Zurich |
| HKIAC | Hong Kong International Arbitration Centre |
| IAI | International Arbitration Institute |
| IBA | International Bar Association |
| IBA Guidelines | IBA Guidelines on Conflicts of Interest in International Arbitration (approved by the IBA Council on 22 May 2004) |

| | |
|---|---|
| IBA Rules | IBA Rules on the Taking of Evidence in International Commercial Arbitration (adopted on 1 June 1999) |
| ibid. | ibidem (in the same place/source) |
| IBL | International Business Lawyer, London |
| ICC | International Chamber of Commerce |
| ICC Arb. Bull. | ICC International Court of Arbitration Bulletin |
| ICC Disp. Res. Bull. | ICC Dispute Resolution Bulletin |
| ICC Rules | Rules of Arbitration of the ICC (Paris) |
| ICCA | International Council for Commercial Arbitration |
| ICCA Handbook | ICCA International Handbook on Commercial Arbitration, The Hague/London/Boston |
| ICCA Yearbook | ICCA Yearbook of Commercial Arbitration, Deventer (NL) |
| ICDR | International Centre for Dispute Resolution, New York |
| ICJ | International Court of Justice, The Hague |
| ICLQ | The International and Comparative Law Quarterly, London |
| ICSID | International Center for the Settlement of Investment Disputes |
| ICSID Convention | Convention on the Settlement of Investment Disputes between States and Nationals of Other States |
| ICSID Review | ICSID Review – Foreign Investment Law Journal, Washington |
| ICSID Rules | ICSID Rules of Procedure for Arbitration Proceedings |
| id. | idem |
| IDR | Journal of International Dispute Resolution |
| i.e. | id est (that is) |
| ILA | International Law Association, London |
| ILR | International Law Reports, Cambridge |
| INCOTERMS | International Commercial Terms |
| Int.A.L.R. | International Arbitration Law Review |
| IP | Intellectual Property |
| IPRax | Praxis des Internationalen Privat- und Verfahrensrechts |
| ISLJ | The International Sports Law Journal |
| ISLR | International Sports Law Review |
| IT | Information Technology |
| JDI | Journal du droit international (Clunet), Paris |
| JdT | Journal des Tribunaux, Lausanne |
| JIDS | Journal of International Dispute Settlement |
| J.Int.Arb. | Journal of International Arbitration |
| J.Int'lL. | Santa Clara Journal of International Law |

| | |
|---|---|
| JPS | Jahrbuch für die Praxis der Schiedsgerichtsbarkeit |
| JWELB | Journal of World Energy Law & Business (Oxford Academic) |
| JWIT | Journal of World Investment & Trade |
| LCIA | London Court of International Arbitration |
| LCIA Rules | Arbitration Rules of the London Court of International Arbitration (LCIA) |
| Lloyd's Rep | Lloyd's Law Reports, London |
| LMCLQ | Lloyd's Maritime and Commercial Law Quarterly |
| LQR | The Law Quarterly Review, London |
| Lugano Convention | Convention on Jurisdiction and the Enforcement of Judgments in Civil and Commercial Matters (Übereinkommen über die gerichtliche Zuständigkeit und die Vollstreckung gerichtlicher Entscheidungen in Zivil- und Handelssachen, abgeschlossen in Lugano am 16. September 1988, SR 0.275.11) |
| Mealey's I.A.R. | Mealey's International Arbitration Report, Wayne (USA) |
| Mich.J.Int'l.Law | Michigan Journal of International Law |
| n. | note |
| NAI Rules | Arbitration Rules of the Netherlands Arbitration Institute |
| N.B. | nota bene (note well) |
| NJW | Neue Juristische Wochenschrift, Frankfurt a.M. |
| No. | Number |
| NSWLR | New South Wales Law Reports, Sydney |
| NYC | New York Convention (Convention on the Recognition and Enforcement of Foreign Arbitral Awards, adopted on 10 June 1958) |
| Ord. | Order |
| OSCE | Organization for Security and Co-operation in Europe |
| p. / pp. | page / pages |
| para. / paras. | paragraph / paragraphs |
| PCA | Permanent Court of Arbitration, The Hague |
| PILS | Private International Law of Switzerland (Bundesgesetz über das Internationale Privatrecht vom 18. December 1987, IPRG, SR 291) |
| PLC Arbitration | Practical Law Company (monthly email) on Arbitration |
| Pra | Die Praxis (des Bundesgerichts), Basel |
| R | Rule |
| RCADI | Collected Courses of the Hague Academy of International Law |
| recht | recht: Zeitschrift für juristische Ausbildung und Praxis, Bern |
| Rev. Arb. | Revue de l'Arbitrage, Paris |

| | |
|---|---|
| Riv.dell'Arb. | Rivista dell'Arbitrato, Milan |
| RIW | Recht der Internationalen Wirtschaft |
| s. / ss. | section / sections |
| SCAI | Swiss Chambers' Arbitration Institution |
| SCC | Arbitration Institute of the Stockholm Chamber of Commerce |
| SCC Rules | Rules of Arbitration of the Arbitration Institute of the Stockholm Chamber of Commerce |
| SchiedsVZ | Zeitschrift für Schiedsverfahren |
| SemJud | La Semaine Judiciaire, Geneva |
| SIA | Swiss Society of Engineers and Architects (Schweizerischer Ingenieur- und Architektenverein) |
| SIAC | Singapore International Arbitration Centre |
| SIAR | Stockholm International Arbitration Review |
| sic! | Zeitschrift für Immaterialgüter-, Informations- und Wetbewerbsrecht, Zurich |
| SIWR | Schweizerisches Immaterialgüter- und Wetbewerbsrecht, Basel |
| SJIR | Schweizerisches Jahrbuch für Internationales Recht |
| SJZ | Schweizerische Juristen-Zeitung (Revue Suisse de Jurisprudence, RSJ), Zurich |
| SKJ | Schweizerische Juristische Kartothek/Fiches Juridiques Suisses (FJS), Geneva |
| SPuRT | Zeitschrift für Sport und Recht |
| SR | Systematic Compilation of Swiss Federal Law (Systematische Sammlung des Bundesrechts, SR / Recueil systématique du droit fédéral, RS / Raccolta sistematica del dirito federale, RS) |
| STEP | Society of Trust and Estate Practitioners |
| STEP Journal | Wealth Structuring Analysis for Trust and Estate Practitioners (official magazine of STEP) |
| subs. / subss. | subsection / subsections |
| successio | Zeitschrift für Erbrecht / Revue de droit des successions |
| Swiss Int'l Arb.L.Rep | The Swiss International Arbitration Law Reports, Huntington |
| Swiss Rules | Swiss Rules of International Arbitration |
| SZIER | Schweizerische Zeitschrift für internationales und europäisches Recht (Revue suisse de droit international et de droit européen, RSDIE) |
| SZW | Schweizerisches Zeitschrift für Wirtschafts- und Finanzmarktrecht (Revue suisse de droit des affaires et du marché financier, RSDA) |

| | |
|---|---|
| SZZP | Schweizerische Zeitschrift für Zivilprozessrecht (Revue suisse de procédure civile, RSPC) |
| TDM | Transnational Dispute Management (Journal) |
| TLR | Tulane Law Review |
| TRIM | Agreement on Trade Related Investment Measures |
| TRIPS | Agreement on Trade related Aspects of Intellectual Property Rights |
| UAA | Uniform Arbitration Act (USA) |
| UCI | Union Cycliste Internationale (International Cycling Union) |
| UK | United Kingdom (of Great Britain and Northern Ireland) |
| UN | United Nations |
| UNCITRAL | United Nations Commission on International Trade Law |
| UNCITRAL Model Law | UNCITRAL Model Law on International Commercial Arbitration |
| UNCITRAL Notes | UNCITRAL Notes on Organizing Arbitral Proceedings |
| UNCITRAL Rules | UNCITRAL Arbitration Rules |
| UNCTAD | United Nations Conference on Trade and Development |
| UNIDROIT Principles | UNIDROIT Principles of International Commercial Contracts |
| US | United States |
| USA | United States of America |
| U.S.C. | United States Code |
| USD | United States dollar(s) / American dollar(s) |
| v. | versus |
| VIAC Rules | Rules of Arbitration and Conciliation of the International Arbitral Centre of the Austrian Federal Economic Chamber, Vienna |
| Vol. | volume |
| VUW Law Review | Victoria University of Wellington Law Review |
| WADA | World Anti-Doping Agency |
| WADA Code | World Anti-Doping Code |
| WAR | World Arbitration Reporter, Huntington NY |
| WIPO | World Intellectual Property Organization |
| WIPO Rules | WIPO Arbitration Rules |
| WTO | World Trade Organization |
| YCA | Yearbook of Commercial Arbitration, Deventer |
| YILJ | Yearbook of International Law and Jurisprudence |
| ZBJV | Zeitschrift des Bernischen Juristenvereins (Revue de la société des juristes bernois, RJB), Bern |

| | |
|---|---|
| ZK | Zivilkammer (Civil Section of the High Court of Bern, Obergericht des Kantons Bern) |
| ZPO | Swiss Code of Civil Procedure of 19 December 2008 (Schweizerische Zivilprozessordnung, ZPO / Code de procédure civile, CPC / Codice di diritto processuale civile svizzero, CPC), in force as from 1 January 2011 |
| ZR | Blätter für Zürcherische Rechtsprechung, Zurich |
| ZSR | Zeitschrift für Schweizerisches Recht, Zurich (Revue de droit suisse, RDS), Zurich |
| ZZZ | Schweizerische Zeitschrift für Zivilprozess- und Zwangsvollstreckungsrecht, Zurich |

# Foreword

*Hao Ma, President of AIPPI*

AIPPI has been improving and promoting the protection of intellectual property (IP) since 1897, celebrating its 120th anniversary in 2017. A truly global association, today it has 70 national and regional groups and over 9000 members from more than 125 countries and regions in Europe, the Americas, Asia, Oceania and Africa.

AIPPI is a non-affiliated, non-profit organization. Its members include lawyers, attorneys and agents working across all fields of IP in corporate and private practice throughout the world, as well as academics, judges, government officials and other persons interested in IP.

AIPPI promotes the value of all types of IP through its comprehensive analyses of existing and proposed IP laws and policies; working with government and nongovernment organisations to develop and improve international and regional treaties and agreements, and national laws; and hosting conferences and seminars. AIPPI is pleased to partner again with Wolters Kluwer to deliver the 'AIPPI Law Series' – another way to promote the protection of IP.

There are important policy considerations underlying the requirement of use of a trademark. It is closely linked to the function of trademarks as signs used in the course of trade to distinguish one proprietor's goods or services from those of others. To exercise that function and merit protection, a trademark must be used in the market. Particularly relevant in the context of this publication, the use requirement is also directly linked to the need to keep trademark registers free from unused trademarks so as not to unduly limit the activities of other traders.

Recognizing the central role of the use requirement, AIPPI conducted a study which resulted in a Resolution – 'The requirement of genuine use of trademarks for maintaining protection' (2011, Hyderabad).This volume builds on that work by providing detailed information about what qualifies as genuine use, how such use may be evidenced and the consequences of non-use of trademarks. For IP professionals, this volume provides practical reference across both prosecution and litigation.

AIPPI expresses its sincere thanks to Wolters Kluwer for its continued cooperation with AIPPI to promote the protection and harmonisation of IP. AIPPI also thanks all chapter authors (many of whom are AIPPI members) for their excellent work which has produced this valuable resource. AIPPI particularly thanks this year's Volume Editor, Eléonore Gaspar, who also Chairs AIPPI's Standing Committee on Trademarks which drives the association's important work on harmonisation of trademark law.

# Argentina

*Graciela Claudia Pérez de Inzaurraga*

## I. INTRODUCING GENUINE USE OF TRADEMARKS

### A. Legal framework

Genuine use is a statutory requirement to maintain protection of a registered trademark. The Argentine Trademark Law[1] foresees use of the mark as a condition to renew a registration as well as to avert cancellation thereof.

In this regard, Article 5 of the Law states that a registration can be renewed indefinitely, provided the mark has been used within the five-year period that precedes the expiry date of said registration.[2]

The Law further provides that property rights over a trademark shall expire upon issuance of a decision that cancels the registration.[3]

The rules on non-use cancellation significantly changed as of January 2018. With the aim of simplifying and expediting proceedings before administrative agencies, President Macri issued an omnibus Decree[4] that deals with a large number of laws and regulations, including the Trademark Law. Besides other amendments to the Trademark Law, the Decree introduces changes to the cancellation proceedings and to the definition of use required to avert cancellation.

1. Law 22362, which entered into force on 2 February 1981 and was partially amended by Decree 27/2018, which entered into force on 12 January 2018.
2. Art. 5, Law 22362: 'The term of duration of the registered trademark shall be of ten (10) years. It can be indefinitely renewed for identical periods if it has been used in the commercialization of a product, in rendering a service or as part of a business name, within five (5) years preceding each expiry date'.
3. Art. 23, Law 22362: 'Property rights over a trademark shall expire, [...] c) upon declaration of voidness or cancellation.'
4. Presidential Decree 27/2018, issued on January 10, 2018, published in the Official Bulletin on January 11, 2018, and entered into force as of 12 January 2018.

Prior to the amendment to the Trademark Law, use of the mark in Argentina on any product or service, or as part of a business or trade name, was sufficient to prevent cancellation in respect of any and all of the products or services covered by the registration for said mark. Thus, use on goods or services not related to those covered by the challenged registration was reckoned sufficient to prevent cancellation. The Law did not provide for partial cancellation of a registration. Cancellation operated upon request from a third party and was decided by the Federal Courts.[5]

As amended, the Law now foresees that to avoid cancellation, the mark should have been used to identify the goods or services covered by the registration, or goods or services similar thereto, or as part of a business or trade name that identifies an activity in a field related to the goods or services covered by the registration for which cancellation is sought.

Moreover, the new provisions state that cancellation can be declared either ex officio or at the request of a third party, and that disputes shall be decided by the National Institute of Industrial Property (hereinafter INPI), with subsequent revision before the Federal Courts.[6]

The changes to the cancellation proceedings under Decree 27/2018 require additional implementing regulations (e.g., to shape the dynamics of the cancellation procedures), which have not been issued at the time this chapter is being written. INPI Rule P-101[7] sorts out which amended sections of the Trademark Law are immediately operative and which are not. Those concerning non-use cancellation proceedings

---

5. The former text of Art. 26 of Law 22362 read: 'On request from a third party, cancellation of the mark shall be declared if it has not been used in the country within five (5) years prior to the filing of the cancellation action, except there were force majeure reasons. A mark registered in a class but not used in said class shall not be cancelled if it has been used in the commercialization of a product or in the rendering of a service included in other classes, or if it is part of a business name.'
6. Art. 26, as amended by Decree 27/2018 now states:

   'The National Institute of Industrial Property, either ex officio or at the request of a third party, and in accordance with the regulations which shall be issued, shall declare the cancellation, including partial cancellation, of the mark which has not been used in the country within five (5) years prior to the filing of the cancellation request, except where there were force majeure reasons.

   The decision issued on a trademark cancellation matter can be appealed within thirty (30) working days from the date notice is served, by means of direct revision before the National Federal Chamber of Appeals on Civil and Commercial Matters; the appeal writ should be filed at the National Institute of Industrial Property.

   A mark registered in a class or for specific goods or services shall not be cancelled if it has been used in the commercialization of a product or in the rendering of a service akin or similar to those covered by the registration, even if included in other classes, or if the mark is part of a tradename related thereto.

   Moreover, after the fifth year of registration and before the expiry of the sixth year, the trademark owner should file a sworn declaration concerning the use of the mark up to said date.'
7. Rule P001/2018, issued on 16 January 2018, published in the Official Bulletin on 23 January 2018 and entered into force on 24 January 2018.

and partial cancellation have been identified as not operative, so that they will not be applicable until the implementing regulation is issued.[8]

### B. Purpose and function of genuine use: policy considerations

Use of the trademark as a requirement to maintain rights thereupon was introduced to the Argentine trademark system in 1981, as of the entry into force of Law 22362. The explanatory recitals to the Law claim that the purpose underlying this requirement is to clean up the Register, so that marks which have been registered for a considerable time but never used no longer block access to registration of junior marks which strive to become players in the marketplace.

As noted by authors, in years immediately preceding the introduction of the use requirement to the Trademark Law, the number of registered trademarks was unreasonably large and not consistent with the country's modest performance in terms of industry and commerce. According to 1976-1978 statistics, Argentina was ranked as the fourth country in terms of number of trademark registrations (preceded only by Japan, United States, and France).[9] The exponential growth of registrations made it increasingly difficult and expensive to register new marks. Before the Law was amended to provide for non-use cancellation, a parasitic and lucrative business was made out of registering marks for the sole purpose of selling them to those actually interested in using them.[10]

The policy considerations underlying the genuine use requirement have been also highlighted in Court decisions and regarded as inherently associated with the basic function of trademarks. It is when actually used that the mark serves the purpose of identifying goods and services.[11]

## II. WHAT QUALIFIES AS GENUINE USE OF A TRADEMARK

### A. Nature of use

#### 1. *Use consistent with the function of the trademark*

In respect of use that suffices to support renewal or avert cancellation, the Law provides a broad concept, which encompasses use of the mark:

(i) in the marketing of a product,
(ii) in rendering a service, or
(iii) as part of the designation of an activity.

---

8. Rule P001/2018, Schedule II.
9. Iván A. Poli, *La caducidad por falta de uso en la nueva ley de marcas*, 3 Revista de Derecho Industrial 55 (Depalma, 1981).
10. Luis E. Bertone and Guillermo Cabanellas de las Cuevas, *Derecho de Marcas/2* (2nd edn, Heliasta, 2003), 339.
11. *Network Beauty & Fashion Cosmetica Ltda. v. Compañía General de Marcas*, Case 10634/05, CCCF, Div. II, 27/03/08.

Based on the reasons which inspired the incorporation of the use requirement into the Argentine trademark system, it may be construed that to comply with said requirement the sign should be used in the market to identify and distinguish goods and services, thus serving its primary function.

However, in stating that use of the sign as part of the designation of an activity – a situation that has been interpreted as use of the mark as part of the name of an enterprise which is actually operating – suffices to meet the use requirement, the Law admits a use of the sign in a function other than that inherent to a trademark.[12]

Use of an umbrella or house mark along with another trademark more prominently displayed has been reckoned as genuine use of the trademark.[13]

### 2. *Use in the course of trade*

Apart from the definition commented in section II.A.1. above, the Law does not provide further guidance to qualify the manner in which a trademark should be used to preserve rights thereupon. Consequently, the standards to meet the use requirement are found in the teachings of doctrine and court decisions.

Local courts have addressed the concept of use on a case by case basis and the decisions are contingent on the specific circumstances and the evidence produced in each case, with the aim of determining whether the use is serious or merely a trick to bypass the legal requirement.[14]

#### *i. Public use*

The issue has been subject to debate in doctrine. In any event, there is consensus in the sense that use should be public and visible and that it is not necessary that the sign should be affixed to the goods or their packaging, containers or tags to comply with the legal requirement, provided the mark is quoted or displayed in a manner which evidences the association between the mark and the product or service.[15]

#### *ii. Commercial purpose*

Beyond such points of consensus, some authors adopt a flexible approach, and understand any manner in which the mark is made known or visible is sufficient to meet the use requirement, to the extent the sign as used serves to identify the product or service or the origin thereof. In this line of reasoning, genuine use would

---

12. Poli, *supra* n. 9, 69-70; *Harrods Limited v. Harrods Buenos Aires Limited* and *Harrods Buenos Aires Limited v. Harrods Limited*, Cases 4112/01 and 11779/02, CCCF, Div. III, 24/02/2011.
13. *Frigorífico Bordon S.A. v. Borden Inc.*, Case 19989/94, CCCF, Div. I, 23/06/1998.
14. *Boehringer Ingelheim KG v. Microsules y* Bernabó SA., Case 15411/91, CCCF, Div. II; *Valydar S.R.L. v. Integración Hogar S.A.* and *Integración Hogar S.A. v. Valydar S.R.L.*, Cases 7288/03 and 3212/03, CCCF, Div. III, 20/05/2010.
15. Poli, *supra* n. 9, 69; Bertone and Cabanellas, *supra* n. 10, 358; Ernesto Aracama Zorraquin, *Sobre el tema del uso obligatorio de la marca*, 97:5423 El Derecho 3, (1982); Jorge Otamendi, *Derecho de marcas* (10th ed., Abeledo Perrot, 2010), 241.

encompass the display of the mark in advertising, leaflets, invoices, catalogues, and also actions in preparation for actual use (e.g., conducting surveys, securing approval from health authorities or procuring the design, manufacture or provision of labels or packaging).[16]

From a stricter perspective, other commentators consider that to comply with the requirement, the products or services should be actually available in the market, i.e., the mark should be a means for the relevant public to get access to the product or service. Activities undertaken in preparation to launch the product or service (preparatory or preliminary uses) might be relevant to avert cancellation provided that subsequent actions by the trademark owner demonstrate a genuine intent to use the mark. Thus, if the products or services were not put into the market within a reasonably short time after the preparatory actions took place, the latter would not satisfy the use requirement.[17]

Stricter positions disregard preparatory actions as sufficient to meet the use requirement, since while they may show the intention to market a product or render a service, only if and when the goods or services are made available to the public is the requirement met. Moreover, such preliminary activities are typically not visible to the public but handled within the realms of the trademark owner's enterprise or organization.[18]

On the other hand, it has been noted that sale, rental or lease is not the only form in which a mark can be genuinely used. Since the rationale underlying the Law and the court decisions is that the trademark should identify either a product or a service actually available in the market, or serve to identify the activities of a company, business or organization, then whether the product is sold or given for free or whether the activities are for profit or not, they would constitute serious or actual use of the trademark.[19] A more restrictive interpretation only considers that there is genuine use where there is for-profit marketing or activity.[20]

Irrespective of doctrinal musings, the decisions handed down by the Federal Courts shed light on the matter and outline the applicable standards. The main relevant considerations which shape what is interpreted as genuine use of the mark in the course of trade can be summarized in the following bullet points:

- Flexible criteria apply in determining whether there has been use of a trademark, so as to encompass any actions which result in making the sign visible in the market. In the context of the Trademark Law, the terms 'use' and 'marketing' ('commercialization') are deemed equivalent, the latter being understood in a broad sense, as including any actions of trade and any other ways to trade the product.[21]

16. Aracama Zorraquin, *supra* n. 15, 3.
17. Otamendi, *supra* n. 15, 214-242.
18. Bertone and Cabanellas, *supra* n. 10, 361.
19. Otamendi, *supra* n. 15, 242.
20. Bertone and Cabanellas, *supra* n. 10, 363.
21. *Bizin, Mario v. Iwch, Inc.*, Case 8197, CCCF, Div. II, 18/06/1991.

- The term 'commercialization' should not be interpreted according to its technical meaning under the Commercial Code, but in a broader sense, so as to include any actions where the trademark is put into the local market.[22]
- References to the trademark on the company's internal documents do not suffice to meet the use requirement unless backed up by evidence of transactions in the marketplace.[23]
- Actions in preparation to the commercialization of the product do not constitute real intent to use where the launch of the product does not take place in a reasonably short time thereafter, in particular if considerable time elapsed since the mark became vulnerable to cancellation on the grounds of lack of use.[24]

### 3. *Use by the owner or with its consent*

To maintain rights over its registration, the sign should have been used by the trademark owner or with its consent. No express authorization from the owner to the third party is required for the use by the latter to inure to the benefit of the registered proprietor.

However, use of the mark by infringers or by third parties who may have adopted identical signs but which are not lawfully related to or authorized by the trademark owner would not qualify to meet the use requirement. Likewise, the use of the trademark by the agent, representative or licensee, once the agreement has been terminated or in breach of the agreement, cannot be invoked by the trademark owner to maintain rights over the registration.[25]

Case law has acknowledged the efficacy of use by an agent[26] or a licensee[27] of the trademark owner or by a third party with consent from the proprietor[28] to prevent cancellation of a registration.

## B. Volume, duration and frequency of use

As commented above, the Law does not set any definite standards in relation to the volume and frequency of the use which would satisfy the requirement to maintain trademark rights.

Doctrine and case law have consistently held that this issue should be analysed on a case-by-case basis and taking into consideration factors such as the nature of

22. *Ibid.*
23. *Uriburu, Miguel P. v. Corporación Argentina de Productores de Carne*, Case 34346/95, CCCF, Div. III, 11/04/2000.
24. *Roemmers S.A. v. Chemotecnica Sintyal*, Case 9630/96, CCCF, Div. I, 18/11/1997; *Citrovita Agro Industrial Ltda. v. Salierno, Juan*, Case 2720/05, CCCF, Div. II, 28/05/09.
25. Bertone and Cabanellas, *supra* n. 10, 372-373.
26. *Bizin v. Iwch, supra* n. 21.
27. *Bordon v. Borden, supra* n. 13.
28. *Atomplast S.A. v. Société d'Étude et d'Application Industrielle de Brevets S.E.A.B.*, Case 8582, CCCF, Div. III, 26/08/1992.

the product or service, the characteristics and turnover of trademark owner and those of the specific business field.[29]

The key issue to determine whether use of the sign meets the legal requirement is neither the volume of the operations nor the frequency thereof, but the serious intention of the owner to put the goods or services bearing the trademark in the market. In the words used by local courts, genuine trademark use is not a matter of quantity but quality.[30]

The Law only requires that the trademark is actually used within the period of five years either preceding the expiry date (to obtain renewal) or the date the cancellation action is lodged (to avert cancellation). Use need not be continuous but should have occurred at some time within such period.[31]

Depending on the circumstances of the case, infrequent or small volume sales have been deemed enough to comply with the use requirement.[32]

If use of the mark is not consistent with the typical operations or turnover of the trademark owner, the nature of the product, or the commercial practices in the field, such use would not suffice to meet the statutory standards and would be deemed a means to merely circumvent the purpose of the Law and therefore a strategy to avert cancellation or indefinitely renew registrations without actual intention of using the mark (the so-called 'token use').[33]

### C. Use of a sign

So that use of the trademark satisfies the requirement to maintain the rights, the sign should be used as registered. Yet, this general rule is not absolute, and use of the mark featuring variations that do not alter its distinctive character would qualify as sufficient use of the registered sign, by virtue of the provisions of Article 5.c.2. of the Paris Convention (to which Argentina is party).

Addition or removal of accessory or secondary elements (such as numbers, acronyms or designs) has been deemed not to alter the distinctive character of the mark. In this line of reasoning, the courts refused cancellation of a composite mark (stylised typeface and device) taking into consideration that the most distinctive element of the mark had been intensively used.[34] In this line of thinking, in a cancellation claim against the registered mark VENECIA (which was being used as VENEZIA), the court held that the variation in spelling was inconsequential, because the meaning of the sign did not change, its sound remained identical – according

29. Otamendi, *supra* n. 15, 243-246; *Exxon Corporation v. Sol Explotación de Petróleo S.A.*, Case 9893, CCCF, Div. II, 22/03/1994; *Valydar v. Integración Hogar, supra* n. 14.
30. *Bizin v. Iwch, supra* n. 21; *Atomplast v. S.E.A.B., supra* n. 28.
31. *Tintas Renner S.A. v. Colorín Industria de Materiales Sintéticos S.A.*, Cases 21778/94 and 1231/97, CCCF, Div. II, 21/12/2000.
32. *Munin, Adolfo v. Bagley S.A.*, Case 5083, CCCF, Div. II, 29/05/1987; *Renner v. Colorín, supra* n. 31.
33. *Toyota Jidosha K.K. v. Deza, Luis E.*, Case 11757, CCCF, Div. I, 29/10/1998.
34. *Exxon v. Sol, supra* n. 29; *Syncro Argentina S.A. v. Dr. Lazar y Cia. S.A.*, Case 7904, CCCF, Div. II, 25/03/1994.

to local pronunciation – and the typeface was significantly close to that of the registered sign.[35]

However, if there are several registrations for a mark that feature different visual or spelling variations thereof, the use of one of said variations would only satisfy the requirement to maintain rights upon the registration which protects the variation actually being used and not the other ones.[36] Yet, if a sign is registered both as a word mark and in stylised script or with a logo, use of the latter is enough to support renewal of both the word mark registration and the stylized one.[37] Similar criteria would apply if there were registrations for the same stylized mark or logo with and without claim for colours. Use of the coloured mark would suffice to keep rights over the registration that does not claim colours. But if there were registrations for the same mark each one displaying and claiming different colours, use of one of them would not amount to use of them all.

Where a sign is only registered as a word mark, use in any typeface or bearing any visual features is regarded sufficient to meet the legal requirement. Also, if there were different registrations for different marks and they were used in combination, such association would amount to use of each of the marks.

In any event, no clear lines can be drawn on this issue, which requires careful consideration of the circumstances and factors in each particular case.

## D. Use in relation to goods and services

In light of the recent amendments to the Trademark Law, different standards will apply to assess whether use on certain goods or services or as part of the name which identifies a business or activity, depending on whether such use is invoked to support renewal or to avert cancellation.

### *1. Use for each of the designated goods and services*

Article 26 of the Trademark Law, in its former version, expressly stated that a mark could not be cancelled if used on any goods or services or as part of the designation of an activity. Therefore, provided the use was serious (as per the standards summarized in the previous section), it was not necessary to use the mark in connection with all goods or services for which it was registered. Even use on goods or services for which it was not registered or as the name of a business or activity unrelated to the goods or services covered by the registration was sufficient to prevent cancellation.

Use limited to a portion of the registered goods or services or use on goods or use on goods or services different to those for which the mark was registered was enough to avoid cancellation of all the registrations for said mark and for all goods and services covered by said registrations. Use of the sign as the name of a commercial

35. *La Veneciana S.C.A. v. Dondero Hnos. y Cia. S.A.*, Case 24989, CCCF, Div. II, 12/10/1992.
36. Otamendi, *supra* n. 15, 247-250; Poli, *supra* n. 9, 70-72; *Cannon S.A. v. Balenciaga S.A.* and *Balenciaga S.A. v. Cannon S.A.*, Cases 11203/95, 901/91 and 6020/93, CCCF, Div. III, 27/09/1995.
37. Poli, *supra* n. 9, 71.

enterprise or non-profit activity, albeit in a field unrelated to the goods or services for which the mark was registered, also met the legal requirement.[38]

The purpose underlying this provision was to allow trademark owners to maintain defensive trademarks, i.e., to secure protection for the marks, in particular for well-known ones, not only in relation to the goods of services they actually identify in the market but also those related and unrelated thereto.

While lack of use in respect of the registered goods or services did not result in cancellation, such circumstance was often taken into account and led to apply a more lenient criteria in determining likelihood of confusion, in conflicts where a senior mark was registered but not used for the goods or services on which a junior mark was used or sought registration.[39]

As mentioned in the introduction to this chapter, this provision of the Law was amended by Decree 27/2018, which came into force on 12 January 2018. Article 26 of the Trademark Law now provides that, in order to avert cancellation, a mark should have been used on the goods or services covered by the registration or on goods or services similar thereto, or as part of the designation of an activity related to the goods and services covered by the registration.

The change involves a substantially stricter standard as to what constitutes sufficient use in terms of the scope of the registrations vis-à-vis the goods, services or activities the sign actually identifies in the marketplace.

Unlike in the former regime, a registration would now be exposed to partial cancellation, depending on the identity or similarity between the goods or services covered by the challenged registration and those on which the mark is being used, or on the relationship between the activities the sign actually designates and the registered goods or services.

In future cancellation proceedings, where a mark is not used for the same goods covered by the challenged registration, it will become a critical issue to establish how close the similarity between the goods or services on which the mark was used and those for which it is registered should be for the use to qualify for averting cancellation.

If construed in accordance to their literal meaning, the words 'akin' and 'similar' used in the new text of Article 26 seem to exclude from qualifying use that of the mark which identifies goods or services that, though not akin or similar to those covered by the registration, are related thereto. Should such interpretation prevail, if a mark is registered for 'suits, coats, and perfumes', use thereof to identify 'suits' might be rendered sufficient to avert cancellation in respect of 'suits and coats' (because these products are somehow similar) but not in respect of 'perfumes' (because while there is a relationship between clothing and fragrances, said goods are not similar in nature or function).

Notably, the Law, as amended, uses the word 'related' in relation to the use of the mark as part of the name of an activity. Consequently, the words 'akin' and 'similar' to qualify use on goods or services different than those listed in the registration seem

38. *Bordon v. Borden*, *supra* n. 13.

39. *Ardu Hnos. S.R.L. v. Daimler-Benz AG*, Case 2613/93, CCCF, Div. II, 23/11/1993; *Zanchetti Hnos. S.A. v. Zanella Hnos.*, Case 7221/94, CCCF, Div. III, 01/03/1995; *Laboratorios Beta S.A. v. Cargos S.R.L.*, Case 18919/96, CCCF, Div. II, 13/07/2000.

to have been purposefully chosen and raise concerns as to whether use on related goods or services might be reckoned sufficient to meet the use requirement under the new Law.

It is too early to predict which criteria will apply to rule on partial cancellation, i.e., how the similarity between the scope of use and that of the registration will be analysed, and whether the wording of the Law will be construed also to encompass use on related goods or services. It is likely that decisions will rely on similar factors to those considered when ruling on opposition or infringement conflicts between marks that do not designate or are not used for the same goods or services. Consequently, the assessment is expected to balance facts and circumstances such as the nature and function of the goods or services, the channels of trade, the relevant public, the commercial practices and trends, the recognition achieved by the mark, the good faith of the parties and the legitimate interest thereof.

When use of the mark as a condition to maintain rights thereupon was incorporated into the Argentine system, it was discussed whether it was possible to immediately apply such provision to cancel registrations or invalidate renewal of marks which were registered under the former Law. At such time, there were divergent opinions and court rulings. While some decisions asserted that the use requirement was applicable to the marks registered under the previous Law only after their renewal under the new Law,[40] the prevailing criterion was that rules on cancellation did not apply until after five years from the date the Law came into force, insofar as genuine use was a new condition imposed upon trademark owners, and it was fair to allow them the time foreseen under the new provisions to commence use of the signs, before exposing the registrations to expungement.[41]

In line with what was ultimately decided when Law 22362 came into force, it is expected that stringent rules concerning use which qualifies to avert cancellation foreseen under Decree 27/2018 will not apply to registrations granted under the previous regime until after five years have elapsed since the coming into force of the amendments, i.e., until after 12 January 2023.

### 2. *Use for a broader category of goods and/or services*

The Law maintains the former broader standard in respect of use required to support renewal of a registration, since Article 5 of the Trademark Law has not been amended. An affidavit of use (or sworn declaration of use) should be submitted along with the renewal application, where the registrant should quote on which goods, services or activities has the mark been used.

Provided the use qualifies as genuine (as per the standards commented on in sections II.A-C above) and takes place within the timeframe and in the territory to be discussed in sections II.E-F below, use of the mark in relation to any goods or services, or as part of the name that identifies an activity, suffices to lawfully renew a registration, irrespective of the identity or relationship between said goods, services or activities, and the goods or services covered by the registration.

40. *Matarazzo S.A. v. D.N.P.I*, Case 2705, CCCF, Div. III, 21/12/1984.
41. *Parfums Rochas S.A. v. Imbrosciano Hnos. S.A.*, Case 7994, CCCF, Div. III, 20/12/1991.

Decree 27/2018 further amends Law 22362 by imposing upon the registrant the burden of filing an affidavit of use between the fifth and the sixth year from the date the original registration was granted and between every fifth and sixth anniversary date in subsequent renewals. The decree has not foreseen the effects resulting from not complying with such obligation. It is unlikely that failure to submit the declaration of use might entail lapse of the registration, inasmuch as the grounds for loss of property rights – and trademark rights are considered property rights – should be expressly stated by the Law. Since Decree 27/2018 did not amend the grounds for lapse of trademark rights, the omission to provide the fifth-year declaration of use should have a different effect, still to be determined by implementing regulations. This provision has been inserted in Article 26 of the Law, which deals with use sufficient to avert cancellation and not to qualify for renewal. Therefore, it is still open to interpretation whether use on any kind of goods or services or to identify any activities (even if not similar or related to the goods or services covered by the registration in respect of which the declaration is filed) would suffice to produce a lawful fifth-year use sworn declaration.

### E. Territorial scope

The Law expressly states that the mark should be used in Argentina to avert cancellation (Article 26 of the Trademark Law) but does not mention where the use should take place for it to be sufficient to validate renewal and issue the requisite affidavit (Articles 5 and 20 of the Trademark Law).

This inconsistency has been construed by some authors in the sense that while use in the country is required to avoid cancellation, any actual use, albeit in foreign jurisdictions would suffice to support renewal.[42] This interpretation would set a double territorial standard of use, namely, a more flexible one to validate renewal and a stringent one to prevent cancellation. Since Law 22362 came into force, there have been no court decisions on this matter, which remains an academic topic.

Where a mark is used to identify products, it is not necessary that they are manufactured in the country. Use on imported products which are commercialized in the Argentine market satisfies the statutory requirement.[43]

Use of the mark in rendering services or as a trade or business name or as the designation of an activity requires the service or the activities to be conducted in Argentina.[44]

Use of the marks on goods which are not to be sold in the country but only for export would qualify to maintain rights to the extent the first sale (i.e., the sale to the exporter) takes place in Argentina.[45]

Use need not extend to the entire territory. Use on a limited geographical area, and even the sale of goods bearing the mark at a single store has been acknowledged

42. Poli, *supra* n. 9, 74. *See* opposite interpretation in Bertone and Cabanellas, *supra* n. 10, 288-289.
43. *Bizin v. Ywch*, *supra* n. 21.
44. Bertone and Cabanellas, *supra* n. 10, 374-378; Otamendi, *supra* n. 15, 243.
45. Bertone and Cabanellas, *supra* n. 10, 375-376; Otamendi, *supra* n. 15, 243.

as genuine, when other circumstances of the case supported the finding of serious use.[46]

Doctrine is consistent in that advertising through any media, including newspapers and magazines which are sold in Argentina, and radio or television commercials which are aired in the country, does not qualify as genuine use if not backed up by the actual presence of the goods in the local market or the rendering of services or the performance of activities in the country.[47] Though this situation has not been raised frequently in cancellation disputes, court decisions support this interpretation.[48]

Similar criteria would rule on use of trademarks on mail-order catalogues or the internet. There will be use if goods bearing the mark are available in the Argentine market, otherwise, the mere circulation of catalogues or the possibility to access web pages or sites where the mark is mentioned would be ineffective to assert genuine use of the mark in the territory.[49]

### F. Relevant time period and its calculation

The Trademark Law states that in order for a registration to be eligible for renewal, the mark should have been used within five years counting backwards from the expiry date of the registration. To avoid cancellation, the mark should have been used within five years counting backwards from the date the cancellation claim is filed.

As from the entry into force of the Law that requires a mandatory mediation process be completed as a preliminary condition to take a dispute before courts,[50] case law has determined that in order to avoid cancellation, use of the mark should have taken place within five years prior to the date notice of the citation to a mediation hearing is served upon the owner of the challenged registration. If a different position were adopted, once alerted of the threat to its registration, the trademark owner would be in the position to cure non-use in the period running as from the date notice of the mediation hearing is served and that of the actual filing of the cancellation complaint (once the mediation proceedings are completed). In such scenario, the plaintiff would be at the mercy of the trademark owner, and the reasons that inspired the introduction of the use requirement would not be served. Use after notice of the mediation proceedings is served would appear as a mere trick to circumvent the legal requirement, since the registrant had plenty of time before such alert to genuinely use the mark.[51]

The requirement to use the mark begins as of the date the first registration is granted. The time while the application for registration is pending is not calculated for the purpose of the use requirement, since cancellation operates against registration.[52]

46. *Prince Manufacturing, Inc. v. Princz, Esteban*, Case 5974, CCCF, Div. II, 16/09/1988.
47. Bertone and Cabanellas, *supra* n. 10, 377-378; Otamendi, *supra* n. 15, 243.
48. *Bayer AG v. Laboratorios Maver S.A.*, Case 8149/06, CCCF, Div. II, 10/08/2011.
49. Otamendi, *supra* n. 15, 243.
50. Law 24573, as amended by Law 26589.
51. *Massalin Particulares S.A. v. Nobleza Piccardo S.A.*, Case 4702/97, CCCF, Div. I, 23/05/2002; *Citrovita v. Salierno*, *supra* n. 24; *Milkaut S.A. v. Industrias Lácteas Buenos Aires S.A.*, Case 8102/07, CCCF, Div. I, 30/11/2010.
52. Poli, *supra* n. 9, 64.

Once the registration is granted, if use of the mark is discontinued for over five years, it becomes vulnerable to cancellation. Renewal of the registration does not reset the timing for using the mark, since it merely extends validity of an existing registration.[53]

As commented in section II.B. above, to satisfy the use requirement it is not necessary that the mark has been continuously used along the whole five-year period. It suffices that use takes place at any time within the five-year period which precedes the renewal deadline or the filing of a cancellation action.[54]

On the other hand, use of the mark which may have occurred before to the five-year period running backwards from the filing date of the cancellation action (or the start of the mediation proceedings) has been considered irrelevant to avert cancellation.[55]

## III. EVIDENCE OF GENUINE USE OF A TRADEMARK

### A. When use should be proved: timing of lack of genuine use claims

#### *1. When filing an application for a trademark*

The Argentine Trademark Law does not require actual use of the mark as a condition to obtain registration. Therefore, it is not mandatory to submit any evidence of use in the course of the registration proceedings. Eventually, evidence may be voluntarily produced with the aim of proving acquired distinctiveness, in case registrability of a sign was objected to on the grounds of lack of distinctiveness.

As mentioned in section I.A. above, genuine use of the mark is required to maintain trademark rights, namely, to validate renewal of the registration and to avert its cancellation.

No evidence of actual use is required when applying for renewal of a registration. The Law only requires a sworn declaration (affidavit) where the applicant quotes on which goods, services or activities the mark has been used within the five-year period which precedes the renewal deadline.

Use of the mark on any goods or services, or as part of the name that identifies an activity suffices to support renewal in respect of all goods or services for which the mark is registered.[56] Therefore, provided the declaration of use is timely submitted, renewal will be granted, regardless of whether use concerns goods, services or activities different or unrelated to those covered by the registration. The affidavit should be filed along with the renewal application. Where the affidavit is not submitted, the Trademark Office shall summon the applicant to produce it, under penalty of refusing renewal of the registration.

The Trademark Office does not review the truthfulness of the declaration of use. Yet, this formal requirement is inherently associated with the condition that there

53. Otamendi, *supra* n. 15, 247.
54. *Banco Francés S.A. v. Najlis de Rosario*, Case 1816/93, CCCF, Div. I, 04/06/1993.
55. *Colorín S.A. v. Cintoplom S.A.*, Case 8415, CCCF, Div. II, 09/08/1981.
56. *See* II.D.2. above.

should be genuine use of the mark to validate its renewal.[57] While the affidavit is not equivalent to proof of use, a false declaration renders the renewed registration null and void, as the renewal is obtained in violation of the provisions of the Trademark Law,[58] and such fraudulent act cannot be validated.[59]

### 2. *In the course of opposition proceedings*

It is not mandatory to submit proof of use in the course of opposition proceedings,[60] unless the applicant brings a cancellation action against the registration invoked by the opposer as a defence to clear the path for acceptance of the junior trademark.

Where there is no cancellation action involved, lack of use of the opposing mark has been reckoned by local courts as one of the factors which may call for more lenient criteria, i.e., a circumstance which may favour the coexistence of otherwise closely similar signs.[61]

### 3. *In cancellation proceedings based on lack of use*

Submission of evidence of use is only required in the course of cancellation proceedings.

Before the recent amendment to the Argentine Trademark Law, cancellation on the grounds of lack of use could only be declared by courts and at the request of a party bearing legitimate interest. As of January 2018, the changes in the Law foresee that cancellation can be declared ex officio, and that INPI will have authority to rule on cancellation disputes, INPI's decision on the matter being subject to direct revision before the Court of Appeals.

In any event, inasmuch as cancellation of a trademark results in revocation of property rights, severe criteria are applied when deciding on the loss of said rights.[62] Both in the previous and the current regime, cancellation on the ground of lack of use shall not operate automatically, but upon a declaration issued as a result of the action taken against the registration by a party and provided cancellation serves a legitimate interest or purpose. Therefore, lack of genuine use is not the only condition for a registration to be cancelled. It is also necessary that a party having legitimate interest takes action to cancel the registration and a declaration is rendered whereby the rights are expunged.

57. Poli, *supra* n. 9, 66; *Laboratorios Andrómaco v. Boehringer Ingelheim KG*, Case 10765, CCCF, Div. III, 28/12/1994.
58. Law 22362 expressly states that trademarks registered in violation of the provisions of the said Law are null and void; *Syncro Argentina S.A. v. Química Montpellier*, CCCF, Div. II, 05/03/1996; *Speedo Holdings v. Metrikis S.A.*, Case 3271/95, CCCF, Div. II, 04/07/1996; *Laboratorios Duncan S.C.A. v. Fada S.R.L.*, Case 2686/94, CCCF, Div. II, 09/09/1999.
59. Aracama Zorraquín, *supra* n. 15, 6; Otamendi, *supra* n. 15, 259, Poli, *supra* n. 9, 67.
60. *Molinos Rio de la Plata S.A. v. Cristina, Luis Alberto*, Case 1201/97, CCCF, Div. II, 16/09/1997.
61. *Sheik Corporation S.A. v. Boehringer Ingelheim KG*, Case 8694/94, CCCF, Div. I, 08/10/1996; *NBA Properties Inc. v. Guevara, Ricardo*, Case 22930/94, CCCF, Div. III, 20/11/1997.
62. *Coppel S.A. de C.V. v. Antonioli, Carlos E., Case* 3269/11, CCCF, Div. II, 04/08/15.

It has been widely accepted that there is legitimate interest to seek cancellation of a registration that has been invoked as grounds for opposition,[63] or raised as a block against the registration of a junior mark, or as a defence to fend off an infringement action.

This rationale inspires the decisions that refuse cancellation of a registration invoked as grounds for opposition, where the opposition is found without merit because the allegedly conflicting signs are deemed not to be confusingly similar. Since the marks are considered different enough to coexist, the applicant for the junior mark lacks legitimate interest in removing the opposing mark, regardless of the fact that the latter may not have been used.[64]

Along the same reasoning, where several registrations were raised against a junior application, and said junior mark is found to be confusingly similar to one of the senior registrations, the latter not being vulnerable to cancellation, no interest is acknowledged for the applicant to pursue cancellation of the other registrations invoked by the opposer, since the block against the junior mark would remain, regardless of any of the other registrations being cancelled.[65]

Legitimate interest has also been acknowledged where a party seeks cancellation of a registration for a mark identical to that the plaintiff had been using for considerable time without objection from the registered proprietor.[66]

#### 4. *As a defence in infringement/cancellation proceedings*

Cancellation on the grounds of lack of use requires that a claim or counterclaim or independent action be lodged, any of these alternatives being admissible. A registration will not be cancelled unless a claim, counterclaim or action seeking cancellation thereof is filed. If in the course of an infringement or opposition dispute there is a finding of lack of use, such circumstance might be taken into consideration for the purpose of applying more lenient criteria, but shall not entail cancellation of the registration.[67]

### B. Burden of proof

The Argentine Trademark Law does not provide for an exception to the general rule, whereby the party pursuing the action is the one who bears the burden of proof.[68] However, since proving negative facts entails an almost impossible mission, doctrine

63. *Farmasa Farmacéutica Argentina S.A. v. Syncro Argentina S.A.*, Case 1585, CCCF, Div. I, 28/02/1991.
64. *Dorsay Industria Farmacéutica Ltda. v. Laboratorios Dr. Gador y Cia. S.A.*, CCCF, Div. III, 17/03/1989.
65. *Molinos Río de la Plata S.A. v. La Arrocera Argentina S.A.*, Case 1594/91, CCCF, Div. III, 17/07/1997.
66. *Duncan v. Fada, supra* n. 58.
67. *Quest Intl. Fragrances Flavours Food Ingredients UK Limited v. Blashig de Fierro, Maria*, Case 4860/97, CCCF, Div. I, 05/02/1998; *Sheik Corporation S.A. v. Boehringer Ingelheim KG*, Case 8694/94, CCCF, Div. I, 08/10/1996.
68. Otamendi, *supra* n. 15, 389.

and case law consistently assert that, insofar as the owner of the challenged trademark is the one who has easy access to the evidence of use, it is said party who should submit the proof to rebuke the alleged lack of use.[69]

In this regard, courts consider that both parties in a cancellation action have the duty to bring forward the evidence for the Judge to reach a fair decision[70] but that the burden should be placed on the party which is in a better position to access and produce the evidence.[71] Even though the Trademark Law has not shifted the burden of proof to the defendant, the latter is not relieved from submitting evidence.[72] Moreover, failure to produce evidence of use by the owner of the mark which cancellation is sought has led courts to presume lack of genuine use.[73] Some decisions have gone even further and directly place the burden of proof on the owner of the mark for which cancellation is pursued.[74]

## C. Relevant evidence

### *1. Use on the product or its packaging*

As commented on in section II.A.2.i. above, for use of a sign to qualify as genuine it is not necessary that the mark be affixed to the products or its packaging. Therefore, while evidence of such use is certainly admissible and helpful, it is neither mandatory to produce it and, ultimately, it may not be the preferred means to prove use.

### *2. Use in advertising*

Use of the mark in advertising or references to the mark in directories or catalogues is deemed a preliminary evidence that products or services have been offered in the market. However, to avoid cancellation, it would be necessary to supplement such evidence with proof that goods or services bearing the mark have actually been available in the local market.[75] Advertisements of the mark in foreign media was regarded not enough to prevent cancellation, since while the commercials were aired in Argentina, the product had not been sold in the country.[76]

---

69. Bertone and Cabanellas, *supra* n. 10, 389-390, Poli, *supra* n. 9, 64-65.
70. *Vivat Holding plc v. Yannit S.A.* and *Yannit S.A. v. Vivat Holding plc*, Cases 21367/95 and 21372/95, CCCF, Div. III, 30/05/1996.
71. *Molson Canada v. Cervecería y Maltería Quilmes S.A.*, Cases 15717/04, 15316/03 and 4577/02, CCCF, Div. III, 29/03/11; Coppel *v. Antonioli*, *supra* n. 62.
72. *Colorin v. Cintoplom*, *supra* n. 55.
73. *Kenzo Société Anonyme v. Vilches, Pedro*, Case 2998/92, CCCF, Div. II, 16/05/1996; *McLean S.A. v. Viviani, Marcelo*, Case 19643/93, CCCF, Div. I, 05/03/2002.
74. *Industria y Comercio de Cosméticos Natura Ltda. v. Losada, Adolfo*, Case 20497/96, CCCF, Div. I, 15/12/1998; *Milkaut S.A. v. Industrias Lácteas Buenos Aires*, Case 8102/07, CCCF, Div. I, 30/11/2010.
75. Bertone and Cabanellas, *supra* n. 10, 359-362.
76. *Bayer v. Maver*, *supra* n. 48.

Along this line of thinking, courts held that the circumstance that a trademark was not listed in a directory did not suffice to prove lack of use thereof.[77]

### 3. *Use on the internet*

Proof that the mark has been used on the internet would be effective to avert cancellation if supported by additional evidence that the goods or services are available in the Argentine market. Supplementary evidence to such purpose may consist in proof of actual transactions with customers in Argentina. If not backed up by actual transactions to local customers, evidence which merely shows that the mark is displayed on a website or webpage accessible from Argentina would not suffice to prove genuine use.

### 4. *Form of evidence*

There is no limitation on the type or form of evidence admissible to prove use or lack thereof.[78] Suitable proof includes, but is not limited to, commercial records and documents, reports and testimonies.[79] In assessing the relevance of the proof, courts have taken into consideration whether the elements (i) are effectively associated with the use of the mark which cancellation is pursued and (ii) convincingly show that the mark has actually identified goods, services or activities.[80]

To effectively prove use of the mark, documentary evidence such as invoices, catalogues, price lists, leaflets or advertising material should be backed up by account records which confirm that commercial transactions actually existed.[81]

Isolated invoices which bear the mark in abbreviated form were disregarded as conclusive evidence of use of the challenged mark, since the trademark owner failed to submit other elements to prove that said abbreviation unequivocally referred to the challenged mark.[82] In a similar line, documents concerning the purchase of labels and tags by the owner of the mark, but where said mark is not mentioned, were deemed insufficient to associate such activities to the intent to use the mark.[83]

### 5. *Distinction between use of a company name and use on the services offered by the company*

As noted in section II.A.1. above, the Argentine Trademark Law expressly states that use of the mark as part of the designation of an activity qualifies to maintain trademark rights. This provision has been interpreted in the sense that use of the

77. *P.I. Rivero y Cia. S.A. v. Laboratorios Andrómaco S.A.*, Case, CCCF, Div. III, 23/05/1996.
78. *Andrés Lagomarsino e Hijos S.A. v. Morixe Hermanos S.A.*, Cases 2163/06 and 5667/06, CCCF, Div. II, 02/07/2015.
79. *Kenzo v. Vilches*, *supra* n. 73.
80. *Lagomarsino v. Morixe*, *supra* n. 78.
81. *Uriburu v. CAP, supra* n. 23; *Toyota Jidosha K.K. v. Deza, Luis*, Case 11757/94, CCCF, Div. I, 29/10/1998.
82. *Network Beauty v. Fashion Cosmetica*, *supra* n. 11.
83. *Valydar v. Integración Hogar*, *supra* n. 14.

company name in internal documents does not satisfy the legal requirement if such use is not visible to the public,[84] i.e., the sign should identify the operations of a party and not merely be a name on its paperwork.

## IV. JUSTIFICATION FOR NON-USE

### A. Obstacles to use

Article 26 of the Trademark Law states that force majeure can be invoked as an excuse to justify lack of use of the trademark and avoid cancellation of the registration.

In the context of Trademark Law, force majeure is understood as encompassing circumstances which prevent use of the mark and are beyond the control and will of the trademark owner, i.e., circumstances which cannot be predicted or, if predictable, cannot be avoided.

To qualify as force majeure reasons, the circumstances should absolutely block use of the mark in the usual course of business.[85] Circumstances which merely make use of the mark more difficult or expensive would not suffice to justify lack of use.[86]

Force majeure has been reckoned as a valid excuse for non-use in cases where laws or regulations ban use of the mark (e.g., restrictions on the importation of certain products, prohibition to use marks which consist of or include words in foreign languages), or commercialization of the products to prior authorization from government authorities.[87]

Case law has not set definite criteria as to whether insolvency or bankruptcy of the trademark owner provides justification for non-use. The fact that the bankrupt is prevented from disposing of its assets has been regarded as an excuse not to use the mark.[88] Conversely, it was held that insolvency does not fall within the notion of force majeure, because the situation which led to the bankruptcy may not have been beyond the bankrupt's control.[89]

Adverse financial or marketing conditions would not suffice to excuse lack of use.[90]

Establishing that non-use can be excused in light of force majeure reasons calls for a case-by-case analysis. In particular, delay in obtaining approval from government agencies has not been deemed to excuse non-use if the trademark owner did not assume a diligent role in securing the authorization or where the delay results from tardiness or non-compliance with requirements on the registrant's side.[91]

---

84. Bertone and Cabanellas, *supra* n. 10, 365-366.
85. Otamendi, *supra* n. 15, 251; Poli, *supra* n. 9, 76.
86. Poli, *supra* n. 9, 75.
87. Bertone and Cabanellas, *supra* n. 10, 382-383; Poli, *supra* n. 9, Otamendi, *supra* n. 15, 252; *Merrell Dow Pharmaceuticals, Inc. v. Gerardo Ramon y Cia. S.A.*, Case 8358, CCCF, Div. II, 30/07/1981.
88. *Porlain S.A. v. Porcelana Americana S.A.*, Case 25031/03, CNC, Div. B, 28/06/2007.
89. *Valydar v. Integración Hogar*, *supra* n. 14.
90. Bertone and Cabanellas, *supra* n. 10, 383; Poli, *supra* n. 9, 75.
91. *Syncro v. Montpellier*, *supra* n. 58; *Behringwerke AG v. Volpino Laboratorios S.A.*, Case 6170, CCCF, Div. II, 28/03/1995; *Syncro Argentina S.A. v. Gerardo Ramon y Cia. S.A.*, Case 3441/93, CCCF, Div. I, 23/11/1995.

Notably, the Law refers to force majeure as a valid excuse to prevent cancellation. However, there is no mention of such justification in relation to the use required as a condition for renewal. Some authors have interpreted this discrepancy in the sense that force majeure excuses do not apply to justify renewal of a mark which has not been used,[92] while others adopt the opposite approach.[93]

### B. Circumstances in which non-use may be cured

Under the Argentine Law, a registration becomes vulnerable to cancellation if use of the mark is discontinued for over five years since the date the original registration is granted. But insofar as cancellation does not operate automatically, lack of use can be cured by resuming genuine use of the mark any time before a cancellation action is lodged.[94]

However, if a false declaration of use was submitted at the time a registration was renewed, subsequent use of the mark would not cure voidness of the renewed registration.[95]

## V. CONSEQUENCES OF NON-USE

### A. Consequences of non-use and relevance of the context

Where a registration is cancelled, there is no provision that limits the ability of a trademark owner or third parties to apply for re-registration of the same trademark. Yet, the specific circumstances of each case should be evaluated to avoid cancellation being used as a means for third parties to unlawfully benefit from the recognition that marks have achieved among consumers.[96]

### B. Extent of any revocation

Before the amendment to the Trademark Law, where a sign was found not to have been genuinely used the whole registration was cancelled. Conversely, actual use of the mark on any goods or services or as the distinctive portion of a trade or business name was sufficient to maintain protection over all goods and services for which the mark was registered, irrespective of the identity or relationship between the goods or services on which the mark was used and those covered by the registration.

As of the changes in legislation in January 2018, the Law provides that to avert cancellation the sign should be used in respect of either the same goods or services covered by the registration, or on goods or services similar or akin thereto, or as part of the designation of an activity related to the registered goods and services. Therefore, a registration will now be vulnerable to total or partial cancellation if the

92. Poli, *supra* n. 9, 74-75.
93. Aracama Zorraquin, *supra* n. 15, 5; Otamendi, *supra* n. 15, 253.
94. *Bristol Myers Squibb Company v. Beiersdorf AG*, Case 9029, CCCF, Div. II, 28/04/1992.
95. *See supra* IV.A.
96. Otamendi, *supra* n. 15, 392.

sign is used on goods or services which differ from the registered ones. However, as the amendments are quite recent, and the Law wisely leaves room for interpretation, it is expected that in the years to come, the decisions from the Courts and the Trademark Office will provide guidelines on how close the relationship between the goods or services on which the sign is used and those for which it is registered should be. [97]

### C. Timing of any revocation

Where a registration is cancelled on the grounds of lack of use, the rights over said registration are deemed expunged as of the date the cancellation action was lodged (or notice of the mandatory mediation proceedings was served upon the trademark owner).[98] Therefore, the trademark owner would still be in a position to pursue infringements that took place before such date, to the extent that no statute of limitations applies.

If a registration is cancelled because renewal thereof was based on a false declaration of use, the effect of the cancellation or voidness of the registration runs as of the date the unlawfully renewed mark expired.

97. See *supra* II.D.1.
98. Poli, *supra* n. 9, 79; Otamendi, *supra* n. 15, 391-392.

# Australia

*Mark Metzeling and Sarah Matheson*

## I. INTRODUCING GENUINE USE OF TRADEMARKS

### A. Legal framework

Australia is a party to the Paris Convention for the Protection of Industrial Property on 20 March 1883 (Paris Convention), the Agreement on Trade-Related Aspects of Intellectual Property Rights (TRIPS Agreement),[1] the Trademark Law Treaty adopted at Geneva on 27 October 1994[2] and the Nice Agreement Concerning the International Classification of Goods and Services for the Purposes of Registration of Marks of 15 June 1957,[3] all of which are concerned to varying degrees with the regulation of trade marks as between the respective Member States.

Under Australian law, an international convention does not by ratification alone become part of Australia's law. International agreements are enforced in Australia to the extent they are implemented by domestic legislation or the rights in question are recognised at common law.

The common law precursor to the legislative system of registration of trade marks in Australia today is passing off. Passing off remains an important and complimentary means of protecting trade marks today.

Australia's first federal trade marks legislation was the Trade Marks Act 1905 (Cth). The current Australian statute relating to trade marks is the Trade Marks Act 1995 (Cth) ('Act').[4]

1. The TRIPS Agreement is annexed to the Marrakesh Agreement Establishing the World Trade Organisation (Marrakesh, 15 April 1994).
2. This treaty was revised by the Singapore Treaty on the Law of Trademarks which came into effect on 16 March 2009.
3. Most recently revised at Geneva on 13 May 1977.
4. References to sections are references to the Act unless otherwise stated.

The Act defines a 'trade mark' as 'a sign *used*, or *intended to be used*, to distinguish goods or services'.[5] A condition of registration of a trade mark is that it must be used or intended to be used.[6]

Notwithstanding this, there is no requirement to lodge proof of use at the time of filing or at any subsequent stage to maintain a trade mark registration, unless a third party brings a non-use action against the trade mark.

Use must be:

(1) in good faith;
(2) as a trade mark;
(3) in the course of trade;
(4) in Australia;
(5) in relation to goods and/or services; and
(6) by the trade mark owner, their predecessor in title or an authorised user.

A significant body of jurisprudence has also been developed in case law and decisions of the Registrar of Trade Marks ('Registrar') that assists in determining when a trade mark has been used.

Whether a trade mark has been *used* is relevant to a number of questions under Australian trade mark law, including:

- whether a trade mark is, or becomes capable of distinguishing;
- whether there has been honest concurrent or prior use of a trade mark;
- who is the owner of a trade mark;
- whether a trade mark should be removed from the Trade Marks Register ('Register') for non-use;
- whether a trade mark has been infringed;
- whether a defensive trade mark should be registered.

The primary focus of this publication is non-use.

## B. Purpose and function of genuine use: policy considerations

In Australia, it is generally accepted that the rationale of the requirement of genuine use is to prevent the Register from becoming cluttered with unused marks, thereby imposing unwarranted impediments on future traders.[7]

5. Sec. 17 (Emphasis added). A 'sign' includes any letter, word, name, signature, numeral, device, brand, heading, label, ticket, aspect of packaging, shape, colour, sound or scent; or any combination of the foregoing: Sec. 6.
6. Sec. 27(1).
7. L Bently and R Burrell, *The Requirement of Trade Mark Use* 13 AIPJ 81, 181-182, 185-188 (2002). More recently, see the Submission to the Australian Government on Productivity Commission Inquiry Report, *Intellectual Property Arrangements* (Report No. 78, 23 September 2016) by M Handler and R Burrell dated 13 February 2017 relating to *inter alia*, 'the inefficiencies caused by cluttering' (p. 1).

In the US, Professor Lee B. Burgunder has described the functions of trade marks as follows:[8]

(1) to enable consumers to discriminate efficiently among similar products in the marketplace with minimal private and social costs ('*Distinguishing Function*');
(2) to preserve the goodwill of traders (and, hence, foster incentives for traders to offer quality goods and services) ('*Goodwill Function*'); and
(3) to allow consumers to associate goods or services in the marketplace with certain forms of information expressed in advertising ('*Advertising Function*').

In the absence of genuine use of a trade mark, these functions will be frustrated, namely the trade mark:

(4) will not serve the Distinguishing Function, due to consumers being unable to develop any relationship between a particular good or service and the trade mark;
(5) will not serve the Goodwill Function, due to traders being unable to develop any goodwill in that trade mark; and
(6) will not serve the Advertising Function, due to consumers being unable to develop any relationship between useful advertising information (cost, origin, composition, quality, value, taste, reputation, characteristics, etc.) and the trade mark.

Australian courts have consistently held that the essential or primary function of a trade mark is to indicate the trade source of the goods or services in relation to which the sign is used (i.e. the Distinguishing Function):

> 'Use "as a trade mark" is use of the mark as a "badge of origin" in the sense that it indicates a connection in the course of trade between goods and the person who applies the mark to the goods [...] that is the concept embodied in the definition of "trade mark" in s 17 – a sign used to distinguish goods dealt with in the course of trade by a person from goods so dealt with by someone else.'[9]

The above statement of the full bench of the Federal Court of Australia ('Federal Court') was subsequently expressly approved by a majority of the High Court of Australia ('High Court'):

> 'the requirement that a trade mark "distinguish" goods encompasses the orthodox understanding that one function of a trade mark is to indicate the origin of "goods to which the trade mark is applied". Distinguishing goods as a registered owner from the goods of others and indicating a connection in the course of trade between the goods and the registered owner are essential characteristics of a trade mark.'[10]

8. Lee B. Burgunder, *An Economic Approach to Trademark Genericism*, 23(3) American Business Law Journal 391, 396 (1985).
9. *Coca-Cola Co v. All-Fect Distributors Ltd*, (1999) 96 FCR 107, 115.
10. *E & J Gallo Winery v. Lion Nathan Australia Pty Ltd*, (2010) 241 CLR 144, 163.

The Goodwill Function has also been recognised,[11] as has the Advertising Function:

> 'Whilst the prime concern of the *Trade Marks Act* assists the capacity of a trade mark to distinguish the goods of the registered owner from those of another trader, trade marks undoubtedly perform other functions. For example [...] it may be accepted that distinctive marks can have a capacity to advertise, and therefore promote sales of products sold under or by reference to them. The advertising function of a trade mark is much more readily appreciated than it once was, and that function may be of great commercial value.'[12]

Further, the absence of a requirement of genuine use of a trade mark provides the potential for trade mark owners to engage in the practice of 'trade mark squatting', i.e. collation of trade marks for on-sale at a profit.

## II. WHAT QUALIFIES AS GENUINE USE OF A TRADEMARK

### A. Nature of use

#### *1. Use consistent with the function of the trademark*

As noted above, the key requirements for use as a trade mark are that:

(1) the use must distinguish the owner's goods or services; and
(2) the goods or services be offered or provided in the course of trade.

Use as a trade mark is a requirement to defend against a non-use action. Not all use *of* a trade mark will constitute use '*as* a trade mark'.

Whether a sign is used as a trade mark is judged objectively.[13] The subjective intention of the user is of secondary importance. Context is all important in indicating whether a sign has been used as a trade mark or for some other purpose.[14]

##### *i. Trade names*

A trade mark is distinguished from a trade name. The latter may be a personal name, or a business or company name, under which a person carries on business. Business names and company names are required by separate legislation to be registered to enable those who deal with a business to ascertain the identity of those trading under a particular name. However, registration gives no exclusive rights.[15]

---

11. *Campomar Sociedad Limitada v. Nike International Ltd*, (2000) 202 CLR 45.
12. *JT International SA v. Commonwealth*, (2012) 250 CLR 1, 102.
13. *Aldi Stores Ltd Partnership v. Frito-Lay Trading Co GmbH*, (2002) 54 IPR 344, para. 30.
14. *Shell Co of Australia v. Esso Oil (Aust) Ltd*, (1963) 109 CLR 407, 422; *Aldi Stores Ltd Partnership v. Frito-Lay Trading Company GmbH*, (2002) 54 IPR 344, para. 30.
15. It is however a defence to an allegation of trade mark infringement that a person (including a body of persons whether incorporated or not) uses in good faith the person's name or the name of the person's place of business: Sec. 122(1)(a).

Registration of a business or company name is not *prima facie* evidence that a sign has been used 'as a trade mark'. It was held in *Shahin Enterprises Pty Ltd v. ExxonMobil Corporation*:[16]

> 'There is a clear distinction [...] between conducting a business under a particular name and using a mark in respect of goods or services. The intention which the applicant needs to establish was an intention to use a mark to distinguish goods or services in the course of the applicant's trade from goods or services provided by any other person. Because it has established that it intended to use the name to brand its businesses, that does not mean, however, it has established that it has used a mark or a sign to distinguish goods or services in the course of trade.'

Where goods or services are provided under a trade name, such name may *also* function as a trade mark if it is used as a badge of origin. It is not uncommon for a business or company name to be used in this way.

*ii. Decorative use*

Trade marks may consist of decorative aspects of products, e.g. colours, patterns and shapes. Use of a sign that is decorative in nature does not preclude its use 'as a trade mark' provided it functions as a 'badge of origin', i.e. something that functions to distinguish trade source.[17]

*iii. Use of a domain name, key word or metatag*

Use of a domain name may constitute use of a trade mark, although mere registration of a domain name will not of itself amount to such use.[18]

16. *Shahin Enterprises Pty Ltd v. ExxonMobil Corporation*, [2005] FCA 1278, para. 73.
17. *Adidas AG v. Pacific Brands Footwear Pty Ltd (No. 3)*, (2013) 103 IPR 521, which concerned an allegation by Adidas that two of its registered trade marks for the well-known three stripes marks were infringed by Pacific Brands shoes having four stripes. Pacific Brands argued that its use of stripes was not use as a trade mark. The Judge accepted, at para. 61, 'that the stripes on the shoes have a decorative element and that the stripes may also act to indicate that each shoe is a sports shoe but that does not mean that those elements constitute the only use of the stripes as a sign: they are not mutually exclusive'.
18. For example, see *Sports Warehouse, Inc v. Fry Consulting Pty Ltd*, (2010) 186 FCR 519 at para. 153:

    'Plainly enough, not all domain names will be used as a sign to distinguish the goods or services of one trader in the course of trade from the goods or services of another trader [...] whether or not a domain name is use of a trade mark will depend on the context in which the domain name was used. In this case the domain name is more than an address for a website, the domain name is also a sign for the Applicant's online retailing service available at the website. In the context of online services, the public is likely to understand the domain name consisting of the trade mark (or something very like it) as a sign for the online services identified by the trade mark as available at the web page to which it carries the internet user.'

In Australia, it had been generally accepted that metatag use was not 'as a trade mark' because it was not visible to the ordinary internet user.[19] However, the Full Federal Court has recently confirmed that the use of a metatag may constitute use of a trade mark for infringement purposes. The reasoning was that although metatags are not actually displayed on a website, the source data of a website is visible to those who know what to look for.[20]

### 2. *Use in the course of trade*

#### *i. Public use*

Australian trade mark law does not draw an express distinction between public and non-public use. The focus for the consideration of whether the use is use 'in the course of trade' looks at the nature of the relevant *dealings* in the goods or services in question.

While some of the instances cited under section (ii) below ('Commercial purpose') might arguably be considered to be non-public use, it is the connection to the supply of goods or services in the market place that is decisive as to whether or not the use is 'use in the course of trade' and therefore 'use as a trade mark'.

#### *ii. Commercial purpose*

While the sign must be used 'in the course of trade',[21] an actual sale is not required.

The authorities establish that use 'in the course of trade' can be demonstrated by:

(1) use through an actual dealing in goods or services;
(2) use through an offering for sale of goods or services; or
(3) use through an intention to offer goods or services which is manifested through preparatory steps showing an objective commitment to using the trade mark.[22]

In general, 'trade' encompasses commercial dealings in goods and services.[23]

*Actual dealings*

This generally means a sale where the mark is used on, or in physical or other relation to, the goods or services in question.[24]

19. *Veda Advantage Ltd v. Malouf Group Enterprises Pty Ltd*, (2016) 241 FCR 161.
20. *Accor Australia & New Zealand Hospitality Pty Ltd v. Liv Pty Ltd*, [2017] FCAFC 56.
21. *Shell v. Esso Standard Oil*, *supra* n. 14; *Christodoulou v. Disney Enterprises Inc*, (2005) 156 FCR 344.
22. *Tonch Pty Ltd v. Christou*, (2008) 78 IPR 358, 364.
23. *Oakley Inc v. Franchise China Pty Ltd*, (2003) 58 IPR 452.
24. In *Thunderbird Products Corp v. Thunderbird Marine Products Pty Ltd*, (1974) 131 CLR 592, a sale of a single boat as a prototype from an American company to an Australian importer was sufficient to constitute a dealing. The trade mark was used on brochures, invoices and shipping documents relating to the sale. It was not material whether the

Use of a trade mark by a non-profit organisation may constitute genuine use, provided that the trade mark is used as an indication of trade origin.[25] A user who does not trade for profit might still be able to show they receive a monetary benefit for providing goods or services under a particular sign, e.g. in the way of donations or sponsorships allowing them to carry out their activities. This may have a sufficiently commercial character to establish that use of the sign is use 'in the course of trade'.

The supply of free samples of goods bearing a trade mark so as to further trade in those goods, may involve a use of the trade mark in the course of trade, but the distribution of a product free of charge does not ordinarily involve trade in that product.[26]

*Offering for sale of goods or services*

Offering goods or services for sale, if genuine, will generally be sufficient for use to have occurred 'in the course of trade'. However, it must be 'possible to identify an actual trade or offer to trade in the goods bearing the mark or an existing intention to offer or supply goods bearing the mark in trade'.[27]

However, an advertisement where the goods or services are not immediately available for sale does not constitute use within the meaning of the Act.[28] In *Settef SpA v. Riv-Oland Marble Co (Vic) Pty Ltd*,[29] provision of brochures and samples was not sufficient to constitute use 'in the course of trade' where the goods were not concurrently available for order or purchase:[30]

> 'it follows from the authorities that a mark is used as a trade mark only if used for the view to facilitating or promoting the operation of a trading channel which in a business sense had already been opened in Australia. [...] The forwarding of samples and brochures is not sufficient to indicate that Settef was ready and willing to fulfil such orders as it received from Australia. The purpose of these items may well have been to ascertain whether there was a market in Australia sufficient for it to be worthwhile for Settef to export here.'

---

trade mark had been used on the boat itself. The other uses were sufficient to conclude that the trade mark had been used *in relation to* the goods.

25. *Herron Pharmaceuticals Pty Ltd v. Brown*, (1998) 45 IPR 321, 327.
26. For example, use of the trade mark 'CSL' on CD-Roms and databases supplied free of charge to promote vaccine/antigen products did not constitute genuine use of the trade mark on class 9 goods, i.e. electronic instruments and computer software and hardware, for the purposes of defeating a non-use action. However, distribution of free classified telephone directories bearing the trade mark constituted use on the basis that the publisher derived income from the sale of advertising in the directories: *Golden Pages Trade Mark* [1985] FSR 27, approved in *Oakley Inc v. Franchise China Ltd*, (2003) 58 IPR 452.
27. *Moorgate Tobacco Co. Ltd v. Philip Morris Limited & Anor*, (1984) 156 CLR 414, which established that it is not necessary that there be an actual dealing in goods bearing the trade mark before being a local use of the mark as a trade mark.
28. See *John Toh v. Paris Croissant Co Ltd*, [2010] ATMO 34.
29. *Settef SpA v. Riv-Oland Marble Co (Vic) Pty Ltd*, (1987) 10 IPR 402.
30. *Ibid.*, at 417.

*Intention to use – preparatory steps manifesting an objective commitment to use*

Use of this nature must go beyond mere preliminary activities, investigating whether to use the mark or planning to use the mark. It must have reached the stage where there is an objective commitment to using the mark.[31] Whether such objective commitment has been demonstrated will depend on all the relevant circumstances. For example:

- in a case concerning a trade mark for magazines, obtaining business cards, having a letterhead printed and using that letterhead to solicit advertisements which were subsequently placed in the magazine was 'sufficient to constitute relevant use of the mark';[32]
- provision of promotional videos, presentation to trade and potential customers and evidence of expenditure on packaging in relation to milk and cheese products was considered 'use as a trade mark'.[33]

However, examples of preliminary activities which were not, by themselves, considered to constitute use 'in the course of trade' include:

- registration of a business or domain name;[34]
- hiring a graphic designer to design a product label;[35]
- use on business cards or letterhead;[36]
- creation of a price list and order form without evidence pertaining as to who had seen the documents and in what circumstances;[37]
- use of the sign on an A-frame board followed by the words 'coming soon'.[38]

It may also be difficult to establish a case for an objective commitment to use the mark at a particular date where the person claiming use has not subsequently sold or offered for sale the goods or services in question.

Some jurisdictions do not recognise the submission of samples to a testing authority as relevant use. However, in Australia, an application to the Therapeutic Goods Administration has been held to be 'a mandatory preparation for marketing pharmaceutical preparations' and was therefore considered to constitute use for the purposes of defending a non-use action.[39]

### 3. *Use by the owner or with its consent*

The Act deems an *authorised use* of a trade mark to be a use of the trade mark *by* its owner.[40] A person is an *authorised user* of the trade mark if they use the trade mark

31. *Woolly Bull Enterprises Pty Ltd v. Reynolds*, (2001) 51 IPR 149.
32. *Buying Systems (Aust) Pty Ltd v. Studio Srl*, (1995) 30 IPR 517, 520.
33. *New South Wales Dairy Corporation v. Murray-Goulburn Co-operative Co Ltd*, (1989) 14 IPR 26, 45.
34. *Tonch Pty Ltd v. Christou*, *supra* n. 22, at 365.
35. *P & T Basile Imports Pty Ltd v. Societe De Produits Nestlé SA*, (2002) 55 IPR 216, 220.
36. *Unilever Australia Limited v. Karounos*, (2001) 113 FCR 322, 339.
37. *Bang on the Door Ltd v. Playgro Australia Pty Ltd*, (2003) 60 IPR 117, 123-124.
38. *Shahin Enterprises Pty Ltd v. ExxonMobil Oil Corporation*, *supra* n. 16, at para. 72.
39. *Sterling Winthrop Pty Ltd v. Herron Pharmaceuticals Pty Ltd*, [1994] ATMO 65.
40. Sec. 7(3).

in relation to goods or services under the *control* of the owner of the trade mark. Such use is an authorised use to the extent of such control.[41]

Control is considered to be a form of connection between the goods or services and the trade mark owner such that it constitutes a relevant dealing *by* the owner and therefore use as a trade mark. It will not be sufficient for the user to use the trade mark in such a way that it exclusively indicates a connection between that user and the goods or services.

The form of control includes quality control *over the goods or services* in relation to which the trade mark is used and financial control by the trade mark owner over *the relevant trading activities*.[42] The control required is not control over the trade mark itself.

### *i. Quality control*

For the purposes of opposing a non-use action, use must be by, or authorised by, the registered owner of the trade mark.[43] The applicant for removal may in turn challenge use said to be authorised use on the basis of the requisite authority to use[44] or insufficient control, notwithstanding the requisite authority.

While there has been some suggestion that a licensor's control could be relatively slight, or that the theoretical possibility of contractual control was sufficient,[45] the Full Federal Court has recently confirmed that the correct interpretation of 'control' is *actual* control as a matter of substance, and not the mere right to exercise control.[46] Actual control is a question of fact and degree. At a minimum, licence agreements should contain explicit quality controls and the licensor should be able to demonstrate that such controls are enforced.

### *ii. Financial control*

The controlling member of a group of companies may be able to demonstrate a relevant dealing with the products of its subsidiaries. It is necessary to demonstrate the precise relationship between the relevant companies. It has been suggested that 'evidence of the exercise of financial control over trading activities, perhaps in the

41. Sec. 8(1) and (2).
42. Sec. 8(3) and (4).
43. Sec. 100(1).
44. For example, in *C A Henschke & Co v. Rosemount Estate Pty Ltd*, (2000) 52 IPR 42, the trade mark was used by a family partnership of wine makers, but following the death of the registered owner, the trade mark was not transferred to his executors for some years, although the trade mark continued to be used by the partnership in the interim period. In a subsequent infringement action brought by the executors of the estate, the alleged infringer sought to have the trade mark removed from the Register on grounds of non-use, arguing that use by the partnership was not authorised use.
45. See *Skyy Spirits LLC v. Lodestar Anstalt*, (2015) 112 IPR 328 in which the trial judge made it clear that he was reluctantly following what he considered to be the *ratio* in *Yau's Entertainment Pty Ltd v. Asia Television Ltd*, (2002) 54 IPR 1, thereby deciding that a perpetual exclusive licence for AU$1 which was 'ceremonial and formulaic' was sufficient.
46. *Lodestar Anstalt v. Campari America LLC*, (2016) 120 IPR 1.

way in which a parent company may exercise control over a wholly owned subsidiary' would be sufficient.[47]

### B. Volume, duration and frequency of use

There are no specific rules relating to quantum of use. In opposing a non-use action, even a small degree of use may be sufficient. The important factor is that the use is in good faith, meaning genuine commercial use.[48]

In general, a relatively small amount of use may be sufficient to constitute 'ordinary and genuine' use judged by commercial standards.[49] A single *bona fide* use of the mark in the relevant period may in fact be sufficient to defeat a non-use action.

However, token use, including use merely designed to maintain trade mark registration, will not be sufficient.[50]

### C. Use of a sign

If a trade mark is used with additions or alterations that do not substantially affect its identity, the Registrar and the courts have a discretion in deciding whether the trade mark has been used, having regard to the circumstances of the particular case.[51]

Examples where an alteration has not been considered to substantially affect the identity of the trade mark include:

- omission of the word 'Service' from the registered trade mark 'Funjet-Service' in the context of travel agency and like services;[52]
- use of the word 'Mindgym' alone was considered use of a composite trade mark consisting of both the word 'Mindgym' and a distinctive device because the word itself was considered to be the key element in the composite mark;[53] and
- different fonts, different positioning of minor decorative elements, additional text in trade marks for beer.[54]

Where a word trade mark is used with a device that is simple or descriptive of the word, the word and the device together may be regarded as substantially identical

47. *Health World Ltd v. Shin-Sun (Aust) Pty Ltd,* (2008) 75 IPR 478.
48. *Woolly Bull v. Reynolds, supra* n. 31.
49. In *E & J Gallo Winery v. Lion Nathan, supra* n. 10, importation and offering for sale of 144 bottles of wine in Australia was deemed genuine commercial use. Further, in *Lodestar Anstalt v. Campari, supra* n. 46, the issue of non-use turned in part on whether a trade mark for wine had been used between 2007 and 2010. One vintage had been sold in small quantities during that period with commercial sales of around AU$10,000, but this was considered sufficient to meet the requirement for use.
50. *Woolly Bull v. Reynolds, supra* n. 31.
51. Sec. 7(1).
52. *QH Tours Ltd v. Mark Travel Corp,* (1999) 45 IPR 553, 559.
53. *Matzka v. The Mind Gym Ltd,* (2006) 68 IPR 339, 344.
54. *Intellectual Property Development Corporation Pty Ltd v. Schindler & Anor,* (2016) 125 IPR 349.

to the word trade mark alone. In the *Gallo* case[55] the High Court considered the use of the word 'BAREFOOT' with a footprint device to be use of the registered trade mark which consisted solely of the word 'BAREFOOT', concluding:[56]

> 'The addition of the device to the registered trade mark is not a feature which separately distinguishes the goods or substantially affects the identity of the registered trade mark because consumers are likely to identify the product under the registered trade mark with the device by reference to the word BAREFOOT. The device is an illustration of the word. The monopoly given by a registration of the word BAREFOOT alone is wide enough to include the word together with a device which does not substantially affect the identity of the trade mark in the word alone.'

Examples of a stricter approach to alternations include:

- the removal of the hyphen from the registered trade mark 'Ugh-Boots' was held to be an alteration which substantially affected the identity of the trade mark, such that the use of 'Ugh Boots' or 'Ugg Boots' did not amount to use of the registered trade mark;[57]
- telephone use of '1800-Flowers' was not considered use of 'FREECALL 1800 FLOWERS AUSTRALIA WIDE' together with a flower device, because the words alone lacked inherent distinctiveness, and the device (that was not used) was considered to confer real distinctiveness.[58]

## D. Use in relation to goods and services

### *1. Use for each of the designated goods and services*

A non-use application may be made in respect of any or all of the goods and/or services in respect of which the trade mark may be, or is, registered.[59] To defeat a non-use action the opponent must rebut the allegation of non-use 'in relation to the goods or services to which the opposed application relates'.[60]

If the non-use action relates to *all* goods or services for which the trade mark is registered, the registration can be preserved only for those goods or services for which use can be shown.

However, the courts have tended to resist drawing too fine a distinction in relation to use for *each* designated good or service. *Pioneer Computers Australia Pty Ltd v. Pioneer KK*[61] concerned the trade mark 'Pioneer' for, amongst other things, 'computers, computer peripheral devices, computer software and computer game programs'. In addressing the application for removal of the registration for these goods, there was considerable discussion as to whether 'computer peripheral devices'

55. *E & J Gallo Winery v. Lion Nathan, supra* n. 10.
56. *E & J Gallo Winery v. Lion Nathan, supra* n. 10, at 170.
57. *McDougall v. Deckers Outdoor Corporation*, (2006) 68 IPR 322.
58. *1-800-Flowers.Com, Inc v. Flowerscorp Pty Ltd*, (2013) 102 IPR 114.
59. Sec. 92(2)(b).
60. Sec. 100(1).
61. *Pioneer Computers Australia Pty Ltd v. Pioneer KK*, (2009) 80 IPR 38.

meant only devices that were externally connected to a computer or whether they included internally connected devices. The removal applicant claimed that the trade mark owner had not used its trade mark in respect of *all* types of computer peripheral devices, so its registration should be restricted to those particular devices that it had sold with its mark. The court rejected the argument that the trade mark owner needed to have used its trade mark in respect of *all* goods within a category of registration in order to retain registration, stating:

> 'It seems artificial to restrict the description of a category of goods on the basis of a distinction drawn by the definition where the distinction is not drawn in use of the term in the marketplace, where it depends upon the way in which the goods are used, sometimes idiosyncratically, and where the use of a particular device as an internal component or an externally connected device may change as the technology changes.'[62]

### 2. *Use for a broader category of goods and/or services*

As a general rule, proof of use of a trade mark on goods and services other than those for which the trade mark is registered will not rebut an allegation of non-use.[63] Proof of use for goods or services of the *same description* will generally also not suffice.[64]

However, Section 101(3) gives the Registrar a discretion to allow a trade mark to remain on the Register, even if no use on the registered goods or services has been shown. The factors that the Registrar may take into account include whether the trade mark has been used by the registered owner in respect of similar or closely related goods or services. The interests of the trade mark owner and the applicant for removal, and the public interest, must be taken into account by the Registrar in exercising such discretion.[65]

62. *Pioneer Computers Australia Pty Ltd v. Pioneer KK*, (2009) 80 IPR 38, 59. However, the Judge acknowledged that where particular computer peripheral devices used by the trade mark owner could be clearly and separately identified, there might be a case that the trade mark owner needed to have used its trade mark in respect of all goods within the category of registration.
63. *Sustainable Agriculture & Food Enterprises Pty Ltd v. Samaritan Pharmaceuticals Inc*, (2008) 81 IPR 122.
64. *McHattan v. Australian Specialised Vehicle Systems Pty Ltd*, (1996) 34 IPR 537. However, in *Trend Windows & Doors Pty Ltd v. Saint-Gobain Glass Fronts*, (2013) 101 IPR 631, the Registrar's delegate indicated that metallic windows may be of the same kind as non-metallic windows, although the actual decision also referred to the fact that if they were not, then the goods were clearly of the same description and the public interest favoured retention for registration for non-metallic windows.
65. Sec. 101(3) and (4). See also *Société Anonyme des Eaux Minérales d'Evian v. The Yoghurt Co Pty Ltd*, (2012) 97 IPR 77. In *General Mills Inc v. Christofis*, (2013) 101 IPR 618, it was found that the trade mark registered for all goods in class 30 had been used in respect of a broad range of such goods, and in respect of those for which use had not been established, discretion was exercised not to remove the mark for various reasons, including the reputation of the trade mark.

Some of the factors which have been considered persuasive in favour of the trade mark owner include:[66]

- evidence of the international profile of the owner's product;
- use in Australia of the trade mark after the period of non-use and the owner's clear intention to use it in the future;
- the period of non-use not being substantially greater than the minimum statutory period of non-use of three years;
- the absence of any evidence concerning the prospect of confusion or public mischief if the trade mark remained on the Register;
- the possibility that removing the trade mark may itself cause confusion to the public as a consequence of its use in Australia and internationally.

## E. Territorial scope

The trade mark must be used in the course of trade in Australia. However, it is not the case that *all* activities connected with the trade must have taken place *in* Australia.

In general, there is no required geographic extent of use in Australia. Use in only one part of Australia is sufficient, unless the provisions of Section 102 apply.[67]

### *1. Goods sold or offered for sale but not present in Australia*

There is conflicting authority as to whether a trade mark used on goods in transit to Australia constitutes use of the trade mark in Australia.[68]

### *2. Services provided outside Australia*

In a case involving cruise services, the Federal Court found that although the actual services of carrying out the cruises did not take place in Australia, 'crucial integers

66. *Austin Nicols & Co Inc v. Lodestar Anstalt,* (2011) 90 IPR 310, affirmed on appeal in *Austin, Nicols & Co Inc v. Lodestar Anstalt,* (2012) 202 FCR 490.
67. Section 102 of the Act allows the Registrar or the court to impose a territorial restriction on the registration of a trade mark where there has been no use in a particular place in Australia (or in a particular export market) for the relevant non-use period. However, its operation is confined to limited circumstances and its practical application has been limited.
68. In *Re Registered Trade Mark 'Yanx'; Ex parte Amalgamated Tobacco Corp Ltd,* (1951) 82 CLR 199, it was held that a trade mark is used in Australia if the goods are offered for sale in Australia but at the time they are offered for sale are actually only in transit to Australia. In *Settef v. RivOland Marble, supra* n. 29, the court drew the inference that, in the absence of any indication that any consignment of goods shipped from Italy to Australia did not arrive, the consignments reached the consignees. However, in *WD&HO Wils (Australia) Ltd v. Rothmans Ltd,* (1956) 94 CLR 182, cigarettes were purchased and paid for in the US and sent direct to consumers in Australia. It was said that there was no use of the mark in Australia as all 'trade' in the goods took place in the US, trading having ended once the goods were purchased or otherwise acquired for consumption, i.e. outside Australia.

in the course of trade took place in Australia', trade having commenced with 'steps taken to encourage inquiries from prospective customers' in Australia.[69]

### 3. *Advertising and foreign publications*

Advertising in foreign publications available 'in Australia' will generally not constitute use 'in the course of trade' in Australia unless it appears to be specific targeting of Australian consumers.[70]

### 4. *Use of the trade mark on the internet*

A trade mark used on a website will generally not be used 'in Australia' unless that website specifically targets Australian consumers.[71] However, later decisions suggest a relatively low bar. Specifically targeting Australian consumers from an overseas website may constitute use of the relevant trade mark in Australia.[72]

---

69. *Carnival Cruise Lines Inc. v. Sitmar Cruises Ltd*, (1994) 31 IPR 375, 390. This may be contrasted to the issue of whether a similar dealing was sufficient to constitute 'use in the course of trade' in the *Settef* case discussed in section A.2. above.
70. For example, in *In Knickerbox Ltd v. Pedita Australia Pty Ltd*, (1995) 31 IPR 108 at 117, the delegate held:

    'Apart from the indication that the goods illustrated in these advertisements originate from the opponent's company, or are available at one of its retail outlets, no information has been made available as to how and where a person in Australia might order the goods, nor what the price of the goods, including their transport, might be in the local currency. I do not think these advertisements can be interpreted as offers for the purpose of soliciting trade in an article of clothing that will be supplied on order.

    [...] given the use of British currency and in the absence of the opponent company's branches in Australia, the advertisements could not be seen to be directed to the Australian consumers, nor intended for the Australian subscribers of the magazines.'
71. In *Ward Group Pty Ltd v. Brodie & Stone plc*, [2005] FCA 471 it was held that 'use of a trade mark on the Internet, uploaded on a website outside of Australia, without more, is not a use by the website proprietor of the mark in each jurisdiction where the mark is downloaded'. This reasoning was subsequently applied in *Deutsche Telekom AG v. E! Entertainment Television* (23 January 2006) (unreported) in Deutsche Telekom owned the mark T-online and used it on the clothing of athletes and their equipment who appeared in sports events broadcast in Europe, on internet services about the Sydney Olympic games which were 'designed for access by audiences in Europe' and on internet services originating in Europe and 'designed for access by European audiences'. The Registrar's delegate concluded:

    'It would logically appear that the upload of the trade mark from Australia to a German website does not constitute use of the trade mark in Australia; neither does accidental viewing by Australian sporting spectators when the trade mark is being used to support, advertise or promote goods or services in a country outside Australia. Absent genuine intention to target and engage the Australian public in trade, the accident [sic] viewing of a trade mark on the internet, on foreign sports teams or on television broadcasts, does not constitute use of a trade mark in Australia.'
72. For example, in *International Hair Cosmetic Groups Pty Ltd v. International Hair Cosmetics Ltd*, (2011) 218 FCR 398, it was found that there had been use of an Australian registered

### 5. *Inadvertent use*

An owner may be able to establish use of the trade mark in Australia even if they had no knowledge of that use. In the *Gallo*[73] case, the High Court held:[74]

> 'The capacity of a trade mark to distinguish a registered owner's goods from those others [...] does not depend on whether the owner knowingly projects the goods into the Australian market. It depends on the goods being in the course of trade in Australia. Each occasion of trade in Australia, whilst goods sold under the trade mark remain in the course of trade, is a use for the purposes of the Trade Marks Act. [...] It is a commonplace of contemporary international trade that prior to consumption goods may be in the course of trade across national boundaries.'

### 6. *Goods produced for export*

If a trade mark is applied in Australia in relation to goods or services to be exported from Australia, the application of the trade mark to the goods or use in relation to the goods or services constitutes use of the trade mark.[75] Similarly, if any act is done in Australia to export goods or services which, if done in relation to goods or services in Australia would constitute use of a trade mark, that act is also use of the trade mark.[76]

## F. Relevant time period and its calculation

Under the current provisions of the Act, the removal applicant in a non-use action must demonstrate that the trade mark owner has not used the registered trade mark:

(1) and never intended to use the registered trade mark (if the removal application is commenced within five years of the lodgement date); or
(2) in a consecutive three-year period (if the removal application is commenced at least five years after the lodgement date).[77]

The Australian Productivity Commission recently completed a public inquiry into Australia's intellectual property system,[78] one recommendation of which was to reduce the grace period for non-use actions from five years to three years. This has

trade mark on an overseas website, even though it was not possible to order and pay for goods directly from the website. The website had a drop down box that listed various countries and a statement that invited users to select Australia amongst other countries. By doing so, a user was taken to information concerning the products in question. While there was no capacity to order goods from the website, there were contact details for the company.

73. *E & J Gallo Winery v. Lion Nathan, supra* n. 10.
74. *E & J Gallo Winery v. Lion Nathan, supra* n. 10, at 165.
75. Sec. 228(1).
76. *Deckers Outdoor Corporation v. Luda Productions Pty Ltd*, (2006) 70 IPR 402.
77. See Secs. 92(4) and 93.
78. See Productivity Commission, *Inquiry into Australia's IP Arrangements*, Final Report, 20 December 2016 (PC Report).

been accepted such that, when enacted, a grace period of three years will run from the date the trade mark becomes registered. It is said that this will reduce the overall period for the vast majority of trade marks, while better aligning Australia's non-use provisions with international standards.[79] The Bill intended to give legislative effect to this change is before Parliament at the time of writing.

See further sections IV and V below.

## III. EVIDENCE OF GENUINE USE OF A TRADEMARK

### A. When use should be proved: timing of lack of genuine use claims

#### 1. *When filing an application for a trademark*

There is no requirement to lodge proof of use at the time of filing an application for a trade mark, or at any subsequent stage to maintain a trade mark registration.

Additionally, there is a five-year grace period from the date of lodging the trade mark application within which a trade mark cannot be removed from the Register on the basis of non-use.[80]

#### 2. *In the course of opposition proceedings*

If opposition proceedings are commenced pursuant to Section 59 ('Applicant not intending to use trade mark')[81] or Section 62A ('Application made in Bad Faith'),[82] the applicant may be required to file evidence of use of the trade mark in its evidence in answer.

Prior to filing evidence in answer, the opponent will be required to file its evidence in support to establish the relevant ground on the balance of probabilities. If the evidence in support indicates that this is likely to be established, it is prudent (but not compulsory) for the applicant to demonstrate genuine use of the trade mark in its evidence in answer. However, a genuine *intent to use the trade mark* will be sufficient to overcome these grounds and as such, it is possible that an applicant will not need to file evidence of genuine use to be successful in the opposition.

79. Explanatory Memorandum to the *Intellectual Property Laws Amendment (Productivity Commission Response Part I and Other Measures) Bill 2018*, Schedule 1, Part 3, paragraph 82.
80. However, it may be removed on the basis that there was never an intention to use the trade mark at the time of filing the trade mark application and there has been no use of the trade mark to date: Sec. 92(4)(a).
81. The registration of a trade mark may be opposed on the ground that the applicant does not intend:
    (a) to use, or authorise the use of, the trade mark in Australia; or
    (b) to assign the trade mark to a body corporate for use by the body corporate in Australia;
    in relation to the goods and/or services specified in the application.
82. The registration of a trade mark may be opposed on the ground that the application was made in bad faith.

### 3. *In cancellation proceedings based on lack of use*

Where a non-use application is filed by a third party, it may be brought before:

(1) a delegate of the Registrar (an administrative proceeding in the Trade Marks Office) by any entity; or
(2) a court (judicial proceeding) if there is an existing court action between the parties to the non-use action (i.e. the trade mark owner and the applicant for removal). See further in section 4 below.

#### *i. Non-use applications in the Trade Marks Office*

Section 92(4) of the Act provides two distinct and alternative grounds for removal. Under paragraph (a) it is necessary to establish:

- first, that the trade mark was registered without the necessary intention and good faith that it should be used; and
- secondly, that there had in fact been no use, or no use in good faith, earlier than one month before the application for removal.

Under paragraph (b), it is necessary to establish only that there had been no use, or no use in good faith, in the three-year period extending up to one month before the application for removal (subject to the grace period referred to above).

A party may make a non-use application to 'clear the way' for their own trade mark application which might otherwise be considered too similar to an existing registered trade mark. Alternatively, where a notice of opposition is filed in respect of a trade mark application already on foot on grounds that it is too similar to a trade mark of the opponent, the trade mark applicant may at that stage file an application for non-use arguing that the opponent has not used their trade mark.

Where a notice of opposition has been filed and a 'defensive' non-use application is made, the opposition proceedings may be suspended until the non-use application is determined. The parties may agree to a cooling-off period in the opposition proceedings, pending the outcome of the non-use application, which must be allowed by the Registrar.[83]

#### *ii. Evidentiary stages of an application for removal for non-use in the Trade Marks Office*

(1) Application for removal for non-use is filed by a third party alleging non-use.
(2) The trade mark owner then has up to two months to oppose the application for removal for non-use. The trade mark owner must file a statement of grounds and opposition which sets out the basis on which the trade mark owner opposes the non-use application. Generally, this merely indicates that the trade mark has been used by the trade mark owner (which includes use by an authorised user), without any evidence being filed at this stage.

83. Regulation 9.19 of the Trade Mark Regulations 1995 (Cth). The cooling-off period is six months, with the possibility of one further extension of six months.

(3) The removal applicant has up to one month to indicate it will defend the opposition to the removal application.
(4) The trade mark owner has three months within which to file evidence in support.
(5) The removal applicant has three months to file evidence in answer.
(6) The trade mark owner has two months to file evidence in reply.

### 4. *As a defence in infringement/cancellation proceedings*

Trade mark infringement proceedings must be brought before a court. It is not a *defence* to infringement that there has been no genuine use of the trade mark that is the subject of the infringement proceeding. However, a defendant may counterclaim in the same court proceeding for revocation of the relevant trade mark on the ground of non-use. The court will determine the revocation issue prior to ruling on the question of infringement.

## B. Burden of proof

(1) In a non-use action under Section 92(4)(a), the trade mark owner bears the onus of showing either:
   - an intention to use the trade mark in good faith, authorise its use by another or assign it to a body corporate for the use of that body corporate; or
   - use in good faith in the period up to one month before the date of filing of the removal application.

(2) In a non-use action under Section 92(4)(b), the trade mark owner bears the onus of demonstrating:
   - use in good faith in the three-year period ending one month before the date of filing of the removal application; or
   - circumstances that were an obstacle to the use of the trade mark during that period.

The trade mark owner must rebut the allegation of non-use on the balance of probabilities.[84] Generally this is done by filing evidence of use (or intention to use if it is a Section 92(4)(a) action).

Once the trade mark owner establishes a *prima facie* case, the evidentiary onus shifts to the removal applicant.[85]

## C. Relevant evidence

The Act provides a definition of use of a trade mark variously in relation to goods and services. These definitions address some of the categories below.

---

84. *Pfizer Products Inc v, Karam*, [2006] FCA 1663.
85. *Television Food Network, G.P. v. Food Channel Network Pty Ltd (No. 2)*, (2009) 80 IPR 314, 323.

(1) 'use of a trade mark in relation to goods' means use of the trade mark upon, or in physical or other relation to, the goods (including second-hand goods);[86]
(2) 'use of a trade mark in relation to services' means use of the trade mark in physical or other relation to the services.[87]

### *1. Use on the product or its packaging*

Evidence of use on the product or its packaging will constitute relevant use.

This can encompass use on the goods themselves, or on packaging, labels or swing tags. Evidence of use can also be in the form of sales revenue and advertising expenditure in the relevant period. This can be supported, for example, by evidence of orders, invoices, receipts or shipping documents.

### *2. Use in advertising*

Advertising and promotional materials, such as print advertisements, and promotional articles or website printouts will also constitute evidence of use.

Use on shop fronts and physical displays where goods or services are offered for sale also constitutes evidence of use.

See also section II.A.2. above.

### *3. Use on the internet*

The *Ward*[88] case concerned infringement, but is considered to have application in non-use actions as well. In a case involving similar issues but decided on another basis,[89] the Registrar's delegate made some comments on what constitutes internet usage targeted at Australia. The website in question made an offer of sale to people in the US, Canada and the United Kingdom, but not Australia. The trade mark owner admitted it had trouble processing Australian credit card orders and so would have found it difficult to fill any orders. In the circumstances, the Delegate concluded that the advertisement did not constitute genuine commercial use.

In contrast, later cases have found that even if the use was only *offering* goods for sale to people in Australia via a website, that can be sufficient evidence of the requisite use.[90]

86. Sec. 7(4).
87. Sec. 7(5).
88. *Ward Group v. Brodie & Stone, supra* n. 71.
89. *McDougall v. Deckers, supra* n. 57.
90. See *Baroque Japan Ltd v. Mad Keen Pty Ltd,* (2015) 113 IPR 383, where the offering of goods was to Australian consumers to be bought with Australian dollars with the website offering delivery exclusively within Australia, even though the use was only found to be the offering of the goods, with the goods themselves being physically located in Hong Kong. See also *International Hair Cosmetics Group Pty Ltd v. International Hair Cosmetics Ltd, supra* n. 72.

### 4. *Form of evidence*

For a non-use action in the Trade Marks Office, a declaration as to evidence of use will be required. In proceedings in the Federal Court, an affidavit will be required.

Survey evidence is not required to establish use of a mark to defeat a non-use action. Generally, survey evidence would not be necessary or even appropriate as evidence of a relatively small amount of use will suffice.

It has, however, been held that 'the fewer the acts [of use] relied on, the more solidly [these acts] ought to be established' or that 'overwhelmingly convincing proof' is required if a single transaction is relied upon.[91]

### 5. *Distinction between use of a company name and use of the services offered by the company*

See 'Trade names' under section II.A.1 above.

## IV. JUSTIFICATION FOR NON-USE

### A. Obstacles to use

Non-use of a trade mark can be excused where 'the opponent has established that the trade mark was not used by its registered owner in relation to those goods and/ or services during that period because of circumstances (whether affecting traders generally or only the registered owner of the trade mark) that were an obstacle to the use of the trade mark during that period'.[92]

At the very least, it was held in *Woolly Bull Enterprises v. Reynolds*[93] that the circumstances justifying non-use must:

(1) rise externally to the registered owner;
(2) be circumstances of a trading nature; and
(3) have a causal link to the mark's non-use.

Examples that have been held *not* to constitute an obstacle to use include:

- product packaging defects;[94]
- the incapacity of the trade mark owner, whether financial or because of illness;[95]
- the issue of cancellation proceedings in relation to the registration;[96]
- transport and distribution problems;
- the delay in the production of product labels;[97]

91. *Converse Inc v. Kavushiki Kaisha Moonstar,* [2010] ATMO 87, para. 18.
92. Sec. 100(3)(c) Trade Marks Act 1995 (Cth).
93. (2001) 51 IPR 149, paras 45, 55.
94. *Nature's Remedy Australia Pty Ltd v. Nature's Remedy Pty Ltd,* [1999] ATMO 47.
95. *Woolly Bull v. Reynolds, supra* n. 31.
96. *Unilever v. Karounas, supra* n. 36.
97. *P & T Basile Imports Pty Ltd v. Societe de Produits Nestle,* [2001] ATMO 16.

- the conduct of clinical trials to investigate possible alternative uses of a pharmaceutical product;[98] and
- lack of manufacturing facilities for the product.[99]

Examples that have been held to constitute an obstacle to use include:

- the imposition of post-war import restrictions preventing importation of milking machines;[100]
- delays in obtaining health department approval for a pharmaceutical product;[101]
- the need to conduct research and trials in relation to pharmaceutical products;[102]
- a long history of litigation in relation to the trade mark.[103]

There is no prescribed time limit for the duration of excusable non-use. However, where the circumstances no longer exist, it is evident that the excuse cannot continue to be used.

The Trade Marks Amendment (Tobacco Plain Packaging) Act 2011 (Cth), which was introduced at the same time as the Tobacco Plain Packaging Act 2011 (Cth) (Tobacco Act), amends the Act to enable tobacco trade mark owners to protect their trade marks (including device marks) from removal for non-use on the basis that an application for removal for non-use can be rebutted by evidence that the registered owner would have used the trade mark, but for the operation of the Tobacco Act.

### B. Circumstances in which non-use may be cured

Non-use can be cured by genuine use, provided no one files a non-use application within one month of the trade mark owner resuming or commencing use of the registered trade mark.

## V. CONSEQUENCES OF NON-USE

### A. Consequences of non-use and relevance of the context

Pursuant to Section 97:

'(1) If:

(a) there is no opposition to an application to the Registrar under Subsection 92(1); or

98. *Kowa Co Ltd v. NV Organon*, (2005) 66 IPR 131.
99. *Angioblast Systems Incorporated v. UCP Gen Pharma AG*, [2010] ATMO 53.
100. *Aktiebologet Manus v. RJ Fullwood and Bland Ltd*, (1949) 66 RPC 71.
101. *Pierre Fabre SA v. Marion Laboratories Inc*, (1986) 7 IPR 387.
102. *Upjohn Co v. Sanofi*, (1998) 42 IPR 576.
103. *Natural Selection Clothing Co Ltd v. Doyle*, [2001] ATMO 23; *Lomas v. WM Productions Pty Ltd*, [2005] ATMO 57 (in obiter).

(b) an opposition to an application under that subsection has been dismissed (see Section 99A);

the Registrar must remove the trade mark from the Register in respect of the goods and/or services specified in the application.

(2) If there is no opposition to an application to a court under Subsection 92(3), the court must order the Registrar to remove the trade mark from the Register in respect of the goods and/or services specified in the application. The court must cause a copy of the order to be served on the Registrar and the Registrar must comply with the order.'

If the removal for non-use application is opposed and non-use is established, then provided the Registrar or the court does not exercise its discretion, the registered trade mark is removed from the Register *in respect of any or all of the goods and/or services to which the* [removal] *application relates.*[104]

If genuine use has been re-established prior to the mark being removed from the Register, there are no consequences for any periods of non-use preceding the same: the owner of the trade mark will be able to enforce its rights against any third parties, including in the period during which the trade mark was not being used.

## B. Extent of any revocation

As set out above, subject to the Registrar or the court exercising its discretion, where it is determined that non-use has been established –

(a) in relation to all of the goods/services, the registered trade mark will be removed from the Register in respect of all goods and/or services; or

(b) in relation to some only of the goods/services, the registered trade mark will be removed from the Register in respect of those goods and/or services in relation to which use hasn't been established,

provided that the goods and/or services in relation to which the use hasn't been established aren't considered to be similar goods or closely related services; or similar services or closely related goods, to those the trade mark has been used (if any) by its registered owner.[105]

## C. Timing of any revocation

An application for the removal of a trade mark from the Register on the basis of non-use may be made at any time after the filing date in respect of the application for the registration of the trade mark.

However, a removal for non-use application brought pursuant to Section 92(4)(b) of the Act, may only be brought after a period of five years has passed from the filing date in respect of the application for the registration of the trade mark.[106]

104. Sec. 101(1).
105. Sec. 101(4).
106. Sec. 93.

# Benelux

*Emmanuel Cornu*

## I. INTRODUCING GENUINE USE OF TRADEMARKS

### A. Legal framework

#### *1. The principles*

Under Benelux law as it currently stands, an obligation to make use of a trademark is enshrined in Articles 2.26 and 2.27 of the Benelux Intellectual Property Convention (or 'BIPC'). Article 2.26(2)(a) BIPC provides that a trademark extinguishes if, for no just cause, once registered and for the ensuing uninterrupted period of five years, no use is made of the trademark in the Benelux territory for the products or services for which it was registered. It goes on to provide that, in the case of a dispute, the court can order that all or part of the onus of proving use will lie with the trademark holder.

In an exception to this rule, Article 2.27(2) BIPC provides that revocation of the trademark can be avoided if a start is made to using the unused trademark or such use is resumed, although such commencement or resumption within three months prior to the tabling in court of the revocation petition will be disregarded where preparations for using it or resuming its use were only embarked on once the holder had learned of the petition's imminent likelihood.

These provisions of the Convention are a transposition into Benelux law of Article 10(1) of Directive 2008/95/EC of the European Parliament and of the Council of 22 October 2008 to approximate the laws of the Member States relating to trade marks.

When the Uniform Benelux Trademarks Act came into force in 1970, the obligation to actually make use of a trademark was one of the most important changes it introduced (now continued under the Benelux Convention), marking as it did a reversal in the roles of use and registration. Previous to the Uniform Benelux Trademarks Act, under the domestic laws of each of Belgium, the Netherlands and Luxembourg, use could form a ground for claiming trademark rights. The Benelux

Trademark Act, on the other hand, made registration an attribute of the trademark right[1] and required use as the means of upholding those rights.[2]

### 2. *Genuine use ('usage sérieux') and normal use ('usage normal')*

Comparing the different language versions of the EU Directive reveals that, in French, the English expression 'genuine use' is rendered as '*usage sérieux*'. Elsewhere, the Benelux Convention talks of 'normal use'. Though worded differently, they all point to the same thing.[3] The notion of '*usage sérieux*' (in this contribution we adopt the Directive's 'genuine use') will incidentally soon replace '*usage normal*', or normal use, as BIPC terminology upon transposition of the new Directive (EU) 2015/2436 of 16 December 2015 to approximate the laws of the Member States relating to trade marks.

## B. Purpose and function of genuine use: policy considerations

The *ratio legis* of the obligation to use the trademark – i.e. genuine use and not merely symbolic – relies on the idea that the exclusive right loses its fundamental purpose if it is not exercised. There are therefore several rationales for the obligation to use a registered trademark. The hoarding of unused distinctive signs leads to a disproportionate proliferation of registrations cluttering the registers which restricts, without any justification, the choice of trademarks by competitors, whereas the number of attractive signs is not unlimited. Furthermore, one can fear that, in the absence of this obligation to use, abusive registrations would be carried out only with a view to block competing companies. The Benelux legislation seeks therefore to tackle the practice of 'reserve trademarks', i.e. trademarks kept too long with any use, and 'defensive trademarks', as well as to tackle another practice involving unduly extending the protection of the registration to goods that registrant never intends to commercialize, a practice known as 'goods reserve'.[4]

The concern vis-à-vis the protection of consumers against confusion regarding the origin of the goods protected by unused registrations plays no role, in our opinion.[5]

1. With the exception of well-known trademarks within the meaning of Art. 6*bis* Paris Convention, which are protected under Benelux law even barring any registration.
2. A. Braun and E. Cornu, *Précis des Marques* (5th edn, Larcier, 2009), p. 579, no. 512; L. Van Bunnen, *Aspects actuels du droit des marques dans le marché commun* (Brussels, Centre interuniversitaire de droit comparé, 1967), p. 385.
3. Braun and Cornu, *supra* n. 2, no. 512, p. 580; J.J. Evrard and P. Péters, *La défense de la marque dans le Benelux* (Larcier, 2nd edn., 2000), no. 196; T. Cohen Jehoram, C.J.J.C. van Nispen and J.L.R.A. Huydecoper, *Industriële eigendom*, Part 2 'Merkenrecht' (Kluwer Deventer, 2008), no. 12.3.3; F. Gotzen and M.-C. Janssens, *Handboek Merkenrecht* (CIR, Bruylant, 2008), p. 249-250.
4. Braun and Cornu, *supra* n. 2, no. 512, p. 579.
5. See also: rapport of the Belgian group of AIPPI on question Q 218, Ing.-Cons., 2011, p. 563.

## II. WHAT QUALIFIES AS GENUINE USE OF A TRADEMARK

### A. Nature of use

#### 1. *Use consistent with the function of the trademark*

The notion of genuine use is part and parcel of the EU's Trademarks Directive and, by dint of that fact, is a term of European Union law. In construing its meaning, reference has to be made to the dicta of the Court of Justice of the European Union and, as the Court itself has stressed, it is the same under both the Trademarks Directive and the EU's Trademarks Regulation.[6]

It is consistently held by the Court of Justice that trademarks are 'genuinely used'

> 'where the mark is used in accordance with its essential function, which is to guarantee the identity of the origin of the goods or services for which it is registered, in order to create or preserve an outlet for those goods or services; genuine use does not include token use for the sole purpose of preserving the rights conferred by the mark. When assessing whether use of the trademark is genuine, regard must be had to all the facts and circumstances relevant to establishing whether there is real commercial exploitation of the mark in the course of trade, particularly the usages regarded as warranted in the economic sector concerned as a means of maintaining or creating market share for the goods or services protected by the mark, the nature of those goods or services, the characteristics of the market and the scale and frequency of use of the mark.'[7]

Generally, using a corporate name or a trade name does not constitute use of the trademark as a means to distinguish products.[8] Thus, a corporate name is intended to identify a legal person, whereas a trade name is aimed at identifying or marking out a business. Their prime aim is not therefore to distinguish products or services,[9] though this is not always the case especially in relation to service marks. *De facto*, a trade name can be deployed such that the general public perceives that use as use of a sign distinguishing from all others the services that are offered.[10]

---

6. Court of Justice, 11 March 2003, C-40/01, *Ansul*, 30, ECLI:EU:C:2003:145.
7. Court of Justice, 19 December 2006, C-149/11, *Leno Merken v. Hagelkruis Beheer*, 29, ECLI:EU:C:2012:816; Court of Justice, 11 May 2006, C-416/04, *The Sunrider Corp. v. OHMI*, 70, ECLI:EU:C:2006:310; Court of Justice (Order), 27 January 2004, C-259/02, *La Mer Technology*, 27, ECLI:EU:C:2004:50; Court of Justice, 11 March 2003, *Ansul*, C-40/01, 43, ECLI:EU:C:2003:145.
8. Antwerp Court of Appeal, 2 February 2007, Annuaire pratiques du commerce, 2007, p. 398.
9. Court of Justice (Grand Chamber), 11 September 2007, C-17/06, *Céline*, ECLI:EU:C:2007:497.
10. Benelux CJ, 20 December 1996, case A95/2, *Europabank*, Ing.-Cons., 1997, p. 9 and note by E. Cornu; on the same lines, see Court of Justice (Grand Chamber), 11 September 2007, C-17/06, *Céline*, *supra* n. 9. See also General Court, 27 September 2007, *La Mer Technology*, T-418/03, 73 and 74, ECLI:EU:T:2007:299; General Court, 21 November 2013, *Recaro Holding GmbH v. OHMI – Certino Mode SL*, T-524/12, ECLI:EU:T:2013:604; appeal dismissed by Court of Justice, order, 14 January 2015, C-57/14 P, ECLI:EU:C:2015:17. Cases where the design on question was held only to have been used as a trade name and not as a trademark: General Court, 14 July 2016, *Modas Cristal SL v. EUIPO – Zorlu Tekstil*

Nor does use of a trademark as a merely description of the characteristics of the products constitute genuine use.[11]

## *2. Use in the course of trade*

### *i. Public use*

Genuine use of the mark entails use of the mark on the market for the goods or services protected by that mark and not just internal use by the undertaking concerned.[12]

In consequence, use of a trademark by a non-profit-making association during purely private ceremonies or events, or for the advertisement or announcement of such ceremonies or events, constitutes an internal use of the trademark and not 'genuine use' of it.[13]

The trademark has to be used publicly and externally and not solely within the undertaking of the trademark owner or within a distribution network owned or controlled by him.[14] This outward use does not necessarily mean that it is use aimed at end consumers. The relevant public to which marks are addressed do not comprise only end consumers, but also specialists, industrial customers and other professional users.[15]

### *ii. Commercial purpose*

#### a. The principle

Genuine use of a trademark requires a 'real commercial exploitation of the mark in the course of trade' or a usage 'regarded as warranted in the economic sector concerned as a means of maintaining or creating market share for the goods or services protected by the mark'.[16]

#### b. Use by an NPO or foundation

Non-profit organizations (NPOs) can also make business use of a trademark. The fact there is no profit motive does not prevent a charity having the intent to create

---

*Ürünleri Pazarlama*, T-345/15, 38 and 39, ECLI:EU:T:2016:405; General Court, 13 May 2009, *Schuhpark Fascies v. OHMI – Leder & Schuh*, T-183/08, 26-41, ECLI:EU:T:2009:156.

11. General Court, 14 April 2016, *Henkel & Co. Sektkellerei KG v. EUIPO – Ciacci Piccolomini d'Aragona di Bianchini*, T-20/15, 39-51, ECLI:EU:T:2016:218.
12. Court of Justice, 11 March 2003, C-40/01, *Ansul*, *supra* n. 6, point 37.
13. Court of Justice, 9 December 2008, C-442/07, *Verein Radetzky-Orden*, ECLI:EU:C:2008:696; point 22.
14. General Court, 10 September 2008, T-325/06, Boston Scientific Ltd/OHIM – *Terumo*, point 47; ECLI:EU:T:2008:338; General Court, 9 September 2015, T-584/14, *Industria de Diseño Textil/OHIM – Zainab Ansell & Roger Ansell*, point 36; ECLI:EU:T:2015:604.
15. General Court, 7 July 2016, *Fruit of the Loom Inc./EUIPO – Takko Holding GmbH*, T-431/15, point 49, ECLI:EU:T:2016:395.
16. Court of Justice, 19 December 2006, C-149/11, *Leno Merken/Hagelkruis Beheer*, *supra* n. 7, point 29 and the case-law referred.

or maintain an outlet for its products or services.[17] Trademarks registered by an NPO can have a *raison d'être* in that they protect the organization against others making commercial use of identical or similar marks.

As long as the NPO uses the trademarks it holds to identify and promote its products or services, that use will be 'genuine'.[18]

c. Free promotional goods

The scholarly view has long been that placing a trademark on merchandising, especially clothing, is not by that simple fact a use intended to designate the clothing.[19]

Thus, affixing the 'Lego' trademark to a T-shirt was dismissed as use for clothing, for the reason that it was an advertisement for the toys bearing that trademark.[20]

It is a position that has now found judicial endorsement from the Court of Justice of the European Union. In *Silberquelle*, it ruled that where a holder of a trademark affixes it to items given out free of charge to the recipient there is no genuine use of the trademark for the class under which the items fall.[21]

d. Use for products already on the market or after-sales services

There are circumstances when use of a trademark can qualify as genuine use for products already on the market and are no longer newly offered for sale, the classic case being where the holder sells spare parts fitting together with or added to products already on the market and for which he makes actual use of the trademark.

The same can happen if a trademark holder uses the mark for items that are not combined with the composition or structure of the products that have already been marketed but that directly relate to those products and are intended to meet the needs of those products' customers, especially after-sales transactions or the provision of maintenance and repair services.[22]

### 3. *Use by the owner or with its consent*

Article 2.26(3)(c) BIPC deems trademark use with the owner's consent to be a use by the owner. The user need not be a registered trademark licensee for the owner

17. Court of Justice (Grand Chamber), 9 December 2008, *Verein Radetzky – Orden v. Bundesvereinigung Kameradschaft ‚Feldmarschall Radetzky'*, C-442/07, 16-29, ECLI:EU:C:2008:696.
18. Report by the AIPPI's Belgian group on question 218, Ing.-Cons., 2011, p. 569.
19. T. van Innis, *Les signes distinctifs* (Larcier, 1997), pp. 500-501; Cohen Jehoram et al., *supra* n. 3, no. 12.3.7.
20. Amsterdam District Court, 26 June 1989, BIE, 1990, p. 43; on the same lines: Amsterdam District Court, 8 November 1990, IER, 1990, p. 59; against this, see: The Hague District Court, 31 March 2009, Bulletin BMM, 2009, Book 2, p. 74.
21. Court of Justice, 15 January 2009, *Silberquelle*, C-495/07, 21 and 22, ECLI:EU:C:2009:10.
22. Court of Justice, 11 March 2003, *Ansul*, C-40/01, *supra* n. 6, 40-43; Cohen Jehoram et al., *supra* n. 3, point 12.3.4.

to assert use by consent. It could therefore simply be use by a wholesaler or retailer of the products and services sold under the trademark.[23]

## B Volume, duration and frequency of use

### *1. The principle*

To meet the genuineness test therefore, use must not be 'token use for the sole purpose of preserving the rights conferred by the mark'.[24]

Both the Court of Justice and the General Court of the EU take the view that the question of 'quantitative' sufficiency of a trademark's use to create or maintain market shares for the products it protects needs answering by reference to various of the aforementioned factors[25] within the bounds of the case in question. Nonetheless the Court of Justice has consistently held that great weight is placed on the features of the relevant products or services, especially the frequency and regularity of the mark's use.[26]

The definition thus given to 'genuine use' by the Court of Justice of the European Union is redolent of that given by the Benelux Court of Justice in its 27 January 1981 judgment on the interpretation of 'normal use'. A reference for a preliminary ruling had been made to it by the *Hoge Raad*, the supreme court of civil jurisdiction in the Netherlands, on use of the 'Whiston' trademark by Turmac, the sole evidence for which was 30 invoices of the same date for sales of 1,000 cigarettes per customer. The Benelux Court of Justice held:

> '1. It is a principle of law that, to prevent revocation of a trademark right under section 5(3) Uniform Benelux Trademarks Act, the mark must at least have been used in the Benelux countries in the course of business by the mark holder or in his behalf during the pertinent period other than inside the business of the user and such use must unequivocally relate to an identified product that is

23. Brussels Court of Appeal, 23 September 2008, IRDI, 2009, p. 161; Brussels Court of Appeal, 24 June 2004, IRDI, 2004, p. 423; Brussels Court of Appeal, 1 October 2003, Ing.-Cons., 2004, p. 237; Brussels Court of Appeal, 12 September 2003, Ing.-Cons., 2004, p. 29; Brussels Court of Appeal, 21 June 2002, Ing.-Cons., 2004, p. 11; President of Ghent Commercial Court, 4 September 2006, TGR-TWVR, 2007, p. 30.
24. Court of Justice, 11 March 2003, *Ansul*, C-40/01, *supra* n. 6, 43.
25. I.e.: 'all the facts and circumstances relevant to establishing whether there is real commercial exploitation of the mark in the course of trade, particularly the usages regarded as warranted in the economic sector concerned as a means of maintaining or creating market share for the goods or services protected by the mark, the nature of those goods or services, the characteristics of the market and the scale and frequency of use of the mark' (Court of Justice, 19 December 2006, C-149/11, *Leno Merken v. Hagelkruis Beheer*, 29, ECLI:EU:C:2012:816; Court of Justice, 11 May 2006, C-416/04, *The Sunrider Corp. v. OHMI*, 70, ECLI:EU:C:2006:310; Court of Justice (Order), 27 January 2004, C-259/02, *La Mer Technology*, 27, ECLI:EU:C:2004:50; Court of Justice, 11 March 2003, *Ansul*, C-40/01, 43, ECLI:EU:C:2003:145).
26. Court of Justice, 11 May 2006, *The Sunrider Corp. v. OHMI*, C-416/04 P, cited above, point 71.

> sold or offered for sale by the user and that, by such use, is distinguished from the products of other parties;
>
> 2. In assessing whether use under (1) qualifies as "normal use", account is taken of the relative importance of all the facts and circumstances of the case, such as: the nature, extent, frequency, regularity, and duration of such use, the nature of the product and the nature and size of the business; the decisive factor is that, provided parties act in a manner that is usual and commercially justified in the business sector in question, the circumstances as a whole point to the intention behind using the mark as being to create or preserve an outlet for the branded products and not to maintain the trademark right.'[27]

It is a dictum reiterated by the Benelux Court of Justice in *Philip Morris v. BAT*, in which it took the view that 'the obligation is founded [...] on the idea that the extended protection conferred by law on the exclusive right of trademarks is only justified if the mark in question is actually used in the course of business to distinguish a company's products or services; hence, if there is no or insufficient use to this end, the mark holder may only exceptionally continue to enjoy the extended protection in cases where, to put it briefly, it isn't his fault it wasn't used.'[28]

For the trademark's use to be genuine, it has 'actually' to have been present 'on the market in question for five years prior to the date of the petition (founding on revocation of the mark)', whereby proof that its use was genuine must be built 'on specific, objective factors evidencing actual, sufficient use of the trademark on the relevant market.'[29] As is moreover clear from a reading of cases decided by the EU's General Court, the more limited the volume of product sold, the more the mark owner must adduce additional evidence to expunge any doubts as to the genuineness of his purported use.[30]

That said, the requirement for genuine use is not a test of a company's commercial success or a yardstick of its business strategy, nor is its aim to make trademark protection the preserve of large-volume trading enterprises,[31] though minimal sales given the nature of the product, the market or the size of the company will obviously raise suspicion. In short, for trademark use be genuine, it cannot be symbolic in order just to simply uphold the rights the mark confers.

Genuine use is what is regarded as justified in the relevant economic sector to maintain or create market share for the branded products or services. Thus, whether use is genuine or not is judged by assessing the nature of the relevant products or services, the characteristics of the market, the extent, frequency and regularity to or with which there is use and the practical abilities of the holder.

---

27. Benelux CJ, 27 January 1981, case 80/1, *Turmac v. Reynold*, Ing.-Cons., 1981, p. 113 and note by F. de Visscher.
28. Benelux CJ, 18 November 1988, case A87/2, *Kim*, Ing.-Cons., 1988, p. 301.
29. Brussels Court of Appeal, 9th Div., 22 December 2005, Ing.-Cons., 2005, p. 401; Brussels Court of Appeal, 9th Div., 4 September 2007, J.T., 2007, p. 523.
30. Along these lines: General Court, 30 April 2008, *Sonia Rykiel v. OHMI – Cuadrado*, T-131/06, 42 and citations, ECLI:EU:T:2008:135.
31. Along these lines: General Court, 27 February 2014, T-226/12, *Lidl Stiftung & Co. KG v. EUIPO – Lidl Music*, 28 and 31, ECLI:EU:T:2014:98; appeal dismissed by the Court of Justice (Order), 6 September 2015, C-224/14 P, ECLI:EU:C:2016:666.

Evidence of use is assessed taking an overall view and bearing in mind all the factors relevant to the individual case. This approach would imply a certain degree of inter-dependence between the factors that are considered. Thus, low product volume sold under the trademark may be offset by high intensity or constancy in the time the trademark was used for and vice versa. Additionally, turnover achieved and quantity of branded products sold cannot be seen as absolutes but have to be viewed relative to other relevant factors like volume of sales activity, manufacturing or sales capacity or degree of diversity in the business operating the trademark plus the characteristics of the products or services on the market under consideration.[32]

### *2. Use sometimes limited in extent that needs to be economically justified*

A limited extent to which there is use may be justified because certain products are produced in short runs or only a few pieces are sold with fairly long intervals between sales. Aircraft are a classic example.

Or low levels of trademark use may be explained by the seasonal nature of the goods, such as Christmas produce.[33]

Genuine use of a trademark does not therefore necessarily mean that the branded products have to be on widespread sale over the entire Benelux territory, but just requires that use should not be symbolic or artificial. Thus, token sales made merely to counter moves to have a trademark revoked will fail the test.

Specious sales that the court views as affected for the sole purpose of the action will therefore be discounted.[34]

Consideration will therefore be given to the nature of the products sold, though also to how the trademark is advertised on Benelux territory.[35] In a case about the 'Sacher' brand for cakes, the Brussels Commercial Court allowed its use as normal or genuine given the size of the company that owned the trademark, which though small was well known, and set greater store in the regularity of sales and the branded products than in their quantity as well as considering online sales via the owner's website and its efforts to promote its products within the Benelux countries by means of advertising.[36] Against this, if there is no or insufficient use having regard to the nature of the products and the size of the company, it will be ruled as having been revoked.[37]

Even though genuine use of a trademark is assessed on the basis of objective, specific factors,[38] there is no minimum standard. The Benelux courts require only

---

32. General Court, 8 November 2007, *Charlott v. OHMI – Charlo*, T-169/06, 36 and 37, ECLI:EU:T:2007:337; General Court, 10 September 2008, *Boston Scientific v. OHMI – Terumo*, T-325/06, 49 and 50, ECLI:EU:T:2008:338.
33. Brussels Commercial Court, 23 November 1995, Ing.-Cons., 1996, p. 154.
34. Braun and Cornu, *supra* n. 2.
35. Brussels Commercial Court, 23 September 2002, Ing.-Cons., 2002, p. 251.
36. *Ibid.*; see also Dutch Supreme Court, 23 December 2005, BIE, 2006, p. 65; BMM Bulletin 2006/4, p. 176.
37. President of Zwolle Court of First Instance, 9 July 1982, BIE, 1984, p. 114; Verviers Commercial Court, 17 March 1980, Ing.-Cons., 1980, p. 156.
38. Brussels Court of Appeal, 22 December 2005, Ing.-Cons., 2005, p. 401.

that the use should have been intended to create outlets for the trademark and that it has been truly utilized.[39]

The courts are especially attentive to the extent, frequency and duration of use and to the nature of the products or services in question.[40] The features of the company in question are also looked to in assessing whether the alleged use is an artifice or otherwise.[41] As indicated above, advertising to promote the trademark will be looked at.[42] However, advertising that is too limited cannot be deemed genuine use.[43]

Merely preparing for use of a trademark and, say, simply trying to lodge a franchise to sell the branded products have been held to fall short of genuine use.[44] Likewise, simple acts preparatory to use are generally regarding as failing the test.[45] And, Belgium's Court of Cassation ruled on 30 June 2000 in a case regarding use of a trademark during clinical trials prior to release for sale, that the Ghent Court of Appeal had not erred when it ruled that sporadic use in correspondence of the 'Amoxypen' trademark during the phase prior to registering a drug under it, without the drug ever having been actually manufactured for sale and never having been in the public domain, did not constitute genuine use of that trademark.[46]

### C. Use of a sign

Article 2.26(3)(a) BIPC extends use of a trademark to use 'in a form differing in aspects not detracting from its distinctive character from that in which it was registered.' This rule accords with Article 5(C)(2°) Paris Convention for the protection of industrial property[47] and is set down in Article 10(1)(a) of Directive 2008/95/EC of 22 October 2008 approximating the laws of the Member States on trade marks, as well as Article 16(5)(a) of Directive 2015/2436 of 16 December 2015 approximating the laws of the Member States on trade marks.

According to the Court of Justice, the purpose of this rule is to allow trademark holders to produce variations in the course of their business that are more flexible

---

39. Brussels Court of Appeal, 2 October 2001, IRDI, 2001, p. 343; Ing.-Cons., 2001, p. 371; Brussels Court of Appeal, 5 May 1999, Ing.-Cons., 1999, p. 361; Brussels Court of Appeal, 19 February 2008, RABG, 2008, p. 1277; Brussels Court of Appeal, 22 March 2005, Bulletin BMM, 2006, p. 39; Brussels Court of Appeal, 22 December 2005, Ing.-Cons., 2005, p. 401; Brussels Court of Appeal, 28 June 1993, Ing.-Cons., 1993, p. 230; Brussels Court of Appeal, 4 June 1986, Ing.-Cons., 1986, p. 278.
40. Brussels Commercial Court, 23 September 2002, Ing.-Cons., 2002, p. 251; Brussels Commercial Court, 23 November 1995, Ing.-Cons., 1996, p. 154.
41. Brussels Commercial Court, 23 September 2002, Ing.-Cons., 2002, p. 251; Oudenaarde Commercial Court, 6 October 1999, Ing.-Cons., 1999, p. 154.
42. Brussels Commercial Court, 23 October 2002, Ing.-Cons., 2002, p. 251.
43. Brussels Commercial Court, 28 June 2007, IRDI, 2007, p. 415; however, against this see: Dinant Commercial Court, 21 April 2000, RDC, 2001, p. 260.
44. Brussels Court of Appeal, 13 January 2000, IRDI, 2000, p. 122.
45. The Hague District Court, 14 October 1998, BIE, 2000, p. 28; The Hague District Court, 27 April 2000, BIE, 2000, p. 99.
46. Court of Cassation, 30 June 2000, Ing.-Cons., 2000, p. 394.
47. Revised in Stockholm on 14 July 1967 and ratified into Belgian law on 26 September 1974, official gazette of Belgium, 29 January 1975.

in terms of the requirements for marketing and promoting the products and services designated by the mark without altering its distinctive character.[48]

It is especially allowed for a word mark to be used in a particular graphic or to be used alongside a logo or other figurative elements provided that, in the configuration in question, the word element continues to function as an indicator of the commercial origin of the designated products or services. The important aspect is therefore, in each case, to be able to clearly identify the registered trademark in the use that is made of it.[49]

It is a trickier situation if word elements are added to a word mark. If the addition forms an indivisible whole together with the registered word mark and puts a different slant on it or gives it a different meaning, the distinctive character of the registered mark will be detracted from. A prime example comes from the General Court's ruling on use of the expression 'Fruit of the Loom' as against the registered mark 'Fruit': whereas the latter had a precise meaning by reference to a plant's produce, the former seems like a fantastical sign, an invented phrase, since there is no fruit in nature that comes from the weaving industry, so that use of that expression has been regarded as altering the distinctive character of the 'Fruit' mark.[50]

Subject to that reservation, the possible variants in use of a trademark are legion. In its *Specsavers* ruling of 18 July 2013, the Court of Justice went a long way in setting the limits to which the owner of a trademark may adapt it to requirements for marketing and promoting its products.[51] In this case, the Court viewed use of a mark made up of a logo incorporating a word sign as capable of constituting genuine use of a trademark incorporating the 'silent' logo i.e. devoid of the word mark in question. The Court ruled that, relative to this use, the trademark as registered has to be clearly recognizable to the relevant audience and that audience has to regard the trademark as still being such as distinguishing the product or service sold under that trademark from those coming from other undertakings.[52]

There is otherwise no rule of trademark law that requires proof of isolated use of trademarks independently from any other mark or sign. It is in fact possible for two or more marks to be used jointly or on their own with or without the name of the product's manufacturer. Joint use of two marks belonging to a single owner does nothing to detract from the identifier function of either of them.[53]

It is perhaps so obvious that we should apologise for pointing it out, but the use of a trademark is not as seen 'in abstracto' from the notional consumer's viewpoint

48. Court of Justice, 25 October 2012, C-553/11, *Rintisch*, Ing.-Cons., 2012, p. 580, 21 and observations by E. Cornu; Court of Justice, 18 July 2013, C-252/12, *Specsavers*, Ing.-Cons., 2013, p. 525, 29; E. Cornu, *L'usage d'une marque en tant qu'élément d'une marque complexe ou en combinaison avec une autre marque*, Ing.-Cons., 2013, pp. 479-485.
49. Cornu, *supra* n. 48, pp. 480 and 481.
50. General Court, 21 June 2012, T-514/10, *Fruit of the Loom Inc. v. OHMI*, ECLI:EU:T:2012:316.
51. Court of Justice, 18 July 2013, C-252/12, *Specsavers*, Ing.-Cons., 2013, p. 525.
52. See numbers 22 and 23 in this decision; on this matter, see Cornu, *supra* n. 48, pp. 179-485.
53. Along these lines, see: General Court, 1 March 2018, T-438/16, *Altunis – Trading, Gestão e Serviços, Lda v. EUIPO and Hotel Cipriani ('CIPRIANI')*, 40, ECLI:EU:T:2018:110; General Court, 8 December 2005, *Castellblanch v. OHMI – Champagne Roederer ('Cristal Castellblanch')*, T-29/04, 33-36, ECLI:EU:T:2005:438; General Court, 14 December 2011, *Völkl v. OHMI – Marker Völkl ('Völkl')*, T-504/09, 100, ECLI:EU:T:2011:739.

but from that of real people in the real world. A trademark always appears in an environment in which it serves to distinguish the product or service in question. In the case of a word mark, it can always be combined with a logo or assume various graphic forms or particular colours and thus quite naturally go hand in hand with other word marks like sub-marks owned by the same entity or with logos and figurative elements that call that entity to mind. Other labelling elements describing the product's composition, where it is from and its characteristics or compulsory legal information can also accompany use of the trademark. Average consumers are in no way taken aback by that and are perfectly used to seeing such information.

All of these specific aspects surrounding use of a trademark do nothing to alter the fact that the people it speaks to will perceive it in terms of its distinctive function, as a guarantee that the product or service bearing it comes from a particular undertaking.

Cases on the merits from courts in the Benelux countries offer a number of examples of how Article 2.26(3)(a) has been applied: the holder of the 'Q' mark could not prevent the same letter being used by the 'Q8' oil company on the ground that the use claimed of its 'Q' mark did not correspond to the form that that letter had when it was registered.[54] The fact that the trademark in question was a single letter might well have been decisive in that assessment.[55]

Similarly, use of the sign 'A CAUSE DES GARCONS' was not deemed normal use of the trademark 'A CAUSE DES GARCONS ON S'HABILLE DE CETTE FACON'[56] and use of the 'BATON DU BERGER' sign was not considered use of the trademark 'LE BATON'.[57] Nor was use of the signs 'VLAM IN DE PAN' judged as forming normal use of the trademark 'FLAMBORJA D'ESPAGNOL – VLAM IN DE PAN'[58] and use of the trademark 'CRISTALINO' was judged insufficient to maintain the trademark 'CRISTALINO JAUME SERRA' as identifying a wine because the signs 'JAUME SERRA' contributed to the distinctive power of the trademark.[59] Nor does use of the trademark 'YELLOW PAGES' count for use of the trademark 'PAGES D'OR'.[60]

By contrast, regarding the 'CHICO' trademark for fruit, it was held that 'neither adopting a different typeface nor deleting certain concentric circles could detract from the distinctive character of the trademark used.'[61] Use of the signs 'SPA ENERGIE VEGETALE' and 'EAU THERMALE SPA' was also ruled by Brussels Appeal Court to constitute normal use of the 'SPA' word mark.[62] All in all, proof of the renown of the

54. Amsterdam District Court, 11 December 1986, BIE, 1987, p. 258; President of Brussels Commercial Court, 18 November 1986, Ing.-Cons., 1986, p. 420.
55. Along these lines, L. Van Bunnen, *Les marques formées d'une seule lettre*, Ing.-Cons., 1987, p. 37.
56. Brussels Commercial Court, 12 July 1991, Ing.-Cons., 1991, p. 486.
57. Brussels Commercial Court, 7 January 1991, BIE, 1994, p. 58.
58. The Hague District Court, 26 May 1989, BIE, 1990, p. 34.
59. Brussels Court of Appeal, 28 June 2011, Ing.-Cons., 2011, p. 430.
60. The Hague District Court, 17 January 2012, BIE, 2012, p. 134.
61. Brussels Commercial Court, 28 December 1989, Ing.-Cons., 1990, p. 36.
62. Brussels Court of Appeal, 4 September 2007, J.T., 2007, p. 724.

figurative 'SPA' trademark accompanied by the image of a leaping Pierrot figure have also been judged as establishing the reputation of the 'SPA' word mark.[63]

The Benelux Court of Justice held elsewhere, in its *Michelin* ruling, that use of a work mark in a configuration possessing distinctive power could also constitute use of that word mark, unless the form in which it was used is so predominant that the general public perceive the sign deployed as merely being a figurative trademark and not as a word mark.[64] It is worth questioning how up-to-date this decision is and whether it accords with cases decide by the Court of Justice of the European Union, particularly its *Specsavers* case, C-252/12, which has since been handed down.

Benelux court decisions at first instance tend to the view that simple differences in the shape given to letters or colours that are used makes no difference when the figurative aspect in question is not of dominant character in the trademark.[65] But, where a graphic element in a complex trademark constitutes a major element of the trademark's distinctive character as registered, use limited to the verbal portion of the complex trademark may sometimes be ruled insufficient to constitute use of the complex trademark.[66] The Hague Court of Appeal nonetheless held that use of the complex 'IKEA' trademark, in which a blue rectangle containing a yellow oval in which is written the name 'IKEA' in blue was use of the same logo but without the word 'IKEA', since the first trademark had a reputation and that renown conferred a distinctive character on the figurative element alone (market research had shown that the trademark devoid of the name 'IKEA' was recognized by a significant portion of the general public as identifying 'IKEA').[67] We cannot but endorse this decision, particularly in light of the Court of Justice's dictum in *Specsavers*, C-252/12, which is referred to above.

By analogy, when a complex trademark is registered in a semi-figurative form but with dominant word elements, its use in a modernised style will generally be deemed use of the trademark that does not detract from the distinctive character of the form in which it was registered.[68]

In a recent judgment, Mons Court of Appeal applied the foregoing principles to conclude, rightly, that use of the figurative sign 'CORA', used to designate a chain of hypermarkets in Belgium, was sufficient to maintain the word mark 'CORA' as well as use of another, semi-figurative mark 'CORA' vaunting the same name in special lettering different from how the trademark was used, since the main distinctive element was still the 'CORA' sign.[69]

---

63. General Court, 17 March 2015, *Spa Monopole v. OHMI and South Pacific Management ('Manea SPA')*, ECLI:EU:T:2015:152; General Court, *Spa Monopole v. OHMI and Orly International ('SPARITUAL')*, ECLI:EU:T:2015:257.
64. Benelux CJ, 16 December 1994, case A-93/7, *Michelin*, Ing.-Cons., 1995, p. 22.
65. Brussels Court of Appeal, 17 November 2009, Ing.-Cons., 2009, p. 636.
66. Antwerp Court of Appeal, 13 April 2006, *Sniffers for Fashion and Sniffers Holding v. Sniffers & FH Dick*, 2004/AR/1978, unreported.
67. The Hague District Court, 26 January 2010, cited by T. de Haan, *Un an de jurisprudence en propriété industrielle dans le Benelux*, Propriété industrielle, Jurisclasseur, January 2011, p. 28.
68. Brussels Commercial Court, 12 December 2005, IRDI, 2006, p. 211.
69. Mons Court of Appeal, 12 June 2017, RDC, 2018, p. 466, and observations by E. Cornu, *A propos de l'intérêt à agir en déchéance d'une marque, à la charge et à l'objet de la preuve.*

## D. Use in relation to goods and services

### *1. Use for each of the designated goods and services*

If a trademark has only been used for one of the parts of products or services for which it was registered, it can be only partially revoked.

### *2. Use for a broader category of goods and/or services*

If a trademark is registered for a category of products of services broad enough for there to be distinguished within it several sub-categories capable of being regarded autonomously, proving genuine use of the trademark for part of these products or services will count only for the sub-categories into which the products or services fall for which the trademark was actually used. By contrast, if a trademark is registered for precisely defined products or services such that it is not possible to make significant sub-divisions within the category concerned, the proof of genuine use of the trademark for those products or services will perforce cover the entire category.[70]

As has been emphasized by Brussels Court of Appeal, 'where a trademark is filed for two products but used only for one of them, the trademark right extinguishes for the product for which it is not used, even if it resembles the product actually marketed.' Therefore, 'the fact that the protection accorded to a valid trademark that has not been revoked extends not only to the products for which it was filed but also all similar products is unaffected since the scope of a trademark's protection is a different question to that concerning maintenance of the trademark right.'[71]

## E. Territorial scope

Maintaining rights over Benelux trademarks requires they be used on Benelux territory, though use need not extend over all three Benelux countries.[72] Use in just one of them or a part of one of them is enough[73] since the Benelux countries no longer have national trademarks and operators in just (part of) one of them have no alternative but to file and register a Benelux trademark.[74]

---

70. General Court, 16 May 2013, *Aleris Holding AB v. OHMI – Carefusion 303 Inc.*, T-353/12, 18 and 19, ECLI:EU:T:2013:257; General Court, 14 July 2005, *Reckitt Benckiser (España) v. OHMI – Aladin ('ALADIN')*, T-126/03, 45, ECLI:EU:T:2005:288; General Court, 13 February 2007, *Mundipharma v. OHMI – Altana Pharma ('Respicur')*, T-256/04, 23, ECLI:EU:T:2007:46; General Court, 23 September 2009, *GlaxoSmithkline, Laboratorios Wellcome The Portugal & The Wellcome Foundation v. OHMI – Sereno Genetics Institute*, T-493/07, T-26/08 and T-27/08, 35-37, ECLI:EU:T:2009:355; appeal dismissed Court of Justice (order), 9 July 2010, C-461/09 P, *The Wellcome Foundation v. OHMI*, ECLI:EU:C:2010:421.
71. Brussels Court of Appeal, 12 October 2001, JLMB, 2002, p. 865; see also Braun and Cornu, *supra* n. 2, p. 605, no. 530.
72. They are Belgium, the Netherlands and Luxembourg.
73. Along these lines, Benelux CJ, 27 January 1981, case 80/1, *Turmac v. Reynold*, *supra* n. 27, and note by F. de Visscher.
74. Along these lines, see report by the AIPPI's Belgian group on this matter 218, Ing.-Cons., 2011, p. 571.

Even if a trademark is representative of high-consumption products, it might perfectly easily be used on the Belgian or Luxembourg market and not on the Dutch one. But sales in a single city of high-consumption products, like foodstuffs or detergent, will not be deemed normal use.

However, it is perfectly feasible for a manufacturer to restrict use of one of its trademarks, even if used for a widely consumed product, to highly specialised clientele though a limited distribution circuit, such as a specially branded whisky for a given club or a trademark applied to products intended only for hospitals.[75]

There must therefore be no confusion between what is genuine use and what is wide use. Limited use will, once it accords with a reasonable sales policy drawn up by the manufacturer, duly qualify as justified if it is compatible with the nature of the product. Thus it was held that sales of products exclusively for export at tax-free shops at Brussels and Luxembourg airports constituted normal or genuine use.[76]

Under Article 2.26(3)(b) of the Benelux Convention, applying a trademark to goods or their packaging solely for export purposes is moreover use of the trademark on Benelux territory.

Similarly, use of trademark in advertising to encourage uptake of services offered outside the Benelux countries may be regarded as constituting use of that mark within the Benelux territory when it involves creating or maintaining an outlet for the services offered under that brand within the Benelux countries. The Dutch Supreme Court or *Hoge Raad* decided this way regarding use of the 'Le Lido' trademark to promote stage shows at the Paris nightspot of that name.[77]

### F. Relevant time period and its calculation

Pursuant to Article 2.26(2)(a) of the Benelux Convention, a Benelux trademark has to be genuinely used once registered during a uninterrupted period of five years. However it is sufficient for genuine use to have been made during part of this relevant period of five years. It is not necessary to prove that the trademark has been used for a continuous and uninterrupted period of five years.[78]

Under Article 10(1) of Directive 2008/95/EC of 22 October 2008 approximating the laws of the Member States and, nowadays, Article 16(1) of Directive 2015/2436 of 16 December 2015 approximating the laws of the Member States on trade marks, the starting point for the period in which a trademark has to be subject to genuine use is the date of completion of the registration procedure. A Benelux trademark has to be used genuinely in the Benelux territory no later than five years after its registration.[79]

In practice, genuine use of a trademark is assessed over the five years preceding presentation of a revocation petition. In this regard, if one compares the different language versions of Article 2.27(2) BIPC and Article 12(1) of the Harmonisation

75. Braun and Cornu, *supra* n. 2, p. 581, no. 513.
76. The Hague District Court, 27 February 1991, IER, 1991, p. 68.
77. Dutch Supreme Court, 13 September 1996, NJ, 1997, p. 312 and note by D.W.F.V.
78. General Court, 8 November 2007, Charlott/OHMI – Charlo, T-169/06, point 41; ECLI:EU:T:2007:337.
79. Braun and Cornu, *supra* n. 2, p. 590, no. 520.

Directive,[80] the conclusion must be that 'presentation of the revocation petition' means the filing of the application for revocation, i.e. the date on which the court is seised of it. It is not, therefore, the date when the trademark owner is informed that a revocation application has been filed against his trademark.

If a registered trademark has not undergone genuine use within the statutory period, the owner may, if no revocation petition has yet been lodged, start or resume use of his trademark. As seen below, the start or resumption of use can in some cases form a solid defence to a revocation petition. The trademark is not therefore automatically revoked.[81]

Nor is there anything stopping the owner of a trademark who has failed to make genuine use of it within the statutory period from filing it again. He will not, however, be able to draw any priority right from his extinguished right, as was stated by the Benelux Court of Justice in *Triomino*.[82] The new filing will be judged in terms of its validity according to the criteria applying to any filing.

## III. EVIDENCE OF GENUINE USE OF A TRADEMARK

### A. When use should be proved: timing of lack of genuine use claims

#### 1. *When filing an application for a trademark*

Under Benelux law, there is no 'intent to use' requirement. It is not required that the trademark has to be used at the time of its filing. A Benelux trademark has to be used genuinely in the Benelux territory no later than five years after its registration.[83]

#### 2. *In the course of opposition proceedings*

Non-use can be cited under current Benelux law as a defence plea in opposition proceedings before the Benelux Office. In such proceedings, defendants are entitled under Rule 1.17(1)(d), (e) and (f) of the BIPC Implementing Regulations to demand proof of use of the mark cited in support of the opposition. In that case, the opposing party has to produce evidence proving genuine use of his trademark during the five years prior to publication of the mark against which the opposition is directed.[84] If the opposing party fails in this, it is enough if he establishes just cause for not making use of the mark. If the opposing party fails within the deadline set by the Office to produce evidence that genuine use was made of his mark during the relevant five-year period and he cannot assert just cause, the opposition is rejected and the case closed.[85]

80. Or Art. 19(2) of the new Directive 2015/2436 of 16 December 2015 approximating the Member States' laws on trade marks.
81. D. Kaesmacher, in *Les droits intellectuels* (2nd edn., Larcier, 2013), 215, at the end of no. 82; Gotzen and Janssen, *supra* n. 3, p. 255.
82. Benelux CJ, 19 December 1996, case A95/3, *Triomino*, Ing.-Cons., 1996, p. 16.
83. See above, section II.F.
84. See rule 1.29(2) BIPC Implementing Regulations.
85. See Art. 2.16(3) BIPC.

If the trademark has been used for only part of the products or services for which it was registered, the Benelux Office will limit its judgement on the opposition to those products and services in respect of which use is proved.[86]

### 3. *In cancellation proceedings based on lack of use*

#### *i. Court proceedings*

Under Article 2.27(1) BIPC, a court action for revocation lies with 'any interested party'. So, who is that?

The expression 'any interested party' was already to be founded Article 14(C) of the Uniform Benelux Trademarks Act, though the statement of reasons for the act remained silent on the point. More detail is nonetheless to be found in the statement of reasons for the Benelux Commission's draft: 'The term is to be interpreted as broadly as possible. It is sufficient if there is a simple non-pecuniary interest on the petitioner's part.'[87]

As was stressed by us in an article that appeared in 2004, the notion of an interested party is to be interpreted according to its common-or-garden meaning:[88] it encompasses any person who might have an interest in having a registration being revoked.[89]

It will generally be a competitor of the trademark owner who seeks to have removed from the register a trademark that could form an obstacle to use or registration of a new trademark. It could also be the defendant in a passing-off action who objects to a registration which has not been subject to normal or genuine use for the products it designates. Or it may be the owner of an anterior right who opts to seek revocation of the previous trademark rather than pleading the existence of a likelihood of confusion. Or it might equally be the owner of a later right who wants to procure the disappearance of a threat to his right. The possible scenarios are every bit as numerous as that.

The plaintiff in the cause has to prove a personal, direct interest to sue, i.e. an interest entitling him to seek the trademark's revocation.

Under the rules of civil procedure, plaintiffs must show a real, current interest.

Those who seek revocation must therefore prove their use or intention to use or request registration of a sign similar to the trademark whose revocation is sought for identical or similar products. It does not require the parties to be competitors, but at least it does mean they carry on business in sectors that are similar in trademark law terms. An assessment of the notion of 'similarity' as used in trademark law goes wider in this respect than just looking for a competitive link. According to the Court of Justice, an assessment of 'similarity' involves that 'all the relevant factors relating to those goods or services themselves should be taken into account. Those

86. See Rule 1.17(e) BIPC Implementing Regulations.
87. Braun and Cornu, *supra* n. 2, p. 618, no. 534.
88. E. Cornu, *L'intérêt à agir en matière de déchéance de marque, aperçu de la situation au Benelux*, Ing.-Cons., 2004, pp. 3-10.
89. Also along these lines, Braun and Cornu, *supra* n. 2, pp. 618-619, no. 534; T. van Innis, *supra* n. 19, 268, no. 345; C. Gielen and L. Wichers Hoeth, *Merkenrecht* (1992), 536.

factors include, inter alia, their nature, their 'intended purpose' and their method of use and whether they are in competition with each other or are complementary.'[90]

A trademark's revocation will therefore generally be sought by one of the owner's competitors and, at least, a party liable to make use of a sign resembling the trademark for similar products or services.

However, a party having an interest to sue for a trademark's revocation can also be the owner of a like trademark to be registered for products or services that are not similar, provided the trademark he owns is 'reputed' in trademark law terms, i.e. known to a significant part of the public affected by the products or services covered by it.[91] The same applies all the more where the trademark in question can be regarded as 'well known' within the meaning of Article 6*bis* Paris Convention, i.e. it is recognized to an extent beyond the branch of industry in question. A 'well-known' trademark is recognizable to a greater audience than a 'reputed trademark.'[92] Protection of a reputed trademark and of a well-known trademark goes beyond the speciality rule. Owners of well-known or reputed trademarks can therefore have a legitimate interest in seeking a trademark's revocation regardless of whether the products or services for which it is registered are identical, similar or non-similar to those for which the trademark is well known or reputed. This will obviously be so if any use of the trademark claimed to be needing to be revoked could occasion harm to the distinctive character or repute of their trademark or would be liable to draw undue profit from it.

The interest possessed by a legal person includes everything to do with its existence, its financial property and its non-pecuniary rights, especially its good standing and its reputation.[93] In the case of the owner of a reputed or well-known trademark, preservation of the image conveyed by the mark will also constitute pursuit of an own interest since, although the prime function of a trademark is that of identification and guaranteed origin, it is no less so that a trademark also functions as a means for sending out other messages concerning, in particular, the particular qualities or characteristics of the products or services it designates or the images and feelings it projects. In this sense, as has been emphasized by the Court of Justice, the trademark has an intrinsic economic worth that is autonomous and separate from that of the products or services for which it is registered. The messages conveyed by a reputed trademark and that are associated with it confer on it a value that is important and worthy of protection.[94]

90. Court of Justice, 29 September 1998, *Canon*, C-39/97, Ing.-Cons., 1998, p. 304, ECLI:EU:C:1998:442, 23.
91. Court of Justice, 14 September 1999, *Chevy*, C-375/97, Ing.-Cons., 1999, p. 555, point 31.
92. Braun and Cornu, *supra* n. 2, p. 232, no. 182.
93. Court of Cassation, 4 April 2014, Pas., 2014, I, p. 911 and opinion of First Advocate General J. F. Leclercq.
94. Court of Justice, 18 June 2009, *L'Oréal v. Bellure*, C-487/07, RDC, 2009, p. 787 and note of observations by E. Cornu, *L'arrêt 'L'Oréal/Bellure' de la Cour de justice: la protection de la marque renommée, la conjonction entre le droit des marques et le droit de la publicité et la consécration des fonctions économiques de la marque*; Court of Justice, 22 September 2001, C-323/09, *Interflora*, 74 *et seq.*, ECLI:EU:C:2011:604; General Court, 22 March 2004, *Sigla v. OHMI – Elleni Holding ('Vips')*, T-215/03, 40; ECLI:EU:T:2007:93; General Court, 19 June 2008, *Mülhens v. OHMI – Spa Monopole, 'Mineral Spa'*, T-93/06, 43; ECLI:EU:T:2008:2015;

Revocation petitions will also be allowed from parties with a less direct interest such as professional associations and consumers' organizations acting to assert a collective interest or the rights of their members to have a sign remain available.[95]

Besides this, the interest to sue must be legitimate. We believe this means that a plaintiff suing for revocation must act in good faith. This requirement for fairness or good faith on the part of a plaintiff is to be found in both Belgian procedural law and Dutch law.[96] We also believe it follows from the dictum of the Benelux Court of Justice in its *Kim* decision of 18 November 1998 in the case between tobacco manufacturers Philips Morris and BAT.[97] In its judgment, the Benelux Court of Justice stressed that 'a party cannot be held to have an interest to sue within the meaning of article 14(C) of the Uniform Benelux Act where, even if the requirements for the trademark of another party to be extinguished are met as laid down in article 5(3) of the Uniform Benelux Act, they act unfairly towards the trademark owner in demanding its extinction given the particular facts of the case.'[98]

Given the requirement for good faith on the part of a plaintiff suing for revocation, it is worth asking whether a party who has been found to have acted in fraud of the rights of a trademark owner and has thus been prohibited from making any use of a trademark for products for which it was registered and for similar products still has any legitimate interest to sue, even in a counterclaim, for revocation of the trademark on the basis of which he was found guilty in relation to certain of the products for which the mark was registered, when those products are similar to those for which the trademark was exploited and to those for which the injunction was handed down or sought. In a case such as that, the view could be taken that, even if the trademark in question was not subject to normal use for certain precise types of products, that does not make it available provided it was genuinely utilized for other, similar products. Although cases turning on this question are rare, reference can be made to a judgment from Brussels Commercial Court of 22 February 2000, which, having found against a Chinese company for passing off goods under the reputed 'Président' trademark for camembert, found that the counterclaim by the defendant in the principal action seeking revocation of the 'Président' trademark for certain specific products in classes 30, 31 and 32 was inadmissible on the ground that the defendant on the principal claim was prohibited from making use of the 'Président' name as a mark for all products in classes 29, 30, 31 and 32, so that it

---

General Court, 25 March 2009, T-21/07, *L'Oréal v. OHMI – Spa Monopole ('Spaline')*, 19; ECLI:EU:T:2009:80; General Court, 16 March 2014, T-201/14, *The Body Shop v. OHMI – Spa Monopole ('Spa Wisdom')*, 56; ECLI:EU:T:2009:80; General Court, 27 October 2016, *Spa Monopole v. EUIPO and YTL ('Spa Village')*, 64 and 65; ECLI:EU:T:2016:631.

95. Cornu, *supra* n. 88, 9.
96. For a recent application in Dutch law, see a case where there was deemed to be no legitimate interest to appeal a decision of the BOIP because it manifested a clear intention to mock the BOIP: The Hague District Court, 29 August 2017, Ing.-Cons., 2017, p. 662 and observations by L. Van Bunnen.
97. Benelux CJ, 18 November 1988, case 87/2, *Kim*, Ing.-Cons., 1988, p. 301.
98. See par. 37 of the judgment. Art. 14(C) of the Uniform Benelux Act is nowadays to be found in Art. 2.27(1) BIPC whilst Art. 5(3) of the Uniform Benelux Act is now enshrined in Art. 2.26(2)(a) BIPC.

lacked standing to raise a claim of averred revocation of the 'Président' trademark for certain specific products in the said classes.[99]

### ii. *Administrative proceedings for revocation before the Benelux Office*

Also new under Directive 2015/2436 of 16 December 2015 to approximate the laws of the Member States relating to trade marks is the introduction of an administrative procedure for revocation, which is intended to be efficient and fast.

Pursuant the Protocols dated 21 May 2014 and 16 December 2014 amending the Benelux Convention, which are prior to the Directive 2015/2436 and which were not intended to transpose this Directive, a new administrative procedure for revocation will already enter into force on 1 June 2018.

This revocation procedure will be raised before the Benelux Office of Intellectual Property. Appeal against its decisions will lie with the Benelux Court of Justice.

The proceeding can be initiated by any natural or legal person and by any grouping or body set up to represent the interests of manufacturers, producers, service providers, traders or consumers that, in terms of the law applicable to him, her or it, has standing to raise proceedings in his, her or its own name.

The current drafts give no indication that an official revocation application would depend on the applicant having an interest of their own in an order revoking registration of the trademark in question. Surprising as it sounds, this is already established practice before the EUIPO, and has been validated by the General Court.[100] There is no requirement to show personal interest or even for good faith on the part of a revocation applicant for the official revocation application to be admissible.[101]

## 4. *As a defence in infringement/cancellation proceedings*

Directive (EU) 2015/2436 of the European Parliament and of the Council of 16 December 2015 to approximate the laws of the Member States relating to trade marks, which is already in force and needs to be transposed on this point no later than 14 January 2019, will make a notable change to the onus of proving non use when it is cited as a defence argument in passing-off proceedings.

Under Article 17 of the new Directive, the owner of a trademark can only prohibit use of a sign insofar as he is not liable to be stripped of his rights for non-use at the time the passing-off action is raised. On a defence motion, the trademark owner has to produce evidence that, during the five years preceding the tabling of the passing-off action, his trademark was subject to genuine use for the products or services for which it is registered and that the owner cites in support of his action. If he fails in proving such use, he at least has to prove 'proper reasons for non-use'.

This facility to cite non-use as a means of defence will thus impose a direct reversal of the onus of proof: it will be the owner of the trademark alleged not to be used who will bear the burden of adducing positive evidence that he made genuine

99. Brussels Commercial Court, 22 February 2000, Ing.-Cons., 2000, p. 122.
100. General Court, 3 December 2009, T-223/08, *Iranian Tobacco Co v. OHMI*, 23, ECLI:EU:T:2009:481.
101. *Ibid.*, 26.

use of it during the relevant period. If he fails to do so, the trademark owner will not forfeit his rights but his passing-off action will be held unproved and thus fail. 'Non-use' as a defence argument in a passing-off case will of course only apply where the trademark cited in support of the passing-off action is under a use obligation, i.e. it had been registered for at least five years on the date the passing-off action tabled.

In the context of a passing-off action of that sort, if evidence of use of the anterior trademark is ordered and the owner of the trademark only falls short in establishing genuine use of his trademark as regards certain of the products or services for which it was registered, we believe his mark ought not to be deemed unregistered for the purposes of judging the merits of his action to the extent of the products and services for which use is established.[102]

Likewise, 'non-use' may be cited as a defence argument in an action to set aside registration of a later mark, just as it can already be cited as a defence argument in an official opposition proceeding, whether before the Benelux Office or the EUIPO.

## B. Burden of proof

### *1. The onus of proving use or lack of use of a trademark in the framework of a court revocation procedure*

Proving a lack of normal or genuine use of a trademark entails proving a negative, which the Benelux legislator recognizes is a difficult onus for a plaintiff in a revocation action to discharge. It is therefore expressly provided that the court before which an action is raised can impose all or part of the onus of proof on the owner of the trademark whose revocation is claimed.[103] The fact is that imposing the onus of proof solely on the plaintiff who seeks revocation would be to impose on him a veritable devil's proof.[104]

In practice, in most litigation, experience would indicate that the owner of the trademark whose revocation is sought will generally, of its own initiative, adduce evidence of its trademark's use, provided, of course, it has been used, so that this procedural question usually poses no difficulties.[105]

On the other hand, even though Belgian judicial procedure requires each party to prove the facts it alleges,[106] the litigants are also required to engage in fair cooperation in the administration of evidence. Under the general rules of procedure, courts can also order a party to produce evidence in its possession.[107]

102. See by analogy Art. 46(4) of Directive 2015/2436 on non-use as a defence argument in an action for a declaration of invalidity.
103. See Art. 2.26(2)(a) BIPC.
104. Incidentally, this is why, under old Dutch law prior to the Uniform Benelux Act, the revocation rules had gone virtually unheeded. On this point, see Braun and Cornu, *supra* n. 2, p. 579, no. 512 and the references cited therein.
105. Along these lines, J. J. Evrard and P. Péters, *La défense de la marque dans le Benelux* (2nd edn, Larcier, 2000), 103, no. 201.
106. See section 870 Belgian Judicial Code.
107. Section 871 Belgian Judicial Code.

Under Belgian procedure, therefore, quite independently from the terms at the end of Article 2.26(2)(a) BIPC, a trademark owner who claims to have made genuine use of it can be obliged to produce evidence of such use and to thereby prove the use thus averred. Although the transposition into Benelux law of Directive 2015/2436 of 16 December 2015 approximating the laws of the Member States on trade marks should soon end up doing away with the last sentence of Article 2.26(2)(a) BIPC, as anticipated by certain draft texts, the procedural law of the Benelux Member States will allow an apportionment of the onus of proving genuine use of a trademark.

Thus, where the defendant's answer to a revocation action is to aver that he makes use of his trademark, it is his duty to adduce the evidence in support of that and the plaintiff cannot be left to prove a negative, i.e. the absence of use.

That does not mean, however, that the plaintiff can remain passive in terms of the production of evidence and purely and simply allege non-use and shoot down the evidence brought forward by the owner of the trademark claimed to be revoked, without producing the slightest evidence to controvert that,[108] since the plaintiff in a revocation action is under a duty to establish liable cause: that the alleged lack of use is probable.

This could be shown, say, by means of a survey or investigative report that convincingly concludes that no products bearing the trademark can be found in large-scale stores or specialist shops, according to the nature of the products in question. If an investigator has been unable to find products marketed under the trademark in the territory in question during the relevant period, that will form a cogent indication of non-use of the trademark, which could be enough to reverse the onus of proof. Also of relevance in reversing the onus of proof would be a report indicating that the trademark in question has not be used as a trademark to designate products marketed by its owner but only as a trade name or domain name. There might also be statements by the trademark owner or its workers acknowledging that the relevant trademark has not been used in the territory in question in relation to the products for which revocation is sought.

Reversal of the onus of proof can therefore be ordered if the court considers non-use to be 'probable'[109] and certainly if there is prima facie evidence of an absence of use.[110]

To order reversal of the onus of proof, it could not reasonably be further required than that the absence of use simple be 'probable'. There can be no question of requiring proof of a negative, which, as indicated above, would be tantamount to demanding a devil's proof, or of taking such a strict stance that Article 2.27(1) should be denuded of any useful effect. It must be remembered in this regard that the possibility it provides for is an apportionment of the onus of proof in as equitable a manner as possible. Once non-use seems probable in light of the facts evidenced

108. Along these lines, Mons Court of Appeal, 12 June 2017, RDC, 2018, p. 466, and observations by E. Cornu.

109. Braun and Cornu, *supra* n. 2, no. 529, p. 603.

110. See also along these lines: The Hague District Court, 18 March 2015, IEF, 1272, 4.3-4.8; Overijssel Court of First Instance, 25 June 2014, IER, 2015, p. 299.

by the plaintiff in the revocation action,[111] the court can require the owner of the trademark in question to adduce positive evidence of its use, which is generally not a difficult task for him to accomplish.[112]

### 2. *The onus of proving use or lack of use in the framework of a administrative revocation*

In the (future) administrative procedure for revocation before the BOIP, it will be up to the defendant (the owner of the trademark whose revocation is sought) to adduce evidence of his trademark's genuine use. The onus of proof is therefore reversed from the very outset.

## C. Relevant evidence

### 1. *Use on the product or its packaging*

Normal or genuine trademark use can be proved by 'any means available in law'. In practice, evidence generally takes the form of invoices for the sale of products bearing the trademark,[113] product or service catalogues showing the trademark (provided they are of provable date), advertising for the relevant products or services displaying the trademark whose revocation is claimed and that might have appeared in the press and media in the relevant territory,[114] press articles that mention the trademark, and such like. But it can also be constituted by testimony from independent parties or even those working for the company holding the allegedly revoked trademark where it is corroborated by other evidence.

As indicated above, the courts are especially attentive to the extent, frequency and duration of use and to the nature of the products or services in question.[115] The features of the company in question are also looked to in assessing whether the alleged use is an artifice or otherwise.[116]

As a pretty rare exception, we might add that, on one occasion, expert testimony was arranged to verify averred genuine use based on accountancy data.[117]

---

111. For example, as already said above, a survey or investigative report or statements made by the trademark holder.
112. Along these lines, Braun and Cornu, *supra* n. 2, no. 529, pp. 603 and 604, which cite the grounds set forth by the Benelux Commission.
113. Brussels Court of Appeal, 23 September 2008, IRDI, 2009, p. 161.
114. Mons Court of Appeal, 25 June 2006, Ing.-Cons., 2006, p. 357; Brussels Commercial Court, 23 September 2002, Ing.-Cons., 2002, p. 251; advertising that is too limited will sometimes not suffice: Brussels Commercial Court, 28 June 2007, IRDI, 2007, p. 415. Nevertheless, one ad can be enough: Dinant Commercial Court, 21 April 2000, RDC, 2001, p. 260.
115. Brussels Commercial Court, 23 September 2002, Ing.-Cons., 2002, p. 251; Brussels Commercial Court, 23 November 1995, Ing.-Cons., 1996, p. 154.
116. Brussels Commercial Court, 23 September 2002, Ing.-Cons., 2002, p. 251; Oudenaarde Commercial Court, 6 October 1999, Ing.-Cons., 1999, p. 154.
117. Brussels Court of Appeal, 12 September 2003, Ing.-Cons., 2004, p. 29.

## 2. *Use in advertising*

As indicated above, advertising to promote the trademark will be looked at.[118] However, advertising that is too limited cannot be deemed genuine use.[119]

## 3. *Use on the internet*

Consideration will be also be given to online sales and to the trademark owner's efforts to promote its products under the trademark within the Benelux countries by means of on online advertising.[120]

## 4. *Form of evidence*

As above mentioned, genuine use can be proven by 'any means available in law', as well invoices, documents, catalogues, advertising, press releases, as testimony, surveys, affidavits. An affidavit from those working for the trademark owner could be taken into consideration where it is corroborated by other evidence.

## 5. *Distinction between use of a company name and use of the services offered by the company*

As already mentioned, in general, using a corporate name or a trade name does not constitute use of the trademark as a means to distinguish products.[121] A corporate name is intended to identify a legal person, and a trade name is aimed at identifying or marking out a business. Their prime aim is not therefore to distinguish products or services,[122] though this is not always the case especially in relation to service marks. *De facto*, a trade name can be deployed such that the general public perceives that use as use of a sign distinguishing from all others the services that are offered.[123]

118. Brussels Commercial Court, 23 October 2002, Ing.-Cons., 2002, p. 251.
119. Brussels Commercial Court, 28 June 2007, IRDI, 2007, p. 415; however, against this see: Dinant Commercial Court, 21 April 2000, RDC, 2001, p. 260.
120. Dutch Supreme Court ('Hoge Raad'), 23 December 2005, BIE, 2006, p. 65; BMM Bulletin 2006/4, p. 176; Brussels Commercial Court, 23 September 2002, Ing.-Cons., 2002, p. 251.
121. Antwerp Court of Appeal, 2 February 2007, Annuaire pratiques du commerce, 2007, p. 398.
122. Court of Justice (Grand Chamber), 11 September 2007, C-17/06, *Céline*, ECLI:EU:C:2007:497.
123. Benelux CJ, 20 December 1996, case A95/2, *Europabank*, Ing.-Cons., 1997, p. 9 and note by E. Cornu; on the same lines, see Court of Justice (Grand Chamber), 11 September 2007, C-17/06, *Céline, supra* n. 122. See also General Court, 27 September 2007, *La Mer Technology*, T-418/03, 73 and 74, ECLI:EU:T:2007:299; General Court, 21 November 2013, *Recaro Holding GmbH v. OHMI – Certino Mode SL*, T-524/12, ECLI:EU:T:2013:604; appeal dismissed by Court of Justice, order, 14 January 2015, C-57/14 P, ECLI:EU:C:2015:17. Cases where the design on question was held only to have been used as a trade name and not as a trademark: General Court, 14 July 2016, *Modas Cristal SL v. EUIPO – Zorlu Tekstil Ürünleri Pazarlama*, T-345/15, 38 and 39, ECLI:EU:T:2016:405; General Court, 13 May 2009, *Schuhpark Fascies v. OHMI – Leder & Schuh*, T-183/08, 26-41, ECLI:EU:T:2009:156.

## IV. JUSTIFICATION FOR NON-USE

### A. Obstacles to use

Want of genuine use of a trademark in the period laid down does not *ipso facto* imply its revocation. The holder must also have failed 'without just cause' or in the English version of the Directive without 'proper reasons for non-use'. This condition only applies to Article 5(C), first paragraph, Paris Convention, under which revocation can only be ordered 'if the party having an interest does not justify the reasons for his lack of action.'

Now contained in the EU Trademarks Directive, the 'proper reasons for non-use' or 'just cause' plea was already previously known in Benelux law.

In *Turmac v. Reynold*, referred to above, the Benelux Court of Justice was required to make a first endeavour to interpret it. It held that

> 'where the owner or someone acting on his behalf has made no genuine use of the trademark on Benelux territory during the relevant period, there is nevertheless no extinction of the trademark right if the owner of the trademark claims and, where challenged, proves the facts and circumstances by dint of which he was powerless (such as do not form part of the normal risks of the business) and made it impossible to make normal use of the trademark or at least constituted such a burden that any such use could not in terms of the usual criteria applying in business life be reasonably demanded from the owner or his licensee.'[124]

It has been inferred from this that 'just cause' or 'proper reasons for non-use' is a notion to be construed restrictively.[125]

The notion of 'proper reasons' is nonetheless enshrined in the Harmonisation Directive and is a term of EU law. It also needs to be interpreted in the light of Article 19(1) of the ADPIC accord, under which valid reasons justifying non-use encompass circumstances outside the trademark owner's control that constitute an obstacle to use of the mark.

In interpreting the notion, the Court of Justice of the European Union has taken the view that 'it follows that only obstacles having a sufficiently direct relationship with a trade mark making its use impossible or unreasonable, and which arise independently of the will of the proprietor of that mark, may be described as "proper reasons for non-use" of that mark'.[126]

---

124. Benelux CJ, 27 January 1981, *Turmac v. Reynold*, *supra* n. 27, and note by F. de Visscher.
125. F. de Visscher, *op. cit.*, Ing.-Cons., 1981, p. 120; Braun and Cornu, *supra* n. 2, p. 589, no. 519; Kaesmacher, *supra* n. 81, no. 82, note 1, p. 214; Gotzen and Janssen, *supra* n. 3, pp. 253-254.
126. Court of Justice, 14 June 2007, *Armin Häupl v. Lidl Stiftung & Co KG*, C-246/05, 54, ECLI:EU:C:2007:340.

Hence, it is only circumstances qualifying as force majeure, such as legal measures prohibiting use of a given trademark, that are capable of constituting 'proper reasons' by which non-use of a trademark can be excused.[127]

Thus, it has been held that the simple fact of court proceedings being raised seeking a trademark's annulment or objecting to its use will not constitute a justifiable reason for not making genuine use of it.[128]

It has in exceptional cases been admitted by Benelux courts that 'proper reasons' existed. One such case involved the 'HUGO BOSS' trademark, the owner of which successfully argued that there was 'proper reasons' for not having used it for tobacco products for which it had been registered. The Amsterdam Court of Appeal took the view that use of its 'HUGO BOSS' trademark for tobacco products in the relevant period would have constituted a threat to its trademark to the extent that it was an identifier for other products such as clothing, perfume and fashion accessories, owing to the coming into force of the European Directive prohibiting tobacco advertising and Hugo Boss's legitimate fear that using its trademark in the tobacco sector might mean it could no longer use it in future to promote its other products, which lay at the heart of its business.[129]

## B. Circumstances in which non-use may be cured: resumption of use, or 'Heilung'

It can no longer be claimed that trademark rights are extinguished if, between expiry of the five-year period and the revocation petition, normal use of the trademark starts or resumes. Nevertheless, commencement or resumption of use within the three months prior to the tabling of the revocation petition is disregarded where the owner has gained knowledge of the fact that a revocation petition might be lodged.[130] The rule is inspired by a practice in Germany known as 'Heilung', or 'cure'.[131]

In practical terms, it is advisable to put the owner of an unused trademark on notice of default and require them to deregister it. If the notice goes unheeded, a revocation action should be lodged within three months of the date of sending the notice of default. Otherwise, subsequent use of the trademark that is genuine could form a defence to the revocation claim.[132]

---

127. Braun and Cornu, *supra* n. 2, p. 590, no. 519; C. Gielen, *Enkele merkenrechtelijke vraagstukken* (1991), p. 137, no. 88.
128. Brussels Court of Appeal, 2 October 2001, Ing.-Cons., 2001, p. 371; The Hague District Court, 19 April 2006, BIE, 2007, p. 30; 's Hertogenbosch District Court, 4 January 1994, BIE, 1994, 112; 's Hertogenbosch District Court, 5 July 1996, BIE, 1999, p. 30.
129. Amsterdam District Court, 18 October 2011, *Hugo Boss v. Reemstsma Cigarettenfabriken GmbH*, IEPT 2011 10 18. This judgment followed on from the Dutch Supreme Court's decision of 11 January 2008 (IER, 2008, p. 19 and note by C.G.), which quashed the decision from The Hague Court of Appeal of 4 August 2005 (BIE, 2006, 27); Bulletin BMM, 2005/4, p. 202.
130. Art. 2.27(2) BIPC; Art. 19 of Directive 2015/2436 of 16 December 2015 approximating the laws of the Member States on trade marks.
131. Cohen Jehoram et al., *supra* n. 3, no. 12.3.13.
132. Braun and Cornu, *supra* n. 2, pp. 591-592, no. 520*ter*.

It follows from the transitional provisions in relation to the 2 December 1992 protocol amending the Uniform Benelux Trademarks Act that the ability to 'resume use' or 'make new use' is also of benefit to old trademarks, i.e. those registered before 1 January 1996, provided the acts relied on by which use is resumed occurred after that date.[133]

Owners of trademarks that are not actionable as revoked due their use having been resumed cannot, however, object to use of a trademark in respect of which there was a filing in the period when their mark could have been declared extinguished for want of normal use.[134] Nor can they claim that the registration of a trademark is void where it was filed during the period when their mark could have been declared revoked for want of normal use. In that case, a situation is created in which the two trademarks coexist.[135]

## V. CONSEQUENCES OF NON-USE

### A. Consequences of non-use and relevance of the context

A registered trademark which has not undergone genuine use within the statutory period of five years is not automatically revoked. If no revocation application has yet been lodged, the owner may start or resume use of his trademark. As mentioned above, the 'Heilung' (start of resumption of use) can in some cases form a defence to a revocation petition.

### B. Extent of any revocation

When ordered by a court, extinction of the trademark right on the ground of revocation for non-use results in deregistration from the Benelux Register.

Furthermore, the trademark becomes *res nullius*, and can be appropriated by any person.

For signs registered as individual marks, re-appropriation can be immediate, given that the two-year waiting period set down in Article 2.4(d) BIPC is required only when the right extinguishes due to the registration not being renewed. The provision goes so far as to explicitly provide that it does not apply where the right extinguishes on grounds of non-use.

By contrast, if the sign was registered as a collective mark, then, as Benelux law stands now, it can only be appropriated by a third party to designate similar products or services, even as an individual mark, once three years have passed starting on the date of extinction of the right over the collective mark for want to genuine use thereof.[136]

---

133. The Hague District Court, 1 October 1998, BIE, 1999, p. 421; President of Amsterdam District Court, 23 January 2003, BIE, 2003, p. 362; B. Van Reepinghen and E. Cornu, *Chronique de jurisprudence*, Droits intellectuels (1997-2004), J.T., 2004, p. 908.
134. See Art. 2.27(3) BIPC.
135. Braun and Cornu, *supra* n. 2, 592, no. 520*ter*.
136. *Ibid.*, 606-607, no. 530bis.

Moreover, a trademark holder whose rights extinguish can always file the trademark again. However, as we saw above and in the Benelux Court of Justice's *Triomino* ruling, the new filing must meet the validity requirements applying to any filing and the holder cannot draw any priority right from his extinguished right.[137]

That being so, re-filing one and the same trademark of which no use is made could be viewed as improper and abusive exercise of the right to file where it has the effect of denying the very purpose of trademark law, as may be so in the case of repeated filings with the sole aim of evading the penalty of revocation for want of use.[138]

### C. Timing of any revocation

A significant difference will be noted between a decision ordering extinction of a right for lack of genuine use during the relevant five-year period and a decision ordering the invalidity of the trademark. A declaration of invalidity of a trademark is retrospective in effect to the time of filing: the trademark's exclusive right is thereby deemed never to have existed. Against this, when the trademark right is ordered extinguished or revoked for lack of genuine use during the relevant five-year period, the order is effective from the time the court deems the conditions leading to extinction to have come into effect.[139]

137. Benelux CJ, 19 December 1996, *Triomino*, Ing.-Cons., 1996, p. 16.
138. Along these lines, T. van Innis, *supra* n.19, no. 306, p. 237.
139. Braun and Cornu, *supra* n. 2, 620-621, no. 537.

# Brazil

*Elisabeth Siemsen do Amaral and Rafael Atab*

## I. INTRODUCING GENUINE USE OF TRADEMARKS

### A. Legal framework

Brazil has traditionally been a first-to-file country. Under the Brazilian Industrial Property Law (Law No. 9.279/96 – 'IP Law'), proprietorship over a trademark derives from a duly granted registration, with few exceptions.[1] Therefore, the rights deriving from the use of an unregistered trademark benefit from more limited protection of the Brazilian IP Law.

Conversely, Industrial Property Laws in Brazil have long considered use of a trademark a paramount factor for preserving the enforceability of a trademark registration.

Brazil's first consolidated Industrial Property Law, Decree 16.264, of 1923, already provided that a trademark registration would be considered forfeit in case the mark had not been in use for a period of three consecutive years.[2] Subsequent IP Laws[3] also provided for the cancellation of a trademark registration in such cases, though providing for an even shorter period of only two years.

Under the current IP Law of 1996, a trademark registration will also be cancelled in case use has not been initiated or has ceased. The new Law has however extended the period for uninterrupted non-use to five years. Such timeframe exceeds the minimum standards provided under the Agreement on Trade-Related Aspects

1. The Brazilian IP Law of 1996 included a new provision, benefiting bona fide senior local users, under section 129. Under such section, any person who had been using a mark in good faith, in Brazil, for at least 6 months prior to the priority date or filing date of a national application, will enjoy preferential rights to the registration of the relevant mark. The INPI has been however recognizing a very narrow interpretation of such possibility, which significantly restricts its availability to local companies.
2. Decree 16264, of 19 December 1923 (Article 115).
3. Enacted in 1945, 1967, 1969 and 1971.

of Intellectual Property Rights ('TRIPS Agreement'), which provides for a minimum standard of three years of non-use.[4]

## B. Purpose and function of genuine use: policy considerations

Trademarks and other distinctive signs are protected under the Brazilian Constitution.[5] However, the Constitution also establishes that protection over any property rights is dependent upon the social and economic role they play in society as a whole.[6]

Since the beginning of the twentieth century, Brazilian scholars such as Bento de Faria have explained that 'the purpose of the industrial mark is to identify and characterize the products of an industry or the goods of a trade, distinguishing them from similar one from competitors [...]. Use consists in the adoption a mark in connection with products, [...] made in a uniform and consistent manner'.[7] Affonso Celso, author of the Brazilian Trademark Law of 1887 (Decree No. 3346, of 1887), adds that 'the trademark is an accessory of the commerce or of the industry'. In his view, if a trademark is not in use, there would be no reasons for its continued protection, given 'the impossibility of criminal competition, which the Law tried to curb, by establishing special guarantees'.[8]

In Gama Cerqueira's classic *Treaty on Industrial Property*, one of the most thorough studies of IP in Brazil, the author outlines that 'the principle of obligatory use of a registered trademark is grounded in the very role [that the marks] play in the field of industry and commerce'.[9] Gama Cerqueira also goes on to explain that:

> 'If the beneficiary [of a trademark registration] does not use the registered mark, which, as a consequence, no longer plays the role for which it was created, and which justifies its legal protection, the *raison d'être* of the exceptional guarantees assured by a registration ceases to exist, reverting the trademark to the public domain. This is required, on the one hand, by [the principle of] freedom of commerce, which would be otherwise curtailed by a useless registration, and, on the other, the absence of any need for a legal guarantee to assure a function that does not exist.'[10]

Therefore, if a trademark is not fulfilling its economic and social roles in the market-place, its protection lacks constitutional support, which should entail its revocation.

This conclusion has been reaffirmed by Brazilian courts several times and can be summarized in a ruling by the Federal Court of Appeals of the 4th Region: 'the maintenance *ad eternum* of a trademark registration with speculative purposes (or

4. Article 19.
5. Article 5, XXIX, of Brazil Constitution of 1988.
6. Article 5, XXIII, of Brazil Constitution of 1988.
7. Antonio Bento de Faria, Das Marcas de Fabrica e Comercio e Do Nome Comercial, 261 (J. Ribeiro dos Santos, 1906) (translated by author).
8. Affonso Celso *apud* Julio de Mello & Fernando de Mello, A Questão da Caducidade do Registro da Marca SAPOLIO. Memorial, 46 (Jornal do Commercio – Rodrigues & C., 1942) (translated by author).
9. João da Gama Cerqueira, *Tratado de Propriedade Industrial* vol. II, tomo II, part III (Revista Forense, 1956), 206 (translated by author).
10. *Ibid.*

those of other nature) is inadequate and violates the principle of freedom of commerce, pursuant to Article 170 of [Brazilian] Constitution of 1988'[11]

## II. WHAT QUALIFIES AS GENUINE USE OF A TRADEMARK

### A. Nature of use

#### *1. Use consistent with the function of the trademark*

Brazilian IP Law does not qualify the type of use which should constitute genuine use for the purpose of preserving a trademark registration. However, scholars, administrative decisions by the Brazilian Patent and Trademark Office (Instituto Nacional de Propriedade Industrial – 'INPI') and judicial precedents have helped set some standards for the analysis of all aspects and assessment of the nature of use that would be considered sufficient to avoid cancellation.

In this regard, José Antonio Faria Correa outlines that the 'legislator does not accept, as an excluding factor to the forfeiture [of a trademark registration] *any use*, but a qualified use: it must be effective. [...] Effective means current, real'. Faria Correa also adds that 'The mark is [...] the mercurial entity that communicates to the public what lies behind the goods being acquired or the services being rendered, their characteristics, their origin, their price range, their reputation, the profile of people to whom they are designed etc.'[12]

In a similar fashion, Denis Borges Barbosa indicates that 'use as a trademark is the use in its distinctive function [...]. Use as a trademark is the one destined to guarantee the creation or continuity of a reputation in relation to the product or service.'[13]

Therefore, the use of a sign merely as a trade name or as a decorative or aesthetic feature in a certain product or service should not constitute, under the Brazilian IP Law, effective use for the purpose of preserving a trademark registration.

#### *2. Use in the course of trade*

##### *i. Public use*

Brazilian IP Law does not explicitly identify 'use in the course of trade' as a requirement for preserving a trademark registration against a non-use forfeiture action.

Nevertheless, it is only in the course of trade that a trademark may fulfil its purpose, which is exactly what supports the special legal guarantees that are afforded to a trademark owner. Naturally, this requires public use of a trademark.

---

11. *Jose Gilmar Michels v. Industrial de Tabacos Zebu Ltda.*, AC. n. 5002459-35.2011.404.7119, TRF4, 3ª Turma, in 15.10.2014.
12. José Antonio B. L. Faria Correa, *O Conceito de Uso da Marca*, 16, Revista da ABPI, 22, 22 (1995) (translated by author).
13. Denis Borges Barbosa, *Ensaios e estudos de Propriedade Intelectual: 2014-2015* vol. 1 (IBPI Europa, 2015), 258 (translated by author).

### *ii.* *Commercial purpose*

Upon reviewing the purposes of a non-use forfeiture action, Federal Judge Marcia Nunes de Barros sums up some of the precedents issued by Brazilian Appeal Courts addressing the matter. In her opinion, 'Court precedents clarify that "the use to which the Law refers is the continued use, able to demonstrate a constant commercial or industrial activity"'. She also adds that, as decided by the courts, 'the exploitation of the mark in the economic circuit of the country is of general interest'.[14]

Therefore, what actually matters must be the purpose of the use and what economic role the trademark plays in marketplace. One may find multiple examples of goods and services that might be available for free, particularly on the internet or other digital channels. This does not mean that the corresponding marks are not in use, as long as the trademark is fulfilling its distinctive purpose and helping its owner develop actual goodwill.

## *3.* ***Use by the owner or with its consent***

In Brazil, a trademark will not be cancelled if the trademark owner is able to prove its own use of the mark or use with its consent. In the past, the INPI would require that a licence agreement be recorded with the Office. Without such recordal, the INPI would not accept use by a third party as valid use for avoiding cancellation of a trademark registration.

Nevertheless, even before current IP Law, courts[15] were already deciding that the recordal of a licence agreement would not be required for considering use by a licensee to be valid and effective. Indeed, while the Brazilian IP Law requires recordal of certain IP contracts in order to preserve third parties' rights that may need to have knowledge of the existence of such contracts, in case of a forfeiture action, it is the right of the owner that should be preserved. If a trademark is in use under authorization of its owner, it is fulfilling its purpose and, therefore, its holder's rights should be preserved.

In any event, the Brazilian IP Law of 1996 has resolved such a discussion by explicitly providing that 'for the purpose of the validity of proof of use, recordal of a licence agreement before the INPI will not be required' (Article 140, § 2o).

Nevertheless, trademark owners should be prepared to prove the existence of an agreement or authorization that would support such use by the third authorized party, particularly if the parties are not related and do not belong to the same economic group.

---

14. *Sie Cosmeticos Ltda. v. Instituto Nacional da Propriedade Industrial and B&K Propaganda Comercial de Artigos para Presentes Ltda.*, Proc. 0017835-75.2013.4.02.5101, 13ª Federal Court of Rio de Janeiro, (28.11.2014)
15. See, e.g., *Cirol Companhia Industrias Royal v. Nabisco Inc. and Instituto Nacional da Propriedade Industrial*, EMBG. INFRG. n. 00005371919914020000, TRF2, (21.06.1994), as well as *Head Sports Inc. and Head Sports AG v. Village Comercio e Assessorios Ltda. and Instituto Nacional da Propriedade Industrial, AC.* n. 0011704-81.2001.4.02.0000, TRF2, 4ª Turma, (08/11/2001).

In case of subsidiaries and companies belonging to the same group, court precedent has already established that a specific 'authorization' should not be required, inasmuch as a subsidiary should not be considered a third party for this particular purpose.[16] All companies belonging to the same group act under the same control with a uniform economic purpose.

## B. Volume, duration and frequency of use

Analysis on what should constitute effective use of a given trademark must address aspects such as volume, duration and frequency of use.

When assessing such aspects, one must necessarily have in mind characteristics related to the product or service, as well as those associated with the owner of the trademark.

In this regard, José Antonio Faria Correa cites some standards that should be adopted when assessing evidence of use.[17]

Faria Correa indicates, for instance, that the 'price range' of a product must be necessarily taken into consideration. A high-end product or a very costly and exclusive service would be expected to meet significantly lower volumes of sales than ordinary goods or essential services.

An additional factor would be the nature of the products and services. One would not expect that the volume of sales and frequency of use of trademarks addressing aeroplane or military helicopters could be similar to those associated with pharmaceuticals or stationery. Similarly, frequency of actual sales would be substantially different in each of these fields of businesses.

In an appeal heard by the Federal Court of Appeals of the 2nd Region involving a trademark registration covering 'eggs', among other food articles, the Court held that while a trademark does not need to be printed on the goods themselves, it would be necessary to show sufficient volume of sales given the nature of the goods. In this particular case, the Court ruled that a small number of packages would not be sufficient to show use during the investigated period, inasmuch as occasional use of a trademark is not sufficient to preserve a trademark registration.[18]

An additional aspect that should be taken into account is the size of the trademark owner. A small undertaking would be expected to have very limited volume of sales vis à vis the amount of sales expected for large multinational companies.

Reviewing such an aspect, Brazil Superior Court of Justice (Superior Tribunal de Justiça - 'STJ') has decided on a case involving the manufacturing of tobacco-related products. While the STJ acknowledged that the local manufacturing of a product for exportation purposes should be considered local use of a trademark, the Court held that the commercialization of a limited number of cigarette packs should not

16. *Smithkline Beecham Corporation and Laboratorio Smith-Kline Enila Ltda v. Laboratorio Zambeletti Ltda. and Instituto Nacional da Propriedade Industrial*, EMBG. INFRG. n. 0000699-14.1991.4.02.0000, TRF2, (01.11.2001).

17. José Antonio B. L. Faria Correa, *O Conceito de Uso da Marca*, 16, Revista da ABPI, 22, 23 (1995) (translated by author).

18. *Cooperativa Agricola de Cotia Cooperativa Central v. Instituto Nacional de Propriedade Industrial*, TRF2, 2ª Truma, (29/11/2005).

be considered sufficient to avoid the cancellation of a trademark, particularly if the owner is a multibillion multinational company. On that matter, STJ considered the commercialization of only 70 packs during the investigated period to be merely occasional.

Another important aspect is the lawfulness of the relevant use. Use of a trademark in products or services that are not offered in a legal fashion may not be considered valid for the purpose of preserving a trademark application.

This analysis is particularly important for products that are subject to government authorizations and regulations, such as pharmaceuticals, veterinary products, cosmetics and even some telecommunication products, which, in Brazil, are subject to prior approval by Brazil Telecommunication Agency (ANATEL).

Use in good faith is also a paramount factor when assessing whether a registration is subject to cancellation. Brazilian Civil Law explicitly provides in its Article 186 that abuse of rights constitutes an illicit act and, therefore, may not validly support the preservation of a trademark registration. Such circumstances are usually exceptional and must be assessed on a case-by-case basis.

### C. Use of a sign

The Brazilian IP Law also stipulates that a trademark registration may be cancelled if the mark has been used in a modified form that implies modification of its original distinctive character, as found on the certificate of registration.

Therefore, if the owner has been using its trademark in a modified version, its trademark rights may be subject to cancellation.

Under INPI's Guidelines for the Examination of Trademark Proceedings[19] ('INPI's Trademark Guidelines'), a word mark can be used in whatever manner its owner seems fit. Therefore, the Office will recognize prove of use showing both the mark featured in its word form or associated with other elements, such as design ones.

A word and design mark, however, must be used exactly as filed, though minor modifications might be accepted. The same goes for a design mark, even though, in this particular case, the Office would accept proof of use of a word and design mark as evidence relating to a design mark in case the relevant design is featured in both marks.

Use of three-dimensional marks requires sufficient evidence of use of the shapes reflected in the depictions filed with the INPI.

In addition, trademarks filed without any colour claims may be used with any combination of colours desired by its owner. If the overall aspect of the mark does not change, the corresponding use will be considered sufficient for the purpose of maintaining the relevant trademark registration.

Brazilian IP Law does not provide clear guidance on the criteria that should be adopted for reviewing whether or not a change on the trademark would impact 'its original distinctive character'. However, under the INPI's Trademark Guidelines, the examiner should not take into account 'minimal modifications to the sign, as long as they refer to ornamental details or secondary elements, especially merely

19. Resolution INPI/PR No. 142/2014, as amended by Resolution INPI/PR No. 177/2017. Full text in Portuguese is available at http://manualdemarcas.inpi.gov.br/

descriptive ones'. Examiners should primarily focus on the main distinctive elements featured in a trademark.

Scholars have long been critical of a very narrow interpretation of the requirements of use of a trademark as registered. According to Gama Cerqueira '[this requirement] must be carefully understood and applied with utmost prudence, so as to avoid unjust harm to the rights of the trademark owner'.[20] Gama Cerqueira also adds the following:

> 'If interpreted in a literal fashion, any modification to a trademark, as small as it might be and even if occasional, would cease *ipso facto* the effects of the registration, [in which case] a trademark would be deprived from its legal protection [...]. Such interpretation, for its absurdity, may not be accepted, so that application of [this interpretation] should be limited to the case that the owner of the registration changes the registered trademark in a substantial manner and ceases using the mark as registered.'[21]

This provision should therefore be interpreted in light of Article 5. C (2) of Paris Convention for the Protection of Industrial Property ('Paris Convention'), of which Brazil is one of the original signing parties. Accordingly, use of a trademark in a form differing in elements which do not alter the distinctive character of the mark in the form in which it was registered shall not entail invalidation of the registration or diminish the protection granted to the mark.

Commenting on the interpretation of this provision, G.H.C. Bodenhausen[22] mentions that 'the purpose of this provision is to allow for unessential differences' in a trademark. He also points out that changes such as 'adaptation' or even 'translation' of certain elements should not be considered significant changes, as long as they do not 'alter the distinctive character of the mark'.

Brazilian scholar Denis Borges Barbosa also criticizes a too stringent approach, indicating the dangers of this formalistic practice by the INPI. Such approach could impose the 'mummification'[23] of trademarks, which does not attend to the protection of either consumers or IP owners interest.

In this regard, Borges Barbosa states that by ignoring the fact that a trademark changes and evolves overtime, the INPI would be missing preserving one of the most important aspects related to a trademark, which is the goodwill that has been associated with a mark. In his view, there should be room to apply, in Brazil, the US doctrine of 'continuing commercial impression' as established by US courts.[24]

---

20. João da Gama Cerqueira, *Tratado de Propriedade Industrial* vol. II, tomo II, part III (Revista Forense, 1956), 165 (translated by author).
21. *Ibid.*
22. G.H.C. Bodenhausen, *Guide to the application of the Paris Convention for the protection of industrial property as revised at Stockholm in 1967*, 76 (BIRPI, 1968), also available at http://www.wipo.int/edocs/pubdocs/en/intproperty/611/wipo_pub_611.pdf
23. Denis Borges Barbosa, *Ensaios e estudos de Propriedade Intelectual: 2014-2015* vol. 1 (IBPI Europa, 2015), 624 (translated by author).
24. *Ibid.*, 639.

He therefore advocates that a review of occasional modifications to a trademark should take into consideration the idea or meaning that a trademark conveys to the public, as explained by Gideon Mark and Jacob Jacoby.[25]

## D. Use in relation to goods and services

### *1. Use for each of the designated goods and services*

In case of a non-use forfeiture action, a titleholder is required to prove use of the mark in connection with all products or services mentioned in the certificate, under penalty of partial forfeiture of the registration with respect to those products or services not similar or akin to those for which use of the mark has been proved (Article 145 IP Law).

Therefore, in case of a wide variety of goods and services covered by the relevant trademarks, the INPI will remove from the description all goods and services that are not considered similar or at least closely related to those that are in actual use. That is, the INPI does not require use in connection with each and every good or service recited in the description but will preserve all those which are at least similar or closely related to those in actual use.

This provision has been considered particularly helpful for those older registrations that are still under the former Brazilian classification system. Brazil only adopted the International Classification System under the Nice Agreement in 2000. The former local system used to adopt very broad class headings. With the availability of a partial cancellation, it has become possible to obtain the restriction of the scope of protection afforded by a registration, allowing similar marks to coexist in different niches.

### *2. Use for a broader category of goods and/or services*

A similar understanding applies if the description of services is somewhat broad or relatively generic, even though the INPI currently requires detailed descriptions. If the goods and services that are in actual use fall within the broader category, the registration will be preserved. On the other hand, in case of very broad categories, a limitation may occur, though this would be exceptional.

On the other hand, the INPI and the courts are usually very attentive as to the correspondence between the actual use of the mark and the products and services described in the trademark registration or at least falling within a broad description adopted when the application was filed.

In a matter involving a trademark covering 'automobile parts and accessories', the Federal Court of Appeals of the 2nd Region ruled that evidence of use relating to the retail service of automobile parts cannot validly support registration covering the goods themselves. According to this unanimous ruling, although the registrant may have mistakenly chosen the wrong class, the INPI must take into consideration

25. *Ibid.*, 640.

the goods and services actually offered by the registrant and whether they match those described in the relevant trademark registration.[26]

## E. Territorial scope

Evidence of use must refer to the Brazilian territory, as clearly indicated in Article 143(I) IP Law. Also, the INPI's Trademark Guidelines indicate that it will only take into consideration, for the purposes of a non-use forfeiture action, evidence referring to the use of a trademark in our country.

In case of foreign IP holders and goods that have been manufactured abroad, INPI's Trademark Guidelines require the company to prove the internalization of the product, which is usually done through customs and/or tax-related documents.

On the other hand, both the INPI and Brazilian courts recognize that the manufacturing of a product in our country exclusively for exportation purposes should be considered local use. Addressing this point, the Brazil Superior Court of Justice explains the rationale behind such an understanding:

> 'If the owner of a trademark registration in Brazil industrializes, manufactures, or develops the product in the national territory, it clearly initiates and makes use of the mark in Brazil, deserving therefore full legal protection, inasmuch as it does business in Brazil, generating production, employments and wealth. It is therefore irrelevant whether goods are locally produced and designed for the local market or exclusively to the foreign market. Local manufacturing of a product bearing the trademark registered here sufficiently complies with the legal requirement of "use initiated in Brazil".'[27]

As this rationale focuses on the economic advantages to the country in recognizing such use, other scenarios could also be potentially covered by such grounds, such as use in duty-free zones in Brazil.

Another important aspect to such a rationale relating to manufacturing in Brazil for exportation purposes is that, under Brazilian IP Law, one will also commit a crime against a trademark registration if one exports or has in stock goods bearing the infringing trademark (Article 190). Therefore, if one may not manufacture and export – or even have in stock – goods bearing a third party's trademark, it becomes clear that local manufacturing for exportation purposes amounts to local use.

Although Brazil is a very large territory, there are no particular provisions establishing requirements on the geographical extension of the use of a mark for the purposes of preserving a trademark registration against its forfeiture.

Much to the contrary, a small business or undertaking is fully entitled to secure a trademark registration before the INPI, even if its activities are limited to a single city or even a neighbourhood. Likewise, if such small businesses are ever required to prove use of their marks, these owners are only expected to submit limited evidence of use, reflecting the size of their business.

---

26. *PST Industria Electronica da Amazonia Ltda. v. PST Veiculos e Peças Ltda. and Instituto Nacional de Propriedade Industrial*, AC 2007.51.01.808584-6, TRF2, 2a Turma, (28.05.2013)
27. *Souza Cruz v. Companhia Industrial de Tabacos Month Paz S/A and Instituto Nacional de Propriedade Industrial*, REsp 1.236.218/RJ, STJ, 4ª Turma, (11.06.2015).

Furthermore, discussions on the territorial scope of the use become highly relevant when it comes to use in a digital environment, as the internet.

In many aspects, use in the digital environment does not necessarily differ significantly from use in the physical environment. This might be particularly true for multi-channel retailing or even online stores that only sells goods through the internet, but clearly targeting local consumers. This would be the case, for instance, for online stores in Portuguese, which follow local regulations on e-commerce (Decree 7.962/2013).

However, questions would arise when it comes to online offering of goods and services that are theoretically available to any consumer in the world, including Brazilian ones, but on websites that are not conceived or designed particularly for the Brazilian consumer (e.g. sites in other languages and showing prices in foreign currencies). This might be the case, for instance, for hotel services. Usually, any consumer around the world may book a hotel room available online with an international credit card.

Brazilian courts have not yet addressed these aspects. However, courts will likely look at the case having in mind the regular criteria to assess genuine use, such as 'volume, frequency and duration' (see section II.B *supra*) and whether such use should be considered 'licit', that is, whether it complies with local laws and regulations.

Nevertheless, in certain cases, it is possible that the main service is not exactly rendered in the country, but other related or ancillary services might be. Reverting to the hotel example, while one may argue that the lodging itself might be in a different country, the reservation services offered online may be targeted at the Brazilian consumer. In that case, as hotel services and reservation of hotel services are related services, the trademark registration would likely be kept in force as a whole (see section V.B *infra*). Therefore, adopting a broad description of goods and services also including ancillary ones might be an important strategy.

## F. Relevant time period and its calculation

Brazilian IP Law stipulates that a trademark registration will become forfeit, upon the request of any person with a legitimate interest, if (1) after five years from its grant, on the date of such request, use of the mark in Brazil has not been initiated; or (2) use of the mark has been interrupted for more than five consecutive years.

As a general rule, the investigated period of five years will be calculated back from the date that a third-party files a request for cancellation based on non-use.

However, there might be differences in the standards of examination, depending on whether the investigated period concerns the first five years following the grant of registration or a period of time much after these five initial years.

The reason for this potential difference lies on the fact that in the first five years the Brazilian IP Law only requires the IP rights holder to show that it has 'initiated' use of the mark.

In an appeal addressing a non-use forfeiture action initiated under the aegis of the former Brazilian Industrial Property Code of 1971 (which set a timeframe of only two years), the Federal Court of Appeals ruled that if the registrant had two years to 'initiate' use, it does not have to show consistent use during the entire period, but only acts that show it has effectively initiated *bona fide* use during such period.[28]

In this particular case, the trademark registration was granted in May 1984, so that registrant had to show it initiated use before May 1986. The registrant was only able to show use during the last five months of the period, which ended up being considered sufficient evidence by the court, given the fact that, at such point, it was only required to show that it initiated using the mark.

Naturally, the factual circumstances of the matter played an important role in such decision, which could have been different if such use was minimal or was merely an attempt to try to artificially preserve a trademark registration, as it seems to have been the case in the aforementioned appeal heard by Brazil Superior Court of Justice in the 'cigarette packs' case (see section II.E *supra*).

## III. EVIDENCE OF GENUINE USE OF A TRADEMARK

### A. When use should be proved: timing of lack of genuine use claims

#### *1. When filing an application for a trademark*

Brazilian IP Law does not require an applicant to submit any statements of use or intent-to-use when filing a trademark registration. The only requirement is that applicants may only apply for the registration of a mark relating to activities in which they are effectively and legally engaged, directly or through undertakings that they control directly or indirectly. Also, a simple declaration of compliance with such requirement is requested at the time of the trademark filing.

Furthermore, use is not a requirement for the grant of registration. Much to the contrary, IP Law provides that a registrant will benefit from a grace period of five years to initiate use and, in any case, evidence of use will only be required upon request by a third party with legitimate interest, in the framework of a specific procedure initiated for such a purpose.

#### *2. In the course of opposition proceedings*

Opponents are not required to provide evidence of use. In addition, applicants are not able to request such evidence from an opponent in the framework of an opposition.

However, if an opposition is filed based on registrations that are already over five years old, opponents should be aware that applicants may counterattack by initiating a separate non-use cancellation action. Should this occur, opposition proceedings are usually stayed pending a final decision on the cancellation action.

28. *Bodega J Edmundo Navarro Correas S.A. v. Navarro Correa Vinhos Finos Ltda. and Instituto Nacional da Propriedade Industrial*, AC. n. 1994.51.01.021842-3, TRF2, 1ª Turma, (07.13.2015)

### 3. *In cancellation proceedings based on lack of use*

The Brazilian IP Law only provides for a single procedure for the review of use requirement, namely, an administrative non-use forfeiture action, to be launched by any third party with a legitimate interest before the INPI. A final decision by the INPI may be reviewed by a Federal Court.

The INPI applies quite broad standards when reviewing legitimate interest. As a general rule, a party will have a legitimate interest in applying for the cancellation of an existing registration based on non-use if such registration is blocking the plaintiff's own trademark application.

However, such an interest is also recognized in other situations, such as the owner of a similar registered mark seeking to avoid dilution of its brand or the owner of other rights based on a similar element (such as trade name or copyright) that failed to timely oppose or challenge the registration acquired by the other party.

### 4. *As a defence in infringement/cancellation proceedings*

As described in section III.A.3 *supra*, the only available route for seeking the cancellation of a registration based on non-use is a non-use forfeiture action brought before the INPI. Therefore, in infringement actions or cancellation proceedings, the defendant will not be able to submit any counterclaims associated with the possible non-use of a trademark by the opponent or plaintiff. In such instances, the defendant will have to initiate a separate and independent administrative procedure which will exclusively entertain the forfeiture claim.

As a general rule, the pendency of a non-use forfeiture action will lead to the suspension of all administrative proceedings based on the trademark that has been challenged due to non-use. A Federal Court however is not required to stay judicial proceedings in the pendency of an administrative action but may do so if the trial Judge so decides, depending on the circumstances.

## B. Burden of proof

In the first IP Laws that established a non-use forfeiture action (namely, those enacted between 1923 and 1969, remaining in force between 1923 and 1970), the burden of proof to demonstrate that a trademark was not in use would lie with the party seeking cancellation.

While such assignment of the burden of proof used to be supported by scholars at the time,[29] it actually created a significantly cumbersome task for those seeking to cancel a trademark registration, as they would have to submit proof of a negative fact (the non-use of a mark by a third party).

The burden of proof was first shifted to the trademark owner in the former Brazilian IP Law of 1971. Hissao Harita,[30] a former President of the INPI, commented

29. João da Gama Cerqueira, *Tratado de Propriedade Industrial* vol. II, tomo II, parte III (Revista Forense, 1956), 212.
30. Hissao Arita, *Obrigatoriedade de uso das Marcas no Brasil*, 2, ABPI. Seminario Nacional da Propriedade Industrial, 74, 78 (1982).

that this was considered at the time one of the most important developments on the subject brought by the 1971 IP Law, together with the possibility for the INPI to initiate ex officio non-use forfeiture actions.

The current Brazilian IP Law of 1996 has continued to require from the trademark owner the burden of proving use of the mark in the framework of such cancellation actions. Happily, though, the current Law no longer allows the INPI to initiate ex officio cancellation actions, limiting the legal standing for so doing to those companies that show legitimate interest in seeking cancellation of a right owned by a third party.

## C. Relevant evidence

### *1. Use on the product and its packaging*

In theory, any type of evidence admissible in court would be admissible before the INPI to prove use of a trademark under a non-use forfeiture action.

Typically, the INPI places a heavy weight on tax-related documents, particularly invoices, showing that goods were effectively sold in Brazil or that services were rendered under the trademark.

If the products identified by the trademark are imported into Brazil, it is necessary to submit copies of the corresponding import declarations issued by the Brazilian Ministry for Financial Affairs. If such copies are not available, it is important to have additional supporting evidence from the importer to demonstrate that the goods were effectively internalized and distributed in Brazil.

Other types of supporting evidence must not only show the mark, but also feature the date they were issued or printed. Without such a date, it is not possible to demonstrate that use of such mark has taken place during the investigated period.

Therefore, the INPI does not usually require evidence or samples of the actual product or of its packaging, though such evidence would certainly be acceptable, if backed up with additional evidence on its sale.

### *2. Use in advertising*

While advertising may be helpful, particularly in case of design marks or word and design marks, the INPI typically would not accept advertisement material without evidence of actual sales, or any evidence without a date that can be objectively verified (such as undated online publicity or undated catalogues).

### *3. Use on the internet*

The INPI may accept any piece of evidence, as long as it shows effective use within the investigated period. Therefore, this can be a difficult issue in connection with internet pages and other online material, which are not always clearly dated.

Also, registrants using the mark on the internet must show that such use was aimed at the Brazilian territory, as detailed in section II.E *supra*.

### 4. *Form of evidence*

As detailed in section III. C.1 *supra*, the INPI accepts any form of evidence admissible in court. While heavier weight is placed on tax documents showing actual sales, other types of proof may also be submitted, such as surveys, affidavits, sales figures, etc.

The usefulness and need for these additional types of evidence may depend on the nature of the product or service covered by the registration, as invoices might not be sufficient to show such use.

On the other hand, the INPI typically would not accept (1) undated evidence, (2) documents issued abroad, including invoices, without tax documents showing that the products were effectively and lawfully internalized in the Brazilian market, and (3) documents issued by third parties without supporting evidence indicating chain of authorization originating from the trademark owner.

Courts are usually more open when reviewing evidence of use, but they are also more careful in checking good faith and possible fraud. Indeed, Federal Courts have reversed a few decisions by the INPI which upheld registrations facing non-use cancellation actions, finding that the invoices and other documents or samples submitted by the IP owner were fraudulent.[31]

### 5. *Distinction between use of a company name and use of the services offered by the company*

As mentioned in section II.A.1. *supra*, the INPI and the courts usually expect evidence showing use consistent with the function of the trademark. Hence, use of a sign merely as a trade name should not be sufficient to prove use of a trademark or service mark.

Naturally, there might be cases in which the registrant uses a certain sign indiscriminately as its trade name or its service mark, not only in tax documents, but service agreements executed with third parties or in advertisements. Any and all supporting evidence should be reviewed individually, and the case should be decided depending on its particular factual circumstances.

## IV. JUSTIFICATION FOR NON-USE

### A. Obstacles to use

In compliance with Article 5, Section C, of the Paris Convention, the Brazilian IP Law also provides, pursuant to its Article 143 § 1, that a trademark will not be cancelled due to non-use if the registrant justifies the lack of use for legitimate reasons.

Such reasons must be presented to the INPI with supporting evidence when the registrant responds to a non-use forfeiture action.

---

31. See, e.g., *N.V. NUTRICIA v. Nutricia S.A. Produtos Dieteticos e Nutricionais and Instituto Nacional de Propriedade Industrial*, AC. n. 0534594-72.2004.4.02.5101, TRF2, 2ª Turma, (08.03.2010) as well as *Emilio Pucci International B.V. v. Amazonas Produtos Para Calçados Ltda. and Instituto Nacional de Propriedade Industrial*, AC. 0805031-13.2011.4.02.5101, TRF2, 2ª Turma, (23.06.2015)

Since the IP Law is silent as to what may constitute 'legitimate reasons' for the purpose of preserving a trademark registration, the INPI and local courts have the task of reviewing claims on a case-by-case basis and setting standards on what such legitimate reasons may be.

A fairly common reason for the non-use of a trademark refers to difficulties in securing government authorization for the commercialization of certain products. This is particularly common in the pharmaceutical area, in which the registrant is required to obtain marketing authorization from Health Authorities (ANVISA, in Brazil) before launching a new product.

Brazil Superior Court of Justice has already recognized this as a legitimate reason for not having initiated use of the mark in Brazil, acknowledging that 'the pursuit of a licence before ANVISA for the commercialization of a pharmaceutical product [the mark of which is] registered at the INPI falls within the concept of legitimate reasons as established by the Law'.[32]

In such cases, however, the registrant is required to demonstrate serious and effective measures that it has undertaken directly or through a licensee for initiating use of the mark in Brazil, as established under INPI's Trademark Guidelines.[33]

For instance, the Federal Court of Appeals of the 2nd Region kept the registration for the trademark FRAPPUCCINO in force after its owner, Starbucks Corporation, showed that it failed to initiate use of the mark solely due to the restrictions that it had faced for the importation of the product. Those were related to the difficulties in securing the required sanitary licences to initiate such importation before local health authorities.[34]

The precedents in Brazil recognizing that hindrances imposed by government authorities should constitute legitimate reasons to avoid cancellation of the trademark registration are closely associated to discussions that took place in the 1980s regarding the closure of Brazilian market to imported goods.

At the time, the Brazilian entity that regulated foreign commerce (CACEX) issued prohibitions on the importation of certain goods that it considered to be 'superfluous', such as cosmetics, perfumery articles, clothing articles, watches etc.[35]

Given such a prohibition, several famous brands, particularly in the fashion industry, ended up excluded from local market. This led to multiple non-use cancellation actions being filed in Brazil against some well-known trademarks, such as PACO RABANE, MOVADO, DAVIDOFF, ELLESSE and HAAGEN-DAZS.

The INPI's initial understanding was that such a prohibition would not constitute a legitimate reason for the absence of use of a trademark, inasmuch as registrants could have manufactured products in Brazil or licensed the mark to a local company. Of course, the Office failed to explain how a Swiss watch manufacturer, for example, could have manufactured goods in Brazil.

32. *Farmoquimica S.A. v. Wyeth Whitehall Ltda. and Instituto Nacional da Propriedade Industrial*, REsp 1377159/RJ, STJ, 3ª Turma (10.05.2016).
33. As per Section 6.5.5, of the Guidelines.
34. *Fran's Café Franchising Ltda. v. Starbucks Corporation and Instituto Nacional da Propriedade Industrial*, AC. n. 0805726-35.2009.4.02.5101, TRF2, 2ª Turma, (29.11.2011).
35. Luiz Leonardos and Maria Cecilia de A. Oswald, *Pequeno Estudo sobre Caducidade de Registro de Marca*, 55, Revista da ABPI, 03, 10 (2001).

Despite some earlier contradictory decisions of Brazilian courts, the understanding that such prohibition constituted a *factum principis* (act of state) that could not harm the rights of a private party ended up prevailing, leading the Office to officially change its understanding in early 1990s.[36]

The STJ's most recent precedents consolidated such an understanding. In one of its recent decisions,[37] STJ outlined that 'the prohibition on importation of goods generates a barrier, which, if not completely unsurmountable, may turn the commercialization of such goods in Brazil either economically or strategically non-viable, in addition to constituting an unexpected measure'.

Another argument that is commonly raised refers to economic problems that a company might be facing. This includes cases referring to bankruptcy and judicial reorganization processes.

Borges Barbosa states that, under Brazilian case law, the mere existence of financial distress is not considered per se legitimate reasons for the non-use of a trademark. Indeed, 'if there is no legal impediment to the development of its activities, there are no reasons to exempt the trademark holder, even under distress, from the duty to use [the mark]'.[38]

As for bankruptcy, while its declaration by a court could theoretically constitute a legitimate reason to support non-use, Borges Barbosa states that one must review whether there were any circumstances that could have allowed the continuance of the activities, for example, through a licence agreement.[39]

Nevertheless, some court precedents have exceptionally recognized that in case of liquidation by bankruptcy, the rights of the insolvency estate should be preserved.[40]

In any case, legitimate reasons for the absence of use must be reviewed on a case by case basis, inasmuch as there are sometimes peculiarities to a trademark that must be taken into account. This has happened, such as with the trademark registration for HORIZON,[41] in the name of French company Eurocopter. The trademark was designed for 'military helicopters and radar systems', which are goods sold exclusively to military forces. The registrant proved that the mark was in use in several different countries and fully available on the marketplace. However, any local sales would exclusively depend on an order by Brazilian military forces, which had not yet occurred. The INPI accepted such explanation as legitimate reason for the non-use of the trademark.

36. *Ibid.*

37. *Hans Schwarzkopf v. Makrofarma Quimica Farmaceutica Ltda. and Instituto Nacional da Propriedade Industrial*, REsp 1071622/RJ, STJ, 3ª Turma, in 03/02/2009.

38. Denis Borges Barbosa, *Ensaios e estudos de Propriedade Intelectual: 2014-2015* vol. 1 (IBPI Europa, 2015), 275 (translated by author).

39. *Ibid.*, 276.

40. *Hasbro Inc. v. Brinquedos Bandeirante S.A. and Instituto Nacional da Propriedade Industrial*, AC. n. 0801060-25.2008.4.02.5101, TRF2, 1ª Turma, in 07.12.2015.

41. Non-use forfeiture action filed by Horizon Desenvolvimento Tecnológico Agroindustrial LTDA against Reg. No. 004092457.

## B. Circumstances in which non-use may be cured

While the Brazilian IP Law does not explicitly address whether non-use can be cured, local practice and precedents indicate that one should only look into the five-year investigated period preceding the filing of a non-use forfeiture action to determine whether or not a trademark is in use.

Hence, if a trademark has not been in use for multiple decades, but use has recommenced in a consistent manner during the five years preceding the action, there would not be any valid reason to have the trademark registration cancelled. Much to the contrary, in that particular case, the Law should preserve the goodwill that has been reestablished by the trademark owner.

This is also the opinion of Gama Cerqueira. In his view, the question of whether or not it would be possible to 'cure' non-use by using the mark would seem 'incongruent', inasmuch as 'use and non-use are mutually exclusive concepts'. In his opinion, 'if a trademark is being used [during the investigated period], fulfilling its own role, the reason that would justify the forfeiture of the registration would then disappear'.[42]

Nevertheless, the author points out two conditions that would have to be met. The first one would be that the non-use forfeiture action may not have been initiated before use was resumed. Use initiated after such challenge cannot harm the rights of the third party that sought the cancellation. Also, use during the investigated period would have to meet the regular requirements for use to be considered genuine, being therefore 'real, effective, made in a regular fashion, uniform and constant'.[43]

Carlos Henrique Froes also shares this point of view. He adds that 'non-use cancellation constitutes a penalty to the inaction of the registration holder, which, by failing to use the mark, turns it into a defensive trademark, which serves merely as an obstacle to the registration of other marks'.[44]

# V. CONSEQUENCES OF NON-USE

## A. Consequences of non-use and relevance of the context

As mentioned above (see section III.A *supra*), any third parties may only challenge a trademark by filing a non-use forfeiture action against the corresponding registration before the INPI. The proceeding will follow its course in an independent fashion and a decision will eventually be issued by an examiner. In the absence of a response by the trademark owner, a decision by default will be issued and the registration will be cancelled.

42. João da Gama Cerqueira, *Tratado de Propriedade Industrial* vol. II, tomo II, part III (Revista Forense, 1956), 215 (translated by author).
43. *Ibid.*, 216.
44. Carlos Henrique de Carvalho Froes, *Caducidade de registro: Efeitos da decisão administrativa*, 12, Revista da ABPI, 1, 12 (1992) (translated by author). Professor Froes also adds that this is also the opinion of other celebrated classic local authors, such as Pontes de Miranda and that this rationale has been adopted by local courts as well.

Any of the parties may appeal against an unfavourable ruling, which will be decided by the President of the INPI. It is possible to submit additional supporting evidence on the use of a trademark at appeal level, even in case of a decision by default.

The decision by the President of the INPI is final within the administrative instance of this Office (closes administrative level), but may be still challenged before Federal Courts.

Until a final decision is rendered, administrative proceedings such as oppositions and nullification procedures that are based on the challenged mark remain suspended. However, infringement actions brought by the trademark owner are usually not suspended by trial Courts.

Hence, one important aspect is the effect of a decision that recognizes the forfeiture of a trademark registration due to the absence of its use. Brazilian scholar Carlos Henrique Froes outlines that 'the effects of the forfeiture, in the silence of the Law, which could have provided for in a different manner (e.g. making them retroactive to the filing of the action), are *ex nunc*, or, more precisely, [such effects] only operate once the administrative decision that orders the forfeiture becomes res judicata'.[45]

In a similar sense, precedents by the Brazil Superior Court of Justice recognize that the effects of the cancellation of a trademark registration based on non-use will only operate once a final decision is rendered.[46]

As Carlos Henrique Froes[47] reminds us, there is a very important practical consequence to establishing the starting date for the effects of the declaration of the forfeiture of a trademark registration. Under current Brazilian case law, until a final decision is rendered in a non-use forfeiture action, the registration remains in force, so that non-authorized use of the relevant trademark may constitute infringement of the rights of the trademark owner.

Some discussions still arise from the position of the Superior Court of Justice, considering that the INPI usually takes a few years to fully decide a non-use cancellation action. Moreover, in case of an administrative decision that favours the registrant, the other side would have to resort to a court action. A lawsuit could then also take a number of years to be ultimately adjudicated. During such period, the third party would not be able to use the desired mark, as it would be otherwise infringing the IP owner's rights, making this company potentially liable for damages based on a trademark that should have been revoked.

Considering that this scenario could potentially allow a trademark registration that had not been genuinely used to remain in force for several years, it might be a matter for Brazil Supreme Court to decide whether or not this rationale fully complies with constitutional requirements related to the social role of a trademark registration and is consonant with the principle of freedom of commerce.

---

45. Carlos Henrique de Carvalho Froes, *Caducidade de registro: Efeitos da decisão administrativa*, 12, Revista da ABPI, 1, 12 (1992) (translated by author).
46. *Lautrec Piublicidade S.A. v. BF Utilidades Domésticas and Instituto Nacional da Propriedade Industrial*, EREsp 964.780/SP, STJ, 2ª Seção, (10/08/2011), DJe 29/08/2011.
47. Carlos Henrique de Carvalho Froes, *Caducidade de registro: Efeitos da decisão administrativa*, 12, Revista da ABPI, 1, 12 (1992) (translated by author).

## B. Extent of any revocation

As detailed in section II.D *supra*, the registrant is required to prove use of the mark in connection with all products or services described in the certificate of registration, under penalty of partial forfeiture of the registration.

However, if the registrant is able to prove use for at least a part of such goods or services, Brazilian IP Law also establishes that those goods and services that are similar or akin to those in use will be kept in the certificate of registration.

The registration will therefore only be revoked for those sufficiently different products or services.

Based on the above rules, in case of broad descriptions, such as 'clothing articles', if the registrant is able to prove use in connection with 'T-shirts' and 'trousers', the INPI will most likely preserve the broad description.

## C. Timing of any revocation

As detailed in section V.B *supra*, a non-use forfeiture action may take a couple of years to be ultimately decided by the INPI and a final decision may take even longer if the matter is brought to a Federal Court. Therefore, since the Brazil Superior Court of Justice has established that the effects of the revocation of a trademark registration based on non-use may depend on a final decision, a junior applicant relying on such an action should carefully consider the best strategies to pursue before initiating use of the mark in Brazil.

# Canada

*Paula Clancy, JD*[1]

'If I have the full use of the thing I am its owner.' – Hegel, *Philosophy of Right* §61

**INTRODUCTION**

The concept of use serves as one of the foundations of Canadian trademark law and is critical for the acquisition, maintenance, and enforcement of trademark rights in Canada.[2] Indeed, use of a mark either in Canada or abroad has traditionally been a precondition to securing a Canadian trademark registration. From a policy perspective, if a trademark is not in use, it is incapable of generating goodwill and performing its role as an indicator of source. Without use there is arguably no rationale for granting exclusive rights in a trademark and removing it from the public domain.

In 2019 significant legislative changes will come into force in Canada which, while not removing use as an essential concept in Canadian trademark law, will permit applicants to obtain a trademark registration without first requiring use of the mark.[3] This shift has proven to be controversial and the Canadian government has faced considerable criticism for eliminating use as a prerequisite to registration.[4]

---

1. The author would like to thank Tania Treciokas, Law Clerk and Trademark Agent Trainee, for her meticulous research assistance, Alfred Macchione (McCarthy Tétrault) and Bill Mayo (Aitken Klee LLP) for their valuable input, and Kathryn Szymczyk, Dr. Renato Cristi and Dr. Marcela Cristi for their keen editorial comments.
2. The other cornerstone of Canadian trademark law is the concept of distinctiveness. The Supreme Court of Canada held in *Kirkbi AG v. Ritvik Holdings Inc.*, 2005 SCC 65 at para. 39, [2005] 3 S.C.R. 302 [*Kirkbi*], 'The foundation of a trade-mark is distinctiveness because only a distinctive mark will allow the consumer to identify the source of the goods'.
3. Bill C-31, Economic Action Plan 2014 Act, No. 1, 2nd Sess, 41st Parl, 2014, which received Royal Assent on June 19, 2014. The amendments to the Trade-marks Act contained in Division 25 of Bill C-31 are the most significant changes to Canada's trademark legislation since the Unfair Competition Act, 1932 was replaced by the *Trade Marks Act* in 1954.
4. Yet Canada is not alone in this approach as several other jurisdictions permit the registration of marks without prior use, including Australia, New Zealand, and the European Union

Notwithstanding these changes, use will continue to be essential for maintaining registered trademark rights in Canada.

This Chapter will examine the concept of use as articulated in the Canadian Trade-marks Act[5] (the 'Act'), as amended by the Economic Action Plan 2014 Act, No. 1 (or 'Bill C-31'), together with legal and judicial commentary on what constitutes genuine trademark use in Canada.

# I. INTRODUCING GENUINE USE OF TRADEMARKS

## A. Legal framework

### *1. Common law*

Canada is a common law jurisdiction in which trademark rights are acquired through *use*. Without use, a trademark cannot generate goodwill, and without goodwill, a trademark has no value.[6]

The Supreme Court of Canada acknowledged the importance of use in creating property rights in a trademark as early as 1888, when it stated:

> 'It is not the registration that makes the party proprietor of a trade mark; he must be proprietor before he can register [...] "proprietor of a trade mark" means a person who has appropriated and acquired a right to the exclusive use of the mark [...]'[7]

More recently, the Supreme Court confirmed that in Canada trademark ownership is established through use. In *Masterpiece Inc. v. Alavida Lifestyles Inc.*, the Supreme

– *See* Paul Tackaberry, *Exploring the Boundaries of 'Use' After the 2014 Amendments to the Canadian Trademarks Act*, 104:6 The Trademark Reporter 1332 at 1339 (2014).

5. Trade-marks Act, RSC 1985, c T-13. All references to sections shall, unless otherwise noted, be references to sections of the Trade-marks Act. Among the many changes introduced by Bill C-31 is the change of spelling of 'trade-mark' to 'trademark'. This is the third spelling change in Canada as the word was previously spelled without a hyphen ('trade mark'). Accordingly, the various iterations will be used throughout depending on whether the word appears in the Act, the jurisprudence or Bill C-31.
6. Daniel R. Bereskin, *Canada's Ill-Conceived New 'Trademark' Law: A Venture into Constitutional Quicksand*, 104:5 The Trademark Reporter 1112 at 1115–1116 (2014); Sheldon Burshtein, *The Basics of Trade-mark Use in Canada: The Who, What, Where, When, Why, How and Future* (Paper delivered at the Twenty-Third Annual Trademark Institute, McGill University, July 2017) at 12 [Burshtein, *Trade-mark Use*].
7. *Partlo v. Todd* (1888), 17 S.C.R. 196 at 200–201, 1888 CarswellOnt 25. *See also*, *Mattel, Inc. v. 3894207 Canada Inc.*, 2006 SCC 22 at para. 5, [2006] 1 S.C.R. 772 [*Mattel, Inc.*], where the Supreme Court of Canada stated:

   'Unlike other forms of intellectual property, the gravaman [sic] of trade-mark entitlement is actual use. By contrast, a Canadian inventor is entitled to his or her patent even if no commercial use of it is made. A playwright retains copyright even if the play remains unperformed. But in trade-marks the watchword is "use it or lose it". In the absence of use, a registered mark can be expunged.'

Court recognized the rights of a prior user in its unregistered marks over those of a registrant who claimed subsequent use. The Court began its analysis as follows:

> 'At the outset, it is important to recall the relationship between use and registration of a trade-mark. ...
>
> Registration itself does not confer priority of title to a trade-mark. At common law, it was use of a trade-mark that conferred the exclusive right to the trade-mark. While the *Trade-marks Act* provides additional rights to a registered trade-mark holder than were available at common law, registration is only available once the right to the trade-mark has been established by use.'[8]

In Canada, it is the first user of a mark who may claim ownership in the mark, and who generally enjoys priority over subsequent users.[9]

## 2. *The Trade-marks Act*

The concept of use is intricately woven throughout the Act. For example, 'use' determines the date of adoption of a trademark,[10] an applicant's entitlement to registration,[11] whether a mark has acquired sufficient distinctiveness to permit its registration,[12] whether a trademark or trade name is confusing with another trademark or trade name,[13] whether a mark has been abandoned,[14] whether a registration ought to be cancelled,[15] and whether one party has infringed the trademark rights of another.[16] Below is a brief overview of the key provisions relating to use found in the Act.

8. *Masterpiece Inc. v. Alavida Lifestyles Inc.*, 2011 SCC 27 at para. 35, [2011] 2 S.C.R. 387 [*Masterpiece*].
9. The first user is entitled to registration *unless* Section 34 or Section 17(2) apply. Section 34 deems the date of Convention priority to be the date of filing in Canada and allows the applicant to claim priority notwithstanding any intervening use in Canada by a third party. Section 17(2) of the Act establishes a five-year period after which a registration may not be challenged on the ground of previous use or making known, unless it is established that the person who adopted the registered mark did so with knowledge of the previous use or making known. In other words, a registration becomes 'incontestable' after five years from the date of registration such that a prior user may not challenge its validity, based on his or her prior use.
10. Section 3 of the Act.
11. Section 16 of the Act.
12. Section 12(2) of the Act.
13. Section 6 of the Act.
14. Section 18 of the Act.
15. Section 45 of the Act.
16. Sections 19 and 20 of the Act.

### *i. Definition of use*

Under the Act, a trademark is not merely a mark or a sign, but rather one that is *used for the purpose of distinguishing.*[17] Section 2 defines 'use' as 'any use that by Section 4 is *deemed to be a use* in association with goods and services'.[18]

The definition of use varies depending on whether the goods are sold within Canada, exported from Canada, or whether the mark is associated with services. Section 4 of the Act provides that a trademark is deemed to be used in association with:

(1) *goods*, if at the time of the transfer of the property in or possession of the goods, in the normal course of trade, it is marked on the goods themselves or on the packages in which they are distributed or it is in any other manner so associated with the goods that notice of the association is then given to the person to whom the property or possession is transferred;[19]
(2) *goods* that are *exported from Canada*, if it is marked in Canada on goods or on the packages in which they are contained when the goods are exported;[20] and
(3) *services*, if it is used or displayed in the performance or advertising of those services.[21]

### *ii. Deemed adoption of a mark*

Section 3 of the Act provides that a mark is deemed to be adopted when it is first used or 'made known' in Canada.[22] If the mark has not been used in Canada, the date of adoption is the application filing date (or the Convention priority date, if applicable).[23]

---

17. Section 2 of the Act – definition of 'trade-mark'. This definition often comes into play in infringement proceedings where the courts will focus on whether the defendant's use of the infringing mark was for the purpose of distinguishing its goods from the plaintiff's goods within the meaning of Section 2: *see Clairol International Corp. v. Thomas Supply & Equipment Co.*, [1968] 2 Ex. C.R. 552, 55 C.P.R. 176 [*Clairol*]. In this case the use by the defendant of the plaintiff's mark on a hair colour comparison chart was not deemed to be infringing since the defendant used the mark to identify the *plaintiff's* products rather than its own.
18. Section 2 of the Act – definition of 'use'.
19. Section 4(1) of the Act. Examples of use of a mark that is 'associated' with the goods when the purchaser takes possession include the use of the mark on documentation accompanying the goods such as invoices or on point of sale materials displayed in close proximity to the goods at the time of purchase.
20. Section 4(3) of the Act.
21. Section 4(2) of the Act.
22. Section 3 of the Act. 'Making known' occurs when a mark becomes well-known in Canada as a result of distribution of the goods in Canada, together with spill-over advertising into Canada. Section 5 of the Act outlines the requirements for establishing that a mark has become well-known.
23. Sections 3 and 34(1) of the Act. Section 34(1) provides that: 'When an application for the registration of a trade-mark has been made in or for any country of the Union other than Canada and an application is subsequently made in Canada for the registration for use in association with the same kind of goods or services of the same or substantially

*iii. Entitlement to registration*

To qualify for a Canadian trademark registration, Section 16 of the Act currently requires one or more of the following preconditions:

(1) that the applicant has *previously used* or *made known* its trademark in Canada;[24]
(2) that the applicant has previously *used* the trademark outside of Canada,[25] and that the applicant has a registration or pending application for the identical mark in its country of origin;[26] or
(3) that the applicant has an *intention to use* the trademark in Canada and that, prior to registration, it declare that *use* of the mark has commenced in Canada.[27]

An applicant is only entitled to secure registration of an otherwise registrable mark if, as of the applicable relevant date,[28] the mark is not confusing with:

---

the same trade-mark by the same applicant or the applicant's successor in title, *the date of filing of the application in or for the other country is deemed to be the date of filing of the application in Canada* [...]' (emphasis added). 'Country of the Union' is defined in Section 2 of the Act as: (a) any country that is a member of the Union for the Protection of Industrial Property constituted under the Convention, or (b) any Member of the World Trade Organization. The priority provision requires that the Canadian application be filed within six months from the date of filing the original application.

24. Section 16(1) of the Act.
25. Provided that the use occurs in a country of the Union, i.e. a member of the Paris Convention or the World Trade Organization.
26. Section 16(2) of the Act. *Thymes, LLC v. Reitmans Canada Limited*, 2013 FC 127, 107 C.P.R. (4th) 319 suggests that in order for a foreign use and registration claim to be valid, the use of the mark outside Canada must have commenced prior to the Canadian filing date. The court held at para. 18: 'There is no doubt a proper reading of [...] section [16(2)] requires that, at the time of filing the application, if an applicant relies on registration or application and use abroad pursuant to that section, there must have been use of the trade-mark *at the time of the application* to rely on this section as a valid basis to obtain registration in Canada' (emphasis added). *But see Coors Brewing Company v. Anheuser-Busch, LLC*, 2014 FC 716, 125 C.P.R. (4th) 15 in which the foreign use and registration claim was added post-filing, and the court found that the claim was valid provided there was foreign use of the mark *when the claim was added*, even if there was no foreign use of the mark at the time of filing.
27. Sections 16(3) and 40(2) of the Act.
28. The relevant date for assessing entitlement depends on the basis for registration claimed by the applicant. If the applicant claims use or making known of its mark in Canada, the date on which the applicant first used or made known its mark is the relevant date; if the applicant claims foreign use and registration of its mark, the relevant date is the date of filing the application in Canada (or, if applicable, the date of priority); and if the applicant claims an intention to use the mark in Canada, the relevant date is the date of filing the application (or date of priority) – *See* Sections 16(1), (2) and (3) of the Act. *See also* Section 30 of the Act which outlines the formal requirements for an application, including identifying the date of first use in Canada, or the country of first use if the mark has been used abroad. If the mark has not been used in Canada, prior to filing, the application must

(1) a trademark that had been *previously used* in Canada or made known in Canada by any other person;[29]
(2) a trademark in respect of which an application for registration had been *previously filed* in Canada by any other person;[30] or
(3) a trade name that had been *previously used* in Canada by any other person.[31]

*iv. Rights and remedies conferred by registration*

Section 19 of the Act confers on the owner of a registered trademark *the exclusive right to use* the mark throughout Canada in relation to the goods and services claimed in the registration.[32] Pursuant to Section 20 of the Act, the exclusive right to use is deemed to be infringed by any person who is not entitled to use the mark and yet who, among other things, sells, distributes or advertises any goods or services in association with a confusing trademark or trade name.[33] Moreover, Section 22 of the Act prohibits the use of a registered trademark by another person in a manner that is likely to have the effect of depreciating the value of the goodwill attaching to the mark.[34]

*v. Test for confusion*

Section 6 of the Act outlines the test for confusion, and provides at 6(2) that:

> 'The use of a trade-mark causes confusion with another trade-mark if the use of both trade-marks in the same area would be likely to lead to the inference that the goods or services associated with those trade-marks are manufactured,

---

include a *bona fide* statement of the applicant's intention to use the mark – *See* Sections 30(b), (c), (d) and (e) of the Act.

29. Sections 16(1)(a), (2)(a), and (3)(a) of the Act.
30. Sections 16(1)(b), (2(b), and (3)(b) of the Act.
31. Sections 16(1)(c), (2)(c), and (3)(c) of the Act.
32. Section 19 of the Act.
33. Section 20 of the Act.
34. Section 22(1) of the Act. Interestingly, the court in *Clairol, supra* note 16, found that Subsection 22(1) *only* requires 'use' within the meaning of Section 4, but not Section 2. The court need not consider whether the use was for the purposes of distinguishing the source of the goods or services (as required by Section 2), but must limit the analysis to whether the criteria for use outlined in Section 4 have been satisfied. Therefore, use of a mark in advertisements for goods would not constitute a violation of Section 22(1) because mere advertising does not qualify as 'use' in relation to goods under Section 4. For further discussion see Tackaberry, *supra* note 4, 1349–1350. In *Veuve Clicquot Ponsardin v. Boutiques Cliquot Ltee.* 2006 SCC 23, [2006] 1 SCR 824, the Supreme Court of Canada also focused its analysis on 'use' within the meaning of Section 4, and clarified that for a claim of depreciation of goodwill the plaintiff does not need to show that use of both marks in the same geographic area will likely lead to confusion. Rather, the plaintiff need only show that the defendant made use of a mark that is sufficiently similar to the plaintiff's mark so as to evoke a mental association, or 'linkage', that is likely to depreciate the plaintiff's goodwill: at para. 49.

sold, leased, hired or performed by the same person, whether or not the goods or services are of the same general class.'[35]

### *vi. Maintenance of registration*

Use is essential to maintaining registered trademark rights. Registrations may be deemed invalid and expunged on the basis of abandonment pursuant to Sections 18(1) and 57 of the Act, or they may be summarily cancelled for non-use under Section 45.

Neither the Act nor Bill C-31 require proof of use in the form of specimens or declarations of continued use during the term of registration or upon renewal.

## *3. Legislative changes coming into force in 2019*

As noted, the Act has been extensively amended by Bill C-31. The amendments are expected to come into force in 2019 and will create the framework necessary for Canada to accede to the Madrid Protocol,[36] the Singapore Treaty[37] and the Nice Agreement.[38]

Perhaps the most significant change introduced by Bill C-31, apart from accession to the Madrid Protocol, is the removal of use as a precondition to registration. Although use will continue to be central to the Canadian trademark system, registration will now be available to applicants whether or not the mark has been previously used in Canada or elsewhere.[39] Any applications based on 'Proposed Use' that are pending as of the date on which Bill C-31 comes into force will be able to proceed to registration without the filing of a declaration of use.

Many Canadian commentators have argued that by abolishing use as an essential precondition to the creation of enforceable trademarks rights, the Canadian government has effectively undermined the very foundation of Canadian trademark

35. Section 6(2) of the Act. Section 6(3) deals with trademark confusion with a trade name and 6(4) with trade name confusion with a trademark. Section 6(5) outlines the factors that must be considered when determining the likelihood of confusion. The Supreme Court of Canada has confirmed that the words, 'if the use of both trademarks in the same area', requires that a confusion analysis be based on a 'hypothetical assumption' that both trademarks are used in the same geographic area, irrespective of whether this is actually the case: *Masterpiece, supra* note 8, para. 30. For a registered owner to enjoy exclusive use of the mark throughout the country, there cannot be a likelihood of confusion with another trademark *anywhere* in Canada.
36. Protocol Relating to the Madrid Agreement Concerning the International Registration of Marks (herein 'Madrid Protocol').
37. Singapore Treaty on the Law of Trademarks (herein 'Singapore Treaty').
38. Nice Agreement Concerning the International Classification of Goods and Services for the Purposes of the Registration of Marks (herein 'Nice Agreement').
39. Specifically, it will no longer be necessary for applicants to identify whether the application is based on prior use or making known, foreign use and registration, or an intention to use the mark. Rather, the applicant's prior use of the mark or its intention to use the mark will be *presumed*. Moreover, applicants will no longer be required to make a declaration attesting that use of a proposed mark has commenced in Canada in order to proceed to registration.

law[40] and negatively impacted Canada's trademark system.[41] They argue that Bill C-31 introduces a registration system that 'confers exclusive rights to the use of a registered mark and the right to prevent the use of confusingly similar trademarks and trade names throughout Canada irrespective whether the registered mark has ever been used'.[42] Concerns have also been expressed that the removal of use as a pre-requisite to registration will encourage trademark squatting and will essentially render Canada a 'first-to-file' regime.[43]

It is certainly true that applicants and rights holders will face increased investigation costs to assess priority and entitlement, as well as increased uncertainty in clearing trademarks for use and registration. There will also no doubt be an increase in opposition and cancellation proceedings in Canada. However, the elimination of use as a precondition to registration will greatly streamline the Canadian trademark application process and may not be as significant as feared.

To begin with, foreign applicants are already able to obtain a Canadian registration without actual use of the mark *in Canada*. Arguably, Bill C-31 has levelled the playing field for Canadian applicants so that they too may obtain registration without actual use in Canada.

Moreover, Canada is already a 'first-to-file' system for the purposes of examination before the Office. In 2005, the Canadian Intellectual Property Office issued a practice notice indicating that when examining entitlement between pending marks, the Office would no longer consider the dates of first use or making known as a relevant consideration.[44] Therefore, when pending marks are confusing, the applicant with the earlier filing date or priority date is already considered to be the person entitled to registration of the trademark. The only option for an applicant with a later filed application who claims an earlier date of first use or making known is to commence opposition proceedings. Accordingly, opposition proceedings will continue to be the forum in which any claims for better title to a mark are considered, including claims based on earlier use of trade names and unregistered marks.

The rights of prior users are also recognized by virtue of Section 18 of the Act which provides that a registration is invalid if the applicant for registration was not the person entitled to secure the registration, and Section 21(1) which permits the concurrent use of confusing marks in cases where a prior user is barred from

---

40. Bereskin, *supra* note 6, 1117–1118.
41. Sheldon Burshtein, *Canada Weakens Trademark Structure by Demolishing Use Foundation*, 105:4 The Trademark Reporter 930 at 933 (2015). *See also* Cynthia Rowden, *A Perfect Storm Confronts Trademark Owners And The Canadian Trademarks Office: 45-Class Applications And The Implications For Trademark Selection, Registration And Enforcement*, Mondaq (5 February 2018), online <http://www.mondaq.com/canada/x/670494/Trademark/A+Perfect+Storm+Confronts+Trademark+Owners+And+The+Canadian+Trademarks+Office+45Class+Applications+And+The+Implications+For+Trademark+Selection+Registration+And+Enforcement>.
42. Bereskin, *supra* note 6, 1118.
43. Benjamin Lai, *Bill C-31, the Trade-marks Act Amendments and Dealing with Trade-mark Squatting*, 28 IPJ 135 at 159 (2015).
44. Canadian Intellectual Property Office, *Entitlement – Confusing Marks*, Practice Notice (Ottawa: CIPO, 19 May 2005). Available at: http://www.ic.gc.ca/eic/site/cipointernet-internetopic.nsf/eng/wr00204.html.

invalidating a registration that has attained incontestability.[45] The Act provides the means to expunge registered marks that are not in use by way of invalidity or cancellation proceedings.

Given the Supreme Court of Canada's clear pronouncements that title to a mark is not established through *registration*, but through *use*, it is unlikely that the courts will uphold a registration if the registrant is unable to establish ownership of the mark through use in Canada.[46] Arguably, non-use of a registered mark on the part of a plaintiff should constitute an absolute defence to an infringement action, which may provide a safeguard following the implementation of Bill C-31.

### B. Purpose and function of genuine use: policy considerations

According to Canadian trademark law, the principal function of a trademark is to act as an indicator of source. A trademark must actually distinguish or be adapted to distinguish the goods or services of one trader from those of another.[47] From a policy perspective, if a mark is incapable of serving this function, it is not worthy of recognition or protection.

Even if one were to take a broader approach to the function of a trademark – one in which the primary role of a trademark is consumer protection, quality assurance, or the prevention of unfair competition – use remains essential. Without actual use in the marketplace, a trademark cannot serve any of these functions.[48]

The Supreme Court of Canada has acknowledged that trademarks are somewhat of an anomaly in intellectual property law. In *Mattel, Inc. v. 3894207 Canada Inc.* the court held that:

> 'Unlike the patent owner or the copyright owner, the owner of a trade-mark is not required to provide the public with some novel benefit in exchange for the monopoly [...] By contrast, a patentee must invent something new and useful. To obtain copyright, a person must add some expressive work to the human repertoire. In each case, the public through Parliament has decided it is worth encouraging such inventions and fostering new expression in exchange for a statutory monopoly (i.e. preventing anyone else from practising the invention or exploiting the copyrighted expression without permission). The trade-mark

45. Sections 18 and 21(1) of the Act. *See also*, Tackaberry, *supra* note 4, 1360, who argues that the main effect of Bill C-31 is 'to shift the requirements for asserting and proving use away from the initial application process and toward the opposition and cancellation process' thereby limiting 'the role of government and transfer[ring] enforcement to the private sector [...]'.
46. This was made clear in *Masterpiece, supra* note 8, where the court confirmed that registration does not confer priority of title to a trademark. *See also* the recent decision in *Micro Matic A/S v. Taizhou TALOS Sanitary Co. Ltd.*, 2017 FC 978, 151 C.P.R. (4th) 433.
47. Section 2 of the Act – definition of 'distinctive'. *See also*, Art. 15 of the World Trade Organization's Agreement on Trade-Related Aspects of Intellectual Property Rights, 1869 *U.N.T.S.* 299, which defines 'trade-mark' in part as 'any sign, or any combination of signs, *capable of distinguishing* the goods or services of one undertaking from those of other undertakings [...]' (emphasis added).
48. David Vaver, *Intellectual Property Law: Copyright, Patents, Trade-marks*, 2nd ed (Toronto: Irwin Law, 2011) at 471–472 [Vaver, *Intellectual Property*].

> owner, by contrast, may simply have used a common name as its "mark" to differentiate its wares from those of its competitors. *Its claim to monopoly rests not on conferring a benefit on the public in the sense of patents or copyrights but on serving an important public interest in assuring consumers that they are buying from the source from whom they think they are buying and receiving the quality which they associate with that particular trade-mark* (emphasis added).'[49]

From a policy perspective, then, the purpose of requiring genuine use of trademarks is that it confirms the economic incentive behind protecting trademarks as a property right: without genuine use in the marketplace, the rationale for preserving an exclusive monopoly in a trademark disappears.

## II. WHAT QUALIFIES AS GENUINE USE OF A TRADEMARK

### A. Nature of use

#### *1. Use consistent with the function of the trademark*

##### *i. Use 'as a trademark'*

For a mark to be used 'as a trademark' it must indicate origin by distinguishing the goods and services of one trader from those of its competitors.[50] Whether a mark is being used 'as a trademark' depends on the message communicated to the public by the display of the mark. In *Tommy Hilfiger Licensing Inc. v. International Clothiers Inc.*, the court found that the plaintiff's mark had not been used by the defendant to distinguish the *defendant's* goods, rather it had been used purely for decorative purposes. In this case the defendant purchased shirts bearing a crest design that was very similar to the plaintiff's crest design and affixed its own label to the shirts. The judge found no evidence that the crest design had been used to distinguish the defendant's goods. Accordingly, the court found that the plaintiff's mark had not been used 'as a trademark' and dismissed a claim of trademark infringement.[51]

---

49. *Mattel, Inc.*, *supra* note 7, para. 21. Similarly, Vaver argues that:

    'Trade-marks are commonly classified as intellectual property, but there is nothing intellectual about them [...] a mark may merely comprise a prosaic word or device – and today many do. The mark may be thought of independently, or it may be someone else's idea. Who originated it, or how, does not matter. No rights to a trade-mark flow from mere creation. Use or its surrogates – public recognition or an intention to use – alone create the rights, and then not in the creator but in the person behind the use, intent, or creation of the public recognition. So advertising agencies that devise trade-marks for their clients do not usually own the mark; it is the client who marks it on his goods who may register it': *ibid.* at 423–424.

50. Bill C-31 simplifies the definition of 'trademark' as: 'a sign or combination of signs that is used or proposed to be used by a person for the purpose of distinguishing or so as to distinguish their goods or services from those of others': Bill C-31, *supra* note 3, s 319.

51. *Tommy Hilfiger Licensing, Inc. v. International Clothiers Inc.*, 2004 FCA 252, [2005] 1 F.C.R. 148. Similarly, if a plaintiff's mark were being used by a defendant to distinguish the *plaintiff's* goods, the defendant would not be using the mark 'as a trademark', and

*ii. Trade names*

The Act defines 'trade name' as 'the name under which any business is carried on, whether or not it is the name of a corporation, partnership or an individual'.[52] While the Act does not permit the registration of a trade name *per se,* if the trade name is also being used 'as a trademark' then it may be registrable. The registrability of a trade name depends entirely on the manner in which it is being used, which in turn is a question of fact. The key consideration is whether the public would perceive use of the trade name merely as a reference to a corporate entity, or whether it would view such use as an indicator of source.[53] In *Sunny Fresh Foods Inc. v. Sunfresh Ltd.*, the Opposition Board found that the use of SUNFRESH did not stand out since the word 'Limited' always appeared immediately following it. The Board concluded that SUNFRESH had been used as a trade name rather than as a trademark.[54]

*iii. Domain names, email addresses, telephone numbers, etc.*

Words or phrases that serve different functions, such as domain names, email addresses, social media handles, telephone numbers, professional designations or titles may not be deemed to function 'as trademarks'. However, if evidence is adduced of acquired distinctiveness through extensive use 'as trademarks', then trademark protection may be available. In *Pizza Pizza v. Canada (Register of Trade Marks)* a telephone number was deemed registrable on the basis that it was highly indicative of Pizza Pizza Limited and it distinguished its products and services from those of others.[55]

*iv. Doctrine of functionality*

Where a trademark is deemed to be functional, it will not be capable of functioning as a trademark. The Supreme Court of Canada upheld the doctrine of functionality in *Kirkbi AG v. Ritvik Holdings Inc.*, confirming that trademark law is not intended

---

infringement would not be found: *see* A. Kelly Gill, *Fox on Canadian Law of Trade-marks and Unfair Competition*, 4th ed (Toronto: Thomson Reuters, 2005) ch 7 at s 7.2(d)(ii) (WLNext Can).

52. Section 2 of the Act.
53. *Road Runner Trailer Manufacturing Ltd. v. Road Runner Trailer Co.* (1984), 1 C.P.R. (3d) 443, 1984 CarswellNat 865 (FCTD). *See also* Gill, *supra* note 50, ch 14 at s 14.4(d).
54. *Sunny Fresh Foods Inc. v. Sunfresh Ltd.* (2003), 30 C.P.R. (4th) 118, 2003 CanLII 67758 [*Sunny Fresh*].
55. *Pizza Pizza Ltd. v. Canada (Registrar of Trade Marks)*, [1989] 3 F.C. 379, 26 C.P.R. (3d) 355 (CA). *See also Star Island Entertainment LLC v. Provent Holdings Ltd.*, 2013 TMOB 84 at para. 22, 112 C.P.R. (4th) 321, where the Trademarks Opposition Board accepted that the display of www.mansioncasino.com qualified as use of the registered mark MANSION and found that 'the addition of "Casino" and ".com" would be viewed as merely descriptive of the type of services offered by Mansion Casino and the online environment through which the [s]ervices are offered [...] [and] would not be perceived as forming part of the trade-mark *per se*'.

to prevent the competitive use of utilitarian features of products.[56] In other words, the courts will not permit trademark law to be used as a means of extending patent or industrial design protection, or to limit the development of any art or industry.[57] However, if the functionality is secondary or peripheral with no essential connection to the goods, it will not bar registration.[58]

*v. Ornamentation*

Where a mark serves a purely decorative or ornamental function it will not be deemed to be used as a trademark.[59] As in the case of functionality, if the ornamentation serves a secondary or incidental function to the primary function of distinguishing source, trademark protection may be available. In *Santana Jeans Ltd. v. Manager Clothing Inc.* the court found that while a cross stitch made of yellow, orange and light blue thread had a decorative effect, its *main purpose* was to distinguish the respondent's jeans.[60] However, in *Adidas (Canada) Ltd. v. Colins Inc.*, the court held that the vertical striping on the sleeves and legs of the applicant's clothing had a 'slenderizing effect' and gave 'an illusion of speed or motion', which made the goods more attractive to consumers. The court concluded that Adidas was attempting to 'convert what is merely ornamental design into a trade mark, which is not permissible'.[61]

### *2. Use in the course of trade*

Use of a mark in relation to goods must be in the normal course of trade.[62] The requirement that use be 'in the normal course of trade' is absent from the definition of use in relation to 'services' and 'export goods'.[63]

56. *Kirkbi, supra* note 2, para. 43. Section 13(2) of the Act articulates the doctrine of functionality in relation to distinguishing guises: 'No registration of a distinguishing guise interferes with the use of any utilitarian feature embodied in the distinguishing guise'. Bill C-31 expands this concept and makes it applicable to all trademarks: 'A trademark is not registrable if, in relation to the goods or services in association with which it is used or proposed to be used, its features are dictated primarily by a utilitarian function': Bill C-31, *supra* note 3, s 326.
57. *Remington Rand Corp. v. Philips Electronics N.V.* (1995), 64 C.P.R. (3d) 467 at 478, 1995 CarswellNat 1846 (FCA); *Kirkbi, supra* note 2 at paras 44–47.
58. *Stork Market Inc v. 1736735 Ontario Inc.*, 2017 FC 779, 149 C.P.R. (4th) 287.
59. *W.J. Hughes & Sons 'Corn Flower' Ltd. v. Morawiec* (1970), 62 C.P.R. 21, 1970 CarswellNat 19 (Ex CR); *Dot Plastics Ltd. v. Gravenhurst Plastics Ltd.* (1988), 22 C.P.R. (3d) 228, 1988 CarswellNat 1430 (TMOB).
60. *Santana Jeans Ltd. v. Manager Clothing Inc.* (1993), 52 C.P.R. (3d) 472, 1993 CarswellNat 388 (FCTD).
61. *Adidas (Canada) Ltd. v. Colins Inc.* (1978), 38 C.P.R. (2d) 145 at 169, 1978 CarswellNat 690 (FCTD).
62. Section 4(1) of the Act.
63. Section 4(2) of the Act. *See Molson Companies Ltd. v. Moosehead Breweries Ltd.* (1990), 32 C.P.R. (3d) 363, 1990 CarswellNat 768 (FCTD); *Shapiro, Cohen, Andrews & Finlayson v. Fireman's Fund Insurance Co.* (1994), 54 C.P.R. (3d) 566, 1994 CarswellNat 3035 (TMOB).

> 'Normal course of trade' is not defined in the Act, but it has been interpreted by the courts as 'requiring that the transfer be a part of a dealing in the wares for the purpose of acquiring goodwill and profits from the marked goods'.[64] The goods do not need to reach the ultimate consumer to establish that a transfer of goods took place in the normal course of trade; a transfer to an intermediary anywhere along the supply chain is sufficient.[65]

Use of a mark on goods that are given away free of charge as promotional items may not qualify as trademark use within the meaning of the Act since there is no element of 'trade';[66] nor would the distribution of product samples for test purposes, or as part of a marketing campaign.[67] Beta testing and clinical trials may qualify as trademark use if this is a common practice within the industry, and an association is made between the trademark and the goods.[68]

Use of a mark in relation to charitable and non-profit services qualify as 'trademark use', provided a public benefit is derived from the performance of the services.[69]

64. *Lin Trading Co. v. CBM Kabushiki Kaisha* (1985), 5 C.P.R. (3d) 27 at 32, 1985 CarswellNat 1110 (TMOB); *Lapointe Rosenstein v. Revlon (Suisse) S.A.* (2002), 26 C.P.R. (4th) 120; 2002 CarswellNat 4658 (TMOB).
65. *Manhattan Industries Inc. v. Princeton Manufacturing Ltd.* (1971), 4 C.P.R. (2d) 6, 1971 CarswellNat 513 (FCTD); *CBM Kabushiki Kaisha v. Lin Trading Co.* (1988), [1989] 1 F.C. 620, 21 C.P.R. (3d) 417 (CA); *Novopharm Ltd. v. Monsanto Canada Inc.* (1998), 80 C.P.R. (3d) 287, 1998 CarswellNat 3214 (TMOB).
66. The concept of 'trade' requires an exchange of consideration or payment for the goods: *Riches, McKenzie & Herbert LLP v. Cosmetic Warriors Limited*, 2018 FC 63, 2018 CarswellNat 131. In this case, Riches McKenzie initiated cancellation proceedings requiring Cosmetic Warriors to evidence use of its mark in association with *t-shirts*. Cosmetic Warriors filed evidence showing that its employees wore the t-shirts as part of their uniform and occasionally purchased the t-shirts. Riches McKenzie argued that this was not use 'in the normal course of trade'. The court found that the free distribution of promotional goods does not qualify as 'use in the ordinary course of trade'. The sale of a small number of t-shirts to employees did not amount to 'use' since they were sold at cost rather than for profit.
67. In some instances, test marketing may be common in the industry and part of the trademark owner's normal business practices: *BFE Inc. v. Broadcast Designs Ltd.* (1993), 48 C.P.R. (3d) 240, 1993 CarswellNat 2491 (TMOB); *Thomas J. Lipton Inc. v. Dial Corp.* (1995), 61 C.P.R. (3d) 123, 1995 CarswellNat 2865 (TMOB); *Fetherstonhaugh & Co. v. ConAgra Inc.* (2002), 23 C.P.R. (4th) 49, 2002 CarswellNat 3549 (FCTD).
68. *See Canadian Olympic Assn. v. Pioneer Kabushiki Kaisha* (1992), 42 C.P.R. (3d) 470, 1992 CarswellNat 1476, where the Canadian Trademarks Opposition Board held that samples shipped for marketing and promotional purposes, followed by a pattern of regular sales did qualify as use. Similarly, in *Estee Lauder Cosmetics Ltd. v. Loveless*, 2017 FC 927, 152 C.P.R. (4th) 188, the Federal Court found in the context of a cancellation proceeding, that the shipment of cosmetic samples during the 'relevant period' which led to subsequent orders for goods, constituted use.
69. *Heritage Canada Foundation v. Tempo Two Native Indian Arts & Crafts Ltd.* (1992), 44 C.P.R. (3d) 533, 1992 CarswellNat 1539 (TMOB).

Use in relation to goods that are manufactured in Canada but destined for export is deemed to be use under the Act, even though the transfer of property in or possession of the goods takes place outside of Canada.[70]

### 3. *Use by the owner or with its consent*

A licence may be differentiated from a 'consent' to use. A consent is an acknowledgement by the party giving the consent that use of a trademark by the party receiving the consent is *not* likely to cause confusion with the former party's trademark.[71] In other words, the parties recognize that concurrent use of their respective marks will not lead consumers to believe there is a connection between the parties, or that the goods or services emanate from the same source.[72] A consent given by one party to another is not deemed to be binding on the Canadian Trademarks Office, and an examiner may find confusion regardless of whether the parties believe that concurrent use is possible.

By contrast, licensing permits the use of another person's mark which, if the licence were not in place, would otherwise create confusion as to source and constitute infringement. Unlicensed use of a mark by a third party could lead to loss of trademark rights for its owner. The owner of a licensed mark must retain direct or indirect control over the character or quality of the licensed goods or services, otherwise any use of the mark by the licensee may not inure to its benefit.[73]

## B. Volume, duration and frequency of use

There are no established pre-requisites as to the volume, duration and frequency of use as this is industry-dependent, and certainly contextual. Nominal sales of products may be construed as being outside of 'the normal course of trade', or possibly fabricated. In highly specialized markets involving the sale of expensive equipment, a single sale may qualify as being 'in the normal course of trade'.[74] Interruptions in use may be deemed to be within the normal course of trade in seasonal industries or in industries

70. Section 4(3) of the Act.
71. Letters of consent are not determinative of the issue of whether two marks are confusing in Canada. The Canadian Intellectual Property Office generally gives limited weight to consents on the basis that 'the Registrar represents the public interest first [...]': *Haw Par Brothers International Ltd. v. Canada (Registrar of Trade Marks)* (1979), 48 C.P.R. (2d) 65 at 68, 1979 CarswellNat 790 (FCTD).
72. Burshtein, *Trade-mark Use, supra* note 6 at 161.
73. Some indicia of control would include the licensor's ability to inspect the licensee's premises and to approve product samples. Where public notice is given of the identity of the trademark owner and the fact that the use of the trademark is licensed, the Act creates a rebuttable presumption that the licensor controls the character and quality of the goods: Section 50(2) of the Act.
74. *Gowling Lafleur Henderson v. Supertex Industrial S.A. de C.V.* (2002), 26 C.P.R. (4th) 251, 2002 CarswellNat 4832 (TMOB); *Miller Thomson LLP v. Terra Equipment Ltd.* (2007), 64 C.P.R. (4th) 53, 2007 CarswellNat 4071 (TMOB); *J.C. Penny Co. v. Gaberdine Clothing Co.* [2001] F.C.J. No. 1845, 16 C.P.R. (4th) 151 (F.C.T.D.)

where products are made by custom order.[75] In other industries, anything short of continuous use of a mark may fall outside the normal course of trade.

### C. Use of a sign

If a trademark is not used in the form in which it was registered, the registration may be cancelled on the basis of deviation if the identity or dominant features of the mark are lost. The Federal Court of Appeal has pointed out that 'with every variation the owner of the trade mark is playing with fire'.[76] While the law permits 'cautious variations' provided the 'same dominant features are maintained, and the differences are so unimportant as not to mislead an unaware purchaser', any changes to the overall commercial impression of the mark could lead to a finding of abandonment or non-use.[77]

### D. Use in relation to goods and services

#### *1. Use for each of the designated goods and services*

Although a trademark application need only include the date of first use for each *general class* of goods and services, the courts have held that the mark must actually have been used in Canada in association with *each of the specific goods and services* identified in the application as of the filing date.[78] This requirement will, of course, be eliminated once Bill C-31 comes into force in 2019.

In the context of Section 45 cancellation proceedings, the Act states that use of the mark must be shown in respect of '*each of the goods or services* specified in the registration', otherwise the registration risks being cancelled or amended.[79] Bill C-31 makes a slight change to the wording so that the new provision will require evidence of use in respect of '*all* the goods or services specified in the registration [...]'.[80]

#### *2. Use for a broader category of goods and/or services*

Specific goods can sometimes be legitimate representatives of a broader grouping or category of goods claimed in the registration. If the registered owner submits an affidavit clearly stating that the mark is used on all goods claimed in the registration

75. *Ivy Lea Shirt Co. v. Muskoka Fine Watercraft & Supply Co.* (2001), 11 C.P.R. (4th) 489, 2001 CarswellNat 5510 (FCTD).
76. *Promafil Canada Ltée v. Munsingwear Inc.* (1992), 44 C.P.R. (3d) 59 at 71, 1992 CarswellNat 181.
77. *Ibid.*
78. *Les Marques Metro/Metro Brands S.E.N.C. v. 1161396 Ontario Inc.*, 2017 FC 806, 2017 CarswellNat 4413 [*Les Marques*].
79. Section 45(1) of the Act.
80. Bill C-31, *supra* note 3, s 349.

and submits representative samples of such use from each category, exhibits may not be necessary for *each* good.[81]

## E. Territorial scope

In the context of filing an application or protecting a registration from cancellation for non-use, the location within Canada where the mark is used is irrelevant: use *anywhere* in Canada is sufficient to base an application on prior use and to evidence use of a registered mark.

Where proof of acquired distinctiveness is required in order to qualify for registration, the Registrar may restrict the registration to the defined territorial area in which the trademark is shown to have become distinctive.[82] Currently, this is the case for marks that consist of distinguishing guises, a name or surname, or words that are clearly descriptive or deceptively misdescriptive, as outlined in Section 13 and Sections 12(1)(a) and (b) of the Act. Under Bill C-31, evidence of acquired distinctiveness will be required for marks consisting of a single colour or combination of colours without delineated contours, or marks consisting of a three-dimensional shape, a mode of packaging goods, a sound, a scent, a taste, a texture, or any other prescribed sign, which means that any of these types of marks may be subject to a territorial restriction if the evidence does not show that the mark has acquired distinctiveness throughout Canada.[83]

## F. Relevant time period and its calculation

If there has been an extended period of non-use following the date of first use in Canada, it is prudent to claim the earliest date from which *continuous* use can be established in the application.[84] Once Bill C-31 comes into force in 2019 a period of non-use prior to filing will no longer be relevant since the date of first use in Canada will not need to be identified in the application.

In the context of cancellation proceedings under Section 45, the relevant period is the three-year period immediately preceding the date of the notice issued by the Registrar. A registrant does not need to show continuous use throughout the entire three-year period but does have to demonstrate that the mark was used in relation to each good and service claimed in the registration during the three-year period.[85]

81. *Saks & Co. v. Canada (Registrar of Trade Marks)* (1989), 24 C.P.R. (3d) 49, 1989 CarswellNat 577 (FCTD) [*Saks*]; *Gowling Lafleur Henderson LLP v. Neutrogena Corp.* (2009), 74 C.P.R. (4th) 153, 2009 CarswellNat 1721 (TMOB); *See also* Cynthia Rowden, *Section 45 Proceedings*, Bereskin & Parr LLP (14 September 2009), online: < http://www.bereskinparr.com/files/file/docs/SECTION45PAPER_0909CR.pdf > at 18.
82. Section 32(2) of the Act.
83. Bill C-31, *supra* note 3, s 339.
84. *Corporativo De Marcas GJB, SA DE CV v. Bacardi & Company Ltd.*, 2014 FC 323, 122 C.P.R. (4th) 389. In this case the court held that a 3½ year gap in use meant non-compliance with Section 30(b) and formed the basis of a successful opposition.
85. *See Carter-Wallace Inc. v. Wampole Canada Inc.* (2000), 8 C.P.R. (4th) 30 at 39, 2000 CarswellNat 1407, where the Federal Court held that the registrant need only provide evidence of use 'at some point within the previous three years'.

Further discussion on cancellation and expungement proceedings is found below at sections IV and V.

## III. EVIDENCE OF GENUINE USE OF A TRADEMARK

### A. When use should be proved: timing of lack of genuine use claims

#### *1. When filing an application for a trademark*

Until Bill C-31 comes into force in 2019, an applicant who wishes to register a trademark that has not yet been used in Canada must include a statement that it intends to use the mark in Canada.[86] Prior to registration of the mark, the applicant must make a declaration that use of the mark has commenced in Canada in association with all of the goods and services claimed in the application. There is no formality associated with such declaration – it need not be legalized, nor are specimens required.[87] However, if a false declaration of use is made, the resulting registration could be declared void *ab initio*, without the need to demonstrate an intent to deceive or fraud.[88]

If the mark has been previously used in Canada, the Act requires that the applicant identify the date from which it, or its predecessors-in-title, have used the mark in Canada in relation to each general class of goods and services claimed in the application.[89] Specimens of use are not required. An application may be opposed on the basis that the applicant did not use the mark from the date of first use claimed in the application, and a registration based on an incorrect date of first use may be invalidated.[90]

If the mark has been 'made known' in Canada, or if the application is based on foreign use and registration of the mark, the applicant must identify the name of the country of the Union in which the mark has been used (see footnote 22 for definition of 'Union'). Neither the date of first use outside of Canada, nor specimens of use, are required.[91]

---

86. Section 30(e) of the Act. The applicant may intend to use the mark by itself and/or through a licensee. An application claiming an intention to use the mark may not be amended to claim prior use of the mark in Canada (Reg 31(d)), and the application may be successfully opposed on the basis that the mark was actually in use in Canada when the application was filed. *See Tone-Craft Paints Ltd. v. Du-Chem Paint Co. Ltd.* (1969), 62 C.P.R. 283, 1969 CarswellNat 426 (Registrar of Trade Marks).
87. In fact, the Office no longer accepts a physical copy of the declaration and requires instead that a box affirming that use has commenced in Canada be checked off on the Canadian Intellectual Property Office's online portal.
88. *Marchands Ro-Na Inc. v. Tefal S.A.* (1981), 55 C.P.R. (2d) 27, 1981 CarswellNat 3 (FCTD); *Unitel Communications Inc. v. Bell Canada* (1995), 61 C.P.R. (3d) 12, 1995 CarswellNat 1004 (FCTD).
89. Section 30(b) of the Act.
90. *Les Marques*, *supra* note 78.
91. Sections 30(c) and (d) of the Act.

Once Bill C-31 comes into force, no use information will be required for the purposes of filing an application; instead, it will be presumed that trademark applicants have either used or intend to use the mark in Canada.

### 2. *In the course of opposition proceedings*

During the course of opposition proceedings, both the applicant and the opponent have the opportunity, though not the obligation, to file evidence of use.[92] If either the applicant or the opponent decides not to file evidence, that party must advise the Opposition Board, otherwise the application will be deemed to be abandoned, or the opposition withdrawn, as the case may be.[93] The likelihood of success decreases substantially in cases where no evidence is filed.

The Registrar only accepts evidence in the form of an affidavit or statutory declaration. Documents or photographs must be attached as exhibits, and oral testimony is not permitted at the hearing stage of the opposition proceedings.[94]

Currently, there are four grounds of opposition available under Canadian law: (a) that the application does not conform to the application filing requirements specified by the Act, (b) that the trademark is not registrable, (c) that the applicant is not the person entitled to registration of the trademark, and (d) that the mark is not distinctive.[95] Bill C-31 introduces two new grounds of opposition: that at the filing date the applicant *was not using and did not propose to use* the trademark in Canada, and that at the filing date the applicant was not *entitled to use* the trademark in Canada in association with those goods or services.[96]

The opponent must adduce evidence to support each ground of opposition or the ground will be dismissed. Conversely, the applicant must adduce evidence to rebut each ground or its application may be rejected. The nature of the evidence depends on the specific ground relied on. It is not sufficient merely to establish the length of time in which a trademark has been in use; it is also necessary to prove that the mark is being used 'as a trademark' for the purposes of 'distinguishing', to show when and how the mark is displayed on the goods or associated with the services (i.e. 'in the normal course of trade', 'at time of transfer of property in or possession of the goods', 'in the performance of the services'), where the use has taken place (i.e. 'in Canada' or elsewhere), who is using the mark (the owner or an authorized licensee), and what mark is actually being used (i.e. the mark at issue or a variant of the mark).

92. Section 38(7) of the Act.
93. Sections 38(7.1) and (7.2) of the Act. *See also* Canadian Intellectual Property Office, *Practice in Trademark Opposition Proceedings*, Practice Notice (Ottawa: CIPO, 31 Mar 2009). Available at: https://www.ic.gc.ca/eic/site/cipointernet-internetopic.nsf/eng/wr01558.html.
94. CIPO, *supra* note 93.
95. Section 38(2) of the Act.
96. Bill C-31, *supra* note 3, s 343.

### 3. *In cancellation proceedings based on lack of use*

'Any person' may initiate cancellation proceedings against a registered mark following the third anniversary from the date of registration.[97] The Registrar will issue a notice to the registrant requiring evidence of use of the mark in Canada during the three-year period immediately preceding the date of the notice (the 'relevant period').[98] Section 45 cancellation proceedings are considered before the Trademarks Opposition Board.[99]

The registered owner must file evidence of use in relation to *each of the goods and services* claimed in the registration, otherwise the registration may be expunged or amended. If the mark is not in use, the registered owner must indicate the date when the mark was last used and the reason for the absence of use.[100]

Cancellation proceedings are meant to be summary in nature, therefore evidentiary overkill is not required.

### 4. *As a defence in infringement/cancellation proceedings*

Invalidity through abandonment may be claimed as a defence in infringement proceedings. Abandonment may be asserted by counter-claim in an action for infringement. However, Section 45 cancellation proceedings must be commenced separately before the Registrar.

The Registrar has no inherent jurisdiction to stay Section 45 cancellation proceedings or to postpone his or her decision pending the outcome of other proceedings, including opposition proceedings or infringement proceedings. Moreover, Section 45 proceedings do not prevent a party from commencing expungement proceedings under Section 18(1) at a later date.[101]

Invalidity is a full defence to an action for infringement or depreciation of goodwill. Similarly, if a registration is cancelled for non-use, its owner no longer has access to a claim of infringement or depreciation of goodwill under the Act. Not surprisingly, defendants in trademark infringement and depreciation of goodwill proceedings often resort to challenging the validity of the plaintiff's trademark registration as a defence tactic.

97. *See* Section 45(1) of the Act and Canadian Intellectual Property Office, *Practice in Section 45 Proceedings*, Practice Notice (Ottawa: CIPO, 14 Sep 2009). Available at: https://www.ic.gc.ca/eic/site/cipointernet-internetopic.nsf/eng/wr01944.html.
98. The registered owner has a period of three months from the date of the notice within which to file its evidence of use and may request a single four-month extension of time to respond. Further extensions are normally not granted as Section 45 cancellation proceedings are intended to be summary in nature. Bill C-31 amends Section 45 to permit the notice to refer to all or a portion of the goods and services contained in the registration.
99. A decision of the Trademarks Opposition Board may be appealed before the Federal Court. *See* Sections 56(1), (2), and (3) of the Act. An appeal must be filed within two months from the date on which notice of the decision was sent by the Registrar. The registered owner may, on appeal, adduce further evidence to support use of its trademark: Section 56(5) of the Act.
100. Section 45(1) of the Act.
101. Gill, *supra* note 51, ch 11 at s 11.3(e).

## B. Burden of proof

An opponent must raise sufficient evidence pertaining to any of the grounds of opposition claimed to substantiate that ground. The evidentiary burden then shifts to the applicant to prove that the ground should be dismissed. In some cases, where the facts are more readily available to the applicant, the opponent's evidentiary burden may be lower.[102]

In cancellation proceedings, the burden of proof falls on the registered owner to show that its mark has been used in relation to *each* of the goods and services claimed in the registration during the relevant period.

In an invalidity proceeding, the burden of proof falls on the party alleging abandonment not only to show that the registered owner is no longer using the mark, but also that it had an *intention* to abandon the mark.

## C. Relevant evidence

### *1. Use on the product or its packaging*

In order to establish use in relation to goods it is necessary to evidence the display of the mark on the products themselves or their packaging. If the nature of the product (e.g. an optical lens) does not permit marking on the goods themselves, there are other ways in which the mark may be associated with the goods, for example, on invoices accompanying the products, on sealing tape applied to the packaging,[103] or on labels or tags affixed to the goods.[104]

### *2. Use in advertising*

Evidence of the display of a mark in advertising will establish use of a mark in relation to services; however, it will not support a finding of use in relation to goods. One exception may be point-of-sale advertising, or 'shelf talkers', provided it can be shown that such advertising is in such close proximity to the goods that an association between the mark and the goods is made by consumers.[105]

---

102. *Seven-up Canada Co. v. Caribbean Ice Cream Company Ltd.*, 2007 CanLII 80903, 2007 CarswellNat 1361 (TMOB); *Brasstech, Inc. v. Elte Carpets Ltd.*, 2014 TMOB 92, 125 C.P.R. (4th) 41.
103. *Gowling, Strathy & Henderson v. Cascades Dominion Inc.*, 1997 CanLII 15708 (TMOB).
104. *Gowling Strathy & Henderson v. The Custom Shop, Inc*, 1997 CanLII 15710 (TMOB).
105. *General Mills Canada Ltd. v. Procter & Gamble Inc.* (1985), 6 C.P.R. (3d) 551, 1985 CarswellNat 1132. In this case the Trademarks Opposition Board found that use of the mark BAKE SOMEONE HAPPY on coupons close to the goods did constitute use in relation to cake mixes since the consumer would have to go to the shelf to find the cake mix, as well as the coupon bearing the mark, and then would have to take both items to the cashier to obtain a discount. However, in *London Drugs Ltd. v. Brooks* (1997), 81 C.P.R. (3d) 540, 1997 CarswellNat 2894, the Opposition Board found that use of the mark on signs, on bags and in newspaper advertisements was insufficient to create an association with the goods in question. There is no definition for 'services' in the Act, however, the courts have held that the term should be given a broad interpretation

### 3. *Use on the internet*

Online use of a mark does not qualify as use of a trademark in relation to goods since the statutory definition of 'use' in relation to goods does not contemplate advertising. By contrast, advertising does constitute use in relation to services and therefore online use may be sufficient to establish use of a service mark in Canada.

Use of a mark on the internet poses several challenges, not the least of which is whether use of the mark can be said to have occurred *in Canada*. Traditionally, Canadian courts have held that advertising of services in Canada did not qualify as 'use' unless the services were also *performed* in Canada.[106] In recent years, however, there has been a move away from this position.

In *HomeAway.com, Inc. v. Hrdlicka*, the Federal Court found that 'a trade-mark which appears on a computer screen website in Canada, regardless [of] where the information may have originated from or be stored, constitutes [...] use and advertising in Canada.'[107] This position was scaled back in *Unicast SA v. South Asian Broadcasting* where the Federal Court clarified that in order for the display of a mark on the internet to constitute 'use' in relation to services, the associated services must be offered to Canadians or performed in Canada.[108] A party based in another country may successfully establish use of a trademark via the internet, provided that it can show that Canadians use the services in question: therefore, 'the current barrier relates to the evidentiary burden that parties must satisfy, not a legal barrier for parties that provide their services in this manner.'[109]

---

such that as long as some members of the public receive a benefit from the activity, it should qualify as a service: *Société Nationale des Chemins de Fer Français SNCF v. Venice Simplon-Orient-Express, Inc.* (1995), 64 C.P.R. (3d) 87, 1995 CarswellNat 2133 (TMOB) [*SNCF v. Venice Simplon-Orient-Express*]. Activities that do not provide a public benefit do not qualify as a service. Therefore, the promotion or advertising of one's own goods and services are not considered to be a 'service' as these activities are merely for the benefit of the trademark owner: *Carling O'Keefe Breweries of Canada Ltd. v. Anheuser-Busch Inc.* (1985), 4 C.P.R. (3d) 216, 1985 CarswellNat 1095 (TMOB); *Ralston Purina Co. v. Effem Foods Ltd.* (1997), 81 C.P.R. (3d) 528, 1997 CarswellNat 2999 (TMOB).

106. *Porter v. Don the Beachcomber* (1966), 48 C.P.R. 280 (Ex. Ct.). In this case the court held that the use of a trademark in advertising, coupled with the performance of the service outside of Canada, did not constitute use of the mark: rather, to qualify as 'use' the service had to be *performed* in Canada. *See also, Motel 6 Inc. v. No. 6 Motel Ltd.* (1981), [1982] 1 F.C. 638, 56 C.P.R. (2d) 44 (TD). More recently in *Supershuttle International, Inc. v. Fetherstonhaugh & Co.*, 2015 FC 1259, 2015 CarswellNat 10798, the Federal Court found that although the registrant had advertised its services in Canada, it had not used its mark SUPERSHUTTLE in connection with 'airport passenger ground transportation services' since the registrant did not operate shuttles in Canada. Allowing Canadians to reserve ground transportation services at airports outside of Canada was insufficient to establish use as the registration did not claim 'reservation services'.
107. *HomeAway.com, Inc. v. Hrdlicka*, 2012 FC 1467, 2012 CarswellNat 5339 [*HomeAway*], at para. 22.
108. *Unicast SA v. South Asian Broadcasting Corporation Inc.*, 2014 FC 295. For a further discussion of online use of trademarks, see David Bowden & Junyi Chen, *Canadian Trademark Law and 'Use' in the Computer and Internet Age*, 33 CIPR 49 (2017).
109. Bowden & Chen, *supra*, note 108, at 60.

Canadian courts have adopted a flexible approach as to what constitutes a 'service' to support a finding of trademark use when a mark is used over the internet. As noted by Bowden and Chen in 'Canadian Trademark Law and "Use" in the Computer and Internet Age':

> 'While use is fundamental to Canadian trademark law, the requirements for use under the *Trade-marks Act* are sometimes at odds with 21st-century commercial practices, particularly in the field of computer technology. Canadian courts and the Trade-marks Opposition Board have shown, to an extent, a willingness to interpret the provisions of the Act in a flexible manner that does not unfairly prejudice the rights of businesses operating in these new commercial spaces.'[110]

The computer software industry is a prime example, given that software is now increasingly delivered over the internet. In *Specialty Software Inc. v. Bewatec Kommunikationstechnik GMBH*, the court held that use in relation to the registrant's *computer software* occurred when the registrant granted access to the software with log in credentials. The registration was maintained even though the registrant was no longer selling the software on disks (as implied by claiming software as a 'good' in the registration) but rather was offering the software as a service.[111]

Careful crafting of the description of the services is necessary when the services will be provided by foreign companies with only an online presence in Canada. In *Saks & Co. v. Canada (Registrar of Trade Marks)*, the Federal Court found that Saks Fifth Avenue was indeed using the mark in Canada in relation to retail store services despite the fact that it did not have any stores located in Canada, by adopting a broad interpretation of what constitutes 'retail department store services'.[112] Yet, while

---

110. *Ibid.*
111. *Specialty Software Inc. v. Bewatec Kommunikationstechnik GMBH*, 2016 FC 223, 2016 CarswellNat 579.
112. *Saks, supra* note 81. The court held at paras 60–61:

> '[T]here is evidence that the company responds to mail and telephone orders from Canada for merchandise, grants credit to Canadian customers, advertises wares in Canada by sending some catalogues to chosen Canadian customers, includes in every statement of account to its Canadian credit customers and to other Canadian customers upon request, monthly focus advertising inserts, provides a toll-free (800) number for purchasing wares by telephone [...] as well as the issuing of gift certificates to Canadian customers. These services are performed without the Canadian customer ever having to leave Canada and the SAKS FIFTH AVENUE mark is used in association with all of these services by marking it on all documents and materials issued by the company relating to same.'

*See also McCarthy Tetrault v. AutoZone Parts Inc.*, 2011 TMOB 73, 93 C.P.R. (4th) 322; *TSA Stores, Inc. v. Registrar of Trademarks*, 2011 FC 273, where the court held that THE SPORTS AUTHORITY mark was used in Canada in association with the operation of a retail sporting goods store even though there were no brick-and-mortal retail locations in Canada, since the registrant's website provided ancillary services to Canadians via its website; *SNCF v. Venice Simplon-Orient-Express, supra*, note 105, where the court held that ORIENT-EXPRESS was in use in Canada in association with 'travel services, namely railway passenger service' (even though no ORIENT-EXPRESS trains operated in Canada) on the basis that the 'ancillary' service of 'train reservation and ticketing services' were indeed provided in Canada.

these services were deemed to be provided in Canada in the *Saks* case, the opposite conclusion was reached when the description of services was 'operating a retail store' even though the client was selling and shipping goods to Canadian consumers.[113] Similar findings have been made when the services have been described as 'hotel reservation services', in which case the services were deemed to have been performed in Canada, vs. 'hotel accommodation services' where the lack of a physical hotel in Canada led to the expungement of the mark for non-use.[114]

#### 4. *Form of evidence*

In the context of opposition and cancellation proceedings, evidence must be submitted by way of affidavit or statutory declaration. It is not sufficient to make mere bald assertions of use.[115] Rather, a statement that the mark is in use must be backed by photographs showing the display of the mark on the products or their packaging, labels or hang tags depicting the mark, purchase orders or invoices confirming sales of the products, aggregate Canadian sales figures, sample advertisements, aggregate advertising expenditures, lists of distributors and/or retail outlets, and so forth. If print, television or radio advertisements are adduced as evidence, the parties also need to include circulation data or viewership figures.

#### 5. *Distinction between use of a company name and use of the services offered by the company*

The presumption that a company name is a trade name rather than a trademark is rebuttable. The key consideration is whether the company name stands out from other identifying information, such as address information, and whether the public would perceive use of the trade name as trademark use for the purpose of identifying the source of the goods and services.[116]

## IV. JUSTIFICATION FOR NON-USE

### A. Obstacles to use

The need to justify non-use arises in the context of summary cancellation and invalidation proceedings. As a general rule, absence of use of a mark is penalized

113. *C.R.A.C. Centre de Recherche et d'Analyses sure les Corp. v. Imco Trading Co.* (1993), 52 C.P.R. (3d) 122, 1993 CarswellNat 2583 (TMOB).
114. *Miller Thomson LLP v. Hilton Worldwide Holding LLP*, 2017 TMOB 19, 2017 CarswellNat 386; *Stikeman Elliott LLP v. Millennium & Copthorne International Ltd.*, 2017 TMOB 34, 2017 CarswellNat 1854.
115. *Aerosol Fillers Inc. v. Plough (Canada) Ltd.* (1980), [1981] 1 F.C. 679, 53 C.P.R. (2d) 62 (CA); *Guido Berlucchi & C. S.r.l. v. Brouilette Kosie Prince*, 2007 FC 245, 56 C.P.R. (4th) 401.
116. *Sunny Fresh*, *supra* note 54.

by expungement.[117] An exception may be made only if the trademark owner can show special circumstances to justify non-use of its mark, as well as an intention to resume use of the mark.

## B. Circumstances in which non-use may be cured

The Registrar will consider the following factors in determining whether the non-use is excusable in the context of cancellation proceedings:

(1) The length of time during which the trademark has not been used;
(2) Whether the reasons for the absence of use were due to circumstances beyond the control of the owner; and
(3) Whether there exists a serious intention to resume use of the trademark.

Canadian courts have excused non-use in the following circumstances: mistake, delay in manufacture and delivery of goods, a misunderstanding of regulatory requirements, damage to place of business, cyclical nature of an industry, serious illness, loss of a distributor, and labour disputes, to name a few.[118]

The Federal Court of Appeal has noted that it is important to distinguish between explaining an absence of use and excusing an absence of use: '[w]hether or not a trade-mark is expunged is a function of the explanation offered for the non-use (the special circumstances) and of the characteristics of the period of non-use'.[119] In other words, the analysis should not only be focused on factors that are beyond the control of the trademark owner, but also on the deliberate decisions of the owner.[120] The Trial Division articulated this distinction in *Lander Co. Canada Ltd. v. Alex E. Macrae & Co.* when considering absence of use due to an economic downturn:

> 'It is obvious that the recession is beyond anyone's control, nevertheless, I fail to see how it can be said that a recession is an uncommon or exceptional situation; it is common knowledge that market conditions fluctuate. Finally, the decision to cease use of a trade mark was entirely voluntary on the part of the appellant.'[121]

Therefore, in order to justify non-use, the special circumstances must affect the trademark owner specifically, rather than all traders, and must not be the result of a voluntary decision of the owner.

117. *Registrar of Trade Marks v. Harris Knitting Mills Ltd.* (1985), 4 C.P.R. (3d) 488, 1985 CarswellNat 534 (FCA).
118. Gill, *supra* note 51, ch 11 at s 11.2(iii).
119. *Scott Paper Ltd. v. Smart & Biggar*, 2008 FCA 129 at para. 31, 65 C.P.R. (4th) 303.
120. Gill, *supra* note 51, ch 11 at s 11.2 (iii).
121. *Lander Co. Canada Ltd. v. Alex E. Macrae & Co.* (1993), 46 C.P.R. (3d) 417 at 421, 1993 CarswellNat 351 (FCTD); *See also Twin River Law LLP v. Continental Capital Investments Inc./Continental Investissements Capital Inc.*, 2017 TMOB 13, 144 C.P.R. (4th) 183.

## V. CONSEQUENCES OF NON-USE

### A. Consequences of non-use and relevance of the context

Trademarks that are no longer in use may be expunged from the Trademarks Register by way of invalidation proceedings or non-use cancellation proceedings.

#### *1. Invalidation proceedings*

Either the Registrar or any person interested may apply to the Federal Court of Canada to invalidate a trademark registration on the basis of abandonment.[122] 'Person interested' is defined as including 'any person who is affected or reasonably apprehends that he may be affected by any entry in the register, or by any act or omission or contemplated act or omission under [the Act] [...]'.[123]

The intention to abandon is a critical evidentiary component in invalidation proceedings, and the burden of proof lies with the person alleging abandonment. Often this can only be shown with circumstantial evidence, such as an extended period of non-use, coupled with widespread infringement, the adoption of a new mark in respect of the same goods, or the discontinuance of the product line.[124]

Although non-use of a mark may be evidence of abandonment, it is not determinative. The registered owner may have valid reasons to excuse the lack of use, which would negate a finding of abandonment, such as where the owner has temporarily parked the mark for future use.

Once a finding of abandonment is made, Section 57 of the Act permits the Federal Court to strike out or amend the registration; however, the court does not have the jurisdiction to assign the mark to the party who commenced the application, or to substitute a new owner for the registered owner.[125]

---

122. Section 18(1) of the Act provides: The registration of a trade-mark is invalid if (a) the trade-mark was not registrable at the date of registration, (b) the trade-mark is not distinctive at the time proceedings bringing the validity of the registration into question are commenced, (c) *the trade-mark has been abandoned*, or (d) subject to section 17, the applicant for registration was not the person entitled to secure the registration. Section 57(1) of the Act gives the Federal Court the exclusive jurisdiction, on the application of the Registrar or of 'any person interested', to order that any entry in the Register be struck out or amended on the ground that the entry does not accurately express the existing rights of the registered owner of the mark. Section 58 of the Act provides that an application for expungement under Section 57(1) must be made either: (a) by filing an originating notice of motion, (b) by way of counter-claim in an action for the infringement of the trademark, or (c) by a statement of claim in an action claiming additional relief under the Act.
123. Section 2 of the Act.
124. David Vaver, *Summary Expungement of Registered Trade Marks on the Ground of Non-Use*, 21:1 Osgoode Hall LJ 17 at 19 (1983).
125. Gill, *supra* note 51, ch 11 at s 11.3(a).

### 2. *Cancellation proceedings based on non-use under Section 45 of the Act*

A mark that is no longer in use may also be expunged by way of summary cancellation proceedings under Section 45 of the Act. These proceedings are intended to be simple and quick; they are not intended to replace *inter partes* expungement proceedings under Section 57 of the Act, where issues such as ownership, distinctiveness or abandonment may be raised.[126]

Section 45 cancellation proceedings may be initiated by any third party, and indeed, law firms often act as the requesting party on behalf of their clients. The Registrar may determine that there are valid reasons for *not* issuing a notice,[127] such as where: (1) the trademark registration is already the subject of a Section 45 proceeding pending before the Registrar or on appeal before the Federal Court of Canada; (2) the request is within three years of the date of issuance of a previous Section 45; (3) the Registrar considers that the request is frivolous or vexatious.

If the registered owner fails to respond to the notice, the mark will automatically be expunged.[128] The burden of proof rests with the registered owner and is relatively light. There is no requirement to show extensive use of the mark, provided that the registered owner can show *some* use. If the registered owner furnishes proof of use only in respect of a portion of the goods and services claimed in the registration, the registration will be amended accordingly.

Bill C-31 clarifies that the Registrar may of his or her own accord initiate cancellation proceedings, and it is anticipated that the Trademarks Office will use this discretion upon implementation as a means to patrol marks that are registered but that are no longer (or never were) in use. Moreover, the Registrar will have discretion to issue a notice in respect of all or a portion of the goods and services claimed in the registration.[129]

Both expungement and cancellation proceedings are only available as against registered marks. These proceedings are not available in respect of unregistered marks at common law.[130]

## B. Extent of any revocation

In Section 45 proceedings, a trademark registration may be cancelled for non-use in whole or in part. Section 45(3) specifically gives the Registrar the power to expunge or amend a registration.[131] The Registrar does not have jurisdiction under Section 45

126. CIPO, *Practice in Section 45 Proceedings*, *supra* note 96.
127. *Ibid.*
128. *Ibid.*
129. Bill C-31, *supra* note 3, s 349.
130. Gill, *supra* note 51, ch 11 at s 11.3(a).
131. It is important to note that in Section 45 proceedings use must be demonstrated in respect of *each* of the goods and services claimed in the registration. For example, if the registration claims 't-shirts, socks and pants' and evidence of use is presented only in respect of socks, the registration will be maintained for socks, but will be cancelled in respect of t-shirts and pants, notwithstanding that these goods fall within the same class, or that they can all be categorized as 'clothing'. By contrast, if the goods are broadly described as 'sunflower seeds', evidence of use in respect of 'roasted sunflower seeds' will suffice.

to limit the territorial ambit of the registration, even if the evidence of use is confined to a specific geographic region.

Section 57(1) gives the Federal Court jurisdiction to strike out or amend a registration in the case of abandonment of a mark.

### C. Timing of any revocation

Once a decision to expunge a registration has been made under Section 45, the Trademarks Opposition Board will generally issue a notice of expungement. The registrant has two months from the date of the notice to commence an appeal before the Federal Court Trial Division, and if no appeal is filed, the Registrar will expunge the mark (generally, within three to four months from the date of the notice).[132]

Where an order for expungement is made by the Federal Court, Section 61 of the Act requires that a certified copy of the order be filed with the Registrar. The Registrar will then update the Canadian Intellectual Property Database to reflect the status as 'expunged' (generally, within one month from the date of filing the certified copy).

## CONCLUSION

Notwithstanding the elimination of use as a precondition to registration in 2019, use will continue to be central to the Canadian trademark system. With the acceptance of non-traditional marks, the explosion of the global online marketplace, the proliferation of user-generated content and social-media platforms, the availability of new generic top-level domains, and the emergence of new technologies, such as 3-D printing, the concept of use will be stretched beyond that which the original statutory framework could have contemplated.

Although Parliament chose not to re-examine the definition of use in Bill C-31, we may nonetheless see an evolution of the concept of use in Canada in the coming years.

---

In *Countryside Canners Co. v. Canada (Registrar of Trade Marks)* (1981), 55 C.P.R. (2d) 25, 1981 CarswellNat 776 (FCTD), a registration for 'canned fruit and vegetables' was maintained upon proof of use of the mark in association with 'canned tomatoes'.

132. Sections 56(1) and (5) of the Act. The standard of review before the Federal Court is one of reasonableness, therefore the court will not substitute its own decision unless the Registrar's decision is unreasonable or wrong. The registrant may adduce additional evidence on appeal before the Federal Court. Appeals from the Federal Court are heard as a matter of right by the Federal Court of Appeal.

# China

*Lena Shen*

## I. INTRODUCING GENUINE USE OF TRADEMARKS

### A. Legal framework

In China, genuine use or intention of use is not a pre-condition to register a trademark but a registered trademark is subject to non-use cancellation if it is not used for over three consecutive years after registration without any legitimate reason.

The current Trademark Law of China took effect on 1 May 2014. For non-use cancellation, there is a different prescription between the current Trademark Law and the previous one.

Article 44 of the previous Trademark Law of China[1] prescribed:

> The Trademark Office *shall order rectification* within a specified time period or cancel a registered trademark in any of the following events in connection with the use of that trademark:
> (1) Where the registered trademark is altered without authorization;
> (2) Where the registrant name or address or any other registration item of the registered trademark is altered without authorization;
> (3) Where the registered trademark is assigned without authorization; or
> (4) *Where the use of the registered trademark has ceased for three consecutive years.*

It was amended into the following in 2013 which took effect on 1 May 2014:

> *Article 49 of the Trademark Law of China*[2]
>
> A trademark registrant that changes, without authorization, the registered trademark, the name or address of the registrant or other registration items during the use of the registered trademark shall be ordered to make correction within

1. Took effect in 2001 and was amended in 2013.
2. Current Trademark Law of China took effect on 1 May 2014.

the prescribed time period by the relevant local administration for industry and commerce, and shall have its registered trademark cancelled by the Trademark Office if it fails to make correction by the prescribed deadline.

Where a registered trademark has become the generic name of the goods for which its use is approved or has not been in use for three consecutive years without justification, any entity or individual may apply to the Trademark Office for cancellation of the registered trademark, and the Trademark Office shall make a decision within nine months upon receipt of the application. Where necessary, the time period may be extended for three months under special circumstances upon approval by the administration for industry and commerce of the State Council.

After the amendment, the Trademark Office will no longer cancel a trademark for non-use on its own initiative and it is only when a third party requests for cancellation that the Trademark Office will start the non-use cancellation procedure. In addition, before the amendment, instead of immediate cancellation, the registrant might be given the chance to rectify the situation of non-use but after the amendment, it will no longer have such a chance.

The other relevant previsions in laws and regulations relating to non-use cancellation are:

*Article 54 of the Trademark Law of China*[3]

A party concerned who has objections to the decision made by the Trademark Office on cancellation or non-cancellation of a registered trademark may apply for review to the Trademark Review and Adjudication Board within 15 days upon receipt of the relevant notice. The Trademark Review and Adjudication Board shall make a decision, and notify the party concerned in writing within nine months upon receipt of the application. Where necessary, the time period may be extended for three months under special circumstances upon approval by the administration for industry and commerce of the State Council. The party concerned who has objections to the decision by the Trademark Review and Adjudication Board may bring a lawsuit to the competent people's court within 30 days upon receipt of the relevant notice.

*Rule 65 of the Implementing Regulations of Trademark Law of China*[4]

Where a registered trademark falls under one of the circumstances prescribed by Article 49 of the Trademark Law and becomes the generic name of the products approved to use the registered trademark, any entity or individual may apply to the Trademark Office for cancellation of the registered trademark, and shall produce evidence materials at the time of application. Upon acceptance of such an application, the Trademark Office shall notify the trademark registrant, and require the latter to submit defence within two months upon receipt of the notice. The failure of the trademark registrant to submit defence by the prescribed deadline shall not affect the decision by the Trademark Office.

3. Current Trademark Law of China took effect on 1 May 2014.
4. Current Implementing Regulations of Trademark Law of China took effect on 1 May 2014.

*Rule 66 of the Implementing Regulations of Trademark Law of China*[5]

Where a registered trademark falls under one of the circumstances prescribed by Article 49 of the Trademark Law and has not been in use for three consecutive years without justification, any entity or individual may apply to the Trademark Office for cancellation of the registered trademark, and shall describe relevant situations at the time of application. Upon acceptance of such an application, the Trademark Office shall notify the trademark registrant, and require the latter to furnish evidence materials proving the use of the registered trademark prior to the cancellation application or explain the justifications for non-use within two months upon receipt of the notice. Where the trademark registrant fails to furnish such evidence materials by the prescribed deadline, or where the evidence materials submitted are invalid and the trademark registrant has no justification, the Trademark Office shall cancel its registered trademark. For the purpose of the preceding Paragraph, evidence materials proving the use of a registered trademark shall include the evidence materials on the use of the registered trademark by the trademark registrant itself and the evidence materials on the licensing granted to others by the trademark registrant to use the registered trademark. An application for cancellation of a registered trademark on the ground of non-use for three consecutive years without justification shall be submitted three years after the date of the registration announcement of the registered trademark.

*Rule 67 of the Implementing Regulations of Trademark Law of China*[6]

The following circumstances shall be considered as justifications prescribed by Article 49 of the Trademark Law:

(1) Force majeure events;
(2) Government policy restrictions;
(3) Bankruptcy liquidation; and
(4) Other legitimate reasons not attributable to a trademark registrant.

## B. Purpose and function of genuine use: policy considerations

According to the Trademark Law of China,[7] the trademark registration system is to serve the need of any natural person, legal person or other organization during their production and business operations. A trademark's basic function is to indicate the origin of goods and services. Genuine business use is a necessary condition to facilitate such a function. A long term idle trademark cannot perform its function and there is no need to give it exclusive rights protection.

In addition, to allow a long-term idle trademark to lie dormant in the trademark Register is a waste of limited resource. It will also encourage unhealthy trademark registration behaviour, such as trademark squatting and bad faith litigation.

Therefore, it is necessary to impose a genuine use requirement on a registered trademark. However, the legislation of the non-use cancellation rule is aimed to

5. *Ibid.*
6. *Ibid.*
7. Article 1 and Article 4 of the Trademark Law of China (2014).

activate the trademark resource and to clean up idle trademarks. Cancellation is the means but not the purpose.[8]

## II. WHAT QUALIFIES AS GENUINE USE OF A TRADEMARK

### A. Nature of use

The use of a trademark in the Trademark Law of China refers to the genuine use of trademarks on the designated goods/services for the commercial purpose. In the use, the trademark performs the function of indicating the origin of goods/services.

#### *1. Use consistent with the function of the trademark*

It should function as a trademark instead of a trade name or pure decoration.

#### *2. Use in the course of trade*

The function of a trademark is to indicate the origin of goods/services and such a function can only be performed during trade. Before the goods/services are put into trade, consumers cannot view them. Here the course of trade should be trade happening on the Chinese mainland. Such a requirement makes it arguable whether the use of a trademark in OEM[9] is valid use to maintain a trademark registration.

Those who holds it is not valid use argue that such products are never put into the Chinese market and therefore it is not use of the trademark in sense of the Trademark Law of China, such an opinion being affirmed by the Supreme Court's decision (2014) Min Ti No. 38 relating to a trademark infringement lawsuit concerning the trademark 'PRETUL'. Those who hold that OEM use is valid use to maintain a trademark registration argue that (1) the policy consideration of non-use cancellation procedure is to encourage use and to avoid idleness of a trademark resource, which decides that a lower requirement of use should be imposed than the use requirement in other previsions; and (2) the manufacture of the products happen in China although such products are not sold in the Chinese market. The Court's attitude on this issue is not consistent. In 2016, the Beijing Intellectual Property Court ruled in the *Soda* trademark cancellation case[10] that the use of the trademark in OEM was valid use to maintain the trademark, while Beijing High Court made an opposite ruling in the *Mango* trademark cancellation case[11] which affirmed the ruling of the Beijing First Intermediate Court. In the *Mango* case, the trademark was only used on OEM

8. China Supreme Court's decision (2010) IP Administrative Decision No. 55.
9. OEM here refers to 'Original Equipment Manufacture'. OEM activity refers to making products for others according to others' requirements to repackage and sell. OEM products manufactured in China are all exported and not sold in the Chinese market.
10. (2015) Beijing IP Administrative First Instance decision No. 5119 made on 23 May 2016 by Beijing Intellectual Property Court.
11. (2015) First Intermediate Court (IP) first instance decision No. 1249 made by Beijing First Intermediate Court in 2016; (2016) Beijing Administrative final decision No. 5003 in December 2017.

products. The Beijing High Court held that such OEM products were all exported for distribution in the market out of the Chinese mainland and the trademark therefore did not perform its function to indicate the origin of goods in the course of trade in China as required by the Trademark Law of China. Thus, the OEM use was denied as valid use to maintain a trademark registration in China.

China is not a case law country so the precedent does not have a binding effect on later cases. The different opinions on OEM use still exist. There may still be conflicting decisions in future cases. However, several recent decisions by Beijing High Court shows that the Court tends to decide that OEM use is not valid use in sense of the Trademark Law of China. However, such a tendency might change in the future.

*i. Public use*

Public use means that the goods/services bearing the trademark enter into circulation and are accessible to the relevant public. The relevant public are consumers and potential consumers of the goods/services. Without being put into trade, the trademark cannot be accessed by the relevant public and cannot perform its function to indicate the origin of goods/services. Using a trademark on goods produced but never put into the market cannot be viewed as public use and therefore cannot be taken as valid use. Also, use of a trademark internally, such as on an internally used notebook, cannot be viewed as public use.

*ii. Commercial purpose*

Commercial use refers to the use of the trademark in commercial operation, which will accumulate the goodwill of the trademark and enable consumers to distinguish the origin of goods with this trademark from others.

The following[12] are not commercial use:

(a) Publication of trademark registration information without other;[13]
(b) Pure trademark licence or assignment without other activities;[14]
(c) Trademark enforcement.

*iii. Lawful use*

It is another arguable issue whether the use to maintain a trademark registration should be lawful. If the use violates certain laws and regulations, can such use maintain the trademark registration?

12. Rules and Cases-Litigations Involving the Authorization and Determination of Trademark Rights (商标授权确权诉讼-规则与判例), author: ZHOU Yunchuan, publisher: Law Press, China, 2014.
13. Article 26 of Provisions of the Supreme People's Court on Several Issues Concerning the Adjudication of Administrative Cases on Granting and Affirming Trademark-related Rights (2017).
14. *Ibid.*

In the Supreme Court's decision (2010) IP Administrative Ruling No. 55 made in 2011, the Court held that whether the use activities violated other laws or regulations (import and export or distribution regulations) was not the issue to be considered in the non-use cancellation case and that the trademark should be maintained as long as its use was public and genuine for the commercial purpose.

Nevertheless, there are different opinions regarding this issue, mainly these three:[15]

(1) there is no need to judge whether the activity is lawful use and the trademark should be maintained as long as it is public, genuine and commercial use;
(2) the use should be completely lawful;
(3) the use should be lawful in respect of the trademark-related laws but there is no need to consider whether the activity violate other laws or regulations.

Personally I agree with opinion (3).

### 3. *Use by the owner or with its consent*

According to Rule 66 of the Implementing Regulations of Trademark Law of China, the use of a trademark can be use by the trademark registrant or ones authorized by the registrant. The Supreme Court of China further expanded it to 'trademark registrant's use of its own, use by others with his consent, and other uses that do not violate the intention of the registrant'.[16]Herein, 'use by others with his consent' refers to use by the licensee. Therefore, there can be three types of user: trademark owner, trademark licensee, and other party.

#### *i. Licence relationship*

The trademark licence required by the Trademark Law of China shall be recorded with the China Trademark Office.[17] However, such a recordal is not a necessary condition for the licence contract to take effect. Article 44 of the Contract Law of China says a contract that is established pursuant to the law shall become effective upon establishment, unless otherwise required by laws and administrative regulations under which a contract may become effective, only after the completion of approval, registration or other procedures. The Trademark Law says 'without recordal, a trademark licence is not effective against bona fide third parties'[18] so the recordal is not a requirement for the licence contract to take effect between the licencor and

15. Rules and Cases-Litigations Involving the Authorization and Determination of Trademark Rights (商标授权确权诉讼-规则与判例), author: ZHOU Yunchuan, publisher: Law Press, China, 2014.
16. Article 20 of Opinions of the Supreme Court on Certain Issues concerning the Trial of Administrative Cases of Trademark Authorization Confirmation (2010) and Article 26 of Provisions of the Supreme People's Court on Several Issues Concerning the Adjudication of Administrative Cases on Granting and Affirming Trademark-related Rights (2017).
17. Article 43 of the Trademark Law of China (2014).
18. *Ibid.*

the licensee. Therefore, even without recordal of the trademark licence, the licence relationship can be established as long as there is a valid licence contract.

#### *ii. Other parties*

In practice, it happens often that the user of the trademark is different from its registrant but there was no written document to prove a licence relationship. In most of such cases, the two parties are associated with each other, such as a mother company and a son company or one being shareholder of the other. In other cases, there is no obvious association between them but the registrant grants a licence afterwards. In such a case, as long as the use of the actual user does not violate the registrant's will, it should be accepted as valid use to maintain the trademark registration.

The trademark licensor shall supervise the quality of goods/services of the licensee for which the licensed trademark is used.[19] The licensee shall guarantee the quality of goods/services. Nevertheless, whether the licensor performed such a duty is not something considered in a non-use cancellation case.

### B. Volume, duration and frequency of use

There is no definite requirement in the laws, regulations or guidelines on the volume, duration and frequency of use in order to maintain a trademark registration. According to the purpose of the non-use cancellation rule, if there is genuine commercial use, it should be maintained even though such a use is only small-scale. Nevertheless, token use is not sufficient. Token use refers very small-scale use in order to meet the requirement for maintaining the trademark registration.

In the review decision on cancellation of TM No. 3703373 made in October 2017 by TRAB (Trademark Review and Adjudication Board),[20] the registrant submitted a licence contract proving the licence relationship between the registrant and the actual user, a distribution contract proving the licensee sold ten items of clothing to another party, one invoice for the said ten items of clothing as well as pictures of clothing bearing the registered trademark. Although it was a complete evidence chain to show commercial use of the registered trademark, it proved only one sale. The TRAB held that only one sale during the three-year period was too little use and it could not meet the minimum requirement on the scale of use. It was reasonable to suspect that such use was only for the purpose of maintaining the registration instead of genuine commercial use. Therefore it was not valid use to maintain a trademark registration.

No court decision gave specific criteria on how much use is sufficient in terms of volume, duration or frequency of use. It is a case-by-case issue and different industries may have different requirements. However, one sale or one advertisement is not sufficient according to a number of decisions.

19. *Ibid.*
20. TRAB decision No. [2017]127281.

## C. Use of a sign

Strictly speaking, use of a registered trademark should be the same as was registered and the registrant should not alter the trademark in use[21] and where the sign contained in a registered trademark requires change, a new registration application shall be filed. In practice minor change is allowed[22] in use in non-use cancellation cases.

In real business operations, the market is changing and the perception of consumers is changing. There will be a need to adjust the trademark on different occasions. Such adjustment may be only minor, such as font change of the word or colour change. It is not realistic nor reasonable to require the registrant to apply for registration for every version. Some business customs also require change on different occasions. For instance, for a registered trademark with combination of a word and a device, it is common practice in business that only the word part is included on the invoice. In such a case, the trademark in use (on the invoice) is different from the registered trademark but it may be viewed as valid use if the word part is the distinctive part in the trademark and consumers are not confused over the origin of goods.

Whether the use of an altered trademark is sufficient to maintain a registered trademark depends on whether such alteration brings substantive change to the distinctive part of the registered trademark.[23] Recent practice shows that the China Trademark Office is not very strict in this aspect. If an altered trademark contains the distinctive part of the registered trademark and evidence shows there is genuine commercial use of the trademark, the Office will normally maintain the registration. Change of font and colour is generally fine as long as the change will not make it substantially different from the registered trademark.

In Beijing High Court's decision (2017) Administrative Ruling No. 4603,[24] the Court held that in trademark No. 662482 '高通 GOTOP', both the Chinese character part '高通' and the English part 'GOTOP' were distinctive parts and the use of Chinese character part '高通' or English part 'GOTOP' separately did not change the distinctive character of the registered trademark so use of these parts separately could be viewed as use of the registered trademark.

## D. Use in relation to goods and services

A trademark registration involves a mark being registered on goods/services. The mark and the goods/services are both indispensable. A mark is merely a sign if there are no goods/services to associate with. Therefore, the use of a registered trademark should be use on the designated goods/services. Use of a mark without being attached to any goods/services cannot be viewed as valid use to maintain a trademark. Beijing High Court supported such a rule in its decision (2010) High Administrative Final Decision No. 309. In that case the registered trademark was a device mark registered on goods in class 25, TM No. 1348732. The registrant submitted a contract to print

---

21. Article 49 of Trademark Law of China (2014).
22. Article 24 of Trademark Law of China (2014).
23. Trademark Examination Guidelines (2016), p. 186.
24. Made on 28 February 2018.

notepaper and envelopes on which the mark was printed. The Court ruled that there was no proof to show use of the mark on the goods, so the trademark was cancelled.

Since 2002, a trademark registration can be partially cancelled in China, i.e., on part of the designated goods/services.

Where the trademark is used only on part of the designated goods/services, can the trademark be maintained on all designated goods/services?

According to the Trademark Examination Guidelines of the Trademark Office and the TRAB, the use of the trademark on designated goods/service is sufficient to maintain its registration on similar goods/services while the use on goods/services that are not designated cannot maintain the registration on any designated goods/services even if such goods/services are similar to each other.[25] Such a prevision comes from Article 23 of Trademark Law of China (2014), which says:

'The holder of a registered trademark shall submit a new registration application if it intends to obtain the exclusive right to use the trademark on goods beyond the approved scope of use.'

This rule was confirmed by a decision made by the Supreme Court of China in a re-trial case of trademark cancellation relating to trademark No. 514091, '萬寶 WANBAO'.[26] In this case, the trademark was used on 'speakers' and a 'power amplifier' but the designated goods were 'TV sets, monitors, etc.' The Court ruled that the trademark should be cancelled since the goods on which the trademark was used were not among the designated goods without considering and commenting on the goods similarity between the real and the designated goods.

However, there are some exceptions in special situations.

### 1. *Use for each of the designated goods and services*

As mentioned above, the use of the trademark on some designated goods/services is not necessarily able to maintain the registration on all goods/services but can maintain it on similar goods/services.

In the review of the cancellation case for TM No. 3906709 'SWEETVALLEY',[27] the designated goods of the registered trademark were: canned fruit, canned vegetables, canned meat, canned aquatic products, canned quail eggs, jam, edible oil, fruit jelly, nuts, dried edible mushrooms. Evidence submitted by the registrant proved the use of the trademark on 'canned yellow peaches'. The TRAB held that 'canned yellow peach' was a type of 'canned fruit' so use on 'canned yellow peaches' was use on 'canned fruit' and meanwhile, 'canned fruit' was similar to 'canned vegetables, canned meat, canned aquatic products, canned quail eggs' so the trademark was maintained on these goods. However, the rest of the designated goods were not similar to 'canned fruit' and therefore the registration should be cancelled on them.

25. Trademark Examination Guidelines (2016), p.186.
26. (2010) IP Administrative decision No. 44.
27. TRAB decision No. [2017] 3185 made by the TRAB in 2017 which was overruled by Beijing IP Court on 26 February 2018 in its decision (2017)BJ 73 Administrative first instance decision No. 3424 due to the validity of evidence.

### 2. *Use for a broader category of goods and/or services*

If the goods/services in actual use is not among the designated goods/services, such use generally cannot be accepted as valid use for the designated goods/services. There have been many such cases.

In the abovementioned '高通GOTOP' case, the goods in actual use were 'word stock card (chip)' while the designated goods were 'Chinese card' to convert the operational system from English into Chinese. The court held that 'word stock card (chip)', compared with the more simply functioned 'Chinese card', had a broader function and use and the two were not the same goods. Therefore, the use of the trademark for 'word stock card (chip)' could not maintain the trademark registration on 'Chinese card'.

However, there is also an exception. In the review of cancellation case 28 of TM No. 3661265'伟迪捷' in class 7, the TRAB accepted the use of the registered trademark for 'machine for spurting codes' as valid use on the designated goods of 'printing machines'. In this case, the registrant had been using its trademark on its machines for spurting codes but 'machines for spurting codes' was not a standard goods name acceptable to the China Trademark Office so the registrant chose the standard 'printing machines' instead.[28]Technologies are developing fast and there are always new goods and goods names appearing. The Trademark Office's updates on the acceptable goods description are not quick enough. So the actual goods wanted by the registrant are not always found in the standard goods list but the Trademark Office of China is very strict with the examination of goods names. Therefore, it often happens that the applicant chooses standard goods descriptions instead of insisting on their real goods names for fear of being required to amend or even being rejected. This was exactly the situation in this case. If this trademark was cancelled, the registrant would lose its trademark and lose its market, which is not fair to the registrant. The TRAB considered the common features, functions and distribution channels of the non-standard goods in real use,'machines for spurting codes', and the standard goods 'printing machines' and decided to maintain this registration.

## E. Territorial scope

Use should be on the Chinese mainland. Hong Kong, Macau and Taiwan are still separate jurisdictions from the Chinese mainland and the territorial scope of protection of a registered Chinese trademark covers only the Chinese mainland. So use in Hong Kong, Macau or Taiwan cannot be accepted as valid use of a registered Chinese trademark. China has a huge geographical coverage and there are 31 provinces in total on the Chinese mainland. It is not necessary to cover the whole country.

---

28. The China Trademark Office is strict with the goods description and prefers the standard goods in the Classification of Goods and Services it publishes.

## F. Relevant time period and its calculation

According to the Trademark Law of China, a trademark registration will be subject to non-use cancellation if use stops for three consecutive years after registration without legitimate reasons. Thus, the grace period for use is three years after registration. Anyone may challenge a trademark registered over three years earlier on the ground of non-use.[29]

For a national registration, it is easy to count the three-year period as it should be three years from the registration date, i.e., the publication date for registration in the trademark gazette. However, for those trademarks registered after opposition, it is more complicated as there will be two publications for registration. The first is an automatic publication at the expiration of the opposition period, from which the ten-year valid term starts and which will be revoked once the opposition is received; the second is the publication of the actual registration, which is done when the opposition fails or partially fails. The three-year period for non-use cancellation starts from the second publication of registration.

For an international registration (IR) to extend to China, the three-year period is more complex to determine since there is no publication of registration for an IR. The three-year period is counted differently for different situations.

(i) For an IR approved for registration without any refusal: the three-year period starts from the expiration of the 12-month or 18-month examination time limit counted from the date of WIPO notifying the China Trademark Office;

(ii) For an IR approved for registration after review of whole refusal: the three-year period starts from the review decision date;

(iii) For an IR partially refused but no review application is raised: the three-year period starts from the expiration of the 12-month or 18-month examination time limit counted from the date of WIPO notifying the China Trademark Office;

(iv) For an IR partially refused and further approved for registration on more goods/services after review of refusal: for the originally approved goods/services (not the refused ones), the three-year period starts from the expiration of the 12-month or 18-month examination time limit counted from the date WIPO notifying the Chinese Trademark Office; for the originally refused but then approved for registration in the review procedure, the three-year period starts from the review decision date.

When a trademark registration is challenged over cancellation for non-use, the registrant shall submit evidence of use for the past three years counted from the date when the non-use cancellation is filed (prescribed period). The registrant has two months to collect, prepare and submit the evidence of use after receiving the notification from the China Trademark Office. New evidence can be accepted in subsequent procedures, including review of cancellation in TRAB and administrative litigation before the court.

If the time of use is before the prescribed period, such use will not be valid to maintain the trademark; if the time of use is after the prescribed period, generally

29. Article 49 of the Trademark Law of China (2014).

speaking it is not valid evidence but if there has already been preparation for use during the prescribed period but the actual use was delayed until afterwards due to some objective reasons, such use may be accepted to maintain the trademark.

## III. EVIDENCE OF GENUINE USE OF A TRADEMARK

### A. When use should be proved: timing of lack of genuine use claims

#### *1. When filing an application for a trademark*

There is no requirement for actual use on filing in China; nor need the applicant claim for intention of use. Use is not a requirement for registration either.

#### *2. In the course of opposition proceedings*

It is not a requirement to demonstrate use of the opposed trademark in response to an opposition unless the opposed party wants to argue the opposed trademark has acquired second meaning in use where the opposition ground is lack of distinctiveness.

Actual use to some extent helps the opposed party to defend an opposition. For instance, actual use may give more confidence to the examiner to believe there is no confusion between the opposed trademark and the cited trademark.

For the opponent, if the opposition ground is earlier trademark filing or registration, there is no need for him to prove the continuous use of his earlier trademark. If he opposes the trademark in question on the following grounds, he needs to prove his use of an identical or similar trademark before the filing date of the opposed trademark:

(1) prior use with certain reputation (Article 32, second paragraph of the Trademark Law of China): on this ground, the opponent needs to provide a big volume of evidence to prove he had been using his trademark before the filing date of the opposed trademark and that such use had made his trademark acquire a high reputation.
(2) contractual or other business relationship (Article 15, second paragraph of the Trademark Law of China): on this ground, the opponent needs to prove, before the filing date of the opposed trademark, (a) there had been contractual or other business relationship between the two parties so the opposed party had been aware of his trademark and (b) he had used his trademark in China on the same or similar goods.

#### *3. In cancellation proceedings based on lack of use*

The procedure of non-use cancellation application against a registered trademark starts from the China Trademark Office; any party may appeal to the TRAB by filing a review application if he is not satisfied with the China Trademark Office decision. The TRAB's decision can be further appealed to the court by initiating administrative litigation. The Beijing Intellectual Property Court is the only competent court to accept such administrative litigation at the first instance. The second (final) instance goes to Beijing High Court. To appeal to the second instance in litigation is a party's right in China. On some occasions, an effective decision may be re-trialled by a higher court but the re-trial is not a right and the higher court may decide whether it takes the case. In the administrative litigation procedure, the court cannot directly decide to cancel or maintain a trademark, it can only decide to order the TRAB to make a new decision if it does not agree with the TRAB's decision.

The Trademark Law of China sets up the time limit of normally nine months for the China Trademark Office to examine a non-use cancellation case[30] counted from the date when the China Trademark Office receives the cancellation application. The time limit for the TRAB to examine a review of non-use cancellation is normally also nine months from when it receives the review case.[31]First instance administrative litigation normally takes six months[32] to conclude, and second instance three months.[33] Below is the flow chart of the cancellation procedure.

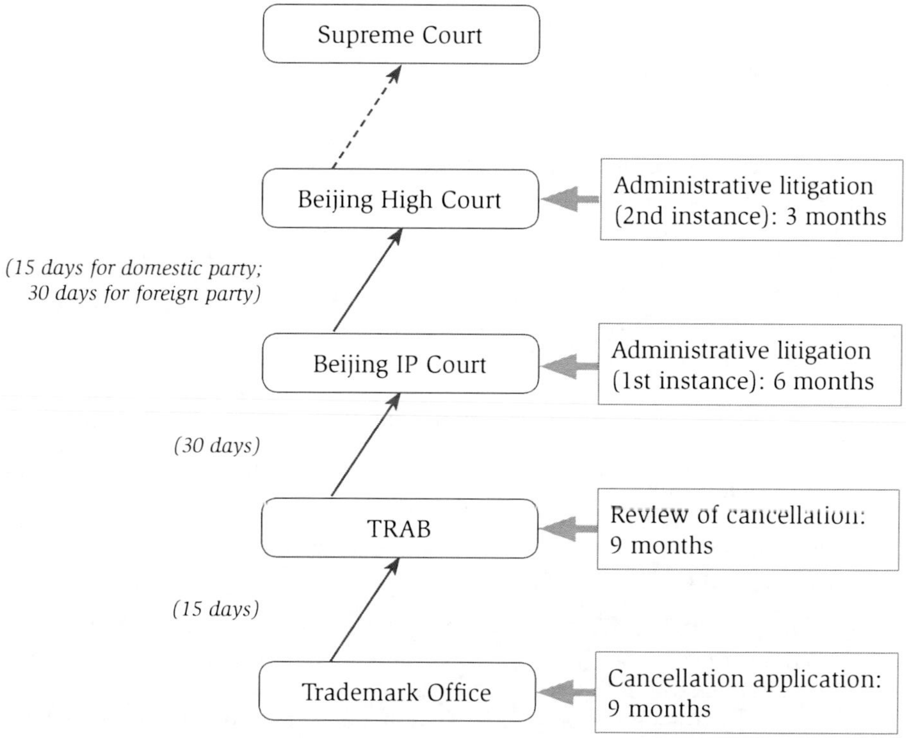


30. Article 49 of the Trademark Law of China (2014).
31. Article 54 of the Trademark Law of China (2014).
32. Article 81 of the Administrative Procedure Law of China (2017).
33. Article 88 of the Administrative Procedure Law of China (2017).

### 4. *As a defence in infringement/cancellation proceedings*

China is a bifurcated country, i.e., the validity of a trademark and the trademark infringement issue are tried by different authorities. Validity issues are decided by the Trademark Office and the TRAB, but infringement issues are handled by the court (judicial means) or the local administrative authority.[34] Non-use of the registered trademark is not a defence for non-infringement but the defendant does not need to pay damages if the registered trademark has been registered for over three years but the plaintiff cannot provide valid evidence of use when it is requested to do so in the infringement litigation.[35]

If the defendant thinks the concerned trademark should be cancelled for non-use and does not want to bear any liability for trademark infringement, he needs to file a separate request with the Trademark Office of China to cancel the trademark. He cannot directly ask the court to decide on non-infringement based on the fact that the trademark is not in use and should be cancelled.

## B. Burden of proof

It is for the registrant to prove his trademark is in valid use. A requester of a non-use cancellation application does not need to provide any evidence in order to prove the concerned trademark is not in use. He just needs to submit a simple application form, asking for the concerned trademark to be cancelled, as long as the concerned trademark has been registered for over three years. This is different from some jurisdictions.

## C. Relevant evidence

The use of a trademark should be use indicating the origin the goods/services.

### 1. *Use on goods*[36]

The following methods can be valid use of a trademark for goods in the sense of the Trademark Law of China:

(1) directly affixing the trademark on the goods, packaging, container, label, etc. by means of sticking, printing, stamping, knitting, and so on, or using the trademark on the goods tags, specifications, manuals, price list, etc.
(2) using the trademark on trading documents relating to the goods, including the goods distribution contract, invoices, receipts, inspection and quarantine

---

34. It used to be Administration of Industry and Commerce (AIC). In April, 2018, China re-organized some government offices and after the re-organization, AIC no longer existed and it is now the new office, State Administration for Market Regulation, that is in charge of trademark administrative enforcement.
35. Article 64 of the Trademark Law of China.
36. Trademark Examination Guidelines (2016).

certificates of imported or exported goods, customs declaration documents, etc.
(3) using the trademark in broadcasting, TV and other media, or in publication materials publicly distributed, or in advertisement activities, such as advertisement boards, mailing advertisement, or other ways of advertisements, to promote the designated goods;
(4) using the trademark at exhibitions and fairs, such as distributing printed or other materials bearing the trademark at exhibitions and fairs, to promote the designated goods;
(5) other means of use.

### 2. *Use on services*[37]

The following methods can be valid use of a trademark for services in sense of the Trademark Law of China:

(1) directly using the trademark in the premises where the services are provided, such as use of the trademark in the introduction manual, service signboard, decoration of the premises, costumes of the staff, posters, menus, price lists, coupons, office stationery, and other articles relating to the designated services;
(2) using the trademark on documents relating to the services, such as service contracts, invoices, receipts, remittance certificates, repair and maintenance certificate, etc.
(3) using the trademark in broadcasting, TV and other media, or in publication materials publicly distributed, or in advertisement activities, such as advertisement boards, mailing advertisements, or other ways of advertisement, to promote the designated services;
(4) using the trademark at exhibitions and fairs, such as distributing printed or other materials bearing the trademark at exhibitions and fairs to promote the designated services;
(5) other means of use.

Its use in a trade name cannot constitute use of a trademark in the sense of the Trademark Law.

Internet use can be an acceptable way of use but the use of the trademark on a foreign website is not acceptable, especially if this website is only in a foreign language.

The form of the evidence shall be either an original copy or notarized duplicate copy. For evidence of internet use, it is advisable to download it with notarization.

The evidence materials submitted to prove the use of the trademark shall meet the following requirements:

(1) the mark specimen in question is demonstrated;
(2) the goods/services on which the trademark is used shall be the designated goods/services;

37. *Ibid.*

(3) the user's name should be demonstrated, which should be either the registrant itself or an authorized party. If it is an authorized party, relevant evidence is needed to prove authorization; and
(4) the geographical scope of use should be demonstrated and it should be on the Chinese mainland;

The above requirements do not need to be demonstrated in one piece of evidence and it is acceptable if they can be proved by a chain of evidence.

The following evidence without other supporting evidence is not sufficient to prove valid use required by the Trademark Law of China:

(1) a goods distribution contract or service agreement;
(2) testimony of witnesses;
(3) material evidence, video, internet information, etc., for which is difficult to judge whether any modification is done on it;
(4) actual products or their reproduction per se, which cannot prove the time and other details.

In practice, the Trademark Office of China is not very strict with the examination of evidence of use submitted in a cancellation application. However, the TRAB and the courts will be stricter.

## IV. JUSTIFICATION FOR NON-USE

### A. Obstacles to use

The Trademark Law of China accepts certain justifications for non-use. Such justifications should be objective reasons. The justification for not using a registered trademark should meet two conditions: (1) the objective reason is a necessary reason for non-use and it is sufficient to stop the use of the trademark; (2) such objective reason shall not attributed to the trademark registrant.

If a trademark registrant fails to actually use or discontinues use of a registered trademark due to objective reasons, such as force majeure, policy restriction, or bankruptcy liquidation, or a trademark registrant has actual intention to use the trademark and has necessary preparation for actual use but has not actually used the registered trademark due to other objective reasons, the cases can be deemed to have justifiable reasons.[38] However, once the event or situation causing such justifiable reasons disappears, the registrant shall resume the use of the trademark within a reasonable period of time.

#### *1. Force majeure*

This mainly includes natural disasters, such as earthquake, flood, drought and disease, and human disaster, such as wars.

38. Article 26 of Provisions of the Supreme People's Court on Several Issues Concerning the Adjudication of Administrative Cases on Granting and Affirming Trademark-related Rights (2017).

## 2. *Policy restriction*

This normally refers to government policies which forbid or restrict the manufacture, sales or import of certain goods, resulting in the inability to use the registered trademark on such goods. Such policies shall not be something that can be overcome or removed by the registrant's efforts and they shall affect the use of the concerned trademark.

In the case of TM No. 3455749 '久保田',[39] government policy was accepted by the Trademark Office as justifiable reason for non-use. The trademark was registered by the Japanese company Asahi-Shuzo Sake Brewing Co.,Ltd. in class 33 on 'Japanese rice wine' etc. It was registered in 2003. The registrant had been exporting the Japanese rice wine to China since 2003 but stopped in 2011 because of an administrative order announced by the Chinese government. In 2011, there was a serious earthquake in Japan which caused the nuclear leak in Fukushima. The areas in and around Fukushima was polluted by nuclear material. So in 2011, the Chinese government banned imports of any food products originating from 12 counties with nuclear pollution risk, including the county where the registrant was located and where they manufactured the products. Hence the registrant stopped the export of its rice wine to China. The import prohibition is not yet lifted. In 2016, a third party requested cancellation of this trademark registration. The Trademark Office accepted the registrant's explanation, determined that such a government order was justification for non-use and therefore maintained the trademark.

In this case, the registrant had used the trademark and had the genuine intention to continue using it but was not able to because of reasons beyond its control.

Acceptable justifiable reasons for non-use may also include administrative requirements of examination and approval or licence. However, such administrative requirements are different from a government prohibition or restriction ban. With efforts, such requirements are likely to be met. So the genuine use intention of the registrant is an important factor to consider when determining whether it can be taken as justifiable reason for non-use. If the registrant never attempted to apply for approval or a licence, nor made any preparation for it, such administrative requirements cannot be accepted as justification for non-use.

The administrative examination and approval requirement was not accepted as a justifiable reason for non-use in TM No. 1129187 'GNC' case.[40] In this case, the concerned goods were fish oil products which should go through administrative examination and approval before being launched in the market. The registrant argued that they did not use the trademark because they did not get such approval. However, the trademark had been registered for seven years and no evidence demonstrated that the registrant ever applied for examination and approval with the relevant authority. Therefore no genuine use intention could be seen. Thus, the Beijing High Court decided there was no justifiable reason of non-use and this trademark should be cancelled.

39. TM Cancellation [2017] No. Y002935.
40. (2006) High Administrative Final No. 78.

### 3. *Bankruptcy liquidation*

This is an internal reason possibly acceptable as justification for non-use. During the bankruptcy liquidation procedure, an enterprise's business activities will be restricted and it therefore can be a justifiable reason for non-use.

### 4. *Other objective reasons*

The purpose of the non-use cancellation procedure is to encourage use. Cancellation is not the aim. The effect of the cancellation of a registered trademark is irretrievable so the authority should be careful in deciding to cancel a registered trademark. In such a consideration, other situations may also be accepted as justifiable reasons for non-use, such as the following:

(1) re-organization or merger: the precondition is that such an act affects the normal use of the trademark.
(2) litigation: this should a) affect the certainty of the validity of the trademark; or b) influence the normal use of the trademark.

## B. Circumstances in which non-use may be cured

Once the decision to cancel a trademark takes effect, there is no procedure to cure it. What the registrant can do is to re-register it. However, the reason for a third party to cancel a prior trademark is normally because this prior trademark obstructs its later application. So it may be too late to re-file the registration application as the new filing will be blocked by this third party's trademark.

Some registrants use repeated registration to solve the problem if they do not use the trademark but want to keep it in the Register, for instance, by re-filing the application every three years.

## V. CONSEQUENCES OF NON-USE

## A. Consequences of non-use and relevance of the context

As China is bifurcated country, the validity action of a trademark and the infringement action are not tried in one case. The defendant in an infringement litigation case cannot defend himself for non-infringement by claiming that the concerned trademark should be cancelled for non-use for over consecutive three years. However, for trademarks registered over three years, if the plaintiff cannot provide valid evidence of use nor prove actual loss caused by the infringing activities, no damage compensation needs to be borne by the defendant even if infringement is established. If the trademark is cancelled before the decision of infringement case is made, there will be no infringement to be established.

There is only one procedure to cancel a trademark registration for non-use, that is, to file a request of non-use cancellation with the Trademark Office of China.

## B. Extent of any revocation

A non-use cancellation procedure can produce three different results:

(1) full cancellation: the concerned trademark has not been used on any of the designated goods and the whole registration is cancelled.
(2) partial cancellation: this can be the result of a partial cancellation request or a partially successful full cancellation request.
(3) whole registration maintained: that means the evidence submitted proves the genuine use of the trademark on all goods, or on part of the goods but with the rest similar to these goods.

## C. Timing of any revocation

The applicant for non-use cancellation may withdraw the request before the decision to cancel the trademark is made and once it is decided to cancel the trademark, the only way to change the situation is that the registrant provides valid evidence of use in the review procedure with the TRAB or during the administrative litigation.

# European Union

*Alexander von Mühlendahl*

## I. INTRODUCING GENUINE USE OF TRADEMARKS

The European Union (EU) is made up of 28 Member States (including the United Kingdom, which is scheduled to depart from the EU after a transition period ending on 31 December 2020). The competence for protecting intellectual property (IP) rights is shared by the EU itself and its Member States. The EU has the competence to provide for unitary, EU-wide protected IP titles, based on Article 118 of the Treaty on the Functioning of the European Union (TFEU), and for adopting measures to harmonize national laws to the extent necessary for the functioning of the internal market, based on Article 114 TFEU.

In the field of trademarks, the EU has exercised both the competence for adopting a unitary title and for harmonizing the laws of the Member States. Based on proposals submitted by the European Commission in 1980, the Council adopted, in 1988, the First Council Directive 89/104/EEC of 21 December 1988 to approximate the laws of the Member States relating to trade marks (OJ 1989 L 40, p. 1), and, in 1993, Council Regulation (EC) No. 40/94 of 20 December 1993 on the Community trade mark (OJ 1994 L 11, p.1). The Directive was codified in 2008, as Directive 2008/95/EC of the European Parliament and of the Council of 22 October 2008 to approximate the laws of the Member States relating to trade marks (OJ 2008 L 299, p. 25), and the Regulation on the Community trade mark was codified in 2009 as Council Regulation (EC) No. 207/2009 of 26 February 2009 on the Community trade mark (OJ 2009 L 78, p. 1).

Following an analysis and review of the functioning of the European trademark system, consisting of the unitary Community trademark system and the harmonized national systems in the Member States, the European Commission proposed in 2013 a new Directive to replace the 1988/2008 Directive, and substantial amendments to the Regulation. Both proposals were finally adopted in December 2015. The Directive, Directive (EU) 2015/2436 of the European Parliament and of the Council of 16 December 2015 to approximate the laws of the Member States relating to trade marks (Recast) (OJ 2015 L 336, p. 1), must be implemented by the Member States by January 2019, with an additional four years for implementing administrative revocation and invalidation proceedings.

The amendments of the Regulation on the Community trade mark and other amendments, adopted by Regulation (EU) 2015/2424 of the European Parliament and of the Council of 16 December 2015 amending Council Regulation (EC) No. 207/2009 on the Community trade mark and Commission Regulation (EC) No. 2868/95 implementing Council Regulation (EC) No. 40/94 on the Community trade mark, and repealing Commission Regulation (EC) No. 2869/95 on the fees payable to the Office for Harmonization in the Internal Market (Trade Marks and Designs) (OJ 2015 L 341, p.21), entered into force in part on 23 March 2016, and in part on 1 October 2017. On the latter date, implementing and delegated acts by the European Commission also entered into force. What was named Regulation on the Community trade mark became the Regulation on the European Union trade mark. That Regulation was codified in 2017 as Regulation (EU) 2017/1001 of the European Parliament and of the Council of 14 June 2017 on the European Union trade mark (OJ 2017 L 154, p.1).

The Regulation is supplemented by the implementing and delegated acts referred to. These are Commission Implementing Regulation(EU) 2017/1431 of 18 May 2017 laying down detailed rules for implementing certain provisions of Council Regulation (EC) No. 207/2009 on the European Union trade mark (OJ 2017 L 205, p. 39), and Commission Delegated Regulation (EU) 2017/1430 of 18 May 2017 supplementing Council Regulation (EC) No. 207/2009 on the European Union trade mark and repealing Commission Regulations (EC) No. 2868/95 and (EC) No. 216/96 (OJ 2017 L 205, p. 1). The Implementing Regulation and the Delegated Regulation were replaced in March 2018 by codified versions, to take into account the codification of the Regulation on the EU trademark in its version of June 2017.

The EU trademark system is administered by the European Union Intellectual Property Office (EUIPO), established in Alicante, Spain, previously known as OHIM, short for the Office for Harmonization in the Internal Market (Trade Marks and Designs). Trade mark applications are examined by the EUIPO *ex officio* for formalities and for absolute grounds of refusal. After publication, they may be opposed by proprietors of earlier EU trademarks or national trademarks in the EU Member States, and of other earlier rights. Third parties may request the revocation or invalidation of registered EU trademarks at the EUIPO. In an infringement action brought by the proprietor of an EU trademark before a competent court in a Member State the defendant may raise absence of genuine use as a defence, and may bring a counterclaim for a declaration of revocation or invalidity.

The presentation in this chapter is based on the 2017 codification of the Regulation on the European Union trade mark, shortened as EU Trade Mark Regulation (EUTMR), and the May 2017 Implementing Regulation and Delegated Regulation, abbreviated as EUTMIR and EUTMDR.

The provisions on the genuine use of trademarks are as to their substance the same in the Directive and in the EUTMR, which means that the legal situation in the Member States of the EU does not differ significantly from that prevailing under the EU trademark system. However, many differences exist in the details of asserting or relying on the absence of genuine use.

## A. Legal framework

The provisions on genuine use in the EU Trade Mark Regulation are organized as follows:

Among the provisions setting out the limitations of the exclusive rights, the basic provision on genuine is the following:

> *Article 18* – Use of an EU trade mark
>
> 1. If, within a period of five years following registration, the proprietor has not put the EU trade mark to genuine use in the Union in connection with the goods or services in respect of which it is registered, or if such use has been suspended during an uninterrupted period of five years, the EU trade mark shall be subject to the sanctions provided for in this Regulation, unless there are proper reasons for non-use.
>
> The following shall also constitute use within the meaning of the first subparagraph:
>
> (a) use of the EU trade mark in a form differing in elements which do not alter the distinctive character of the mark in the form in which it was registered, regardless of whether or not the trade mark in the form as used is also registered in the name of the proprietor;
>
> (b) affixing of the EU trade mark to goods or to the packaging thereof in the Union solely for export purposes.
>
> 2. Use of the EU trade mark with the consent of the proprietor shall be deemed to constitute use by the proprietor.

The sanctions referred to in Article 18(1) EUTMR are the following:

A registered EU trademark is liable to be revoked on grounds of non-use, pursuant to Article 58(1)(a) EUTMR, upon an application brought by a third party *in revocation proceedings*_before the EUIPO. Article 58(1)(a) EUTMR provides as follows:

> *Article 58* – Grounds for revocation
>
> 1. The rights of the proprietor of the EU trade mark shall be declared to be revoked on application to the Office or on the basis of a counterclaim in infringement proceedings:
>
> (a) if, within a continuous period of five years, the trade mark has not been put to genuine use in the Union in connection with the goods or services in respect of which it is registered, and there are no proper reasons for non- use; however, no person may claim that the proprietor's rights in an EU trade mark should be revoked where, during the interval between expiry of the five-year period and filing of the application or counterclaim, genuine use of the trade mark has been started or resumed; the commencement or resumption of use within a period of three months preceding the filing of the application or counterclaim which began at the earliest on expiry of the continuous period of five years of non-use shall, however, be disregarded where preparations for the commencement or resumption occur only after the proprietor becomes aware that the application or counterclaim may be filed;

*In opposition proceedings* before the EUIPO, the applicant, if confronted by an earlier EU trademark or by an earlier national mark, may request that the opposing party provides proof of genuine use of the earlier mark if the earlier mark, at the filing or priority date of the opposed mark, was registered for more than five years. Article 47(2) and (3) EUTMR provide as follows:

> *Article 47* – Examination of opposition
>
> 1. [...]
>
> 2. If the applicant so requests, the proprietor of an earlier EU trade mark who has given notice of opposition shall furnish proof that, during the five-year period preceding the date of filing or the date of priority of the EU trade mark application, the earlier EU trade mark has been put to genuine use in the Union in connection with the goods or services in respect of which it is registered and which he cites as justification for his opposition, or that there are proper reasons for non-use, provided that the earlier EU trade mark has at that date been registered for not less than five years. In the absence of proof to this effect, the opposition shall be rejected. If the earlier EU trade mark has been used in relation to only part of the goods or services for which it is registered it shall, for the purposes of the examination of the opposition, be deemed to be registered in respect only of that part of the goods or services.
>
> 3. Paragraph 2 shall apply to earlier national trade marks referred to in Article 8(2)(a), by substituting use in the Member State in which the earlier national trade mark is protected for use in the Union.

*In invalidation proceedings* before the EUIPO seeking to invalidate a registered EU trademark based on an earlier EU trademark or an earlier national trademark, the proprietor of the EU trademark may request that the proprietor of the earlier mark provides proof of use. Article 64(2) and (3) EUTMR provide as follows:

> *Article 64* – Examination of the application
>
> 1. [...]
>
> 2. If the proprietor of the EU trade mark so requests, the proprietor of an earlier EU trade mark, being a party to the invalidity proceedings, shall furnish proof that, during the period of five years preceding the date of the application for a declaration of invalidity, the earlier EU trade mark has been put to genuine use in the Union in connection with the goods or services in respect of which it is registered and which the proprietor of that earlier trade mark cites as justification for his application, or that there are proper reasons for non-use, provided that the earlier EU trade mark has at that date been registered for not less than five years. If, at the date on which the EU trade mark application was filed or at the priority date of the EU trade mark application, the earlier EU trade mark had been registered for not less than five years, the proprietor of the earlier EU trade mark shall furnish proof that, in addition, the conditions set out in Article 47(2) were satisfied at that date. In the absence of proof to this effect, the application for a declaration of invalidity shall be rejected. If the earlier EU trade mark has been used only in relation to part of the goods or services for which it is registered, it shall, for the purpose of the examination of the application for a declaration of invalidity, be deemed to be registered in respect of that part of the goods or services only.

> 3. Paragraph 2 shall apply to earlier national trade marks referred to in Article 8(2)(a), by substituting use in the Member State in which the earlier national trade mark is protected for use in the Union.

*In infringement proceedings* based on a registered EU trademark, the defendant may raise absence of genuine use of the claimant's mark as a defence, as provided for in Article 127(3) EUTMR, and may also, as provided for in Article 128 EUTMR, bring a counterclaim against the proprietor of the EU trademark seeking to obtain a declaration of revocation on grounds of absence of genuine use. If such a counterclaim is brought, the counterclaimant must prove, if so requested by the other party, that the earlier mark relied on as the basis of the counterclaim has been genuinely used as provided for in Article 64(2) and (3) EUTMR.

Article 127(3) EUTMR reads as follows:

> *Article 127* – Presumption of validity – Defence as to the merits
>
> 1. [...]
>
> 2. [...]
>
> 3. In the actions referred to in points (a) and (c) of Article 124, a plea relating to revocation of the EU trade mark submitted otherwise than by way of a counterclaim shall be admissible where the defendant claims that the EU trade mark could be revoked for lack of genuine use at the time the infringement action was brought.

Article 128 EUTMR reads, in relevant parts, as follows:

> *Article 128* – Counterclaims
>
> 1. A counterclaim for revocation or for a declaration of invalidity may only be based on the grounds for revocation or invalidity mentioned in this Regulation.
>
> 2. [...]
>
> 3. [...]
>
> 4. [...]
>
> 5. Article 64(2) to (5) shall apply.
>
> 6. [...]
>
> 7. [...]

## B. Purpose and function of genuine use: policy considerations

The EU trademark system is based on a first-to-file approach to trademark registration – in case of conflicts the mark with the earlier filing (or priority) date prevails over later marks. There is no requirement to use the mark prior to application of registration. However, once registered, and subject to a 'grace period' of five years from the date of registration, a registered EU trademark must be genuinely used – in the absence of genuine use the registered mark may be revoked, and it may not be enforced in opposition, invalidation and infringement proceedings. At the time the original Regulation on the Community trademark was adopted, most EU Member States already had adopted the requirement of genuine use as part of their national legislation. Thus, it was not difficult to obtain agreement on the provisions governing use, set out in detail above.

The justification for requiring that registered trademarks must be put to genuine use in order to be enforceable is set out in the Preamble to the EU Trade Mark Regulation as follows:

> (24) There is no justification for protecting EU trade marks or, as against them, any trade mark which has been registered before them, except where the trade marks are actually used.
>
> (25) For reasons of equity and legal certainty, the use of an EU trade mark in a form that differs in elements which do not alter the distinctive character of that mark in the form in which it is registered should be sufficient to preserve the rights conferred regardless of whether the trade mark in the form as used is also registered.

A somewhat more extensive justification was provided by the European Commission when it presented the original proposal for a Community trademark in 1980 (Bulletin EG Supplement 5/80, p. 60):

> 'One of the basic ideas in Community trade-mark law is that ownership can be maintained only by using the trade-mark. The obligation to use the trade-mark is part of the price paid for the right to obtain a trade-mark simply by registration. There is no justification nowadays, when it, is becoming increasingly difficult to find, an attractive trade-mark, for allowing a proprietor to monopolize a trade-mark that he neither uses himself nor allows to be used by others.'

## II. WHAT QUALIFIES AS GENUINE USE OF A TRADEMARK

### A. Nature of use

#### *1. Use consistent with the function of the trademark*

In order to be recognized as genuine, the use must be made 'as a trademark'. This requirement is central to trademark law in general: the mark must be used for the purposes of distinguishing the goods or services for which it is registered from goods or services having a different commercial origin. The same concept applies for determining whether the challenged use of a sign is infringing use: in these situations as well, the use must be made 'as a trademark', which means that the use must be in relation to goods or services and so as to distinguish these goods or services from goods or services of a different commercial origin. This was expressed by the Court of Justice as follows (Judgment of 11 March 2003, C-40/01, *Ansul BV v. Ajax Brandbeveiliging BV (MINIMAX)*, at para. 36): '"Genuine use" must therefore [...] be consistent with the essential function of a trademark, which is to guarantee the identity of the origin of goods or services to the consumer or end user by enabling him, without any possibility of confusion, to distinguish the product or service from others which have another origin.'

No trademark use is use of the sign as a corporate or company name or store name without any relation to goods or services. Similarly, use of sign as decoration or illustration without any relation to goods or services will not qualify as genuine use.

The use must be made 'in relation' to goods or services. Thus, while in many situations the mark is affixed to the goods or to their packaging, this is not an absolute requirement. For example, use in a catalogue may suffice, even if the goods themselves do not bear the trademark. Use may also be made on business papers, such as invoices, and in advertisements depicting the goods. The use of marks protected for services will always be 'in relation' to such services, as such a mark cannot be used 'on' services. Service marks will always be on some support, such a stationery, or brochures. Overall, in accordance with the case law of the Court of Justice, all relevant circumstances must be taken into account, and the customs or usages regarding the manner in which a trademark is used in the particular branch of business are among those circumstances.

Whether advertising the availability of goods or services prior to their actual appearance on the market is sufficient to constitute 'genuine use' has not been the subject of a decision of the General Court of the Court of Justice. If such a case should arise, such advertisements or promotion activities, as they are common when announcing a new model of a product, or the release of a television show or film, should be accepted as sufficient to maintain and enforce an EU trademark.

Mere preparations for use are not 'genuine' use for purposes of Article 18 EUTMR. Thus, plans to produce a particular product, even if announced to the public, are not 'use'. This also follows from Article 58(1)(a) EUTMR, according to which preparations of use are not in themselves 'use' which may prevent revocation. However, such preparations must be distinguished from use in advertising, which may indeed be sufficient to constitute genuine use (see above).

### *2. Use in the course of trade*

The use, in order to be judged as 'genuine' within the meaning of Article 18 EUTMR, must be use 'in the course of trade'. This excludes use which is private, i.e. outside of the course of commerce.

#### *i. Public use*

The use must also be 'public'. This excludes use which is entirely within a business. Thus, goods which have been produced and marked with the protected mark, but which have not yet been delivered to the trade, do not fulfil the requirement of 'public' use. Similarly, intra-company transport (delivery from the factory to a storage unit, delivery from parent to subsidiary) are not 'public' use and thus do not count as evidence of genuine use.

#### *ii. Commercial purpose*

While as a general proposition the public use made of a trademark in the course of trade will be made for purposes of obtaining a commercial benefit, the 'commercial benefit' does not necessarily mean that the use must be made by a commercial enterprise. Rather, also entities without any profit motive, such as charitable institutions, may obtain trademarks and may also make genuine use of such a trademark. This was confirmed by the Court of Justice in the judgment of 9 December 2008, Case C-442/07, *Verein Radetzky-Orden v. Bundesvereinigung Kameradschaft 'Feldmarschall Radetzky'*, where the Court held (at para. 24): '[A] trade mark is put to genuine use where a non-profit-making association uses the trade mark, in its relations with the public, in announcements of forthcoming events, on business papers and on advertising material and where the association's members wear badges featuring that trade mark when collecting and distributing donations.'

On the other hand, where the use is made for a product which does not actually look for a position on the market, in the actual case use for a product which was given for free with the purchase of a principal product, this was held not to be 'genuine' (Judgment of 15 January 2009, C-495/07, *Silberquelle GmbH v. Maselli-Strickmode GmbH (WELLNESS)*). The Court held (at para. 21): 'In such a situation, those items are not at all distributed with the aim of penetrating the market for goods in the same class. Under those circumstances, affixing the mark to those items does not contribute to creating an outlet for those items or to distinguishing, in the interest of the customer, those items from the goods of other undertakings.' That case should be seen as an exception to the general rule that goods which are available to customers as part of the overall 'service' – such as accessories in hotel rooms etc., internet services available without payment, etc. – even if no separate charge is made for them, are still genuinely used within the meaning of Article 18 EUTMR.

### *3. Use by the owner or with its consent*

According to Article 18(2) EUTMR, genuine use of a registered trademark may be made by the proprietor or by a person having the consent of the proprietor. The consent may be given expressly or may be implied from the circumstances. A proprietor who relies on use with his consent has the burden of proof. In normal situations, such proof should not be difficult even in the absence of an express agreement because in the normal course of events a trademark proprietor will not allow third parties to use his trademark unless he has given his consent. Typical 'consent' cases are use of a trademark by a licensee (Article 25 EUTMR) or use by one or more of several companies belong to a group of companies. In order to attribute use made with the consent of the proprietor, there is no requirement that the proprietor actually has the power to or actually exercises control over the quality of the goods or services placed on the market. On the other hand, use which is actually infringing cannot be converted ex post to use made with the consent of the proprietor so as to avoid revocation or unenforceability as a result of non-use.

With regard to EU collective marks and certification marks, Articles 78 and 87 EUTMR respectively provide that use by any person authorized to use the mark suffices.

## B. Volume, duration and frequency of use

The EU Court of Justice has on a number of occasions been called upon to set out the principles for determining whether use is genuine or not. Initially the cases were decided with regard to national trademarks, but the Court has applied the same principles to EU trademarks as well. Thus, 'there is "genuine use" of a trade mark where the mark is used in accordance with its essential function, which is to guarantee the identity of the origin of the goods or services for which it is registered, in order to create or preserve an outlet for those goods or services; genuine use does not include token use for the sole purpose of preserving the rights conferred by the mark. When assessing whether use of the trade mark is genuine, regard must be had to all the facts and circumstances relevant to establishing whether there is real commercial exploitation of the mark in the course of trade, particularly the usages regarded as warranted in the economic sector concerned as a means of maintaining or creating market share for the goods or services protected by the mark, the nature of those goods or services, the characteristics of the market and the scale and frequency of use of the mark' (Judgment of 19 December 2012, C-149/11, *Leno Merken BV v. Hagelkruis Beheer BV (ONEL/OMEL)*, para. 30). The earlier cases relied on, which are still cited as the 'leading cases' today, are Cases C 40/01, *Ansul*, [2003] ECR I 2439, C 416/04 P, *Sunrider v. OHIM*, [2006] ECR I 4237, and the order in Case C 259/02, *La Mer Technology*, [2004] ECR I 1159.

Under the 'all circumstances' rule applied by the Court, there is no *de minimis* rule, there are no absolute requirements as regards length of use, frequency of use, and intensity of use – what is relevant that the use must be 'real' use of the mark on the market, rather than 'token' use made to preserve trademark rights.

## C. Use of a sign

According to the applicable legal provisions, the use that must be assessed must have been use of the registered EU trademark. This raises the issue to which degree the mark as used is allowed to differ from the mark as registered. Article 18 EUTMR, to this extent in entire agreement with Article 5C(2) of the Paris Convention, provides that differences between the mark as used and the mark as registered do not constitute an obstacle to accepting genuine use if the use made was 'in a form differing in elements which do not alter the distinctive character of the mark in the form in which it was registered, regardless of whether or not the trade mark in the form as used is also registered in the name of the proprietor [...]'.

### *1. Use of a variation*

Case law of the General Court and of Court of Justice of the European Union has established that Article 18 EUTMR pursues the following objective (Judgment of 18 July 2013, C-252/12, *Specsavers International Healthcare Ltd v. Asda Stores Ltd*, at para. 29): 'avoiding imposing a requirement for strict conformity between the form used in trade and the form in which the trade mark was registered, [...] allow[s] the proprietor of the mark, in the commercial exploitation of the sign, to make variations in the sign, which, without altering its distinctive character, enable it to be better adapted to the marketing and promotion requirements of the goods or services concerned.' This statement focusses correctly on commercial realities, but – nevertheless – requires that the distinctive character should not be altered, without providing a standard as to how the 'alteration' of the distinctive character should be assessed.

It should be recognized that any 'alteration' necessarily also has an effect on the distinctive character of the mark; what matters is whether the effect is sufficiently small or insignificant so as to allow the enforcement of the mark as registered, even though it is actually not used in exactly that form. The *Specsavers* court required that the form as registered (not the form as used) 'always refers in that form to the goods of the proprietor covered by the registration, which is to be determined by the referring court.' It is submitted that this is a rather irrelevant argument, insofar as the mark as registered is not before the consumer. The proper analysis would (or should) focus on accepted commercial practices, taking into account the interests of the trademark proprietor and those of third parties. A useful analysis is provided by the case law of the General Court which considers that it is acceptable if 'the sign used in trade differs from the form in which it is registered only in negligible elements, so that the two signs can be regarded as broadly equivalent.' Obviously, if only the mark as used causes confusion, but not the mark as registered, the alteration should be judged to go beyond the acceptable variation.

In deciding individual cases where the question is whether the variation stays within acceptable bounds or not, reasonable minds may differ whether the outcome actually is supported by the general principles. Thus, in a recent judgment of the Court of Justice (Judgment of 11 October 2017, C-501/15 P, *EUIPO v. Cactus SA & Isabel Del Rio Rodriguez (CACTUS)*) a mark was judged to have been used as registered even though it was composed of a word and a figurative element and the use was made only of the word element. The Court justified its decision as follows (at para. 69):

> 'In paragraph 61 of the judgment under appeal, the General Court found – and that finding has not been disputed by EUIPO in the present appeal – that the two elements of the earlier figurative mark, namely a stylised cactus and the word element "Cactus", conveyed, in their respective forms, the same semantic content. It follows from that finding, however, that the General Court considered that the word element "Cactus" could not be regarded as possessing a distinctive character that was different from that of the stylised cactus and that the absence of that word element in the abbreviated version of the earlier figurative mark was not sufficiently important in the perception of that mark as a whole as to alter its distinctive character.' This outcome is considered to be unacceptable, but should be taken as an isolated instance without effects beyond the actual decision.'

## *2. Registration of the variation*

Pursuant to Article 18(1)(a) EUTMR, the separate registration of the variation of an EU trademark already registered does not constitute an obstacle to recognizing the use of the (registered) variation as sufficient to maintain rights in the original mark. While this is now part of the legislation, there was a period in the past where there were doubts whether the registration of the variation would constitute an obstacle to accepting the use of that variation as sufficient to support a registered trademark. Previously, the case law of the Court of Justice had already rejected the notion that the separate registration of the variation precluded attribution of its use to another mark.

### 3. *Use of more than one trademark*

The use of a variation – where the sign as used differs from the mark as registered – must be distinguished from the concurrent use of more than one trademark. No problems or difficulties arise in situations where several marks are used on or in relation to a single product, such as a 'house' mark and a product mark. More problematic are cases where the signs are used in close proximity, and it is not clear whether there are two or more marks or a single combination sign. Such situations arise for example when the issue is acquired distinctiveness, and the sign alleged to have acquired distinctiveness is used as part of a larger sign. That issue arose in the case 'Have a Break – Have a Kit-Kat', where it was claimed that 'Have a Break' was itself a trademark that had acquired distinctiveness through use (Judgment of 7 July 2005, C-353/03, *Société des produits Nestlé SA v. Mars UK Ltd*). The Court of Justice decided that such combinations did not exclude that one element thereof was itself a mark.

The Court applied the same principle in a case where the issue was genuine use and the mark contested on grounds of non-use was a figurative (position) mark, a little textile flag on the side of jeans pockets, which was however used together with the word 'LEVI'S' (Judgment of 18 April 2013, C-12/12, *Colloseum Holding AG v. Levi Strauss & Co.*). The Court held as follows (at para. 36): '[T]he condition of genuine use of a trade mark [...] may be satisfied where a registered trade mark, which has become distinctive as a result of the use of another composite mark of which it constitutes one of the elements, is used only through that other composite mark, or where it is used only in conjunction with another mark, and the combination of those two marks is, furthermore, itself registered as a trade mark.' The Court added (para. 35): '[A] registered trade mark that is used only as part of a composite mark or in conjunction with another mark must continue to be perceived as indicative of the origin of the product at issue for that use to be covered by the term "genuine use".' Interestingly, in the *Specsavers* judgment (see above), the Court came to an opposite conclusion when a figurative mark was used with a word mark superimposed – the Court concluded that this altered the figurative mark.

## D. Use in relation to goods and services

According to Article 18(1) EUTMR, in order to avoid the sanctions of non-use, the mark must be used for the goods or services for which it is registered. In revocation proceedings, the mark will be invalidated for those goods or services for which it has not been genuinely used, whereas it will remain registered for goods or services for which it has been used (Article 58(2) EUTMR). In opposition or invalidation proceedings, when the proprietor of the earlier mark has to prove use, only those goods or services will be taken into account for which such use has been proven (Article 47(2), second sentence EUTMR and Article 64 (2), fourth sentence EUTMR).

### 1. *Use for each of the designated goods and services*

In practice, cases arise where the mark is actually used, but not for the goods or services for which the mark is registered. For example, where a mark is registered for footwear in class 25, but only used for orthopaedic shoes, which are classified in class 10, there was no use for the goods for which the mark was registered, and the respective sanctions apply, even though class 10 shoes and class 25 shoes may be considered to be similar. The rule previously applied in some jurisdictions that use for similar goods or services may be sufficient to maintain the mark has not become part of EU law.

## 2. *Use for a broader category of goods and/or services*

What is required is that the mark is used for goods or services which are actually the goods or services for which the mark is registered, or which fall under a broader category of goods or services for which the mark is registered. It is common practice that trademarks are registered for broadly defined categories of goods and services but are subsequently used in relation to a much narrower range of goods and services. Failure to use the trademark in respect of all the goods and services covered by the registration does not mean that the proprietor risks losing his rights entirely. On the contrary, it is clear from the provisions cited above that rights will be maintained or may be enforced in respect of those goods and services for which he has been able to prove genuine use of the trademark.

The principal difficulty in this respect arises when determining for which specification the mark may be maintained or will be deemed to be registered – at the one extreme, only the actual, specific goods or services for which the mark was used, or, at the other extreme, for a broad, generic category within which the specific goods or services fall.

The leading case of what was then the Court of First Instance (now the General Court) is the judgment of 14 July 2005, T-126/03, *Reckitt Benckiser (España), SL. OHIM/EUIPO – Aladin Gesellschaft für innovative mikrobiologische Systeme GmbH (ALADDIN/ALADIN)*, where the Court stated (at para. 51) that the provisions on genuine use must be applied by taking into account 'the legitimate interest of the proprietor in being able in the future to extend his range of goods or services, within the confines of the terms describing the goods or services for which the trade mark was registered, by using the protection which registration of the trade mark confers on him. That is particularly so when, as here, the goods and services for which the trade mark has been registered form a sufficiently narrowly-defined category.' According to current case law of the General Court and current practice of the EUIPO, a broader category of goods or services may continue to be registered if the goods or services for which the mark was used fall under this category. However, where the category allows itself to be divided into more specific subcategories, only a properly defined subcategory will remain registered or will be taken into account. Good examples are marks registered for pharmaceutical products in class 5 or software in class 9, broad terms accepted for registration purposes. In most situations such marks are used only for a specific product. The practice of the General Court and of the EUIPO in these situations allows the mark to be enforced (or maintained) for an appropriate subcategory formed by the indication (e.g. for the treatment of allergies) under which the product falls, while other criteria, such as dosage form (injection, oral) or whether the product is available only on prescription, are not relevant. These same principles are applied in relation to software and other broad categories of products as well.

## E. Territorial scope

### *1. Use in the European Union*

According to Article 18(1) EUTMR, an EU trademark must be used 'in the Union' (previously: 'in the Community'). From the beginning it was subject to debate whether the use must be made in more than one EU Member State. For the 'creators' of the system it was self-evident that such an 'interstate' requirement was incompatible with the notion of a single market. At the time the original Regulation on the Community trademark was adopted, Commission and Council agreed on an interpretative statement (published in the Official Journal of OHIM, 1996, p. 613), according to which '[t]he Council and the Commission consider that use which is genuine within the meaning of Article 15 in one country constitutes genuine use in the Community'. It is well-established that such statements are not binding, but they may still have an influence on the approach that courts take to the issue.

The leading Court of Justice judgment on the territorial scope of use – more accurately: the territory where an EU trademark must be used in order for the use to be recognized as genuine – is the *ONEL* judgment (Judgment of 19 December 2012, C-149/11, *Leno Merken BV v. Hagelkruis Beheer BV (ONEL/OMEL)*), where the Court concluded as follows: 'Article 15(1) of Regulation No. 207/2009 [now Article 18 (1)] must be interpreted as meaning that the territorial borders of the Member States should be disregarded in the assessment of whether a trade mark has been put to "genuine use in the Community" within the meaning of that provision.' Rather, the issue of 'genuine use' must be assessed by taking into account all of the circumstances of the case, among which also the territory where the mark has been used. According the Court (at para. 55): 'Since the assessment of whether the use of the trade mark is genuine is carried out by reference to all the facts and circumstances relevant to establishing whether the commercial exploitation of the mark serves to create or maintain market shares for the goods or services for which it was registered, it is impossible to determine a priori, and in the abstract, what territorial scope should be chosen in order to determine whether the use of the mark is genuine or not. A *de minimis* rule, which would not allow the national court to appraise all the circumstances of the dispute before it, cannot therefore be laid down.' The Court also added (at para. 50):

> 'Whilst there is admittedly some justification for thinking that a Community trade mark should – because it enjoys more extensive territorial protection than a national trade mark – be used in a larger area than the territory of a single Member State in order for the use to be regarded as "genuine use", it cannot be ruled out that, in certain circumstances, the market for the goods or services for which a Community trade mark has been registered is in fact restricted to the territory of a single Member State. In such a case, use of the Community trade mark on that territory might satisfy the conditions both for genuine use of a Community trade mark and for genuine use of a national trade mark.'

In conclusion, 'territory' is relevant, whereas the borders between Member States are not. Use in a small territory may – depending on the circumstances – still fail to meet the 'genuine use' test.

### 2. *Export use*

Under Article 18(1)(b) EUTMR, a mark shall be deemed to have been used in the EU when it is affixed to goods or their packaging within the EU solely for export outside the EU. In this way, an EU trademark may continue to be registered even if the goods are placed on the market outside, where the same mark is likely to be protected as well. For example, if the EU trademark provides the basis for a network of International Registrations, and the products made in the EU are made for export only, the EU trademark can be maintained, and thus also the network of International Registrations.

## F. Relevant time period and its calculation

In order to avoid the sanctions of non-use, as provided in Article 18(1) EUTMR, genuine use must have been made no later than after five years from the date of registration. If genuine use is begun within this period, the registered EU trademark is not subject to revocation and may be enforced vis-à-vis later marks in opposition, invalidation and infringement proceedings. This is the initial 'grace period', which is counted from the registration date.

The same time period of five years applies not only for the initial period after the date of registration, but for any subsequent period of absence of genuine use after a period of genuine use. Thus, if genuine use had been made for two years, and thereafter genuine use was discontinued, the EU trademark would be subject to the sanctions outlined above if the use was discontinued for more than five years.

# III. EVIDENCE OF GENUINE USE OF A TRADEMARK

## A. When use should be proved: timing of lack of genuine use claims

### 1. *When filing an application for a trademark*

Under EU trademark law, the applicant is under no obligation to claim – and even less to provide any evidence – of any intent to use the mark which is the subject of the application. Rather, a registered EU trademark is subject to revocation if not used within a continuous period of five years and may not be enforced against later EU or national trademarks if not used within a period of five year prior to the priority or filing date of the later EU or national trademark.

EU trademark registrations are valid for ten years from the filing date. They may be renewed for additional periods of ten years. There is no requirement of proving use when renewing an EU trademark. Renewal is effected merely by requesting renewal and paying the renewal fee.

### 2. *In the course of opposition proceedings*

As provided for in Article 47(2) and (3) EUTMR, in opposition proceedings brought against the registration of a later EU trademark on the basis of an earlier EU trademark or an earlier national trademark, the opposing party, if so requested by the EU trademark applicant, must prove genuine use of the earlier mark in the period of five years before the application or priority date of the later EU trademark, provided (as follows from the 'grace period' previously referred to) that the earlier mark at that point in time was registered for more than five years. Procedurally, the request for proving genuine use must be made by the EU trademark applicant at the first opportunity of replying to the substantiated opposition. The request must be made in a separate document. If the request to prove use is not made at this point in time and in this manner, the opponent is under no obligation to prove genuine use.

As regards earlier EU trademarks relied upon in opposition proceedings, the five-year grace period is calculated from the registration date of the earlier mark. Where the earlier mark is a national trademark, the five-year period is also calculated from the date of registration unless the national trademark system applies a post-registration opposition, in which case the five years are calculated from the end of the opposition period or, if an opposition was brought against the national trademark, from the end of that proceedings.

### 3. *In cancellation proceedings based on lack of use*

In revocation proceedings before the EUIPO based on the absence of genuine use under Article 58(1)(a) EUTMR, the EU trademark proprietor is under an obligation to prove genuine use of the challenged mark in the five years prior to the filing of the revocation action if, at that time, the earlier mark was registered for more than five years. Any use made by the proprietor after such expiration of five years and before the action for revocation was brought must be disregarded if it was made within three months before the revocation action was brought and the preparations for such use were initiated by the proprietor only while it was aware that the revocation action could be brought.

The request for a declaration may be brought by any person. There is no requirement of 'interest' or any other 'standing' requirement.

The same rule applies for revocation actions brought by way of a counterclaim before a national EU trademark court as a reply to an infringement action brought by the proprietor of the EU trademark. In such a situation, the action for a declaration is brought by the defendant in the infringement action, and the defendant will be in compliance with any standing requirements which may exist under the applicable national procedural or substantive law.

### 4. *As a defence in infringement/cancellation proceedings*

Pursuant to Article 64(2) and (3) EUTMR, in an invalidation proceedings brought against a later EU trademark on the basis of an earlier EU trademark or an earlier national trademark, the proprietor of the earlier mark, if so requested by the proprietor of the contested mark, must prove genuine use of the earlier mark in the period of five years before the filing of the request for a declaration of invalidity, provided (as follows from the 'grace period' previously referred to) that the earlier mark at that point in time was registered for more than five years. In addition, the proprietor of the earlier mark must also prove genuine use of the earlier mark at the filing or priority date of the later mark if the earlier mark, at that point in time, was already registered for more than five years.

Procedurally, the request for proving genuine use must be made by the proprietor of the contested EU trademark at the first opportunity of replying to the substantiated invalidity request. The request must be made in a separate document. If the request to prove use is not made at this point in time and in this manner, the opponent is under no obligation to prove genuine use.

As regards earlier EU trademarks relied upon in an invalidation proceedings, the five-year grace period is calculated from the registration date of the earlier mark. Where the earlier mark is a national trademark, the five-year period is also calculated from the date of registration unless the national trademark system applies a post-registration opposition, in which case the five years are calculated from the end of the opposition period or, if an opposition was brought against the national trademark, from the end of that proceedings.

Pursuant to Article 127(3) EUTMR, in an infringement action brought by the proprietor of an EU trademark, the defendant may raise the absence of genuine use of the asserted EU trademark as a defence. The defence will succeed if the asserted trademark was not genuinely used in the five years preceding the initiation of the infringement action, provided the asserted EU trademark was, at that point in time, already registered for more than five years.

As an alternative to the defence of absence of use, the defendant may bring an action for a declaration of revocation before the EUIPO (see above) or bring a counterclaim in the infringement action pursuant to Article 128 EUTMR requesting a declaration of revocation of the asserted mark.

If a separate action is brought before the EUIPO, the court before which the infringement action is pending may suspend or stay the action until the revocation action is terminated. Whether such suspension or stay will be ordered will depend on the procedural situation and the relevant procedural rules applied by the infringement court.

## B. Burden of proof

In all actions before the EUIPO – opposition, revocation, invalidation – the proprietor of the earlier mark, if properly challenged, bears the burden of proof.

If the issue of genuine use is raised as a defence in an infringement action (see above section A.5) or in a counterclaim seeking a declaration of revocation of the asserted EU trademark, the procedural rules of the Member State where the infringement court is located apply. The EU Trade Mark Regulation does not prescribe what the defendant must allege or prove as a requirement for raising the defence of non-use or requesting a declaration of revocation.

## C. Relevant evidence

According to Article 97(1) EUTMR, in proceedings before EUIPO, including opposition, revocation, and invalidation proceedings, the means of giving or obtaining evidence shall include the following:

(a) hearing the parties;
(b) requests for information;
(c) the production of documents and items of evidence;
(d) hearing witnesses;
(e) opinions by experts;
(f) statements in writing sworn or affirmed or having a similar effect under the law of the state in which the statement is drawn up.

In practice, almost all proceedings before the EUIPO are in writing, without any oral hearings or hearing of witnesses. Thus, all the evidence that is provided, including evidence provided to prove genuine use in revocation, opposition and invalidation proceedings, consists of documents and other items of evidence available in writing and statements provided on behalf of the parties, sworn or affirmed, but also just as written witness statements.

When genuine use must be proved, the evidence must show that all the elements of genuine use are satisfied – time of use, i.e. within the required periods prior to bringing the action; nature of use, i.e. use of the mark as a trademark, or use of an acceptable variation of the mark as registered; use of the mark for the goods or services for which it is registered and on which the party relies; use in the relevant territory (in the Union for EU trademarks, and in the respective Member State for national trademarks); extent of use, such as volume of sales, amount and nature of advertising, so as to establish the genuineness of the use.

As regards proceedings before national courts where genuine use issues may also arise the rules of the respective national jurisdiction on evidence and proof apply.

In what follows we will focus on evidence of genuine use in proceedings before the EUIPO.

## *1. Use on the product or its packaging*

### *i. Use for goods*

It is undisputed that use of the mark on the product itself or on the packaging of the product constitutes trademark use. 'On the product' includes use of the mark on labels or tags or similar identifying means. The answer to the question of whether the use must be visually prominent or otherwise apparent depends on the circumstances of the case: on a small item of jewellery the mark will obviously not be prominently displayed, whereas a mention of a mark in small type on a product displaying a primary trademark may not suffice as genuine use.

### *ii. Use for services*

Following from the nature of services as incorporeal products, trademarks cannot be directly used 'on' services, and obviously not on their packaging. Therefore, as regards service marks, evidence of use will consist of use on business papers and in advertising or displays.

## 2. *Use in advertising*

The Guidelines of the EUIPO contain the following very helpful statement: 'Trade marks fulfil their function of indicating the commercial origin of goods or services and as symbols of the goodwill of their owner not only when they are actually used on or for goods or services, but also when they are used in advertising. In fact, the advertising or market communication function of trademarks is one of their most important functions. Therefore, use in advertising will generally be considered as amounting to genuine use: (1) if the volume of advertising is sufficient to constitute genuine public use of the mark; and (2) if a relation can be established between the mark and the goods or services for which the mark is registered.' The CJEU followed this analysis in the *Minimax* case, Judgment of 11 March 2003, C-40/01, *Ansul BV v. Ajax Brandbeveiliging BV (MINIMAX)* (see above section II.A.1), where the Court held (at para. 37):

> 'It follows that "genuine use" of the mark entails use of the mark on the market for the goods or services protected by that mark and not just internal use by the undertaking concerned. The protection the mark confers and the consequences of registering it in terms of enforceability vis-à-vis third parties cannot continue to operate if the mark loses its commercial *raison d'être*, which is to create or preserve an outlet for the goods or services that bear the sign of which it is composed, as distinct from the goods or services of other undertakings. Use of the mark must therefore relate to goods or services already marketed or about to be marketed and for which preparations by the undertaking to secure customers are under way, particularly in the form of advertising campaigns.'

Evidence of use of a mark in advertising is relevant evidence, primarily so if the use in advertising is in addition to the use of the mark on goods or their packaging. As regards service marks, advertising use is direct evidence of use. Also, advertising prior to the actual introduction and presence of goods or services on the market, if made to establish a market, is relevant evidence for proving genuine use. There is currently no case law on advertising use without any subsequent marketing of goods or services. Like in most situations, the outcome will depend on the circumstances of each case – such as the nature of the goods or services, availability in third countries, etc. For example, advertising hotel services offered in a third country in the EU may be sufficient to constitute proof of genuine use.

## 3. *Use on the internet*

Evidence of use of a mark on the internet is relevant evidence, provided it is connected with the use of the mark for goods or services offered or available within the European Union. Thus, the evidence must show, in particular, place, time, and extent of use. For example, evidence of the visits to a particular website and a particular product may be particularly relevant. Internet evidence may often be the only reliable evidence when the sale of goods is made only via the internet.

In particular, the value in terms of evidence of internet extracts can be strengthened by submitting evidence that the specific website has been visited and, in particular, that orders for the relevant goods and services have been made through the website by a certain number of customers in the relevant period. For instance, useful evidence in this regard could be records that are generally kept when operating a business web page, for example records relating to the hits attained at various points in time or, in some cases, the countries from which the web page has been accessed. As to the relevant period, information on the internet or in online databases is considered to be of the date on which the information was posted. Internet websites often contain highly relevant information. Certain information may even be available only on the internet from such websites. This includes, for example, online catalogues or other information that is actually not available in paper form.

#### 4. *Form of evidence*

In proceedings before the EUIPO, the means of proving facts, including proving genuine use, are set out above. Sworn or affirmed statements are a frequently used means of evidence. Their value depends on who is making the statement and how that person is related to the parties. Statements made by persons close to the party having the burden of proof are given less weight, and statements unsupported by other evidence are less convincing than those supported by relevant documentary evidence.

#### 5. *Distinction between use of a company name and use of the services offered by the company*

Use of a company name as such, without any relation to any goods or services, is only use of a company name, and not use of a trademark which may be identical to or contained within the company name.

## IV. JUSTIFICATION FOR NON-USE

### A. Obstacles to use

The leading case on the concept of proper reasons for non-use is the *Häupl* judgment of the Court of Justice (Judgment of 14 June 2007, C-246/05, *Armin Häupl v. Lidl Stiftung & Co. KG (LE CHEF DE CUISINE)*). In that case, the claimant had sought to revoke a mark registered in Austria on the first day after the end of the 'grace period'. The proprietor invoked 'proper reasons', relying on bureaucratic obstacles which had prevented the use of the mark, in particular delays in the issue of operating licences. The Court took account of Article 19(1) of the TRIPS Agreement, according to which circumstances arising independently of the will of the owner of the trademark which constitute an obstacle to the use of the trademark, such as import restrictions or other government requirements, are valid reasons for non-use. The Court concluded as follows (at para. 54 and 55):

> 'It follows that only obstacles having a sufficiently direct relationship with a trade mark making its use impossible or unreasonable, and which arise independently of the will of the proprietor of that mark, may be described as "proper reasons for non-use" of that mark. It must be assessed on a case-by-case basis whether

> a change in the strategy of the undertaking to circumvent the obstacle under consideration would make the use of that mark unreasonable. [...] Article 12(1) of the Directive must be interpreted as meaning that obstacles having a direct relationship with a trade mark which make its use impossible or unreasonable and which are independent of the will of the proprietor of that mark constitute "proper reasons for non-use" of the mark.'

This judgment, while rendered in the case of a national mark, is applicable, *mutatis mutandis*, also for EU trademarks.

Typical instances where there may be 'proper reasons' are pending certification or authorization procedures, such as for new pharmaceuticals, or for operating licences. On the other hand, mere economic difficulties – such as absence of funds etc. – are not 'proper reasons'. Similarly, absence of use for a 'defensive' trademark may not be excused on grounds of 'proper reasons'.

During the period of 'proper reasons' for the absence of genuine use, the respective 'grace period' of five years does not continue to run. Once the obstacles disappear, the 'grace period' continues. Thus, while resumption of use sets into motion a new 'grace period' if use is interrupted, 'proper reasons' merely suspend the running of the 'grace period'.

### B. Circumstances in which non-use may be cured

Once use is begun, after the end of the initial grace period or after the end of a continuous period of five years of non-use, the EU trademark is no longer subject to revocation, and it may be enforced vis-à-vis later EU or national marks. Thus, the mark – even though previously potentially invalid – is 'revalidated', the defect is 'cured'. This 'curing' effect is not absolute, however. Rather, as was explained earlier (under section III.A), a mark which was not in genuine use at the filing or priority date of later mark may not be enforced in opposition or subsequent invalidation proceedings. Also, the proprietor of a 'cured' EU trademark may not enforce this mark in infringement actions against proprietors of trademarks (national or EU) (so-called 'intervening rights') acquired through filing at a time when the earlier mark would have been subject to revocation on grounds of non-use.

## V. CONSEQUENCES OF NON-USE

### A. Consequences of non-use and relevance of the context

#### 1. *Consequences of non-use in revocation proceedings*

In a revocation action before EUIPO, it is for the revocation applicant to identify the extent to which he seeks a declaration of revocation. The request may extend to all the goods or services or be more restricted and cover only some of the goods or services for which the mark is registered. In the absence of proof of genuine use for any goods or services which are the subject of the proceedings, the contested EU trademark will be revoked for all those goods or services. To the extent that the proprietor is able to prove use, the revocation request will be rejected. In cases of partial use, the principles explained above (under section II.D.2) on partial use will be followed. This means that the mark may remain registered for a properly defined category or subcategory of the goods for which use has been demonstrated. The same principles apply when revocation is requested by way of a counterclaim pursuant to Article 128 EUTMR.

### 2. *Consequences of non-use in opposition or invalidation proceedings*

In opposition proceedings or in invalidation proceedings based on an earlier EU or national mark, failure to prove any use of the earlier mark for the goods or services on which the opposition or invalidation request is based, if genuine use has been properly put in issue, leads to the rejection of the opposition or the invalidation action as unfounded. If the opposing party or the invalidation applicant can show partial use, the earlier mark will be deemed to be registered for the properly defined group or sub-group of goods or services, defined in accordance with what has been explained regarding partial use (see section II.D.2). The conflict between the earlier and the later mark will then be determined on the basis of the goods or services for which the earlier mark is deemed to be registered.

The same principles apply when invalidation is requested by way of a counterclaim pursuant to Article 128 EUTMR.

### 3. *Consequences of non-use in infringement proceedings*

When the defendant in an infringement action brought on the basis of an EU trademark raises the defence of absence of genuine use, as provided for in Article 127 (3) EUTMR, the infringement action will be dismissed as unfounded if the claimant cannot show any use for the goods or services on which the infringement action is based. If he can show partial use, the rules referred to earlier on the effects of partial use will apply, and the infringement action will be judged on the basis of the goods or services for which the EU trademark is deemed to be registered.

### 4. *Intervening rights*

When a national or EU trademark has been filed at a point in time (or has a priority date) when an earlier EU (or national) mark would have been liable to be declared revoked on grounds of absence of genuine use, the proprietor of the earlier mark is precluded from prevailing in an infringement action (Article 16 EUTMR). The later mark – a so-called 'intervening right' – will coexist with the earlier mark.

## B. Extent of any revocation

In the absence of proof of genuine use for any goods or services which are the subject of the proceedings, the contested EU trademark will be revoked for all those goods or services. To the extent that the proprietor is able to prove use for all the goods and services which are contested, the revocation request will be rejected. To the extent that the proprietor is able to prove use for some but not for all the contested goods or services, the mark will be revoked in part and remain registered in part. In deciding whether the mark will remain registered for the part for which it has been used, the principles explained above (under II.D.2.) on partial use will be followed. This means that the mark may remain registered for a properly defined category or subcategory of the goods for which use has been demonstrated. The same principles apply when revocation is requested by way of a counterclaim pursuant to Article 128 EUTMR.

## C. Timing of any revocation

According to Article 62(1) EUTMR, the revocation of an EU trademark registration shall have the effect that the mark 'shall be deemed not to have had, as from the date of the application for revocation or of the counterclaim, the effects specified in this Regulation, to the extent that the rights of the proprietor have been revoked.' Thus, revocation has effect *ex nunc*. Nevertheless, a party to the revocation action may request that 'an earlier date, on which one of the grounds for revocation occurred, may be fixed in the decision'. This latter provision may be useful or even necessary when a party has been involved in infringement litigation, or attacked for an alleged infringement, and the revocation for non-use could have occurred even at date earlier or much earlier than the filing date of the request or counterclaim for revocation. In this way possible sanctions for infringement in the period up to the filing of the request or the counterclaim can be avoided.

# France

*Eléonore Gaspar and Laurine Janin-Reynaud*

## I. INTRODUCING GENUINE USE OF TRADEMARKS

### A. Legal framework

France is a party to international agreements on trademark protection, in particular the Paris Convention for the Protection of Industrial Property signed in 1883, the Agreement on Trade-Related Aspects of Intellectual Property Rights (TRIPS), the Madrid Agreement Concerning the International Registration of Marks and the Protocol relating to that Agreement, and also the Singapore Treaty on the Law of Trademarks.

As an EU member state, France is also part of the EU trademark system and will implement the latest EU Trademark Directive (2015/2436) in its national trademark law.

Since 1964, genuine use has been required in France to maintain a trademark. Article 11 §1 of Law No. 64-1360 of 31 December 1964, on manufacturer's marks, trademarks and service marks thus provided that the rights of a trademark owner 'who did not, without a legitimate reason, use the mark or cause it to be used in a public and unequivocal way during the five-year period preceding the revocation claim' were to be revoked.

Article L.714-5 of the Intellectual Property Code (IPC) (stemming from the law of 4 January 1991) reiterated the requirement of the Law of 1964 by taking account of the terms of the of the First Council Directive 89/104/EEC of 21 December 1988:

> 'An owner who has not, without good reason, put his mark to genuine use in connection with the goods or services referred to in the registration during an uninterrupted period of five years shall be liable to revocation of his rights.
>
> The following qualifies as such a use:
>
> a) Use made with the consent of the owner of the mark or, for collective marks, under the conditions of the regulation;
>
> b) Use of the mark in a different form that does not alter the distinctive character;

> c) Affixing of the mark to goods or to their packaging solely for export purposes.
>
> Revocation may be requested in court by any interested party. If the request bears only on part of the goods or services referred to in the registration, the revocation shall extend only to the goods or services concerned.
>
> Genuine use of the mark commenced or resumed after the five-year period set out in the first paragraph of this Article shall not preclude revocation if it was commenced within the three months preceding the request for revocation and after the owner became aware of the possibility of such a request.
>
> Proof of use shall rest with the owner of the mark that is the subject matter of the revocation request. It can be brought by any means.
>
> The revocation shall take effect on the date of expiry of the five-year period set out in the first paragraph of this Article. It shall have an absolute effect.'

On 16 December 2015, the European Parliament and Council adopted Directive 2015/2436 to approximate the laws of the Member States relating to trademarks. Such Directive contains provisions that will introduce new rules, authorities and practices in French national law, once implemented, and in particular the administrative proceedings for revocation because of non-use which are presently absent from French law.

Two dates of implementation are provided:

- 14 January 2019 for substantial provisions
- 14 January 2023 for administrative proceedings

Therefore, the present chapter will present, wherever it is relevant, the present rules as provided by the IPC, as well as the new mandatory rules provided by the Directive that will enter into force and modify French law when the Directive is implemented.

It has been announced that the Directive should be transposed into French law by the future law entitled 'PACTE' that the French government presented shortly on June 18, 2018. However, to date, the draft law has not been made available for consultation.

## B. Purpose and function of genuine use: policy considerations

Genuine use is an effective use of the trademark on the market, for goods or services, in accordance with the essential function of the trademark which is to guarantee the origin of these goods or services in order to create or preserve an outlet for them.

The French Supreme Court gives the following definition:

> 'A trademark is used genuinely when it is used in accordance with its essential function, which is to guarantee the identity of the origin of the goods or services for which it is registered.'[1]

The Supreme Court also refers to the requirement of the creation of market share as laid down by community case law: 'The use of a trademark, even minimal, can be

1. Supreme Court, Commercial Chamber, 30 November 2004, Appeal No. 02-18731; Supreme Court, Commercial Chamber, 9 March 2010, Appeal No. B 09-11.330.

sufficient for it to be qualified as genuine provided it is considered justified, in the relevant economic sector, to maintain or create a share in the market for the goods or services protected by the trademark'.[2]

According to the economic sense of the word, genuine use seems to be a compromise between the contradictory objectives of worldwide freedom of trade, on the one hand, and the special privilege granted to the trademark owner to create temporary barriers against competition, on the other hand. The obligation to use a trademark is therefore the consideration of the monopoly granted to its owner. Indeed, the right to assert a trademark against a third party is only conceivable if said trademark fulfils its function, that is to say if it is used. Legal commentators thus recall that the obligation of use is 'so to speak of the very nature of the trademark', since the function of trademarks, which is to indicate an origin, is only fulfilled through use. As a result, a trademark that is not used does not perform its function, and actually ceases to be a trademark in the legal and economic sense of the word.[3] Recital 31 of Directive 2015/2436 also provides that: 'Trade marks fulfil their purpose of distinguishing goods or services and allowing consumers to make informed choices only when they are actually used on the market.'

Another objective of genuine use is also to 'reduce the total number of trademarks registered and, consequently, the number of conflicts which arise between them.' The objective of uncluttering the Trademark Register is also to give new applicants a broader range from which to choose when they adopt new trademarks.

## II. WHAT QUALIFIES AS GENUINE USE OF A TRADEMARK

### A. Nature of use

#### *1. Use consistent with the function of the trademark*

Genuine use implies a use of the sign that is consistent with the function of the trademark, which is to guarantee the identity of the origin of a good or a service to consumers by allowing them to distinguish it, with no possible confusion, from those that have another origin.[4]

This was already what the Supreme Court asserted in a consistent way.[5] Thus, to avoid revocation, the use of the trademark must be real and effective. The sign filed must perform its function, which is to distinguish the goods offered for sale, either because they are affixed to such goods or because they are provided to the public together. Use 'as a mark' is required to maintain protection.

This means that uses that are not consistent with the function of a trademark cannot constitute genuine use that would avoid cancellation because of non-use.

2. Supreme Court, Commercial Chamber, 9 November 2010, *Sony – UMD trade mark*, Appeal No. D 09-999.
3. Paul Mathély, *Le Droit Français des Signes Distinctifs* (1984), 417.
4. Court of Justice of the European Union (CJEU), 11 March 2003, *Ansul*, C-40/01, paragraphs 43 and 36.
5. For instance Supreme Court, Commercial Chamber, 29 January 2013.

Use as a corporate name, trade name or business name is not adequate to prevent revocation, except if such use also aims to distinguish goods or services from those of other undertakings in the minds of consumers.

The affixing of the title of a work, or of the name of a character, on merchandise may or may not constitute use as a trademark, depending on whether it is considered that the sign is affixed to designate the work or, on the contrary, to identify the origin of the merchandise. In France, the Supreme Court overturned a decision that had ordered the revocation of the GOLDORAK trademark affixed to video cassettes and figurines on the grounds that GOLDORAK was used as the title of a work.[6] Indeed, the Supreme Court considered that the GOLDORAK denomination was affixed to identify goods on the market 'regardless of the fact that this denomination also designates the work and its main character.' The Supreme Court thus considered that the denomination was used as a trademark. It can be concluded that even when a sign is the title of a work or the name of a character, it can be used as a trademark to identify products based on this work (such as figurines). For the trademark rights to be maintained through such use, it must comply with the function of the trademark to indicate the origin of the good. Such a solution makes it possible to preserve the rights of owners who filed as a trademark the title of a work or the name of a character for goods other than the classic physical media of the work (books, DVDs, etc.). As a consequence, the fact that a sign is found to be a title is not sufficient to rule out its use as a trademark for merchandise products. This is consistent with the General Court of the European Union (EGC) case law.[7]

Courts have also ruled out descriptive uses that were not considered as trademark use. Thus, by way of example, the Supreme Court ruled that since the sign BOTOX was commonly used to designate the botulinum toxin itself, the Court of Appeals had rightly considered that this word had not been used as a trademark to designate the origin of the goods and ordered the revocation of the rights in this trademark.[8] Likewise, exclusively decorative uses should be ruled out. This is the position that the Supreme Court seems to have adopted. It thus rejected an appeal lodged against the decision of the Court of Appeals ordering the revocation of DPAM's 86 trademark on the grounds that the 86 signs affixed to the products and used with many variations over the years were perceived by the relevant public exclusively as a decorative pattern, and that the public did not associate it with the registered trademark.[9]

All the circumstances of use should be assessed on a case-by-case basis and assessment of whether a sign is used as a trademark depends on the perception of the consumer. It is necessary to make a concrete analysis of the function fulfilled by the sign in light of the products to which it is affixed.

6. Supreme Court, 1st Civil Chamber, 30 October 2007.
7. EGC, 30 June 2009, T-435/05 Danjaq.
8. Supreme Court, 27 January 2009, *Allergan v. Jouve.*
9. Supreme Court, 15 December 2009, *DPAM v. Zara France.*

## 2. *Use in the course of trade*

### *i. Public use*

Use of a trademark that is not visible or intended for the public shall not qualify as genuine use. Therefore, internal use or that exclusively made within the private or company sphere should not be taken into account to maintain trademark rights.

Genuine use results from a public and outward use of the trademark.

### *ii. Commercial purpose*

Only use in the course of trade is relevant but use in the course of trade does not necessarily imply a profit motive behind it. Nevertheless, genuine use shall be motivated by a commercial purpose, i.e., the aim of creating and preserving an outlet for the goods on the concerned market in order to remain consistent with the essential function of the mark and qualify as genuine use.

This raises the question of some specific uses, such as promotional use or effective preparation of commercial use.

In accordance with EU case law, use of a mark on promotional material for other goods shall not, in principle, be sufficient to constitute genuine use of the trademark. Therefore, distributing tee-shirts, mugs, accessories or other goodies in the context of promotional events in order to promote the goods and activities that are offered under the mark, will not be considered as genuine use for said promotional products. Indeed, the Court of Justice of the European Union (CJEU) has ruled that the use of a trademark in the form of free promotional goods which are given to purchaser of other goods of the trademark owner does not constitute genuine use.[10] However, the EGC also specified that the fact that promotional goods were distributed free of charge was not sufficient to exclude genuine use and that the real criteria lay in the creation of an outlet for those items or in distinguishing those items from the goods of other undertakings.[11]

In France however, the Paris Court of Appeals admitted a solution that seems a bit more flexible, but it predates the above-cited decisions of the CJEU and EGC. Indeed, the Court of Appeals considered that 'even if they are promotional items, such use also constitutes a use as a trademark, namely a use consistent with the essential function of the trademark which is to guarantee to the consumer or the end user the identity of origin of a good or a service, by allowing it to distinguish this good or service from those having another origin, with no possible confusion'.[12] To our knowledge, the Paris Court of Appeals has not ruled again on this issue since the *Silberquelle* decision.

Can use be genuine when the goods or services are not yet available on the market, but when preparations for placing them on the market are underway, such as advertising campaigns or public launch announcements?

10. CJEU, 15 January 2009, C-495/07 *Silberquelle.*
11. EGC, 9 September 2011, T-289/09, *Omnicare Inc. v. OHIM.*
12. Paris Court of Appeals, 23 May 2008.

French courts consider that genuine use implies an effective exploitation, and not a mere project. The Supreme Court has held that preparations made within the owner's company or with service providers or subcontractors were not sufficient to evidence genuine use.[13] However, under certain circumstances, French Courts could be more flexible, in light of European case law, which considers that the use of the mark could relate to goods or services about to be marketed and for which preparations by the undertaking to secure customers are underway, particularly in the form of advertising campaigns.[14] Indeed, in a decision of 16 April 2010, the Paris First Instance Court admitted genuine use of a trademark even though the goods were not on the market, as they had been presented to professionals and to the public under the trademark and commercialization was underway.[15]

As regards use on goods designed for export, there is a specific provision in the IPC.

Indeed, in application of Article L.714-5 c) of the IPC, 'the affixing of the trademark to products or their packaging with the sole purpose of exporting them' is likened to a genuine use of the trademark allowing the avoidance of revocation. In this regard, Article 16.5b) of Directive 2015/2436 includes, in potential genuine use, the 'affixing of the trademark to goods or to the packaging thereof in the Member State concerned solely for export purposes.'

The use of the trademark will be considered as genuine only if it can be evidenced that the trademark was affixed in France to the products or their packaging, and that the products to which the trademark is affixed are being sold for export.[16]

The mere manufacture of labels in France is not sufficient to demonstrate that the affixing of the trademark took place in France.[17]

The export must 'be real and not limited to mere preparations not followed by implementation'.[18]

### 3. *Use by the owner or with its consent*

In application of the provisions of Article L.714-5-a of the IPC, 'use made with the trademark owner's consent' is considered as genuine use of the trademark. The use of the trademark by a licensee, a distributor or any other market player, with the trademark owner's consent (franchisee, sales representative, agent, etc.), can therefore constitute genuine use for the maintenance of protection. Such is naturally not the case for acts of use carried out by an infringer.

The acts of the licensee can in themselves constitute genuine use of the trademark 'regardless of the lack of a duly registered license agreement'.[19] Nor is it important to evidence the existence of a written contract authorizing the use of the

13. Supreme Court, 19 March 2013, *Voyager*.
14. Decision of 11 March 2003, C-40/01, *Ansul*, para. 37.
15. Paris First Instance Court, 16 April 2010, *Oxygène*.
16. Supreme Court, Commercial Chamber, 12 November 1992; Paris First Instance Court, 23 February 2016.
17. Paris Court of Appeals, 20 September 2002.
18. Paris First Instance Court, 10 January 2001, Docket No. 99/1130.
19. Supreme Court, Commercial Chamber, 31 January 2006, Appeal No. 04-12945.

trademark since, according to case law, it is not required for the trademark owner's authorization to be given in writing,[20] provided however that it is 'revealed by a positive act, as mere tolerance cannot be admitted as valid as an authorization of the trademark owner'.[21]

In case of trademark use by a duly authorized licensee, the trademark owner does not need to satisfy any specific requirements (such as quality control, inspections) for the maintenance of the rights in the trademark. However, it may risk the revocation of its rights if the licensee does not use the trademark seriously, or if it does not use it 'as a trademark.'

Article 16 of Directive 2015/2436 confirms that '6. Use of the trademark with the consent of the proprietor shall be deemed to constitute use by the proprietor.'

### B. Volume, duration and frequency of use

The degree of use to maintain protection is to be assessed on a case-by-case basis, in a factual way, taking account of all the relevant factors of the case likely to evidence the reality of the commercial use. It is thus necessary to take account of the nature of the goods or services, the frequency or regularity of the use, the characteristics of the relevant market, the volume of the business activity, the production and marketing capacities of the company using the trademark, etc. A global assessment is required, which implies some interdependency between the various factors (for instance, small volume can be offset by high intensity or a certain consistency over time).

There is no quantitative threshold that could be taken into account abstractly. Therefore, no *de minimis* rule can be set.

However, it is necessary to determine whether the use is quantitatively sufficient to maintain or create market shares for the goods or services protected by the trademark.[22]

Token use whose sole purpose is to preserve the rights conferred by the trademark is not sufficient. Thus, fictitious use, which does not aim to ensure that the goods or services find their place on the market, is not sufficient.

By way of example, the Versailles Court of Appeals thus ordered the revocation of a LES CROISICAISES trademark for sardines because the owner only provided a small leaflet with the representation of a single tin of sardines bearing the trademark in a barely visible way when fish-based preserved foods are goods designed for the general public that can be manufactured or distributed in great quantities. On this occasion, the Court of Appeals ruled that 'genuine use shall mean a use which is not a token use for the sole purpose of preserving the rights conferred by the trademark, but one that complies with its function, which is to guarantee for the consumer the identity of the products for which it was registered'.[23] Similarly, the Supreme Court ruled that for mass markets, which are highly competitive, a trademark owner must

20. Paris Court of Appeals, 18 February 2005, Docket No. 03/00832, '*SOLEVITA*', PIBD 2005, III, p.266.
21. Supreme Court, Commercial Chamber, 24 September 2003, Appeal No. 01-14934.
22. Supreme Court, 9 November 2010.
23. Versailles Court of Appeals, 9 March 2010, *Saupiquet v. Chancerelle.*

prove its intent to be part of such market and cannot solely provide token use for the sole purpose of maintaining its rights.[24]

However, even minimal use can be sufficient when it serves a true commercial purpose. The Supreme Court ruled along these lines in the following terms: 'Even minimal use of a trademark can be sufficient to be considered genuine, provided it is considered justified in the relevant economic field to maintain or create market share for the goods or services protected by the trademark'.[25]

### C. Use of a sign

Use in the exact form in which the trademark is registered is not required.

Article 16 of Directive 2015/2436, which is identical to Article 5-C-2 of the Paris Union Convention, provides that 'use of the trademark in a form differing in elements which do not alter the distinctive character of the mark in the form in which it was registered' constitutes genuine use. This means that, in spite of a slight difference between the registered sign and the sign that is used, the trademark must be considered as used when this difference does not alter its distinctive character.

Under French law, this tolerance is set forth in paragraph 2b) of Article L.714-5 of the Intellectual Property Code, which provides that 'the use of the trademark in a different form that does not alter the distinctive character' constitutes a genuine use.

The owner may therefore change its trademark slightly, particularly for commercial reasons, and use a slightly different sign without incurring revocation.

Whether this rule also applies when the use of a trademark in a different form is also filed as a trademark is a question that has been widely discussed in French case law within the past twenty years.

After many reversals, French case law today is aligned on the EU position settled by the CJEU in the *Rintisch* decision of 25 October 2012 (C-553/11) and to which the French Supreme Court expressly refers:

> 'Article 10(2)(a) of Directive 89/104 must be interpreted as meaning that, in order to establish use of the trademark for the purposes of that provision, the proprietor of a registered trademark is not precluded from relying on the fact that it is used in a form which differs from the form in which it was registered, without the differences between the two forms altering the distinctive character of that trademark, even though that different form is itself registered as a trademark.'[26]

However, the Supreme Court also ruled that, in the specific context of a family of trademarks, the use of a trademark cannot be claimed to justify the use of another trademark.[27]

The rules enacted by the CJEU were incorporated in Article 16 of Directive 2015/2436 which provides that the following constitutes genuine use: 'use of the trade mark in a form differing in elements which do not alter the distinctive character

24. Paris First Instance Court, 6 March 2009.
25. Supreme Court, Commercial Chamber, 9 November 2010, *Sony – UMD trade mark*, Appeal No. D 09999.
26. Supreme Court, 3 June 2014, and 12 May 2015.
27. Supreme Court, Commercial Chamber, 19 January 2016.

of the mark in the form in which it was registered, regardless of whether or not the trademark in the form as used is also registered in the name of the proprietor.'

In principle, only minimal differences, namely those that do not alter the distinctiveness of the trademark, are permissible. Case law assesses on a case-by-case basis the impact of these differences on the distinctiveness of the trademark and on the public's perception thereof.

By way of example, it has been ruled that:

- The use of the YVRESSE denomination constitutes a use of the word mark claimed IVRESSE in a modified form which could be likened to a genuine use of this trademark.[28]
- The substitution of a cross of Lorraine for the dragonfly as represented in the sign (also composed of the word element *'le verre français'*) alters the distinctive character of the mark and cannot qualify as genuine use.[29]

- The use of the sole word element LOLLIPOPS without the specific presentation and calligraphy as registered does not alter the distinctive character of the trademark and qualifies as genuine use.[30]

- The addition of common words to the trademark composed of the sole Roman numeral alters the distinctive character of the mark and the use of 'Parfum VII Bonheur Triomphant' cannot qualify as genuine use of the trademark.[31]

- The use of the sole word element LA FUETTE (alone or in combination with other descriptive elements such as *'croustillante de Plaisir', 'gourmande', 'céréale',*) but without the figurative elements as registered could qualify as genuine use (the Court of Appeals will have to determine whether the word element LA FUETTE constitutes the dominant distinctive element).[32]

Concerning signs filed in black and white, as per French case law, their use in colour constitutes valid use. It is thus necessary to analyse a trademark application that claims no colour as an application allowing uses in many colours, provided that the words and lines of the trademark as registered are reproduced in the sign used. Indeed, with a black and white application, the owner wants the distinctiveness of the trademark to reside in elements other than colours.

As an example, in a decision relating to a revocation claim against the figurative 'Crumpler' trademark registered in black and white, the Paris First Instance Court

28. Paris Court of Appeals, 10 September 2008.
29. Supreme Court, Commercial Chamber, 9 March 2010.
30. Supreme Court, Commercial Chamber, 15 March 2011.
31. Supreme Court, Commercial Chamber, 15 September 2015.
32. Supreme Court, 16 February 2016.

considered that although the trademark was used in a modified form, the figurative and verbal signs were reproduced in the exact same way as they were filed. The only difference lies in the fact that, in order for the trademark to be visible on the products to which it was affixed, it was used in white on a black background instead of black on a white background. The First Instance Court admitted that 'this inversion of the basic colours of the trademark does not at all alter its distinctiveness because the determining elements were identically reproduced.' As a result, it dismissed the revocation claim.[33]

The matter is different for trademarks filed in colour: it is necessary to assess whether or not the owner wants the distinctiveness of the trademark to reside in its colour.

## D. Use in relation to goods and services

### *1. Use for each of the designated goods and services*

The trademark should be used for each of the goods or services listed in the trademark registration. Otherwise, the trademark can be partially revoked for the goods or services of its listing for which it has not been used. Partial use therefore only allows the maintenance of trademark rights for the goods and services used.

Indeed, where grounds for revocation of a trademark exist in respect of only some of the goods or services for which that trademark has been registered, revocation shall cover those goods or services only. On the other hand, revocation only extends to goods and services that have not been put to genuine use. However, when a trademark is partially revoked for certain goods or services, it can still be claimed against a later trademark for goods or services that are similar to those for which it remains valid.

### *2. Use for a broader category of goods and/or services*

As a principle, use for a similar product cannot be claimed as a use of the product listed in the trademark registration.[34]

When the use of the trademark is limited to a specific good, EU case law has introduced a distinction as to whether the good is listed in the trademark registration or whether only the general category containing the good is designated:

- If both the specific good (e.g., chair) and the general category (e.g., furniture) are designated, the use for the specific good allows the maintenance of the rights in the good (chair) but not in the general category (furniture).
- If the specific good (e.g., champagne/fishing rod) is not designated in the trademark registration, its use validates only the general category (e.g., alcoholic beverages/sports articles), designated in the trademark registration.[35]

33. Paris First Instance Court, 2 June 2009.
34. Supreme Court, Commercial Chamber, 9 March 2010, Appeal No. T 09-13231.
35. Paris First Instance Court, 20 January 2014 and December 2014.

Indeed, it was ruled that use for fishing equipment could qualify as genuine use of sports articles, as a subcategory thereof.[36]

However, EU case law has stated, in terms of proof of use in the context of an opposition, that

> 'if a trade mark has been registered for a category of goods or services that is sufficiently broad for a distinction to be made, within it, between several subcategories that can be considered autonomously, the proof of genuine use of the trade mark for part of these goods or services only provides protection, in opposition proceedings, for the subcategory or subcategories for which the trade mark has actually been used. However, if a trade mark was registered for goods or services defined in such a precise and circumscribed manner as to make it impossible to make significant divisions within the category at issue, then proof of genuine use of the trade mark for said goods or services will necessarily cover the entire category for the purposes of the opposition'.[37]

Thus, in opposition proceedings, although it is possible to contemplate autonomous subcategories within a general category designated in the trademark registration, the trademark is deemed registered only for the subcategory used when claimed against a third party. At present time, this is not applied in France as INPI does not assess in detail evidence of use in opposition proceedings.

In addition, as case law currently stands, even if a single specific product of a general category is used, the courts cannot limit the goods and services listed in the trademark only to this specific product not designated in the registration.

Indeed, the Paris First Instance Court has ruled that:

> 'Although the similarity between the goods or services that have been used and those that have not been used is traditionally inoperative with regard to the action for trademark cancellation covering them, it should nevertheless be remembered that although the principle of partial use operates to ensure that trademarks which have not been used for a given category of goods are not rendered unavailable, it must not, however, result in the proprietor of the earlier trademark being stripped of all protection for goods which, although not strictly identical to those in respect of which he has succeeded in proving genuine use, are not in essence different from them and belong to a single group which cannot be divided other than in an arbitrary manner.'[38]

## E. Territorial scope

Article 16 of Directive 2015/2436 provides that the trademark must be put to 'genuine use in the Member State in connection with the goods or services in respect of which it is registered.' The rule is therefore to use it in the national territory, which is the protected territory.

36. Paris First Instance Court, 5 December 2014.
37. EGC, July 14, 2005, and more recently General Court, 24 January 2017.
38. Paris First Instance Court, 30 September 2011.

It is clearly stated by the courts: 'In addition, the genuine use of the trademark must be assessed according to the principle of territoriality which implies only taking account of the use made in the national territory, excluding use abroad.'[39]

However, the texts in force do not define the expanse of the geographical area of trademark use in the national territory for it to be qualified as genuine.

Case law admits that use in a limited part of the French territory can constitute genuine use and the use of the trademark does not have to cover the entire territory of France. Thus, the Paris Court of Appeals ruled on 14 September 2001, that the use of a trademark exclusively in Martinique was genuine use.[40] The territorial scope is only one of the elements taken into account to assess whether the use is genuine.

### F. Relevant time period and its calculation

Article L.714-5 of the IPC only provides that non-use of a trademark registration during an uninterrupted period of five years can lead to the revocation of the trademark. It gives no indication on the starting point of the five-year period for genuine use but, in application of Article R.712-23 of the IPC, this time period begins upon the completion of the registration procedure.

Indeed, as per Article R.712-23 of the IPC, the date on which a mark is deemed registered shall be:

- for French marks: the date of the Official Bulletin of Industrial Property in which the registration is published
- for international marks: i) the date when the international trademark can no longer be provisionally refused in accordance with the Madrid Agreement or Protocol; or ii) the date when the international trademark can no longer be opposed in accordance with the IPL; or iii) the date when a decision ruling on the grant of protection, following either provisional refusal based on absolute grounds or an opposition, becomes final.

As a consequence, French trademarks and international registrations designating France benefit from a grace period of five years as of their registration during which non-use of the trademark will have no impact on their validity. In other words, the owner of a trademark registration has five years following registration to start using it genuinely and will not have to justify non-use of the trademark during this grace period.

Non-use of a trademark during an uninterrupted period of five years can lead to its revocation only if an action is brought against the trademark in this regard. As a result, resumption of use of a trademark, even after a period of non-use exceeding five years, can allow its owner to maintain its rights in the trademark (provided it proves genuine use once use is resumed).

However, the French IPC provides that the start or the resumption of the use of the trademark after the expiration of the five-year period allows the trademark owner to avoid the revocation of its trademark except:

---

39. Paris Court of Appeals, 3 November 2010; Supreme Court, Commercial Chamber, 9 March 2010, Appeal No. 0911813.
40. Paris Court of Appeals, 14 September 2001, Docket No. 1999/19667, *Danone v. Sodiaal*.

- if it was initiated within three months prior to the revocation claim (suspect period)
- and after the trademark owner became aware of the possibility of such a claim.[41]

This is consistent with Article 19 of Directive 2015/2436 which states that:

> '3. The commencement or resumption of use within the three-month period preceding the filing of the application for revocation which began at the earliest on expiry of the continuous five-year period of non-use shall be disregarded where preparations for the commencement or resumption occur only after the proprietor becomes aware that the application for revocation may be filed.'

## III. EVIDENCE OF GENUINE USE OF A TRADEMARK

### A. When use should be proved: timing of lack of genuine use claims

#### *1. When filing an application for a trademark*

French law does not require that the sign to be registered as a trademark be already used at the time of its filing nor does it require any intent of use. The French trademark office (i.e., INPI) will therefore not ask for evidence of use or declaration of use of the sign to proceed with its application and registration.

#### *2. In the course of opposition proceedings*

In opposition proceedings currently, Article R.712-17 of IPC provides that the owner of the trademark registration can, in its first observations in reply, invite the opponent to submit evidence capable of evidencing the fact that it is not subject to the revocation of its rights for non-use. The evidence submitted must prove the use of the earlier trademark during the five years preceding the request for proof of use, for at least one of the goods or services on which the opposition is based or provide proper reasons for non-use. The INPI will then give the opponent a deadline to submit such evidence.

As a result, INPI shall carry out a very limited examination of the proof of use submitted in the proceedings. Indeed, INPI 'cannot, except in case of obvious lack of relevance, make any pronouncement as to the exact scope of the evidence submitted [...] as only the court before which a revocation action has been submitted can make such an assessment'.[42]

Moreover, the opponent only has to prove the use of its trademark for one of the goods claimed. INPI does not have the power to reject the other goods and services claimed and for which use has not been proven, nor obviously to order their revocation.

41. Reims Court of Appeals, 13 November 2002; Supreme Court, 9 June 2009, appeal U/2008/16333.
42. Paris Court of Appeals, 19 October 1999, Jurisclasseur PI September 2002 p. 17.

Lastly, the applicant can request the suspension of the opposition proceedings when a lawsuit before judicial courts has been filed claiming the revocation of the trademark on which the opposition is based (Article L.712-4 of the IPC).

However, these rules regarding evidence of use in opposition proceeding will be modified when implementing Directive 2015/2436. Indeed, Article 44 of Directive 2015/2436 provides that

> '1. In opposition proceedings [...], where at the filing date or date of priority of the later trademark, the five-year period within which the earlier trademark must have been put to genuine use [...] had expired, at the request of the applicant, the proprietor of the earlier trademark who has given notice of opposition shall furnish proof that the earlier trademark has been put to genuine use [...] during the five-year period preceding the filing date or date of priority of the later trademark, or that proper reasons for non-use existed. In the absence of proof to this effect, the opposition shall be rejected.'

Once implemented, this provision will therefore change the starting point of the time period within which genuine use must be proven in the context of opposition proceedings. The relevant period to be taken into account will be the five years preceding the filing/priority date of the trademark application opposed and no longer the five years preceding the request of proof of use.

After implementation of the Directive, INPI will also have to examine in detail evidence of use provided as the earlier trademark will be deemed to be registered only in respect of the goods and services that were genuinely used.

### 3. *In cancellation proceedings based on lack of use*

At the present time, judicial courts have exclusive jurisdiction on revocation actions based on non-use of a trademark and could have to deal with the question of genuine use in a principal action or as a counterclaim to an infringement action.

Before French courts, it is only admissible for 'interested parties' to request proof of genuine use, which means that as per civil procedure rules, the plaintiff will have to demonstrate standing to sue (Article L.714-5 of the IPC).

Case law has refused to interpret the expression 'any interested party' as meaning that it would be admissible for anyone having some interest to file a revocation action. On the contrary, it has considered that only a revocation action initiated by a person having a legitimate interest, as defined in Article 31 of the Code of Civil Procedure, could be admissible.

There is a standing to sue when a claimant has shown a legitimate interest when its claim aims at removing an obstacle to the use of the sign in the context of its commercial activity.[43] For instance, the owner of an identical corporate name registered in the same field of business as the trademark owner[44] or the owner of a similar trademark who wishes to use its trademark without risking an infringement

43. Supreme Court, 24 June 2014.
44. Paris First Instance Court, 12 May 1993.

action,[45] all had their cancellation action found admissible based on non-use. This also applies to counterclaims.

Conversely, the claimant who did not establish that the goods for which it claimed the cancellation of the trademark for non-use were not part of his own field of business saw his claims declared inadmissible since he had failed to demonstrate that he was barred from commercializing this kind of goods.[46]

The same was decided about a claimant that did not show any actual plans to use the trademarks and whose cancellation action against them only reflected his intent to appropriate trademarks already known on the relevant market.[47]

In practice, in order to maintain the rights in the trademark, its owner must show genuine use during the period of five years preceding the action (principal) or the counterclaim (defence). However, the plaintiff can ask for the revocation to be pronounced as of the end date of an uninterrupted five-year period of non-use, if use has never been resumed.

Directive 2015/2436 introduces administrative proceedings. At the present time, no draft law has been made publicly available. According to Article 45 and Recital 38 of the Directive, Member States shall provide for an efficient and expeditious administrative procedure before their offices, particularly for the revocation of trademarks based on non-use. Such procedural changes enjoy a longer transposition period of seven years after the entry into force of this Directive (i.e., until 14 January 2023).

This is completely new in France and raises many questions in terms of double jurisdiction, INPI's scope of power, grounds of opposition and implementation of the procedural requirements imposed on the parties before the judicial courts.

After implementation of the Directive, introduction of administrative proceedings in revocation for lack of use will also raise issues as to where such actions should be brought and possible stays of the proceedings.

### 4. *As a defence in infringement/cancellation proceedings*

At the present time, it is possible to claim non-use of a trademark as a defence in infringement and cancellation proceedings, by requesting revocation because of non-use as a counterclaim in the same proceedings. A counterclaim seeking revocation based on non-use can even be invoked for the first time before the court of appeal.[48] Such a counterclaim will be examined by the court before ruling on the infringement or cancellation of the subsequent trademark or use.

However, Directive 2015/2436 introduces the possibility of claiming 'non-use' as a defence in cancellation proceedings or infringement actions even if the trademark at issue can no longer be revoked. Indeed, in such cases, the proprietor of a trademark shall be entitled to prohibit the use of a sign or to cancel a subsequent trademark only to the extent that the proprietor's rights are not liable to be revoked at the time the infringement action is brought (Article 17 of the Directive) or a subsequent trademark is filed (Article 46 of the Directive). Therefore, even if the prior trademark cannot

45. Paris First Instance Court, 16 June 2000.
46. Supreme Court, 13 January 2009.
47. Supreme Court, 1 March 2006.
48. Supreme Court, 31 January 2018.

be revoked because use was validly resumed, it cannot be opposed if it was liable to be revoked at the time the infringement action was brought, or to a subsequent trademark registration if it was liable to be revoked at the filing date or the date of priority of the later trademark.

This will drastically change the current practice before the judicial courts and the trademark strategy in France.

### B. Burden of proof

The burden of proof of genuine use or of the justification for non-use lies with the trademark owner wishing to avoid revocation.

### C. Relevant evidence

Proof of genuine use can be brought by any means.

It is necessary to make a global assessment of the use made, taking account of all the relevant factors of the case in point. Thus, the documents should be analysed globally and combined with each other.[49]

In French law, the IPC contains no indications on the relevant evidence. However, case law shows that, like EU law, the documents filed in the proceedings must, at least when combined together, establish the following facts:

- trademark used
- date of use
- territory of use
- goods or services at issue
- extent of use

#### *1. Use on the product or its packaging*

The best way to evidence genuine use consists in submitting pictures or samples of products bearing the trademark and proof of its presence on the French market. Use on the product or on its packaging is indeed the classic form of use of the trademark.

However, the trademark is not always affixed to the goods and when it comes to evidencing use of a mark in relation to services, the owner must rely on other kinds of evidence.

#### *2. Use in advertising*

Use in advertising can serve to avoid revocation. The advertisements must prove use in the territory for the designated goods and/or services. In general, the advertisements published are corroborated by other proof such as invoices or catalogues.

In this regard, the Paris Court of Appeals ruled that

49. EGC, 30 April 2008, T131/06, *Sonia Rykiel v. OHIM*.

> 'among the functions of the trademark there is not only the essential function of the trademark, which is to guarantee for consumers the origin of the goods or services, but also its other functions, particularly that of guaranteeing the quality of the goods or services in question, and also those of communication, investment and advertising. The proof of use submitted by [the appellant] Queguiner cannot be dismissed on the sole ground that certain elements of such proof are advertisements, store signboards, posters and advertising displays. It should be noted that on such communication media, the trademark is used as an element to attract clients in order to maintain or create market share relative to the goods listed in the trademark, which can establish the genuineness of the use'.[50]

Considering the nature of the goods and services, and the specificities of the market concerned, it is therefore possible to refer to advertising and other types of communication material.[51]

However, use in advertising is not sufficient if the goods or services are not sold on the market, or at least about to be sold.

Thus, in an old decision, it was ruled that an advertising campaign made without actually putting the product on the market did not constitute genuine use.[52] Indeed, the use of the trademark must bear on goods or services that are already being sold or whose sale is imminent and has been prepared by the company in order to create its clientele, particularly by means of advertising campaigns.[53] The solution of the *Ansul* decision is transposable to French trademark matters.

The Supreme Court also overturned a decision of the Rennes Court of Appeals which had considered that the affixing of a trademark on advertising and commercial documents constituted sufficient use, even though the trademark owner had not affixed it to the products (manufactured by third parties but selected and sold by him). The Supreme Court pointed out that genuine use of a trademark implies use on the market to designate the protected goods or services.[54]

However, a more recent decision of the Paris Court of Appeals of 23 November 2012, (referenced above) admitted that advertisements shall not be automatically set aside if they comply with the function of the trademark. Case law is therefore unsettled in such matters and the assessment of genuine use shall clearly be conducted on a case-by-case basis.

### 3. *Use on the internet*

Use on the internet can only serve to avoid revocation in case of trademark use to designate the goods or services designated by the trademark, in the relevant territory, to acquire or maintain market shares with the public of the relevant territory.

If the website is clearly intended for the French public, the use of the trademark on the internet to designate goods or services should be considered genuine. However,

50. Paris Court of Appeals, 23 November 2012.
51. Supreme Court, Commercial Chamber, 16 February 2016.
52. Paris First Instance Court, 31 October 1973, D 1974 763.
53. CJEU, 11 March 2003, *Ansul*, C-40/01, paragraphs 43 and 36.
54. Supreme Court, Commercial Chamber, 30 November 2004.

if no good or service has ever been ordered on the website for shipping to France, it is not certain that such use alone will be sufficient. It will be necessary to assess *inter alia* whether the offer for sale to the French public is real and serious or if a sale is imminent.

If products bearing the trademark have been ordered from France on a website clearly not intended for French consumers, such sales can be used to demonstrate the existence of genuine use. It will then be necessary to demonstrate that the trademarks at issue are truly affixed to the products purchased by French consumers via this site.[55] This also requires that the trademark be visible on the site, that it is used to designate the goods or services, and that goods are actually available for order by French consumers.

If a website is clearly not intended for the French public and if there is no significant number of orders made from France (or no serious proof of sales), this cannot qualify as serious use of the trademark.[56]

### 4. *Form of evidence*

The trademark owner may submit all kinds of evidence and does not need to provide an affidavit.

The most effective method would be to submit documents such as catalogues, products, packaging, orders, invoices, advertising. All evidence should be designed to show use of the trademark in France for the goods at issue, on a specific date, and the extent of such use.

Sales figures cannot in themselves prove that a trademark has been put to genuine use. Indeed, it is necessary to submit in the proceedings exhibits to supplement the sales figures to identify the products sold and prove that the trademark at issue was duly affixed to the products.

It is not mandatory to carry out a survey to evidence the use of a trademark. In fact, this is not common practice in France to evidence the use of a trademark.

### 5. *Distinction between use of a company name and use of the services offered by the company*

Classically, use as a corporate name, trade name or business name is not adequate to prevent revocation.[57] Nevertheless, even though the purpose of a corporate name, a trade name or a business name is not ultimately to distinguish goods or services, it can be found that use has been made for goods or services when the owner affixes the sign constituting its corporate name, trade name or business name to the goods it sells or if there is a connection between the sign and the goods or services. In such circumstances, use of a sign can constitute use as a trademark if the sign actually

55. Supreme Court, 9 November 2010.
56. Paris Court of Appeals, 24 November 2006, Supreme Court, Commercial Chamber, 20 March 2012.
57. Supreme Court, Commercial Chamber, 25 April 2001, Paris Court of Appeals, 25 April 2017, and Douai Court of Appeals, 29 March 2018.

designates goods and services in the minds of consumers.[58] As a consequence, the fact that a sign is found to be a corporate name or trade name is not sufficient to rule out its use as a trademark.

## IV. JUSTIFICATION FOR NON-USE

### A. Obstacles to use

Even if the trademark was not used genuinely, if the owner can evidence that there are 'proper reasons' for the non-use of the trademark (i.e. justifications for non-use), the owner can avoid the revocation of its trademark rights.

Article 19 of Directive 2015/2436 also refers to 'proper reasons for non-use'.

For a more specific definition, reference can be made to the TRIPS Agreement which refers to 'proper reasons based on the existence of obstacles to such use are shown by the trade mark owner', and specifies that 'circumstances arising independently of the will of the owner of the trade mark which constitute an obstacle to the use of the trade mark, such as import restrictions on or other government requirements for goods or services protected by the trade mark, shall be recognized as proper reasons for non-use.'

In a decision dated 13 September 2007 (case C-234-06P), pertaining to an opposition against the registration and providing of proof of use of the earlier trademark claimed, the ECJ also ruled that the notion of proper reason 'refers essentially to circumstances unconnected with the proprietor of a trademark which prevent him from using the mark, rather than to national legislation which makes an exception to the rule that a trade mark that has not been used for a period of five years must be revoked, even where such lack of use is intentional on the part of the proprietor of the trademark' (paragraph 102).

The exception to genuine use for proper reasons is not subject to any time limit.

Article L.714-5 of the IPC does not specify exactly the starting point of the five-year period for genuine use.

### B. Circumstances in which non-use may be cured

Generally speaking, French case law therefore admits that a 'proper reason' is:

- an obstacle of fact or law beyond the trademark owner's control, and
- one that is directly connected with the trademark to such an extent as to make the use thereof unreasonable, if not impossible.[59]

French case law has admitted that, for the assessment of proper reasons, the circumstances must clearly show the trademark owner's intent to use it for the goods

58. Toulouse Court of Appeals, 27 January 1992.
59. Paris Court of Appeals, 9 November 2007, *SA Chronopost v. DHL EXPRESS France*, Paris First Instance Court, 6 July 2017.

and services listed in the registration[60] . However, mere intent is not sufficient to avoid revocation.

The reasons for non-use must stem from an obstacle of fact or law.[61]

For example, the following situations have been considered by French case law as qualifying as proper reasons:

- the need to obtain an administrative authorization, particularly a drug marketing authorization;[62]
- the existence or threat of legal proceedings.[63]

## V. CONSEQUENCES OF NON-USE

### A. Consequences of non-use and relevance of the context

In theory and in application of the provisions of Article L.714-5 of the IPC, an owner who has not put its trademark to genuine use during an uninterrupted period of five years shall be liable to revocation of its rights. The revocation shall take effect upon the expiration of the period of five years of non-use (starting from the date of registration) and its effects are absolute.

However, the consequences of non-use depend on the context in which proof of use was requested as, at the present time, revocation because of non-use can be ordered only judicially.

As a result, at the present time, French courts have jurisdiction to revoke trademarks on the basis of non-use (either as a main claim or as a counterclaim) and, in this context, their decisions have absolute effect on the validity of the trademark.

It shall be borne in mind that use can be resumed at any time and that it can be relied upon to maintain validity of the trademark valid as long as the trademark is not challenged on the basis of non-use.

In the context of opposition proceedings, at the present time, INPI carries out a very limited examination of the proof of use submitted in the proceedings. Indeed, INPI cannot, except in case of obvious lack of relevance, make any assessment as to the exact scope of the evidence submitted. In any case, INPI is not entitled to revoke a trademark on the basis of non-use, it only looks into this in order to decide whether the opposition is admissible. This will change with the implementation of the Directive 2015/2436 that will give broader powers to INPI. The opposition will be examined only for the products and services that have been used.

With the implementation of Directive 2015/2436, non-use will also be a possible defence, even if the trademark cannot be revoked. Indeed, the proprietor of a trademark shall not be entitled to prohibit the use of a sign if the proprietor's rights were liable to be revoked because of non-use at the time the infringement action is brought. The same defence can be brought in cancellation proceedings if

60. Supreme Court, Commercial Chamber, 18 May 1993.
61. Supreme Court, Commercial Chamber, 2 February 2016.
62. Paris First Instance Court, 1 June 1999, and Orleans Court of Appeals, 13 February 2003.
63. Paris First Instance Court, 27 January 2017; Strasbourg First Instance Court, 13 January 2004; Versailles Court of Appeals, 29 April 2004).

the prior rights claimed were liable to be revoked because of non-use at the time the subsequent application is filed, or at the priority date.

This means that resumption of use (after a period of non-use) will not allow the trademark owner to oppose its rights to a subsequent user if this occurs after the introduction of the infringement proceedings or to a subsequent trademark owner if it occurs after the filing date or priority date of the subsequent trademark.

### B. Extent of any revocation

The trademark can be partially revoked for the goods or services of its listing for which it has not been used. Partial use therefore only allows the maintenance of trademark rights for the goods and services used.

When the use of the trademark is limited to a specific good, EU case law has introduced a distinction as to whether the good is listed in the trademark registration or whether only the general category containing the good is designated which seems to be applied by French courts:

- If both the specific good (e.g., chair) and the general category (e.g., furniture) are designated, the use for the specific good allows the maintenance of the rights in the good (chair) but not in the general category (furniture).
- If the specific good (e.g., champagne/fishing rod) is not designated in the trademark registration, its use validates only the general category (e.g., alcoholic beverages/sports articles), designated in the trademark registration.[64]

### C. Timing of any revocation

The date on which revocation of a trademark takes effect is determined by the courts.

As a principle, a trademark can be revoked with effect on the date five years after registration or after the expiration of an uninterrupted period of five years, provided use has not been resumed.

Said five-year period of uninterrupted non-use therefore starts either on the trademark registration date or on the date of last use of the trademark.

The revocation can therefore be pronounced as of a date that is prior to the revocation claim.

However, in practice, revocation because of non-use is often requested as of the date of the revocation claim and courts do therefore pronounce revocation at this date.

64. Paris First Instance Court, 20 January 2014 and 3 December 2014.

# Germany

*Rainer Böhm and Sandra Müller*

## I. INTRODUCING GENUINE USE OF TRADEMARKS

### A. Legal framework

In line with the European provisions, the owner of a German trademark is granted a grace period of five years from the date of registration, or in case of an opposition proceeding, the date of the final decision. The requirements of genuine use and the legal consequences of non-use are governed by several provisions in the German Trademark Act. The revised European Trademark Directive had no material impact on the legal framework.

### B. Purpose and function of genuine use: policy considerations

A registered trademark is the only registered IP right which potentially lasts forever. Following registration and subject to payment of fees a German trademark can be extended every ten years (cf. Section 47(2) German Trademark Act). In order to avoid the Trademark Register being cluttered with unused marks, German trademark law provides for instruments to terminate trademark protection once a registered trademark is no longer in use. As will be discussed further below, for removing a trademark from the German Register which is no longer in the grace period and has not been taken into genuine use, any third party may file a request for revocation of such a mark based on non-use. Further, non-use may be invoked by way of defence in infringement, opposition and cancellation proceedings. In all such proceedings the use requirement will be examined by the competent court or the German Patent and Trademark Office (GPTO).

## II. WHAT QUALIFIES AS GENUINE USE OF A TRADEMARK

### A. Nature of use

#### *1. Use consistent with the function of the trademark*

Genuine use in the sense of Section 26(1) of the German Trademark Act requires use that is consistent with the basic function of the trademark which is to guarantee the identity of the origin of goods or services to the relevant public by allowing its members to distinguish the goods or services from others which originate from a different undertaking.[1] It is thus crucial that the mark is (at least also) used as an indication of origin for certain goods or services.

Mere decorative use, for example on merchandising products or promotional material for other goods, will not be regarded as sufficient.[2]

The same applies if a designation is exclusively used as a company symbol in the sense of Section 5(2) German Trademark Act, for example at the top of order forms or invoices or as part of a domain name. Drawing a clear distinction is often impossible; in many cases, a designation will be perceived by the relevant public as a company symbol and as an indication of origin for certain products at the same time.[3]

Given that work titles, which are protected under Section 5(3) of the German Trademark Act, are intended to enable the relevant public to distinguish between different works, namely printed publications, cinematic works, music works, stage works and comparable works, they will often also be understood as an indication of origin. This applies, for example, if the right owner attaches the ® symbol or uses the designation for a whole series of works.[4]

#### *2. Use in the course of trade*

*i. Public use*

To fulfil its essential function as an indication of origin a trademark must be used externally. Use in the private sphere or mere internal use within a company does not qualify as genuine use. This principle has been well-established under German case law in a series of decisions at a rather early stage. For example, the German Federal Court of Justice (*Bundesgerichtshof* – BGH) ruled in 1979 that mere preparations to use a mark in the course of product testing within a rather short period of time will, in general, not be considered use of the mark in trade.[5] Similar considerations apply if a mark is used in a pharmaceutical approval process, however, in this case the requirement of passing such procedure may be regarded as a proper reason

1. BGH, judgment of 10 April 2008, I ZR 167/05 para. 22 – *LOTTOCARD*
2. Munich Appellate Court, judgment of 6 July 1995, 6 U 4543/94 – *The Beatles*.
3. BGH, judgment of 13 September 2007, I ZR 33/05 para. 28– *THE HOME STORE*; BGH, judgment of 9 October 2003, I ZR 65/00 – *Leysieffer*.
4. See, for example, Hamburg Appellate Court, judgment of 18 January 2012, 5 U 147/09.
5. BGH, decision of 19 December 1979, I ZB 4/78 – *Trend*.

for non-use.[6] If a subsidiary uses a trademark only when trading with its holding company and the products are not designated with the mark when they are finally brought on the market, such use within the group of companies is, in general, also be considered mere internal use under German case law.[7]

*ii. Commercial purpose*

In more recent decisions, the German courts have relied on the guidelines provided by the European Court of Justice (ECJ) when interpreting the concept of genuine use and in particular use with commercial purpose.

As follows from the *Ansul* case (*Ansul BV v. Ajax Brandbeveligung BV*, Case C-40/01),[8] genuine use of a mark entails use of the mark on the market for the goods or services protected by that mark. This is because the protection the mark confers and the consequences of registering it in terms of enforceability against third parties cannot continue to subsist if the mark loses is commercial purpose as an indication of origin. The ECJ also confirmed in this decision that token use, in order to preserve the rights conferred by the mark, does not qualify as genuine use, but a small amount of use might be sufficient as long as it serves a real commercial purpose.

In the decision *The Sunrider Corp*, Case C-416/04 P,[9] the ECJ noted the following on the issue of what constituted genuine use:

- Genuine use of a trademark means use of a mark in accordance with its essential function, namely to guarantee the identity of the origin of the relevant goods.
- Genuine use does not include token use for the sole purpose of preserving the rights conferred by the mark.
- When assessing whether use of the trademark is genuine, the court should consider all the facts and circumstances relevant to establishing whether the commercial exploitation of the mark is real.

In this judgment, the ECJ also held that the contested aspects of the judgment of the European General Court (CFI) which was under appeal were issues of fact, not law, and therefore should not be the subject of a review by the ECJ.

This principle was developed in the following cases, among others:

- In a case where a trademark was used by a non-profit organisation, the ECJ ruled that using a trademark for activities carried out for non-profit purposes such as announcing public fund-raising events, collecting donations from the public and distributing donations can constitute genuine use even where no trade is involved, provided that the mark is registered for such services (*Verein Radetzky-Orden v. Bundesvereinigung Kameradschaft "Feldmarschall Radetzky"*, Case C-442/07).[10] In particular, the ECJ ruled that use by a

6. BGH, decision of 24 November 1999, I ZB 17/97 – *IMMUNINE/IMUKIN*.
7. BGH, decision of 2 March 1979, I ZB 3/77 – *lamod*.
8. EU:C:2003:145 para. 37.
9. EU:C:2006:310 para. 70 et seq.
10. EU:C:2008:696.

non-profit organisation of its trademark in its relations with the public, in announcements of forthcoming events, on business papers and advertising material, and where the organisation's members wore badges featuring that trademark when collecting and distributing donations, amounted to genuine use of a mark. However, use of a trademark for internal purposes, such as on correspondence with existing members of the organisation, would not constitute genuine use of the mark. This distinction between external and internal uses of the trademark is consistent with principles set out by the ECJ in the *Ansul* case, but can become difficult to apply in practice in this sector, as it can be in the commercial sector.

- The ECJ further ruled that the requirement of genuine use was not fulfilled where promotional items were handed out as a reward for, and to encourage, the purchase of other goods as those items were not distributed with the aim of creating or maintaining the market for goods in the same class. This situation can be distinguished from that where the trademark owner sells promotional items, for example T-shirts, to promote goods sold under its mark (*Silberquelle GmbH v. Maselli Strickmode GmbH*, Case C-495/07).[11]

### 3. *Use by the owner or with its consent*

Section 26(1) German Trademark Act requires that the trademark has been used by its owner. Use of the trademark by a third party with the owner's consent may be attributed to the trademark owner provided that the consent has been declared prior to the respective act of use, for example by way of conclusion of a licence agreement. If consent is declared only after commencement of use, such use may only be attributed to the trademark owner from the moment of giving consent. A subsequent approval of use may thus not constitute genuine use in relation to a period in the past.[12] Merely accepting third party use is not considered consent in the sense of the provision.[13]

## B. Volume, duration and frequency of use

According to the ECJ's *Sunrider* decision[14] the question of whether use is sufficient to maintain or create market share for the goods or services protected by the mark depends on several factors and is to be assessed on a case-by-case basis taking into account all the relevant factors of the particular case (such as the characteristics of the goods and services, the frequency of use of the mark, and the commercial volumes involved). As the ECJ specifies under para. 72 of the decision, this means that it is not possible to determine *a priori*, and in the abstract, a quantitative threshold in order to determine whether use is genuine or not. A *de minimis* rule could thus

11. EU:C:2009:10.
12. BGH, judgment of 6 February 2013 – I ZR 106/11- *VOODOO*; BGH, decision of 11 October 1984, I ZB 14/83 – *FLUOSOL*.
13. Hamburg Appellate Court, decision of 3 July 1997, U 193/96 – *MATADOR*.
14. EU:C:2006:310 para. 71.

not be laid down. Instead, even minimal use of the trademark could be sufficient to establish genuine use as long as it served a real commercial purpose.

From the perspective of previous German case law, the ECJ applied a rather generous approach in this respect. It has thus been clarified by the German courts in more recent decisions that quantity of use is a factor which has to be handled comparatively subordinately and that a particular threshold of turnover would not be required. In line with the ECJ's findings according to which genuine use does not include mere token use it would be sufficient if it could be ruled out that the trademark owner was using its trademark merely formally, namely for the sole purpose of preserving the rights.[15]

As far as duration and frequency of use are concerned, corresponding to the European requirements any use before or after the five-year term will be considered irrelevant. Nevertheless, such use may be taken into account in the overall assessment of whether any use which has taken place in the relevant time frame is considered sufficiently serious. To qualify as sufficiently serious use must not necessarily cover the entire five-year-term. In fact, if use is only commenced shortly before the grace period expires this does not count in each and every case against the trademark owner seriously trying to acquire a commercial position on the relevant market, as it shall be conceded to the trademark owner to make use of the statutory time frames.[16]

Also the frequency of use is a factor which has to be assessed according to the circumstances of the individual case. Here again it is necessary that in consideration of the trademark owner's business concept the use in question can be deemed significant enough to exclude mere token use. For example, the Appellate Court Frankfurt considered use of a mark for fair hosting services every two years sufficient in the light of the two-yearly rhythm of the respective fair.[17]

## C. Use of a sign

Section 26(3) sentence 1 German Trademark Act provides that 'genuine use' includes use of the trademark in a shape which deviates from the registration if the deviations do not change the distinctive character of the trademark. According to Section 26(3) sentence 2, this also applies if the trademark is also registered in the shape in which it has been used.

Whether the distinctive character of the mark is altered in the sense of this provision depends on whether the relevant public, when taking the standards of use of trademarks in the relevant sector into account, would still recognise 'the same trademark', irrespective of the differences between the signs.[18] According to the BGH, this is a question of law.[19]

When evaluating use of a mark in a different form to which it has been registered, the German courts very much focus on the case law of the ECJ and the

15. See, for example, Cologne Appellate Court, judgment of 20 May 2009, 6 U 195/08 - *Protipower/Protifit*.
16. BGH, judgment of 17 January 1985, I ZR 107/83 - *topfitz/topfit*.
17. Frankfurt Appellate Court, judgement of 9 November 2006, 6 U 18/06 - *ISH*.
18. BGH, judgment of 10 January 2013, I ZR 84/09 - *PROTI II*.
19. BGH, judgment of 8 January 2014, I ZR 38/13 - *Probiotik*.

guidelines and practice of the EUIPO. This applies *inter alia* to the EUIPO's guidelines in relation to use of a monochrome (black-and-white) version of a colour mark for the purpose of establishing genuine use of a colour mark, and vice versa.

## D. Use in relation to goods and services

Pursuant to Section 26(1) of the German Trademark Act the mark has to be used for the goods and services for which it has been registered.

### *1. Use for each of the designated goods and services*

The requirements of genuine use must be fulfilled for each designated good and service claimed by the mark at issue. Whether the goods or services branded with the mark actually fall under a term contained in the description of goods and services has to be assessed from the perspective of the relevant public.[20] Also the edition of the Nice Classification which applied when the mark was registered may be taken into account.[21]

From the European perspective, it has to be considered that the ECJ ruled in *Ansul*[22] that use of a mark in relation to component parts for goods and related repair or maintenance services could constitute genuine use of the mark in relation to the main goods. In this case, use of the mark MINIMAX in relation to spare parts and ancillary services for fire extinguishers could, in principle, maintain a registration for fire extinguishers, despite there being no use on fire extinguishers themselves within the relevant period

### *2. Use for a broader category of goods and/or services*

If a mark has been registered for a certain category of goods and/or services, but has only been genuinely used for a specific good or service from that category, the question arises whether such use may nevertheless qualify as use for all goods and/or services contained in this category. Under German case law this question has been primarily discussed in the procedural context of revocation proceedings. We will therefore comment on that question further below.[23]

## E. Territorial scope

Section 26(1) of the German Trademark Act explicitly provides for the trademark being used in the territory of Germany. Use in the whole of Germany is not required in each and every case, though. Depending on the kind of goods and/or services as well as industry practice involved, regional use may still qualify as sufficiently serious.[24]

20. BGH, judgment of 13 July 1989, I ZR 157/87 – *Taurus*.
21. BGH, judgment of 8 January 2014, I ZR 38/13 – *Probiotik*.
22. EU:C:2003:145.
23. Please refer to section V.B.
24. BGH, judgment of 6 February 2013 – I ZR 106/11- *VOODOO*.

Use on the internet may only be deemed relevant if it has a commercial effect in the territory of Germany. There may be various indications speaking in favour of such commercial effect, for example, if the website content is in the German language, if there is a German flag displayed on the website and/or the currency information is set out in euros. In any event, mere accessibility of the website to German users will not amount to a commercial effect.

### F. Relevant time period and its calculation

The actual period for which use must be demonstrated is calculated back from the relevant date which is specified in the statutory provisions governing the respective procedural scenario in which non-use is invoked. We will therefore discuss these time periods in the context of our observations below on the different procedural scenarios.[25] The same applies with respect to the requirements that have to be fulfilled for non-use being cured.[26]

## III. EVIDENCE OF GENUINE USE OF A TRADEMARK

The genuine use requirement may become the subject of various procedural scenarios before the national courts and/or the GPTO. In all scenarios, German procedural law applies. In this section, we provide an overview of the major aspects of these procedural scenarios: timing, burden of proof and relevant evidence.

### A. When use should be proved: timing of lack of genuine use claims

The aspect of timing involves two perspectives: from the perspective of the party challenging the trademark the question arises as to when it should attack the mark and/or raise the defence of non-use. The trademark owner, on the other hand, has to consider at which point in the proceedings evidence of use has to be furnished.

#### *1. When filing an application for a trademark*

The GPTO does not assess whether the trademark that has been applied for has been put into genuine use. Correspondingly, no statements or documentation on use have to be filed by the applicant.

Unless the registration is challenged by a third party, a German trademark which has not been put into genuine use within the grace period of five years will remain in the Register. The GPTO will not request evidence of use. Any third party may apply for revocation of the mark at any time, though (Section 49 German Trademark Act). Further, the trademark owner shall not be entitled to enforce its rights in the trademark anymore. The adversary party may raise the defence of non-use (*Einrede mangelnder Benutzung*) in opposition proceedings (Section 43(1) German Trademark Act), infringement proceedings (Section 25(2) German Trademark Act)

25. Please see under section III.A.
26. Please refer to section IV.B.

and invalidation proceedings (Section 55(3) German Trademark Act). Non-use may also be asserted by way of a counter-claim action.

### 2. *In the course of opposition proceedings*

Pursuant to Section 43(1) German Trademark Act, if the trademark applicant disputes the use of the opposition mark, the opponent has to make a plausible case that the opposition mark has been genuinely used within the last five years prior to the publication of the registration of the objected-to trademark. If the period of five years of non-use ends after the publication of the registration, the opponent, if the applicant disputes the use, must make a plausible case that the trademark has been genuinely used within the past five years prior to the decision on the opposition.

The defence of non-use does not have to be raised at the earliest possible point in time before the GPTO. It may still be raised in the appeal proceedings.[27] However, given that with respect to genuine use the principle according to which the GPTO shall investigate the facts of the case *ex officio* is displaced by the principle of party presentation (*Beibringungsgrundsatz*) the civil procedural rules on preclusion apply.[28] In particular, the defence of use may be dismissed, if admitting it would delay the proceedings and the delay is the result of gross negligence by the opponent. This will only rarely be the case, though, as in German opposition proceedings, there is, in general, no specific date for a decision so that it is hardly ascertainable whether and if so, to which extent proceedings are delayed by the defence of non-use. Only where the GPTO in an appeal proceeding schedules oral hearings may the provision become relevant. Furthermore, when determining whether the opponent acted in a grossly negligent manner, the general rule, under which the defence of non-use must not be raised at the earliest possible point in time, has to be taken into account. The trademark applicant is rather to be granted a reasonable period of time for gathering information about the use of the opposition mark before raising a defence of non-use which is unfounded.

After the defence of non-use has been raised the opponent has to furnish all evidence which is required to make a plausible case that the opposition mark has been genuinely used. There are no statutory exclusion periods, however the opponent has to render evidence in such due time that the applicant is able to comment on such evidence. It is suggested to apply the standard of Section 132(2) of the German Code of Civil Procedure according to which preparatory written pleadings submitted in response to new submissions are to be filed in such due time that it may be served at least three days prior to the oral hearing.[29] Regarding the risk of preclusion, the above considerations basically apply accordingly. In particular, there will not be a preclusion in a written proceeding and if evidence of use is furnished shortly before or in an oral hearing, a delay may be avoided by the court determining a time limit within which the opponent may submit its submissions in response.[30]

---

27. BPatGE 17, 151, 153 – *Angifant*; BPatG 23, 168, 161 – *FLUDEX*; BPatG, decision of 16. February 1994, 26 W (pat) 63/92 – *Duotherm*.
28. BGH, decision of 4 May 1998, I ZB 9/96 – *DRAGON*.
29. Ströbele/Hacker/Thiering, § 43 para. 54.
30. Ströbele/Hacker,/Thiering § 43 para. 56.

### 3. *In cancellation proceedings based on lack of use*

Under Section 49(1) German Trademark Act any third party may file an application for revocation based on non-use. Two procedural options are available: according to Section 55(1) German Trademark Act, non-use may be asserted by way of an action with a national court. Without prejudice to this option, Section 53(1) German Trademark Act provides for the possibility to file the request for cancellation with the GPTO. If in this case the trademark owner does not object to the cancellation request within two months from being notified thereof by the GPTO, the registration shall be cancelled. If the trademark owner objects to the cancellation, the applicant has to proceed with an action in accordance with Section 55 German Trademark Act.

The decision between these two options is a strategic one. If the applicant does not expect the trademark owner to object to the revocation request, the office procedure provides for the possibility to obtain a revocation of the mark in a cost- and time-efficient manner, as the question of genuine use will not be examined. If it is likely that the trademark owner defends the mark, the two-step procedure may only consume additional time.

Apart from the general rules on forfeiture there is no specific time limit for filing a cancellation request based on non-use. However, in order to keep chances at bay that non-use is cured by commencement or resumption of use by the trademark owner, applicants who first lodged the request with the GPTO are well-advised to file the court action within three months after being notified by the GPTO that the trademark owner objects to the cancellation request. We shall comment on this requirement in more detail in relation to the conditions under which non-use may be cured.

From the perspective of the trademark owner, it is crucial to object to a cancellation request served by the GPTO within the given deadline. If the applicant follows up with a cancellation action before court, the general rules of German civil procedural law on preclusion apply.

### 4. *As a defence in infringement/cancellation proceedings*

In an infringement proceeding before a German court, the defendant may raise the defence of use under Section 25(2) German Trademark Act. In this case, the plaintiff must demonstrate that the trademark in suit has been genuinely used within the last five years prior to filing the action. If the five-year period of non-use ends subsequent to the filing of the action, the plaintiff shall demonstrate that the mark has been genuinely used within the last five years prior to the conclusion of the oral hearing.

Likewise, under Section 55(3) German Trademark Act the defence of use may also be raised by a defendant in a cancellation proceeding filed by the owner of a trademark with an earlier priority. If the contested mark was published when the earlier mark in suit had been eligible for cancellation, it has the status of an intermediate right and is thus allowed to coexist with the earlier mark and the cancellation action is to be dismissed.

In addition to raising the non-use defence, the defendant in an infringement or cancellation action may also file a counterclaim for revocation based on non-use of the plaintiff's mark in suit. Unlike cancellation requests based on absolute grounds of refusal, for which the GPTO is exclusively competent, applications for revocation based on non-use fall into the competence of the German courts. The additional option to file such a request with the GPTO only qualifies as a pre-procedure and amounts to a revocation of the mark only if the trademark owner does not defend itself. Against this background, a court proceeding would usually not be suspended until a decision on the counterclaim is issued, but the court would usually decide upon both actions at the same time.

For both parties the general rules on preclusion under German civil procedural law apply. This means, in particular, that submissions submitted after a deadline imposed by the court are only to be admitted at the court's discretion and only if admitting them would not delay the process of proceedings, or if the party provides sufficient excuse for such delay. The same considerations apply to submissions which are not submitted or communicated in due time, if the court finds at its discretion that admitting them to the proceedings would delay the process of proceedings and that the delay is the result of gross negligence. Submissions following close of the oral hearings, subsequent to which a judgment is handed down, are only admitted in exceptional cases.

## B. Burden of proof

As always under German civil procedural law, which would apply in opposition proceedings before the GPTO as well as far as genuine use is concerned, the required depth of substantiation and proof very much depends on the submissions of the other party, i.e. on which facts the counterparty contests. In German proceedings, what is not contested is deemed admitted. As a result, only facts which are disputed by the other party have to be proven.

The required standard of proof varies in the abovementioned procedural scenarios. The standard of proof in revocation, infringement and cancellation proceedings before a German court is that the judge must be fully convinced of the truth of the factual allegations of the respective party. In opposition proceedings before the GPTO as well as in preliminary injunction proceedings, a lower standard of prima facie evidence (*Glaubhaftmachung*) applies. This standard, which is set out in Section 294 German Code of Civil Procedure, is met if the factual allegations supporting the claim are in all probability true. In addition to the means of evidence admitted in proceedings where the standard of full conviction of the judge applies, Section 294 German Code of Civil Procedure allows the party to prove the facts by presenting an affidavit.

In opposition proceedings, it is, according to the wording of Section 43(1) sentence 1 German Trademark Act, the opponent who has to render prima facie evidence that the opposition mark has been genuinely used. The opponent's duty to furnish such evidence according to the standard of Section 294 German Code of Civil Procedure only extends to allegations of fact which require proof, though this may not be the case if the respective facts are undisputed or known by the GPTO.

In revocation proceedings before a German court, the plaintiff who challenges the mark based on non-use has the burden of substantiation and proof for the underlying facts. Given that non-use of a mark is a negative fact, i.e. the plaintiff has to demonstrate something which the defendant has *not* done, the burden of substantiation and proof may be facilitated or even shift. This applies in particular if the plaintiff had no chance to obtain knowledge of the circumstances of use and has no reasonable options to investigate by itself. In this case, the court may in particular assume a secondary burden of substantiation (*sekundäre Darlegungslast*) of the defendant to provide for sufficient information for the plaintiff to substantiate its claim.[31]

Pursuant to Section 25(2) German Trademark Act, in an infringement proceeding it is the plaintiff who must demonstrate genuine use in response to a defence of non-use raised by the defendant. As this explicit wording was not in line with Article 11 of EU Directive 2008/95/EG, which provided for the defendant to render proof that the mark in suit was eligible for revocation due to non-use, it had been discussed in Germany whether Section 25(2) German Trademark Act was in accordance with this provision or whether this provision was of mere optional nature based on the assumption that the burden of proof was a procedural question not to be harmonised. This dispute has been resolved by Article 17 of EU Directive 2015/2436 which now expressly provides that the trademark owner has to furnish proof for the genuine use of the mark in suit.

## C. Relevant evidence

Whether a trademark has been 'seriously used' in the sense of Section 26(1) German Trademark Act is deemed a question of law.[32] The response to that question requires finding of facts, though, which have to be presented by the parties in accordance with the respective procedural framework.

Evidence of genuine use shall consist of documentation concerning the place, time, extent and nature of use of the mark for the claimed goods and services. Such documentation may *inter alia* consist of catalogues and further advertising material, both online and offline, product packaging, labels, price lists, invoices and turnover figures. In line with the European requirements these pieces of evidence have to be assessed in conjunction with each other. The GPTO or the competent court has to evaluate the evidence by way of an overall assessment that takes all the circumstances of the individual case into account. In line with this principle, although certain pieces of evidence may be regarded insufficient by themselves to prove use of the mark in question, they may contribute to the finding of genuine use when combined with other materials.

### *1. Use on the product or its packaging*

Which kind of use is considered sufficient very much depends on the standard practice in the respective industry. If trademarks are customarily used on the goods themselves or their packaging, demonstration of such use will, in general, be considered essential by the German courts in order to assume genuine use. In particular, it will not be considered sufficient if the trademark merely appears on at the seller's premises, on

31. BGH, judgment of 10 April 2008, I ZR 167/05 para. 19– *LOTTOCARD*; BGH, judgment of 19 September 1996, I ZR 124/94 – *Beratungskompetenz*; BGH, judgment of 26 October 2006, I ZR 33/04 para. 31– *Regenwaldprojekt II*.
32. BGH, decision of 16 March 2000, I ZB 43/97 – *Bayer/BeiChem*.

its website, catalogues or shipping material. German courts have tended to apply this rather strict standard in a more flexible way recently, however, always provided that a sufficiently close connection between the mark and the goods in relation to which it shall be used can be assumed.

### 2. *Use in advertising*

Just like use on the goods or their packaging, use in advertising material will, in general, be considered a suitable means of evidence. This applies at least if such material shows the mark either on the goods displayed or in such direct context that a functional connection between the mark and the respective goods is established in the mind of the relevant public.

When furnishing proof of use by presenting advertising material trademark owners are well-advised to additionally substantiate in what quantities such material has been distributed, to which target public it has been handed out and within which timeframe.

### 3. *Use on the internet*

Use on the internet is, in general, not assessed differently as compared to other forms of advertising.

Print-outs from the trademark owner's website or social media sites may be considered a valuable piece of evidence, always provided that they are addressed to the German public (which will usually be assumed if the website content is in German with currency information in euros) and document use within the relevant time frame.

It is thus crucial to file print-outs with website time-stamps. Apart from internal website archives, which are operated by many trademark owners in view of the increasing importance of online marketing, the internet archive 'wayback machine' is frequently used before German courts in order to present proof of use on the internet. Depending on the respective industry, press clippings from online newspapers and magazines, postings in discussion boards on consumer websites or on social media channels such as Facebook, Instagram, Pinterest or YouTube may also provide valuable documentation of use, in particular when considering that they usually set out the date when the information was published.

### 4. *Form of evidence*

The GPTO and the German courts will also consider other forms of evidence such as invoices, orders, turnover figures, information on money spent in advertising and declarations by the trademark owner, its employees or third parties.

When presenting invoices and sales figures the trademark owner has to indicate the time period, frequency and territory of sales as well as why these sales are significant for proving genuine use. It also has to be set out what the respective goods and/or their packaging actually looked like, in particular, where the mark exactly appeared.

Such additional explanations will often be contained in other pieces of evidence. In many cases, the trademark owner will refer to an affidavit by itself or by an employee as the basis of its submissions. This applies at least in opposition and preliminary injunction proceedings where affidavits are deemed an admissible means of demonstrating a *prima facie* case. In a revocation, infringement or cancellation proceeding before a German court affidavits are not considered a proper means of evidence. In these proceedings, evidence may be provided by witness statements in a witness hearing or by an oral or written expert opinion.

Affidavits and witness statements by the trademark owner or its employees are often given less evidential value as compared to statements from independent sources, as the party issuing such statement will often have a personal interest in the matter. The German courts do not tend to generalise, though. A very detailed and specific statement by the director of the trademark owner which matches with the other pieces of evidence may, for example, be awarded more evidential value than a rather abstract and general statement of an independent party. Nevertheless, it is often recommendable to file additional third party declarations. The same applies for market surveys.

#### 5. *Distinction between use of a company name and use of the services offered by the company*

Under European and German case law where use of a designation is limited to identifying a company or designating a business such use does not qualify as use in relation to goods or services and is thus not suitable to establish genuine use.

It has been acknowledged under German case law, however, that it is rather difficult to draw a line between use as a trademark and use as a company identifier and that there may be quite some overlap. This applies in particular in cases in which use of the mark for services is claimed, as the mark may not be applied on the services as such. In such cases, use of the mark at the company's premises, on the website or on items which are used in connection with the supply of services (e.g. business papers, advertising material, professional apparel or price lists) will often be deemed sufficient to contribute to the assumption of genuine use.

In fact, German courts have found that the relevant public is accustomed to the parallel use of a designation as a trademark and a company identifier. Nevertheless, this does not relieve the trademark owner of the burden to present sustainable facts which confirm that the relevant public perceives the designation not only as a company identifier, but as an indication of origin for the designated goods/services in the individual case, and to submit evidence for such facts, if disputed by the other party.

## IV. JUSTIFICATION FOR NON-USE

### A. Obstacles to use

Section 26(1) German Trademark Act states that where the assertion of rights from a registered trademark or the upholding of the registration depends on the trademark having being used, it is required that the mark has been seriously used by its owner in Germany for the goods or services in respect of which it is registered unless the trademark owner can demonstrate proper reasons for non-use. As follows from the wording of this provision, reasonable justifications for non-use may be asserted in any kind of proceeding in which non-use may have legal consequences.

In the decision *Armin Häupl/Lidl*,[33] the ECJ provided some guidance about the concept of 'proper reasons for non-use'. In particular, the ECJ ruled that since the objective of proper reasons was to make it possible to justify situations in which there is no genuine use of the mark in order to avoid revocation of the mark, the function of such reasons was thus closely linked to that of genuine use, with the result that the term 'proper reasons for non-use' as contained in Article 16(1) of the EU Trade Mark Directive had to be interpreted in a uniform way, just as the concept of 'genuine use' of the mark.[34] In light of Article 19(1) of the TRIPS Agreement, which would be a factor to be taken into account in the interpretation, circumstances arising independently of the will of the trademark owner and constituting an obstacle to the use of the trademark were to be recognised as valid reasons for non-use.[35] Such obstacles would not necessarily need to make the use of the trademark impossible. Non-use could also be justified in cases where it would be unreasonable for the trademark owner to change its commercial strategy to overcome the obstacle.[36]

Although the concept of 'proper reasons for non-use' is a uniform one, the ECJ makes it clear that the national courts should apply this concept on a case-by-case basis. The German courts tend to apply rather strict standards. Mere economic obstacles such as unprofitability, difficulties in finding suitable licensees or shortage of raw material or capital will not be regarded as sufficient justifications for non-use. 'Proper reasons for non-use' have rather been assumed in cases where regulatory approvals make use of the mark impossible or at least unreasonable. This may, for example, apply for registration procedures for pharmaceutical products which may not be offered under the trademark before the administrative procedure is concluded.[37] In such cases, the regulatory procedure has to be directly linked to the envisaged use of the trademark and not a mere bureaucratic obstacle which is not be considered sufficient under the decision of the ECJ.[38]

## B. Circumstances in which non-use may be cured

If genuine use of a mark for which the grace period has expired is commenced or resumed before a cancellation action or revocation request based on non-use is filed, non-use may still be cured, although the trademark owner may have to accept the coexistence of intermediate rights acquired by third parties within the time frame in which the trademark was eligible for revocation.

33. EU:C:2007:340 para. 42 et seqq.
34. EU:C:2007:340 para. 44.
35. EU:C:2007:340 para. 48.
36. EU:C:2007:340 para. 53.
37. See already BGH, decision of 24 November 1999, I ZB 17/97 – *IMMUNINE/IMUKIN*; District Court Munich, judgment of 22 December 2015, 33 O 18890/14.
38. EU:C:2007:340 para. 52.

Section 49(1) sentence 3 German Trademark Act provides for an exception, however, where genuine use of the mark was commenced or resumed within three months preceding the date on which the action or request for revocation was filed. In this case, such use will be disregarded, if the trademark owner started preparations for the commencement or resumption of use only after becoming aware of the fact that a cancellation action or revocation request may be filed. The burden of proof for this exception is on the applicant for revocation. It is thus the applicant who has to demonstrate to the court that the trademark owner was aware of its intention to attack the mark.

If the application for cancellation is filed with the GPTO, according to Section 49(1) sentence 4 German Trademark Act the date of this application is considered material for the calculation of the three-month period only if the action for cancellation with court is lodged within three months after service of the notice that the trademark owner has objected to the cancellation request. If the plaintiff fails to meet this deadline the exception will not apply and non-use may, again, be cured.

## V. CONSEQUENCES OF NON-USE

### A. Consequences of non-use and relevance of the context

The consequences of non-use depend on the procedural scenario:

If the trademark owner does not object to the revocation request a third party filed with the GPTO, the Office will order the revocation of the mark without even having examined the question of genuine use. If the trademark owner objects to the revocation request or if the third party directly filed an action for revocation of the mark with court, it has to be noted that the court does not have the competence to order the revocation of the mark by itself. Instead, the defendant is sentenced to approve to the revocation of the mark towards the GPTO.

If the opponent in an opposition proceeding is not able to furnish sufficient evidence of use after the trademark applicant has raised the defence of non-use, the opposition will be dismissed.

Basically the same considerations apply to infringement and cancellation actions before a German court. If the defendant raises the non-use defence and the plaintiff fails to substantiate and, if disputed by the defendant, prove genuine use, the action will be dismissed. If the defendant additionally invoked non-use by way of a counterclaim action, such action will result in the plaintiff and counter-defendant being sentenced to approve the revocation of the mark towards the GPTO. There is no settled case law on the question of whether the trademark owner may nevertheless claim damages for infringing acts which occurred before the mark in suit became vulnerable to revocation. What counts in favour of this is that a trademark owner, who has obtained such claims, must have the economic freedom to cease using the mark anytime without the risk of losing its damage claims.[39] Considering that is does not seem justified to preserve damage claims which are merely based on formal

39. Ingerl/Rohnke, § 25 para. 29.

registration, it seems justified, though, to limit such time to the timeframe in which the mark in suit was no longer in the grace period.[40]

Non-use may be cured under the conditions set out under Section 49(1) sentences 2-4 German Trademark Act. In this case, trademark protection revives with its original priority. As follows from Section 22(1) no. 2 German Trademark Act, however, third parties may have acquired intermediate rights in the period of time during which the trademark was vulnerable to revocation. The trademark owner is not entitled to enforce its revived right against the owner of such intermediate mark or commercial designation, neither in an infringement proceeding (Section 22(1) no. 2 German Trademark Act) nor in a cancellation proceeding based on earlier rights (Section 55(3) sentence 3 German Trademark Act). Conversely, the owner of the intermediate right may not enforce such right against the owner of the mark for which non-use has been cured (Section 22(2) German Trademark Act), in other words, both parties have to accept a coexistence of their respective rights.

## B. Extent of any revocation

Section 49(4) German Trademark Act explicitly clarifies that if the trademark owner was not able to prove genuine use for a part of the goods or services for which the mark has been registered, the registration shall only be revoked for these goods and services.

At the time of the trademark application it is often not entirely clear for which goods and services the mark shall actually be used, which is the reason why trademark applicants will often opt for general terms used in the headings of the Nice Classification. If it later turns out that the mark has only been genuinely used for a part of the goods covered by the general term (say, for example, 'children's shirts' as part of the general term 'clothing') the question arises as to whether the general term may remain on the Register or whether this term has to be limited by way of partial revocation.

According to the BGH, if a trademark is used only for certain goods or services falling under a general term for which the mark is registered, the trademark is to be partially revoked. It is not necessarily just the goods/services for which genuine use has been assumed that remain on the Register, though. In order to preserve some economic mobility for the trademark owner, the goods/services which can be attributed to the same sector of products (*gleicher Warenbereich*) may also be maintained.[41]

The assessment of which goods/services fall under the same section of products depends on findings of facts, which for example involves the question of whether the relevant public distinguishes between certain sub-groups of products. Given that the demarcation criterion of the same product sector is rather indefinite applying this standard developed by the BGH may be difficult. In a more recent decision, the District Court Düsseldorf assumed that the general term 'clothing' was to be limited

40. Ingerl/Rohnke, § 25 para. 29.
41. BGH, judgment of 10 April 2008, I ZR 167/05 – *LOTTOCARD*.

to 'men's outerwear',[42] whereas a limitation to 'menswear' would have also been an option given that also other types of menswear may fall into the same product sector.

## C. Timing of any revocation

Under Section 52(1) sentence 1 German Trademark Act the effects of the registration of a trademark shall be deemed not to have taken place from the point in time when the action for revocation was lodged. The provision thus provides for a revocation *ex nunc*. If the revocation request was filed with the GPTO the provision applies accordingly, provided that the court action was filed within a time frame of three months according to the rationale of Section 49(1) sentence 4 German Trademark Act.[43]

Pursuant to Section 52(2) sentence 2 German Trademark Act an earlier point in time may be determined at which one of the grounds for revocation commenced to apply, if one of the parties requests so. It has been discussed whether such request requires a specific interest in such declaratory judgment, which will in most cases of such request be given.[44]

Correspondingly, if the contested trademark becomes eligible for revocation only after the action has been filed the court will have to consider this later point in time. In this case the plaintiff has not entirely prevailed, which may have consequences for the court's decision on costs.

42. District Court Düsseldorf, judgment of 26. November 2014 – 2a O 357/13 – *Anson's*.
43. Ströbele/Hacker/Thiering, § 52 para. 3; Ingerl/Rohnke, § 52 para. 11.
44. See, for example, Munich Appellate Court, judgment of 16 June 2005, 29 U 5456/04 – 800-Flowers in relation to a case where the point in time when the contested trademark became vulnerable for cancellation due to non-use was considered relevant in an opposition proceeding based on that mark before the EUIPO.

# Israel

*Avi Ordo and Moran Katz*

## I. INTRODUCING GENUINE USE OF TRADEMARKS

### A. Legal framework

Israeli trademark law is largely based on statutes and case law – i.e. judgments and decisions rendered by the civil courts and by the Registrar of Trademarks ('the Registrar') or Adjudicators (Registrar's and Adjudicators' decision shall be referred to as 'Registrar decisions').

The principal pieces of legislation regulating trademarks are:

- Trademarks Ordinance (new version) 5732 – 1972 ('the Ordinance');
- Trademarks Regulations, 1940 ('the Regulations');

In addition, Israel is party to the following multi-national conventions (among others):

- Protocol Relating to the Madrid Agreement Concerning the International Registration of Marks 1989 (Madrid Protocol);
- Nice Agreement Concerning the International Classification of Goods and Services for the Purpose of the Regulation of Marks 1957;
- Paris Convention for the Protection of Industrial Property 1883 (Paris Convention); and
- Agreement on Trade-Related Aspects of Intellectual Property Rights (TRIPS).

'Genuine Use' is regulated in Section 41 of the Ordinance ('Section 41') titled 'cancellation of registration on grounds of non-use'. According to Section 41 any interested person may submit an application for cancellation of the registration of a trademark, on the following grounds:

1. there was no *bona fide*[1] intention to use the trademark in connection with the goods in respect of which cancellation of the registration is requested and there has in fact been no *bona fide* use of the trademark in connection with said goods; or
2. there has been no such use as aforesaid during three years preceding the application for cancellation.

However, the above provision shall not apply if it is proven that the non-use is due to special circumstances in the trade, and not to any intention not to use, or to abandon, the registered trademark.

Accordingly, usually, three questions are considered in proceedings involving the cancellation of the registration of a trademark on grounds of non-use:

1. whether the trademark was in use during the relevant time period (three years prior to the application for cancellation); if not –
2. is there special circumstances in the trade that prevented use of the trademark; if not –
3. whether, despite non-use of the mark, the Registrar should exercise his discretion and not remove the trademark from the Registry of Trademarks ('Registry').

### B. Purpose and function of genuine use: policy considerations

The objects and purposes of the genuine use provision are: (i) to ensure that the Registry truly and accurately reflects the commercial reality, that all registered trademarks should be used on a regular and ongoing basis; (ii) to prevent the possibility of registering defensive trademarks, with no intention or desire of using them for commercial purposes; and (iii) to prevent the survival of trademarks that have become obsolete.

In the absence of the Registrar's authority to revoke trademarks on his own initiative, and in so far as the Ordinance does not require use of a mark prior to its registration, it is important to allow the public to 'clean out' the Registry by seeking for the removal of marks that have not been used for inordinate periods of time.

## II. WHAT QUALIFIES AS GENUINE USE OF A TRADEMARK

According to Section 41, there shall be deemed *not* to have been *bona fide* use of the trademark in the event of either of the following:

1. Use of the trademark in Israel in advertising only, whether in the local press or in foreign newspapers reaching Israel, unless there are special circumstances which, in the opinion of the court or the Registrar justify the non-use of the mark on goods manufactured or sold in Israel;
2. Cancellation of a licence to use the mark, given to a manufacturer in Israel under Section 50 of the Ordinance (which deals with a licence to use a mark

---

1. In one Registrar's decision it was clarified that the term '*bona fide* use' in Section 41 is the equivalent of the term 'genuine use' used in the EU Directive.

and the registration thereof), unless the licence is cancelled following an infringement of conditions, or because the person who granted the licence intends himself to manufacture the goods for which the mark is to be used or to grant the licence to another manufacturer in Israel.

According to Section 41, 'use' of a trademark, includes:

1. Use of a registered trademark by its proprietor or by a registered licensee as provided in Section 50 of the Ordinance, even if such use is of a different form from that entered in the Register in relation to details that do not alter the distinctive character of the mark as registered;
2. Use by a registered licensee as provided in Section 50 of the Ordinance, on condition that use thereof has been made subject to control by the proprietor of the mark.

## A. Nature of use

Genuine use must be examined having regard to all the circumstances of the case.

Particular emphasis when examining whether there has been genuine use of a trademark is placed on the specific goods or services and on the specific market in which the proprietor of the mark operates, as well as the trademark in question. In appropriate cases even single use of the trademark can sometimes be considered as genuine use. Nonetheless, the single use of a trademark must to be reasonable, according to the commercial circumstances related to the goods or service in question.

### *1. Use consistent with the function of the trademark*

According to case law, the mark should be used as a trademark, i.e., as a mark indicating the owner(s) of the mark as the source of the goods. Therefore, genuine use must be consistent with the purpose of the trademark. When there is no use of the trademark on its own, but merely use of the trademark as a descriptive addition alongside other trademarks – such use shall not be deemed 'genuine use' for the purpose of Section 41.

The term 'use' is interpreted broadly under Israeli case law. For example, based on rulings rendered by the Registrar, use of a registered trademark as part of a corporate name may also be considered as use for the purposes of Section 41, provided that the name of the corporation (which includes the mark) is impressed on the product or service falling within the specification thereof.

### *2. Use in the course of trade*

#### *i. Public use*

The object of a trademark is to distinguish the product or service offered by the trademark owner from other products or services; if the use is not public this object will not be fulfilled. Accordingly, use of a trademark entails offering products bearing the mark to the general public. In one Registrar's decision it was held that private sale to individuals, which is not made public, does not constitute 'use' of a trademark for the purpose of Section 41.

In another case, tax invoices bearing the mark (in words and not in the stylized form in which the mark was registered) issued for the payment of rent and not for the goods included in the mark's specification of goods, were not considered use of the mark.

#### *ii. Commercial purpose*

'Use' for the purpose of Section 41 is use of the mark with respect to the goods in respect of which it was registered. According to Section 41, use of the trademark in Israel in advertising only, whether in the local press or in foreign newspapers reaching Israel, will not be considered *bona fide* use of the trademark, unless there are special circumstances which, in the opinion of the court or the Registrar justify the non-use of the mark on goods manufactured or sold in Israel.

In accordance with the aforesaid provision the Registrar has ruled that using the mark in connection with promotional efforts alone, such as contacting potential customers, will not suffice to be considered as genuine use of the mark.

The court has also held that defensive actions, such as the sending of 'cease and desist' letters to infringers, customs seizures and the like, will not be considered 'use' for the purposes of proving genuine use.

### *3. Use by the owner or with its consent*

In accordance with Sections 41 and 50 of the Ordinance, 'use' includes use by the owner of the trademark or by a licensee on his behalf, provided that the licence to use has been duly registered in the Registry and on condition that the use has been made subject to control by the proprietor of the mark.

Therefore, if the licence has not been duly registered, use thereof by a licensee will not be deemed as genuine use by the owner of the trademark.

However, use by a corporation that is not the owner of the trademark, may be deemed genuine use by the proprietor of the trademark, even if the licence has not been duly registered in the Registry, where the corporations are affiliates, or connected in such a manner that any separation between them is insignificant or artificial.

## B. Volume, duration and frequency of use

The requirement of Section 41 includes two cumulative conditions which should exist for the trademark owner: (1) intent to use the mark, and (2) actual use of the mark. In one case the Registrar ruled that intent to use the trademark, as manifested in a distribution agreement to distribute products bearing the trademark in Israel, does not, in and of itself, constitute genuine use. In order to satisfy the requirements of Section 41, the intent to sell must be realized. In another case, the Registrar ruled that conducting legal proceedings around the world in relation to the challenged trademark, for example, cannot be construed as substantial intent to use the trademark.

It has been held in Israeli case law that the term 'use' should be interpreted broadly and leniently, and includes even minimal use. The requirement to use the trade mark does not entail extensive use of the trademark, and it is sufficient that it be used in a limited manner. However, the content of 'use' should not be exhausted, and it is necessary that actual use be made of the registered mark. In this context, the frequency of use of a trademark should be measured with reference to the specific product, and with emphasis being placed on normal use during the ordinary course of business, and taking into account the specific product and the specific market in which the proprietor of the trademark operates. In one Registrar's decision it was held that use should be continual. However, under certain circumstances even single use of a mark might amount to genuine use, provided that the commercial circumstances related to the specific product indicate so. For example, according to these principles, the Registrar ruled that a single sale of relatively simple products such as shoes is insufficient to prove use. Likewise, products such as newspapers require a certain frequency and rate of distribution: therefore, a publication of a single issue after a break of few years, cannot be considered normal use during the ordinary course of business. Further, it was held that a single act, such as submitting a preparation for registration with the Ministry of Health, does not constitute genuine use: use will commence only upon importation of the products (even if in small quantities).

## C. Use of a sign

According to the Ordinance, and as noted above, use of a trademark includes use of a registered trademark by its proprietor or by a proprietor of a registered trademark as provided in Section 50 of the Ordinance, 'in a different form from that entered in the register in relation to details that do not alter the distinctive character of the mark as registered'.

The question as to whether certain use of a trademark led to the distinctive character of the mark being altered, or whether the changes are minor, depends largely on the circumstances of the given case. The more the use deviates or differs from the format recorded in the Registry, the greater the possibility of the use not being considered as use of the registered trademark. Merely altering the trademark format from black and white to colour would not be deemed as altering the distinctive character of the trademark, since, according to guidelines published by the Registrar on the issue, registration in black and white by default covers colours as well.

The Registrar has discussed in a few cases the distinction between minor alterations and changes which affect the distinctive character of the mark. According to one Registrar's decision, in the event that a stylized registered logo is used in the format of printed (block) letters only, and not in the format of the stylized logo as recorded in the Registry, the use by the trademark owner in such a manner would not constitute genuine use of the stylized logo.

When the issue concerns a translated trademark (e.g. use in English of a trademark registered in Hebrew and vice versa) the pivotal question is whether there is an associative connection or phonetic similarity between the marks. Use of the trademark in one language does not automatically infer use of the trademark when translated into a different language, if there is no immediate associative connection between the marks. For example: in one Registrar's decision it was held that the use of the trademark 'Pesek Zman', in Hebrew (which can be translated as 'time-out' in English) does not constitute use of the mark 'Time-Out', since there is no immediate associative connection between the marks.

By contrast, it was held that using a mark which is a transliteration into Hebrew of a trademark originally registered in Latin (English) letters, will still constitute 'use' of the original trademark, provided that there is a phonetic similarity between the registered trademark and its transliteration.

Using a mark alongside trademarks owned by different entities will be considered genuine use, when it is clear that the source of the publication in which the trademark appears alongside such other trademarks is the trademark owner.

## D. Use in relation to goods and services

### 1. *Use for each of the designated goods and services*

With regard to the type of use, reference should be made to use of mark in relation to the goods or services falling within the specifications thereof. Use relating to goods or services which are not included in the specification has very little relevance, if any.

Use should be made in relation to each of the goods and services and cancellation due to non-use may be sought with respect to only a portion of the goods/services included in the specification. However, in one case, cancellation of the registered trademark on grounds of non-use was requested only for a portion of the goods included in the trademark specification, and the Registrar held that the interest of avoiding confusion among consumers dictates that the challenged goods which are related to goods for which cancellation was not sought, will also be kept on the Registry, notwithstanding non-use with respect to such challenged related goods.

With regard to the cancellation of a registered well-known trademark which is not used in the trade in Israel, the status conferred on a registered well-known trademark is higher than that conferred on a 'regular' registered trademark that is not well-known. In one Registrar's decision it was stated (in obiter dictum) that such a mark is not exposed to the same level of scrutiny regarding the extent of use with respect to each of the goods and services detailed in the specification. Nevertheless, even for a well-known trademark the Registrar requires 'some level of proof of use' (however, the Registrar does not specify the exact level of proof required, or how it will be measured) in the local market, in order to reject an application under Section 41.

### 2. *Use for a broader category of goods and/or services*

Use relating to goods or services which are not included in the specification has very little relevance, if any. Use for a broader category of goods and/or services may, however, be relevant to the issue of classification of the trademark as a well-known mark, which may, in turn, be taken into account by the Registrar when hearing an application to cancel the trademark registration on ground of non-use (see also section D.1. above).

## E. Territorial scope

Generally, the use should be territorial, i.e. the use should occur in Israel. It was determined that use of the mark abroad, and even long-term use, cannot replace the territorial use requirement.

However, the emphasis should be on the consumers' point of view, and the degree of their familiarity with the trademark. Thus, in certain cases, even if the product is sold over the internet, use of a trademark on a website will still constitute use in Israel. Use of a trademark on the internet may be relevant as long as it is made available to Israelis. The case law is unclear however, on whether the website on which the trademark is used must specifically be intended for and used by consumers in Israel.

## F. Relevant time period and its calculation

The relevant time period during which use should have taken place is the three-year period preceding the date on which an application is filed for cancellation of a trademark based on grounds of non-use. The relevant period does not take into account, in addition, the period preceding the date of registration of the trademark.

Examination is done according to the status of the trademark and the use made thereof as at the date of submission of an application seeking its cancellation. When rebutting such application, evidence concerning the period following the date of submission of the application is irrelevant. The reason for this stems from the need to prevent 'artificial resuscitation' of a trademark after receipt by the proprietor of an application seeking cancellation of the mark on grounds of non-use.

# III. EVIDENCE OF GENUINE USE OF A TRADEMARK

## A. When use should be proved: timing of lack of genuine use claims

### *1. When filing an application for a trademark*

There is no requirement under the Israeli law to show either actual use or intent to use, at the time of filing the application or even the registration.

Therefore, genuine use should be proven by the proprietor of the mark only where cancellation proceedings are instituted based on grounds of non-use.

### *2. In the course of opposition proceedings*

There are a few Registrar decisions expressing the view that registration of a defensive trademark (i.e. with no intention to use the trademark) for the purpose of blocking competitor or other third parties from using the trademark amounts to bad faith and/or unfair competition and as such may be the basis for a successful opposition. Accordingly, while there is no general requirement to provide proof of use in opposition proceedings, proof of use may be relevant in opposition proceedings in order to establish the secondary meaning of the trademark or where the basis for the opposition is a claim whereby there was no *bona fide* intention to use the trademark in connection with the goods in respect of which registration is requested and there has in fact been no *bona fide* use of the trademark in connection with said goods.

### 3. *In cancellation proceedings based on lack of use*

According to the language and purpose of Section 41, after a grace period of three years from the date of registration of the trademark, the right to file an application for cancellation of a trademark on grounds of non-use during the three years preceding the application is not allocated in time and is not obsolete. An application to cancel the registration of a trademark on the basis of a claim whereby there was no *bona fide* intention to use the trademark in connection with the goods in respect of which registration is requested and there has in fact been no *bona fide* use of the trademark in connection with said goods may be filed at any time after the trademark was registered.

Cancellation proceedings are available and are conducted in the Israeli Trademarks Office ('ITMO') before the Registrar or an Adjudicator.

The process is as follows:

- A cancellation notice must include a statement detailing the grounds of the application.
- The proprietor of the trademark must then submit to the ITMO, within two months of receipt of the application, a counter-notice, setting out the grounds for its submission.
- The applicant must then submit his evidence within two months.
- The proprietor can submit counter-evidence within two months after that.
- The applicant can also file, within an additional two-month period, evidence-in-reply to the evidence filed by the proprietor.
- After all evidence is filed, a hearing is set at which the respective attorneys can cross-examine each party's witnesses.
- Upon completion of the hearing, the parties must file their respective summations and the Registrar (or an Adjudicator) will then render his decision.

The Registrar's decision in an application for cancellation is subject to appeal to the District Court, within 30 days from the date of the Registrar's decision.

Any interested person can submit an application for cancellation of a registered trademark on the basis of non-use; the 'any interested person' standard is more lenient than the 'aggrieved person' standard required in other sections of the Ordinance (such as Section 38 of the Ordinance which deals with the rectification of the Register on the application of non-proprietor). It is immaterial whether the applicant in this regard wishes to use the contested trademark or to be registered as the proprietor of the trademark. A person wishing to use a trademark and who is prevented from so doing due to prior registration, may try to establish that no genuine use was made of the registered trademark. The Registrar may, however, ascribe weight to the applicant's motives and behaviour, for example when the applicant's behaviour is tainted with bad faith, when the Registrar considers whether he should exercise his power to keep the trademark on the Register notwithstanding the non-use thereof (see section IV.B below).

### 4. *As a defence in infringement/cancellation proceedings*

Where infringement proceedings are filed with the court and the defendant wishes to raise a defence claim based on grounds of non-use, he may choose between either of the following courses of action:

1. To apply to the ITMO and initiate separate proceedings for the cancellation of the mark, while in the meantime filing an application with the court seeking a stay of the main action.
2. To raise the argument of non-use as a defence argument in the main action itself.

According to case law, when the court hears a motion for a stay of proceedings, it must take into account various considerations, including the efficiency of the hearing and the judicial system, saving time and expenses, preventing conflicting decisions, a quick decision, the balance of convenience of the parties and the like.

## B. Burden of proof

In civil proceedings conducted in Israel the burden of proof is preponderance of the evidence. The onus of proving non-use lies with the applicant seeking cancellation of the trademark registration. The applicant must therefore file initial evidence showing lack of use of the trademark during the three-year period preceding the filing of the application. The extent of evidence required depends on the specific circumstances of the case, including the size of the relevant market and the variety and type of the goods. Results of enquiries conducted in a few stores will usually not suffice. If the evidence filed by the applicant is deemed sufficient, the burden shifts to the trademark proprietor, who will then be required to provide evidence contradicting the applicant's evidence, i.e. evidence demonstrating the use made by him of his trademark during the relevant three-year period. If after all the evidence is filed doubt still exists, such doubt will tend to favour the proprietor of the trademark.

## C. Relevant evidence

### 1. *Use on the product or its packaging*

Evidence with respect to use of a contested trademark on the packaging of products, even as a means for identifying the proprietor as the manufacturer of the products, can assist in proving genuine use of the trademark.

### 2. *Use in advertising*

As mentioned above, use of a trademark in Israel in advertising only would not be considered genuine use. However, use of a trademark in such a manner may actually strengthen a claim of genuine use, provided that other kinds of use were made of the trademark in Israel, or special circumstances exist justifying non-use of the mark.

Accordingly, evidence demonstrating use of a trademark in advertisements is submitted in trademark cancellation proceedings, even though, as noted above, such evidence has limited weight. Such evidence can merely show that many people were exposed to the trademark and that the proprietor had invested considerable resources in it.

### 3. *Use on the internet*

The question of whether use of a trademark on the internet can serve as evidence of genuine use depends on the specific circumstances of the case. Use of a trademark on the internet may be relevant as long as it is made available to Israelis. The case law is unclear however, on whether the website, in which the trademark is used, must specifically be intended for and used by consumers in Israel. Nevertheless, even if use is made of a trademark on the internet, use still needs to be made of the goods or services for which the mark is registered. For example, in one case it was held that use of a mark on a newspaper published on the internet, as oppose to the printed edition during the relevant period, was not considered as use because the trademark was registered in relation to paper products.

According to the Registrar's decisions, taking the evidence aspects into consideration, internet search results should be treated with caution and the veracity and significance of the search results must be established by the evidence. In one Registrar's decision, the Registrar did not ascribe any weight to an affidavit filed by the applicant's attorney with respect to internet searches performed by him. It was held that a representing attorney may testify with respect to the technical issues, such as the furnishing of documents or other procedural matters, but not with respect to facts giving rise to the cause of action. internet searches may be considered a technical issue, but not in cases where the search results have substantial weight. Based on the same logic, in another Registrar's decision it was held that the affidavit filed by the applicant himself claiming non-use, has limited, if any, weight. In another Registrar's decision it was held that the meaning of the searches conducted using the various search engines should be established by an expert opinion. In any event, the evidence should refer to the method of search and relevant circumstances.

### 4. *Form of evidence*

Evidence submitted to the ITMO must be given in the form of a written affidavit. Additional supporting evidence such as pictures, advertising, sales data and the like should be submitted through an affiant.

Since the applicant bears the initial burden of proof, and is required to prove a negative fact, the applicant usually submits affidavits given by private investigators who have carried out the necessary investigation to determine whether the mark is being used in commerce, such as when products bearing the mark are offered for sale in shops and in other prominent places. Issues requiring specific expertise must be addressed through expert's written opinion, which can be submitted as evidence, and the court/Registrar assesses its credibility. The regular rules which apply in civil proceedings with regard to the inadmissibility of hearsay evidence and the like, also apply in proceedings for trademark cancellation on grounds of non-use.

It is important that a comprehensive private investigator's report be submitted, in order for the Registrar to formulate an objective impression of the extent to which the proprietor of the mark has, or has not, made genuine use of the mark, with respect to the goods for which it was registered. Failure by the applicant to submit evidence in the form of a private investigator's report would usually stand against him, since it establishes a presumption that if the report had been presented, its results would have adversely affected the applicant.

Where the affiant is an interested party, the content of his affidavit would be examined with caution, and when the affiant is the applicant himself, a lesser weight will be given to the affidavit.

The proprietor will usually submit an affidavit in response, to which he will attach evidence of use, such as: product orders, receipts, invoices, photographs of products, advertisements, etc.

Each party is entitled to cross-examine the affiants and experts of the counter-party in order to corroborate and verify the facts and statements included in their affidavits/expert opinions.

#### 5. *Distinction between use of a company name and use of the services offered by the company*

See section II.A.1. above – use of a corporate name may be considered use for the purposes of Section 41, provided that the name of the corporation is impressed on the product or service in relation to goods or services falling within the specification thereof.

## IV. JUSTIFICATION FOR NON-USE

### A. Obstacles to use

If the Registrar or the court decide that no genuine use was made during the three years preceding the filing of the application for cancellation, they will proceed to examine whether (1) the non-use is due to special circumstances in the trade, and not due to any intention not to use, or to abandon, the mark (Section 41(b)); (2) there are special circumstances which, in the opinion of the court or the Registrar, justify non-use of the mark (Section 41(c)(1)). The timing for such special circumstances is the relevant period during which non-use is examined, i.e., during the three-year period prior to the filing of the cancellation application

#### 1. *Special circumstances in trade*

Section 41(b) sets out two cumulative conditions justifying non-use, as follows:

1. special circumstances in trade prevented use of the trademark; and
2. the trademark proprietor intends to use the mark.

Special circumstances in trade will occur where the proprietor is prevented from using his trademark due to external elements, and not due to his own failings.

Considerations mentioned in the case law as relevant to the issue of whether special circumstances in trade exist, are as follows:

1. The circumstances in question cannot be ordinary circumstances or ordinary difficulties.
2. They should be special circumstances not falling within the ordinary course of trade or the market, or severe interferences, even if temporary, in the proprietor's business.
3. Delays due to regulations can constitute 'special circumstances', however, routine delays in operation and marketing and delays due to the proprietor's conduct would not.
4. Circumstances falling within in the proprietor's control would not constitute 'special circumstances'.

### 2. *Special circumstances justifying non-use in the opinion of the court or the Registrar*

The Registrar will exercise his discretion only in exceptional circumstances, particularly where the proprietor did not have any control over the circumstances. When exercising his discretion the Registrar should balance between the special circumstances justifying keeping the registration of the trademark, and the purpose of Section 41, which is to enable all merchants engaging in a certain area to make use of marks still registered in the registry despite the lack of commercial use or intention to use them.

Examples of situations that have been recognized in case law as special circumstances in trade:

- import or legal restrictions imposed on all manufacturers importing and manufacturing certain types of goods.
- extensive unauthorized use of the trademark (e.g. by a licensee, whose licence was not registered, which is inconsistent with the terms of the licence granted, or by a person not authorized to use the mark), may constitute special circumstances in trade. However, the proprietor's failure to institute infringement proceedings against a third party for unlawful use of the mark may indicate that such third party's conduct was not the cause for the proprietor's non-use.

By contrast, market competition or lack of commercial profitability does not amount to special circumstances in trade.

With regard to the second enumerated condition set out in Section 41(b) (trademark proprietor intends to use the mark), it was held that an intention to grant a licence to use the trademark will not suffice, since this would not be deemed use in connection with the goods for which the mark was registered. Additionally, defending against legal proceedings (against the applicant), including trademark infringement proceedings, do not demonstrate a proprietor's intent to use the mark, because, according to the Registrar, such act is merely defensive.

### B. Circumstances in which non-use may be cured

As stated above, a trademark shall not be cancelled notwithstanding non-use if the proprietor thereof is able to establish that there are special circumstances in trade which prevented use of the trademark which, in the opinion of the court or the Registrar justify the non-use.

In addition, even if no such special circumstances exist which prevented use of the trademark, the Registrar may use his discretion and decide to keep the trademark on the Registry.

Examples of circumstances where the Registrar may exercise his said authority:

- A well know trademark: if the trademark is a well-known trademark in Israel, the object of preventing consumers' confusion may justify keeping the trademark on the Registry, even if the trademark owner has lost any commercial interest in it and has stopped manufacturing products bearing the trademark. The reason for this is due to the vast goodwill attached to the trademark, which is not completely obliterated.
- Applicant's bad faith: for example: (i) when a licensee, following a dispute with the trademark proprietor and the termination of their business relationship, applies for cancellation of the trademark on the basis that the licence was not registered; (ii) when cancelling the trademark registration will amount to an ex post facto validation of the infringement, e.g. the mark has become more associated with the applicant for cancellation rather than with the trademark proprietor, due to trademark infringement.

## V. CONSEQUENCES OF NON-USE

### A. Consequences of non-use and relevance of the context

If an application for cancellation of a trademark due to non-use is accepted, the trademark will be cancelled in respect of the goods for which the cancellation was requested and the non-use was proven. There is no curing effect of resumption of use after an application to cancel a trademark due to non-use was filed.

### B. Extent of any revocation

Revocation of the mark would only apply to goods that have not been used.

### C. Timing of any revocation

The trademark will be deemed revoked from the date the ruling revoking it on grounds of non-use is rendered, unless the Registrar or the court has stated otherwise in the ruling.

# Italy

*Gualtiero Dragotti and Elena Varese*

## I. INTRODUCING GENUINE USE OF TRADEMARKS

### A. Legal framework

The basic principles of trademark law in Italy are set forth in the Civil Code, namely in Volume V, Title VIII, Section III, Articles 2569-2574 and expanded in Title II, Section I of the Industrial Property Code ('IPC'), which contains the substantive provisions concerning trademarks, including the discipline concerning use requirements.

Further regulations are set forth in other sections of IPC, in particular those related to the enforcement of IP rights and those related to opposition and other PTO proceedings.

According to the general principles, trademark rights stem mainly from registration; however, the use of a trademark plays a very significant role in connection with the establishment of the exclusive rights, their scope and their maintenance.

For example, Article 19 IPC provides that trademark registrations can be obtained only by persons using or intending to use the mark; Article 24 IPC provides that lacking genuine use within certain terms lapses the mark; Article 28 IPC requires that a mark is actually and genuinely used to safeguard its validity, precluding novelty attacks for marks that have been genuinely used in commerce in good faith for a certain number of years.

### B. Purpose and function of genuine use: policy considerations

Most commentators link the genuine use requirements to the need to leave unused signs available to third parties.[1] This construction is based on the assumption that

1. See, *inter alia*: L.C. Ubertazzi, *Commentario breve alle leggi su proprietà intellettuale e concorrenza* (2016), 296; A. Vanzetti, *Codice della Proprietà Industriale* (2013), 495; M. Ricolfi, *I segni distintivi* (1999), 108.

(a) the number of available signs is limited;[2] and (b) the lack of genuine use makes it unjustified the grant of an exclusive right.[3] Both these assumptions can be discussed: the expanding notion of distinctive signs, not limited to words or devices, increases the number of theoretically available signs; the exclusive rights can be justified by commercial reasons independent of the actual use of the mark by its owner.

A further and possibly more convincing rationale justifying the genuine use requirements is focused on the need to look at the use of the mark in order to properly understand and design the scope of protection to be attributed to the mark.

Indeed, most of the provisions concerning the validity and extension of exclusive rights (think, for example, of the notion and discipline of reputed trademarks) are based on the assumption that the mark is actually used and cannot be easily construed if registration without use can attribute unconditional exclusive rights.

## II. WHAT QUALIFIES AS GENUINE USE OF A TRADEMARK

### A. Nature of use

#### 1. *Use consistent with the function of the trademark*

Generally speaking, genuine use should be assessed with respect to a trademark that has been used in a distinctive function and it is perceived as such by consumers.[4] An effective use, aimed to link a given product or service to a certain undertaking qualifies for genuine use. On the other hand, a merely token use or an use exclusively aimed to preserve exclusive rights over the trademark would not meet the threshold. Italian scholars and case-law have set out some guiding principles when assessing genuine use, highlighting that the actual capacity of the company to be present on the market, in light of its structure, dimensions and penetration and of the nature of the product as well as the targeted public should be taken into account.[5] For example, the small or local dimensions of a company (and consequently a minor penetration of the market) would not necessarily entail that genuine use is not met. On the other hand, the subjective attitude of the trademark holder is not a significant factor, so that the intention to prevent lapse for non-use would not be considered *per se*.[6]

2. See A. Vanzetti – V. Di Cataldo, *Manuale di diritto industriale* (2012), 282-284; C. Galli – A.M. Gambino, *Codice commentato della Proprietà Industriale e Intellettuale* (2011), 390 and M. Ricolfi, *I segni distintivi* (1999), 109.
3. See, among others, A. Vanzetti, *Codice della Proprietà Industriale* (2013), 495 and C. Galli, *Funzione del marchio ed ampiezza della tutela* (1996), 120-121. See also the ruling in Court of Bologna, 28 August 2008, in *GADI* 2008, 5304, p. 1050.
4. See, among others, A. Vanzetti – V. Di Cataldo, *Manuale di diritto industriale* (2012), 283 and M.S. Spolidoro, in *Commento tematico* (1998), 298. The same link between genuine use and the indication of origin of the trademark has been highlighted also by Italian case-law (see *inter alia* Court of Milan, 5 February 2014, in *GADI* 2014, 6132, p. 771; Court of Bologna, 13 June 2011, in *GADI* 2012, 5816, p. 284 and Milan Court of Appeal, 16 December 1994, in *GADI* 1995, 3278, p. 753).
5. See A. Vanzetti – V. Di Cataldo, *Manuale di diritto industriale* (2012), 233.
6. M.S. Spolidoro, in *Commento tematico* (1998), 299.

There is a certain interdependence between trademarks and other distinctive signs, so that use in diverse contexts may still account for genuine use. The Italian system has, in fact, allowed the possibility to prohibit trademark infringement against use which is made of a sign for purposes other than distinguishing goods or services.[7] In fact, Article 22 IPC – establishing the so called 'principle of unity of distinctive signs' – provides that the holder of a trademark has the right to prohibit the use of such trademark by third parties not only if it is affixed on products or used in connection with the services offered, but also if it is included in domain names, company names, shop-signs or other distinctive signs.

According to the majority of scholars and case-law,[8] titles (mainly of literary works, magazines – including their columns – and movies) also fall within the category of 'other distinctive signs' mentioned in Article 22, which adds to the protection granted by copyright and unfair competition under Articles 100 and 102 of Italian Copyright Law. On this point, the Court of Milan has stressed that titles may have the capacity to distinguish the commercial origin of a publication (and not only to identify the content of a copyrighted work), thus it is feasible to cumulate different IP protections.[9]

In light of the principle of unity of distinctive signs, use of a sign as a domain name, company name, shop-sign or title would account for genuine use,[10] although some practical problems might arise in assessing effectivity of use with respect to distinctive signs often not attached to physical goods.

For instance, with regard to company names, the mere presence on the Company Register of a formally active entity would not account for genuine use, if not followed by actual presence on the market. According to the most recent case-law, genuine use of a company name is recognized if (i) such name is registered in the Italian Company Register; and (ii) the entity characterized by such company name is actually and actively present on the Italian market.[11] Further, the comparison between products and services will take place between the business objective of the concerned company (provided that the company is actually active in each of the activities listed in the Company Register) and the goods/services covered by the trademark.

As to domain names, they should be continuously and consistently used, therefore a page under construction or merely embedding promotional links in third parties' websites would not reasonably establish genuine use. As to the comparison

7. The possibility to do so is also expressly recognized for each Member State by the CJEU, 11 September 2007, case C-17/06, *Céline*, par. 20.
8. The Italian system allows protection of titles through different cumulative means: trademarks, copyright and unfair competition. See C. Galli – A.M. Gambino, in *Codice commentato della Proprietà Industriale e Intellettuale* (2011), 387, Court of Milan, 23 July 2003, in *GADI* 2005, 4793, p. 186 and Court of Rome, 20 December 2000, in *GADI* 2001, 4256, p. 529.
9. Court of Milan, 2 September 2003, in *GADI* 2004, 4683, p. 550.
10. See P. Auteri, *Segni distintivi dell'impresa*, in *Enc. giur. Treccani* (1992), 9 and M. Ricolfi, in *I segni distintivi* (1991), 108.
11. See, for instance, Court of Milan, 7 July 2017, in www.giurisprudenzadelleimprese.it.

between goods and services, this should take place with respect to the activities actually carried out through the website corresponding with the domain name.[12]

Lastly, Italian scholars have expressly admitted that use of a sign as an ornamentation on the surface of a product would account for genuine use, if consumers are able to perceive the sign also (and not only) as an indicator of origin.[13] To this end, it is to be noted that the debated issue of ornamental use might have a more relevant impact when assessing distinctiveness of the concerned trademark, rather than when evaluating evidence of use.

## *2. Use in the course of trade*

### *i. Public use*

Genuine use shall be directed to the market and the mere inclusion in internal communications of the company or in materials confined to employees has been considered insufficient with the lack of proper exposure of the product before the public. For instance, the Court of Milan has denied genuine use, as a concerned trademark distinguishing cigarettes was neither included in the company's website, nor in important independent sources including a list of all the cigarettes present in the Italian market, nor in the tariffs of the cigarettes marketed in Italy issued by the Italian Authority in charge of controlling the production and marketing of tobacco products in the national market.[14]

The issue of use in advertisements has been widely debated among Italian scholars: some authors point out that such use may account for genuine use, even if the products are not (yet) available on the market,[15] whilst other authors criticize this opinion that would unduly circumvent the operation of lapse for non-use and restrict commercial freedom, by granting, for instance, prior rights to passive websites, merely accessible from a given territory and not allowing ordering of the concerned products.[16]

Also, case-law has not reached conclusive results: use within advertisements has been – in principle – recognized as a satisfactory evidence of use by most national courts.[17] On the other hand, the same decisions highlight that if the trademark had never been included in advertisements, this does not necessarily entail a lack of genuine use. In a case before the Court of Cagliari, proof of an intention to put the product back on the market was established through the declaration on the

12. See, for instance, Court of Rome, 2 August 1997, in *GADI* 1998, 3770, p. 381; Court of Brescia, 10 October 2000, in *GADI* 2001, 4241, p. 380
13. See M.S. Spolidoro, in *Commento tematico* (1998), 298 and C.E. Mayr, in *L'onere di utilizzazione del marchio d'impresa* (1991), 162-164.
14. Court of Milan, 5 February 2014, in *GADI* 2014, 6132, p. 771.
15. See, among others, A. Vanzetti - V. Di Cataldo, *Manuale di diritto industriale* (2009), 267 and M.S. Spolidoro, in *La decadenza della registrazione*, in Aa. Vv. *Commento tematico della legge marchi* (1998), 296-298.
16. M. Ricolfi, *Trattato dei marchi. Diritto europeo e nazionale* (2015), 944.
17. Court of Bologna, 28 August 2008, in *GADI* 2008, 5304, p. 1050; Court of Milan, 13 June 2011, in *GADI* 2012, 5816, p. 284 and Court of Milan, 10 October 1996, in *GADI* 1997, 3617, p. 377.

company's website that the concerned product (wine) was destined to 'reappear in the sale points and on the finest tables in 2000'.[18] In the same case, the Court found that use in advertisements amounts to a presumption that the product has anyhow remained on the market and was still available to consumers, even if such product was not commercialized by the company for a certain period.

Nonetheless, the link between advertisements and the actual presence of the products on the market has not been frequently addressed by national decisions. An interesting precedent by the Court of Bologna held that 'effective use could be established also through advertisements, if aimed to maintain or create an actual way out of the products on the market'.[19] This decision seems to uphold the more restrictive approach that goods should be about to be marketed or already present in the Italian market and the level of advertisement should not be nominal. As a consequence, genuine preparatory acts of marketing and advertising the product in Italy might also be taken into account, provided that its subsequent entrance into the Italian market is proved at a later stage.

### *ii. Commercial purpose*

Traditionally, the purpose of making a profit is not relevant for assessing genuine use, which could also be established through the distribution of promotional items bearing the trademark.[20] Some scholars point out that this approach would create an undue monopoly over a sign which is not used to gain a market share for the goods and services for which the trademark has been registered.[21]

Along the same lines, some decisions tried to draw a line between promotional gadgets that merely support the sale of ultimately different products, and items distributed for free, bearing a given trademark (which was perceived as distinctive by consumers), stating that only the latter case would account for genuine use.[22] For instance, use of promotional flyers to advertise a foodstuff would not account for genuine use of paper materials within class 16, whilst a calendar or a pen bearing the trademark of a bank might also account for genuine use of stationery products, insofar as those items are not perceived as mere ancillary advertising tools of different products and services. More recently, the Court of Bologna has apparently adopted a stricter approach, in line with the case-law of the CJEU, holding that the distribution by a non-profit organization of a negligible quantity of promotional gadgets on the occasion of amateur sailing events was not aimed at gaining a market share for the concerned products (clothing) nor for sailing events in general.[23]

18. Court of Cagliari, 17 May 2007, in *GADI* 2008, 5241, p. 462.
19. Court of Bologna, 13 June 2011, in *GADI* 2012, 5816, p. 284.
20. See, among many, Milan Court of Appeal, 16 December 1994, in *GADI* 1995, 3278, p. 753 and Court of Modena, 19 February 1994, in *GADI* 1994, 3108, p. 648.
21. M. Ricolfi, *Trattato dei marchi. Diritto europeo e nazionale* (2015), 946.
22. See, for instance, Court of Milan, 10 October 1996, in *GADI* 1997, 3617, pp. 378-379.
23. Court of Bologna, 22 September 2015, in *Ius Explorer*.

### 3. *Use by the owner or with its consent*

Article 24(1) IPC provides that genuine use could also be established through use by the trademark holder or with its consent, thus including use of authorized third parties, including licensees.

The prevalent opinion is that the consent of the trademark holder could also be proved through presumption, regardless of the existence of a written licence agreement.[24] The same conclusion is supported by the fact that specific formal requirements are not necessary to enter into a licence agreement under Italian law. However, it is questionable whether a revocable and provisional authorization to use a trademark would still be considered as effective use by the trademark holder.

On a different note, it is disputed whether a valid licence would be requested, in order to prove effective use by the licensee. A first opinion argues that the existence of a licence should be considered just as a factual element without looking at the validity of the underlying contract.[25] This opinion seems to be preferable, although it goes against the principle that void agreements are radically and retroactively unproductive of effects. There is, however, one exception provided by Article 19 TRIPS whereby 'when subject to the control of its owner, use of a trademark by another person shall be recognized as use of the trademark for the purpose of maintaining the registration'. In light of this, a licensor could benefit from use of the trademark by its licensee, provided that the underlying agreement is not void, due to lack of coordination and control by the licensor. In this respect, quality control of the licensor does not seem to be requested by Italian law to prove genuine use, while this issue might be relevant in assessing the deceptiveness of a trademark.

The relevance of control has also arisen with respect to use of the trademark by companies belonging to the same group, when the trademark is owned by one of them, but used in different markets by various affiliate companies. Likewise, it is reasonable to assume that this would establish genuine use if made by the licensee under the coordination and control of the company holding the trademark, based on two grounds: the unitary nature of group companies offered by Italian Company Law and the fact that, under Italian law, licence agreements could also be proved by conclusive behaviour.[26]

On the other hand, use by the infringer does not traditionally account for genuine use, as it is made without the consent of the trademark holder.[27] It is unclear whether and to what extent the same would apply if the trademark holder subsequently enters into a licence agreement with the former infringer.

## B. Volume, duration and frequency of use

The traditional principle whereby a minimum use of the trademark would also account for genuine use has been apparently overcome by the most recent case-law, holding that use of the trademark should be aimed at least to gain continuous and consistent

24. M. Ricolfi, *Trattato dei marchi. Diritto europeo e nazionale* (2015), 964.
25. See, for instance, M.S. Spolidoro, in *Commento tematico* (1998), 301.
26. See, among others, Milan Court of Appeal, 10 January 1997, in *GADI* 1997, 3638, p. 521.
27. See, among others, Court of Milan, 4 March 2014, in *GADI* 2015, 6238, p. 401.

market recognition (so-called 'qualitative criteria of use'), without taking into account specific quantitative thresholds as to volumes, duration and frequency of use.

In application of the above principles, Italian case-law considered sufficient use of a trademark (i) the sale of one million ice-creams in the course of the summer season;[28] (ii) the sale of some products awarding the company a turnover of EUR 200,000, also when the concerned sign was used in a smaller font on the back of the packaging;[29] (iii) orders and requests for information placed by the clients over the years, indirectly showing that the product was on the market and known among consumers;[30] (iv) the inclusion of the trademark for few years on paper promotional material, catalogues and the company's letterhead;[31] (v) the sale of a batch of pharmaceutical products over a year, shown by invoices and data collected by distributors, prescriptions from doctors and the market data collected by an independent organization;[32] (vi) the consistent offering for some years of cigarettes, albeit only in duty free shops;[33] (vii) the purchase of 3,000 wine labels in September 2010 and the purchase of another 10,000 in September 2005, together with an advertising publication showing the label in 2010 and the company's records, attesting to the selling of wine bottles bearing such labels;[34] (viii) e-mails, invoices, price lists, advertising material, menus and wine lists used over the years as well as shop windows and the internal furnishing of the restaurants showing the trademark;[35] (ix) the licence agreement, the invoices relating to the selling of 1.600 bags, the bank statement attesting payment of royalties to the trademark owner by the licensee, an original sample of the bag bearing the product and the promotional offering of the bags along with beverages;[36] and (x) in the case of a hair salon, the purchase of t-shirts between 2012 and 2013, copies of an Italian newspaper dated 27 March 2010 and 16 April 2011 and pictures of gloves, boxes of combs and nail polish displays, all of which show the trademark.[37]

On the other hand, it was found to be not sufficient use of a trademark: (i) invoices issued by two distributors, evidence of the selling of the product at an Italian shop and copies of some marketplace websites, where the products were actually meant to be sold only abroad;[38] (ii) the dispatch of cease and desist letters to defend the trademark and the renewal of the relevant application, pending bankruptcy proceedings of the trademark holder;[39] (iii) a single and isolated act of the sale of a product;[40] (iv) use of the trademark on few dozens of scooters, when importation has ceased, regardless of the presence on the market of some merchandising products,

28. Court of Rome, 22 May 2003, in *GADI* 2004, 4676, p. 479.
29. Court of Turin, 8 April 2011, in *GADI* 2011, 5713, p. 950.
30. Supreme Court, 19 June 2008, No. 6647, in *GADI* 2008, 5222, p. 202.
31. Court of Milan, 23 May 2011, in *GADI* 2011, 5737, p. 1175.
32. Court of Milan, 7 May 2015, in www.giurisprudenzadelleimprese.it.
33. Court of Rome, 1 April 2013, in *IDI* 2014, 5, p. 477.
34. Court of Turin, 18 April 2016, in *DartsIP*.
35. Court of Rome, 10 March 2017, in *DartsIP*.
36. Court of Venice, 17 June 2016, in *DartsIP*.
37. Court of Turin, 9 January 2018, in www.giurisprudenzadelleimprese.it.
38. Court of Milan, 16 May 2017, in *DartsIP*.
39. Court of Milan, 23 December 2013, in *GADI* 2014, 6117, p. 585.
40. Court of Milan, 28 September 1972, in *GADI* 1972, 172, p. 1197.

the recognition by consumers and presence within the second-hand market of the product, as this use was not enough to modify the market situation and the competitors' position;[41] (v) the devising of the packaging of the product and the relevant advertising campaign limited to the internal organization of the trademark holder, in the absence of the products in the market,[42] and (vi) the inclusion of the trademark on an internet webpage without any reference to the availability of products and services.[43]

### C. Use of a sign

Article 24(2) IPC expressly admits that use of a sign showing modifications to the trademark as registered also accounts for genuine use, as long as such modifications do not alter the distinctive character of the trademark. The rationale of this provision is to grant a certain degree of flexibility to the trademark owner, who is allowed to restyle and update its trademark.[44]

The criteria to establish the non-alteration of the distinctive character are generally quite strict, in order to avoid the potentially anti-competitive effects of the provision. To this end, case-law has denied lapse for non-use only when the signs were almost identical. For instance, it accounted for genuine use: (i) usage of an inverted 'C', as it was a mere positioning of the sign, which did not alter the individual features of the trademark;[45] (ii) use of the mark 'Teominale' when the registered mark was 'Theominale';[46] (iii) use of the mark 'OE' when the registered mark was 'OE Deodorant Fraicheur';[47] (iv) use of the mark 'Denthoral' when the registered mark was 'Dent-Horal';[48] (v) use of a mark in colour, when the trademark was registered in black and white;[49] and (vi) use of the mark 'Sevenbell' or the alternative '7 bell'.[50]

In case of complex trademarks (including trademarks composed of word and figurative elements), use of the distinctive core of such trademarks is sufficient to avoid lapse for non-use. For example, use of the word 'Selex' in a family of registered trademarks, combining the same wording with different distinctive components, was sufficient to avoid lapse for non-use.[51] Further, use of the mark 'DiG' was found to have an autonomous role within the complex trademark 'DiG Di Gregorio', which prevented the lapse for non-use.[52]

---

41. See the 'Lambretta' line of precedents: Supreme Court, 28 March 2017, No. 7970, in *Foro it.* 2017, 5, I, p. 1553; Milan Court of Appeal, 12 April 2013, in *GADI* 2014, 6088, p. 300 and Court of Milan, 26 November 2008, in *GADI* 2008, 5317, p. 1178.
42. Court of Bologna, 13 June 2011, in *GADI* 2012, 5816, p. 284.
43. Court of Turin, 13 July 2012, in *Ius Explorer*.
44. The same allowance was also formerly granted when renewing the trademark under the old Italian Trademark Law.
45. Court of Milan, 24 February 2003, in *GADI* 2004, 4664, p. 402.
46. Supreme Court, 22 January 1974, n. 176, in *GADI* 1974, p. 20.
47. Court of Rome, 14 June 2011, in *GADI* 2012, 5817, p. 293.
48. Court of Milan, 14 April 2011, in *GADI* 2011, 5718, p. 1013.
49. Court of Milan, 23 November 1961, in *RDI* 1963, II, p. 91.
50. Court of Florence, 8 May 2017, in www.giurisprudenzadelleimprese.it.
51. Court of Rome, 8 June 2010, in *GADI* 2010, 5560, p. 648.
52. Court of Rome, 12 January 2015, in *GADI* 2015, 6266, p. 721.

## D. Use in relation to goods and services

### *1. Use for each of the designated goods and services*

The Italian system does not require that proof of use is submitted for each designated good and service. This is confirmed, on the one hand, by some precedents that denied lapse for non-use of reputed trademarks, even if they were not used for each designated good and service; on the other hand, by the narrower scope of the products covered by lapse for non-use, which is limited to products which are dissimilar to those of effective use.

As to the first issue, Italian case-law is divided on the possibility of admitting that reputed trademarks would not be subject to lapse for non-use with respect to all the designated goods and services. For instance, the Court of Rome denied that the fashion trademark 'Hugo Boss' lapsed for non-use with respect to magazines and newspapers;[53] contrarily, the Court of Turin declared lapsed for non-use the trademark 'Martini' for lighting systems, as proof was submitted only for spirits, glasses and clothing.[54] The second approach seems preferable and more in line with the EU provisions, which admit a different protection of reputed trademarks for the respective enforcement and not with respect to lapse for non-use.

As to the second issue, notwithstanding the literal formulation of Article 27 IPC,[55] use for some designated goods accounts for genuine use also with respect to a broader category goods and services, provided that such goods and services are similar to those of actual use. The creation of this 'safe harbour' has been criticized, as it would prevent lapse for non-use with respect to products that are in turn similar to the 'similar products' and so forth, until virtually all designated products and services are covered. Some authors have then proposed to limit the operation of the safe harbour only to the products and services that are similar to those that are effectively used within the first five-year period from trademark registration.[56]

Italian case-law has not apparently dealt with the issue of a possible time limitation of the safe harbour, but was conversely quite strict in limiting the broader category of similar goods not affected by lapse for non-use. For example, the Court of Bologna recognized genuine use in relation to usage of hand-cutting devices also for mechanical cutting devices, as these responded to the same need and destination of use, even if they belonged to different classes.[57] The Court of Turin held that use for panettone (a traditional cake) also accounted for genuine use for biscuits, as these products were somehow complementary.[58] Along the same lines, the Court of Milan found that use for female clothing also prevented lapse for non-use for male-clothing, as the destination of men or women did not allow establishment of

53. Rome Court of Appeal, 9 December 2008, in *IDI* 2009, p. 241.
54. Court of Turin, 26 January 2007 in *DartsIP*.
55. Article 27 IPC provides that 'if the reasons for lapse for non-use or invalidity of a trademark are present only for a part of the products or services for which the trademark has been registered, lapse for non-use or invalidity concern only this part of products or services'.
56. M. Ricolfi, *Trattato dei marchi. Diritto europeo e nazionale* (2015), 948.
57. Court of Bologna, 15 May 2012, in *GADI* 2013, 5977, p. 402.
58. Court of Turin, 4 April 1996, in *GADI* 1996, 3478, p. 738.

a separate category of goods within garments, belonging to class 25.[59] Again, use of the mark for foodstuff destined for cats and dogs determined use within the broader category of foodstuff for animals.[60]

However, it should be pointed out that few decisions have broadened the concept of similarity too much, holding that the trademark registration would not lapse with respect to all the products falling within the sphere of a potential expansion of the business. With this reasoning, a decision of the Court of Appeal of Milan recognized that use of a mark for oil was able to prevent lapse for non-use of a mark within classes 32 and 33 (drinks, wine and spirits).[61] Similarly, the Court of Bologna found that use of the mark 'Essere Benessere' for beauty and para-pharmaceutical products accounted for genuine use of fitness and wellness products.[62]

### 2. *Use for a broader category of goods and/or services*

Use for a broader category of goods and/or services has not been frequently addressed by Italian case-law in relation to genuine use. This could be due to the fact that the Italian system provides for unregistered trademarks, thus use for a broader category of goods exceeding the designated goods would be either covered by the trademark registration (insofar as the goods of use are similar) or by the unregistered trademark (insofar as the goods of use are dissimilar).

## E. Territorial scope

There are no specific prescriptions relating to the extent of the Italian territory where the trademark should be used, in order to prove genuine use. Nonetheless, it is well-established that use only within a reasonable part of the territory would meet the threshold. In this respect, the Court of Milan considered as proof of genuine use the submission of ten invoices issued between 2003 and 2004, showing the selling of margarine to few resellers located in four important Italian regions (namely Lombardy, Liguria, Veneto and Sicily).[63]

Use shall not be necessarily in connection with products that are present on the national market, as use of the trademark on goods that are exclusively destined for exportation is able to prevent lapse for non-use, as expressly admitted by Article 24(2) IPC. The same applies to the affixing of the trademark in Italy for exported products, as well as to the sale in Italy of products imported from abroad.

Applying the above principles when the former Trademark Law was in force, the Supreme Court held that affixing the trademark 'Zara' by some Italian third-party weavers on products subsequently exported in Spain and sold thereto accounted for genuine use, even if the products were not present in the Italian shops, but only offered by minor stockists.[64]

59. Court of Milan, 19 November 2012, in *GADI* 2012, 1184, p. 1184.
60. Court of Verona, 23 July 2001, in *GADI* 2001, 4308, p. 1013.
61. Milan Court of Appeal, 22 May 1998, in *GADI* 1999, 3923, p. 465.
62. Court of Milan, 7 October 2010, in *GADI* 2011, 5657, p. 465.
63. Court of Milan, 20 June 2017, in *DartsIP*.
64. Supreme Court, 12 August 2010, No. 18647, in *RDI* 2012, II, p. 178.

On the other hand, limited episodes of displaying the products abroad cannot prevent lapse for non-use, in the absence of proof of a certain notoriety among Italian consumers. For instance, the Court of Florence denied genuine use of the trademark 'Cinque Terre' in connection with a perfume, as the concerned product was only shown in a trade fair in Cannes in 2009. This single episode, in fact, was not accompanied by the sale abroad and did not show a serious intention to market the product in Italy. In addition, three non-signed orders for limited quantity of products with a date subsequent to the relevant five-year period did not show market recognition or goodwill of the trademark acquired with the Italian public.[65]

Use of the mark not attached to physical goods (as in case of use through websites) must be directed to the Italian public. For instance, the Court of Turin declared that the display of the trademark 'Heraklith' in a foreign website written in a foreign language did not account for genuine use, as such website was not specifically targeting the Italian speaking public and received a negligible number of visits (1,700 in three years):[66] both circumstances revealed that the mark was not known among national consumers. Consistent with this, use of the trademark 'Radio Pane e Pomodoro – La radio dei tipi da spiaggia' in the commercial breaks of a small Italian radio via internet streaming (formerly operating over the airwaves) was found sufficient to avoid lapse for non-use, as it was directed to and known by the local public.[67]

More generous criteria were apparently applied to goods sold in Italy, but destined for foreigners, by virtue of the limited number of consumers to which the product was intentionally directed: this was the case for 10-15 copies of books in English which were available each month in a few but important point of sales based in Rome and Milan[68] and for 48 bottles of a US drink only sold within the NATO bases and at the American Market in Livorno.[69]

## F. Relevant time period and its calculation

Article 24(3) IPC provides that lapse for non-use shall not take place if between the expiration of the five-year period of non-use and the proposition of a claim, counterclaim or exception of revocation for non-use, effective use of the lapsed trademark has started or resumed. The rationale of the provision is to cure the so-called 'errors of youth of a trademark now solidly redeemed'.[70] Calculation of lapse for non-use runs from the registration date of the trademark for the following five years[71] and, after this period, each national trademark is subject to proof of use within court proceedings. In light of the possibility to resume use of the lapsed mark, Italian judges tend to make an overall assessment of use, looking whether the trademark has not

65. Court of Florence, 22 July 2010, in *GADI* 2010, 5575, p. 804.
66. Court of Turin, 23 June 2014, in *GADI* 2015, 6246, p. 484.
67. Court of Bari, 12 March 2012, in *Ius Explorer*.
68. Court of Milan, 24 February 2000, in *GADI* 2000, 4141, p. 819.
69. Court of Rome, 7 march 1973, in *GADI* 1973, 4141, p. 488.
70. See Vanzetti - V. Di Cataldo, *Manuale di diritto industriale* (2012), 218.
71. Court of Milan, 28 January 2016, in www.giurisprudenzadelleimprese.it.

been used for five consecutive years in the period between trademark registration and the proposition of the claim, counterclaim or exception of revocation for non-use.

Resumption of use by the trademark holder after the expiry of the five-year period operates for the future and does not prevent trademark rights acquired in the meantime in good faith by third parties through registration or use (even with a filing or registration prior to the occurring of lapse for non-use). This provision is peculiar to the Italian system and allows third parties to 'book' a similar or identical trademark, in case the earlier trademark is subsequently revoked for non-use.

Another exception to the possibility to cure a lapse for non-use is when the trademark holder makes serious and effective preparations for the start or resumption of use of the mark only after knowing that a claim, counterclaim or exception of revocation for non-use was about to be proposed, unless such start or resumption take place at least three months before the filing of the concerned claims, counterclaims and exceptions. In other words, only preparatory acts started at least three months before the proposition of revocation for non-use would prevent the operation of lapse for non-use. This three-month grace period is only relevant if the five-year period of non-use has fully expired.

## III. EVIDENCE OF GENUINE USE OF A TRADEMARK

### A. When use should be proved: timing of lack of genuine use claims

The above principles and comments help explain why genuine use is required. With regard to timing and circumstances to provide the related evidence, the following principles and rules apply.

#### *1. When filing an application for a trademark*

According to Article 19 IPC, trademark registrations can only be obtained by persons using or intending to use the mark, directly or indirectly; this provision does not translate into the need to submit, at the time of the application, evidence of the actual use of the mark or an intent-to-use declaration.

However, if a mark is filed for purposes different from the abovementioned ones the resulting registration could be null and void. Of course, such a consequence is mainly theoretical, as the applicant is usually in the position to claim that at the time of the application it intended to use the mark, at least indirectly.

#### *2. In the course of opposition proceedings*

Opposition proceedings were introduced in Italy only in 2011, so that the number of decisions is still limited.

Article 178 IPC provides that any opposition based on a trademark registered more than five years before the opposition date and therefore subject to genuine use requirements shall be rejected if, upon request by the opposed trademark applicant, the opponent does not submit to the opposition division adequate evidence of genuine use of the mark on which the opposition is based.

### 3. *In cancellation proceedings based on lack of use*

In judicial proceedings the lack of genuine use could be invoked by any party interested in claiming the invalidity of a trademark registration. Such a claim follows the usual rules on burden of proof, so that the claimant must substantiate and support the non-use claim. However Article 121 IPC provides that non-use evidence can be given by any means, including presumptions.

This provision acknowledges that providing evidence that a mark has been never used could be extremely difficult or even impossible, while for the trademark owner it is pretty easy to provide evidence of use (of course, if the mark has been actually used).

Hence the concession granted to the party claiming the lack of genuine use. This concession is not, however, a full inversion of the burden of proof: it is still necessary to substantiate the non-use claim with elements suggesting that the mark was not actually or sufficiently used in commerce.

### 4. *As a defence in infringement/cancellation proceedings*

In infringement or cancellation proceedings the lapse for lack of genuine use of the infringed/prior mark can be brought both as a counterclaim and as an exception.

In the first case, the revocation decision would be *erga omnes*, i.e. valid for everyone. In the latter case, the decision will be valid only *inter partes*: third parties will not be in the position to invoke the decision and would need to independently ask for the cancellation of the mark.

In both cases the claim based on lack of genuine use does not need to be brought in separate proceedings, and it is indeed common to bring it as a counterclaim/defence within the infringement/cancellation proceedings.

If the two sets of claims (infringement or cancellation on the one hand, revocation for lack of genuine use on the other) are brought in separate proceedings, those proceedings can be joined, provided that they are pending before the same court and at substantially the same stage. Otherwise the two proceedings continue separately and it is not mandatory (nor it is common) that one of them is stayed.

## B. Burden of proof

Article 121(1) IPC provides that lapse for non-use of a trademark shall be proved by the party claiming revocation through any evidentiary means (including documentary and witness evidence as well as presumptions of law). The Supreme Court has clarified that 'the burden of proof, pursuant to Article 2697 of the Italian Civil Code, lies on the party who intends to assert a right in court [...] This principle does not suffer derogation even when it regards "negative facts" [...] However, since the material demonstration of a fact that has not been made is not possible, the relative proof can be given by demonstration of a specific contrary positive fact or even by means of presumptions from which the negative fact can be inferred'.[72]

72. Supreme Court, 9 June 2008, No. 15162, in *Giust. Civ. Mass.* 2008, 6, p. 897.

Italian scholars and case-law have frequently highlighted the difficulty for the party claiming lapse for non-use to provide negative evidence of a circumstance that is ultimately out of its control. This is only partially mitigated by the possibility to prove lapse for non-use through the request of discovery of the trademark holder's documentation, as well as through presumption.

In order to be considered as a fool-proof, presumptions should have the following three requirements: (i) conclusiveness (they assume an unknown fact as a logical consequence of a known fact, based on common experience); (ii) preciseness (they are not vague, but well-determined); and (iii) consistency (they start from a series of known and non-contradictory facts). Once the party claiming lapse for non-use has submitted evidence to determine its claim, the trademark holder shall submit counter-evidence. If the party claiming lapse for non-use is not able to provide conclusive evidence, its claims should be rejected even without considering the evidence provided by the other side, based on the general principle of the burden of proof in the Italian system.

## C. Relevant evidence

### *1. Use on the product or its packaging*

The usual proof to determine use is by submitting a sample of the product bearing the trademark. Equally reliable evidence concerns sales data, price lists, catalogues and invoices, showing the market presence and penetration of the product bearing the trademark. There is no specific requirement as to the form of submission of such evidence: notarized excerpts of the company's accounting documents or spreadsheets including sales data combined with witness statements are both acceptable. One main concern is to attribute a date to each piece of evidence, in order to establish continuous use over a reasonable time-frame, as well as to tackle which part of the company's turnover is linked with the sale of the specific product.

In this respect, for instance, the Court of Milan held that six invoices issued by two distributors in the course of five years and some tags and labels allegedly belonging to the existing production of clothing (not including the relevant dates) did not account for genuine use.[73] On the other hand, the Court of Florence held that the inclusion of the trademark in paper tags hanging from jeans, invoices dating from 2009 to 2015, advertising materials and campaigns, packaging, shoppers, ribbons and boxes, as well as the investments made to implement the e-commerce channel and an invitation to a 'Pitti Immagine Uomo 2013' event was to be considered as an effective use in class 25. In the same case, documentary evidence was submitted to prove use of the trademark also as a domain name, with proof of all the visits received from April 2013 to January 2015.[74]

73. Court of Rome, 23 November 2016, in *DartsIP*.
74. Court of Florence, 8 May 2017, in www.giurisprudenzadelleimprese.it.

## 2. *Use in advertising*

Investment made to promote the trademark, its appearances on TV and in the press, as well as the prizes won by the relevant products are also a suitable proof of use along with the evidence of the advertisements by the trademark holder. The same applies also to the company's catalogues and price-lists including the trademark.

For example, the Court of Rome recognized as effective the use of the trademark 'Fortuny' for lighting instruments bearing the trademark, samples of which were included in catalogues and price lists. In such case, the mere lack of any reference to cosmetic products in the submitted invoices was considered not enough to determine lapse for non-use (this only showed that the relevant production was ancillary, but not completely non-existent).[75] The same Court considered as satisfactory evidence of use of the trademark its inclusion on the company's letterhead and correspondence, on invoices, price lists, promotional items, shop windows and their interiors, menus, wine-lists and on the website.[76]

## 3. *Use on the internet*

Websites are frequently used in order to determine genuine use of a trademark together with web archives showing use of the trademark over time. It should be noted, however, that older decisions have pointed out that this kind of evidence alone is not always conclusive to determine lapse for non-use. For example, the Court of Naples held that the absence of the trademark in the internet page of the trademark owner was not to be considered *per se* as a proof of lapse for non-use, but as a mere presumption. In that case, the Court observed that the excerpts of the concerned internet pages had no date, have not been notarized and no witness was called to confirm the genuineness and date of the webpages. On the other hand, it was not admissible to request the judge to autonomously verify the content of such pages outside the proceedings.[77]

More recent decisions, however, considered excerpts from websites also taken over time through web archives showing the absence of the trademark 'FMR' in connection with clothing, perfumes and cosmetics as conclusive evidence of non-use. Such evidence was completed by witness statements, who confirmed the outcome of said internet search.[78]

## 4. *Form of evidence*

Under Italian practice, the presence of some of the following is considered enough to prove lapse for non-use: (i) the absence of the trademark in the company's catalogues and price-lists; (ii) the absence of the trademark in the list of the relevant trade association; (iii) the fact that the product classes in which the trademark has been registered are unrelated to the company's main activities, as described in the

75. Court of Rome, 18 July 2017, in *DartsIP*.
76. Court of Rome, 10 March 2017, in *DartsIP*.
77. Court of Naples, 2 February 2006 in *Pluris*.
78. Court of Milan, 6 April 2017 in www.giurisprudenzadelleimprese.it.

Company Register; (iv) the lack of the trademark on the website of the trademark holder; and (v) the absence of results from inserting the trademark as a keyword in search engines and web archives.

Italian courts were quite reluctant in the past to rely on trademark surveys, on the basis that this evidence could be easily manipulated. Now trademark surveys (whether carried out by the parties' expert or by a court-appointed expert) are more commonly used in Italian proceedings and taken into account by Italian judges, also with respect to trademark use. The judge has, however, the last word and is of course free to decide whether to determine his assessment (and decision) on such surveys.

As to further evidence, some cases recognized as a valid proof of non-use the witness statements and affidavits of market operators, wholesalers and retailers in the field stating that they had no knowledge of the concerned trademark.[79]

There are some precedents that adopted a particularly strict approach and denied any relevance to an investigative report on use coming from a private investigator[80] or to technical expertise commissioned by the party,[81] as in both cases such proof did not come from reliable sources and was created unilaterally, without the other side being heard.

### 5. *Distinction between use of a company name and use of the services offered by the company*

As mentioned above, there is a certain interdependence between use of a company name and use of the services offered by the company and the same applies to the relevant evidence to determine use, which does not differ significantly from the evidence just outlined.

In this respect, a recent decision of the Court of Milan held lapse for non-use, based on the fact that the company's purpose, as included in the Company Register, did not include the product and services for which the trademark was registered. Similarly, there was no registration of any licence before the Italian Trademark and Patent Office, from which one could infer genuine use of the concerned trademark.[82]

## IV. JUSTIFICATION FOR NON-USE

### A. Obstacles to use

Article 24 IPC provides that revocation for non-use is excluded when the lack of use is due to legitimate reasons.

These legitimate reasons include events and causes independent of the will and control of the trademark owner, such as the need to obtain an administrative

79. Court of Rome, 5 June 1986, in *GADI* 1986, 2046, p. 522.
80. Court of Naples, 2 February 2006, in *Pluris*.
81. Court of Milan, 23 May 1991, in *RDI* 1993, II, p. 174.
82. Court of Milan, 23 May 2017 in www.giurisprudenzadelleimprese.it.

authorization,[83] a war,[84] a prohibition to sell the product bearing the mark,[85] etc. These cases are listed by most commentators as uncontroversial. It is however debated if a decision not to use or to discontinue use due to a (legitimate) commercial policy may be deemed justified and, therefore, qualify as a legitimate reason for the purposes of Article 24 IPC. Assuming a negative solution seems advisable, lacking any established case-law on the point and in view of the restrictive approach adopted by the European Court of Justice since the *Haupl* decision.[86]

The former Italian Trademark Law, superseded by the IPC, explicitly stated that lack of financial resources did not amount to a justification for non-use. This provision is no longer included in the law, but most commentators are inclined to exclude lack of financial resources from the list of legitimate reasons for the purposes of Article 24 IPC.[87] A different solution has been reached by some (older) decisions concerning bankruptcy, which could justify an extended non-use.[88] As noted by some commentators, this conclusion is now less and less relevant, as the foremost concern of the receiver is usually to assign or license the trademark(s) in order to avoid them losing market value.[89]

Another debated issue is the possibility that an extended infringement could be deemed a legitimate reason for postponing the beginning of the use of a mark, or to suspend it.[90]

## B. Circumstances in which non-use may be cured

As noted in section III.F above, Article 24(3) IPC provides that non-use is cured, i.e. lapse does not take place, if between the expiration of the five-year period and the revocation action, effective use of the trademark has started or resumed.

However trademark rights acquired meanwhile in good faith by third parties through registration or use are not affected by this mechanism.

It is also worthwhile mentioning that some marks, namely the so-called defensive marks, are not subject to use requirements. According to Article 24(4) revocation for non-use does take place if the trademark owner actually uses similar trademarks in connection with the same goods.

---

83. Court of Milan, 27 May 2015, in *IDI* 2016, I, p. 70; Court of Milan, 2 September 2003, in *GADI* 2004, 4683, p. 550.
84. C. Galli – A.M. Gambino, *Codice commentato della Proprietà Industriale e Intellettuale* (2011), 397.
85. M. Ricolfi, *I segni distintivi* (1999), 110.
86. CJEU, 14 June 2007, case C-246-05, *Armin Häupl v. Lidl Stiftung & Co. KG.*
87. See, among others, G. SENA, *Il nuovo diritto dei marchi* (2007), 148; C.E. Mayr, in *L'onere di utilizzazione del marchio d'impresa* (1991), 137.
88. Court of Milan, 15 March 2006, unedited, which appears to be the most – and only – recent one ruling in this sense.
89. C. Galli – A.M. Gambino, *Codice commentato della Proprietà Industriale e Intellettuale* (2011), 398.
90. According to some decisions of the Court of Milan, extensive infringement or multiple claims against the validity or ownership of a trademark may justify the discontinuance of its use until the end of the dispute. This position is not shared by the Court of Rome.

This provision has the purpose of allowing the trademark owner to register families or groups of similar trademarks and keep all these registrations in force (i.e. not be exposed to revocation for non-use) provided that at least one of the marks is genuinely used in commerce.

## V. CONSEQUENCES OF NON-USE

### A. Consequences of non-use and relevance of the context

In opposition proceedings before the Italian PTO the non-use of the trademark on which the opposition is based would lead to the rejection of the opposition claim. However, the concerned trademark registration will remain formally in force (albeit exposed to a possible revocation action brought by the owner of the opposed trademark or by other interested third parties).

In court proceedings a successful non-use revocation claim would lead to a decision *erga omnes* and to the cancellation of the mark; for this purpose the decision is communicated by the Court to the PTO.

The revocation for non-use and the cancellation of a registered trademark does not prevent, *per se*, the use of the mark by the (former) trademark owner or by any other party, unless of course such a use interferes with any exclusive right; this is often the case when the non-use revocation claim is brought within an infringement proceeding.

### B. Extent of any revocation

Article 27 IPC provides that invalidity and revocation, including revocation for non-use, can be limited to certain goods or services when the invalidity or revocation grounds are related to said goods or services.

In the past this principle usually led courts to tailor revocation or invalidity decisions at class level.[91] This approach will be superseded by a more granular one, whereby invalidity and revocation are linked to specific goods and services.[92]

### C. Timing of any revocation

The general principles of civil procedure law provide that invalidity declarations operate *ex tunc*, i.e. go back in time to the origin of the (invalid) right; revocation declarations operate *ex nunc* (for the future), so that the revoked right remains and must be deemed valid until the decision date.

According to one author,[93] revocation for non-use decisions extend their effects backwards to the time when the non-use period matured. According to an approach

91. See, among others, Court of Turin, 26 January 2007, in *DartsIP*.
92. Court of Venice, 29 January 2015, in *RDI* 2016, 1, II, p. 3; Bologna Court of Appeal, 13 April 2004, in *GADI* 2004, 4739, p. 1028.
93. P. AUTERI, *Territorialità del diritto di marchio e circolazione di prodotti originali* (1973), 170.

shared by most commentators, revocation decisions follow the normal rules; the issues related to the rights of third parties which may come to life after the non-use period, but before the commencement of the revocation action, are satisfactorily dealt with by Article 24(3)[94] commented in section II.F above.

94. In this connection see, in particular, G. SENA, *Il nuovo diritto dei marchi* (2007), 148.

# Japan

*Masako Nishimura*

## I. INTRODUCING GENUINE USE OF TRADEMARKS

### A. Legal framework

Article 50 of the Trademark Act on non-use cancellation mainly relates to 'genuine use' or 'legitimate use' under Japanese law:

> *Article 50*:
>
> Any person may request a trial against a registered trademark that is not used continuously for three years or more to cancel the registration. Cancellation can be requested for unused goods/services.
>
> Where a request for non-use cancellation is filed, unless the registrant proves that the holder of trademark right, of the exclusive right to use or of the non-exclusive right to use has used the registered trademark in Japan in connection with any of the designated goods or designated services pertaining to the request within three years prior to the registration of the request for the trial, the registrant may not prevent the rescission of the trademark registration in connection with the relevant designated goods or designated services. However, this shall not apply where the registrant shows just cause for non-use of the registered trademark in connection with the relevant designated goods or designated services.

'Legitimate use' is also considered in a revocation based on unfair use, as to whether use of a registered mark is appropriate, that is, whether a mark in use is regarded as identical to a registered mark. A registered mark which is likely to cause confusion with another's brand by use of a similar mark to the registered one, or mislead over the quality of goods or services, is to be cancelled under Article 51 of the Trademark Act. Similar Articles are provided as to use of an assigned trademark (Article 52-2) and also as to use by a licensee (Article 53). An example of use of a similar mark is as follows. In this case, the similar mark was not considered as causing confusion with the brand ELLE.

*ELLEGARDEN case (IP High Court decision, 24 February 2009)*

| Subject mark (registered mark)<br>Designated goods: compact disc etc.<br>Designated services: music playing | Trademark in use (similar mark) | Cited mark<br>In Japan, its brand has widely penetrated through the publication of the magazine 'ELLE' and many licences. |
|---|---|---|
| ELLEGARDEN (with Japanese katakana) | ELLE/GARDEN (logo in two lines) | ELLE (logo) |

> 'In conclusion, although the mark usage is similar to the cited mark, from the usage form of the compact disc, for the consumer who knows that "ELLE-GARDEN" is the name of this rock band, as well as the consumer who does not know that when contacting this Compact Disc, it was judged that there was no likelihood that the mark usage could be misunderstood and confused as having any relationship with the "ELLE" brand.'

It was therefore concluded that this use was regarded as legitimate and the revocation was prevented.

## B. Purpose and function of genuine use: policy considerations

Under Japanese law, while there is no definition of 'genuine use' or 'legitimate use', 'legitimate use' is considered as being required for revoking registered trademarks not in use for the following reason.

In Japan, since the use of a trademark is neither a registration requirement nor a renewal requirement, there is currently a large number of unused registered trademarks. Even if a trademark is used for part of designated goods/services, it is very common that greater part of the goods/services covered by the trademark are actually not used.

According to 'Outline of the Results of the Investigation of Intellectual Property Activities' (Japan Patent Office (JPO), 2014), the utilization rate of domestic trademark rights is 68.0%. However, according to '2009 Trademark Application Trend Survey Report' (JPO, March 2010), in addition to trademarks that could not be confirmed as being used at all, the proportion of trademarks with designated goods/services not in use is high, at 87.9%. The reasons for designating goods/services that are not intended to be used is as follows:

> 'In addition to the possibility of future use and defensive measures, when actually developing business, they do not want to violate the rights of others.'[1]

1. *2013 Trademark Application Trend Survey Report (Summary) – Effects after Non-use Trademark Countermeasures, Application for Analysis, Registration Status Survey* (JPO, February 2014).

Under the circumstances, it is more important in Japan to check 'genuine use' not to allow 'nominal use' for goods/services which are not substantially used.

## II. WHAT QUALIFIES AS GENUINE USE OF A TRADEMARK

### A. Nature of use

#### *1. Use consistent with the function of the trademark*

In the case of infringement of a trademark right, in accordance with the revision of the Trademark Act in 2014, it is stipulated in Article 26 (Limitations of Effects of Trademark Rights) that 'a trademark which is not used in the mode that consumers may recognize the goods or services as those pertaining to the business of a particular person.' (paragraph 1, item 6) is outside the effect of a trademark right. Before the revision of the Trademark Act, the Courts already have ruled that a use not consistent with the function of the trademark does not constitute trademark use. However, even after the Trademark Act revision, the qualification of trademark use is still assessed on a case-by-case basis.

On the other hand, regarding non-use cancellation, views and court decisions are divided as to whether trademark use or use consistent with the function of the trademark is required or not. One opinion is that, if it formally falls under one of 'use' definitions provided under Article 2(3), it is considered as use of the trademark, even if it does not function as a source indicator or it is not trademark use; another is that trademark use is necessary to maintain a trademark registration. Since the definition of a trademark under the Trademark Act does not include 'distinctiveness', from the interpretation of the current law, the former opinion is considered as appropriate. The following is a case demonstrating the former opinion.

*LE MANS case (IP High Court decision, 14 September 2016)*

> 'The main purpose of Article 50 of the Trademark Act is as follows. Exclusive monopoly rights will occur in registered trademarks regardless of their use or not. To keep registered trademarks that are not used at all for a long period of time may narrow down the scope of choice of trademarks by persons other than the right holder pertaining to the trademark and may cause harmful effects of unjustly infringing the interests of the public. For these reasons, it is possible to cancel registration of a trademark that has not been used for a certain period of time.
>
> In view of the above purpose, "use" prescribed by Article 50 of the Trademark Act is sufficient as long as the trademark is used in any form with respect to the designated goods or designated services (Article 2(3)) and is not limited to the manner in which the trademark functions as a source designator.'

Regarding a corporate/trade name etc., Article 26 (Limitations of Effects of Trademark Rights) provides that use of a trademark indicating, in a common manner, one's likeness, name, known pseudonym, professional name, or well-known abbreviation thereof, without the purpose of unfair competition (referring to the purpose of gaining unfair profits, using another person's goodwill), that is, provided it complies with the principle of fairness of commercial relations, is outside the protection of a trademark right.

Regarding use as a title, in the case of indicating the contents of the book, since it is only to indicate the quality of the book, it is considered that the trademark right does not extend to such descriptive use under Article 26(1)(ii). Or, it is considered that a title cannot be used as a trademark because it does not function as a source indicator (Article 26 (6)). (However, in my personal opinion, the title of a book can function as a trademark because, recently in Japan, creation of a title determines whether the book sells well.)

Regarding decorative use, a court has ruled that a trademark and a design are not exclusive or alternative relationships, and decorative patterns that appear designs could be trademark use, as long as the patterns function as a source indicator.

*Louis Vuitton case (Osaka District Court decision, 18 March 1987)*

> 'it is clear that the plaintiff and the defendant use the marks (1) and (2) for the function of self-identification of the product' (mark (1) is a Louis Vuitton logo and (2) is a Louis Vuitton monogram).

## 2. *Use in the course of trade*

### *i. Public use*

While there is no public use doctrine in Japanese practice, abbreviations or nicknames of brands used by the public might give rise to protectable trademark rights, even if the brand owners themselves do not use those abbreviations. The following is a nickname case, and the registrant is SUNTORY, one of large liquor companies in Japan.

*'角瓶' (meaning 'Square Bottle') case (Tokyo High Court decision, 30 January 2002)*

'角瓶' (square bottle) did not appear on the label of the square bottle, but the nickname, derived from the shape of the bottle, had been established by consumers.

'角瓶' is not inherently distinctive, but a secondary meaning was recognized by the registrant's use for a long time in large quantities.

### *ii. Commercial purpose*

According to Article 2(3)(viii), 'use' of a trademark includes:

> 'To display or distribute advertisement materials, price lists or transaction documents relating to goods or services to which a mark is affixed, or to provide information on this content, to which a mark is affixed by electronic or magnetic means'.

In order to decide legitimate use, it is not decisive whether or not goods are provided free, though this is taken into consideration.

Regarding use on goods destined for export, if affixing a mark on the goods or packaging of goods etc. takes place in Japan, then 'use' is in Japan. At any rate, 'Export' by itself is provided as 'use' under Japanese law.

### 3. *Use by the owner or with its consent*

According to court decisions, an implied licence without a written agreement could be considered as consent to use or a licence, so that such use might prevent a registered mark from being cancelled due to non-use.

Quality control, although not required in principle, is necessary for a registrant to avoid cancellation based on unfair use under Article 53 as above.

## B. Volume, duration and frequency of use

Under Japanese practice, volume, duration and frequency of use are not much of a problem for non-use cancellation, and it is rare that a trademark registration is cancelled because of nominal use.

## C. Use of a sign

Under Article 50(1), 'a registered trademark' includes a trademark consisting of characters identical with the registered trademark but in different fonts, a trademark that is written in different characters, hiragana characters, katakana characters, or Latin characters, but identical to the registered trademark in terms of pronunciation and concept, and a trademark consisting of figures that are considered identical in terms of appearance as those of the registered trademark.

When another word is added to a registered mark, there is the possibility that the mark in use is not considered as identical to the registered one, as in the following case.

*Pearl Filter case (IP High Court decision, 25 December 2013)*

Registered mark: PEARL (with Japanese katakana)
Used mark: Pearl Filter (in Japanese katakana)
Designated goods: Class 34 – Tobacco

In this case, the JPO recognized use of the registered mark in the use of 'Pearl Filter' because 'Filter' is not distinctive in the mark, the distinctive part being 'Pearl'. However, the IP High Court denied the use of the trademark because of the fact that 'Pearl Filter' as a whole is used as a sub brand of the goods, which is not identical to the registered trademark.

Regarding the problem of alternative forms of use, relating to existence of other registered trademark, please refer to the above *ELLEGARDEN* case.

In general, a registered mark in black and white is considered identical to a colour mark having the same shape, and vice versa, provided that both marks would be regarded as being identical with each other if they used the same colour(s), according to Article 70.

## D. Use in relation to goods and services

### 1. *Use for each of the designated goods and services*

In order to have a non-use cancellation request dismissed, it is enough to establish use of the mark for one of the designated goods or services against which the cancellation is requested.

It can be hard to determine whether the goods in use are identical to the designated ones against which a cancellation request is filed. If we consider the range of identity of goods with emphasis on the secondary use – 'it can also be used as [...]' – in a broad manner, the determination of which goods are covered becomes less clear. On the other hand, depending on the nature of goods, considering the possibility of confusion, it may be possible to judge the range of identity widely and consequently maintain registration. Whether or not to judge the identity of the goods in use strictly is related to whether to raise hurdles to maintain rights, which is the same kind of problems as trademark use whether a registered mark should function as a source indicator. The following is a court decision which judged the identity of goods broadly with respect to 'charm'.

*SAVOY case (IP High Court decision, 5 December 2012)*

In light of the name and composition of the goods, the advertisement mode at the time of sale, the mode or manner of use of the goods or of the same type of goods, and the recognition by traders and consumers to be inferred from these, these goods are a decoration attached to the chain part of ornaments such as a watch, bracelet, or necklace. While a 'charm' is attached to a bag, it can be also attached to clothes or used for 'accessories' like a bracelet or a necklace. Since a single product may have multiple functions or uses, it is not necessary to decide that one product always belongs to one category of goods and cannot have other intended uses. Accordingly, it cannot be said that there is no possibility that products used as key holders may not be used for different purposes.

It can also be said that 'charms' are 'personal ornaments' which are decorative items used for fashion purposes.

As to the nature of goods, it may be judged that they belong to two or more product groups because they do not correspond to a single similar product group, but have a multiple purpose. See the INTEGRAN case (Tokyo High Court decision, 12 July 2001) on 'medical supplies' and 'medicines'.

In the following case, proof of use for 'providing information concerning commercial sales' in class 35 was denied for the reason that it is merely to introduce the products sold by other companies to consumers, since introducing products to consumers who are the end users of goods does not fall under the business-to-business services in class 35. Although 'providing information concerning commercial sales' can be literally understood as including the services of providing product information to end consumers, the services should be corporate services according to the following Explanatory Note of the International Classification:

> 'Class 35 includes mainly services rendered by persons or organizations principally with the object of: (1) help in the working or management of a commercial

undertaking, or (2) help in the management of the business affairs or commercial functions of an industrial or commercial enterprise, as well as services rendered by advertising establishments primarily undertaking communications to the public, declarations or announcements by all means of diffusion and concerning all kinds of goods or services.'

*ARIKA case (Supreme Court decision, 20 December 2011)*

As for 'providing information concerning commercial sales' in class 35, it was decided as follows and clarified that the service in this class is not directed to end consumers.

> 'It is reasonable to assume that it is a service providing information to a company engaged in commerce, etc. to aid in management, operations and the like. Then, providing information on sales performance of products and information on statistical analysis concerning product sales to companies engaged in commerce etc. falls under this service.'

### 2. *Use for a broader category of goods and/or services*

Japanese practice includes broad categories of goods/services such as 'telecommunication machines and apparatus' and 'electronic machines, apparatus and their parts', which are not acceptable in most foreign countries. When a non-use cancellation is requested against such a category, it is enough to establish use of the mark for one of the designated goods in the category.

If a cancellation request is filed against one item (e.g. 'computer software') among 'electronic machines, apparatus and their parts', and if the registrant fails to prove use of the mark for 'computer software', registration of the goods is to be cancelled. However, as the remaining goods in the same category are regarded as similar to 'computer software', a party demanding cancellation still cannot register its mark for the goods because of the prior mark for the remaining goods.

## E. Territorial scope

If a trademark owner or a licensee uses a registered trademark only in a foreign country, it is not possible to have a cancellation request dismissed, while export and import of goods are considered 'use' (Article 2(3)(ii)). As for whether foreign corporations could be considered to carry out 'use' in Japan might be problematic, such as when there are no licensees in Japan, when there are only few sales, when there is no continuity as a business, when there is only a sale by personal import, and when there is a website only in English.

Although advertising (Article 2(3)(viii)) is 'use' of a trademark, in the case of the website of a foreign registrant who advertises goods only in English, though many Japanese consumers will understand the contents, it is necessary to consider whether it corresponds to advertisement in Japan. While there are many people who can trade with Japanese customers in English, in the case of another foreign language, there will be very few. However, depending on the nature of the product, it is necessary to consider if there are circumstances that a trader can purchase it only from the site of the specific language.

For English-language websites, when the server is in a foreign country (1), the website does not have a Japanese domain, the price indication is in a foreign currency such as US dollars (2), and there are no indications that the website is targeted to Japanese consumers, it would be doubtful if the website can be considered as functioning as an advertisement in Japan.

If one Japanese consumer purchases goods once from such an English-language website, could the English website be deemed an 'advertisement' in Japan? In addition, it might be necessary for the website to be recognized as intended for Japanese consumers by such indications as 'delivery to Japan', or a price display automatically switching to yen.

Regarding points (1) and (2) above, there are the following JPO trial decisions.

(1) Cancellation Trial 2007-300017
As the subject web page is provided on a US server and its contents are all displayed in English, it is not recognized as intended for Japanese consumers. The subject web page cannot be considered as corresponding to use in Japan simply because it is accessible from Japan.

(2) Cancellation Trial 2009-300585
For computer software provided on a website, in order to be recognized as used in Japan, in addition to the explanation of the purchase method, the advertisement, the descriptions of the goods, and the manner of payment all being in Japanese, the price must be displayed in yen or dollars with an exchange rate, and the trademark must be used for this purpose.

Regarding a foreign magazine also read in the country of the mark at issue, as long as Japanese consumers also read it, use in the magazine can be regarded as use in Japan.

Even if a trademark is used in only part of Japan, it will establish use as long as it is considered as legitimate use.

### F. Relevant time period and its calculation

As for non-use cancellation, since it is a requirement that a registered trademark remains unused for three years or more continuously, it is necessary for three years or more to have passed since the registration of the trademark as a prerequisite for a request for a non-use cancellation trial. Even when a registered trademark right is transferred in the three years, the non-use period is totalled despite the respective registrants' periods. Accordingly, if a trademark is used every three years, it will not be cancelled.

## III. EVIDENCE OF GENUINE USE OF A TRADEMARK

### A. When use should be proved: timing of lack of genuine use claims

#### *1. When filing an application for a trademark*

Under Japanese law, there is an intent-to-use requirement, but actual use at the time of the trademark application is not required. Unless an applicant claims too many categories of goods/services in one class, it is not necessary to prove intent to use. If too many categories (more than 22 according to the Examination Guidelines as of April 2018) are claimed, the examiner requests proof of such a wide range of business activity by advertising materials, etc.

### 2. *In the course of opposition proceedings*

Under Japanese law, there is no system that proof of use is to be provided in opposition proceedings upon request by the applicant or upon request of the opponent, nor that the opponent should submit evidence of use of the trademark.

### 3. *In cancellation proceedings based on lack of use*

Cancellation proceedings based on non-use are primarily administrative proceedings at the JPO trial division. It is possible to appeal to IP High Court against a trial decision, and then to the Supreme Court.

With respect to the claim of use and the timing of submitting proof of use, according to a Supreme Court decision of 23 April 1991, it is understood that it is permissible to submit proof of use until the end of oral pleadings at the IP High Court, even if the registrant had not proved any use at the JPO trial.

Also according to the Supreme Court decision, it is not contrary to the provisions of the Trademark Act to hear facts regarding use of the registered trademark not claimed by the registrant and to recognize the use from evidence submitted by the official authority (IP High Court decision, 7 November 2016).

Where a trial decision to the effect that the trademark registration is to be rescinded becomes final and binding under Article 50(1), the trademark right shall be deemed to have become extinguished on the date of registration of the request for a trial under the said paragraph (Article 54(2)).

### 4. *As a defence in infringement/cancellation proceedings*

In principle, a trademark right is effective until it is cancelled or invalidated. However, as a defence in infringement proceedings, the exercise of a trademark right which is not used might be considered as an abuse of rights in infringement proceedings, that is, it might not be possible to exercise a trademark right based on the mark of non-use.

Where an action is instituted or a motion for the order of a provisional seizure or an order of a provisional disposition is filed, the court may, if it considers it necessary, suspend the court proceedings until the trial decision becomes final and binding (Patent Act 168(2) applicable to Trademark Act).

## B. Burden of proof

When a non-use cancellation is requested against a registered mark, the registrant must prove use of the mark.

## C. Relevant evidence

'Use' with respect to a mark as used in the Trademark Act means any of the following acts (Article 2(3)):

(i) to affix a mark to goods or packages of goods;
(ii) to assign, deliver, display for the purpose of assignment or delivery, export, import, provide through an electric telecommunication line, goods or packages of goods to which a mark is affixed;
(iii) in the course of the provision of services, to affix a mark to articles to be used by a person who receives the said services (including articles to be assigned or loaned;
(iv) in the course of the provision of services, to provide the said services by using articles to which a mark is affixed and which are to be used by a person who receives the said services;
(v) for the purpose of providing services, to display articles to be used for the provision of the services (including articles to be used by a person who receives the services in the course of the provision of services) to which a mark is affixed;
(vi) in the course of the provision of services, to affix a mark to articles pertaining to the provision of the said services belonging to a person who receives the services;
(vii) in the course of the provision of services through an image viewer, by using electromagnetic means (electromagnetic means shall refer to any electronic, magnetic or other method that is not recognizable by human perception), to provide the said services by displaying a mark on the image viewer; or
(viii) to display or distribute advertisement materials, price lists or transaction documents relating to goods or services to which a mark is affixed, or to provide information on such content to which a mark is affixed by using electromagnetic means.

### 1. *Form of evidence*

Examples of evidence of use include:

- Business documents showing sales and advertisements of the goods/services against which a cancellation request is filed, in Japan
- Copies of articles or advertisements relating to goods/services in magazines or newspapers distributed in Japan, and/or records of advertisements through any media during the period of three years, establishing dates of use.
- Photos of goods/services showing use during the period.
- Affidavits are not necessarily requested.

### 2. *Distinction between use of a company name and use of the services offered by the company*

A company name can also function as a trademark for the services offered by the company.

## IV. JUSTIFICATION FOR NON-USE

If there is a justifiable reason for not using the registered trademark, the registration should not be cancelled. 'Justifiable reason' may be when production of a product scheduled to be used for the trademark is being prepared, a factory was destroyed as a result of natural disaster, or the use of the trademark is prohibited for more than three years by legislation.

## V. CONSEQUENCES OF NON-USE

### A. Consequences of non-use and relevance of the context

When cancellation of the registration of a trademark is finalized by the decision of the trial for cancellation due to non-use, the trademark right is deemed to have been extinguished on the day of registration of the request of the trial (Article 54 (2)). As a result, if a person, who has received the exercise of a trademark right pertaining to an unused registered trademark, immediately requests a non-use cancellation, even if it takes time for the decision, the trademark owner in the meantime cannot claim damages if the registration is cancelled.

### B. Extent of any revocation

Nominal use of a trademark is not sufficient to avoid cancellation on the grounds of non-use. If an extremely small quantity of goods has been sold, or if advertisement of goods is not followed by any sales, this will not be considered as substantial use and will not qualify as trademark use.

For example, even if an advertisement for a course pertaining to a registered trademark exists, but use of the name has ceased and the course did not take place, this does not correspond to use (FRM – Financial Risk Manager case, IP High Court decision, 25 August 2016).

However, as for very expensive foreign-made cars, even if there are only a few sales per year, use cannot be said to be nominal, taking into account the nature of the product, the demand, the sales route, etc.

Since it is normal to pre-advertise products that are scheduled, it cannot be said to be nominal use if the mark has only been used for advertisements at that point in time. For an example of an advertisement judged to be nominal use, see the VUITTON case (Tokyo High Court decision, 30 November 1993), showing use of a plurality of marks (LOUIS VUITTON, VUITTON, LV logo and LV monogram) for many items of merchandise.

### C. Timing of any revocation

The trademark right is deemed to have been extinguished on the date on which the cancellation action based on non-use is recorded (once the final decision is issued).

Where the demandant of a trial conspires with the demandee to cause a trial decision to be rendered for the purpose of harming the rights or interests of a third party, the third party may file a request for a retrial against the final and binding trial decision (Article 58(1)).

# The Republic of Korea

*Seung Yong Lee and Doug Jay Lee*

## I. INTRODUCING GENUINE USE OF TRADEMARKS

### A. Legal framework

A trademark needs to be used to function as a source identifier for goods. It only makes sense that a person using or intending to use a mark should be bestowed with rights to a trademark that is subject to use. If a mark remains unused, the proprietor's rights to the mark should be subject to cancellation so as to allow the mark to be available for use by another person as a source indicator of goods. The inextricable connection between a trademark and its use can be inferred from provisions of the Korean Trademark Act governing the definition of a trademark, a person entitled to register a mark, cancellation of a trademark registration, and the like.

The Korean Trademark Act basically adopts a first-to-register system that allows a person to file and register a mark without proof of its use or intention of use. The Korean trademark system does not provide a specific procedure for confirming use or intent to use when applying for a trademark registration. The first-to-register system has its merits in reducing the burden placed on the trademark office reviewing trademark applications through its omission of a confirmation procedure for use of a mark. On the other hand, the same system has the problem of limiting another person's opportunity to select a trademark by increasing the number of unused registered marks. The Korean Trademark Act has been evolving in the direction of maintaining the first-to-register system and at the same time strengthening the use requirement to supplement the inherent problems of this system. That is, efforts are being made to have a trademark application that is filed without a bona fide intent to use to be refused at the examination stage and ease the requirements for the cancellation of unused registered marks so as to ensure more fairness in the trademark system.

On a related note, the Korean Trademark Act uses the language 'use' without classifying it into specific types of use, such as genuine use, normal use, actual use, or token use. The 'use' referenced in the Korean Trademark Act means a legal and valid use and is always understood to refer to genuine use. It may be of interest to know that the courts rely on the concept of token use – the opposite concept of genuine use that refers to use made to merely avoid cancellation due to non-use – to determine that token use does not constitute the 'use' meant in the Korean Trademark Act.

## B. Purpose and function of genuine use: policy considerations

The number of trademarks filed and registered in Korea is continuously rising. Trademark applications filed in Korea increased from 122,384 in 2006 to 181,606 in 2016, showing a 48% increase in ten years. The number of trademark registrations increased from 65,825 in 2006 to 119,255 in 2016, showing an 81% increase during the same period.[1] An increase in the number of existing trademark registrations may interfere with another person's opportunity to select a trademark as it may also mean an increase in the number of 'unused registered marks', which may not only limit another person's freedom of choice in selecting a trademark, but also contributes to negligence of trademark preoccupancy by parties without a legitimate right. Accordingly, the non-use cancellation system should remain fully functional for easy cancellation of unused registered marks. Also, it is very important to clarify what constitutes genuine use of a trademark as the determination of whether or not a trademark registration should be cancelled is based on this interpretation.

The Korean Supreme Court explains the intention behind a non-use cancellation proceeding as follows:

> 'The proprietor or licensee of a registered mark has a duty to use the mark on the designated goods, and the cancellation of the registered mark is available as a countermeasure against non-use of the mark for a specified period of time, so as to mitigate the negative consequences of the first-to-register system under which a trademark satisfying certain requirements is allowed to be registered irrespective of its use and to also expand another person's freedom of choice in selecting a trademark.'[2]

| | *2010* | *2011* | *2012* | *2013* | *2014* |
|---|---|---|---|---|---|
| Number of new trademark registrations | 53,136 | 71,255 | 77,903 | 100,094 | 99,791 |
| Number of trademark rights in effect at the end of the year | 720,709 | 768,019 | 817,862 | 886,273 | 950,145 |
| Number of non-renewed trademark registrations | 22,809 | 23,945 | 28,060 | 31,683 | 35,919 |

As shown in the table above, the number of new trademark registrations each year increased by 88% in five years – from 53,136 in 2010 to 99,791 in 2014. The number

1. Intellectual Property Statistics for 2016 published by the Korean Intellectual Property Office.
2. Supreme Court ruling on Case No. 2012 HU 2463, delivered on 26 September 2013.

of trademark rights in effect increased by 32% during the same period – from 720,709 at the end of 2010 to 950,145 in 2014.[3]

Theoretically, the number of new trademark registrations in a year plus the number of trademark rights in force at the end of the previous year should equal the number of trademark rights in effect at the end of the year. However, the number of actual trademark registrations in effect at the end of each year is less than such sum. The differences between the numbers may be accounted for by the number of non-renewed trademark registrations, which may be categorized into unused registered marks.[4] The number of unrenewed trademark registrations showed a 57% increase in five years, from 22,809 in 2010 to 35,919 in 2014.

As shown above, the number of unused registered marks are increasing along with increases in the number of trademark rights in effect. Such registrations may be cancelled through a non-use cancellation trial. Alternatively, such registrations may be prevented at the examination stage by confirming lack of bona fide intent to use. Genuine use or bona fide intent to use a trademark provides the cornerstone for a non-use cancellation trial and refusal of a trademark registration as trademarks not being genuinely used nor intended for genuine use are not conducive to the maintenance of goodwill and fair competition and may even interfere with them.

## II. WHAT QUALIFIES AS GENUINE USE OF A TRADEMARK

### A. Nature of use

#### 1. *Use consistent with the function of the trademark*

##### *i. Use of a trademark under the Korean Trademark Act*

The Korean Trademark Act defines 'use of a trademark' as any of the following acts.[5]

(i) indicating a trademark on goods or packages of goods;
(ii) transferring or delivering goods or packages of goods on which a trademark is indicated, or displaying, exporting, or importing them for such purpose;
(iii) indicating a trademark on advertisements on goods, price lists, transaction documents, signboards, or tags and displaying or distributing them.

The above provision merely enumerates types of use that may be acknowledged as valid use. The determination of valid use in real cases is based on whether a used

3. Intellectual Property Statistics published by the Korean Intellectual Property Office: http://www.kipo.go.kr/kpo/user.tdf?a = user.html.HtmlApp&c = 3043&catmenu = m04_05_02_02.
4. The number of the registrations that actually expired due to non-renewal are estimated to be slightly less than the number of the reported non-renewed registrations as registrations that are invalidated, cancelled due to non-use, or abandoned are included in the reported total number of the non-renewed registrations. However, such discrepancies have been disregarded here as the number of non-renewed registrations due to invalidation, non-use cancellation, or abandonment would be relatively scant and difficult to trace.
5. Korean Trademark Act, Article 2(1)(xi).

mark can be perceived as a designation of the source of the goods the petitioner is seeking to cancel.[6]

*ii. Use of a personal name/business name*

A personal name or business name indicated on products is highly likely to be perceived as a source identifier for the products, and the chances are likely that such usage is just what the user had in mind. Accordingly, a personal name or business name, even if jointly used with another mark on the product, is likely to be acknowledged as valid use of a trademark. However, in the event the personal name or business name is not at all likely to be perceived as a source identifier for goods, such use cannot be regarded as valid use of a trademark.

On a related note, use of a personal name or business name that conflicts with another person's trademark right is not deemed to amount to infringement of the trademark right if it had been used to signify one's personal identity in a commonplace way (Korean Trademark Act, Article 90(1)). However, the same use for the purpose of unfair competition after registration of the trademark right amounts to trademark infringement.

*iii. Titles of books, magazines, newspapers, movies and other copyrighted works*

Use of a trademark as the title of a copyrighted work is not recognized as valid use unless there are special circumstances for the trademark to be perceived as indicating the source of the copyrighted work. The use of a trademark as the title of periodical literature, however, will be recognized as valid use since the sources of periodicals are easily identified through their titles.

Supreme Court ruling on Case No. 2000 HU 3395, delivered on 10 December 2002:

> 'Titles of books are merely descriptive of the contents of the books and do not identify the publishers from which the books originate. That is, titles of books are not used to distinguish the books of one publisher from those of another publisher. Accordingly, effects of registered trademark rights do not extend to identical or confusingly similar marks that are being used as the title of a book.'

*iv. Use as a design*

If a mark is not perceived as a source identifier for goods but is instead perceived as a design, it cannot be considered to have been validly used.

---

6. There can be a gap between the intention of a trademark owner and perception of consumers as to whether the concerned mark is perceived as a source identifier for goods. In Korea, a non-use cancellation trial tends to be determined with a focus on whether the trademark owner (or a licensee) used the mark to indicate the sources of the goods. However, in infringement claims, even if a trademark had been used to indicate the sources of the goods, if that is not how the mark would be perceived amongst consumers, there is tendency to deny usage of the mark as a trademark.

Supreme Court ruling on Case No. 2011 DA 18802, delivered on 24 January 2013: In an infringement claim brought by the owner of the trademark

designating necklaces of precious metal and the like, for relief against distribution of pendant necklaces looking like

,

the court ruled that 'in addition to not looking similar to the cited trademark,

was merely used as a design and would not be perceived as a source identifier for goods. So, it cannot be considered to have been used as a trademark, leading to the conclusion that its use did not amount to infringement of the registered mark.'

### *2. Use in the course of trade*

#### *i. Public use*

A trademark is used to indicate the source of the goods or services in commerce. Accordingly, use of a trademark on goods that are not distributed through normal transactions cannot be considered valid. Valid use of a trademark on the concerned goods can be acknowledged in a non-use cancellation proceeding or a trademark infringement litigation, insofar as the goods are the subject of an independent commercial transaction. Valid use will be recognized only if use of the trademark was public.

#### *ii. Commercial purpose*

a. The goods against which a cancellation trial was filed should be the subject of an independent commercial transaction

Use of a trademark on articles merely serving as an advertising medium is not acknowledged as valid use. The gist of the Supreme Court ruling on Case No. 2012 HU 1415, delivered on 26 December 2013, found the goods at issue to be the subject of an independent commercial transaction as follows:

The trademark owner, whose principal goods included bags, ordered 500 bottles of perfume be manufactured and made available for mileage purchase by his VIP clients. A mark considered identical to the subject mark was used on the containers and packaging of the perfume, which was offered to his clients. In light of the aforesaid transactions, size, and packaging of the perfume, the perfume is deemed to have been an article for independent commercial transactions with exchange value. Accordingly, use of the concerned mark on packaging of the perfume or sale of such perfume is considered to constitute valid use of the subject mark.

b. Advertisement

The Korean Trademark Act recognizes the act of using, exhibiting, or advertising a mark in or on product advertisements, price tags, transaction documents, or other materials as amounting to valid use. Use of a mark in advertisements prior to actual manufacturing/distribution of the goods may be recognized as valid use provided that the goods are scheduled for manufacture or distribution sometime soon.

c. Non-monetary transactions

The Supreme Court reached the same conclusion in Case No. 2010 HUH 9668, delivered on 29 April 2011:

> 'The owner of the concerned trademark registration is a manufacturer and distributor of "Elite School Uniform," a leading brand of school uniforms in Korea. It hosted a model contest with sponsorship of Fashionable Communications (a local educational TV program). Although the contest was sponsored by Fashionable Communications, since the trademark owner prepared and hosted the model contest for the purpose of recruiting models to endorse their Elite school uniforms, the mark cannot be considered to have been used for the services "organizing of model contests" or "organizing of beauty contests" in a pure sense. Accordingly, unless there is evidence indicating that the trademark owner used the concerned mark in the performance of services for benefit of consumers and was compensated with income for such services, the trademark cannot be considered to have been validly used.'

d. Promotional use

Use of a mark on Goods B to promote sales of Goods A constitutes valid use of the mark in respect of Goods A, but not in respect of Goods B.

In this regard, the Patent Court ruled as follows on Case No. 2003 HUH 5408 on 28 May 2004:

> 'In light of the descriptions "long wallets for promotional use, wallets for July promotion" entered in the relevant internal order sheet, it seems that the concerned goods were manufactured and distributed merely as a free gift for customers or to promote sales of other products rather than as an article for an independent commercial transaction. Unless the long wallets and other articles at issue on which the subject mark is used were intended for normal distribution in the relevant market, they are deemed to be merely articles serving as an advertising medium not intended for normal distribution in the relevant

> market. Accordingly, the subject mark, though indicated on the concerned long wallets, cannot be considered to have been validly used for the long wallets.'

e. Token use

According to the Korean Supreme Court's ruling in Case No. 89 HU 1257, delivered on 10 July 1990, and Case No. 2011 HU 354, delivered on 30 June 2011, respectively, 'In light of the intent behind the non-use cancellation system, even if a mark was used in a local advertisement/promotion, such use is valid only if the advertisement/promotion was run with general and normal distribution of the concerned goods in Korea in mind (the goods should be being currently distributed or at least being scheduled/prepared for distribution).'

In this regard, the Korean Supreme Court ruled as follows in Case No. 2015 HU 2006, delivered on June 29, 2017:

> 'Even if the registered mark at issue was indicated in a blank space in a newspaper advertisement of another brand of soju (a Korean alcoholic drink distilled from rice), produced and distributed by the trademark owner, for about one month before the filing of the subject non-use cancellation proceeding, such use cannot be considered to have been made to indicate the source of soju (the designated goods) but to merely protect the trademark registration from cancellation, which constitutes token use; therefore, the mark cannot be considered to have been validly used.'

### 3. *Use by the owner or with its consent*

Under the Korean Trademark Act, right to use a mark comes into existence upon the conclusion of a licence agreement between the trademark owner and a licensee, and such right to use does not need to be recorded with the Korean Intellectual Property Office (KIPO). A trademark registration can be protected from cancellation upon proof of its use by a licensee. Right to use is acknowledged upon proof of the fact that the trademark owner granted right to use the registered mark to a licensee, and quality control does not need to be embodied in a licence agreement.

## B. Volume, duration and frequency of use

There are no legal provisions or precedents available as of yet that set forth certain standards for volume, duration, or frequency of use required to protect a registered mark from being cancelled due to non-use. Whether or not use is token use made to merely avoid cancellation or genuine use is determined based on a review of the evidence of use submitted in each case. If the volume of use, the duration of use, or the frequency of use is too low, the higher the chances of a conclusion that such use is token use that was made merely to prevent cancellation due to non-use. In the case of types of goods that need to be produced in certain minimum quantities to be distributed, production below the minimum level may lead to a determination of non-use.

## C. Use of a sign

### *1. Identicalness to the registered mark*

According to standards established by local precedents, use of a registered mark in an altered form is recognized as valid use provided that the altered form still falls within the scope of what is considered identical to the registered mark. Use of a registered mark in a different font or colour is generally acknowledged as constituting use of the registered mark. There is no default font suggested by KIPO.

### *2. Use of a registered mark after deleting its distinctive element*

In Case No., 2004 HU 1588, decided on 15 July 2005, the Supreme Court concluded that of the portions comprising the registered mark 'GUESS BY MAURICE MARCIANO,' 'GUESS,' and 'MAURICE MARCIANO,', or 'MAURICE' and 'MARCIANO,' each constituted the distinctive essence and had the same weight as 'GUESS'. Accordingly, the mark

GUESS
BY MARCIANO

in which the portion 'MAURICE' was deleted did not fall within the scope of what is considered identical to the registered mark.

### *3. Use of a registered mark consisting of an English portion and a Korean transliteration*

In the case of a registered mark consisting of an English portion and a Korean transliteration thereof, use only of the English portion is acknowledged as amounting to use of the registered mark.

If a registered mark is used after one of its distinctive essences is deleted, it is no longer considered identical to its registered form. However, if a registered mark consists of English letters and a Korean transliteration thereof, use only of the English portion is considered to amount to use of the registered mark.

The Korean Supreme Court unanimously reached just such a conclusion in Case No. 2012 HU 2463, delivered on 26 September 2013, as follows, and repealed previous decisions to the contrary:

> 'In regard to the registered mark, which consists of English letters and a Korean transliteration thereof, use only of the English letter portion or the Korean transliteration portion amounts use of a mark that falls within the scope of what is considered identical to the registered mark, provided that the two portions combined do not give rise to any new concept other than what is derivable from the meaning of the English letter portion and that even with either the English letter portion or the Korean transliteration portion deleted, the remaining portion will be pronounced the same way the registered mark will be generally pronounced amongst consumers or traders. Accordingly, we are invariably led to the conclusion that use only of the English letter portion

of the registered mark amounts to use of the registered mark the petitioner is seeking to cancel.'

## D. Use in relation to goods and services

### *1. Use for each of the designated goods and services*

In Korea, a cancellation trial against only some of the goods covered by a trademark registration is allowed. A trademark owner can defend his registration by establishing use of the registered mark for only one of the designated goods the petitioner is seeking to cancel in a non-use cancellation proceeding. In real cases, it can be difficult to determine whether the registered mark was actually used on goods or services that are considered identical to the designated goods or services of the registered mark. The trademark owner needs to provide a clear definition of the designated goods by considering the dictionary definition of the designated goods, the goods classification system, explanatory notes on goods that were published in the relevant industry, or the name by which the designated goods are actually referenced in the relevant industry and show that the designated goods and the goods on which the registered mark was actually used are identical. The registered mark is required to have been used on goods identical to the designated goods and cannot be defended from attack if it was used only on similar goods.

### *2. Use for a broader category of goods and/or services*

In Korea, 'broad goods description' is allowed in respect of certain goods such as clothing, footwear, headwear, playthings, toys, musical instruments, tea, cosmetics, and kimchi (food). In the event the designated goods are 'clothing,' use of the registered mark on 'Tee-shirts' is valid enough. That is, use of a registered mark for any one of the goods encompassed in a broad goods description is sufficient to preserve the trademark right for the broad goods description, and the trademark right will remain valid for the entire scope of the goods covered by the broad goods description.

## E. Territorial scope

### *1. Distribution of foreign magazines containing advertisements of the concerned mark in Korea*

There is a court precedent acknowledging the import and distribution of a magazine published abroad, which contains an advertisement of the relevant mark, as amounting to valid use, as follows:

Patent Court Case No. 98 HUH 867, delivered on 4 September 1998:

> 'In regard to the method of advertising through publications, publications are not limited to the ones issued in Korea. Even if publications were issued abroad, if they are being imported and distributed in Korea, the chances are likely that the advertisement of goods in the publications would be seen by local consumers.

Accordingly, advertising/promotion of a mark in Korea through publications that are issued abroad amounts to valid use of the mark.'

## 2. *Distribution of foreign magazines containing an advertisement of the designated services in Korea in the absence of the facilities required to provide the designated services in Korea*

In principle, valid use of a service mark is acknowledged when the mark was used to indicate the source of the services through the process of providing the services to consumers in Korea. In the case of a registered mark whose designated services are available to consumers in Korea from abroad, use of the mark in the provision of the services from abroad to consumers in Korea may be acknowledged as valid use.

However, the Patent Court ruling on Case No. 2010 HUH 1480, decided on 21 October 2010 (not appealed and subsequently affirmed) shows a somewhat different view. A US company registered 'Sea World' for 'dolphin shows, animals shows, etc.' in Class 41 with KIPO. However, the designated services could not be provided in Korea due to a lack of supporting facilities. Nonetheless, the Patent Court concluded that the mark had been validly used for the designated services based on the fact that the trademark owner distributed a travel guide issued by California Tourism Office that contained descriptions and advertisements of the Sea World theme park (operated by the trademark owner) to guests (and later to a number of local travel agencies) during the opening ceremony of a Korean branch of California Tourism Office, held in the Chosun Hotel in Seoul, Korea.[7]

## 3. *Sale of goods, which were purchased by a local distributor directly from the trademark owner, though they do not have a business relationship, constitutes valid use*[8]

Import and sales of goods in Korea, on which a registered mark is used, that are purchased by a local distributor from the trademark owner amount to valid use of the registered mark in Korea. The local distributor does not need to be under any business agreement with the foreign trademark owner. Under the Korean Trademark Act, import of the concerned goods into Korea by the trademark owner or a licensee is seen as amounting to use by the trademark owner. In this case, import and distribution of the goods by an unrelated third party was acknowledged as amounting to valid use of the registered mark in Korea by the trademark owner.

---

7. This decision acknowledged valid use of the registered mark solely based on advertising activity in Korea even though the concerned services remained unavailable to consumers in Korea. Such decision is contrary to the intent of the non-use cancellation system, which is aimed at promoting commercial use of the registered mark by the trademark owner in Korea.
8. Case No. 2002 HU 2020 *The Manhattan Passage* case, decided by the Korean Supreme Court on 26 December 2003.

### 4. *Purchase by Korean consumers in duty-free shops in Fukuoka Airport in Japan is not acknowledged as valid use*

In Case No. 2004 HUH 4919, decided on 7 January 2004, the Patent Court did not acknowledge valid use of

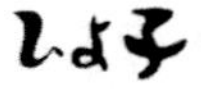

,

a registered mark in Korea:

> 'It is acknowledged that the trademark owner sold confectionery in duty-free shops in Fukuoka, one of which was located at a Fukuoka port where a passenger liner would make as many as four round trips a day between Fukuoka and Busan, Korea, and the other which was located in the Fukuoka Airport. However, the said duty free shops were not operating in Korea and it could not be supported with sufficient evidence that the mark used on the said confectionery had come to signal to consumers and traders in Korea the person whose goods are indicated by the mark through transfer and display of the goods in the normal course of trade in Korea nor that the relevant publications had been distributed to a certain extent in Korea. By the trademark owner's admission, the places where the said confectionery or publications were sold served final consumers rather than intermediaries or retailers. Also, considering that the persons who purchased the goods are passengers of a passenger liner or flights, the said confectionery could not be expected to be distributed in Korea through transfer or exhibition in the normal course of transactions. Nor could the said publications be expected to be distributed in Korea in sufficient quantities. Accordingly, the aforesaid use of the registered mark could not be considered valid.'

### 5. *Trademark use on overseas-based e-commerce sites and in Korea*[9]

The Patent Court's decision on Case No. 2011 HUH 4868 (an appeal against cancellation of a 'Victoria's Secret' trademark registration) involves a mark used on goods that could only be ordered from an overseas-based shopping site due to no physical sales outlets in Korea at the time:

> 'Even if the trademark owner did not have stores selling the concerned goods in Korea at the time, local consumers ordered the goods bearing the registered mark online via the website of the trademark owner and received their order at

9. Our opinion as to what activities can be performed by an operator of an overseas-based e-commerce site to make the goods available to local consumers is as follows: 1) the availability of the goods to local consumers alone is not sufficient; 2) although there are precedents in which the availability of the goods to local consumers and actual transactions with local consumers alone were found to be sufficient enough to reach a conclusion of valid use, the same conditions may not be considered sufficient enough depending on the facts of the matter; 3) the availability of the goods *and* advertising of the same to local consumers are sufficient enough to draw the conclusion of valid use; and 4) the availability of the goods and advertising of the same to local consumers, plus actual transactions will, without a doubt, be sufficient to draw the conclusion of valid use.

their addresses in Korea, which, unless there is particular reason to conclude differently, amounts to transfer or delivery of the goods on which the mark has been used and therefore constitutes valid use. [...] Also, if local consumers received catalogues carrying pictures of models in products bearing the registered mark together with their orders, this amounts to use of the trademark in advertising and distribution of the goods and constitutes valid use [...]. The petitioner contends that purchase of the goods by local consumers via an overseas-based online site and receiving the orders at their addresses in Korea cannot be seen as amounting to transfer of the goods on which the trademark has been used in Korea, since in doing so, local consumers merely received the goods that were purchased outside of Korea. Where local consumers made an overseas purchase through online transactions, as was done in this case, the sales may be seen as having been completed abroad through the plaintiff's overseas-based website. However, it is reasonable to conclude that delivery of the order, a process through which the ownership of the ordered goods changes hands, was completed in Korea when the customer received the orders.'

### 6. *OEM/use on goods destined for export*

Use of a mark on goods produced only for export in Korea is recognized as valid use. This remains the same even if the goods are not at all likely to be distributed in Korea. The trademark owner is seen as having fulfilled his obligation to use a registered mark in Korea by using it in the aforesaid manner.

Regarding use of a registered mark on OEM products, the Korean Supreme Court ruled as follows in Case No. 2012 HU 740, decided on 12 July 2012:

'In the case of OEM (original equipment manufacturer) products that were manufactured and exported under the brand of the requesting party instead of the manufacturer's, the requesting party maintains actual control over the manufacture of the goods, including quality control. The exporter's production exclusively relies on orders of the requesting party and all the manufactured products are normally delivered to the requesting party. So, when determining the party that has used the mark in a non-use cancellation proceeding, unless there are particular reasons to conclude differently, the mark should be deemed to have been used by the trademark owner or a licensee thereof who requested the manufacture of the goods. It is reasonable to conclude that even when the goods on which a registered mark was used were OEM-manufactured in Korea for export and were unavailable to consumers in Korea, the trademark owner fulfilled his duty to use the registered mark in Korea.'

### 7. *Use in limited regions*

A registered mark does not need to be used nationwide in Korea to be protected from cancellation; use in a limited area is valid. A trademark right used in a limited area has nationwide effects.

## F. Relevant time period and its calculation

It is not required that a registered trademark be used continuously without a non-use period to be protected from non-use cancellation. Even if the registered mark was used only once in the three years before the filing of the non-use cancellation proceeding, if it was not token use made solely for the purpose of preserving the registration, the registration will not be subject to cancellation.

# III. EVIDENCE OF GENUINE USE OF A TRADEMARK

## A. When use should be proved: timing of lack of genuine use claims

### *1. When filing an application for a trademark*

Proof of use or intent to use is not required when applying for a trademark registration in Korea.[10] However, the provision that only the person using or intending to use a trademark may obtain a registration for the mark[11] may be cited to refuse a trademark application. When the trademark examiner has reasonable suspicion to doubt the intent to use, he may issue a preliminary rejection of the application to confirm the applicant's bona fide intentions.

According to the trademark examination guidelines, cases in which it is reasonable to doubt a bona fide intent to use include the following:

- A trademark application filed by an individual designates goods/services requiring large capital and capabilities, e.g., satellites, railway vehicles, department stores, banking.
- A trademark application designates a large number of unrelated, non-similar goods
- Intention of preoccupying a trademark registration can be inferred from the applicant's trademark filing history.

If a preliminary rejection of an application due to a lack of bona fide intent to use is issued, the applicant may submit a statement explaining how he plans to use the mark. If the trademark examiner is not convinced, he may issue a final rejection of the trademark application. However, such statement does not tend to be reviewed strictly.

### *2. In the course of opposition proceedings*

An opposition can be filed by anyone within two months from the publication of a trademark application. In the case of an opposition grounded on a likelihood of confusion with a prior registration of the opponent, the opponent does not have duty to prove use of the cited registration. On a related note, the opponent's cited registration will be vulnerable to a cancellation attack based on non-use by the

10. A Korean trademark application form does not have a section for declaring intent to use.
11. Article 3(1) (Persons Entitled to Registration of Trademark): Any person who uses or intends to use a trademark in the Republic of Korea may obtain registration of his/her trademark.

applicant if it had not been used for more than three years before the filing of the cancellation proceeding.

### 3. *In cancellation proceedings based on lack of use*

A trademark registration that has not been used by the owner or an exclusive/non-exclusive licensee thereof for more than three years before the filing of a non-use cancellation proceeding in Korea for the designated goods may be cancelled via a non-use cancellation proceeding. Such non-use cancellation proceeding may only be filed with the Intellectual Property Tribunal and Appeal Board (IPTAB), an administrative body under KIPO. A panel of three IPTAB trial judges reviews and decide cases independently, without supervision of the Chief Officer of KIPO or IPTAB. It took an average of eight to nine months from the date of request for appeal up to the issuance of the decision in IPTAB in 2016. However, a case in IPTAB can take beyond a year depending on how aggressively the concerned parties respond.[12] An IPTAB decision can be challenged in the Patent Court, and if it is not successful, to the Supreme Court. Patent Court cases are decided more quickly than IPTAB cases, taking about four to five months on average.

### 4. *As a defence in infringement/cancellation proceedings*

In Korea, whether or not a trademark should be subject to cancellation is determined with the IPTAB (an administrative panel under KIPO) and can be challenged up to the Patent and Supreme Court levels. Trademark infringement cases are determined in district courts and can be challenged in the Patent Court and the Supreme Court. In Korea, a non-use cancellation cannot be directly filed in a court, and this rule remains the same even if there is a trademark infringement case pending in a district court or the Patent Court. The infringer is not allowed to file a non-use cancellation proceeding in the court as a counter-action. Whether or not the court in which a trademark infringement litigation is pending should wait until a related non-use cancellation proceeding is decided in IPTAB is determined at the discretion of the relevant panel of judges as there is no relevant provision governing this aspect.

## B. Burden of proof

The burden of proof is placed on the trademark owner. If the trademark owner establishes valid use of the registered mark for any one of the designated goods the petitioner is seeking to cancel, the registration will be protected from cancellation for all of the targeted goods. The petitioner may seek to cancel only some of the goods covered by a target registration.

---

12. Intellectual Property Statistics for 2016 published by the Korean Intellectual Property Office: http://www.index.go.kr/potal/main/EachDtlPageDetail.do?idx_cd = 2786. A total of 2,122 non-use cancellation were filed with IPTAB in 2016. Among the 1,666 cases decided in the same year, 1,219 were allowed, 99 cases were denied, 141 cases were dismissed and 207 cases were withdrawn. A substantial number of the 207 withdrawn cases may have been settled between the concerned parties outside of IPTAB.

## C. Relevant evidence

### *1. Use on the product or its packaging*

Materials showing use of the concerned mark on packaging is considered strong evidence of use. In real cases, pictures of products or packaging on which the concerned mark is used are submitted with other evidence in support of the valid use of the concerned mark in the three years before the filing of a non-use cancellation proceeding.

### *2. Use in advertising*

The relevant page(s) of a book or magazine in which the concerned mark or products featuring the mark were advertised and the cover page thereof with the title, issue date, and source of the books or the magazines printed thereon serve as strong evidence of use. In the case of online advertisements or TV commercials, captured images of the advertisements may be submitted, preferably accompanied by a statement of fact by the operator on the advertised medium or records of ad spending.

### *3. Use on the internet*

Web screens showing use of the concerned mark as a source identifier for the relevant goods or services may be captured and printed for submission as evidence of use. When it is difficult to establish that the web screenshots are from a date falling within the three-year period before the cancellation proceeding, the date of the web screenshots needs to be established with additional evidence such as an affidavit.

### *4. Form of evidence*

An affidavit attesting to the performance of manufacturing, distribution, import, or advertising of the goods on which the registered mark was used by the trademark owner or a licensee in the three years before the cancellation proceeding may be drafted by a third party, the trademark owner, a licensee, or a related party for submission as evidence. Affidavits are often relied upon to substantiate use in real cases and need to be signed and notarized to be considered valid.

Use of a service mark is generally established with evidence showing use on the means by which the services are provided, or goods supplied to offer the services or their packaging or exhibition thereof, or use of the service mark in advertisements, price lists, transaction records, signs or labels, or exhibition or distribution thereof. Due to the particular nature of the services, the said materials may be seen as transaction records as well as the means of services or goods supplied to provide the services. Hence, use of a service mark on the materials without exhibition or circulation of the materials is deemed sufficient use of a service mark (Patent Court Case No. 2008 HUH 10207, delivered on 2 April 2009).

### 5. *Distinction between use of a company name and use of the services offered by the company*

In regard to a service mark registration consisting of a business name, if evidence shows its use solely as a business name, the registered mark will not be considered to have been validly used for the designated services. However, if the business name was used in such a way so as to identify the party providing the designated services, valid use of the registered mark for the designated services may be acknowledged. These differences determine whether or not use of a business name will be deemed valid use of a service mark. A corporate image advertisement from which the goods or services offered cannot be identified is deemed insufficient as evidence to establish use of a registered mark for certain goods or services in a cancellation proceeding.

## IV. JUSTIFICATION FOR NON-USE

### A. Obstacles to use

Even if a trademark registration had not been used by the owner or a licensee for more than three years, it may still be protected from cancellation if there is a legitimate reason for such non-use. The legitimate reason meant herein includes both force majeure events and events not due to fault or negligence of the trademark owner. Force majeure events include epidemics or contagious diseases, and natural disasters. Events not due to fault or negligence of the trademark owner include situations in which the goods of a registered mark cannot be traded through general, normal transactions in the local market due to prevention by law, prohibition against sales, or import restrictions placed by the government.

In Case No. 98 HU 8144, decided on 29 October 1998, the Patent Court dismissed a petition for cancellation of a trademark registration, concluding that there was a justifiable reason for such non-use as follows:

> 'The plaintiff, The Cartoon Network Inc., a foreign corporation, was forbidden by law to either directly manage or partake in management of television broadcasting or cable channel services covered by the subject registered mark "CARTOON NETWORK" through financing of the CEO in Korea. Therefore, non-use of the subject mark for the designated services was not due to the fault of the plaintiff, and constituted a satisfactory reason for non-use.'

On a related note, the burden of proving a legitimate reason for non-use of a trademark registration as well as proving that such non-use did not occur through fault of the trademark owner is placed on the trademark owner. In Case No. 89 HU 599, decided on 26 June 1990, the Korean Supreme Court found that non-use as a trademark registration was not justified as follows:

> 'It cannot be asserted that a justifiable reason for non-use of the concerned trademark registration exists just because the "cigarettes" covered by the registration requires a licence or consignment to be imported (into Korea).

> Non-use lasting over one year[13] would be justified only if there were such circumstances so as to make it impossible to receive permission, designation, or consignment required for import through no fault of the trademark owner. Also, the trademark owner would have the burden of proving such justifiable reason for non-use.'

### B. Circumstances in which non-use may be cured

If a trademark registration had not been used for more than three years before a cancellation proceeding is initiated, it becomes vulnerable to cancellation via a trial based on non-use. If no such cancellation trial is filed, a trademark registration will continue to remain valid even if it had not been used for more than three years. Also, non-use lasting over three years can be avoided if the registered mark begins to be used prior to a cancellation proceeding. On a related note, three years of continuous non-use of a trademark registration should have elapsed before the filing of a cancellation proceeding. If a cancellation action is filed before the three-year non-use period elapses, the registration will not be subject to cancellation.

## V. CONSEQUENCES OF NON-USE

### A. Consequences of non-use and relevance of the context

Once a decision ordering cancellation of a trademark registration is confirmed, the registration will not only be cancelled, but the trademark owner or a licensee will be enjoined from filing a mark identical/confusingly similar to the cancelled mark for identical/similar goods until three years have passed from the confirmation date of the cancellation decision. (Korean Trademark Act Article 34(3)(iii)) Non-compliance with this provision is a ground for rejection of a trademark application, an opposition, or an invalidation trial.

Non-use of a registered mark is inured to the advantage of the infringer in claims for damage. If no business operations, such as manufacturing and distribution of products, were performed by the trademark owner in which the registered mark was used, there would be no loss suffered by the trademark owner's business, and generally, an infringer would not be found to have an obligation to compensate the trademark owner for damages in such a case.

The Supreme Court denied any obligation on the part of infringer to compensate a trademark owner as follows in Case No. 2003 DA 62910, decided on 22 July 2004.

> 'As the plaintiff did not conduct any business operations, such as manufacturing or distribution of products using the "X-GIRL" trademark in Korea, there is no business loss deemed to have been caused to the plaintiff's business. Even if the plaintiff registered "X-GIRL" as a trademark and carried out surveillance on possible trademark infringement, such activities cannot be seen as amounting

13. The period of non-use required for cancellation of a trademark registration was one year up until 1 September 1990, when it was extended to three years as a result of the 1 September 1990 amendment to the Korean Trademark Act.

to manufacturing or distribution of products or similar commercial activity. Considering that a claim for damages only stands insofar as that there is damage caused to the plaintiff's business through the defendant's commercial activities, the plaintiff's claim for damage compensation, based on the plaintiff's rights to a Korean trademark registration, is not reasonable.'

Use of a registered mark means use of a mark identical/confusingly similar to the registered mark on goods which are identical or fall within the scope of what is considered identical to the designated goods of the registered mark. Use of a mark that is not within the scope of what is considered identical to the registered mark or on goods merely similar to the designated goods of the registered mark cannot be considered valid use of the registered mark. In such cases, provisions relied upon to calculate the amount of losses and statutory damages, which inure to the benefit of the trademark owner, if enforceable, cannot be applied.[14]

## B. Extent of any revocation

In a case in which a trademark registration designates at least two goods, a cancellation proceeding may be filed against specified designated goods. It should be noted that KIPO classifies goods and services under local similarity codes and deems the goods that fall under the same similarity code similar and vice versa. Accordingly, the petitioner should seek to cancel all of the goods that fall in the same similarity code as the goods he is specifically seeking to cancel to accomplish the goal of his cancellation proceeding.

The use of a registered mark on goods that are seen as falling within the scope of what is considered identical to the designated goods or any of the goods encompassed under a broader goods description is considered valid use. The Patent Court concluded that use of a registered mark on 'CDs, Video tapes featuring music' amounted to use of the registered mark on the goods 'recorded tapes' covered by the registration,[15] and that a simple bag for carrying swimming suits, either hand-held or worn over the shoulder, was similar to the goods 'beach bags' and' travel bags' covered by the concerned registration.[16]

## C. Timing of any revocation

Once a cancellation proceeding is decided in favour of the petitioner by IPTAB, the registrant has an opportunity to file an appeal in the Patent Court within 30 days from its receipt of the decision. In the event the Patent Court decides against the trademark owner's appeal, the owner has an opportunity to file an appeal in the Supreme Court within two weeks from its receipt of the Patent Court decision. If the Supreme Court decides against the owner, the cancellation of the trademark

14. According to Article 111 (claim for legal damages) of the Korean Trademark Act, a court may grant compensation for a reasonable amount not exceeding KRW 50 million based on its review of the entire pleadings and evidence in a case.
15. Patent Court Ruling on Case No. 98 HUJH 4328, decided on 11 February 1999.
16. Patent Court Case No. 2008 HUH 2350, decided on 5 September 2008.

registration will be affirmed. In the event the trademark owner does not challenge the IPTAB decision ordering the cancellation of his registration in the Patent Court or the Supreme Court, the cancellation of the trademark registration will be confirmed upon expiration of the deadline for appeal.

Once the cancellation of a trademark registration is confirmed, the registration will be deemed to have been cancelled from the filing date of the cancellation proceeding.

# Malaysia

*Karen Abraham and Janet Toh Yoong San*

## I. INTRODUCING GENUINE USE OF TRADEMARKS

### A. Legal framework

#### *1. Overview of statutory requirements (national/regional laws)*

The sources of Malaysian trademark law are found in primary and secondary legislation and common law. The Trade Marks Act 1976 ('TMA 1976') forms the primary legislative framework for the registration of trademarks in Malaysia, with the Trade Marks Regulations 1997 ('TMR') being the subsidiary legislation made thereunder. The Trade Descriptions Act 2011 ('TDA 2011') came into force on 1 November 2011 and provides for criminal enforcement against trademark infringement.

The key legislation in regulating the genuine use of a trademark is the TMA 1976. Section 46 of the TMA 1976 provides as follows.

#### *2. Provisions as to non-use of trade mark*

Subject to this section and to section 57, the Court may, on application by a person aggrieved, order a trade mark to be removed from the Register in respect of any of the goods or services in respect of which it is registered on the ground –

(a) That the trade mark was registered without an intention in good faith, on the part of the applicant for registration or, if it was registered under subsection (1) of section 26, on the part of the body corporate or registered user concerned, to use the trade mark in relation to those goods or services and that there has in fact been no use in good faith of the trade mark in relation to those goods or services by the registered proprietor or registered user of the trade mark for the time being up to the date one month before the date of the application; *or*

(b) That up to one month before the date of the application a continuous period of not less that three years had elapsed during which the trade mark was a registered trade mark and during which there was no use in good faith of the trade mark in relation to those goods or services by the registered proprietor or registered user of the trade mark for the time being'.

There are therefore two limbs to this provision of rectification in relation to non-use.

The first limb relates to the registration of a trademark without a genuine intention to use it as such, i.e., in the course of trade to indicate a connection between the goods or services and the proprietor or a registered user, and where no *bona fide* use has in fact been made since the date of registration.

In the High Court case of *E-Toyo Global Stationery*,[1] Ramly Ali J (as he then was) interpreted 'good faith' as follows:

> '43. In both sub-paras (a) and (b) of s 46(1), the words 'good faith' play a role. In this context, the *Law of Intellectual Property* by Staniforth Ricketson, at para. 37.13 at p 727 states as follows:
>
> In this context, "good faith" means "genuine", rather than "honest" or "without deceit or intention to defraud". Thus, the intended or actual use must not be colourable, as would be the case where the mark is used in advertisements without the goods or services in respect of which it is used being made available.'

The second limb of non-use on the other hand relates to non-use for a continuous period of three years. If the applicant for rectification can establish that there has been no use in good faith of the mark for a continuous period of at least three years ending one month before the date of the application, it may be struck from the register. This limb is mainly relied on for the cancellation of a trademark based on the grounds of non-use as it is often difficult to prove that there was no intention to use a trademark at the time of its registration.[2]

## B. Purpose and function of genuine use: policy considerations

It is a fundamental principle of trademark law that a registered trademark requires a certain use to maintain its protection. Malaysian legislation on trademarks establishes an 'obligation' for the owner of a registered trademark to use that mark in good faith. The obligation of use is not applicable immediately after registration. Instead, the owner of a registered mark has a so-called 'grace period' of three years during which it is not necessary to demonstrate use of the mark in order to rely upon it. After this grace period, the owner may be required to demonstrate use of the registered mark for the relevant goods and services. Before this period lapses, the mere formal registration gives the mark its full protection.

The following are possible reasons behind the requirement that registered marks are required to be used in good faith.

1. *E-Toyo Global Stationery Sdn Bhd v. Toyo Ink Sdn Bhd & Ors, The Registrar of Trade Marks (Interested Party)* [2005] 1 MLJ 445, at para. 43.
2. *E-Toyo Global Stationery Sdn Bhd v. Toyo Ink Sdn Bhd & Ors* [2005] 1 MLJ 459.

### *1. Unfair competition*

A trademark registration potentially grants a registered proprietor a monopoly on the use of the trademark in relation to the goods and services registered for an unlimited period of time. Under normal circumstances, once an individual obtains these exclusive rights, other individuals are not able to use an identical or similar mark in the same trade. This may be seen as a method to restrict competition. Placing a 'genuine use' requirement therefore ensures that a proprietor does not simply 'squat' on the Register or 'hijack' the mark of a competitor. In the absence of such a statutory provision, there is a risk that some registrants may only register a trademark to prevent others from using it, without any actual intent to use the trademark themselves.

The purpose of such a statutory requirement therefore ensures that along with such monopoly comes the obligation to put the trademark to genuine use, failing which the owner of a registered trademark would run the risk of losing the trademark registration which he would have spent considerable time, cost and effort securing.

### *2. Prevent trafficking of trademarks*

Another purpose of the 'genuine use' requirement could possibly be also to prevent trademark trafficking, which is against the public interest.

### *3. Affecting the rights of other merchants in the same trade*

Trademarks which remain on the Register and are not used for a prescribed period of time should not be allowed to remain on the Register if they affect the rights of other merchants in the same trade to use these marks or similar marks.

This was recognised by the court in the case of *Re Arnold D. Palmer in the Matter of Trade Mark Registration No. 63249* (Singapore), in which Chan Sek Keong JC referred to Romer J in *Batt's Trade Marks*, where he said 'I take it to be clear that if persons register new trademarks, which they say they intend to use, they must, at the date of registration, have a bona fide intention of using those trademarks in respect of the goods for which they register them'. Cotton LJ pointed that out very clearly in the case of *Edwards v. Dennis*, when he said:

> 'Even if a trade-mark can be registered which is not in actual use (and, I may add, which has since been used) it *ought to be restricted to those goods in connection* with which it is going to be used; and, indeed, one cannot help seeing the evils that may result from allowing trademarks to be registered broadcast, if I may use the expression, there being no real intention of using them, or only an intention possibly of using them in respect of a *few articles*. The inconvenience it occasions, the costs it occasions, is very large, and beyond that I cannot help seeing that it would lead in some cases to absolute oppression, and to persons using the position they have obtained as registered owners of trade-marks (which are not really bona fide trade marks) for the purpose of trafficking in them and using them as a weapon to obtain money from subsequent persons who may want to use bona fide trademarks in respect of some classes in respect of which they find those bogus trade marks registered.' (Emphasis added)

Therefore, on the above basis, registered trademarks should either be used or cleared from the Register to allow for new registrants to make use of the mark in question.

#### 4. *Reduce the number of conflicts between marks*

Another possible reason behind the requirement that registered marks must be put to genuine use is to restrict the number of trademarks registered and protected and, consequently, the number of conflicts that may occur. This would also have the effect of reducing the likelihood of a crowded Register as unused marks could be cancelled upon application.

## II. WHAT QUALIFIES AS GENUINE USE OF A TRADEMARK

### A. Nature of use

A 'trademark' is defined under section 3 of the TMA 1976 to mean 'a mark *used or proposed* to be used in relation to goods or services for the *purpose of indicating or so as to indicate a connection in the course of trade between the goods or services* and a person having the right either as proprietor or as registered user to use the mark whether with or without an indication of the identity of that person.' (Emphasis added)

Based on the above definition of a trademark, it is pertinent to note that the required 'nature of use' of a trademark refers to:

- its use as a trademark which is to indicate a connection between the goods or services registered for and the registered proprietor;
- its use in the course of trade; and
- its use in relation to goods and services.

#### 1. *Use consistent with the function of the trademark*

The use of a trademark must be consistent with the essential function of the trademark, which is to guarantee the identity of origin of the marked goods or services to the consumer or end user by enabling him, without any possibility of confusion, to distinguish the goods and services from others which have another origin.[3] Section 3(2) of the TMA 1976 requires proof of use in connection with goods or services.

The High Court in the case of *Godrej Sara Lee*[4] held that if the provisions of section 46 were to be considered for use to be constituted and taken into account, *the use must be in relation to the goods claimed in the registration*. The High Court further held in that case that merely seeking approval and obtaining it did not translate to actual use of the mark in the course of trade.

As a trademark has, inter alia, the function of operating as a connection between the goods and services and the person responsible for their marketing, the proof of use must establish a *clear connection between the use of the mark and the relevant*

3. *Tan Kim Hock Tong Seng Food Industry Sdn Bhd v. Tan Kim Hock Product Centre Sdn Bhd & Anor* [2016] 7 MLJ 561.
4. *Godrej Sara Lee Ltd v. Siah Teong Teck & Anor (Part 2)* [2007] 7 MLJ 164; [2008] 7 CLJ 24.

*goods and services*. In deciding whether there was genuine use, the Court in *Jost Cranes*[5] referred to the English case of *Daimler AG*[6] and laid down the following principles of genuine use:

(a) Genuine use is a qualitative and not a quantitative criterion.
(b) The use has to be consistent with the essential function of the trademark, which is to guarantee the identity of the origin of the goods or services to the customer or end users.
(c) The use has to be on market by the proprietor or an authorized third party and not just internal use by the undertaking concerned.
(d) The use has to maintain or create a share in the market for the goods or services protected by the mark.
(e) The court has to take into consideration all the facts and circumstances relevant to establishing whether the commercial exploitation of the mark has been real, and in particular, whether such use would be viewed as warranted in the trade sector concerned to maintain or create a share in the market for the relevant goods or services protected by the mark.
(f) And finally, the nature of the goods at issue, the characteristics of the market concerned and the scale and frequency of the use of the mark have to be taken into account.

Depending on the circumstances, the following situations may be suitable for supporting genuine use of the registered trademark. That is because the use of the mark can serve more than one purpose at the same time. Consequently, the following uses can also be use of the mark as a trademark. However, the purpose for which a mark is used needs to be assessed individually.

(i) The use of a mark as a business, company or trade name can be regarded as trademark use provided that the use of the mark can represent and has the likelihood of inducing members of the buying public to believe that goods bearing the mark, name, description or other indication belong to 'the distinctive group of goods' under the particular trade name with the ‹recognisable and distinctive qualities that has acquired a certain goodwill and reputation'.[7] If however the mark was only used as a company's name and no products were associated with it, then this is unlikely to be regarded use as a mark.[8]
(ii) Today a domain name is no longer just an address to locate or identify a particular website. With the advent of the internet in e-commerce and retail, the domain name has become an invaluable asset and has been considered the equivalent of a trademark or trade name used to identify a commercial service or product on the internet. In *Petroliam Nasional Berhad*,[9] the first reported Malaysian case involving domain name disputes, the High Court

5. *Jost Cranes GmbH & Co KG v. Jost Cranes Sdn Bhd* [2010] 4 MLJ 191.
6. *Daimler AG v. Sany Group Co. Ltd* [2009] All ER (D) 128.
7. *Maestro Swiss Chocolate Sdn Bhd & Ors v. Chocosuisse Union Des Fabricants Suisses De Chocolat & Ors And Another Appeal* [2016] 3 CLJ 345.
8. *H & R Johnson Tiles Ltd & Anor v. H & R Johnson (M) Bhd* [1998] 4 MLJ 13.
9. *Petroliam Nasional Berhad & Others v. Khoo Nee Kiong* [2003] 4 MLJ 16.

granted prohibitory injunctions against the defendant from registering any domain names containing the word 'Petronas', as it was a household name in Malaysia belonging to the plaintiff. Similarly, in the decision of *Intel Corporation*,[10] the first defendant had registered the domain name www.intelcardsystems.com. The High Court granted an interim injunction against the defendant from using the name 'INTELCARD', whilst affording protection to the plaintiff who was the proprietor of the 'Intel' trademark. Therefore these cases suggest that use of a domain name can amount to use of a trademark, particularly where the use of the domain name impliedly represents such goods offered on the relevant website to be the goods of the proprietor or the goods of the proprietor of a particular class or quality.

(iii) On the other hand, the genuineness of use can be put into doubt in cases, when used together with another mark, the registered mark is perceived as a mere decorative element.

## 2 *Use in the course of trade*

### *i. Public use*

The use must be on the market by the proprietor or an authorised third party and not just internal use by the undertaking concerned (*Jost Cranes*,[11] Azahar Mohamed J (as he then was) following the decision of Susan Prevezer QC (sitting as a Deputy Judge) in the English High Court case of *Daimler AG*[12]).

In determining what constitutes use in the course of trade, the High Court in *Lam Soon Edible*[13] referred to *HERMES Trade Marks*,[14] where it was held that the phrase 'in the course of trade' was wide enough to cover the steps necessary for the production of goods as well as their actual placing on the market: 'If the registered proprietor should commence a series of advertisements featuring his mark as part of an introductory campaign, prior to putting his goods on the market under the mark, but before they were actually on the market, in my judgment such use would clearly be use of the mark in the course of trade, not upon the goods or in physical relation thereto, but it would be in other relation thereto, the point being that it would be use in the course of trade in those goods, albeit in advertisement.'

### *ii. Commercial purpose*

The TMA 1976 defines the terms 'course of trade' to mean in relation to the provision of services, in the course of business. In the case of *Hai-O Enterprise*,[15] it was

10. *Intel Corporation v. Intelcard Systems Sdn. Bhd. & Others* [2004] 1 MLJ 595.
11. *Jost Cranes GmbH & Co KG v. Jost Cranes Sdn Bhd* [2010] 1 LNS 331; [2010] 4 MLJ 191, at para. 20.
12. *Daimler AG v. Sany Group Ltd* [2009] All ER (D) 37.
13. *Lam Soon Edible Sdn. Bhd. Hup Seng Perusahaan Sdn Bhd*. (2010) 5 CLJ 975.
14. *HERMES Trade Marks* [1982] RPC 425 at 426.
15. *Hai-O Enterprise v. Nguang Chan @ Nguang Chan Liquor Trader (a firm, intervening)* 1 BLJ 53.

decided that goods remain the course of trade so long as they are on a market for sale. However, when these goods are bought for consumption, they cease to be in the course of trade.

In the case of *Doretti Resources*,[16] the court held that for the purpose of deciding the first user of the mark, courts may consider the four ways in which a mark may be used in the course of trade; namely:

(a) An invoice bearing a mark may be considered as use of the mark in the course of trade;[17]
(b) The mark is used 'upon' the goods pursuant to section 3(2)(b) TMA 1976;
(c) The mark is used 'in physical' relation to goods pursuant to section 3(2)(b) TMA 1976; and
(d) The mark is used in 'other relation to' goods pursuant to section 3(2) (b) TMA 1976.

*iii.* *Use of trademark on goods that are free*

Where the mark is protecting goods or services of a not-for-profit organisation, and the mark has been used, the fact that the organisation is a non-trading entity is irrelevant:

> 'The plaintiff is not, in the normal sense, a trader. The plaintiff was created by *Act 29* and is entrusted with various functions and duties thereunder. However, the plaintiff being a sports organisation with the statutory duties under *s. 4(1)(d) of Act 29* to promote sports locally and internationally, as well as to participate in the development of sports in Malaysia was fully empowered and entitled pursuant to s. 4(2)(g) to acquire, hold and enjoy movable and immovable assets and properties of every description including those over the trade marks namely of the tiger stripes design and to further deal with such rights including to deal with or to develop such rights belonging to the plaintiff. [...] It is plain from the reading of the relevant provisions of *Act 29* that the plaintiff was authorised to engage in commercial business that was incidental to its primary statutory objects; the plaintiff need not be a "trading entity" before it can be said that the plaintiff had used the tiger stripes design "in the course of trade".'[18]

*iv.* *Use of trademark on goods to be exported*

Section 71 of the TMA 1976 provides that the application of a trademark to goods to be exported from Malaysia constitutes use of the trademark in relation to those goods. The High Court in the case of *Hiu Kuan Hoe*[19] held that the use of the mark by associated companies in Malaysia is use by and attributable to the respondent. The Court considered random copies of shipping invoices and sales invoices showing substantial sales and distribution of the respondent's goods bearing the mark in

16. *Doretti Resources Sdn Bhd v. Fitters Marketing Sdn Bhd & Anor* [2017] 1 LNS 737.
17. *Acushnet Company v. Metro Golf Manufacturing Sdn Bhd* [2006] 7 CLJ 557; *Ooi Siew Bee & Ors v. Zhu Ge Kong Ming Sdn Bhd & Anor* [2017] 2 AMR 161.
18. *Mesuma Sports Sdn Bhd v. Majlis Sukan Negara (Malaysia) (Pendaftar Cap Dagangan Malaysia, interested party)* [2015] MLJU 556.
19. *Hiu Kuan Hoe v. Societe Des Produits Nestle SA* [2011] 7 CLJ 179.

question by the respondent's associated companies in Malaysia. The Court also noted that the associated companies were authorized by the respondent to use the mark including to apply the mark onto packaging of goods, to manufacture, pack, export, commercially distribute or to otherwise deal by way of trade in goods bearing the trademark. Therefore, the Court found that the respondent had used the mark in Malaysia and was the first and prior user of the mark in Malaysia.

### 3. *Use by the owner or with its consent*

#### *i. Use by the owner*

One of the conditions that a trademark must fulfil if it is to remain on the Register is that it must be 'used' either by the proprietor or 'registered users'. The case of *Godrej Sara Lee* illustrates the importance that the trademark must be used by the registered proprietor, and/or its registered user and not by anyone else. In the case of *Godrej Sara Lee*, the respondent relied on the evidence of an application made by a certain company to the Pesticides Board for approval to show that there has been use of the disputed trademark. The court held that the use during the material period has to be by the registered proprietor or a registered user. The court held that the company that had applied to the Pesticides Board for the approval was not the registered proprietor and therefore, even in the hypothetical situation that the application to apply for an approval was considered use in the course of trade, such use was not recognized as use within the context of section 46 TMA 1976.

#### *ii. Use with its consent*

Section 48 (5) TMA 1976 provides that the use of a trademark by a registered user is deemed to be use by the registered proprietor of the trademark. This was confirmed by the Federal Court in the case of *LB (Lian Bee) Confectionery*.[20] The Federal Court further elaborated that the effective period of use of a mark by a registered user may pre-date the date of application for a registered user. The Federal Court agreed with the Court of Appeal in finding that there is no provision in section 46 TMA 1976 or the TMR which will preclude the use of a trademark by a registered user to be in force retrospectively before the date of the application to register the registered user. It follows from this that a registered user status, including the recognized period of use, does not take effect only on the date of registration of the user status. The court in *Lian Bee* further highlighted that section 48(1) evidently recognizes the pre-existence of the licensing agreement between the proprietor and the intended user pursuant to the wording of the provision 'where the registered proprietor of a trademark grants, by lawful contract, a right to any person to use the trademark, for all or any of the goods or services in respect of which the trademark is registered, that person may be entered on the Register as a registered user of the said trademark whether with or without any conditions or restrictions, provided that it shall be a condition of any such registration that the registered proprietor shall retain and exercise control over

20. *LB (Lian Bee) Confectionery Sdn Bhd v. QAF Limited* [2012] MLJU 70.

the use of the trademark and over the quality of the goods or services provided by the registered user in connection with that trademark'.

Further, contractual activities would amount to use of the trademark. In the case of *McLaren International*,[21] shoes bearing the trademark 'McLaren' were contracted to two companies by the registered proprietor of the trademark for manufacture. These companies were granted lawful contracts providing the right to use the trademark by the registered proprietor. The Court of Appeal held that the contractual activities amounted to use of the trademark and could be attributed to the respondent as the registered proprietor of the trademark.

### B. Volume, duration and frequency of use

The duration of use is an important factor which the Registrar considers before exercising its discretion whether to remove the registered trademark in situations where the validity of the registration of a trademark is challenged. This is because the duration and frequency of use of the trademark is inevitably woven together with the public interest when considering the removal of a registered trademark based on non-use.

This is illustrated in the case of *Yong Sze Fun & Anor v. Sharifah binti Mohd Tamin & Ors*, where the judge had explained that in the event where a registered trademark has been used for a substantial period of time to the extent where the public has come to recognize goods bearing that trademark as having originated from the registered proprietor, it would be in the public interest for the trademark to remain on the Register.

Yet, the case of *Tokai Corporation*[22] would suggest that 'substantial use' of a trademark is not a requirement under section 46(1)(b) TMA 1976 to remove the mark from the Register based on non-use. In reaching this conclusion, Wong Kian Kheong JC explained the following:

- Section 46(1)(b) TMA 1976 only required a 'bona fide' use of the trademark by the registered proprietor.
  The cases of *Nodoz Trade Mark* and *E-Toyo Global Stationery* have recognized a single and bona fide use of the registered trademark as being sufficient.

Having said that, it is important to bear in mind that in the case of *Nodoz*, Wilberforce J had held that one single act of user of the trademark may be sufficient, if that one single act which is relied upon can be established by conclusive or overwhelmingly convincing proof.

Further, the cases of *Daimler AG* and *Jost Cranes GmbH* have applied a qualitative and not a quantitative assessment.

21. *McLaren International Ltd v. Lim Yat Meen* [2007] 7 MLJ 581.
22. *Tokai Corporation v. DKSH Malaysia Sdn Bhd* [2016] MLJU 621.

## C. Use of a sign

Generally, a trademark must be used in the form as registered. Use of a registered trademark becomes an issue in an action for expungement based on non-use. Pursuant to section 23(2) of the TMA 1976, where use of a trademark is required to be proved for any purpose, the Registrar may, if he thinks is right, only accept use of a trademark with additions and alterations which do not substantially affect the identity of the registered trademark.

This can be seen in the case of *Lam Soon*[23] where the court held that section 23(2) of the TMA 1976 permits the use of a trademark with additions and alterations provided that such addition and alterations do not substantially affect the identity of the registered trademark. The court in this case cited the English decision *In The Matter of Morny Ld's Trade Marks*[24] where it was held that the law has recognized the undesirability of requiring proprietors of registered trademarks to maintain the precise form of embellishments or representations which at the time of application were included in the form of the mark submitted for registration. The court also cited the case of *Levi Strauss*[25] which stated the same principle in relation to evidence of use of a trademark with additions and alterations that do not affect substantially the identity of the trademark.

## D. Use in relation to goods and services

### *1. Use for each designated good and service*

Malaysian trademarks can be vulnerable to revocation actions if the trademarks are not appropriately used in the market for each of the relevant goods and services. In the case of *Leo Pharmaceutical*,[26] the defendant contended that the plaintiff had not used its trademark for veterinary preparations and, in view of this, attempted to persuade the court to expunge the mark.

The High Court emphasized that the discretion to expunge a trademark is a principled discretion which must be exercised judicially after considering all facts and circumstances of the case, including public interest. In the present case, the High Court held that the expungement of 'veterinary preparations' from the plaintiff's trademark registration would lead to the possibility that the defendant or some third parties may begin to use the mark in respect of the expunged goods. This would create confusion in the market as to the origin of the pharmaceutical products. Therefore, the Court held that it is in the public interest that this confusion should be eliminated or minimized. Accordingly, the defendant was unable to expunge the plaintiff's mark in relation to veterinary preparations.

---

23. *Lam Soon (M) Bhd v. Forward Supreme Sdn Bhd & Ors* [2001] 4 CLJ 673.
24. *In The Matter of Morny Ld's Trade Marks* [1951] 68 RPC 55.
25. *Levi Strauss & Co v. Shah & Another* [1985] RPC 371.
26. *Leo Pharmaceutical Products Ltd A/S (Lovens kemiske Fabrik Prodktionsaktieselskab) v. Kotra Pharma (M) Sdn Bhd* [2009] MLJU.

### 2. *Use for a broader category of goods and/or services*

Pursuant to section 46(2) of the TMA 1976, non-use of trademark on goods/services for which it is being registered may be 'cured' if it can be shown that during the relevant period of non-use, the registered proprietor has used the trademark on goods/services which are of the *same description*. Section 46(2) of the TMA 1976 provides as follows:

> 'Except where an applicant has been permitted under section 20 to register an identical or a nearly resembling trade mark in respect of the goods or services to which the application relates or the Court is of the opinion that the applicant can properly be permitted to register the trade mark, the Court may refuse an application made under subsection (1) –
> (a) in relation to any goods, if there has been, before the relevant date or during the relevant period, as the case may be, use in good faith of the trade mark by the registered proprietor of the trade mark for the time being in relation to goods of the same description, being goods in respect of which the trade mark is registered; and
> (b) in relation to any services, if there has been, before the relevant date or during the relevant period, as the case may be, use in good faith of the trade mark by the registered proprietor of the trade mark for the time being in relation to services of the same description, being services in respect of which the trade mark is registered.'

## E. Territorial scope

It is pertinent to bear in mind that trademark law is generally territorial in nature. Appellate cases have held in the past that the mark should be used within Malaysian territory.[27]

### 1. *Well-known marks*

Pursuant to section 70B of the TMA 1976, the proprietor of a well-known trademark which is entitled to protection under the Paris Convention or the TRIPS Agreement may, through an injunction, prevent the use of a trademark by others in Malaysia, if the said trademark is identical or nearly resembles the proprietor's trademark and the use is likely to deceive or cause confusion. In determining whether a mark is well-known or not, the following criteria may be taken into account:

(a) degree of knowledge or recognition of the mark in the relevant sector of the public;
(b) the duration, extent and geographical area of any use of the mark;

---

27. *Yong Teng Hing B/S Hong Kong Trading Co & Anor v. Walton International Ltd* [2012] 6 CLJ 337; *Mesuma Sports*; *Lim Yew Sing v. Hummel International Sports & Leisure A/S* [1996] 3 MLJ 7; *Meidi (M) Sdn Bhd v. Meidi-Ya Co Ltd, Japan & Anor* [2008] 6 MLJ 433.

(c) the duration, extent and geographical area of any promotion of the mark, including advertising or publicity and the presentation, at any fairs or exhibitions, of the goods or services to which the mark applies;
(d) the duration and geographical area of any registrations, or any applications for registration, of the mark to the extent that they reflect use or recognition of the mark;
(e) the record of successful enforcement of rights in the mark, in particular, the extent to which the mark was recognized as well-known by competent authorities;
(f) the value associated with the mark.

### 2. *Use on foreign websites*

In *Yong Teng Hing*, the Federal Court held that modern technology and communications have improved to such an extent that the public in Malaysia would be aware of foreign marks even though such marks had not been previously used in Malaysia. The Court held in this case that limited use of the 'GIORDANO' trademark by the appellant was sufficient to prevent the respondent or any other party from claiming proprietorship in the said mark if its use by the respondent was likely to cause confusion to the public. The appellant in the present case had, through its licensees, used the 'GIORDANO' trademark in Malaysia prior to the filing date of the respondent's application for registration.

In the case of *Abercrombie*,[28] the plaintiff's products including articles of clothing bearing the second plaintiff's Registered Trade Marks were available for purchase over the internet. In that case, the plaintiffs had maintained, *inter alia*, websites offered for sale the plaintiff's products bearing the 'ABERCROMBIE & FITCH' and 'ABERCROMBIE' trademarks: *www.abercrombie.com* and *www.abercrombiekids.com*. All these websites were accessible to all customers who had access to the internet including customers and/or potential customers in Malaysia.

The High Court in *Abercrombie* referred to *1-800*,[29] where Buxton LJ had provided obiter guidelines on whether use of a mark on the internet might be sufficient to constitute use of a mark in the United Kingdom for the purpose of trademark law. Placing a mark on the internet from a non-United Kingdom location would not suffice. An active move, such as direct encouragement to purchase the product or advertisement, would be necessary in order to establish the necessary use. In the United States personal jurisdiction cases can be divided into three categories:

(1) Active websites: cases where defendants actively conduct business on the internet in the forum state and personal jurisdiction is found.
(2) Intermediate websites: interactive website cases where a user exchanges data with the host computer. In such cases, jurisdiction is determined by examining the level of interactivity and commercial nature of the information exchange.

---

28. *Abercrombie & Fitch Co & Anor v. Fashion Factory Outlet Kl Sdn Bhd & 4 Ors* [2007] 1 LNS 486.
29. *1-800 Flowers Inc v. Phonenames Ltd* [2002] FSR 12 CA.

(3) Passive websites: sites that merely provide information or advertisements. Decisions refuse to find personal jurisdiction in such cases.

The Court emphasized that in considering whether there was actual use of the mark, the intention and purpose for putting up the website must be taken into account. In the present instance, the website was intended to be used to seek worldwide trade with the view to commercial gain, and therefore, its activities were held to be 'doing business over the internet' and may constitute 'use' for the purpose of trademark proceedings. The High Court therefore held that the sale of goods bearing a trademark over the internet or via a website was recognized as a use of the trademark because the plaintiffs had in fact sold and shipped products bearing the second plaintiff's Registered Trade Marks to Malaysian customers in Malaysia.

Therefore online use may be recognized as genuine use of a trademark where the goods bearing the trademark are made available for purchase by customers or potential customers in Malaysia. Pursuant to the decision in *Abercombie*, it would appear that merely placing a mark on the internet from a non-Malaysian location without more will likely not be sufficient. There must be an active move, for example, direct encouragement to purchase the product or advertisement in order to establish the necessary use.

### 3. *Foreign publications*

Advertisements of the goods bearing a trademark in an internationally circulated/ distributed magazine which is available in Malaysia will constitute use of the trademark in Malaysia. Advertisement of a product in a magazine that is available for purchase in Malaysia is possibly a direct encouragement to purchase the product and constitutes use of the trademark in Malaysia. In the Malaysian case of *Godrej Sara Lee*,[30] it was held that: 'For actual use to arise, there must be use of the mark to the goods or in relation to the goods. This would include affixing the mark to the goods or in an advertisement, circular or a catalogue.'

### 4. *Use in part of the territory*

In *Degem*,[31] the High Court recognized that the defendant's goodwill was only within Johor Bahru or at most the State of Johor and allowed the first plaintiff an injunction restraining the defendant from expanding beyond the State of Johor.

## F. Relevant time period and its calculation

Malaysian registrations are vulnerable to non-use cancellation actions from the date of the actual registration of the trademark (i.e the date when the mark is entered on the Register) and *not* the date of the application: *Industria De Diseno*.[32] The period of time of three years would be the period of time up to one month preceding the

30. *Godrej Sara Lee Ltd v. Siah Teong Teck & Anor (No. 2)* [2008] 7 CLJ 24.
31. *Degem Bhd & Anor v. De Gem Goldsmith & Jewellery Sdn Bhd & Ors* [2013] 3 CLJ 209.
32. *Industria De Diseno, SA v. Edition Concept Sdn Bhd* [2005] 3 MLJ 347.

application to remove or expunge the trademark.[33] For example, where the application to expunge a trademark for non-use is filed on 10 April 2008, the relevant three-year period of non-use that has to be shown would be between 10 March 2005 and 10 March 2008. Further, it was confirmed by Ramly Ali J in the case of *Godrej Sara Lee* that use outside the relevant period cannot be considered to overcome the assertion of non-use.

## III. EVIDENCE OF GENUINE USE OF A TRADEMARK

### A. When use should be proved: timing of lack of genuine use claims

#### 1. *When filing an application for a trademark*

A mark that is applied for does not have to be in use during the time of the application. Based on the definition of a trademark, an intention to use the mark would be sufficient to register a mark. In Malaysia, it is not a requirement to enquire whether the applicant has used his mark at the application stage.

Section 25 of the TMA 1976 provides that 'any person claiming to be the proprietor of a trademark used or proposed to be used by him may make application to the Registrar for the registration of that mark in the Register in the prescribed manner'. Notwithstanding the above, an exception may be made if the Registrar is of the view that the applicant does not carry on trade. However, the applicant may provide evidence of use to show that the mark has acquired distinctiveness through such use in the event where the Registrar raises an objection on the non-distinctiveness of the mark. If evidence of use is intended to be submitted by the applicant in support of his application, such evidence must be by way of a Statutory Declaration, i.e. a statement on oath.

#### 2. *In the course of opposition proceedings*

Within two months from the publication in the Government Gazette, the opponent may file a Notice of Opposition to the Registrar along with the grounds of opposition. Within two months from the receipt of the Notice of Opposition, the applicant may file and serve a counter-statement, which includes a statement setting out the grounds relied upon in support of the application. During these preliminary stages, no evidence of use is adduced yet.

Within two months from the receipt of the counter-statement, the opponent may file and serve a statutory declaration to adduce evidence in support of his opposition. The applicant then has his turn to file evidence in support of the application, also by way of a Statutory Declaration, and this is also to be done within a two-month time period. The opponent then has a right to reply, and is to file his evidence in reply, by way of Statutory Declaration in Reply, within two months from the date of receipt of the applicant's Statutory Declaration.

---

33. *Godrej Sara Lee*, *supra* n. 4.

Upon completion of the evidence filed, the Registrar will set a date for both parties to file their respective written submissions. The Registrar will consider the evidence before him and any written submissions, and deliver a written decision and notify the parties accordingly. All of the time periods stated above are subject to an application for an extension of time. The opposition is usually disposed of after consideration of documents filed, without the need for an oral hearing.

The decision of the Registrar can be appealed to the High Court.

### 3. *In cancellation proceedings based on lack of use*

An application for revocation of a trademark shall be filed with the High Court by way of originating summons, supported by affidavit evidence pursuant to Order 87 r2 of the Rules of Court 2012. An aggrieved person who has used the mark or has a genuine intention to use the mark is entitled to commence the relevant proceedings. In deciding what an aggrieved person is, the Federal Court explained in *McLaren*,[34] as follows:

> '[22] We understand that passage as laying down the principle that a person aggrieved is a person who has used his mark as a trade mark – or who has a genuine and present intention to use his mark as a trade mark – in the course of a trade which is the same as or similar to the trade of the owner of the registered trade mark that the person wants to have removed from the register.'

In the case of *Lam Soon Edible*,[35] the High Court referred to the case of *Lim Yew Sing* which provides that if a market survey is to be relied on to prove non-use, it must fairly reflect the factual position at the time the application to expunge is made.

### 4. *As a defence in infringement/cancellation proceedings*

A cancellation proceeding based on non-use can be brought by a counterclaim as a defence in infringement proceedings. It is important for an individual who brings an infringement and/or cancellation proceeding to first ensure that his trademark registration could not be challenged. In the case of *Colliers*,[36] the Learned Judge held the following:

Whilst the Plaintiff's claim against the Defendant is for trademark infringement and the tort of passing off, the Defendant's counterclaim is for rectification under section 45 and section 46 of the Trade Marks Act 1976. As such, I will deal with the Defendant's application to expunge the Plaintiff's registered mark from the Register, then there is no longer any issue of the Plaintiff's claim for infringement under section 38 TMA 1976.

It is noted from the above that the court will first consider the issue of whether it is possible to expunge the mark from the Register based on non-use before considering the issue of whether there is any infringement.

---

34. *McLaren International Ltd v. Lim Yat Meen* [2009] 4 CLJ 749, at para. 22.
35. *Lam Soon Edible, supra* n. 13.
36. *Colliers International Property Consultants Inc v. Colliers International Property Consultants Sdn Bhd* [2015] 1 LNS 252.

## B. Burden of proof

When an aggrieved person brings a cancellation action based on non-use, the applicant who has applied for the action has to prove at the outset that there is a prima facie case of non-use by the registered proprietor. In order to determine whether there was non-use, a market survey or an investigation can be conducted.

The High Court in *Lam Soon Edible*[37] held that in relying on survey evidence for a non-use action, the applicant must be able to show that the reports complied with the 'minimum criteria' governing admissibility of such evidence. In arriving at this decision, the High Court in *Lam Soon* applied the decision of the Court of Appeal in *Lim Yew Sing*[38] which held that that survey evidence – to have validity – must meet certain minimum criteria which have been spelt out in *Imperial Group*.[39] The 'minimum criteria' referred to by the Court of Appeal in *Lim Yew Sing* as laid down in *Imperial Group PLC* are as follows:

(a) the interviewees must be selected so as to represent a relevant cross-section of the public;
(b) the size must be statistically significant;
(c) it must be conducted fairly;
(d) all the surveys carried out must be disclosed including the number carried out, how they were conducted, and the total number of persons involved;
(e) the totality of the answers given must be disclosed and made available to the defendant;
(f) the questions must not be leading nor should they lead the person answering into a field of speculation he would never have embarked upon had the question not been put;
(g) the exact answers and not some abbreviated form must be recorded;
(h) the instructions to the interviewers as to how to carry out the survey must be disclosed; and
(i) where the answers are coded for computer input, the coding instructions must be disclosed.

In the case of *USA PRO IP*,[40] the Court held that a prima facie case of non-use was established based on the investigation of a private investigator who had found out that the first defendant company had been dormant for at least the last five years and that the first defendant had not been using its trademarks in relation to any of its goods.

Nonetheless, it is worth noting that the Federal Court in the case of *Liwayway*[41] held that the compliance or fulfilment of the Whitford guidelines (laid down by Whitford J in *Imperial Group* for the conduct of surveys for trademark matters) does not automatically translate into a case where the prima facie burden of proof has

37. *Lam Soon Edible, supra* n. 13.
38. *Lim Yew Sing v. Hummel International Sports & Leisure A/S* [1996] 4 CLJ 784.
39. *Imperial Group PLC & Another v. Philip Morris Limited & Another* [1984] RFC 293.
40. *USA PRO IP Ltd v. Montfort Services Sdn Bhd & Anor* [2016] 9 MLJ 258.
41. *Liwayway Marketing Corporation v. Oishi Group Public Company Limited* [2017] MLJU 464.

been discharged. The Federal Court emphasized the sufficiency and the weight to be attached to the content of the market survey.

Once the prima facie case of non-use is established, the burden then shifts to the registered proprietor to show that there has been use in relation to the goods or services applied for or show that non-use was due to special circumstances pursuant to section 46(4) of the TMA 1976. This was confirmed by Ramly Ali J in the case of *Godrej Sara Lee* where he referred to the case of *Trina Trade Mark* (1977) RPC 131. It is pertinent to note that unless and until such burden of proof is discharged, there is no requirement for the registered proprietor to show that it had used the mark in question upon any of the goods or services it has been registered for.[42]

## C. Relevant evidence

### *1. Use on the product or its packaging*

The High Court in the case of *Godrej Sara Lee* held that there must be use of the mark on or in relation to goods for actual use to arise. This would include affixing the mark to the goods or in an advertisement, circular or a catalogue. An application made by the company to the Pesticides Board does not amount to actual use as it only indicates a mere intention to use the mark.

### *2. Use in advertising*

In the case of *Lam Soon Edible*,[43] the court referred to *HERMES Trade Marks* [1982] RPC 425 where it was held that the phrase 'in the course of trade' is wide enough to cover the steps necessary for the production of goods as well as their actual placing on the market. The court observed the following: 'If the registered proprietor should commence a series of advertisements featuring his mark as part of an introductory campaign, prior to putting his goods on the market under the mark, but before they were actually on the market, in my judgment such use would clearly be use of the mark in the course of trade, not upon the goods or in physical relation thereto, but it would be in other relation thereto, the point being that it would be use in the course of trade in those goods, albeit in advertisement.'

### *3. Use on the internet*

As discussed above, the sale of goods over the internet bearing a trademark is recognized as use of the trademark as held in the case of *Abercrombie*.[44] The Court recognized that mere access of a website which carries a particular trademark is not sufficient to constitute use of the mark and held that the website must be intended to be used to seek worldwide trade with a view towards commercial gain.

42. *E-Toyo Global Stationery Sdn Bhd v. Toyo Ink Sdn Bhd & Anor* [2004] 7 CLJ 368.
43. *Lam Soon Edible, supra* n. 13.
44. *Abercrombie & Fitch Co & Anor v. Fashion Factory Outlet KL Sdn Bhd* [2008] 7 CLJ 413.

### 4. *Form of evidence*

In the case of *Mesuma*, the court took into account the invoices as evidence of the first use. Sales and market surveys are usually taken into account to show use of the mark.

### 5. *Distinction between use of a company name and use of the services offered by the company*

In principle, the use of the sign as a company name or trade name, is not, of itself, intended to distinguish goods or services. The purpose of a company name is to identify a company, whereas the purpose of a trade name or a shop name is to designate a business which is being run. Accordingly, having a company name alone does not necessarily mean that the company is using the name as a mark.

The use of a business, company or trade name can be regarded as use 'in relation to goods' where:

(a) there is continual use of these names on its products, letterhead, invoices; or
(b) there are promotional activities and advertisement, which have led to significant growth in sales.

Such activities shall have led the products to be identified with the plaintiff's company name.[45]

## IV. JUSTIFICATION FOR NON-USE

### A. Obstacles to use

Section 46(4) of TMA 1976 provides an excuse for the non-use of a trademark. The registered proprietor must show that failure to use a trademark must have been because of special circumstances in the trade and not to an intention not to use or to abandon the trademark in relation to the goods to which the application relates.

The High Court in the case of *Godrej Sara Lee* made reference to the English case of *BULOVA Trade Mark* (1967) RPC 29. Ramly Ali J clarified that to fall within 'special circumstances' to excuse non-use, two conditions must be satisfied, namely: (1) the special circumstances must be peculiar or abnormal and must be external in nature and (2) the non-use must be as a result of those circumstances.

For example, in the case of *BULOVA*, the failure to use the mark was a result of a quota imposed by a Government Order. This was not under the control of the proprietor and was recognized by the court to be a 'special circumstance' to excuse the non-use of a trademark. Other situations that may amount to 'special circumstances' are import restrictions, war or saturation of the goods in the market place.[46]

---

45. *Tan Kim Hock Tong Seng Food Industry Sdn Bhd v. Tan Kim Hock Product Centre Sdn Bhd & Anor* [2017] 1 LNS 2113.
46. Tay Pek San, *Intellectual Property Law In Malaysia* (2013 Sweet & Maxwell Asia).

### B. Circumstances in which non-use may be cured

Non-use of a trademark can be cured through use. As discussed above, the relevant period is three years up to one month preceding the application to remove or expunge the trademark (*Godrej Sara Lee*). The purpose of the one-month provision is to prevent a proprietor circumventing the cancellation of his mark by using the mark solely for the purpose of circumventing the non-use criteria and without genuine intention to trade under the mark. Therefore, those who intend to launch rectification actions must do so without delay as non-use cancellation can be cured through the resumption of use.

## V. CONSEQUENCES OF NON-USE

### A. Consequences of non-use and relevance of the context

Section 46 TMA 1976 provides the court with the discretion to order for a removal of a trademark from the Register in respect of any goods or services based on non-use. When an application under section 46(1) is successful, one of the two consequences may follow: (a) removal of the trademark from the Register; or (b) the court may impose conditions, amendments, modifications or limitations as it thinks proper to the registration of the trademark. This is provided for under section 46(3) and section 46(3A).

The question then arises as to whether a party who has infringed a trademark may bring an action for cancellation based on the ground of non-use as a defense. The answer is in the negative. The case of *LB (Lian Bee) Confectionery*[47] provides a clear illustration of this. To bring an application for a removal of a trademark under section 46 TMA 1976, the applicant must establish himself to be an aggrieved person. In the case of *LB (Lian Bee)*, the Federal Court agreed with the High Court Judge that the *conduct* of the applicant must be taken into account when deciding whether the appellant was a 'person aggrieved'. In considering the conduct of the appellant, the High Court Judge had found the appellant to be an infringer of the respondent's registered trademark.

The Federal Court further explained that as a matter of policy, the appellant in this instance, could not be regarded as a 'person aggrieved' for the purposes of section 46(1) of the TMA 1976 as it was essential for the courts to apply the principle of construction *in bonam partem* as a matter of statutory interpretation. The Federal Court further quoted *Bennion on Statutory Interpretation* (fifth edition, Lexis Nexis, 2015) as follows:

> 'In pursuance of the principle that law should serve the public interest, the courts have evolved the important technique known as construction *in bonam partem* (good faith). If a statutory benefit is given on a specified condition being satisfied, it is presumed that Parliament intended the benefit to operate only where the required act is performed in a lawful manner.

47. *LB (Lian Bee) Confectionery*, *supra* n. 20.

> Construction *in bonam partem* is related to three specific legal principles. The first is that a person should not benefit from his own wrong. Next is the principle *allegans suam turpitudinem non est audiendus*. If a person had to prove an unlawful act in order to claim the statutory benefit, this maxim would preclude him from succeeding. The third related principle is stated by Coke in the words *ubi quid generaliter conceditur inest haec exceptio si non aliquid sit contra jus fasque* (where a grant is in general terms there is always an implied provision that it shall not include anything which is unlawful or immoral)'.

The appellant in the present case had, before the application for expungement, used a mark which was confusingly similar to the respondent's registered trademark. Based on this, the Federal Court held that the appellant is not a 'bona fide person aggrieved' under section 46(1) TMA 1976. In arriving at its conclusion, the Federal Court highlighted that to hold the otherwise would allow all infringers to expunge the very trademark they have been infringing and that this would contradict the principle of not allowing wrongdoers to benefit from their own wrong or unlawful act.

### B. Extent of any revocation

In the case of *McLaren International*,[48] the appellant sought to have certain goods removed from the respondent's trademark entry on the grounds that the respondent had never used the mark on those goods. Amongst the questions that were posed to the Federal Court, was whether the Court can expunge certain registered goods from the Register based on the ground of non-use. In the present case, the respondent's trademark was registered for 'articles of clothing, including boots, shoes and slippers' in class 25. However, the respondent's trademark had only been used in respect of executive men's shoes. The Court held that it is empowered to order the removal of the respondent's trademark in respect if such articles of clothing i.e. to grant the appellant's plea for deletion of the words 'articles of clothing, including' from the description of the goods. Nevertheless, it is pertinent to note that the appellants did not succeed in their appeal as they did not establish that they have the necessary *locus standi*. The appeal was accordingly dismissed with costs.

### C. Timing of any revocation

Malaysian registrations are vulnerable to non-use cancellation actions from the date of the actual registration of the trademark (i.e the date when the mark is entered on the Register) and *not* the date of the application.[49]

It is pertinent to note that the date when the mark is actually entered on the Malaysian Register of Trade Marks is the date of the issuance of the certificate of registration. The relevant period of 'non-use' is calculated three years and one month prior to the application for a non-use cancellation action.

---

48. *McLaren International, supra* n. 34.
49. *Industria De Diseno, SA v. Edition Concept Sdn Bhd* [2005] 3 MLJ 347.

# Mexico

*Eduardo Kleinberg*

## I. INTRODUCING GENUINE USE OF TRADEMARKS

### A. Legal framework

The Industrial Property Law is the basis of the use requirements. As per the Mexican Constitution, the Federal Congress has the exclusive authority to legislate on trademark matters. While the executive branch of the Mexican Government is allowed to issue regulations and other administrative rulings on the subject, they cannot exceed the scope of the Industrial Property Law nor contradict it in any way

While use is not a requirement to file a trademark application or to obtain registration, Article 113(III) of the Industrial Property Law dictates that the applicant must indicate the date of first use of the trademark or that it has not been used. When such a date is claimed, Article 56 of the Regulations to the Industrial Property Law provides that the applicant must indicate the address of the establishment or premises where the trademark is used.

The lack of any indication as to use, Article 113 provides, will be understood as a presumption that the trademark is not in use. For instance, all designations made from the Madrid System for Mexico lack an indication about use of the trademark and thus they have a legal presumption that the trademark is not in use at the time of the filing.

For their part, Articles 128, 130, 131, 134 and 152 of the Industrial Property Law set out the basic requirements of using a registered trademark and outline the consequences of failing to carry out such use, namely the impossibility to renew and the possibility of having the registration cancelled by a third party. Article 62 of the Regulations of the Industrial Property Law exemplifies what is to be considered use of a trademark for the purposes of the Industrial Property Law. Articles 92 and 151 of the Industrial Property Law set out the possibility of raising prior use as a defence and as a basis for challenging a registration, so as long as the use precedes the filing date of the registration or its claimed date of first use.

In addition, Article 19 of the Agreement on Trade-Related Aspects of Intellectual Property Rights (TRIPS) answers the question as to who may use the registered trademark for fulfilling the requirements of use set forth by the Industrial Property Law, namely any use under the control of the trademark owner can be considered.

As per Articles 1 and 133 of the Mexican Constitution, and their interpretation, international treaties signed by the executive branch of government and ratified by the Senate, one of the two chambers of the Federal Congress of Mexico, are considered to be higher in hierarchy than any law or regulation, usually only the Federal Constitution will have a higher hierarchy. Legal dispositions contravening an international treaty can often be challenged and judges are constrained to apply the treaty rather than the law against it. Thus, for example, TRIPS has been interpreted as barring Mexican authorities from refusing to acknowledge use made by a third party on the basis that such use was not made by a recorded licensee.

Also, there is a wide body of binding and merely persuasive opinions issued by the Federal Judiciary and the Federal Administrative Justice Tribunal interpreting the requirements of use, how to fulfil them and how to demonstrate compliance within litigation. Final decisions issued by the Mexican Institute of Industrial Property, including those cancelling a trademark registration due to lack of use or prior use can be appealed with the Federal Tribunal of Administrative Justice, a court independent from the judiciary that solves appeals on intellectual property matters through a specialized three-judge panel aptly named the Specialized Chamber on Intellectual Property Matters. When appeals involve novel interpretations of the law or international treaties, the Superior Chamber of the Tribunal, composed of eleven judges, may demand that the Specialized Chamber review the case so that those eleven judges may issue the decision. Both chambers can issue opinions and have done so in a variety of topics regarding use of trademarks.

Decisions from the Federal Administrative Justice Tribunal can be challenged through an '*amparo directo*' a challenge that may comprise both legal and constitutional aspects of the decision and the law of the case, to be solved by one of the currently 21 Federal Circuit Courts, each composed of three judges, located in Mexico City. These Courts also issue opinions and they supersede other courts, and all lower courts including the Federal Administrative Justice Tribunal can be bound by them.

Finally, in very rare cases generally involving constitutional issues considered extremely important, the Supreme Court of Mexico can further review the case. Their opinions can bind all lower courts including those detailed in previous paragraphs.

## B. Purpose and function of genuine use: policy considerations

Trademarks are defined in Article 88 of the Industrial Property Law as visible signs capable of identifying products or services from others of the same species or class in the market. The courts have consistently interpreted the Industrial Property Law and emphasized the importance of trademarks in assisting consumers in identifying products or services. To a certain extent, the exclusivity over a trademark is justified by this purpose. However, cases from the Federal Judiciary and the Federal Tax and Administrative Justice Tribunal also point out to the importance of trademarks in maintaining a system of free competition. On the one hand the system seeks to protect consumers and allow them to identify the goods or services of their preference, and on the other it seeks to protect free competition by allowing merchants to identify their goods and services.

In such a context, use of a registered trademark is necessary for justifying the exclusivity. A trademark that is unused is not identifying products or services for the benefit of consumers, merchants and free competition. Quite the contrary, it may bar the use of distinctive signs that could potentially fulfil such roles. For these reasons the Mexican Institute of Industrial Property and the courts have consistently ruled that once registered, the owner of a registered trademark has the duty to use it at least in one of the products or services covered by the respective registration. Modifications are only allowed to the extent they do not alter the distinctive character of the mark as filed by the owner of the registration and the use must comprise at least one of the products covered, that over the years has been interpreted to refer only to products expressly indicated by the applicant.

Further, because only the owner of the trademark will usually have the documents necessary to prove use and know the extent of such use, the courts have considered it desirable to interpret Articles 81 and 82 of the Federal Code of Civil Procedure as imposing upon the defendant to an action based on lack of use the burden of proving use of the trademark.

In such manners, the legislative branch of government along with the Mexican Institute of Industrial Property and the judiciary have sought to avoid the maintenance of registrations whose coverage does not reflect what is actually in the market.

## II. WHAT QUALIFIES AS GENUINE USE OF A TRADEMARK

### A. Nature of use

#### *1. Use consistent with the function of the trademark*

Article 62 of the Industrial Property Law provides that a trademark will be considered to be in use when the products or services identified by the mark have been put into commerce or are available in the market or are destined for export. This is consistent with the concept and function of a trademark; the use of the trademark must identify the products or services covered. To exemplify, cases have been decided where the owner of the registration showed use of a trademark in connection with the sales services for certain products, but the products themselves were identified by other trademarks. In such a scenario, use of the first trademark in connection with the sale service could be demonstrated but not with the products being sold.

It is important to differentiate trademarks from trade names and slogans for the purposes of use. In Mexico a trade name, which may be protected through use alone or a publication on the Industrial Property Gazette is the name by which an establishment is identified. For its part a slogan is a phrase used to identify products, services or an establishment and which can be protected through registration. Use of a trademark as a trade name, namely as the name by which an establishment is known may not qualify for the purposes of maintaining a trademark registration if the mark is not used to identify the products or services.

For instance, in a case regarding a challenge based on lack of use against both a trademark and a trade name used in connection with apparel, the owner submitted invoices and the authority concluded that although invoices can prove use of a trademark they cannot show that a certain establishment is known by a certain name and declared its cancellation.

Equally, in the case of slogans the Federal Administrative Justice Tribunal has interpreted that because of their nature, proving use of a slogan implies that the same must be used in media such as print or television or on the internet to truly advertise products, services or an establishment. Failure to announce, even if the slogan were identifying the products or services, can result in a cancellation due to lack of use. By contrast a trademark only needs to identify the products or services in order for use to be established.

## 2. *Use in the course of trade*

Use of a trademark in the course of trade is usually sufficient to maintain a trademark registration and defeat potential actions based on lack of use against it. This does not mean that actual sales need to have taken place and demonstrated but only that the products or services identified by the mark must be available to consumers in Mexico or destined for export.

Promotional use can be sufficient, at least if performed with sufficient frequency. For example, in a case involving a burger restaurant where their registration was challenged based on lack of use, their defence was that although they did not have restaurants in Mexico they did perform several promotional events in Mexican territory. The Mexican Institute of Industrial Property agreed and refused to declare the cancellation of their mark.

## 3. *Use by the owner or with its consent*

Use controlled by the owner of the registration is acceptable. The courts of Mexico have established that failing to acknowledge such use would breach international treaties.

Control can be demonstrated by a written agreement between the owner of the trademark and the third party whose use is being asserted. While a licence agreement, recorded or not, would usually be the most common and usual kind of evidence, it need not be. Other kinds of agreements can be acceptable as well, such as those pertaining the advertising or transport of the products bearing the mark, if they mention it or otherwise permit the authority to infer that the products bearing the mark were part of the respective transaction. Agreements that do not permit the authority to make such inference, such as agreements for the leasing of premises in which the products are allegedly sold, can be considered insufficient. Regarding use of a trademark made by an entity, the articles of incorporation, records of shareholders and other similar documents can be enough to prove control over the entity. In the event of bankruptcy, use made or controlled by the administrator of the assets will be considered.

Between individuals, control can be harder to establish and will typically demand the existence of a written agreement. For deceased individuals, use made or controlled by the administrator of the respective estate will be considered at least until the registrations pass on to the heirs.

## B. Volume, duration and frequency of use

No specific volume of sales is set forth by the law or its regulations and some opinions of the Federal Administrative Justice Tribunal are lax in providing that proving use in any time during the three years preceding the filing of a cancellation action based on lack use can be enough to defeat it. In most cases, however, it will be difficult to persuade the Mexican Institute of Industrial Property or a judge that a trademark has been used with one or a few instances of use.

## C. Use of a sign

Article 128 of the Industrial Property Law expressly provides trademarks should be used as registered or with modifications that do not alter the distinctive character. Such modifications are not defined by the law and both the Mexican Institute of Industrial Property and the courts will decide on a case-by-case basis if a trademark has been modified past that point. A criterion that is usually followed is that the first step to determine if a modification is permissible is to consider the kind of trademark registered under the notion that if a trademark registration grants exclusivity over certain word and/or design elements, exclusivity should not be maintained over them if they are not used.

In most cases, for word marks, the owner can use the mark in any font and size and this will not jeopardize the registration as the regulations of the Industrial Property Law expressly provide that a registration for a word mark comprises a right of exclusive use over any size and font. The inclusion of additional words can usually be tolerated, especially for descriptive words, as long as all the registered letters and numbers are present and their order is not altered. The inclusion of design elements will not usually jeopardize a registration for a word mark.

For composite marks, the exclusion of a design element can alter the distinctive character of the mark. A typical example would be a trademark registered with a distinctive design but using only the word elements. Usually, slight changes in colour and shape can be tolerated, but if the changes alter the most visually distinctive elements, which may be determined from their location, size, etc., the changes can alter the distinctive character of the mark.

## D. Use in relation to goods and services

### *1. Use for each of the designated goods and services*

Use must comprise at least one of the products or services covered by the respective trademark registration. Use in other products or services will provide no defence against an action based on lack of use. The authority will usually consider the products or services expressly pointed out by the applicant as the starting point for determining if the use has comprised the right products or services. Because in Mexico we do not have multi-class registrations, the classification of the products of a challenged registration at the time it was issued can determine whether a particular product is protected or not.

### 2. *Use for a broader category of goods and/or services*

Over time, the criteria regarding the interpretation of descriptions of products and services have become stricter. Important decisions from the Federal Administrative Justice Tribunal and the Federal Circuit Courts point out that the authority should in principle limit the description to the products or services expressly covered and that even very similar products not mentioned in the description are not included. Also, precedents point out that the description should be read as referring to 'finished products and services'. This means that if the description reads, for example, 'nails and screws' the owner of the registration cannot defeat an action based on lack of use by showing use of the mark in products incorporating the nails and the screws.

### E. Territorial scope

Use of the mark must include Mexico for the purposes of defending the registration from actions based on lack of use. Use on the internet will be considered to have been made in Mexico if it can be inferred that the product was available for Mexican consumers in Mexico. Indications that the product is shipped to Mexico, Spanish text, invoices and receipts for the delivery of products in Mexico and other elements may be considered to show that the product bearing the mark is in fact available in Mexico through the website.

It is important to keep in mind, however, that there are some evidentiary issues regarding websites and specifically proving that their contents existed prior to the litigation. Also, simply having a website may be insufficient to prove use if the site does not offer the products or services for consumers in Mexico.

### F. Relevant time period and its calculation

From the registration date, the owner of a registered mark will have three years to begin use at latest, after which third parties may attempt actions based on lack of use against the registration.

## III. EVIDENCE OF GENUINE USE OF A TRADEMARK

### A. When use should be proved: timing of lack of genuine use claims

### 1. *When filing an application for a trademark*

There is no requirement for using the trademark at the time that the application is filed or to submit any proof thereof. However, the Industrial Property Law dictates that the applicant must indicate the date of first use of the trademark or the lack of use and the omission thereof will be understood as a presumption that the trademark is not in use. This will be the case for all designations made through the Madrid System. Once the application is filed, the date of prior use cannot be changed. The claiming of a date of prior use will not exempt the party asserting prior use, which can be done either as a claim or as a defence, from submitting evidence that proves such use in litigation.

Although claiming a date of first use does not demand that the applicant file any evidence along with the application to support such claim, it is important to consider that Article 151(III) of the Industrial Property Law allows a trademark registration to be cancelled if granted based on false information contained in the application such as the date of first use. The statute of limitations for challenging a registration on the basis that its date of first use is false is five years, meaning evidence of use when claiming a date of first use should be kept for at least such a time period. This should not be confused with the evidence of use that the owner of a registration has to file in the event of an action based on lack of use, which should prove use for three years immediately preceding the filing of the action.

It is important to note that the claimed date of first use does not have any effect on the filing date of the application. The application with the earlier filing date still is the one that obtains registration over others filed afterwards, even if they have a claimed date of first use. However, claiming the date implies that third parties attempting to assert prior use against the resulting registration will not only need to precede the filing date, but also the date of first use.

For example, if A seeks to challenge B's registration based on prior use, A will need to prove use that not only precedes the filing date or priority date of B's registration but also any date of prior use that B may have claimed. The same applies when asserting prior use as a defence or counterclaim. For example, if A claims infringement against B and B wishes to assert prior use as a defence, B will need to prove use that precedes a date of prior use claimed in A's registration.

Finally, it is important to note that although use is not a requirement for obtaining registration, to obtain the renewal of the registration, the owner has to sign a form declaring under oath that use of the trademark has not been interrupted for a period of time equal or greater than three consecutive years. So while evidence of use is not requested to renew, the owner of the registration cannot usually renew a registration for an unused trademark without committing perjury. A workaround to this issue is to re-file the application at the time that renewal is due as applications do not need to be filed on a basis of use.

### 2. *In the course of opposition proceedings*

Mexico's new opposition system has peculiarities that must be given proper consideration. The applications are published before being examined, meaning that potential opponents should know that certain aspects of the trademarks they oppose such as the description of goods or services may change afterwards. The examiner may still cite trademarks *ex officio* and will ultimately decide the case. Neither the opposition or the lack thereof will force any outcome as the examiner is not bound to agree with the opposition. It is merely informative and persuasive. Finally, the opposition does not interrupt the registration process and replying to an opposition, while generally advisable, is not mandatory.

Generally, neither the applicant nor the opponent would need to prove that their marks are in use. Cases from the Federal Administrative Justice Tribunal point out that the examination of a trademark must be generally delimited to the trademark as requested in the official form and all the products or services indicated therein. Use of the trademark is not generally relevant for determining if an opposition based on likelihood of confusion will succeed and will not usually provide a valid basis for filing an opposition.

For example, if an application is filed and an opposition is submitted arguing likelihood of confusion with a prior registered trademark, the examiner will only consider the products or services covered by the application and the prior registration and the goods in which the marks are used will not be considered.

However, it is important to consider that if an opposition succeeds in persuading the examiner not to grant registration based on a prior mark, an office action citing the trademark will be issued and from that moment in time, the applicant has the legal standing necessary to challenge the cited mark, based on prior use or lack of use. Therefore, before filing an opposition based on a prior trademark, it is important to determine if such trademark is being used in Mexico.

### *3. In cancellation proceedings based on lack of use*

Article 130 of the Industrial Property Law provides that a trademark registration may be cancelled based on lack of use if no use of the trademark is made in Mexico for at least one of the products or services covered by the respective registration in a period equal or greater than three years. This term has been interpreted as forbidding actions based on lack of use against trademarks whose registrations have not been issued three or more years ago.

This is an important upside for re-filing an unused or scarcely used trademark. By re-filing, the resulting registration will enjoy an additional three years counted from its registration date, in which it will not be vulnerable to actions based on lack of use. Given that in Mexico, use of a trademark is not a condition for obtaining registration, this can be a viable strategy to take when the time to renew an unused trademark approaches.

To determine if the trademark will be cancelled due to lack of use, the Mexican Institute of Industrial Property must, either *ex officio*, which is hypothetically possible but extremely rare, or as per the complaint of a party with legal standing, go through the proceeding set forth by Chapter II of the Tenth Title of the Industrial Property Law, which includes giving the owner of the challenged registration a term of one month to reply and produce evidence that demonstrates use of the trademark.

The standing necessary to challenge a trademark registration on the basis of lack of use is defined by the law as legal standing. This is defined in turn in case-law as standing that is only held by those suffering an injury in a right that is legally protected or susceptible of being protected. In other words, the plaintiff must have a right that is recognized by the law or susceptible of being recognized by the law. This severely limits the population of potential plaintiffs.

If a trademark registration is being used as the basis of a legal action, for example when claiming infringement, then the defendant in that action will have the legal standing necessary to file the non-use cancellation action as a counterclaim as it is understood that the registration will affect the alleged infringer's rights to participate in commerce, employment, etc. It should be mentioned that the standing is not limited to infringement but to any legal action based on the trademark registration, such as a cancellation action based on a prior registration.

The legal standing necessary to challenge a registration based on lack of use is also understood to arise when the registration is cited against a pending trademark application. In such case, the applicant is understood to have the standing to challenge the cited registration, but the action needs to be filed before the examiner issues a final refusal of the application. In other words, the applicant needs to file the action against a cited registration while the application is still pending the final decision by the examiner.

The action based on lack of use must meet certain very basic statutory requirements. It must clearly indicate who the plaintiff and defendant are, a domicile where notice may be given to the defendant, narrate the facts that justify the legal standing and the basis for the action and cite the relevant legal dispositions. If the plaintiff files the action through legal counsel or another representative, a valid power of attorney must be filed along with the complaint.

### 4. *As a defence in infringement/cancellation proceedings*

Prior use of a trademark in Mexico can be raised as a complete defence against infringement or cancellation proceedings based on a trademark registration, so if the use precedes the filing date and the claimed date of first use of the registration(s) in which the action is based. For example, if A claims trademark infringement against B, B can escape liability if he submits evidence to prove that his use of the mark in Mexico precedes the filing date and claimed date of first use of A's registration. It is important to consider that the owner of the registration will not be able to defeat a defence based on prior use by proving an even earlier use. So in the previous example if A claims infringement against B and B raises a defence of prior use, A cannot defeat the defence by proving an even earlier use of the mark. So as long as B's use precedes the filing date or the claimed date of first use, the defence will work.

The prior use that is needed must have been uninterrupted and comprise goods or services identical or similar to those covered by the registration. Of course, the trademark used must be identical or confusingly similar to the trademark on which the action is based. These two aspects are rarely an issue when prior use is raised as a counterclaim or defence in infringement proceedings as the plaintiff in an infringement action will almost always assert that the trademark used by the infringer is identical or confusingly similar and that the products/services are the same.

While it is possible to assert prior use in Mexico as a defence, in some instances it is also viable to assert it as a counterclaim. In such manner, the defendant will not only escape liability but may also destroy the plaintiff's registration. For this purpose, the counterclaim may be asserted if no more than three years have passed since the publication of the registration in the Industrial Property Gazette. Because the effects of a successful action or counterclaim based on prior use are understood to be the annulment, not just the cancellation of the registration, it can be argued that the effects of the registration are retroactively destroyed and thus that a successful counterclaim bars the finding of liability against the alleged infringer.

Lack of use, while the law does not foresee it as a defence, can be asserted as a counterclaim against any action based on a trademark registration. However, it should be mentioned that a cancellation based on lack of use will generally be understood as a taking effect when the decision cancelling the registration is issued and not retroactively like an action based on prior use. Therefore, it is usual for parties to argue that a cancellation based on lack of use is not a complete defence to infringement and that the authority could find that in spite of the cancellation, at the moment in time the infringement took place the registration was still valid and thus infringed.

In addition, it shall be mentioned that one important aspect in which use can also come into play is when determining the penalty imposed upon an infringer. As per Article 220 of the Industrial Property Law, when determining the fine or other sanctions imposed for infringing a trademark registration, intentionality plays a relevant role. Such intentionality is easier to infer when infringing a widely used trademark that when infringing one that has not been used and may result in a larger fine. Thus, a defendant may attempt to point out to the lack of use of a trademark as a basis for arguing that the use of the mark was not intentional and that the corresponding sanction should be less severe. Alternatively, a plaintiff in an infringement action may try to demonstrate the use of the mark in order to show the defendant's knowledge and thus that the sanction should be more severe.

Also, use of the mark can be relevant for other aspects of infringement. Article 229 provides that in order for damages to be awarded, for criminal prosecutions to be attempted and for the imposition of preliminary injunctions, the plaintiff must show that the trademark has been used with indications that it is registered such as the initials 'M.R.' ('TM'), the 'R' symbol or the expression '*marca registrada*'. In practice, workaround that has been used in instances where the mark is not used with any indication that it is registered is to publish an announcement on a newspaper indicating the existence of the registration. The Mexican Institute of Industrial Property often has allowed preliminary injunctions on such a basis.

### B. Burden of proof

For actions based on lack of use, the burden of proof is placed on the defendant as per Articles 81 and 82 of the Federal Code of Civil Procedure, which provide that a party making a denial must prove such denial if it involves an affirmation. Thus by denying the action, defendant is affirming use and must prove it. Another theme typical of burden of proof issues on this kind of action is that only the defendant may have evidence of use available and knowledge of how the mark is used and thus could prove use, while the plaintiff could not prove a total lack of use. The defendant must demonstrate that the mark has been used during the three years immediately preceding the filing of the action. If the defendant fails to answer the action a default judgement declaring the cancellation due to lack of use will be declared. If the defendant answers but fails to submit evidence of use, the defendant will typically lose unless the defendant successfully challenges plaintiff's standing and authority to file the action.

The defendant must offer and submit all available evidence when the action is answered, typically a month from the date in which defendant receives notice. However, a defendant can obtain an extension of time if the evidence is in a foreign jurisdiction, so if the defendant makes the request in the response to the action and clearly identifies each item of evidence that will be filed it can be submitted afterwards. Evidence that does not exist at the time of the filing of the response or whose existence is unknown at that time can also be submitted afterwards.. Failure to timely offer and submit evidence will preclude defendant from doing so afterwards. Challenges to the usual burden of proof established herein routinely fail.

While the plaintiff does not have the burden to prove lack of use, the plaintiff can and often will submit evidence to try to dispute the evidence filed by the defendant or destroy its credibility. Usually, the plaintiff has a three-day term to do so after receiving notice of the reply. A typical example would be that if the defendant submits invoices pointing out an establishment where the mark is used, the plaintiff could submit evidence to prove that the establishment belongs to a different entity, no longer operates, etc. The plaintiff can also object to the defendant's evidence as stated below.

For actions or defences based on prior use, it is the party asserting such use who has the burden to prove such use. Like the defendant in an action based on lack of use, the party asserting prior use as a claim or a defence must offer and file all available evidence when their complaint or response is filed. Otherwise, they will be precluded from doing so afterwards. Failure by the party against whom prior use is being asserted to timely object or refute the evidence submitted can result in a tacit recognition of the facts the other party attempts to prove through such evidence.

### C. Relevant evidence

Given that a trademark is in use when the products or services are available, the most typical evidence for proving use is invoices or sales receipts showing that the products or services bearing the mark have in fact been sold. Several invoices or sales receipts clearly indicating the trademark and the goods/services covered along with evidence of advertising such as advertisings in magazines or newspapers would usually suffice to prove use.

#### *1. Use on the product or its packaging*

Physical evidence consisting of the product and its packaging is usually insufficient by itself to prove use as the date of use is usually very relevant for both actions based on lack of use and actions or defences based on prior use and it is usually impossible to put such physical evidence within a specific time period. However, this evidence can be used in support of and in conjunction with other evidence.

#### *2. Use in advertising*

Advertising in publications such as newspapers or magazines can be very relevant, especially if regularly published by a third party. For catalogues, flyers, brochures and other documents published by the owner of the trademark to advertise it, the authority will usually consider them sufficient to prove use by themselves only if some proof of their distribution or delivery is provided. For example, if a catalogue is filed to prove that a trademark has been used, the party filing the catalogue should also file the agreement with the distributor of the catalogue or at least a letter from them attesting to the fact that they were hired to distribute it.

Further, when using catalogues and other advertising as evidence, it is usually desirable to be able to provide and demonstrate the volume of advertising printed and distributed to show that the products or services were indeed available. If the advertising points to a telephone number or website on which to purchase the products or services, showing the volume of traffic can also be helpful.

### 3. *Use on the internet*

Use on the internet can be considered. However, because the party asserting use must usually prove it took place within a certain timeframe, use on the internet is rarely used by itself to prove use. Usually, the party trying to prove use through a website will either request that an appropriate government officer such as a court clerk review the site or hire a notary public to review it and attest to the contents and dates mentioned therein.

### 4. *Form of evidence*

It is important to consider that as per our evidentiary rules non-public documents (as invoices) can be objected to, even without expressing a reason, and if so their contents can be corroborated through other evidence attesting to the same facts. Non-certified copies can be easily objected to and thus, it is usually advisable to file the original document instead or certified and legalized copies.

Witness testimony is not admissible to prove use of a trademark during these proceeding and while the constitutionality of such limitation has been questioned, to date it stands. The only possibility to have such testimony admitted is to have it in writing, usually by having the witness declare before a notary. However, it is advisable to have the facts declared by the witness supported by documents or other evidence as the testimony by itself will not suffice to support a finding that the trademark is in use.

In addition, the parties may request that the authority solving the dispute, either the Mexican Institute of Industrial Property or the court clerk of the judge in charge of the case, inspect websites, stores premises, etc. However, timing is again very important. Present use of a trademark cannot by itself support a finding that the trademark has been used in the same manner in the past. Thus, these inspections must usually be supported by other evidence to be effective.

In sum, proving use is not usually done by a single overwhelming item of evidence, but by combining several.

### 5. **Distinction between use of a company name and use of the services offered by the company**

Though the difference is subtle, the owner of a trademark should clearly differentiate that the mark is not being used as a trade name but as an identifier for the service.

As stated above, use of a trade name and use of trademark are similar but not interchangeable.

## IV. JUSTIFICATION FOR NON-USE

### A. Obstacles to use

Lack of use of a trademark can be excused only in a very limited set of circumstances. As stated above, use is a precondition for renewal. However, for renewal purposes only, the law can excuse lack of use for a trademark registration if use is made for the same mark but for products and services of a different class covered by another registration. For example, if a company owns two registrations for the same mark in connection with different products or services, use of the mark in connection with the products covered by the first registration suffices for renewing the second registration as well. Keep in mind, this exception only applies to renewals and is no defence whatsoever against actions based on lack of use.

Regarding actions based on lack of use, use can only be excused if the defendant proves that there are government restrictions or other circumstances that barred use and that are independent of the will of the owner. Bankruptcy is not considered as an excuse. Instead, courts have held that the administrators of the assets during the bankruptcy proceeding would have the duty to ensure it is in use and may be liable to the creditors if they allow the registrations to be cancelled due to lack of use.

Equally, the assignee who obtains a trademark registration, even in good faith and for consideration, is not excused from demonstrating that the prior trademark owner used the mark. So if A assigns a trademark registration to B and B receives notice of an action based on lack of use in the immediately next instant, B is in no manner whatsoever excused from proving use.

For individuals, though there is no binding precedent on the subject, death is unlikely to be an excuse for non-use. Instead, the administrator of the estate of the deceased individual would have the duty to keep using the registered mark until assigned to the heir.

### B. Circumstances in which non-use may be cured

While the law does not expressly estate how to 'cure' the lack of use of a registered mark, for the purposes of cancelling a registration it is only the three years immediately preceding the filing of the action that matter, so if use of the trademark begins within that three-year period, it could be hypothetically possible to effectively 'cure' the lack of use, so as long as the owner can persuade the authorities about such use.

Again, it is not necessary to prove use within the three whole years preceding the filing of a cancellation action based on lack of use. Instead if the owner of the registration can persuade the authorities that the mark has been in use somewhere in that period, that might be enough. Of course, the more use that can be demonstrated the more likely it will be that the authorities will be persuaded. In any case, the use should commence with enough anticipation to a cancellation action based on lack of use as to avoid the risk that the authority will consider the use to be a simulation prepared for litigation rather than a real use in the Mexican market.

Re-filing is also possible to 'cure' the lack of use. Because use is not a requirement to obtain registration, the trademark owner may simply re-file for registration and obtain a new registration that will not be vulnerable to actions based on lack of use until three years elapse from the registration. Precedents from the Federal Circuit Courts acknowledge that even if the registration is cancelled based on lack of use, it is perfectly lawful for the owner of the cancelled registration to file again for the same trademark.

There are some downsides to re-filing however. Given that the new application will not enjoy the filing date and priority of the prior registration it is possible that the applicant will run into prior trademarks. Also, if the cancellation action was pursued at the request of a third party who received a citation, the applicant will have to deal with the fact that such third party may obtain registration and that it will not be possible based on lack of use until three years have passed.

## V. CONSEQUENCES OF NON-USE

### A. Consequences of non-use and relevance of the context

The consequences of not using the mark can be summarized as follows:

(1) The vulnerability of the registration to actions based on lack of use, or to counterclaims based on such lack of use when attempting to enforce the registration in infringement, annulment or opposition proceedings, which can result in a termination of the exclusivity conferred by the registration.
(2) The inability to file for renewal, unless the exception outlined above applies.

However, neither of these consequences can keep the owner of the mark from re-filing a trademark application to obtain registration again and this is a commonly used remedy. Because in Mexico use is not a requirement to obtain registration, the owner of the registration may simply decide to file a fresh application. In such manner, the new registration will not be vulnerable to actions based on lack of use until three years elapse from its issuance.

The effect of a cancellation due to lack of use is understood to take place at the moment in time when the decision declaring the cancellation due to lack of use is issued, even if such decision is challenged afterwards, given that the Federal Administrative Procedure Law provides that decisions from authorities such as the Mexican Institute of Industrial Property generally have legal effects from the date in which notice is given and until an administrative or judicial authority rules otherwise. Thus, a decision subject to an appeal for example declaring the cancellation for lack of use is valid until the appeal is decided and rules otherwise.

The law does not differentiate between proving use as a defence in a cancellation action and proving use as a defence to a counterclaim based on lack of use. If the owner of the registration can prove that use was made in the three years preceding the claim or counterclaim that forces the proving use, the registration will survive.

## B. Extent of any revocation

As stated above, use may comprise any of the products covered by the registration to survive an action based on prior use. The whole registration will survive. No partial cancellations exist, even for unused products or services. It is thus possible that trademark exclusivity can subsist over unused products so as long as other products covered by the registration are in use.

## C. Timing of any revocation

The cancellation based on lack of use will take effect from the issuance and notice of the decision. As the Federal Administrative Procedure provides that resolutions from most federal administrative authorities including the Mexican Institute of Industrial Property will have effect from notice and until a contrary ruling is issued, there are grounds to argue that the filing of an appeal or other legal challenge to a cancellation based on lack of use will not interrupt its effect until the decision regarding the respective legal remedy comes into effect.

Also, for this reason, a cancellation based on lack of use can be asserted as a counterclaim against actions based on a trademark registration such as an infringement action, but that does not mean that if the counterclaim succeeds the infringer may escape all liability. After all, if the registration is valid until the cancellation is issued and notice is given to its owner, then it may be considered that it was valid when infringed.

In practice when a counterclaim is filed against a trademark registration, what the Mexican Institute of Industrial Property will do when deciding the case is that it will 'accumulate' the proceedings so that both the claim and the counterclaim are simultaneously solved in a single resolution. In this manner, the authority will try to avoid problems derived from the timing of the cancellation. Courts hearing appeals regarding such cases will usually do the same.

Claims or counterclaims based on prior use on the other hand are understood to produce an annulment of the registration rather than a cancellation, which is why prior use can provide a better defence against infringement, for example, as it can be argued that the effects of a legal nullity are retroactively destroyed. Again, such effect will take place once the decision declaring the annulment due to prior use is issued and notice is given and it is understood such effect can persist even if such a decision is challenged and until a ruling to the contrary is issued.

The lapsing of a registration based on lack of renewal, whether or not such lack of renewal is due to lack of use, is understood to operate *ipso iure*, without need for any declaration, notice or resolution from the moment in time when the time to file for renewal has elapsed.

# The Philippines

*Aleli Angela G. Quirino and Victor Basilio n. De Leon*[1]

## I. INTRODUCING GENUINE USE OF TRADEMARKS

### A. Legal framework

Under the Philippine Intellectual Property Code (IP Code),[2] showing actual or genuine use of a trademark is not required before one can apply for registration. In fact, a trademark can be registered even without prior use. This is markedly different from the old Trademark Law (Republic Act No. 166), which adopted the first-to-use system. Under the old Trademark Law, a mark must have been commercially used in the Philippines for at least two months before it can be the subject of a trademark application.[3]

Under the present system, trademark rights are acquired through registration:[4]

> 'How Marks are Acquired. – The rights in a mark shall be acquired through registration made validly in accordance with the provisions of this law.'

However, despite the clear language of the IP Code, the Philippines does not wholly embrace the first-to-file rule. Use is not completely dispensed with. In fact, the IP Code requires a mark to be used within three years from the filing date of the application. Otherwise, the trademark application will be refused or the registration removed.[5] This means that registration is insufficient to perfect trademark ownership. As the

1. The authors would like to thank Kurt Glen Yeung, a senior associate in the Intellectual Property Department of Angara Abello Concepcion Regala & Cruz Law Offices, and Danielle S. Cadiz, an associate in the Intellectual Property Department of Angara Abello Concepcion Regala & Cruz Law Offices.
2. Intellectual Prop. Code, which took effect on 1 January 1998.
3. Rep. Act No. 166, (1947), sec. 2-A.
4. Intellectual Prop. Code, sec. 122.
5. *Ibid.*, sec. 124.2.

Philippine Supreme Court held in *Berris Agricultural Co., Inc. v. Abyadang*,[6] 'ownership of a trademark is acquired by its registration and actual use'. This is because a 'trademark is a creation of use and belongs to one who first used it in commerce.'

Thus, registration is only prima facie evidence of one's ownership over a particular mark.[7] This presumption of ownership can be disputed and overcome by anyone who could show prior use and, therefore, a superior right.

Apart from showing use within three years from the filing date,[8] the IP Code and the Rules and Regulations on Trademarks, Service Marks, Trade Names, and Market or Stamped Containers of 2017 (Trademark Regulations) mandate showing proof of actual use during the following periods:

- Within one year from the fifth anniversary of the trademark registration,
- Within one year from the date of renewal, and
- Within one year from the fifth anniversary of each renewal.[9]

If actual use is not established by evidence during any of these periods, the registered mark will be cancelled and removed from the Trademark Register.[10]

### B. Purpose and function of genuine use: policy considerations

It is essential that a trademark be actually used in commerce because allowing registered marks to remain valid without the requirement of actual use will lead to a cluttered Trademark Register and an accumulation of registered marks that do not perform any actual function in business.[11] This is contrary to the State policy of having 'an effective intellectual and industrial property system'.[12]

Moreover, because a trademark indicates the origin or ownership of the articles to which it is attached and guarantees that those articles have a certain standard of quality,[13] an unused mark serves no purpose. The government, therefore, has no reason to protect it.

## II. WHAT QUALIFIES AS GENUINE USE OF A TRADEMARK

### A. Nature of use

#### *1. Use consistent with the function of the trademark*

Genuine or actual use requires the application of the mark to the goods, containers, or documents associated with the goods involved, such as receipts or invoices of

6. G.R. No. 183404, 13 October 2014.
7. Intellectual Prop. Code, sec. 138.
8. Or International Registration date for trademarks that went through the Madrid Protocol.
9. IPO Memorandum Circular No. 17-010, rule 204 (2017).
10. *Ibid.*
11. R. Aquino, *Intellectual Property Law Comments and Annotations* (2006), 153.
12. Intellectual Prop. Code, sec. 2.
13. *Mirpuri v. Court of Appeals*, 318 SCRA 516, 532 (1999).

sale.[14] For service marks, actual use may be shown through contracts, brochures, or advertising materials, both online or offline, which are used in the promotion and sale of services.[15] Actual use must also be consistent with the function of a trademark. Otherwise, any instance of the mark's public appearance (for example, as part of an article reviewing a particular brand) will be considered trademark use.

The case of *Mirpuri v. Court of Appeals* describes the function of a trademark this way:

> 'In Philippine jurisprudence, the function of a trademark is to point out distinctly the origin or ownership of the goods to which it is affixed; to secure to whoever has been instrumental in bringing into the market a superior article of merchandise, the fruit of his industry and skill; to assure the public that they are procuring the genuine article; to prevent fraud and imposition; and to protect the manufacturer against substitution and sale of an inferior and different article as his product.'[16]

In other words, for trademark use to be genuine or actual it must be use in public in connection with sale or offering of sale of a particular product or service.

### 2. *Use in the course of trade*

#### *i. Public use*

The goods or services bearing the mark must be for sale or made available to the public. Internal use does not constitute genuine or actual use. In *Asari Yoko Co., Ltd. v. Kee Boc*,[17] the Supreme Court sustained the opposition of Asari Yoko Co., Ltd. (Asari), a Japanese corporation, against Kee Boc's application for the registration of the mark 'RACE' and design to be used on shirts and undershirts. In sustaining the opposition, the Supreme Court found that there was continuous use by Asari as evidenced by its importation of the shirts and undershirts bearing the mark sought to be registered from Japan to the Philippines since 1949.[18] Shipping documents also supported Asari's importation activities.[19] Thus, in this case, the importation of products bearing the mark to the public as evidenced by shipping documents amounted to use in commerce.

#### *ii. Commercial purpose*

Advertising alone is generally not considered as actual use within the meaning of Philippine jurisprudence with respect to the promotional items on which the mark is or has been used. Accordingly, issuing circulars, advertising, and giving out price

14. IPO Memorandum Circular No. 17-010, rule 210 (2017).
15. *Ibid.*
16. *Mirpuri v. Court of Appeals*, *supra* n. 13, at 532.
17. *Asari Yoko Co., Ltd. v. Kee Boc*, 1 SCRA 1 (1998).
18. *Ibid.*, at 6.
19. *Ibid.*

lists on certain goods alone do not give rise to any rights over a mark.[20] In *Pag-asa Industrial Corporation v. Court of Appeals,* the Supreme Court ruled that sending free samples of zippers to the Philippines are not sufficient to establish use, since the samples have no commercial value.[21]

A contradiction therefore arises between case law and the 2017 Trademark Regulations. Under the Trademark Regulations, advertising materials are acceptable proof of use.[22] Thus, it appears that to maintain a trademark registration with the Intellectual Property Office (IPO), advertising materials are sufficient to prove use. However, for purposes of proving ownership before the courts, advertising materials may not be enough.

With respect to goods bearing the mark destined for export outside the Philippines, a distinction may be made between use for the purpose of maintaining a trademark registration with the IPO and use for the purpose of proving ownership of the mark. Under the Trademark Regulations, receipts or invoices of sale of the goods or services or other evidence of use and copies of contracts for services showing the use of the mark are accepted as proof of use. Thus, it appears that as long as the trademark owner is able to furnish the IPO with any of the aforementioned documents, the IPO will consider the use by the trademark owner as sufficient despite the fact that the products bearing the mark are destined for sale outside of the Philippines. However, it is unclear whether the mere activity of exporting goods outside of the Philippines can be considered as actual use in commerce in the Philippines for proving ownership over the mark. Since the mark is used on goods that are being exported outside the Philippines, it may be argued that the goodwill generated from the use of such mark cannot be said to be coming from within the territory of the Philippines.

This rule may be likened to the well-established principle in Philippine Corporate Law that Philippine exporters, by the mere act alone of exporting their products, are not doing or transacting business in the Philippines for purposes of Section 133 of the Corporation Code.[23]

### 3. *Use by the owner or with its consent*

In general, it is the trademark owner who is required under Philippine law to put the mark in actual use in commerce. However, the trademark owner is free to authorize third parties such as related companies, importers, distributors, and affiliates to use the mark. Such use by authorized third parties inure to the benefit of the trademark owner in accordance with Rule 902 of the Trademark Regulations, provided that such use does not affect the validity of the mark or of the mark's registration.[24] If the trademark owner controls the use of a mark by such third party with respect to

---

20. *Sterling Products International, Inc. v. Farbenfabriken Bayer Aktiengesellschaft*, 27 SCRA 1214, 1224 (1969).
21. *Pag-asa Industrial Corporation v. Court of Appeals*, 118 SCRA 526, 533 (1982).
22. IPO Memorandum Circular No. 17-010, rule 210 (d) (2017).
23. *Cargill, Inc. v. Intra Strata Assurance Corporation*, 615 SCRA 304, 316 (2010).
24. IPO Memorandum Circular No. 17-010, rule 902 (2017).

the nature and quality of the goods or services, such use shall likewise inure to the benefit of the trademark owner.[25]

### B. Volume, duration and frequency of use

Philippine law is silent as regards volume, duration, and frequency of actual use. As long as the mark is used in commerce and the applicant or registrant is able to submit a Declaration of Actual Use (DAU) evidencing use in the Philippines during the periods prescribed by law, protection over the mark is maintained. A DAU is a statement under oath, filed by the trademark applicant or registrant, or their duly authorized representative stating that the mark is in use in commerce. It must be supported by evidence showing use.

In *Converse Rubber Corporation v. Universal Rubber Products*, the Supreme Court considered as actual use in commerce the petitioner's sale of 12 to 20 pairs a month of rubber shoes.[26] As to whether the goods or services must be directly sold by the trademark owner, the Court noted that:

> 'It is a matter of public knowledge that all brands of goods filter into the market, indiscriminately sold by jobbers, dealers, and merchants not necessarily with the knowledge or consent of the manufacturer. Such actual sale of goods in the local market establishes trademark use which serves as a basis for any action aimed at trademark pre-emption.'[27]

Therefore, no matter how insignificant the quantity of goods sold in the country may be, or how clear is the lack of direct or indirect participation by the trademark owner in the transaction involved, still any sale of the trademarked goods is sufficient to constitute genuine or actual use.

### C. Use of a sign

The IP Code and the Trademark Regulations expressly provide that the use of the mark in a form different from the form in which it is registered is not a ground for cancellation or removal of the mark and shall not diminish the protection granted to it, provided that the distinctive feature or character is not altered.[28] However, there are no specific criteria by which to judge if a revision to the mark alters its distinctive feature or character.

25. *Ibid.*
26. *Converse Rubber Corporation v. Universal Rubber Products*, 147 SCRA 154, 162 (1987).
27. *Ibid.*
28. IPO Memorandum Circular No. 17-010, rule 900 (2017); Intellectual Prop. Code, sec. 152.2.

## D. Use in relation to goods and services

### *1. Use for each of the designated goods and services*

Philippine law does not require the trademark owner to use the mark on each and every product or service listed in a particular class. The IP Code and the Trademark Regulations provide that when a mark is used for some of the goods or services under the same class, such use is considered use for the entire class to which those goods or services belong.[29] Consequently, the use of the mark for some of the goods or services will prevent the mark's cancellation or removal in respect of all other goods and services belonging to the same class.[30]

### *2. Use for a broader category of goods and/or services*

The Trademark Regulations prohibit the use of broad terms in identifying the goods or services intended to be covered by a trademark.[31] However, what term may be deemed broad is subject to the examiner's discretion. Thus, there may be registrations that have a broad listing of goods and services. In any case, use of a mark for a broader category of goods and services is acceptable. In fact, under the Trademark Regulations, trademark use on goods belonging to a different class may be considered use for another class, provided that the classes are related.[32]

## E. Territorial scope

Use of the mark must be in the Philippines.[33]

The territorial scope of a trademark is evident in *Kabushi Kaisha Isetan v. Intermediate Appellate Court*. In the afore-mentioned case, Kabushi Kaisha Isetan, a Japanese corporation operating a retail outlet, petitioned for the cancellation of the similar 'ISETANN' trademark claimed by private respondent Isetann Department Store, a Philippine registrant.[34] In ruling against the position of Kabushi Kaisha Isetan, the Supreme Court explained the rationale for the concept of territoriality under Philippine Trademark Law:

> 'What is to be secured from unfair competition in a given territory is the trade which one has in that particular territory. There is where his business is carried on where the goodwill symbolized by the trademark has immediate value; where the infringer may profit by infringement.'[35]

Petitioner Kabushi Kaisha Isetan-catered solely to the Japanese market. It had no business activity whatsoever in the Philippines during the period of controversy.

29. *Ibid.*, at rule 901 and at sec. 152.3.
30. *Ibid.*, at rule 901 and at sec. 152.3.
31. *Ibid.*, at rule 406.
32. IPO Memorandum Circular No. 17-010, rule 209 (2017).
33. Rule 208, Trademark Regulations
34. *Kabushi Kaisha Isetan v. Intermediate Appellate Court*, 203 SCRA 583 (1991).
35. *Ibid.*, at 592.

Hence, it was ultimately held by the Supreme Court that Kabushi Kaisha Isetan had trademark rights solely within the territory where it engaged in the sale of its goods. Moreover, since it failed to present any evidence to prove that it also sold its items to Filipino buyers, the Supreme Court ruled that there was no reason to recognize Kabushi Kaisha Isetan's ownership over the same trademark within the Philippine archipelago:

> 'The records show that the petitioner has never conducted any business in the Philippines. It has never promoted its tradename or trademark in the Philippines. It has absolutely no business goodwill in the Philippines. It is unknown to Filipinos except the very few who may have noticed it while travelling abroad. It has never paid a single centavo of tax to the Philippine government. Under the law, it has no right to the remedy it seeks.'[36]

In sum, there was no need to protect the petitioner Kabushi Kaisha Isetan from any potential unfair competition, considering the fact that the Philippine market was not exposed to its sales operations. The Supreme Court also recognized that any goodwill that might have sprung in the Philippines was solely the work of Isetann Department Store, in view of the fact that it registered its trademark in 1979, and had continuously used the same in commerce:

> 'The respondent registered its trademark in 1979. It has continuously used that name in commerce. It has established goodwill through extensive advertising. The people who buy at Isetann Store do so because of Isetann's efforts. There is no showing that the Japanese firm's registration in Japan or Hong Kong has any influence whatsoever on the Filipino buying public.'[37]

The exception to this general rule that actual use of a mark must be 'use in the Philippines' is if the mark is an internationally well-known trademark.[38] The Trademark Regulations provide the general mechanics by which a trademark may be considered to be well known.[39] If it is proven that a foreign mark has attained the status of a well-known mark, then that mark is granted protection in the Philippines, even if the mark is not used in the country.

In *Fredco Manufacturing Corporation v. President and Fellows of Harvard College*,[40] Fredco Manufacturing Corporation (Fredco), a local proprietor, wanted to cancel Harvard's registration of the mark 'HARVARD' in the Philippines. Fredco based its alleged superior claim of ownership on prior registration in this manner:

> 'Fredco alleged that it was formed and registered with the Securities and Exchange Commission on 9 November 1995 and had since then handled the manufacture, promotion and marketing of "Harvard" clothing articles. Fredco alleged that at the time of issuance of Registration No. 56561 to Harvard University, New York Garments had already registered the mark "Harvard" for

36. *Ibid.*, at 591.
37. *Ibid.*, at 593.
38. D. Funa, *Trademark Law of the Philippines* (2012), 91.
39. IPO Memorandum Circular No. 17-010, rule 103 (2017).
40. *Fredco Manufacturing Corporation v. President and Fellows of Harvard College (Harvard University)* 650 SCRA 232 (2011).

goods under Class 25. Fredco alleged that the registration was cancelled on 30 July 1998 when New York Garments inadvertently failed to file an affidavit of use/non-use on the fifth anniversary of the registration but the right to the mark "Harvard" remained with its predecessor New York Garments and now with Fredco.'[41]

The Supreme Court noted, however, that notwithstanding Fredco's prior registration of the 'HARVARD' mark, respondent Harvard nevertheless had prior use thereof in multiple countries abroad, securing registration in multiple countries.[42] In this case, the 'HARVARD' mark was officially recognized to be an internationally well-known mark, thus entitling it to protection despite having no business in the Philippines:

> 'There is no question then, and this Court so declares, that "Harvard" is a well-known name and mark not only in the United States but also internationally, including the Philippines. The mark "Harvard" is rated as one of the most famous marks in the world.'[43]

The Court in the *Fredco* case summarized the exception to the territorial scope of trademark use:

> 'Indeed, Section 123.1 (e) of R.A. No. 8293 now categorically states that a mark which is considered by the competent authority of the Philippines to be well-known internationally and in the Philippines, whether or not it is registered here, cannot be registered by another in the Philippines.
>
> Section 123.1 (e) does not require that the well-known mark be used in commerce in the Philippines but only that it be well-known in the Philippines.'[44]

This case should not be read as an exemption for owners of well-known marks not to file a DAU during the periods mandated by the Trademark Regulations. The DAU requirement still stands, but for purposes of determining ownership of a trademark, the owner of a well-known mark need not show use in the Philippines.

## F. Relevant time period and its calculation

In the Philippines, there is no minimum period provided by the IP Code and the Trademark Regulations for use to qualify as actual use in commerce. While the trademark applicant or registrant is required to submit a DAU evidencing the fact of use during the periods prescribed under Rule 204 of the Trademark Regulations, use is not required to be continuous during the relevant periods. It shall be sufficient for purposes of determining whether the mark is actually used in commerce if use of the mark has been made at the beginning, middle, or end of such period.

41. *Ibid.*, at 236.
42. *Ibid.*, at 237.
43. *Ibid.*, at 255.
44. *Ibid.*, at 252.

## III. EVIDENCE OF GENUINE USE OF A TRADEMARK

### A. When use should be proved: timing of lack of genuine use claims

#### *1. When filing an application for a trademark*

When filing an application for a trademark with the IPO, the intention to use the mark applied for is not required to be stated. It is also not required to show that the mark being applied for is currently in use.[45] However, the trademark applicant must submit a DAU with evidence of use within three years counted from the filing date of the application.[46]

#### *2. In the course of opposition proceedings*

If an opposition is being filed on the ground that the opposing party is the true owner of the mark based on prior and continuous use, then the opposer must attached to the Verified Notice of Opposition pieces of evidence proving such use. Without evidence of use, the opposition will fail.

Normally, if the trademark applicant feels that he has superior right over the mark, he would try to provide evidence showing that his use of the same or confusingly similar mark is earlier than the opposer's.

It is worth noting that 'the determination of priority of use of a mark is a question of fact',[47] and in arriving at a decision in an opposition or cancellation case, the Hearing Officer will consider the evidence presented by both parties.

#### *3. In cancellation proceedings based on lack of use*

In a Petition for Cancellation based on lack of use, the petitioner may submit proof evidencing non-use of the mark by the registered trademark owner for an uninterrupted period of at least three years.[48]

#### *4. As a defence in infringement/cancellation action proceedings*

In trademark infringement and cancellation proceedings, the parties are allowed to submit proof of prior use of the mark to the proper court or the Bureau of Legal Affairs (BLA) by way of defence.

In cancellation proceedings based on non-use, the petitioner must allege ultimate facts constituting non-use and may attach to the petition affidavits of witnesses, documentary or object evidence, and other supporting documents which may prove

45. IPO Memorandum Circular No. 17-010, rule 204 (2017).
46. *Ibid.*
47. *Berris Agricultural Co., Inc. v. Abyadang, supra* n. 6.
48. Intellectual Prop. Code, 151.1. (c).

the trademark owner's non-use.[49] In this instance, the burden of proof lies with the trademark owner to prove actual use of the mark in the Philippines.

Note also that the prior institution of a cancellation case will not bar the institution of a subsequent action to enforce trademark rights:

> 'The filing of a suit to enforce the registered mark with the proper court or Bureau shall exclude any other court or agency from assuming jurisdiction over a subsequently filed petition to cancel the same mark. On the other hand, the earlier filing of petition to cancel the mark with the Bureau shall not constitute a prejudicial question that must be resolved before an action to enforce the rights to the same mark may be decided.'

The case of *Superior Commercial Enterprises, Inc. v. Kunnan Enterprises Ltd.*[50] highlights the effect of a final and executory decision in a cancellation case to a subsequently instituted trademark infringement proceeding. In this case, petitioner Superior Commercial Enterprises, Inc. (Superior) filed a complaint for trademark infringement and unfair competition against respondent Kunnan and Sports Concept (Kunnan). Superior claimed that it was the owner and prior user of the marks 'KENNEX', 'KENNEX & DEVICE', 'PRO KENNEX' and 'PRO-KENNEX'. Respondent Kunnan disputed Superior's claim of ownership, claiming that it created and first used the 'KENNEX' and 'PRO KENNEX' trademarks. It further claimed that Superior was a mere distributor as evidenced under a Distributorship Agreement. Prior to and during the pendency of the infringement and unfair competition case, Kunnan filed Petitions for Cancellation of the 'KENNEX' and 'PRO KENNEX' marks with the BLA.[51] The BLA found that respondent Kunnan was able to establish its rights as the true and lawful owner of the marks in dispute and ordered the cancellation of Superior's registrations. The Supreme Court, in dismissing Superior's complaint for trademark infringement and declaring that Kunnan was the true and lawful owner of the 'KENNEX' and 'PRO KENNEX' marks, held that the finality of the BLA decision in the cancellation case rendered the trademark infringement aspect of Superior's case moot and academic.[52]

### B. Burden of proof

Generally, the burden of proof for actual use of a mark in commerce lies with the trademark owner. To prove that a mark is actually used in commerce, the trademark owner must submit a DAU along with proof of actual use during the periods prescribed by the Trademark Regulations. For purposes of showing proof of actual use, the DAU requires a declaration under oath including the following information:

- The name and address of the applicant or registrant;
- Declaration that the mark is in actual use in the Philippines;
- The list of goods or services for which the mark is used;

49. IPO Memorandum Circular No. 7, rule 2, sec. 7 (2016).
50. *Superior Commercial Enterprises, Inc. v. Kunnan Enterprises Ltd.*, 618 SCRA 531, 534 (2010).
51. *Ibid.*, at 539.
52. *Ibid.*, at 550.

- The name(s) of the establishment(s) and address(es) where the products are being sold or where the services or being rendered; and
- If the goods or services are available by online purchase, the website must be indicated on the form in lieu of the name and address of the establishment or outlet.[53]

The role of the IPO is to determine whether there was actual use by the trademark owner based on the evidence submitted to it. The IPO is not required to collect evidence on the trademark owner's behalf to determine whether the mark is actually being used in commerce.

In a cancellation proceeding instituted by another party against the registrant based on non-use, the burden of proof for genuine use also rests upon the trademark registrant. It is not enough for the registrant to merely assert that the mark that is the subject of the cancellation case is actually being used in commerce in the Philippines. Since a cancellation case is administrative in nature, Rule 2, Section 17 of the Inter Partes Rules provides that the quantum of evidence needed is substantial evidence.[54] Thereafter, the BLA shall decide the cancellation case on the basis of the pleadings, records, and evidence submitted as well as matters which may be taken up by judicial notice.[55]

## C. Relevant evidence

### *1. Use on the product or its packaging*

For goods, trademarks are traditionally used on the product itself, its container and/or its packaging. The IPO allows the submission of labels, tags, packaging, or photographs of the product clearly showing use of the mark on the packaging or on the product itself as evidence of actual use.[56] In view of the general rule that the mark must be used in the Philippines, the evidence of actual use must pertain to the same labels, tags, packaging being used in the Philippines.

For services, it is not possible to physically place or affix a mark by virtue of their intangible nature. Forms of evidence of actual use accepted by the IPO for services include literature, brochures, ads, downloaded pages from websites, and advertising materials showing that the services are being rendered in the Philippines.[57]

### *2. Use in advertising*

As discussed earlier, advertising by itself may not give rise to the acquisition of rights over a mark and is generally not acceptable as evidence of actual use to prove ownership. However, for the purposes of maintaining a trademark registration and

53. IPO Memorandum Circular No. 17-010, rule 208 (2017).
54. IPO Memorandum Circular No. 7, rule 2, sec. 17 (2016).
55. *Ibid.*
56. IPO Memorandum Circular No. 17-010, rule 210 (2017).
57. *Ibid.*

for submitting the DAU, brochures or advertising materials can serve as evidence of actual use.[58]

### 3. *Use on the internet*

In light of the developments in the field of information and communication technology, the IPO allows a trademark owner to submit evidence of use pertaining to online purchases or services. To be able to prove actual use online, the trademark owner must indicate in the DAU the website where the goods or services are available in lieu of the name or address of the establishment or outlet.

The Supreme Court in the case of *W Land Holdings, Inc. v. Starwood Hotels and Resorts Worldwide, Inc.*, declared that the 'use of a registered mark representing the owner's goods or services by means of an interactive website may constitute as proof of actual use that is sufficient to maintain the registration of the same.'[59] However, the Court qualified its pronouncement in this case by stating that the operation of an interactive website alone will not suffice as actual use; such use must be 'genuine', such that the trademark owner has transacted or has the intention of transacting with targeted customers of a particular jurisdiction in order to show that the mark is actually being used in the course of business in that country.[60] Ultimately, the Court ruled that the trademark owner was able to prove genuine or actual use of the mark in the Philippines based on the following facts and circumstances:

> 'In this case, Starwood has proven that it owns Philippine registered domain names, i.e., www.whotels.ph, www.wreservations.ph, www.whotel.ph, www.wreservation.ph, for its website that showcase its mark. The website is readily accessible to Philippine citizens and residents, where they can avail and book amenities and other services in any of Starwood's W Hotels worldwide. Its website also readily provides a phone number for Philippine consumers to call for information or other concerns. The website further uses the English language – considered as an official language in this country – which the relevant market in the Philippines understands and often uses in the daily conduct of affairs. In addition, the prices for its hotel accommodations and/or services can be converted into the local currency or the Philippine Peso. Amidst all of these features, Starwood's "W" mark is prominently displayed in the website through which consumers in the Philippines can instantaneously book and pay for their accommodations, with immediate confirmation, in any of its W Hotels. Furthermore, it has presented data showing a considerably growing number of internet users in the Philippines visiting its website since 2003, which is enough to conclude that Starwood has established commercially-motivated relationships with Philippine consumers.'[61]

---

58. *Ibid.*
59. *W Land Holdings, Inc. v. Starwood Hotels and Resorts Worldwide, Inc.*, G.R. No. 22366, 04 December 2017.
60. *Ibid.*
61. *Ibid.*

### 4. *Form of evidence*

The IPO accepts the following as actual proof of use of the mark:

- Labels of the mark as these are actually used in the Philippines;
- Downloaded pages from the website clearly showing that the goods are being sold or the services are rendered in the Philippines;
- Photographs of goods bearing the marks as these are actually used or of the stamped or marked container of goods and of the establishment(s) where the services are being rendered;
- Brochures or advertising materials showing the actual use of the mark on the goods being sold or services being rendered in the Philippines;
- Receipts or invoices of sale of the goods or services rendered or other similar evidence of use, showing that the goods are placed on the market or the services are available in the Philippines; and
- Copies of contracts for services showing use of the mark in the Philippines.[62]

### 5. *Distinction between use of a corporate name and use of the services offered by the company*

A trademark that appears only as a corporate or trade name can be also considered trademark use. In *Shangri-la v. Developers,*[63] one of the issues raised was that the marks SHANGRI-LA and the S logo were only being used as a corporate or trade name. The Philippine Supreme Court did not give much weight to this by ruling that the two concepts (corporate or business name and trademark) are not mutually exclusive. Thus, a company's use of a corporate name or trade name may be considered trademark use, if the name can be associated to a specific product or service. For example, evidence showing use of a hotel name (which is also a trademark) on the hotel website where clients can book their stay, will be accepted as evidence of actual use.

## IV. JUSTIFICATION FOR NON-USE

### A. Obstacles to use

If the trademark owner is unable to use the mark because of circumstances independent of his will, the owner may submit a Declaration of Non-Use (DNU) indicating therein the ground(s) and justification for the non-use of the mark. The DNU must state the facts and circumstances beyond the owner's control that prevent him from using the mark in commerce. Lack of funds is not an acceptable excuse.[64]

The Trademark Regulations gives the following grounds that justify the non-use of a mark:

62. IPO Memorandum Circular No. 17-010, rule 210 (2017).
63. *Shangri-la International Hotel Management, Ltd. et. al v. Developers Group of Companies, Inc.*, G.R. No.159938, 31 March 2006.
64. IPO Memorandum Circular No. 17-010, rule 211 (2017).

- The owner is prohibited from using the mark in commerce due to a requirement imposed by a government agency prior to putting the goods on the market or rendering of the services;
- A restraining order or injunction was issued by the BLA, the courts, or quasi-judicial bodies prohibiting the use of the mark; or
- The mark is subject of an opposition or cancellation case.[65]

The DNU must also indicate the steps being taken by the trademark owner to put the mark back in commerce as well as the date when use in commerce is expected to resume. The periods prescribed under the Trademark Regulations for filing the DNU are the same as the periods for filing a DAU:

- For pending applications and existing registrations, a DNU may be filed within (3) years from the filing date of the application,[66] extendible for another six (6) months upon the filing by the applicant or the registrant of a request of extension of time to file the DNU prior to the expiration of the three-year period.[67]
- When the use of a registered mark or a mark that is the subject of an active application has been interrupted or discontinued by a pending litigation, the DNU must be filed within the periods prescribed for filing the DAU.[68]

The existence of legitimate or justified reasons for the non-use of a mark does not imply that non-use can extend indefinitely. The Trademark Regulations expressly provide that except when the use of a mark is interrupted or discontinued by pending litigation, the non-use by the trademark owner shall not be allowed to exceed a period of more than six years. If the trademark owner fails to use the mark within one year from the fifth anniversary of registration, the IPO shall remove the mark from the Trademark Register.

For other reasons of non-use not stated in the Trademark Regulations, these will be examined by a trademark examiner who will make a determination whether the reason is independent of the will of the trademark owner. If the examiner determines that the facts and circumstances stated in the DNU do not show excusable non-use, the trademark examiner will deny the acceptance of the DNU and recommend the cancellation of the trademark registration.

The following cases illustrate acceptable reasons for not using a trademark:

*Romero v. Maiden Form Brassiere Co., Inc.*[69] In this case, the Supreme Court held that the non-use of the mark by the respondent Maiden Form Brassieres Co., Inc. did not affect its rights over the mark since the respondent's non-use was by reason of the Philippine Government's restrictions on the importation of the respondent's goods bearing the mark.[70] In considering the respondent's non-use as excusable, the Court

65. *Ibid.*
66. *Ibid.*, at rule 212.
67. *Ibid.*
68. *Ibid.*
69. *Romero v. Maiden Form Brassiere Co., Inc.*, 10 SCRA 556, 562 (1964)
70. *Ibid.*

observed that the mark's non-use was not permanent, voluntary, and intentional on the part of the respondent.[71]

*Philippine Nut Industry, Inc., v. Standard Brands Incorporated.*[72] In this case, one of the contentions of the petitioner Philippine Nut Industry, Inc. was that the respondent's use of the mark 'PLANTERS' was 'interrupted during the Japanese occupation and in fact was discontinued when the importation of peanuts was prohibited by Central Bank Regulations effective July 1, 1953, hence it cannot be presumed that it has acquired a secondary meaning.'[73] The Court reiterated the rule established in the *Romero* case that 'the non-use of a trademark on an article or merchandise due to legal restrictions or circumstances beyond one's control is not to be considered as an abandonment'.[74] The events that transpired during the respondent's non-use of the mark, specifically the Japanese occupation as well as the issuance by the Central Bank of a regulation prohibiting the importation of peanuts, were considered by the Court to be circumstances beyond the respondent's control.

### B. Circumstances in which non-use may be cured

The failure of the trademark owner to file the DAU or the DNU together with the evidence of actual use during the prescribed periods and the owner's inexcusable non-use of the mark are not curable under Philippine law. The effect of the trademark owner's failure to file the required DAU or DNU is summarized in the case of *Birkenstock* in this manner:

> 'Failure to file the declaration of actual use (DAU) within the requisite period results in the automatic cancellation of registration of a trademark. In turn, such failure is tantamount to the abandonment or withdrawal of any right or interest the registrant has over his trademark.'[75]

The IP Code[76] and the Inter Partes Rules[77] provide that the continuous and uninterrupted non-use of a mark by a registered owner for three years or longer is a ground for the cancellation of its registration. If the trademark owner has not used the mark for three years or longer, but begins using the mark prior to the institution by another party of a cancellation case against said mark, the law does not allow the trademark owner to cure his previous non-use of the mark. However, both the IP Code and the Inter Partes Rules do not prohibit the owner from filing a new application for registration of the abandoned or cancelled mark for non-use, provided that no conflicting marks have been registered during the interim period.

71. *Ibid.*
72. *Philippine Nut Industry, Inc. v. Standard Brands Incorporated*, 65 SCRA 575, 586 (1975).
73. *Ibid.*
74. *Ibid.*
75. *Birkenstock Orthopaedie GmbH and Co. KG (formerly Birkenstock Orthopaedie GmbH) v. Philippine Shoe Expo Marketing Corporation*, 710 SCRA 474, 485 (2013).
76. Intellectual Prop. Code, sec. 151.1 (c).
77. IPO Memorandum Circular No. 7, rule 8, sec. 2 (c) (2016).

## V. CONSEQUENCES OF NON-USE

### A. Consequences of non-use and relevance of the context

Should a trademark owner fail to use the mark within the prescribed periods under the Trademark Regulations, the following are the possible consequences provided by law:

- The trademark application shall be refused registration by the IPO, in case the application for registration of the mark is still pending by the time the DAU or DNU becomes due and the trademark applicant fails to submit evidence of actual use or a DNU.
- The trademark registration shall be deemed automatically cancelled by the IPO, in case the mark is already registered with the IPO and the trademark owner fails to file the DAU or DNU.
- The trademark registration shall be susceptible to a cancellation action by interested third parties, if the mark has not been used by the trademark owner for an uninterrupted period of three years or longer without legitimate reason.[78]

Should non-use of the mark occur prior to, or at the time of, filing the trademark application and before the DAU falls due, such non-use does not amount to an abandonment of the mark. The IP Code does not require actual use of the mark prior to the filing of a trademark application, and allows the submission of a trademark application based on intent to use. Thus, as long as the trademark applicant is able to submit evidence of use during the period of three years counted from the filing date of the application (or within the extension period), prior non-use would not result in the cancellation or removal of the mark from the Trademark Register.

In opposition proceedings, the party opposing the mark is not allowed to oppose an application for registration of a mark based on non-use since it is not a valid ground for filing an opposition case. Consequently, non-use during the institution of opposition proceedings is immaterial to the resolution of the opposition case.

In cancellation proceedings, if the registrant fails to use the mark for an uninterrupted period of three years or longer without legitimate reason, any interested third party may file a cancellation case against the registration of the mark. Should the cancellation case succeed, the BLA shall order the cancellation of the registration of the mark. Upon finality of the order or judgment, the registrant loses any right conferred by such registration.[79]

### B. Extent of any revocation

If a trademark is not used at all, then it will be removed from the Trademark Register. However, if a trademark covers several classes, some of which have use, then only those classes that have no use will be cancelled. This can be inferred from Rule 209 of the Trademark Regulations, which states that a subsequent DAU is allowed in case the first submitted DAU fails to cover all the classes. However, if no subsequent DAU is filed,

78. Intellectual Prop. Code, sec. 151.1 (c).
79. *Ibid.*, at sec. 154.

then the classes that are not supported by the first DAU will automatically be dropped from the application or registration without need of notice to the trademark owner.

## C. Timing for any revocation

A petition for cancellation based on non-use of a mark for three years can be filed at any time.[80] As for those trademarks where no DAU was filed, removal from the Trademark Register or refusal of the trademark application is automatic and occurs the day after the deadline has lapsed.

80. *Ibid.*, at sec. 151(c).

# Poland

*Bartosz Krakowiak and Klaudia Błach-Morysińska*[1]

## I. INTRODUCING GENUINE USE OF TRADEMARKS

### A. Legal framework

Poland is a signatory to all of the major international agreements on protection of trademarks, including the Paris Convention for the Protection of Industrial Property, the Agreement on Trade-Related Aspects of Intellectual Property Rights (TRIPS), the Madrid Agreement Concerning the International Registration of Marks and the Protocol relating to that Agreement, as well as the Singapore Treaty on the Law of Trademarks. As an EU Member State, Poland is also part of the EU trademark system and is required to harmonize its national trademark laws with the relevant EU directives, including the recast EU Trademark Directive (2015/2436).

The most important Polish Act related to trademarks is the Industrial Property Law of 2000 ('the IPL').[2] Under said Act genuine use is a requirement for maintaining protection for a trademark. However, the obligation of using a trademark, as such, is not specified in the Act. It results, indirectly, from the provisions of the IPL which provide for certain sanctions and limits, i.e. legal consequences of a trademark non-use.[3]

To avoid being exposed to said consequences, a registered trademark must be put to genuine use for the goods or services covered by its registration within a period of five years following the registration date, unless there are proper reasons for non-use. What is understood as use of a trademark is defined in the IPL, however, there is no definition of genuine (or actual) use as such. As far as this requirement

1. I–II.A, III.A and V by Bartosz Krakowiak; II.B–F, III.B–C and IV by Klaudia Błach-Morysińska, PhD candidate at Polish Academy of Science.
2. The Act of 30 June 2000 – Industrial Property Law (O.J. of 2017, item 776).
3. See also the Polish Group's report for the AIPPI Study Question Q218, *The requirement of genuine use of trademarks for maintaining protection* (2011), co-authored by M. Fijałkowski, J. Fiołka and B. Krakowiak (available at www.aippi.org).

is concerned, the Polish Patent Office and the competent Polish courts follow the principles resulting from the well-established EU case law, though there are also some Poland-specific nuances.

Some important changes in respect of the provisions of the IPL related to the consequences of non-use are envisaged within the framework of transposing the recast EU Trademark Directive (2015/2436) into Polish national law. The draft law amending the IPL was first made publicly available in December 2017 and, following an extensive consultation process, it is expected to be adopted by the Parliament by the end of 2018.[4]

## B. Purpose and function of genuine use: policy considerations

Regarding justification of the genuine use requirement, Polish law follows the EU law, where use of trademarks is considered the essential justification for their protection. It is also stressed that the ability to remove unused trademarks from the Register is necessary in order to facilitate the acquisition of new marks and reduce the number of potential conflicts. Therefore, one of the main purposes of the genuine use requirement is to keep the Register uncluttered and to thereby allow for new proprietors (including competitors) to make use of a 'limited' supply of possible marks.[5] Making non-used trademarks available to competitors is necessary for maintaining the freedom of business activity and free competition on the market.

It is further argued that the genuine use requirement and the corresponding sanctions provided for in law aim at eliminating the risk of 'simulated' performance of the exclusive rights resulting from trademark registration or even the risk of abusing these exclusive rights, contrary to their actual function (by blocking competitors' access to certain non-used designations).[6] It is stressed that the exclusive rights resulting from trademark registration, if not supported with genuine use of the mark, are devoid of any economic sense.[7] The sanctions for non-use of a trademark also aim at eliminating legal uncertainty as to the status and protection of trademarks which are kept in the Register but which are not used for a considerable period of time.

4. Comments in this chapter are based on the official draft law published for the consultation purposes on 6 December 2017 and further discussed at the governmental level (with an active participation of the Polish Group of AIPPI).
5. Judgment of the Supreme Administrative Court (further: NSA) of 21.10.2012, II GSK 1646/11, where NSA refers to the 'cleansing' function of the genuine use requirement.
6. Judgment of the Administrative Court of Warsaw (further: WSA) of 7.02.2006, VI SA/Wa 1749/05.
7. M. Trzebiatowski, *Obowiązek używania znaku towarowego – Studium z prawa polskiego na tle prawnoporównawczym* (Warszawa 2007), 51 and following (with the summary of opinions expressed by M. Kępiński, R. Skubisz, U. Promińska, A. Kowalczyk, M. Modrzejewska and S. Sołtysiński); M. Trzebiatowski, in R. Skubisz (ed.), *System Prawa Prywatnego 14B – Prawo własności przemysłowej* (Warszawa 2017), 961-962; K. Jasińska, in P. Kostański (ed.), *Prawo własności przemysłowej – Komentarz* (Warszawa 2010), 858-859.

## II. WHAT QUALIFIES AS GENUINE USE OF A TRADEMARK

### A. Nature of use

#### 1. *Use consistent with the function of the trademark*

In accordance with the well-established EU case law, also in Poland genuine use of a trademark means that the mark is used in accordance with its essential function, which is to guarantee the identity of the origin of the goods or services for which it is registered, in order to preserve an outlet for those goods or services. The commercial exploitation of the mark must be real and the use must be viewed as warranted to maintain or create a share in the market for the goods or services protected by the mark.

Polish administrative courts also interpret genuine use in connection with the essential function of trademarks (i.e. to identify the origin of the goods or services), which can be fulfilled only where the mark is put on goods and where such goods are directed to their consumers on the market.[8] A trademark is capable of guaranteeing the identity of the origin of the goods only where these goods are actually put into trade and consumers have actual contact with the trademark (it means that trademark use must always have a connection – physical or at least conceptual – with certain product or service).[9]

According to the above, only use 'as a mark' can constitute genuine use as required by Polish law. This means that a trademark must be used, in the course of trade, as a sign distinguishing the goods or services of the proprietor from those of others, i.e. in accordance with its essential function. Consequently, use of a mark merely as a business/trade name (i.e. to identify the entrepreneur rather than the goods or services) is not sufficient to establish genuine trademark use.[10] This is so because genuine use of a trademark requires putting the trademark on the goods or packaging thereof and not (not only) its use in order to identify the entrepreneur and his business activity.[11] Therefore, for example, the sale of various goods in a store the name of which corresponds with the trademark cannot, as such, constitute genuine use of that trademark (if registered only for goods and not for retail services). Further, if the trademark is used in invoices only as a part of the seller's business/trade name, this is not considered use 'as a mark' and, consequently, this is not sufficient to establish genuine trademark use. A more careful approach is taken in relation to service marks, due to the 'immaterial' character of services.[12] It is

8. WSA 21.03.2006, VI SA/Wa 1653/05 and 21.02.2007, VI SA/Wa 1166/06.
9. NSA 24.05.2006, II GSK 70/06.
10. NSA 8.05.2014, II GSK 375/13; WSA 27.02.2015, VI SA/Wa 2015/14; WSA 18.01.2016, VI SA/Wa 1857/15 and WSA 18.05.2016, VI SA/Wa 3272/15.
11. M. Trzebiatowski, *Obowiązek używania znaku towarowego*, supra n. 7, 104-105; M. Trzebiatowski, in *System Prawa Prywatnego*, supra n. 7, 970.
12. Accordingly, it is noted that service marks are, by their nature, used in the manner attributed to business names (NSA 10.01.2002, II SA 3390/01); however, use of a service mark only in its owner's business name does not alone meet the genuine trademark use requirement (WSA 15.10.2010, VI SA/Wa 1425/10).

believed, however, that genuine use of a service mark also requires actual provision of services under that mark and use of the mark in order to identify the services, not only the service provider.

As far as the use of a trademark as a title is concerned, where the title (e.g. press title) corresponds with registered trademark, mere appearance of that title on a product (e.g. newspaper) is not sufficient to establish genuine use of the corresponding trademark. In such situations it is always necessary to indicate that such use of the title can be simultaneously perceived by the public as use aimed at designating the product's commercial origin.[13]

Also mere decorative use of a trademark, i.e. not aimed at distinguishing certain goods or services as to their commercial origin but at making a product visually more attractive, despite its possible promotional impact, does not fulfil the essential function of trademarks and consequently cannot amount to genuine trademark use.

## 2. *Use in the course of trade*

### *i. Public use*

Genuine use means actual and serious use in the course of trade, which has unambiguous character. A trademark must be objectively used on the market in a manner which is real, durable and stable. Such use should have more or less serious impact on the market position of the trademark owner, from the perspective of the goods or services under his trademark.

It results from the above that only use which has public character, i.e. which is made outside the trademark owner's enterprise, is legally relevant. Otherwise, without the consumers' contact with the trademark (physical or at least conceptual), its use could not have any commercial significance required to deem such use genuine.[14] However, said use does not always need to be directed to final consumers. It is also acceptable if it is directed to the trademark owner's outside clients, including other (separate and independent) entrepreneurs (e.g. wholesalers, trade agents).[15] Use of a trademark solely within internal activities of the trademark owner and in connection with his own purposes (including in relations with the trademark owner's subsidiaries, affiliates, representatives, etc., even where such relations involve sale of goods) is not considered use in the course of trade and, consequently, it cannot be treated as genuine trademark use.[16]

---

13. M. Trzebiatowski, *Obowiązek używania znaku towarowego*, supra n. 7, 105 .
14. Judgment of the Supreme Court (further: SN) of 17.02.2005, I CK 626/04.
15. M. Kępiński in his commentary to the judgment of SN 17.02.2005, I CK 626/04; M. Trzebiatowski, *Obowiązek używania znaku towarowego*, supra n. 7, 94-96.
16. M. Trzebiatowski, *Obowiązek używania znaku towarowego*, supra n. 7, 95 and 100; M. Trzebiatowski, in *System Prawa Prywatnego*, supra n. 7, 971; WSA 11.12.2012, VI SA/ Wa 1732/12.

### *ii. Commercial purpose*

Only the activities taken in the course of trade, in connection with certain goods or services made available on the market, are relevant for the genuine use requirement.[17] This also implies the commercial nature of such activities, in alignment with the trademark owner's business, which however does not require that they are in each case taken for profit.[18]

As far as trademark use in advertising is concerned, there is a provision in the IPL which stipulates that use of a trademark in advertising of the product (or service) which is neither available on the Polish market nor is manufactured in Poland for export purposes shall not be deemed to constitute genuine use.[19] It means that mere advertising is not sufficient to establish genuine trademark use if the mark is not put on the goods effectively present on the market (or at least manufactured for export purposes). This is irrespective of the fact that use in advertising is listed in the IPL as one of the forms of trademark use covered by the proprietor's exclusive rights. Accordingly, Polish administrative courts confirmed several times that use of a trademark only for advertising or promotional purposes, e.g. in newspapers, leaflets or promotional lotteries, and not in order to identify the goods as originating from certain entrepreneur (e.g. by putting the mark on the goods available on the market or packaging thereof), cannot constitute genuine use of the trademark in the course of trade.[20]

Regarding use of trademarks in merchandising, if such use serves only promotional purposes, the same approach should be taken as with the use in advertising. By way of example, if various articles bearing a trademark (pencils, T-shirts, candles, etc.) are given on some occasions to commercial partners or clients as gifts only, this does not constitute genuine trademark use. On the other hand, where various merchandising products are actually offered and put on the market, the marks used on them – even if they correspond with movie titles, names or logos of pop bands, sports teams or events, etc. – can be deemed to be genuinely used 'as a mark', i.e. (among some other functions) in order to guarantee the identity of the origin of the goods for which they are registered.[21] It is, however, not decisive whether the goods are distributed for free or only against payment, as long as the goods are genuinely present on the market.

As far as use of trademarks on the internet is concerned, it is believed that the assessment should always depend on the function in which the trademark is actually used on the internet, which can only be found after careful analysis of the respective website's purpose and content.[22] In particular, it is necessary to establish

17. NSA 10.10.2006, II GSK 173/06.
18. The respective provisions of the IPL, when defining the scope of the exclusive trademark owner's right, refer to the term "use for profit or for professional purposes".
19. Article 169(5) IPL.
20. NSA 24.05.2006, II GSK 70/06; WSA 12.10.2005, VI SA/Wa 937/05.
21. It is, however, important that the goods for which genuine use is to be established, as per the trademark registration, are the merchandising products as such (e.g. pencils, T-shirts, pendants) and not other goods or services to which the merchandising products only relate or refer to (e.g. by promoting them).
22. M. Trzebiatowski, *Obowiązek używania znaku towarowego*, supra n. 7, 137 and 153-155.

that certain goods or services are offered or rendered under the trademark used on the website and that such goods or services are directed to consumers in Poland. It is not sufficient where the website has only promotional character – in such situation the use of a mark should be treated as use in advertising only. Further, it is not sufficient where the mark is used only in various offers or tariffs available on the website, where no goods or services (under the trademark) are actually available to consumers in Poland (e.g. through the website). Finally, mere use of a trademark in domain names, keywords or meta-tags, not corresponding with actually offered/sold/rendered goods or services, is not considered genuine use 'as a mark', even if such use has commercial and promotional character (i.e. is aimed at attracting the consumers' attention to certain website). On the other hand, it is believed that use of a trademark merely on the internet can constitute genuine use and the same criteria should be applied as to the 'traditional' forms of use. In particular, establishing a direct connection between the trademark and certain goods or services offered/sold/rendered through the website is a must. Such goods or services must be available to consumers in Poland (commercial effect in Poland), who must have an actual possibility of acquiring them through 'real' transactions (even if carried out virtually).

Regarding use of a trademark only in relation to goods to be exported, the IPL clearly provides that genuine use of a trademark shall also mean the use by affixing the mark to goods or the packaging thereof solely for export purposes.[23] It results from that provision that the genuine use requirement is satisfied only when the mark is actually affixed to exported goods, or the packaging thereof, on the territory of Poland ('domestic' use of the mark in Poland, without putting goods on the Polish market), even if the export as such is carried out by another entity, not necessarily by the trademark owner himself.

As far as various actions taken in preparation for actual use of a trademark in the course of trade are concerned (not only those related to promotion), it is stressed that only putting goods with a trademark on the market or rendering services under a trademark can constitute genuine trademark use. Therefore, as a principle, mere preparatory actions cannot be deemed to meet the genuine use requirement.[24] As long as the products or services are not available to consumers in Poland and consumers have no actual contact with them (even non-physical), there is no commercial effect and, consequently, no genuine use of a trademark in commerce can be established (with the exception of goods solely for export purposes).

23. Article 169(4)(2) IPL.
24. M. Trzebiatowski, *Obowiązek używania znaku towarowego*, supra n. 7, 142 and following – the author suggests that, under certain circumstances, the situations where various actions have been taken in preparation to actual trademark use (including those of obligatory character, e.g. in order to meet regulatory or other administrative requirements), but no actual use has been started yet, can be assessed from the perspective of proper reasons for non-use.

### 3. *Use by the owner or with its consent*

Under the IPL, genuine use of a trademark shall also mean the use of a mark by a third party with the right holder's consent.[25] There is no difference made between the use of a trademark by its proprietor, licensees, distributors or other third parties having appropriate (i.e. prior and explicit) consent from the proprietor. There are no additional requirements for such parties, as long as their use – as agreed with the trademark owner – can be deemed genuine.[26] Mere entering into a licence or similar agreement is not sufficient to establish genuine use if such use has not been effectively initiated by the third party.[27] Recording of a licence in the trademark Register is not required for the purpose of attributing the trademark use by the licensee to the trademark owner, though it is always advisable to have such licence recorded for evidence purposes.[28] Also quality control by the trademark owner over the goods or services offered by a third party under his trademark and with his consent does not appear to be necessary from this perspective, though it is believed that such third party's use cannot be absolutely independent of the trademark owner.[29]

## B. Volume, duration and frequency of use

It is not possible to assume in advance which quantitative threshold should be taken to recognize that the use was of genuine character. The genuine use should always be examined in the context of a particular case. For this reason, in various cases, due to all circumstances, the actual nature of the use of a trademark can be determined in different ways. The turnover, volume of sales of the product, frequency or regularity of the use of the trademark shall always be assessed in relation to all the other relevant factors, such as: (i) the volume of business; (ii) production or

25. Article 169(4)(3) IPL.
26. M. Trzebiatowski, *Obowiązek używania znaku towarowego*, supra n. 7, 254 and following; M. Trzebiatowski, in *System Prawa Prywatnego*, supra n. 7, 989-991 – the author proposes to apply very strict criteria for assessing the trademark owner's consent to use his trademark by a third party, which, according to the author, must always have a form of written agreement (e.g. licence, distributorship, agency, dealership, franchising, leasing), so no implied consent (*per facta concludentia*), for example resulting from mere tolerating of a third party's use, should be acceptable (the same concerns *ex post* declarations confirming the trademark owner's consent – in practice, acceptable by the Polish Patent Office); WSA 29.02.2016, VI SA/Wa 2048/15.
27. WSA 21.12.2005, VI SA/Wa 1315/05.
28. WSA 26.01.2005, VI SA/Wa 285/04; a different view presented in M. Trzebiatowski, *Obowiązek używania znaku towarowego*, supra n. 7, 258 – the author considers entry of a licence to the trademark Register necessary, with the exception of situations which involve various structural and corporate relations between the parties (holdings or other groups of companies).
29. M. Trzebiatowski, *Obowiązek używania znaku towarowego*, supra n. 7, 258-259; M. Trzebiatowski, in *System Prawa Prywatnego*, supra n. 7, 990 – the author, rightly making a reference to Article 19(2) of TRIPS (explicitly providing for the control requirement), considers the trademark owner's control over third party's trademark use, based on an appropriate written agreement, a must.

marketing capacity; (iii) the degree of diversification of the undertaking using the trademark; (iv) characteristics of the products or services on the relevant market.

Use need not always be quantitatively significant for it to be deemed genuine, as that depends on the characteristics of the goods or services concerned on the corresponding market. The same refers to the frequency of use. In a case regarding a trademark registered for cosmetics by small entity, the court said that sales of 4000 goods within 10 months is enough to prove genuine use.[30] Presenting 42 invoices and offer of inquiries within the term of three years was enough to prove the genuine use for transport services served by one-person entity.[31] On the other hand, the sale of dozen of shirts, few skirts and few jackets covered with trademark by the entity who sells a large amount of goods to department stores does not constitute genuine use.[32]

## C. Use of a sign

To bring the law into the line with business, the IPL allows use of the trademark in trade in a form different from the form in which it was registered on condition that the differences refer to elements, which do not alter the distinctive character of the mark in the form in which it was registered.[33] The purpose of this regulation is to avoid imposing strict conformity between the used form of the trademark and the form in which the mark was registered. It allows the proprietor to make variations in the mark, which, without altering its distinctive character, enable it to be better adapted to the marketing and promotional requirements of the goods or services concerned.

The comparison shall be made between trademark as used and the registered trademark, while the trademark as registered shall be the reference point. First of all the distinctive element in the registered trademark shall be defined. Then the distinctive or dominant character of the added or omitted elements in the used trademark shall be assessed. As the final step, the interdependence between the strength of the distinctive character of the trademark and the effect of alterations shall be examined.

As a general rule, an alteration or omission of the distinctive element shall be treated as an alteration of the trademark's distinctive character. The same refers to replacement of the distinctive element by another element. Considering word trademarks, the alteration of typeface, use of upper/lower case or colour shall not alter the trademark.[34] As to the figurative trademarks with no word elements, any alteration may lead to the change of the distinctive character. If the figurative trademark includes additional word element, changes to certain non-distinctive figurative elements do not normally affect the distinctive character of the marks. The use of a 3D mark in varying sizes usually amounts to use of the mark as registered. The addition of a word/figurative element to such a mark does not generally alter the distinctive character of the sign. Colour marks must be used with the colours as

30. WSA 02.02.11, VI SA/Wa 1794/10.
31. WSA 18.01.2016, VI SA/Wa 1875/15.
32. WSA 26.03.2015, VI SA/Wa 3375/14.
33. Article 169(1)(4) IPL.
34. WSA confirmed that word trademarks may be used in any form (WSA 18.01.2016, VI SA/Wa 1875/15);

registered. Insignificant variations in the colour shade and strength shall not change the distinctive character of the trademark.

Historically, in the case of two registered trademarks, the use of one trademark could not constitute a genuine use of the other trademark.[35] Nowadays, this statement has been changed to reflect EU case law. It is generally accepted that addition of elements to mark or use of a mark in conjunction with another registered trademark or as part of a complex registered trademark may constitute the genuine use.[36]

Historically, black and white registrations were deemed to covered all colour variations of the trademark. Currently, change only in colour does not alter the distinctive character of the trademark, as long as: a) the word/figurative elements coincide and are the main distinctive elements; b) the contrast of shades is respected; c) colour or combination of colours does not possess distinctive character in itself; and d) colour is not one of the main contributors to the overall distinctiveness of the mark.[37]

## D. Use in relation to goods and services

### 1. *Use for each of the designated goods and services*

In accordance with the principle of specialty, a trademark shall be used for the goods and services for which it was registered. To keep the protection for all goods and services for which trademark was registered it is necessary to use the trademark for all of them.[38]

The jurisprudence admits that if the trademark is registered for a wide category of goods or services, which is defined with one category and it is used just for some subcategory of these goods or services, it is acceptable to revoke trademark protection due to non-use for the other subcategories. Doubts may arise with regard to the extent to which the subcategories can be subdivided and revoked from the protection.[39]

35. NSA 24.05.2006, II GSK 70/06 – court decided that use of the registered word trademark 'Naomi Campbell' was not treated as use of registered word trademark 'Naomi'; NSA 26.10.2006, II GSK 190/06, where use of the registered word trademark 'Twoje Dziecko' was not treated as use of registered word trademark 'Dziecko'.
36. WSA 23.09.2014, VI SA/Wa 3386/13.
37. Poland participates in EUIPO Common Practice including CP4 regarding 'Scope of Protection of Black and White (B&W) Marks'.
38. This obligation arises from Article 171(1) IPL which says that where the reason for the lapse of a trademark protection right refers solely to certain goods, the lapse of the right refers exclusively to those goods.
39. NSA approved the partial cancellation of the trademark registered for alcohol beverages except wines, as the trademark owner provided proof of use for wine only. The trademark was registered in class 33 for alcohol beverages. The trademark owner was the wine producer and presented proof of use for wine. The Polish Patent Office made distinction between the alcoholic beverages on high, middle and low percentage and decided that wine is middle percentage alcohol. As no evidence was provided for other alcoholic drink categories, the Polish Patent Office decided to cancel the trademark in this scope (16.04.2009, II GSK 708/08). In another case NSA decided to divide between spices in terms of multi-component and one-component. Presented proof of use referring just to

## 2. *Use for a broader category of goods and/or services*

The trademark shall be used for the goods for which it was registered. It is not possible to fulfill the trademark use obligation for the registered goods and services by using the trademark for the goods or services, which are not included in the list of goods and services. Use of the trademark for a broader category of goods or services than registered cannot be treated as use of the trademark. The main argument for the strict interpretation is that, unlike the regulation on use of the trademark in different form than it was registered, the IPL provisions do not allow use of the trademark for the goods or services different from those for which the trademark was registered.

The trademark owner defines the list of goods and services in the trademark application and it sets the limits of trademark protection. If he uses the trademark for other goods than registered, even if similar to that registered, he uses it beyond the trademark protection. It cannot be treated as fulfilling the obligation of use of the trademark.[40] There is a concept of the extended use effect where the use of the trademark for some goods may be treated as use of the trademark for similar goods from the same Nice class. Nevertheless Polish courts do not approve it.

## E. Territorial scope

Historically[41] it was demanded that the trademark be used on the territory of Poland. If the trademark was used in an advertisement, it was required that it was an advertisement published in Polish mass media. The production of goods in Poland for exportation purposes was not treated as use of the trademark. It was crucial that consumers became familiar with a trademark in Poland.

Nowadays, the IPL provisions do not stress that the use of trademark must take place in the territory of Poland, but the territorial character of the trademark determines that trademark must be used in the territory where it is protected. The use of the trademark does not have to cover the entire territory of country.[42] The territorial scope of use of the trademark is only one of the aspects of genuine use. It is enough if a trademark is used in a part of Poland.[43] Nevertheless, due to the economic aspects, the use of the trademark on partial territory of Poland is not enough to prove genuine use. It is important that the use is clear in the light of the other aspects of genuine use.

---

one-component spices does not entitle the trademark owner to keep the registration for other subcategories (31.01.2012, II GSK 1502/10). Similarly, WSA decided that the Polish Patent Office shall separate bags with wheels from bags for which the protection right was issued (WSA, 16.12.2015, VI SA/Wa 1936/15).

40. NSA, 08.05.2013, II GSK 375/13.
41. Based on the Act on Trademarks of 31 January 1985 (O.J. of 1997, item 554 with amendments).
42. Nevertheless, in a case regarding the development and maintaining of the website, the Polish Patent Office and WSA underlined that offering services in different regions of Poland was important for the genuine use (WSA, 18.10.2016, VI SA/Wa 2125/16).
43. In a case regarding home building investment services, WSA confirmed that local use of the trademark shall be treated as the genuine use (WSA, 18.05.2016, VI SA/Wa 3272/15).

The goods and services with the trademark shall reach consumers in Poland.[44] The only exception is for goods solely for export purposes. The exception shall be interpreted narrowly. The goods with the trademark shall be provided to the foreign consumers, but affixing of the trademark to the goods must take place in Poland.

No special rules shall be applied for the use of a trademark on the internet. As to the use of a trademark on the internet or in the foreign mass media, it is crucial that goods are available for Polish consumers. There are many aspects that may be examined in this kind of use of the trademark, but the most important is that goods and services offered or advertised on the internet or in the foreign press are the subject of the transaction with Polish consumers. Important aspects are for example: the language of the offer, currency, information on delivery to Poland, etc.

## F. Relevant time period and its calculation

Registered trademarks shall be put to genuine use for the goods covered by the protection right within five years of the date of issuing the decision on granting a trademark protection right, unless proper reasons for not using the trademark exist.[45] For international registrations the five-year term counts from the date of advertising the recognition of said protection in the Patent Office's Official Gazette.[46] Nevertheless, the planned amendments to the IPL will change the relevant date for international registrations. The five-year period will be counted from: i) the date when the international trademark can no longer be provisionally refused in accordance with the Madrid Agreement or Protocol; ii) the date when the international trademark can no longer be opposed in accordance with the IPL; or iii) the date when a decision ruling on the grant of protection, following either a provisional refusal based on absolute grounds or an opposition, became final. Use does not have to be made throughout the period of five years, but rather within the five years. The five-year period of non-use shall be continuous.

The revocation request of the trademark due to non-use shall be rejected when it is filed before the expiration of the five-year period after the registration of the trademark. The Polish Patent Office shall also refuse a request to declare the revocation of a trademark, if before making the request the genuine use of the trademark was commenced or resumed. However, if the trademark owner started or resumed the use of the trademark after the expiry of a continuous five-year period of not using the mark and within a three-month period preceding the making of the request to declare the lapse of a trademark, the Polish Patent Office shall not take it into consideration if the preparations to commence or resume use were started after the proprietor had become aware that the request would be made.[47]

The three-month period and awareness of the upcoming revocation request conditions shall be fulfilled jointly. If any of the above conditions is not fulfilled the

44. It is confirmed by Article 169(5) IPL – 'The genuine use of a trademark does not mean the using of the trademark to advertise goods that are not available on a domestic market or are not made in the country for exportation purposes.'
45. Article 161(1)1 IPL.
46. Article $152^{15}$ IPL.
47. Article 170(1-2) IPL.

commencement of the use shall break the five-year non-use period. Awareness of an upcoming revocation request may come from the owner of later trademark suggesting the clash between trademarks. A revocation request filed by third person that has not warned the trademark owner or after three-month period from commencement or resumption of the trademark use shall not cancel the period of non-use.

The commencement or resumption of use of the trademark may take place after several five-year non-use periods. In such a situation, the Polish Patent Office shall not consider the previous five-year non-use periods, as the commencement of use cancels the period of non-use.[48]

Justified non-use of the trademark does not restart the five-year non-use period. It only suspends the terM. Once the justifiable reasons are gone, the five-year non-use period resumes. The sale of the trademark does not break the non-use period.[49]

As for the date of expiry of the protection right, please see section V.C below.

## III. EVIDENCE OF GENUINE USE OF A TRADEMARK

### A. When use should be proved: timing of lack of genuine use claims

#### *1. When filing an application for a trademark*

Polish law does not provide for an intention-of-use or actual use requirement applicable at the time of filing or registration of a trademark. Further, no declarations of use are required after trademark registration, e.g. upon renewal of protection. Accordingly, a trademark owner can be obliged to submit a proof of use only at the request of another party, in the course of cancellation, opposition, invalidation or enforcement proceedings. The problem whether filing of a trademark without any intention to use it (including so-called repeated applications, filed in order to circumvent the genuine use requirement) can be treated as an application made in bad faith has not been ultimately decided in Poland.[50]

#### *2. In the course of opposition proceedings*

The existing pre-grant opposition procedure was introduced to the IPL with the effect as from April 2016. So far, there has been no case law related to the genuine use requirement in connection with the 'new' opposition proceedings. The IPL clearly states that, in defence to the opposition, the trademark applicant may bring up an objection (claim) concerning the lack of genuine use of the trademark on which the opposition is based, within an uninterrupted five-year period preceding the application date of the opposed trademark, for the goods or services covered by the

48. M. Trzebiatowski, *Obowiązek używania znaku towarowego*, supra n. 7, 345.
49. NSA 25.02.2015, II GSK 117/14; NSA 15.06.2005, II GSK 84/05.
50. According to some commentators, under certain circumstances, such filings – especially if intended to circumvent the genuine use requirement or solely to block competitors' access to certain designation – can be treated as made in bad faith (M. Trzebiatowski, *Obowiązek używania znaku towarowego*, supra n. 7, 97 and 442-444).

opposition, unless there are proper reasons for non-use or the earlier trademark has been registered for less than five years. In case the non-use objection (claim) is found by the Patent Office to be justified, the opposition shall be rejected.[51]

It is believed that, despite its literal meaning, the expression 'for the goods (services) covered by the opposition' (i.e. the goods (services) contained in the opposed trademark application) should be logically understood as the goods (services) covered by the registration of the earlier trademark, to the extent on which the opposition is based on theM. It is also believed (though not explicitly stated in the respective provision) that in case the non-use objection (claim) is found only partially justified (i.e. in respect of some goods or services only), the opposition shall be rejected only in part, depending on the assessment of similarity (and other relevant factors) with the goods or services for which the use has been duly proven. In other words, in such cases the earlier trademark shall, for the purpose of the examination of the opposition, be deemed registered in respect of only that part of goods or services in relation to which it has been genuinely used.[52]

The above principles are also applied where the earlier trademark is an EU trademark. In such a case, the genuine use of the EU trademark shall be determined in accordance with the EU Trademark Regulation (1001/2017).[53] For the time being, this is the only situation provided for in the law where the Polish Patent Office is entitled to examine the genuine use of an EU trademark.

### 3. *In cancellation proceedings based on lack of use*

Following a third party's cancellation request filed with the Patent Office, trademark protection can be revoked due to lack of genuine use of a registered trademark for the goods or services covered by its registration for an uninterrupted term of five years after the registration date, unless there are proper reasons for non-use. If such cancellation proceedings before the Patent Office are initiated, the obligation to demonstrate use of the trademark or the existence of proper reasons substantiating non-use lies with the trademark owner.[54] As far as international trademark registrations designating Poland are concerned, the aforementioned five-year period starts running from the date of publication of the grant of protection for such trademark in the Patent Office's Official Gazette.[55]

Until April 2016, under the IPL there existed one additional requirement to effectively file a cancellation application based on non-use, which had a tremendous practical significance. Namely, one had to demonstrate its legitimate interest in filing

51. Article 15219(4) IPL.
52. A. Szewc/M. Mazurek, in R. Skubisz (ed.), *System Prawa Prywatnego 14B – Prawo własności przemysłowej* (Warszawa 2017), 607.
53. This means that it is not required to establish the genuine use of the earlier EU trademark in the territory of Poland – according to the relevant EU case law, the assessment of whether an EU trademark has been put to genuine use in the EU must be made irrespective of the territorial borders of the Member States (CoJ EU judgment in the case C-149/11, *Leno Merken*).
54. Articles 169(1)(1), 169(2) and 169(6) IPL.
55. Article $152^{15}$ IPL (as to the planned amendments to the IPL in this respect please see section II.2.F above).

such application. Said interest was understood as legal and not only factual (but needed to be reflected in facts of the case); it had to be material and objectively justified, as well as individual, specific, current, direct and real.[56] If the legitimate interest of the cancellation applicant was not properly established, the Patent Office was under the obligation to reject the cancellation application even if the use of the attacked trademark had not been proven whatsoever.

### 4. *As a defence in infringement/cancellation proceedings*

A non-use defence (claim) in the event of invalidation proceedings does exist under the IPL. Namely, a trademark registration may not be invalidated on the sole ground that the trademark is similar to an earlier trademark where the latter has not been genuinely used, for the goods or services covered by its registration, for a period of five successive years. In such case, a claim that the earlier trademark has not been used may be raised only together with an application for cancellation of that mark, which shall be examined together with the application for invalidation.[57] In other words, a defendant in invalidation proceedings is entitled to initiate cancellation action due to non-use of the earlier trademark, assuming the earlier trademark has been registered for over five years. However, such cancellation action is not dealt with separately but is formally treated as part of the invalidation proceedings.[58] In practice, the Patent Office needs to first examine the non-use claim and only then the invalidity grounds, provided that the genuine use of the earlier trademark has been properly established. If there are grounds to revoke the earlier trademark registration due to non-use, the Patent Office revokes it (or its respective part) and rejects the application for invalidation.[59]

The initial draft law aimed at implementation of the recast EU Trademark Directive (2015/2436) did not provide for modification of the aforesaid rule according to which a non-use claim in the invalidation proceedings could only be raised together with an application for cancellation of the earlier trademark. However, in the course of discussions following the initial draft presentation, it was made clear that such solution is not in line with Article 46 of the Directive, which in consequence led to further changes in the respective provisions of the draft. As a result, when said changes are finally adopted, a non-use claim will be treated as a defence raised in the course of invalidation proceedings, as is now the case with a non-use objection in the opposition proceedings (see above).[60]

---

56. M. Trzebiatowski, *Obowiązek używania znaku towarowego*, supra n. 7, 465 and following, with the extensive case law referred to.
57. Article 166 IPL.
58. WSA 20.03.2006, VI SA/Wa 1747/05; WSA 21.02.2008, VI SA/Wa 2018/07; although this is not made entirely clear in the respective provision, both aspects (non-use and invalidity) should be decided in one decision (M. Trzebiatowski, *Obowiązek używania znaku towarowego*, supra n. 7, 499).
59. K. Marciniszyn, in A. Michalak (ed.), *Prawo własności przemysłowej – Komentarz* (Warszawa 2016), 519.
60. The respective new provision of the IPL, according to the updated draft law made available on 8 June 2018, appears to mirror Article 46 of the Directive.

A non-use defence does also exist under the IPL in the event of court proceedings related to trademark infringement. Namely, a trademark owner may not prohibit another person from using his mark or a similar mark in the course of trade if he has refrained from using that mark genuinely, for the goods or services covered by the registration, for a period of five successive years.[61] Here, the IPL does not require the defendant to initiate separate cancellation proceedings due to non-use before the Patent Office.[62] This means that, upon raising of said non-use defence by the alleged infringer, the court examining the main lawsuit should on its own be entitled to examine genuine use of the trademark constituting the ground for action (just as a defence and not counter-claim).[63] However, unlike the EU Trademark Court competent to deal with EU trademarks, Polish courts have no powers to revoke protection of a non-used trademark – to achieve such a result, the defendant must always initiate separate cancellation proceedings before the Patent Office, which appears to be the most typical scenario (with the civil court staying its proceedings related to the alleged infringement until the Patent Office's decision related to cancellation becomes final).[64]

Under the draft law aimed at implementation of the recast EU Trademark Directive (2015/2436), it is going to be clearly stated that the trademark owner shall be entitled to prohibit the use of a sign only to the extent that his rights are not liable to be revoked due to non-use at the time the infringement action is brought. If the defendant so requests, the trademark owner shall furnish proof that, during the five-year period preceding the date of bringing the action, the trademark has been put to genuine use for the goods or services in respect of which it is registered or that there are proper reasons for non-use, provided the trademark at issue has been registered for not less than five years.

## B. Burden of proof

The general rule of Polish law is that the burden of proving a fact shall lie with the person who asserts legal consequences arising from this fact.[65] In proceedings for the revocation of a trademark the burden of proof is reversed.[66] In such proceedings the obligation to prove the use of a trademark or the existence of proper reasons for not using the trademark rests with the proprietor of the trademark. The trademark owner is obliged to furnish proof of use of the trademark or prove that there was a justification for non-use. Such construction can create a non-use presumption that can be rebutted by the trademark owner's action. The trademark owner is the only person who knows how and where the trademark was used and who has proof of use. Even if the presented proofs seem weak, the Polish Patent Office must not call the party to present additional evidence as such action may be treated as violence

61. Article 157 IPL.
62. M. Mazurek, in R. Skubisz (ed.), *System Prawa Prywatnego 14B – Prawo własności przemysłowej* (Warszawa 2017), 1262; differently: K. Marciniszyn, in *Prawo własności przemysłowej – Komentarz*, 492-493.
63. M. Trzebiatowski, *Obowiązek używania znaku towarowego*, supra n. 7, 503.
64. M. Trzebiatowski, in *System Prawa Prywatnego*, supra n. 7, 1008.
65. Article 6 Polish Civil Code.
66. Article 169(6) IPL.

of principle of equality of parties.[67] If the trademark owner takes no action after receiving the motion on revocation due to non-use, the Patent Office will cancel the trademark.[68]

## C. Relevant evidence

### *1. Use on the product or its packaging*

Use of the trademark on the product or its packaging is the most common form of trademark use.[69] The trademark shall be used on the product in the manner that product and trademark are treated as an inseparable whole which is visible to the consumer. The way the trademark is placed on the product or its durability are irrelevant. Connection of the trademark with the product may be temporary and does not have to last for the whole time of use.[70]

It should be stressed that placement of the trademark itself on goods or packaging is not sufficient to prove genuine use. Putting the trademark on elements, especially the packaging, which is required by the law shall not be treated as genuine use. This refers to streamers, stamps with security marks or certificates or stamps. Also putting the trademark in a hardly visible place or just on the bands, bottom, cap or the seal shall not be treated as genuine use.[71]

The trademark must be used on the products or its packaging to identify the origin of goods.[72] Moreover, goods or packaging must be available for the Polish consumers in Poland.[73] As for services, the trademark shall be used on the documentation, such as agreements, invoices, letters of reference or offers.[74]

### *2. Use in advertising*

Advertising is one of the most important trademark functions. Nevertheless, the use of a trademark solely in an advertisement cannot be treated as genuine use. It is necessary to prove that the advertised goods or services were offered on the Polish market. Where advertising is carried out in parallel to the marketing of goods or services and there is proof of both activities, advertising will support the genuineness

67. WSA, 07.11.2014, VI SA/Wa 252/14.
68. WSA confirmed that the Polish Patent Office is not obliged to look for the evidence as in a reverse burden proof system the trademark owner is the only party who is responsible for presenting the proof of use (11.01.2017, VI AS/Wa 1740/16).
69. It is listed as one of the typical trademark uses in Articles 154(1) IPL and 169(4)2 IPL.
70. K. Szczepanowska-Kozłowska, in E. Nowińska, K. Szczepanowska-Kozłowska (ed.), *System Prawa Handlowego. Tom II. Prawo własności przemysłowej* (Warszawa 2015), 692-693.
71. WSA 07.11.2017, VI/Wa 252/14.
72. Putting the trademark on the back of the goods or packaging marked with other entities' trademarks among informative marks was not treated as using the mark for an identity of origin function (NSA 17.11.2016, II GSK 872/15).
73. The use of the trademark in the advertisement of goods which were not available in Poland was not treated as genuine use (NSA 24.05.2006, II GSK 70/06).
74. WSA 18.01.2016, VI SA/Wa 1875/15.

of the use. The marketing materials such as calendars, logo pictures, cards, bags, etc. may only support other evidence.[75]

The advanced advertising of goods or services before introducing them to the Polish market shall be treated as preparation to use the trademark. This use shall be supported by other evidence proving that proper steps to use trademark on the market were taken, and advertising is just one of them.

### 3. *Use on the internet*

Considering use of a trademark on the internet, the trademark is usually used in marketing or in identity of origin function. To define in which function the trademark is used on the internet, the content and the aim of the website where it was used shall be defined. As a general rule, the website registration, running commercial activity and advertising this activity through the website shall be deemed as trademark use.[76]

If the trademark owner can prove that he offers goods or services through the website, this shall be deemed use of the trademark. If the website has only an advertising purpose, the use of the trademark on such website will have advertising function. If the domain name is inactive and is not used to advertise, offer or sell goods or services, it cannot be seen as use of the trademark.[77] The same refers to use of the trademark on private, non-profit or administrative bodies' websites.

All comments dealing with the issue of trademark use on the Polish market, as discussed in section II.E above, as well as the issue of trademark use for commercial purposes, as discussed in section II.A.2(ii) above, are relevant also for this section.

### 4. *Form of evidence*

The trademark owner may present all kinds of evidence that may prove use of the trademark.[78] To prove use of the trademark, most effective are documents such as invoices, orders, delivery confirmations, sale confirmations. Additionally catalogues, trade information, advertisements and marketing materials, archival prints from the internet, court decisions regarding trademark infringement may be helpful, as are brochures, flyers, stickers used in relation to the goods or services. Even though the evidence catalogue is wide, the analysis of the court decisions leads to the conclusion

75. Polish administrative courts stated that use of trademark only for advertising purposes, e.g. in newspapers, leaflets or promotional lotteries, and not in order to identify the goods as originating from certain entrepreneur, cannot constitute genuine use of the trademark in the course of trade (WSA 22.02.2017, VI SA/Wa 2049/16; WSA 15.10.2010, VI SA/Wa 1425/10); M. Trzebiatowski, *Obowiązek używania znaku towarowego*, supra n. 7, 472.

76. Judgement of Appeal Court in Poznań 26.04.2006, I ACa 1228/05.

77. J. R. Antoniuk, *Ochrona znaków towarowych w domenach internetowych w świetle prawa własności przemysłowej* (Warszawa 2006), 12-13; R. Chmura, W. Włodarczyk, in R. Skubisz (ed.) *Bezprawne używanie zarejestrowanych znaków towarowych w adresach domenowych w świetle ustawy o znakach towarowych, Internet – problemy prawne* (Lublin 1999), 102; A. Kobylańska, *Ochrona znaków towarowych w Internecie* (Warszawa 2005), 97.

78. Pursuant to Article 75 of Polish Administrative Proceeding Code (APC), anything that can clarify the case and is not against the law shall be allowed as a proof, in particular documents, witness testimonies, expert opinions or inspections.

that invoices seem to be the most important evidence. In most cases marketing materials indicate the intention to use of the trademark which does not constitute genuine use. As to catalogues or labels, it is stressed that they do not prove the scope of use and without additional evidence cannot prove genuine use.

The evidence supplied by the witnesses[79] shall be supported by documents.[80] Courts will examine the value of the witness testimony in the light of provided documents.[81] In administration proceedings the Polish Patent Office may take a statement of a party made under penalty of perjury. In civil proceedings the statements and affidavits shall be treated as private documents and prove that the person who signed the document made the statement in the wording as presented.[82]

All evidence: (i) shall refer to the use of the trademark which is to be cancelled due to non-use; (ii) must include information allowing definition of the date of use, form of the trademark which was used, the goods or services for which trademark was used and the territorial scope of use.

## 5. *Distinction between use of a company name and use of the services offered by the company*

Identity between the business name and the trademark is a special case. In administrative proceedings the rule is that the trademark shall be used in connection with the goods and services for which it was registered and use of trademark merely as a business name (i.e. to identify the entrepreneur rather than the goods or services) is not sufficient to establish genuine trademark use.[83] In each case, it shall be examined if evidence provided, in particular invoices, prove the use of the trademark in relation to the goods or services, or just as a company name. Putting the registered form of the trademark on the invoices shall be treated as use of the trademark. It is helpful if the registered form of the trademark is used next to the company's name with the symbol ®, especially with word trademarks.[84] The use of the trademark just as a

79. The evidence from witness testimony is commonly used in contentious proceedings like revocation and invalidation proceedings. In new opposition proceedings it is possible to have witness testimony, nevertheless, bearing in mind that the IPL emphasizes the written character of opposition procedures, it shall be assumed that statements and affidavits will replace witness testimony.
80. In the doctrine there is an opinion that genuine use of the trademark should be proved only by documents. The parties testimony and witness testimony shall have no evidence value (*see* M. Trzebiatowski, *Obowiązek używania znaku towarowego*, supra n. 7, 472). It seems that in practice such strict approach could cause evidential difficulties and such opinion shall not be accepted.
81. The statements or the testimonies without any additional documents proving the use may be viewed as untrustworthy; *see* WSA, 11.01.2017, VI SA/Wa 1740/16.
82. The Polish Patent Office will accept the statements and affidavits but in civil procedure they will have small value.
83. Nevertheless, in the civil case regarding the trademark infringement, the court underlined that Article 154 IPL gives just examples of use of the trademark and use of the trademark as a business name can be treated as trademark use in context of trademark infringement (Supreme Court 07.10.2008, III CSK 192/08).
84. WSA, 15.10.2010, VI SA/Wa 1425/10; WSA, 18.01.2016, VI SA/Wa 1857/15.

business name shall not be treated as a use of the trademark.[85] Additionally, evidence such as an extract from a commercial register including the name of the company was treated by the WSA as preparation to use the mark as a company name, and not as a trademark.[86]

## IV. JUSTIFICATION FOR NON-USE

### A. Obstacles to use

Proper reasons for non-use may justify non-use of the trademark and save it from revocation. The IPL does not define justification for non-use. In doctrine and jurisprudence, justifiable reasons are defined as material, legal or factual reasons, not caused by carelessness, helplessness or negligence of the trademark owner. Such reasons are independent of the trademark proprietor's will. The extraordinary and sudden circumstances of an external nature, impossible to foresee and prevent (*force majeure*), shall justify non-use of the trademark. The same refers to reasons, which cannot be foreseen, stopped and are beyond trademark owner's control.[87] The obstacles shall be material to the trademark owner's business. They must have direct relationship with a trademark and make its use impossible or unreasonable.[88]

### B. Circumstances in which non-use may be cured

When examining whether there are justifiable reasons, the behavior of trademark owner shall be examined in the light of the criteria mentioned in section IV.A above. Below are selected examples from the jurisprudence regarding justifiable reasons for non-use:

(i) non-use of the trademark due to decision on the interim injunction prohibiting the use of the trademark;[89]
(ii) pending court proceedings regarding the trademark;[90]
(iii) government requirements implying a ban on importing or placing goods on the market;[91]

85. If the trademark is used in invoices only as a part of the seller's business name, it is not considered to be use as a trademark and it is not sufficient to establish genuine trademark use (vide: NSA, 08.05.2013, II GSK 375/13).
86. WSA 11.01.2017, VI SA/Wa 1740/16.
87. This correlates with TRIPS, *see* NSA 9.10.2018, II GSK 1389/15; NSA 20.09.2007, II GSK 127/07.
88. K. Jasińska, in P. Kostański (ed.), *Prawo Własności Przemysłowej. Komentarz* (Warszawa 2010), 862.
89. But receiving the cease and desist letter shall not be treated as justifiable reason.
90. NSA 20.09.2007, II GSK 127/07.
91. Nevertheless, the WSA said that if law includes the ban on use of the trademark on the trademark application date, the applicant should be aware of the ban, especially if he is professional. Such circumstances cannot be treated as justifiable reason for non-use (22.02.2017, VI Sa/Wa 2049/16).

(iv) legal obstacles, such as the prohibition of placing goods on the market for which the trademark is registered, resulting from individual acts of administration or normative acts.[92]

On the other hand, the following shall not justify non-use of the trademark:

(i) non-use of the trademark by the licensee;
(ii) pending cancellation proceedings before the Patent Office and agreement talks between the parties;[93]
(iii) demands barriers;[94]
(iv) the commercial, economic, organizational or negligent difficulties in running the economic activity of the trademark owner as well as failure to comply with the requirements of the entity, which result in the inability to produce goods or provide services.

## V. CONSEQUENCES OF NON-USE

### A. Consequences of non-use and relevance of the context

The lack of genuine use of a trademark, as required by the respective provisions of the IPL, makes the trademark subject to various limits and sanctions, which depend on the context and type of procedure in the course of which the use of trademark has been examined. The most far-reaching consequence of non-use (provided there are no proper reasons for it) is revocation of the trademark protection – as a result of either an independent request for cancellation or a separate counter-request for cancellation lodged in the course of the invalidity proceedings. If non-use of the earlier trademark is successfully raised as a defence in the opposition proceedings or in the infringement-related proceedings, it may lead to rejection of the opposition or the court action, total or partial (depending on the extent to which the use of the earlier trademark, cited as a ground for the opposition or the court action, has been properly proven).

### B. Extent of any revocation

As a principle, if the reason for trademark protection being revoked concerns only some of the goods or services, protection is revoked only for those goods or services.[95] Accordingly, if genuine use of the trademark is limited to certain goods or services only, the trademark shall be subject to revocation with respect to those goods or services for which no genuine use has been proven. Partial revocation means that the

92. U. Promińska, *Ustawa o znakach towarowych, Komentarz* (Warszawa 1998), 107; E. Nowińska, U. Promińska, M. du Vall, *Prawo własności przemysłowej, Przepisy i omówienie* (Warszawa 2003), 245.
93. The NSA said that a trademark owner can control the agreement re: trademark's coexistence talks. As a result it cannot be treated as justification of non-use (29.11.2016, II GSK 919/15).
94. WSA 15.12.2004, II SA 3166/03.
95. Article 171 IPL.

protection right stemming from the trademark's registration is limited by appropriate modification of the list of goods and services but it continues to exist.[96]

Where the use of a trademark has been demonstrated in respect of only specific goods or services falling within a broader category of goods or services as covered by the registration, the use of the trademark for that broader category of goods or services should be accepted, unless that category would be perceived as composed of a number of subcategories. In the latter case, if it is established that the trademark has been genuinely used in respect of such subcategory of goods or services only, the trademark protection should be revoked for all other (non-used) subcategories falling within the registered broad category.[97]

### C. Timing of any revocation

Under the existing provisions of the IPL, as a principle, trademark protection is revoked with effect as from the date on which an event occurred leading by force of law to revocation. The date of revocation should be confirmed in a revocation decision.[98] According to the Supreme Administrative Court's interpretation, this means that in case the trademark owner does not appropriately establish use of his trademark in the non-use cancellation proceedings, the trademark protection shall be revoked with effect as from the last day of the five-year non-use grace period (where the lapse of the non-use grace period is treated as 'an event leading by force of law to revocation').[99] Said five-year period of uninterrupted non-use starts either on the day of the trademark's registration (if there was no genuine use whatsoever) or on the day of the last effective use (if the use had been started and ceased after some time).

From the above, it is possible to have trademark protection revoked due to non-use with effect earlier than the date of the cancellation application (but not earlier than five years after the registration), provided that in the meantime (that is between the last day of the five-year non-use period and the day of filing the cancellation application) no genuine use has been started or resumed. This results from the provision of the IPL which stipulates that in case – prior to filling a cancellation application by a third party – the trademark owner started or resumed genuine use

96. K. Szczepanowska-Kozłowska, in R. Skubisz (ed.), *System Prawa Prywatnego 14B – Prawo własności przemysłowej* (Warszawa 2017), 1293.
97. NSA 16.06.2009, II GSK 709/08 – in the case at issue the trademark was registered for 'alcoholic beverages' in class 33 and genuinely used only for wine, so the protection was revoked in respect of 'alcoholic beverages with the exception of wine'.
98. Article 172 IPL.
99. NSA resolution of 23.04.2008, II GPS 1/08 – taken by seven judges due to discrepancies in the decisions of the Patent Office and the administrative courts, where the prevailing view had been to deem a trademark revoked due to non-use with effect as from the date of the cancellation application (as referred to and criticized in M. Trzebiatowski, *Obowiązek używania znaku towarowego*, supra n. 7, 412 and following). The resolution, or its respective parts, was also subject to criticism (K. Szczepanowska-Kozłowska, in *System Prawa Prywatnego*, supra n. 7, 1332 – the author argues that it should still be the filing of the cancellation application which determines the effective date of revocation; R. Skubisz/M. Trzebiatowski, commentary to the NSA resolution II GPS 1/08; K. Jasińska, in *Prawo własności przemysłowej*, supra n. 7, 873).

of his trademark that has not been in use within the five-year non-use grace period, the cancellation application shall be dismissed (with some exceptions).[100] The party filing a cancellation application, interested in revocation with effect earlier than the date of the cancellation application, should explicitly request this, either in the application or in the course of cancellation proceedings (once it has been found that the trademark had not been genuinely used for the respective period prior to the date earlier than the date of the cancellation application).[101]

According to the draft law aimed at implementation of the recast EU Trademark Directive (2015/2436), protection of a registered trademark shall be deemed revoked (to the extent that genuine use has not been proven) as from the date of the application for cancellation due to non-use. However, another date (namely, the date on which an event occurred leading by force of law to revocation) may be fixed in the Patent Office's decision at the request of one of the parties.

100. Article 170(1) and (2) IPL.
101. M. Trzebiatowski, in *System Prawa Prywatnego*, supra n. 7, 997.

# Russia

*Denis Voevodin and Valeria Larionova*

## I. INTRODUCING GENUINE USE OF TRADEMARKS

### A. Legal framework

The trademark legislation of the Russian Federation consists of (1) international agreements, (2) national laws and regulations, (3) by-laws and (4) judicial practice.

Russia belongs to the continental legal system, which has a registration-based system for the establishment of trademark priority. At the same time, matters relating to the genuine use of trademarks registered in the Russian Federation are covered in the following legal acts:

- the Paris Convention for the Protection of Industrial Property, 1883 (Paris Convention);
- the Agreement on Trade-Related Aspects of Intellectual Property Rights, 1994 (TRIPS Agreement);
- the Singapore Treaty on the Law of Trademarks, 2006 (Singapore Treaty);
- the Civil Code of the Russian Federation, Part IV, 2008 (CC RF).

Even though judicial practice in Russia is not a direct source of law, acts of judicial authorities, in particular, acts of the specialized Court for Intellectual Property Rights (IPC), are instrumental in terms of determining the genuine use of trademarks.

### B. Purpose and function of genuine use: policy considerations

In implementing the provisions of the TRIPS Agreement and the Paris Convention, the lawmaker established the principle of mandatory use of a registered trademark as a condition for preservation of the right thereto.

Commenting on the Trademark Law back in 2003,[1] the Constitutional Court of the Russian Federation stated that 'the registration of a trademark not only has the effect of establishing a right, it also implies a duty for the holder to exercise the given exclusive right'.[2]

The following are among the reasons why the lawmaker introduced a trademark use requirement:[3]

(1) the trademark use requirement is designed to ensure both intensive and informal trademark operation in business.
(2) the trademark use requirement makes it possible to 'purge' the register of 'dead' marks that are not being used by their holders to mark goods they are producing, and to remove obstacles for manufacturers that are prepared to make and sell goods under similar trademarks.
(3) the trademark use requirement can in some cases facilitate efforts to combat piracy – actions by persons that register trademarks in their own name in order to block the real makers of goods (including foreign ones) from entering the market.

## II. WHAT QUALIFIES AS GENUINE USE OF A TRADEMARK

### A. Nature of use

#### *1. Use consistent with the function of the trademark*

Based on the definition of a trademark that is enshrined in Article 1447 CC RF, it should be noted that a trademark (service mark) functions to differentiate goods or services: it enables the consumer to distinguish one manufacturer's goods from another manufacturer's similar goods.

At the same time, legal names and trade names are granted separate legal protection as intellectual property designations. To wit, a legal name is intended to help the consumer (and the contractual partner) distinguish one legal entity from another that is operating in a similar field (Article 1473 CC RF). Trade names, in turn, identify businesses as a portfolio of assets.

The practical difference between the aforementioned intellectual property designations can be pinpointed in a panel decision by the Chamber for Patent Disputes (CPD) of the Russian Intellectual Property Service (Rospatent) denying an attempt by the well-known company Mail.Ru to register the word trademark 'MAIL'. The applicant's argument that the designation reproduces part of its legal name in

1. RF Law on Trademarks, Service Marks and Geographical Indications dated 23 September 1992.
2. RF Constitutional Court Decision No. 393-O of 2 October 2003, 'Dismissing the request for a hearing of the complaint by Samy SMAK Production Firm limited liability company that Article 22(3) of the Russian Federation Law on Trademarks, Service Marks and Geographical Indications violates constitutional rights and freedoms';
3. L.A. Trakhtengerts, S.A. Gorlenko, V.O. Kalyatin and L.L. Kiry, *Commentary on the Civil Code of the Russian Federation (Part IV): in 2 volumes* vol. 2, (2nd edn, INFRA-M, 2016).

English ('LLC Mail.Ru') was rejected by the CPD because 'the distinguishing feature of the given legal name is the "Mail.Ru" element. At the same time, a legal name and trademark are different intellectual property designations that perform different functions, are governed by different institutions of law and, accordingly, are subject to different requirements. The fact that a particular word is contained within the legal name of a legal entity is not a basis for granting legal protection to that word as a trademark on behalf of the given entity.'[4]

A similar line of argument is contained in the CPD decisions concerning the registration to Rusmedia, LLC of a trademark including the verbal element 'Rusmedia,'[5] and concerning the registration to Belstar Trading Company, LLC of a trademark containing the verbal element 'BELSTAR.'[6]

In denying the state registration of the word trademark '*Dengi do zarplaty*' ('Paycheck advance') to Tsentrozaym (Dengi do zarplaty), LLC, the CPD noted that the legal name and the trade name identify the applicant and its business, but they do not attest to the designations used therein being capable of further differentiating services, or of distinguishing between certain services of the applicant and other entities' similar services, i.e., of functioning as a trademark (service mark).[7]

Meanwhile, the use of a trademark without seeking to differentiate a product and with no intent to use the trademark to promote these products is not regarded as genuine use. In particular, advertising and promoting non-alcoholic beverages at trade shows in order to promote alcoholic beverages does not mean that the trademark is being used in relation to the non-alcoholic beverages. The CPD's argument for this conclusion was that advertising nonalcoholic beverages the trademark holder intended to promote alcoholic beverages the advertising of which is prohibited in Russia. Therefore advertising of drinking water was directed to promote alcoholic beverages instead of commercialization of nonalcoholic beverages, and consequently did not constitute the use of the trademark in relation to nonalcoholic beverages.[8]

At the same time, using someone else's trademark without intent to differentiate one's own goods is nevertheless regarded as trademark infringement, even though the lack of such intent could serve as an argument for reduced liability, as is confirmed by judicial practice. To wit, Lina, LLC filed suit against McDonalds, LLC's in connection with the latter's use of the trademark '*S pylu s zharu*' ('Piping hot' or 'Fresh out of the oven') in order to describe the properties of its product as a hot and freshly made dish, without intending to use someone else's trademark to promote its own product. The courts acknowledged that McDonalds was not intending to use someone

4. CPD Conclusion dated 26 Dec. 2013 (Appendix to the Rospatent decision of 31 Mar. 2014, on application No. 2012701954), 'Denying the state registration of a trademark'.
5. CPD Conclusion dated 20 Dec. 2013 (Appendix to the Rospatent decision of 5 Mar. 2014, on application No. 2012707768), 'Reversing a Rospatent decision and revoking the state registration of a trademark'.
6. CPD Conclusion dated 13 Aug. 2013 (Appendix to the Rospatent decision of 18 Oct. 2013, on application No. 2011714522), 'Upholding a decision on the state registration of a trademark'.
7. CPD Conclusion dated 25 June 2012 (Appendix to the Rospatent decision of 28 Aug. 2012, on application No. 2010720950), 'Denying the state registration of a trademark'.
8. CPD Conclusion dated 21 Feb. 2011 (Appendix to the Rospatent decision of 23 Apr. 2011, on application No. 2006726037/50), 'On revocation of a trademark'.

else's trademark to differentiate its own product, but they still held that such actions constitute a violation.[9]

## 2. *Use in the course of trade*

### *i. Public use*

Article 1483(3) CC RF stipulates: 'no one may use designations that are similar to a trademark without its holder's permission in relation to the goods for the differentiation of which the trademark was registered or similar goods, if such use is likely to result in confusion'.

At the same time, the legislation does not specify when persons other than the trademark holder can freely use a registered trademark.

There are cases in judicial practice involving the use of trademarks for informational purposes. For example, the display of trademarks in printed propaganda materials was regarded as use for informational purposes, inasmuch as the use was designed to inform voters about institutions and businesses without intending to differentiate any goods, work or services.[10]

Free use includes a verbal mention of someone else's trademark in an interview.[11] Mentioning a word (despite someone having registered it as a trademark) in articles for descriptive or informational purposes is regarded as fair use (that being said, the websites and other resources on which the articles are posted should not be being used for commercial purposes).[12]

Therefore, an analysis of judicial practice leads to the conclusion that only a *word* trademark is subject to free use. Any reference to a design and/or combined trademark in their registered form would more than likely be regarded as misuse and, consequently, as infringement of the trademark holder's rights.

At the same time, the courts point out that using someone else's registered trademark directly on product packaging cannot be regarded as the use for informational purposes and such use constitutes trademark infringement in any case.[13]

As for use of trademark directed to the public, the Article 1484 requires that the trademark has to be distributed to the end consumer.

Analysis of court practice shows that the usage cannot be internal and has to be directed to the community outside the organization. Thus, the court rejected such evidence of use as a service agreement between ZIL- IP LLC, the trademark owner, and ZIL Consulting, LLC, the licensee of trademark and principal under the

9. Decision of the Commercial Court of the City of Moscow dated 17 Oct. 2012, in case No. A40-2569/11-27-22.
10. Appellate decision of the RF Supreme Court, No. 91-APG16-6 dated 17 Sept. 2016.
11. RF Supreme Commercial Court Presidium Ruling No. 10852/09 of 1 Dec. 2009, in case No. A45-15761/2008-8/270.
12. IPC Ruling No. S01-377/2016 of 22 July 2016, in case No. A40-128923/2015.
13. Tenth Commercial Court of Appeals Ruling No. 10AP-13368/2017 of 29 Nov. 2017, in case No. A41-50224/14.

service agreement because the use of a trademark by an affiliated person cannot constitute genuine use.[14]

Similarly, the delivery of goods by one affiliated person of a trademark owner to another affiliated person does not prove that these affiliated companies introduced goods into civil circulation and distributed them to final consumers.[15]

In view of the above, public use of trademark is an essential part of trademark genuine use in Russia.

*ii. Commercial purpose*

According to Article 1484 CC RF, a trademark holder can use the mark by any legitimate means, including by displaying the trademark:

(1) on goods, including on the labelling and packaging of goods that are being made, offered for sale, sold, exhibited at trade shows and fairs or otherwise commercialized in the Russian Federation, or else are being stored or transported for that purpose, or else are being imported into the Russian Federation;
(2) when performing work or providing services;
(3) on documentation related to the commercialization of goods;
(4) in offers for the sale of goods, performance of work or provision of services, and also in notices, on signs and in advertising;
(5) on the internet, including in a domain name and in other addressing methods.

In deciding whether trademark use is genuine, the courts proceed on the basis that any use must be associated with the commercialization of goods, i.e., distributing them to the end consumer.

For example, unilaterally produced fact sheets on goods for sale,[16] agreements between affiliates[17] and leasing domain names that are similar to a trademark[18] do not themselves demonstrate the commercialization of goods.

Using a trademark on the packaging of goods also does not demonstrate the use of the trademark for that packaging as a standalone product. It is the product that is inside the packaging and which the consumer intends to acquire that has consumer value. If the separate packaging (without its contents) has not been commercialized, has not been offered for sale and has not been advertised, then the trademark use cannot be regarded as genuine use.[19]

Promotional materials will demonstrate genuine use of a trademark if they have been distributed to the consumer. For example, a product advertisement using a trademark in the trademark holder's corporate newsletter, which has not been distributed to consumers, does not demonstrate genuine use of the trademark.[20]

14. IPC Presidium Ruling No. S01-296/2016 of 30 May 2016, in case No. SIP-451/2015.
15. IPC Decision of 21 Mar. 2016, in case No. SIP-252/2015.
16. IPC Decision of 17 Aug. 2017, in case No. SIP-105/2017.
17. IPC Decision of 24 Oct. 2017, in case No. SIP-288/2016.
18. IPC Decision of 21 Mar. 2016, in case No. SIP-252/2015.
19. IPC Presidium Ruling No. S01-352/2016 of 8 July 2016, in case No. SIP-448/2015.
20. IPC Presidium Ruling No. S01-902/2016 of 16 Jan. 2017, in case No. SIP-185/2016.

At the same time, the court holds that exhibiting goods at trade shows (fairs) does serve as evidence of trademark use. The legislation 'does not include any reservations or limitations vis-à-vis the range of consumers or visitors to those trade shows (fairs), which means that they can be conducted for a narrow group of professionals'.[21]

Giving property as a gift generally constitutes commercialization of goods. However, for the purposes of determining genuine use of a trademark, the receiver of the gift should not have ties to the company giving the gift; among other things, the gift receivers should not be employees of the giver.[22] Giving out trademarked prizes[23] or distributing trademarked goods free of charge to promote another product[24] is regarded as the commercialization of goods.

### 3. *Use by the owner or with its consent*

Article 1486 CC RF stipulates that a trademark can be used by its holder, and by a licensee or other person controlled by the trademark holder.

That being said, a trademark holder's intention for a third party to use the trademark can be expressed as follows:

(1) In an agreement with the third party:

- in a franchise agreement;
- in a simple partnership agreement (e.g., if the trademark is contributed by its holder to the partners' communal property[25] or in the event that persons have undertaken to combine efforts, professional and other knowledge, skills and expertise, goodwill or business ties and to use them to promote and develop a trademark);[26]
- in a contractor agreement;
- in a paid services agreement,[27] etc.

Declaration of the granting of the right under an agreement null and void, as well as the invalidation or reversal of the agreement, has no effect on the trademark holder's intention to use the mark, and so the actual use of the trademark is deemed established.

At the same time, the mere existence of an agreement without evidence of its actual performance is not confirmation of trademark use.[28]

(2) The existence of a corporate relationship between the trademark holder and another person that is using the mark, which may be supported either by one legal

21. IPC Decision of 31 Jan. 2017, in case No. SIP-638/2016.
22. IPC Presidium Ruling No. S01-902/2016 of 16 Jan. 2017, in case No. SIP-185/2016.
23. IPC Decision of 11 Feb. 2016, in case No. SIP-448/2015.
24. IPC Presidium Ruling No. S01-58/2015 of 2 Mar. 2015, in case No. SIP-601/2014.
25. IPC Presidium Ruling No. S01-596/2014 of 29 Oct. 2014, in case No. SIP-19/2014.
26. IPC Presidium Ruling No. S01-451/2016 of 22 July 2016, in case No. SIP-643/2015.
27. IPC Presidium Ruling No. S01-992/2016 of 12 Dec. 2016, in case No. SIP-206/2016.
28. IPC Decision of 26 Apr. 2016, in case No. SIP-30/2016.

entity having a dominant participation in another, or by constitutional documents. That being said, consent to the use does not necessarily have to be formalized.[29]

(3) In documents evidencing the trademark holder's application to the customs authorities to have the mark included in the customs register.[30]

### B. Volume, duration and frequency of use

Generally, in order to maintain trademark protection, the mark has to be used continuously within a three-year period before the application for trademark cancellation was filed.

The CPD's position on the quantity of goods that is sufficient for the recognition of genuine trademark use comes down to the fact that the legislation has no stipulation as to the number of uses, and so this criterion is irrelevant.[31] For example, releasing a batch of 800 bottles of mineral water is sufficient to confirm actual use as far as the CPD is concerned.[32]

The IPC, in turn, sees the free-of-charge distribution of a total of six wooden surface cleaners,[33] a single sale of a total of 41 sets of Christmas decorations,[34] and the production of a 'small' quantity of notepads, which were then distributed to a limited audience – *VOGUE* magazine readers (subscribers) as sufficient evidence of actual trademark use.[35]

However, there is an opposing viewpoint. For example, a small number of sales of expensive (luxury) goods could be regarded by the court as sufficient evidence of trademark use, while the same number of sales of less expensive goods, due to their qualities (more accessible to the ordinary consumer due to their cost and/or abundance and/or their being household items, everyday items and the like), cannot demonstrate a trademark holder's *bona fide* use of its mark and cannot serve as sufficient evidence of use.[36]

As for frequency of use, in one case the court stated that a single instance of providing gala event planning services cannot be regarded as sufficient evidence of trademark use.[37] In another case, the one-time publication of a newspaper in the three years prior to the date when the statement of claim was filed with the court was deemed to be sufficient evidence of trademark use.

29. IPC Presidium Ruling No. S01-184/2014 of 2 Apr. 2014, in case No. SIP-247/2013.
30. IPC Presidium Ruling No. S01-330/2014 of 27 June 2014, in case No. SIP-193/2013.
31. CPD Conclusion dated 20 June 2011 (Appendix to the Rospatent decision of 19 Aug. 2011, on application No. 2005732763), 'On revocation of a trademark'; CPD Conclusion dated 6 Feb. 2009 (Appendix to the Rospatent decision of 30 Mar. 2009, on application No. 2000709106/50), 'On revocation of a trademark'.
32. CPD Conclusion dated 8 Oct. 2008 (Appendix to the Rospatent decision of 2 Dec. 2008, on application No. 2001727390/50), 'On revocation of a trademark'.
33. IPC Presidium Ruling No. S01-902/2016 of 16 Jan. 2017, in case No. SIP-185/2016.
34. IPC Presidium Ruling No. S01-186/2016 of 4 July 2016, in case No. SIP-438/2015.
35. IPC Presidium Ruling No. S01-58/2015 of 2 Mar. 2015, in case No. SIP-601/2014.
36. IPC Decision of 18 Jan. 2018, in case No. SIP-597/2017.
37. IPC Presidium Ruling No. S01-724/2016 of 9 Sept. 2016, in case No. SIP-626/2015.

Therefore, as can be seen from the examples cited above, there is no established practice in Russia concerning the frequency and intensity of trademark use: in each instance the court decides the matter by analysing the totality of the various evidence and other circumstances of the case.

## C. Use of a sign

In Russia, the use of a trademark in its registered form or in a modified form, provided that the mark is still recognizable (distinguishable) as a particular trademark in the eyes of consumers, is regarded as trademark use.[38]

The following trademark use scenarios have been examined in judicial practice.

*Using a standard typeface instead of a logotype does not demonstrate use*

1

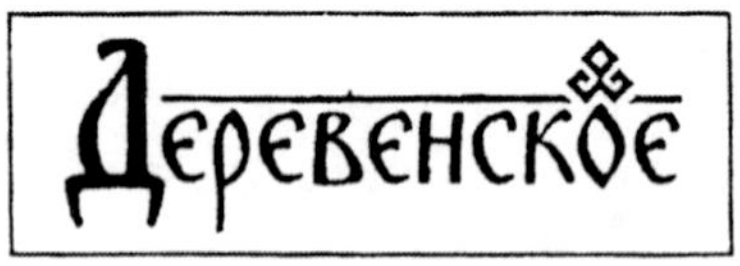


2

The court concluded that using the verbal element 'Деревенское' ('Derevenskoye') (Country-style) within designation (1) does not demonstrate use of trademark (2), because legal protection was granted to the 'Деревенское' ('Derevenskoye)' logotype, which is integral to the trademark's distinctiveness. For this reason, using the word 'Деревенское' ('Derevenskoye)' written in a standard typeface affects the distinctiveness of the designation.

3

4

ЦАРЬ

5

38. IPC Presidium Ruling No. S01-840/2017 of 17 Nov. 2017, in case No. SIP-103/2017.

The court came to a similar conclusion when hearing cases on the use of trademarks (3)[39] and (4).[40] The trademark holders used the verbal designation 'Царь' ('Tsar') (5) in a standard typeface, which according to the court could not be characterized as use of the contested trademarks with certain elements thereof modified without changing the substance of the mark.

*Using one version of a verbal designation instead of the registered combination of verbal elements, written with letters from various alphabets, does not demonstrate trademark use*

**С ПЫЛУ С ЖАРУ**
**S PILY S ZHARY**
**S PILU S ZHARU**
**С ПЫЛУ S ЖARU**
**С ПIЛУ S ЖARU**
**S PIЛУ S ЖАРУ**

5

С пылу с жару

6

The trademark holder contended that using designation (6) demonstrates use of trademark (5). According to the court, the trademark holder failed to prove that the verbal designation 'С пылу с жару' ('S pylu s zharu' – 'Piping hot' or 'Fresh out of the oven') used by it in documentation does not change the substantive composition of verbal elements in the trademark that was recognized by Rospatent as a distinctive mark at the time of its registration.[41]

*Using a registered word trademark other than in its registered form can lead to infringement of third-party rights*

TESORO/
TECOPO

7/8

MARQUES DEL
REALTESORO

9

10

39. IPC Presidium Ruling No. S01-842/2017 of 17 Nov. 2017, in case No. SIP-105/2017.
40. IPC Presidium Ruling No. S01-841/2017 of 17 Nov. 2017, in case No. SIP-104/2017.
41. IPC Decision of 3 Apr. 2017, in case No. SIP-749/2016.

The owner of word trademarks (7/8) filed an infringement suit against the owner of trademark (9), because the latter used its trademark in the form shown in picture (10). The court found that the designation 'REAL TESORO' occupies a dominant position on the defendant's label, which increases the similarity to trademarks (7/8).[42]

*Using a colour version of a registered black-and-white image constitutes trademark adaptation and is regarded as genuine use*

**MARKIES**

11 (black-and-white) 12 (colour)

In a case regarding the revocation of word trademark (11), the CPD concluded that the trademark holder's changes (12) to the registered mark (using a colour version of the black-and-white designation, using upper or lower case letters in a nonstandard typeface) constitute adaptation of the mark to market conditions and do not change the distinguishing features of the mark itself, because the distinctiveness of the mark is contained in the phonetics and semantics of the 'Markies' element.[43]

*When using a trademark that was registered with Latin-script letters, it may be transliterated into Cyrillic script*

**Dr. Theiss Angi Sept**

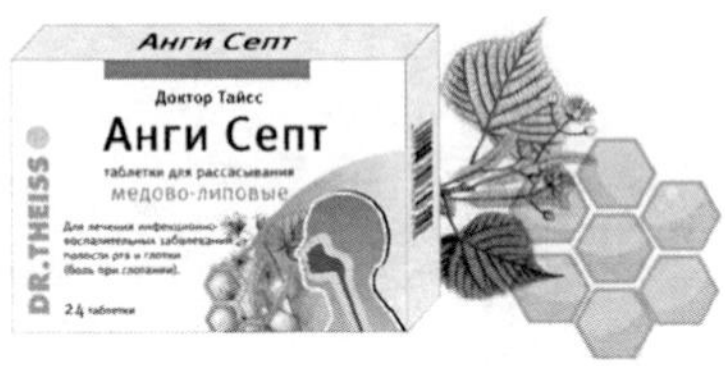


13 14

In a case regarding the revocation of trademark (13), the IPC concluded that the trademark can be used by transliterating the Latin-script letters into Cyrillic script (14),

42. RF Supreme Commercial Court Presidium Ruling No. 14567/13 of 4 Feb. 2014, in case No. A40-27210/12-19-247.
43. CPD Conclusion dated 24 Feb. 2010 (Appendix to the Rospatent decision of 26 Mar. 2010, on application No. 129129/50), 'On partial revocation of a certified trademark'.

inasmuch as this change does not affect its distinctiveness and does not limit the protection granted to the trademark.[44]

## D. Use in relation to goods and services

### *1. Use for each of the designated goods and services*

Russian legislation requires a trademark holder to use the mark in relation to all of the goods for the differentiation of which it was registered.

To wit, in the hearing of a case regarding the revocation of a trademark in relation to meat, pork and liver, the defendant produced evidence relating to goods such as 'frankfurters,' which was not recognized by the court as evidence demonstrating the use of the trademark in relation to other goods.[45]

In another case, the court found that using a trademark in relation to dietary supplements does not demonstrate the use of the trademark in relation to pharmaceutical drugs, because dietary supplements are a separate type of product with a different purpose from pharmaceutical drugs.[46]

### *2. Use for a broader category of goods and/or services*

In most cases, judicial practice proceeds on the basis that a trademark has to be used to differentiate the goods/services indicated in the certificate.

Generally in Russia, the use of a trademark in relation to goods similar to those indicated in the certificate is not recognized. However, the similarity criterion becomes more important when 'strong' (well-known, distinctive) trademarks are in need of protection.[47]

For example, the court found that the actual use of the widely recognized trademark '*Dinastiya*' (Dynasty) in relation to chocolate and candy demonstrates use of the trademark in relation to goods such as 'frozen yogurt and ice cream'.[48]

## E. Territorial scope

In accordance with Article 1484 CC RF, the exclusive right to a registered trademark belongs to its holder and is effective throughout the entire territory of the Russian Federation, irrespective of that trademark being used within the boundaries of a particular geographic area or administrative division.[49]

In one court case, a trademark holder asserted that it disseminates information in the Russian Federation on the holding of trade shows and conducts advertising campaigns in the Russian Federation. The court concurred that taking such actions

44. IPC Decision of 9 Dec. 2014, in case No. SIP-530/2014.
45. IPC Presidium Ruling No. S01-842/2017 of 17 Nov. 2017, in case No. SIP-105/2017.
46. RF Supreme Commercial Court Presidium Ruling No. 8817/11 of 29 Nov. 2011, in case No. A40-103027/10-12-654.
47. RF Supreme Court Decision of 11 Jan. 2016, in case No. 300-ES15-10765, SIP-530/2014.
48. IPC Decision of 12 Dec. 2016, in case No. SIP-124/2016.
49. IPC Ruling No. S01-657/2015 of 21 Aug. 2015, in case No. A73-14762/2014.

in order to attract potential clients and guests is an essential preparatory part of holding trade shows and fairs. At the same time, it has no legal bearing that the trade show itself (with, among others, Russian businesses and organizations participating therein) is actually held outside of the Russian Federation.[50]

### F. Relevant time period and its calculation

Generally, in order to maintain trademark protection, it has to be used continuously within any three-year period after the date of its registration. The three-year period is counted back from the date when an interested party files an application to revoke the trademark.

There is no prohibition against resuming the use of a trademark after a certain period of non-use.[51]

## III. EVIDENCE OF GENUINE USE OF A TRADEMARK

### A. When use should be proved: timing of lack of genuine use claims

#### *1. When filing an application for a trademark*

Russia has a registration-based system in that, when a trademark is registered with the patent authority, protection is granted from the time when the application is filed. That being said, the applicant does not have to prove actual intent to use the trademark. Moreover, using a trademark in the examination stage before it is registered could be regarded as infringement of someone else's trademark that is confusingly similar to the applied mark.

#### *2. In the course of opposition proceedings*

Russia has no procedure in place for opposing the use of a trademark. The trademark holder has an exclusive right to use the trademark only from the date of its registration.

#### *3. In cancellation proceedings based on lack of use*

Russia has a judicial procedure in place for cancelling a trademark due to non-use. The IPC serves as the trial court.

Only an interested party can file an application to revoke a trademark. Meanwhile, the absence of interest is independent grounds for refusal to revoke a trademark.

A producer of similar goods (work, services)[52] that actually intends to use the contested designation in its own business and which has taken the necessary steps in preparation for such use (e.g., has filed an application to register an identical or

50. IPC Decision of 20 Oct. 2015, in case No. SIP-263/2015.
51. IPC Presidium Ruling No. S01-502/2017 of 21 Aug. 2017, in case No. SIP-233/2016.
52. IPC Decision of 3 Apr. 2017, in case No. SIP-749/2016.

similar mark[53]) may be recognized as an interested party. That being said, the mere fact of filing an application cannot serve as sole and sufficient evidence of interest.[54]

A person whose rights and legitimate interests are affected by the existence of legal protection of a trademark can also be recognized as an interested party (e.g., if there is a dispute between a plaintiff and a defendant over protection of the exclusive rights to a disputed trademark). In this context, interest is contingent on the need to defend against the filing of claims by a trademark holder that is not using the trademark.[55]

### *4. As a defence in infringement/cancellation proceedings*

In the event that an infringement action is filed, the defendant can file a statement to claim for revocation of the disputed trademark. That being said, the infringement matter and the trademark revocation matter will be heard under different court cases, because trademark infringement cases are under the jurisdiction of the territorial commercial courts where the defendant has its registered office, while revocation actions are heard by the IPC.

In the event of parallel trademark revocation proceedings, the defendant can file a motion to stay proceedings in the infringement case until the IPC passes judgment.

That being said, the decision to grant the motion is taken by the judge independently, inasmuch as the cancellation of a trademark means the termination of legal protection at a future time. Until that time, the person to which the disputed trademark is registered cannot be denied protection of the exclusive right to the trademark.[56]

## B. Burden of proof

In the hearing of a demand for revocation of a trademark due to its non-use, the burden of proving trademark use lies with the trademark holder. The applicant, in turn, has to prove that there is a legitimate interest in the revocation of the trademark.

## C. Relevant evidence

### *1. Use on the product or its packaging*

Using a trademark on the packaging of a specific product demonstrates the use of that mark on the product itself, but not on its packaging or on the raw materials from which the packaging or the product itself is made. At the same time, packaging on which a trademark is placed forms an integral part of the product.

53. IPC Decision of 17 Aug. 2017, in case No. SIP-105/2017.
54. IPC Presidium Ruling No. S01-479/2015 of 26 June 2015, in case No. SIP-1021/2014.
55. IPC Presidium Ruling No. S01-806/2017 of 2 Nov. 2017, in case No. SIP-121/2017.
56. Tenth Commercial Court of Appeals Ruling No. 10AP-10918/2016 of 31 Oct. 2016, in case No. A41-12080/16.

To wit, the court found that, in and of itself, ordering packaging paper, placing a trademark thereon and using it to package other products (paints, modelling clay, etc.) does not denote use of the trademark for that paper, inasmuch as the defendant was commercializing not the packaging paper as a standalone product, but rather the other products – modelling clay, paint, chalk.[57]

In the hearing of a case for the revocation of the '*Yuniton*' trademark, the court stated that 'using a trademark on the packaging of a product does not demonstrate use of the trademark for that packaging as a standalone product. It is the product that is inside the packaging and which the consumer intends to acquire that has consumer value, not its packaging'.[58]

In another case, the court admitted as evidence a contract for the manufacture of packaging for floral arrangements using the disputed mark 'maison des fleur' and stated that, considering the proven sale of products and ordering of a large quantity of packaging for floral arrangements, the court finds that the trademark was used by the defendant in the sale of floral arrangements by being placed on the packaging.[59]

## *2. Use in advertising*

When a trademark is used in advertising, it is necessary to ensure that the advertising has been distributed to the consumer.[60] To wit, a defendant-provided copy of a promotional brochure depicting tin cans using a contested trademark, in the court's opinion, cannot demonstrate the trademark's use in the provision of class 41 services under the International Classification of Goods and Services (ICGS). The defendant also failed to produce evidence that the said promotional brochure had been distributed to consumers.[61]

In addition, the advertising has to be aimed at promoting the particular goods and services for the differentiation of which the trademark was registered. To wit, in the hearing of a dispute over the revocation of the '*Alleya*' ('Avenue') trademark, promotional materials and an advertising card were not admitted by the court as evidence confirming trademark use, because they are advertising a shop, but not the actual goods in relation to which the trademark was registered.[62]

In another case, using the

trademark on products for promotional purposes served as sufficient evidence of use of the trademark in relation to some goods.

57. IPC Decision of 22 Dec. 2016, in case No. SIP-631/2016.
58. IPC Presidium Ruling No. S01-352/2016 of 8 July 2016, in case No. SIP-448/2015.
59. IPC Decision of 25 Jan. 2018, in case No. SIP 614/2017.
60. IPC Decision of 17 Aug. 2017, in case No. SIP-105/2017.
61. IPC Decision of 31 May 2016, in case No. SIP-711/2015.
62. IPC Presidium Ruling No. S01-30/2016 of 14 July 2016, in case No. SIP-372/2015.

### 3. *Use on the internet*

Like the aforementioned uses, the use of a trademark on the internet has to be directly related to the commercialization of a product.

To wit, in the court's opinion having a reference to an internet address (domain name) and the email address 'bversia@yandex.ru' does not confirm trademark use.

In documents provided by the defendant, a domain name and email address containing elements that are a transliteration of the disputed trademark were given as the contact details for an Office (organization) and were not directly related to the commercialization of the plaintiff's applied services. According to the panel of judges, the domain name and email address are not perceived as the service brand that is the disputed trademark.[63]

At the same time, placing a trademark and information about an offer to provide accounting and auditing services on the internet demonstrates both distributing the trademark and information to the consumer, and use of the trademark in relation to the services indicated in the certificate.[64]

### 4. *Form of evidence*

Analysis of judicial practice has shown that the courts accept the following as evidence of trademark use:

- a contract to manufacture goods under the client's trademark;[65]
- payment orders and goods consignment notes under contracts demonstrating the commercialization of goods;
- printouts from the State Register of Trademarks;
- printouts of internet webpages about goods;
- advertising, promoting and offering to sell goods using the disputed trademark (in documents accompanying the goods);
- pictures of products for sale in store displays, advertisements on transport vehicles and products on offer on the defendant's internet website;[66]
- a cargo customs declaration for goods being delivered to Russia from foreign territory.

At the same time, in accordance with the Commercial Procedure Code of the Russian Federation, a court can reject evidence as irrelevant or inadmissible. For example, having noted that the defendant-provided copies of pictures of calamari packaging labelled with the mark '*S PYLU S ZHARU*' ('PIPING HOT' or 'FRESH OUT OF THE OVEN') do not have a date of manufacture, the court rejected this evidence as irrelevant to the case considering the period of trademark use under investigation.[67]

63. IPC Decision of 3 Oct. 2016, in case No. SIP-385/2016.
64. IPC Presidium Ruling No. S01-814/2016 of 30 Sept. 2016, in case No. SIP-157/2016.
65. IPC Decision of 3 Apr. 2017, in case No. SIP-749/2016.
66. IPC Decision of 7 Dec. 2016, in case No. SIP-476/2016.
67. IPC Decision of 3 Apr. 2017, in case No. SIP-749/2016.

If the court is unable to establish the origin of copies of pictures, then it also rejects them as irrelevant to the case, because it cannot be determined what period of trademark use they are confirming.[68]

## IV. JUSTIFICATION FOR NON-USE

In deciding on the revocation of a trademark due to its non-use, consideration may be given to evidence produced by the trademark holder that the trademark was not being used due to circumstances beyond its control.

To wit, if there is a court judgment confirming that a trademark holder has been declared bankrupt and that bankruptcy proceedings have been initiated against it, it is effectively prevented from using the trademark.[69]

At the same time, while recognizing the initiation of bankruptcy proceedings against a trademark holder as a circumstance beyond its control, the court holds that it is unreasonable to count the three-year period of non-use prior to the date of the legal action excluding the time during which the bankruptcy proceedings against the trademark holder were under way, because the evidence of trademark use or of the trademark holder's inability to use it due to circumstances beyond its control must cover the three-year timeframe preceding the filing of the claim.[70]

Meanwhile, in hearing a case for the revocation of the 'KREMONA' trademark, the court stated that the ban on imports into the Russian Federation of confectioneries made at the Roshen confectionery corporation's plants due to the trademark holder's violation of health and disease-control standards (inadequate information on the product's nutritional value given on the label) is a circumstance within the trademark holder's indirect control.[71]

The court also rejected a reference to non-use of a trademark due to circumstances beyond the trademark holder's control in light of the introduction of the Russian food embargo on agricultural goods (including alcohol) produced in Moldova, because the trademark holder failed to provide any evidence whatsoever corroborating these circumstances (alcoholic product supply contracts with entities located in Moldova, evidence of taking steps to prepare for the supply of alcoholic products from the specified territory to the Russian Federation).[72]

Lack of demand for products also cannot be counted as a circumstance that is completely beyond the control of the manufacturer of those goods.[73]

68. IPC Decision of 14 Aug. 2017, in case No. SIP-103/2017.
69. IPC Decision of 12 July 2016, in case No. SIP-225/2016.
70. IPC Decision of 12 Sept. 2016, in case No. SIP-240/2016.
71. IPC Decision of 20 Jan. 2018, in case No. SIP-602/2017.
72. IPC Decision of 11 Apr. 2016, in case No. SIP-702/2015.
73. IPC Presidium Ruling No. S01-800/2015 of 30 Sept. 2015, in case No. SIP-151/2015.

## V. CONSEQUENCES OF NON-USE

### A. Consequences of non-use and relevance of the context

Non-use of a trademark may result in a court judgment revoking the trademark. The respective judgment is forwarded to Rospatent, which in turn makes an entry in the Register on the revocation. Legal protection is terminated from the effective date of the court judgment.

### B. Extent of any revocation

The trademark is only cancelled in relation to those goods/services for which the trademark holder has not been able to demonstrate actual use.

### C. Timing of any revocation

Revocation of a trademark due to non-use means the cancellation of the trademark at a future time.

Therefore, in Russia protection claims filed by trademark holders before the termination of trademark protection are to be granted.[74]

74. Review of judicial practice in cases involving the resolution of disputes over the protection of intellectual property rights (approved by the RF Supreme Court Presidium on 23 Sept. 2015).

# Singapore

*Stanley Lai, SC*

## I. INTRODUCING GENUINE USE OF TRADEMARKS

The statutory regime governing the use of registered trade marks in Singapore is the Trade Marks Act (Cap 332) ('TMA').

### A. Legal framework

Under the TMA, the first observation that should be made is that the requirement of 'genuine use' of a trade mark per se is absent from the infringement provisions of the TMA. Section 27 TMA is the main infringement section, and (by paraphrase) it envisages that infringement will occur by the unauthorized use of a sign that is identical or similar to a registered trade mark and is applied to identical or similar goods and services. If 'use' is to be distilled further, Section 27(4) TMA provides that a person uses a sign if, in particular, he:

(a) applies it to goods or the packaging thereof;
(b) offers or exposes goods for sale, puts them on the market or stocks them for those purposes under the sign, or offers or supplies services under the sign;
(c) imports or exports goods under the sign;
(d) uses the sign on an invoice, wine list, catalogue, business letter, business paper, price list or other commercial document, including any such document in any medium; or
(e) uses the sign in advertising.

There is an important qualification in Section 27(5) TMA, which provides that a person who applies a sign to any material used or intended to be used for labelling or packaging goods, or uses a sign on any document described in Section 27(4)(d) or in advertising is not deemed to use the sign if, at the time of such application or use, he does not know or has no reason to believe that the proprietor or a licensee of the registered trade mark did not consent to such application or use of the sign. The *mens rea* behind the use of a sign may disqualify the use of the sign if it is absent at the point of application or use. Implicit in this requirement is that obviously the state of mind (in the manner provided by Section 27(5) TMA) will also impact 'genuine use' of a sign or absent *mens rea*, an infringement action would also not succeed, on the basis that 'use' has not been properly established.

The view has been taken in Singapore that Section 27(4) 'use' is equally applicable to determining the issue of genuine use. In *Bluestar Exchange (Singapore) Pte Ltd v. Teoh Keng Long and others (trading as Polykwan Trading Co)* [2003] 4 SLR(R) 92 at [53] it was said:

> 'Mr Gill also relied on Kerley's Law of Trade Marks and Trade Names (13th Ed), at p 281, which opined that although (the UK equivalent of our) s 27(4) was expressed to be for the purposes of s 27 (only), the types of use set out therein are *equally applicable* when considering the question of use under (the UK equivalent of our) s 22. Mr Wong did not contest this submission.' (emphasis added)

Section 5(2)(e) TMA arguably captures the essence of genuine use. It requires that a trade mark application contains a statement from the applicant stating that:

> '(a) that the trade mark *is being used in the course of trade*, by the applicant or with his consent, in relation to those goods or services; or
> (b) that the applicant has a *bona fide intention* that the trade mark should be so used.' (emphasis added)

There is also the 'non-use' revocation section of Section 22(1) TMA, which exposes a trade mark to a non-use revocation if it has not been put to 'genuine use' in the course of trade in Singapore within a period of five years following the completion of the registration procedure (see discussion below). It is also worth remembering that under the TMA, use is not confined to use by graphical representation. Section 2(4) TMA states that:

> 'References in this Act to use (or any particular description of use) of a trade mark, or of a sign identical with, similar to, or likely to be mistaken for a trade mark, include use (or that description of use) otherwise than by means of a graphic representation.'

In other words, 'oral use' of a trade mark can also suffice.

## B. Purpose and function of genuine use: policy considerations

The genuine or bona fide use of a trade mark clearly falls within the proper exercise of trade mark rights under the TMA. The Singapore Court of Appeal has clarified that 'genuine use means bona fide use as a trade mark'.[1]

As seen from the requirements of Section 5(2)(e) TMA, *bona fide* use of a trade mark undergirds the requirement of genuine use. Applicants who aspire to protect

1. *Societe des Produits Nestle SA and anor v. Petra Foods Ltd and anor* [2017] 1 SLR 35 at [121]-[123].

a trade mark to cover all classes of goods or services should also note that if they should fail to trade in any one class, any aggrieved person may apply under Section 22(6) TMA to revoke the registered mark in relation to the classes of goods or services for which the trade mark has not been genuinely used. There is also potential for an allegation that a wide-scoped application is made in bad faith (contrary to Section 7(6) TMA). This is a basis for potential revocation under Section 23 TMA.

Unused or abandoned marks are also vulnerable to revocation under Section 22(1) TMA (e.g. the absence of 'genuine use' for a period of five years from the completion of the registration procedure or suspended use for a continuous period of five years) for policy reasons. These were articulated by the Singapore High Court in *Weir Warman*:

(a) Trade marks serve to indicate the origin of goods and the registration system facilitates and protects this function. This protection given by registration is no longer justified if the mark is not used in the capacity as a badge of origin.
(b) The Register serves as a notice to rival traders of trade marks which are already in use. To allow unused marks to remain on the Register is to jeopardise this function of the Register as an accurate record.
(c) This ground for revocation prevents traders from 'hoarding' trade marks or 'squatting' on trade marks which have been abandoned. The process of revocation releases trade marks back to the public domain for the use by others.[2]

## II. WHAT QUALIFIES AS GENUINE USE OF A TRADEMARK

### A. Nature of use

The evaluation of genuine use is a factual inquiry. In *Société des Produits Nestlé*, the Court of Appeal stated (at [122]):

> 'The question of whether there has been genuine use of a registered trade mark entails a factual inquiry. In *La Mer Technology Inc v Office for Harmonisation in the Internal market (Trade Marks and Designs) (OHIM)* [2008] ETMR 9 at [57], which was cited in Wing Joo Loong at [41], the CFI elaborated on how an examination of whether a trade mark had been put to genuine use should be carried out:
>
> > "To examine whether [...] [a] trade mark has been put to genuine use, an overall assessment must be carried out, which takes into account all the relevant factors of the particular case. That assessment entails a degree of interdependence between the factors taken into account. Thus, the fact that [the] commercial volume achieved under the mark was not high may be offset by the fact that use of the mark was extensive or very regular, and vice versa. In addition, the turnover and the volume of sales

2. *Weir Warman Ltd v. Research & Development Pty Ltd* [2007] 2 SLR(R) 1073, at [42], [99]; recently affirmed by the Singapore Court of Appeal in *Société des Produits Nestlé and another v. Petra Foods Ltd and another* [2017] 1 SLR 35 at [119].

> of the product under the [...] trade mark cannot be assessed in absolute terms but must be looked at in relation to other relevant factors, such as the volume of business, [the] production or marketing capacity or the degree of diversification of the undertaking using the trade mark and the characteristics of the products or services on the relevant market."'

## *1. Use consistent with the function of the trademark*

The requirement of visual perceptibility was removed from the current definition of 'trade mark' within the TMA by the Trade Marks (Amendment) Act (No. 20 of 2004). This paved the way for non-conventional signs such as smells, taste, touch, gestures or moving images, musical tunes and sounds to be potentially registrable as trade marks in Singapore.

The registered trade mark performs the function as an indicator of origin of goods and services. No other functions are contemplated. The requirement of use in a 'trade mark sense' is well seeded in Singapore case law.[3]

Other tolerated uses of a sign include, inter alia, (a) the use of an own name in accordance with honest practices in industrial and commercial matters (Section 28(1)(a) TMA); (b) use of a sign to indicate certain descriptive characteristics of goods or services, in accordance with honest practices in industrial and commercial matters; (c) use of the registered trade mark to indicate the intended purpose of goods (in particular accessories or spare parts) or services in accordance with honest practices in industrial and commercial matters.

## *2. Use in the course of trade*

The requirement that a trade mark be 'used in the course of trade' arises under Section 5(2)(e) TMA (see above). Inherent in the definition of a 'trade mark' lies the requirement of a sign which is capable of graphical representation and which is capable of distinguishing goods and services dealt with or provided *in the course of trade* by a person from goods or services provided by any other person (Section 2(1) TMA).

### *i. Public use*

Genuine use does not necessarily require public use to the extent that there does not need to be communication of the mark to the end user or ultimate consumer of the goods. However, the mark should be communicated to a third party, e.g. to a potential agent or distributor, in such a way as can be said to be 'consistent with the essential function of a trade mark' (i.e. as a badge of origin).[4]

3. See *Nation Fittings (M) Sdn Bhd v. Oystertec plc* [2006] 1 SLR(R) 712 at [74]; *City Chain Stores (S) Pte Ltd v. Louis Vuitton Malletier* [2010] 1 SLR 382 at [20]
4. *Weir Warman, supra* n. 2, [109].

*ii.* *Commercial purpose*

The requirement of use in the course of trade traverses a broad range of commercial uses by the trader who owns the trade mark. The application of a trade mark on promotional merchandise[5] or leftover stocks[6] have been held to be sufficient to constitute use in the course of trade. Other uses in the course of trade are set out in Section 27(4) TMA (outlined above).

### 3. *Use by the owner or with its consent*

Unauthorized use within meaning of the TMA gives rise to potential infringement. Under the predecessor to the TMA (the 1939 Act), the use of a registered trade mark by licensees was generally regarded as 'permitted use' and such use was regarded as use by the proprietor. The licensee could be registered as a 'registered user'.

The trade marks regime generally envisages the proprietor maintaining control over the use of the trade mark. Under the repealed Trade Marks Act 1939, the licensee who used a trade mark with permission, was accountable for quality of goods and services. This control over quality of the use of the mark in relation to goods and services undergirded the trade connection between the proprietor and the 'registered user'. Under the TMA, there is no reference to the registered user. Licences are registrable transactions (Section 39(2)(b) TMA) but registration is not mandatory (Section 39(5) TMA).

Whist the TMA does not grant any express restrictions over the grant of trade mark licences, the Singapore courts still require proprietors to exercise some supervisory control over the use of the trade mark by the licensee.[7] The trade connection for the purposes of a valid trade mark licence is established by this means.

One may indeed venture to say that the absence of quality control over the licensed use of a trade mark may expose the mark to a potential revocation under Section 22(1)(d) TMA. Section 22(1)(d) provides that a trade mark may be revoked if, 'in consequence of the use made of it by the proprietor or with his consent, in relation to goods and services for which it is registered, it is liable to mislead the public, particularly as to the nature, quality or geographical origin of those goods or services.'

The potential for deception arising out of the failure to supervise nature, quality or geographical origin of goods or services in question under Section 22(1)(d) TMA resonates strongly.

The UK (House of Lords) decision of *Scandecor*[8] may have suggested a shift of paradigm and the question was whether this shift equally applies in Singapore. In *Scandecor*, a claim for revocation of the registered trade mark was made on the basis that the trade mark had ceased to be distinctive because of an absence of quality control over the licensee's use of the trade mark. The House of Lords clarified that the absence of quality control in a bare exclusive licence does not, in and of itself, render the trade mark inherently deceptive so as to result in its revocation pursuant

5. *Nation Fittings (M) Sdn Bhd v. Oystertec plc* [2006] 1 SLR(R) 712.
6. *Swanfu Trading Pte td v. Beyer Electrical Enterprise Pte Ltd* [1994] 1 SLR(R) 330
7. *Weir Warman, supra* n. 2, [65].
8. *Scandecor Developments AB v. Scandecor Marketing AB* [2002] FSR 7.

to Section 46(1)(d) of the UK Trade Marks Act 1994 (the equivalent of Section 22(1)(d) of the TMA). Lord Nicholls, in particular, opined that the UK 1994 Act did away with the traditional view that the only trade source permissible resided in the trade mark proprietor alone. With the removal of the *registered user* requirement, as well as the need for a connection in the course of trade between the trade mark proprietor and the goods of the licensee, the UK 1994 Act envisaged that the source of the goods under the mark can be either the owner or the licensee, depending on the circumstances. The modern approach to trade mark licences was one where the business source was either the proprietor or the licensee, depending on the circumstances. Lord Nicholls observed:

> 'What is the message which a trade mark conveys today? What does a trade mark denote? It denotes that goods bearing the mark come from one business source: the goods of one undertaking, in the words of section 1(1) of the 1994 Act. That much is clear. But what does the mark denote about that source? Must the source be the proprietor of the trade mark? On this the Act is silent. But so to read the Act would accord ill with the statutory power to grant licences. And this interpretation would turn the clock back to 1905. Even under the 1938 Act the source was not so confined. Under the 1938 Act the source could be a registered user.
>
> [...]
>
> Or is the business source the person who is for the time being entitled to use the mark, whether as proprietor or exclusive licensee? I prefer this view. The commercial background has changed greatly since 1938, and even more so since 1905.
>
> [...]
>
> Thus, the wider interpretation, according to which the source may be either the proprietor or an exclusive licensee, would not be at variance with customers' perceptions. Customers are well used to the practice of licensing of trade marks. When they see goods to which a mark has been affixed, they understand that the goods have been produced either by the owner of the mark or by someone else acting with his consent.'[9]

Applying the above approach, the House of Lords held that the absence of quality control over the licensee's use of the trade mark in a bare exclusive licence does not in itself render the mark 'deceptive' such that it is capable of being revoked within Section 46(1)(d) of the English Trade Marks Act 1994. The matter was sent on reference for a ruling from the (then) European Court of Justice, but the action was settled before a ruling was dispensed.

In *Kickapoo (Malaysia) Sdn Bhd v. The Monarch Beverage Co (Europe) Ltd* [2010] 1 SLR 1212 at [44]-[53], Andrew Phang JA expressed some concerns over accepting the approach that was taken in *Scandecor*. In particular:

(a) there was the uncertainty as to what would happen to the distinctiveness of the trade mark once an exclusive licence came to an end;

9. *Scandecor, supra* n. 8, [36]-[39].

(b) it was unclear if such a concept would apply to non-exclusive (as opposed to exclusive) licences;
(c) the concept of an evolving trade source was not irreconcilable with the traditional approach to the effect that trade source resided in the trade mark proprietor.

The learned Judge also made a passing observation which was an approach that adopted the traditional approach by way of a rebuttable presumption as a starting point, which presumption could be rebutted by clear evidence that the public identified the licensee as the trade source. To address rebuttal evidence, factors such as the duration of the licence as well as efforts spent on advertising and marketing would also be relevant.

Insofar as evidence is to be proffered to rebut the presumption which the Court in Kickapoo outlined, that is another facet of genuine use of a trade mark which may also be recognized in future.

### B. Volume, duration and frequency of use

In terms of the duration of use that should inform the decision as to whether a trade mark is vulnerable to revocation, Section 22(1) TMA provides that a trade mark may be revoked on any of the following grounds:

(a) that, within the period of five years, following the date of completion of the registration procedure, it has not been put to *genuine use* in the course of trade in Singapore, by the proprietor or with his consent, in relation to the goods or services for which it is registered, and there are no proper reasons for non-use;
(b) that such use has been suspended for an uninterrupted period of five years, and there are no proper reasons for non-use;
(c) that, in consequence of acts or inactivity of the proprietor, it has become the common name in the trade for the product or service for which it is registered;
(d) that, in consequence of the use made of it by the proprietor or with his consent in relation to the goods or services for which it is registered, it is liable to mislead the public, particularly as to the nature, quality or geographical origin of those goods or services.

Trade mark owners should be mindful of commencing *use* of a trade mark within five years of the completion of the registration procedure, or periods of suspension of use lasting up to five years.

Quite apart from volume and frequency of use, the requirement of genuine use also mandates that use must be *in* Singapore, and in the course of trade. This is read from Section 22(1)(a) TMA. Singapore Courts have held that for use to be considered genuine, it does not have to be significant in the quantitative sense provided that it was in accordance with the essential trade mark (i.e. as a guarantee of origin). A token use for the sole purpose of preserving rights conferred by the registration is not genuine use.[10]

Use of a trade mark outside Singapore would also not satisfy the requirement of genuine use within the scope of Section 22(1) TMA. It should be noted, however, that if the trade mark is being used on goods for export, this would also be considered as 'use' under Section 22(2) TMA. Use also includes applying the trade mark to goods or to materials for the labelling or packaging of goods in Singapore.

Is a single use of a trade mark within the relevant five-year period sufficient to satisfy the requirement of 'genuine use'? The Singapore Court of Appeal in *Nike International Ltd v. Campomar SL* [2006] 1 SLR(R) 919 held that one single use of the trade mark could satisfy the use requirement provided that overwhelmingly convincing proof of the act is adduced.

From the discussion above it would be apparent that there is no rule that *de minimis* use cannot constitute genuine use.

### C. Use of a sign

As for the use of trade marks in a different form, Section 22(2) TMA provides that use of a trade mark 'includes use in a form differing in elements which do not alter the distinctive character of the mark in the form in which it was registered'. Instructive guidance can be taken from the judgment of Simon Thorley QC in *BUD and BUDWEISER BUDBRAU Trade Marks* [2002] RPC 38 where it was identified that the Court starts by determining what constitutes the distinctive character of the mark. For word marks, the distinctive character must surely reside in the word itself and any pictorial variations would be allowed as these differing elements do not alter the basic distinctive character of the trade mark.

In *Richement International SA v. Da Vinci Collections Pte Ltd* [2006] 4 SLR(R) 369, the Singapore High Court held that the registered mark 'DA VINCI' in plain block capitals was not an 'out of ordinary form' from the stylised works 'Da Vinci'. Use of the latter sufficiently constituted use of the registered mark.

Where a registered trade mark is surrounded by graphic elements which render it unrecognizable, the trade mark may no longer be considered to be 'used'. In *Nation Fittings*, Andrew Phang J (as he then was) commented on whether a mark remained visually recognizable:

> 'There must, in other words, be a certain point in time when the registered trade mark in question, *whilst being used literally, is subsumed within other elements to the extent that it is no longer reasonably recognizable as such.* It seems to me that when that particular situation arises, the registered trade mark concerned can no longer, in law, be said to have been used within the meaning of ss 22(1)(a) and 22(1)(b) of the TMA.' (emphasis added)

10. See *Wing Joo Loong Ginseng Hong (Singapore) Co Pte Ltd v. Qinghai Xinyuan Foreign Trade Co Ltd* [2009] 2 SLR(R) 814 at [38]-[39].

## D. Use in relation to goods and services

### *1. Use for each of the designated goods and services*

This is a necessary requirement, otherwise a proprietor faces the risk of complete or partial revocation of the trade mark for non-use, pursuant to Section 22(1) TMA.

### *2. Use for a broader category of goods and/or services*

The potential revocation of a trade mark on the basis of non-use depends on the extent to which use has been made in relation to each specified good or service (see below). If use is in respect of a broader category of goods and/or services, this may not expose a trade mark to a non-use revocation if such broader use includes and covers use of the existing specifications under the class in question.

## E. Territorial scope

A question arises over use in relation to use on a foreign website: is availability of goods over the internet considered to be use of the trade mark? The answer would be 'yes', provided that there is some other 'active step' that goes beyond the passive offering of goods on a website. In *Weir Warman* this issue arose in the context of revocation for non-use. The registered proprietor's separate fax communication that offered goods bearing the registered trade mark to a prospective customer was, *inter alia*, held to constitute an 'active step' to amount to genuine use of a trade mark. In *Weir Warman*, it was also held that the actual purchase of goods was not a mandatory requirement for genuine use to be established.

Even if goods under a registered trade mark are made available on the internet to customers of another jurisdiction, to establish genuine use the trade mark owner must go beyond the mere provision of facilities for download. Additional steps such as direct encouragement, solicitation or advertisement of goods or services must be evidently present before the courts are prepared to accept that a trade mark has been used in a particular jurisdiction, via the internet.[11]

## F. Relevant time period and its calculation

As explained above, the relevant time period for the calculation of genuine use is within a period of five years from the completion of the registration procedure, and alternatively, any continuous period of five years' suspended use during the life of a registered trade mark without proper reasons can also potentially expose a trade mark to a risk of revocation proceedings under Section 22 TMA on the grounds of non-use.

In addition, the TMA also provides that the registration of a trade mark shall not be revoked on the above grounds if use is commenced or resumed after the expiry of the five-year period and before the application for revocation is made (Section 22(3)). This is a further grace period provided by the TMA.

11. See *1-800 Flowers Inc v. Phonenames Ltd* [2001] EWCA Civ 721.

## III. EVIDENCE OF GENUINE USE OF A TRADEMARK

### A. When use should be proved: timing of lack of genuine use claims

#### *1. When filing an application for a trademark*

As mentioned above, under Section 5(2)(e) TMA, a trade mark application must be made in a prescribed manner, and *inter alia*, contain a statement from the applicant stating:

(i) that the trade mark is being used in the course of trade by the applicant or with his consent, in relation to those goods or services; or
(ii) that the applicant has a *bona fide* intention that the trade mark should be so used.

At the point of filing an application for a trade mark, evidence of use may also be tendered to assist in overcoming potential objections for refusal under Section 7(1)(b), (c) or (d) TMA. These objections apply to trade marks that are devoid of distinctive character; consist of signs and indications which designate quality or geographical origin and other characteristic of goods; or consist exclusively of signs or indications which have become customary in current language or in the *bona fide* or established practices of trade. Such evidence of use should demonstrate that prior to the date of application, the trade mark has in fact acquired a distinctive character as a result of use made of it (Section 7(2) TMA).

Chapter 6 of the Singapore Trade Mark Work Manual ('Manual') provides further guidance as to how evidence of use (to prove acquired distinctiveness) is to be proffered (see below).

Section 9(1) TMA provides that notwithstanding an earlier trade mark, an applicant can adduce evidence of honest concurrent use in support of an application, (note that 'honest concurrent user' does not apply in a conflict scenario in Section 8(4) TMA). The other limitation is that the 'honest concurrent user' provision will not be accepted for registration if there are opposition proceedings made under Section 8(1)-(3) or (7) TMA. The evidence should give indications that the public is used to distinguishing the two trade marks and there is no confusion arising from their respective use.

#### *2. In the course of opposition proceedings*

For trade mark opposition proceedings, both opponent and applicant are required to file Statutory Declarations setting out the evidence that they rely on in support of the opposition/application (see Rule 31A Trade Mark Rules S635/98 ('TMR'), subject to strict rules on extensions (Rules 32-34 TMR)).

As a general indication, evidence in support of an opposition/application would traverse the following:

(a) first use of the trade mark in Singapore;
(b) first use of the trade mark globally;
(c) countries where the trade mark has been used;

(d) details of global trade mark applications/registrations;
(e) turnover of business in Singapore in relation to the use of the trade mark;
(f) turnover of business globally in relation to the use of the trade mark;
(g) advertising and promotion in Singapore in relation to the use of the trade mark;
(h) advertising and promotion globally in relation to the use of the trade mark;
(i) domain name registrations;
(j) third party licensing practices.

Where a 'well known trade mark in Singapore' ('WKTMS') is raised as a basis to prevent an application mark from being registered, Section 2(7) TMA sets out additional criteria to establish that the earlier trade mark is a WKTMS:

(a) the degree to which the trade mark is known to or recognized by any relevant sector of the public in Singapore;
(b) the duration, extent and geographical area of –
   (i) any use of the trade mark; or
   (ii) any promotion of the trade mark, including any advertising of, any publicity given to, or any presentation at any fair or exhibition of, the goods or services to which the trade mark is applied.
(c) any registration or application for the registration of the trade mark in any country or territory in which the trade mark is used or recognized, and the duration of such registration or application;
(d) any successful enforcement of any right in the trade mark in any country or territory, and the extent to which the trade mark was recognized as well known by the competent authorities of that country or territory;
(e) any value associated with the trade mark.

Section 2(8) TMA also provides that if a trade mark is determined to be well known to any 'relevant sector of the public' in Singapore, that trade mark is also potentially a WKTMS. Under Section 2(9) TMA, the phrase 'relevant sector of the public' is defined to include any of the following:

(i) all actual consumers and potential consumers in Singapore of the goods or services to which the trade mark is applied;
(ii) all persons in Singapore involved in the distribution of the goods or services to which the trade mark is applied;
(iii) all businesses and companies in Singapore dealing in the goods or services to which the trade mark is applied.

Evidence of use of a trade mark will also be relevant to establishing the above criteria.

### 3. *In cancellation proceedings based on lack of use*

Proceedings for revocation under Section 22 TMA are typically brought before the Intellectual Property Office of Singapore ('IPOS') although Section 22(5) TMA allows an aggrieved person to bring proceedings for revocation to the High Court. Proceedings before IPOS commence by the filing of Form TM28, accompanied by a statement of grounds on which the application is made. Within two months after the receipt of this application the trade mark proprietor may file a counter-statement (Form HC6) setting out the grounds in support of the registration. In the case of an application on the grounds of Section 22(1)(a) or (b) TMA, the counter-statement should be accompanied by a statutory declaration setting out, *inter alia*:

(1) evidence of the use by the proprietor of the trade mark in relation to goods or services for which it is registered during the period of non-use alleged by the applicant (in Form TM28);
(2) evidence supporting proper reasons for non-use during the period of non-use alleged by the application;
(3) evidence of commencement or resumption of use of the trade mark in relation to the goods or services for which it is registered on a date which falls after the period of non-use alleged by the applicant in Form TM28, and before the three-month period immediately preceding the date of the application.[12]

#### 4. *As a defence in infringement/cancellation proceedings*

In infringement/cancellation (i.e. invalidity) proceedings, it is possible to claim a revocation of the earlier mark on the basis of non-use. As infringement proceedings are commenced in the High Court, a claim for revocation of the registered trade mark that is alleged to be infringed on the basis of non-use would typically be made by way of a counterclaim. If the claim for revocation is made by way of a counterclaim in an infringement action, the claim for infringement and counterclaim for revocation will be heard together.

Where the proceedings for invalidity of a registered trade mark on the basis of similarity to an earlier mark are brought before IPOS, a claim for revocation of the earlier mark will have to be made by way of a separate action. Where two separate actions are initiated at the IPOS level, it is possible that IPOS may, on the application of the parties, stay the action for invalidation pending the outcome of the revocation action.

### B. Burden of proof

In the case of revocation of a registered trade mark for non-use, the burden of proof rests on the applicant to prove non-use. To defeat an application for non-use, the proprietor of the registered trade mark bears the burden to prove genuine use within the relevant five-year period.[13] Section 105 TMA provides that where a question arises as to the 'use' to which a registered trade mark has been put, it is for the proprietor to show what use has been made of it.

The Singapore courts have noted that where a proprietor cannot show a single sale during the relevant five-year period, the court may infer that this is due to non-use of the trade mark.[14]

12. Rule 58 TMR.
13. Confirmed by the Singapore Court of Appeal in *Nike International Ltd v. Campomar SL* [2006] 1 SLR(R) 919 at [21]
14. *Reemtsma Cigarettenfabriken GmbH v. Hugo Boss* [2003] 4 SLR(R) 155.

## C. Relevant evidence

### *1. Use on the product or its packaging*

As set out above, Section 27(4) TMA provides a non-exhaustive list of 'uses' for the infringement provisions that is relevant for determining whether there has been use for the purposes of infringement and revocation proceedings. Section 27(4)(a) TMA specifically covers use of the sign on goods or the packaging thereof.

The use must also strictly pertain to the relevant goods and services for which the trade mark is registered. This is well illustrated by *Intuition Publishing Ltd v. Intuition Consulting Pte Ltd* [2012] SGHC 149. The proprietor held a word mark registration in class 9 for computers and computer programmes. The proprietor used the word mark on its website, which was principally used to market technology-enabled learning services. The Court ordered revocation of the class 9 registration on the basis that the use in question pertained to technology-enabled learning services as opposed to use for computers and computer programmes. Hence use of a trade mark per se is clearly insufficient and use must pertain to the relevant goods and services designated under the trade mark registration.

### *2. Use in advertising*

Use in advertising can constitute relevant evidence of use of a trade mark as Section 27(4)(e) specifically covers use of the sign in advertising.

### *3. Use on the internet*

As set out above, the availability of goods over the internet can be considered to be use of the trade mark in Singapore, provided that there is some other 'active step' that goes beyond the passive offering of goods on a website.[15]

Even if goods under a registered trade mark are made available on the internet to customers of another jurisdiction, to establish genuine use, the trade mark owner must go beyond the mere provision of facilities for download. Additional steps such as direct encouragement, solicitation or advertisement of goods or services to consumers in a particular jurisdiction must be evidently present before the courts are prepared to accept that a trade mark has been used in a particular jurisdiction via the internet.[16]

### *4. Form of evidence*

Guidance on the evidence of acquired distinctiveness to support the operation of a Section 7(2) TMA proviso are set out in Chapter 6 of the Manual. The evidence that is submitted should answer the following questions in the affirmative (page 5 of the Manual):

15. *Weir Warman, supra* n. 2, [106] to [108].
16. See *1-800 Flowers Inc v. Phonenames Ltd* [2001] EWCA Civ 721.

(i) Has the trade mark been used as a trade mark, i.e. as a means of identifying trade origin of the goods?
(ii) Has the applicant promoted the trade mark as a trade mark?
(iii) Does the evidence show, as a matter of fact, that the trade mark is operating in the market place as an indicator of origin?
(iv) Has the relevant public (or a significant proportion thereof) come to rely upon the mark, in the course of trade, as a means of identifying the trade origin of the goods?
(v) Although the applicant need not necessarily have used the mark as the only means of identifying the trade origin of the product as there is no rule that two or more trade marks cannot operate alongside each other, the trade mark applied for must, by itself, come to foster a concrete expectation amongst the relevant public that the goods bearing the mark originate from, or are under the control of, a single undertaking.

The evaluation of evidence of use is conducted by reference to the period of use (page 5 of the Manual), extent of use (territory) (page 6 of the Manual), advertising expenditure (page 6 of the Manual) and nature of exhibits (page 7 of the Manual). For exhibits, the Registrar will have regard to materials such as articles, brochures, catalogues, balance sheets, statements of account, annual reports, newsletters, magazines, stationery, invoices, receipts, sales advices, samples, staff business cards, delivery vans, advertising, posters, menus, bar lists, wine lists (for restaurants), room folders, guest information cards (for hotels), and timetables (travel services). (page 9 of the Manual).

### 5. *Distinction between use of a company name and use of the services offered by the company*

As mentioned above, genuine use must be as a trade mark, i.e. so as to indicate origin. Further, the use relied upon must be use in the course of trade and in relation to goods or services falling within the specification. Hence, mere use of a company name that is not in relation to the goods or services falling within the trade mark specification will not suffice as an instance of genuine use of the trade mark.

## IV. JUSTIFICATION FOR NON-USE

### A. Obstacles to use

Under Section 22(1)(a) and (b) TMA, a trade mark is susceptible to revocation for non-use unless there are proper reasons for the non-use. In terms of legitimate reasons to justify non-use, the courts are likely to have regard to Article 19(1) TRIPS Agreement, and the reference to 'circumstances arising independently of the will of the owner of the trade mark which constitute an obstacle to the use of the trade mark, such as import restrictions on or other government requirements for goods or services protected by the trade mark.'

The Singapore courts may also have regard to exceptional delays giving rise to proper reasons for non-use. In *Magic Ball Trade Mark* [2000] RPC 439, unexpected delays that arose in developing a new manufacturing technique for a new type of lollipop were accepted as reasons for non-use. The activities of an independent manufacturer fell outside the control of the proprietor, who at all times evinced the intention to use the mark, but for the unexpected delays.

In *Nation Fittings* the High Court prescribed helpful guidance as to what may constitute 'proper reasons' for non-use (at [129]):

(a) The onus is on the proprietor to show that he has proper reasons for not using the trade mark during the relevant five-year period.
(b) The word 'proper' means 'apt, acceptable, reasonable, justifiable in all the circumstances', and it is intended to cover abnormal situations in the industry or the market, or even some temporary but serious disruption affecting the registered proprietor's business.
(c) To constitute proper reasons for non-use, the obstacle to use must have arisen independently of the will of the proprietor and ordinary delays in producing a new product bearing the trade mark would not constitute a valid reason.

### B. Circumstances in which non-use may be cured

Under Singapore law, non-use of a registered trade mark, in the context of Section 22(1)(a) or (b) TMA cannot be cured generally unless use is commenced or resumed after the expiry of the five-year period *and* before the application for revocation is made.[17]

Section 22(4) TMA provides that any commencement or resumption of use referred to in Subsection (3) after the expiry of the five-year period but within the period of three months before the making of the application for revocation shall be disregarded *unless* preparations for the commencement or resumption began *before* the proprietor became aware that the application might be made.

## V. CONSEQUENCES OF NON-USE

### A. Consequences of non-use and relevance of the context

When the trade mark is revoked (completely or partially) for non-use under the TMA, the rights of the proprietor shall be deemed to have ceased to that extent as from:

(a) the date of the application for revocation; or
(b) if the Registrar or court is satisfied that the grounds for revocation existed at an earlier date, that date.[18]

### B. Extent of any revocation

Based on Section 22(7) TMA, a trade mark registration continues to have effect prior to an application for revocation. Liabilities for trade mark infringement in respect of

17. Section 22(2) TMA.
18. Section 22(7) TMA.

past activity prior to revocation would be preserved, and the trade mark owner still has *locus standi* to bring proceedings.

Partial revocation is provided for under Section 22(7) TMA. The Singapore High Court has held that its task is 'to limit the specification so that it reflects the circumstances of the particular trade and the way that the public would perceive the use'.[19]

Last year, IPOS partially revoked a registered trade mark for 'watches', and not 'jewellery', on the basis that the proprietor only demonstrated use within the five-year period following the completion of registration for the latter.[20]

### C. Timing of any revocation

As explained above the relevant time period for commencing genuine use to avoid a non-use revocation action is within a period of five years from the completion of the registration procedure. Thereafter, any continuous period of five years' suspended use during the term of registration of a registered trade mark without proper reasons can also potentially expose a trade mark to a risk of revocation proceedings under Section 22 TMA on the grounds of non-use.

19. *Bluestar Exchange (Singapore) Pte Ltd v. Teoh Keng Long* [2003] 4 SLR(R) 92; *Weir Warman, supra* n. 2, [112].
20. *Van Cleef & Arpels SA v. FMTM Distribution Limited* [2017] SGIPOS 06.

# Spain

*David Gómez and Pedro Merino Baylos*[1]

## I. INTRODUCING GENUINE USE OF TRADEMARKS

### A. Legal framework

The obligatory use of trademarks stands out as the fundamental principle in current trademark systems. Indeed, the European Union law and national laws of Member States set out the general obligation to use a registered trademark. The establishing of this uniform fundamental principle entails the end of a period in which the majority of the national systems had rather fragmented and partial regulation.

The former Spanish Trademarks Act 32/1988 dedicated three articles to regulation of the obligation of use: in Article 4 it established the obligation of use of the trademark and its structure; in Article 53(c) it configured the infringement of the obligation of use as a ground for revocation; and finally, in Article 7.2 it addressed a very specific aspect of use, which is the requirement for the trademark owner to file a declaration of use when applying for renewal.

This regulation did not met the criteria of the EU legislation on trademarks, resulting in serious differences, not only terminological, but also substantive. In particular, the Spanish Trademarks Act of 1988 contradicted the Directive 89/104EEC (hereinafter referred to as 'the Directive'), as it did not admit the lack of use in the course of the cancellation proceedings (Article 11.1 of the Directive). Neither did it limit the effects of partial use of the products and services for which the trademark had been effectively used (Article 11.4 of the Directive) and the temporal aspects of the procedure of cancellation for lack of use (Article 12.1 of the Directive), nor did it include the faculty established in the Directive for the Member States to include, either in the course of the registration procedure or in the course of the trademark infringement procedure, the possibility to invoke a lack of use of the opponent's or the claimant's trademark (Articles 11.2 and 3 of the Directive).

---

1. This work has been written with the help of Joanna Rozanska.

The current Spanish Trademarks Act 17/2001 has introduced relevant amendments to the regulation of the obligatory use of a registered trademark, adapting the Spanish legislation to EU law. First of all, it eliminated the intervention *ex officio* of the Trademark and Patent Office, complying with the rule contained in Article 13.4(iii) of the Trademark Law Treaty of 27 October 1994, which prohibits requiring declaration of use of the trademark when applying for renewal. Furthermore, interested third parties can invoke infringement of the legal burden of use of the trademark in three scenarios. The first scenario is infringement proceedings: according to Article 41.2 of the Spanish Trademarks Act 17/2001 (which transposes Article 11.3 of the Directive), the defendant can argue that the claimant has not used the allegedly infringed trademark within the five years preceding the claim. The second scenario is the cancellation proceedings: according to Article 52.2 of the Spanish Trademarks Act (that transposes Article 11.1 of the Directive), the defendant proprietor of the trademark can argue that the claimant has not fulfilled the legal burden of use of the earlier trademark on which the cancellation action is based. The third scenario is revocation proceedings of registered trademarks: according to Article 58 of the Spanish Trademarks Act (which transposes Article 12.1 of the Directive), interested third parties can bring an action requesting the declaration of revocation of the trademark for lack of use.

In this regard, the obligation of genuine use of a trademark in the current Spanish legislation is reflected fundamentally in two articles, namely Articles 39 and 58 of the Spanish Trademarks Act 17/2001. The first establishes the obligation of use of a trademark within a period of five years from the publication of its grant, in relation to products and services for which it has been registered. The article states that, if a trademark has not been put to genuine use in the legally established period or if the use has been suspended for a period exceeding five years, then the trademark will be subject to the sanctions set out by the Trademarks Act, unless the proprietor provides justifiable reasons for non-use. Secondly, according to Article 58, the burden of proof of the genuine use or the provision of justifiable reasons for non-use, is on the proprietor of the trademark. However, cancellation for lack of use will not be declared when, during the interval between the expiry of the five-year period and the filing of the application for cancellation, genuine use of the trade mark has been started or resumed. Finally, the commencement or resumption of use within a period of three months preceding the filing of the application or counterclaim which will not begin before the expiry of the continuous period of five years of non-use, shall be disregarded when preparations for the commencement or resumption occur only after the proprietor becomes aware that the application or counterclaim may be filed.

In light of all of the above, there can be no doubt that the obligatory use has been reinforced substantially in the context of the Spanish Trademarks Act 17/2001.

### B. Purpose and function of genuine use: policy considerations

As Lobato[2] (the outstanding scholar) states, the use of trademark is a burden, in its technical-legal sense, for the trademark owner. In other words, it is not a mere

2. Manuel Lobato, *Commentary on the Spanish Trademarks Act 17/2001* (Civitas, Madrid, 2002).

legally enforceable obligation, but an imperative in his own interests: the owner of the trademark must use it in order to prevent revocation.

Thus, the case-law justifies the burden of use with the active character of the trademark as the mere inscription in the Register is not enough. Accordingly, the General Court of the European Union in its Judgment of 15 September 2011 declared (emphasis added):

> 'The rationale for the requirement that a mark must have been put to genuine use in order to be protected under EU law is that *OHIM's register cannot be compared to a strategic and static depository granting an inactive proprietor a legal monopoly for an unlimited period.*'

The aim is to prevent the legal system from protecting unnecessary registrations that provide unfounded monopolies. In this regard, the unquestionable result is that the trademark system should not give legal protection to a trademark proprietor who does not use a trademark registration. In this respect, the recent Judgment of Chamber 28 of the Appeal Court of Madrid of 22 March 2017 stated (emphasis added):

> 'A trademark right has predominantly registry character, providing its proprietor with a monopoly regime in relation to using the distinctive sign that, if it wasn't for the legal configuration of the right, would remain naturally in the public domain, in such a manner that *the attribution of this exclusivity is justified solely because, when distinguishing products and services of the company on the market, the trademark benefits consumers and competitors, strengthening the economic and concurrence system.*'

Moreover, obligatory use fulfils an important functional purpose. In particular, it significantly facilitates the granting of new trademark applications, cleaning up the Trademark Registry which would otherwise be saturated with unused distinctive signs. Therefore, it seems clear that, as we revoke unused registered trademarks, the saturation will decrease, benefitting the position of new trademark applicants enormously. We can refer to the Conclusions issued in the *Ansul* case (Case C-40/01), by Advocate General Ruiz-Jacobo Colomer (emphasis added):

> 'Trade mark registers *cannot simply be repositories for signs hidden away, lying in wait for the moment when an unsuspecting party might attempt to put them to use, only then to be brandished with an intent that is at best speculative.* The opposite is true; *they must faithfully reflect the reality of indications used by undertakings in the market to distinguish their goods and services. Only marks that are used in commercial life should be registered by offices with responsibility for industrial property matters.*'

## II. WHAT QUALIFIES AS GENUINE USE OF A TRADEMARK

### A. Nature of use

#### *1. Use consistent with the function of the trademark*

The first paragraph of Article 39 of the current Spanish Trademarks Act of 2001 qualifies the obligatory use of the trademark with a double adjective: the use of the trademark must be 'genuine' and 'real', whereas Article 10.1 of the Directive refers to the 'genuine use' of the trademark. In this regard, in the Conclusions issued in the case *Ansul* case, Advocate General Ruiz-Jacobo Colomer highlighted as follows (emphasis added):

> 'Examination of the various language versions of the Directive leads me to the conclusion that *the kind of use intended by the Community legislature is what may be described as sufficient in relation to the function performed by a trade mark.*'

Neither the Directive nor the Spanish Trademarks Act 2001 list which acts count as genuine use. In this regard, as highlighted by the Judgment of 17 June 2014, Madrid Commercial Court Number 12 (Juzgado de lo Mercantil) stated:

> 'Article 39 of the Spanish Trademarks Act requires a genuine and real use. The double legal requirement, which is actually redundant, implies the requirement of public and external use, entailing an objective presence on the market for which the sign has been conferred, which configuration should remain stable, so that the commercial impression of the consumers would perceive a trademark as the corporate source and provenance of the products.'

Additionally, the sign must be used as a trademark which, according to Article 4 of the Spanish Trademarks Act 2001, means that it should be capable of distinguishing the goods or services of one undertaking from those of other undertakings.

The use of the trademark to distinguish products and services requires establishing an adequate connection between the distinctive sign and the referred products and services. In a concrete situation, in order to determine if the connection is there, it should be taken into account whether consumers perceive that the sign identifies the products and services for which it has been registered.

On the contrary, the use of a sign does not qualify as trademark use when it fulfils functions beyond the scope of the Trademark Law, by the way of example, when a trademark is used for academic, informative or decorative purposes, or in the course of the administrative procedures to obtain marketing authorization. Nor is there a genuine use of trademark when the sign identifies elements other than products and services, for instance the name of individuals or companies, establishment emblems or domain names.

In that context, the Judgment of the General Court of the European Union in *Völkl GmbH & Co. KG Office for Harmonisation in the Internal Market (Trade Marks and Designs)* (OHIM) ECJ 2011/406, stated (emphasis added):

> 'There is genuine use of a trade mark where the mark is used in accordance with its essential function, which is to guarantee the identity of origin of the goods or services for which it is registered, *in order to create or preserve an outlet for those goods or services*; genuine use does not include token use for the sole purpose of preserving the rights conferred by the mark.'

Furthermore, it is to be noted that when assessing the concurrence of the requirement of genuine and real use, it is irrelevant whether the products or services covered by the trademark have achieved commercial success on the market, as highlighted by Chamber 28 of the Appeal Court of Madrid in the Judgment of 27 February 2012 (emphasis added):

> 'The requirement of genuine use *does not claim to evaluate the commercial success, nor control the economic strategy of the company, nor reserve the protection of trademark exclusively to quantitatively important commercial exploitations* [...] Even minimal use can therefore be sufficient to qualify as genuine, on condition that it is deemed to be justified, in the economic sector concerned, for the purpose of preserving or creating market share for the goods or services protected by the trademark.'

Additionally, the use must be public, meaning it shall be apparent to actual or potential customers of the goods or services. As stated by the Supreme Court in the Judgment of 23 of June 2006 (emphasis added):

> 'In order to appreciate the genuine and real use, relevant to the trademark, *it shall be apparent in the commercial sector for which it has been registered*, and, therefore, fulfil the purpose for which its proprietor has been granted an exclusive right.'

Products bearing the trademark shall be introduced on the market through distribution channels and commercial networks. The use in the private sphere or purely internally, within the company or a group of companies, is not use in the course of trade.[3] That occurs when, for example, the proprietor of the registered trademark limits his activities to preparations for use, such as designing a sign that constitutes a trademark, printing off labels or packaging that incorporate the sign, producing containers without a real intention to launch the product into the market with the aim of getting a market share. These are the acts that constitute internal use, and consequently not use in the course of trade for the present purposes, unless preparations are already under way to secure customers, particularly in the form of advertising campaigns for the imminent commercialization.[4]

In this regard, in the Conclusions issued in the *Ansul* case, Advocate General Ruiz-Jacobo Colomer points out as follows (emphasis added):

> 'On the other hand, *private use that does not extend beyond the internal province of the proprietor's undertaking does not count, in so far as it is not directed at securing a place in the market*. Preparations for the marketing of goods or services do not, therefore, constitute sufficient or effective use, nor does getting them shop-ready or storing them where they do not leave the undertaking's premises.'

### 2. *Use in the course of trade*

#### *i. Public use*

Effective protection of the trademark depends on it being effectively used in the course of trade. The meaning of 'course of trade' for the purpose of genuine use is

3. Ansul, ECJ C-40/01, nr. 37.
4. Ansul, TJCE C-40/01, ap. 37

not the same as for the purpose of trademark infringement. By way of example, mere storage of products constitutes a trademark infringement when it is carried out in order to be offered for sale or commercialized.[5] However, this use is not 'public' as required in the present context.

#### *ii. Commercial purpose*

The term 'course of trade' should be put in the context of commercial or industrial continuous activity. For example, a trademark used by a non-profit organization amounts to a genuine and real use if it is in the market to sell or commercialize products and services.

In this regard, the Judgment of the European Court of Justice of 27 January 2004, case C-259/02, Le Mer stated as follows (emphasis added):

> 'When assessing whether use of the trade mark is genuine, regard must be had to *all the facts and circumstances relevant to establishing whether the commercial use of the mark is real in the course of trade, particularly whether such use is viewed as warranted in the economic sector concerned to maintain or create a share in the market for the goods or services protected by the mark, the nature of those goods or services, the characteristics of the market and the scale and frequency of use of the mark. When it serves a real commercial purpose, in the circumstances cited above, even minimal use of the mark or use by only a single importer in the Member State concerned can be sufficient to establish genuine use within the meaning of the Directive.*'

It is worth noting the Judgment of Chamber 28 of the Appeal Court of Madrid of 24 February 2012 which stated (emphasis added):

> 'The concept of use "in the course of trade" from Article 34 of the Trademark Law and the "genuine and real" use of the trademark from Articles 39 and 55 of the Trademark Law, the lack of which may determine revocation, do not coincide.
>
> As a consequence, *it is not contradictory that giving gifts consisting of products with the infringing sign could constitute use in the course of trade within Article 34 of the Trademark Law and that this activity does not imply genuine and real use of the trademark* that satisfies the legal burden imposed by Article 39 of the Trademark Act which, in case of non-compliance within the statutory period, causes its revocation according to Article 55 of the above-mentioned legal text.'

### 3. *Use by the owner or with its consent*

Furthermore, the proprietor of the trademark can comply with the use requirement either directly or indirectly. The proprietor fulfils the condition directly when he uses the trademark to differentiate products manufactured or distributed by himself. The use will be fulfilled indirectly when the trademark is used by a third party with the consent of the proprietor.

5. Opposition Guidelines, part 6, *Proof of Use*.

The proprietor of the registered trademark can authorize a third party to use a trademark expressly or implicitly. Although the consent granted by the proprietor can derive from other legal transactions, in most cases this authorization will be reflected in licence agreements that confer on the licensee the faculty to use the trademark in trade. It is irrelevant whether the licence granted to a third party is exclusive or non-exclusive. We should bear in mind that a licence that extends exclusively to some of the products or services covered by the trademark, could lead to a partial use of the corresponding trademark. It should be highlighted that, in order to consider if the trademark is used by the licensee, it is not essential that the licence be in the Trademark Register. Additionally, when analysing the use carried out by the licensee we shall take into consideration the characteristics of the company of the trademark licensee, such as its size.

As regards the implicit consent of the trademark proprietor, the issue has been interpreted in the *Vitafruit* case in the Judgment of the Court of First Instance of 8 July 2002 (Case T-203/02). In this case, the applicant for the EU trademark argued that the proprietor of the Spanish trademark 'Vitafruit' had failed to show that the trademark had been used with his consent. The Court rejected this argumentation, stating:

> 'Where an opposing party maintains that the use of an earlier trade mark by a third party constitutes genuine use [...] *he claims, by implication that he consented to that use.*'

Consequently, as a general principle the use of the registered trademark by a third party equates to the use made by the registered owner provided the third party has the consent or authorisation of the owner. However, it is understood that the consent has not been given when the owner did not have knowledge of the use made by the third party or when it was clear that the third party is infringing the registered trademark. The consent should not be considered has been given either when granting such a consent lacks any relevance because the third party could continue using his trademark in Spain without the consent of the registered owner, as happens in cases of exhaustion of trademark rights.

Finally, in case of collective and warranty trademarks, the requirement of use can be fulfilled by any person entitled to use them (Article 75 of the Spanish Trademarks Act 2001).

### B. Volume, duration and frequency of use

European case-law and doctrine state that a trademark is genuinely used when it is used in accordance with its essential function, which is to guarantee the identity of the origin of the goods or services for which it is registered in order to create or preserve an outlet for those goods or services which does not include token use for the sole purpose of preserving the rights conferred by the trademark. In this regard, we shall recall some of the criteria for sufficient use to create or preserve an outlet for goods or services. The Judgment of the Court of Justice of 27 January 2004 in *La Mer Technology Inc. v. Laboratoires Goemar SA* (C-259/02) pointed out (emphasis added):

'The question of whether use is sufficient to preserve or create market share for those products or services depends on several factors and on a case-by-case assessment which is for the national court to carry out. *The characteristics of those products and services, the frequency or regularity of the use of the mark, whether the mark is used for the purpose of marketing all the identical products or services of the proprietor or merely some of them, or evidence which the proprietor is able to provide, are among the factors which may be taken into account.*'

Additionally, Chamber 28 of the Appeal Court of Madrid in its Judgment of 23 November 2012 highlighted that the factors for obligation of use shall be analysed globally (emphasis added):

'Thus, by the way of example, information such as the commercial volume of all the use on the one hand and the duration of the period in which that use occurred, and the frequency of those results are highly important. *When we consider them, we can conclude that a low volume of goods marketed under that trademark may be compensated for by a high intensity or a certain constancy in time of the use of that trade mark or vice versa. In the same way, the broad and continued market introduction of a certain general consumer good under a particular trademark, through an extensive distribution network reinforced by intense advertising campaigns, does not have to result in high sales volume, but in this scenario we shall not deny the existence of a genuine and real use of a trademark.*'

Assuming that a commercialization of products is a perfect example of act of genuine use, we shall define when the commercialization of products covered by the trademark should be considered sufficient to fulfil the requirement of use. First of all, we should stress that commercialization of products bearing the trademark must be continuous. As pointed out in the Conclusions issued by Advocate General Ruiz-Jacobo Colomer in the *Ansul* case (Case C-40/01), (emphasis added):

'In any event, irrespective of the volume of transactions carried out under the mark or their frequency, *the use must be consistent, and not sporadic or occasional.*'

Secondly, we shall indicate that the sales turnover of the products covered by the trademark must achieve a certain level. Nevertheless, the sales turnover must be balanced considering two factors: (i) the nature of products and (ii) the dimension of the company proprietor of the trademark.

When analysing the factor of the nature of products covered by the trademark, we shall apply different criteria, depending on whether the trademark is used to identify mass consumption goods or to distinguish specialized or high-cost goods. If goods covered by the trademark are mass consumption products, there shall be a relatively high sales turnover. On the contrary, if goods covered by the trademark are specialized or high-cost products, the sales turnover could be relatively low. The relevance of this factor was highlighted by Advocate General Ruiz-Jacobo Colomer in the *Ansul* case (Case C-40/01), (emphasis added):

'The point beyond which commercial use of the trade mark may be considered appropriate and genuine is directly related to the type of goods or category of service. As the Commission noted in its written observations, *the degree of use required of a mark employed in relation to luxury goods of limited distribution cannot be the same as for a mark employed in relation to mass consumption goods*.'

In the context of the nature of the product covered by the trademark, European case-law tempers the requirement of the sales turnover. Accordingly, in the Judgment of 11 March 2003 the Court of Justice stated as follows:

> 'Assessing the circumstances of the case may thus include giving consideration, inter alia, to the nature of the goods or service at issue, the characteristics of the market concerned and the scale and frequency of use of the mark. Use of the mark need not, therefore, always be quantitatively significant for it to be deemed genuine, as that depends on the characteristics of the goods or service concerned on the corresponding market.'

This approach has been confirmed by the Judgment of the Court of Justice of 27 January 2004 in *La Mer Technology Inc. v. Laboratoires Goemar SA* (C-259/02), (emphasis added):

> '*Even minimal use can therefore be sufficient to qualify as genuine, on condition that it is deemed to be justified, in the economic sector concerned*, for the purpose of preserving or creating market share for the goods or services protected by the mark.'

The size of the company owning the trademark is the second factor to take into consideration when analysing the sales turnover of products covered by the trademark. This circumstance was mentioned expressly in the Preamble to the German Act of 4 September 1967, which implemented the principle of obligatory use of the registered trademark in the German system. The German doctrine maintains that we shall not require the same sales turnover from a large company as from small or medium-sized ones.

In summary, when analysing the intensity of use of a trademark, we shall take into account the nature of the products and services. However, it does not mean there is no correlation between the size of the company and the sales required to appreciate genuine use of a trademark. This correlation is clear when a large company and a small one distribute comparable products and services in the same market. In this scenario, the same sales turnover may be considered insufficient with respect to a large company, and sufficient in relation to a small one.

## C. Use of a sign

Over time, the proprietor of a trademark could modify the form in which the trademark is actually used in the course of trade in order to adapt it to market trends. For this reason, the legal system establishes a mechanism that allows the proprietor to use – within certain limits – the trademark in a form that differs from the form in which it was registered. Indeed, the Judgment of Chamber 28 of the Appeal Court of Madrid of 8 May 2008 declared (emphasis added):

> 'Article 39.2(a) of the Spanish Trademarks Act admits, in fact (and being more accurate than Article 4.2 of the Spanish Trademarks Act 32/1988) *the possibility that use of a trademark in a form different from registered could satisfy the legal burden of use of the registered trademark*. For this, *it is required to use a trademark in a form differing in elements which do not alter the distinctive character of the trademark in the form in which it was registered*. This is the case when the trademark evolves due to its updating, as long as it keeps the distinctive character of the trademark as registered, *which would imply maintaining its distinctive elements, even if the proprietor introduces some non-substantial variations or changes in its ancillary elements*. In consequence, *a sign would fulfil the burden of use of a registered trademark if in the course of trade the signs are perceived as the same trademark, as the commercial impression they produce in the consumer has not be altered*.'

The admissibility of the divergence between the registered trademark and the way it is used depends on a relatively complex assessment. In this regard, the Supreme Court in the Judgment of 22 September 1999 applied the following rationale (emphasis added):

> 'Now is the moment to determine when the registered trademark is used in the form that is different but not incorrect and consequently unlawful, and in this regard we shall invoke the North American doctrine of the "commercial impression", also known in the German doctrine as the "*Unterscheidungskraft*", establishing that *the change of form of the trademark will not be considered relevant if it doesn't alter the commercial impression of the registered trademark*.'

In view of the above, what is decisive is determining whether the version of the trademark that is genuinely used in the market provokes in the consumers the same commercial impression generated by the form in which the trademark is registered. To this end, we should carry out two basic operations: first, determining which element attributes a distinctive character to the trademark; and secondly, examining if this element has been altered while using the trademark. In doing this exercise, we should rely on the perception of the average consumer interested in acquiring the goods or services in question. In this regard, according to the Spanish scholar Aréan Lalín, we should establish if the consumers 'contemplate the old and the modern version of the correspondent sign as one and the same trademark.[6]'

## D. Use in relation to goods and services

### 1. *Use for each of the designated goods and services*

The proprietor of the trademark has the obligation to genuinely use the trademark in respect of the goods or services for which it is registered. For the partial use of the trademark, Article 60 of the Spanish Trademarks Act 2001 states that if the trademark has been used only for part of the goods or services for which the trademark is registered, the declaration of revocation shall extend exclusively to the goods or services affected by the non-use. The previous Spanish Trademarks Act 32/1988

6. Manuel Areán Lalín (auth.-ed.), *The change of the form of the trademark* (Santiago de Compostela, 1985), 135.

contained a provision that had not been foreseen by the Directive, regulating the importance of the partial use in terms of fulfilling the obligation of use (Article 4.4 of the Spanish Trademarks Act 1988). In accordance with this provision, the use of the trademark for a certain good or service could serve to prove the use in relation to other goods or services from the same class of the International Nomenclature, in relation to similar goods or services that did not belong to the same class, or in relation to goods and services that could create an association with consumers with the corporate origin. Undoubtedly, this was beneficial for the trademark proprietor, as in many classes he was not exposed to cancellation of trademarks which were not used, or partial revocation of the trademark with respect to goods and services that had not been put to use covered by the trademark. The reason that probably justified this divergence between the UE regulation and the Spanish Trademarks Act 32/1988 was that the Spanish trademark legislation was based on the *uniclass* registration system. The *multiclass* cancellation, endorsed by the Directive, has special importance in the *multiclass* system.

### 2. *Use for a broader category of goods and/or services*

If a trademark has been registered for a category of goods or services which is sufficiently broad for it to identify within it a number of subcategories capable of being perceived as being independent from each other, and if it can be shown that the trademark has been genuinely used in relation to only part of the initial broad specification, then protection is available only for the subcategory or subcategories to which the used goods or services belong.

## E. Territorial scope

Article 39(1) of the Spanish Trademarks Act 2001 states that a registered trademark has to be used in Spain in a genuine and real manner. Obligatory use is related to the territoriality principle of trademarks. This implies that products covered by the corresponding trademark must be marketed within Spanish commercial channels. This condition will be fulfilled when products identified with the trademark are commercialized in any part of the Spanish territory, even in cases of export. According to Article 39.2(b), genuine use is:

> 'the use of a trademark in Spain, as it is applied to goods or services or its packaging, exclusively for export purposes.'

The case of export is an exception to the territoriality principle, as the proprietor, instead of using the trademark, conducts on the national territory just a particular act of use: placing the sign on goods or on its appearance. Undoubtedly, in this case products covered by the trademark are not introduced in Spanish commercial channels, but the commercial channels of countries to which the products are exported. Nonetheless, the rule contained in Article 39.2(b) is perfectly justifiable given the legitimate interests of companies that export products covered by the trademark to certain third countries, without introducing them into the Spanish market.

### F. Relevant time period and its calculation

As stated before, a trademark must be used. If it is not used during a period of five years from the publication of its grant, it can then be cancelled for lack of use. The same applies if use of the trademark has been suspended for the same length of time. The cancellation does not apply if the lack of use is due to justified reasons. This is established in Article 39.1 of the Spanish Trademarks Act 2011. The Supreme Court in the Judgment of 28 March 2005 (*Nike*) recalled, in the context of the calculation of the period of five years that determines the application of the sanction of cancellation, that the publication of the grant of new products and services covered by the registered trademark does not constitute a *dies a quo* regarding that calculation of time nor interrupts the running of the same if it has already started. An application for new products and services covered by the trademark has no relevance in the context of the calculation of the said time period, as the publication of the grant of the new products and services is not a factor for considering the time period for lack of use, nor the said application amounts to real and effective use.

Finally, we should highlight that lapse of the abovementioned period does not imply *ipso facto* cancellation of the registered trademark, as this has to be declared by the courts (Article 55.1, last paragraph of the Spanish Trademarks Act).

## III. EVIDENCE OF GENUINE USE OF A TRADEMARK

### A. When use should be proved: timing of lack of genuine use claims

#### *1. When filing an application for a trademark*

The Spanish Trademarks Act 2001 does not establish a requirement of intent-to-use when filling an application. Therefore, when registering a trademark, it is irrelevant if the proprietor puts the trademark to genuine and real use, as this is only relevant when the period of five years since the registration expires. In this respect, the Judgment of the Court of Justice of 21 December 2016 established (emphasis added):

> 'It is apparent from the wording and purpose of Article 15(1), (1)(a) and (2) of Regulation No. 207/2009 that, until the period of five years following registration of the EU trade mark has expired, the rights of the proprietor cannot be declared to be revoked in respect of either some or all of the goods or services for which the mark is registered. *Those provisions thus confer on the proprietor a grace period for beginning genuine use of his mark, during which he may rely on the exclusive rights which the mark confers, pursuant to Article 9(1) of the regulation, in respect of all those goods and services, without having to demonstrate such use.*'

In view of the above, once the abovementioned period of five years lapses, then third parties are entitled to make use of the legally established defence mechanisms, invoking cancellation of the trademark for non-use.

## 2. *In the course of opposition proceedings*

The Spanish Trademarks Act 2001 has not transposed into national law the facultative rule established under Article 11.2 of the Directive that allows a trademark applicant to argue that the trademark of the opponent has not been used. According to the Spanish scholar Fernandéz Novoa,[7] the main reason behind this omission is the enormous difficulties that would entail having to provide the Spanish Trademarks and Patents Office with the necessary means to resolve complex issues regarding the requirement of genuine use of registered trademarks. Consequently, the second paragraph of Article 55.1 of the Spanish Trademarks Act 2001 establishes that cancellation for lack of use will be declared by the courts.

Notwithstanding the foregoing, and with respect to the applicant of the trademark bringing a legal action against the opposing party, according to Article 59 of the Spanish Trademarks Act 2001, apart from the Spanish Trademarks and Patents Office, individuals and entities entitled to file an action for cancellation for lack of use are:

> 'any natural or legal person or any legally formed organization for representation of interests of manufacturers, producers, services providers, traders or consumers that are affected or hold any subjective right or legitimate interest.'

In this regard, as pointed out by the Spanish scholar García Vidal,[8] it is clear that when a party applies for a trademark and the proprietor of the earlier trademark opposes, invoking the risk of confusion between them, the applicant for the new trademark is entitled to file an action for cancellation of the earlier trademark. This is because his interest in the loss of the trademark by the opposing party is unquestionable, as it would remove the obstacle to the registration of the trademark applied for.

In addition, the opponent trademark may be registered for different products and services and the risk of confusion with the trademark applied for is assessed only in relation to some specific products and services from all the products and services for which the earlier trademark has been registered. In this case, is the applicant for the subsequent trademark entitled to file a cancellation action for lack of use of the earlier trademark regarding all the products and services in relation to which the sign has not been used or only regarding specific products or services that generate a risk of confusion with the subsequent trademark and that constitute an obstacle to registration of his trademark?

This issue has been resolved in the recent Judgment from Chamber 28 of the Appeal Court of Madrid of 18 December 2017, (emphasis added):

> 'The loss of the title of the claimant, as a consequence of the success of the action for revocation produces per se the same consequences favourable to the interests of the plaintiff. As a consequence, *his legitimate interest is unquestionable, and so is his legal standing to file not only for a revocation in relation to some concrete products but all the products for which it has been registered.*'

7. Carlos Fernandéz Novoa, José Manuel Otero Lastres and Manuel Botana Agra, *Handbook of the Industrial Property* (Marcial Pons, Madrid, 2009).
8. Ángel García Vidal, *The obligatory use of registered trademark for pharmaceutical products* (2018), http://www.gomezacebo-pombo.com.

### 3. *In cancellation proceedings based on lack of use*

The second paragraph of Article 55.1 of the Spanish Trademarks Act 2001 establishes that cancellation for lack of use will be declared by the courts when the proprietor of the trademark does not prove use of the trademark. The claim for cancellation shall be filed with the Judge of the First Instance Commercial Court of the defendant's domicile. However, the courts of Madrid, Valencia and Barcelona have exclusive jurisdiction in this matter. In case the defendant does not have a domicile in these jurisdictions, then the plaintiff can forum shop among the three of them.

Article 58 sets out, in cancellation actions for lack of use, a reversal of the burden of proof that in the legal system usually corresponds to the plaintiff. Due to this reversal it is the proprietor of the trademark (defendant) who must prove that it has been used or that there are justified reasons for non-use.

In the current legal system lack of use of the trademark for an uninterrupted period of five years does not automatically and definitively provoke its cancellation. Nevertheless, a trademark that is not used during this period would be in danger of a possible action for revocation.

Finally, as regards the effects of the resumption of use, if the use of a trademark resumes, the cause for cancellation is remedied, as long as the use is resumed in good faith. Article 58 of the Spanish Trademarks Act 2001 opts for reproducing the Directive, exclusively regulating the issue of the disputable period of three months:

> 'the commencement or resumption of use within a period of three months preceding the filing of the application or counterclaim [...] which began at the earliest on expiry of the continuous period of five years of non-use shall, however, be disregarded where preparations for the commencement or resumption occur only after the proprietor becomes aware that the application or counterclaim may be filed.'

### 4. *As a defence in infringement/cancellation proceedings*

According to Article 41 of the Spanish Trademarks Act 2001, the proprietor of a trademark registered for at least five years, who initiate an infringement action should prove (if the defendant requires so in his defence), that within the period of five years prior to the date of filing the claim, the trademark has been put to genuine and real use in connection with the goods or services for which it is registered, or that there exist justifiable reasons for non-use.

According to Article 41.2 of the Spanish Trademarks Act 2001, the said exception should be invoked explicitly in the statement of defence by requesting the claimant to prove that within the period of five years prior to issuing the claim, it has been put to genuine use in connection with the goods or services for which it is registered. In this regard, the Judgment of Chamber 28 of the Appeal Court of Madrid of 20 May 2011 stated (emphasis added):

> '*the exception of non-use cannot have the form of the generic criticism made against the claimant*, based on the fact that, instead of using the trademark he has registered, he uses another trademark to commercialize his products. In the same way, we cannot consider alleged, by way of example, the exception of

the lapse of the legal term to exercise an action, based exclusively on providing information about the *dies a quo* of the limitation period.'

Therefore, the trademark will be considered registered exclusively for products and services in relation to which it has been used. The defendant can contest its validity by way of an exception or with a counterclaim for a declaration of cancellation for non-use, which, as well as the exception, must be brought expressly, as highlighted in the Judgment from Chamber 28 of the Appeal Court of Madrid from 19 June 2009, (emphasis added) :

> 'It is clear that the defendants did not file a counterclaim in order to obtain a declaration of cancellation of the trademark and its removal from the register, as a consequence of non-use, neither had they opposed the lack of use in their defence. *We cannot declare ex officio the cancellation of the trademark for non-use, which requires a counterclaim, neither can we admit the exception of non-use in the defence (Article 41.2 of the Spanish Trademarks Act), since it must be invoked expressly by the defendant.*'

Additionally, it should be noted that the effect of cancellation of a trademark will vary depending on whether the defendant invokes its validity by way of a defence or with a counterclaim. That being so, if the cancellation is brought as a successful counterclaim, then the court will declare it so and, once the decision is final, it will take effect against all third parties or *erga omnes* (Article 61 of the Spanish Trademarks Act 2001). Whereas, if it is brought as a successful exception, it will exclusively affect the parties to the proceedings or *inter parte*.

## B. Burden of proof

Article 58 sets out the reversal of the burden of proof. Given that proving a negative fact (non-use) constitutes an enormously difficult proof, the Spanish Trademarks Act 2001 has placed the burden of proof on the proprietor of the trademark. In this respect, Chamber 28 of the Appeal Court of Madrid declared the following in its Judgement of 29 March 2007 (emphasis added):

> 'Article 58 of the Spanish Trademarks Act refers to the onus probandi, meaning that the proprietor of the trademark shall demonstrate that it has been put to use according to Article 39. The judgement of the Supreme Court from 29 January 2000, quoting the judgement from 30 October 1986, points out that *the burden of proof is on the proprietor of the trademark, who invokes its use and application, as it is the positive proof, clear and accessible (invoices, catalogues, testimonies of traders acquires, etc.), as invoking lack of use has the difficulty always inherent in every negative proof, which is impossible to be shown (diabolical proof).*'

## C. Relevant evidence

### *1. Use on the product or its packaging*

Along these same lines, as explained above, the use of a trademark has to be external, public and in the course of trade, for products bearing the trademark placed on the market through the distribution channels and commercial networks. In consequence, the perfect example of genuine use is the commercialization of the products and services bearing the trademark. This has been accurately highlighted in the Conclusions issued in the *Ansul* case (Case C-40/01) by Advocate General Ruiz-Jacobo Colomer (emphasis added):

> 'It is necessary to provide a purposive interpretation, again by reference to the function performed by a trademark, and to look at whether the use by the proprietor is directed at distinguishing his goods or services in the market so as to create an outlet by free, open and fair competition [...] *If the goods or services are to be placed in the market, the paradigm for genuine use is their sale or supply bearing the trademark.*'

Furthermore, the Judgment of the Court of First Instance of 12 December 2002 (*Hiwatt*) stated as follows (emphasis added):

> 'It must be noted first of all that genuine use implies real use of the mark on the market concerned for the purpose of identifying the goods or services [...] *In that regard, even if it is the owner's intention to make real use of his trade mark, if the trade mark is not objectively present on the market in a manner that is effective, consistent over time and stable in terms of the configuration of the sign, so that it cannot be perceived by consumers as an indication of the origin of the goods or services in question, there is no genuine use of the trade mark.*'

## 2. *Use in advertising*

Additionally, as regards advertising use, the burden of use has been fulfilled when the advertisement of the trademark is endorsed by a potential supply of products and services covered by the trademark. For this reason, there will be a genuine use of the trademark if it appears in advertisements that precede the initiation of a sales campaign for the products. Moreover, there will be genuine use of the trademark if, even though the advertisement does not connect directly with the sales campaign, it can be showed that the proprietor makes effective and serious preparations to initiate the procedure of distribution of the products or services covered by the trademark. In this regard, the Judgment of the Supreme Court of 29 March 2006 pointed out (emphasis added):

> 'the use for advertising purposes can be relevant, *but only when it involves the effective introduction on the market of the articles covered by the trademark*. Other preparations for this introduction through distribution channels and commercial networks are not relevant, as happens with designs or the preparation of labels or containers that include the sign.'

Accordingly, the Judgment of the Commercial Court of Palma de Mallorca (Juzgado de lo Mercantil de Palma de Mallorca) of 13 June 2014 stated (emphasis added):

> 'The advertising also constitutes an external and public act, as it is a marketing instrument especially suitable for facilitating the recognition of the trademark among consumers, as well as reinforcing the image of products and services of the trademark. The question of whether the advertising itself can be considered

sufficient to fulfil the requirement of use has been much debated in the doctrine. In my opinion, given the informative purpose of the advertisement, its sufficiency in this context will depend on whether it has been organized in good faith, that is, without disappointing the expectations of consumers if its only purpose is avoiding revocation. The good faith in this matter is typically translated as a requirement for using the trademark to enable consumers to identify products and services and facilitate them taking present and future decisions on the market. *For this reason, the advertisement shall be considered a suitable means to fulfil the requirement of obligatory use of the trademark, as long as it is complemented or followed by the commercialization of products or services.*'

### 3. *Use on the internet*

Moreover, in connection with the use of trademarks solely on the internet, following the arguments of the Judgement of the General Court of the European Union from 25 March 2016, it does not result sufficient to prove the genuine use of trademark[9] in the following context, (emphasis added):

> 'Regarding internet screenshots of companies using a denomination that includes the term "ocean" or "océano" in relation to similar or identical services to those we are analysing, if they do not include relevant information such as web site internet users, business volume and advertising budget that would allow the conclusion that there has been an intensive use of those signs [...] *Mere presence on the internet gives no information about the impact on the public.*'

In other cases, the Supreme Court has interpreted that participation in car rallies, fairs and exhibitions or purely occasional distribution of spare parts when the product has not been manufactured in decades are not acts that constitute genuine use of a trademark. Additionally, there are multiple resolutions that reject mere printing of labels as evidence of use.[10]

### 4. *Form of evidence*

With regard to concrete forms of evidence, the proprietor of the trademark can provide all type of evidence, the most common ones being packaging, labels, price lists, catalogues, invoices, newspaper advertising and buyers testimonies.

### 5. *Distinction between use of a company name and use of the services offered by the company*

Finally, we should highlight that the effective use of the trademark required is the use consistent with its essential function, which is to guarantee the identity of the origin of goods or services to the consumer or end users by enabling them, without

9. Judgment of the Supreme Court of 1 March 1999 (*Hispano-Suiza*).
10. Judgement of the Appeal Court of Madrid (Audienicia Provincial de Madrid) from 24 March 1998 (*Cumarey*). Judgement of the Appeal Court of Madrid (Audiencia Provincial de Madrid) from 18 July 1999 (*Gaudi*).

any possibility of confusion, to distinguish the product or service from others which have another origin. In consequence, the use of the company's name is not sufficient. In this regard, the Judgment of the General Court of the European Union of 14 July 2016 points out, (emphasis added):

> 'even though the tickets mention Modal Cristal, it is used as the company name of the taxpayer, in order to identify the name of the legal person subject to payment of the IAE. Given that *the denomination Modas Cristal is not used on any of the tickets to identify products or services, the tickets provided by the claimant do not prove the genuine use of Modas Cristal as a trademark that identifies the corporate origin of the products and services*.'

## IV. JUSTIFICATION FOR NON-USE

### A. Obstacles to use

In order to reduce the burden of proof, the legal system contemplates the possibility to exonerate the proprietor of a trademark from fulfilling the requirement of use by showing justified reasons for non-use. Thus, Article 39.4 of the Spanish Trademarks Act 2001, when describing reasons for non-use due to exceptional circumstances independent of the will of the proprietor, implicitly makes allusion to the existence of *force majeure*. In this regard, according to Articles 41.2 and 52.3 of the Spanish Trademarks Act 2001, the proprietor of a registered trademark should prove the existence of justifiable reasons for non-use (e.g. export restrictions).

### B. Circumstances in which non-use may be cured

The proprietor of the trademark can invoke as justifiable reasons for non-use the traditional cases of *force majeure*: war and natural catastrophes such as floods, earthquakes or fires. In this regard the proprietor can plead, for example, that a trademark stopped being in use as a consequence of a fire that destroyed the factory where the products covered by the trademark were manufactured. For trademarks designed to identify pharmaceutical products, a delay in granting or withdrawal of the administrative authorisation to commercialize a product could constitute a justifiable reasons for non-use.[11]

On the other hand, among the circumstances that cannot justify lack of use are the lack or insufficiency of financial resources to manufacture and distribute the products. Thus, financial difficulties encountered by a company as a result of an economic recession or due to its own financial problems, such as insolvency proceedings, bankruptcy or temporary stoppage of activities are not considered proper reasons for non-use. In this regard, the Judgment of the Supreme Court of 22 September 2012 stated that such circumstances (emphasis added):

> 'are not covered by Article 39.4 of the Spanish Trademarks Act, *as they constitute ordinary business risks, such as commercial factors* (time required to launch a

11. Judgement from Chamber 15 of the Court of Appeal of Barcelona of 5 October 1995.

> new product range, main client requirements) or *economic factors* (bankruptcy of the previous proprietor of the trademark) or *they do not justify the impossibility to use the trademark* (the existence of the competitor that uses the trademark inappropriately).'

Finally, according to the majority of Spanish doctrine the justifiable reason for non-use will result only in suspension of the period of five years, but not in its interruption. Accordingly, when a justifiable reason for non-use disappears, the calculation of the period of five years resumes from the moment the obstacle emerged.

## V. CONSEQUENCES OF NON-USE

### A. Consequences of non-use and relevance of the context

If a trademark has not been used within the period of five years since the day of the publication of the grant in the Official Bulletin or if the use has been interrupted during the same period, the trademark will be subject to the sanctions established in the Spanish Trademarks Act 2001 (Article 39.1), the principal one being its cancellation.

Thus, the most relevant effect is that the trademark is weakened. Thus, its registered owner has a limited or null capacity of asserting the same against infringers. The reason is simple: the alleged infringer may counterclaim asking the cancellation for lack of use.

The proprietor of a trademark in this situation can bring an infringement action against a third party or an invalidity action on relative grounds. Nevertheless, at the request of the defendant – by way of exception or with a counterclaim – it can be declared that his trademark has not been put to use, and, consequently, the application is dismissed. In this regard, if a proprietor fails to demonstrate that the trademark has been put to genuine and real use, the effects will vary, depending on how the issue was raised – as an action (counterclaim) or as a defence. In case the issue of revocation is brought as an action or as a counterclaim, the court will declare a revocation and, once the decision is finalized, it will take effect against all third parties (Article 61). Whereas, if it is brought as an exception, it will only affect the parties to the proceedings.

If the use of the trademark is resumed, then there is no basis for its cancellation for non-use provided that it is resumed in good faith. Article 58 of the Spanish Trademarks Act 2001 opts for reproducing the Directive, exclusively regulating the issue of the disputable period of three months:

> 'the commencement or resumption of use within a period of three months preceding the filing of the application or counterclaim [...] which began at the earliest on expiry of the continuous period of five years of non-use shall, however, be disregarded where preparations for the commencement or resumption occur only after the proprietor becomes aware that the application or counterclaim may be filed.'

## B. Extent of any revocation

In the context of extension of the cancellation, Article 60 of the Spanish Trademarks Act 2001 provides that when a trademark has been used exclusively for part of products or services for which it has been registered, the cancellation will be partial, extending exclusively to the affected goods and services. As a consequence, the cancellation will not affect those goods and services in relation to which the proprietor proves genuine use.

## C. Timing of any revocation

The cancellation for non-use of the trademark is not automatic, thus it will be declared by the courts. The Spanish Trademarks Act 2001 in Article 55.2 establishes a general principle of retroactivity of the effects of the cancellation to the moment when the acts or omissions which gave rise to the revocation occurred:

> 'Trademarks which have been cancelled shall cease to have legal effect from the time when the events or omissions that gave rise to the cancellation occurred, irrespective of the date on which they were published in the Official Bulletin.'

# Sweden

*Karin Cederlund, Dan Eklöf and Anders Kylhammar*

## I. INTRODUCING GENUINE USE OF TRADEMARKS

### A. Legal framework

The current Swedish Trademark Act (2010:1877, below 'the Trademark Act', 'the Act') came into force in 2011 as a result of the implementation of Directive 2008/95/EC. The implementation of Directive 2015/2436 is currently being prepared. A Government Bill is expected in the first half of 2018 and is planned to be adopted by the Parliament later in 2018. The amended Act is expected to enter into force by 1 January 2019. This chapter is based on the assumption that a proposal of the amendments, presented by a Government Inquiry in late 2016, will be followed by the Government and the Parliament. The following is thus subject to possible changes to the proposal that may be decided when the Act is finally adopted.

The central provision on genuine use is found in the Trademark Act, chapter 3, section 2. This provision will not be amended in the process of implementing Directive 2015/2436. The provision reads (Unofficial translation, Justice Department, March 2013):

> 'A trade mark registration may be revoked, if the proprietor has not, within five years from the final decision on the registration matter, or within a consecutive period of five years, made genuine use of the mark in Sweden in respect of the goods or services for which it has been registered. The registration may not, however, be revoked on grounds of failure to use, if there are proper reasons for the non-use.
>
> To be "use" pursuant to the first paragraph is also deemed
> 1. when the trade mark is used in a form differing in elements which do not alter the distinctive character of the mark in the form in which it was registered, and
> 2. when the trade mark is affixed to goods or the packaging thereof in this country solely for export purposes.

> To be trade mark use by the proprietor is deemed trade mark use by someone else with the proprietor's consent.
>
> The registration may not be revoked, if the trade mark has been used during the interval between the expiry of the five-year period and the filing of the application for revocation. Use that is commenced or resumed after the expiry of the five-year period and within three months preceding the filing of the application for revocation shall, however, be disregarded, if the preparations for the commencement or resumption occurred only after the proprietor becoming aware that the application for revocation might be filed.'

### B. Purpose and function of genuine use: policy considerations

The historical policy considerations behind the provisions on genuine use in Sweden – dating from the 1950s – were primarily that the (then) purely national Register was packed and that competition could be limited if trademark registrations could be upheld in the absence of use. It was pointed out that through a strategy with 'defensive' registrations a proprietor could achieve an unduly wide *de facto* scope of protection, thereby raising costs for other operators to enter into a market where the registered trademark was not used but nonetheless would enjoy an infinite protection. Such a 'defensive' registration strategy could *inter alia* be to register trademarks for products and services outside the scope of the proprietor's business or to register many similar trademarks of which only one was actually used. Another policy consideration historically set forth was a perception that the access to 'good trademarks' – i.e. trademarks that are easy to remember and pronounce, and that are capable of wide penetration and creative in appeal – was somewhat limited and should be as available as possible to various operators on the market.

It is fair to say that there has been no pronounced policy discussion in Sweden on genuine use. Since Sweden's entry into the European Union in 1995, Swedish law appears to be in all material respects faithful to the policy considerations on genuine use underlying the adopted EU instruments.

## II. WHAT QUALIFIES AS GENUINE USE OF A TRADEMARK

### A. Nature of use

#### *1. Use consistent with the function of the trademark*

In order to qualify as 'genuine use', the use should be consistent with the function of the trademark. It follows that use of a trademark as a title or for purely descriptive purposes cannot be invoked as a defence in a genuine use challenge.

In terms of company names Swedish law is in line with EU legislation and jurisprudence, with an added layer of protection for trade names.

As clarified by the CJEU in C-17/06, *Céline*, a trademark proprietor may prevent a third party from using a company name where the use is made in such a way as to affect or be liable to affect the functions of the trademark, e.g. when the use of the company name is made for the purposes of distinguishing goods or services. For the sake of clarification a provision to this effect will be included in the forthcoming amended Trademark Act. It will also be clarified that using a part of the company name, e.g. the dominant feature of it, may constitute such infringing use that the trademark proprietor may prevent.

Swedish law contains the somewhat unusual concept of a so-called cross-protection of trademarks and company names. In short, the proprietor of a company name has the exclusive right to use its company name as a trademark and vice versa. Against this backdrop it can be assumed that the use of a company name for the purposes of distinguishing goods or services would qualify as genuine use of a trademark. Thus such use of a company name may be invoked as being of genuine character in proceedings against a challenge based on lack of genuine use.

## 2. *Use in the course of trade*

### *i. Public use*

In order to qualify as genuine the use shall be real, i.e. the sole purpose of the use cannot be to maintain the registration of the trademark. However, as clarified by the CJEU in C-259/02, *La Mer*, use of a trademark can be considered genuine even if the use has been limited. One has to take into account all relevant circumstances and although actions taken after the relevant time period should not be considered, such actions can give some guidance on how to interpret actions that have been taken during the relevant period. This was also discussed in a ruling from the Swedish Supreme Court (NJA 1987 s. 22, *Budvar*). The requirements of genuine use of a national trademark is probably lower than for a Community trademark since there is a significant difference between having exclusive rights in *one* Member State compared to having it in *all* Member States.[1]

Taking the above into account, there is no specific requirement of the use being 'public' but it is a circumstance among others that should be taken into account when deciding whether the use has been genuine or not.

### *ii. Commercial purpose*

The use shall be commercial in order to qualify as genuine use. Swedish law is in line with EU jurisprudence and no deviation is intended. National case-law is sparse and to the extent the CJEU has not addressed the issue the legal state of affairs remains uncertain. As far as use on goods destined for exports is concerned the prevailing opinion is that such use qualifies as genuine. Preparation for commercial activities may, in principle, qualify as genuine use. The threshold is unclear. It has been debated whether measures such as the display of a trademark in market surveys would qualify. Another debated issue is whether the use has to target consumers or other relevant end-users on the relevant market, or if it is sufficient that the use has been carried out in a pure B2B setting before launching the product to end-users,

1. Richard Wessman, *The Swedish Trademark Act* (1 September 2017, zeteo.nj.se) comment to chapter 3, section 2.

such as use during a demonstration at a trade fair or that the trademark proprietor has entered into various contracts with distributors and retailers. The prevailing opinion appears to be that the use in a pure B2B setting may suffice as genuine but that an assessment on a case-by-case basis needs to be conducted. There also seems to be a consensus that there must be a reasonable time frame within which such preparatory activities must be followed by an actual market launch in order for the preparatory activities to qualify as genuine use, although this timeframe is unclear and needs to be defined on a case-by-case basis.

### 3. *Use by the owner or with its consent*

The trademark shall be used by the trademark proprietor or with its consent in order to qualify as genuine use. A licence agreement would constitute the typical form of consent, although other actions from the trademark proprietor may qualify as consent under general principles of contract law. No requirement of quality control set up by the proprietor exists in Swedish law.

## B. Volume, duration and frequency of use

Swedish law adheres to EU law and jurisprudence. There is thus no *de minimis* rule with regard to volume, duration or frequency of use. Whether the use is genuine shall be assessed on a case by case basis, taking into account all relevant circumstances. Only one Swedish authoritative ruling can be noted. In that case from 2005 (NJA 2005 s. 643, *ENTELCARD*) the Swedish Supreme Court revoked the registrations of two trademarks based on a finding that the use was too insignificant to meet the bar of genuine use.

The trademark proprietor sold telephone cards for long-distance calling from Sweden to Chile (Sweden has a fairly large Chilean community) and the market concerned was thus of a niche character. The activities of the trademark proprietor consisted, put somewhat simply, of an insignificant (and in terms of content unsubstantiated) promotional campaign on commercial radio, a contract between the trademark proprietor and a private individual who undertook to promote the sale of the telephone cards, and a list of a few telephone calls made between Sweden and Chile using the service. The Swedish Supreme Court held in a brief statement that the use had been too insignificant and unsubstantiated to qualify as genuine.

## C. Use of a sign

The trademark shall be used in its registered form in order to qualify as genuine. No deviation from EU law and jurisprudence is intended. The relevance of use in a different form than the registered form has however been debated in Sweden. The Swedish Supreme Court, in a case from 1991 (NJA 1991 s. 23, *GATT*) and hence before the Swedish accession to the EU, applied a rather generous view in favour of the proprietor of the trademark. A trademark proprietor had registered a combined word and figure mark consisting of the word 'GATT', but had used the same word in another typeface and with different figurative elements. In a brief judgment the

Swedish Supreme Court held that this met the threshold of genuine use, since the word 'GATT' in itself qualified for registration as a trademark. It thus seems as if the Supreme Court considered that the distinctive part of the trademark was the word 'GATT' rather than the figurative elements and this played a decisive role for the outcome.

In a subsequent case from 2001 (NJA 2001 s. 265, *Spotlight*), i.e. after the Swedish accession to the EU, the Swedish Supreme Court relied on the first Trademark Directive (as implemented) when facing the question of whether, again put somewhat simply, a registration of 'SPOTLIGHTSERIEN' – with the letters in different typefaces and size and additional arrangements in terms of layout (a combined word and figure mark) – should be revoked on the ground that the trademark proprietor merely had used the word 'SPOTLIGHT' in a different typeface. The Swedish Supreme Court's point of departure, again in a brief judgment, was that the registered trademark contained three word elements in different typefaces and sizes and arranged in a distinctive manner. Further, the Supreme Court concluded that the trademark proprietor's legal protection, presumably the scope of protection acquired by the registration, included the whole combination together with the form, and not only the words taken separately. The use during the past five years consisted, as the Supreme Court phrased it, not of the registered form but merely of the form 'SPOTLIGHT', in uniform typeface and in a sequence. This was determined to be such a deviation from the registered form that the distinctive character of the registered form had been 'altered'. Hence the trademark was revoked.

It appears as if the Swedish Supreme Court took into account the scope of protection of the registered trademark in terms of an assessment of confusion and other parts of trademark jurisprudence where the scope of protection is relevant, when determining whether the use of the trademark in a different form was to be considered as genuine use of the registered trademark. The Supreme Court may also have been influenced by a provision in the Trademark Act enabling a trademark proprietor to register non-essential changes of the registered trademark, in so far as this does not affect the overall impression.

## D. Use in relation to goods and services

With regard to use in relation to goods and services, Swedish law is in accordance with EU law and no deviation is intended. Lower courts have underlined the adherence to EU law principles. Thus it has been recognized that if a trademark has been registered for goods or services defined so precisely and narrowly that it is not possible to make any significant sub-divisions within the product or service category concerned, then the proof of genuine use of the mark for the registered goods or services necessarily covers the entire product or service category. On the other hand, if a trademark has been registered for a broader category of goods and services, and this category is considered sufficiently broad to cover several subcategories that may exist independently of the broader category, the trademark proprietor will only enjoy protection with regard to the goods and services for which the trademark actually has been used. This assessment is to be made on a case-by-case basis starting out from the phrasing of the specific category and how precise this phrasing is, including the purpose and intended use of the goods or services. Concepts are to be defined from a general linguistic understanding of their meaning bearing in mind that they should be interpreted in a uniform way across the EU.

For instance, a lower court faced with the concept of 'banking services' did not attribute to this concept the legal meaning it has in Swedish banking law regulations but rather held, under trademark law considerations, that various services – such as issuing credit cards – constituted a relevant subdivision which could be carried out independently of banking services. An autonomous genuine use within the subdivision of the category was therefore required.

## E. Territorial scope

Use in any part in Sweden would qualify as genuine; no territorial limitations or the like apply. With regard to use on foreign websites and foreign magazines there is no case-law and the issue has not been treated in the legal literature. Foreign websites have been subject to infringement and/or injunction proceedings in older case-law, but it has to be assumed that such situations would today be assessed under EU law and jurisprudence. It appears that use on foreign websites targeting the Swedish territory, would – *mutatis mutandis* – be assessed under some influence of the principles governing infringement and/or injunction actions, but the state of the law remains uncertain.

## F. Relevant time period and its calculation

A trademark registration may be revoked, if the proprietor has not, within five years from the final decision on the registration matter, or within a consecutive period of five years, made genuine use of the mark in Sweden in respect of the goods or services for which it has been registered.

The legislative committee proposing amendments to the Trademark Act has suggested a new provision (chapter 2, section 30) based on Article 46 in Directive 2015/2436. Accordingly, in proceedings for a declaration of invalidity based on a registered trademark with an earlier filing date or priority date, the proprietor of the earlier trademark shall furnish proof that, during the five-year period preceding the date of the application for a declaration of invalidity, the earlier trademark has been put to genuine use, if the proprietor of the later trademark so requests. In order for the later trademark to be revoked due to an earlier trademark, the following have to be fulfilled:

- at the time of filing the application, or in the event of a priority date, more than five years must have passed since the registration of the earlier trademark was finally terminated;
- the earlier trademark must have been genuinely used in accordance with the Trademark Act, chapter 3, section 2 (or if a Community trademark, following the EU Regulation); and
- the genuine use must have occurred within the five years preceding the date of filing the application, or the priority date, for the later trademark.

Should these requirements not be fulfilled, the later trademark can only be revoked should there be proper reasons for the non-use of the earlier trademark.[2]

The legislative committee has also suggested that the registration authority shall make a note in the Register which day the five-year-period begins based on Article 16.4 in Directive 2015/2436.[3]

A registration may not be revoked if the trademark has been used between the end of the five-year period and the submission of the application for revocation of the registration. However, use commencing or resumed after the expiration of the five-year period and within three months before the application for revocation is submitted is disregarded if the preparation for commencement or resumption was taken after the trademark proprietor was informed that a revocation request could be made.

## III. EVIDENCE OF GENUINE USE OF A TRADEMARK

### A. When use should be proved: timing of lack of genuine use claims

#### *1. When filing an application for a trademark*

Use should be proven by the trademark proprietor when it has been contested by an opponent in administrative proceedings or before judicial courts. There is no intent-to-use requirement, nor is use required, at the time of the trademark filing.

#### *2. In the course of opposition proceedings*

Proof of use shall be provided by the opponent upon request by the trademark applicant. The opponent is not required to present any evidence of use, if there is no such request from the trademark applicant.

#### *3. In cancellation proceedings based on lack of use*

Use should be proven in revocation proceedings in both administrative proceedings and before judicial courts.

#### *4. As a defence in infringement/cancellation proceedings*

A claim for revocation of a trademark can be made as part of a defence in infringement and cancellation proceedings. Such claim can be made by way of a counterclaim or in a separate action.

According to the Swedish Code of Judicial Procedure (Sw. *rättegångsbalken*) chapter 32, section 5, an action can be stayed pending the outcome of another proceeding if that outcome is of extraordinary importance for the adjudication of the first action. However, in practice it is rather unusual that infringement proceedings are stayed pending the outcome of revocation proceedings.

2. Legislative Committee Report SOU 2016:79, p. 285.
3. Legislative Committee Report SOU 2016:79, p. 284.

## B. Burden of proof

The trademark proprietor has the burden of proof to show genuine use when it has been contested.

## C. Relevant evidence

### *1. Use on the product or its packaging, in advertising and on the internet; form of evidence*

Swedish courts apply the principle of free production of evidence (Sw. *fri bevisföring*) and the principle of free evaluation of evidence (Sw. *fri bevisvärdering*) in accordance with the Swedish Code of Judicial Proceedings (Sw. *rättegångsbalken*) chapter 35, section 1. This means that the parties can refer to whatever evidence they want and the courts will independently evaluate all the evidence. Relevant evidence may for example be packaging, sales statistics, advertisements and commercials – both printed and digital. Depending on the circumstances of the case, different types of evidence may be referred to and may be considered more or less relevant by the court.

Nonetheless, one should bear in mind that Swedish courts also apply the principle of oral proceedings and the principle of immediateness. The principle of oral proceedings means that, as a general rule, the parties and the witnesses shall testify at the main hearing. Depositions and affidavits are only allowed if the court finds that such documentary evidence would facilitate the understanding or in any other way be beneficial for the handling of the case.

The principle of immediateness means that, as a main rule, only evidence presented or at least referred to during the main hearing will be taken into account by the court. Any evidence presented during the preparatory stage of the case has therefore to be presented or at least referred to during the main hearing.

### *2. Distinction between use of a company name and use of the services offered by the company*

Sweden is rather unusual in that there is a cross-protection regulation between trademarks and company names (the Trademark Act, chapter 1, section 8 and the Trade Names Act, section 2a). The consequence of the cross-protection is that a trademark proprietor whose trademark is infringed by a company name may claim protection under the Trademark Act and, vice versa, the proprietor of a company name may claim protection under the Trade Names Act.

The provision in Directive 2015/2436 stating that a trademark proprietor has the exclusive right to the registered trademark and that it includes using the sign as a company name is already covered in Sweden by the cross-protection. According to the legislative committee proposing amendments to the Trademark Act this should however be clarified by an amendment in the Act making it possible for a trademark proprietor whose trademark is infringed by a third party's use of a company name to claim protection both under the Trademark Act and under the Trade Names Act.

This amendment will however not have any substantial practical impact since the sanctions are the same under the two acts.[4]

Against this backdrop it can be assumed that the use of a company name for the purposes of distinguishing goods or services would qualify as genuine use of a trademark. Thus such use of a company name may be invoked as being of genuine character in proceedings against a challenge based on lack of genuine use.

## IV. JUSTIFICATION FOR NON-USE

### A. Obstacles to use

If the trademark proprietor can show that the reasons for the non-use are proper, the trademark registration cannot be revoked. The CJEU has provided examples of such reasons in e.g. C-246/05, *Le Chef de Cuisine*. In general, proper reasons for non-use shall be circumstances that are not within the trademark proprietor's control such as administrative decisions or regulations.

The Swedish Supreme Court has ruled that a trademark proprietor's financial considerations are not proper reasons for non-use (NJA 1987 s. 22, *Budvar*). In that specific case, Budvar (as the defendant) stated that it could not use its trademark in Sweden due to Swedish alcohol legislation which prohibited Budvar from selling beer with a certain alcohol percentage. This reason was not accepted by the Swedish Supreme Court which stated that even though it would be an adjustment made by Budvar only for the Swedish market, is would be possible for Budvar to sell beer with a lower alcohol percentage and use its registered trademark for such beer.

### B. Circumstances in which non-use may be cured

The non-use of a trademark can be cured by starting to use it. A trademark cannot be revoked should the trademark proprietor start using it within the period after the expiration of the five-year period but before the filing of the application of revocation. However, use commencing or resumed after the expiration of the five-year period and within three months before the application for revocation is submitted is disregarded if the preparation for commencement or resumption was taken after the trademark proprietor was informed that a revocation request could be made. Consequently, a party that intends to apply for revocation of a trademark should not wait longer than three months between sending the first warning letter to the trademark proprietor and filing the application of summons.

## V. CONSEQUENCES OF NON-USE

### A. Consequences of non-use and relevance of the context

The consequences of non-use is the risk of the registration being revoked and that an opponent can claim non-use as a defence in an infringement proceeding.

4. Legislative Committee Report SOU 2016:79, p. 232.

There are two possible ways to revoke a registration. First, the party intending to revoke a trademark may initiate proceedings in the Swedish Patent and Market Court.

Secondly, it is possibly to apply for revocation of a trademark with the Swedish Patent and Registration Office, a so-called administrative revocation. To the extent the trademark proprietor agrees to the registration being revoked, the Swedish Patent and Registration Office will revoke the registration. If the trademark proprietor contests the application in full or in part, the applicant has the opportunity to request that the case is referred to the Swedish Patent and Market Court.

## B. Extent of any revocation

A registered trademark can be revoked partially or in full depending on how it has been used. The registration of the trademark will only be revoked for goods and services for which the trademark has not been used. If the same trademark has been used for other types of goods and services the registration will be upheld for such goods and services.[5]

## C. Timing of any revocation

When a trademark registration has been revoked, wholly or in part, by the Patent and Registration Office or by a court, the registration shall, to the same extent, be removed from Trademark Register when the decision has acquired legal force. The decision shall be published.

5. Richard Wessman, *The Swedish Trademark Act* (1 September 2017, zeteo.nj.se) comment to chapter 3, section 4.

# Switzerland

*Thierry Calame*

## I. INTRODUCING GENUINE USE OF TRADEMARKS

### A. Legal framework

The Trade Mark Act of 28 August 1992 (as amended on 1 January 2017) sets out the national basis for trademark law in Switzerland. Subsidiary provisions concerning the registry and administrative proceedings are set out in the Trade Mark Ordinance of 23 December 1992 (as amended on 1 January 2017).

Switzerland is also party to a number of international treaties and conventions in relation to trademarks, including the Paris Convention for the Protection of Industrial Property of 1883 (Paris Convention), the WTO Agreement on Trade-Related Aspects of Intellectual Property Rights of 1994 (TRIPS Agreement), the Nice Agreement concerning the International Classification of Goods and Services of 1957 (Nice Classification) and the Madrid Agreement Concerning the International Registration of Marks of 1891 (Madrid Agreement).

Under the Paris Convention and the TRIPS Agreement, there is no mandatory requirement to use a registered trademark to maintain its protection, it is only optional. However, the Paris Convention makes clear that if in any country use of the registered mark is compulsory, the registration may be cancelled only after a reasonable period. The TRIPS Agreement specifies that this period should be at least three years. The Paris Convention further provides that if the proprietor uses his trademark in a form which is not significantly different from the form in which it was registered, such use shall be sufficient to maintain protection. Finally, the TRIPS Agreement provides that use of the trademark with the consent of the proprietor is considered to be use by the proprietor himself.

Switzerland provides for a mandatory use requirement. The relevant provisions in the Trade Mark Act with regard to genuine use read as follows:

*Art. 11 Use of the trade mark*

1 A trade mark is protected if it is used in relation to the goods or services for which it is claimed.

2 Use in a manner not significantly different from the registered trade mark and use for export purposes also constitute use of the trade mark.

3 Use of the trade mark with the consent of the proprietor is considered to be use by the proprietor himself.

*Art. 12 Consequences of non-use*

1 Where the proprietor has not used the trade mark in relation to the goods or services for which it is claimed for an uninterrupted period of five years following the expiry of the opposition period with no opposition having been filed or upon conclusion of opposition proceedings, he may no longer assert his right to the trade mark, unless there are proper reasons for non-use.

2 If use of the trade mark is commenced or resumed after more than five years, the right to the trade mark is restored with effect from the original priority date, unless non-use of the trade mark has been invoked under paragraph 1 prior to its commencement or resumption of use.

3 Any person who invokes non-use of a trade mark is required to substantiate his claim; evidence of use is required to be provided by the proprietor of the trade mark.

### B. Purpose and function of genuine use: policy considerations

There are a number of justifications for the genuine use requirement. One reason is that the number of useful marks is limited. If registered trademarks are not used, they should be cleared from the Register to make them available again. Another aspect is that clearance of unused trademarks from the Register prevents the trademark system from clogging up and it keeps the system efficient and manageable. Indeed, the genuine use requirement serves to reduce the total number of trademarks registered and, consequently, the number of conflicts which arise between them. Finally, some believe that trademarks can only fulfil their function, i.e., distinguish goods and services of one undertaking from those of other undertakings, if they are used.

## II. WHAT QUALIFIES AS GENUINE USE OF A TRADEMARK

### A. Nature of use

#### *1. Use consistent with the function of the trademark*

Only use which is consistent with the function of the trademark qualifies as genuine use. In other words, the mark must be used 'as a mark' to maintain protection, i.e., it must identify specific goods or services originating from a particular undertaking.[1] Mere decorative use will not qualify as genuine use. However, the use of a mark in

1. *See* BGE 139 III 424, 431 (*M-Watch II*).

merchandising will generally qualify as use 'as a mark' since also in merchandising consumers usually perceive the mark as a sign distinguishing goods or services rather than a decorative element.[2]

If the mark is used solely to identify an undertaking rather than a specific product, the use will not qualify as genuine use. In other words, use as a corporate or trade name is not sufficient to maintain protection.[3] Similarly, use of the mark in company-related (rather than product specific) advertising or sponsoring activities will not qualify as genuine use.

Use of the mark as an internet domain name or as part of a domain name will generally not be considered as use 'as a mark'. An exception may apply in case the website actively offers the products and the proprietor can demonstrate that the website is not only consulted by customers, but also products have been sold.[4] Clearly product related use of the mark on the website may also qualify as genuine use.

## 2. *Use in the course of trade*

### *i. Public use*

Use in the course of trade, i.e., in commerce, is required. The prevailing view is that mere public use does not constitute genuine use. However, to the extent publicly owned companies are active in commerce, their use of a mark also qualifies as genuine use.

### *ii. Commercial purpose*

The notion 'in the course of trade' is very broad. Any kind of commercial use is considered use in the course of trade. This includes activities which are not directed at achieving commercial profit. Therefore, use by not-for-profit associations equally qualifies as genuine use.

Whether use of the mark on free promotional goods or in connection with test sales qualifies as genuine use, is not settled. Some think that if it is sufficiently serious use, it should count as genuine use. Others stress that use of the mark on free samples or in connection with test sales should only qualify as genuine use if the handing out of free samples or test sale is followed by a commercial sale.

Internal use of a mark within a company or a group of companies (i.e., intra-company and intra-group sales) will not constitute genuine use. It is required for the mark to be used outside the sphere of the trademark proprietor, i.e., on products that are sold to external customers.[5]

---

2. *See* AIPPI Question Q218, The requirement of genuine use of trademarks for maintaining the protection, Swiss Group Report, page 2.
3. BGer of 25 September 2008, No. 4A.309/2008, cons. 6.1 (*Eva Astrologia*); BGE 88 II 32 ff. (*Au Bûcheron*).
4. BVGer of 20 January 2014, No. B-2910/2012, cons. 9.1.1 (*Artelier*).
5. *See* BVGer, sic! 2008, 638, 640 (*K Swiss*).

### 3. *Use by the owner or with its consent*

According to Article 11(3) Trade Mark Act use of the trademark with the consent of the proprietor is considered to be use by the proprietor himself. Therefore, use of the mark by a licensor or a distributor constitutes genuine use for maintaining protection. There is no legal requirement to be met by the trademark proprietor (such as quality control) to maintain protection. The consent does not have to be explicit, it can also be implicit (e.g., in the case of affiliated companies).[6] Consent has to be given prior to use. If the mark is used without the proprietor's consent, the use will not be considered genuine.[7] This even applies in case the proprietor approves of the use afterwards.

## B. Volume, duration and frequency of use

Only serious use will constitute genuine use, token use is not sufficient to maintain protection. There are no strict requirements in terms of volume. According to the long standing case law of the Federal Supreme Court even a marginal turnover is sufficient for maintaining protection if the trademark owner has the genuine will to satisfy the demand of the market.[8] For instance, in an older case the delivery of 24 bottles of port wine within three years was considered sufficient.[9] However, turnover is nevertheless relevant. The higher the turnover the more likely genuine use will be accepted. The required degree of use also depends on the type of goods. In case of more expensive goods a lower threshold applies, while a higher degree of use is required for mass products. For instance, genuine use was denied in a case involving cookies where a turnover of about € 8000 was generated per year.[10]

There are no strict requirements in terms of duration and frequency of use. Typically, use over a certain period is required. Single deliveries and temporary offers are usually not sufficient. For instance, a single delivery of 567 kg of Swiss cheese[11] and two deliveries of 4329 bottles of fruit juice within 8 months[12] were considered insufficient. On the other hand, genuine use was accepted in cases where 5000 copies of a quarterly magazine were sold within two years[13] and 3260 ties were sold within five years.[14]

---

6. *See* BVGer, sic! 2008, 638, 640 (*K Swiss*).
7. *See* RGKE, sic! 1997, 577, 579 (*Canal Plus*).
8. BGE 102 II 111 (*Silva II*). It is noteworthy that the more recent case law of the Federal Administrative Tribunal appears to be stricter than the older case law of the Federal Supreme Court.
9. BGE 102 II 111 (*Silva II*).
10. BVGer, GRUR Int. 2014, 840 cons. 6.4 (*Koala*).
11. BVGer of 19 July 2010, No. B-892/2009, cons. 6.9 (*Heidiland*).
12. BVGer of 8 April 2010, No. B-1791/2009, cons. 3.3.2 (*Yo*).
13. RGKE, sic! 2004, 783 cons. 8 (*IP*).
14. BVGer of 19 December 2007, No. B-7500/2006, cons. 5.4 (*Diva Cravatte*).

## C. Use of a sign

According to Article 11(2) Trade Mark Act, use in a manner not significantly different from the registered trademark also constitutes use of the trademark. On the other hand, if a mark is used in a significantly different form, such use will no longer be sufficient for maintaining protection.

What does 'significantly different' mean? Typically, a form is considered significantly different if the distinctive character of the mark has been altered to an extent that the relevant consumers do not recognize the same sign anymore. In a case involving the Tripp Trapp high chair, the Federal Supreme Court said that use is significantly different if the distinctive core of the mark has lost its identity.[15] The more distinctive a mark is the more the mark can be changed without losing 'the identity of its distinctive core'.

In practice, it is recognized that certain changes like modernizing are permissible as long as the trademark is still recognized. For instance, no significant change was found when small letters were converted into capital letters, single letters without phonetic significance were omitted, the spelling of single words was modernized and graphic drawings were used in slightly different proportions. Equally, using a black and white trademark registration in colours or written inversely with white writing on black background was considered acceptable. On the other hand, replacing an old-fashioned script with modern characters and rephrasing important word elements were considered significant changes.[16]

Adding or omitting non-distinctive (e.g., descriptive) elements is generally possible. On the other hand, adding or omitting distinctive elements is generally not acceptable. For instance, it is usually not possible to only use the word element of a figurative mark and omit the figurative element.[17]

If several trademark registrations exist for marks which deviate from one another only in an insignificant manner, genuine use for one mark will be sufficient to maintain all marks.[18]

## D. Use in relation to goods and services

### *1. Use for each of the designated goods and services*

As a general rule, genuine use will only be recognized with regard to the goods and services the mark is actually used for. If the mark is not used for each designated good and service, the mark will only be maintained with regard to those goods and services which are actually offered and sold on the market. On the other hand, if the

15. BGE 130 III 267 cons. 2.4 (*Tripp Trapp*).
16. Examples taken from Noth/Bühler/Thouvenin-Markus Wang, Art. 11 MSchG N 81 et seqq. and BSK MSchG-Bernard Volken, Art. 11 MSchG N 106 et seqq. *See* also AIPPI Question Q218, *The requirement of genuine use of trademarks for maintaining the protection*, Swiss Group Report, p. 3.
17. *See* BGE 130 III 267 cons. 2.4 (*Tripp Trapp*).
18. *See* BGE 99 II 104, 119 (*Silva I*).

mark is used for products other than the designated ones, such use will not qualify as genuine use.

Use of the mark for one product does not qualify as genuine use for similar products. For instance, the use of a mark for cheese is genuine use for cheese, but not for milk even though cheese and milk are similar goods.

If a designated component or constituent has a recognizable self-contained function within an overall product (e.g., a processor within a personal computer), use of the mark for the overall product (personal computer) will qualify as genuine use also for the component (processor).[19]

If goods and services have no independent economic function, but merely serve to promote designated goods and services, use of the mark for these auxiliary goods qualifies as use for the designated goods and services. For instance, in a case where beer glasses were used to promote a beer, use of the mark for beer glasses was deemed use of the mark for the beer.[20]

### *2. Use for a broader category of goods and/or services*

Use of a mark for a broader category of goods and services is discussed in connection with the headings of the Nice Classification. The use of one specific product of a certain class does not qualify as genuine use for all products that fall under the relevant class heading. However, it is recognized that use for one specific product of a certain class *may* qualify as use for the class heading. It is not settled when this is the case. As a general rule, it is safe to assume that use for typical products falling under the class heading will qualify as use for the class heading as a whole. For instance, use for sports clothing was qualified as genuine use for clothing,[21] use for cigars was considered sufficient use for tobacco products[22] and use for cheese was deemed sufficient use for dairy products.[23] On the other hand, use for a contrasting agent was not considered genuine use for pharmaceutical products.[24]

## E. Territorial scope

Based on the longstanding principle of territoriality, the use of the mark must take place in Switzerland in order to qualify as genuine use. This means the mark must be used on goods or services which are sold or furnished in Switzerland. There are two exceptions to this principle.

*First*, according to Article 11(2) Trade Mark Act, use for export purposes also constitutes use of the trademark. In other words, if the mark is used on products which are not sold in Switzerland, but exported from Switzerland, this use will be sufficient.

*Secondly*, Switzerland and Germany are parties to a bilateral treaty of 13 April 1892 according to which use of a mark in Germany qualifies as genuine use

19. Example taken from Noth/Bühler/Thouvenin-Markus Wang, Art. 11 MSchG N 45.
20. BGE 58 II 167, 170 (*Urhell*).
21. BVGer of 25 July 2016, No. B-6249/2014 (*Campagnolo*).
22. RKGE, sic! 2003, 138 (*Boss*).
23. HGer Berne, sic! 2007, 632 (*Vichy check pattern*).
24. BVGer of 4 March 2013, No. B-5871/2011 (*Gadovist*).

in Switzerland and vice versa. This means that a trademark can be maintained in Switzerland on the basis of use of the parallel trademark in Germany[25] only and vice versa. Only citizens and residents of Germany and Switzerland are eligible to benefit from this bilateral treaty.[26]

Use of the mark in a foreign magazine which is also read in Switzerland or use of mark on a 'foreign' website (e.g., a website which is accessible under a gTDL) only constitutes genuine use if there is a link to Switzerland. This is only the case when the magazine or the website specifically target the Swiss market, e.g., if on the website product is specifically offered for sale to Swiss residents, or if the websites contains local reference points such as local contact persons, local currency or local language.

Use in part of Switzerland only is sufficient. The Federal Supreme Court even considered sales in one city (the city of Berne) to be sufficient.[27] In a case involving the sale of cosmetic products of Chanel, the Federal Administrative Tribunal left the question open as to whether the sale of products in airport shops constitutes sufficient use in the territory of Switzerland, as it was known to the court that the Chanel products are sold elsewhere in shops in Switzerland.[28] However, in scholarly writing it is held that even sales in airport shops would constitute sufficient use in view of the fact that Swiss residents equally buy products there.[29]

## F. Relevant time period and its calculation

Swiss law has a grace period of five years. According to Article 12(1) Trade Mark Act, where the proprietor has not used the trademark in relation to the goods or services for which it is claimed for an uninterrupted period of five years, he may no longer assert his right to the trademark, unless there are proper reasons for non-use. For Swiss trademarks, the time period starts running on the day following the conclusion of opposition proceedings or, if no opposition has been filed, on the day following the expiry of the opposition period.

The same grace period applies in case the mark is initially used, but use is later given up. The proprietor again has five years within which to resume use of the mark in order to avoid the mark being vulnerable to cancellation. If use of the trademark is then resumed, the right to the trademark is restored with effect from the original priority date, unless non-use of the trade mark had been invoked prior to resumption of use.[30]

---

25. The trademark needs to be registered both in Switzerland and Germany. However, it is not required that both registrations are identical. It is sufficient if the two registrations are identical with regard to the essential parts of the trademark; see BGE 96 II 243, 254 (*Blauer Bock*).
26. BGE 124 III 277, 283 (*Nike*).
27. BGE 102 II 111 (*Silva II*).
28. BVGer of 16 March 2011, No. B-3126/2010 (*CC*).
29. Noth/Bühler/Thouvenin-Markus Wang, Art. 11 MSchG N 52.
30. In order to invoke the defence of non-use the third party asserting non-use usually sends a formal letter to the trademark proprietor asserting that the mark in question has not been used. This leads to a reversal of the burden of proof: it is now up to the trademark proprietor to prove genuine use of the mark prior to receipt of the formal letter.

## III. EVIDENCE OF GENUINE USE OF A TRADEMARK

### A. When use should be proved: timing of lack of genuine use claims

#### *1. When filing an application for a trademark*

There is no intent-to-use requirement in Swiss trademark law. Equally, there is no requirement of actual use at the time of the trademark filing. As more fully explained above, Swiss law provides for a five-year grace period.[31]

#### *2. In the course of opposition proceedings*

In opposition proceedings, the defendant may invoke the defence of non-use and demand proof of genuine use of the earlier mark. The non-use must be asserted in the defendant's first submission.[32] In case the defence of non-use is invoked, the Intellectual Property Office requires the opponent to either submit prima facie evidence of genuine use of the mark or proper grounds for non-use. If the opponent fails to submit sufficient evidence of use or grounds for non-use, the opposition will be dismissed.

#### *3. In cancellation proceedings based on lack of use*

A trademark can be cancelled based on non-use in two ways: (1) in administrative cancellation proceedings before the Intellectual Property Office and (2) in cancellation proceedings before a judicial court. While administrative cancellation proceedings only require prima facie evidence and are relatively simple, expedient and inexpensive, cancellation proceedings in court require full proof and are more elaborate and expensive.

##### *i. Administrative cancellation proceedings*

Administrative cancellation proceedings before the Intellectual Property Office can be initiated at the earliest five years after the conclusion of opposition proceedings or, if no opposition has been filed, five years after the expiry of the opposition period.

As clearing the Register from marks which are not used is also in the interest of the public, anyone can file a request for cancellation of a trademark in administrative proceedings based on non-use.

The request will be rejected if the plaintiff fails to submit prima facie evidence of non-use of the mark or if the defendant submits prima facie evidence of use of the

31. By way of exception, an applicant who registers a defensive trademark with no intention to use it in Switzerland will generally not be able to rely on the grace period.
32. Article 22 (3) Trade Mark Ordinance. In case the defendant fails to invoke the defence of non-use in his first submission, he must initiate separate cancellation proceedings (see below, III.A.3.) and ask for the opposition proceedings to be stayed pending the resolution of the cancellation proceedings.

mark or proper grounds for non-use. On the other hand, if the plaintiff successfully establishes a prima facie case of non-use and the defendant fails to submit sufficient prima facie evidence, the trademark will be cancelled.

Where non-use is shown only for some of the designated goods or services, the Intellectual Property Office will only approve the request to the extent thereof.

*ii. Cancellation proceedings before a judicial court*

Cancellation proceedings before a judicial court can also be initiated at the earliest five years after the conclusion of opposition proceedings or, if no opposition has been filed, five years after the expiry of the opposition period.

As to standing to sue, Article 52 Trade Mark Act provides that anyone proving a legal interest may introduce a cancellation action before a judicial court and request to declare a trademark null and void. Such interest is given if the plaintiff is affected in his economic activity. However, if the cancellation of a trademark is requested on the basis of non-use, the courts accept any legally relevant interest, including the interest to clear the Register from marks which are not used. However, unlike administrative cancellation proceedings, the plaintiff has no standing to sue if for other reasons he is not able or not entitled to use the trademark in question or a similar mark.[33] The plaintiff has to submit prima facie evidence of non-use of the mark.[34] This leads to a reversal of the burden of proof: it is now up to the trademark proprietor to provide full proof of genuine use of the mark prior to the filing date of the cancellation request. If the plaintiff successfully establishes a prima facie case of non-use and the defendant fails to submit full evidence of genuine use, the trademark will be cancelled.

Where non-use is established only for some of the designated goods or services, the court will only approve the request to the extent thereof.

### 4. *As a defence in infringement/cancellation proceedings*

In infringement proceedings, the defendant may invoke the defence of non-use or even file a counterclaim seeking the cancellation of the plaintiff's trademark. If non-use of plaintiff's mark is claimed by way of defence, the court will first determine the validity of the plaintiff's mark as a preliminary question. If the defendant successfully establishes a prima facie case of non-use and the plaintiff fails to submit full evidence of genuine use, the court will find the plaintiff's trademark invalid and, as a result, reject the plaintiff's infringement action. The finding of invalidity due to non-use only has *inter partes* effect and does not result in the cancellation of the plaintiff's trademark.

On the other hand, if the defendant files a counterclaim seeking the cancellation of the plaintiff's trademark, the court will deal with the infringement action and the counterclaim together. The court will typically first decide the counterclaim for

33. BGE 125 III 193, 206 (Bud); BGE 136 III 102, 104 (*Yello/Yallo*).
34. In practice, prima facie evidence is usually furnished by submitting an in-use report prepared by one of the standard commercial providers.

cancellation. If the defendant successfully establishes a prima facie case of non-use and the plaintiff fails to submit full evidence of genuine use, the court will find the plaintiff's trademark invalid and, as a result, approve the defendant's counterclaim and reject the plaintiff's infringement action. The finding of invalidity due to non-use has *erga omnes* effect and results in the cancellation of the plaintiff's trademark.

## B. Burden of proof

In judicial proceedings before a court the general rule of Article 8 Civil Code regarding burden of proof applies. According to this rule, unless the law provides otherwise, the burden of proving the existence of an alleged fact shall rest on the person who derives rights from that fact. Given that non-use is a negative fact and, thus, difficult to prove, Article 12(3) Trade Mark Act provides for a reversal of the burden of proof, as already mentioned above:[35] the party asserting non-use has to submit prima facie evidence of non-use of the mark, and it is then up to the trademark proprietor to provide full proof of genuine use of the mark or proper grounds for non-use.

In administrative proceedings, different rules apply. In administrative cancellation proceedings, the lower level of prima facie proof applies to both parties. While the plaintiff asserting non-use must submit prima facie evidence of non-use of the mark, the defendant (i.e., the trademark proprietor) must submit prima facie evidence of use of the mark or proper grounds for non-use. In opposition proceedings, the party invoking the defence of non-use does not even have to furnish prima facie evidence, a substantiated allegation of non-use is sufficient. The opponent on the other hand has to furnish prima facie evidence of use of the mark or proper grounds for non-use.[36]

This rather confusing situation may be summarized as follows:

| | *Cancellation proceedings before judicial court* | *Administrative cancellation proceedings* | *Opposition proceedings* |
|---|---|---|---|
| Party asserting non-use | Prima facie evidence of non-use | Prima facie evidence of non-use | Substantiated allegation of non-use |
| Trademark proprietor | Full proof of use of mark or proper grounds for non-use | Prima facie evidence of use of mark or proper grounds for non-use | Prima facie evidence of use of mark or proper grounds for non-use |

## C. Relevant evidence

### 1. *Use on the product or its packaging*

Samples of the product or of the product packaging and labels from the relevant time are helpful means of providing evidence as they show the use of the mark on the

35. *See* III.A.3.
36. Article 32 Trade Mark Act.

product itself and the packaging. However, samples of product and packaging alone will not be sufficient to demonstrate genuine use. Additional evidence is required to show the volume, duration and frequency of the use of the mark.

### 2. *Use in advertising*

Sales brochures, catalogues and other advertising materials from the relevant five year period are equally helpful means of providing evidence to the extent they show the product and the use of the mark on the product or its packaging. Sales brochures and catalogues and other advertising materials can also be helpful to show the volume and duration of the use of the mark, especially if the trademark proprietor is able to submit multiple catalogues (showing the product and the use of the mark on the product or its packaging) for a number of consecutive years during the relevant time period.

### 3. *Use on the internet*

The internet can be an important source of evidence to prove genuine use as long as the product and the use of the mark on the product or its packaging are visible on the relevant screenshots. The internet can also be helpful to show the volume and especially the duration of the use. To this end, the trademark proprietor must provide multiple screen shots (showing the product and the use of the mark on the product or its packaging) for a number of consecutive years during the relevant time period.[37]

The internet can also be a source of evidence for a third party seeking to challenge the proprietor's trademark on the basis of non-use. For instance, if a standard Google search does not reveal any hits with regard to a specific product, respective screen shots may be filed to establish a *prima facie* case of non-use.[38]

### 4. *Form of evidence*

In judicial proceedings the trademark owner may rely on all admissible forms of evidence, including documentary evidence and witness testimony, among others. Documentary evidence includes delivery notes or invoices, sales brochures, catalogues or other advertising materials. Sales figures should be externally audited to qualify as proper evidence for demonstrating genuine use. Mere internal figures will not be considered proper evidence.

Survey evidence is not required and not common practice. Affidavits may be helpful to provide background to the court, but they are not common practice in Switzerland and will unlikely be taken into account unless they are backed up with witness testimony.

---

37. In practice, it is often helpful to rely on the Wayback Machine (web.archive.org) which is a digital archive of the World Wide Web and other information on the internet.
38. In practice, prima facie evidence is often furnished by submitting an in-use report prepared by one of the standard commercial providers (*see* also BGer, sic! 2018, 59 (*Abanca (fig.)/ Abanka (fig.)*)).

For all of the documentary evidence, it is essential that they derive from the relevant five-year-period. Therefore, the date of the evidence should be apparent either by virtue of a date on the evidence itself or according to the circumstances. Undated evidence may also be helpful in combination with dated evidence.

### 5. *Distinction between use of a company name and use of the mark for services offered by the company*

This distinction between use of the mark as a company name and use of the mark for services offered is difficult in practice. In addition, it is more difficult in practice to prove genuine use for services than for goods. Therefore, the courts tend to be more generous in assessing genuine use of a service mark. Any kind of documentary evidence shall be taken into account, including catalogues, advertising materials, brochures, insurance policies, business letters, and even auxiliary documents such as uniforms, tools and vehicles. [39] In one case, the court took into account licence agreements as evidence to prove genuine use.[40]

## IV. JUSTIFICATION FOR NON-USE

### A. Obstacles to use

Where the trademark proprietor has proper reasons for not having used his trademark, it will not be revoked, according to Article 12(1) Trade Mark Act. But what constitutes a proper reason for non-use? Article 19(1) of the TRIPS Agreement recognizes the following reasons for non-use: 'circumstances arising independently of the will of the owner of the trademark which constitute an obstacle to the use of the trademark, such as import restrictions on or other government requirements for goods or services protected by the trademark'.

In line with this legal requirement, it is recognized in Switzerland that events which lie outside the will, or at least the control, of the trademark proprietor will be excused. Such events include acts of force majeure and acts of god, including natural catastrophes, war, riots, boycotts and import bans due to outbreak of a disease. In an older case involving the pencil manufacturer Koh-i-noor, the expropriation of a production facility abroad was also found to constitute a proper reason for non-use.[41]

The impossibility to distribute a product in Switzerland before completion of the market authorization procedure is equally recognized as valid excuse for non-use. This applies, for instance, to pharmaceuticals. However, the trademark proprietor cannot invoke circumstances for which he alone is responsible. Equally, he cannot rely on delays caused by technical or economic difficulties that were predictable. For instance, in one case the court noted that the difficulty of selling foreign butter in

39. *See* Noth/Bühler/Thouvenin-Markus Wang, Art. 11 MSchG N 24 and BSK MSchG-Bernard Volken, Art. 11 MSchG N 28.
40. *See* BVGer of 10 February 2011, decision B-3686/2010 (*Heidiland II*).
41. BGE 83 II 312, 331 (*Koh-i-noor*).

Switzerland was well known and could have been foreseen by the trademark owner. As a result, the non-use of the mark was not excused.[42]

Imminent or pending legal procedures between private parties may also justify non-use. In particular a pending action against the validity or use of a trademark may excuse non-use. A trademark owner cannot be expected to use his mark as long as he runs the risk that he will be injuncted or his mark even cancelled. However, if only a cease and desist letter has been sent to the trademark owner and no litigation initiated, non-use will not be justified, as he can be expected to either ignore the cease and desist letter or clarify the situation by seeking a negative declaratory judgment.

It is not settled whether infringement proceedings commenced by the trademark owner during the grace period may justify non-use. The issue arises in cases where the grace period expires during pending proceedings. Some believe that the trademark owner should not be required to start use during pending proceedings while infringement is ongoing. Others rightly think that the fact that the trademark owner may end up having to tolerate another mark he considers infringing does not justify non-use.

In a case involving the German magazine *Focus*, the Federal Supreme Court held that a pending legal procedure against the foreign basis of an international registration (so-called central attack[43]) also justifies non-use.[44] This is because the extension for Switzerland depends entirely on the foreign basis of the international registration, whose cancellation lies outside the control of the trademark proprietor.

If a proper reason for non-use exists, what is the consequence in terms of timing? The five -year grace period is not interrupted. Rather, it is suspended as long as the reason for non-use exists. In other words, the grace period is extended by the time period during which the justification for non-use exists.

### B. Circumstances in which non-use may be cured

As long as no one has invoked non-use, the proprietor may at any time resume use with the effect that the right to the trademark is restored (see above, II.E.).

## V. CONSEQUENCES OF NON-USE

### A. Consequences of non-use and relevance of the context

The consequence of non-use is that the proprietor can 'no longer assert his right to the trademark', according to the wording of Article 12(1) Trade Mark Act. However,

42. RKGE, sic! 1998, 406 (*Anchor/Ancora*).
43. *See* Article 6 (3) Madrid Agreement.
44. BGE 130 III 371, 373 (*Focus*). The situation is different when an opposition is pending against the foreign basis of an international registration. Rather than justifying non-use for the entire duration of the foreign procedure, in case of an opposition the postponement of the grace period ends at the latest when the five year period of Article 6(2) Madrid Agreement expires because at this stage the pending opposition no longer carries the risk of causing a loss of protection in all countries of extension of the international registration; *see* BGer, sic! 2008, 896, 897 (*Focus II*).

this wording is not entirely precise. The consequences of non-use only enter into effect once a third party claims non-use. As long as no third party claims non-use or invokes the defence of non-use the trademark owner is still able to assert his right to the trademark.

If a third party claims non-use by way of cancellation proceedings (either before a judicial court or in administrative proceedings before the Intellectual Property Office) the consequence of non-use is the revocation of the trademark with *erga omnes* effect.

If a third party invokes the defence of non-use in opposition or trademark infringement proceedings, the consequence of non-use is not the revocation of the mark. Rather, the finding of invalidity has *inter partes* effect only, and the trademark owner can no longer assert his right to the trademark against such counterparty. This *inter partes* loss of trademark rights is permanent. This means that even if the trademark proprietor resumes use of the mark for the designated goods and services, this will not have any effect towards the counterparty who successfully invoked non-use by way of defence prior to resumption of use. The trademark will only be restored in relation to everyone else.

## B. Extent of any revocation

Where non-use is established only for some of the designated goods or services, the court will only approve the request to the extent thereof.

## C. Timing of any revocation

The timing of the consequences of non-use is not entirely clear. There are three possibilities, but the prevailing view is that the consequences of non-use occur with *ex tunc* effect from the date of expiry of the grace period.

# Thailand

*Say Sujintaya*[1]

## I. INTRODUCING GENUINE USE OF TRADEMARKS

### A. Legal framework

In Thailand, genuine use is not a requirement for trademark registration or renewal. The applicant or registrant does not have to prove that the mark has been used to secure or maintain a trademark registration. However, according to Section 63 of the Trademark Act B.E. 2534,[2] any interested party or the Registrar may submit a petition to the Trademark Board to cancel a trademark registration on the grounds of non-use, if it is proved that (a) the trademark has not been used for three years prior to the petition date, and (b) that the registrant has no intention to use the mark. Under Thai practice, this action is referred to as *non-use cancellation*.

In cases where non-use cancellation petitions are filed against registered trademarks, genuine use would be pertinent. Nevertheless, there is no legal requirement in the Trademark Act for trademark owners to prove genuine use of their trademarks.

---

1. The Author would like to thank Tirasha Punjabutrachai, an intellectual property associate at Baker McKenzie in Bangkok, for her assistance with this chapter.

2. Section 63 of the Trademark Act reads as follows:

> 'Any interested person or the Registrar may petition the Board to cancel a trademark registration if it is proved that at the time of registration the owner of the trademark had no bona fide intention to use the trademark with the goods for which it was registered and in fact there was no bona fide use whatsoever of the trademark for such goods or that during the three years prior to the petition for cancellation there was no bona fide use of the trademark for the goods for which it was registered unless the owner can prove that such non-use was due to special circumstances in the trade and not to an intention not to use or to abandon the trademark for the goods for which it was registered.'

### B. Purpose and function of genuine use: policy considerations

An interested party may use non-use cancellation as an option in trying to remove third party obstacles to their trademark applications. While the Registrar is entitled to file a non-use cancellation petition to the Trademark Board, this chapter does not discuss non-use cancellation initiated by Registrars, as under current practice in Thailand, it appears that only trademark owners have been submitting non-use cancellation petitions, not the authorities.

According to the trademark law, if any interested party submits a non-use cancellation petition to the Trademark Board, the burden of proof falls on such interested party to prove to the Trademark Board that the registered mark has not been used for three years before the petition date and that the trademark owner no longer has a real intention to use the registered mark with the goods or services for which the mark is registered.

In practice, when a petitioner submits a non-use cancellation petition to the Trademark Board, the Trademark Board will serve a notice to the registration owner providing the opportunity for the registrant to submit evidence of use of the mark and to defend against the non-use cancellation within 60 days from the date of receipt of the Trademark Board's notice. If the registration owner can submit evidence of use or claims that he still has an intention to use the registered mark, the Trademark Board will likely dismiss the non-use cancellation petition. A party who is not satisfied with the Trademark Board's decision can bring a lawsuit to the Court within 90 days of receipt of the Trademark Board's decision.

It is interesting to note that, under current practice in Thailand, the burden of proving non-use and the trademark owner's intention not to use the mark are so unfairly onerous as to make it exceedingly difficult for a petitioner to succeed in a non-use cancellation action. As a result, a non-use cancellation action may not be an appropriate option for interested parties to pursue to remove third party obstacles to their trademark applications. In other words, non-use cancellation actions in Thailand, unfortunately, do not serve their intended purpose.

## II. WHAT QUALIFIES AS GENUINE USE OF A TRADEMARK

The Trademark Act does not define the words *use* or *genuine use*. Given this, in order for trademark owners to determine if their use would be sufficient to constitute genuine use, the Trademark Board's decisions and its current practice as well as court's judgments need to be studied and taken into consideration.

### A. Nature of use

#### *1. Use consistent with the function of the trademark*

The Trademark Board tends to consider that genuine use of the mark must be use of the mark in association with the goods or services applied for. That is, it must be used as a badge of origin of the goods or services which can allow the public to distinguish the goods or services under the mark from goods or services of the others. In this regard, use of marks as decorative elements which cannot identify the origin of goods or services would not qualify as genuine use. For example, use of the picture of the Incredible Hulk as a t-shirt design would unlikely be regarded as use as a trademark.

### 2. *Use in the course of trade*

To constitute genuine use, the mark should be used in the course of trade of goods or services sought for registration. However, there is no indication under current practice in Thailand that public use or use of the mark in a non-commercial manner will not be considered as genuine use.

### 3. *Use by the owner or with its consent*

It is not mandatory that use of the mark has to be carried out by the trademark owner alone. The trademark owner can claim use of the mark by his authorized licensee as genuine use of the mark. However, to allow the licensee's use to qualify as genuine use, the trademark licence agreement between the trademark owner and the licensee must fulfil formality requirements, i.e. such licence must be in writing and registered with the trademark Registrar.

It is worth noting that the Trademark Act also requires that the trademark licence agreement must contain a clause on quality control of goods under the licensed mark. If there is no clause on quality control, the Registrar could not approve registration of such licence agreement. As a consequence, if the trademark licence is not registered, the trademark owner cannot claim use of the mark by the licensee as genuine use of the mark.

## B. Volume, duration and frequency of use

The Trademark Board has not clearly identified volume, duration and frequency of use to qualify for genuine use. In practice, only a small token use, e.g. product catalogues showing that goods under the mark are available for sale, or a one-time shipment of a small number of products to a distributor, would be accepted by the Trademark Board as evidence of use.

## C. Use of a sign

Based on precedents, the Trademark Board allows slight differences between the marks as used and the marks as registered when defending against non-use. That is, if the marks as used are slightly different from the registered marks, such use could still be used to successfully defend against third party non-use cancellation petitions. However, if the marks as used are significantly different from the registered marks then such use would not be regarded as genuine use of the registered marks.

It is worth noting that the Supreme Court adopted a conservative view on use of a mark in a different form from that as registered in Case No. 2082/2557, where use of the marks VITARAL S.M. or VITANA-EZ was unacceptable as genuine use of the registered mark VITANA -SM.

### D. Use in relation to goods and services

To qualify as genuine use, the mark must be used with goods or services sought for registration.

### E. Territorial scope

While the law does not stipulate that genuine use of the mark must take place within Thailand, it is an established practice of the Trademark Board that evidence to prove genuine use must be for use of the mark in Thailand.

### F. Relevant time period and its calculation

The period of non-use after which a mark will become vulnerable to non-use cancellation is three years counting back from the date when the petition is filed. If use of the mark is resumed after a non-use period of three years, such resumption would reset the non-use period.

## III. EVIDENCE OF GENUINE USE OF A TRADEMARK

### A. When use should be proved: timing of lack of genuine use claims

#### *1. When filing an application for a trademark*

At the time of filing an application, an applicant can file a trademark application without being required to submit evidence of genuine use of the mark or state that he has an intention to use the mark within a specific period in the future.

#### *2. In the course of opposition proceedings*

A party who files an opposition against a pending application does not have to submit evidence of genuine use to support the opposition. Non-use of the mark cannot be used as a ground to oppose a pending application in Thailand.

#### *3. In cancellation proceedings based on lack of use*

In cancellation proceedings in Thailand, any interested party who files a non-use cancellation petition with the Trademark Board, which is at the administrative level, has to prove that the mark has not been used in good faith and that the trademark owner has no intention of using the mark in good faith.

#### *4. As a defence in infringement/cancellation proceedings*

In response to the non-use cancellation, the trademark owner has the right to submit a counterclaim and evidence of genuine use to the Trademark Board in the same cancellation action in order to undermine a non-use cancellation raised by the petitioner within 60 days from receiving notice of the cancellation petition from the Trademark Board.

## B. Burden of proof

As discussed above, the burden of proof of the non-use of the trademark falls on the party who files a non-use cancellation action with the Trademark Board. Any interested party who files a non-use cancellation has to prove the following elements in order to satisfy the Trademark Board:

- the mark has not been used in good faith for three years prior to the non-use cancellation petition being filed; and
- the trademark owner does not have an intention of using the mark in good faith.

It is interesting to note that the non-use cancellation petitioner's burden of proof under the Thai law has long been seen to put a huge burden on the petitioner, as proving that something does not exist (non-use) is so much more difficult than proving that something does exist (use). If the mark has not been used, such non-use would not leave any footprint which the petitioner could use as provable evidence to submit to the Trademark Board.

Moreover, proving the trademark owner no longer has an intention of using the mark in good faith appears to be more difficult as, in practice, the Trademark Board would consider that renewal of trademark can be deemed an intent to use and dismiss the non-use cancellation petition. As such, if the continuous period of non-use of three years is interrupted by the filing of a renewal, it would, perhaps, be better to wait until three years after the renewal before filing for cancellation on the basis of non-use.

## C. Relevant evidence

(1) Use on the product or its packaging – Evidence of use of the mark on the product or its packaging would be acceptable by the Trademark Board.
(2) Use in advertising – Evidence of use of the mark in advertising clearly targeting consumers in Thailand, for example advertisements in Thai or showing Thai baht currency, would be acceptable by the Trademark Board.
(3) Use on the internet – Evidence of use of the mark on the internet should be in Thai language to prove to the Trademark Board that such use is targeted at consumers in the country.
(4) Form of evidence – Evidence of use of the mark could be in the form of invoices (containing references to the mark), receipts (containing references to the mark), expenses on advertising, market surveys, etc.
(5) Distinction between use of a company name and use of the services offered by the company – If the company name is only used as a trade name, such use would not be regarded as genuine use of the mark. However, if the company name is registered as a trademark or service mark, use of the company name

as a badge of origin of goods or services sought for registration would then be deemed as use of the mark.

## IV. JUSTIFICATION FOR NON-USE

### A. Obstacles to use

According to the Trademark Act, it is for the trademark owner to prove that there is a valid reason for such non-use i.e. special circumstances in trade. In practice, the Trademark Board would accept a shortage of supply, business hardship, economic crisis, inability to appoint a distributor, etc. as special circumstances in trade which would justify the non-use.

### B. Circumstances in which non-use may be cured

The trademark owner will not be given the opportunity to rectify a situation of non-use by starting to use the mark within a fixed period after the non-use cancellation action has been raised.

## V. CONSEQUENCES OF NON-USE

### A. Consequences of non-use and relevance of the context

The consequence of non-use of the mark is that the trademark registration could be vulnerable to non-use cancellation. In Thailand, an actionable non-use cancellation is a different proceeding from trademark opposition proceedings or enforcement of right against trademark infringement. If any interested party tries to cancel a trademark registration on the ground of non-use, he has to initiate a separate cancellation action before the Trademark Board.

### B. Extent of any revocation

Partial non-use cancellation is not applicable in Thailand. If the mark is registered with many different goods or services, evidence of genuine use of the mark with some applied goods or services under the registration would be sufficient to defend against a non-use cancellation of the whole registration.-

### C. Timing of any revocation

There is no fixed period of time in which an interested person needs to bring a non-use cancellation action to the Trademark Board. An interested person can file a non-use cancellation petition to the Trademark Board at any time if he can prove that such registered mark has not been used for three consecutive years prior to the filing of the cancellation petition.

# Turkey

*E. Kerim Yardimci and Sarper Sever*[1]

## I. INTRODUCING GENUINE USE OF TRADEMARKS

### A. Legal framework

Protecting the trademark domestically and globally is very important for international and national companies for the sustainability of market share. Furthermore, the competitiveness between entities and presenting goods by the entities in a new market in Europe are one of the aims of European Union (hereinafter 'EU'), especially the logic behind the free trade zone.[2] In brief, the registered trademark shall be used continuously in accordance with these aims of European Union.

Since Turkey is one of the candidate countries for EU membership, the laws are being harmonized with EU laws and regulations to reach the common standards of the EU across the internal market. One of the most recent examples of such harmonization acts is in the Turkish Industrial Property Law (entered into force in 10 January 2017) area; in other words, the Articles regarding 'use', namely 'genuine use' of trademarks now comply with the EU laws, especially with the following Articles of Directive (EU) 2015/2436 of the European Parliament and of the Council of 16 December 2015[3] to approximate the laws of the Member States relating to trademarks:

> (31) Trademarks fulfil their purpose of distinguishing goods or services and allowing consumers to make informed choices only when they are actually used on the market. A requirement of use is also necessary in order to reduce the total

---

1. Many thanks to Feride Utku, an experienced intellectual property lawyer at the law firm Deris, for her valuable help.
2. Esen Cam, *Avrupa Birligi Hukuku Uyarinca Markanin 'Gercek Kullanim' Problemi*, (2016), 2 TFM 27, at 28.
3. Directive (EU) 2015/2436 of the European Parliament and of the Council of 16 December 2015 available at <https://eur-lex.europa.eu/legal-content/EN/TXT/?uri=CELEX:32015L2436> (last accessed on 14 March 2018).

> number of trademark registered and protected in the Union and, consequently, the number of conflicts arising between them. It is therefore essential to require that registered trademarks actually be used in connection with the goods or services for which they are registered, or, if not used in that connection within five years of the date of the completion of the registration procedure, be liable to be revoked.
>
> (32) Consequently, a registered trade mark should only be protected in so far as it is actually used and a registered earlier trade mark should not enable its proprietor to oppose or invalidate a later trade mark if that proprietor has not put his trade mark to genuine use. Furthermore, Member States should provide that a trade mark may not be successfully invoked in infringement proceedings if it is established, as a result of a plea, that the trade mark could be revoked or, when the action is brought against a later right, could have been revoked at the time when the later right was acquired.

Article 9 of the Turkish Industrial Property Law,[4] which regulates 'trademark usage' has been harmonized with the above provisions and if the registered trademark is not being genuinely used in Turkey within five years from the date of registration with regard to registered goods and services, the subject trademark registration will become vulnerable to a cancellation action in regard to the goods and services[5] not being used. On the other hand, if the registered trademark is genuinely used only for certain goods and services, these goods and services will not be affected by the cancellation.

For example, according to the wording of the law, the use of a registered trademark with various elements without altering its distinctive character only for goods or packages for export purposes and the use of a trademark by third parties with the consent of the trademark owner shall be regarded as genuine/valid use of the mark. Under these circumstances, the registered trademark cannot be subject to a cancellation decision.

For this reason, it can be said that the new law aims to harmonize its applications with EU legislation.[6]

This new enactment by Turkey also complies with the Paris Convention for the Protection of Industrial Property of 20 March 1883:[7]

> *Article 5*
>
> *C. Failure to Use; Different Forms; Use by Co-proprietors*
>
> (1) If, in any country, use of the registered mark is compulsory, the registration may be cancelled only after a reasonable period, and then only if the person concerned does not justify his inaction.

4. Industrial Property Law, Law No. 6769 enactment date: 22 December 2016.
5. Onder Erol Unsal, *Markanin Kullanimi Kavrami, Kullanmama Savunmasi Muessesesi ve Avrupa Birligi Kurumlarinin Bu Hususlardaki Degerlendirmeleri*, 14 Fikri Gündem 30, at 31 (2017).
6. *Ibid*, at 34.
7. Paris Convention for the Protection of Industrial Property of 20 March 1883, as revised at Brussels on 14 December 1900, at Washington on 2 June 1911, at The Hague on 6 November 1925, at London on 2 June 1934, at Lisbon on 31 October 1958, and at Stockholm on 14 July 1967, and as amended on 28 September 1979.

> (2) Use of a trademark by the proprietor in a form differing in elements which do not alter the distinctive character of the mark in the form in which it was registered in one of the countries of the Union shall not entail invalidation of the registration and shall not diminish the protection granted to the mark.

The wording 'a reasonable period' means that every signatory country shall designate the proper period according to their practice or internal needs.

In addition to the Paris Convention, this new enactment by Turkey also complies with the Agreement on Trade-Related Aspects of Intellectual Property Rights:[8]

> *Article 19*
>
> *Requirement of Use*
>
> 1. If use is required to maintain a registration, the registration may be cancelled only after an uninterrupted period of at least three years of non-use, unless valid reasons based on the existence of obstacles to such use are shown by the trademark owner. Circumstances arising independently of the will of the owner of the trademark which constitute an obstacle to the use of the trademark, such as import restrictions on or other government requirements for goods or services protected by the trademark, shall be recognized as valid reasons for non-use.
>
> 2. When subject to the control of its owner, use of a trademark by another person shall be recognized as use of the trademark for the purpose of maintaining the registration.

It can be seen below that the wording 'at least three years of non-use' has been designated as five years in Turkey which therefore complies with this provision.

## B. Purpose and function of genuine use: policy considerations

In general, the trademark is the face of the company and it is one of the most valuable assets of the company. However, registering trademarks and not using them may block the economic system. So, one of the purposes of the genuine use necessity is to remove unused trademarks from the Register and give new opportunities to applicants. Therefore, if the Register is kept uncluttered and clean, new parties may use an unlimited supply of possible marks.

According to the preamble to the Turkish Industrial Property Law, the main rationale for the provisions of genuine use can be determined as enhancing industrial growth. For instance, if owners register the trademarks and do not use them, the use or registration of the trademark by third parties can be blocked. In terms of industrial growth, this cannot be accepted. Therefore, the function of genuine use is to prevent the Register of the Turkish Patent and Trademark Office[9] becoming a 'junkyard for trademarks'.

Furthermore, the Turkish legislation also aims to ensure that registered trademarks are used effectively on the market to meet the essential function of the trademark by distinguishing goods and services from others and enabling entrepreneurs

8. Agreement on Trade-Related Aspects of Intellectual Property Rights (TRIPS).
9. The establishment of the Turkish Patent and Trademark Office has hereby been decided by the Council of Ministers on 19 June 1994.

to register new trademarks without receiving any opposition due to the unused trademarks. In other words, the idea is avoiding obstacles to entering a market and for the competition in the market.[10]

Another purpose of genuine use is to provide fair competition and to prevent registering of trademarks in bad faith, namely without the intention of using the mark but with the actual intention of preventing others from using it. Usually in practice not only are some trademarks registered without intention of use, such trademarks are even renewed many times, even though their owners have no intention of using them at all.

According to the statistics of the Office, 121,108 of trademark applications were filed before the Office in 2017, a 13% increase compared to 2016.[11]

According to officers, the main reason for this increase is that most owners re-applied for their trademark which was registered more than five years ago. The main idea behind that is to overcome provisions of Turkish Industrial Property Code regarding non-use. No trademark application has been rejected due to the 're-filing' to date.

## II. WHAT QUALIFIES AS GENUINE USE OF A TRADEMARK

### A. Nature of use

#### *1. Use consistent with the function of the trademark*

The trademark should be used with respect to its basic (in fact essential) function and has to have a direct connection in relation to the goods and services it is registered.

The usage of a trademark as a trade name, business name, title or only online may not be accepted as sufficient or genuine use at all times. To evaluate the sufficiency of such use concrete case must be considered.

If such usage is sufficient to clearly distinguish the goods and services of the trademark owner from the goods and services of other competitors in the market and makes a physical connection between the mark and the goods and services of the owner by being perceived by the consumers in the same manner, it will be deemed as genuine use.

#### *2. Use in the course of trade*

##### *i. Public use*

Trademarks, which are used to distinguish the goods or services of one undertaking from those of other undertakings in the market, should be put on the market.

10. Ilhami Gunes, *Kullanilmayan Markanin Iptali Sorunu*, available at <http://www.ankarabarosu.org.tr/siteler/ankarabarosu/frmmakale/2015-1/2.pdf> (last accessed on 1 March 2018), at 42.
11. Trademark – Breakdown of Trademark Applications by Years, available at <http://www.turkpatent.gov.tr/TURKPATENT/statistics/> (last accessed on 20 March 2018).

Otherwise, using the trademark internally regardless of the size of the company will not meet the requirements for the genuine use of the trademark.

*ii. Commercial purpose*

As the essential function of a trademark is to distinguish the goods and services of the trademark from others in the market and to establish a direct connection with the goods and services it is registered for; genuine use in the course of trade is required.

That being said, use in form of free promotional goods which are given to consumers as gifts does not constitute genuine use since the essential function of a mark is not fulfilled. Promotional goods are not given to consumers with the purpose of penetrating the market for those goods. The preparation for commercial activities or internal use does not constitute genuine use. Genuine use of the trademark requires the use of the trademark within trading activities.

On the other hand, use of a trademark for merchandising may be deemed to constitute genuine use if the trademark used together is perceived by the consumer as distinguishing the goods, such use will then be considered as genuine use for the original products.

Use only in relation to goods to be exported is accepted as sufficient according to the clear wording of the Turkish Industrial Property Law.

### *3. Use by the owner or with its consent*

According to the Turkish Industrial Property Law, use of the trademark with the consent of the proprietor shall be deemed as use by the proprietor of the trademark.

A common example is granting a licence; therefore, use by the licensee will be deemed as genuine/valid use. Furthermore, the use of a trademark by an entity which is connected to the parent company shall be deemed as a genuine/valid use as well.[12]

The licensor, however, shall undertake to take the necessary measures to ensure the quality of the goods to be produced or the services to be rendered by the licensee.

Use of the trademark by the licensee inures to the benefit of the owner. There is no requirement for the licence agreement to be recorded with the Turkish Patent and Trademark Office.

## B. Volume, duration and frequency of use

Evaluating the genuine use of trademark based on objective criteria is essential.

In other words, the registered goods and services, the way they are used and the scope of this use are important for the evaluation of genuine use of the trademark.

For instance, there will be fewer buyers of expensive goods or services compared to cheaper goods or services and therefore, the sales of expensive goods or services will naturally be lower. Accordingly, this does not mean that the fewer sales do not prove the genuine use. However, it is true that designating genuine use of the

12. Basak Ozkok, *Marka Basvurusunun Yayıma Itiraz Asamasında Kullanim Ispatı Talebi*, 12 Terazi Law Journal 43, at 50.

everyday products, such as goods with high volumes of sale, or services will be easier. At this point, the size of the company can be also an objective indicator to designate genuine use of its products.

For instance, the evaluation of genuine use of the goods or services of a large company can be stricter than for a smaller company.[13] Furthermore, it is significant whether the trademark is used in accordance with the basic function of the goods or services registered for, whether the trademark is able to create a market share. In addition to this, concerned costumer group is also important for the evaluation the volume of genuine use of trademark.[14]Yet, needless to say, token use is not accepted as genuine use.

In summary, minimal use is sufficient as long as it qualifies as genuine use, since the use of the trademark does not always have to be extensive for it to be deemed as genuine, since the use or sales of the trademark depends on the characteristics of the goods or service concerned on the corresponding market.

Concerning court action for cancellation, the defendant is expected to present evidence to prove genuine use within the last five years from the date of the cancellation action. It should be noted that such usage falling within the three months prior to the court action effected with the expectation that cancellation would be requested may be ignored by the court.

In addition to the date, the place of the use of the trademark should be evaluated in detail.[15]

| *Goods/Services* | *Quantity of Use* | *Sufficiency* |
|---|---|---|
| Clothes | 2 invoices and 100 shirts | No |
| Retail services of cosmetics products | Sales amount: 1117 perfumes, 1000 deodorants, 2500 lipsticks in 2015 | Yes |
| Dairy products | Sales value: TRY 5000 in total in 2014–2015–2016 | No |
| Music instruments | 30 handmade guitars, 15 drums, 10 pianos | Yes |
| Toy car | 1000 miniature cars for the purpose of collection | Yes |
| Bottle of water | Sales amount: 5000 bottles of water in total in 2014–2015–2016 | No |

## C. Use of a sign

Article 9 of IP Code clearly states that 'use of the trademark in a form differing in elements which do not alter the distinctive character of the trademark' will be deemed as the use of trademark. In other words, as long as the distinctive character of the

13. Hanife Dirikkan, *Tescilli Markayi Kullanma Kulfeti* (Vedat Kitapçılık 1998), 244.
14. The Guidelines of Proving of Use of Turkish Patent and Trademark Office, available at <https://www.turkpatent.gov.tr/TURKPATENT/resources/temp/06BC31D4-EDDC-411A-9E47-332F344B27F6.pdf> (last accessed on 20 March 2018).
15. The Guidelines of Proving of Use, *supra* n. 14.

trademark remains, a different form of the use of trademark will also be deemed as use of the trademark.

Undoubtedly, there is logic behind this statement being that due to the dynamics of business life and due to the changes in business custom, the owner can modernize the trademark without changing its distinctive character.

In brief, even though the trademark is modernized, without changing its distinctive character it will be accepted as valid use of the trademark. Additions or subtractions made in the character of the trademark, or a simple change of motif, capitalization of a lower case branding mark, addition of relative denominations which are descriptive to the goods or services, use with a colour other than the one for which it is registered will not be deemed as changing the distinctive character of the trademark.

It is difficult to designate definite objective rules for which conditions will be deemed as changing the distinctive character of the trademark. At this point, the following examples have been published by the Turkish Patent and Trademark Office. These examples will give a general idea as to whether a change will be deemed as use or not.[16]

### 1. *Uses that do not change the distinctive character of trademarks*

Use of the registered trademark with another registered trademark such as umbrella mark:

| *Registered trademark* | *Use* | *Is it acceptable?* |
|---|---|---|
| GALAXY S6 | SAMSUNG<br>Galaxy S6 | YES |

Use of the registered trademark with the addition of descriptive components:

| *Registered trademark* | *Use* | *Is it acceptable?* |
|---|---|---|
| AINOA | AINOA BIO<br>AINOA SPA WORLD<br>AINOA CHOCOLATE | YES |

Use of the registered trademark with joint denominations to create a impression of serial trademark:

| *Registered trademark* | *Use* | *Is it acceptable?* |
|---|---|---|
| ELISIO | ELISIO15M<br>ELISIO15H<br>ELISIO·25H | YES |

16. *Ibid.*

Use of the registered trademark with auxiliary denominations which are not dominant:

| *Registered trademark* | *Use* | *Is it acceptable?* |
|---|---|---|
| COLORIS | COLORIS Global Coloring Concept | YES |

Use of the registered trademark with a distinctive device:

| *Registered trademark* | *Use* | *Is it acceptable?* |
|---|---|---|
| SEMBELLA | sembella® | YES |

Use of the registered device mark with an additional word mark:

| *Registered trademark* | *Use* | *Is it acceptable?* |
|---|---|---|
| X | X BIONIC® | YES |

Use of the registered trademark with a prefix or suffix:

| *Registered trademark* | *Use* | *Is it acceptable?* |
|---|---|---|
| TENTATION | TENTATIONS | YES |

Use of the registered trademark removing a word which is not distinctive nor descriptive:

| *Registered trademark* | *Use* | *Is it acceptable?* |
|---|---|---|
| GarantiBank | Garanti | YES |

Use of the registered trademark removing the device:

| *Registered trademark* | *Use* | *Is it acceptable?* |
|---|---|---|
| evaGROUP | EVA | YES |

Use of the registered trademark removing a prefix/suffix or denomination which is not important:

| *Registered trademark* | *Use* | *Is it acceptable?* |
|---|---|---|
| CASTILLO DE PERELADA | CASTILLO PERELADA | YES |

Use of the registered trademark with a different font or adding the device:

| *Registered trademark* | *Use* | *Is it acceptable?* |
|---|---|---|
| FOL'EPI | Fol Epi | YES |

## *2. Uses that change the distinctive character of trademarks*

Use of the registered trademark adding certain denominations:

| *Registered trademark* | *Use* | *Is it acceptable?* |
|---|---|---|
| CAPTAIN | CAPTAIN BIRDS EYE | NO |
| YILDIZ | AYYILDIZ | NO |

Use of the registered trademark removing certain denominations:

| *Registered trademark* | *Use* | *Is it acceptable?* |
|---|---|---|
| TONY HAWK | HAWK | NO |

| *Registered trademark* | *Use* | *Is it acceptable?* |
|---|---|---|
| RED KID | RED KIT  | NO |
| D'MEX | DIMEX | NO |

Apart from these, some changes such as registering a trademark with a circle, square, or triangle and using it without these elements, or registering a trademark which is composed of two words separated by a hyphen or broken line and using it without the hyphen/broken line will be deemed as use.[17]

17. Unal Tekinalp, *Fikri Mulkiyet Hukuku* (5th edn, Vedat Kitapcilik 2005), 429.

Furthermore, as mentioned above, the addition of descriptive denominations will not affect the distinctive character of the trademark; however, adding another word which itself has a distinctive character will definitely change the distinctive character of the trademark and it will not deemed as use.[18]

## D. Use in relation to goods and services

### 1. *Use for each of the designated goods and services*

The registered trademark has to be used for all goods and services covered by the registration. The non-use of the registered trademark regarding some of the goods and services indicated in the Register means that the obligation to use the mark for all the goods and services is not fulfilled.

It is important to note that the genuine use of goods or services will be evaluated per individual good or service. If genuine use is proved only for one good or service, it will be accepted that the other registered goods and services are unused. In other words, partial cancellation of trademarks is applicable in Turkey.

For instance, if an opposition is filed based on a trademark registered for the following goods in class 25:

- 'inner-outer clothing made from all kinds of material, except clothing for protective purposes, socks, foulards, shawls, bandannas, scarves, belts; (sub-class 1)
- footwear: shoes, slippers, sandals (sub-class 2)
- head gear: hats, flat caps, berets, skull caps, caps (sub-class 3)'

and the trademark owner has been invited to prove the usage thereof and yet can only prove that the subject trademark is being used only for 'caps'; other items will be available for the opposition trademark application and it will be registered for the goods which are not proved as used.

It should be borne in mind that the situation in cancellation actions will be different. For instance, if the trademark is registered for the abovementioned goods and during the cancellation action the owner could prove use only for 'caps'; the trademark registration will be cancelled for those other goods not being used and shall remain registered for 'caps' only. However, the critical point is that since the registration will be valid for 'caps', the whole subclass will be blocked by the owner anyway. In other words, the applicant of a new trademark application will not be able to register its trademark for 'head gear: hats, flat caps, berets, skull caps, caps' because of the national sub-classification system in Turkey.

### 2. *Use for a broader category of goods and/or services*

Turkey has adopted the Nice Classification system; furthermore, the system of sub-classes is available in Turkey.

18. Sabih Arkan, *Marka Hukuku* (Vedat Kitapcılık, 1998), 151.

During the process of filing a trademark application, the trademark can only be registered for all the goods in a sub-class.

If there is only one good that it is used within the sub-class, all the other goods in this sub-class will be cancelled and deleted from the registration. That being said, this one good will be enough to block a third party from registering other goods in the same sub-class.

According to the IP Code, identical or almost identical goods will be ex officio rejected by the Office and the evaluation of the identicalness of the goods and services is judged simply by examining whether the goods fall within the same sub-class or not.

For instance, taking again the example of sub-class 25.1, if only 'belt' is used then a third party cannot register the same or a very similar trademark for other goods in sub-class of 25.1.

Briefly, the use should be examined for each good or service, but if a trademark is used for a certain sub-class, then a third party cannot obtain a trademark for other goods within the same sub-class.

### E. Territorial scope

According to the related articles of the Turkish Industrial Property Law, genuine use of the trademark shall take place in Turkey. In other words, evidence for proving genuine use from outside of Turkey cannot be accepted.

This condition complies with the territoriality of Turkish Industrial Property Law.[19] On the other hand, this does not mean that the trademark must be used throughout Turkey. As mentioned above, locality is also important for evaluating genuine use of the trademark. This occasion rises especially when goods are local. If the product is used widely in that specific area, genuine use can be accepted. For instance, the burden of proof of genuine use for a local business cannot be the same as the burden of proof of genuine use for a country-wide company. Furthermore, the use of the trademark in free zones that are beyond customs frontiers can be accepted as a genuine use.[20]

The only exception is the use of the trademark on goods or their packaging solely for export purposes, which can be accepted for proving genuine use. Here, it should be proven that the trademark is affixed to the goods or packaging within Turkey and thus the use of the trademark can be deemed genuine use.

### F. Relevant time period and its calculation

The applicant of the opposed trademark may require from the opponent to evidence the use of its trademark in Turkey if the trademark is registered for more than five years. The five-year period will be calculated as of the date of registration of the subject trademark.

---

19. The Guidelines of Proving of Use, *supra* n. 14.
20. Hamide Erdogdu, *Markanin Hükümsüzlük Halleri ve Marka Hakkinin Sona Ermesi* available at <http://www.teknolojitransferi.gov.tr/TeknolojiTransferPlatformu/resources/temp/7A2F2FAB-C98E-422B-9199-A17C75B64262.pdf;jsessionid=6E94FC5D164CADE795013850143A00A4> (last accessed on 18 March 2018), at 20.

In cancellation actions due to non-use, the calculation is made starting from the date of the institution of the cancellation action. Therefore, the defendant must prove the use of trademark within five years prior to the institution of the cancellation action. It should be noted that such usage shall be ignored if the trademark was used within three months before such a request was submitted to the Court with the expectation that a cancellation would be requested.

According to the Turkish Industrial Property Law, resumption of use of the trademark before the opposing party institutes a cancellation action due to non-use will halt the cancellation action and avoid cancellation of the registered trademark.

However, such usage shall be ignored if the trademark was used within three months before such request was submitted to the Court (or the Office as of 10 January 2024) with the expectation that a cancellation request would be filed.

In other words, the resumption of use will not be accepted for very particular circumstances for a very short period of time and only if the Judge comes to the conclusion that the defendant used the trademark just to avoid cancellation.

## III. EVIDENCE OF GENUINE USE OF A TRADEMARK

### A. When use should be proved: timing of lack of genuine use claims

#### *1. When filing an application for a trademark*

According to Turkish Industrial Property Law, there is no intent to use requirement or necessity to prove actual use when filing the trademark. Neither is there any further grace period to submit such use documents or intention of usage declarations.

#### *2. In the course of opposition proceedings*

For oppositions procedures filed on the ground of risk of confusion and risk of association, the opponent, upon the request of the applicant, may be requested to prove that the opposing trademark has been genuinely used in Turkey for the relevant goods or services provided that it has been registered in Turkey for more than five years as from the registration date.

If the opponent fails to submit any documents to prove genuine use within one month as of the notification the administrative decision, the opposition shall be refused.

If it is proven that the opposing trademark has only been used for part of the goods or services covered by the registration, the opposition shall be examined only with respect to the goods or services for which use has been proven.

It should be noted that the requirement of use is valid only for oppositions filed on the ground of risk of confusion. In other words, if the opposition is filed based on other grounds such as prior use, notoriety or bad faith, the non-use counterclaim cannot be requested.

It should be noted that the request for submitting evidence of use shall be filed during the first administrative instance (involving one examiner from Trademark Department to examine the opposition). It is critical to mention that the applicant will not be able to request the opponent to submit evidence regarding the use of trademark during the second administrative instance (involving Council of Re-Examination and Evaluation i.e. the appeal body consisting of three very experienced examiners).

At this point, it is very important for holders of trademark applications extended through WIPO by way of international registration to note that the Turkish Patent and Trademark Office does not inform International Registration applicants of opposition filed by third parties in Turkey.

In other words, the opposition is examined without the applicant's knowledge of the opposition proceedings. As applicants will not have knowledge of opposition proceedings they will not be able to request the opponent proves use of its registered trademarks.

For this reason, it is most important to monitor the status of trademark applications which are extended through WIPO in order to prevent any loss of rights.

### 3. *In cancellation proceedings based on lack of use*

According to the Turkish Industrial Property Law, the Turkish Patent and Trademark Office is entitled to cancel the trademark application if, within a period of five years following the registration, a trademark has not without a justifiable reason been put to genuine use for the goods or services for which it has been registered, or if the use has been suspended for an uninterrupted period of five years, the trademark shall be repealed.

Interested parties may file for the cancellation of a trademark with the Office.

The request for the cancellation of a trademark shall be filed against the persons registered as the proprietor of the trademark at the time of the request, or against their successors.

The requests for cancellation in relation to non-use shall be refused if the trademark was substantially used in terms of the goods or services for which it had been registered in the five-year period prior to the date of submission of the request for cancellation before the Office.

It should be highlighted that such usage shall be ignored if the trademark was used within three months before such request was submitted to the Office with the expectation that a cancellation request would be filed.

Where the cancellation is involved for only some of the goods or services for which the trademark has been registered, partial cancellation shall be ruled only with respect to those goods or services.

In any case, the cancellation cannot be ruled so as to change the representation of a trademark.

All requests for cancellation shall be notified to the proprietor of the trademark for which cancellation is requested. If the proprietor changes during the examination of cancellation, procedures shall continue against the registered proprietor.

The proprietor shall submit to the Office any relevant evidence and counterclaims against the request within one month. The Office shall grant a further one-month extension if requested within the said one-month period. The Office, if it deems necessary, may request further information and documents to be submitted. The Office shall grant a decision on the case within the framework of the claims and counterclaims as well as the evidence submitted.

It should be noted that the Office will examine the cancellation requests up to 10 January 2024 as according to the IP Code the Office will be entitled to examine the cancellation request for seven years from the entry in force of the IP code on 10 January 2017.

Until 10 January 2024, the cancellation request should be made before the court through institution of a cancellation action due to non-use. Court actions instituted before 10 January 2024 shall be finalized by the courts. The decisions rendered by the courts shall be ex officio notified to the Office once they are finalized.

This provision also complies with the Directive (EU) No. 2015/2436 of the European Parliament and of the Council of 16 December 2015 to approximate the laws of the Member States relating to trade marks.

> 'Article 45
>
> *Procedure for revocation or declaration of invalidity*
>
> Without prejudice to the right of the parties to appeal to the courts, Member States shall provide for an efficient and expeditious administrative procedure before their offices for the revocation or declaration of invalidity of a trade mark.
>
> The administrative procedure for revocation shall provide that the trade mark is to be revoked on the grounds provided for in Articles 19 and 20.
>
> The administrative procedure for invalidity shall provide that the trade mark is to be declared invalid at least on the following grounds:
>
> (a) the trade mark should not have been registered because it does not comply with the requirements provided for in Article 4;
>
> (b) the trade mark should not have been registered because of the existence of an earlier right within the meaning of Article 5(1) to (3).'

According to the Directive, Member States will be obliged to constitute the administrative procedure until 14 January 2024.[21] The seven years were designated in order to have an enough time to build sufficient and efficient administrative proceedings.

## 4. *As a defence in infringement/cancellation proceedings*

In infringement proceedings, the non-use claim may be provided as a defence by way of counterclaim. If the plaintiff cannot evidence use of the trademarks on the goods and services, the infringement action will simply not succeed but the non-used trademark will not be affected by this decision in the sense that it will continue to exist in the Register as a valid trademark.

It is important to note that the date of the proceedings shall be taken as the basis for determining the period of five years regarding use.

21. The Article states that 'Member States shall bring into force the laws, regulations and administrative provisions necessary to comply with Articles 3 to 6, Articles 8 to 14, Articles 16, 17 and 18, Articles 22 to 39, Article 41, Articles 43 to 50 by 14 January 2019. Member States shall bring into force the laws, regulations and administrative provisions to comply with Article 45 by 14 January 2023. They shall immediately communicate the text of those measures to the Commission.'

It is also possible to institute a counter-action against an infringement action at the same the same court. The court will not stay but first examine the extent of use of the trademark upon which the infringement action is based and there enter into the matter of infringement. The advantage of the counter-action is that the non-use registered trademark will be cancelled and deleted from the Register.

This opportunity is a significant breakthrough in the Turkish Industrial Property Law; claiming this defence will prevent 'compensation hunters' who register trademarks and do not use them, waiting for infringements by third parties.[22]

## B. Burden of proof

According to Turkish Procedural Law the burden of proof is on the plaintiff who is the owner of the claim.[23] However, in non-use actions before the courts, considering the impossibility of submitting evidence which proves non-use of the trademark, the burden of proof will be switched to the defendant where defendant will submit evidence regarding its use.[24]

It should be noted that the shift in burden of proof is confirmed by longstanding jurisprudence of the Supreme Court.

## C. Relevant evidence

To proceed smoothly with regard to proving usage through opposition cases, the Turkish Patent and Trademark Office publish a specific form; therefore, submission of evidence should follow the requirement of this form. There is no official form to submit evidence for a cancellation action. The Office strictly adheres to this official form, and therefore, the parties shall use this form. All the evidence shall be indicated separately on the form.

The same principle applies for a cancellation action before the Courts. Indeed, the defendant will be obliged to prove that he is using the trademark on specific goods by clearly classifying the all relevant evidence for each of the goods and/or services.

Acceptable evidence for proving use varies according to the current practice, in the form of invoices, catalogues, price lists, product codes, products, packages, signage samples, advertisements, promotions, promotions, market research, public opinion research, advertising, promotional and promotional products images, videos and so on. It is very important that the trademark itself and the type of products in line with the goods and/or services registered for should clearly appear in the evidence. The date, and the appearance of the trademark on the evidence should be clearly demonstrated.

22. Hayrettin Caglar, *6769 Sayili Sinaî Mulkiyet Kanununa Gore Tescilli Markanin Kullanilmamasinin Hukuki Sonuçlari ve Anayasa Mahkemesinin 556 Sayili KHK M. 14 Hukmunu* İptal *Eden Kararinin Etkileri*, available at < http://webftp.gazi.edu.tr/hukuk/dergi/21_1_1.pdf > (last accessed on 11 March 2018), at 11.
23. Sami Karahan, Cahit Suluk, Tahir Saraç, Temel Nal, *Fikri Mulkiyet Hukukunun Esasları* (3rd edn, SeckinYayincilik2011), 154.
24. Unal Tekinalp, *Fikri Mulkiyet Hukuk* (5th edn, VedatKitapcilik 2012), 420.

It should be borne in mind that when the Office will be entitled to examine cancellation action due to non-use, it is likely that the parties will be obliged to follow the same or similar form.

### 1. *Use on the product or its packaging*

Photos of the product and its packaging or photos of signboards can be submitted as evidence. According to the practice of the Office, these photos should be A4 size clearly showing the trademark in use.

### 2. *Use in advertising*

Advertising documents will be accepted as evidence. According to the common practice, it is a very good indicator of use as it makes sense that the owner will be not willing to make investment for advertising for the trademark which is not in use. It is beneficial to submit these kinds of evidence along with their invoices.

### 3. *Use on the internet*

As other evidence, use of trademark on the internet can be submitted as evidence, such as captions from online shopping websites showing the goods. It may be beneficial to show and indicate the URL of the website which offers the goods or services in order to demonstrate the accuracy of the evidence.

### 4. **Form of evidence**

Any form of evidence such as sales, surveys, and affidavits will be accepted during opposition or court action proceedings. It should be noted that it is beneficial to submit these kinds of documents with a notarized affidavit.

### 5. *Distinction between use of a company name and use of the services offered by the company*

The functions of commercial name, company name and trademark are different. However, the distinction between the use of a company name and the use of the services offered by the company can be difficult to make. If the company does not produce goods, the trademark may be used on the signboard such as for services of hospitality, hotel, merchandising, restaurants. Yet, it is possible to provide the services without confusion with others' trademarks and company names. Therefore, the infringement can be avoided.[25]

25. Mehmet Emin Bilge, *Marka ve Ticaret Unvani Arasinda Iltibas,* 2TFHD 7, at 13 (2015).

## IV. JUSTIFICATION FOR NON-USE

### A. Obstacles to use

If the opponent presents good reasons why it has not been possible to use the trademark, it will not be necessary to prove of use of this trademark.

According to well established Supreme Court decisions, a good reason will be an excuse, which will not depend on the will of the proprietor of the trademark and which will be impossible to avoid, such as an act of god or force majeure.

Namely, justification for non-use means that the conditions arising will prevent him from using his trademark. Therefore, these conditions can be any conditions which are factual or judicial. In other words, it does not mean that these conditions are only force majeure conditions. The important point here is that the hurdle should be serious and originate from reasons beyond the will of owner, his control or his area of influence. In other words, the owner shall act as a prudent merchant; however, the conditions cannot be eliminated even though he acts as a prudent merchant.

Valid reasons why the trademark owner has not used the trademark for the last five years can include war, economic crises, natural disasters, changes in customs legislation, import restrictions and embargoes.[26]

Beside these, reasons can arise because of the actions of a competent authority. For instance, the following reasons can also be evaluated as justification for non-use: the owner not being able to produce as a result of the precautions taken by the authorities or the authorities not giving the necessary permission for him to sell the product.

Other acceptable reasons for non-use can be an invalidation action instituted against the registered trademark. Considering the legal dispute, the owner of the trademark will have valid reasons for not using its trademark due to risk of invalidation with *ex tunc* effect. In fact, invalidation can give rise to compensation as the trademark will be considered not to have existed from the beginning.

Worsening of the trademark owner's financial situation is not a valid reason for not using the trademark. The business owner's being out of action (due to disease or injury) is also not a valid reason. Yet, this hurdle can be overcome by replacing the owner with a commercial representative.

However, the situation such as the owner's personal labour being important for production (for instance; instruments the owner produces are high quality) or the owner having a chronic disease shall be evaluated objectively. On the same lines, situations where the goods can be modified consistent with changes in the market cannot be accepted as a justification for non-use; it is always possible to overcome the hurdle by changing the product.

In the case of granting a licence to a licensee, a trademark not being used by the licensee cannot be put forward as a reason by the licensor and therefore cannot be deemed as a justification for non-use. The logic behind this is that in non-exclusive licence agreements, the licensor also has the right to use his trademark.

26. Hasan Oguzhan Karsli, *Markanin Kullanim Zorunlulugu* <http://www.teknolojitransferi.gov.tr/TeknolojiTransferPlatformu/resources/temp/A15BC11B-8E82-4E84-8D76-E9AAE-758B2AB.pdf> (last accessed on 15 March 2018), at 118.

On the other hand, in exclusive licence agreements, the licensor can prevent his trademark from being cancelled due to non-use by cancellation of the licence agreement. Furthermore, the owner of the trademark may use the trademark by entering into a new licence agreement with the third party.

As a rule, bankruptcy of the owner cannot be accepted as a valid reason since the owner has a chance to grant an assignment in order to use his trademark. However, if the owner goes bankrupt because of a global economic crisis or his manufacturing site is condemned, such bankruptcy may be accepted as a valid reason for justification of non-use.[27]

### B. Circumstances in which non-use may be cured

According to the Turkish Industrial Property Law, resumption of use of the trademark before the opposing party institutes cancellation due to non-use will stop cancellation action and avoid cancellation of registered trademark.

However, such usage shall be ignored if the trademark was used within three months before such request was submitted to the Court (or the Office as of 10 January 2024) with the expectation that a cancellation request would be filed.

In other words, the resumption of use will not be accepted for very particular circumstances for a very short period of time and only if the Judge comes to the conclusion that the defendant has used the trademark just to avoid cancellation.

## V. CONSEQUENCES OF NON-USE

### A. Consequences of non-use and relevance of the context

Once a decision for cancellation becomes final and definitive, the rights regarding the trademark shall lapse in the Register as of the date of the cancellation action (*ex nunc*) and such decision will be published in the Official Trademark Bulletin.

Exceptionally, where the conditions for cancellation have arisen at an earlier date, it may be decided, upon request, that decision of cancellation shall take retroactive effect as of that date – (earlier than the date of the cancellation action).

The retroactive effect a decision for cancellation due to non-use shall not extend to any final decisions for infringement of the trademark reached and enforced prior to the decision or contracts concluded and executed prior to the decision. That being said, partial or total reimbursement of sums paid under these contracts may be claimed on grounds of equity.

The retroactive effect will be without prejudice to claims for compensation of third parties for damage caused by gross negligence or bad faith on the part of the proprietor of the trademark,

27. Ozkok, *supra* n. 11, 57.

## B. Extent of any revocation

Evaluation of use during the cancellation action process will be carried out item by item. Therefore, if the defendant proves genuine use for only one item, the other goods or services covered will be removed from the trademark application.

At the end of the proceeding, the list of goods and services for the trademark can be modified by the court in accordance with proof of use of the trademark. In other words, if the evidence attesting the use of goods is submitted for only one of the goods, the others can be removed from the list. Therefore, the trademark will remain valid only for the goods which are proven in use.

## C. Timing of any revocation

The timing of the cancellation decision will be the date of filing of the cancellation request. However, where the conditions for cancellation have arisen at an earlier date, it may be decided, upon request, that decision shall take effect as of that date.

# United Kingdom

*Paul A. Harris and Jan M. Vleck*

## I. INTRODUCING GENUINE USE OF TRADEMARKS

### A. Legal framework

The United Kingdom is a member of the European Union (EU) so the law in the United Kingdom builds upon the EU legal framework. The comments made in the chapter on the EU are therefore also applicable to the United Kingdom.

#### *1. United Kingdom Trade Marks Act 1994*

The relevant provision in the Trade Marks Act 1994 ('TMA 1994'), which was brought in pursuant to the TMD 1988, is section 46. This is as follows:

> *Revocation of registration.*
>
> (1) The registration of a trade mark may be revoked on any of the following grounds –
>
> (a) that within the period of five years following the date of completion of the registration procedure it has not been put to genuine use in the United Kingdom, by the proprietor or with his consent, in relation to the goods or services for which it is registered, and there are no proper reasons for non-use;
>
> (b) that such use has been suspended for an uninterrupted period of five years, and there are no proper reasons for non-use;
>
> (c) [...]
>
> (d) [...]
>
> (2) For the purposes of subsection (1) use of a trade mark includes use in a form differing in elements which do not alter the distinctive character of the mark in the form in which it was registered, and use in the United Kingdom includes affixing the trade mark to goods or to the packaging of goods in the United Kingdom solely for export purposes.

(3) The registration of a trade mark shall not be revoked on the ground mentioned in subsection (1)(a) or (b) if such use as is referred to in that paragraph is commenced or resumed after the expiry of the five year period and before the application for revocation is made:

Provided that, any such commencement or resumption of use after the expiry of the five year period but within the period of three months before the making of the application shall be disregarded unless preparations for the commencement or resumption began before the proprietor became aware that the application might be made.

(4) An application for revocation may be made by any person, and may be made either to the registrar or to the court, except that –

(a) if proceedings concerning the trade mark in question are pending in the court, the application must be made to the court; and
(b) if in any other case the application is made to the registrar, he may at any stage of the proceedings refer the application to the court.

(5) Where grounds for revocation exist in respect of only some of the goods or services for which the trade mark is registered, revocation shall relate to those goods or services only.

(6) Where the registration of a trade mark is revoked to any extent, the rights of the proprietor shall be deemed to have ceased to that extent as from –

(a) the date of the application for revocation, or
(b) if the registrar or court is satisfied that the grounds for revocation existed at an earlier date, that date.

As only European Courts can interpret EU law, what this means is that where the General Court (which handles appeals from the tribunals of EUIPO, namely the Boards of Appeal) or the full Court of Justice of the European Union ('CJEU'), give judgement (and whether on either the TMD 1988 – now TMD 2015 – or the EUTMR), that decision will impact on both sets of legislation because of the identical wording. However, it also impacts on the domestic legislation, and how that must be interpreted. This is because the domestic legislation must comply with the TMD 1988 (and now the latest version, TMD 2015), and so the meaning should be the same.

For those not familiar with the function of the various legal tribunals within the EU, it is perhaps worth pausing to gain a better understanding of them, and the interplay with each Member States' domestic courts. Insofar as all revocation actions of EUTMs commenced in the Cancellation Division of EUIPO, any appeal is to one of the Boards of Appeal. Thereafter, any further appeal is to the General Court and potentially (on issues of law) to the CJEU. Where, however, issues have arisen as a result of a counterclaim seeking to revoke either a EUTM or a national mark in an infringement action before a domestic court, questions are referred directly to the CJEU to seek guidance on the interpretation of the law. The issues relating to the facts are for the national courts only to decide.

TMA 1994 requires that an applicant for a UK registered trade mark has an intent to use the mark in respect of the listed goods and/or services when filing a trade mark application.

Section 32(3) TMA 1994 Act states:

> 'The application shall state that the trade mark is being used, by the applicant or with his consent, in relation to those goods or services, or that he has a *bona fide* intention that it should be so used.'

Applying for a UK registered trade mark (which necessarily involves a statement that the mark is being used or there is an intention to use it) without that use or intention to use is considered to be bad faith, and is therefore a ground for refusal of an application and/or revocation of a registered trade mark. However, it is important to note that there is a presumption of good faith so bad faith must be proven by whoever is attacking the registration or application.

### B. Purpose and function of genuine use: policy considerations

There are several purposes for requiring genuine use. The 9th recital of the Codified Directive says that the purpose is to 'reduce the total number of trade marks registered and protected in the Community and, consequently, the number of conflicts which arise between them'.

In *Laboratoires Goemar v. La Mer Technology* [2005] EWCA Civ 978, Mummery LJ in the English Court of Appeal considered Articles 10 and 12 of the Directive and said at paragraph 14 of his judgment that it was evident that these purposes are 'to prevent the Register from being cluttered up with unused marks, which would obstruct later traders wishing to use the marks and would create unnecessary conflict and confusion between the registered trade marks of traders'. This is consistent with the position stated in the 10th recital of the EUTMR 207/2009 (now recital 24 in EUTMR 2017/1001), that there is no justification for protecting trade marks 'except where the trade marks are actually used'. The Court of Appeal took into account the ruling of the ECJ (C- 259/02) on an earlier reference in the case by Jacob J (as he then was).

The prevention of trade mark 'trafficking' or 'trolling' may be considered to be another purpose. The use requirement limits the ability of a non-user to hold a stockpile of trade mark registrations for an indefinite period.

## II. WHAT QUALIFIES AS GENUINE USE OF A TRADEMARK

### A. Nature of use

#### *1. Use consistent with the function of the trade mark*

The issue of what constituted 'genuine use' was considered by the ECJ (now renamed Court of Justice of the European Union - CJEU) and the Court of First Instance (now renamed the General Court) within one day of each other in Case C-40/01 *Ansul BV v. Ajax Brandbeveiliging BV* (ECRI - 2439, [2003] RPC 40), following a referral by the Hoge Raad (in relation to wording used in TMD 1988), and in Case T-174/01 *Jean Goulbourn v. OHIM* respectively (in relation to an opposition in which the opponent's mark was put to proof of use).

In *Jean Goulbourn*, the CFI held that the condition of genuine use required that the mark, as protected in the relevant territory, be used publicly and outwardly for the purpose of ensuring an outlet for the goods or services which it represents.

Genuine use, the Court held, is real use of the earlier mark on the marketplace, so as to draw attention of potential consumers to the goods or services effectively offered under the mark.

However, the ECJ decision in *Ansul* gave far more guidance on the meaning of 'genuine use'. In that case it was held that 'genuine use' means actual use of the mark and not merely token use, serving solely to preserve the rights conferred by the mark, and such use must be consistent with the essential function of a trade mark. Genuine use entails use of the mark on the market for the goods or services protected by the mark, and not just internal use by the undertaking concerned. Use must relate to goods/services already marketed or about to be marketed and for which preparations by the undertaking to secure customers are under way (i.e. in the form of advertising campaigns). When assessing genuine use, regard should be had to all the facts and circumstances relevant to establishing whether the commercial exploitation of the mark is real. Use of the mark need not always be quantitatively significant to be deemed genuine. Use of the mark could also be genuine for goods in respect of which it is registered that were sold at one time but are no longer available, provided the proprietor makes actual use of the same mark for component parts that are integral to the make-up or structure of such goods, or for goods or services directly connected with the goods previously sold and intended to meet the needs of customers.

The ECJ's decision in *Ansul* did not deal with whether there should be a *de minimis* rule for 'genuine use' and, if so, what it should be. Nor did it give guidance as to the position when the use is tiny but has only a genuine commercial motive.

However, in considering 'genuine use' under the TMA 1994, in the case of *Laboratoires Goemar v. La Mer Technology* [2005] EWCA Civ 978, applying the principles laid down in *Ansul*, the Court of Appeal found that limited sales to an agent were sufficient (only £800 during the relevant period) to defeat a claim that a mark should be revoked on the basis of non-use. Allowing an appeal by Goemar, the Court of Appeal held that the first instance judge had misapplied *Ansul* by adding further requirements beyond those envisaged by the ECJ. First, Mummery LJ noted that the judge had characterised the use by Goemar as being 'akin to internal use', on the basis that there were no sales to the ultimate consumer, which he considered incorrect as those sales were at 'arm's length' and could only be considered as external. In this sense, the Court considered it was not necessary to show 'public and external use', in the sense of communication of the mark to the end-user, to show genuine use. Commenting on the quantity of sales, Mummery LJ stated that there was no qualitative threshold for sales to constitute genuine use of a mark. In his view, the modest amounts did not prevent the use from being genuine use on the market. He added that the subjective reasons for the sale were not relevant, but that the existence of the sale between the companies proved the existence of a market for the goods. Adding to Mummery LJ's decision, Neuberger LJ stated that any use of a trade mark, which is consistent with its essential function, and which is not token or internal, would constitute genuine use, and stressed that minimal use will not disqualify the use from being genuine (although it may raise the questions as to whether it was token). Arguably, this is a departure from what was initially thought the *Ansul* case required, namely that a certain level of use was required in order for that use to be classed as 'genuine'.

Where the use is internal (for example internal emails), this is not use, or not use 'in the course of trade' and is likely to constitute use as a 'private matter' (see paragraph 66 *Och-Ziff Management Europe Limited and ors v. OCH Capital LLP and ors* [2010] EWHC 2599 (Ch)).

However, the principles applicable to genuine use in English domestic trade mark law have been set out most recently in *The London Taxi Corporation v. Frazer-Nash Research Ltd* [2016] EWHC 52 (Ch), and reiterated (by the same judge) in *W3 Ltd v. Easygroup Ltd and anor* [2018] EWHC 7 (Ch). The judge, Arnold J, reviewed the authorities and the judgements of other tribunals, before setting out those principles. It is considered his comments are useful in this context:

> '*Use of the CTM itself*
>
> 217. *The law with respect to genuine use.* In *Stichting BDO v BDO Unibank Inc* [2013] EWHC 418 (Ch), [2013] FSR 35 I set out at [51] a helpful summary by Anna Carboni sitting as the Appointed Person in *SANT AMBROEUS Trade Mark* [2010] RPC 28 at [42] of the jurisprudence of the CJEU in Case C-40/01 *Ansul BV v Ajax Brandbeveiliging BV* [2003] ECR I-2439, Case C-259/02 *La Mer Technology Inc v Laboratories Goemar SA* [2004] ECR I-1159 and Case C-495/07 *Silberquelle GmbH v Maselli-Strickmode GmbH* [2009] ECR I-2759 (to which I added references to Case C-416/04 P *Sunrider Corp v Office for Harmonisation in the Internal Market (Trade Marks and Designs)* [2006] ECR I-4237). I also referred at [52] to the judgment of the CJEU in Case C-149/11 *Leno Merken BV v Hagelkruis Beheer BV* [EU:C:2012:816], [2013] ETMR 16 on the question of the territorial extent of the use. Since then the CJEU has issued a reasoned Order in Case C-141/13 *P Reber Holding & Co KG v Office for Harmonisation in the Internal Market (Trade Marks and Designs)* [EU:C:2014:2089] and that Order has been persuasively analysed by Professor Ruth Annand sitting as the Appointed Person in *SdS InvestCorp AG v Memory Opticians Ltd* (O/528/15).
>
> [...]
>
> 219. I would now summarise the principles for the assessment of whether there has been genuine use of a trade mark established by the case law of the Court of Justice, which also includes Case C-442/07 *Verein Radetsky-Order v Bundervsvereinigung Kamaradschaft 'Feldmarschall Radetsky'* [2008] ECR I-9223 and Case C-609/11 *Centrotherm Systemtechnik GmbH v Centrotherm Clean Solutions GmbH & Co KG* [EU:C:2013:592], [2014] ETMR 7, as follows:
>
> > "(1) Genuine use means actual use of the trade mark by the proprietor or by a third party with authority to use the mark: *Ansul* at [35] and [37].
> >
> > (2) The use must be more than merely token, that is to say, serving solely to preserve the rights conferred by the registration of the mark: *Ansul* at [36]; *Sunrider* at [70]; *Verein* at [13]; *Centrotherm* at [71]; *Leno* at [29].
> >
> > (3) The use must be consistent with the essential function of a trade mark, which is to guarantee the identity of the origin of the goods or services to the consumer or end user by enabling him to distinguish the goods or services from others which have another origin: *Ansul* at [36]; *Sunrider* at [70]; *Verein* at [13]; *Silberquelle* at [17]; *Centrotherm* at [71]; *Leno* at [29].
> >
> > (4) Use of the mark must relate to goods or services which are already marketed or which are about to be marketed and for which preparations to secure customers are under way, particularly in the form of advertising

campaigns: *Ansul* at [37]. Internal use by the proprietor does not suffice: *Ansul* at [37]; *Verein* at [14]. Nor does the distribution of promotional items as a reward for the purchase of other goods and to encourage the sale of the latter: *Silberquelle* at [20]-[21]. But use by a non-profit making association can constitute genuine use: *Verein* at [16]-[23].

(5) The use must be by way of real commercial exploitation of the mark on the market for the relevant goods or services, that is to say, use in accordance with the commercial *raison d'être* of the mark, which is to create or preserve an outlet for the goods or services that bear the mark: *Ansul* at [37]-[38]; *Verein* at [14]; *Silberquelle* at [18]; *Centrotherm* at [71].

(6) All the relevant facts and circumstances must be taken into account in determining whether there is real commercial exploitation of the mark, including: (a) whether such use is viewed as warranted in the economic sector concerned to maintain or create a share in the market for the goods and services in question; (b) the nature of the goods or services; (c) the characteristics of the market concerned; (d) the scale and frequency of use of the mark; (e) whether the mark is used for the purpose of marketing all the goods and services covered by the mark or just some of them; (f) the evidence that the proprietor is able to provide; and (g) the territorial extent of the use: *Ansul* at [38] and [39]; *La Mer* at [22]-[23]; *Sunrider* at [70]-[71], [76]; *Centrotherm* at [72]-[76]; *Reber* at [29], [32]-[34]; *Leno* at [29]-[30], [56].

(7) Use of the mark need not always be quantitatively significant for it to be deemed genuine. Even minimal use may qualify as genuine use if it is deemed to be justified in the economic sector concerned for the purpose of creating or preserving market share for the relevant goods or services. For example, use of the mark by a single client which imports the relevant goods can be sufficient to demonstrate that such use is genuine, if it appears that the import operation has a genuine commercial justification for the proprietor. Thus there is no *de minimis* rule: *Ansul* at [39]; *La Mer* at [21], [24] and [25]; *Sunrider* at [72]; *Leno* at [55].

(8) It is not the case that every proven commercial use of the mark may automatically be deemed to constitute genuine use: *Reber* at [32]."'

## 2. *Use in the course of trade*

### i. *Public use*

As pointed out in the *Jean Goulbourn* decision above, the use must be publicly and outwardly. However, the degree of 'public' use has to be viewed in the context of the cases, in particular *Ansul* and more importantly for the UK, the *Laboratoires Goemar v. La Mer Technology* decision, since the sale was to a single distributor.

### ii. *Commercial purpose*

Although *Ansul* requires that for 'genuine use' the mark must be used on the market, in Case C-442/07 *Verein Radetzky-Orden v. Bundesvereinigung Kameradschaft*

*'Feldmarschall Radetzky'*, the ECJ held that use of a trade mark by a non-profit organisation in its relations with the public was genuine use of the trade mark.

Use on give-away promotion items is not considered genuine use of the mark on those goods. In Case C-495/07 *Silberquelle GmbH v. Maselli-Strickmode GmbH* a clothing manufacturer that gave away drinks bearing its trade mark in order to promote its clothing brand, was not considered to have used the mark with respect to drinks. This was because the drinks were not distributed with an intention to penetrate the drinks market.

This point had, in fact, already been considered in the case of *Premier Brands UK Ltd v. Typhoon Europe Ltd* [2000] EWHC 1557 (Ch). In this case, the mark in issue was TY.PHOO, registered in relation to 'tea' in class 30, but also for 'domestic utensils' in class 21, and cutlery in class 8. These marks having been registered since 1928 and 1945, TY.PHOO is considered to be a well-known brand in the UK. The defendants wished to use the mark 'TYPHOON' in relation to oriental kitchen hardware made in the Far East.

The case involved consideration of some peculiarities of English law that arose in relation to the two 'defensive' marks in classes 8 and 17 registered in 1945 at the time when the Trade Marks Act 1938 was in force. However, with the TMA 1994 in force as well as the TMD 1988 applying, one of the main issues revolved around whether there had been genuine use of the 1945 registrations, which the defendants had sought to revoke on the basis of non-use. The specific items which Premier Brands claimed use of and which fell within the classes (referred to as the 'Goods') were tea canisters, tea pots, mugs, tea towels, milk jugs, sugar bowls, biscuit barrels, cups, mug trees, aprons, storage jars, memo boards, tea cosies, trivets, tea pot stands, trays, cafetieres and drinking glasses. However, it was accepted that these items were given away as promotional items in order to promote the tea sold under the same brand ie TY.PHOO.

In concluding that there was genuine use, the judge gave his reasons as follows:

> 'In my judgment, it brings this case near to the borderline (as in ELLE Trade Marks, albeit for somewhat different reasons) but in the end it is not sufficient to bring it over the line. First, although it pales in comparison with the use of the mark in relation to tea, the use of the mark TY.PHOO in relation to the Goods over the past five years does appear to have been reasonably substantial; secondly, the use has been genuinely in relation to the Goods, in the sense that they are supplied in part for their own intrinsic merit, unlike an advertising hoarding or a shelf; thirdly, "giving away" involves a slightly narrow view, in the sense that Premier were clearly not indulging in charitable activities, and hoped therefore to profit indirectly from the manufacture and distribution of the Goods; fourthly, although very limited, there is evidence of two types of actual sale of the Goods, one of which was by Premier itself; fifthly, paragraph 28 of the judgment of the ECJ in Canon [1999] RPC 117 at 134 seems to me to suggest that trade mark use involves use in connection with goods in the course of trade, and, while in the main given away, the way in which, and the purpose for which, the Goods are distributed by Premier appears to me to fall within that concept; sixthly, the breadth of the sort of activities which fall within the requisite use seems to me to be underlined by the reference in Section 46(1) to the "consent" of the proprietor being sufficient.

> In the event, therefore, at least in so far as use of the mark TY.PHOO on the Goods is concerned, TEL's case on non-use of the two defensive marks fails.'

Given the date of this case, and that *Ansul* was decided some 3 years later, it is likely never to be followed, and the conclusion of the judge regarded as wrong. In particular, the fact that Premier Brands did not seek to secure or preserve a market in relation to those actual goods, and so the use (being promotional items) would not be consistent with the essential function of the mark.

### 3. *Use by the owner or with its consent*

Section 46(1)(a) provides that there must not have been use by the proprietor 'or with his consent'. Use by a licensee whether or not recorded on the trade marks register pursuant to section 25 TMA 1994 would seem to be sufficient. However, uncertainties remain:

(1) there is no definition of consent either in the TMD or the TMA 1994;
(2) it is by no means clear whether the 'consent' must be in the form of a written document (licence), albeit that it may be unregistered, or whether this can be oral consent or even implied;
(3) if the consent to use is contained in a licence, which either does not contain quality control provisions or the proprietor does not seek to enforce them, does this create a problem? Arguably, it may do under English domestic law, since the proprietor potentially faces the problem that the mark becomes vulnerable to removal, on the basis that the use has become misleading (or 'deceptive' as was used in the 1938 Act) – see *BOSTITCH TM* [1963] RPC 183, *Bowden Wire Ltd v. Bowden Brake* (1913) 30 RPC 45, and also *Scandecor Development v. Scandecor Marketing and ors* [2001] HL. In the latter case, Lord Nicholl's view was that it is no longer the position under the 1994 Act that unless the trade mark proprietor exercised quality control over the goods put out under that trade mark it was rendered invalid.

There is some guidance from the ECJ decision in Case C-416/04 *The Sunrider Corporation v. OHIM* where the proof of use partly consisted of invoices and orders showing sales of the products had been made by a third party. On appeal to the ECJ, the Sunrider Corporation argued that the CFI should not have taken account of a third party's use of the earlier trade mark. In doing so they argued that the CFI had taken inconclusive evidence from the opposing party into account and relied on presumption rather than firm evidence. This plea was rejected by the ECJ. They considered that the proprietor (who had the burden of proving genuine use of his earlier mark) could only have provided the third party's invoices and orders showing sales if he had consented to the use of the trade mark by a third party. The ECJ found that it was correct to rely on the presumption there was consent to the use of the mark on the basis it was unlikely that the proprietor would have been in a position to submit the evidence of use by that third party otherwise.

## B. Volume, duration and frequency of use

The scale and frequency of the use of the mark, together with territorial extent, etc., are factors to be taken into consideration in the determination of what constitutes genuine use. Even before the helpful laying out of the principles for assessment of genuine use by Arnold J in *The London Taxi Corporation* decision (supra), it was long recognised that the main area of contention would be the scale and frequency of use. Inevitably, this was always going to be tested when there was perceived to be little use over the relevant period.

In *The Sunrider Corporation* decision notwithstanding the probity of the invoices, they showed that only EUR 4,800 of product were sold and only to one customer over a period of ten months. Here the ECJ refused to overturn the CFIs decision that the use was more than 'token'. In their view, the CFI did not make any error in its assessment of genuine use. First, the CFI analysed the place, time, extent and nature of use of the mark. Secondly, it considered whether the mark had been used to create or preserve an outlet for 'concentrated fruit juices' (the products in issue). Thirdly, the fact that it was only sold to a single customer and so was not present in a substantial part of Spain was only one of many factors and by no means fatal to that determination.

However, in the UK, the Appointed Person in *Time Machine* (o/306/08), on appeal from a Trade Mark Registry decision, held that there was genuine use of the trade mark, the registration of which was only acquired by the defendant three weeks before the end of the five year non-use period. Following the approach in *Laboratoires Goemar*, the Appointed Person found that a small number of sales within the three week period were sufficient to constitute genuine use under the circumstances. The Appointed Person took into consideration the pre-acquisition use of the mark as being relevant because it enabled the defendant to begin use of the mark as soon as it was acquired and was consistent with the *Ansul* decision.

## C. Use of a sign

A proprietor is not precluded from seeking to rely on the fact that his mark was used in a form differing from that which was registered, provided that the differences do not alter the distinctive character of the mark, even where the later mark is itself registered. In Case C-533/11 *Bernhard Rintisch v. Klaus Eder*, Mr Rintisch had not used the mark relied on ('PROTI'), but had used 'PROTIPLUS' (also registered) and 'Proti Power'. The German Court considered that the words 'PLUS' and 'Power' respectively, did not alter the distinctive character of the 'PROTI' mark, and so use of 'PROTIPLUS' and/or 'Proti Power' was considered to be genuine use of the mark 'PROTI'. The CJEU agreed, and made the proposition stated above.

Since then, the CJEU have ruled that the principles that apply to marks, which are either part of another mark or are only ever used with another mark (as set out in case C-353/03 *NESTLÉ* [2005] ECR I-6135) in acquiring distinctiveness through use, apply equally to verifying genuine use: see Case C-12/12 *Colloseum Holding AG v. Levi Strauss & Co.* In this case, Levi Strauss argued that the use of their red tab could be genuine use, even though the word 'LEVI'S' always appeared on the tab. They argued the situation was analogous to the NESTLÉ case in establishing distinctiveness through use. The CJEU agreed.

Insofar as English domestic law cases are concerned, the CJEU have also ruled on the questions referred to it in Case C-252/12 *Specsavers International Healthcare Ltd v. ASDA Stores Ltd*. These questions stemmed from a trade mark infringement action in the UK. In the infringement case, the proprietor's mark was a figurative mark, which consisted of two overlapping ovals always used in conjunction with the word mark 'SPECSAVERS' (that composite combination was also registered). The CJEU considered there is genuine use of the figurative mark as part of the combination, provided the differences in form between the combination mark and figurative mark do not alter the distinctive character of the figurative mark, as registered.

On returning to the Court of Appeal, it was considered that the use of the Shaded logo (which is a black and white version of the well-known Specsavers' green overlapping ovals logo with 'Specsavers' in white script superimposed on it), constituted use of a mark consisting only of overlapping ovals in black (which was called the Wordless logo). The Court of Appeal considered that the evidence supported their conclusion that the Wordless logo had a distinctive character and could signify Specsavers' goods on its own, to the average consumer. It did, however, make clear that each case depended on its own facts.

## D. Use in relation to goods and services

### *1. Use for each of the designated goods and services*

As pointed out above, section 32(3) TMA 1994 Act requires that there is a *bona fide* intention to use the mark. That has to be for all the goods and services the subject of any trade mark application. Accordingly, there must be genuine use of the trade mark for each of the goods and services specified.

### *2. Use for a broader category of goods and/or services*

As Mr Justice Arnold commented in the *W3 Ltd* decision (supra):

> 'It is often the case that, when challenged to prove genuine use of a trade mark, the proprietor shows use in relation to some goods or services covered by the specification, but not others. In these circumstances Article 51(2) of the Regulation requires the competent authority only to revoke the trade mark to the extent that it has not been genuinely used. This can lead to difficulties where the trade mark is registered in respect of a broad class of goods or services, but it is said that the proprietor has only established use of a narrower sub-class within that broad class.'

The applicable principles have been set out by the Court of Appeal in *Merck KGaG v. Merck Sharp & Dohme Corp* [2017] EWCA CIV 1834 as:

> '245. First, it is necessary to identify the goods or services in relation to which the mark has been used during the relevant period.

> 246. Secondly, the goods or services for which the mark is registered must be considered. If the mark is registered for a category of goods or services which is sufficiently broad that it is possible to identify within it a number of subcategories capable of being viewed independently, use of the mark in relation to one or more of the subcategories will not constitute use of the mark in relation to all of the other subcategories.
>
> 247. Thirdly, it is not possible for a proprietor to use the mark in relation to all possible variations of a product or service. So care must be taken to ensure this exercise does not result in the proprietor being stripped of protection for goods or services which, though not the same as those for which use has been proved, are not in essence different from them and cannot be distinguished from them other than in an arbitrary way.
>
> 248. Fourthly, these issues are to be considered from the viewpoint of the average consumer and the purpose and intended use of the products or services in issue. Ultimately it is the task of the tribunal to arrive at a fair specification of goods or services having regard to the use which has been made of the mark.'

What the fourth point above highlights is the practice in the UK of the relevant tribunal being able to amend the specification of goods or services in order to achieve fair protection for the proprietor, having considered what genuine use there has been of the goods or services, with certainty for third parties as to the scope of that protection. This practice is considered more fully below in relation to section V.

## E. Territorial scope

Territorial scope will depend on whether the mark in issue is a national mark, or a EUTM.

Insofar as a UK national mark is concerned, in *Euromarket Designs v. Peters (CRATE & BARRELL)* [2000] ETMR 1025, Jacob Errant J stated that 'genuine use must involve that which a trader or consumer would regard as real or genuine trade in this country. This involves quantity as well as the nature of the use.' This case involved three cases of very minor use:

- Items bought in America by American customers and sent as presents to people in the UK;
- UK residents sending orders to the US via a wedding present scheme to be supplied in the US;
- 'Spillover' circulation of US magazines.

This was not found to be use in the UK.

Use on the internet was also considered in *Richard Dearlove v. Shaun Combs* [2007] EWHC 375 (Ch) by Kitchin J (occasionally referred to as the *P. Diddy* case) who at paragraph 25 said:

> 'The fundamental question is whether or not the average consumer of the goods or services in issue within the UK would regard the advertisement and site as being aimed and directed at him. All material circumstances must be considered and these would include the nature of the goods or services, the appearance of the website, whether it is possible to buy goods or services from the website,

whether or not the advertiser has in fact sold goods or services in the UK through the website or otherwise, and any other evidence of the advertiser's intention.'

This case was considered in the decision of HHJ Colin Birss QC in *Yell Limited v. Louis Giboin and ors* [2011] EWPCC 9 (a trade mark infringement case). On the facts, there was found to be use in the UK, even though the domain names were not .co.uk.

However, use of the mark (where it is a EUTM) in only a single territory of the EU may not be sufficient (see Case C-140/11 *Leno Merken BV v. Hagelkruis Beheer BV*) to constitute genuine use. The same principles apply as to creating/keeping the market, but the national court is left to consider all the circumstances and facts.

This issue of 'genuine use in the Community, and whether such use must be in more than one Member State' has been considered in the IPEC by HHJ Hacon in *The Sofa Workshop Ltd v. Sofaworks Limited* [2015] EWHC 1773 (IPEC)). The judge considered carefully the *Leno Merken* case together with other CJEU decisions. Analysing the *Leno Merken* case decision the judge gave his guidance and interpretation on this specific issue as follows:

(1) In considering 'genuine use in the Community' territorial borders are to be disregarded;
(2) A mark is put to genuine use where it is used in accordance with its essential function, namely, to guarantee the identity of the origin of the goods or services, and used for the purpose of maintaining or creating market share;
(3) To determine use all relevant facts and circumstances are to be considered, including characteristics of the market, nature of the relevant goods/services, territorial extent, scale of use and frequency and regularity of use;
(4) In relation to territorial extent of use, use in more than one Member State will generally be required; and
(5) An exception is where the market for the relevant goods/services is restricted to a single Member State.

The principles were more concisely set out in the *The London Taxi Corporation* decision and again repeated in the *W3 Ltd* decision (supra), as follows:

- territorial borders of the Member States should be disregarded in the assessment of whether a trade mark has been put to genuine use in the Community;
- while it is reasonable to expect that a Community trade mark should be used in a larger area than a national trade mark, it is not necessary that the mark should be used in an extensive geographical area for the use to be deemed genuine, since this depends on the characteristics of the goods or services and the market for them;
- it cannot be ruled out that, in certain circumstances, the market for the goods or services in question is in fact restricted to the territory of a single Member State, and in such a case use of the Community trade mark in that territory might satisfy the conditions for genuine use of a Community trade mark.

## F. Relevant time period and its calculation

The period of time is always the same: five years. However, from when that time period is counted will depend on the proceedings in which the issue of genuine use is raised. Where there is an application for revocation brought at the EUIPO, then the five-year period will run backwards from the date on which the application was lodged with EUIPO.

Where, however, the issue is raised by way of a counterclaim, and as part of a defence, then the five years is counted back from the date of the counterclaim (see *The London Taxi Corporation* decision supra).

## III. EVIDENCE OF GENUINE USE OF A TRADEMARK

### A. When use should be proved: timing of lack of genuine use claims

#### *1. When filing an application for a trade mark*

As mentioned above in section I.A.3, UK domestic law requires that, on filing, the mark in question is in use or there is a *bona fide* intention to use that trade mark, in respect of the goods and services applied for.

The requirement for use or an intention to use on filing is not examined by the UK IPO. Failure to have the necessary use or intention to use is a possible ground for opposition or revocation on the ground of bad faith.

Arnold J. summarised the law relating to bad faith filing at paragraphs 130 to 138 of his judgment in the *Red Bull* case in the following terms:

> '130. A number of general principles concerning bad faith for the purposes of section 3(6) of the 1994 Act/Article 3(2)((d) of the Directive/Article 52(1)(d) of the Regulation are now fairly well established. (For a helpful discussion of many of these points, see N.M. Dawson, 'Bad faith in European trade mark law' [2011] IPQ 229.)
>
> 131. First, the relevant date for assessing whether an application to register a trade mark was made in bad faith is the application date: see Case C-529/07 Chocoladenfabriken Lindt & Sprüngli AG v Franz Hauswirth GmbH (2009) ECR 1-4893 at [35].
>
> 132. Secondly, although the relevant date is the application date, later evidence is relevant if it casts light backwards on the position as at the application date: see Hotel Cipriani Srl v Cipriani (Grosvenor Street) Ltd [2009] EHWC 3032 (Ch), [2009] RPC 9 at [167] and cf. Case C-259/02 La Mer Technology Inc v Laboratoires Goemar SA [2004] ECR I-1159 at [31] and Case C-192/03 Alcon Inc v OHIM [2004] ECR 1-8993 at [41].
>
> 133. Thirdly, a person is presumed to have acted in good faith unless the contrary is proved. An allegation of bad faith is a serious allegation which must be distinctly proved. The standard of proof is on the balance of probabilities but cogent evidence is required due to the seriousness of the allegation. It is not enough to prove facts which are also consistent with good faith: see BRUTT Trade Marks [2007] RPC 19 at [29], von Rossum v Heinrich Mack Nachf. GmbH & Co KG (Case R 336/207-2, OHIM Second Board of Appeal, 13 November 2007 at [22] and Funke Kunststoffe GmbH v Astral Property Pty Ltd (Case R 1621/2006-4, OHIM Fourth Board of Appeal, 21 December 2009 at [22].

134. Fourthly, bad faith includes not only dishonesty, but also 'some dealings which fall short of the standards of acceptable commercial behaviour observed by reasonable and experienced men in the particular area being examined': see Gromax Plasticulture Ltd v Don & Low Nonwovens Ltd [1999] RPC 367 at 379 and DAAWAT Trade Mark (Case C000659037/1, OHIM Cancellation Division, 28 June 2004 at [8].

135. Fifthly, section 3(6) of the 1994 Act, Article 3(2)(d) of the Directive and Article 52(1)(b) of the Regulation are intended toprevent abuse of the trade mark system: see Melly's Trade Mark Application [2008] RPC 20 at [51] and CHOOSI Trade Mark (Case R 633/2007-2, OHIM Second Board of Appeal, 29 February 2008 at [21]. As the case law makes clear, there are two main classes of abuse. The first concerns abuse vis-á-vis the relevant office, for example where the applicant knowingly supplies untrue or misleading information in support of his application; and the second concerns abuse vis-á-vis third parties: see Cipriani at [185].

136. Sixthly, in order to determine whether the applicant acted in bad faith, the tribunal must take an overall assessment, taking into account all the factors relevant to the particular case: see Lindt v Hauswirth at [37].

137. Seventhly, the tribunal must first ascertain what the defendant knew about the matters in question and then decide whether, in the light of that knowledge, the defendant's conduct is dishonest (or otherwise falls short of the standards of acceptable commercial behaviour) judged by ordinary standards of honest people. The applicant's own standards of honesty (or acceptable commercial behaviour) are irrelevant to the enquiry: see AJIT WEEKLY Trade Mark [2006] RPC 25 at [35]-[41], GERSON Trade Mark (Case R 916/2004-1, OHIM First Board of Appeal, 4 June 2009 at [53] and Campbell v Hughes [2011] RPC 21 at [36].

138. Eighthly, consideration must be given to the applicant's intention. As the CJEU stated in Lindt v Hauswirth:

> "41. [...] in order to determine whether there was bad faith, consideration must also be given to the applicant's intention at the time when he files the application for registration.
>
> 42. It must be observed in that regard that, as the Advocate General states in point 58 of her Opinion, the applicant's intention at the relevant time is a subjective factor which must be determined by reference to the objective circumstances of the particular case.
>
> 43. Accordingly, the intention to prevent a third party from marketing a product may, in certain circumstances, be an element of bad faith on the part of the applicant.
>
> 44. That is in particular the case when it becomes apparent, subsequently, that the applicant applied for registration of a sign as a Community trade mark without intending to use it, his sole objective being to prevent a third party from entering the market.
>
> 45. In such a case, the mark does not fulfil its essential function, namely, that of ensuring that the consumer or end-user can identify the origin of the product or service concerned by allowing him to distinguish that product or service from those of different origin, without any confusion (see, inter

alia, Joined Cases C-456/01 P and C-457/01 P Henkel v OHIM [2004] ECR 1-5089, paragraph 48)."'

## 2. *In the course of opposition proceedings*

Where in opposition proceedings before the UKIPO in relation to a domestic UK application a proprietor of an earlier mark relies on that mark in order to oppose, Section 100 of the 1994 Act states:

> 'If in any civil proceedings under this Act a question arises as to the use to which a registered trade mark has been put, it is for the proprietor to show what use has been made of it.
>
> If an opposition is based on an earlier registered trade mark which is more than five years old at the date of publication of the application under opposition, the Applicant of the opposed mark may ask the Opponent to prove that its earlier mark has been used or there are good reasons for its non-use. The Opponent will only be asked to prove use of registrations which are more than five years old if the Applicant asks him to do so. The Tribunal will not inquire of its own volition as to whether the earlier mark invoked in an opposition has been used or not.
>
> If an opposition is based on the ground of bad faith by virtue of the Applicant not having used or had a bona fide intention to use the mark on the date of filing the application, then the Applicant could overcome such a ground by either simply filing evidence of use, or evidence that it did have a bona fide intention to use the mark on the goods and services applied for.'

## 3. *In cancellation proceedings based on lack of use*

Cancellation proceedings are known in the UK as Revocation proceedings. Proceedings for revocation on the ground of non-use may be brought before either the courts or UK Intellectual Property Office, save where infringement proceedings are already before a Court, in which case, any defence by way of counterclaim has to be before that Court.

An application for revocation on the grounds of non-use may be brought by anyone. There is no need for a particular standing to sue.

The relevant sections of TMA 1994 for calculating the dates for revocation are set out above in section II.F.

The UK IPO Trade Marks Manual helpfully explains:

> 'By way of example, if the registration procedure was completed on 18 March 2000, the five year period would expire on 18 March 2005. An application for revocation for non-use could only be made after the latter date. The earliest date that an application for revocation could be made would therefore be 19 March 2005.
>
> [...]
>
> In the case of a revocation action founded upon section 46(1)(b) of the Act, where the applicant claims that the mark has not been used in the five years prior to the application, section 46(6)(a) of the Act will also apply to a successful revocation. In this instance, if the application is made on 18 March 2005 and

> this is the revocation date that is sought, the five year period during which there was no use will run from 18 March 2000 to 17 March 2005.
>
> [...]
>
> Revocation actions may also be brought on the grounds of section 46(1)(b), but for a five year period which does not coincide with the date of the application (that is, that section 46(6)(b) will apply). For example, if the applicant claims a revocation date of 18 March 2005, this means that the five year period for which there has been no genuine use of the mark will run from 18 March 2000 until 17 March 2005, with the rights of the proprietor ceasing to exist on 18 March 2005: the revocation date.'

It is possible for an application for revocation to be made for more than one date.

Section 46(3) provides that commencement or resumption of use after a five year period of non-use is a defence against an application for revocation on the ground of non-use for that period of five years, provided that the commencement or resumption was done without any knowledge of a possible action for revocation on the ground of non-use, or within three months of the proprietor being aware that such an application for revocation might be filed. This three month period is to allow a third party to threaten to file such a revocation action, with a view to coming to a settled agreement with the proprietor of the registration without having to immediately file a revocation action to protect its position.

#### 4. *As a defence in infringement/cancellation proceedings*

The old legal adage 'attack is often the best form of defence' means that the majority of trade mark infringement cases brought before the courts in the UK (primarily the English and Welsh Courts), have an associated counterclaim for invalidity. Depending on the age and uses the mark have been put to, that will often include a claim for revocation on the ground of non-use.

### B. Burden of proof

As mentioned in section I.C.2 above, Section 100 TMA 1994 puts the burden of proving use on the proprietor, whether that is of an earlier mark relied on in opposition proceedings or in a straightforward revocation action for non-use. The standard of proof to establish that there has been genuine use of the mark in relation to the goods for which it is registered in the relevant period is the ordinary civil standard of proof upon the balance of probabilities.

### C. Relevant evidence

As set out above in section II.A, one of the principles applied to genuine use is that use must be consistent with the essential function of a trade mark, which is to indicate source or origin of the goods or services.

Evidence of use in UK legal proceedings (whether UK IPO or court proceedings) typically takes the form of a witness statement from someone ('the declarant') from or associated with the entity or business attempting to prove use. The witness statement identifies its declarant and why they are able to make the declaration of use. This often includes a statement that the evidence comes from the records of the business making the use and that the declarant has full access to those records.

The witness statement should include full details of the scale and nature of the use which has been made. This typically includes statements confirming:

(1) the form in which the mark has been used;
(2) where the mark has been used;
(3) the volume of sales bearing the mark;
(4) the geographical scale of the use and sales;
(5) the nature and extent of advertising and marketing activities (including advertising and marketing expenditure).

The witness statement should have supporting annexes or exhibits which might include copies or pictures of:

(1) specimens of use of the mark;
(2) sales invoices;
(3) price lists;
(4) sample marketing materials.

### *1. Use on the product or its packaging*

As mentioned immediately above, use is required in relation to the goods and/or services in respect of which the trade mark is registered. Such use may be directly on the product or its packaging: the test is whether it is genuine use on the market in connection with the relevant goods and/or services.

Genuine use 'entails use of the mark on the market and not just internal use by the undertaking concerned', noting that the relevant market may be re-sellers, e.g. UK importers, and not necessarily end users.

Genuine use must relate to 'goods or services already marketed or about to be marketed and for which preparations by the undertaking to secure customers are under way, particularly in the form of advertising campaigns'.

### *2. Use in advertising*

Use in advertising will usually constitute genuine use, if it amounts to external use of the mark that enables consumers to make a connexion between the mark and the goods or services.

Use of a mark in merchandising will not necessarily constitute genuine use. However, it would appear that provided the use on the goods - whether promotional or otherwise - penetrates the market for those particular goods and functions as a badge of origin in relation to those particular goods, such use will be sufficient to constitute genuine use. If the distribution of promotional items is as merely a reward for the purchase of other goods or to encourage the sale of other goods without creating a market for those other goods under the mark, it will not be sufficient.

### 3. *Use on the internet*

Use on a website may constitute genuine use. However, the mere presence of the mark on a website accessible in the UK is insufficient to establish genuine use in the UK. In 1-800 Flowers [2000] ETMR 369 at 378 it was held that: 'the mere fact that websites could be accessed anywhere in the world did not mean, for trade mark purposes, that the law should regard them as being used everywhere in the world. However, use of the mark on a website which a UK customer would understand to be directed at him should be sufficient to constitute genuine use in the UK.'

### 4. *Form of evidence*

As mentioned above, evidence of use usually takes the form of a witness statement by someone able to attest to the use which has been made.

Appointed Person Arnold QC (as he then was) in *Pan World Brands v. Tripp Ltd (Extreme Trade Mark)* [2008] RPC 2 at 31, said that in relation to UK trade marks, whilst a bare assertion by a witness is insufficient evidence, a statement by a witness with knowledge of the facts setting out in narrative form when, where, in what manner and in relation to which goods or services the trade mark has been used would not in his view constitute a bare assertion and could be sufficient evidence of genuine use. This was followed in *Daimler AG v. Sany Group Co Ltd* [2009] EWHC 2581 (Ch) where it was held that a statement from in-house counsel, together with an extract from its parts database, marketing date and packaging, was sufficient evidence to establish genuine use. Daimler's in-house counsel had the requisite knowledge and experience to give the evidence that he did. It was not fatal that Daimler had not produced sample invoices in respect of sales and/or that in some instances there was only an uncorroborated statement from its counsel.

This is different to the practice of the EU IPO which gives little weight to witness statements from representatives of the proprietor which are unsupported by documentary records.

Survey evidence is not required and not generally appropriate to show genuine use.

The acceptable specimens for proving genuine use for goods and services are not necessarily different, except to the extent necessary to reflect the fact that marks cannot be used directly 'on' services. Invoices, price lists and advertising samples will normally be relied upon.

### 5. *Distinction between use of a company name and use of services offered by the company*

The use of a sign as a business or trade name may be regarded as trade mark use, provided that the relevant goods or services themselves are identified and offered on the market under that sign, so that it functions as an indication of origin and that it enables consumers to make a connection between the mark and the goods or services sold.

## IV. JUSTIFICATION FOR NON-USE

### A. Obstacles to use

Not long after the enactment of the TMA 1994, a decision of the Trade Mark Registry in *INVERMONT TM* [1997] RPC at 125 provided an observation as to what 'proper' meant:

> 'Apt, acceptable, reasonable, justifiable in all the circumstances.'

The following passage of the Registrar gives some guidance as to what constitutes a proper reason for non-use:

> 'I do not think the term "proper" was intended to cover normal situations or routine difficulties. I think it much more likely that it is intended to cover abnormal situations in the industry or the market or even perhaps some temporary but serious disruption affecting the registered proprietor's business. Normal delays occasioned by some unavoidable regulatory requirement, such as the approval of a medicine, might be acceptable but not, I think, the normal delays found in the marketing function. These are matters within the businessman's own control and I think he should plan accordingly.'

In *Nestle UK Ltd v. Zeta Espacial SA* [2000] RPC 439, Park J held that technical difficulties affecting production of an intended product could be a proper reason for non-use. He considered that the problems of product development experienced by Zeta were exceptional and it would be a travesty to describe them as 'normal or routine'.

The ECJ and now the CJEU have also considered proper reasons for non-use in the *Armin Haupl v. Lidl Stiftung & Co KG* decision (Case C-246/05). The ECJ considered that use of the mark did not have to be impossible for there to be a genuine reason for use. In this case, Lidl claimed the bureaucratic reasons relating to the opening of stores in Austria caused the delay in being able to use the mark. The ECJ suggests that whatever prevents the use, there must be a direct link with the reason, and it must be independent of the will of the trade mark properties. This proposition has recently been repeated in the joined cases T- 427/16 to T – 429/16 *Martin Osete v. EUIPO (Rey – AN IDEAL WIFE)*.

### B. Circumstances in which non-use may be cured

There are no specific circumstances relating to the curing of non-use, save that any use by the proprietor or with its consent in the 3 months prior to any revocation proceedings for non-use will be disregarded, unless the proprietor can show he was preparing to commence or resume use. Although as yet untested, it is likely a tribunal would require the proprietor to show 'serious and effective' preparations for such commencement or resumption, as opposed to simply stating an intention to do so.

The commencement or resumption must still constitute 'genuine use', and so seek to establish a market for the goods or services, or to preserve such. Accordingly, all of the points highlighted above concerning genuine use will apply to any such commencement or resumption.

## V. CONSEQUENCES OF NON-USE

### A. Consequences of non-use and relevance of the context

The consequences of a failure to prove genuine use of a trade mark differ, depending on the nature of the proceedings in which proof has been requested.

If proof of use has been requested in the context of opposition proceedings (see section I.C.2 above) failure to prove use or that there were proper reasons for that non-use means that the Tribunal will when considering the opposition act as though the earlier mark only covers the goods and/or services for which use or a proper reason for non-use has been proven. The Tribunal will act for the purposes of the opposition and the comparison of the later mark under opposition with the earlier registration for which use had to be proven, as though the earlier registration has been revoked or partially revoked but will not revoke or partially revoke the registration. A formal revocation action is necessary for the registration to be revoked.

If proof of use has been requested in the context of cancellation or revocation proceedings (see section I.C.3 above), failure to prove use, or that there were proper reasons for that non-use, will result in revocation or partial revocation of the registration subject to the cancellation or revocation action.

### B. Extent of any revocation

Section 46(5)(a) of the Trade Marks Act 1994 says that 'where grounds for revocation exist in respect of only some of the goods or services for which the trade mark is registered, revocation shall relate to those goods or services only.' Hence the hearing officer or court in the revocation action has the discretion to limit the specification of the registered mark to those goods for which the mark has actually been used. The specification will be limited to a fair description which would be used by the average consumer of the products on which the mark has actually been used. This approach was formulated by Pumfrey J in *Decon* [2001] RPC 17 and *Merc* [2001] RPC 42.

Aldous LJ in *Thomson Holidays v. The Norwegian Cruise Line* [2003] RPC 32, considered Pumfrey's judgments in *Decon* and *Merc* and said, at 31, that this

'task should be carried out so as to limit the specification so that it reflects the circumstances of the particular trade and the way that the public would perceive the use.'

He said that the court adopts the attitude of the 'average reasonably informed consumer of the products' when assessing infringement. He considered that the court should do likewise when deciding the fair way to describe the use that the proprietor has made of his mark. Thus the court will inform itself of the nature of the trade and then decide how the notional consumer would describe such use, and limit the specification accordingly.

In *Animal Trade Mark* [2004] FSR 19 at 20, Jacob J (as he then was) said:

> 'The reason for bringing the public perception in this way is because it is the public which uses and relies upon trade marks. I do not think there is anything technical about this: the consumer is not expected to think in a pernickety way because the average consumer does not do so [...] Thus, for instance, if there

> has only been use for three-holed razor blades imported from Venezuela [...] "threeholed razor blades imported from Venezuela" is an accurate description of the goods. But it is not one which an average consumer would pick for trade mark purposes. He would surely say "razor blades" or just "razors".'

This approach has recently been affirmed in *Galileo International Technology LLC v. European Union* [2011] EWHC 35 (Ch). In that case, Floyd J also affirmed Laddie J's comments in *Mercury Communications v. Mercury Interactive (UK) Ltd* [1995] FSR 850 that there is a strong argument that a registration of a mark simply for 'computer software' will be too wide. Floyd J held that, among other limitations, an overbroad specification for computer software should be limited to 'computer software for travel and accommodation reservations'. In so doing, he rejected Galileo's argument that use of the mark for document management and calendar software should also have been reflected in the specification, since the average consumer would not have perceived Galileo as selling software for such applications.

### C. Timing of any revocation

Section 46(6) provides that: 'where the registration of a trade mark is revoked to any extent, the rights of the proprietor shall be deemed to have ceased to that extent as from:

(a) the date of the application for revocation, or
(b) if the Registrar or Court is satisfied that the grounds for revocation existed at an earlier date, that date.'

The significance of this is, of course, if the application for revocation is made in a counterclaim to infringement proceedings, the applicant/infringer will not avoid liability unless there are grounds for showing that revocation should be from a date which predates the first infringing act.

# United States

*Dennis Prahl*

## I. INTRODUCING GENUINE USE OF TRADEMARKS

### A. Legal framework

In contrast to its treatment of patents and copyrights, the United States Constitution makes no express provision for the protection of trademarks. Instead, trademark rights in the United States arose first under English common law rules, which are based solely on use of trademarks, and secondly by statute. Each of the 50 United States and the Commonwealth of Puerto Rico have their own trademark laws that govern trademarks essentially within the boundary of that state or territory.[1] However, the primary trademark statutory system exists under US federal law.

The commerce clause of the United States Constitution gives the United States Congress the power 'to regulate Commerce with foreign Nations, and among the several States, and with the Indian Tribes'. ('US Commerce'). The first Trademark Act under the commerce clause was enacted in 1881 and replaced in 1905. The current federal Trademark Act, known as the Lanham Act, is the primary trademark law of the United States. In turn, the Lanham Act is interpreted by the decisions of the Trademark Trial and Appeal Board of the USPTO, as well as the federal courts, resulting in a large body of decisions that augment the interpretation of the Lanham Act's provisions.

Because the Lanham Act derives from Congressional power to regulate commerce, the concept of trademark use is essential to the creation and maintenance of federal trademark rights in the United States. As a general rule, absent use in US Commerce, a trademark cannot be granted rights under the Lanham Act. The Lanham Act does not apply to trademarks that are used solely within a single US state or solely outside of the United States and its territories.

1. This chapter will not address issues of trademark use under the separate laws of the 50 states or the Commonwealth of Puerto Rico.

As a result of the legal history and underpinnings of trademark law in the United States, there are essentially two regimes under which trademark rights arise in the United States: rights under common law based upon use of a trademark and rights under federal law based on use of a trademark in US Commerce. The US system differs from non-common law countries where rights are generally created only upon registration of a trademark.

The Lanham Act, codified at 15 U.S.C. § 1051 et seq., is the primary federal trademark law in the United States. In order for a trademark to be federally registered in the United States, the Lanham Act requires:

(1) the trademark must be used in commerce which the US Congress can regulate (interstate commerce, commerce with US territories or foreign commerce with the United States); and
(2) the use in commerce must be '*bona fide* use of a mark in the ordinary course of trade, and not made merely to reserve a right in a mark'.

Failure to satisfy either of these statutory requirements means that the trademark is not eligible for federal trademark protection.

The first requirement indicates that, generally, use solely outside the United States and its territories, or use solely within a US state, will not qualify for federal protection. The use in commerce requirement requires that the mark must be used as a trademark in connection with an entity offering for sale certain goods and/or services.

## B. Purpose and function of genuine use: policy considerations

Since use of a trademark is so essential to trademark protection in the United States, the concept of genuine or *bona fide* use is also essential.

The requirements for use of the mark in US Commerce serve dual purposes: first, the requirements qualify trademark use to be of relevance to, and therefore capable of regulation by, the federal government, and second the requirements ensure that trademark rights mirror commercial realities in the marketplace.

Originally, trademark rights were protected only by state common law and were geographic in nature. The exclusive right to use a mark was limited to the specific geographic area within which the mark was used in commerce. This allowed similar marks for similar goods to exist in different areas of the country without infringing upon the other. However, with the expansion of transportation and therefore commerce, the concern for consumer protection nationwide grew. The federal government began playing a bigger role in the regulation of rights, like trademarks, that impacted interstate commerce. Marks that were once only seen by consumers in small towns, single states, or at most certain regions of the country, began to travel coast to coast. After a number of previous federal laws, Congress enacted the Lanham Act in July 1946 pursuant to its constitutional power to regulate interstate and international commerce in the US. As a federal agency, the United States Patent and Trademark Office's power extends only to those trademarks that are operating within and impacting interstate or international commerce. Federal registration of a trademark grants exclusive rights in the trademark for the identified goods and services nationwide.

The feature that distinguishes US registrations from most of the world is the use requirement. By granting registered protection only for marks that are already in use, the USPTO's grant of exclusive rights is limited to those marks that have been deemed unique enough to distinguish the goods and services they represent from others in the marketplace with similar or related goods or services. This exclusive use allows trademark owners to stop others from obtaining registrations that would be confusing to theirs, and also allows owners to stop others in the marketplace that are using a mark that is confusing to theirs.

## II. WHAT QUALIFIES AS GENUINE USE OF A TRADEMARK

### A. Nature of use

#### *1. Use consistent with the function of the trademark*

Under US trademark law, use of a trademark must be consistent with the function of a trademark. The definition of a trademark encompasses any word, name, symbol, or device, or any combination thereof (1) used by a person, or (2) which a person has a *bona fide* intention to use in US Commerce and applies to register on the Principal Register established by the Lanham Act, to identify and distinguish his or her goods, including a unique product, from those manufactured or sold by others and to indicate the source of the goods, even if that source is unknown.

There are numerous types of uses that will not be considered consistent with the function of a trademark such as mere trade name use, matter that is purely functional, purely ornamental or decorative matter, slogans or words that are not likely to have source-indicating significance, merely informational matter, and marks that consist *solely* of one or more colours used on particular objects. Likewise, a mark applied to goods that are not goods in trade will not be registrable (e.g., a mark is not registrable for packaging where the packaging is merely incidental to the actual goods sold by the trademark owner and the trademark owner does not sell packaging by itself). Titles of a single creative work, names and pseudonyms of authors and artists of a single work, and names and designs of characters in a creative work are not considered to function as trademarks. Repeating pattern marks, social media hashtags, and model or grade designations are also types of uses that are not considered to function as trademarks. It is possible that some of the above types of uses could be considered trademarks if sufficient use has been made of the marks so that they have acquired distinctiveness or secondary meaning.

#### *2. Use in the course of trade*

There are two important factors that determine if use of a mark is genuine for purposes of federal trademark protection. First, the use must be in commerce which the US Congress can regulate (interstate commerce, commerce with US territories or foreign commerce with the United States). Second, the definition of use in commerce includes the phrase 'the *bona fide* use of a mark in the ordinary course of trade, and not made merely to reserve a right in a mark', thereby requiring legitimate use.

*i. Public use*

*Bona fide* use should be a public use, not a purely private use or use between related parties simply to show use. Thus an intracompany sale, sales merely to friends and family, etc. would be considered secret use and may not ordinarily qualify as *bona fide* use.

Although usually it is the trademark owner that must make use of the mark, there are occasions where use of a mark by the public, even though the trademark owner has not made use of the mark, can create trademark rights. The so-called public use doctrine provides that the public's use of a term to describe a company or its products may give the company trademark rights in that term, just as if the company had adopted the term itself. The Trademark Trial and Appeal Board has stated that '[i]t is well-settled that in order to establish rights in a mark, a party need not have actually used a mark if the public nevertheless associates the mark with the goods or services of that party'. This doctrine has been used to give Coca-Cola rights in COKE, American Express in AMEX, and Volkswagen in BUG even though none of these parties actually used those terms at the time.

*ii. Commercial purpose*

The definition of use in commerce includes the phrase 'the *bona fide* use of a mark in the ordinary course of trade, and not made merely to reserve a right in a mark'. The primary aim of this phrase is to eliminate the practice of 'token use', or use made solely to reserve rights in a mark. The meaning of the phrase can vary from one industry to another. Some industries may make infrequent sales due to the seasonal nature of their business or the high cost of their products. Other industries may make few sales in the ordinary course of trade yet ship products to investigators during a regulatory approval process. The definition is interpreted flexibly in order to encompass genuine, yet less traditional, trademark use.

As a result of the flexibility of the definition of *bona fide* use, various factors will determine whether the use claimed satisfies the definition, including the amount of use, the nature or quality of the transaction and what is typical use in the relevant industry.

So long as Congress can regulate the transaction, the use should be considered use in commerce. There is no actual requirement that the goods or services be sold, only that the goods or services under the mark are used in US Commerce. Therefore, promotional use of goods shipped or services offered across state lines may qualify, as could use on goods destined only for export from the United States. Mere promotion in the United States of goods or services not available in the United States (e.g., a restaurant in England) would not meet the test.

When use of a trademark in intrastate commerce, taken as a whole, has a direct effect on interstate commerce, it qualifies as sufficient use for registration. For example, transactions with out-of-state customers qualify as use in commerce that has a substantial effect on interstate commerce. As little as a single sale to an out-of-state customer has qualified as sufficient to impact interstate commerce. Intrastate services may qualify as having a direct impact on interstate commerce when they are advertised, using the trademark, in at least two states or if the services are utilized by consumers in interstate transportation (e.g., hotels or restaurants). Offering goods and services on the internet has also been seen as constituting use in US Commerce if the goods or services are available nationwide. Although the Trademark Office will not ordinarily enquire into the substance of the use on its own, it is in the best interest of the trademark owner to ensure its use qualifies as *bona fide* use in US Commerce, because insufficient use in US Commerce is a ground for a third-party challenge to the mark.

Certain seemingly commercial uses of a mark may not qualify as use in US Commerce for trademark registration purposes. Promotional use, advertising and contests to incentivize the sale of the goods or services covered by a mark do not qualify as a service under the mark. Trademark owners are expected to promote their own products or services through advertising so using a mark in advertising would not qualify it as a service mark for advertising services. However, any unique or additional services offered with the trademark, beyond traditional advertising or promotion, may qualify as commercial use of the mark for such services. For example, holding a sporting event to promote accounting services is beyond the expected form of promoting the services and may qualify those services as commercial use of the mark for sporting events.

Mere use of a mark in connection with preparation efforts for commercial activity also does not constitute adequate use in US Commerce. Preparation efforts for commercial activity include creating a business plan, having a mark designed or ordering sample products. Until the mark is used in connection with the goods or services in the ordinary course of trade, the efforts to provide the goods and services do not constitute use in US Commerce.

The ordinary course of trade does not always mean 'for sale'. The ordinary course of trade varies by industry and can encompass goods or services that are offered free of charge. For example, a mark used to indicate the source of *pro bono* legal services still qualifies as a service being offered in the ordinary course of trade. Many of the software applications that are downloaded on computers, tablets and smart phones are free of charge, yet the trademark used to indicate the source of the application may still qualify as use on downloadable software. As long as the goods or services meet the US Commerce requirement, the price tag or frequency of transactions does not disqualify the use as not operating within the ordinary course of trade.

Trademarks that identify the source of goods that are manufactured and destined exclusively for export to a foreign country or a territory of the United States still qualify for federal registration. This type of territorial or international commerce between the United States and a foreign country is among the types of commerce that may be regulated by the United States Congress, pursuant to its authority under the Constitution and the Commerce Clause. Unless the commerce involves the United States, Congress does not have the authority to regulate it and therefore cannot provide protection for a mark that is used only in a foreign country.

### 3. *Use by the owner or with its consent*

The Trademark Act provides that a trademark or service mark application based on use in US Commerce must be filed by the owner of the mark. If anyone other than the owner files the application, it becomes a void application. The applicant must provide a verified statement that the applicant believes it is the owner of the mark. However, the applicant may claim use of the mark in ways other than its own. Valid use may be the owner's own exclusive use of the mark; use of the mark solely by a related company where such use is to the benefit of the owner; or use of the mark by both the owner and the related company. The owner is the entity that controls the quality of the goods or services being offered in connection with the mark and also controls the nature of the use of the mark. A related company is defined as 'any person whose use of a mark is controlled by the owner of the mark with respect to the nature and quality of the goods or services on or in connection with which the mark is used'. Control may take many forms ranging from a licence, a relationship between a parent company and subsidiary, or a general understanding between the owner and the related company. When the owner files an application and bases its ownership upon use by a related company, the USPTO rarely enquires into this relationship as long as there is no contradicting information about who the owner is and who controls the nature and quality of the goods or services being offered with the mark. The issue can be raised, however, in *inter partes* proceedings concerning the use of the mark.

## B. Volume, duration and frequency of use

The Lanham Act defines use in US Commerce as 'the *bona fide* use of a mark in the ordinary course of trade, and not made merely to reserve a right in a mark'. Therefore mere token use made only to obtain or maintain a right in a mark does not meet the statutory definition.

There is no bright line test that would necessitate a certain volume of sales, value of sales, duration of sales or frequency of sales to qualify as *bona fide* use. On the other hand, decisions have held that limited use will not qualify, such as: sales of a single low-cost product over a period of years; monthly shipments to wholesalers with no indication that the goods ever reached end consumers; minimal sales over a lengthy period for minimal profit absent extraordinary circumstances; sales only to friends and family, etc.

On the other hand a single sale or infrequent sales under some circumstances could qualify as use in US Commerce. It is important that continuing use, or continuing efforts at use, are shown in the case of a registered mark. As discussed in greater detail below, failure to use a mark in US Commerce for a period of three years constitutes *prima facie* abandonment of the mark.

## C. Use of a sign

It is important that any use of a mark that is sought to be registered, or that is registered, be used in the form of the mark that is (sought to be) registered. The USPTO will reject attempts to satisfy the use requirement with specimens of use where the mark differs from the applied-for or registered mark. If the mark as used is not the same as the applied-for mark or the registered mark, the use may not support the mark. However, it is possible for the applicant, or registrant, of the mark to amend the form of the mark on the Register provided that the amended form of the mark is not a 'material alteration' of the original mark. In determining whether an amendment is a material alteration, the controlling question is always whether the new and old forms of the mark create essentially the same commercial impression. Whether the alteration is 'material' is a highly fact-specific and subjective decision. If the USPTO finds that the new form of the mark is a material alteration of the former mark, the application or registration may be refused.

Unlike the current position in the European Union, use of a trademark in colours will support the application or registration for the mark in black and white since US trademark law considers that the black and white version of a registered trademark protects the mark in all colours.

## D. Use in relation to goods and services

### *1. Use for each of the designated goods and services*

Since use of a trademark is integrally connected with the ability to register the trademark and maintain it, it is essential that use of the trademark is required on each of the goods and services for which registration is sought or maintained. If the trademark owner is seeking registration, or to maintain or defend a registration, for goods A, B and C, it is necessary that the trademark owner use the trademark in US Commerce in connection with each of goods A, B and C. Where use on less than all of these goods is made, the trademark should be registrable for only the goods in use.

Although usually only one specimen of use per class is required to obtain and maintain a trademark registration, if an application or registration includes multiple items in a class, the mark must actually be in use on or in connection with every identified good and service. If the mark is not used on some of the goods or services, those goods and services should be deleted from the application or registration. The USPTO may request a substitute or additional specimen to show the mark in use on another identified good or service in the same class.

### *2. Use for a broader category of goods and services*

The USPTO requires that identifications of goods and services in trademark registrations be specific, definite, clear, accurate, and concise. As a result, with limited justifiable exceptions in the case of house marks or in the case of a trademark used for a full line of products (e.g., a full line of clothing), it is not possible to register broad categories of goods and services. The use of the trademark must be in connection with the identified goods and services. Use on other goods and services, even in the same general category of identified goods and services, is not acceptable to support the identified goods and services.

As US law has traditionally not allowed amendment of a trademark registration to expand the scope of the goods or services it originally covered, registrants must file new applications for updated goods or services. However, the USPTO recently implemented a pilot program that, under limited circumstances, allows trademark owners to amend the goods and/or services identified in their trademark registrations when the technology has evolved such that the originally registered goods were rendered obsolete. For example, a registration covering motion picture films could be amended to include streaming of motion pictures.

## E. Territorial scope

Trademark use prior to federal regulation granted exclusive rights only in the territory where the mark was used to offer goods or services. If goods or services were only available in the northeastern United States, that is where the rights were protected. Other parties on the west coast could offer the same goods or services under the same mark without infringing the rights in the northeast. Although that rule still holds true for marks not protected by a federal trademark registration, today, a federal registration does not limit rights to a mark in one geographic region. Instead, a federal registration grants exclusive rights to a mark for the identified goods and services across the entire United States and its territories. The acquisition and maintenance of a federal registration depends on use in interstate commerce, that is commerce between as few as two states, or use in foreign commerce with the United States. It is not necessary that a trademark be used in many or all states to qualify for federal trademark protection. Use of a trademark solely outside the United States is not capable of supporting a US federal trademark registration. Likewise advertising in the United States either in print, radio, television or online of a good or service that is not available in the United States would not support a US federal trademark registration.

## F. Relevant time period and its calculation

The relevant time period for trademark use will vary depending upon whether the use is being submitted to obtain registration, to establish acquired distinctiveness, or secondary meaning, of a descriptive/non-distinctive mark to obtain registration, to maintain/renew a registration, or to defend against an allegation of non-use.

For trademark applications based on an intent-to-use, it is necessary for the applicant to commence use in interstate commerce, or foreign commerce with the United States, within six months from the date the USPTO issues a Notice of Allowance. As long as there is a continued *bona fide* intention to use the mark for the specified goods and services, up to five six-month extensions of time are allowed from this deadline, namely up to three years from the date of the Notice of Allowance, to commence use and to file a sworn statement of use with an appropriate specimen or specimens. If the requirements are not satisfied by the final deadline, the USPTO will issue a Notice of Abandonment giving the applicant two months to petition to revive the application on certain grounds and comply with the requirements. All applications based on use or intent to use must include the date of first use anywhere and the date of first use in US Commerce. This does not apply to applications based solely on foreign registrations or applications that are designations of International Registrations under the Madrid Protocol, however, such resulting registrations are always subject to attack on the ground of non-use in the same way as other registrations.

If a mark is refused for being descriptive, misdescriptive, geographically descriptive, or primarily a surname, an applicant may be able to overcome the refusal by submitting evidence that the applied-for mark has acquired distinctiveness based upon five years of exclusive and continuous use of the mark in connection with the applicant's goods or services. This extensive use and promotion of the mark means consumers likely associate the mark with the applicant as the source of the goods, and therefore the mark has a secondary meaning for consumers beyond its descriptive characteristics. Evidence to support a claim of acquired distinctiveness includes: money spent on advertising; promotional and marketing materials for the goods that show the mark being used as a trademark and source-identifier; and consumer statements regarding the mark. Long term use to support an acquired distinctiveness claim must be in US Commerce; use only in a foreign country does not qualify for the five-year period. A mark owner may also argue a mark has acquired distinctiveness based upon prior active registrations for the mark or portions of the mark where the new application adds goods or services, or because it is a new member of a family of registered marks that shares a common characteristic that is recognized by the public.

Registered marks must be maintained between the fifth and sixth years of registration and must be renewed between the ninth and tenth anniversary of the registration date and every ten years thereafter to keep the registration active. The renewal must be signed by the registrant, include a fee for each class to be renewed, include a list of goods or services to be renewed if the renewal and current use does not cover the entire registration, and provide acceptable evidence of such use for at least one item in each class of goods and services. There is a six-month grace period following expiration of the above deadlines within which to comply with the requirements for an additional fee.

A registration owner has the option to file an affidavit after five years of a mark being registered to declare its exclusive right to use a mark incontestable on certain grounds. The affidavit must state the mark has been used in US Commerce on or in connection with the identified goods and services continuously for five years. The period of continuous use may begin no earlier than the date of issuance of the registration, unless the holder owns a prior US registration of the same mark for the same goods, or services, or collective membership organization. The USPTO does not determine incontestability. If the incontestability affidavit meets the requirements, the status of a registration will be updated to notify the public of the acknowledgment of incontestability, but a court is the only proper venue to determine whether the registration is incontestable if there is a proceeding involving the mark.

In a proceeding or counter-claim for a determination of non-use, a period of non-use of a trademark for three years is considered *prima facie* evidence of abandonment of the trademark. However, this presumption may be rebutted. If there is evidence that the trademark owner intended to cease use of and surrender the trademark earlier than the three-year period, then a shorter period of non-use will apply. On the other hand, if the trademark owner intended to resume use of the trademark after the three-year period, and had documented plans and efforts to do so, the presumption may not apply. Likewise, excusable non-use, discussed below, may also rebut the presumption.

If the owner of a registered trademark has not used the trademark for some or all of the registered goods, they may cure the non-use situation by resuming use of the trademark in US Commerce, provided that, in the interim, no third party has filed an action for cancellation of the trademark on the ground of non-use.

## III. EVIDENCE OF GENUINE USE OF A TRADEMARK

### A. When use should be proved: timing of lack of genuine use claims

#### *1. When filing an application for a trademark*

The issue of use arises at the time of filing a trademark application with the USPTO. A US trademark application for registration on the Principal Register may be based either on (a) use in US Commerce, (b) a *bona fide* intent to use in US Commerce, (c) a foreign application/registration or (d) an International Registration, or (e) any combination of (a), (b) and (c).

An application based on use requires: (1) a declaration swearing to the use of the mark in US Commerce signed by the owner of the application or a person properly authorized to sign on behalf of the owner; (2) the date of first use anywhere and the date of first use in US Commerce; and (3) a specimen that shows how the mark is being used in connection with the identified goods and services in the application.

The use of the mark in US Commerce must be in good faith and must be more than 'token use' to merely reserve rights in the mark. The mark should be used in the ordinary course of trade, beyond sporadic or nominal transactions. The USPTO requires a specimen that shows how the mark is being used in US Commerce. The specimen must reflect how consumers see the mark in the marketplace on or in connection with the goods or services. A specimen is required for each class of goods or services for which the mark is applied. Appropriate specimens for goods can include the mark on the outside of the goods or packaging, normally on a label or tag, or at a point of sale display associated with the goods as long as those goods are sold or transported in US Commerce. Appropriate specimens for services may include the mark as shown in advertising or promotional materials for the services, or may demonstrate how the mark is used in providing the services, such as on a website. Service mark specimens must reference the services and create a connection between the trademark and those services.

When a trademark is not yet in use in US Commerce, an owner may file an Intent to Use (ITU) application. ITU applications must be filed with a good faith intent to use the mark in US Commerce on the identified goods and services within a reasonable time frame. To prove *bona fide* intent, ITU applications include a sworn statement that the applicant has a *bona fide* intention to use the mark. However, until the owner files a Statement of Use swearing to use of the mark in US Commerce and provides a use specimen for each applied-for class of goods or services, the mark will remain unregistered. Although only one specimen of use per class is required, if an application includes multiple items in a class, the mark must actually be in use on or in connection with every identified good and service. If the mark is not used on some of the goods or services, those goods and services should be deleted from the application. In some cases the Examining Attorney may request a substitute or additional specimen to show the mark in use on other identified goods or services in the same class.

While a trademark that is the subject of an ITU application is not registered and therefore does not enjoy the federal protections of a registered trademark, the application does stop junior filers from obtaining a registration of that mark for the same of similar trademark for the same or similar goods and services, unless their actual use predates the ITU application date. Within six months after the USPTO issues a Notice of Allowance, indicating the mark is 'allowed' but not yet registered, the owner must file a Statement of Use. If the mark will not yet be in use within the allowed six-month time frame, the applicant may file a Request for Extension of time, plus the required fee, to file a Statement of Use for an additional six months. Five total Requests for Extension of time to file a statement of use may be granted for an application, up to three years from the allowance date. Each new request must include a verified statement of the applicant's continued *bona fide* intent to use the applied-for mark for the identified goods and services. Once registered, the application date of the registration is considered the constructive first use date in the United States for purposes of priority over third party marks.

An applicant may also apply for a US registration by relying on a registration in another country, without showing use in the US. However, to obtain the US registration, the foreign registration must cover the applied-for goods and services in the US application. The applicant must also meet the following criteria:

(1) Submit a true copy, photocopy, certification, or a certified copy of the registration in the applicant's country of origin, *and* if the foreign registration or certification is not in English, provide a translation of the document;
(2) Include the applicant's verified statement that the applicant has a *bona fide* intention to use the mark in US Commerce, which is required even if use in US Commerce is asserted in the application; and
(3) The applicant's foreign country must either be a party to a convention or treaty relating to trademarks to which the US is also a party or the foreign country must extend reciprocal registration rights to US applications by law.

If the foreign registration does not adequately cover the applied-for goods or services identified in the US application, the applicant must submit a use specimen for the goods or services not covered by the foreign registration. Only one specimen per class is required.

A single trademark application may also be based on a combination of either ITU, use in US Commerce or foreign application/registration.

Evidence of use of a registered trademark in US Commerce is also required between the fifth and sixth years following registration of the mark, between the ninth and tenth years following registration of the mark and every ten years thereafter in order to maintain the registration. The requirements for a verified statement and evidence of use are the same as for obtaining a trademark registration.

Under a recent audit program started in November 2017, the USPTO is randomly selecting maintenance and renewal applications to request that trademark owners provide additional specimens of use for two additional goods or services covered by the registration being maintained or renewed. A timely filed response in which use is not acceptably demonstrated in relation to the selected goods may result in a subsequent Office Action requiring evidence of use in relation to each and every individual good or service identified in the registration. Likewise, deletion of one or more of the additional goods will result in a similar Office Action. Failure to reply to the Office Action will result in the cancellation of the entire registration, regardless of whether valid evidence was initially submitted. This is an effort to clear what is considered to be an increasing amount of dead-wood registrations or overly broad registrations for which there is no use.

Sufficient use in US Commerce and the first-use date provided for an application should be determined with careful consideration. The use evidence and claimed date(s) of first use can be challenged on several different fronts and if successfully challenged, the application or resulting registration can be refused, cancelled, or limited in its applicability.

### 2. *In the course of opposition proceedings*

Once an in-use application is approved by the USPTO, it is published for opposition. A third-party who believes they will be damaged by the registration of the proposed mark may file a Notice of Opposition with the Trademark Trial and Appeal Board (TTAB) within thirty days of publication; the term is extendible on certain grounds. If the application is based on intent-to-use, a foreign application/registration or a designation of an International Registration, there is no basis on which the opponent can challenge the applicant's use although the opponent may challenge the applicant's claim to a *bona fide* intent to use in certain circumstances. If the application is based on use, an opponent may challenge any use claimed by the applicant, who will then usually need to offer additional evidence concerning the claimed use during the opposition. If the TTAB finds the allegations to be true, the application can be cancelled, or limited to the goods or services on which the mark is being used or for which there was a *bona fide* intent to use.

In most cases the opposition will be based on earlier rights of the opponent, whose use of such rights, whether registered rights or rights at common law based on prior use, will need to be proven and can also be challenged by the applicant in a timely counterclaim.

### 3. *In cancellation proceedings based on lack of use*

A third party may challenge a registered trademark by commencing a cancellation proceeding before the TTAB on the ground of abandonment. Under some circumstances, a party can include a claim for cancellation in federal district court proceedings if it is combined with other causes of action. The principles are the same under cancellation proceedings as under opposition proceedings, discussed above. The petitioner must have standing to petition to cancel, which is satisfied by a belief that it is damaged by the existence of the registration that is to be cancelled. Standing usually exists where there is a likelihood of confusion with petitioner's mark or when the petitioner's application was rejected on the basis of the challenged mark's prior right. The cancellation action can be brought either by attacking the claims of use of the registrant or alleging abandonment of the trademark based on non-use, usually for a period of three years or longer. In order to survive a petition to cancel for non-use, the registrant must prove use of its claims of use or prove continuing use or reasons for non-use, discussed below. If the petitioner is successful, the TTAB will issue an order granting the cancellation of the registered mark, either in whole or in part, as appropriate. The cancellation of a trademark registration does not affect any common law rights the registrant may have based on use of its mark.

#### *4. As a defence in infringement/cancellation proceedings*

Infringement actions, cancellation actions and oppositions may all be defended with a counterclaim of non-use of the initiating party's mark.

In defending an opposition or cancellation action before the TTAB it is necessary to raise this defence either by bringing a timely counterclaim or by filing a separate petition to cancel the opponent's/petitioner's registration. The better practice is to assert the defence by way of counterclaim. If a separate petition to cancel is filed, the TTAB will usually consolidate the proceedings. It is important to note that if the counterclaim is not timely made, it is not possible to make the counterclaim at a later date or to petition to cancel at a later date.

In an infringement action in federal court, it is also necessary to raise the defence of non-use in a timely counterclaim. Alternatively, the defendant may bring a separate petition to cancel before the TTAB. Typically, when parties are engaged in civil litigation that may affect the outcome of proceedings before the TTAB, a request to suspend the TTAB proceedings until a final determination is made in the civil action is entered. The TTAB may suspend the proceeding at its discretion but nearly always does so.

### B. Burden of proof

In the context of obtaining a trademark registration, maintaining or renewing a trademark registration, the burden of proof to evidence genuine use of a trademark is on the trademark owner. In *inter partes* proceedings regarding use of a trademark, the party asserting the non-use claim initially has the burden of proof to demonstrate that there was no use. This is done primarily during the course of discovery in the proceeding when obtaining documentation and testimony from the trademark owner. Once the *prima facie* case of non-use is established, the burden shifts to the trademark owner to prove that a registered trademark is in use for the identified goods and services.

### C. Relevant evidence

#### *1. Use on the product or its packaging*

Appropriate specimens to prove use of a trademark on goods can include labels or tags, product packaging showing the mark, a photograph of the product showing the mark directly on the goods, catalogues, or a webpage. Acceptable specimens to prove use of a trademark for services include marketing materials, print and digital advertising, menus for restaurants or catering, or a photograph with the trademark displayed while the services are being rendered (e.g., a band performing).

There are a number of materials that will not be sufficient to evidence use. Drawings or pictures of marks will not suffice, including artist's renderings, printer's proofs, computer graphics that only show what the mark looks like, or digitally created or altered images of the goods or packaging. Likewise, package inserts that are merely advertising material, or invoices, order forms, bills of lading and press releases will not be able to serve as a specimen of use.

## 2. *Use in advertising*

Generally mere use of a trademark in advertising material would not be considered an appropriate specimen of use of the mark on goods. Therefore, advertising circulars and brochures, price lists, press releases, business cards, online and social media advertising are not sufficient to evidence use of a trademark on goods. Such evidence may be sufficient if it is a point of sale material use in close proximity to the goods for sale or if it includes a photograph of the mark appearing on the goods or on their packaging.

A trademark owner may also show use of the mark on a display associated with the goods. Such displays essentially comprise point-of-sale material such as banners, shelf-talkers, window displays, menus, and similar media. They must display the trademark prominently and associate it with the goods and must be designed to catch a consumer's attention. In contrast, folders, brochures and materials that serve merely as advertising will not be sufficient to evidence use.

The rules are more relaxed for service marks, so that advertising use can evidence service mark use so long as the specimen shows proper use of the mark in US Commerce, namely showing the mark as used or displayed as a service mark during the sale of the services or showing the mark used or displayed in advertising or promoting the services. It is essential that the evidence show the mark in association with the services for the specimen to be suitable; a specimen merely showing the mark without reference to the services will not be acceptable.

## 3. *Use on the internet*

A webpage may be used as a specimen of use for goods, provided that the webpage must: (1) contain a picture or textual description of the identified goods; (2) show the mark in association with the goods; and (3) provide a means of ordering the identified goods. The mark must be displayed in a way in which consumers will recognize it as a mark to identify the source of goods. The more prominently a mark appears on a webpage, the more likely consumers will perceive the mark as a trademark. Displaying the trademark as part of the URL, internet address, email address, or domain of the website is generally not viewed by consumers as trademark use. Just as a mark would be used in a retail store to identify the source of goods, the mark on a webpage must do the same. A webpage of a third party that satisfies the requirements can also qualify as an acceptable display of the mark. A webpage that does not show the goods or services are available in US Commerce will not be sufficient.

The rules for webpages also apply to printed catalogue pages, which can serve as evidence of use on goods provided the above three criteria are met.

Webpages are often used as evidence of use of a trademark on services. The webpage must show the mark in a manner that indicates a direct association with the relevant services so that consumers will perceive it as identifying the services and indicating their source. Acceptable evidence of use will be determined by the facts and the relevant commercial context in which the mark is used.

### 4. *Form of evidence*

For purposes of obtaining a trademark registration or maintaining or renewing a trademark registration, it is normally sufficient to submit a valid specimen of use together with an affidavit signed by an authorized signatory of the trademark holder attesting that the trademark is in use in US Commerce.

Additional evidence is required for purposes of defending a claim of abandonment, for establishing that a descriptive or non-distinctive trademark has acquired distinctiveness, known as secondary meaning, and for establishing rights based on common law use of a trademark.

In proceedings before the Trademark Office, a trademark owner may submit affidavits, verified statements, deposition testimony of relevant parties, letters or statements from the trade or public, or other appropriate evidence showing duration, extent, and nature of the use in US Commerce and advertising expenditures in connection therewith. Advertising evidence should identify types of media and include examples of typical advertisements. Sales figures should indicate dates of sales, volume and value of sales.

In *inter partes* proceedings, a party may submit any evidence that would be admissible under the Federal Rules of Civil Procedure, including documents, affidavits, declarations, expert reports or expert testimony, survey evidence, government records or fact witness testimony. Any such evidence will need to demonstrate the duration, extent and nature of the use.

### 5. *Distinction between use of a company name and use of the services offered by the company*

A term used only as a company name, or trade name, is not capable of functioning or of registration as a service mark. Trade names are defined separately in the Lanham Act as any name used by a person to identify his business or vocation and there is no provision for their registration. A trade name may be used as a trademark or service mark and depends on the manner of use and the probably impact of such use on consumers, as indicated in the evidence of use. For example, if the trade name is used separately or in a distinctive font or logo in materials in the manner of a trademark or service mark, it may function as such. On the other hand, if the trade name is used in proximity to a company address, this may be an indication of mere trade name use.

## IV. JUSTIFICATION FOR NON-USE

### A. Obstacles to use

In general, a registered trademark must be continuously used on all the identified goods and services in a registration in the ordinary course of trade.

If a registered trademark is not being used due to circumstances beyond the control of the owner, the owner may be able to maintain its rights in the registration without evidence of use under a claim of 'excusable non-use'. At the appropriate time to maintain/renew the registered mark the owner must file an affidavit that includes a statement of the facts and the special circumstances that it considers excuses its non-use of the mark in relation to the goods and services. The affidavit must include the following: the factual basis for the excuse; the date when use of the mark in US Commerce stopped; the approximate date when use in US Commerce is expected to resume; specific steps being taken to put the mark back in use for all identified goods and services; and the specific goods and services to which the claimed excusable non-use applies, including a request to delete goods or services for which there is no use and to which the excuse does not apply. The facts provided in the declaration must clearly demonstrate that non-use is beyond the owner's control or forced by outside conditions. The above also applies when the trademark owner is defending a claim of non-use.

There are few excuses that justify non-use. Examples of excusable non-use include: trade embargos, the sale of the business, acts of God, illnesses impacting the operation of the business, or deficiency in supply for necessary materials, equipment, or labour. Examples of reasons that are not acceptable to excuse non-use include: business decisions, decreased demand, negotiations with suppliers or distributors, use outside the United States, use on goods or services not listed on the registration, or use of a materially different mark.

### B. Circumstances in which non-use may be cured

If the owner of a registered trademark has not used the trademark for some or all of the registered goods, they may cure the non-use situation by resuming use of the trademark in US Commerce, provided that, in the interim, no third party has filed an action for cancellation of the trademark on the ground of non-use.

## V. CONSEQUENCES OF NON-USE

### A. Consequences of non-use and relevance of the context

There are three key moments in the life of a trademark registration where non-use has consequences: at the application stage, at the renewal/maintenance stage and when asserting/defending a registered trademark.

An application based on a claim of use in US Commerce must be in use on the identified goods and services at the time the application is filed. If the trademark is not in use on some or all of the goods and services, the application or resulting registration may be revoked in part for the non-used goods and services or in whole if fraud is shown. The same is true for an application based on a *bona fide* intent to use where the Statement of Use filed by the applicant claims use on goods and services for which there is no use.

In the same way, when a trademark registrant seeks to maintain a registration during the fifth and sixth anniversary following registration, or seeks to renew a registration every ten years following registration or renewal, if the trademark is not in use on some or all of the goods and services claimed in the registration, the registration may be revoked in part for the non-used goods and services or in whole if fraud is shown.

In the above cases of applications and registrations, as long as the minimum requirements are met, namely a verified Statement of Use and a suitable specimen per class, the Trademark Office will generally not make further enquiries. The issue of non-use in these cases arises therefore in the context of oppositions and petitions to cancel (or counterclaims for non-use cancellation) brought by third parties against the application/registration. In oppositions, non-use cancellation actions and counterclaims of non-use brought by third parties, an application or registration may be revoked in part for any goods or services which were not in use when claimed or have not been in use where abandonment is evident. In cases where fraud is established, which is a high burden, the entire application or registration may be revoked.

## B. Extent of any revocation

Revocation of a trademark registration is generally granted only in respect of the goods and services for which the trademark has not been used. In rare cases where fraud by the trademark owner is established, the entire registration may be revoked.

## C. Timing of any revocation

In the context of claiming use for the purposes of obtaining or maintaining a trademark registration, claims of use must be accurate as of the date they are made. If use is claimed prior to the filing date of an application, appropriate evidence of use from the date of claimed first use should be available if challenged. In the context of a petition to cancel a registration for abandonment or a counterclaim for non-use, it is important to establish through evidence that the registered trademark has been used within the last three years before the petition or claim was made or that there was an intention to resume use or that the non-use was excusable.

A petition to cancel a registration or a counterclaim for non-use may be brought at any time during the life of a trademark registration provided that there is sufficient evidence to indicate that the trademark owner either did not have the required use or *bona fide* intent to use as claimed, or that the trademark has been abandoned by the trademark owner, as discussed earlier. This applies equally to registrations based on use and to registrations based on foreign trademark registrations or International Registration designations to the United States.

ExCo Hyderabad 2011
Adopted Resolution
October 18, 2011

**RESOLUTION**

**QUESTION Q218**

AIPPI

## THE REQUIREMENT OF GENUINE USE OF TRADEMARKS FOR MAINTAINING PROTECTION

AIPPI

*Noting that:*

(1) AIPPI has previously adopted Resolutions regarding questions of use of trademarks, notably Resolutions on the following Questions: Q70, Q92A, Q95, Q164, Q168 and Q210.

*Considering that:*

(1) The purpose of the use requirement is closely linked with the function of trademarks as signs used in the course of trade to distinguish goods or services of the proprietor of the mark from those of other undertakings: in order to exercise its function and merit protection, a mark must actually be present on the market. Additional purposes are also relevant, notably the need to keep the trademark registries free from unused trademarks so as to facilitate the acquisition of new trademarks.

(2) In countries where trademark rights are acquired through registration, the issue of use to maintain protection arises after registration, whereas in countries where trademark rights are acquired through use rather than through registration, the requirement of use is inextricably linked with the acquisition of rights.

(3) In order to maintain rights the mark must be used 'as a mark'. It should not matter that the same sign may also be protected as a trade name or by some other right, in particular because many trade names are also registered and used as marks, and many marks are used also as trade names.

(4) In view of Article 5C of the Paris Convention a mark shall be deemed to be used even if the actual form of use departs from the form of the mark as registered, provided that the distinctive character of the mark is not altered.

(5) Account must be taken of situations where the proprietor of the mark holds other similar registered marks, such as variations adopted over time.

(6) According to Article 19 (2) of the TRIPS Agreement, use by third parties subject to the control of the trademark proprietor shall be sufficient to maintain trademark rights. More generally, use by third parties is sufficient to maintain trademark rights if the use is authorized by the proprietor; typical cases of such authorization are the use by a licensee or by related companies.

(7) Due consideration must be given to jurisdiction where the use by third parties in order to be recognized requires the exercise of control of the quality of the goods or services by the proprietor.

(8) Where a mark is used only for some of the goods or services for which it is registered, account must be taken of the interests of the proprietor in not unreasonably restricting a further extension of the use, as well as of third parties, which should not be unduly restrained in acquiring new marks.

(9) The territory includes all areas within the jurisdiction of the country or region, without prejudice to any applicable international agreements.

(10) Use in the course of trade includes situations where goods or services are provided free of charge or not-for-profit, such as by charitable organizations.

(11) The nature of use differs between trademarks and service marks. Whereas for the former use must be in relation to the goods, use of service marks is of a different kind because of the intangible character of services.

(12) The burden of proof of use should ultimately be on the proprietor of the trademark, without prejudice to the procedural requirements applicable to the raising of absence of genuine use as a defense or as a claim when seeking to invalidate a registered trademark.

(13) Where registration of trademarks is available without prior use, a 'grace period' applies after the expiry of which the sanctions in case of absence of genuine use apply. The same grace period applies when the use has been interrupted. The minimum duration of such grace period must be three years, in accordance with Article 19 (1) of the TRIPS Agreement, and should not be longer than five years, which provides a proper balance between the interests of proprietors of registered trademarks and those seeking to obtain new trademarks.

(14) Where the grace period has expired and use has begun or was resumed thereafter, such use has the effect that the mark can no longer be challenged on grounds of absence of use. This is without prejudice to the application of rules on abandonment.

(15) This 'curing' does not apply when the mark has been properly challenged on grounds of absence of genuine use. The applicable law should determine the notion of 'proper challenge', such as the need to initiate formal proceedings to obtain cancellation of the mark.

(16) Due regard should be given to the rights acquired by third parties, for example through registration of a trademark, in a period when an earlier trademark was subject to cancellation on grounds of absence of genuine use.

(17) Justification of absence of genuine use must be available in accordance with Article 5C (1) of the Paris Convention and Article 19 (1) of the TRIPS Agreement. A typical example of 'valid reasons' are ongoing administrative authorization or certification procedures, extending beyond the grace period, for example those necessary before bringing medicinal products or chemicals or any other product needing government authorization on the market.

(18) Persons asserting the absence of genuine use must be able to bring an action for the removal of the mark from the register or challenge the absence of use in the respective administrative and judicial proceedings. The defendant in an infringement action should be entitled to raise the issue of non-use, either as defense or by way of a counterclaim or by way of a separate claim before the competent authority.

*Resolves:*

***(1) Genuine use: Principle***

The rights to a trademark should only be maintained if the mark is put to genuine use, subject to a reasonable grace period.

***(2) Use as a mark***

In order for the use to be recognized, the use must be made as a mark distinguishing goods or services as to their origin. Whether the actual use is use as a trademark should, inter alia, depend on the perception of the public to which the mark is addressed. To assess whether a sign is used as a mark, it should not matter whether or not this sign is also used as another distinctive sign, such as a trade name.

***(3) Use of variations***

The use of a mark which differs from the registered mark must be recognized as sufficient use of the registered mark if the differences between the used and the registered mark do not alter the distinctive character of the mark. The evaluation of whether or not the differences alter the distinctive character will depend on a case by case analysis. Where the proprietor has registrations for a number of similar marks, and is using a variation mark, the maintenance of rights in each of the registered marks must be evaluated by comparing each to the variation mark that is in use, regardless of whether the variation itself is also registered.

***(4) Use by the proprietor or with his consent***

The mark must be used by the proprietor or by third parties with his consent, such as by licensees or related companies.

Where, under applicable law, use by a licensee or any other third party may be authorized by the proprietor without the requirement of control over the nature and quality of the goods or services of the user, the use made by the third party should be attributed to the proprietor. Where, under applicable law, use by third parties is authorized only if the proprietor exercises control over the nature and quality of the goods or services of the user, use by the third party should be attributed to the proprietor only when such control is actually carried out.

Use by a licensee should be attributed to the proprietor regardless of whether or not the license is registered.

### *(5) Use for the goods or services for which the mark is registered*

The mark must be used for the goods or services for which it is registered.

Where the mark is used only for some of the registered goods or services, the mark should not be maintained for the goods or services for which it has not been used. However, where there has been use for a specific product or service in a broader category of goods or services, a properly restricted broader category should be recognized.

### *(6) Territory*

The mark must be used in the territory for which it is protected. The territory includes customs-free zones. No minimum territorial extent should be required.

Export of goods or rendering of services from the territory should be recognized as use in the territory.

### *(7) Use in the course of trade*

The use must be made in the course of trade, which excludes private use and purely internal use.

Making available goods or services for free or not for profit should not exclude the use from being recognized as use in the course of trade.

### *(8) Genuine use*

Only use which is genuine is sufficient to maintain trademark rights. The genuineness of use must be judged by taking into account all the relevant circumstances of the case, which include inter alia the nature of the goods or services, the size of the undertaking using the mark, and the market for the goods or services. Where it is established, in view of all the facts, that the use is made to establish or maintain a presence on the market, it should be held to be genuine. In this respect, no minimum threshold should be established. However, if the use is made merely to protect or obtain a registration ('token' or 'sham' use), it should be disregarded.

Genuine use may be found when a trademark is used on or in relation to the goods for which it is registered. In judging the genuineness of use of a service mark, the nature of the services should be taken into account. Use in advertising should also be recognized if a sufficient relation with the goods or services is established.

### *(9) Proof of use*

When proof of use is required, the trademark proprietor should bear the burden of proof.

All available evidence admissible in administrative or judicial proceedings should be admitted to prove use in the respective proceedings.

### *(10) Grace period*

A reasonable grace period should be provided before a mark becomes liable to sanctions on grounds of non-use. The grace period should be at least three years and should not be longer than five years.

Beginning or resuming use after the expiry of the grace period and before validity of the mark is properly challenged should 'cure' the defect arising from the absence of use.

The re-filing of application for registration of unused marks, during or after the expiry of a grace period, should be allowed, provided that it is not in bad faith.

### *(11) Justification of non-use*

The trademark proprietor must be allowed to justify the absence of genuine use. In judging whether the grounds invoked by the proprietor constitute 'valid reasons' for the absence of use, all relevant circumstances should be taken into account, for example the need to obtain authorizations for the marketing, to the extent that the circumstances are beyond the control or sphere of the trademark proprietor.

### *(12) Challenge*

Any person, subject to the requirements of the respective legal system, must be entitled to seek cancellation of a mark that has not been genuinely used.

Where a jurisdiction provides for an administrative opposition or invalidation procedure on the basis of earlier trademarks, the applicant or proprietor of the contested mark should be entitled to invoke the absence of genuine use of the earlier mark, which should lead to the proprietor of the earlier mark being required to prove genuine use.

In infringement proceedings the defendant should be able to invoke the absence of genuine use, either as a defense or by way of a counterclaim.

## AIPPI LAW SERIES

1. Sanna Wolk, Kacper Szkalej (eds.), *Employees' Intellectual Property Rights*, second edition 2017 (ISBN 978-90-411-9230-1).
2. Jochen Bühling (ed.), *Patent Protection for Second Medical Uses*, 2016 (ISBN 978-90-411-8253-1).
3. Christopher V. Carani (ed.), *Design Rights: Functionality and Scope of Protection*, 2017 (ISBN 978-90-411-8864-9).
4. Eléonore Gaspar (ed.), *Genuine Use of Trademarks*, 2018 (ISBN 978-94-035-0563-3).